PHYSICAL CONSTANTS

Speed of light	c	$= 3.00 \times 10^8$ m/sec
Gravitational constant	G	$= 6.67 \times 10^{-11}$ N $\cdot$ m^2/kg^2
Avogadro's number	N_A	$= 6.023 \times 10^{23}$ particles/ g $\cdot$ atom
Boltzmann's constant	k	$= 1.3806 \times 10^{-23}$ J/K
Gas constant	R	$= 8.314$ J/g $\cdot$ mole $\cdot$ K
		$= 1.9872$ cal/g $\cdot$ mole $\cdot$ K
Planck's constant	h	$= 6.63 \times 10^{-34}$ J $\cdot$ sec
Electron charge	e	$= 1.60 \times 10^{-19}$ C
Electron rest mass	m_e	$= 9.11 \times 10^{-31}$ kg
		$= 5.4858 \times 10^{-4}$ u
Proton rest mass	m_p	$= 1.67 \times 10^{-27}$ kg
		$= 1.00727$ u
Neutron rest mass	m_n	$= 1.67 \times 10^{-27}$ kg
		$= 1.00866$ u
Coulomb's law constant	k	$= 9.0 \times 10^9$ N $\cdot$ m^2/C^2
Permeability constant	μ_0	$= 4\pi \times 10^{-7}$ N/A^2
Standard gravitational acceleration	g	$= 9.81$ m/s^2 $= 32.17$ ft/s^2
Mass of earth		5.98×10^{24} kg
Average radius of earth		6.38×10^6 m
Average density of earth		5570 kg/m^3
Average earth-moon distance		3.84×10^8 m
Average earth-sun distance		1.496×10^{11} m
Mass of sun		1.99×10^{30} kg
Radius of sun		7×10^8 m

PROBLEM-SOLVING TECHNIQUES

PHYSICS

PHYSICS

A General Introduction

Second Edition

Alan Van Heuvelen

New Mexico State University

Little, Brown and Company

BOSTON TORONTO

Library of Congress Cataloging-in-Publication Data

Van Heuvelen, Alan.
   Physics, a general introduction.

   Includes index.
   1. Physics. I. Title.
QC23.V25 1985   530   85-18186
ISBN 0-316-89716-7

Library of Congress Catalog Card No. 85-18186

ISBN 0-316-89716-7

9 8 7 6 5 4 3 2 1

MU

Published simultaneously in Canada
by Little, Brown & Company (Canada) Limited

Printed in the United States of America

Cover photo: The image was created by the Lissajous orbit of a light pendulum. The orbit is controlled by the direction and force of the push of the swing. At the end of the motion the pendulum will come to rest in the center. Photo by Jerome Kresch—Peter Arnold, Inc.

Color insert: Facing page 452, photos by Dave Vasquez. From Paul G. Hewitt, *Conceptual Physics,* 5th ed. Copyright © 1985 by Paul G. Hewitt. Reprinted by permission of Little, Brown and Company. *Facing page 453,* Courtesy of Bausch and Lomb.

To my parents,
teachers by example

PREFACE

Physics: A General Introduction is intended for use in the introductory physics course with an algebra and trigonometry prerequisite taken by students in biology, the health professions, geology, engineering technology, and other fields of science. We have been pleased with the success and wide usage of the first edition and by the favorable comments the first edition received from users, reviewers, and even students. Thus, our task in the second edition has been to preserve, and in some cases to improve and refine, the qualities that made the first edition such a success—a friendly and understandable presentation of physics, interesting worked examples and problems, and a useful approach to problem solving.

In the second edition, as in the first, four goals have guided the writing of the text: (1) to provide a clear, understandable introduction to the basic principles of physics; (2) to develop students' ability to use physics principles in solving problems; (3) to bring life to physics and problem solving by integrating into the presentation a wide variety of interesting, real-world examples and problems; and (4) to develop the general analytical skills of students in a way that will aid their future scientific and personal endeavors, not just their physics problem solving. Let me mention briefly how each goal has shaped the text.

1. *A clear and understandable introduction to the basic principles of physics.* I have used two strategies to attain this goal. First, new principles are carefully developed through a series of logical steps ending with a statement of the principle in a general form. Whenever possible, the presentation and development of a new principle is related to the students' experiences.

Second, the text is organized to help students systematically learn and store in their minds the *basic concepts* of physics rather than a plethora of unrelated facts. Novice problem solvers often store knowledge quite randomly; they scramble through large quantities of diverse information in search of the "right equation" to solve a problem. On the other hand, more experienced problem solvers store information in a small number of fundamental blocks—like the sections of a well-organized library. We should help students store their knowledge in a similar manner. To this end, Part I, "Newtonian Mechanics," emphasizes one basic idea of cause and effect: forces, impulses, and torques cause objects to accelerate, change momentum, and experience an angular acceleration. Newton's second law and equations derived from it are combined with kinematics in the Newtonian approach to problem solving.

The unifying idea in Part II, "Energy and Its Transformations," is the conservation-of-energy principle. This conservation principle, as expressed in the work-energy equation and the first law of thermodynamics, can be introduced and easily understood as a second fundamental principle and problem-solving technique.

Parts I and II, then, show two fundamental concepts and two ways that physicists perform calculations when dealing with natural phenomena. Each subsequent part of the book not only increases the students' repertoire of concepts but also combines new analytical tools with those learned earlier. By emphasizing the basic principles, the presentation of new concepts unfolds in a logical manner from a relatively small number of important ideas, key equations, and central analytical techniques.

In the second edition the chapters in Part I have been rearranged in a more conventional order: vectors, kinematics, dynamics, momentum, circular motion, statics, and rotational dynamics.

2. *The development of students' abilities to use physics principles to solve problems.* First, as just mentioned, the book is organized to help students store their knowledge of physics in the form of basic principles and the associated problem-solving procedures.

Second, general problem-solving techniques are outlined in seven critical places throughout the text (see examples on pages 62 and 119), with a step-by-step illustration of each procedure. These techniques are used in over 300 worked examples. Also included are 360 questions and 1700 end-of-chapter problems. The problems are labeled for difficulty by squares beside the problem: easier, one-step substitution problems have no squares, and challenging problems that combine different concepts and techniques have two squares. Special care has been taken in the second edition to achieve a balance of difficulty in the problems, with the majority being of intermediate difficulty (one square). About one-third of the problems in the second edition are new and include a wider variety of problems that integrate concepts from different sections and chapters. (The General Problems of the first edition appear in this edition with two squares.)

The SI system of units is used throughout except for a small number of English-system problems in Part I.

3. *Physics concepts and problem-solving procedures brought to life by applying them to interesting, realistic situations.* Teaching and learning are much more enjoyable and effective if the principles can be illustrated with relevant, real-world examples and problems. In recent years, textbooks have attempted to motivate students in physics by providing applications to other subjects, notably to biology. These applications have usually been presented in separate sections or in essays that qualitatively discuss the ways in which physics relates to other subjects. But because there is seldom enough time to learn all of the important physics concepts and how to use them, optional sections, no matter now interesting, are often omitted. Thus, I have integrated many applications into the presentation of the physics principles themselves. Although a large number of standard physics examples appear, physics is also applied to everyday life, biology, technology, the earth and its climate, and other fields of science. All of these examples involve quantitative calculations and are presented with physical analogues that are easily understood by both students and physicists. The integration of these examples in the presentation allows the student to learn physics principles while applying them to a variety of interesting situations.

4. *Development of analytical skills.* Students should be encouraged to develop analytical thinking skills that carry over into their personal and professional lives. As a means to this end, I have included a number of estimation problems in which students are not given all the information they need to work the problem. Instead, they must decide what quantities are needed to make the estimate, the approximate values for these quantities, and what principles and equations must be used. Interesting situations have been chosen for these estimation problems so that students will be curious to learn the results of their estimates.

Also included are seven short sections called Interludes that encourage analytical thinking by exposing students to specific techniques such as proportional reasoning, dimensional analysis, scaling, and the exponential function. These Interludes include worked examples and problems that are applied to real-world situations. Each Interlude can serve as the subject for one lecture period.

Ancillary Aids

Several useful items have been prepared to accompany the second edition of *Physics: A General Introduction.* Professor Lois M. Kieffaber's (Whitworth College) *Study Guide and Workbook,* Second Edition, provides an exceptional problem-solving aid for students. For each chapter of the text, it includes a list of important terms and equations; about ten worked examples and three or four self-test problems (with complete solutions in the appendix); and a section on avoiding pitfalls that indicates common problem-solving mistakes that students make. Also, I have prepared for the instructor a *Test Bank* available to adopters of the text by writing Little, Brown and Company. The *Test Bank* contains over 500 questions, available both in booklet form and on diskette for Apple II and IBM microcomputers. A *Solutions Manual* with brief solutions to all problems accompanies the text. *Transparency Masters* for use in demonstrating the problem-solving procedures in the text are also available to adopters of the second edition. Finally, a math pretest written by Professor Thomas Hudson of the University of Houston is available. Students who do poorly on this test may be encouraged to use a self-paced *Math Review* written by Professor Hudson and available through Little, Brown and Company.

Acknowledgments

I have incurred more debts in writing this book than I can hope to acknowledge. I am very grateful for the many useful comments and suggestions by the following professors who reviewed parts of the manuscripts for the first and/or second editions: Angelo Armenti (Villanova University), Robert Bearse (University of Kansas), John Botke (Effects Technology), Bruce Brackenridge (Lawrence University), Frank O. Clark (University of Kentucky), Robert Beck Clark (Texas A&M), Peter G. Debrunner (University of Illinois–Urbana), Dewey Dykstra (Boise State University), W. M. Hartmann (Michigan State University), Stanley Hertzback (University of Massachusetts, Amherst), Verner Jensen (University of Northern Iowa), F. M. Kelly (University of Manitoba), Lois M. Kieffaber (Whitworth College), Elie Lowy (Queensborough Community College), David Markowitz (University of Connecticut, Storrs), Konrad Mauersberger (University of Minnesota), Joseph McCauley (University of Houston), H. W. Norton (University of Western Florida), Mildred O'Donnel (Wentworth Institute), W. C. Parke (George Washington University), Neil Peek (University of California, Davis), Alan Peltzer (Nassau Community College), Robert B. Prigo (Middlebury College), Annette Rappleyea (City College of San Francisco), John W. Robson (University of Arizona), John Rohrs (Kearney State College), Lawrence Rowan (University of North Carolina, Chapel Hill), Radah R. Roy (Arizona State University), D. L. Rutledge (Oklahoma State University), Melvin Schwartz (St. John's University), Carol Spader (Wentworth Institute), Fred

Thatcher (Indianapolis University, Purdue University–Indianapolis), Ernest Urvater, Richard Whitlock (University of North Carolina, Greensboro), Charles A. Whitten, Jr. (University of California, Los Angeles), George Williams (University of Utah), Gordon G. Wiseman (University of Kansas), and Noel Yeh (State University of New York, Binghamton). A special thanks to Professor Lois M. Kieffaber for her many helpful suggestions in addition to her work in the preparation of an exceptional *Study Guide and Workbook*. I was also aided by personal discussions with my colleagues at New Mexico State University, including Bob Armstrong, Ramish Bhandari, George Burleson, Alex Burr, Horace Coburn, Twan Chen, George Goedecke, Roger Greensfelder, Richard Ingraham, Gary Kyle, Robert Liefeld, August Miller, Dave Mott, Jim Ni, Lee Radziemski, Budh Ram, and Thor Stromberg.

Several graduate and undergraduate students, including Clifford Flint, Assad Fotovatian, and Bill Maloney, checked the answers and the wording for the new problems for the second edition. A special thanks to Nancy Denzler, Jan Lane, Jodi Van Heuvelen, and Josie Vigil for their expert typing and editing.

I am grateful to the very competent people at Little, Brown and Company for their professional yet personable help in all stages of the project. In preparing the first edition, I am especially indebted to Ian Irvine, Jane Tufts, and Elizabeth Schaaf for their guidance, review, and production. In preparation of the second edition, the cast has changed but the quality has continued with the able assistance of Ron Pullins, John Covell, Sue Warne, Virginia Shine, and their colleagues.

Finally, I would like to thank Linda, Jodi, and Scot for helping to make each day special.

SUMMARY CONTENTS

CONTENTS

PART IV
Vibrations and Waves 333

PART V
Light and Optics   397

PART VI
Electricity and Magnetism   473

PART VII
Modern Physics 633

30 Special Relativity *635*

31 Wave-Particle Duality *655*

INTRODUCTION

Physics is a fundamental science encompassing subjects ranging from atoms and the subatomic particles of which atoms are made to galaxies and their constituents—pulsars, black holes, neutron stars, white dwarfs, and even the sun and our own earth. Physicists search for general rules or laws that bring understanding to the chaotic behavior of our surroundings. The laws, once discovered, often seem obvious, yet their discovery usually requires years of theorizing and experimentation.

For example, Apollonius of Perega in 200 B.C. adopted the concept that the earth occupied the center of a revolving universe. Three hundred years later Ptolemy provided a theory to explain the complicated motion of the planets in that earth-centered universe. Ptolemy's theory, which predicted with surprising accuracy the changing positions of the planets, was accepted for the next fourteen hundred years. Then Copernicus, who studied astronomy at the time Columbus sailed to America, developed a theory of motion for the heavenly bodies; the sun resided at the center of the universe and the earth moved in orbit about it just as did the other planets. More than a hundred years later the theory was confirmed by careful observation by Galileo Galilei. Finally, fifty years after Galileo's death, Isaac Newton formulated three simple laws of motion and the universal law of gravitation, which together provided the basic understanding needed to explain the orbital motion of the earth and the other planets.

Newton's simple laws also give us the understanding needed to guide rockets to the moon, to build skyscrapers, and to realize why we should not lift heavy weights when in a bent position. Newton's inspiration provided not only the basic resolution of an eighteen-hundred-year-old problem but also a general framework for analyzing the mechanical properties of nature. Now, three hundred years after Newton's death, we are able to learn these basic laws of motion and how to use them during the first few weeks of our physics course. It is hard to appreciate the great struggle of our predecessors who developed the new understanding that now seems routine.

Today, those same struggles occur in most branches of science. Only the subject matter has changed. How does the brain work? What causes the earth's magnetic field? What is the nature of the pulsating sources of x-ray radiation in our galaxy? Are protons and neutrons really made of three smaller particles called quarks held together by exchanging other particles called gluons—or is this quark theory a misconception such as Ptolemy's complicated model of an earth-centered universe?

The pursuit of basic understanding often seems greatly removed from the activities of daily living. How could knowing that a proton is made of three smaller particles called quarks possibly affect our lives? Contemporaries of J. J. Thomson could just as easily have asked him a similar question in 1897 after his discovery of the electron. He probably could not have provided a satisfactory answer. Yet, less than a century later, the electron plays an integral part in our

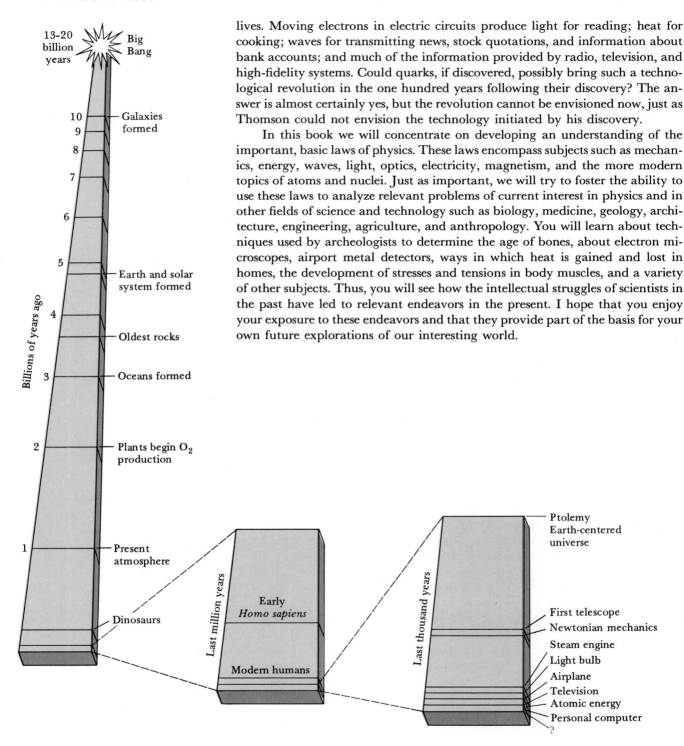

lives. Moving electrons in electric circuits produce light for reading; heat for cooking; waves for transmitting news, stock quotations, and information about bank accounts; and much of the information provided by radio, television, and high-fidelity systems. Could quarks, if discovered, possibly bring such a technological revolution in the one hundred years following their discovery? The answer is almost certainly yes, but the revolution cannot be envisioned now, just as Thomson could not envision the technology initiated by his discovery.

In this book we will concentrate on developing an understanding of the important, basic laws of physics. These laws encompass subjects such as mechanics, energy, waves, light, optics, electricity, magnetism, and the more modern topics of atoms and nuclei. Just as important, we will try to foster the ability to use these laws to analyze relevant problems of current interest in physics and in other fields of science and technology such as biology, medicine, geology, architecture, engineering, agriculture, and anthropology. You will learn about techniques used by archeologists to determine the age of bones, about electron microscopes, airport metal detectors, ways in which heat is gained and lost in homes, the development of stresses and tensions in body muscles, and a variety of other subjects. Thus, you will see how the intellectual struggles of scientists in the past have led to relevant endeavors in the present. I hope that you enjoy your exposure to these endeavors and that they provide part of the basis for your own future explorations of our interesting world.

"It's an exciting time to be alive." – Albert Einstein

PART I

Newtonian Mechanics

We often hear that knowledge in science increases rapidly. We should note, however, that many of the principles used in physics today were developed over a hundred years ago. Our present subject, Newtonian mechanics, was introduced over three hundred years ago by men such as Galileo Galilei and Isaac Newton.

In Newtonian mechanics we consider the way in which forces affect the motion of objects. Under what conditions do objects move faster or slower, and under what conditions do they remain at rest? Newtonian mechanics helps us answer many down-to-earth questions. For example: What is the compression in the discs of your backbone when you lift a 50-lb weight? What is the minimum stopping distance in a car collision that allows the passengers to avoid injury? How does a ballerina spin so fast during a pirouette? How far from the earth must a communications satellite orbit if it is to remain above the same point on the earth's rotating surface? Besides helping us analyze such questions, Newtonian mechanics provides the foundation for much of the rest of our development of physics.

CHAPTER 1

Scalars and Vectors

Physicists seek to describe nature in terms of a small number of basic laws. These laws involve equations that relate quantities such as force, mass, acceleration, momentum, and energy. Many of these quantities are said to be **scalars**—quantities having only a magnitude and a unit of measure. Examples are time, distance, and energy. The normal rules of algebra are used to add, subtract, multiply, and divide scalar quantities.

Other quantities, such as force and velocity, have a magnitude, unit of measure, and a direction. These quantities with direction are called **vectors.** To add and subtract them we must use special rules, which are developed in this chapter. This mathematical background plays an essential role in subsequent chapters, where we develop and use some of the important laws of physics.

1.1   Physical Quantities and Standards

In the physical sciences, nature is described using a variety of precisely defined quantities. These quantities differ from the descriptive words used in everyday language—a person is *nice* or a flower is *pretty*. The meanings of such everyday words depend on the previous experiences of the persons using them. The word *nice* may be used by one person to rate physical appearance and by another to rate personality. For this reason we cannot rate a person exactly using such words as *nice*.

In physics and in some other branches of science, the quantities used to describe nature are defined so that they have the same meaning for any person using them. Quantities such as time, length, speed, and pressure can be measured and given specific values. Each quantity, when measured, has a magnitude (a number) and a unit of measure. For example, a basketball player 2.1 m tall is 2.1 times longer than a unit of length called the meter. If we all agree on the definition of a meter, then the player's height is described unambiguously.

An essential component of measurement is precise standards that define units such as the meter. Standards have not always been defined precisely. At one time a "foot" was the length of the king's foot in England. Early in the present century the "foot" was defined in several different ways. Property evaluations in Brooklyn were made using the U.S. foot, the Bushwick foot, the Williamsburg foot, and the foot of the 26th Ward. Some strips of Brooklyn property were untaxable because after surveys made using different "feet," these strips of land legally did not exist!

We can see that it is important to have standards that are precisely defined and that are used in common by people involved in trade, science, and industry. The unit of length presently used in science is called the *meter*. It was originally defined as the distance between two lines inscribed on a bar of metal stored in a special vault in a small town near Paris, France. Replicas made from the standard were distributed around the world. A meter in France had the same length as a meter in Canada, the United States, Peru, or any other country.

A metal bar such as used for the length standard can deteriorate with time and is accessible to only a few. In 1960, therefore, the meter was redefined in terms of the wavelength of light emitted by a particular atomic transition. The meter was redefined in 1983 as the distance traveled by light in a vacuum in a time interval of 1/299,792,458 second. This new standard equals the length of the old but is more accurate than the old, is accessible to scientists in any laboratory in the world, and does not deteriorate. The current standard for time also depends on an atomic phenomenon.

1.2 Systems of Units

We might ask ourselves how many quantities such as distance and time need to have standards. For example, if we have standards for length and for time, then we do not need a separate standard for speed, since its unit (meters per second) is already defined by the length and time standards. At the present time, we can describe most scientific observations using the units of the seven basic quantities listed in Table 1.1. They are called **basic quantities** because each unit has a well-defined standard.

Scientists use many more than seven quantities to describe their observations, but the units of the other quantities, called **derived quantities,** are just combinations of basic units. Speed is an example of a derived quantity; its unit is the ratio of meters and seconds. Often, a special name is given to a frequently used derived unit, usually to honor a noteworthy physicist. For example, the derived unit of force—the kg m/s² —is called a *newton* in honor of Isaac Newton.

A *system of units* consists of a set of standards for some or all of the basic units. In a system of units that evolved in English-speaking countries, the so-called English system, the units for length, mass, and time are the foot, slug, and second, respectively. In the metric system the units of length, mass, and time are the meter, kilogram, and second, respectively. The modernized version of the metric system, based on atomic standards, is called the **SI system,** from the French *Système International d'Unités* (international system of units).

Today, SI units are used in most world trade and in science, although units from other systems still retain favor in some fields of study. One big advantage of the metric system is that large and small units are all related to the basic units by powers of ten. For example, 1 kilometer = 1000 meters, and 1 centimeter = 1/100 meter. Thus, it is much easier to convert meters to kilometers than to convert feet to miles. The prefixes used for these multiples are listed in a table on the inside back cover.

TABLE 1.1 Basic Units of the SI System

Quantity	Unit Name	Unit Symbol
Length	meter	m
Mass	kilogram	kg
Time	second	s
Electric current	ampere	A
Temperature	degree kelvin	K
Amount of substance	mole	mol
Luminous intensity	candela	cd

1.3 Scalar and Vector Quantities

We have said that quantities in science can be measured and specific values given to them. Each quantity can be represented by a symbol indicating its numerical value and unit of measure. For example, the symbol E could represent an energy of 100 calories, or t could represent a time of 3.5 seconds. These quantities are examples of what are called scalars.

A **scalar** is a quantity that has a magnitude and usually a unit of measure. Scalars are given algebraic symbols such as E, t, and v. The familiar rules of algebraic addition, subtraction, multiplication, and division are used to mathematically manipulate scalars.

Often we need to indicate the direction as well as the magnitude of a quantity. Perhaps you have been told that a hidden treasure has a "displacement" from you of 100 miles in a direction 50° north of west. **Displacement** is a vector quantity indicating (1) the distance from one point to another and (2) the direction of an arrow pointing from the first to the second point. **Distance** is the corresponding scalar quantity; it indicates only the separation of the points, not the direction from one to the other.

A **vector** is a quantity with magnitude, direction, and usually a unit of measure. Vectors are represented by boldface algebraic symbols such as **F**, **d**, and **a**.* Special rules will be developed to add, subtract, and multiply vectors.

The **magnitude** of a vector is a scalar that indicates only how large or small the vector is and its unit of measure. The magnitude is always either a positive number or zero and is represented by the same symbol as the vector, except that it is not boldface. For example, a 30-m displacement vector **A** that points west has a magnitude $A = 30$ m.

Several examples of scalar and vector quantities are listed in Table 1.2. Notice that each vector in Table 1.2 is specified by a magnitude and a direction, whereas scalars are specified only by magnitudes.

TABLE 1.2 Examples of Scalar and Vector Quantities

	Scalars	
Symbol	Name	Example
d	distance	30 m
v	speed	50 m/s
t	time	15 s
E	energy	2000 cal

	Vectors	
Symbol	Name	Example
d	displacement	30 m north
v	velocity	50 m/s west
F	force	100 lb up
a	acceleration	12 m/s² down

Which of the signs are vectors and which are scalars? (Photos by David Conklin and Peter Menzel/Black Star)

*In handwritten work, vectors are usually represented by the symbol with an arrow over it ($\vec{F}$, $\vec{d}$, and $\vec{a}$).

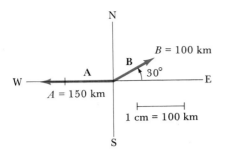

FIG. 1.1. Two displacement vectors. **A** is 1.5 cm long and represents a 150-km displacement toward the west. **B** is 1.0 cm long and represents a 100-km displacement 30° north of east.

1.4 Graphical Representation of Vectors

Vector quantities are often represented graphically by arrows. The arrow's direction indicates the direction of the vector, and the arrow's length indicates the vector's magnitude. An appropriate scale must be used when drawing the length of the arrow. For example, in Fig. 1.1, 1 cm represents 100 km. Thus, vector **A**, which is 1.5 cm long, represents a 150-km displacement toward the west. Vector **B** represents a 100-km displacement 30° north of east.*

Forces are another familiar quantity represented by vectors. **Forces** are often caused by a push or pull of one object on another. Sometimes, but not always, the objects are clearly in contact with each other. A book pushing down on a table or a boxer's glove punching an opponent are forces, but so too is the earth's gravitational pull on a skydiver (see Fig. 1.2). The unit of force in the English system of units is the *pound*. In the metric system, the unit of force is the *newton*. One newton is approximately the weight of a medium-sized apple, a little less than one-fourth of a pound. Force will be defined more precisely in Chapter 3.

1.5 Graphical Addition and Subtraction of Vectors

How are vector quantities added and subtracted? The normal rules of algebraic addition and subtraction are not satisfactory. Suppose, for example, that you take a two-day trip. The first day you travel 500 km toward the east along a straight road. The second day involves another 500-km displacement, but not necessarily in the same direction. In what direction and how far from your starting position are you now located? The answer, of course, depends on the direction of the second day's trip. You could be 1000 km east of your home if you continued traveling east during the second day (Fig. 1.3a). However, if you traveled west on the second day, your net displacement would be zero—you would be back where you started (Fig. 1.3b). If you traveled north during the

(a)

(b)

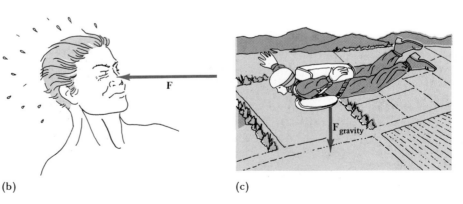

(c)

FIG. 1.2. (a) A boxer receiving a punch; (b) the vector representation of the force caused by the punch on the boxer's head. (c) The gravitational force of the earth pulling down on a skydiver.

*A displacement whose direction is 30° north of east makes a 30° angle with an axis pointing toward the east. The vector is oriented above the axis toward the north rather than below the axis toward the south.

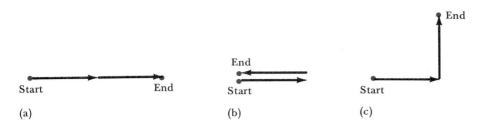

FIG. 1.3. Two separate 500-km trips are each represented by a displacement vector. The net displacement from the starting position depends on the direction of the second day's trip.

second day, your net displacement would be a little over 700 km to the northeast of your home (Fig. 1.3c). When adding displacement vectors or vectors of any type, we are concerned only with their net result. For displacement vectors, we are concerned only with the magnitude and direction of the final displacement from the starting position. If we travel around the world and end back where we started, our net displacement is zero. We get no credit for the manner in which we reached that destination.

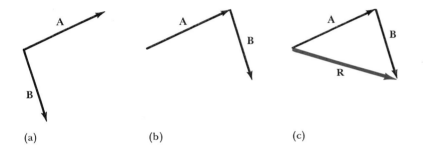

FIG. 1.4. (a) Two vectors. (b) To add the vectors, the tail of one vector is placed at the head of the other. (c) The resultant, $R = A + B$, goes from the tail of the first to the head of the last vector.

Vector addition is visualized most easily using a graphical technique. Suppose we want to add the two vectors **A** and **B** shown in Fig. 1.4a. These could be displacement vectors, force vectors, or any other type of vector. To add them graphically, we place the tail of **B** at the head of **A,** as in Fig. 1.4b. We can move vectors from one place to another for addition, but we cannot change the magnitude or direction of a vector while moving it. Having moved vector **B,** we draw another vector, **R,** from the tail of **A** to the head of **B,** as in Fig. 1.4c. This vector **R** represents the net displacement (or net force if **A** and **B** are force vectors) and is called the **resultant** of vectors **A** and **B.** We can write the resultant as a mathematical equation: $R = A + B$. Because **A** and **B** are vectors, we must use a technique such as illustrated in Fig. 1.4c to find **R.** *We cannot simply add the magnitudes of* **A** *and* **B** *to find* **R.** The graphical addition of vectors is summarized as follows with reference to the three vectors shown in Fig. 1.5:

FIG. 1.5. Three vectors shown in (a) are added by joining them head to tail in (b). (c) The resultant, $R = A + B + C$, goes from the tail of the first to the head of the last vector. (d) Joining the vectors in a different order produces the same resultant, $R = C + B + A.$

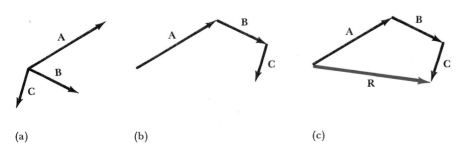

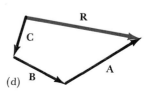

FIG. 1.6. (a) Three displacement vectors; (b) the resultant displacement.

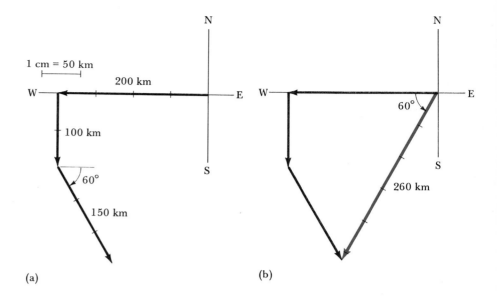

(a)

(b)

Graphical addition of vectors: To add two or more vectors graphically, place the vectors head to tail one at a time, as illustrated in Fig. 1.5b. The magnitudes and directions of the vectors must not be changed as they are moved about. The resultant vector **R** goes from the tail of the first vector to the head of the last vector, as illustrated in Fig. 1.5c, and equals the sum of these vectors (**R** = **A** + **B** + **C**). The order in which we add the vectors makes no difference (for example, **A** + **B** + **C** = **C** + **B** + **A,** as can be seen by comparing Figs. 1.5c and d).

Usually, the graphical technique for adding vectors is used as a rough check on another vector addition technique introduced later. However, if done with care using a ruler and protractor, the graphical vector addition technique is a fairly accurate method for determining the resultant of several vectors.

EXAMPLE 1.1 A car travels 200 km west, 100 km south, and finally 150 km at an angle 60° south of east. What is the net displacement of the car?

SOLUTION A vector diagram representing the three displacements drawn head to tail is shown in Fig. 1.6a. To find the resultant displacement **R** we draw an arrow from the tail of the first displacement to the head of the last (Fig. 1.6b). We measure the magnitude of the resultant with a ruler and find that its length is 5.2 cm. Since each centimeter represents 50 km, the magnitude of the resultant displacement is 5.2 × 50 km = 260 km. Using a protractor, we find that the direction of the resultant is 60° south of west. ∎

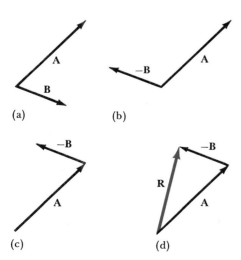

FIG. 1.7. (a) Two vectors **A** and **B;** (b) **A** and −**B;** (c) the addition of **A** and −**B;** (d) the resultant **R** = **A** + (−**B**) = **A** − **B.**

Occasionally we will need to subtract two vectors. A general case of vector subtraction is summarized as follows and is illustrated in Fig. 1.7, where vector **B** is subtracted from **A.**

Graphical subtraction of vectors: To graphically subtract vector **B** from vector **A**, first reverse the direction of **B**, thus producing −**B** (Fig. 1.7b). Then add **A** and −**B** (Fig. 1.7c). The resultant goes from the tail of **A** to the head of −**B** (Fig. 1.7d) and is the difference of **A** and **B**:

$$R = A + (-B) = A - B.$$

1.6 Components of a Vector

The graphical method of adding and subtracting vectors takes considerable time if done accurately. A different vector addition technique developed in this and the next section is more accurate and is also faster and more convenient for solving a variety of interesting problems involving vectors. The *component method*, as it is called, depends on our ability to resolve a vector into components, as illustrated in Fig. 1.8.

Consider the displacement vector in Fig. 1.8a. A coordinate system has been drawn that has an *x* axis pointing east and a *y* axis pointing north. The *x* **component** of **A** is a vector A_x that points along the *x* axis and whose length equals the projection of **A** on that axis, as shown in Fig. 1.8b. The *y* **component** of **A** is the projection of **A** on the *y* axis and is labeled A_y in Fig. 1.8b. It is important to notice that the vector sum of A_x and A_y equals **A**, as shown in Fig. 1.8c.

In the component method of adding vectors, each vector being added is first resolved into its components. Then, instead of adding the vectors, we add their components. This seems reasonable since the components of a vector, when added, equal the vector itself. Thus, it should not matter whether we add the vectors or their components.

How do we determine the value of the components of a vector of known magnitude and direction? First, consider a component's direction. The subscript on a component—for example, the *x* on A_x—indicates the axis along which the component lies. In Fig. 1.9, the *x* and *y* axes each have a positive direction, with arrows at their heads (to the right for the *x* axis and pointing upward for the *y* axis). The opposite directions are the negative directions. If a component points in the positive direction of an axis, it is given a positive sign. If it points toward the negative direction of the axis, it is given a negative sign. For instance, the *x* component of the force **F** shown in Fig. 1.9 is negative because F_x points in the negative *x* direction, whereas the *y* component is positive because it points parallel to the positive *y* direction. We do not need to use a vector symbol (boldface print) to specify a component because its direction is implied by its subscript and sign (positive or negative).

Next we need to determine the magnitude of the components. For this, a brief review of trigonometry is helpful. Consider the right triangle shown in Fig. 1.10. Recall the definitions of the sine, cosine, and tangent functions:

$$\sin \theta = \frac{\text{Opposite side}}{\text{Hypotenuse}} = \frac{o}{h},$$

$$\cos \theta = \frac{\text{Adjacent side}}{\text{Hypotenuse}} = \frac{a}{h},$$

$$\tan \theta = \frac{\text{Opposite side}}{\text{Adjacent side}} = \frac{o}{a}.$$

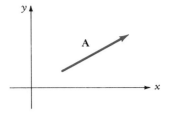

(a)

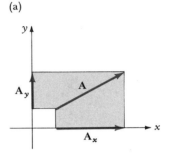

(b)

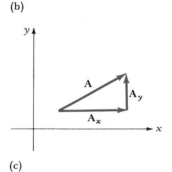

(c)

FIG. 1.8. (a) A displacement vector **A** and (b) its *x* and *y* components (A_x and A_y). (c) The components when added equal **A**.

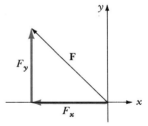

FIG. 1.9. F_x is negative, F_y positive.

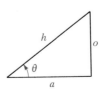

FIG. 1.10. The sides of the triangle are labeled *h* (hypotenuse), *o* (side opposite the angle θ), and *a* (side adjacent to the angle θ).

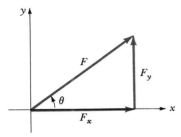

FIG. 1.11. If F and θ are known, F_x and F_y can be found using trigonometry.

Using these relationships, we can find the magnitude of the force components shown in Fig. 1.11. The hypotenuse is the magnitude F of the force and the sides are the magnitudes of the x and y components of **F**. Thus

$$\sin \theta = \frac{F_y}{F} \quad \text{or} \quad F_y = F \sin \theta, \tag{1.1}$$

$$\cos \theta = \frac{F_x}{F} \quad \text{or} \quad F_x = F \cos \theta, \tag{1.2}$$

$$\tan \theta = \frac{F_y}{F_x}.$$

Equations (1.1) and (1.2) are used to calculate the magnitudes of F_x and F_y if F and θ are known. A table of sine, cosine, and tangent functions appears on the inside back cover. Most calculators also give the sines and cosines of different angles.

The technique for finding the components of a vector may be summarized as follows.

The x and y **components** of a vector **F** are

$$F_x = \pm F \cos \theta, \tag{1.3}$$
$$F_y = \pm F \sin \theta, \tag{1.4}$$

where F is the magnitude of the vector and θ is the angle (90° or less) that **F** makes with the x axis. F_x is positive if it points in the positive x direction and negative if it points in the negative x direction. F_y is either positive or negative, depending on its direction relative to the y axis.*

Be sure to note that the angle θ appearing in Eqs. (1.3) and (1.4) is the angle that the vector makes with the positive or negative x axis and not the angle it makes with the y axis.

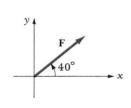

(a)

EXAMPLE 1.2 A carpenter pulls a load of lumber on a cart with a 60-lb force directed 40° above the horizontal (Fig. 1.12a). Calculate the horizontal and vertical components of the force.

SOLUTION The force and a set of axes used to describe it are shown in Fig. 1.12b. Both force components shown in Fig. 1.12c are positive. Since $F = 60$ lb and $\theta = 40°$, we find using Eqs. (1.3) and (1.4) that

$$F_x = F \cos \theta = (60 \text{ lb})(\cos 40°) = \underline{46 \text{ lb}},$$
$$F_y = F \sin \theta = (60 \text{ lb})(\sin 40°) = \underline{39 \text{ lb}}.$$

It is important to realize—and it may surprise you—that the 60-lb force pulling the wagon in the direction shown in Fig. 1.12b produces exactly the same effect on the wagon as if its components, a 46-lb horizontal force and a 39-lb vertical force, pulled the wagon. ∎

(b)

(c)

FIG. 1.12. (a) A carpenter pulling a load of lumber. (b) The force of the rope on the cart. (c) The horizontal and vertical components of the force.

*Different conventions are used for calculating components of vectors. You may find in another course or book that the positive and negative signs are omitted when calculating F_x and F_y. In that case, the angle θ is the counterclockwise angle from the positive x axis to the vector and can assume a value between 0 and 360°. If you use this convention, then $F_x = F \cos \theta$ and $F_y = F \sin \theta$, and the signs of F_x and F_y depend on the signs of the trigonometric functions. Either convention works fine, but do not mix them. Choose one and use it consistently.

EXAMPLE 1.3   You fly 4200 km in a direction 47° north of west while traveling from Minneapolis to Fairbanks. Using a coordinate system such as shown in Fig. 1.13a with the x axis pointing east and the y axis north, calculate the x and y components of your displacement.

SOLUTION   The x component of the trip is negative because it points west in the negative x direction, whereas the y component is positive. The components are then

$$A_x = -A \cos \theta = -(4200 \text{ km})(\cos 47°) = \underline{-2900 \text{ km}},$$
$$A_y = +A \sin \theta = +(4200 \text{ km})(\sin 47°) = \underline{3100 \text{ km}}.$$ ∎

A Note About Significant Figures

If you repeat the calculations in Example 1.3 using a calculator, your answers will differ from those stated. For example, your calculator will indicate that $A_y = +(4200 \text{ km})(\sin 47°) = 3071.6856$ km. Why is the answer in the example rounded off to 3100 km? By convention, unless otherwise stated in the problem, a distance of 4200 km is assumed to be known only within about 100 km, that is, $A = (4200 \pm 100)$ km. The last nonzero digit (the two in 4200) is considered uncertain by ± 1. We cannot know the value of A_y with more precision than the value of A used to calculate A_y. Thus, our answer was rounded off to two significant digits, the same as the number of significant digits in the value of A. (If we leave the answer as 3071.6856 km, we imply a precision of ± 0.0001 km or about ± 4 in, certainly not justified for the problem.) A detailed discussion of significant figures appears in Appendix A. You should get used to rounding off the answers to your problems to the appropriate number of significant figures.

1.7   Vector Addition by Components

Having learned to calculate the components of a vector, let us next see how vectors can be added by adding their components. In Fig. 1.14a, vectors **A** and **B** are added graphically to produce a resultant vector **R** (the dashed line). In Fig.

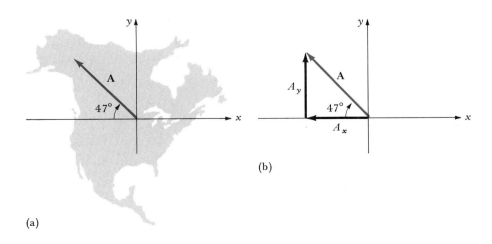

(b)

(a)

FIG. 1.13   (a) A 4200-km trip from Minneapolis to Fairbanks. (b) The x component is negative and the y component positive.

FIG. 1.14. (a) Vectors **A** and **B** are added graphically to produce **R**. (b) **A** and **B** are replaced by their components, and then the components are added to produce **R**. (c) The order of component addition is rearranged to determine the components R_x and R_y. (d) The magnitude R and direction θ of **R** can be determined from R_x and R_y.

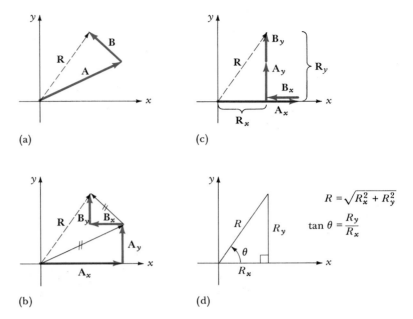

(a)

(c)

(b)

(d)

$$R = \sqrt{R_x^2 + R_y^2}$$

$$\tan \theta = \frac{R_y}{R_x}$$

1.14b, the double-crossed lines through vectors **A** and **B** indicate that they have been replaced by their components. Instead of adding **A** and **B**, we add their components—A_x, A_y, B_x, and B_y—as illustrated in Fig. 1.14b. The resultant is the same as when we add **A** and **B** directly, as in Fig. 1.14a.

We can, if we wish, change the order of addition of vectors. This has been done in Fig. 1.14c, where A_x and B_x are added first to find the x component of the resultant vector, called R_x. Note that since B_x points to the left, it cancels some of the effect of A_x, which points to the right. After finding R_x, we then add A_y and B_y to find the y component of the resultant, R_y. Then, when we add R_x and R_y, as shown in Fig. 1.14d, we once again get the same resultant vector **R**.

Notice that the triangle made of R, R_x, and R_y, shown in Fig. 1.14d, is a right triangle. According to the Pythagorean theorem, the sum of the squares of its sides $(R_x^2 + R_y^2)$ equals the square of the triangle's hypotenuse (R^2), or

$$R = \sqrt{R_x^2 + R_y^2}.$$

Thus, we can determine the magnitude of R if R_x and R_y are known.

The direction of **R** is determined from a picture of **R** constructed from R_x and R_y. The exact angle, θ, that R makes with the x axis is also determined from the values of R_x and R_y by taking the tangent of θ for the triangle shown in Fig. 1.14d. Note that

$$\tan \theta = \left| \frac{R_y}{R_x} \right|,$$

or

$$\theta = \text{invtan} \left| \frac{R_y}{R_x} \right|.$$

The lines on either side of R_y/R_x in the above equations are absolute value signs and mean that only the positive values of R_x and R_y are substituted into these equations. When solving for θ, we then determine the angle of 90° or less that the resultant vector makes with either the positive or negative *x* axis. The quadrant in which the vector points depends on the signs of R_x and R_y.

This vector addition technique can be used for any type of vector; it is summarized as follows for force vectors.

Vector addition technique using components: To add a group of vectors to find the resultant $\mathbf{R} = \mathbf{F}_1 + \mathbf{F}_2 + \mathbf{F}_3 + \cdots$, we first add the *x* components of the vectors to find R_x and then add the *y* components to find R_y:

$$R_x = F_{1x} + F_{2x} + F_{3x} + \cdots, \tag{1.5}$$

$$R_y = F_{1y} + F_{2y} + F_{3y} + \cdots. \tag{1.6}$$

Equations (1.3) and (1.4) are used to find the components of each vector. Having found R_x and R_y, the magnitude of the resultant is

$$R = \sqrt{R_x^2 + R_y^2}. \tag{1.7}$$

The quadrant in which **R** points is determined from a picture of **R** constructed from R_x and R_y. The exact angle, θ, that **R** makes with the positive or negative *x* axis is determined from the equation

$$\tan \theta = \left| \frac{R_y}{R_x} \right|, \tag{1.8}$$

or

$$\theta = \text{invtan} \left| \frac{R_y}{R_x} \right|. \tag{1.9}$$

EXAMPLE 1.4 Three people pull ropes attached to a trailer hitch on the bumper of a car stuck in the mud, as shown in Fig. 1.15a. Calculate the resultant of the three forces.*

SOLUTION Check carefully the sign and angle used to calculate the components of each force. The resultant components are as follows:

$$R_x = F_{1x} + F_{2x} + F_{3x} = -F_1 \cos 30° - F_2 \cos 0° - F_3 \cos 53°$$
$$= -(180 \text{ N})0.87 - (520 \text{ N})1.00 - (380 \text{ N})0.60 = -905 \text{ N},$$
$$R_y = F_{1y} + F_{2y} + F_{3y} = F_1 \sin 30° + F_2 \sin 0° - F_3 \sin 53°$$
$$= (180 \text{ N})0.50 + (520 \text{ N})0.00 - (380 \text{ N})0.80 = -213 \text{ N}.$$

We see that the three ropes together exert a net or resultant force whose *x* component equals -905 N and whose *y* component equals -213 N. (We will round off to two significant digits when we get our final answer.)

You may prefer to add components by constructing a table, such as the one below. Compare the values of the components listed in the table with those in the preceding equations to see how the table was completed. Either procedure, the equations or a table, works well for adding vectors.

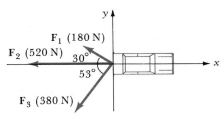

(a)

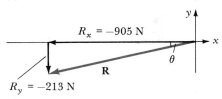

(b)

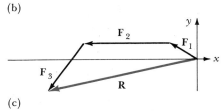

(c)

FIG. 1.15. (a) Three ropes pull on a car bumper with the forces shown. (b) The resultant **R** is constructed from its components R_x and R_y. (c) A graphical addition of the three forces.

*For purposes of comparison, a force of magnitude 380 N ($\mathbf{F}_3$ in Fig. 1.15a) equals a little more than the weight of a bag of cement (380 N = 85 lb).

Vector	x Component (N)	y Component (N)
F_1	-156	90
F_2	-520	0
F_3	-229	-303
R	-905 N	-213 N

Having found R_x and R_y, we construct **R** as shown in Fig. 1.15b. We see that **R** points into the third quadrant. The rough graphical addition of F_1, F_2, and F_3 shown in Fig. 1.15c confirms that **R** points into the third quadrant (a good check on our math). The magnitude of **R** and the exact angle that it makes with the x axis are calculated using Eqs. (1.7) and (1.8):

$$R = \sqrt{905^2 + 213^2} = 930 \text{ N},$$

$$\tan \theta = \frac{213 \text{ N}}{905 \text{ N}} = 0.235.$$

According to our table of tangents, a $13°$ angle has a tangent of 0.235. Since **R** points into the third quadrant, its direction must be $13°$ below the negative x axis and it has a magnitude of 930 N (210 lb).

We should note that this single resultant force **R** would have exactly the same effect on the car as the three forces shown in Fig. 1.15a. The resultant force equals the vector sum of the three forces both mathematically and in terms of its effect on the car. ∎

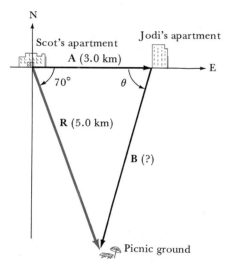

N
Scot's apartment
A (3.0 km)
Jodi's apartment
E
70°
θ
R (5.0 km)
B (?)
Picnic ground

FIG. 1.16. A + B = R or B = R − A.

EXAMPLE 1.5 The displacement from Scot's to Jodi's apartment is 3.0 km east. Scot tells Jodi that he wants to meet her at a picnic ground that is 5.0 km $70°$ south of east from his apartment. What is the displacement of the picnic ground from Jodi's apartment?

SOLUTION An important part of physics problem solving involves drawing pictures or diagrams of situations described in the problems. (We will discuss this more in the next chapter.) Such a drawing for this example appears in Fig. 1.16. Notice that Scot, to get to the picnic ground, could first go to Jodi's apartment (displacement **A**) and then from Jodi's apartment to the picnic ground (displacement **B**), or he could go directly to the picnic ground (displacement **R**). Since he starts and ends at the same place, his resultant displacement is the same for each route:

$$\mathbf{A} + \mathbf{B} = \mathbf{R},$$

or, rearranging,

$$\mathbf{B} = \mathbf{R} - \mathbf{A}.$$

Thus, we can find **B**, the displacement from Jodi's apartment to the picnic ground by subtracting **A** from **R**. We first determine the x and y components of **B** by subtracting the x and y components of **R** and **A**:

$$B_x = R_x - A_x = R \cos 70° - A \cos 0°$$
$$= (5.0 \text{ km})0.34 - (3.0 \text{ km})1.0 = -1.3 \text{ km},$$
$$B_y = R_y - A_y = -R \sin 70° - A \sin 0°$$
$$= -(5.0 \text{ km})0.94 - (3.0 \text{ km})0.00 = -4.7 \text{ km}.$$

Displacement **B** points toward the southwest since both the x and y components are negative.

The magnitude of **B** is determined using Eq. (1.7):

$$B = \sqrt{B_x^2 + B_y^2} = \sqrt{1.3^2 + 4.7^2} = 4.9 \text{ km}.$$

The angle that **B** makes below the x axis in the negative direction is given by Eq. (1.9):

$$\theta = \text{invtan} \left| \frac{R_y}{R_x} \right| = \text{invtan} \frac{4.7}{1.3} = 75°.$$

Thus, **B** points 75° below the x axis in the negative direction (toward the south-west) and has a magnitude of 4.9 km. ∎

Summary and Additional Readings

1. **Measurable quantities:** In physics we describe nature in terms of measurable quantities. Most quantities are given in a unit of measure based on a standard or a unit derived from standardized units. The SI metric system provides most of the standards for scientific measurement. There are two types of measurable quantities: scalars and vectors.

Scalars are quantities that have only a magnitude, including a unit of measure. The normal rules of algebraic addition, subtraction, multiplication, and division are used with scalars.

Vectors are quantities that have a magnitude and a direction. Special mathematical rules are needed to manipulate vectors.

2. **Vector addition and subtraction by the graphical technique:** To add vectors graphically, we join the vectors head to tail. The resultant vector goes from the tail of the first vector to the head of the last and is the sum of these vectors ($\mathbf{R} = \mathbf{A} + \mathbf{B} + \mathbf{C} + \cdots$). Vectors may be moved from one place to another for this type of addition, but their magnitudes and directions cannot be changed. Vectors are subtracted graphically by adding the negative of the vector to be subtracted; that is, $\mathbf{R} = \mathbf{A} + (-\mathbf{B}) = \mathbf{A} - \mathbf{B}$. A negative vector, for example $-\mathbf{B}$, is the vector with its direction reversed.

3. **Vector addition by components:** To add vectors by the component technique, we must first find the x and y **components** of each vector being added. They are given by the equations

$$F_x = \pm F \cos \theta, \qquad (1.3)$$
$$F_y = \pm F \sin \theta, \qquad (1.4)$$

where F is the magnitude of the vector and θ is the angle (90° or less) that **F** makes with either the positive or negative x axis. The signs of F_x and F_y depend on the direction of the projection of **F** on the x and y axes, respectively. We then calculate the x and y components of the resultant vector by adding the x and y components of each vector being added:

$$R_x = F_{1x} + F_{2x} + F_{3x} + \cdots, \qquad (1.5)$$
$$R_y = F_{1y} + F_{2y} + F_{3y} + \cdots. \qquad (1.6)$$

Having determined R_x and R_y, we calculate the magnitude and direction of the resultant vector using the equations

$$R = \sqrt{R_x^2 + R_y^2} \qquad (1.7)$$
$$\tan \theta = \left| \frac{R_y}{R_z} \right|, \quad \text{or} \quad \theta = \text{invtan} \left| \frac{R_y}{R_x} \right|, \quad \textbf{(1.8) or (1.9)}$$

where θ is the angle, 90° or less, that **R** makes with either the positive or negative x axis. The quadrant in which **R** points is determined from a rough drawing of **R** based on its component values.

Larry D. Kirkpatrick and Adele S. Pittendrigh, "A Writing Teacher in the Physics Classroom," *The Physics Teacher*, March (1984). This article provides excellent guidelines for writing answers to the qualitative questions at the end of each chapter.

Robert A. Nelson, "Foundations of the International System of Units (SI)," *The Physics Teacher*, p. 596, December (1981).

Banesh Hoffman, *About Vectors*, Prentice-Hall, Englewood Cliffs, N.J. (1966).

Questions

1. Many years ago in England, the standard of weight was based on the weight of 32 wheat kernels. In what ways does this seem like a good standard compared to our present one—a metal cylinder stored near Paris? In what ways does it seem like a bad standard?

2. Most quantities used in physics can be measured in units with well-defined standards. Why are there not similar standards for descriptive words used in everyday language such as *pretty, nice, lovely,* and *bad*?

3. Choose some word, such as mentioned in the previous question, and invent a standard for it. For example, develop a standard for the word *friendly*. Your standard for this or some other everyday word should allow us to use the word with a more precise meaning. We should, for instance, be able to use your standard to assign numerical values of friendliness for different people. Indicate the reasons why your definition may be deficient.

4. A medium-sized apple weighs 1 N. (a) Name five other objects that weigh about 1 N ($\pm$0.5 N). (b) Name several objects that you estimate weigh between 5 and 15 N; (c) between 50 and 150 N.

5. Which of the following quantities are scalars and which are vectors: (a) wind velocity, (b) population of a town, (c) force of a cable pulling an elevator, (d) weight of a book, (e) number of leaves on a tree, (f) age of the universe, (g) flow rate of blood in the circulatory system? Explain your choice in each case.

6. (a) Displacement **A** is 10 km west. Describe displacement $-$**A**. (b) What two conditions must be true for two vectors **C** and **D** to add to zero?

7. Can two forces of unequal magnitude add to zero? Explain.

8. Can a nonzero displacement vector pointing north and one of equal magnitude pointing east add to zero? Explain.

9. If $\mathbf{F}_1$ has a magnitude of 100 lb and $\mathbf{F}_2$ a magnitude of 200 lb, does $\mathbf{R} = \mathbf{F}_1 + \mathbf{F}_2$ necessarily have a magnitude of 300 lb? If not, construct a graphical example for which **R** does not equal 300 lb.

10. Can the magnitude of a vector ever be less than the value of one of its components? Explain.

Problems

1.5 Graphical Addition and Subtraction of Vectors

All problems in this section must be worked using graphical techniques.

1. To reach your first class walking on sidewalks, you must go 500 m west, 200 m south, and then 100 m east. Estimate your resultant displacement.

2. A mule is pulled by ropes with forces of 100 lb in a direction 30° north of west and 120 lb in a direction 37° south of west. Estimate the resultant force on the mule.

3. A Frisbee is thrown three times: 20 m in a direction 30° south of east, 30 m toward the north, and then 40 m in a direction 15° south of west. Estimate its resultant displacement.

4. An airplane flies 1000 km in a direction 25° west of north from Miami to Atlanta and then 1200 km in a direction 5° south of west to Dallas–Fort Worth. Estimate the displacement from Miami to Dallas–Fort Worth.

5. On January 16, 1954, a South African sharp-nosed frog named Leaping Lena (later found to be a male) set the world's record distance for three consecutive jumps: 4.0 m 30° north of east, 3.2 m 15° south of east, and 3.6 m 25° south of east. Estimate his record displacement from the starting position after the three jumps (magnitude and direction).

6. A hot air balloon drifts 12 km in a direction 70° south of east. An abrupt change in air movement then carries the balloon 8.0 km in a direction 40° south of west. Estimate the balloon's resultant displacement.

7. Determine the sum of forces $\mathbf{F}_1$ and $\mathbf{F}_2$ shown in Fig. 1.17.

8. Determine the sum of forces $\mathbf{F}_1$, $\mathbf{F}_3$, and $\mathbf{F}_4$ shown in Fig. 1.17.

■ 9. Determine (a) $\mathbf{F}_1 + \mathbf{F}_2$; (b) $\mathbf{F}_2 + \mathbf{F}_1$; (c) $\mathbf{F}_1 - \mathbf{F}_2$; (d) $\mathbf{F}_2 - \mathbf{F}_1$ for the forces shown in Fig. 1.17.

■ 10. Determine (a) $\mathbf{A} + \mathbf{C}$; (b) $\mathbf{A} - \mathbf{C}$ for the displacements shown in Fig. 1.18.

■ 11. Use graphical addition to show that $\mathbf{A} + \mathbf{B} + \mathbf{D} = \mathbf{B} + \mathbf{A} + \mathbf{D} = \mathbf{D} + \mathbf{B} + \mathbf{A}$ for the displacement vectors in Fig. 1.18.

■ 12. Determine the magnitude and direction of a displacement **E** that can be added to **D** in Fig. 1.18 to produce a resultant displacement 25 km west.

■ 13. A man hikes 10 km south and then an unknown distance west. He is now 14 km as the crow flies from where he started. Determine the distance that he walked toward the west.

■ 14. A force of magnitude 100 N is directed west, and a second force of magnitude 200 N is directed 30° east of north (or 60° north of east). Find the magnitude and direction of a third force that, when added to the first two, gives a resultant whose magnitude is zero.

■ 15. Repeat Problem 14 for the case where the resultant force is not zero but instead has a magnitude of 150 N and points 30° north of east.

1.6 Components of a Vector

16. Calculate the x and y components of displacements **A** and **C** shown in Fig. 1.18.

17. Calculate the x and y components of displacements **B** and **D** shown in Fig. 1.18.

18. Calculate the x and y components of $\mathbf{F}_1$ and $\mathbf{F}_3$ shown in Fig. 1.17.

19. Calculate the x and y components of $\mathbf{F}_2$ and $\mathbf{F}_4$ shown in Fig. 1.17.

20. Determine the components of each force acting on the mule in Problem 2.

21. Determine the components of each displacement made by Leaping Lena as described in Problem 5.

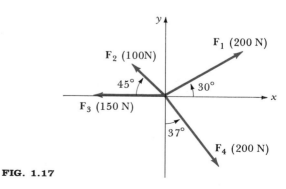

FIG. 1.17

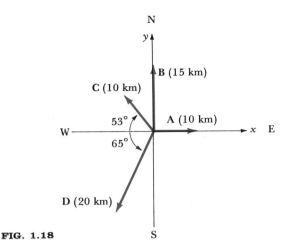

FIG. 1.18

■ 22. A woman has hiked 12 km in a straight line and is 7 km west of where she started. Determine two possible directions in which she hiked and her distance north or south of her starting position for each case.

■ 23. A force F of magnitude 500 N has a y component of -200 N. Determine two possible directions for F and the corresponding x component for each situation.

1.7 Vector Addition by Components

All vector addition in this section must be done by the component technique.

24. Find the magnitudes and directions of forces whose x and y components are (a) 10 N and -10 N; (b) -20 N and -30 N; (c) -5 N and 10 N; (d) 20 N and -10 N.

25. Find the magnitudes and directions of displacements whose x and y components are (a) -7 km and $+4$ km; (b) $+20$ km and -30 km; (c) 0 and $+10$ km; (d) -20 km and 0.

■ 26. Calculate the resultant force on the mule described in Problem 2.

■ 27. Calculate your resultant displacement in Problem 1.

■ 28. Calculate the resultant displacement of the balloon described in Problem 6.

■ 29. Calculate the resultant displacement of the Frisbee described in Problem 3.

■ 30. Calculate the resultant displacement of Leaping Lena in Problem 5.

■ 31. Add forces F_1 and F_2 shown in Fig. 1.17.

■ 32. Add forces F_1, F_3, and F_4 shown in Fig. 1.17.

■ 33. Add displacements A, B, and C shown in Fig. 1.18.

■ 34. Use the component method to calculate (a) B + C; (b) B − C; (c) C − B for displacements B and C shown in Fig. 1.18. Check your results using graphical addition or subtraction of vectors.

■ 35. For forces F_3 and F_4 shown in Fig. 1.17, use the component method to calculate (a) $F_3 + F_4$; (b) $F_3 − F_4$; (c) $F_4 − F_3$. Check your results using graphical addition or subtraction of vectors.

■ 36. Two displacement vectors when added produce a resultant vector of magnitude 100 km pointing 37° south of east. One displacement vector is 200 km in a direction 60° north of east. Determine the magnitude and direction of the other displacement vector.

■ 37. A group of backpackers, after three days hiking, are 30 km north of their starting position. On the first day they hiked 20 km east. On the second day they hiked 30 km in a direction 53° north of west. Use the component method to calculate their displacement on the third day.

■ 38. Three forces when added produce a resultant force that is 100 N pointing 63° above the positive x axis. If one of the forces is 50 N in the positive y direction and another is 100 N in a direction 45° below the positive x axis, what are the magnitude and direction of the third force?

■■ 39. Two friends leave the same point and hike in different directions. The displacement components for one hiker are $A_x = 15$ km and $A_y = -4$ km, and for the other hiker $B_x = 2$ km and $B_y = 8$ km. Determine the smallest angle between their paths.

■■ 40. Suppose that hiker A in Problem 39 remains at his or her present location while B takes the shortest route to A. In what direction relative to the x axis and how far must B travel?

CHAPTER 2

Describing Motion: Kinematics

Our first use of physical quantities in the analysis of natural phenomena involves a subject called **kinematics,** the quantitative description of an object's motion. The usual intent of a kinematic analysis is to determine from known information some unknown property concerning that motion. For example, kinematics can be used to estimate the time needed for a rapid-transit train to travel between stations, the time for a skydiver to fall a certain distance, or the deceleration of a person stopped abruptly during an automobile collision or stopped while landing on a cushion after a pole vault. These calculations rely on careful definitions of quantities such as velocity and acceleration. We will formulate equations that relate an object's acceleration to its changing position and velocity at different times. These kinematical equations appear frequently in our future discussions of physics.

2.1 Coordinate Systems

To describe an object's motion, we must first construct a coordinate system consisting of one or more axes used to indicate an object's position, velocity, and acceleration at different times. Each axis must have (1) a direction called positive, the opposite direction being negative, and (2) an origin, the zero point, that serves as a reference for indicating distances along each axis.

The way in which an object's motion is described depends very much on the choice of axes for the coordinate system. Suppose, for example, that we wish to describe the position and velocity of a person who has jumped from the window of a burning building into a safety net below. We need only one axis, oriented in the vertical direction, to describe the motion. We call this **one-dimensional** or **linear motion** since the person moves along a line. If we choose the ground as the origin and the positive direction pointing up, as in Fig. 2.1a, then the person's position is always positive when above the ground, and the person's velocity, which is downward, is negative because it points in the negative direction. We could equally well choose the origin of the vertical axis to be at the window, with the positive direction down, as in Fig. 2.1b. In that case, the person's position and velocity during the fall are both positive.

How can we use a time lapse photograph such as this to determine an object's changing velocity and acceleration? Read on. (The flashes for this photograph occurred every 0.01 s.) (Courtesy of Harold Edgerton)

The fact that the velocity was negative in one coordinate system and positive in the other does not mean that one description is wrong. Each description is correct relative to the coordinate axis used. A coordinate system is in some ways similar to a language; the words used to describe an object, such as a rose, differ from one language to another. This does not mean that the rose is different or that one description is wrong. All we ask is that a person uses the language of choice correctly. The same is true for coordinate systems in kinematics. Once a particular coordinate system is chosen, the description must be consistent with that system.

2.2 Displacement and Distance

Having specified a coordinate system, we can now start our quantitative description of motion. The description involves careful definitions of vector quantities such as displacement, velocity, and acceleration. Qualitatively, these three quantities are part of our everyday experience and should be quite familiar.

Displacement, **d,** a vector quantity, indicates the separation of one point from another and the direction of that separation. Suppose that at time t_1 a boat is northeast of Bermuda at position r_1, as shown in Fig. 2.2. Two hours later at time t_2, the boat is directly north of Bermuda at position r_2. The boat's displacement **d** during that time is represented by a vector pointing from its initial position to its final position.

Using the graphical vector addition technique developed in Chapter 1, we see from Fig. 2.2 that $r_1 + d = r_2$, or, rearranging,

$$d = r_2 - r_1. \tag{2.1}$$

The boat's displacement is its final position minus its initial position.

Displacement is defined as a vector that points from an object's initial position to its final position and whose magnitude equals the distance separating the points. Displacement can be calculated by subtracting a vector from the origin to the object's final position and a vector from the origin to its final position:

$$\text{Displacement} = \text{Final position} - \text{Initial position.} \tag{2.2}$$

The units of displacement are any units of length—meter, kilometer, mile, foot, and so on.

For linear motion along a single axis, such as the x axis, we can represent an object's initial position at time t_1 by the coordinate x_1 and its final position at time t_2 by the coordinate x_2. The object's displacement between those times is then

$$d = x_2 - x_1. \tag{2.3}$$

We need not use vector symbols for displacements in one dimension. Positive and negative signs suffice. In the last equation, for example, if x_2 is greater than x_1, then the object's displacement is in the positive direction and d is positive. If x_2 is less than x_1, the displacement is in the negative direction and d is negative.

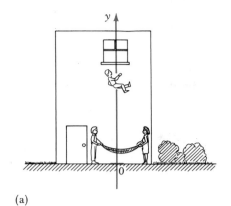

(a)

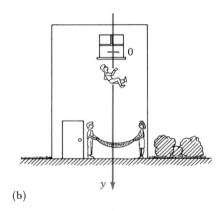

(b)

FIG. 2.1. Different coordinate axes used to describe the same one-dimensional, or linear, motion. (a) The origin is at the ground and the positive direction points upward. (b) The origin is above the ground and the positive direction points downward.

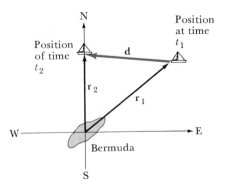

FIG. 2.2. A boat's displacement between times t_1 and t_2 is $d = r_2 - r_1$.

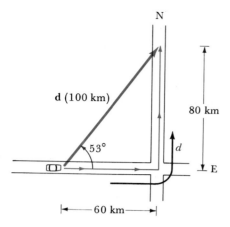

FIG. 2.3

2.3 Speed and Velocity

Speed and velocity are similar quantities in that they both describe how fast an object moves. There is, however, a significant difference. Velocity is a vector quantity that specifies both how fast and in what direction an object moves, whereas speed specifies only how fast the object moves—not its direction.

Average Speed

The difference between speed and velocity becomes apparent if we examine a two-hour trip taken by the car shown in Fig. 2.3. The car travels 60 km east and then 80 km north, or a total distance of 140 km in a time of two hours. The car's average speed is the distance traveled divided by the elapsed time: 140 km/ 2 h = 70 km/h. The speed does not reflect in any way the direction of motion.

Average speed, $\bar{v}$, is defined as the distance traveled divided by the time required to travel that distance:

$$\bar{v} = \frac{\text{Distance traveled}}{\text{Elapsed time}} = \frac{d}{t}. \qquad (2.4)$$

A line over a symbol, as in $\bar{v}$, indicates the average value of the quantity under the line. Units of speed include km/h, m/s, mi/h, or any unit of distance divided by time.

EXAMPLE 2.1 In 1926, Johnny Weissmuller set the men's world 400-m swimming record in 4 min, 57.0 s. In 1966, Martha Randall set the women's record in 4 min, 38.0 s. By how many meters would Martha have beaten Johnny if they had raced each other?

SOLUTION When thinking of a solution to this problem, try to visualize the situation at the end of the race. Martha finishes the race 19.0 s before Johnny. If we know Johnny's average speed, we can rearrange Eq. (2.4) to determine the distance he travels in the last 19.0 s, that is, his distance from the finish line when Martha completes the race. His average speed for the whole race is

$$\bar{v} = \frac{d}{t} = \frac{400 \text{ m}}{(4 \text{ min } 57 \text{ s})} = \frac{400 \text{ m}}{297 \text{ s}} = 1.35 \text{ m/s}.$$

In the last 19.0 s while traveling at this speed, he moves a distance

$$d = \bar{v}t = \left(1.35 \frac{\text{m}}{\text{s}}\right)(19.0 \text{ s}) = 25.6 \text{ m}.$$

Thus, he would have been beaten by 25.6 m. ∎

Average Velocity

The car's average velocity during the trip shown in Fig. 2.3 is quite different from its average speed. When calculating velocity, we are interested in an object's displacement—its final position minus its initial position. The displacement **d** of the car shown in Fig. 2.3 is a vector of magnitude 100 km that points 53° north of east—from the car's initial position to its final position. The *magnitude* of the car's average velocity during the two-hour trip is the magnitude of its displacement (100 km) divided by the time needed for that displacement (2 h). Thus, the magnitude of the velocity is 100 km/2 h = 50 km/h. The velocity points in the same *direction* as the displacement—53° north of east. In calculating average velocity, we give the car no credit for the distance it traveled, only for its final displacement from its starting position.

The **average velocity** $\bar{v}$ of an object during some time period is the object's displacement (a vector) during that time divided by the time:

$$\bar{v} = \text{Average velocity} = \frac{\text{Displacement}}{\text{Elapsed time}} \qquad (2.5)$$

The units of velocity are the same as the units of speed—m/s, km/h, mi/h, and so forth.

For two- or three-dimensional motion, Eq. (2.5) can be written as

$$\mathbf{v} = \frac{\text{Displacement in two or three dimensions}}{\text{Elapsed time}} = \frac{\mathbf{r}_2 - \mathbf{r}_1}{t_2 - t_1},$$

where $\mathbf{r}_1$ is the object's position at time t_1 and $\mathbf{r}_2$ is its position at time t_2 (its displacement during that time interval is $\mathbf{r}_2 - \mathbf{r}_1$, as illustrated in Fig. 2.2). For one-dimensional motion, we need only one axis (for example, an x axis or a y axis) to indicate an object's position. Equation (2.5) then becomes

$$v = \frac{\text{Displacement along one axis}}{\text{Elapsed time}} = \frac{x_2 - x_1}{t_2 - t_1},$$

where x_1 is the object's position along the axis at time t_1 and x_2 is its position at time t_2.

EXAMPLE 2.2 An automobile moves along a straight track beside which are markers indicating the car's position at different times. If the car is at position $x_1 = 51.2$ m at time $t_1 = 5.3$ s and is at position $x_2 = 49.8$ m at time 5.4 s, what is the car's velocity during that time interval?

SOLUTION For this linear motion problem, the change in position of the car is $x_2 - x_1 = 49.8$ m $- 51.2$ m $= -1.4$ m. The negative sign means that the car has moved to the left, in the negative x direction. The time interval during which this displacement occurred is $t_2 - t_1 = 5.4$ s $- 5.3$ s $= 0.1$ s. Hence, the car's average velocity during that time is

$$v = \frac{x_2 - x_1}{t_2 - t_1} = \frac{-1.4 \text{ m}}{0.1 \text{ s}} = -14 \text{ m/s}.$$

The velocity is negative because the car is moving in the negative x direction.
    Vector symbols are not needed for linear motion. The sign (+ or −) indicates the direction of motion along the axis. ■

TABLE 2.1 Position and Speed of a Drag Racer at Different Times

t (s)	x (m)	Δx (m)	$\Delta x/\Delta t = \bar{v}$ (m/s)
0.0	0		
		3	3/0.5 = 6
0.5	3		
		10	10/0.5 = 20
1.0	13		
		16	32
1.5	29		
		22	44
2.0	51		
		27	54
2.5	78		
		34	68
3.0	112		
		41	82
3.5	153		
		47	94
4.0	200		

Instantaneous Speed and Velocity

We found earlier that the average speed of a car that traveled 140 km in a time of 2 h was 70 km/h. It is unlikely that the car traveled at a constant 70 km/h during the whole trip because most driving involves many starts and stops. The car probably traveled slower than 70 km/h for part of the trip and faster during other portions. The car may even have moved backward for part of the trip. To have complete knowledge of how the trip was made, we need to know the car's velocity at each instant of time during the trip.

Instantaneous velocity indicates how fast an object moves at each instant of time and the direction of that motion. **Instantaneous speed** indicates only how fast an object is moving at each instant of time. (The speedometer of an automobile indicates its instantaneous speed.) For most of this chapter, we will consider only one-dimensional motion. In this case we need not use vector symbols for velocity. If an object moves in the positive direction along the axis, its velocity is positive; if it moves in the negative direction, its velocity is negative.

To develop a more precise definition for instantaneous velocity and speed in one dimension, consider the motion of a drag racer, shown in Fig. 2.4. The car starts from rest and, when given a signal, accelerates in the positive direction along a 200-m track, arriving at the finish line 4.0 s later. Hence, the average velocity of the car during the race is $\bar{v} = +200$ m/4.0 s $= +50.0$ m/s.

To obtain a more complete profile of the changing velocity of the car, suppose we photograph the racer at 0.5-s time intervals from the start of the race to the finish. Distance markers beside the track will show us where the car is located at the end of each successive time interval (Fig. 2.4). Table 2.1 gives values of x for different times. Notice that at time 0.0 s the race car is at the starting line at position 0 m. During the next 0.5 s, the car moves to position $x = 3$ m. Its average velocity during that time is its *change in position* Δx, divided by the change in time Δt required for the position change:*

$$\bar{v} = \frac{\Delta x}{\Delta t} = \frac{3\text{ m} - 0\text{ m}}{0.5\text{ s} - 0\text{ s}} = +6\text{ m/s}.$$

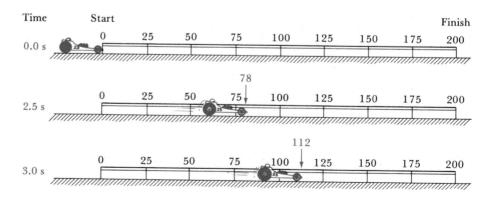

FIG. 2.4. The position of a drag racer at different times is determined by photographs showing the racer's position relative to markers in the background.

*The symbol Δ, delta, means a change in the quantity following the Δ. Thus, Δx (read "delta x") is a change in the position coordinate x; Δt (read "delta t") is a change in time; Δx and Δt are single symbols, not the product of Δ times x or Δ times t. The units of Δx are the units of x (for example, meters) and the units of Δt are the units of t.

During the next 0.5 s, from $t = 0.5$ s to $t = 1.0$ s, the car moves from position 3 m to 13 m. Thus, its average velocity during the second time interval is

$$\bar{v} = \frac{\Delta x}{\Delta t} = \frac{13 \text{ m} - 3 \text{ m}}{1.0 \text{ s} - 0.5 \text{ s}} = \frac{+10 \text{ m}}{0.5 \text{ s}} = +20 \text{ m/s}.$$

A list of the calculated values of average velocity for each successive 0.5-s time interval appears in the last column of Table 2.1. The velocities are positive because the car moves in the positive x direction.

Since the car's velocity is continually increasing, it almost certainly moves faster at the end of each 0.5-s time interval than at the beginning. To calculate its instantaneous velocity, we must divide the trip into much smaller time intervals. We might require, for example, that pictures be taken so often that the change in velocity between successive time intervals can be no more than 0.1 percent. If this condition is satisfied, we consider the velocities we calculate from these frequent pictures to be instantaneous velocities. In practice, such accuracy is seldom necessary. We see that instantaneous velocity can be thought of as the average velocity for a very short time interval. For one-dimensional motion, instantaneous velocity and speed are defined as follows:

The **instantaneous velocity** of an object moving along the x axis is its change in position Δx divided by the very short time Δt required for this change:

$$\text{Instantaneous one-dimensional velocity} = v = \frac{\Delta x}{\Delta t}. \qquad \textbf{(2.6)}$$

If the object moves in the positive x direction, Δx is positive, as is the velocity. If the object moves in the negative x direction, Δx and v are both negative. The **instantaneous speed** of the object is just the magnitude of v and is always a positive number with the appropriate unit of measure.*

For the more general case where the object's motion is not restricted to one dimension, its instantaneous velocity is defined as its displacement $\Delta \mathbf{d}$ (a vector) divided by the very short time Δt needed for this displacement:

$$\text{Instantaneous velocity} = \mathbf{v} = \frac{\Delta \mathbf{d}}{\Delta t}.$$

Suppose that a car undergoes a displacement of 2.3 m in a direction 30° north of east during a 0.1-s time interval; the car's velocity is 2.3 m/0.1 s = 23 m/s and is directed 30° north of east. The car's speed is simply the magnitude of the velocity or, in this case, 23 m/s.

Hereafter, we use the word *speed* to mean "instantaneous speed" and *velocity* to mean "instantaneous velocity."

---

*The formal definition of one-dimensional instantaneous velocity relies on the use of calculus and is defined as

$$v = \lim_{\Delta t \to 0} \frac{\Delta x}{\Delta t} \equiv \frac{dx}{dt}.$$

The instantaneous velocity equals the limiting value of the change in x divided by the change in t as the time interval approaches zero. As Δt approaches zero, so does Δx. However, the ratio of the two small numbers, which is the velocity, can be quite large, depending on how fast the object moves. This limiting value of $\Delta x / \Delta t$ is called the **derivative** of x with respect to t and is written as $v = dx/dt$.

2.4 Unit Conversions

Occasionally, we need to convert a quantity from one system of units to another. Suppose that we wish to convert a speed of 55 mi/h to a speed in units of m/s. The conversion table inside the front cover of the book indicates that 1 mi = 1609 m. Also, we know that 3600 s = 1 h. We rewrite these equalities as 1 = (1609 m/1 mi) and 1 = (1 h/3600 s). Now to convert 55 mi/h to m/s, we multiply by 1 twice, substituting the conversion factors that equal 1, as follows:

$$55\,\frac{\text{mi}}{\text{h}} = \left(55\,\frac{\text{mi}}{\text{h}}\right)(1)(1) = \left(55\,\frac{\cancel{\text{mi}}}{\cancel{\text{h}}}\right)\left(\frac{1609\text{ m}}{1\,\cancel{\text{mi}}}\right)\left(\frac{1\,\cancel{\text{h}}}{3600\text{ s}}\right) = 25\text{ m/s}.$$

The mile in the numerator cancels the mile in the denominator, just as algebraic quantities cancel. The hours also cancel.

When multiplying by the conversion factor that equals 1, you will have a choice of placing one quantity in the numerator and the other in the denominator. The arrangement of the two equal quantities must be made so that the desired units cancel. For example, to convert the miles in the numerator of 55 mi/h to meters, we place the miles of the conversion factor (1609 m/1 mi) in the denominator (the miles then cancel) leaving us with the new unit, meters, in the numerator. Similarly, the hour in the denominator of 55 mi/h was cancelled by the hour in the numerator of the time conversion factor (1 h/3600 s). If you try the other arrangement, you will see that the units do not cancel.

If a unit to be converted appears to the second or a higher power, then each power must be converted. Suppose that the cross-sectional area of the bone in your ankle is 0.50 in^2, and you need to know its area in units of cm^2. According to the conversion table inside the front cover, 1 in = 2.54 cm, or 1 = (2.54 cm/ 1 in). Each inch of the 0.50-in^2 area must be converted, so we multiply by 1 twice:

$$0.50\text{ in}^2 = (0.50\text{ in}^2)(1)(1) = (0.50\text{ in}^2)\left(\frac{2.54\text{ cm}}{1\text{ in}}\right)\left(\frac{2.54\text{ cm}}{1\text{ in}}\right) = 3.2\text{ cm}^2.$$

2.5 Acceleration

Whenever the velocity of an object changes, we say that the object is **accelerating.** A car waiting at a red light accelerates after the stoplight changes to green. Its velocity increases from zero as it moves through the intersection. A car also accelerates when it comes to a stop at a stop sign. We often call this a **deceleration**—a decrease in the speed of an object. The words *change in velocity* are especially important in understanding the meaning of acceleration. An object moving at a constant high speed in a straight line is not accelerating because its velocity is not changing.

Acceleration can be defined more precisely with reference to Fig. 2.5. Suppose that at time t_1 an object is moving with velocity $\mathbf{v}_1$ and at a later time t_2 its velocity has changed to $\mathbf{v}_2$.

The **average acceleration,** $\overline{\mathbf{a}}$, of the object is its change in velocity divided by the time required for that change:

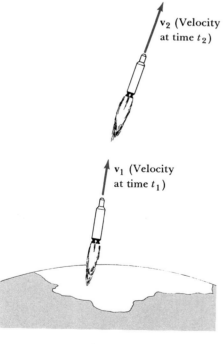

$\mathbf{v}_2$ (Velocity at time t_2)

$\mathbf{v}_1$ (Velocity at time t_1)

FIG. 2.5. The acceleration of a rocket during time $t_2 - t_1$ is the ratio of its change in velocity $(\mathbf{v}_2 - \mathbf{v}_1)$ and its change in time $(t_2 - t_1)$.

$$\overline{\mathbf{a}} = \frac{\text{Change in velocity}}{\text{Elapsed time}} = \frac{\mathbf{v}_2 - \mathbf{v}_1}{t_2 - t_1}. \qquad (2.7)$$

The units of acceleration are the units of velocity (for example, m/s or ft/s) divided by the unit of time (s). Thus, the metric unit for acceleration is $(m/s)/s = m/s^2 = m \cdot s^{-2}$ (ft/s² in the English system), that is, distance divided by time squared.

To get a better sense of the units of acceleration, you might think of an acceleration of 1 m/s² as being a *change* in velocity of 1 m/s in a time of 1 s, a *change* in velocity of 2 m/s in 2 s, or a *change* in velocity of 10 m/s in 10 s. In each case, the velocity changed on the average by 1 m/s during each second and the average acceleration was the same—(1 m/s)/1 s = (2 m/s)/2 s = (10 m/s)/10 s = 1 m/s².

Three factors affect acceleration: (1) the change in the magnitude of the object's velocity—that is, its change in speed; (2) the change in the direction of the object's velocity; and (3) the time required for these changes. The second factor is considered in Chapter 5 when we discuss circular motion. The following examples involving motion in one direction illustrate the first and third factors. For one-dimensional motion we do not need to use vector symbols if we indicate directions by positive and negative signs. A positive velocity indicates motion in the positive x direction; a negative velocity, motion in the negative x direction.

EXAMPLE 2.3 An automobile travels east in the positive x direction at $+20$ m/s and needs to pass a truck (Fig. 2.6). The automobile's velocity is increased to $+25$ m/s in 2 s; it still moves toward the east. Calculate its acceleration.

SOLUTION Since the motion is in the positive x direction, both the initial and final velocities are positive. The change in velocity is $v_2 - v_1 = (+25 \text{ m/s}) - (+20 \text{ m/s}) = +5$ m/s. This change occurs in time $t_2 - t_1 = 2$ s. The average acceleration is

$$\bar{a} = \frac{v_2 - v_1}{t_2 - t_1} = \frac{+5 \text{ m/s}}{2 \text{ s}} = +2.5 \text{ m/s}^2. \qquad \blacksquare$$

EXAMPLE 2.4 A car traveling east in the positive x direction at $+20$ m/s comes up behind a large truck and is unable to pass (Fig. 2.7). The car slows to $+15$ m/s in a time of 2 s. Calculate the car's acceleration.

SOLUTION The change in the car's velocity is $v_2 - v_1 = (+15 \text{ m/s}) - (+20 \text{ m/s}) = -5$ m/s. Note that the change in velocity is negative because the car's velocity in the positive x direction decreased by 5 m/s. This decrease occurred in time $t_2 - t_1 = 2$ s. The average acceleration is

$$\bar{a} = \frac{v_2 - v_1}{t_2 - t_1} = \frac{-5 \text{ m/s}}{2 \text{ s}} = -2.5 \text{ m/s}^2.$$

The acceleration is negative because the car's velocity along the positive x direction decreased. It lost 2.5 m/s of speed in this direction each second. $\blacksquare$

Acceleration also depends on the time required for the change in velocity. If the change in velocity had occurred in one second instead of two, the car's acceleration would have been twice as great.

Time (t_1): 0
Velocity (v_1): 20 m/s, east

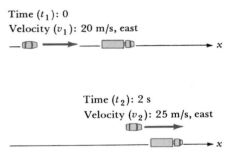

FIG. 2.6. The acceleration of the car is positive since its velocity in the positive x direction increases.

Time (t_1): 0
Velocity (v_1): 20 m/s, east

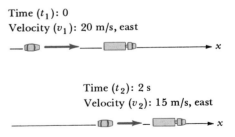

FIG. 2.7. The acceleration of the car is negative since its velocity in the positive x direction decreases.

As with speed and velocity, we are often interested in an object's instantaneous acceleration.

An object's **instantaneous acceleration a** is defined as its change in velocity $\Delta\mathbf{v}$ divided by the very short time Δt needed for that change:

$$\mathbf{a} = \frac{\Delta\mathbf{v}}{\Delta t}.$$

(2.8)

The time interval Δt should be so short that the acceleration varies little from one time interval to the next.*

2.6 Linear Motion with Constant Acceleration

The techniques described in previous sections for calculating the velocity and acceleration of an object require that we know an object's position at many different times. In this section we derive four simple equations that relate the position, velocity, and acceleration of an object at different times. With one of these equations we can calculate an object's acceleration if its position and velocity at two different times are known. By using other equations, we can calculate an object's future position and velocity if its acceleration is known.

Although these simple equations are easier to use than the techniques developed in the last sections, they have one major disadvantage: *the equations apply only to linear motion for objects whose acceleration is constant.* Few situations in life involve constant acceleration, but many come close (for example, the acceleration of a falling object or of a car skidding to a stop). These equations are often used in such situations even though the acceleration is not strictly constant.

The four equations are derived using the definitions of average velocity and acceleration. The object whose motion we consider is initially (that is, at time zero) at position x_0 moving at velocity v_0. The object experiences constant acceleration (or deceleration), a, and at some later time t is at position x moving at velocity v. Because the derivations are done for linear motion, we omit the vector signs and keep track of directions using positive and negative signs only.

EQUATION 1 Since the object's velocity has changed by an amount $v - v_0$ during a time $t - 0$, its acceleration during that time, according to Eq. (2.7), is

$$a = \frac{v - v_0}{t - 0}.$$

Rearranging this equation to solve for v gives us our first equation of linear motion for constant acceleration:

$$v = v_0 + at.$$

(2.9)

---

*One-dimensional instantaneous acceleration is defined using calculus as

$$a = \lim_{\Delta t \to 0} \frac{\Delta v}{\Delta t} \equiv \frac{dv}{dt}.$$

The instantaneous acceleration equals the derivative of v with respect to t.

EQUATION 2 Our next equation is derived by equating two different expressions for the average velocity of an object that starts at time zero with velocity v_0 and ends at time t with velocity v. To develop the first expression, we plot velocity on the ordinate of a graph versus time on the abscissa, as shown in Fig. 2.8. The straight line represents the equation $v = v_0 + at$. At $t = 0$, the velocity equals v_0. Since the acceleration is constant, the velocity increases in direct proportion to t and is represented by the straight line shown in the graph. The average value of the velocity during that time interval is just the arithmetic average of the initial velocity and the final velocity:

$$\bar{v} = \frac{v + v_0}{2}.$$

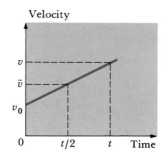

FIG. 2.8. A velocity-versus-time graph is a straight line for constant acceleration.

The average velocity also equals the displacement of the object divided by the time required for the displacement. If the object starts at position x_0 at time zero and ends at position x at time t, its displacement is $x - x_0$. Thus, its average velocity is

$$\bar{v} = \frac{\text{Displacement}}{\text{Time}} = \frac{x - x_0}{t}.$$

We can now equate the two expressions for average velocity:

$$\frac{x - x_0}{t} = \frac{v + v_0}{2}.$$

Rearranging the above produces our second equation of motion for constant acceleration:

$$x - x_0 = \left(\frac{v + v_0}{2}\right)t. \tag{2.10}$$

EQUATION 3 The third equation is obtained by substituting $v = v_0 + at$ [Eq. (2.9)] for v in Eq. (2.10):

$$x - x_0 = \left(\frac{v + v_0}{2}\right)t = \left[\frac{(v_0 + at) + v_0}{2}\right]t.$$

When rearranged, we have our third equation:

$$x - x_0 = v_0 t + \frac{1}{2}at^2. \tag{2.11}$$

EQUATION 4 The last equation is obtained by solving Eq. (2.9) for t; that is, $t = (v - v_0)/a$. This expression for t is then inserted into Eq. (2.10):

$$x - x_0 = \left(\frac{v + v_0}{2}\right)t = \left(\frac{v + v_0}{2}\right)\left(\frac{v - v_0}{a}\right) = \frac{v^2 - v_0^2}{2a}.$$

Rearranging, we obtain the fourth equation of motion:

$$2a(x - x_0) = v^2 - v_0^2. \tag{2.12}$$

In summary, then, we have four equations of motion for constant acceleration that relate the object's acceleration a, initial position x_0 and initial velocity v_0 at time zero to the object's final position x and final velocity v at time t:

EQUATIONS FOR LINEAR MOTION WITH CONSTANT ACCELERATION

$$v = v_0 + at, \tag{2.9}$$

$$x - x_0 = \left(\frac{v + v_0}{2}\right)t, \tag{2.10}$$

$$x - x_0 = v_0 t + \frac{1}{2}at^2, \tag{2.11}$$

$$2a(x - x_0) = v^2 - v_0^2. \tag{2.12}$$

How do we use these equations to solve problems in kinematics? We outline next a useful problem-solving procedure and illustrate its use with a worked example.

Solving Problems in Kinematics

This kinematics problem-solving procedure involves six steps. The first two steps and the last step are especially important because they apply to most problem solving in physics. Practice using all the steps now, and you will develop a habit that will help you throughout your study of physics.

Each step of the procedure is illustrated for the following problem: Determine the average acceleration of Shirley Muldowney's 1979 record-setting drag racecar. The initial velocity was zero, and the final velocity, after the car had traveled 402 m (one-quarter mile), was 114 m/s (256 mi/h).

1. The first step in solving most physics problems involves drawing a picture or diagram of the situation described in the problem. Include in the diagram (a) a drawing that represents your interpretation of what is happening, (b) a coordinate axis or axes, (c) the values of known quantities represented in terms of appropriate symbols, and (d) a symbol for the unknown quantity you wish to determine. The picture puts all the problem information in front of you in an easily accessible form and also gives you a better intuitive grasp of the problem.

   The diagram for Shirley Muldowney's racecar is shown in the margin.

2. Divide and conquer. Some of the more complex and interesting problems of real life must first be divided into parts. Each part can often be solved with relative ease even though the original problem might have seemed impossibly complicated.

   The problem illustrated in the figure for Step 1 does not need to be broken in parts. However, we use the "divide and conquer" procedure later in Examples 2.7 and 2.8.

3. Find one of the equations listed earlier in which the only unknown quantity in the equation is the one whose value you wish to calculate.

   For the problem illustrated here, the only unknown in Eq. (2.12) is acceleration—$(x - x_0)$, v, and v_0 are known.

   $$2a(x - x_0) = v^2 - v_0^2$$

4. Rearrange the equation so that the unknown appears alone on the left and

Shirley Muldowney holds the second fastest time of 255.58 mi/h in drag racing history which she earned in the 1979 Winter Nationals in Pomona, CA. She is the only person to have gone over 250 mi/h more than twice. (Courtesy of General Motors, Corp.)

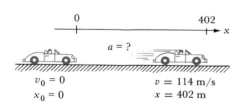

$v_0 = 0$
$x_0 = 0$
$v = 114$ m/s
$x = 402$ m

the known quantities are on the right. You have much less chance of error if you rearrange the equation before you substitute known quantities.

$$a = \frac{v^2 - v_0^2}{2(x - x_0)}$$

5. Substitute the values of the known quantities and calculate the value of the unknown, including its units.

$$a = \frac{(114 \text{ m/s})^2 - 0^2}{2(402 \text{ m} - 0)}$$
$$= +16.2 \text{ m/s}^2$$

6. Check your work. Does your answer have the correct units? If not, you may have made a mistake in your algebraic manipulations. Does the magnitude of the answer seem reasonable?

For this problem, we found that the answer had units of m/s^2, the correct units for acceleration. The drag racer's average acceleration was 16.2 m/s^2—the velocity increased by 16.2 m/s (about 36 mi/h) during each second. This seems reasonable—a drag racer's velocity should be able to increase from zero to 36 mi/h in one second or from 36 to 72 mi/h in the next second. However, if your answer had been 1000 m/s^2, you had better check your work—the car's velocity could certainly not increase by 1000 m/s (over 2000 mi/h) in one second!

Now let's try these procedures on some problems.

EXAMPLE 2.5   The driver of a car traveling at 25 m/s (55 mi/h) needs to quickly reduce the car's speed to 16 m/s (35 mi/h) to round a curve in the highway. If the car's deceleration has a magnitude of 6.0 m/s^2, how far does it travel while reducing its speed?

SOLUTION   The situation described in the problem is pictured in Fig. 2.9. Note that the acceleration has a negative sign because the car's velocity in the positive direction decreases. Equation (2.12) can be used to calculate the car's displacement $x - x_0$ while its velocity changes because all other quantities (a, v_0, and v) in that equation are known. After rearranging, we find the car's displacement:

$$x - x_0 = \frac{v^2 - v_0^2}{2a} = \frac{(16 \text{ m/s})^2 - (25 \text{ m/s})^2}{2(-6.0 \text{ m/s}^2)}$$
$$= \underline{31 \text{ m}}.$$

The answer has the correct units for displacement and seems reasonable—the car needs to travel about three-tenths the length of a football field to make the change in speed. ∎

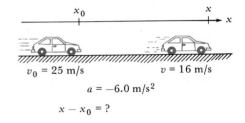

$v_0 = 25 \text{ m/s}$ $v = 16 \text{ m/s}$

$a = -6.0 \text{ m/s}^2$

$x - x_0 = ?$

FIG. 2.9

EXAMPLE 2.6   A pole vaulter after crossing the bar lands on a foam cushion. His downward speed as he first touches the cushion is 6.8 m/s, and he sinks 25 cm into the cushion before stopping. Estimate his acceleration, assumed constant, while being stopped by the cushion.

FIG. 2.10. A pole vaulter's downward velocity decreases as he sinks into a cushion.

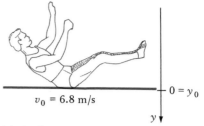

$v_0 = 6.8$ m/s

$0 = y_0$

(a) At time zero

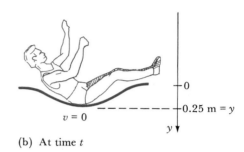

$v = 0$

0

0.25 m $= y$

(b) At time t

SOLUTION A diagram illustrating the situation described in the problem is shown in Fig. 2.10. To express the given information quantitatively, we must first choose a coordinate system. For this problem, we choose the y-axis pointing down, with the origin at the point where the vaulter first touches the cushion (Fig. 2.10a). With this coordinate system, the known information is as follows:

At time zero (as vaulter touches the cushion)	At time t (as vaulter stops after sinking into cushion)
$y_0 = 0$	$y = +0.25$ m
$v_0 = +6.8$ m/s	$v = 0$

Equation (2.12) can be rearranged to determine the vaulter's acceleration:

$$a = \frac{v^2 - v_0^2}{2(y - y_0)} = \frac{0^2 - (6.8 \text{ m/s})^2}{2(0.25 \text{ m} - 0)} = \underline{-92 \text{ m/s}^2}.$$

The minus sign indicates that the acceleration, a vector, points in the negative direction—up in this case. The vaulter's velocity in the positive (downward) direction decreased.

The answer has the correct units, but the magnitude seems fairly large—a velocity change of 92 m/s in one second. The velocity does not of course change by 92 m/s since the vaulter's initial velocity was only 6.8 m/s. However, we can get an acceleration of magnitude 92 m/s^2 from a small velocity change if the change occurs over a very short time—the time needed for the vaulter to stop. For example, if the stopping time was 0.074 s (less than one-tenth of a second), the magnitude of the average acceleration $\Delta v/\Delta t$ during that short time would be $(6.8 \text{ m/s})/0.074 \text{ s} = 92 \text{ m/s}^2$. Thus, the deceleration can be large in magnitude if the stopping time is short—which it is in this problem and in many problems that involve the abrupt impact of one object with another. ∎

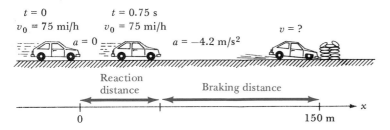

EXAMPLE 2.7 A bank robber traveling at 75 mi/h in an escape car suddenly sees a barricade 150 m ahead. The robber slams on the brakes following a 0.75-s delay (the driver's reaction time). After the brakes are applied, the car decelerates at 4.2 m/s². Will the car hit the barricade and, if so, what is its speed?

SOLUTION The problem is pictured schematically in Fig. 2.11. Notice that the picture contains a coordinate axis and the known information given with the appropriate kinematic symbols.

Our next task is to restate the problem (does the car hit the barricade?) in a form that relates more directly to some kinematic quantity. There are two ways we might do this. We could try to calculate the car's velocity as it reaches the barricade. If the velocity at that point is positive (toward the right), the car will hit the barricade. Or we might try to determine at what position the car would stop if the barricade were not present. If the stopping position is to the right of the barricade, the car will hit it before stopping. Let us try this last approach—to see where the car stops if the barricade is not present.

The calculation of this stopping distance requires a *divide and conquer approach*. We divide the car's motion in two parts: (1) It moves at constant speed (zero acceleration) while the driver reacts to get his or her foot on the brakes, and (2) the car's brakes are applied, causing it to decelerate and eventually stop. The distance traveled in the first part is called the *reaction or thinking distance* and the distance in the second part is called the *braking distance*. The total stopping distance is the sum of the reaction and braking distances.

Reaction distance: The known information for this part of the trip is as follows: $v_0 = \left(75 \dfrac{\text{mi}}{\text{h}}\right)\left(\dfrac{1609 \text{ m}}{\text{mi}}\right)\left(\dfrac{1 \text{ h}}{3600 \text{ s}}\right) = 33.5 \text{ m/s}$, $a = 0$, $t = 0.75$ s. Substituting into Eq. (2.11), we find the reaction distance:

$$x - x_0 = v_0 t + \frac{1}{2}at^2 = (33.5 \text{ m/s})(0.75 \text{ s}) + \frac{1}{2}(0)(0.75 \text{ s})^2$$

$$= 25.1 \text{ m}.$$

Braking distance: This represents a separate part of the problem in which the known conditions are as follows: $v_0 = 33.5$ m/s (initial velocity when braking starts), $v = 0$ (final velocity), $a = -4.2$ m/s².

The minus sign must be included with the acceleration because the car's velocity in the positive direction decreases. Equation (2.12) can be rearranged to find the braking distance:

$$x - x_0 = \frac{v^2 - v_0^2}{2a} = \frac{0^2 - (33.5 \text{ m/s})^2}{2(-4.2 \text{ m/s}^2)}$$

$$= 133.6 \text{ m}.$$

The stopping distance is the sum of the reaction and braking distances:

Stopping distance = 25.1 m + 133.6 m = 158.7 m.

Since the barricade is only 150 m from where the robber sees it, the <u>car will hit the barricade</u>.

To determine the car's speed at the barricade, note that the barricade is 150 m − 25.1 m = 124.9 m from where the brakes are first applied. Using Eq. (2.12), we can solve for the car's speed when it reaches the barricade after decelerating for 124.9 m:

$$v^2 = 2a(x - x_0) + v_0^2$$
$$= 2(-4.2 \text{ m/s}^2)(124.9 \text{ m}) + (33.5 \text{ m/s})^2 = 73.1 \text{ m}^2/\text{s}^2$$

or

$$v = \underline{8.5 \text{ m/s}}. \qquad \blacksquare$$

EXAMPLE 2.8 A hobo wishes to catch a freight train that stands in front of a train station. He waits 50 ft behind the train, whose last car is 50 ft from the station. As the train starts to leave the station with an acceleration of 2 ft/s^2, the hobo runs toward the train at a constant speed of 15 ft/s. Will he catch the train before the last car reaches the station, or will he have to run past the station to catch the train?

SOLUTION The situation described in the problem is pictured in Fig. 2.12. To express the known information quantitatively, we must first choose a coordinate system. For convenience, the axis points in the direction of motion. The origin can be placed anywhere, but once the choice is made, all of our positions must be expressed relative to that choice. We have placed the origin of the x axis at the initial position of the hobo. Thus, the initial conditions at time zero are as follows:

Hobo	Train
$x_0 = 0$ ft	$x_0' = 50$ ft
$v_0 = 15$ ft/s	$v_0' = 0$
$a = 0$	$a' = 2$ ft/s^2

|←— 50 ft —→|←— 50 ft —→|

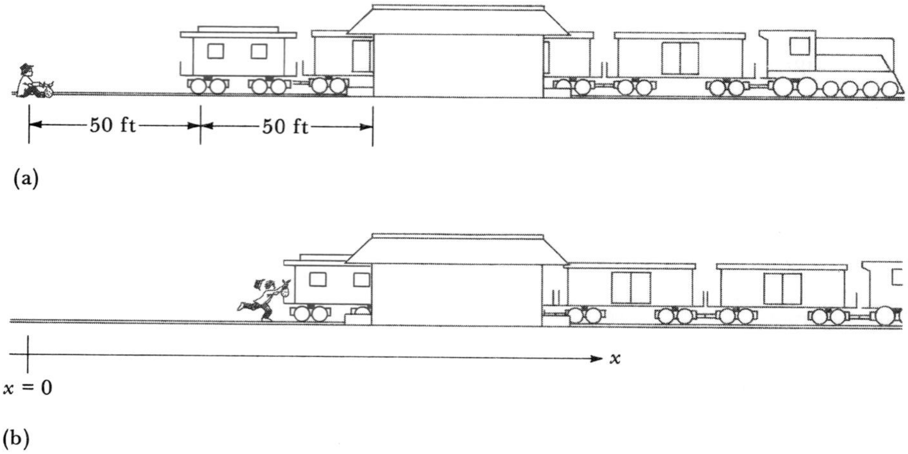

(a)

$x = 0$

(b)

FIG. 2.12. (a) A hobo waits to catch a train. (b) Later, after the train has started, the hobo and the last car reach the same position at some unknown time.

Our problem is to find the position where the hobo reaches the train. We use Eq. (2.11) to write general expressions for the position of the hobo (x) and of the train (x') at some future time:

Hobo: $\quad x = x_0 + v_0 t + \frac{1}{2} a t^2 = 0 + 15t + \left(\frac{1}{2}\right)(0)t^2 = 15t,$ **(2.13)**

Train: $\quad x' = x'_0 + v'_0 t + \frac{1}{2} a' t^2 = 50 + (0)t + \left(\frac{1}{2}\right)(2)t^2 = 50 + t^2.$ **(2.14)**

We have omitted units to keep the equations simple. These two equations give the position of the hobo and of the train at any time in the future. However, we are interested in the specific time when the hobo and the train are at the same position; that is, when $x = x'$. If we set Eqs. (2.13) and (2.14) equal to each other, we find that

$$15t = 50 + t^2,$$

or

$$t^2 - 15t + 50 = 0. \qquad \textbf{(2.15)}$$

This equation, when solved for t, will tell us the time when the hobo and the back of the train are in the same position, that is, when $x = x'$. After finding that time, we substitute back into either Eq. (2.13) or Eq. (2.14) to find the position where this occurs.

Equation (2.15) is a quadratic equation of the form

$$at^2 + bt + c = 0. \qquad \textbf{(2.16)}$$

Two values of t are solutions of this equation:

$$t = \frac{-b \pm \sqrt{b^2 - 4ac}}{2a}. \qquad \textbf{(2.17)}$$

For our particular problem, $a = 1$, $b = -15$, and $c = 50$. Thus, the times when the hobo and train are in the same position are

$$t = \frac{-(-15) \pm \sqrt{(-15)^2 - (4 \times 1 \times 50)}}{2} = 5 \text{ s or } 10 \text{ s}.$$

The hobo and the last car of the train will be at the same position 5 s after the train starts and also 10 s after the train starts. The first of these solutions corresponds to the time the hobo first reaches the train. Evidently he is moving fast enough to run past the last car and farther up along the train. By 10 s, the train has accelerated to such a speed that the last car catches up to and passes him.

If the hobo joins the train at the earlier time, he will be at position $x = 15t = 15(5 \text{ s}) = 75$ ft, or 25 ft before the station. ■

2.7 Gravitational Acceleration

One important example of linear motion with approximately constant acceleration is the vertical motion of objects thrown into the air or of objects falling through the air. In fact, if it were not for the resistive force of air on these objects, their acceleration would be constant. This acceleration is caused by the gravitational force of the earth pulling down on the objects. This force—their weight—causes them to accelerate toward the center of the earth with what is called **gravitational acceleration.** The magnitude of this acceleration, which we give the symbol g, has a value at the earth's surface

$$g = 9.8 \text{ m/s}^2 = 32.2 \text{ ft/s}^2. \tag{2.18}$$

The value of g varies slightly with elevation and latitude and also depends on the geological formations at different parts of the earth's surface. In all of our calculations, we will use the rounded-off value $g = 9.8 \text{ m/s}^2$ or 32 ft/s^2.

We have said that this gravitational acceleration is a constant only if we ignore air resistance. For what situations can air resistance be ignored? The answer to this question depends on several factors:

1. If you desire great accuracy, then you must consider air resistance.
2. How fast is the object moving? The faster the object moves, the more air resistance affects its acceleration.
3. The shape and surface area of an object greatly affect air resistance. Air resistance on an open parachute is much greater than on a parachute still enclosed in its bag.
4. Air resistance is much more important for a light object, such as a Ping-Pong ball, than for a heavier object of the same size, such as a golf ball.

Obviously, no general rule tells us when air resistance can be ignored. In our examples and problems the objects will be moving through the air at moderate speeds and will be relatively heavy and/or streamlined. We will ignore air resistance, even though its effect may be significant.

We will use the set of four equations derived in the previous section to relate the position, velocity, and acceleration, at different times, of objects moving in the vertical direction under the influence of the gravitational acceleration. We will use a y coordinate to indicate the vertical position of the object. For this vertical motion, Eqs. (2.9)–(2.12) become

$$v = v_0 + at, \tag{2.9y}$$

$$y - y_0 = \left(\frac{v + v_0}{2}\right)t, \tag{2.10y}$$

$$y - y_0 = v_0 t + \frac{1}{2}at^2, \tag{2.11y}$$

$$2a(y - y_0) = v^2 - v^2_0 \tag{2.12y}$$

These equations relate the object's acceleration a, initial vertical position y_0, and initial vertical velocity v_0 at time zero to its final vertical position y and final vertical velocity v at time t. If the positive y axis points up, then the acceleration due to gravity is down; that is, $a = -g$. If the y axis points down, then $a = +g$ because the acceleration is downward in the positive y direction. Let us consider some examples.

EXAMPLE 2.9 A high-jumper leaps into the air with an initial speed of 5.8 m/s. How high will she be at the top of her flight, when $v = 0$? *

SOLUTION If we choose the ground level as the origin of our y-coordinate axis and the y axis points up, then the known quantities for this problem are:

---

*At the top of a person's jump, her velocity is neither up nor down. The person has stopped for just an instant while changing directions.

$$y_0 = 0 \text{ m},$$
$$v_0 = 5.8 \text{ m/s},$$
$$a = -g = -9.8 \text{ m/s}^2.$$

We wish to know the value of y when $v = 0$. Equation (2.12y) can be used to calculate the jumper's height at the top of the jump:

$$y - y_0 = \frac{v^2 - v_0^2}{2a} = \frac{0^2 - (5.8 \text{ m/s})^2}{2(-9.8 \text{ m/s}^2)} = \underline{1.7 \text{ m}}. \quad \blacksquare$$

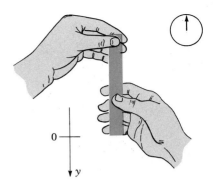

EXAMPLE 2.10 A person can test your reaction time by dropping a ruler between your fingers, as shown in Fig. 2.13. The distance the ruler falls before you catch it indicates the time the ruler falls and hence your reaction time. What is your reaction time if the ruler falls 20 cm before you catch it?

SOLUTION Let us choose the initial position of the lower edge of the ruler as the origin of the coordinate system. We will follow the motion of this lower edge. If we orient the y axis so that it points down (Fig. 2.13b), then the known quantities are:

$$y_0 = 0,$$
$$v_0 = 0,$$
$$a = g = 9.8 \text{ m/s}^2.$$

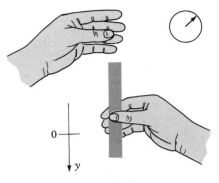

FIG. 2.13. (a) A person drops a meter stick. (b) A second person catches it. The distance the stick falls is related to the second person's reaction time.

We wish to find the time t when the bottom edge of the ruler has fallen to $y = 20 \text{ cm} = 0.20 \text{ m}$. It is important to use units all from the same system of units. Thus, if g is in m/s^2, y should be in meters and not in centimeters. Equation (2.11y) can be used to find t:

$$y - y_0 = v_0 t + \frac{1}{2}at^2,$$

or

$$(0.20 \text{ m}) - (0) = (0)t + \frac{1}{2}(9.8 \text{ m/s}^2)t^2.$$

Thus,

$$t^2 = \frac{2(0.20 \text{ m})}{9.8 \text{ m/s}^2} = 0.041 \text{ s}^2,$$

or

$$t = \underline{0.20 \text{ s}}.$$

This is about the amount of time needed for information to travel from your eye to your brain and then to your hand so that it can grab the ruler. Do not be discouraged if your time is longer than this; 0.2 s is a fast reaction time. $\blacksquare$

2.8 Projectile Motion

A **projectile** is an object that moves through space under the influence of the earth's gravitational force. Two coordinates must be used to describe the projectile's motion, since it moves horizontally as well as vertically. Examples of pro-

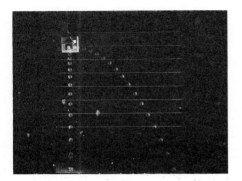

How does the vertical motion of the projectile shot toward the right compare to that of the ball that falls straight down? Does the projectile have a constant horizontal component of velocity? (Courtesy of Educational Development Corp.)

jectile motion are the path followed by a baseball having been hit into the air by a bat or by an ice skater jumping over some barrels. Often a strong, abrupt force initiates the motion of a projectile. Following this force, the projectile moves through the air and is influenced only by the gravitational force of the earth pulling down on it and by air resistance. If the effect of air resistance is ignored, it is quite easy to use equations, such as those developed in the previous sections, to analyze the motion of a projectile—how far it will travel, how high it will go, and so forth.

When the four linear motion equations are used to analyze projectile motion, the horizontal and vertical motions of the projectile are treated independently, as illustrated in Figs. 2.14 and 2.15. An ejector seat is placed in a car. When the car is at rest, the seat is adjusted so that, when released, the driver shoots straight up into the air and returns again to the seat (Fig. 2.14). Without readjusting the seat, the driver next travels at constant speed toward an underpass (Fig. 2.15). Just before the car reaches the underpass, the ejector seat shoots the driver straight up into the air. Since the driver and car are both moving horizontally at the same speed before the driver is released, both continue this horizontal motion as though nothing had happened. Thus, when the driver comes back down to the level of the car, he again finds himself back in the front seat, having flown over the underpass. The point is that both objects had the same horizontal motion, which was unaffected by the vertical flight of the driver.

Projectile motion, then, is described in terms of motion in the horizontal or x direction and independently by motion in the vertical or y direction. If air resistance can be ignored, the projectile's horizontal motion continues at constant velocity because its horizontal acceleration is zero. The vertical motion, however, will have acceleration due to gravity. We can therefore restate the four linear motion equations as two separate sets of equations: One set describes the horizontal or x motion of the projectile, for which $a_x = 0$; the other set describes the vertical or y motion, for which $a_y = -g$ (the y axis is oriented upward).

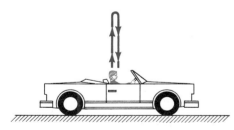

FIG. 2.14. An ejector seat shoots a person straight up into the air. The seat is adjusted so that, when the car is at rest, the person returns to the car seat.

FIG. 2.15. Without readjusting the seat, a person ejected from the moving car also returns to the seat. Since the car is moving, the driver has an initial x component of velocity equal to that of the car. This velocity is unaffected by the person's vertical motion.

PROJECTILE MOTION

x Equations ($a_x = 0$)		y Equations ($a_y = -g$)	
$v_x = v_{x_0}$	$(2.9x)$	$v_y = v_{y_0} - gt$	$(2.9y)$
$x - x_0 = \left(\dfrac{v_{x_0} + v_x}{2} \right) t$	$(2.10x)$	$y - y_0 = \left(\dfrac{v_{y_0} + v_y}{2} \right) t$	$(2.10y)$
$x - x_0 = v_{x_0} t$	$(2.11x)$	$y - y_0 = v_{y_0} - \dfrac{1}{2} g t^2$	$(2.11y)$
$0 = v_x^2 - v_{x_0}^2$	$(2.12x)$		
		$-2g(y - y_0) = v_y^2 - v_{y_0}^2$	$(2.12y)$

Most often a projectile leaves the ground at some angle θ with a speed v_0, as shown in Fig. 2.16. In such a case, the initial speed in the x direction is the x component of $\mathbf{v}_0$, or $v_{x_0} = v_0 \cos \theta$. The initial speed in the y direction is the y component of $\mathbf{v}_0$, or $v_{y_0} = v_0 \sin \theta$ [see Eqs. (1.3) and (1.4)].

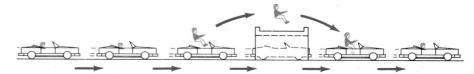

Often, the y equations are used to determine the length of time the projectile remains in the air and the x equations to determine how far the projectile will have traveled horizontally during that time. For example, if a projectile leaves ground level and returns to ground level, Eq. (2.11y) with $y = 0$ and $y_0 = 0$ can be solved for t. Equation (2.11x) can then be used to calculate the distance the projectile travels in that time.

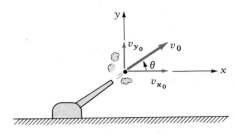

FIG. 2.16. When fired at an angle, a projectile has an initial motion in both the x and the y directions.

EXAMPLE 2.11 A football player kicks off from the 40-yd line. How far will the ball travel before hitting the ground if its initial speed v_0 is 80 ft/s and the ball leaves the ground at an angle of 30°? (Assume that air resistance can be ignored, a poor assumption in this problem.)

SOLUTION First, we solve the y equations to find the time of flight of the ball. The known conditions for the y motion are

$$y_0 = 0 \quad \text{and} \quad v_{y_0} = v_0 \sin 30°, \quad \text{where } v_0 = 80 \text{ ft/s.}$$

We need to find the time t when $y = 0$, that is, when the ball returns to ground level. Equation (2.11y) can be used:

$$0 - 0 = v_0 \sin 30° \, t - \frac{1}{2}gt^2,$$

or

$$t = \frac{2v_0 \sin 30°}{g} = \frac{2(80 \text{ ft/s})(0.5)}{(32 \text{ ft/s}^2)} = 2.5 \text{ s.}$$

Next we find the horizontal distance $(x - x_0)$ the ball travels during this time. Equation (2.11x) provides this information:

$$x - x_0 = v_{x_0}t = (v_0 \cos 30°)t = (80 \text{ ft/s})(0.87)(2.5 \text{ s})$$
$$= \underline{170 \text{ ft.}} \qquad \blacksquare$$

EXAMPLE 2.12 An airplane traveling horizontally at 78 m/s at an elevation of 210 m drops bales of hay to cows stranded in a snowstorm. (a) At what horizontal distance from the cows should the bales be dropped? Ignore air resistance. (b) Calculate the velocity of a bale as it reaches the ground.

SOLUTION (a) The situation described in the problem is pictured in Fig. 2.17. We have placed the origin of the coordinate system at the ground directly under the position of the plane when it drops the bales of hay. The bales after leaving the plane move horizontally at a constant speed $v_x = v_{x_0} = 78$ m/s. At the same time, they accelerate downward ($a_y = -g = -9.8$ m/s^2) starting with an initial downward velocity of zero ($v_{y_0} = 0$).

The divide and conquer strategy is essential for most projectile problems. For this problem, we first rearrange Eq. (2.11y) to determine the time needed for the bales to fall a distance of 210 m (remember that $v_{y_0} = 0$):

$$t = \left[\frac{2(y - y_0)}{-g}\right]^{1/2} = \left[\frac{2(0 - 210 \text{ m})}{-9.8 \text{ m/s}^2}\right]^{1/2} = 6.5 \text{ s.}$$

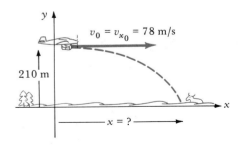

FIG. 2.17. A bale of hay moves forward at constant speed as its downward velocity increases because of the gravitational acceleration. The combination of the horizontal and vertical motion causes it to fall along the path of the dashed line.

Next we calculate the horizontal distance that the bales travel during the 6.5 s needed to reach the ground. We can use Eq. (2.11x):

$$x - x_0 = v_{x_0} t = (78 \text{ m/s})(6.5 \text{ s}) = \underline{510 \text{ m}}.$$

The bales should be dropped when the plane is a horizontal distance of 510 m from the cows.

(b) The bale's horizontal component of velocity v_x remains constant at 78 m/s. The vertical component of velocity v_y as the bale reaches the ground can be determined using Eq. (2.9y), with $v_{y_0} = 0$ and $t = 6.5$ s:

$$v_y = v_{y_0} - gt = 0 - (9.8 \text{ m/s}^2)(6.5 \text{ s}) = -64 \text{ m/s}.$$

The magnitude of the bale's velocity is given by Eq. (1.7):

$$v = \sqrt{v_x^2 + v_y^2} = \sqrt{(78 \text{ m/s})^2 + (-64 \text{ m/s})^2} = 100 \text{ m/s}.$$

The angle of the velocity relative to the x axis is determined using Eq. (1.9):

$$\theta = \text{invtan} \left| \frac{v_y}{v_x} \right| = \text{invtan} \frac{64 \text{ m/s}}{78 \text{ m/s}} = \text{invtan } 0.82 = 39°.$$

The bale's velocity is $\underline{100 \text{ m/s in a direction } 39° \text{ below the positive } x \text{ axis}}$. ∎

2.9 Graphical Analysis of Velocity and Acceleration

In the last three sections we solved problems using kinematic equations that related an object's constant acceleration to its changing position and velocity at different times. In this section, we develop methods to analyze nonconstant accelerated motion.

Usually, the basic data for this analysis will be provided by a set of measurements of position or velocity as a function of time. The data, often recorded in a table, are easier to interpret when displayed in the form of a graph. But even graphs require some skill to interpret. We will find that the slopes of graphs and the area under a curve plotted in the graph provide especially useful information.

Slope of a Graph

As an example of the use of the slope of a graph, consider the changing position of a San Francisco Bay Area Rapid Transit (BART) commuter train as it leaves a station. Its position at different times is tabulated in Fig. 2.18a. These measured data are plotted in a graph in Fig. 2.18b. The BART train's position appears on the vertical axis and the time on the horizontal axis. Each point on the graph corresponds to a measurement of position at one particular time. A smooth curve runs through these data points.

The slope of the curve shown in Fig. 2.18b—that is, how fast the curve rises or falls—indicates the velocity of the object. We can estimate the slope at a point by drawing a triangle, such as shown in Fig. 2.18b. The horizontal side of the triangle is called the *run* of the curve, and in this case it equals a change in time

Time (s)	0	10	20	30	40	50	60
Position (m)	0	50	180	370	600	870	1200

(a)

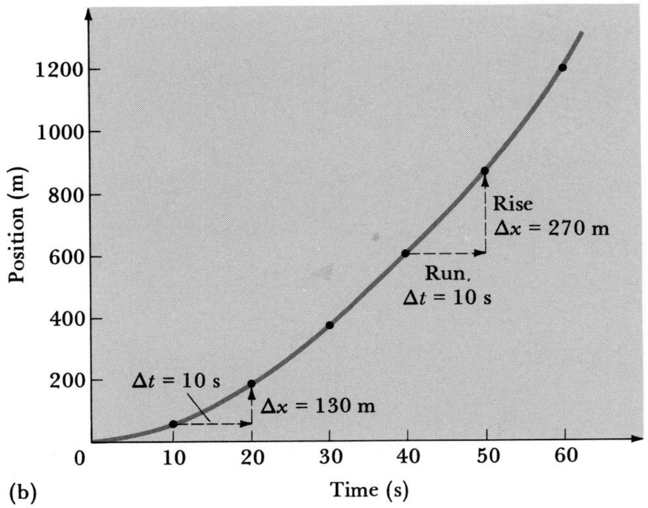

FIG. 2.18. (a) A table showing the position of a BART train at different times. (b) The same data displayed in a position-versus-time graph.

(b)

(10 s). The vertical line is called the *rise* of the curve; in this graph it equals the change in position (270 m) that occurs during the run time. The ratio of these two changes—the rise divided by the run—equals approximately the slope of the curve:

$$\text{Slope} \cong \frac{\text{Rise}}{\text{Run}}.$$

For the graph plotted in Fig. 2.18b, the slope is simply the velocity of the train:

$$\text{Slope} = \frac{\text{Rise}}{\text{Run}} = \frac{\Delta x}{\Delta t} = \text{Velocity}.$$

In the last paragraph we used the word *approximately*. The slope equals the ratio of the rise and run only if the run is infinitesimally small. To calculate exactly the slope of the curve at a particular point, we draw a line tangent to the curve at that point, as shown in Fig. 2.19a. The slope equals exactly the rise divided by the run of the tangent line.

For a position-versus-time graph, the slope of the tangent line at a particular time is the object's velocity at that time.

By taking the slopes of the tangent lines of the position-versus-time graph at different times, we can construct a velocity-versus-time graph, as shown in Fig. 2.19b. With this graph we can construct an acceleration-versus-time graph, shown in Fig. 2.19c. Notice that the slope of the velocity-versus-time graph in Fig. 2.19b is given by

$$\text{Slope} = \frac{\text{Rise}}{\text{Run}} = \frac{\Delta v}{\Delta t},$$

which is the acceleration.

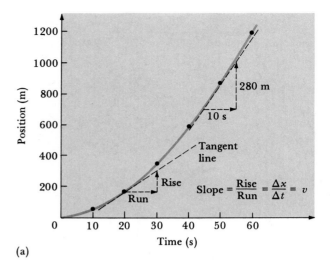

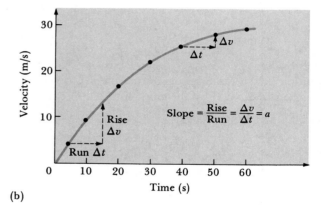

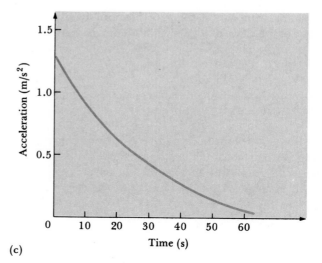

FIG. 2.19. (a) The slope of a position-versus-time graph is the velocity. (b) The slope of a velocity-versus-time graph is the acceleration. (c) An acceleration-versus-time graph.

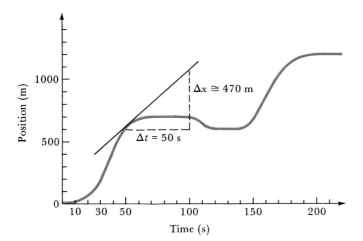

FIG. 2.20. A graph showing the position of a bus at different times.

For linear motion, the acceleration of an object at a particular time equals the slope of the tangent line of its velocity-versus-time graph at that time.

EXAMPLE 2.13 The position of a bus as a function of time is shown in Fig. 2.20. (a) At what time or during what time intervals is the bus at rest? (b) During what time intervals is the bus moving in the positive x direction? (c) In the negative x direction? (d) At what time is the bus moving fastest in the positive direction? (e) Fastest in the negative direction? (f) Estimate the velocity of the bus at $t = 50$ s.

SOLUTION (a) If the bus is at rest, its position change Δx during a short time Δt is zero, as is the rise of the tangent line. Thus, the tangent line is horizontal. This occurs at 0 s, when the bus is starting its trip and during the time intervals from 70 s to 90 s, 120 s to 140 s, and 200 s and later.

(b) If the bus moves forward, in the positive direction, its position change Δx during a short time Δt is positive, as is the rise of the tangent line. Thus, the tangent line tilts upward. This occurs from a little after 0 s to 70 s and from 140 s to 200 s.

(c) If the bus moves in the negative direction, its position change Δx during a short time Δt is negative—the tangent line falls or tilts downward. This occurs during the time interval from about 90 s to 120 s.

(d) The bus moves fastest in the positive direction when its upward slope is greatest—during the time from about 30 s to 40 s.

(e) The bus moves fastest in the negative direction when its downward slope is greatest, at about 110 s.

(f) The velocity at $t = 50$ s is the slope of a line tangent to the

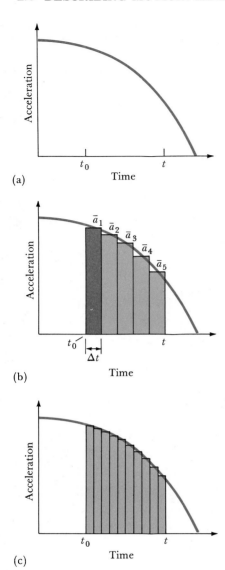

FIG. 2.21. (a) An acceleration-versus-time graph. (b) The graph in (a) is divided into small time intervals Δt to determine the velocity change during each interval. (c) If the intervals are even smaller, the area of the narrow acceleration-versus-time rectangles equals the area under the curve between t_0 and t.

position-versus-time graph at that time. As we see from Fig. 2.20, at $t = 50$ s, the slope is

$$v = \frac{\Delta x}{\Delta t} \cong \frac{+470 \text{ m}}{50 \text{ s}} = \underline{+9.4 \text{ m/s}}.$$

Area Under a Graph

We have seen that a position-versus-time graph can be used to construct a velocity-versus-time graph, which can in turn be used to construct an acceleration-versus-time graph. Sometimes it may be desirable to work in the opposite direction. Suppose that we know the time variation of a person's acceleration while landing on a trampoline. Could we then work backward to determine the person's velocity and position at different times as he or she sinks into the trampoline while being stopped? The answer is yes.

Let us assume that an object's acceleration as a functon of time is as shown in Fig. 2.21a and that the object's initial velocity at time t_0 is v_0. What is the object's velocity v at some later time t? To determine the change in velocity, we divide the graph into short time intervals each of width Δt, as shown in Fig. 2.21b. The darkened rectangle above the first time interval has a height equal to the average acceleration $\bar{a}_1$ of the object during that time. The velocity change caused by this acceleration can be determined using the equation that defines acceleration ($a = \Delta v/\Delta t$). Rearranging, we find that during the first short time interval the velocity changes by $\Delta v_1 = \bar{a}_1 \Delta t$. If the initial velocity was v_0, then at the end of the first time interval, the velocity has become $v_1 = v_0 + \bar{a}_1 \Delta t$.

It is interesting to note that *this velocity change equals the area of the tall, shaded rectangle in Fig. 2.21b.* Its height $\bar{a}_1$ times its width Δt has area $\bar{a}_1 \Delta t$, which equals the change in velocity Δv_1.

During the second short time interval after t_0, the average acceleration is reduced to $\bar{a}_2$, and the velocity changes by an amount $\Delta v_2 = \bar{a}_2 \Delta t$. This velocity change equals the area of the second tall rectangle, whose height $\bar{a}_2$ times width Δt equals $\bar{a}_2 \Delta t$. At the end of the second time interval the velocity has increased to $v_2 = v_0 + \bar{a}_1 \Delta t + \bar{a}_2 \Delta t$.

By continuing this process, we can determine the object's velocity at the end of each time interval. From this information, we can construct a velocity-versus-time graph. Also, the net velocity change between any two times is the sum of the areas of all rectangles of width Δt between those times. If the width Δt of the rectangles is very narrow, as in Fig. 2.21c, the sum of their areas between times t_0 and t equals the area under the acceleration-versus-time graph between those same times.

The area under an acceleration-versus-time graph between times t_0 and t equals an object's change in velocity $v - v_0$ between those same times. *

A similar process can be followed using a velocity-versus-time graph to determine an object's position at different times. Consider the

---

*In calculus this area under an acceleration-versus-time graph is determined by taking the integral of a with respect to t between times t_0 and t:

$$v - v_0 = \int_{t_0}^{t} a \cdot dt.$$

velocity-versus-time graph shown in Fig. 2.22. It has been divided into rectangles of width Δt (a time interval) and height $\overline{v}$ (the average velocity of the object during the time interval). Recalling that $\overline{v} = \Delta x/\Delta t$ and rearranging, we see that the object's displacement Δx during a short time interval Δt while traveling with velocity $\overline{v}$ is $\Delta x = \overline{v} \, \Delta t$. If the object's initial position was x_0 at time t_0, then after the first time interval its position has changed to $x_1 = x_0 + \overline{v}_1 \, \Delta t$. During the second time interval, the object undergoes another displacement $\Delta x_2 = \overline{v}_2 \, \Delta t$, and its position changes to $x_2 = x_0 + \overline{v}_1 \, \Delta t + \overline{v}_2 \, \Delta t$. We can use this same procedure to determine the object's changing position by simply adding the $\overline{v} \, \Delta t$ displacements for successive time intervals. If we wish to determine the object's net change in position (its displacement) from an initial time t_0 to a final time t, we add the $\overline{v} \, \Delta t$ displacements for all the short time intervals between those times.

Since each $\overline{v} \, \Delta t$ rectangle has height $\overline{v}$ and width Δt, the product of the two quantities equals the area of the rectangle and also equals the object's displacement during that time. By adding the areas of all the rectangles of width Δt between t_0 and t, we have the object's net displacement between those times.

The area under a velocity-versus-time graph between times t_0 and t equals an object's displacement $x - x_0$ during that same time period.*

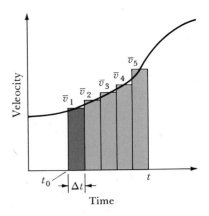

FIG. 2.22. The displacement during each time interval Δt is the area of each narrow velocity-versus-time rectangle.

EXAMPLE 2.14 The graph shown in Fig. 2.23 indicates the downward velocity of a skydiver at different times. Approximately how far does the diver fall in the first 3.0 s?

SOLUTION To estimate the distance fallen from 0.0 to 3.0 s, we simply estimate the area under the velocity-versus-time graph. The simplest way to do this is to count the number of squares under the curve and multiply by the area of each square. The area of the large, dark square of width 1.0 s and height 5 m/s shown in Fig. 2.23 is $(1.0 \text{ s})(5 \text{ m/s}) = 5$ m. Thus, each large square accounts for a displacement of 5 m. To find the total distance fallen, we estimate the number of squares under the curve and multiply by 5 m per square. Squares 1 and 2 in Fig. 2.23 together have an area of about one full square, as do squares 5 and 8. Since squares 3, 4, 6, and 7 are almost entirely under the curve, the total area under the curve is approximately 6 squares times 5 m per square, or 30 m, our estimate of the skydiver's displacement in 3.0 s. ■

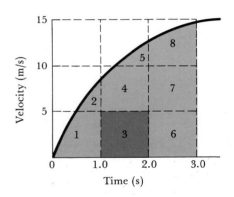

FIG. 2.23. The area under a skydiver's velocity-versus-time graph is divided into squares, the areas of which are added to determine the diver's displacement.

Summary and Additional Readings

1. **Speed, velocity, and acceleration: Kinematics** is the quantitative description of motion used to determine some unknown property of the motion, such as an object's acceleration or its location or velocity at some future time. The vector quantities displacement **d**, velocity **v**, and acceleration **a** are represented in slightly different ways, depending on whether we are interested in their average values over a long period of time t or their instanta-

neous values during a very short time Δt. The symbols used for motion along a line, called **linear** or **one-dimensional motion**, are different from the symbols used for motion in two or three dimensions. However, the basic idea underlying the definitions for these different situations is the same: **Displacement** is a vector quantity pointing from the initial position to the final position, **velocity** is the displacement of an object during a certain time divided by

*In calculus this area under a velocity-versus-time graph is determined by taking the integral of v with respect to t:

$$x - x_0 = \int_{t_0}^{t} v \cdot dt.$$

TABLE 2.2   Kinematic Quantities

	One-Dimensional Motion		Two- or Three-Dimensional Motion	
	Average Values	Instantaneous Values*	Average Values	Instantaneous Values*
Displacement	$x_2 - x_1$	Δx	$\mathbf{d} = \mathbf{r}_2 - \mathbf{r}_1$	$\Delta \mathbf{d}$
Velocity	$v = \dfrac{x_2 - x_1}{t_2 - t_1}$	$\dfrac{\Delta x}{\Delta t}$	$\mathbf{v} = \dfrac{\mathbf{r}_2 - \mathbf{r}_1}{t_2 - t_1}$	$\dfrac{\Delta \mathbf{d}}{\Delta t}$
Acceleration	$a = \dfrac{v_2 - v_1}{t_2 - t_1}$	$\dfrac{\Delta v}{\Delta t}$	$\mathbf{a} = \dfrac{\mathbf{v}_2 - \mathbf{v}_1}{t_2 - t_1}$	$\dfrac{\Delta \mathbf{v}}{\Delta t}$

*For very short time intervals Δt.

that time, and **acceleration** is the change in an object's velocity divided by the time required for that change. The symbolic representations of these vector quantities for different situations are summarized in Table 2.2. **Speed** is the scalar magnitude of the velocity.

For one-dimensional motion, x indicates the position of an object along a coordinate axis. The vector nature of the motion is indicated using positive and negative signs depending on the object's position along the axis and on the direction of the motion relative to the axis.

2. **Equations of linear motion with constant acceleration:** When acceleration is constant during linear motion, a set of four simple equations can be used to relate position, velocity, and acceleration at different times:

$$v = v_0 + at, \tag{2.9}$$

$$x - x_0 = \left(\frac{v + v_0}{2}\right) t, \tag{2.10}$$

$$x - x_0 = v_0 t + \frac{1}{2} a t^2, \tag{2.11}$$

$$2a(x - x_0) = v^2 - v_0^2. \tag{2.12}$$

In these equations, x_0 and v_0 represent the object's position and velocity at time zero; x and v represent its position and velocity at a later time t; and a represents the object's constant acceleration. These equations can be used to calculate an object's acceleration if its changing position and velocity at different times are known. If the acceleration is known, the equations can be used to calculate the future position and velocity of the object.

3. **Gravitational acceleration:** The vertical motion of objects moving freely in the air comes close to being linear motion with constant acceleration. If we ignore the effects of wind and air resistance, objects experience constant acceleration caused by the earth's gravitational force. The acceleration is directed toward the center of the earth and has a magnitude

Gravitational acceleration = g = 9.8 m/s^2 = 32 ft/s^2.

4. **Projectile motion:** Projectiles are objects moving both horizontally and vertically through the air. If we ignore the resistive force of the air on the projectiles, then the horizontal and vertical components of their motion can be described independently using two separate sets of equations derived from Eqs. (2.9)–(2.12). For the horizontal motion equations, the acceleration is zero ($a_x = 0$). For the vertical motion equations, the acceleration is that caused by gravity ($a_x = -g$), where, for the vertical component of motion, the y axis points up.

5. **Graphical analysis of velocity and acceleration:** For linear motion, the slope of a position-versus-time graph at a particular time equals an object's velocity at that time. The slope of a velocity-versus-time graph at a particular time equals the object's acceleration at that time. Also, the area under a velocity-versus-time graph between two times equals an object's displacement between those times, and the area under an acceleration-versus-time graph between two times equals an object's change in velocity between those times.

Bernard Cohen, "Galileo," *Scientific American* **181,** 40 (1949). A biography about the person called the father of kinematics.

Seville Chapman, "Catching a Baseball," *American Journal of Physics* **36,** 868 (1968).

P. Kirkpatrick, "Bad Physics in Athletic Measurements," *American Journal of Physics* **12,** 7 (1944).

Several excellent papers discuss general problem-solving procedures:

Jill Larkin, John McDermott, Dorthea P. Simon, and Herbert A. Simon, "Expert and Novice Performance in Solving Physics Problems," *Science*, pp 1335–1342, June 20 (1980).

Frederick Reif, "Teaching Problem Solving—A Scientific Approach," *The Physics Teacher*, p 310, May (1981).

Arnold B. Arons, "Student Patterns of Thinking and Reasoning," *The Physics Teacher*, p 576, December (1983); p 21, January (1984); and p 88, February (1984).

Questions

1. Design and describe an experiment to measure the speed of sound.

2. Roger Bannister ran the mile on an oval track in 3 min, 59.4 s. His average speed and average velocity were quite different. Explain.

3. How much velocity does your body gain in 2 s when it undergoes an acceleration of $+2$ m/s^2? Of 4 m/s^2? What is its final velocity in each case if its initial velocity is $+1$ m/s? -1 m/s?

4. A car's initial velocity is $+10$ m/s. What is the car's velocity 1 s later if its acceleration is (a) $+2$ m/s^2; (b) $+4$ m/s^2? (c) and (d) Repeat, but determine the velocity 3 s later for the accelerations given in (a) and (b). (e)–(h) Repeat all four calculations from (a)–(d) for the case where the accelerations are -2 m/s^2 and -4 m/s^2.

5. (a) Can a car traveling at 30 m/s have the same acceleration as a bicycle rider traveling at 2 m/s? Explain. (b) Give an example of an object with zero speed that has a nonzero acceleration.

6. Give an example of an object that has a nonzero acceleration but moves with a constant speed.

7. Suppose that the positive axis of a one-dimensional coordinate system points east. (a) How does the motion of a car traveling east change if its acceleration is positive? (b) If its acceleration is negative? (c) How does the motion of a car traveling west change if its acceleration is positive? (d) If its acceleration is negative? Calculate numerical examples to help justify each answer.

8. A ball thrown straight up has an initial upward speed v_0. The ball's speed decreases as it rises until it stops for an instant at the apex of its flight. The ball then returns to the hand that threw it. Draw a series of seven pictures indicating the ball's position at roughly equal intervals of time for its complete flight up and down. Draw arrows from each picture that represent the velocity and the acceleration of the ball at each position.

9. Is it true that a rock dropped from rest falls three times farther in the second second than in the first? Explain.

10. For novice tennis players who have trouble getting their serve into the service court, it is advantageous to stand toward the corner of the court (B in Fig. 2.24) rather than near the center line (A). The ball is in the air longer when it reaches the net when served from B than from A. Why does this help get it into the service court?

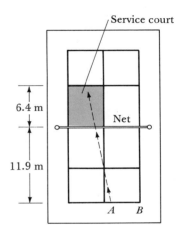

FIG. 2.24

Problems

2.3 Speed and Velocity

1. If the continents are drifting apart at a constant speed of 3 cm/yr and Africa and South America are separated by 6.4×10^8 cm, how long has this drift been occurring?

2. Sound caused by an explosion at the ocean's surface travels vertically down to the bottom, is reflected, and returns to the surface in 11.6 s. Calculate the depth of the ocean if sound travels at 1450 m/s in water.

3. One person can run 400 m in 58 s and another can run the same distance in 60 s. If they run together, by what distance does the faster runner beat the slower runner?

4. A person drives for 50 min at a speed of 28 m/s, then stops for 20 min, and finally drives for 30 min at 36 m/s. Calculate (a) the total distance traveled and (b) the average speed.

5. A person runs 800 m at an average speed of 6.0 m/s and then the next 800 m at an average speed of 8.0 m/s. Calculate (a) the total time required and (b) the average speed.

■ 6. A bat flying toward a wall in a cave sends sound waves toward the wall. The waves travel at 330 m/s, and the wall is 2 m from the bat. If the bat is flying at 20 m/s, approximately how far will the bat travel during the time required for the sound waves to make a round trip from the bat to the wall?

■ 7. A student in Tallahassee drives to Fort Lauderdale for spring vacation by first going 280 km east to pick up a friend in Jacksonville and then 510 km south to Fort Lauderdale. (We will ignore the small eastward displacement from Jacksonville to Fort Lauderdale.) The whole trip takes 9.6 hr. Calculate (a) the student's average speed and (b) average velocity.

■ 8. An airplane flies over a farm 110 km north and 42 km west of Winnipeg. Fifteen minutes later, the plane passes over a second farm 50 km north and 38 km east of Winnipeg. Calculate (a) the plane's displacement during that time and (b) its average velocity.

■ 9. On the way to class you walk east and then south such that your resultant displacement is 360 m in a direction 53° south of east. Your average velocity is 2.0 m/s in the same direction as the displacement. Calculate (a) your walking time and (b) your average speed. [*Hint:* The distance traveled is greater than 360 m.]

10. A skydiver falls from an airplane. His positions below the plane at different times are given in the following table.

(a) Calculate the diver's average speed during the first 5 s. (b) Calculate the diver's average speed during each 1-s interval.

Time (s)	0	1	2	3	4	5	6
Position (m)	0	5	19	42	74	112	154

■ 11.   A weather balloon continuously radios its elevation back to a base station. Its elevations at several different times are given as follows:

Time (min)	0	5	10	15	20	25
Elevation (km)	1.2	1.6	1.9	2.0	1.8	1.1

Calculate its average velocity for the time interval (a) from 5 to 10 min; (b) from 20 to 25 min; (c) from 0 to 25 min. (d) If the balloon's velocity after 25 min is the same as it was from 20 to 25 min, when will it hit the ground?

■ 12.   A young woman wishes to run 400 m in 60 s. Her positions on the track at 3-s intervals for the first 15 s are listed in the following table. (a) Calculate her average speed during each time interval. (b) If she continues to run for the last 298 m at the same speed as she ran during the time from 12 to 15 s, what will be her time for the full 400 m?

Time (s)	0	3	6	9	12	15
Distance (m)	0	16	35	57	79	102

2.4   Unit Conversions

13.   What is the speed in meters per second of a car traveling (a) 65 mi/h? (b) 25 mi/h? (c) What is the speed in miles per hour of a car traveling at 100 km/h? (d) At 30 m/s?

14.   Four seconds after seeing the lightning in a distant cloud, you hear thunder. How far is the cloud in kilometers and in miles? Sound travels at about 330 m/s.

15.   The surface area of a person's body is 1.6 m². Calculate the area (a) in square centimeters and (b) in square feet.

16.   The volume of a 130-lb person is about 1.8 ft³. Calculate the person's weight in newtons and volume in cubic meters and cubic centimeters.

17.   (a) An electron involved in energy conversion in photosynthesis travels about 100 μm in 3 ms. What is the average speed of the electron in meters per second? (b) What is the electron's speed in miles per hour?

2.5   Acceleration

■ 18.   A truck traveling east has a velocity of $+26$ m/s. (a) Calculate its velocity after accelerating at $+1.0$ m/s² for 5.0 s. (b) The truck then decelerates at -2.0 m/s² for 3.0 s. Now what is its velocity? (c) The truck then takes 8.0 s to change its velocity to $+21$ m/s. What was its final acceleration?

■ 19.   A jogger running at $+4.0$ m/s undergoes the same acceleration during 5.0 s as a bus whose velocity increases from $+16.0$ m/s to $+20.0$ m/s in 8.0 s. Calculate the jogger's final velocity.

■ 20.   (a) Calculate the acceleration of a car whose velocity changes from -10 m/s to -20 m/s in 4.0 s. (b) Calculate the car's acceleration if its velocity changes from -20 m/s to -18 m/s in 2.0 s. (c) Provide some physical insight for why the sign of the acceleration is different in (a) and (b). (d) Calculate the

car's final velocity if its acceleration for the next 2.0 s is -3.0 m/s².

■ 21.   A car traveling at 16 m/s crashes into an embankment and stops in 0.080 s. Calculate its average acceleration and indicate the direction of the acceleration relative to the initial velocity.

■ 22.   An airport runway is 1000 m long. An airplane must attain a speed of 66 m/s to take off. (a) If the average speed of the airplane down the runway is 33 m/s, for what length of time is the plane on the runway? (b) What is the average acceleration of the plane?

23.   The velocity of a drag racer at different times after the start of a race is given in the following table. Calculate (a) the racer's acceleration during each time interval and (b) the average acceleration for the first 4 s.

t (s)	0.0	0.5	1.0	1.5	2.0	2.5	3.0	3.5	4.0
v (m/s)	0.0	5.4	14.3	25.7	38.0	49.2	58.1	67.3	74.4

■ 24.   The light turns yellow as you approach a stoplight in a car moving at 18 m/s. After a 1.0-s reaction-time delay, the brakes are applied causing the car to decelerate at 3.0 m/s² until it stops (the acceleration is opposite the initial velocity). (a) Calculate the distance the car travels after the light turns yellow and before the brakes are applied. (b) Calculate the time needed to stop the car after the brakes are applied and (c) the distance the car travels during that time. (Assume that the car's average speed during that time is the average of its initial and final velocities.) (d) Calculate the total stopping distance, the sum of the answers to (a) and (c).

■ 25.   A rock on Mars falls from rest. Its vertical positions below its starting point at different times are listed in the following table. Calculate the rock's average speed (a) from 0 s to 1 s and (b) from 2 s to 3 s. (c) Estimate the rock's average acceleration from 0.5 s to 2.5 s.

t (s)	0	1	2	3
x (m)	0	1.9	7.7	17.3

2.6   Linear Motion with Constant Acceleration

26.   A Triumph TR-7 starts at rest and accelerates to a speed of 27 m/s (equivalent to 60 mi/h) in 11.8 s. (a) Calculate the car's average acceleration. (b) Calculate the distance traveled during that acceleration.

27.   A soapbox derby race car starts at rest and moves 300 m down a track in 22.4 s. Calculate its average acceleration (assumed constant) and its speed at the end of the track.

28.   A car traveling at 35 mi/h is stopped in a distance of 2.5 ft when it hits a tree. (a) Calculate the average deceleration of the car. (b) Calculate the time needed to stop the car (assuming constant deceleration).

29.   A skydiver slows from 52 m/s to 8 m/s in 0.80 s as her parachute opens. Calculate the magnitude of her deceleration (assumed constant) and the distance she falls while the parachute is opening.

30.   To make a long voyage, a spaceship must travel as fast as possible. Relativistic considerations limit the ship's speed to about one-tenth the speed of light (the speed of light is

3.0×10^8 m/s). To avoid physiological problems for the ship's occupants, the ship can accelerate at only 10 m/s². (a) What length of time in years is required for the ship to attain its final speed? (b) How far does it travel in that time?

■ 31. A bus driver forgets to stop at a bus stop. Just as she passes the stop, she decelerates. Five seconds later the bus is 80 m beyond the stop and has slowed to 10 m/s. (a) Calculate the speed of the bus as it passed the bus stop. (b) What is the deceleration of the bus (assumed constant)?

■ 32. The driver of a car traveling at 24 m/s sees a cow on the road. (a) Calculate the stopping distance if the driver's reaction time before applying the brakes is 0.75 s and the car's deceleration after the brakes are applied is 3.6 m/s². (b) Suppose the cow is 92 m from the car when seen by the driver. Calculate the deceleration needed so that the car stops just as it reaches the cow.

■ 33. Most driver's manuals show the distances required to stop a car when traveling at various speeds. The following table is taken from the 1976 *California Driver's Handbook*:

STOPPING DISTANCE: From Eye to Brain to Foot to Wheel to Road (with Perfect 4-Wheel Brakes and Ideal Conditions)

MPH	Thinking Distance	Braking Distance	Total Distance Traveled
25	27 ft	34.4 ft	61.4 ft
35	38 ft	67.0 ft	105.0 ft
45	49 ft	110.0 ft	159.0 ft
55	60 ft	165.0 ft	225.0 ft
65	71 ft	231.0 ft	302.0 ft

(a) Calculate the reaction time associated with the "thinking distance" for two different speeds. Assume that the car travels at a constant speed during that time. (b) Calculate the deceleration associated with the "braking distance" for two different speeds. (c) Using the averages of your numbers for (a) and (b), calculate the total stopping distance for a speed of 40 mi/h.

■■ 34. The maximum acceleration of a subway train is ±2.0 m/s² and its maximum speed is 22 m/s. Calculate the minimum time needed for it to travel 990 m between stops at adjacent stations.

■ 35. The average of any quantity that changes uniformly is the sum of its initial and final values divided by 2. We used this fact to derive Eq. (2.10). Try it for the following examples: (a) Your semester grade-point averages increase uniformly as follows: 2.4, 2.7, 3.0, 3.3. Calculate your average for the four semesters, first by taking the average of all four numbers, second by taking the average of the first and last numbers. (b) Your weight decreases uniformly each month: 150, 143, 136, 129, 122 lb. Calculate your average weight during the five months using both techniques. (c) Invent some other example of a uniformly changing quantity and present similar calculations for it.

■■ 36. A herd of bulls is stampeding at a speed of 7 m/s. A boy at rest 20 m in front of the bulls accelerates at 3 m/s². Will the herd trample the boy? Explain.

■■ 37. A truck moving at a constant speed of 20 m/s passes a car just as the car starts from rest with an acceleration of 2.0 m/s². How far does the car travel before it overtakes the truck?

■■ 38. A subway train accelerates from rest at 1.0 m/s². At the same instant as the train starts, another train is 1000 m from the first train and moving toward it on another track at a constant speed of 20 m/s. How far from the starting point of the first train are the two trains when they meet?

2.7 Gravitational Acceleration

39. A stone that is initially at rest is dropped from the top of a building. It falls for 5.0 s before hitting the ground. How tall is the building?

40. Divers in Acapulco fall from a cliff that is 36 m above the water. *Estimate* their speed when they hit the water.

41. A basketball player can jump 0.72 m off the floor from a standing position. Calculate the player's speed when leaving the floor.

■ 42. A rock is thrown down into a well with an initial downward speed of 3.0 m/s. The rock hits bottom 1.8 s later. (a) How deep is the well? (b) What is the rock's speed as it hits the bottom?

■ 43. Do you recall Galileo's famous experiment at the Leaning Tower of Pisa (the tower is 55 m high)? If Galileo had accidentally dropped one of the rocks 0.50 s before the other, with what velocity would the second rock have had to be thrown downward to reach the ground at the same time as the first rock?

■ 44. A building is 50 m tall. A jealous lover wishes to drop a tomato from the top of the building onto the head of his girlfriend's suitor. If the suitor walks at a speed of 1.0 m/s, how many meters should the suitor be from the position below the tomato when it is dropped? (See Fig. 2.25.)

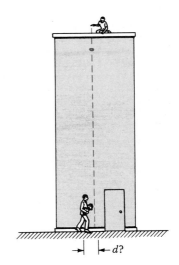

FIG. 2.25

■ 45. A construction worker ascending at 6.0 m/s in an open elevator 42 m above the ground accidentally drops a hammer. Calculate the time needed for the hammer to reach the ground and (b) its velocity just before touching the ground.

■ 46. A person in a hot air balloon that is 25.0 m above the ground and descending at 2.5 m/s drops a bag of ballast. How much time does a 2.0-m tall crew member standing below the balloon have to jump out of the way of the falling ballast?

■■ 47. A burglar drops a bag of loot from a window in a hotel. The bag takes 0.15 s to pass the 1.6-m tall window of your room as it falls toward the ground. How far above the top of your window is the burglar who dropped the bag? (The bag's initial speed was zero.)

2.8 Projectile Motion

■ 48. An airplane traveling horizontally at 120 m/s at an elevation of 240 m drops a load of flame retardant on a forest fire. At what horizontal distance from the flames should the retardant be dropped?

■ 49. A car drives off a horizontal embankment and lands 11 m from the edge of the embankment in a field that is 3 m lower than the embankment. With what speed was the car traveling when it left the embankment?

■ 50. The unofficial record for a single jump by a frog was set by a South African sharp-nosed frog that leaped 5 m when being retrieved for placement in its container. If this frog left the ground at an angle of 40°, what was its initial speed?

■ 51. Daring Darless wishes to cross the Grand Canyon of the Snake River by being shot from a cannon. She wishes to be launched at 60° to the horizontal so that she can spend more time in the air waving to the crowd. With what speed must she be launched to cross the 520-m canyon?

■ 52. A football punter wants to kick the ball so that it hits the ground 50 m from where it is kicked and so that the ball is in the air 4 s. At what angle and with what initial speed should the ball be kicked? Assume that the ball leaves the kicker's foot from an elevation of 1 m.

■ 53. A baseball leaves the bat of Henry Aaron with a speed of 34 m/s at an angle of 37° above the horizontal. The ball is 1.2 m off the ground when it leaves the bat. To be a home run, the ball must clear a fence that is 3.0 m high and 106 m from the batter. (a) At what times after being hit will it reach the height of the fence? (b) How far from the batter will the ball be at these times? (c) Will Henry have a home run? Explain.

■ 54. A tennis ball is served from position *A* in Fig. 2.24 such that it leaves the racket 2.4 m above the ground in a horizontal direction with a speed of 22 m/s (50 mi/h). (a) Will the ball be able to cross a 0.91-m-high net 11.9 m in front of the server? (b) Will the ball land in the service court, which is within 6.4 m of the net on the other side?

2.9 Graphical Analysis of Velocity and Acceleration

■ 55. A car's changing position as a function of time is shown in Fig. 2.26. During what time period(s) was the car (a) at rest,

(b) moving in the positive direction, and (c) moving in the negative direction? Estimate the car's greatest speed in (d) the positive direction and (e) the negative direction and the times at which these speeds occurred.

■ 56. A car's velocity as a function of time is shown in Fig. 2.27. Calculate the car's acceleration during each segment of the trip.

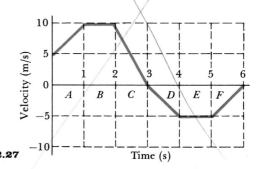

FIG. 2.27

■ 57. Estimate the acceleration of the skydiver whose velocity-versus-time graph is shown in Fig. 2.23 at (a) 0.0 s, (b) 1.0 s, (c) 2.0 s, and (d) 3.0 s. (e) Using these estimates, construct an acceleration-versus-time graph and draw a smooth line through the points on the graph.

■ 58. (a) Construct a position-versus-time graph for the rock dropped on Mars using the data from Problem 25. Draw a smooth line through the data points and estimate the slope of the graph (the rock's velocity) at (b) 0.5 s, (c) 1.5 s, and (d) 2.5 s. (e) Using this information, construct a velocity-versus-time graph with a smooth line passing through the data points and estimate the rock's acceleration as it falls.

■ 59. Estimate the displacement of the skydiver whose velocity-versus-time graph is shown in Fig. 2.23 during the time interval from (a) 0.0 to 1.0 s, (b) 1.0 to 2.0 s, and (c) 2.0 to 3.0 s. (d) If the skydiver's position at time zero is 0 m, roughly what is the diver's location at 1.0 s, 2.0 s, and 3.0 s?

■ 60. An acceleration-versus-time graph of the head of a dummy hit by a heavyweight boxer is shown in Fig. 2.28. Estimate the final velocity of the head if it starts at rest.

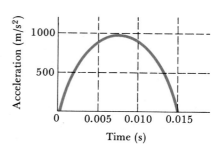

FIG. 2.28

■ 61. The initial velocity of a baseball just before it hits a bat is −34 m/s (the positive direction points away from the bat opposite the initial velocity). If the ball's acceleration as a function of time while in contact with the bat is as shown in Fig. 2.29, estimate the ball's final velocity.

■ 62. The brakes of a car traveling at an initial speed of 18 m/s (40 mi/h) are applied to stop for a washed-out bridge 80 m

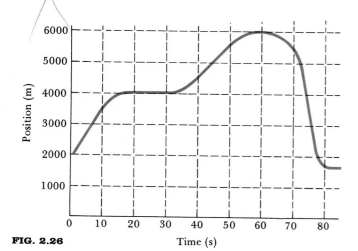

FIG. 2.26

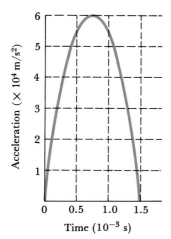

FIG. 2.29

ahead. If the car's velocity changes as shown in Fig. 2.30, will the car stop before reaching the bridge or will the car and driver take a dip? Explain.

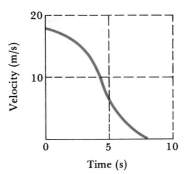

FIG. 2.30

A proportion is defined as the equality of two ratios; for instance, $a/b = c/d$ is a proportion. Proportions can be used to determine the expected change in one quantity when another quantity changes. Suppose that the cable for an elevator is replaced by a new cable whose diameter is 50 percent larger. By what percentage does the weight that can be lifted increase? (It is 125 percent, not 50 percent, as we learn later.) Proportions can be used in everyday life to answer questions such as whether a large container of a product is a better buy on a cost-per-unit-weight basis than a small container.

To illustrate the use of proportions, consider the pricing of pizza. Suppose that a small pizza costs a certain amount. How much should a larger pizza cost? If the cost depends on the amount of ingredients used, then the cost should increase in proportion to the pizza's area, not its diameter:

$$\text{Cost} = k(\text{Area}) = k(\pi r^2), \tag{I.1}$$

where r is the radius of the pizza and k is a constant that depends on the price of ingredients per unit area. If the area of the pizza doubles, the cost should double, but k remains unchanged.

Let us rearrange Eq. (I.1) so that the two variable quantities (cost and radius) are on the right side of the equation, and the constants are on the left:

$$k\pi = \frac{\text{Cost}}{r^2}.$$

This equation should apply to any size pizza. If r increases, the cost should increase in such a way that the ratio Cost/r^2 remains constant. Thus, we can write a proportion for pizzas of two different sizes:

$$k\pi = \frac{\text{Cost}}{r^2} = \frac{\text{Cost}'}{r'^2}. \tag{I.2}$$

EXAMPLE I.1   A 7-in pizza costs $2.50. How much should a 10-in pizza cost?

SOLUTION   Substituting Cost = $2.50 for a pizza of radius $r = 7\text{ in}/2 = 3.5$ in, we can then use Eq. (I.2) to calculate the cost of a pizza having a radius $r' = 10\text{ in}/2 = 5$ in:

$$\text{Cost}' = \frac{r'^2}{r^2}(\text{Cost}) = \frac{(5\text{ in})^2}{(3.5\text{ in})^2}(\$2.50) = \underline{\$5.10}. \qquad \blacksquare$$

This procedure can be used for most equations relating two quantities that change while all other quantities remain constant. We simply place all the constant quantities on one side of the equation and the variables on the other side. If one variable changes, the other variable must change in a manner that keeps the variable side of the equation equal to the side with constants.

EXAMPLE I.2 The resistive drag force of air on a moving car F_D depends on the car's speed v according to the following equation:

$$F_D = \frac{1}{2} C_D \rho A v^2, \qquad \text{(I.3)}$$

where A is the car's cross-sectional area, ρ is the density of air, and C_D is a unitless coefficient called the drag coefficient whose value depends on the car's shape and is approximately equal to 1. For a particular car, C_D, ρ, and A remain nearly constant as the car's speed changes. If the drag force is 130 N (29 lb) when the car travels at 15 m/s, what is the drag force at 25 m/s?

SOLUTION We first rearrange Eq. (I.3) so that the constants are on one side and the variables (F_D and v) are on the other:

$$\frac{1}{2} C_D \rho A = \frac{F_D}{v^2}.$$

If v changes, F_D must change so that the ratio F_D/v^2 remains equal to the constants on the left. Thus, for two different speeds, v and v', the corresponding drag forces, F_D and F'_D, must be related by the proportion

$$\frac{1}{2} C_D \rho A = \frac{F_D}{v^2} = \frac{F'_D}{v'^2},$$

or

$$F'_D = \frac{v'^2}{v^2} F_D = \left[\frac{(25 \text{ m/s})^2}{(15 \text{ m/s})^2} \right] 130 \text{ N} = \underline{360 \text{ N}} \text{ (81 lb).} \qquad \blacksquare$$

Sometimes we are told that a quantity changes by a certain percent. Suppose that the actual length L of an object is 5 percent less than the desired length L_0. In that case the actual length L is related to the desired length L_0 by the equation $L = 0.95L_0$. If the length is 10 percent greater than it should be, then $L = 1.10L_0$. A small percent change in one quantity can sometimes cause a much larger change in another quantity, such as the volume of a cord of wood.

EXAMPLE I.3 A cord of wood costs $120 and has the following dimensions: length $L_0 = 8.0$ ft, width $W_0 = 4.0$ ft, and height $H_0 = 4.0$ ft. By what percent should the price of the cord be reduced if each dimension is reduced by 10 percent? What is the reduced price?

SOLUTION The cost of the wood should be proportional to its volume V:

$$\text{Cost}_0 = kV_0$$

or

$$k = \frac{\text{Cost}_0}{V_0}.$$

The constant k depends on the cost per unit volume and should be the same for a large volume as for a small volume. As volume changes, the cost should also change so that the ratio remains constant. Thus,

$$k = \frac{\text{Cost}_0}{V_0} = \frac{\text{Cost}}{V}.$$

Rearranging and substituting for the volume of a cord ($V = LWH$), we find that

$$\text{Cost} = \left(\frac{V}{V_0}\right)\text{Cost}_0 = \left(\frac{L\,W\,H}{L_0 W_0 H_0}\right)\text{Cost}_0.$$

Since each dimension is 10 percent less than it should be, we substitute $L = 0.90L_0$, $W = 0.90W_0$, and $H = 0.90H_0$ to get

$$\text{Cost} = \left[\frac{(0.90L_0)(0.90W_0)(0.90H_0)}{L_0 W_0 H_0}\right]\text{Cost}_0$$
$$= (0.90^3)\,\text{Cost}_0 = (0.73)\,\text{Cost}_0.$$

Thus, the cost should be reduced by 27 percent from the cost of a full cord, or

$$\text{Cost} = 0.73(\$120) = \underline{\$87}. \qquad \blacksquare$$

Problems

■ 1. The distance y that an object falls in time t if starting at rest is given by the equation $y = 1/2at^2$. On the moon a rock falls 10 m in 3.50 s. How much time does it take to fall 15 m, assuming that the acceleration is constant?

■ 2. A circular wool quilt of 1.2-m diameter costs $82. What should the price of a 1.6-m-diameter quilt be if it is to cost the same per unit area?

■ 3. A 0.80-cm-diameter rope will support a 5000-N weight. If the strength of a rope increases in proportion to its cross-sectional area, how much weight will a 1.00-cm-diameter rope support?

■ 4. A box of wood 1.0 m on each side costs $56. What should be the price of a box of wood that is 1.8 m on each side if its cost per unit volume equals that of the smaller box?

■ 5. The distance in which a car stops after the brakes are applied (the braking distance) depends on the square of the car's speed. Suppose that the braking distance when traveling at 18 m/s (40 mi/h) is 26 m. Calculate the braking distance when traveling at 27 m/s (60 mi/h).

■ 6. By what fraction should the driver of a car reduce its speed to decrease the drag force by 30 percent? (See Example I.2 for a discussion of drag force.)

■ 7. You decide to open a pizza parlor. The ingredients and cost factors require that you charge $3.10 for a 7-in-diameter pizza. How large should you make a pizza whose price is 50 percent more than that of the 7-in pizza?

■ 8. By what percent does the braking distance change if the speed of a car increases by 20 percent? (See Problem 5 for a discussion of braking distance.)

CHAPTER 3

Newton's Laws of Motion

In the summer of 1665 the Great (bubonic) Plague was devastating England. Ten percent of London's population died. When the universities were closed to prevent its spread, Isaac Newton, a 22-year-old student at Trinity College, Cambridge, returned to his mother's farm in Woolsthorpe, Lincolnshire. The next eighteen months of his life, spent at the farm, initiated one of the most productive periods of achievement by an individual in the history of science. The period culminated about 22 years later with the publication of Newton's major work, a book entitled *Philosophiae Naturalis Principia Mathematica*. During this relatively short time, Newton had invented differential and integral calculus, discovered the universal law of gravitation, developed a new theory of light, and put together the ideas for his three laws of motion, the subject of this chapter.

Three hundred years after their formulation, these three laws still serve as the basis of **dynamics,** a branch of physics that analyzes the effect of force in causing accelerated motion. How much force is required to stop your body during a car collision? To catch a child jumping from a wall? To hit a baseball for a 360-ft home run? Every dynamic process in nature in one way or another involves these three laws.

3.1 Newton's First Law of Motion

Newton's description of motion represented a dramatic change from the ideas of Aristotle, the Greek scientist who two thousand years earlier had written extensively about the mechanical behavior of nature. At Newton's birth, Aristotle's work was still considered the ultimate authority on the subject.

Aristotle distinguished two types of motion: natural and nonnatural. The motion of heavenly bodies in a circle was considered natural, as was the motion of an apple falling toward the earth. These objects seemed to move without a visible source of force pulling or pushing them. The motion was considered natural because it continued unaided by human activity.

The motion of an object being pulled or pushed by a visible source of force, such as a rope pulling a sled, was considered nonnatural. The object's motion

The arrows represent forces acting on the skateboard and its rider. Acceleration occurs if the resultant force is not zero, as we learn shortly.

continued only if acted on by the force. For example, if you push a book lying on a table, it will slide. When the force stops, so does the book. According to Aristotle, it was nonnatural for the book to slide; hence it stopped when the force was removed.

Newton saw the situation differently. A moving book slides to a stop not because it is involved in a nonnatural motion but because a friction force opposes its motion. Without friction, the book would continue sliding in a straight line until it fell off the table. The motion that Aristotle considered natural because it continued unaided by human assistance was also caused in many situations, according to Newton, by a force. For example, the acceleration of a falling apple was caused by the gravitational force of the earth pulling down on the apple.

As we see, force plays a crucial role in Newton's description of nature. **Force** can be considered qualitatively as the push or pull of one object on another. A car pulls a trailer; a cable lifts an elevator; a shopper pushes a grocery cart. Newton's first law of motion describes situations in which the net force on an object is zero.

Newton's first law of motion: Every body continues in its state of rest or of constant speed in a straight line unless it is compelled to change that state because of forces acting on it.

According to Newton's first law, an object sitting at rest will remain at rest if the sum of all forces acting on the object is zero. Similarly, if an object is moving and if the sum of the forces acting on the object is zero, then the object will continue to move in the same direction with the same speed. This tendency of objects to remain at rest or to continue in motion is called **inertia**. To change the motion of an object with inertia, unbalanced forces must act on the object—the subject of Newton's second law.

3.2 Newton's Second Law of Motion and Mass

According to Newton, an object accelerates if acted on by an unbalanced force. Consider the situations depicted in Fig. 3.1. Each object is subjected to an unbalanced force. In Fig. 3.1a, a person steps off a high diving board. The force labeled **w** is the person's weight and is caused by the gravitational pull of the earth on the person. Since no other force balances **w,** that is, pushes up in the opposite direction with equal magnitude, the person accelerates toward the water. In Fig. 3.1b, a moving car hits a wall, and the unbalanced force of the wall causes the car to slow down and stop—that is, to decelerate.

These examples point to a general rule: An unbalanced force causes an object's motion to change. If the force is in the direction of motion, the object moves faster; it accelerates. If the force opposes the motion, the object slows down; it decelerates.

The acceleration, according to Newton, depends on two factors: (1) the resultant force acting on the object and (2) the mass or amount of matter of which the object is made. Figure 3.2 illustrates the relation between force and acceleration. In Fig. 3.2a, a force **F** causes a car to have an acceleration **a.** If the force is doubled, as in Fig. 3.2b, the acceleration doubles. If the force is tripled,

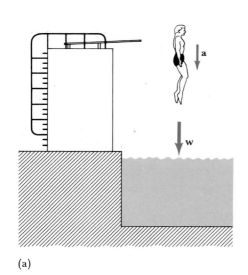

(a)

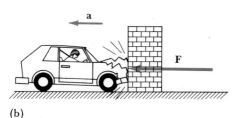

(b)

FIG. 3.1. Accelerated or decelerated motion caused by an unbalanced force.

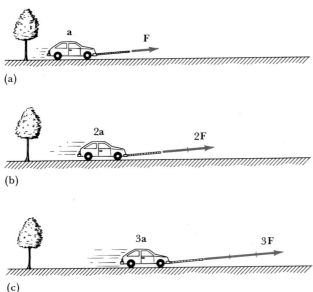

(a)

(b)

(c)

as in Fig. 3.2c, the acceleration triples. The acceleration increases in direct proportion to the resultant force applied to the object:

$$\mathbf{a} \propto \mathbf{F}.$$

The symbol $\propto$ means "is proportional to."

An object's acceleration also depends on its mass.

Mass is often defined qualitatively as the amount of matter of which an object is made. Mass is also defined as a measure of an object's inertia. The greater an object's mass, the greater its inertia and the less its motion changes when pushed or pulled by a force.

The idea of inertia can be illustrated by considering defensive linemen on a football team. Most of these young men have relatively large mass, and so they are difficult to move. A large force causes little acceleration. A large mass is also advantageous for a running back on a football team. The larger his mass, the less he decelerates or slows down when a force is applied.

The quantitative relation between mass and acceleration is illustrated in Fig. 3.3. In Fig. 3.3a, a force exerted on a single cart having mass m causes the cart to have an acceleration **a**. In Fig. 3.3b, two identical carts attached together and having a total mass $2m$ are pulled by the same force. However, their acceleration is only $\mathbf{a}/2$ instead of **a**. Four carts of total mass $4m$, shown in Fig. 3.3c, experience an acceleration $\mathbf{a}/4$ when pulled by the same force. The acceleration caused by a force is inversely proportional to the mass of an object or objects on which the force acts. This inverse relation between acceleration and mass is expressed as follows:

$$\mathbf{a} \propto \frac{1}{m}.$$

By choosing the units of force in an appropriate fashion, we can combine the two proportionality equations ($\mathbf{a} \propto \mathbf{F}$ and $\mathbf{a} \propto 1/m$) into a single equation

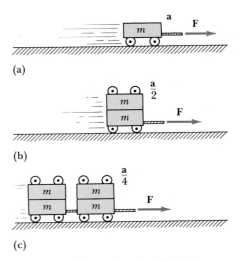

(a)

(b)

(c)

FIG. 3.3. The acceleration of objects on which the same force acts is less for objects with greater mass.

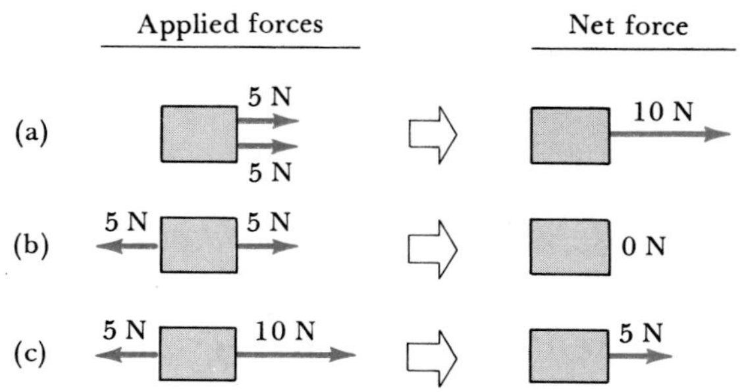

FIG. 3.4. An object's acceleration depends on the net force acting on it.

$(\mathbf{a} = \mathbf{F}/m)$ called Newton's second law. The units for this equation are discussed in the next section.

Newton's second law of motion: An unbalanced force acting on an object will cause the object to accelerate in the direction of the force. The acceleration is directly proportional to the resultant force acting on the object and inversely proportional to the object's mass. If a group of forces act on an object of mass m, the vector sum of these forces $(\Sigma \mathbf{F})$* causes the object to have an acceleration $\mathbf{a}$, given by

$$\mathbf{a} = \frac{\Sigma \mathbf{F}}{m}, \quad \text{or} \quad \Sigma \mathbf{F} = m\mathbf{a}. \tag{3.1}$$

The direction of $\mathbf{a}$ is in the direction of the resultant force.

Newton's second law is often written as $\mathbf{F} = m\mathbf{a}$. The force $\mathbf{F}$ on the left side of the equation, however, represents the net or resultant force acting on the object. For example, if three forces $\mathbf{F}_1$, $\mathbf{F}_2$, and $\mathbf{F}_3$ act on an object, then

$$\Sigma \mathbf{F} = \mathbf{F}_1 + \mathbf{F}_2 + \mathbf{F}_3 = m\mathbf{a}.$$

This idea is illustrated in Fig. 3.4. Note that if the forces acting on an object add to zero, as in Fig. 3.4b, the object's acceleration is zero: if the object is at rest, it remains at rest; if it is moving, its velocity remains constant.

3.3 Units for Newton's Second Law

When using Newton's second law, the units of force, mass, and acceleration must all be from the same system of units. In this text SI metric units are used almost exclusively. However, English system units will be introduced and used occasionally because of their widespread use in the past.

SI Metric System

The unit of acceleration in the SI system is the **meter/second²** (m/s^2). Both meters and seconds are basic units with well-defined standards. The SI unit of mass is the **kilogram** (kg). One kg is a basic unit defined as the mass of a solid

*The Greek letter sigma, Σ, means summation. Thus, $\Sigma \mathbf{F}$ represents the sum of all forces acting on the object.

platinum-iridium cylinder stored in a special vault in the small town of Sèvres near Paris, France.

The SI unit of force is called the **newton** (N). A medium-sized apple weighs 1 newton; it is a little less than one-fourth of a pound. The newton is defined not in terms of a basic standard but by using Newton's second law. A force of 1 newton is an unbalanced force that will cause a 1-kg mass to experience an acceleration of 1 m/s². Substituting these units into Newton's second law ($\Sigma \mathbf{F} = m\mathbf{a}$) produces an expression relating newtons to the other units:

$$1 \text{ N} = (1 \text{ kg})(1 \text{ m/s}^2). \tag{3.2}$$

This expression is used often in the text for making unit conversions. To remember it, just remember $\mathbf{F} = m\mathbf{a}$ and the units for each quantity.

English System

The unit of acceleration in the English system is the **foot/second²** (ft/s²) and the unit of force is the **pound** (lb), which is defined in terms of the newton:

$$1 \text{ lb} = 4.45 \text{ N},$$

or

$$1 \text{ N} = 0.225 \text{ lb} \cong \frac{1}{4} \text{ lb}.$$

The unit of mass in the English system is called the **slug** and is defined using Newton's second law in the form $m = \mathbf{F}/\mathbf{a}$. One slug is that mass which experiences an acceleration of 1 ft/s² when acted on by an unbalanced force of 1 lb:

$$1 \text{ slug} = \frac{1 \text{ lb}}{1 \text{ ft/s}^2}. \tag{3.3}$$

A one-slug mass is equal to a 14.6-kg mass:

$$1 \text{ slug} = 14.6 \text{ kg},$$

or

$$1 \text{ kg} = 0.0685 \text{ slug}.$$

One slug equals approximately the mass of four gallons of milk.

Warning: When using any of the laws or equations of physics, express all quantities in the same system of units. For example, do not substitute mass in units of kg and acceleration in units of ft/s² into Newton's second law to calculate force. The ft/s² must first be converted to m/s². Also, each quantity must be expressed using the appropriate unit for the system of units being used. For example, when using the metric system, mass must be in kilograms and not in grams or milligrams. If mass is given in grams, first convert to kilograms before substituting into an equation.

In the examples that follow in this section, Newton's second law is applied only to one-dimensional motion, in which case we need not use vector notation. In Section 3.6 we consider more carefully the vector nature of Newton's second law.

EXAMPLE 3.1 A 0.145-kg baseball thrown by fastball pitcher Bob Feller was measured to have left his hand at a speed of 44 m/s (98 mi/h). (a) Calculate the average acceleration of the ball while pushed forward in his hand a distance of 2.8 m. (b) Calculate the average force of his hand on the ball during this acceleration.

SOLUTION (a) You should recognize that this part of the problem involves kinematics, an analysis of the ball's motion to determine its acceleration. Since we know the distance the ball moved while being accelerated ($x - x_0 = 2.8$ m) and the ball's velocity at the start ($v_0 = 0$) and at the end when it left his hand ($v = 44$ m/s), we can use Eq. (2.12) to determine the ball's average acceleration while in his hand:

$$a = \frac{v^2 - v_0^2}{2(x - x_0)} = \frac{(44 \text{ m/s})^2 - 0^2}{2(2.8 \text{ m})} = 346 \text{ m/s}^2.$$

Rounding off to two significant figures, the ball's average acceleration was 350 m/s².

(b) We can now use Newton's second law to determine the force needed to produce this acceleration:

$$F = ma = (0.145 \text{ kg})(346 \text{ m/s}^2)$$
$$= 50 \frac{\text{kg} \cdot \text{m}}{\text{s}^2} = 50 \text{ N}.$$

In Example 3.1 we first used kinematics to determine acceleration and then Newton's second law to determine an unknown force. The procedure is reversed in Example 3.2.

EXAMPLE 3.2 A Boeing 707 jet with a takeoff mass of 1.2×10^5 kg has four engines each producing an average net thrust of 7.5×10^4 N during take-off. (a) Calculate the plane's average acceleration down the runway and (b) the distance it must travel on the runway to attain its 73 m/s liftoff speed. Ignore air resistance and other friction-type forces acting on the plane.

SOLUTION (a) The net force acting on the plane from its four engines is $4(7.5 \times 10^4 \text{ N}) = 3.0 \times 10^5$ N. Hence, according to Newton's second law, its acceleration is

$$a = \frac{F}{m} = \frac{3.0 \times 10^5 \text{ N}}{1.2 \times 10^5 \text{ kg}} = 2.5 \text{ m/s}^2.$$

(b) Next we use kinematics to determine the displacement $x - x_0$ needed for the plane's velocity to change from an initial value $v_0 = 0$ to its liftoff velocity $v = 73$ m/s. The plane's acceleration is $a = 2.5$ m/s². We can rearrange Eq. (2.12) for this purpose:

$$x - x_0 = \frac{v^2 - v_0^2}{2a} = \frac{(73 \text{ m/s})^2 - 0^2}{2(2.5 \text{ m/s}^2)}$$
$$= 1066 \text{ m}.$$

Rounding off to two significant figures, the plane will need to travel 1100 m before reaching its liftoff velocity.

3.4 Weight

The force of gravity affects almost all of nature—from the energy needed to climb stairs or rise from a chair to the tides caused by the gravitational attraction of the sun and moon on the oceans. Weight is just one example of a gravitational force. When you are standing on the earth's surface, your weight is the gravitational force of the earth's mass pulling on your body. If you were standing on the moon, your weight would be reduced by one-sixth because the moon's mass is smaller and exerts less force on your body. (The force also depends on the radius of the moon, as we will learn in Section 5.6.)

Weight is defined as the gravitational force exerted on an object because of its attraction to some other mass such as the earth.

An object's weight when on the earth's surface is related by a simple equation to its mass. Suppose that an object with mass m is falling freely (we ignore air resistance). The cause of its acceleration is the object's weight w—the force of gravity pulling down on it. The object experiences a downward acceleration of magnitude g equal to 9.8 m/s² or 32 ft/s². Substituting weight for force and g for acceleration into Newton's second law ($\Sigma F = ma$), we see that

$$w = mg. \tag{3.4}$$

If an object's mass is known, Eq. (3.4) can be used to calculate the object's weight when near the earth's surface, and vice versa.

Weight and mass are often incorrectly substituted for each other in Newton's second law. *Weight, the force of gravity, is in units of newtons or pounds, and it appears with other forces on the left side of Newton's second law ($\Sigma F = ma$).* Mass represents an inertial property of an object—how effective forces are in changing the object's motion. *An object's mass appears in units of kilograms or slugs on the right side of Newton's second law.*

How does the astronaut's mass and weight when on the moon compare to his mass and weight when on earth? (One quantity is the same both places.)

EXAMPLE 3.3 Calculate the weight of a 0.510-kg jar of Smucker's natural peanut butter.

SOLUTION Mass is in units of kg. Thus, the peanut butter's mass must be 0.510 kg. Its weight is

$$w = mg = (0.510 \text{ kg})(9.8 \text{ m/s}^2) = 5.0 \, \frac{\text{kg} \cdot \text{m}}{\text{s}^2} = \underline{5.0 \text{ N}}. \qquad \blacksquare$$

3.5 Newton's Third Law

The great potential for analyzing interesting and complex situations by using Newton's second law is not apparent from the relatively simple examples presented so far. As situations become more complex and interesting, we must use greater care to identify which forces act on an object to cause its acceleration. The words *act on* are especially important.

Newton indicated in his third law of motion that forces always come in pairs—an action force and its reaction force. The action force is the force exerted by one object on another. The reaction force is the force exerted by the other object on the first. For example, rest your forearm against the edge of the pages

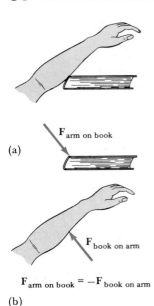

$$\mathbf{F}_{\text{arm on book}} = -\mathbf{F}_{\text{book on arm}}$$

(b)

FIG. 3.5. The force (a) of the arm on the book is equal in magnitude but opposite in direction to the reaction force (b) of the book on the arm.

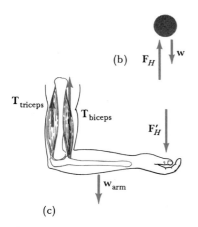

FIG. 3.6. (a) A woman throws a ball. (b) A force diagram of the ball and (c) of the woman's forearm.

of this book, as illustrated in Fig. 3.5. Your arm exerts a force into and against the book. But the book also exerts a reaction force into and against your arm, as is apparent from the indentation of your skin and muscle at the region of contact. According to Newton's third law of motion, these forces have equal magnitudes but point in opposite directions and act on different objects. (For this example, one force acts on the book, the other on the arm.)

Newton's third law of motion: Whenever one object exerts a force on a second object, the second object exerts a reaction force of equal magnitude but opposite direction on the first object.

When using Newton's second law of motion to calculate an object's acceleration, we must consider only the forces acting on the object. The reaction forces that it exerts on other objects do not affect its acceleration. These reaction forces, however, affect the other objects and can be considered if we wish to study their acceleration.

Suppose, for example, that a person throws a ball in the air, as shown in Fig. 3.6a. If we are interested in the speed of the ball as it leaves the person's hand, we might construct what is called a **force diagram** for the ball. (Force diagrams are sometimes called **free-body diagrams.**) In the diagram, we draw arrows representing all forces acting on the object. For the ball shown in Fig. 3.6b, a force $\mathbf{F}_H$ represents the force of the hand pushing up on the ball and $\mathbf{w}$ represents the ball's weight, a downward force. However, if our concern is with the muscle tension in the thrower's arm, we construct a force diagram of the forearm, as shown in Fig. 3.6c. Notice that the ball exerts a reaction force $\mathbf{F}'_H$ down on the thrower's hand. The magnitude of the reaction force equals $\mathbf{F}_H$ but points in the opposite direction.

We see that we must define carefully the object whose motion we are considering. Then, when constructing a force diagram for that object, we consider only the forces acting on it and not the reaction forces that it exerts on other objects. Choosing an object for a force diagram helps us isolate our attention on one part of a complicated situation.

3.6 Applications of Newton's Second Law in Component Form

In this section we add two new features to our Newton's law analysis of dynamics problems. First, the objects of our interest will usually have several forces acting on them. Second, the forces in many cases are *not* directed along a single axis but instead point in different directions in a plane. For these problems, it is most convenient to use Newton's second law in component form.

In the component method each force that acts on an object is broken into its x and y components. Because force is a vector quantity, the component forces produce the same effect on the object as the force itself does. The x components of forces cause acceleration along the x axis whereas the y components cause acceleration along the y axis. The component form of Newton's second law is summarized as follows:

Suppose that all forces $(\mathbf{F}_1, \mathbf{F}_2, \mathbf{F}_3, \ldots)$ acting on an object of mass m cause it to have an acceleration $\mathbf{a}$. The sum of the x components of these forces

$(F_{1x}, F_{2x}, F_{3x}, \ldots)$ equals the product of the object's mass and x component of acceleration a_x. Similarly, the sum of the y components of these forces $(F_{1y}, F_{2y}, F_{3y}, \ldots)$ equals the product of the object's mass and y component of acceleration a_y. **Newton's second law** rewritten in component form becomes:

$$\Sigma F_x = F_{1x} + F_{2x} + F_{3x} + \cdots = ma_x \qquad (3.5)$$
$$\Sigma F_y = F_{1y} + F_{2y} + F_{3y} + \cdots = ma_y \qquad (3.6)$$

When working problems, we can usually choose our coordinate axes so that the acceleration along one of the axes is zero. Consider the young man holding the flowerpot shown in Fig. 3.7a. Experience tells us that if the man lets go of the pot, it falls to the ground. If he pushes or lifts hard in an upward direction, the pot rises. Evidently, two forces act on the pot: one pushing up and one pulling down. If the two forces balance, the pot remains stationary. If the upward force becomes greater than the downward force, the pot rises, and vice versa.

The two forces are represented by vectors $\mathbf{F}_N$ and $\mathbf{w}$ in Fig. 3.7b. $\mathbf{F}_N$ is called a **normal force** and represents the force of the man's hand pushing up on the pot. Two objects exert normal forces on each other whenever they touch. The direction of a normal force is always perpendicular to the surface of contact. (In physics the word *normal* means perpendicular.) The force labeled $\mathbf{w}$ is the pot's weight and is caused by the gravitational pull of the earth on the pot.

In Fig. 3.7c we have added a coordinate system with the positive y axis pointing up. The two forces have been moved so that their tails are at the origin of the coordinate system. We can ignore the x-component form of Newton's second law since the forces do not push the pot in the horizontal direction; the pot's x component of acceleration is zero. However, the vertical component of acceleration may not be zero if $\mathbf{F}_N$ and $\mathbf{w}$ do not balance. The y-component form of Newton's second law [Eq. (3.6)] is written as follows:

$$\Sigma F_y = F_{N_y} + w_y = ma_y$$

The symbols F_{N_y} and w_y represent the y components of the normal force and the weight force, respectively. We have not yet indicated the magnitude or sign of these components. The y components are determined using Eq. (1.4), in which case the previous equation becomes

$$F_N \sin 90° - w \sin 90° = ma_y$$

Substituting $\sin 90° = 1$, we have

$$F_N - w = ma_y$$

The F_N and w in the previous equation are not force vectors but are instead the *magnitudes* of these forces. (Remember that magnitudes of vectors are scalar quantities with positive values, or zero.) Notice that if F_N is greater than w, the pot's acceleration is positive—it accelerates up. If F_N is less than w, the pot's acceleration is negative—it accelerates down.

Let us next apply the second law in component form to the situation shown in Fig. 3.8a on the next page. How large a tension force $\mathbf{T}$ must the rope exert on the wagon in order to cause the wagon to accelerate up the hill? A **tension force** $\mathbf{T}$ is the force exerted by a string, rope, or cable on an object to which it is attached. A tension force pulls in the direction of the rope and is exerted uniformly along its entire length.

(a)

(b)

(c)

FIG. 3.7. (a) A man holding a flowerpot. (b) The forces acting on the pot. (c) The forces with their tails at the origin of a coordinate system.

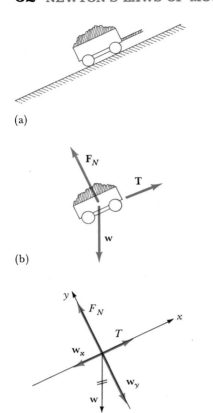

FIG. 3.8. (a) A rope exerts a tension force on a wagon. (b) The forces acting on the wagon. (c) The forces resolved into components.

Two other forces in addition to **T** act on the wagon, as shown in Fig. 3.8b. First, the normal force **F**$_N$ of the ground pushes up on the tires. The normal force is directed perpendicular to the surface of contact between the tires and the road. You might note that the tires also exert a normal force on the ground. Since we are interested only in forces acting on the wagon, we do not show the force of the wagon on the ground.

The third force acting on the wagon is the downward force of gravity **w** equal in magnitude to the wagon's weight.

In Fig. 3.8c, we have added x and y coordinate axes with the x axis drawn parallel to the incline and the y axis perpendicular. This choice of axes is useful since no motion occurs perpendicular to the plane; thus, $a_y = 0$.

In Fig. 3.8c, we move the force vectors so that their tails are at the origin of a coordinate system. Also, the weight force **w** is resolved into its components, w_x and w_y. The x component form of Newton's second law is:

$$\Sigma F_x = T_x + w_x + F_{N_x} = ma_x$$

Using Eq. (1.3) and Fig. 3.8c, we find that $T_x = T \cos 0°$, $w_x = -w \cos \theta$, and $F_{N_x} = F_N \cos 90° = 0$. Substituting these components with $\cos 0° = 1$ into the previous equation, we find that

$$T - w \cos \theta = ma_x$$

If T is greater than $w \cos \theta$, the wagon accelerates up the hill in the positive x direction. If T is less than $w \cos \theta$, the wagon accelerates down the hill in the negative x direction. If T and $w \cos \theta$ have the same magnitude, the acceleration is zero.

We need not worry about the y component equation of Newton's second law for this problem, as all motion is along the x axis. However, it is clear that the normal force **F**$_N$ and the y component of the weight force w_y must balance each other. If they do not, the wagon "sinks" into the hill ($w_y > F_N$) or rises into the air ($F_N > w_y$). We know from experience that this does not happen, which implies that F_N and w_y balance; the acceleration perpendicular to the plane in the y direction is zero.

As we see, much can be learned by considering the forces that act on an object and the components of these forces along well-chosen coordinate axes. Newton's second law and our equations of kinematics from Chapter 2 provide a powerful tool for analyzing a variety of interesting problems. For many of these problems it is essential to combine the techniques introduced earlier into a procedure that is outlined next and illustrated by a worked example in the margin.

Solving Problems in Dynamics

Our procedure for solving problems in dynamics involves five steps described in general below and illustrated for the following worked example: Determine the tension in a cable needed to lift a 2000-kg elevator with an acceleration of 1.5 m/s².

1. Make a drawing of the whole situation being considered. Identify as precisely as possible the known information and the unknown quantity you wish to determine.

The diagram for the elevator in our problem is shown in the margin.

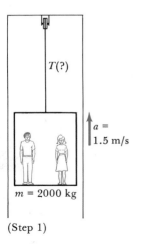

$T(?)$

$a = 1.5$ m/s

$m = 2000$ kg

(Step 1)

2. Construct a force diagram for a particular object in your drawing. The force diagram allows you to focus your attention on one small part of a complicated situation. Usually all, or all but one, of the forces acting on this object are known. Draw arrows to represent each known and unknown force acting on the object. The object's acceleration is a consequence of the forces shown in the diagram and is not itself a force. Hence, the $m\mathbf{a}$ (or ma_x and ma_y) of Newton's second law should not appear as a force in the diagram, although you should indicate the magnitude and direction of the acceleration if known.

The force diagram for the elevator is shown in the margin.

3. Superimpose a coordinate system on your force diagram. The coordinate system provides a set of axes along which you calculate the force components when solving Eqs. (3.5) and (3.6). You may wish to move your vectors so that their tails are at the origin. This makes it easier to calculate their components. You can reduce the difficulty of solving a problem by orienting your coordinate axes in particular directions, as illustrated in the examples.

The diagram in the margin shows the coordinate system for the elevator.

4. Apply Newton's second law in component form to the problem.

$$\Sigma F_x = ma_x,$$
$$\Sigma F_y = ma_y.$$

Determine the components of the forces acting on the object in your force diagram using Eqs. (1.3) and (1.4).

$$\Sigma F_x = T_x + w_x = 0 + 0 = ma_x$$
$$\Sigma F_y = T_y + w_y = T - w = ma_y$$

5. Solve the equations for the unknowns. Frequently, you will also use kinematic equations such as Eqs. (2.9)–(2.12) along with Newton's second law when solving for an unknown.

$$T = ma_y + w = ma_y + mg$$
$$= 2000 \text{ kg } (1.5 \text{ m/s}^2 + 9.8 \text{ m/s}^2)$$
$$= \underline{2.26 \times 10^4 \text{ N}}$$

The last step in this problem-solving procedure usually proceeds in one of two ways—another divide and conquer approach. If all the forces acting on an object are known, we use Newton's second law to calculate the object's acceleration. Having found the acceleration, we can then use kinematics to determine the future velocity and position of the object. This procedure is summarized as follows:

$$\frac{\Sigma \mathbf{F}}{m} \xrightarrow{\text{second law}} \mathbf{a} \xrightarrow{\text{kinematics}} \text{Changing velocity and position}.$$

On the other hand, if the changing position and velocity of an object are known, we can use kinematic equations to calculate the acceleration. Newton's

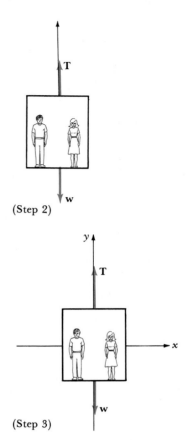

(Step 2)

(Step 3)

second law is then used to determine an unknown force responsible for the acceleration:

$$\text{Changing velocity} \atop \text{and position} \xrightarrow{\text{kinematics}} \mathbf{a} \xrightarrow{\overset{\text{second}}{\text{law}}} \frac{\Sigma \mathbf{F}}{m}.$$

The latter procedure is used in Example 3.4.

EXAMPLE 3.4 A car with an 80-kg driver runs into the back end of a parked cement mixer. The car and its driver, held in place by a seat belt and shoulder strap, stop 1.0 m from the point of impact. If the car's original speed was 18 m/s, calculate the average force of the seat belt and shoulder strap on the driver.

SOLUTION The situation is pictured in Fig. 3.9a. A force diagram for the person is shown in Fig. 3.9b. In this problem we are concerned primarily with the forward motion in the positive x direction, and we ignore the y-component equation. The x-component equation of Newton's second law is

$$\Sigma F_x = F_{N_x} + w_x + F_{B_x} \tag{3.7}$$
$$= F_N \underset{0}{\cos 90°} + w \underset{0}{\cos 90°} - F_B \underset{1}{\cos 0°} = ma_x, \tag{3.8}$$

or

$$-F_B = ma_x, \tag{3.9}$$

where F_B is the unknown magnitude of the force of the seat belt on the person. The negative sign appears in this equation because $\mathbf{F}_B$ points in the negative x direction opposite the car's motion. Without this opposing force of the seat belt, the person would continue moving forward, through the car's window, during the collision.

The above equation can be used to calculate F_B after a_x is calculated using kinematics. Remember that the car's initial speed $v_0 = 18$ m/s and its final speed is zero after decelerating to a stop in a distance of 1.0 m. Thus, using Eq. (2.12), we find that

$$a_x = \frac{v^2 - v_0^2}{2(x - x_0)} = \frac{0^2 - (18 \text{ m/s})^2}{2(1.0 \text{ m})} = -160 \text{ m/s}^2.$$

Substituting for m and a_x in Eq. (3.9), we find that

$$-F_B = ma_x = (80 \text{ kg})(-160 \text{ m/s}^2) = \underline{-13,000 \text{ N}}.$$

The magnitude of the force of the seat belt and shoulder strap on the person is 13,000 N, or about 1.4 tons of force. This force is great enough to break the person's ribs or cause other chest injuries.

In the solution of Example 3.4, we used three steps, Eqs. (3.7)–(3.9), to get Eq. (3.9), $-F_B = ma_x$. Equation (3.8) involved a rather formal evaluation of the x components of the three forces acting on the person. The x component of the person's weight was $w_x = w \cos 90°$, which equaled zero because $\cos 90° = 0$. We could have written Eq. (3.9) immediately in one step by simply inspecting Fig. 3.9b.

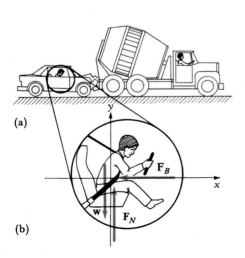

(a)

(b)

FIG. 3.9. (a) A car crashes into a cement mixer. (b) A force diagram of the car's driver.

To use the inspection method, you must remember that the *x* component of a force is its projection on the *x* axis. Notice that for the person shown in Fig. 3.9b, forces F_N and **w** are perpendicular to the *x* axis, and their projections on that axis are zero, as are their *x* components. On the other hand, force F_B is parallel to the *x* axis. Hence, its projection on that axis equals its magnitude F_B. Since F_B points in the negative *x* direction, its *x* component has a negative sign. Thus, the sum of the *x* components of all forces acting on the man is $-F_B$. We will often use this projection-inspection approach to write the component equations for Newton's second law.

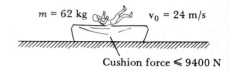

$m = 62$ kg $v_0 = 24$ m/s

Cushion force $\leqslant 9400$ N

(a)

EXAMPLE 3.5 A 62-kg stunt diver is to jump from a helicopter and land on an air cushion (Fig. 3.10a). Her speed as she reaches the cushion will be 24 m/s. To avoid injury, the average force of the cushion on her body while she is being stopped should be no more than 9400 N. Calculate the distance she must be able to sink into the cushion while being stopped by a cushion force of this magnitude.

(b)

FIG. 3.10. (a) A stunt diver lands on an air cushion after jumping from a helicopter. (b) A force diagram of the diver while being stopped by the cushion.

SOLUTION The problem is worked in two parts: (a) We first use Newton's second law to calculate the diver's deceleration while stopping, and (b) we then use kinematics to determine her stopping distance.

(a) A force diagram for the diver is shown in Fig. 3.10b. Two forces act on her: a downward weight force **w** and the upward force **F** of the cushion that stops her fall. (*Beware:* Students often forget to include an object's weight in problems involving vertical acceleration or deceleration of an object.)

If we choose the *y* axis pointing up, then the *y*-component form of Newton's second law becomes

$$\Sigma F_y = F - w = ma_y.$$

(We ignore the *x*-component equation because no forces act on the diver in the horizontal direction.) We are given that the magnitude of the stopping force **F** is 9400 N. The diver's weight $w = mg = (62 \text{ kg})(9.8 \text{ m/s}^2) = 608$ N. Rearranging the equation for Newton's second law and substituting for *F*, *w*, and *m*, we find the diver's acceleration:

$$a_y = \frac{F - w}{m} = \frac{9400 \text{ N} - 608 \text{ N}}{62 \text{ kg}} = +142 \text{ m/s}^2,$$

about 15 times the acceleration of a freely falling object.

(b) We can now use Eq. (2.12) from kinematics [$2a_y(y - y_0) = v^2 - v_0^2$] to calculate the stopping distance $y - y_0$ needed to cause this acceleration. The diver's initial velocity as she first touches the cushion is $v_0 = -24$ m/s (the negative sign is included because she moves in the negative *y* direction) and her final velocity after stopping is zero ($v = 0$). Thus,

$$y - y_0 = \frac{v^2 - v_0^2}{2a_y} = \frac{0^2 - (24 \text{ m/s})^2}{2(142 \text{ m/s}^2)} = -2.0 \text{ m}.$$

The negative sign indicates that her displacement $y - y_0$ while being stopped is in the negative *y* direction. ∎

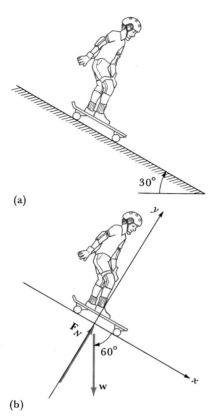

(a)

(b)

FIG. 3.11. (a) A person on a skateboard rolling down a hill. (b) A force diagram of the person. The *x* axis is parallel to the incline and the *y* axis is perpendicular to it.

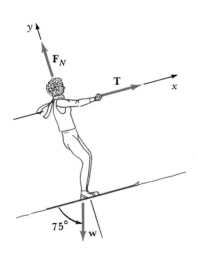

FIG. 3.12. A force diagram for a skier being pulled up a hill.

Inclines

We next consider problems in which an object moves up or down an incline such as a hill, ski slope, or ramp. Problems involving inclines are solved more easily if the *x* axis points parallel to the incline and the *y* axis points perpendicular to the incline. In this coordinate system, all motion is along the *x* axis parallel to the incline. No motion occurs perpendicular to the incline, and $a_y = 0$.

EXAMPLE 3.6 A 50-kg person on a skateboard rolls down a short hill inclined at 30° with the horizontal. Calculate the person's acceleration if friction is ignored.

SOLUTION Figure 3.11a shows the situation described in the problem. Figure 3.11b is a force diagram of the person. The *x* axis points parallel to the plane of the incline and the *y* axis points perpendicular to the plane of the incline. The weight force **w** makes an angle of 60° with the *x* axis when the plane is inclined at 30°.*

For the forces and axes shown in Fig. 3.11b, the *x*-component equation for Newton's second law is

$$\Sigma F_x = F_{N_x} + w_x = F_N \cos 90° + w \cos 60°$$
$$= F_N(0) + w(0.5) = ma_x,$$

or

$$a_x = \frac{0.5w}{m}.$$

But since the magnitude of the weight force equals mg, we find that

$$a_x = \frac{0.5w}{m} = \frac{0.5(mg)}{m} = 0.5(9.8 \text{ m/s}^2) = \underline{4.9 \text{ m/s}^2}.$$

The *y*-component equation is not needed for this problem. ∎

EXAMPLE 3.7 A skier wishes to build a rope tow to pull herself up a ski hill that is inclined at 15° with the horizontal. Calculate the tension needed in the rope to give the skier's 54-kg body (including the skis on which she stands) a 1.2-m/s² acceleration. Ignore friction.

SOLUTION A force diagram for the skier is shown in Fig. 3.12. The *x* axis points up the hill parallel to its surface. The *x*-component form of Newton's second law is

$$\Sigma F_x = T - w \cos 75° = ma_x.$$

---

*To see that the weight force makes a 60° angle with respect to the inclined plane, construct a right triangle whose three sides are parallel to the ground, the inclined plane, and the weight force. The right angle lies at the intersection of the ground and the side parallel to the weight force; the inclined plane is the hypotenuse of the triangle. The angle of the plane relative to the ground (30° in Example 3.6) is the complement of the angle that the weight force makes with respect to the plane (60° in Example 3.6). If the plane is inclined at 10°, the weight force makes an 80° angle with respect to the plane.

Substituting $w = mg$ for the skier's weight and rearranging, we find the tension needed in the rope:

$$T = ma_x + mg \cos 75°$$
$$= (54 \text{ kg})(1.2 \text{ m/s}^2) + (54 \text{ kg})(9.8 \text{ m/s}^2)(0.26)$$
$$= \underline{200 \text{ N}}.$$ ■

Multiple Force Diagrams

In some problems, the procedures we have been using must be applied to more than one object. The results of each application are then combined to get the desired result.

In many of these problems, a rope or cable connects two different objects. If one object moves, the rope causes the other object to move. A rope tension force must be included in the force diagram for each object. Consider the situation pictured in Fig. 3.13a. A rope, tied to a heavy barrel on one end, passes over a pulley and down to a person whose weight is less than that of the barrel. Since the barrel is heavier than the person, the barrel accelerates down while the person accelerates up.

If the rope's mass is small compared to the masses on its end and if there is little friction

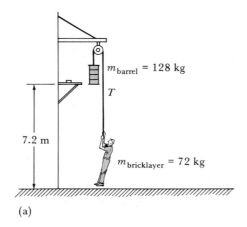

$m_{\text{barrel}} = 128 \text{ kg}$

T

7.2 m

$m_{\text{bricklayer}} = 72 \text{ kg}$

(a)

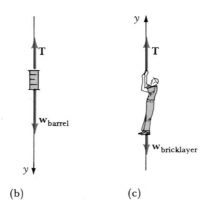

(b)   (c)

FIG. 3.13. (a) A bricklayer being lifted as a barrel falls. (b) A force diagram for the barrel; (c) a force diagram for the bricklayer.

in the pulley, then the tension is uniform along the entire length of the rope. The magnitude of the force it exerts on the person is the same as the magnitude of the force it exerts on the barrel.

Now let us consider in detail the unfortunate problem pictured in Fig. 3.13a.

EXAMPLE 3.8   *The Guardian of London and Manchester* described the perils of a bricklayer who filed the following report:

> I was asked to bring down some excess bricks from the third floor, so I rigged up a beam and pulley, hoisted up a barrel, and tied it in place. After filling the barrel with bricks, I returned to the ground and untied the rope, intending to lower the barrel to the ground.
>
> Unfortunately, I had misjudged the weight of the bricks. As the barrel started down, it jerked me off the ground so fast and so far that I was afraid to let go. Halfway up, I met the barrel coming down and received a severe blow on the shoulder.
>
> I then continued to the top banging my head against the beam and getting my fingers jammed in the pulley. When the barrel hit the ground it burst its bottom, allowing the bricks to spill out. I was now heavier than the barrel and so I started down again at high speed.
>
> Halfway down I met the barrel coming up and received severe injuries to my shins. When I hit the ground I landed on the bricks, getting several painful cuts from the sharp edges. At this point, I must have lost my presence of mind because I let go of the line; the barrel then came down, giving me another heavy blow on the head and putting me in the hospital.
>
> I respectfully request sick leave!

Let us suppose that the mass of the barrel when loaded with bricks is 128 kg, the mass of the bricklayer is 72 kg, and the vertical distance traversed by the barrel and bricklayer is 7.2 m, as shown in Fig. 3.13a. Calculate the bricklayer's speed as he hits the overhead beam that holds the pulley.

SOLUTION   Separate force diagrams are drawn for the filled barrel and for the bricklayer in Figs. 3.13b and 3.13c. We have oriented the positive y axis down for the barrel and up for the bricklayer. The downward acceleration of the barrel will then equal the upward acceleration of the bricklayer.

The y-component form of Newton's second law for the objects in the force diagrams are

$$\text{Bricklayer:} \quad T - w_{\text{bricklayer}} = m_{\text{bricklayer}}a,$$
$$\text{Barrel:} \quad w_{\text{barrel}} - T = m_{\text{barrel}}a.$$

If we add these two equations, the tension cancels from the left side leaving an equation with only the acceleration as an unknown:

$$w_{\text{barrel}} - w_{\text{bricklayer}} = (m_{\text{barrel}} + m_{\text{bricklayer}})a.$$

This equation makes sense. The barrel's weight produces a downward pull on the barrel and rope to which it is attached. But the force of the bricklayer's weight opposes that motion, so the net force in the direction of motion is the difference of these weights. This net force has to accelerate both masses, that of the barrel and that of the bricklayer. So both masses are added in the inertial term on the right.

If we now substitute $w = mg$ for the weights [Eq. (3.4)] in the last equation and rearrange, we can calculate the acceleration:

$$a = \frac{(m_{\text{barrel}}g) - (m_{\text{bricklayer}}g)}{m_{\text{barrel}} + m_{\text{bricklayer}}}$$

$$= \left(\frac{m_{\text{barrel}} - m_{\text{bricklayer}}}{m_{\text{barrel}} + m_{\text{bricklayer}}}\right)g$$

$$= \left(\frac{128\text{ kg} - 72\text{ kg}}{128\text{ kg} + 72\text{ kg}}\right)9.8\text{ m/s}^2$$

$$= 2.74\text{ m/s}^2.$$

Now we use Eq. (2.12) from kinematics to calculate the bricklayer's speed after traveling 7.2 m with this acceleration:

$$v^2 = 2a(y - y_0) + v_0^2$$

$$= 2(2.74\text{ m/s}^2)(7.2\text{ m}) + 0^2 = 39.5\text{ m}^2/\text{s}^2,$$

or $\qquad v = \underline{6.3\text{ m/s}}$, about 14 mi/h. ∎

EXAMPLE 3.9 Two lovers are parked 10 m from the edge of a cliff in a sports car whose mass, including that of the occupants, is 1000 kg. A jealous suitor ties a rope to the car's bumper and a 50-kg rock to the other end of the rope. He then lowers the rock over the cliff, and the car, in neutral, accelerates toward the cliff (we ignore all friction forces). (a) Calculate the car's acceleration. (b) How much time do the lovers have to leap from the car before it is pulled over the cliff?

SOLUTION (a) The situation is shown in Fig. 3.14a. In this problem, it is necessary to make separate force diagrams for the car and for the rock (Figs. 3.14b and 3.14c). We have chosen the x axis in each figure to be the direction of motion because the car's acceleration a_x will then equal the acceleration of the rock along its x axis. The x-component equation of Newton's second law for the car is

$$\Sigma F_x = T = m_{\text{car}}a_x, \tag{3.10}$$

where T is the x component of the rope tension force pulling on the car. The x components of the weight and contact forces are zero.

Next we write the x-component equation for the rock:

$$\Sigma F_x = w_{\text{rock}} - T = m_{\text{rock}}a_x,$$

or, substituting $w_{\text{rock}} = m_{\text{rock}}g$, we obtain

$$T = m_{\text{rock}}g - m_{\text{rock}}a_x. \tag{3.11}$$

If we substitute Eq. (3.10) into Eq. (3.11) and rearrange to solve for a_x, we find that

$$a_x = \left(\frac{m_{\text{rock}}}{m_{\text{car}} + m_{\text{rock}}}\right)g = \left(\frac{50\text{ kg}}{1050\text{ kg}}\right)9.8\text{ m/s}^2 = \underline{0.47\text{ m/s}^2}.$$

(b) We now use kinematics to calculate the time before the car is pulled over the cliff. We know the following information about the car: $v_{x_0} = 0$, $a_x = 0.47$ m/s^2,

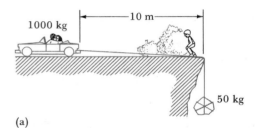
(a)

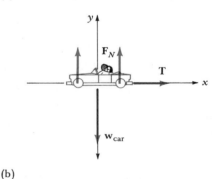
(b)

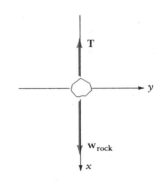
(c)

FIG. 3.14. (a) A car being pulled toward a cliff by a rope attached to a hanging rock. (b) A force diagram for the car; (c) a force diagram for the rock.

and $x - x_0 = 10$ m. We can use Eq. (2.11) $(x - x_0 = v_{x_0}t + 1/2a_xt^2)$ to calculate the time taken by the car to travel 10 m.

Since $v_{x_0} = 0$, we find that

$$t^2 = \frac{2(x - x_0)}{a_x} = \frac{2(10 \text{ m})}{0.47 \text{ m/s}^2} = 42.6 \text{ s}^2,$$

or

$$t = \underline{6.5 \text{ s.}}$$

This should be sufficient time for the lovers to leap from the car! They may even have time to apply the brakes and stop the car. ∎

3.7 Friction Forces

So far, our examples in this chapter have involved three kinds of force: (1) a tension force transmitted from one point to another along the length of a rope or cord; (2) a normal force exerted perpendicular to the surface of an object where it touches another object; and (3) a weight force caused by the gravitational pull of the earth on an object. In this section we examine another kind of force—friction force. A **friction force** opposes the motion of an object across a surface on which the object rests or slides and is directed parallel to the surface of contact. We consider here two common types of friction: static friction and kinetic friction.

Static Friction

Static friction exists when an object does not slide along a surface on which it rests even though a force is exerted to make it slide. If we push a large crate but it does not slide, we say that a static friction force resists our efforts. This friction force is static because the crate remains at rest, or "static," even though we push it.

The static friction force is illustrated in Fig. 3.15. In Fig. 3.15a, a large crate rests on a floor. No force pushes the crate; hence, there is no static friction force opposing a push. In Fig. 3.15b, a person exerts a gentle push $\mathbf{P}$ against the crate. However, the crate does not move because a static friction force $\mathbf{F}_s$ equal in

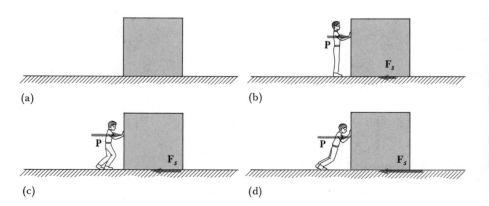

FIG. 3.15. A static friction force $\mathbf{F}_s$ resists the tendency of an object to start sliding with whatever force is needed up to a maximum value given by $F_{s \text{ (max)}} = \mu_s F_N$.

magnitude to the push opposes the motion. In Figs. 3.15c and 3.15d, even larger pushing forces are unable to make the crate slide. As **P** increases, so does the static friction force resisting the motion.

If several people pushed the crate, the force **P** might become so large that the static friction force could no longer prevent the crate from sliding. The maximum force that a static friction force can exert depends on two factors:

1. The relative roughness of the two surfaces. A measure of the roughness of two surfaces is indicated by the **coefficient of static friction** μ_s of the surfaces. The larger the value of μ_s, the rougher the surfaces and the harder it is to make the object start sliding.

2. The magnitude of the normal (perpendicular) force between the object and the surface on which it rests. The larger the normal force, the harder it is to make the object start sliding. Often, the magnitude of a normal force between an object and the surface on which it rests equals the object's weight. Thus, the greater its weight, the harder it is to make the object slide. However, the magnitude of the normal force does not always equal an object's weight, as we will see in some of the examples and problems.

Figure 3.16, which shows an enlarged view of two surfaces in contact with each other, may make these two factors clearer. The figure shows that there are many indentations on any surface; if viewed through a microscope, no surface is perfectly smooth. The microscopic indentations and bulges in one surface tend to interlock with bulges and indentations in the other surface. If we try to pull one of the objects across the other, the bulges and indentations "hook on" to each other and resist the motion. The coefficient of static friction μ_s describes this relative roughness of surfaces. The coefficient of static friction between a rubber car tire and a concrete highway ranges from about 0.6 to 1.0, depending on the condition of the road; between two pieces of Teflon it is 0.04, and it is even less between the cartilage and bone in a healthy joint. The smaller the value of μ_s, the easier it is for the surfaces to slide across each other. Table 3.1 gives the values of μ_s for several surfaces.

Figure 3.16 also indicates why the maximum value of the static friction force depends on the normal force between the surfaces. As the normal force increases, the tiny bulges and indentations of one surface press more closely into those of the other surface, making it harder for the object to slide. A heavy crate is therefore harder to slide than a light crate.

As a result of many experiments, it has been determined that the magnitude of the static friction force F_s for relatively smooth surfaces is less than or equal to the product of the coefficient of static friction μ_s and the magnitude of the normal force F_N between the two surfaces:

$$F_s \leq \mu_s F_N. \qquad (3.12)$$

The direction of the static friction force F_s is parallel to the surface of contact and opposes the tendency to move. Note that an object sitting at rest begins to slide if pulled or pushed by a force directed parallel to the surface whose magnitude exceeds the maximum possible static friction force.

EXAMPLE 3.10 A 45-kg box sits on a floor. The coefficient of static friction between the block and the floor is 0.40. What magnitude of horizontal pushing force **P** is needed to make the box slide?

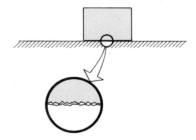

FIG. 3.16. An enlargement of a section of contact between a block and a floor shows many indentations on the two surfaces, which hook and bond to each other.

TABLE 3.1 Coefficients of Static (μ_s) and Kinetic (μ_k) Friction

Surfaces	μ_s	μ_k
Rubber on concrete	0.6–1.0	0.8
Steel on steel	0.74	0.57
Aluminum on steel	0.61	0.47
Glass on glass	0.94	0.4
Wood on wood	0.25–0.5	
Ski wax on wet snow		0.1
Ski wax on dry snow		0.04
Teflon on Teflon	0.04	0.04
Greased metals	0.1	0.06
Surfaces in healthy joint		0.015

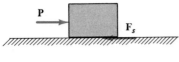

(a)

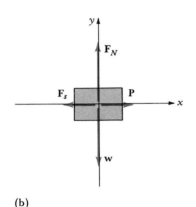

(b)

FIG. 3.17. (a) A pushing force **P** on a block is resisted by a static friction force **F**$_s$. (b) A force diagram for the block.

SOLUTION The situation is shown in Fig. 3.17a. A force diagram of the box is shown in Fig. 3.17b. We have moved all the vector arrows so that their tails are at the origin of the coordinate system.

For the box to start sliding, the magnitude of the pushing force **P** must be greater than the maximum possible magnitude of the static friction force **F**$_s$:

$$P > F_{s(max)} = \mu_s F_N.$$

We can see from the force diagram that the normal force **F**$_N$ must balance the box's weight **w**. Thus, $F_N = w = mg = (45 \text{ kg})(9.8 \text{ m/s}^2) = 440 \text{ N}$. Therefore,

$$P > \mu_s F_N = (0.40)440 \text{ N} = \underline{180 \text{ N}}.$$

Notice that we rounded off the answer to two significant figures. ∎

Kinetic Friction

If an object slides across a surface, a **kinetic friction force** opposes its sliding. The word *kinetic* implies that the object is moving. A road, for example, exerts a kinetic friction force on a car that skids to a stop.

After an object that is pushed or pulled starts to slide, less force is usually required to keep it sliding than was needed to start it sliding; that is, the kinetic friction force is less than the maximum static friction force. This is probably because the two surfaces cannot settle into each other as easily when the object moves as when it is at rest.

This effect of the motion of an object on the friction force is accounted for by defining a **coefficient of kinetic friction,** μ_k, which is usually less than the coefficient of static friction. In addition to values for the coefficient of static friction for various surfaces, Table 3.1 gives values of μ_k that have been determined experimentally for the same surfaces. Notice that μ_s and μ_k are dimensionless numbers; they have no unit of measure.

Actually, the coefficient of kinetic friction μ_k is not constant for a surface. The value of the coefficient can depend on the condition of the surface and also on the speed with which one object moves across the other. Usually, the value of μ_k decreases slightly as the speed of the object increases. This effect of velocity on μ_k is small, though, and will be ignored in our examples and problems.

The sliding or kinetic friction force **F**$_k$ depends not only on μ_k but also on the magnitude of the normal force between the two surfaces, just as the static friction force does. From many experiments with relatively smooth surfaces, it has been found that the magnitude of the kinetic friction force **F**$_k$ acting on an object sliding across a surface is given by

$$F_k = \mu_k F_N. \tag{3.13}$$

The friction force **F**$_k$ is directed parallel to the surface of contact and opposes the motion.

Surprisingly, kinetic and static friction forces usually do not depend on the area of contact between the surfaces. When the surfaces are hard and cannot fold into the little bulges and indentations of each other, friction forces are independent of area. If the two surfaces can conform to each other's shape, however, the friction force does depend on area. Drag racers underinflate the rear tires of their cars so that the tires can fold into the bumps and valleys in a road. The back tires are also very wide to increase the area of contact between the tires and the road. This folding of wide tires into the indentations of the road causes the friction between the road and the tire to increase, enabling the racer to start faster. In our examples we will ignore the effect of contact area.

EXAMPLE 3.11 Police, examining the scene of an accident, observe skid marks 30 m long left by a 1200-kg car. The car skidded to a stop on a concrete highway having a coefficient of kinetic friction with the tires of 0.80. Estimate the car's speed at the beginning of the skid.

SOLUTION The situation described in the problem is shown in Fig. 3.18a, and a force diagram of the car appears in Fig. 3.18b. The kinetic friction force $\mathbf{F}_k$ is parallel to the surface of contact between the road and tires, and it points opposite the direction of motion. Before calculating the magnitude of $\mathbf{F}_k$, we need to calculate the normal force between the road and the tires. To find F_N, we use the y-component equation of Newton's second law:

$$\Sigma F_y = F_N - w = ma_y.$$

But since no vertical motion occurs, $a_y = 0$ and $F_N - w = 0$. Thus, for this problem

$$F_N = w = mg = (1200 \text{ kg})(9.8 \text{ m/s}^2) = 1.18 \times 10^4 \text{ N}.$$

We can now calculate the kinetic friction force using Eq. (3.13):

$$F_k = \mu_k F_N = (0.80)(1.18 \times 10^4 \text{ N}) = 0.94 \times 10^4 \text{ N}.$$

We can then use the x-component equation of Newton's second law to determine the car's acceleration:

$$\Sigma F_x = -F_k = ma_x,$$

or

$$a_x = -\frac{F_k}{m} = -\frac{0.94 \times 10^4 \text{ N}}{1200 \text{ kg}} = -7.8 \text{ m/s}^2.$$

Finally, we can use kinematics Eq. (2.12) to calculate the car's initial speed v_0 if its final speed is zero after skidding 30 m:

$$v^2 - v_0^2 = 2a_x(x - x_0),$$

or

$$0^2 - v_0^2 = 2(-7.8 \text{ m/s}^2)(30 \text{ m}).$$

Thus,

$$v_0 = \underline{22 \text{ m/s}}, \text{ or } 48 \text{ mi/h}. \qquad \blacksquare$$

EXAMPLE 3.12 A 42-kg toboggan, including its load, is pulled along a horizontal surface covered with snow by the 120-N tension force of a rope directed 37° above the horizontal. If the kinetic coefficient of friction between the toboggan and the snow is 0.20, what time is required to increase the toboggan's speed from zero to 4.0 m/s?

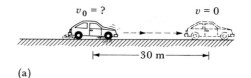

(a)

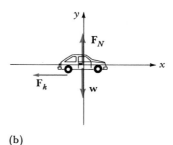

(b)

FIG. 3.18. (a) A car skidding to a stop. (b) A force diagram for the car.

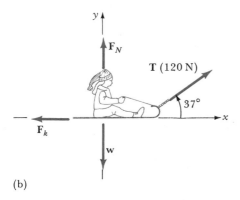

(a)

(b)

FIG. 3.19. (a) A toboggan pulled by a rope. (b) A force diagram for the toboggan.

SOLUTION The situation described in the problem is shown in Fig. 3.19a, and Fig. 3.19b shows a force diagram for the toboggan. We first calculate the acceleration of the toboggan and then use kinematics to determine the time needed to increase its speed.

The toboggan's acceleration in the x direction depends on the horizontal component of the tension force and on the kinetic friction force. To find the magnitude of the friction force, we must first find the normal force F_N. We can use the y-component equation of Newton's second law to calculate F_N:

$$\Sigma F_y = (T \sin 37°) + F_N - w = ma_y.$$

Since there is no motion along the y axis, $a_y = 0$, and the above equation, after rearranging, becomes

$$F_N = w - 0.6T = mg - 0.6T$$
$$= [(42 \text{ kg})(9.8 \text{ m/s}^2)] - [0.6(120 \text{ N})] = 340 \text{ N}.$$

We can now calculate the magnitude of the kinetic friction force using Eq. (3.13):

$$F_k = \mu_k F_N = 0.20(340 \text{ N}) = 68 \text{ N}.$$

The x-component equation of Newton's second law allows us to calculate the toboggan's acceleration:

$$\Sigma F_x = (T \cos 37°) - F_k = ma_x,$$

or
$$a_x = \frac{0.8T - F_k}{m} = \frac{0.8(120 \text{ N}) - 68 \text{ N}}{42 \text{ kg}}$$
$$= 0.67 \text{ m/s}^2.$$

Having determined the toboggan's acceleration, we can use kinematics to calculate the time needed to increase its velocity from zero ($v_0 = 0$) to a final value $v = 4.0$ m/s. Equation (2.9), after rearranging, provides the answer:

$$t = \frac{v - v_0}{a_x} = \frac{(4.0 \text{ m/s}) - 0}{0.67 \text{ m/s}^2} = \underline{6.0 \text{ s}}. \quad \blacksquare$$

Summary and Additional Readings

1. **Newton's laws of motion:** The study called dynamics considers the way in which forces affect the motion of objects. An understanding of Newton's three laws of motion is fundamental to dynamics.

   Newton's first law: Every body continues in its state of rest or of constant speed in a straight line unless it is compelled to change that state because of forces acting on it.

   Newton's second law: An unbalanced force acting on an object will cause the object to accelerate in the direction of the force. The acceleration is proportional to the vector sum of the forces acting on the object and inversely proportional to the object's mass:

$$\Sigma \mathbf{F} = m\mathbf{a} \quad \text{or} \quad \begin{array}{l} \Sigma F_x = ma_x \\ \Sigma F_y = ma_y \end{array} \quad (3.1) \quad \text{or} \quad \begin{array}{l} (3.5) \\ (3.6) \end{array}$$

   Newton's third law: Whenever one object exerts a force on a second object, the second object exerts a reaction force of equal magnitude but opposite direction on the first object.

   Newton's second law relates the forces acting on an object to its acceleration. Kinematics is often used to relate an object's acceleration to its changing velocity and position. Procedures for using the second law and kinematics to solve problems in dynamics are outlined in Section 3.6.

2. **Mass and weight:** Mass is a measure of an object's inertia. The greater its inertia, the less its acceleration when pushed or pulled by a force. Mass is also an indication of the amount of matter of which the object is made.

   An object's **weight** is the gravitational force exerted on the object due to its gravitational attraction to other masses such as the earth. Near the earth's surface, an object's weight and mass are related by the equation

$$w = mg, \tag{3.4}$$

where $g = 9.8 \text{ m/s}^2$.

3. **Friction forces:** Friction forces oppose the motion of one object across another surface. If an object, when pushed or pulled, does not move because of the opposing friction force, the friction is said to be static friction. If the object slides across a surface, it is opposed by a kinetic friction force.

The **static friction force** $\mathbf{F}_s$ has whatever value is needed to keep an object from sliding but does not exceed a limiting value given by the equation

$$F_s \leq \mu_s F_N, \tag{3.12}$$

where μ_s is the coefficient of static friction between the object and the surface on which it rests and F_N is the normal force between the two surfaces.

The **kinetic friction force** $\mathbf{F}_k$ has a magnitude

$$F_k = \mu_k F_N, \tag{3.13}$$

where μ_k is the coefficient of kinetic friction. Both static and kinetic friction forces are directed parallel to the surface of contact and oppose the motion.

I. B. Cohen, "Newton," *Scientific American* **193**, 73 (1955). A biography of Isaac Newton.

C. G. Adler and B. L. Coulter, "Aristotle: Villain or Victim?" *The Physics Teacher* **13**, 35 (1975). Discusses Aristotle's ideas about motion.

Robert Weinstock, "The Laws of Classical Motion: What's *F*? What's *m*? What's *a*?" *American Journal of Physics* **29**, 698 (1961).

Jean Bratlin, "Action and Reaction? What's That?" *The Physics Teacher* **19**, 326 (1981).

Ernest Rabinowicz, "Stick and Slip," *Scientific American*, p 109, May (1956). A microscopic look at friction.

Questions

1. (a) *Estimate* the weight of this book in newtons, using the fact that one medium-sized apple weighs 1 N. (b) Using this estimate, calculate the book's weight in pounds and its mass in kilograms and slugs.

2. (a) A force is seen acting on an object, but the object does not move. Explain. (b) A 200-N force acts on a 10-kg block but causes it to have an acceleration of only 5 m/s². Explain.

3. Two balls of the same material, one of mass m and the other of mass $2m$, are dropped simultaneously from the Leaning Tower of Pisa. Will the balls hit the ground at the same time? Explain carefully, using Newton's second law. Ignore air resistance.

4. A box full of lead and a box of the same size full of feathers are floating weightless near a spaceship. Develop an experiment to estimate the mass of one compared to the mass of the other.

5. A horse pulls on a wagon, and the wagon pulls back on the horse with an equal reaction force. How can the horse accelerate?

6. A person jumps from a wall and lands stiff-legged. Use kinematics and Newton's second law to explain why the person is less likely to be injured when landing in soft sand than on concrete.

7. A friend drops a lead ball on your hand while your hand rests on a table. Why is your hand more prone to injury than if you catch the ball in the air, even if it is dropped from the same height?

8. A person lifts a box from a bent position. Why is there more chance of back injury if the box is lifted quickly than if it is lifted slowly?

9. A person is able to push down harder on a bicycle pedal when pulling up hard on the handlebars. Explain, using a force diagram.

10. Why is it easier to pull a lawn mower than to push it?

11. A large mass is suspended from the ceiling by a string, and another identical string hangs from the mass. Why does the upper string break first if you exert a firm, steady pull on the lower string whereas the lower string breaks first if it is given a jerk?

12. Why do your books slide off the front seat of a car if you stop abruptly?

13. A car accelerates along a road. In a force diagram, (a) show the forces of the car on the road; (b) show the reaction forces of the road on the car. (c) Which force is responsible for the acceleration?

Problems

3.2 and 3.3 Newton's Second Law and Units

1. An earth satellite is being constructed in orbit above the earth. A 200-N force is used to push a steel beam into place. The beam accelerates at 0.20 m/s² when acted on by this force. Calculate the mass of the beam.

2. A farmer pushes a 500-kg wagon with a horizontal force of 125 N. (a) Calculate the acceleration of the wagon (ignore friction forces). (b) If the wagon starts at rest, how fast is it moving after being pushed for a time of 5.0 s?

3. The lower left chamber of the heart (the left ventricle) pumps blood into the aorta. The left ventricular contraction lasts 0.20 s, during which time a mass of 88 g of blood is accelerated

from rest to a speed of about 0.45 m/s. (a) What is the average acceleration of the blood during this time? (b) What is the average net force of the left ventricle on the blood during the contraction?

4. A 90-kg baseball player running at 9.0 m/s slides into home plate and is stopped by the slide in 3.0 m. Calculate (a) the average acceleration of the player while being stopped and (b) the average net force causing the change in motion. Assume that all motion is in the horizontal direction.

■ 5. The speed of a 60-kg dolphin increases uniformly from 2.0 m/s to 6.0 m/s in 5.0 s. Calculate the average net force acting on the dolphin while this speed change occurs.

■ 6. A 1000-kg sports car accelerates from rest to 60 mi/h in 11 s. Calculate the net force acting on the car.

■ 7. During a head-on collision, the person in the front seat of a car decelerates from 13.3 m/s (30 mi/h) to rest in 0.10 s. If the person's mass is 75 kg (165 lb), calculate the average force of the seat belt and shoulder strap on the person.

■ 8. A 40-g arrow initially traveling at 50 m/s passes through a 16-cm-thick grapefruit hanging from a tree. If the arrow leaves the fruit at 40 m/s, calculate the average resisting force of the fruit on the arrow.

■ 9. A 0.145-kg baseball enters the glove of a catcher at a speed of 92 mi/h. The baseball is stopped in 21 cm. Calculate the average force of the glove on the ball, assuming constant deceleration.

■ 10. A rope will break if its tension exceeds 3000 N. Calculate the minimum time needed to pull with this rope a 1400-kg car from rest to a speed of 15 m/s. Ignore friction.

3.4 Weight

11. (a) The mass of a gallon of milk is 3.8 kg. Calculate its weight in newtons. (b) A cement truck when filled weighs 4.2×10^5 N. Calculate its mass in kilograms.

12. The liquid in a can of Coca-Cola has a mass of 0.35 kg. Calculate your change in weight in newtons and pounds immediately after drinking the contents of one can.

13. Calculate the mass of a 3-lb bag of oranges (a) in slugs; (b) in kilograms. (c) Calculate its weight in newtons.

14. (a) Indicate your weight in pounds and calculate your mass in slugs. (b) If a 50-lb unbalanced force acted on you, what would be your acceleration?

3.5 Newton's Third Law

■ 15. One book lies on top of a second book, which rests on a table. Construct a force diagram for each book and identify in words the cause of each force. Indicate which, if any, of the forces shown in your diagrams are action-reaction pairs.

■ 16. A car exerts a forward force on a trailer, and the trailer exerts an equal-magnitude backward force on the car. Construct a force diagram for each vehicle and explain what force causes the car to accelerate when pulling the trailer.

■ 17. Two blocks of equal mass are connected by a string and rest on a frictionless table. A free string is attached to the right block. You pull the free string, which causes both blocks to accelerate to the right. Construct force diagrams for each block and explain in words the cause of each force.

■ 18. A car pulls a trailer up a road inclined at 10° with the horizontal. Construct a force diagram for the car and for the trailer and identify in words the cause of each force.

■ 19. Construct force diagrams for each block shown in Fig.

3.20 if the right-hand rope pulls at an angle of 30° above the horizontal.

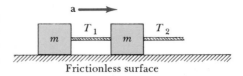

FIG. 3.20

3.6 Applications of Newton's Second Law

20. A rope connected to a 50-kg sled pulls it along a frictionless sheet of ice. The tension in the rope is 100 N, and the rope is oriented at an angle 30° above a line drawn parallel to the ice. Calculate the horizontal acceleration of the sled.

21. A 20-kg block is pulled on a horizontal frictionless surface by a rope that makes an angle of 37° with the horizontal. Calculate the tension in the rope if the block accelerates at 2.0 m/s².

■ 22. A tennis ball of mass 0.058 kg traveling at 25 m/s in the horizontal direction hits a practice wall and rebounds in the opposite direction at 20 m/s. Calculate (a) the ball's average acceleration and (b) the magnitude and direction of the average force of the wall on the ball if the force acts for 0.060 s.

■ 23. An 80-kg woman jumps from a wall and lands stiff-legged in soft sand. The woman's downward speed just as she touches the sand is 6.0 m/s. She is stopped after sinking 0.050 m into the sand. Calculate the force, assumed constant, of the sand on the woman while she is being stopped.

■ 24. A 3.1-g penny thrown from the top of the Sears building in Chicago lands on the head of a passerby 460 m below. Just before the penny hits the person's head, it is falling at a speed of 25 m/s. It is stopped by the head in a distance of 1.0 mm (10^{-3} m). (a) Calculate the deceleration of the penny. (b) Calculate the force of the penny on the head. (c) Parts of the skull bones will be fractured by forces of about 900 N (200 lb) or more. Is there a chance of fracturing the skull with the penny?

■ 25. An 8000-kg elevator is pulled by a cable. (a) Construct a force diagram for the elevator. (b) Write Newton's second law for motion along the vertical direction. (c) Determine the tension in the cable if the elevator accelerates up at 1.2 m/s², and (d) down at 1.2 m/s².

■ 26. A 130-lb rock climber must be rescued from a mountain ledge by a helicopter. After a rope from the helicopter is attached to the climber, the helicopter accelerates upward at 24 ft/s². What is the force of the rope on the climber?

■ 27. A string breaks if a mass larger than 40 kg hangs from it. Calculate the maximum vertical acceleration of a 24-kg mass pulled upward at the end of the string.

■ 28. A 60-kg skier is pulled by a rope up a frictionless slope that is inclined at an angle of 20° with the horizontal. The rope pulls parallel to the slope and has a tension of 230 N. Calculate the skier's acceleration.

■ 29. A 52-kg skier starting from rest slides 30 m down a short hill inclined 12° above the horizontal. Calculate (a) the time needed to reach the bottom and (b) the skier's speed at the bottom. Ignore friction.

■ 30. Two blocks of equal mass are attached as shown in Fig. 3.20. Calculate the ratio T_2/T_1. [*Hint:* Use the problem-solving procedure on pages 62–63 for each block separately to get two independent equations that can be combined.]

■ 31. Suppose that rope 2 in Fig. 3.20 pulls at an angle 37° with the horizontal rather than horizontally. Calculate the ratio T_2/T_1.

■ 32. Calculate tensions T_1 and T_2 for the situation shown in Fig. 3.20 if the acceleration is 2.0 m/s² and the mass of each block is 4.0 kg.

■ 33. Suppose that T_2 in Fig. 3.20 pulls at an angle 30° above the horizontal with a tension of 16 N. Calculate the acceleration of the blocks and the tension in rope 1 if the mass of each block is 6.0 kg.

■ 34. (a) Determine the downward acceleration of the 72-kg bricklayer described in Example 3.8 as he falls from the position of the pulley while hanging on to the rope with the 28-kg empty barrel at the other end. (b) How much time is needed for the bricklayer to fall 7.2 m if he starts at rest?

■ 35. A 4.0-kg can of paint rests on a ledge 20 m above the ground. A taut rope attached to the can passes over a pulley and straight down to a 3.0-kg can of nails. If the can of paint is accidentally knocked off the ledge, how much time does a carpenter have to catch the can of paint before it smashes into the floor?

■ 36. A 10-kg monkey hangs from the tail of a 15-kg monkey that holds a rope hanging from an outdoor elevator. Calculate the tension in the rope and in the 15-kg monkey's tail if (a) the elevator is at rest; (b) the elevator accelerates up at 2.0 m/s²; (c) the elevator accelerates down at 2.0 m/s²; (d) the elevator falls freely.

■ 37. A car sitting at rest is hit from the rear by a truck moving at 13 m/s. The car lurches forward with an acceleration of about 300 m/s². Figure 3.21 shows the neck muscle that causes the head to accelerate forward with the body. If the head has a mass of 4.5 kg, what is the horizontal component of **F** required to accelerate the head? If **F** is directed 37° below the horizontal, what is the magnitude of **F**? [*Note:* Whiplash victims often experience sore neck muscles because of these unusually large tension forces.]

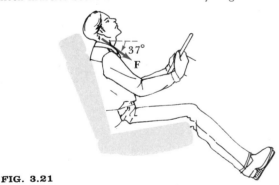

FIG. 3.21

■ 38. A string with one 10-g die on the end is attached to the rear-view mirror of a car. When the car leaves an intersection, the string makes an angle of 5° with the vertical. What is the acceleration of the car? [*Hint:* Choose the die as the object for your force diagram. Two forces act on the die, which is accelerating in the horizontal direction but not vertically. First use the vertical-component equation of Newton's second law to find the tension in the string. The horizontal component of **T** causes the die to accelerate.]

■ 39. A 20-kg block rests on a frictionless table. A cord attached to the block extends horizontally to a pulley at the edge of the table. A 10-kg mass hangs at the end of the cord after it passes over the pulley. Calculate the acceleration of the block and mass and the tension in the cable.

■ 40. A 20-kg block rests on a frictionless table. A cord attached to the block extends horizontally to a pulley at the edge of the table. When another block of unknown mass is hung at the end of the cord after it passes over the pulley, the hanging block accelerates downward at 2.7 m/s² and pulls the other block with it. Calculate the mass of the hanging block and the tension in the cord.

■■ 41. The system shown in Fig. 3.22 accelerates to the left. Find the magnitude of this acceleration and the tension in the cable. There is no friction between the block and the inclined plane.

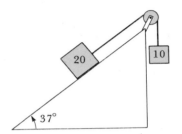

FIG. 3.22

■■ 42. Suppose that the 10-kg block in Fig. 3.22, instead of hanging, lies on a horizontal, frictionless surface to the right of the pulley so that the rope between the pulley and the 10-kg mass is horizontal. Calculate (a) the acceleration of the blocks and (b) the tension in the cable.

3.7 Friction Forces

43. A 200-lb crate lies on the floor. The coefficient of static friction between the crate and floor is 0.60. What is the minimum force required to start the crate sliding?

44. A 60-kg student sitting on a hardwood floor does not slide until pulled by a 240-N horizontal force. Calculate the coefficient of static friction between the student and floor.

■ 45. A block rests on an inclined plane. Its angle with the horizontal is increased. When the angle becomes 40°, the block starts to slide. Calculate the coefficient of static friction between the block and the plane.

■ 46. A 20-kg crate sitting on a horizontal floor is attached to a rope that pulls 37° above the horizontal. The coefficient of static friction between the crate and floor is 0.50. (a) Construct a force diagram for the crate. (b) Determine the least rope tension that will cause the crate to start sliding. [*Note:* The normal force is not 196 N.]

■ 47. A 50-kg block rests on the floor. The coefficients of static and kinetic friction are 0.70 and 0.50, respectively. (a) What is the minimum force needed to start the box sliding? (b) If the same force continues to push the block after it starts sliding, what is the acceleration of the block?

■ 48. A 900-kg car traveling at 20 m/s skids to a stop on a wet surface after the brakes are applied. How far does the car skid if the coefficient of kinetic friction is 0.40?

■ 49. A 20-kg wagon accelerates on a horizontal surface at 0.50 m/s² when pulled by the 90-N horizontal force of a rope. Calculate the magnitude of the friction force acting on the wagon and the coefficient of friction that would cause this force.

■ 50. A 3000-lb auto traveling at 59 ft/s (40 mi/h) coasts to a stop in 2200 ft when in gear and in 3000 ft when in neutral. Calculate the net resistive force slowing the auto in each case. The difference in the two situations is the extra friction force caused by the transmission and drive shaft when in gear.

■ 51. A 20-kg block moves along a horizontal surface with a 0.61 coefficient of kinetic friction. The block is pulled by a rope that exerts a 162-N tension force at a 37° angle above the horizontal. (a) Construct a force diagram for the block. (b) Write Newton's second law in component form for each axis. (c) Solve the equations to find the crate's acceleration.

■ 52. A 70-kg skier slides down a hill inclined at 12° with the horizontal. Calculate the skier's acceleration if opposed by a 100-N friction force.

■ 53. A 20-kg crate is lowered down a plane inclined at an angle of 37° with the horizontal by a rope that exerts on the crate a 50-N force directed parallel to the plane. A 40-N friction force, also parallel to the plane, opposes the downward motion. What is the acceleration of the crate?

■ ■ 54. A 60-kg skier slides down a hill inclined at 12° with the horizontal. Calculate the skier's acceleration if the coefficient of kinetic friction between the person and the hill is 0.16.

■ ■ 55. A 3.0-kg box sits on a horizontal surface of your car seat as you drive at a speed of 20 m/s. The coefficient of friction between the box and the seat is 0.50. You apply the brakes to stop the car. Calculate the shortest possible stopping distance so that the box does not start to slide off the seat.

■ ■ 56. Show that the minimum distance needed to stop a car traveling at speed v is $v^2/2\mu g$, where μ is the coefficient of friction between the car and the road and g is the acceleration of gravity.

■ ■ 57. Use the proportion technique and the results of Problem 56 to determine the percent increase in the distance needed to stop a car if its initial speed is increased by 20 percent.

■ ■ 58. In designing a rope tow at a ski resort, an engineer needs a motor that will cause a 100-kg skier to be accelerated up a 10° incline at 0.80 m/s². Calculate the tension needed in the rope if the coefficient of friction between the skier and the snow is 0.10.

CHAPTER 4

Momentum

In Chapter 3 Newton's second law ($\Sigma\,\mathbf{F} = m\mathbf{a}$) summarized a cause-effect relation between the forces acting on an object and the acceleration caused by those forces. Force, the cause, produced an acceleration, the effect.

Another cause-effect relation, the impulse-momentum equation, will be developed in this chapter. We will find that the impulse of a force causes an object's momentum to change. Impulse, the cause, produces a change in momentum, the effect. The impulse-momentum equation is useful for studying collisions of automobiles, impacts that occur during athletic events, demolition of buildings, and many less dramatic examples involving the changing motion of objects.

Momentum is a measure of the difficulty encountered in bringing an object to rest. The greater the object's mass and velocity, the greater is its momentum. **Impulse** is the product of a force acting on an object and the time that the force acts. An impulse, such as the force of a bat on a ball, causes the object's momentum to change.

Momentum has a special significance: It is one of a small number of quantities that supposedly does not change, when considering the universe as a whole, during the many interactions of life. It is called a *conserved quantity*. The total momentum of two cars involved in a crash is the same just after the crash as it was just before the crash. The fact that momentum is conserved provides a useful tool for analyzing a variety of physical processes ranging from the study of rocket propulsion to the discovery of new particles produced by high-energy accelerators.

4.1 Momentum

Momentum is a property related to an object's motion and mass. The faster the object moves and the larger its mass, the greater is its momentum. A truck moving at 30 mi/h has more momentum than a Ping-Pong ball flying at the same speed or than another truck of the same mass moving at a slower speed. Qualitatively, *momentum reflects the tendency of a moving object to continue moving and the difficulty encountered in reducing that motion.* Momentum is defined quantitatively as follows:

Physics professor Paul Hewitt gives a friend (?) an unusual impulse. (Courtesy, Paul Hewitt)

The **momentum p** of an object is defined as the product of its mass m and velocity **v**:

$$\text{Momentum} = \mathbf{p} = m\mathbf{v}. \tag{4.1}$$

The units of momentum are kg·m/s.

Since velocity is a vector quantity having magnitude and direction, momentum is also a vector quantity with magnitude and direction. The direction of an object's momentum equals the direction of its velocity.

EXAMPLE 4.1 A cement truck full of cement has a mass of 42,000 kg. It travels north at a speed of 18 m/s, or about 40 mi/h. (a) Calculate the truck's momentum. (b) How fast must a 750-kg Chevy Sprint travel to have the same momentum?

SOLUTION (a) The truck's momentum points north in the same direction as its velocity. The magnitude of its momentum is

$$p = mv = (4.2 \times 10^4 \text{ kg})(18 \text{ m/s}) = 7.6 \times 10^5 \text{ kg·m/s}.$$

(b) For the momentum of the Sprint to equal that of the truck, it must drive north in the same direction as the truck. The magnitude of the Sprint's momentum must also equal that of the truck. Thus,

$$m_{\text{Sprint}} v_{\text{Sprint}} = 7.6 \times 10^5 \text{ kg·m/s},$$

or

$$v_{\text{Sprint}} = \frac{7.6 \times 10^5 \text{ kg·m/s}}{m_{\text{Sprint}}} = \frac{7.6 \times 10^5 \text{ kg·m/s}}{750 \text{ kg}} = 1000 \text{ m/s}.$$

This is approximately 2300 mi/h! It is unlikely that the momentum of the Sprint or of any small car will equal that of a cement truck moving at 40 mi/h. ∎

4.2 The Impulse-Momentum Equation

What causes an object's momentum to change? Force must be involved. Force is needed to start or stop a car. The length of time that the force acts also affects the change in momentum. For instance, a small force acting for a long time can stop a car (Fig. 4.1a) just as a large force acting for a short time can stop the car (Fig. 4.1b). These two quantities, the force and the time the force acts, determine the impulse of the force.

The **impulse** of a force **F** acting on an object for a time t is defined as the product of the force and time:

$$\text{Impulse} = \mathbf{F}t. \tag{4.2}$$

Impulse has units of newton-seconds (N·s) and is the cause of an object's change in momentum.

An equation relating the impulse of a force and the subsequent change in

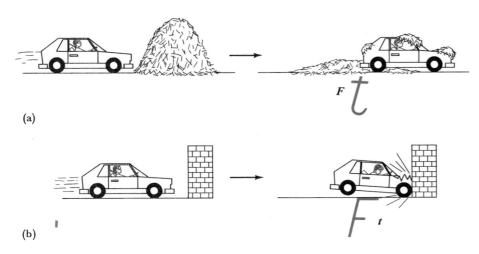

(a)

(b)

FIG. 4.1. (a) A small force acting for a long time can cause a large change in momentum. (b) A large force acting for a short time also can cause a large change in momentum.

momentum of the object on which thé force acts is derived easily using Newton's second law of motion ($\Sigma \mathbf{F} = m\mathbf{a}$) and kinematics. We replace the acceleration in Newton's second law by the acceleration given by Eq. (2.9) from kinematics,

$$\mathbf{a} = \frac{\mathbf{v} - \mathbf{v}_0}{t},$$

to obtain

$$\Sigma \mathbf{F} = m\mathbf{a} = m\frac{\mathbf{v} - \mathbf{v}_0}{t}.$$

Rearranging the above equation produces the **impulse-momentum equation:**

$$(\Sigma \mathbf{F})t = m\mathbf{v} - m\mathbf{v}_0. \tag{4.3}$$

Notice that the right side of the equation equals the change in an object's momentum. The left side is the cause of the change in momentum and equals the sum of the impulses of all forces acting on the object during the time of the momentum change.

The impulse-momentum equation provides a nice interpretation for a variety of phenomena ranging from impulses occurring in athletics and car collisions to the impulse of a molecule of air bouncing off the inside wall of a balloon or off your skin. Most of our examples and problems are restricted to one-dimensional motion, in which case we will not use the vector symbols in Eq. (4.3). Instead, the directions of forces and velocities are indicated by positive and negative signs, depending on the orientation of the force or velocity relative to the axis along which motion occurs.

EXAMPLE 4.2 A golf club strikes a golf ball (Fig. 4.2). The club and ball remain in contact for 0.60×10^{-3} s. The 45-g ball leaves the club with a speed of 70 m/s. Calculate the average force of the club on the ball.

SOLUTION The force of the club on the ball is directed along the line of motion of the club and is the only force acting on the ball in that direction. We omit vector signs and consider only the magnitude of this force and the ball's

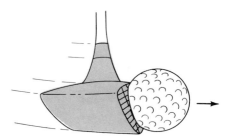

FIG. 4.2. A golf club transferring momentum to a golf ball.

change in momentum in that direction. The ball's initial speed v_0 is zero and its final speed v equals 70 m/s. Rearranging Eq. (4.3), we find that

$$F_{club} = \frac{mv - mv_0}{t} = \frac{(4.5 \times 10^{-2}\,\text{kg})(70\,\text{m/s}) - 0}{0.60 \times 10^{-3}\,\text{s}} = \underline{5.3 \times 10^3\,\text{N}.} \quad \blacksquare$$

The force calculated in Example 4.2 is the average force exerted by the club on the ball during its contact time with the ball. The actual force varies—being less than the average force during portions of the contact time and greater at other times. Later in this section we will use Eq. (4.3) with variable forces.

In many applications of the impulse-momentum equation, the impulse time lasts less than 1 s. For example, during a head-on car collision, the car may stop in less than 0.1 s. To calculate the force exerted on an object during an impulse, we must know the impulse time.

Kinematics provides a moderately reliable method for estimating the impulse time. Suppose that a car crashes into a large tree. Examination of the car after the collision indicates that its front end was smashed in a distance of about 0.5 m during the collision. This 0.5 m represents the distance the car traveled from the beginning to the end of the impact and is called the car's stopping distance.

Equation (2.10) from kinematics can be used to relate this stopping distance $x - x_0$ to the stopping or impulse time t:

$$t = \frac{2(x - x_0)}{v + v_0}, \quad (2.10)$$

where v is the car's final speed (zero for our example) and v_0 is its initial speed. This equation is exact only if the car's deceleration is constant—a very improbable collision. However, the equation does provide us with a technique for estimating the stopping time if the stopping distance is known.

EXAMPLE 4.3   A car and its driver traveling at a speed of 18 m/s (40 mi/h) are stopped abruptly by a collision with the rear end of a parked moving van (Fig. 4.3a). The car's front end is pushed in 0.55 m by the collision; the stopping distance of the car, then, is also 0.55 m. A 60-kg driver, who is restrained by a seat belt and shoulder strap, is also stopped in 0.55 m. Calculate (a) the time required to stop the car, (b) the average force acting on the driver, and (c) the deceleration of the driver.

SOLUTION   (a) Choosing the positive x axis to point in the direction of initial motion, the driver's initial velocity in the x direction is 18 m/s and the final velocity is zero. The time required to stop the car is calculated using Eq. (2.10):

$$t = \frac{2(x - x_0)}{v + v_0} = \frac{2(0.55\,\text{m})}{0 + 18\,\text{m/s}} = 0.061\,\text{s} = \underline{61\,\text{ms},}$$

where 1 ms = 1 millisecond = 10^{-3} s.

(b) The average force of the seat belt and shoulder strap on the driver is determined from the impulse-momentum equation:

$$F = \frac{m(v - v_0)}{t} = \frac{(60\,\text{kg})(0 - 18\,\text{m/s})}{0.061\,\text{s}} = \underline{-1.8 \times 10^4\,\text{N}.}$$

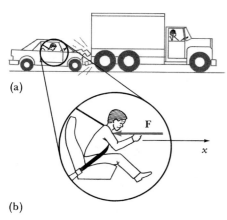

(a)

(b)

FIG. 4.3. (a) A car crashes into the rear of a moving van. (b) The force of the seat belt and shoulder strap restrains the driver of the car as it stops abruptly.

The negative sign implies that the force points in the negative x direction—opposite the direction of motion. The magnitude of the force is equivalent to 3600 lb (almost the weight of a rhinoceros).

(c) The driver's deceleration can be calculated using Eq. (2.9):

$$a = \frac{v - v_0}{t} = \frac{0 - (18 \text{ m/s})}{0.061 \text{ s}} = \underline{-300 \text{ m/s}^2}.$$

It is common to write acceleration or deceleration during impact in terms of gravitational acceleration—that is, as so many g's. Since $g = 9.8 \text{ m/s}^2$, we find that

$$a = (-300 \text{ m/s}^2)\left(\frac{1g}{9.8 \text{ m/s}^2}\right) = -30g.$$

The deceleration is 30 times greater than that of a freely falling body. ∎

Will the driver in Example 4.3 be injured? In October 1976 a preliminary report of the European Experimental Vehicle Committee (EEVC)* estimated that chest injuries were improbable if shoulder strap tension was less than about 8000 N. If one-half of the 18,000-N force on the driver in Example 4.3, or 9000 N, is caused by the shoulder strap, then there is a reasonable chance of injury. Common chest injuries in collisions include broken ribs, lacerations of the lungs, bruises of the heart, and even rupture of the aorta. This latter injury can occur if the heart is pushed or pulled away from the large vessels that normally anchor it to the body. To avoid such injuries, shoulder straps can be constructed that extend or loosen slightly if the tension reaches some maximum value (for example, 8000 N). This increases the stopping time and thus decreases the magnitude of the force during the impulse.

EXAMPLE 4.4 A 30-kg girl jumps from a ledge that is 2.5 m high. Over what distance must the force that stops her at the bottom of the jump be applied so that the average stopping force is 1.0×10^4 N or less, small enough to avoid injury.

SOLUTION The solution requires a divide and conquer technique involving three steps: (1) Calculate the girl's velocity as she reaches the ground. (2) Use the impulse-momentum equation to calculate her stopping time when opposed by a 1.0×10^4-N force. (3) Use kinematics to determine the downward distance that she moves during that time.

The velocity and forces can be described using a coordinate system with the y axis pointing down (Fig. 4.4). Since the motion and force occur along the y axis, we can denote directions using positive and negative signs (+ for down and − for up).

We use Eq. (2.12) from kinematics to calculate the girl's velocity as she reaches the ground after falling freely for 2.5 m (her acceleration while falling is

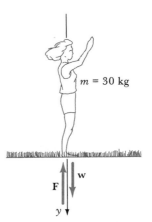

FIG. 4.4. A force diagram for a girl jumping onto the ground.

$m = 30 \text{ kg}$

*Report on the Sixth International Technical Conference on Experimental Safety Vehicles (Washington, D.C.: U.S. Department of Transportation, 1976), pp. 20–40.

the acceleration of gravity, $g = 9.8$ m/s²):

$$v^2 = v_0^2 + 2a(y - y_0)$$
$$= 0^2 + 2(9.8 \text{ m/s}^2)(2.5 \text{ m}) = 49 \text{ m}^2/\text{s}^2,$$

or

$$v = 7.0 \text{ m/s}.$$

The rest of the problem involves the stopping force at the bottom of the fall. Two forces act on the girl (Fig. 4.4): the stopping force **F** caused by the ground on which she lands points up in the negative y direction and the gravitational force due to her weight **w** points down in the positive y direction (do not forget to include weight in problems involving vertical motion). Both these forces appear on the left side of the impulse momentum equation: $(w - F)t = mv - mv_0$. For this part of the problem, the girl's initial velocity as the force begins to stop her fall is $v_0 = 7.0$ m/s, and her final velocity after she has stopped is $v = 0$. Rearranging the impulse-momentum equation and substituting $w = mg$, we find her stopping time:

$$t = \frac{m(v - v_0)}{mg - F} = \frac{30 \text{ kg}(0 - 7.0 \text{ m/s})}{(30 \text{ kg})(9.8 \text{ m/s}^2) - (1.0 \times 10^4 \text{ N})}$$
$$= 0.022 \text{ s}.$$

Now we can rearrange kinematic Eq. (2.10) to calculate the stopping distance:

$$y - y_0 = \left(\frac{v + v_0}{2}\right)t = \left(\frac{0 + 7.0 \text{ m/s}}{2}\right)0.022 \text{ s}$$
$$= 0.076 \text{ m} = \underline{7.6 \text{ cm}}.$$

It is unlikely that the girl will sink so far into the ground while stopping. However, if she bends her knees as she stops, the force is spread over a longer time, thus helping to reduce the magnitude of the stopping force. ∎

Impulse from a Variable Force

If the net force acting on an object varies with time, such as shown in Fig. 4.5a, the impulse caused by the force can be calculated by dividing the time into many small intervals Δt, such as we did in Section 2.10. During one of these time intervals, for example, the first dark-shaded interval shown in Fig. 4.5b, the impulse-momentum equation becomes

$$\overline{F}_1 \Delta t = m \Delta v_1,$$

where $\overline{F}_1$ is the average net force acting on the object during the first short time Δt, and Δv_1 is the object's change in velocity during that time. For the next time interval, the average force acting on the object has changed to $\overline{F}_2$, and the impulse-momentum equation becomes

$$\overline{F}_2 \Delta t = m \Delta v_2.$$

During each interval Δt, we calculate a different momentum change. To calculate the total change in momentum from an initial time t_0 to a final time t,

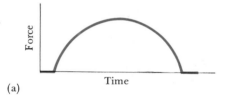

(a)

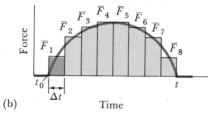

(b)

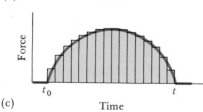

(c)

FIG. 4.5. (a) The time variation of a force acting on an object. (b) The area of each narrow force-versus-time rectangle equals the impulse during the short time Δt. (c) The sum of the impulses is the total impulse and equals the area under the force-versus-time graph.

we simply add the changes for each small interval between those times:

$$\overline{F}_1 \, \Delta t + \overline{F}_2 \, \Delta t + \overline{F}_3 \, \Delta t + \cdots = m(\Delta v_1 + \Delta v_2 + \Delta v_3 + \cdots)$$
$$= m(v - v_0),$$

where $m(v - v_0)$ is the total momentum change from t_0 to t.

The left side of the preceding equation (the impulse side) can be evaluated by determining the area under the force-versus-time graph from the initial time t_0 to the final time t. Note, for example, that the area of the first shaded rectangle in Fig. 4.5b is the product of its height $\overline{F}_1$ and width Δt. Similarly, the area of the second rectangle is the product of $\overline{F}_2$ and Δt. Thus, the sum of these products $(\overline{F}_1 \, \Delta t + \overline{F}_2 \, \Delta t + \overline{F}_3 \, \Delta t + \cdots)$ between times t_0 and t is the sum of the areas of all the rectangles between those times. If the time intervals are made small, as in Fig. 4.5c, the area of the rectangles equals the area under the force-versus-time graph.

The impulse caused by the resultant force F acting on an object between times t_0 and t equals the area under the force-versus-time graph between those times. The impulse causes a change in momentum equal to $m(v - v_0)$, where v_0 is the object's velocity at time t_0 and v is its velocity at time t:

$$\text{Impulse} = \frac{\text{Area under force-versus-time graph}}{\text{between times } t_0 \text{ and } t} = m(v - v_0).^* \qquad \textbf{(4.4)}$$

EXAMPLE 4.5 The force of an automobile's seat belts on a 29-kg child sitting in the front seat of the car when it collides with a barrier is shown in Fig. 4.6. Determine (a) the impulse of the force and (b) the car and child's initial velocity (their final velocity is zero). (c) Suppose that the child was not in a seat belt but was stopped instead by hitting the dashboard. Estimate the maximum force of the dashboard on the child's head if the force had the same shape as in Fig. 4.6 but took only about one-fifth the time (a more abrupt impulse).

SOLUTION (a) To calculate the impulse, which equals the area under the force-versus-time graph, we add the number of squares under the graph. A rough estimate, in which we combine squares that are partly under the graph and partly over the graph, indicates that there are approximately 58 squares under the graph, with an uncertainty of about 3 or 4. Each square is $0.10 \times 10^4 \, \text{N} = 1000 \, \text{N}$ high and $10 \, \text{ms} = 0.010 \, \text{s}$ wide. Thus, the area of one square, its height times its width, is $(1000 \, \text{N})(0.010 \, \text{s}) = 10 \, \text{N·s}$. Since there are approximately 58 squares under the curve, the impulse is $58(10 \, \text{N·s}) = \underline{580 \, \text{N·s}}$. (b) Using Eq. (4.3), with the final velocity $v = 0$, we find that

$$v_0 = -\frac{\text{Impulse}}{m} = -\frac{580 \, \text{N·s}}{29 \, \text{kg}}$$
$$= -20 \, \text{N·s/kg} = \underline{-20 \, \text{m/s}}, \text{ or } 45 \, \text{mi/h}.$$

*In calculus, the area under a force-versus-time graph is the integral of F as a function of t between the times of interest:

$$Impulse = \int_{t_0}^{t} F \, dt = m(v - v_0).$$

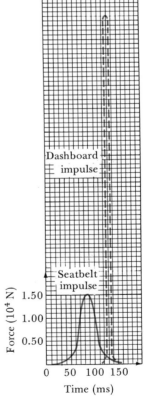

FIG. 4.6. The time variation of the force on a child when stopped abruptly by a seat belt (solid curve) and by hitting the dashboard (dashed curve).

This unit conversion was possible because $1\,N = 1\,kg\cdot m/s^2$. The negative sign means that the initial velocity was opposite the direction of the impulse. Since the impulse had a positive sign (the restraining force was in the positive direction), the car must have been moving initially in the negative direction.

(c) If the child is stopped more abruptly by hitting the dashboard, the momentum change and area under the curve still remain the same since the child's initial and final momenta are the same as before. But if the dashboard force-versus-time graph is one-fifth the width of the seat belt graph, the dashboard force will need to be five times greater for the areas to be the same. The dashed-line graph from about 120 to 130 ms in Fig. 4.6 represents the impulse caused by the dashboard. Its height is five times the 1.60×10^4-N height of the seat belt curve—that is, $8.0 \times 10^4\,N$, about 18,000 lb (9 tons)! Clearly, it is important to extend the impulse time by restraining a child (adults, too) by using seat belts and shoulder straps. Airbags are even better. ∎

4.3 Conserved Quantities

Most scientists think that the amount of certain quantities in the universe never changes. These quantities are called **conserved quantities.** Their amount in the universe at a particular time should be the same as at any time in the future. Momentum is one example of a conserved quantity, as are electric charge and energy.

It may seem presumptuous to talk about quantities that are conserved in a universe, most of which we have never seen. The idea of conserved quantities has evolved from experiments done with very small, isolated parts of the universe. For example, chemists can study groups of chemicals reacting in closed vessels. The chemicals can be isolated from their environment by ensuring that no heat or matter can cross the walls of the vessels. After studying many such isolated reactions, chemists have found that the number of atoms of a certain type (for example, carbon atoms or hydrogen atoms) is always the same before the reaction as after the reaction. Out of these observations has grown a conservation principle: The number of atoms of a certain type is the same in the molecules present before a reaction as after the reaction. This conservation law is the basis for the technique we learn in elementary chemistry courses—the technique of balancing chemical reactions. For example, when sugar and oxygen are "burned" to form carbon dioxide and water, we write the chemical reaction as

$$C_6H_{12}O_6 + 6\,O_2 \longrightarrow 6\,CO_2 + 6\,H_2O.$$

The number of C, O, and H atoms is the same on each side of the chemical equation. The number of atoms of each type is conserved in the reaction.*

Many other physical quantities seem to be conserved in processes that take place in the isolated confines of a laboratory apparatus. Since no exception to the conservation of these quantities in isolated systems has ever been found, we assume that they are conserved in any isolated system, the universe being the largest such system.

*The conservation law is true for most chemical processes that occur on the earth. Exceptions to this rule involve radioactive nuclei and nuclear reactions that occur in power plants, on the sun, and in particle accelerators.

But what is an isolated system? Fig. 4.7 is a schematic representation of the universe. We have drawn dashed lines around one part of the universe, which we call the system. The system can be any part of the universe we choose—a human body, the whole earth, a single atom. All the rest of the universe outside the system is called the **environment.** The dashed lines can be an actual boundary between the system and its environment, or it can be an imaginary boundary that we use to help identify the system. The system is said to be **isolated** if nothing (energy, matter, light) crosses the boundary of the system to or from the environment and if the net sum of all external forces acting on the system is zero. Within an isolated system, the amount of a conserved quantity such as momentum will not change.

Often, an isolated system is constructed in the laboratory. The walls containing the system prevent energy and matter from entering or leaving it. At other times, we can consider a system approximately isolated by simply choosing its boundaries in an appropriate manner. Then we can apply conservation laws to learn more about elements within the system. When two cars collide, for instance, we can choose the system to be the region of space that includes both cars. The interaction of each car with the other is much greater than any other force affecting the cars. As a result, we consider the system as being approximately isolated, and we can use conservation principles to learn more about the motions of the cars during their collision. In the next section we introduce the law of conservation of momentum and see how it can be used.

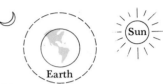

(a)

(b)

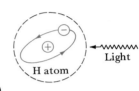

(c)

FIG. 4.7. (a) A system is some part of the universe, such as a person, in which we have a special interest. The rest of the universe not in the system is called the environment. (b) The earth is the system, and the sun and moon are its immediate environment. (c) A hydrogen atom is the system, and it is being approached by light from the environment.

4.4 Conservation of Momentum

We can derive the conservation of momentum principle by looking at the simple case of a collision between two objects. Figure 4.8 shows two objects before, during, and after they collide. The objects could be balls, spherically shaped molecules, or whatever. Let us suppose that they are balls.

First consider ball 1. What is its change in momentum? Before the collision, its velocity is v_{1_0}; following the collision, its velocity is v_1. Its change in momentum is then $mv_1 - mv_{1_0}$. This change in momentum is caused by the impulsive force of ball 2 on ball 1, $\mathbf{F}_{2\,\text{on}\,1}$, which we assume is constant and lasts for time t. Using Eq. (4.3), we can then write for ball 1 that

$$\mathbf{F}_{2\,\text{on}\,1}t = m_1\mathbf{v}_1 - m_1\mathbf{v}_{1_0}.$$

Similarly, the momentum of ball 2 is changed because of the impulsive force of ball 1 on ball 2, $\mathbf{F}_{1\,\text{on}\,2}$. Applying Eq. (4.3) for ball 2, we obtain

$$\mathbf{F}_{1\,\text{on}\,2}t = m_2\mathbf{v}_2 - m_2\mathbf{v}_{2_0}.$$

Adding these two equations, we find that

$$(\mathbf{F}_{2\,\text{on}\,1} + \mathbf{F}_{1\,\text{on}\,2})t = (m_1\mathbf{v}_1 - m_1\mathbf{v}_{1_0}) + (m_2\mathbf{v}_2 - m_2\mathbf{v}_{2_0}). \tag{4.5}$$

Newton's third law, however, says that the action-reaction pair of forces, $\mathbf{F}_{2\,\text{on}\,1}$ and $\mathbf{F}_{1\,\text{on}\,2}$, must be equal in magnitude but opposite in direction. Thus their sum must be zero:

$$\mathbf{F}_{1\,\text{on}\,2} = -\mathbf{F}_{2\,\text{on}\,1},$$

or

$$\mathbf{F}_{2\,\text{on}\,1} + \mathbf{F}_{1\,\text{on}\,2} = 0.$$

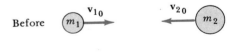

Before

During

After

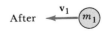

FIG. 4.8. Two balls before, during, and after a collision.

The push that transfers forward momentum to the skater on the left is balanced by the backward momentum of the pusher. Momentum, a vector quantity, is conserved. (Photo by David Schaefer)

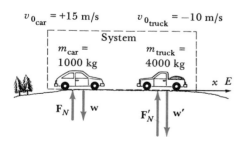

FIG. 4.9. The external forces acting on the vehicles in a system.

As a result, the left side of Eq. (4.5) must be zero. We find, then, that the change in momentum of ball 1 plus the change in momentum of ball 2 must add to zero:

$$0 = (m_1\mathbf{v}_1 - m_1\mathbf{v}_{1_0}) + (m_2\mathbf{v}_2 - m_2\mathbf{v}_{2_0}). \tag{4.6}$$

Equation (4.6) can be rewritten as

$$(m_1\mathbf{v}_1 + m_2\mathbf{v}_2) = (m_1\mathbf{v}_{1_0} + m_2\mathbf{v}_{2_0}). \tag{4.7}$$

The term on the left side of the equation is the total momentum of the two balls after the collision. The term on the right side is their momentum before the collision. There has been no change in the total momentum during the collision, although each ball has changed momentum.

This is an example of the general principle called the **conservation of momentum principle:**

When the net sum of the external forces acting on a system of objects is zero, the vector sum of the momenta of these objects is a constant, both in magnitude and direction. The conservation of momentum principle is expressed quantitatively for two objects using either Eq. (4.6) or Eq. (4.7).

External forces are forces originating from outside the system that act on objects in the system. If no such external forces act on the system, or if the external forces add to zero, we can say that the system is isolated. In such cases, the total momentum of objects in the system will not change. There can, of course, be changes in the momentum of individual particles caused by the forces they exert on each other, but when all these momentum changes are added, the net change is zero.

To apply the conservation of momentum principle to problems, we must first carefully identify the system being considered. This can be done by drawing a dashed line around the objects in the system, as in Fig. 4.9. Second, we superimpose a coordinate system on the system. The coordinate system provides a reference frame for expressing quantitatively the known momenta of objects in the system. Third, if the system is isolated, we substitute the known information into the conservation of momentum equation [either Eq. (4.6) or (4.7) for systems with two objects]. We must be sure to include the appropriate signs for the velocities of the objects. Finally, we solve this equation or equations for the unknown quantity.

EXAMPLE 4.6 A 1000-kg car traveling east at a speed of 15 m/s has a head-on collision with a 4000-kg truck traveling west at 10 m/s. If the vehicles are locked together after the collision, what is their final velocity v? (This is a one-dimensional problem, so directions are indicated using positive and negative signs.)

SOLUTION Figure 4.9 shows the system we have chosen. A downward gravitational force $\mathbf{w}$ and an upward normal force $\mathbf{F}_N$ act on each vehicle. These forces originate from outside the system and are thus external forces. But since $\mathbf{w} = -\mathbf{F}_N$ and $\mathbf{w}' = -\mathbf{F}'_N$, the net sum of these external forces is zero. We can therefore consider the system isolated.

Next we choose a coordinate system. If we choose the positive x axis as pointing toward the east, then $v_{0_{car}} = 15$ m/s and $v_{0_{truck}} = -10$ m/s. We are also given that $m_{car} = 1000$ kg and $m_{truck} = 4000$ kg. The final velocity v of the car and truck in the x direction is an unknown. We do not yet know whether v is positive or negative. Equation (4.6) becomes

$$0 = (1000 \text{ kg})(v - 15 \text{ m/s}) + (4000 \text{ kg})[v - (-10 \text{ m/s})],$$

or

$$(5000 \text{ kg})v = -25,000 \text{ kg} \cdot \text{m/s}.$$

Thus,

$$v = \underline{-5 \text{ m/s}}.$$

The negative sign means that the car and truck move in the negative x direction after the collision—that is, toward the west.

We can look at these results from another point of view. Initially, the total momentum of the car and the truck was the sum of their individual momenta, or

$$\text{Initial momentum} = (1000 \text{ kg})(+15 \text{ m/s}) + (4000 \text{ kg})(-10 \text{ m/s})$$
$$= -25,000 \text{ kg} \cdot \text{m/s}.$$

Following the collision, the car and truck are locked together and have a mass of 5000 kg. They are moving to the west at 5 m/s (in the negative x direction). Thus,

$$\text{Final momentum} = (5000 \text{ kg})(-5 \text{ m/s}) = -25,000 \text{ kg} \cdot \text{m/s}.$$

There has been no change in momentum. ■

Spaceship propulsion is another interesting phenomenon involving conservation of momentum. It is often mistakenly thought that fuel ejected from the spaceship pushes against the air outside the ship and thereby propels the ship. But a force exerted against the air would propel the air, not the spaceship. Also, when flying beyond the earth's atmosphere, there is no air for the ejected fuel to push against. Spaceship propulsion can be most easily understood as an example of conservation of momentum. Since no outside forces act on the system that consists of the spaceship and its fuel, the momentum gained by fuel ejected in the backward direction must be balanced by forward momentum gained by the spaceship.

The situation is similar to that of a person standing on roller skates on a hardwood floor and holding lead weights. To move east without skating, the person could throw the weights toward the west. The momentum gained by the weights flying west is equal in magnitude to the momentum the skater gains in the opposite direction, toward the east. The ability to gain forward momentum by ejecting mass in the backward direction is illustrated in Example 4.7.

EXAMPLE 4.7 The 5000-kg final stage of a spaceship ejects the 10,000-kg second stage. Before ejection, the two stages together travel with a velocity of 200 m/s. If after ejection the second stage travels at 50 m/s opposite the direction of its initial motion, what is the velocity of the final stage? (See Fig. 4.10.)

Before ejection

$v_0 = 200$ m/s

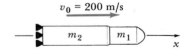

After ejection

$v_2 = -50$ m/s $\qquad$ $v_1 = ?$

FIG. 4.10. The ejection of the second stage of a rocket helps the first stage move faster.

SOLUTION The momentum of the two stages together before ejection must equal the sum of the momenta of the two stages separately after ejection:

$$(m_1 + m_2)v_0 = m_1v_1 + m_2v_2,$$

$$\begin{array}{ccc} \text{Initial} & = & \text{Final} \\ \text{momentum} & & \text{momentum}, \end{array}$$

where m_1 and m_2 are the masses of the first and second stages, respectively; v_0 is their initial velocity (200 m/s); v_1 is the unknown final velocity of the first stage; and v_2 is the final velocity of the second stage (-50 m/s). Solving for v_1, we find

$$v_1 = \frac{(m_1 + m_2)v_0 - m_2v_2}{m_1}$$

$$= \frac{(15{,}000 \text{ kg})(200 \text{ m/s}) - (10{,}000 \text{ kg})(-50 \text{ m/s})}{5000 \text{ kg}}$$

$$= \underline{700 \text{ m/s}}. \qquad\blacksquare$$

The previous examples involved changes in momenta occurring along only one axis. Often, the collision of one object with another results in changes in direction and must be described using two or more coordinates. We must then rewrite Eqs. (4.6) and (4.7) in their component forms and consider conservation of momentum along each axis separately. Equation (4.7) becomes

$$m_1v_{1_{0x}} + m_2v_{2_{0x}} = m_1v_{1_x} + m_2v_{2_x}, \qquad (4.7x)$$

$$m_1v_{1_{0y}} + m_2v_{2_{0y}} = m_1v_{1_y} + m_2v_{2_y}. \qquad (4.7y)$$

The components of momentum along the x axis must be conserved, as must the components along the y axis.

EXAMPLE 4.8 A cue ball traveling with a speed of 2 m/s along the x axis (see Fig. 4.11) hits the eight ball. After the collision, the cue ball travels with a speed of 1.6 m/s in a direction 37° below the positive x axis. Will the eight ball fall in the side pocket, which is oriented along a line 53° above the positive x axis? The two balls have equal mass.

SOLUTION We are given that $m_1 = m_2 = m$; $v_{1_0} = 2$ m/s; $v_{2_0} = 0$; $v_1 = 1.6$ m/s; and $\theta_1 = 37°$, where θ_1 is the direction of the cue ball after the collision. We wish to calculate the direction of the eight ball after the collision. This is done most easily by using the two equations for the conservation of the x and y components of momentum to solve for v_{2_x} and v_{2_y}, the x and y components of the velocity of the eight ball after the collision. We have

$$\text{Momentum before} = \text{Momentum after},$$

$$x \text{ component:} \qquad m_1v_{1_0} = m_1v_1 \cos 37° + m_2v_{2_x},$$

$$y \text{ component:} \qquad 0 = -m_1v_1 \sin 37° + m_2v_{2_y}.$$

Substituting $m_1 = m_2 = m$ and the values of v_{1_0}, v_1, cos 37°, and sin 37° into the preceding equations and rearranging, we find that

$$v_{2_x} = v_{1_0} - v_1 \cos 37° = (2 \text{ m/s}) - (1.6 \text{ m/s})0.8 = 0.72 \text{ m/s},$$

$$v_{2_y} = v_1 \sin 37° = (1.6 \text{ m/s})0.6 = 0.96 \text{ m/s}.$$

Before collision

Side pocket

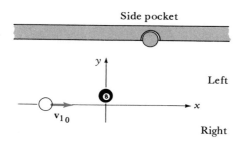

Left

Right

After collision

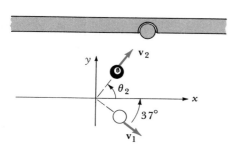

FIG. 4.11. A coordinate system for describing the motion of pool balls. (a) The balls before the collision and (b) after the collision.

We can determine the direction of v_2 using Eq. (1.8):

$$\tan \theta_2 = \frac{v_{2_y}}{v_{2_x}} = \frac{0.96 \text{ m/s}}{0.72 \text{ m/s}} = 1.33,$$

or

$$\theta_2 = \underline{53°}.$$

The ball falls in the side pocket.

■

Summary and Additional Readings

1. **Momentum and impulse:** The momentum of an object is defined as the product of its mass and its velocity **v**:

$$\text{Momentum} = \mathbf{p} = m\mathbf{v}. \qquad (4.1)$$

A force **F** acting on an object for time t is said to exert an impulse on the object. The impulse of the force is the product of **F** and t, or

$$\text{Impulse} = \mathbf{F}t. \qquad (4.2)$$

The **F** in Eq. (4.2) represents the average force during time t. For forces that vary with time, the impulse can be determined by calculating the area under a force-versus-time graph during the time of interest.

The sum of the impulses of all forces acting on an object for a certain time is equal to the change in momentum of the object during that time. This is called the **impulse-momentum equation** and is written as

$$(\Sigma \mathbf{F})t = m\mathbf{v} - m\mathbf{v}_0. \qquad (4.3)$$

When using the impulse-momentum equation to analyze one-dimensional collisions, the impulse time is often estimated using Eq. (2.10) from kinematics:

$$t = \frac{2(x - x_0)}{v + v_0},$$

where $x - x_0$ is the distance the object moves during the impulse, v_0 is its initial velocity at the start of the impulse, and v is its velocity at the end of the impulse.

2. **Conservation of momentum:** Momentum is one of a group of conserved quantities. The amount of a conserved quantity in the universe never changes. When a small part of the universe, which we call the **system,** is isolated from the rest of the universe, the total momentum of objects in the system is constant. For a system to be **isolated,** the sum of external forces (those originating from outside the system) that act on objects in the system must add to zero. Also, no matter, energy, or light should enter or leave the system.

For an isolated system with two objects, the conservation of momentum equation can be written as

$$0 = (m_1\mathbf{v}_1 - m_1\mathbf{v}_{1_0}) + (m_2\mathbf{v}_2 - m_2\mathbf{v}_{2_0}). \qquad (4.6)$$

The change in momentum of object 1 plus the change in momentum of object 2 must add to zero. Equation (4.6) can also be written as

$$(m_1\mathbf{v}_1 + m_2\mathbf{v}_2) = (m_1\mathbf{v}_{1_0} + m_2\mathbf{v}_{2_0}) \qquad (4.7)$$

or

$$\begin{array}{c} \text{Final momentum} \\ \text{of two objects} \end{array} = \begin{array}{c} \text{Initial momentum} \\ \text{of two objects} \end{array}$$

In an isolated system, the initial and final momenta are equal.

Martin S. Tiersten, "Force, Momentum Change, and Motion," *American Journal of Physics* **37**, 82 (1969).

F. I. Ordway, "Principles of Rocket Engines," *Sky and Telescope* **14**, 48 (1954).

P. Kirkpatrick, "Batting the Ball," *American Journal of Physics* **31**, 606 (1963).

Jearl D. Walker, "Karate Strikes," *American Journal of Physics* **43**, 845 (1975).

Questions

Three of these questions require rough estimates. Before working them, you might read Interlude II following this chapter.

1. A 0.4-kg hammerhead hits a nail and pounds it partway into a board. (a) *Estimate* the speed of the hammerhead before it hits the nail. (b) Estimate the stopping distance. (c) Calculate the stopping time. (d) Use these numbers to estimate the average force of the hammerhead on the nail (which will be equal in magnitude but opposite in direction to the force of the nail on the hammerhead).

2. A hand lies palm up on a tabletop. A 1.0-kg ball strikes the palm with a speed of 5.0 m/s after falling from a height of 1.25 m. (a) *Estimate* the stopping distance—that is, the extra distance the ball falls after first touching the hand. (b) Use this distance to estimate the stopping time. (c) Use these numbers and Eq. (4.3) to

estimate the force of the hand on the ball (which is equal in magnitude to the force of the ball on the hand). (d) Why would the force be somewhat less if the hand caught the ball in midair?

3. A 30-kg girl jumps from a garage roof 3.5 m high onto a concrete driveway. *Estimate* the force on the girl's leg (a) if she lands on her heels with her legs stiff and (b) if she lands on her toes and lets her knees bend as she stops.

4. Why is an impact-absorbing bumper that retracts during a collision preferable to a rigid, hard-steel bumper?

5. Explain why an inflated balloon shoots across a room when its valve stem is opened so that the air can escape.

6. The car of a toy train rests on a level track. Will its speed

increase more if (a) hit from the rear by a wad of clay that sticks to the car or (b) hit by a superball of equal mass and velocity that rebounds in the opposite direction after hitting the car? Explain.

7. A meteorite falls through the earth's atmosphere and forms a crater after striking the earth. If momentum is supposed to be conserved, what happened to the meteorite's momentum?

8. Two cars of unequal mass moving at the same speed collide head-on. Explain why a passenger in the smaller car is more likely to be injured than one in the larger car. Justify your reasoning with physical principles. [*Hint:* Which passenger has the greatest change in momentum? Why?]

Problems

4.1 Momentum

1. A 1000-kg car travels at a speed of 20 m/s. (a) At what speed must a 3000-kg truck travel to have the same momentum? (b) At what speed must a 7.3-kg bowling ball travel to have the same momentum?

2. (a) Calculate the momentum of a 30-kg dog running at a speed of 8.0 m/s toward the west. (b) How fast must a 70-kg person run to have the same momentum as the dog?

3. *Estimate* the magnitude of your momentum in units of kg·m/s when walking to physics class. Justify verbally the numbers used in your estimate.

■ 4. What is the change in momentum of a 0.0567-kg tennis ball traveling at 25 m/s if it hits a practice wall and rebounds in the opposite direction with the same speed? Is the vector change in momentum in the direction of the initial velocity, the direction of the final velocity, or neither? Explain.

■ 5. (a) A 145-g baseball traveling at 35 m/s is hit by a bat and rebounds in the opposite direction at 40 m/s. What is the change in momentum of the ball? (b) A 0.055-kg golf ball starts at rest and leaves the tee with a speed of 50 m/s. What is the change in momentum of the golf ball?

■ 6. (a) The momentum of a 900-kg car initially moving north at 10 m/s increases by 10,800 kg·m/s. Calculate the car's final velocity (a vector). (b) Calculate the car's final velocity if its momentum decreases by 10,800 kg·m/s.

4.2 The Impulse-Momentum Equation

7. An impulse of 7.5×10^4 N·s stops a car in 0.10 s. (a) Calculate the average force on the car. (b) If the impulse time is tripled, what is the average force on the car?

8. A 2.0×10^5-N friction force caused by the brakes of a subway train stops the train in 50 s. (a) Calculate the force needed to stop the train in 25 s. (b) Calculate the stopping time if a 1.0×10^5-N friction force acts on the train.

9. About 80 g of blood is pumped from the heart during each heartbeat. The blood starts at rest and has a speed of about 0.6 m/s in the aorta. If the pumping takes 0.17 s, what is the magnitude of the average force on the blood?

■ 10. A 0.60-kg basketball falling vertically at a speed of 6.0 m/s hits a floor and rebounds at a speed of 5.2 m/s. (a) Calculate the ball's change in momentum (magnitude and direction).

(b) Calculate the average net force on the ball if the collision lasts 0.12 s.

■ 11. A person will experience potentially serious chest injuries when stopped during a collision if the combined force of a seat belt and shoulder strap on the person's body exceeds 16,000 N. Calculate (a) the minimum stopping time and (b) the corresponding stopping distance for an 80-kg person moving initially at 16 m/s in order to avoid serious injury.

■ 12. A bat contacts a 0.145-kg baseball for 1.3×10^{-3} s. The average force of the bat on the ball is 8900 N. If the ball has an initial velocity of 36 m/s toward the bat and the force of the bat causes the ball's motion to reverse direction, what is the ball's final speed?

■ 13. A 100-kg astronaut pushes against the inside back wall of a 2000-kg spaceship, causing her speed toward the front to increase from zero to 0.80 m/s. (a) If the push lasts 0.30 s, what is the average force of the astronaut on the spaceship? (b) With what speed does the spaceship recoil if it is initially at rest? [*Note:* A reaction force acts on the spaceship, the magnitude of which equals the force on the astronaut (Newton's third law).]

■ 14. The bone in the upper part of the cheek (the zygomatic bone) can be fractured by a 900-N force lasting 6.0 ms or longer. A hockey puck can easily provide such a force when hitting an unprotected face. (a) What change in velocity of an 0.11-kg hockey puck is needed to provide that impulsive force? (b) A padded face mask doubles the impact time. By how much does it change the force on the head behind the mask? Explain.

■ 15. An impulse of 150 N·s is required to stop a person's head in a car collision. (a) If the head stops in 0.020 s when the cheekbone hits the steering wheel, what is the average force on the cheekbone? (b) Will the cheekbone be fractured (see Problem 14)? (c) What is the shortest impact time to avoid breaking the bone?

■ 16. A 0.10-kg apple will bruise if subjected to a force greater than 8.0 N. A 0.10-kg apple falls from a tree and reaches the ground at a speed of 5.0 m/s where it stops after sinking 8.0 cm into the grass below. Is the average force of the grass enough to bruise the apple? Explain.

■ 17. An 80-kg man stands stiff-legged in an elevator whose cable has just broken. The elevator is falling at 20 m/s when it starts to hit a shock-absorbing device at the bottom of the elevator shaft. To avoid injuring the man, the upward force of the elevator floor on the man while stopping should be no more than 40,000 N. Calculate the minimum stopping distance (do not forget to include the man's weight in your calculations) to avoid injury.

■ 18. A 50-kg woman jumps from a window and lands in a safety net that stops her in a distance of 1.0 m. Calculate the average force on her if she enters the net at a speed of 24 m/s.

■ 19. The majority of the European Experimental Vehicle Committee has recommended that the knee, thigh bone (femur), and hip should be subjected to a force no greater than 4000 N during car collisions. Greater forces could cause dislocations or fractures. Assume that a knee is stopped when it hits the instrument panel of the car. Assume also that the effective mass carried along with the knee is 20 percent of the body mass. (a) What is the minimum stopping time to avoid injury to the knee of an 80-kg person traveling at 15 m/s (34 mi/h)? (b) What is the minimum stopping distance?

■ 20. Water sprays from a firehose against a burning wall at a rate of 60 kg/s. The water's speed before hitting the wall is 40 m/s, and its speed after hitting the wall is zero. Calculate the force of the water against the wall.

■ 21. Air blowing with a speed of 10 m/s hits a sign, causing the air to stop and move to the side past the sign. The mass of air hitting the sign each second is 13 kg. (a) Calculate the change in momentum of the air each second. (b) Calculate the steady force of the air on the sign.

■ 22. A 6.0-m/s wind blows against one side of one of the twin towers of the World Trade Center in New York City. The area of the side is 63 m × 411 m, and 2.0×10^5 kg of air hits the wall each second. If the air's speed is reduced by one-half when hitting the wall (part of the air is deflected and its speed is changed little), calculate the average force of the air on the wall.

■ 23. An impact-absorbing bumper is built to stop a car of mass m moving with an initial speed v_0. The bumper exerts an average force F on the car as it retracts a distance Δx during a collision. (a) Using Eqs. (2.10) and (4.3), show that the magnitude of the force and retraction distance are related by the equation

$$F = \frac{mv_0^2}{2\,\Delta x}.$$

This general equation applies to all collisions that stop an object with initial speed v_0. Notice that the net force required to stop the car in a fixed distance increases in proportion to the square of the car's initial velocity. (b) What force is required to stop a 1500-kg car moving at 5 mi/h in a distance of 0.20 m? (c) Repeat (b) for a car moving at 10 mi/h.

■■ 24. You are asked to suggest methods to reduce by 20 percent the average force stopping a car with the impact-absorbing bumper discussed in Problem 23. Using proportional reasoning and the equation in Problem 23, determine (a) the percent change in the retraction distance Δx that would produce the desired reduction in force and (b) the percent change in initial speed that would produce the desired reduction in force.

■■ 25. A 0.40-kg beach ball flying toward the east at a speed of 8.0 m/s is bumped toward the north by a person's hand. The force has an average magnitude of 84 N and acts on the ball for 0.010 s. Calculate the ball's velocity (magnitude and direction) following the bump.

■ 26. The graph in Fig. 4.12 indicates force as a function of time on a dummy's 6.0-kg head when hit by a heavyweight prizefighter. Estimate (a) the impulse of the force and (b) the head's speed the instant following the punch, assuming it started at rest.

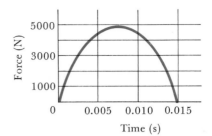

FIG. 4.12

■■ 27. The graph in Fig. 4.13 shows the time variation of the force of an automobile's airbag on the 9.1-kg effective mass of a person's head while being stopped during a collision. Determine (a) the total impulse of the airbag's force on the person's head and (b) the person's speed just before the collision occurred. (c) Draw a rough graph of force-versus-time if the head is stopped not by the airbag but by hitting the car's steering wheel. The impulse time is about one-tenth the time shown in Fig. 4.13 and has a similar shape.

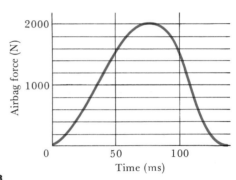

FIG. 4.13

4.4 Conservation of Momentum

28. A 3.0-kg gun mounted on frictionless wheels shoots a 5.0-g bullet in the horizontal direction at a speed of 300 m/s. What is the recoil speed of the gun?

29. A cannon mounted on the back of a ship fires a 50-kg cannonball in the horizontal direction at a speed of 150 m/s. If the cannon and ship to which it is firmly attached have a mass of 4000 kg and are initially at rest, what is the speed of the ship just after shooting the cannon? Ignore water resistance.

■ 30. A 90-kg space mechanic is drifting away from a spaceship at a speed of 0.20 m/s. The mechanic has no rocket to help him get back to the ship—only a 0.50-kg wrench. With what speed and in what direction relative to the spaceship must he throw the wrench to acquire a speed of 0.10 m/s toward the spaceship? Explain.

■ 31. An 80-kg astronaut is floating at rest a distance of 10 m from a spaceship when she runs out of oxygen and fuel to power her back to the spaceship. She removes her oxygen tank (3.0 kg) and flings it into space away from the ship with a speed of 15 m/s. (a) At what speed does she recoil toward the spaceship? (b) How long must she hold her breath before reaching the ship?

■ 32. A radioactive nucleus at rest decays to form an alpha particle and a product nucleus. With what speed does the product nucleus move following decay if the alpha particle moves at

2.0×10^7 m/s? The mass of the product nucleus is 54 times greater than the mass of the alpha particle.

■ 33. There are approximately 240 million people in the United States, each with an average mass of about 60 kg. If all simultaneously jumped upward from the earth with an initial speed of 6 m/s, *estimate* the recoil velocity of the earth.

■ 34. A group of people in Quebec are playing ice baseball. A 72-kg person initially at rest catches the 145-g ball traveling at 18 m/s. If the person's skates are frictionless, how much time is required for the person to slide 5.0 m after catching the ball?

■ 35. A 10-kg sled carries a 30-kg girl on a horizontal, frictionless surface at a speed of 6.0 m/s toward the east. The girl jumps off the back of the sled, and the sled shoots forward toward the east at 20 m/s. What was the girl's velocity (magnitude and direction) relative to the ground at the instant she left the sled?

■ 36. A 10,000-kg coal car coasts under a coal storage bin with a speed of 2.0 m/s. As it goes under the bin, 1000 kg of coal is dropped into the car. What is the final speed of the loaded car?

■ 37. An 80-g arrow shot from a bow passes through an apple balanced on William Tell's head. The arrow has a speed of 50 m/s before passing through the apple and 40 m/s after. The apple has a mass of 100 g. Calculate the final speed of the apple.

■ 38. Two carts on a frictionless air track are forced together with a spring of force constant 1000 N/m between them. After the carts, initially at rest, are released, the 1.6-kg cart moves right at a speed of 5.0 m/s. Calculate the velocity of the other cart, whose mass is 1.2 kg.

■ ■ 39. A 2100-lb sports car traveling east at 20 mi/h has a head-on collision with a 4000-lb station wagon traveling west at 30 mi/h. (a) If the vehicles remain locked together, what is their final velocity? (b) If the collision time is 0.10 s, what is the force needed to restrain a 150-lb person in the sports car? (c) In the station wagon?

■ ■ 40. A 3.0-g bullet traveling at 300 m/s hits and embeds itself in a 1.0-kg wooden block resting on a frictionless surface. (a) What is the final speed of the block and bullet? (b) If the block after being hit slides on a surface that exerts a 3.0-N friction force on the block, how far will it slide?

■ ■ 41. A 1200-kg car traveling south at 24 m/s collides with and attaches itself to a 2000-kg truck traveling east at 16 m/s. Calculate the velocity (magnitude and direction) of the two vehicles when locked together just after the collision.

■ ■ 42. A 3000-kg spaceship travels toward the moon at a speed of 1.5×10^4 m/s. The captain of the ship wishes to change direction by 5°. Rockets eject a short burst of fuel at a speed of 3×10^6 m/s perpendicular to the ship's initial direction. Calculate the mass of fuel that must be ejected to change course as the captain wishes.

■ ■ 43. An 1100-kg car at rest at a stop sign is hit from the rear by a 2200-kg truck. The car and truck remain locked together and skid 4.6 m before stopping. If the coefficient of friction between the vehicles and the road is 0.70, what was the truck's initial velocity?

INTERLUDE II Making Rough Estimates

The ability to make estimates and rough calculations is useful for a variety of reasons. (1) You need to decide whether a goal is worth pursuing; for example, can you make a living as a piano tuner in a town that already has a certain number of tuners? (2) You need to know roughly the amount of material needed for some activity—for instance, the food needed for a party or the number of bags of fertilizer needed for your lawn. (3) You want to estimate a number before it is measured; for example, what scale should be used when measuring a voltage or how rapid a time-measuring device is needed to detect laser light reflected from a distant object? (4) You wish to check whether a measurement you have made is reasonable—for instance, the measurement of the time for light to travel to a mountain and back or the oxygen consumed by a humming-bird. (5) You can determine an unknown quantity only by an estimation proce-dure; for example, you estimate the number of cats in the United States or the compression force on the discs in your back when lifting a weight in different ways.*

The procedure for making rough estimates usually means selecting some basic numbers whose values are known or can be estimated and then combining the numbers in some mathematical framework that leads to the desired answer. In Chapter 4 the mathematical framework involved the use of the impulse-momentum equation and equations of kinematics. Sometimes no physics equa-tion is needed; basic numbers are simply multiplied and divided to determine the desired result. This procedure is used in Example II.1 and in problems at the end of this interlude.

EXAMPLE II.1 Estimate the fraction of the U.S. population that eats at McDonald's each day. [*Hint:* In 1985 a sign outside one of McDonald's build-ings said that the company had served over 50 billion meals. You might try making a wild guess before examining our estimation technique.]

SOLUTION For this type of problem, we need a scheme that allows us to calculate the unknown by using other numbers or quantities that we can esti-mate reliably. Suppose we know the number of meals served by McDonald's each day. This number equals roughly the number of people eating at McDonald's each day. Then, dividing by the U.S. population (about 230 mil-lion) would give us the fraction of the U.S. population eating at McDonald's each day:

$$\frac{\text{Number of meals served each day}}{\text{U.S. population}} \cong \frac{\text{Fraction of population}}{\text{eating at McDonald's each day}}.$$

---

*The subject of making rough estimates has been considered in two interesting articles in *The Physics Teacher:* Mark St. John, "Thinking Like a Physicist in the Laboratory," *The Physics Teacher* **18**, 436–443 (1980), and J. D. Memory and A. W. Jenkins, Jr., "Estimating Orders of Magnitude," *The Physics Teacher* **15**, 43–44 (1977).

To estimate the number of meals served by McDonald's each day, we divide the total number of meals served since the company started business (over 50 billion according to the sign) by the total number of days McDonald's has served meals. We can probably make a rough estimate of the latter number. If you are in your twenties, then you probably remember McDonald's for most of your life. My first recollection of McDonald's goes back to the early 1960s. Let us assume that the company has been in business for 25 years as of 1985. Since there are 365 days per year, McDonald's has served meals for about

$$25 \text{ years} \times \frac{365 \text{ days}}{\text{year}} = 9125 \text{ days}.$$

We could easily be in error by 1000 or 2000 days. Thus, we will round our estimate of 9125 days to 9000 days. Retaining the 125 in 9125 days implies greater accuracy than we can justify.

Next we divide the total number of meals served by the number of days in business to get an estimate of the average number of meals served per day:

$$\frac{50,000,000,000 \text{ meals}}{9000 \text{ days}} = 5,500,000 \text{ meals/day}.$$

It seems likely that more meals are served per day now than during the earlier days in McDonald's history. Thus, we should probably fudge this number up to 6 or 7 million per day. We will settle for 6 million.

Now then, we can return to our first equation to determine the fraction of people eating at McDonald's:

$$\frac{6 \text{ million meals/day}}{230 \text{ million people}} \cong 0.026 = \frac{1}{38},$$

or about one of every 40 people eats at McDonald's each day.

This whole procedure can be summarized in a chart as follows:

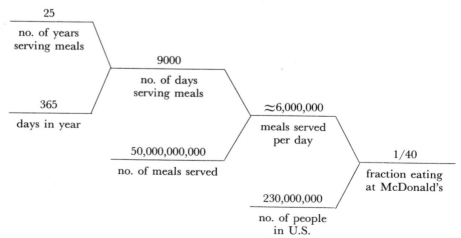

TABLE II.1 Percent Uncertainty of Numbers Used in McDonald's Estimate

Quantity	Number and Uncertainty	Percent Uncertainty
Years in service	25 ± 10	$\dfrac{10}{25} \times 100 = 40\%$
Days in year	365 ± 0	$\dfrac{0}{365} \times 100 = 0\%$
No. meals served	$50 \text{ bil} \pm 10 \text{ bil}$	$\dfrac{10}{50} \times 100 = 20\%$
People in U.S.	$230 \text{ mil} \pm 20 \text{ mil}$	$\dfrac{20}{230} \times 100 = 10\%$

In the chart, two quantities on the left are combined by multiplying or dividing to get the next quantity on the right. Each number on the right is rounded off.

All four numbers on the left side of the chart are numbers we supplied to make the estimate. Some of them are known with considerable accuracy—for example, the number of days in the year. Others are known with less accuracy—the number of years that McDonald's has served meals. The uncertainty in our final answer can be estimated using the **weakest-link rule:** The calculated value is no more precise than the least-precise value used to make the calculation. We first estimate the percent uncertainty in each of the numbers used to make the estimate. The final answer is assigned the same percent uncertainty as the least uncertain of the estimated numbers. We estimate the uncertainties for the McDonald's example in Table II.1. Since the largest percent uncertainty is 40 percent—for the number of years in service—we assign our final answer a 40 percent uncertainty:

Fraction eating at McDonald's $= 0.026 \pm 0.4(0.026) = 0.026 \pm 0.010.$

Our fraction could vary from about 0.016 to 0.036, or from 1/63 to 1/28. We round these numbers and estimate that somewhere between <u>1/60</u> and <u>1/30</u> of the population eats at McDonald's each day.

A check with McDonald's national organization indicates that the company serves each day about 1 out of every 30 or 40 U.S. citizens. ■

Problems assigned in most science and mathematics courses usually include "input information" that allows you to calculate a precise answer. Yet in much of real-life problem solving you must provide the input information yourself, and this information is often estimated or measured imprecisely. It should not upset you to find answers that are 50 or 100 percent uncertain. You still know more about the desired solution than if you made a wild guess.

The following problems should help you develop the ability to make rough estimates. To determine a solution for each problem, use a technique similar to the one we used in the McDonald's example. Also, using the weakest-link rule, estimate the uncertainty in your answer. Solutions provided to your instructor are either values determined by contacting an appropriate reference or the author's estimate. The latter may be significantly in error, so do not be upset if your answer differs somewhat from that estimated by another person. You should, however, justify your estimation procedure with a reasonable plan and your input numbers with moderate accuracy.

Problems

1. How long would it take to sail around the earth nonstop?

2. How many times does the average person's heart beat during his or her lifetime? By comparison, a four-stroke car engine traveling at 60 mi/h completes about 10^7 strokes in its lifetime.

3. Estimate the average number of people in the United States who ride a bicycle each day.

4. Estimate the volume of your body in cubic centimeters.

5. Estimate the number of piano tuners that work in Atlanta, Georgia.

6. Estimate the cost of gasoline used for all purposes by people who work at and attend your college or university each day.

7. Estimate the average number of miles jogged by students at your college or university each day.

8. Estimate the volume of water consumed during the day by people in your physics class, and estimate the three greatest uses of the water and the amount used for each purpose.

CHAPTER 5

Circular Motion at Constant Speed

Kinematics and Newton's second law of motion provided the basis in the first four chapters for analyzing the motion of objects from one place to another (called translational motion) and the forces that cause the motion to change. Similar techniques are used in this chapter and Chapter 7 to analyze circular and rotational motion. In this chapter we restrict our attention to objects moving in a circular path at constant speed—a ball swinging in a horizontal circle at the end of a string, the moon circling the earth, a race car moving at constant speed around a circular track.

We will find, using Newton's second law, that a force is needed to keep an object in a circular path, even if the object moves at constant speed. The force pulls the object toward the center of the circle; it does not change the object's speed, only its direction of motion as it moves in the circle. Examples of these forces that exert an inward force on an object as it moves in a circle are the earth's gravitational pull on the moon and the tension force of a string pulling on a ball swung at its end.

If an object's speed while in orbit increases or decreases, then another force is responsible for the change in speed. The forces that cause changing speed are directed tangent to the circle of motion; we consider them in Chapter 7.

5.1 Angular-Position Coordinate

When describing linear motion in Chapter 2, we used an x coordinate or a y coordinate to indicate an object's position at different times. To describe circular motion, such as the movement of a satellite about the earth, we need to use some other type of coordinate to indicate the object's position.

One way of doing this might be to place markers along the circumference of the circular path. The position of the object would then be given by its distance along the circumference from some reference point—such as the furlong markers indicating the position of a horse from the starting line of a horse race.* Usually, in physics, the reference point for one of these circular coordinate systems is the

---

*One furlong is a unit of length equal to 220 yd.

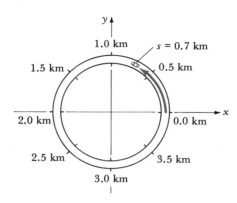

FIG. 5.1. The arc length *s* is used to indicate the location of an object on a circular path along which it moves. The race car is at the 0.7-km position—that is, *s* = 0.7 km.

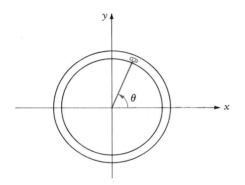

FIG. 5.2. The angular position coordinate *θ* is the angle from the positive *x* axis to a radial line drawn from the center of the circle to the object moving in a circle.

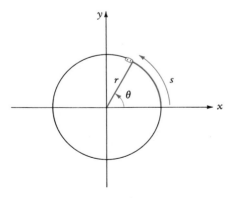

FIG. 5.3. The angle *θ* in units of radians is the ratio of *s*, the arc length, and *r*, the radius of the circle.

position where the positive *x* axis crosses the circle (see Fig. 5.1). This position coordinate along the arc of the circle is called the **arc-length coordinate** and is given the symbol *s*. By convention, *s* is positive in the counterclockwise direction from the reference point and negative in the clockwise direction.

While the arc-length coordinate *s* is convenient for describing the motion of some objects, such as the position of horses during a race, it is not useful for describing the position of an earth satellite. We cannot place markers along the satellite's orbit to show its position. We can, however, indicate its angular position above the earth using telemetry. This angular coordinate technique is illustrated in Fig. 5.2. A race car travels around a circular track. To indicate the car's position by an angle, we draw a radial line from the center of the circle to the car. The angle from the *x* axis to the radial line, as in Fig. 5.2, is called the **angular-position coordinate** and is given the symbol *θ* (theta). The angular position for the race car shown in Fig. 5.2 is about 75 degrees—that is, *θ* = 75°. *θ* is positive when measured in a counterclockwise direction from the positive *x* axis and is negative in a clockwise direction.

5.2 Units to Describe Angular Position

Three different units are used to measure the angular position of an object: (1) degrees, (2) revolutions, and (3) radians. The unit degree is the most familiar. There are 360 degrees in a circle. If an object is at the top of the circle, its position is 90°. When at the bottom, its position is 270°.

Revolution is also a familiar unit. One revolution (rev) corresponds to a complete rotation about the circle and equals 360°. An object that had made one-fourth of a revolution would be at the top of the circle. Revolutions are usually used to indicate the change in angular position per unit time. For example, an object that rotates in a complete circle 120 times each minute is said to have an angular speed of 120 revolutions per minute (120 rpm). Motorcycle engines rotate at about 500 rpm. The turntable of a stereo system usually rotates with an angular speed of 33.3 rpm.

The unit of angular position most useful in physics is the radian. It is defined in terms of the two lengths shown in Fig. 5.3. One of these, the arc length *s* defined earlier, is the distance along the circumference of the circle from the positive *x* axis to the position of the object. The other length used in defining the radian is the radius *r* of the circle. The angle *θ* in units of **radians** (rad) is the ratio of *s* and *r*:

$$\theta \text{ (in radians)} = \frac{s}{r}. \tag{5.1}$$

Since the circumference of a circle is $2\pi r$, one complete rotation around a circle corresponds to a change in arc length *s* of $2\pi r$ and a change in angular coordinate of

$$\theta = \frac{s}{r} = \frac{2\pi r}{r} = 2\pi.$$

Thus, in a complete circle, there are 2π radians.

We can now relate the three units of angular position as follows:

$$360° = 1 \text{ rev} = 2\pi \text{ rad.} \qquad (5.2)$$

This equation is useful when converting between different units for angles.

It is important to note that the radian unit has no dimensions; it is the ratio of two lengths. Consider, for instance, an angle of 1 rad, shown in Fig. 5.4. Its arc length equals 2 cm, as does the radius of the circle. Thus,

$$\theta = \frac{s}{r} = \frac{2 \text{ cm}}{2 \text{ cm}} = 1 \text{ rad.}$$

But $2 \text{ cm}/2 \text{ cm} = 1$ and is unitless. We see that one radian is just the number 1. We can remove the word *radian* from an expression or multiply an expression by 1 radian with no adverse consequences. The unit radian is usually kept in an expression that involves angles to remind us that we are using units of radians and not degrees or revolutions.

The radian is useful in science because it can be used to relate angle positions or changes in angular position to arc lengths. For example, if a car travels 2.0 rad around a circular racetrack of radius 300 m, the distance along the arc that the car has traveled is determined by rearranging Eq. (5.1):

$$s = r\theta = (300 \text{ m})(2.0 \text{ rad}) = 600 \text{ m.}$$

We dropped the rad unit in the answer because we are calculating a length, and the radian is therefore not needed as a reminder. Note that we could not have calculated this arc length using degrees or revolutions unless we first converted those units to radians.

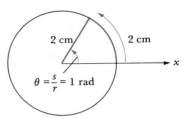

FIG. 5.4. An angle of one radian has an arc length that equals the radius of the circle.

EXAMPLE 5.1 Ocean vessels can sail up the Amazon River from Belem, Brazil, at its mouth to Iquitos, Peru—a twisting, curved trip of 3700 km along the river. Calculate the distance as the crow flies between the two cities, both of which are near the equator, Belem at 48° longitude and Iquitos at 73° longitude (Fig. 5.5).

SOLUTION The arc length s of a portion of a circle is given by Eq. (5.1)—that is, $s = r\theta$. For this problem, r is the radius of the earth (6.38×10^6 m). The angular separation of the two cities on the equator in units of degrees is the difference in their longitude: $73° - 48° = 25°$. To use Eq. (5.1), we must first convert 25° to units of radians:

$$\theta = 25° \left(\frac{2\pi \text{ rad}}{360°}\right) = 0.44 \text{ rad.}$$

Then, according to Eq. (5.1),

$$s = r\theta = (6.38 \times 10^6 \text{ m})(0.44 \text{ rad})$$
$$= 2.8 \times 10^6 \text{ m} = \underline{2800 \text{ km}}.$$

■

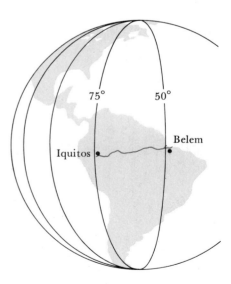

FIG. 5.5. The angular separation of cities on the equator is the difference in their longitudes.

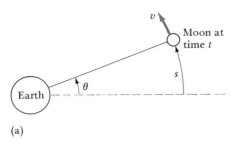

(a)

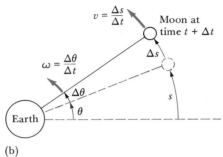

(b)

FIG. 5.6. (a) At time t the moon is at angular position θ and arc-length position s. (b) A short time, Δt, later its angular position has changed by $\Delta\theta$ and the moon has moved a distance Δs along the arc.

5.3 Tangential Speed and Angular Velocity

In Section 5.1, we used two ways to indicate the location of an object moving in a circular path: (1) the distance s along the circumference of the circle from some reference point to the object and (2) the angular-position coordinate θ of the object. Figure 5.6a shows both ways of describing an object's position.

We can also use two ways to describe how fast an object moves. Using s, we define the object's **tangential speed** v along the circumference of the circle as the ratio of its change in position Δs and the time Δt required for the change (see Fig. 5.6b). Thus,

$$\text{Tangential speed} = v = \frac{\Delta s}{\Delta t}. \tag{5.3}$$

Tangential speed indicates how fast the object actually moves along the arc of the circle. In the case of a horse race, the tangential speed is the actual speed of a horse along the track. The word *tangential* is used because the motion is directed tangent to the arc of the circle.

If an object is moving, then its angular position θ is changing. We say that the object has an angular velocity. **Angular velocity** is defined as the ratio of the change in angular position $\Delta\theta$ and the time Δt required for this change and is given the symbol ω (omega):

$$\text{Angular velocity} = \omega = \frac{\Delta\theta}{\Delta t}. \tag{5.4}$$

If, for example, a race car travels $90°$ around a circular track in 30 s, then its angular velocity is

$$\omega = \frac{90°}{30\text{ s}} = \frac{3°}{\text{s}} = 3° \text{ s}^{-1}.$$

Notice that angular velocity indicates a change in angular position per unit time and not a change in distance per unit time. You might think of the radial line from the center of the circle to the moving object as the spoke of a bicycle tire that rotates in a circle. The angular velocity is the spoke's change in angular position or orientation per unit time. The word *velocity* is used rather than *angular speed* because ω is actually a vector quantity. The direction of angular velocity is an axis that is perpendicular to the plane in which the circular motion occurs. For example, the angular velocity of the valve stem of a rotating bicycle tire points along the axis of rotation perpendicular to the plane of the tire. If an object moves counterclockwise in the plane of the paper, the direction of $\boldsymbol{\omega}$ is out of the paper perpendicular to the plane of motion. If an object moves clockwise, $\boldsymbol{\omega}$ is directed into the paper. We will not concern ourselves in the future with the direction of $\boldsymbol{\omega}$.

In many examples of circular motion, we will need to determine the tangential speed of an object from its angular velocity, or vice versa. For example, how fast does the groove on a stereo record travel under the needle when the record turns with a certain angular velocity? To derive an expression relating v and ω, we note that when θ is in units of radians, the tangential distance s and the angular position θ are related by the equation $s = r\theta$ [Eq. (5.1)]. We can use this equation to relate v and ω. Consider the motion of the moon depicted in Fig. 5.6b. It rotates a distance Δs along its circular path in a time Δt. During the same time the moon's angular position coordinate has changed by $\Delta \theta$. From Eq. (5.1) we see that Δs and $\Delta \theta$ are also related by

$$\Delta s = r\,\Delta\theta.$$

If we divide both sides of this equation by Δt, we find that

$$\frac{\Delta s}{\Delta t} = r\,\frac{\Delta\theta}{\Delta t}.$$

But $\Delta s/\Delta t$ is just the tangential speed v of the moon, and $\Delta\theta/\Delta t$ is the moon's angular velocity ω. Thus, we find that tangential speed and angular velocity are related by the equation

$$v = r\omega. \tag{5.5}$$

When using this equation, the angular velocity must be in units of radians per second because $\Delta s = r\,\Delta\theta$ applies only for $\Delta\theta$ in units of radians.

We should note that $v = \Delta s/\Delta t$ and $\omega = \Delta\theta/\Delta t$ are instantaneous values if Δt is very short and are average values if Δt is long. Equation (5.5) relates v and ω in either case.

EXAMPLE 5.2 The earth makes one rotation about its axis in 24 hours. (a) Calculate its angular velocity in units of rad/s. (b) Calculate the tangential speed of an object at the equator due to this rotational motion. The radius of the earth is 6.38×10^6 m.

SOLUTION (a) According to Eq. (5.4), we find that the earth's angular velocity is

$$\omega = \frac{\Delta\theta}{\Delta t} = \frac{1\ \text{rev}}{24\ \text{h}} = \left(\frac{1\ \text{rev}}{24\ \text{h}}\right)\left(\frac{2\pi\ \text{rad}}{1\ \text{rev}}\right)\left(\frac{1\ \text{h}}{3600\ \text{s}}\right) = \underline{7.3 \times 10^{-5}\ \text{rad/s}}.$$

(b) Using Eq. (5.5), we find that the tangential speed of an object on the earth's equator is

$$v = r\omega = (6.38 \times 10^6\ \text{m})(7.3 \times 10^{-5}\ \text{rad/s}) = \underline{460\ \text{m/s}}.$$

Thus, a person standing on the equator travels in a circular path at a speed of 460 m/s, or a little over 1000 mi/h. Fortunately, the earth's atmosphere rotates at about the same speed; otherwise the earth would be very windy! ∎

5.4 Centripetal Acceleration

We might think that an object moving with constant speed in a circular path is not accelerating. However, an object that is not accelerating has constant velocity, not constant speed. Notice in Fig. 5.7 that the velocity changes as a ball swings around a circle at the end of a string. The speed may not change, but the direction of the velocity changes. Since the velocity changes direction, the ball is accelerating.

We can derive an expression for this acceleration by considering the motion

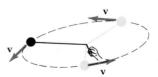

FIG. 5.7. The velocity of an object moving at constant speed in a circular path is continually changing because its direction of motion is changing.

FIG. 5.8. (a) As the moon moves from position A to position B, its velocity changes from $\mathbf{v}_A$ to $\mathbf{v}_B$. (b) The change in velocity $\Delta\mathbf{v} = \mathbf{v}_B - \mathbf{v}_A$. (c) A triangle showing the distance Δs traveled by the moon. (d) A similar velocity triangle.

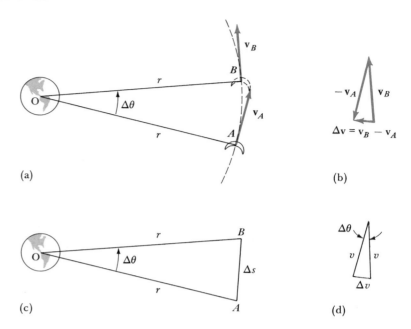

of the moon in its orbit about the earth (Fig. 5.8a). During a time Δt the moon moves from position A to position B, and the moon's velocity changes from $\mathbf{v}_A$ to $\mathbf{v}_B$. The magnitude of the velocity remains the same; only the direction changes. The acceleration of the moon equals the change in velocity divided by the change in time:

$$\mathbf{a} = \frac{\Delta\mathbf{v}}{\Delta t} = \frac{\mathbf{v}_B - \mathbf{v}_A}{\Delta t}.$$

The change in velocity $\Delta\mathbf{v} = \mathbf{v}_B - \mathbf{v}_A$ is shown in Fig. 5.8b and points toward the center of the circle.* The acceleration must also be toward the center of the circle. This acceleration of objects moving in circular paths at constant speed is called **centripetal acceleration,** after the Latin *centripetus* meaning "toward the center."

Centripetal acceleration is always a little difficult to understand. If the moon is accelerating toward the earth, why does it remain a constant distance away rather than "falling" into the earth? In fact, if the moon did not accelerate, it would move straight ahead, as in Fig. 5.9, and leave its orbit around the earth. Because of its centripetal acceleration, the moon does, in a sense, fall toward the earth, but while it falls, it continues to move forward. The combination of the two motions keeps it on a circular path of constant radius.

To find the magnitude of the centripetal acceleration, we apply some basic geometry. In Fig. 5.8c, a triangle is constructed whose sides are radial lines from the earth to the moon when the moon is at positions A and B. The third side of the triangle is the distance Δs that the moon travels from position A to position B.† In Fig. 5.8d, another triangle is constructed from $\mathbf{v}_B$, $-\mathbf{v}_A$, and

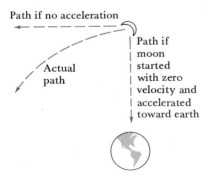

FIG. 5.9. The earth pulls the moon down toward the earth. While this happens, the moon travels to the left. The net result is a circular path of constant speed.

*The graphical subtraction of vectors was discussed at the end of Section 1.5.

†This straight line is slightly shorter than the actual curved arc along which the moon travels. The error introduced is negligible when $\Delta\theta$ is small.

$\Delta \mathbf{v} = \mathbf{v}_B - \mathbf{v}_A$. Both $\mathbf{v}_B$ and $-\mathbf{v}_A$ have the same magnitude, equal to the moon's speed v.

Comparing these triangles and the vectors shown in Fig. 5.8a, we see that $-\mathbf{v}_A$ is perpendicular to line OA and $\mathbf{v}_B$ is perpendicular to line OB. Thus, the angle $\Delta\theta$ between OA and OB equals the angle between $-\mathbf{v}_A$ and $\mathbf{v}_B$ (Fig. 5.8b). The corresponding triangles shown in Figs. 5.8c and 5.8d must have equal angles and are similar triangles. Because they are similar triangles, the ratio of the length of the side opposite $\Delta\theta$ and the length of the side adjacent to $\Delta\theta$ for one triangle must equal the same ratio for the other triangle:

$$\Delta\theta = \frac{\Delta s}{r} = \frac{\Delta v}{v}. \tag{5.6}$$

Since the moon travels with speed v, the distance it travels in time Δt is

$$\Delta s = v\,\Delta t.$$

When we substitute this into Eq. (5.6), we find that

$$\frac{v\,\Delta t}{r} = \frac{\Delta v}{v}.$$

After rearranging, we have

$$\frac{\Delta v}{\Delta t} = \frac{v^2}{r}.$$

But $\Delta v/\Delta t$ is just the magnitude of the acceleration and equals v^2/r. Since v and r are the same at each point on the circle, the magnitude of the acceleration is also the same at each point. We can summarize what we have learned about centripetal acceleration by the following statements:

An object moving in a circular path of radius r with constant speed v has an acceleration, called **centripetal acceleration a_c**, of magnitude

$$a_c = \frac{v^2}{r} \tag{5.7}$$

$$= r\omega^2. \tag{5.8}$$

The acceleration is directed toward the center of the circle. The expression for centripetal acceleration $a_c = r\omega^2$ was obtained by substituting $v = r\omega$ [Eq. (5.5)] for v in the first expression. An object's centripetal acceleration can be calculated using the most convenient of these two expressions. The angular velocity must be in units of rad/s when using Eq. (5.8).

At what minimum speed must the roller coaster travel at the top of the loop in order for passengers to remain in their seats? We will learn about problems like this in Section 5.5. (Peter Arnold, Inc./Vic Cox, Photographer)

EXAMPLE 5.3 Calculate the centripetal acceleration of a car traveling on a circular race track of radius 1000 m at a speed of 180 km/h.

SOLUTION We first convert the speed to units of meters and seconds:

$$v = 180\,\frac{\text{km}}{\text{h}}\left(\frac{1000\ \text{m}}{1\ \text{km}}\right)\left(\frac{1\ \text{h}}{3600\ \text{s}}\right) = 50\ \text{m/s}.$$

Then, using Eq. (5.7), we find that

$$a_c = \frac{v^2}{r} = \frac{(50\ \text{m/s})^2}{1000\ \text{m}} = \underline{2.5\ \text{m/s}^2}.\qquad\blacksquare$$

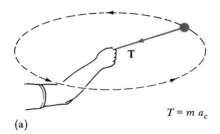

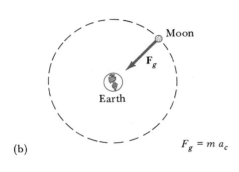

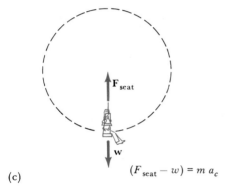

FIG. 5.10. Examples of forces responsible for the centripetal acceleration of different objects moving in circular paths: (a) a string tension force pulling on a ball; (b) the gravitational force of the earth on the moon; (c) the seat of a Ferris wheel pushing opposite the weight of a person on the seat.

5.5 Centripetal Force

We have seen that an object moving in a circular path at constant speed experiences an acceleration toward the center of the circle—a centripetal acceleration. But, according to Newton's second law ($\Sigma \mathbf{F} = m\mathbf{a}$), acceleration occurs only if a net nonzero force acts on an object in the direction of the acceleration. For the case of circular motion, the net force must be exerted radially inward to cause the centripetal acceleration. Any force responsible for centripetal acceleration is said to act as a **centripetal force.** For example, the gravitational force of the earth on the moon is a centripetal force that keeps the moon in its circular orbit.

Newton's second law, when applied to the centripetal forces directed along the radial direction and the corresponding centripetal acceleration they cause, is written as follows:

$$\Sigma F \text{ (in radial direction)} = ma_c, \qquad (5.9)$$

where $a_c = v^2/r = r\omega^2$ and ΣF (in radial direction) is the sum of the radial components of all forces acting on the object that moves in a circle. If a force points toward the center of the circle, its radial component has a positive sign; if it points outward, the radial component is negative. For circular motion to occur, the inward components must be greater in magnitude than the outward components so that the net force is inward.

Several examples of forces responsible for centripetal acceleration are shown in Fig. 5.10. The application of the radial form of Newton's second law for each situation is also shown. Notice that many different types of forces can act as the centripetal forces that keep objects in circular paths.

EXAMPLE 5.4 A 1.2-kg block slides on a horizontal frictionless surface in a circular path at the end of a 0.50-m-long string. (a) Calculate the block's speed if the tension in the string is 86 N. (b) By what percent does the tension change if the block's speed decreases by 10 percent?

SOLUTION (a) A force diagram for the block is shown in Fig. 5.11. The upward normal force balances the block's weight. The tension force of the string on the block provides the centripetal force that keeps the block moving in a circle. Newton's second law for forces along the radial direction is

$$\Sigma F \text{ (in radial direction)} = T = \frac{mv^2}{r},$$

or

$$v = \sqrt{\frac{Tr}{m}} = \sqrt{\frac{(86 \text{ N})(0.50 \text{ m})}{1.2 \text{ kg}}} = \underline{6.0 \text{ m/s}}.$$

(b) A 10 percent reduction in the speed results in a speed $v' = 5.4$ m/s. The new tension is

$$T' = \frac{mv'^2}{r} = \frac{(1.2 \text{ kg})(5.4 \text{ m/s})^2}{0.50 \text{ m}} = 70 \text{ N}.$$

Thus,

$$\frac{T'}{T} = \frac{70 \text{ N}}{86 \text{ N}} = 0.81,$$

or a <u>19 percent reduction in the tension</u>.

The same result is obtained using a proportionality method:

$$\frac{T'}{T} = \frac{(mv'^2/r)}{(mv^2/r)} = \left(\frac{v'}{v}\right)^2 = \left(\frac{0.90v}{v}\right)^2 = 0.81.$$

∎

EXAMPLE 5.5 An airplane moves at 64 m/s in a vertical loop of radius 120 m, as shown in Fig. 5.12a. Calculate the force of the plane's seat on a 72-kg pilot while passing through the bottom part of the loop.

SOLUTION Two forces act on the pilot (Fig. 5.12b): his downward weight force **w** and the upward force of the airplane's seat F_{seat}. Because the pilot moves in a circular path, these forces along the radial direction must, according to Newton's second law ($\Sigma \mathbf{F} = m\mathbf{a}$), equal the pilot's mass times his centripetal acceleration, where $a_c = v^2/r$. We find that

$$\Sigma F \text{ (in radial direction)} = F_{seat} - w = m\frac{v^2}{r}.$$

Remember that forces pointing toward the center of the circle ($\mathbf{F}_{seat}$) are positive and those pointing away from the center (**w**) are negative.

Substituting $w = mg$ and rearranging, we find that the force of the airplane seat on the pilot is

$$\begin{aligned}
F_{seat} &= m\left(\frac{v^2}{r} + g\right) \\
&= 72 \text{ kg}\left[\frac{(64 \text{ m/s})^2}{120 \text{ m}} + 9.8 \text{ m/s}^2\right] = 72 \text{ kg}\left(34.1 \text{ m/s}^2 + 9.8 \text{ m/s}^2\right) \\
&= \underline{3200 \text{ N}}.
\end{aligned}$$

∎

The pilot in Example 5.5 feels very heavy. To keep him in the circular path, the seat must push up on the pilot with a force of 3200 N, 4.5 times his normal weight. He is said to experience an acceleration of 4.5 g, that is, 4.5 times the acceleration of gravity.

Blood in the pilot's body may be "left behind" in his legs. The vessels in the lower body can easily absorb a large amount of excess blood. If an upward centripetal acceleration of three or four times that caused by gravity lasts for 20 seconds or more, blood will be unable to reach the pilot's brain. The pilot will lose consciousness, and "blackout" will occur. Special flight suits are made that exert considerable pressure on the legs during such accelerations. This pressure prevents blood from draining into the veins of the legs.

If a pilot initiates a power dive from a horizontal position, the opposite effect can occur. The pilot accelerates downward and blood rushes to his or her head. Vision may be blurred. Blood rushing to the head fills the vessels in the retina of the eye and the pilot literally "sees red." This phenomenon is called a "redout."

EXAMPLE 5.6 A 62-kg woman is a passenger in a "rotor ride" at an amusement park. A drum of radius 5.0 m is spun with an angular velocity of 25 rpm. The woman is pressed against the wall of the rotating drum as shown in Fig. 5.13a. (a) Calculate the normal force of the drum on the woman (the centripetal force that prevents her from leaving her circular path). (b) While the drum

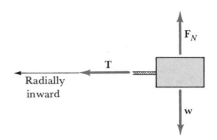

FIG. 5.11. Side view of a block rotating in a circle on a frictionless surface. The string tension force provides the centripetal force to keep the block moving in a circle.

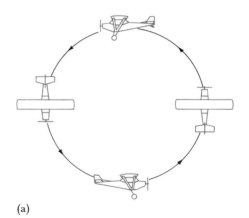

(a)

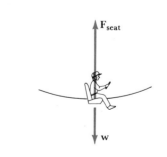

(b)

FIG. 5.12. (a) An airplane doing a loop. (b) A force diagram of the pilot at the bottom of the loop.

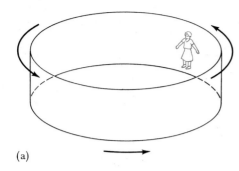

(a)

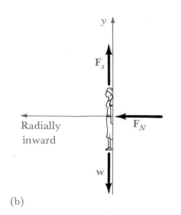

(b)

FIG. 5.13. (a) A passenger on a "rotor ride." (b) A force diagram for the person.

rotates, the floor is lowered. A vertical static friction force supports the woman's weight. What must the coefficient of friction be to support her weight?

SOLUTION (a) A force diagram for the woman is shown in Fig. 5.13b. The drum exerts a normal force directed toward the center of the circle of motion. Using Eq. (5.9), we find that

$$F_N = ma_c = mr\omega^2$$

$$= (62 \text{ kg})(5.0 \text{ m})\left(25 \frac{\text{rev}}{\text{min}} \times \frac{2\pi \text{ rad}}{1 \text{ rev}} \times \frac{1 \text{ min}}{60 \text{ s}}\right)^2$$

$$= \underline{2100 \text{ N}}.$$

(b) Next we consider forces that have vertical components acting along the y axis. Two forces have nonzero vertical components (see Fig. 5.13b): The woman's weight force **w** points down, and the static friction force of the drum wall on her back points up and prevents her from falling. The magnitude of the static friction force is given by Eq. (3.12): $F_s = \mu_s F_N$, where μ_s is the coefficient of static friction between the woman and the wall and F_N is the normal force of the wall on her back, calculated in (a). Since the woman does not move up or down, her vertical acceleration a_y is zero. Thus, according to Newton's second law, the sum of the vertical components of the forces along the y axis must be zero:

$$\Sigma F_y = F_s - w = 0.$$

Substituting $F_s = \mu_s F_N$ and $w = mg$ into this equation, we can determine the minimum coefficient of friction needed for the static friction force to balance the woman's weight:

$$\mu_s F_N - mg = 0,$$

or

$$\mu_s = \frac{mg}{F_N} = \frac{(62 \text{ kg})(9.8 \text{ m/s}^2)}{2100 \text{ N}} = \underline{0.29}. \qquad \blacksquare$$

5.6   Newton's Law of Gravitation

Another important force that causes centripetal acceleration is the force of gravity. The quantitative description of this force was discovered by Isaac Newton. As he put together the ideas for his three laws of motion, he also wondered about the circular motion of the moon about the earth and the circular motion of the planets about the sun.* What prevented these bodies from leaving their paths?

According to one of his biographers, Newton was sitting under an apple tree one day when an apple fell. Newton was struck by the idea that the force causing the apple to fall might also cause the moon to fall toward the earth. However, since the moon moved forward as it "fell," the moon followed a circular path rather than falling into the earth as the apple had done. According to Newton, he "began to think of gravity extending to the orb of the moon. . . ." Using observations about planetary motion made by Kepler, Newton deduced that the force that kept planets in orbit could also account for the earth's gravi-

*The orbits of the moon and planets are elliptical but are close enough to circular that we can apply the techniques of this chapter with little error.

tational force on an apple. Newton had developed a logical and basic understanding of an important aspect of the structure of our universe.

This gravitational force that one mass exerts on another and that is responsible for the circular motion of the moon, earth satellites, and the planets is described as follows:

Newton's universal law of gravitation: The gravitational force exerted by a mass m_1 on another mass m_2 is proportional to the products of their masses and inversely proportional to the square of their separation r:

$$F = \frac{Gm_1 m_2}{r^2}, \qquad (5.10)$$

where G is a constant called the universal gravitational constant and has a value

$$G = 6.67 \times 10^{-11}\,\text{N}\cdot\text{m}^2/\text{kg}^2.$$

Figure 5.14 shows the direction of the forces that the two masses exert on each other. $\mathbf{F}_{1\ \text{on}\ 2}$ is the force that m_1 exerts on m_2. $\mathbf{F}_{2\ \text{on}\ 1}$ is the force that m_2 exerts on m_1. As you can see, each mass exerts an attractive force on the other. The magnitude of each force is the same and can be calculated using Eq. (5.10). The separation of the masses used in these calculations must be from the center of each object (assuming the mass of each object is of uniform density) and not from the edge of the object.

Newton's gravitational force accounts for the weight of an apple or of any object at the earth's surface. The gravitational force caused by the earth on an object of mass m at the earth's surface is

$$F = \frac{Gm_e m}{r_e^2},$$

where r_e is the radius of the earth and m_e is its mass. Substituting for G, r_e, and m_e, we find that

$$F = \left[\frac{(6.67 \times 10^{-11}\,\text{N}\cdot\text{m}^2/\text{kg}^2)(5.98 \times 10^{24}\,\text{kg})}{(6.38 \times 10^6\,\text{m})^2}\right] m$$

$$= (9.8\,\text{m/s}^2)m = mg.$$

Note that $9.8\,\text{m/s}^2$ is the value of g, the earth's gravitational acceleration at its surface, which is given by

$$g = \frac{Gm_e}{r_e^2}. \qquad (5.11)$$

An object's weight is the gravitational force of the earth on the object's mass m and equals

$$w = \frac{Gm_e m}{r_e^2} = \left(\frac{Gm_e}{r_e^2}\right) m = mg.$$

A similar procedure could be used to calculate an object's weight when on the moon or on any of the planets. We simply substitute into Eq. (5.11) the mass and radius of the planet and calculate the gravitational acceleration g on that planet. An object of mass m when on the planet has a weight $w = mg$. Table 5.1 gives the radii and masses of the planets. Other astronomical constants, such as the sun's mass, are on the inside front flyleaf of the book.

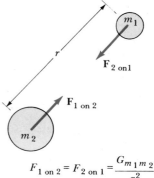

FIG. 5.14. Two masses exert attractive gravitational forces on each other. The forces are equal in magnitude but opposite in direction.

TABLE 5.1 Planetary Constants

Planet	Equatorial Radius (km)	Planet Mass ÷ Earth Mass*	Mean Density (kg/m³)	Mean Distance from Sun (AU)**
Mercury	2,439	0.0553	5440	0.39
Venus	6,052	0.8150	5240	0.72
Earth	6,378.140	1	5497	1.00
Mars	3,397.2	0.1074	3900	1.52
Jupiter	71,398	317.89	1300	5.20
Saturn	60,000	95.17	700	9.54
Uranus	27,900	14.56	1000	19.2
Neptune	24,300	17.24	1700	30.0
Pluto	2,500 (?)	0.0017 (?)	700 (?)	39.4

*The earth's mass is 5.98×10^{24} kg.
**1 AU (astronomical unit) is 1.496×10^{11} m.

EXAMPLE 5.7 (a) Calculate the gravitational acceleration on the surface of Mars (Fig. 5.15) and (b) the weight of a 0.50-kg rock on its surface.

SOLUTION (a) From Table 5.1, we see that the radius of Mars is 3.40×10^6 m (about one-half the earth's radius) and that its mass is $0.1074(5.98 \times 10^{24}$ kg$) = 6.42 \times 10^{23}$ kg. The gravitational acceleration on Mars is calculated using Eq. (5.11):

$$g_M = \frac{Gm_M}{r_M^2} = \frac{(6.67 \times 10^{-11} \, \text{N} \cdot \text{m}^2/\text{kg}^2)(6.42 \times 10^{23} \, \text{kg})}{(3.40 \times 10^6 \, \text{m})^2}$$
$$= 3.70 \, \text{N/kg} = 3.70 \, (\text{kg} \cdot \text{m/s}^2)/\text{kg}$$
$$= \underline{3.70 \, \text{m/s}^2}.$$

(b) The rock's weight on Mars is

$$w_M = mg_M = (0.50 \, \text{kg})(3.70 \, \text{m/s}^2) = \underline{1.9 \, \text{N}}.$$

By comparison, the same rock weighs $w = (0.50 \, \text{kg})(9.8 \, \text{m/s}^2) = 4.9$ N on the earth's surface. ∎

Another important application of Newton's law of gravitation and of our ideas in the last section concerning circular motion involves the motion of satellites. Communication satellites are used routinely to transmit information around the globe. Most communication satellites are synchronous with the earth's rotation about its axis because they remain in orbit above the same point on the earth. This requires that they circle the earth once in 24 hours—that is, their motion is synchronous with the earth's rotation.

FIG. 5.15. A Viking I photograph of the crater Yuty, taken June 22, 1976 from 1877 km above the surface of Mars. Yuty is 18 km in diameter and was probably formed by the collision of a meteorite. The whole area has been worn down by wind and possibly water erosion. Yuty was named for a village in Honduras. (NASA)

EXAMPLE 5.8 (a) At what distance from the earth should we place a satellite so that it remains synchronous with the earth's rotation? (b) What is the speed of the satellite?

SOLUTION (a) The earth's gravitational force keeps the satellite moving in a circular path. Substituting this gravitational force and the expression for centripetal acceleration, given by Eq. (5.8), into Newton's second law for circular motion [Eq. (5.9)], we find that

$$\frac{Gm_em}{r^2} = m(r\omega^2), \tag{5.12}$$

where m is the mass of the satellite, m_e the mass of the earth, r the separation of the satellite from the center of the earth, and ω the satellite's angular velocity.

Since the satellite is to make one revolution about the earth each 24 hours, its angular velocity is

$$\omega = \frac{1 \text{ rev}}{24 \text{ h}} = \left(\frac{1 \text{ rev}}{24 \text{ h}}\right)\left(\frac{2\pi \text{ rad}}{1 \text{ rev}}\right)\left(\frac{1 \text{ h}}{3600 \text{ s}}\right) = 7.27 \times 10^{-5} \text{ rad/s}.$$

Solving Eq. (5.12) for r, we find that

$$r^3 = \frac{Gm_e}{\omega^2} = \frac{(6.67 \times 10^{-11} \text{ N} \cdot \text{m}^2/\text{kg}^2)(5.98 \times 10^{24} \text{ kg})}{(7.27 \times 10^{-5} \text{ rad/s})^2} = 7.55 \times 10^{22} \text{ m}^3, *$$

or

$$r = \sqrt[3]{7.55 \times 10^{22} \text{ m}^3} = \underline{4.23 \times 10^7 \text{ m}}.$$

This is almost seven times the radius of the earth.

It is interesting that our result did not depend on the mass of the satellite. Thus, any satellite placed at this distance from the earth's center will remain synchronous with the earth; it rotates with the earth and always stays above the same part of the earth's surface. In fact, the time required to circle the earth one time depends only on the separation of the satellite from the earth's center and on the earth's mass, but not on the mass of the orbiting object.

(b) The satellite's speed, according to Eq. (5.5), is

$$v = r\omega = (4.23 \times 10^7 \text{ m})(7.27 \times 10^{-5} \text{ rad/s}) = \underline{3070 \text{ m/s}},$$

or about 7000 mi/h. ∎

5.7 Weightlessness

An interesting application of centripetal force and acceleration is the experience called weightlessness. This phenomenon is associated most often with the motion of astronauts floating in spaceships. Let us consider a more down-to-earth situation.

In Fig. 5.16a, a person on the surface of the earth supports a metal sphere on his hand. Two forces act on the sphere, as shown in Fig. 5.16b. The downward weight force **w** is the actual weight of the sphere and is caused by its gravitational attraction to the earth. The second force acting on the sphere, F_{hand}, is the upward force of the person's hand.

In Fig. 5.16a, we see that the sphere moves in a circular path as the earth rotates. For any object to move in a circular path, there must be a net force pulling the object toward the center of the circle. This centripetal force causes the object's path to bend in a circle rather than move straight ahead. For the

*We have used the fact that $1 \text{ N} = 1 \text{ kg} \cdot \text{m/s}^2$ to convert units.

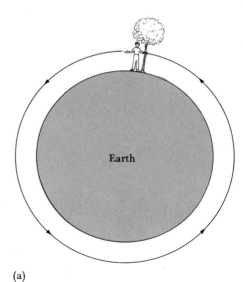

(a)

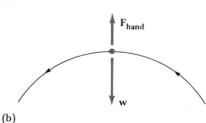

(b)

FIG. 5.16. (a) A person holds a metal sphere; since the earth is rotating, the sphere moves in a circle. (b) The forces acting on the sphere.

sphere, the net centripetal force is merely the difference of **w** and $\mathbf{F}_{\text{hand}}$. Newton's second law becomes

$$w - F_{\text{hand}} = ma_c$$
$$= m(r\omega^2).$$

(5.13)

We see that F_{hand} must be less than w if the mass experiences a centripetal acceleration.

The earth's angular velocity is so small that, for practical purposes, $r\omega^2$ can be ignored for objects on the earth's surface. If $r\omega^2$ is very small, then according to Eq. (5.13) the object's weight w is equal in magnitude to the force F_{hand} needed to support that weight. However, if the earth's angular velocity is greater, or if the object circles the earth in an orbit of greater radius, then its centripetal acceleration is not negligible. All of its weight force may be needed to keep it in orbit about the earth, and no force F_{hand} is needed to support its weight. The object seems to have no weight.

Suppose that the earth's angular velocity could be increased. Imagine it rotating so fast that a mass would fly out of your hand, much like a drop of water being spun out of wet clothes in the spin cycle of a washing machine. At this point, a mass would appear to be weightless. It would require no force to support its weight.

EXAMPLE 5.9 At what angular velocity must the earth and a metal ball rotate so that the ball seems weightless?

SOLUTION The condition of weightlessness occurs when the force F_{hand} needed to support the sphere's weight is zero. Substituting $F_{\text{hand}} = 0$ into Eq. (5.13), we find that

$$w - 0 = mr\omega^2,$$

or

$$\omega = \sqrt{\frac{w}{mr}} = \sqrt{\frac{mg}{mr}} = \sqrt{\frac{g}{r}}$$
$$= \sqrt{\frac{9.8 \text{ m/s}^2}{6.38 \times 10^6 \text{ m}}} = \underline{1.2 \times 10^{-3} \text{ rad/s}}.$$ ∎

We see from Example 5.9 that if the earth rotated with an angular velocity of 1.2×10^{-3} rad/s, or about 17 times the earth's actual angular velocity, we would need no force to support the ball's weight. If we stood on a bathroom scale, the scale would read zero. If the earth's angular velocity were even greater, we would have to grab hold of a tree or something else attached to the earth just to stay on its surface. The metal ball we were holding would rise up out of our hand and we would rise up with it if we did not anchor ourselves.

An astronaut in orbit about the earth is in a similar situation. The whole gravitational weight force caused by the attraction between the earth's mass and that of the astronaut is needed to keep the astronaut in the circular orbit. If the astronaut stood on a bathroom scale, it would exert no outward normal force on the astronaut. The astronaut would appear to have no weight.

We see that the expression *weightlessness* is inappropriate: A weight force does act on the astronaut. It would be better to say that the astronaut has no apparent weight because no outward force is needed to support the astronaut's actual weight when moving in the circular orbit about the earth.

Summary and Additional Readings

1. **Angular position:** The angular position of an object moving in a circular path is specified by an **angular-position coordinate, θ.** θ is the angle from the positive x axis to a radial line drawn from the center of the circle to the object. The units of θ can be degrees, revolutions, or radians. These units are related by the equation

$$360° = 1 \text{ rev} = 2\pi \text{ rad}. \tag{5.2}$$

The angle θ in units of **radians** is defined as the arc length s along the circumference of the circle from the x axis to the object divided by the radius r of the circle:

$$\theta \text{ (in radians)} = \frac{s}{r}. \tag{5.1}$$

2. **Angular velocity and tangential speed:** The **angular velocity** ω of an object moving in a circular path is the ratio of the change in the object's angular position $\Delta\theta$ and the time Δt required for this change:

$$\text{Angular velocity} = \omega = \frac{\Delta\theta}{\Delta t}. \tag{5.4}$$

The object's **tangential speed** is the ratio of its change in position Δs along the arc of the circle and the time Δt required for this change:

$$\text{Tangential speed} = v = \frac{\Delta s}{\Delta t}. \tag{5.3}$$

The two quantities angular velocity and tangential speed are related by the equation

$$v = r\omega, \tag{5.5}$$

where r is the radius of the circle.

3. **Centripetal acceleration:** When an object moves in a circular path at constant speed, the direction of its velocity changes continually even though the magnitude of its velocity is constant. The object experiences a **centripetal acceleration** $\mathbf{a}_c$ because of this changing direction of motion. The magnitude of the acceleration is given by

$$a_c = v^2/r \tag{5.7}$$
$$= r\omega^2. \tag{5.8}$$

The direction of $\mathbf{a}_c$ is toward the center of the circle.

Forces cause acceleration. When an object moves in a circular path, a **centripetal force** is needed to pull in on the object and keep it from leaving the circle. Any force responsible for keeping the object in its circular path is said to be a centripetal force. Newton's second law relates these forces that act along the radial direction to the centripetal acceleration they cause:

$$\Sigma F \text{ (in radial direction)} = ma_c. \tag{5.9}$$

A force directed radially inward is given a positive sign and one directed radially outward a negative sign.

4. **Gravitational force:** The gravitational force of attraction exerted by mass m_1 on mass m_2 is proportional to the product of their masses and inversely proportional to the square of their separation r:

$$F = \frac{Gm_1m_2}{r^2}, \tag{5.10}$$

where $G = 6.67 \times 10^{-11} \text{ N·m}^2/\text{kg}^2$ is the universal gravitational constant.

Van E. Neie, "Analysis of Running on Banked and Unbanked Curves," *The Physics Teacher* **19**, 321 (1981).

John L. Roeder, "Physics and the Amusement Park," *The Physics Teacher* **13**, 327 (1975).

Curt Covey, "The Earth's Orbit and the Ice Ages," *Scientific American*, p. 58, February (1984).

T. C. Van Flandern, "Is Gravity Getting Weaker?," *Scientific American* **234**, p. 44 (1976).

Questions

1. In the following examples, identify the centripetal force that causes the object to move in a circular path: (a) a ball rotating in a horizontal circle at the end of a string; (b) a ball rotating in a vertical circle at the end of a string when at the top of the circle; (c) a car rounding a curve on a level highway; (d) the moon circling the earth.

2. A pilot sits upright in an airplane at the top and bottom of a vertical loop traversed at constant speed. Why is it more likely that the pilot will black out at the bottom of the loop than at the top?

3. *Estimate* the force needed to prevent a penny from sliding off the edge of a long-playing record while rotating on the turntable of a stereo system. What is the cause of this force?

4. *Estimate* the centripetal force needed to hold your foot on your leg when swinging the leg to kick a football. Justify any numbers used in your estimate.

5. A rope is tied to the handle of a pail of water. Why does the water stay in the pail when the pail is swung in a vertical circle?

6. Explain how the spin-dry cycle on a washing machine causes water to be removed from clothes. Be specific.

7. Why do we feel we are being thrown upward, out of our seats, when going over a hump on a roller coaster?

8. Where on the earth's surface would you expect to experience the greatest centripetal acceleration as a result of the earth's rotation? Explain.

9. How would you estimate the earth's mass using a knowledge of the orbital motion of the space shuttle? Indicate all information you would need for your estimate.

Problems

5.1 and 5.2 Angular Position Coordinate and Units

1. Perform the following conversions: 50° to rad, 160° to rad, 1.5π rad to degrees, 0.4 rad to degrees, 2.1 rev to rad, 3 rev to degrees.

2. Make the following conversions: 500 rpm to rad/s, 1 rev/27 days to rad/s, 37° to rad.

3. The Space Needle in Seattle is 185 m tall. From where you stand, it subtends an angle of 6.62°. Calculate the distance from you to the Space Needle.

4. The beam of a laser used in student laboratories diverges, or spreads, at an angle of 1.0×10^{-3} rad. The laser light illuminates the wall of a distant building, producing a bright spot of diameter 0.25 m. How far is the wall from the laser?

■ 5. The headlights on a car have an angular separation of 0.40° when viewed by a distant observer. *Estimate* the distance of the car from the observer. Justify your estimation technique.

■ 6. An insect on the edge of a 12-in record completes one rotation in 0.56 s. (a) Calculate the change in angular position of the insect in 1.0 s. (b) Calculate the distance traveled by the insect in 1.0 s.

■ 7. A bicycle with a 0.68-m-diameter tire travels 2000 km before the tire becomes worn and must be replaced. Calculate the number of revolutions made by the tire in traveling that distance.

■ 8. The moon is 3.8×10^8 m from the earth. Outline and carry out a procedure for estimating the moon's diameter. [*Hint:* You might want to invent a method using a ruler or meter stick to estimate the angle subtended by the moon.]

5.3 Tangential Speed and Angular Velocity

9. The earth circles the sun in a nearly circular orbit of radius 1.5×10^{11} m. (a) Calculate the angular velocity of the earth about the sun. (b) Calculate the earth's speed while moving in this orbit.

10. Calculate the angular velocity of a passenger car traveling at 50 mi/h on the circular turn of a highway. The radius of the turn is 400 m.

11. A bicycle tire of radius 0.68 m rotates 1000 times in 600 s. Calculate (a) the distance traveled by the valve stem, (b) its average tangential speed, and (c) its average angular velocity.

12. The moon is an average distance of 3.8×10^8 m from the earth. It circles the earth once each 27.3 days. (a) Calculate its angular velocity in rad/s. (b) Calculate the distance it travels in its orbit in one day.

13. The electron in a hydrogen atom rotates about its proton nucleus with an angular velocity of 4.15×10^{16} rad/s. (a) Calculate the number of revolutions the electron makes per second. (b) Calculate the electron's speed. The average distance of the electron from the nucleus is 0.53×10^{-10} m.

5.4 Centripetal Acceleration

14. Calculate the centripetal acceleration of the electron in the hydrogen atom in Problem 13.

15. An airplane makes a loop of radius 500 m. The centripetal acceleration of the airplane is 40 m/s². Calculate the angular velocity and tangential speed of the airplane.

16. Jupiter rotates once about its axis each 9 h, 56 min. Its radius is 7.15×10^4 km. Calculate the centripetal acceleration in m/s² of a hypothetical person standing at the equator of Jupiter.

17. Extremely dense neutron stars rotate with an angular velocity of about 1 rev/s. (a) If the radius of a star is 2.5×10^4 m, what is the speed of a mass on the star's equator? (b) Calculate the centripetal acceleration in m/s² and in g's of a point on the equator.

■ 18. Compute the centripetal acceleration due to the earth's rotation about its axis for a person standing (a) at the equator, (b) at the North Pole, and (c) halfway between the equator and the North Pole. Justify your answers to (b) and (c) carefully.

■ 19. *Estimate* the centripetal acceleration of a person riding on a Ferris wheel at an amusement park. Justify your choice of numbers used in the estimate.

5.5 Centripetal Force

20. Trout weighing less than 4.5 N (16 oz) must be returned to the stream. A disappointed angler, after catching a 3.4-N trout, decides to weigh it by swinging it at the end of a string attached to a scale in a circle of 1.5-m radius. With what angular velocity must the fish be swung to get the scale to read 4.5 N? Assume that it swings in a horizontal circle.

21. Calculate the centripetal force acting on an 1100-kg car that travels around a highway curve of radius 150 m at 27 m/s.

22. A 2.0-kg block slides on a horizontal, frictionless surface in a circular path at the end of a string. Calculate the maximum speed of the block if the string's radius is 0.70 m and the string breaks when its tension exceeds 40 N.

23. A car speeds around the 80-m-radius curved exit ramp of a freeway. A 70-kg passenger holds the armrest of the car door with a 220-N force to prevent sliding across the front seat of the car (ignore friction). Calculate the car's speed in km/hr.

■ 24. A 2100-kg truck travels around a highway curve whose radius is 250 m. The truck travels at 30 m/s. (a) Calculate the centripetal force acting on the truck. (b) This centripetal force is provided by static friction, which prevents the truck from sliding off the highway. What is the minimum coefficient of static friction needed to keep the truck from sliding? The road is flat.

■ 25. A 28-kg child sits on the edge of a merry-go-round 3.0 m from its center. If the coefficient of static friction between the child and merry-go-round is 0.70, what is the maximum angular velocity at which the merry-go-round can rotate before the child slides off?

26. A 40-kg person sitting on the seat of a swing whose ropes are 5.0 m long passes the vertical position under the supports at a speed of 7.0 m/s. (a) Construct a force diagram for the person and identify all forces acting on the person. (b) Calculate the force of the swing's seat on the person.

27. The fire department attaches a 10-m-long rope to a tree and the other end to a dormitory window to serve as a fire escape. A 68-kg student while swinging in a circular arc from the window can hold the rope with a force no greater than 1200 N. What is the maximum speed of the swinger at the bottom of the swing? Do not forget to include the swinger's weight in your force diagram and equations.

28. A 70-kg person sits on the seat of a Ferris wheel having a radius of 24 m. The wheel rotates with an angular velocity of 0.50 rad/s. (a) Construct a force diagram showing and identifying all the forces acting on the person when at the top of the Ferris wheel. (b) Calculate the force (magnitude and direction) of the seat on the person when at that position.

29. A 74-kg man rides in a carnival car that rotates at the end of a beam in a vertical circle of 20-m radius with an angular velocity of 0.20 rev/s. (a) Draw a force diagram for the man at the top of the circle (he is inverted). (b) Use Newton's second law to determine the magnitude of the force of the car seat on the man at the top of the circle.

30. A roller coaster travels around a vertical loop of radius 9.0 m in which the cars are inverted at the top of the loop. Calculate the minimum speed at the top if passengers are not to fall away from their seats.

31. A 64-kg person stands 5.0 m from the center of a merry-go-round that rotates with an angular velocity of 0.50 rad/s. (a) Construct a force diagram for the person. (b) Calculate the static friction force $\mathbf{F}_s$ needed to prevent the person from sliding off the merry-go-round. (c) Calculate the magnitude of the vertical normal force $\mathbf{F}_N$ that balances the person's weight (remember that the normal force is perpendicular to the surface of contact). (d) The net force of the merry-go-round on the person is the vector sum $\mathbf{F}_s + \mathbf{F}_N$ and points in the same direction that the person leans toward the center of the merry-go-round. Calculate this leaning angle θ relative to the horizontal.

32. A person revolves at the end of a swing, as in Fig. 5.17. (a) Draw a force diagram of the swing and person, whose combined mass is 80 kg. (b) Use the vertical-component equation of Newton's second law to find the tension in the cable attached to the swing (note that $a_y = 0$). (c) The horizontal component of this tension force provides the centripetal force. Use this fact to calculate the angular velocity of the swing.

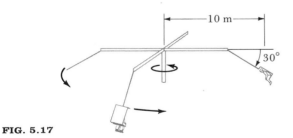

FIG. 5.17

33. A highway curve of 240-m radius is banked at 16°. At what speed must a car travel around the curve if the banking alone, and no friction force, keeps the car on the road?

34. A highway engineer, to help decide the speed limits for unbanked curves of highways, desires a general formula for expressing the maximum speed v that a car can travel around a curve without sliding off the road. Derive this equation for the engineer in terms of the following quantities: the estimated coefficient of static friction μ_s between the car tires and road, the radius r of the curved road, and the acceleration of gravity g. (Note that the formula does not depend on the car's mass and thus applies to all cars with the same μ_s.)

35. A battery-operated toy car attached to the end of a string rotates in a circle on a smooth surface. The radius of the string is then changed, and the car's speed is also changed so that the string tension remains constant. By what percent must the speed change if the radius of the string increases by 50 percent? Show your solution so that a fellow student could understand it.

5.6 Newton's Law of Gravitation

36. Calculate the gravitational force (a) of the sun on the moon and (b) of the earth on the moon.

37. The acceleration of gravity on the surface of Jupiter, the most massive planet, is 6.47 m/s^2, and Jupiter's radius is 7.1×10^4 km. Calculate Jupiter's mass.

38. Calculate the weight and mass of a 50-kg person standing on the north pole of Mercury.

39. A person weighing 760 N on the earth moves to Saturn. (a) What is the person's mass on Saturn? (b) Calculate the person's weight on Saturn.

40. The average radius of the earth's orbit around the sun is 1.5×10^8 km. The mass of the earth is 5.98×10^{24} kg, and it makes one revolution in 365 days. (a) Calculate the speed of the earth. (b) Estimate the sun's mass, using Newton's law of gravitation and his second law of motion.

41. Calculate the radius of the orbit of an earth satellite that makes two revolutions per day.

42. The moon travels in an orbit about the earth that is nearly circular. Calculate the time needed for one moon orbit about the earth using Newton's law of gravitation and his second law of motion.

43. Calculate the time needed for one orbit of an earth satellite that is just above the earth's surface. Use a force diagram and Newton's second law as the starting point for your calculations.

44. Calculate the ratio of the earth's gravitational force on an 80-kg person when at the earth's surface and when 1000 km above the earth's surface. The radius of the earth is 6370 km.

45. A black hole exerts a 50-N gravitational force on a spaceship. The black hole is 10^{14} m from the ship. Calculate the force on the ship when it moves to one-half that distance from the black hole. (*Note:* One-half of 10^{14} m is not 10^7 m.)

46. At what distance from the earth's surface, in terms of earth radii, should a satellite's orbit be for the gravitational force of the earth on the satellite to be half what it is at the earth's surface.

47. (a) If the mass of both the earth and the moon were doubled, by what factor would the radius of the moon's orbit about the earth change if it had the same velocity? (b) By what factor would its velocity have to change if it had the same radius? Explain.

48. A moon makes one complete orbit of radius r around a planet of mass M in time T. Use Newton's law of gravitation and

his second law of motion to show that $M = (4\pi^2 r^3)/(GT^2)$, where G is the universal gravitational constant.

5.7 Weightlessness

■ 49. A 50-kg person stands on a bathroom scale on the earth's surface at the equator. (a) Draw a force diagram for the person. (b) With the assistance of this diagram write an expression using Newton's second law for the forces and acceleration along the radial direction. (Remember that the person rotates daily at a constant speed in a circular path, so that $a_c = r\omega^2$.) (c) If the earth's rotational speed about its axis increased by a factor of 10, what would the bathroom scale read?

■ 50. With what angular velocity should Mars rotate about its axis so that a person standing on the equator of Mars would feel weightless?

■ 51. People living on a circular space platform in outer space would normally feel weightless because there are no large masses nearby exerting gravitational forces on them. To simulate the effect of gravitation, people could live on the edge of the platform with their heads toward the center and their feet on an outside rim (see Fig. 5.18). If the platform rotates, the outside rim exerts an inward centripetal normal force on the space people. At what angular velocity must a platform of radius 100 m rotate to provide for the space people an apparent weight of mg? Explain.

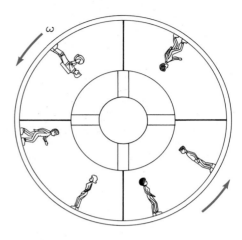

FIG. 5.18. A space platform.

CHAPTER 6

Statics

In Chapter 5 we considered objects moving at constant speed in a circular path. In Chapter 7 we will study the rotational motion of extended bodies—bodies whose mass is distributed in space. In preparation for that analysis, this chapter considers the conditions that are necessary for an extended body to remain stationary, both in terms of its position in space and its rotational stability.

The subject of the chapter, **statics,** involves an analysis of the forces acting on objects that are either at rest or moving at constant speed in a straight path. Both groups of objects are said to be in equilibrium. Objects at rest are in static equilibrium, and those moving in a straight path at constant speed are in dynamic equilibrium. In this chapter we consider only objects at rest.

For a rigid object to remain in equilibrium, it must satisfy two conditions: (1) The vector sum of all forces acting on the object must be zero. Then, according to Newton's second law of motion, the object's acceleration is zero: If at rest, the object remains at rest; if moving, its motion continues unchanged. (2) For the object to be rotationally stable, the forces must act on it in a way that prevents it from rotating, tipping, or falling over. This condition leads to the introduction of a new quantity called *torque*—a measure of the ability of a force to cause an object to rotate. For an object to remain rotationally stable, the sum of the torques caused by all forces acting on the object must be zero.

Each of these conditions can be written in the form of a mathematical equation, and the equations can be used to analyze a variety of situations. For example, an engineer can estimate the safe load that the girders of a building can carry. Or we can estimate the tension force in back muscles when a person bends over to lift a weight.

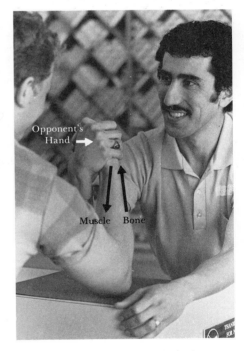

6.1 First Condition of Equilibrium

The first condition of equilibrium is a special case of Newton's second law ($\Sigma \mathbf{F} = m\mathbf{a}$) for situations in which objects are not accelerating, for example objects sitting at rest. If the acceleration is zero, then the forces acting on the object must add to zero.

First condition of equilibrium: If an object is in equilibrium, then the vector sum of all forces acting on the object must be zero:

$$\Sigma \mathbf{F} = \mathbf{F}_1 + \mathbf{F}_2 + \mathbf{F}_3 + \cdots = 0. \qquad (6.1)$$

To solve dynamics problems using Newton's second law, it is convenient to

Three forces act on the hand of the man with his back toward the camera. If the hand is stationary, the forces add to zero. This fact allows us to estimate the tension in the muscle of the man's forearm. (Photo by David Conklin)

117

(a)

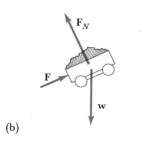

(b)

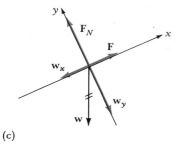

(c)

FIG. 6.1. (a) A man pushing a wagon to hold it at rest. (b) The forces acting on the wagon; (c) the forces resolved into components.

use the law in component form. The same is true in statics; the first condition of equilibrium is usually written in component form.

Alternative form of the first condition of equilibrium: For an object in equilibrium, the sum of the x components (F_{1x}, F_{2x}, F_{3x}, . . .) of all forces acting on the object must be zero, and the sum of the y components (F_{1y}, F_{2y}, F_{3y}, . . .) must also be zero:

$$\Sigma \, F_x = F_{1x} + F_{2x} + F_{3x} + \cdots = 0. \tag{6.2a}$$

$$\Sigma \, F_y = F_{1y} + F_{2y} + F_{3y} + \cdots = 0. \tag{6.2b}$$

The component form of the first condition of equilibrium is illustrated in Fig. 6.1 where we use it to determine the force **F** with which a person must push a wagon to hold it stationary on the side of a hill. Two other forces in addition to **F** act on the wagon, as shown in Fig. 6.1b. First, the normal force $\mathbf{F}_N$ of the ground pushes up on the tires. The normal force is directed perpendicular to the surface of contact between the tires and the ground. The third force acting on the wagon is the downward force of gravity **w**, equal in magnitude to the wagon's weight.

In Fig. 6.1c all of the force vectors have been moved so that their tails are at the origin of a coordinate system. Also, the weight force **w** has been resolved into its components, $\mathbf{w}_x$ and $\mathbf{w}_y$. If $\mathbf{w}_x$ is greater in magnitude than **F**, the wagon will start to roll down the hill in the negative x direction. For the wagon to remain stationary, **F** must just balance $\mathbf{w}_x$. Also, $\mathbf{F}_N$ and $\mathbf{w}_y$ must balance each other. If they do not, the wagon "sinks" into the hill ($\mathbf{w}_y > \mathbf{F}_N$) or rises into the air ($\mathbf{F}_N > \mathbf{w}_y$). We know from experience that this does not happen, which implies that $\mathbf{F}_N$ and $\mathbf{w}_y$ balance each other. As we see, for the wagon to remain stationary, the sum of the x components and the sum of the y components of the forces acting on the wagon must independently add to zero.

6.2 Problem-Solving Technique and Examples

Often we use statics to calculate one or two unknown forces acting on an object if all other forces acting on it are known. Suppose that you bend over to lift a box of books. We can identify many of the forces acting on your body: the force of gravity, the books pulling down on your arms, and the floor pushing up on your feet. Statics enables us to estimate with reasonable accuracy two difficult-to-measure forces acting inside your body: the tension in your back muscles and the compression force on discs that separate vertebrae in your backbone. If we can calculate these forces for different lifting techniques, perhaps we can learn what is the best lifting technique to avoid back injury.

How do we use statics and the equations of equilibrium to calculate these unknown forces? The procedure is similar to that used to solve problems in dynamics. We take one or two small parts of a complex situation and apply the equations of equilibrium to these parts. If the following technique is mastered, complex statics problems can be solved routinely.

Solving Problems in Statics

Each step of the problem-solving procedure is outlined in general and illustrated for the following problem: Determine the tension in each cord that helps support the 100-N hanging block shown in the margin.

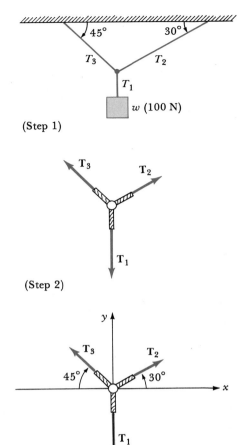

(Step 1)

(Step 2)

(Step 3)

1. Make a drawing of the whole structure being considered. Include all of the known information in the drawing and identify in symbols unknown quantities you wish to determine.

   The diagram is shown in the margin.

2. Draw a separate force diagram (also called a free-body diagram) for one part or object of the structure. The part of the structure chosen is usually one on which many of the known forces and at least one of the unknown forces act. The conditions of equilibrium must be satisfied by this object because it is at rest. After choosing an object on which to focus attention, draw arrows to represent all forces acting on that object. These forces should be identified symbolically (for example, as $\mathbf{P}$, $\mathbf{w}$, $\mathbf{T}$, $\mathbf{F}_N$).

   For the problem illustrated here, we choose the knot at the junction of the three cords as the object for a force diagram. The knot is the only part of the whole structure on which the three tension forces ($\mathbf{T}_1$, $\mathbf{T}_2$, and $\mathbf{T}_3$) act. Note that the object for the force diagram does not have to be separate from other objects but can be attached to adjacent objects, such as the three cords.

3. Superimpose an x axis and a y axis on the force diagram. These axes may be oriented in any direction. Once you decide on the placement of the axes and the origin of the coordinates, all future calculations must be consistent with that choice.

   The diagram in the margin shows the coordinates for our problem.

4. Write the two equations for the first condition of equilibrium [Eqs. (6.2a) and (6.2b)]. To do this, you must first resolve the known forces into their components as described in Section 1.6. You will also have to include in your equations symbols for the unknown components. When finished, you will have two equations that you can use to solve for unknowns. (Later in the chapter, the second condition of equilibrium provides a third equation to use in solving for unknowns.)

   The x and y equilibrium equations for the force diagram shown in Step 3 are as shown here. Try to reproduce these equations on your own.

   $\Sigma F_x = 0$, or

   $T_{1x} + T_{2x} + T_{3x} = +T_1 \cos 90° + T_2 \cos 30° - T_3 \cos 45° = 0$

   $\Sigma F_y = 0$, or

   $T_{1y} + T_{2y} + T_{3y} = -T_1 \sin 90° + T_2 \sin 30° + T_3 \sin 45° = 0$

5. Substitute known information into the equations from Step 4 and perform the necessary algebraic manipulations to solve for the unknowns.

   For the problem being illustrated, we substitute $T_1 = 100$ N and the values of the sines and cosines into the equations from Step 4 to get the first two equations here. We then solve these two equations for the unknown tensions T_2 and T_3.

$$0.87T_2 - 0.71T_3 = 0,$$
$$-100 + 0.50T_2 + 0.71T_3 = 0$$

From the first equation, we find that $T_2 = (0.71/0.87)T_3 = 0.82T_3$. Substituting for T_2 in the second equation, we find that $-100 + 0.50(0.82T_3) + 0.71T_3 = 0$. Rearranging to solve for T_3, we have

$$T_3 = \frac{100}{0.5(0.82) + 0.71} = \underline{89\text{ N}}.$$

Since $T_2 = 0.82T_3$, we find that $T_2 = 0.82 \times 89 = \underline{73\text{ N}}.$

As you can see, the object considered in statics problems (such as the knot) is often some small part of the whole structure shown in the original picture. Note carefully the object chosen in Example 6.1. It is the one part of the whole structure on which the known and unknown forces act.

EXAMPLE 6.1   Calculate the magnitude of the tension force **T** pulling on the foot in the traction device shown in Fig. 6.2a. The weight w hanging from the cable attached to the wall is 40 N, or about 9 lb.

SOLUTION   When several ropes or cables are joined at a junction or a knot, the tension in each cable is usually different. However, the tension along a single cable is the same at every point. This is true even if the cable changes directions as it passes around a pulley. Thus, the tension at all points along the hanging cable shown in Fig. 6.2a is the same and equals the 40-N weight hanging at its end.*

We choose the pulley attached to the foot as the object for a force diagram (Fig. 6.2b). To visualize more easily the forces acting on the pulley, we include in the force diagram the small sections of rope attached to and passing around the pulley. In this way it becomes apparent that the hanging cable exerts two forces on the pulley, each of magnitude w and oriented as shown in Fig. 6.2b. The cable attached to the foot exerts an unknown tension force **T** on the pulley. The equations of equilibrium for the pulley are as follows:

$$\Sigma F_x = 0, \quad \text{or} \quad -T + w\cos 30° + w\cos 30° = 0,$$
$$\Sigma F_y = 0, \quad \text{or} \quad 0 + w\sin 30° - w\sin 30° = 0.$$

Since $w = 40$ N, the x equation allows us to solve for T:

$$T = 2w\cos 30° = 2(40\text{ N})0.87 = \underline{69\text{ N}}. \qquad \blacksquare$$

EXAMPLE 6.2   A couple parked in a car on a moonlit evening find that their car is stuck in the sand. The woman, having studied physics, ties a rope to the car's front bumper and stretches it around a tree in front of the car as shown in Fig. 6.3a. She then pushes with a 290-N (65-lb) force against the center of the

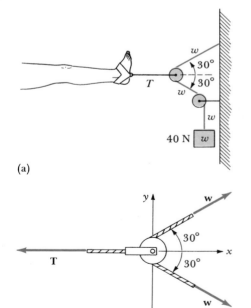

(a)

(b)

FIG. 6.2.   (a) A traction device for the leg. (b) A force diagram of the pulley of the traction device.

*This is true only if the cable has negligible weight compared to the weight hanging at its end and if the pulleys around which it passes are frictionless. In all our problems we will assume that these two conditions hold.

rope, causing it to deflect as shown. The car starts to move. Calculate the force of the rope on the car.

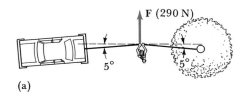

(a)

SOLUTION We choose as the object for a force diagram the short section of rope being pushed by the woman (Fig. 6.3b). This is the one part of the whole situation shown in Fig. 6.3a on which the known force acts. Three forces act on this center section of rope: The woman pushes it with a 290-N force, and the rope is pulled from each side with an unknown tension force caused by the adjacent pieces of rope.

Using the coordinate axes shown in Fig. 6.3b, we find that the equations of equilibrium for the center section of rope are:

$$\Sigma F_x = T_{1x} + F_x + T_{2x} = T_1 \cos 5° + 0 - T_2 \cos 5° = 0,$$
$$\Sigma F_y = T_{1y} + F_y + T_{2y} = -T_1 \sin 5° + F - T_2 \sin 5° = 0.$$

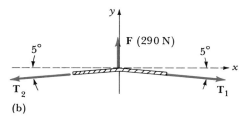

(b)

Note that since **F** is perpendicular to the x axis, its projection on that axis—its x component—is zero. Its y component is the magnitude of **F** since **F** points along the y axis.

The first equation above implies that $T_1 = T_2$; that is, the tension is the same on each side of the rope. Substituting $T_1 = T_2 = T$ into the second equation and solving for T produces the equation

$$T = \frac{F}{2 \sin 5°} = \frac{(290 \text{ N})}{2(0.087)} = \underline{1700 \text{ N}}.$$

FIG. 6.3. (a) A technique for pulling a car out of the sand. (b) A force diagram for the section of rope being pushed by the woman.

The woman, using ingenuity, has succeeded in applying a force of almost 400 lb to the car. ∎

6.3 Torques

When the sum of all forces acting on an object is zero, the object does not accelerate. The first condition of equilibrium guarantees that the object, if at rest, will not start moving to a new location. However, this condition does not guarantee that the object will not tip over or rotate. Another condition is needed to ensure rotational equilibrium.

Consider a seesaw in equilibrium (Fig. 6.4a). Since it is in equilibrium, the sum of the forces acting on it must be zero. We can disturb this equilibrium by applying two forces of equal magnitude, one up, the other down, at opposite ends of the seesaw (Fig. 6.4b). The seesaw begins to rotate and is no longer in rotational equilibrium. However, the sum of all forces acting on the seesaw is still zero, since we have added a net force of zero [**F** + (−**F**) = 0]. Because the forces are applied at opposite ends of the seesaw, they cause the seesaw to rotate about the fulcrum. The fact that the forces add to zero does not guarantee that the seesaw is rotationally stable. Evidently, another condition besides balanced forces is needed to maintain rotational equilibrium. This condition concerns the torques caused by the forces, which depends on where the forces are applied to the object.

Torque is a measure of the tendency of a force to cause an object, such as the wrench shown in Fig. 6.5, to turn or rotate. When the hand applies a force near the head of the wrench, as in Fig. 6.5a, the wrench does not turn. The torque is small even though the force may be large. If the hand applies the same

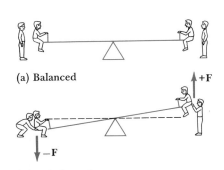

(b) Unbalanced

FIG. 6.4. A seesaw rotationally unbalanced by two forces that add to zero.

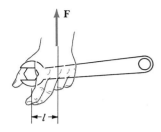

(a)

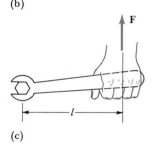

(b)

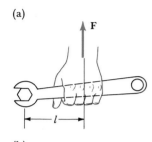

(c)

FIG. 6.5. The ability of a force to turn a wrench depends on where the force is applied: the farther from the head, the better the turning ability, or torque.

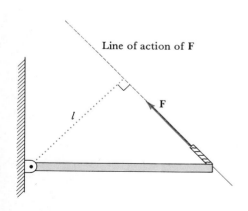

FIG. 6.6. The line of action of a force pulling on a beam, shown together with the moment arm *l* of the force.

force halfway along the handle, as in Fig. 6.5b, the torque is greater and the wrench may turn the bolt, although with difficulty. If the hand applies the same force at the end of the handle, as in Fig. 6.5c, the torque is much greater and the wrench and bolt will turn more easily. We see that the place where a force is applied to an object is important in determining the magnitude of the torque that causes the object to rotate.

Torque, then, is a measure of the ability of a force to cause rotation. Torque is given the symbol τ (the Greek letter tau). In the simplest cases, when the force is directed perpendicular to an object such as a wrench, torque depends on the magnitude of the force and on the distance between the point of application of the force and the point about which the object rotates, called the **fulcrum**. This distance is called the **moment arm** or lever arm of the force and is given the symbol l (see Fig. 6.5). The magnitude of the torque τ of a force is the product of the magnitude of the force **F** and the moment arm l; that is, $\tau = (F)(l)$.

Often the forces that produce rotation are not directed perpendicular to the object. Consider the force **F** of the rope pulling on the beam in Fig. 6.6. The beam can rotate about a bearing placed at its left end. We will choose this point of rotation as the origin of coordinates. To calculate the torque caused by the force **F** about this origin, we first need to find the moment arm.

To determine the **moment arm** of a force, first draw an extended, dashed line passing through the point where the force acts on the object and pointing in the direction of **F** (Fig. 6.6). This line is called the **line of action** of **F**. Next draw a dotted line from the origin of coordinates to the line of action. Make the second line perpendicular to the line of action, forming a right angle at their intersection. The distance from the origin to the intersection of the two lines is the moment arm l of the force. Thus, a general definition of **moment arm** is the perpendicular distance from the origin of the coordinates to the line of action of the force.

Having found the moment arm, we can then define the torque:

The **torque** τ caused by a force **F** is the product of the magnitude of the force and its moment arm l:

$$\tau = \pm(F)(l). \tag{6.3}$$

The torque is positive if the force tends to rotate the object counterclockwise about the origin and negative if it tends to rotate the object clockwise.

EXAMPLE 6.3 Calculate the torque caused by the 12-kg mass hanging from the beam shown in Fig. 6.7a. For this problem place the origin of coordinates at the pin on the left side of the beam.

SOLUTION The line of action and the moment arm of this force are shown in Fig. 6.7b. We see that $l = 1.5$ m, $w = mg = (12 \text{ kg})(9.8 \text{ m/s}^2) = 118$ N, and the torque is negative, since it tends to rotate the beam clockwise about the origin. Thus, the torque caused by the weight is

$$\tau = -(118 \text{ N})(1.5 \text{ m}) = \underline{-180 \text{ N} \cdot \text{m}}. \quad \blacksquare$$

EXAMPLE 6.4 A rope pulls on the end of a beam as pictured in Fig. 6.7a. The tension in the rope is 800 N. Calculate the torque about the pin at the end of the beam due to this tension force.

SOLUTION The line of action of **T** passes directly through the pin where the origin of coordinates is located (Fig. 6.7c). Consequently, the perpendicular distance (the shortest distance) from the origin to the line of action is zero, as is the moment arm. Thus, the torque is zero:

$$\tau = (l)(T) = (0)(800 \text{ N}) = \underline{0}.$$ ∎

EXAMPLE 6.5 Calculate the torque on the beam shown in Fig. 6.8a due to the 540-N rope tension force **F**. The length of the beam L is 3.0 m.

SOLUTION We are given that $F = 540$ N and $L = 3.0$ m. The moment arm of **F** is shown in Fig. 6.8b and can be calculated by taking the sine of the 30° angle:

$$\sin 30° = \frac{\text{Opposite side}}{\text{Hypotenuse}} = \frac{l}{L},$$

or $l = L \sin 30°$. (There is no magic equation for calculating moment arms. You will often have to make a drawing such as Fig. 6.8b and use a trigonometric function to relate l to another known length and angle.) The torque due to **F** is positive, since it tends to rotate the beam counterclockwise, and is calculated as follows:

$$\tau = +(F)(l) = (F)(L \sin 30°) = (540 \text{ N})(3.0 \text{ m} \times 0.50) = \underline{810 \text{ N·m}}.$$

Another way to calculate the torque caused by **F** is shown in Fig. 6.8c. The force is first resolved into its x and y components, $\mathbf{F}_x$ and $\mathbf{F}_y$. We then calculate separately the torque caused by each component. As you can see in Fig. 6.8c, the moment arm of $\mathbf{F}_y$ is L and that of $\mathbf{F}_x$ is zero because its line of action passes through the origin. The total torque caused by **F** is the sum of the torques of its components:

$$\tau = (F_y)(l_y) + (F_x)(l_x) = (F \sin 30°)(L) + (F \cos 30°)(0)$$
$$= (540 \text{ N} \times 0.50)(3.0 \text{ m}) + 0 = \underline{810 \text{ N·m}}.$$

This is the same answer as calculated using the technique shown in Fig. 6.8b. Often it is easier to calculate a torque by replacing a force with its components and then adding the torques caused by each component. ∎

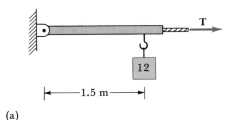
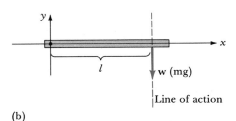
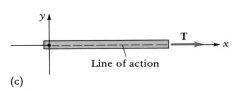

(a)

(b)

(c)

FIG. 6.7. (a) A beam being pulled from its end by a rope tension force and supporting a 12-kg mass. (b) The line of action and moment arm of the weight force. (c) The line of action of the tension force passes through the origin, so its torque is zero. Another force (not shown) is needed to keep the beam in equilibrium.

FIG. 6.8. (a) A force acting on a beam; (b) the moment arm l of the force; (c) resolution of the force into components to calculate the torque by the component method.

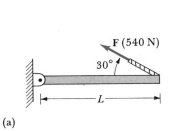

(a)

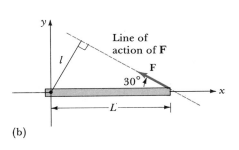

(b)

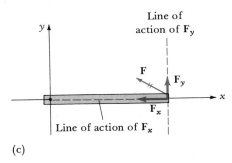

(c)

6.4 Second Condition of Equilibrium

We learned in Section 6.3 that the torque exerted by a force is a measure of its ability to cause an object to rotate or turn. For an object to remain in rotational equilibrium, the positive torques that tend to rotate it counterclockwise must be balanced by equal-magnitude negative torques that tend to rotate it clockwise. This is the second condition of equilibrium.

Second condition of equilibrium: For an object to remain in rotational equilibrium, the sum of all torques acting on the object must be zero:

$$\Sigma \tau = \tau_1 + \tau_2 + \tau_3 + \cdots = 0, \tag{6.4}$$

where $\tau_1, \tau_2, \tau_3, \ldots$ represent the torques caused by all forces $\mathbf{F}_1, \mathbf{F}_2, \mathbf{F}_3, \ldots$ acting on the object.

The two conditions that must be satisfied for an object to remain in equilibrium are summarized as follows:

$$\text{First condition:} \quad \Sigma F_x = 0 \tag{6.2a}$$
$$\Sigma F_y = 0 \tag{6.2b}$$
$$\text{Second condition:} \quad \Sigma \tau = 0 \tag{6.4}$$

In the following examples these conditions are used to calculate unknown forces acting on objects in equilibrium.

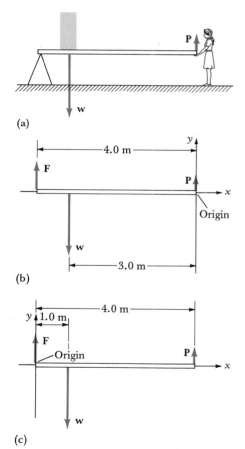

(a)

(b)

(c)

FIG. 6.9. (a) A woman applying a small force to the end of a beam to support a large weight near the beam's fulcrum. (b) and (c) Force diagrams of the beam for two different coordinate systems.

EXAMPLE 6.6 The woman shown in Fig. 6.9a supports the weight of an 840-N crate resting on a lever. Calculate the upward force of her hand on the lever and the force of the fulcrum on the lever. Her hand is 4.0 m from the fulcrum, and the crate is 1.0 m from the fulcrum.

SOLUTION The lever is a logical object for a force diagram, since both unknown forces act on it, as does the known weight force of the crate pushing down on it (see Fig. 6.9b).

The origin of the x and y coordinates may be placed anywhere. However, once the origin is chosen, the torques must then be calculated relative to that origin even though it may not be the real fulcrum of the lever. If the object is in equilibrium, the torques must add to zero no matter what origin is used. For example, in Fig. 6.9b, the origin is located at the point where the woman's hand pushes on the beam. We calculate the tendency of forces to cause rotation about that point. The weight force **w** tends to rotate the beam counterclockwise about this point and causes a torque equal to $\tau_w = +(w)(l) = +(840 \text{ N})(3.0 \text{ m}) = 2520 \text{ N} \cdot \text{m}$. The actual fulcrum tends to rotate the beam clockwise about the origin shown in Fig. 6.9b and causes a torque given by $\tau_F = -(F)(l_F) = -(F)(4.0 \text{ m})$. Since **P** has a zero moment arm about the origin shown in Fig. 6.9b, its torque is zero. The sum of these three torques must be zero if the beam is in rotational equilibrium:

$$\Sigma \tau = \tau_F + \tau_w + \tau_P = -(4.0 \text{ m})F + 2520 \text{ N} \cdot \text{m} + 0 = 0.$$

Rearranging this equation, we find that

$$F = \frac{2520 \text{ N} \cdot \text{m}}{4.0 \text{ m}} = \underline{630 \text{ N}}.$$

Let us rework this problem, this time with the origin of coordinates located at the real fulcrum, as shown in Fig. 6.9c. The moment arms of the forces are different about that origin, as are the torques. But since the beam is in equilibrium, the torques must still add to zero:

$$\Sigma \tau = \tau_F + \tau_w + \tau_P = +(F)(0\text{ m}) - (840\text{ N})(1.0\text{ m}) + (P)(4.0\text{ m}) = 0.$$

Rearranging this equation, we find that

$$P = \frac{(840\text{ N})(1.0\text{ m})}{(4.0\text{ m})} = \underline{210\text{ N}}.$$

We have been able to solve for both unknown forces by simply applying the torque equation for two different coordinate systems. ∎

When using the torque equation to solve for unknown forces, it is often useful to place the origin of coordinates at the point where one of the unknown forces acts on the object, as we did in Figs. 6.9b and 6.9c. At that point the torque caused by the unknown force is zero because it has a zero moment arm. Consequently, that particular unknown force does not appear in the torque equation, and the equation can be used to solve for the other unknown force.

Finally, we should note that the results of these calculations using the second condition of equilibrium are consistent with those using the first condition of equilibrium. Referring to Fig. 6.9b, we see that Eq. (6.2b) should be written

$$\Sigma F_y = F_y + w_y + P_y = F - w + P = 0.$$

Substituting $w = 840$ N as well as the results of our previous calculations using the torque equations, $F = 630$ N and $P = 210$ N, in the above equation produces a net force of zero on the left side of the equation. Our torque calculations are consistent with the first condition of equilibrium.

EXAMPLE 6.7 The foot and the Achilles tendon muscle provide a nice example of a body lever (Fig. 6.10a). The ball of the foot touching the ground is the fulcrum, the bones in the foot are the arm of the lever, and the Achilles tendon and its associated muscles provide the tension force needed to lift the load (the calf bone in this case). The ground pushes up on the ball of the foot with a force of half the person's weight. The heel of the foot is just off the ground. Find the tension **T** in the Achilles tendon and the compression force **C** at the joint between the leg and foot for a person weighing 800 N (180 lb).

SOLUTION The object of interest is the lever consisting of the foot bones. The coordinates used are shown in Fig. 6.10b. The conditions of equilibrium for this lever are as follows:

$$\Sigma F_x = 0, \quad \text{or} \quad w_x + C_x + T_x = 0 + 0 + 0 = 0,$$
$$\Sigma F_y = 0, \quad \text{or} \quad w_y + C_y + T_y = (800\text{ N})/2 - C + T = 0,$$
$$\Sigma \tau = 0, \quad \text{or} \quad -\frac{w}{2}(l_w) + C(l_c) + T(l_T) = 0.$$

Substituting the known information in the last equation, we have

$$-[(800\text{ N})/2](0.14\text{ m}) + C(0) + T(0.04\text{ m}) = 0.$$

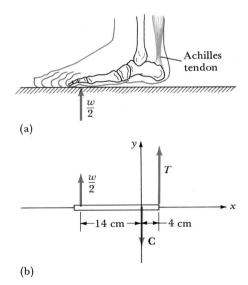

(a)

(b)

FIG. 6.10. (a) The foot as a lever. (b) A force diagram of the foot as a lever while supporting the leg by tension in the Achilles tendon.

T can be calculated using the preceding equation:

$$T = \frac{[(800 \text{ N})/2](0.14 \text{ m})}{0.04 \text{ m}} = \underline{1400 \text{ N}} \ (310 \text{ lb}).$$

Using the y-force equation, we find that

$$C = \frac{800 \text{ N}}{2} + T = 400 \text{ N} + 1400 \text{ N} = \underline{1800 \text{ N}} \ (400 \text{ lb}).$$

A person's Achilles tendon becomes ripped or torn if the tension exceeds about 7000 N (1500 lb). This tension is exceeded occasionally when athletes with powerful muscles in their lower legs run or jump.

EXAMPLE 6.8 Calculate the tension in the cable supporting the beam shown in Fig. 6.11a if a plant weighing 100 N hangs from the beam as shown. Also calculate the force of the wall support on the beam. The beam's weight can be ignored.

SOLUTION A force diagram for the beam is shown in Fig. 6.11b. The origin of the coordinates is placed at the location of the unknown force **F** of the wall support on the beam. The horizontal component of **F** balances the horizontal component of **T**. The vertical component of **F** is needed to prevent the left edge of the beam from falling. To calculate **F**, it is easiest to consider its components F_x and F_y as unknowns (see Fig. 6.11c). These components plus the unknown cable tension T can be found using the three equations of equilibrium:

$$\Sigma F_x = F_x + 0 - T\cos 37° = 0,$$
$$\Sigma F_y = F_y - w + T\sin 37° = 0,$$
$$\Sigma \tau = (F)(0) - (w)(0.60 \text{ m}) + \underbrace{(T\sin 37°)(0.90 \text{ m})}_{T_y} = 0.$$

Note that the torques due to $\mathbf{F}_x$, $\mathbf{F}_y$, and $\mathbf{T}_x$ (shown in Fig. 6.11c) are zero because these component forces have zero moment arms.

The last equation can be solved for T after substituting $w = 100$ N:

$$T = \frac{w(0.60 \text{ m})}{\sin 37° \ (0.90 \text{ m})} = \frac{(100 \text{ N})(0.60 \text{ m})}{0.60(0.90 \text{ m})} = \underline{110 \text{ N}}.$$

Using the x-force equation, we find that

$$F_x = T\cos 37° = (110 \text{ N})(0.80) = 88 \text{ N}.$$

Using the y-force equation, we find that

$$F_y = w - T\sin 37° = 100 \text{ N} - (110 \text{ N})(0.60) = 34 \text{ N}.$$

The magnitude of the force of the wall support is determined using Eq. (1.7):

$$F = \sqrt{F_x^2 + F_y^2} = \sqrt{88^2 + 34^2} = 94 \text{ N}.$$

F is drawn from its component values in Fig. 6.11d. It points into the first quadrant. The tangent of the angle that it makes with the x axis is given by Eq. (1.8):

$$\tan \theta = \left| \frac{F_y}{F_x} \right| = \frac{34 \text{ N}}{88 \text{ N}} = 0.39.$$

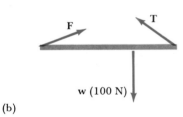

(a)

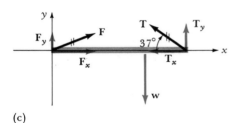

(b)

(c)

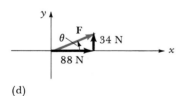

(d)

FIG. 6.11. (a) A beam supported by a cable; (b) a force diagram of the beam; (c) a force diagram of the beam with the force of the wall support and the cable tension force resolved into components; (d) the wall support force **F** constructed from its components.

A 21° angle has a tangent of 0.39. Thus, **F** has a magnitude of 94 N and points 21° above the positive x axis. ∎

6.5 Gravitational Torques

The force of gravity often produces a torque on an object. For example, when a person bends over to pick up something from the ground, the weight of the upper body creates a torque that tends to rotate the upper body about the hips, as shown in Fig. 6.12. To prevent this, the back muscles must exert an opposing torque. The line of action of this muscle force is along the muscle itself. However, it is not obvious what the line of action of the weight force is, since it is exerted approximately uniformly over the whole upper body. Gravity acts on the head, shoulders, arms, chest, and so forth. Each weight force on each body part has a different moment arm. This is true for any **extended body**—that is, any body whose mass is distributed in space.

It would be cumbersome to calculate separately the torques caused by the weights of each small part of an extended body. Fortunately, we do not need to do this. Every extended body has what is called a **center of gravity**—a point at or near the center of the body at which we consider all of the weight force as acting when calculating torques. A variety of experimental and theoretical techniques can be used to measure or calculate the center of gravity of different objects, but we will not consider these techniques here. In most of the problems in this book the position of the center of gravity of an object is given. To calculate the gravitational torque acting on the object, we assume that all of its weight force acts at the center of gravity, and we use the usual rules for calculating torques.

EXAMPLE 6.9 A 68-kg rock climber hangs out from a ledge, as shown in Fig. 6.13a. Calculate the force **F** of the ledge on the climber's feet and the horizontal tension force **T** on the climber's arms. The climber's center of gravity is 0.50 m out from the side of the cliff.

SOLUTION A force diagram for the climber is shown in Fig. 6.13b. We place the origin of coordinates at the climber's feet so that the unknown force **F** does not enter into the torque equation. The equations of equilibrium are as follows:

$$\Sigma F_x = 0, \quad \text{or} \quad F_x - T + 0 = 0,$$
$$\Sigma F_y = 0, \quad \text{or} \quad F_y + 0 - w = 0,$$
$$\Sigma \tau = 0, \quad \text{or} \quad \tau_F + \tau_T + \tau_w = (F)(0) + (T)(1.50 \text{ m}) - (w)(0.50 \text{ m}) = 0.$$

Notice that when we calculate the weight torque, the whole weight force **w** is assumed to act at the center of gravity. Substituting $w = mg = (68 \text{ kg})(9.8 \text{ m/s}^2) = 666 \text{ N}$ into the last equation, we find that

$$T = \frac{(666 \text{ N})(0.50 \text{ m})}{1.50 \text{ m}} = \underline{222 \text{ N}}.$$

We find from the x-component equation that $F_x = T = 222$ N and from the y-component equation that $F_y = w = 666$ N. The magnitude of **F** is determined using Eq. (1.7):

$$F = \sqrt{F_x^2 + F_y^2} = \sqrt{222^2 + 666^2} = 700 \text{ N}.$$

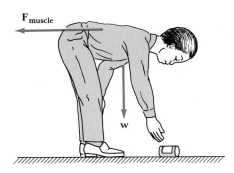

FIG. 6.12. A man picking up an empty can. The weight of his upper body creates a torque about the hips that must be balanced by the torque produced by his back muscles.

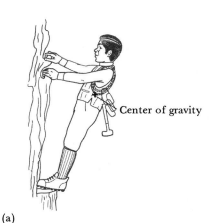

(a)

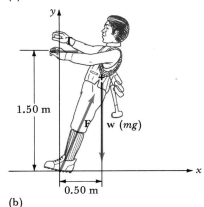

(b)

FIG. 6.13. (a) A rock climber hanging on a ledge. (b) A force diagram for the climber.

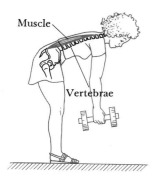

FIG. 6.14. Lifting a weight incorrectly by bending the back rather than the legs.

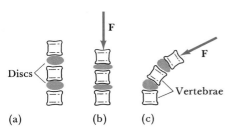

FIG. 6.15. (a) Vertebrae separated by discs under normal load; (b) larger compression forces causing the discs to flatten; (c) change of shape of the disc to accommodate changes in direction of the spine.

The components of **F** are both positive. Thus, **F** must point into the first quadrant. The tangent of the angle it makes with the x axis is

$$\tan \theta = \frac{666}{222} = 3.0, \quad \text{or} \quad \theta = 72°.$$

Thus, **F** has a magnitude of 700 N and points 72° above the positive x axis. ∎

The final example in this chapter illustrates an important problem related to muscle tensions and to the compression forces on body joints. The actual calculations are a little difficult, but the background gained from other examples in this chapter should allow us to understand the principles.*

EXAMPLE 6.10 Lower-back problems have been with humans since our prehistoric ancestors assumed an upright position. In this example we calculate the large compression force produced on the backbone when improper lifting techniques are used and the large tension forces in back muscles needed to do this lifting.

Figure 6.14 shows a person lifting a 50-lb weight from a leaning position. One group of back muscles is the primary "cable" holding the person in this position. The backbone serves as a "beam" to support the lifting activity. The backbone is made of many small bones called vertebrae, which are separated by discs. A disc is similar to a flexible, fluid-filled bag. As the back flexes, the discs change shape (Fig. 6.15).

Figures 6.16a and 6.16b show a mechanical model of the back of a 180-lb person lifting a 50-lb weight. We want to solve for the tension T in the cable (the back muscle) and the compression force **C** on the rotary joint at the bottom of the beam (this compression is present in much of the spine and can cause serious disc problems, particularly in the lower spine). The cable supporting the beam attaches to the beam (spine) one-third of the way from its end. The weight of the trunk w_1 at its center of gravity is 72 lb; the combined weight w_2 of the head, arms, and 50-lb object being lifted is 86 lb and acts at the center of gravity of these three objects.

FIG. 6.16. (a) A force diagram of the back of a person lifting a weight. The beam is the spine, the cable is the spinal extensor muscles, w_1 is the weight of the trunk, and w_2 is the total weight of the head, arms, and the 50-lb weight being lifted. (b) The forces on the beam.

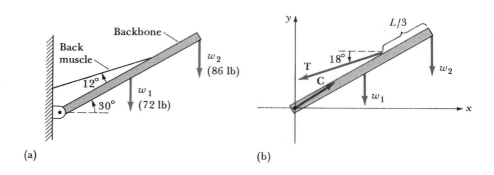

(a)   (b)

*This example is adapted from L. A. Strait, V. T. Inman, and H. J. Ralston, *American Journal of Physics* **15**, 375–382 (1947).

SOLUTION The equations of equilibrium for this system are as follows:

$$\Sigma F_x = -T\cos 18° + C_x = 0,$$
$$\Sigma F_y = -T\sin 18° - 72 - 86 + C_y = 0,$$
$$\Sigma \tau = -72\left(\frac{L}{2}\cos 30°\right) - 86(L\cos 30°) + T\left(\frac{2L}{3}\sin 12°\right) = 0.$$

C_x and C_y are the unknown components of the compression force **C**. Solving the last equation, we find that $T = \underline{760\ \text{lb}}$! We then find from the x-force equation that $C_x = 730$ lb and from the y-force equation that $C_y = 390$ lb. Thus, the magnitude of C is

$$C = \sqrt{730^2 + 390^2} = \underline{830\ \text{lb}}.$$

The direction of **C** is determined from Eq. (1.8):

$$\tan\theta = \frac{C_y}{C_x} = \frac{390}{730} = 0.53, \quad \text{or} \quad \theta = \underline{28°}.$$

We see that the back muscles and backbone experience forces greater than one-third ton when lifting from a bent position (Fig. 6.14). The person's 1-in-diameter fluid-filled discs in the lower back are compressed by an 830-lb force—like resting a grand piano on the disc. The forces are reduced to less than half these values if the person lifts the 50-lb weight by bending his or her knees while keeping the back vertical. ∎

Summary and Additional Readings

1. **Conditions for equilibrium:** Objects in equilibrium are either at rest or moving with constant speed in a straight line. We have considered only objects at rest. For an object to be in equilibrium, all the forces and torques acting on the object must add to zero.

First condition: $\Sigma \mathbf{F} = 0$ (6.1), or $\begin{aligned}&\Sigma F_x = 0. \quad (6.2\text{a})\\ &\Sigma F_y = 0. \quad (6.2\text{b})\end{aligned}$

Second condition: $\Sigma \tau = 0.$ (6.4)

If the first condition of equilibrium is satisfied, an object at rest remains at rest; if the object is moving, its state of motion is unchanged. If the second condition is satisfied, an object is rotationally stable.

2. **Torques:** Torque indicates the ability of a force to cause a rotation. The torque τ of a force **F** acting on an object is

$$\tau = \pm(F)(l), \qquad (6.3)$$

where l is the moment arm of the force. The moment arm is the perpendicular distance from the line of action of the force to the origin of coordinates. The torque is positive if the force tends to rotate the object counterclockwise about the origin and is negative if it tends to rotate the object clockwise.

3. **Gravitational torques:** The gravitational force is exerted on every part of an extended body. To calculate the gravitational torque, however, we assume that all the weight force acts at a point near the center of the body, called the center of gravity.

M. Williams and H. R. Lissner, *Biomechanics of Humans,* Saunders, Philadelphia (1962). Contains many examples of static body levers.

R. M. Alexander, *Animal Mechanics,* University of Washington Press, Seattle (1968).

Jim Minstrell, "Explaining the 'At Rest' Condition of an Object," *The Physics Teacher* **20**, 10 (1982).

G. Stroink, "Center of Gravity of a Student," *The Physics Teacher* **17**, 254 (1979).

Questions

1. A hammock is tied with ropes between two trees. Are its ropes more or less likely to break if stretched tightly or loosely between the trees? Explain.

2. Something is wrong with the orientation of the ropes shown in

Fig. 6.17. Using the first condition of equilibrium for the hanging pulley, explain this error and redraw the system as you would expect to see it.

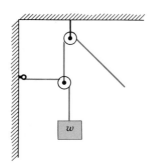

FIG. 6.17

3. Is it possible for an object to be in equilibrium when only one force acts on it? Explain.

4. Two books lie on top of each other on a table. Construct a force diagram showing the forces acting on the bottom book. Identify each force.

5. Explain the meaning of torque so that a friend not taking physics can understand.

6. Can two forces of equal magnitude and direction exert different torques on an object? Explain and illustrate with an example.

7. Can two forces of equal magnitude applied to the same point on an object exert different torques? Explain and illustrate with an example.

8. Give three examples of an object that is not in rotational equilibrium even though the sum of the forces acting on the object is zero.

9. The tension in body muscles needed to lift various objects is usually five to ten times greater than the weight being lifted. Explain. A simple example would be helpful.

10. A ladder leans against a wall. Construct a force diagram showing the direction of all forces acting on the ladder. Be sure to identify the cause of each force.

11. Using a crowbar, a person can remove a nail with little force, whereas pulling directly on the nail requires a large force to remove it. Explain and illustrate with a figure so that a friend can understand.

12. Is it more difficult to do a sit-up with your hands stretched in front of you or with them behind your head? Explain.

Problems

6.1 and 6.2 First Condition of Equilibrium and Problem-Solving Technique

■ 1. Three ropes pull on a knot that holds them together. Rope 1 exerts a 20-N force in the positive x direction and rope 2 pulls with a 40-N force at an angle 53° above the negative x axis. Rope 3 pulls with a force that balances the first two so that the net force on the knot is zero. (a) Construct a force diagram for the knot. (b) Use Eqs. (6.2a) and (6.2b) to write equations that can be used to solve for F_{3x} and F_{3y}. (c) Solve for F_{3x} and F_{3y} and for the magnitude and direction of $\mathbf{F}_3$.

■ 2. Repeat Problem 1 for the case where $\mathbf{F}_1$ has magnitude 100 N and points 30° below the positive x axis and $\mathbf{F}_2$ has magnitude 150 N and points in the negative y direction.

■ 3. Three ropes tied to a ring pull on it with the following forces: $\mathbf{T}_1$ (50 N in the positive y direction), $\mathbf{T}_2$ (20 N, 25° above the negative x axis), and $\mathbf{T}_3$ (70 N, 70° below the negative x axis). A fourth rope exerts a force $\mathbf{T}_4$ that allows the ring to remain at equilibrium. (a) Construct a force diagram for the ring. (b) Use Eqs. (6.2a) and (6.2b) to write two equations that can be used to solve for T_{4x} and T_{4y}. (c) Solve these equations and determine the magnitude and direction of $\mathbf{T}_4$.

■ 4. Calculate the tensions in the three ropes shown in Fig. 6.18 for the case where $\theta_2 = 37°$, $\theta_3 = 0°$, and $w = 12$ N.

■ 5. Calculate the tensions T_1, T_2, and T_3 in the ropes shown in Fig. 6.18 for the case where $\theta_2 = 63°$, $\theta_3 = 45°$, and $w = mg = 120$ N.

■ 6. Calculate the tensions in the three ropes shown in Fig. 6.18 for the case where $\theta_2 = 30°$, $\theta_3 = 50°$, and $w = mg = 240$ N.

■ 7. Redraw Fig. 6.18 with $\theta_2 = 50°$ and $\theta_3 = 0$. Rope 2 is found to have a tension T_2 equal to 100 N. Calculate the values of T_1, T_3, and m.

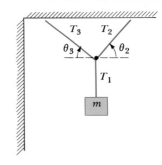

FIG. 6.18

■ 8. Rope 3 shown in Fig. 6.18 has a tension of 120 N and $\theta_3 = 37°$. Determine T_2 and θ_2 if $w = mg = 180$ N.

■ 9. Find the tension T and weight w_1 needed to keep the 120-N weight w_2 shown in Fig. 6.19 in equilibrium.

10. Determine weights w_1 and w_2, shown in Fig. 6.19, that cause the tension T in the horizontal cable to be 64 N.

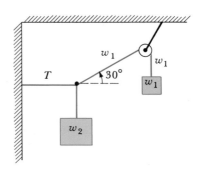

FIG. 6.19

■ 11. A 640-kg crate hangs by a single rope from the ceiling. A second rope attached to a pulley pulls the hanging rope to the side as shown in Fig. 6.20. If $\theta_1 = 60°$, what is the magnitude and direction of the tension force **T**?

■■ 12. The 640-kg hanging crate shown in Fig. 6.20 is pulled to the side by a 630-N tension force **T**. Write the two equations for the first condition of equilibrium using the pulley as the object for a force diagram. Calculate θ_1 and θ_2. You may need to use the identity $\sin^2\theta + \cos^2\theta = 1$.

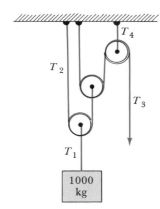

FIG. 6.20

■■ 13. (a) Construct a force diagram for each weightless pulley shown in Fig. 6.21. (b) Use the equations of equilibrium and the force diagrams to determine the tension in each rope.

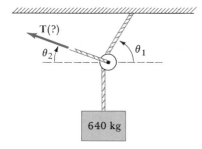

FIG. 6.21

■ 14. A 150-lb tightrope walker stands at the center of a tightrope that is 20 ft long. The tightrope will break if its tension exceeds 1500 lb. Calculate the smallest angle at which the rope can bend up from the horizontal on either side of the walker to avoid breaking.

■ 15. Calculate the tension in the cable that supports the 3000-kg gondola as it moves across the canyon shown in Fig. 6.22.

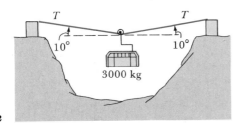

FIG. 6.22

■ 16. At what angle above the horizontal does the rope shown in Fig. 6.23 bend while supporting the pulley from which the 78-kg man hangs?

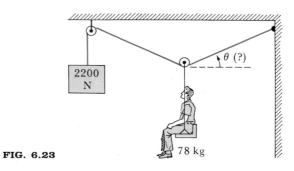

FIG. 6.23

6.3 Torques

17. A 260-lb mutineer on Captain Bligh's ship is made to "walk the plank." The plank, which extends 9 ft beyond its support, will break if subjected to a torque greater than 2200 lb·ft. Will the sailor break the plank before stepping off its end? Explain.

18. A 68-kg high diver stands at the end of a diving board 3.5 m from its support. Calculate the magnitude of the torque caused by the diver on the board about the point of support.

■ 19. Calculate the torques about point O caused by forces $\mathbf{F}_1$, $\mathbf{F}_2$, $\mathbf{F}_3$, and $\mathbf{F}_4$ in Fig. 6.24. All forces have magnitudes of 120 N and are applied a distance of 2.0 m from O.

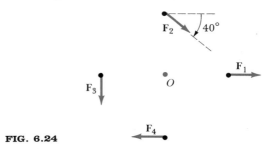

FIG. 6.24

■ 20. Your hand holds a quart of milk while your arm is bent at the elbow in a 90° angle. *Estimate* the torque caused by the milk on your arm about the elbow joint. Indicate all numbers used in your calculations. This is an estimate, and your answer may differ by 10 to 50 percent from the answers of others.

■ 21. A person's bent arm is shown in Fig. 6.25. The person holds a ball whose weight is 40 N. Calculate the torques on the forearm about the elbow joint caused by the downward force of the ball and the upward forces of the biceps and triceps muscles. $F_{triceps} = 100$ N and $F_{biceps} = 340$ N.

■ 22. A 24-kg girl sits 1.2 m from the fulcrum of a seesaw. (a) Calculate the magnitude of the torque exerted by the girl on the seesaw. (b) At what distance from the fulcrum on the other side should the girl's 34-kg friend sit so that his torque is equal in magnitude to her torque?

■ 23. Three forces are applied to the beam shown in Fig. 6.26. (a) Calculate the torques about the pivot point on the left caused by forces $\mathbf{F}_1$ and $\mathbf{F}_2$. (b) At what distance must $\mathbf{F}_3$ be applied to cause a torque that balances those produced by $\mathbf{F}_1$ and $\mathbf{F}_2$? All forces have magnitudes of 200 N.

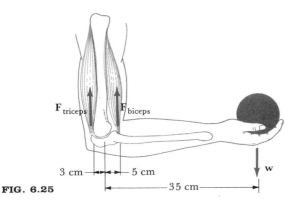

FIG. 6.25

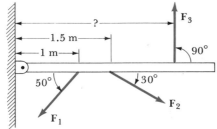

FIG. 6.26

■ 24. The torque caused by the tension force **T** shown in Fig. 6.16 is $T[(2L \sin 12°)/3]$. Justify this statement.

■ 25. A 2.4-m-long weightless beam, such as shown in Fig. 6.8, is supported on the right by a cable that makes an angle of 53° with the beam. A 32-kg mass hangs down from the beam 1.5 m from its support on the left. (a) Calculate the torque caused by the hanging mass on the beam. (b) Determine the cable tension needed to produce a torque that will balance the torque caused by the mass.

6.4 Second Condition of Equilibrium

26. The fulcrum of a seesaw 4.0 m long is located 2.5 m from one end. A person weighing 300 N sits on the long end. What must the weight of a person sitting at the short end be to balance the seesaw? Ignore the weight of the seesaw.

■ 27. A 1.0-m-long horizontal, weightless rod hangs from the ceiling by a string tied to its middle. A 20-kg mass hangs from one end of the rod and a 10-kg mass hangs from the other end. Where should a third 15-kg mass hang relative to the center string to balance the rod?

■ 28. A 3.0-m-long weightless beam is supported at each end by cables. A painter weighing 900 N stands 1.0 m from the left cable. Calculate the tension in each cable.

■ 29. A 3.0-m-long weightless beam has an 82-kg mass resting on one end and a 64-kg mass on the other end. How far from the 64-kg mass should a fulcrum be located to balance the beam?

■ 30. The object marked A in the mobile in Fig. 6.27 weighs 10 N. Calculate the weight of object B. The numbers in Fig. 6.27 indicate the relative lengths of the rods on each side of their supporting cord. Ignore the weight of the rods.

■ 31. A plank 10 m long is supported by cables from each end. The left cable has a tension of 300 N and the right cable, 400 N. Calculate the weight and location of a person sitting on the plank. Ignore the weight of the plank.

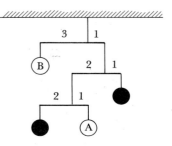

FIG. 6.27

■ 32. The biceps muscle of the person's arm shown in Fig. 6.25 provides the positive torque needed to support the 5.2-kg ball held in the person's hand. Calculate the force of the biceps muscle on the arm. Ignore the triceps muscle and the arm's weight.

■ 33. The person's arm shown in Fig. 6.25, instead of holding a ball, pushes down on a table with a force of 20 N at the fingers. Calculate the force of the triceps muscle needed to balance the torque caused by the upward force of the table on the fingers. Ignore the biceps muscle and the arm's weight.

■ 34. Find the force of the biceps muscle on the arm in Fig. 6.25 when a person lifts a 15-kg mass with the hand. Also calculate the force of the bone in the upper arm on the bone in the forearm at the elbow joint. Ignore the triceps muscle and the weight of the arm.

■ 35. Calculate the magnitude of the tension force **T** exerted by the hamstring muscles (in the back of the thigh) and the compression force **C** at the joint between the thigh bone and calf bone when a 20-lb force pulls on the foot as shown in Fig. 6.28.

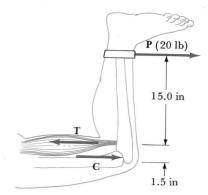

FIG. 6.28

■ 36. A 62-kg person stands on a diving board (Fig. 6.29). Calculate the forces of the two supports on the board. Ignore the weight of the board.

■ 37. The weightless beam shown in Fig. 6.30a is analogous to a human body when lifting a weight in a bent position (Fig. 6.30b). Calculate the tension in the cable and the force of the wall support on the beam when the hanging weight is 420 N (94 lb).

■ 38. Calculate the tension in a system such as that shown in Fig. 6.30a, but with a 620-N hanging weight and a beam tilted at 30° rather than 15°. The cable remains horizontal.

■ 39. Find the tension T in the cable and the force **F** (magnitude and direction) of the wall support on the weightless beam shown in Fig. 6.31.

■ 40. Determine the tension in the cable shown in Fig. 6.31 if

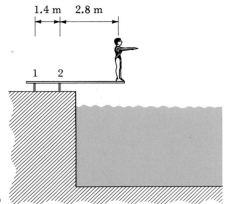

FIG. 6.29

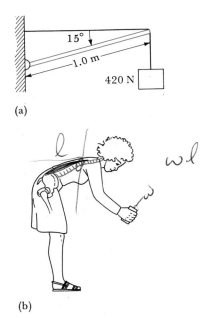

(a)

(b)

FIG. 6.30. (a) A beam with hanging weight; (b) its biological analogue.

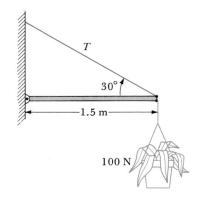

FIG. 6.31

it is attached to the center of the weightless beam and makes a 53° angle with the beam. A plant weighing 100 N still hangs at the end.

■ 41. Determine the tension in the cable shown in Fig. 6.30a if the cable is moved down so that it attaches to the beam 0.60 m from its bottom left end. The cable remains horizontal.

6.5 Gravitational Torques

■ 42. The fulcrum of a uniform seesaw that is 4.0 m long and weighs 200 N is located 2.5 m from one end. A person weighing 300 N sits on the long end. Determine the weight of a person at the other end if the seesaw is balanced.

■ 43. A 4.0-m-long uniform board of mass 21 kg is supported at each end by cables. A 62-kg painter stands 1.0 m from the left cable. Calculate the tension in each cable.

■ 44. The diving board shown in Fig. 6.29 has mass of 28 kg, and its center of gravity is at the board's geometrical center. Calculate the forces of the two supports on the board when a 56-kg person stands on the end.

■ 45. The person's hand shown in Fig. 6.25 supports a 100-N weight. The forearm weighs 10 N and its center of gravity is 18 cm from the elbow joint. Calculate the force of the biceps muscle on the person's arm. Ignore the tension in the triceps muscle.

■ 46. The person's arm shown in Fig. 6.25, instead of holding a ball, pushes down on a table with a 30-N force at the fingers. The forearm weighs 12 N and its center of gravity is 18 cm from the elbow joint. Calculate the tension in the triceps muscle. Ignore the tension in the biceps muscle.

■ 47. A 2.0-m-long uniform beam of mass 8.0 kg supports a 12.0-kg bag of vegetables at one end and a 6.0-kg bag of fruit at the other end. At what distance from the vegetables should the beam rest on your shoulder to balance?

■ 48. Determine the tension in the cable in Fig. 6.31 if the beam weighs 150 N and is uniform.

■ 49. Determine the tension T in the cable in Fig. 6.31 and the force **F** of the wall support on the beam, which weighs 150 N and is uniform.

■ 50. The beam shown in Fig. 6.30 is uniform and weighs 210 N. Determine the tension in the cable.

■ 51. The beam shown in Fig. 6.30 is uniform and weighs 210 N. Determine the tension in the cable and the force of the wall support on the beam.

■■ 52. A uniform cubical box weighing 2000 N rests on the floor with its bottom left edge pressing against a ridge as shown in Fig. 6.32. The length L of a side of the box is 1.2 m. Determine the least force **F** applied horizontally at the top right edge of the box that will cause its bottom right edge to be slightly off the floor. [*Note:* With the right edge slightly off the floor, the ground and ridge exert their forces on the bottom left side of the box.]

■■ 53. If the force **F** shown in Fig. 6.32 is 840 N and the bottom right edge of the box is slightly off the ground, calculate the box's weight.

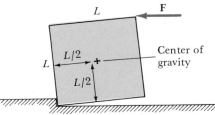

FIG. 6.32

■ 54. A football player's head, shown in Fig. 6.33, can be considered as a lever. The vertebra at the bottom of the skull serves as the fulcrum. The torque caused by the weight w of the head and helmet (18 lb) is balanced by the torque caused by the trapezius muscle tension T in the neck. (a) Calculate T. (b) If a 40-lb downward force is applied to the face mask, what muscle tension T is required to keep the head in equilibrium?

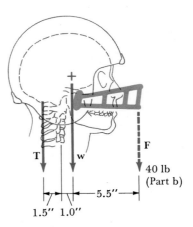

FIG. 6.33

CHAPTER 7

Rotational Motion with Angular Acceleration

We learned in Chapter 5 that a centripetal force is needed to keep an object moving in a circular path. The force points toward the center of the circle, perpendicular to the object's motion, and does not change the object's speed, only its direction of motion. To make the object move faster or slower, a force must be exerted parallel to its velocity—that is, tangent to the circle along which the object moves (Fig. 7.1). Tangential forces can give the object a push, causing it to move faster, or can slow it by exerting an opposing force. This chapter examines these tangential forces and their effect on objects moving in circular paths.

Most of our examples in Chapter 5 had to do with motion in which a single object moved in a circle whose radius was much larger than the object: A race-track is much larger than a car moving around it, and the moon's orbit about the earth is much larger than the moon. This motion in circles becomes slightly more complicated when we consider the rotational motion of **extended bodies**—objects whose mass is distributed continuously in a certain shape. Examples of the rotation of extended bodies are the earth's rotation about its north-south axis or the rotation of your leg about the hip joint when kicking a ball. To understand the motion of extended bodies, we must consider the torques caused by forces and the subsequent change in angular motion caused by the torques. The treatment is very similar to that used in Chapter 3, but Newton's second law and the kinematic equations of that chapter are replaced by similar equations appropriate for describing rotational motion. We will also introduce a rotational analogue of linear momentum. These rotational equations should help you understand phenomena such as the torque needed by a motor to turn the wheels of a car and the technique used by ballerinas to perform a pirouette.

FIG. 7.1. A person's tangential pushing force causes a merry-go-round to rotate faster. (Photo by David Conklin)

135

7.1 Angular Acceleration

Much of our analysis in this chapter relies on a quantity called angular acceleration. If an object's speed around a circle increases, then its angular velocity also increases. We say that the object experiences an angular acceleration.

The **angular acceleration** α of an object moving in a circle or rotating about an axis is defined as the ratio of the object's change in angular velocity $\Delta\omega$ and the time Δt required for that change:

$$\alpha = \frac{\Delta\omega}{\Delta t}. \tag{7.1}$$

The units of angular acceleration are rad/s^2.

If the time interval Δt is very small, then α, as defined above, is an instantaneous angular acceleration. Often we are interested only in the average angular acceleration over a longer time interval:

$$\bar{\alpha} = \frac{\omega - \omega_0}{t}, \tag{7.2}$$

where ω_0 is the initial angular velocity at time zero and ω is the final angular velocity at time t.

EXAMPLE 7.1 A bicycle tire starts from rest and 10 s later has an angular velocity of 2.5 rev/s. Calculate its average angular acceleration during those 10 s.

SOLUTION We know that the initial angular velocity $\omega_0 = 0$, and the final angular velocity $\omega = 2.5$ rev/s at $t = 10$ s. Using Eq. (7.2), we find that

$$\bar{\alpha} = \frac{\omega - \omega_0}{t} = \frac{2.5 \text{ rev/s} - 0}{10 \text{ s}} = 0.25 \frac{\text{rev}}{\text{s}^2}$$

$$= 0.25 \frac{\text{rev}}{\text{s}^2} \left(\frac{2\pi \text{ rad}}{1 \text{ rev}}\right) = \underline{1.6 \text{ rad/s}^2}.$$

The angular velocity changed, on the average, 1.6 rad/s during each second of the 10-s interval. ■

Angular acceleration is a vector quantity. We will restrict our attention in this chapter to rotation about a single axis and will not adhere strictly to conventions about the direction and sign of angular acceleration and angular velocity. For our problems and examples, α is positive if the magnitude of the angular velocity increases and negative if it decreases.

7.2 Tangential Acceleration

If an object rotates faster—if its angular velocity increases—then each point on the object also moves faster. The object experiences a tangential acceleration; its speed increases.

Tangential acceleration a_t is defined as the change in speed Δv of an object moving in a circle, or of a point on a rotating object, divided by the time Δt required for that change in speed:

$$a_t = \frac{\Delta v}{\Delta t} \tag{7.3}$$

The units of tangential acceleration are m/s².

If Δt in the above definition is small, then an instantaneous tangential acceleration has been defined. The average tangential acceleration can be defined as

$$\bar{a}_t = \frac{v - v_0}{t}, \tag{7.4}$$

where v_0 is the object's tangential speed at time zero, and v is its speed at time t.

Angular acceleration and tangential acceleration can be related easily. Recall from Section 5.3 that tangential speed and angular velocity are related by the equation $v = r\omega$. Substituting for v and v_0 into Eq. (7.4), we find that

$$\bar{a}_t = \frac{r\omega - r\omega_0}{t} = r\left(\frac{\omega - \omega_0}{t}\right). \tag{7.5}$$

But $(\omega - \omega_0)/t$ is the average angular acceleration [Eq. (7.2)]. Thus,

$$a_t = r\alpha. \tag{7.6}$$

The tangential acceleration (change in speed) is the product of an object's angular acceleration and the distance r of the object from its center of rotation. The average signs are dropped because Eq. (7.6) applies either to average values of a_t and α or to instantaneous values.

EXAMPLE 7.2 A bicycle tire undergoes an average angular acceleration of 1.50 rad/s²; that is, its angular velocity increases by 1.50 rad/s each second. (a) Calculate the average tangential acceleration of its valve stem, which is 0.68 m from the axis of rotation. (b) Calculate the valve stem's tangential speed after 10 s, if it starts at rest.

SOLUTION (a) We are given that $r = 0.68$ m and $\alpha = 1.50$ rad/s². Thus,

$$a_t = r\alpha = (0.68)(1.50 \text{ rad/s}^2) = \underline{1.0 \text{ m/s}^2}.$$

(b) Rearranging Eq. (7.4), we find that

$$v = v_0 + a_t t$$
$$= 0 + (1.0 \text{ m/s}^2)(10 \text{ s}) = \underline{10 \text{ m/s}}. \quad\blacksquare$$

We now have equations relating angular position, velocity, and acceleration (θ, ω, and α, respectively) to an object's position, speed, and acceleration (s, v, and a_t) along its circular path:

$$s = r\theta,$$
$$v = r\omega,$$
$$a_t = r\alpha.$$

Several kinematic equations can be derived that relate these angular and tangential quantities to each other at different times for the special case when the angular acceleration is constant.

7.3 Kinematic Equations for Constant Angular Acceleration

In Chapter 2 we derived four equations (2.9–2.12) that relate an object's position, velocity, and acceleration at two different times. The equations apply to linear motion with constant acceleration. They apply also to the tangential motion of an object along a circular path with constant tangential acceleration. The position coordinate x is replaced by an arc-length coordinate s, the velocity v by tangential speed v, and the acceleration a by tangential acceleration a_t.

A similar set of equations can be derived to relate the angular position coordinate θ, the angular velocity ω, and the angular acceleration α at time t to its angular position θ_0 and angular velocity ω_0 at time zero. *These equations apply only if angular acceleration is constant.* The two sets of equations are given in the following table.

Tangential	Angular	
$v = v_0 + a_t t$	$\omega = \omega_0 + \alpha t$	**(7.7)**
$s - s_0 = \left(\dfrac{v + v_0}{2}\right)t$	$\theta - \theta_0 = \left(\dfrac{\omega + \omega_0}{2}\right)t$	**(7.8)**
$s - s_0 = v_0 t + \dfrac{1}{2}a_t t^2$	$\theta - \theta_0 = \omega_0 t + \dfrac{1}{2}\alpha t^2$	**(7.9)**
$2(s - s_0)a_t = v^2 - v_0^2$	$2\alpha(\theta - \theta_0) = \omega^2 - \omega_0^2$	**(7.10)**

EXAMPLE 7.3 The switch for the turntable on a 33-rpm stereo system is turned off. The table makes 20 complete rotations before slowing to a stop. Calculate its angular acceleration, assumed constant.

SOLUTION The turntable makes 20 complete turns, revolutions, while stopping. Thus,

$$\theta - \theta_0 = 20 \text{ rev} = 20 \text{ rev}\left(\frac{2\pi \text{ rad}}{1 \text{ rev}}\right) = 126 \text{ rad}.$$

Its initial angular velocity is

$$\omega_0 = 33\frac{\text{rev}}{\text{min}} = 33\frac{\text{rev}}{\text{min}}\left(\frac{2\pi \text{ rad}}{1 \text{ rev}}\right)\left(\frac{1 \text{ min}}{60 \text{ s}}\right) = 3.5 \text{ rad/s},$$

and its final angular velocity $\omega = 0$. Equation (7.10) can be used to determine the turntable's angular acceleration:

$$\alpha = \frac{\omega^2 - \omega_0^2}{2(\theta - \theta_0)} = \frac{0^2 - (3.5 \text{ rad/s})^2}{2(126 \text{ rad})} = -4.9 \times 10^{-2} \text{ rad/s}^2.$$

The negative sign implies that the turntable's angular velocity decreases. ■

7.4 Torque and Angular Acceleration

What causes an object to experience an angular acceleration? Recall from Chapter 6 that for an object to remain in rotational equilibrium, the vector sum of the torques acting on it must be zero. When the torques do not add to zero, the rotational equilibrium is disturbed; the object undergoes an angular acceleration. The angular acceleration depends on the magnitude and sign of the torques exerted by forces acting on the object.

Figure 7.2 provides an example. A metal ball is attached to the end of a rod, and the other end of the rod pivots freely about a pin. The ball and the rod can rotate without friction in a horizontal plane. For the ball's speed to increase and to have greater angular velocity, a force must be applied. As we saw in Chapter 6, the magnitude of the torque τ caused by the force is the product of the magnitude of the force F and its moment arm l, which is the perpendicular distance from the line of action of the force to the pivot point or origin of coordinates. Thus,

$$\tau = lF. \tag{6.3}$$

The nearer the force is applied to the axis about which rotation occurs, the smaller the torque and the less effective it will be in causing angular acceleration of the rod and ball.

Newton's second law of motion can be used to find a precise relation between the torque of a force and the resulting angular acceleration. Consider the simple example shown in Fig. 7.3. The ball's mass is m; the rod's mass is assumed small compared to that of the ball and can be ignored. A single force $\mathbf{F}$ acts on the ball in the tangential direction. Thus, Newton's second law applied to the tangential direction becomes

$$F = ma_t.$$

The moment arm of the force is just r, the distance of the ball and force from the pivot point, as shown in the figure. Multiplying both sides of the preceding equation by r, we obtain

$$rF = mra_t.$$

The left side of this equation equals the torque τ caused by $\mathbf{F}$. Substituting for a_t, using Eq. (7.6), $a_t = r\alpha$, we find that

$$rF = mr\,(r\alpha), \quad \text{or} \quad \tau = (mr^2)\alpha.$$

Often several forces act on a rotating object. In these cases we must add the torques caused by all forces acting on the object to find the angular acceleration:

$$\Sigma\,\tau = \tau_1 + \tau_2 + \cdots = (mr^2)\alpha, \tag{7.11}$$

where $\tau_1, \tau_2, \ldots$ are the torques caused by forces $\mathbf{F}_1, \mathbf{F}_2, \ldots$. Notice that if the torques add to zero, the angular acceleration is also zero. The object is in rotational equilibrium.

EXAMPLE 7.4 A sledgehammer with a 7.0-kg head is used to pound a stake. The hammerhead is 0.80 m from the body of the person swinging it. The person exerts a 50-N·m torque on the hammer (see Fig. 7.4a). (a) Calculate the angular acceleration of the hammerhead when in the position shown in Fig.

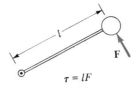

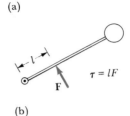

$\tau = lF$

(a)

$\tau = lF$

(b)

FIG. 7.2. The torque caused by a force depends on where the force acts on an object. The greater the moment arm l of the force, the greater the torque.

FIG. 7.3. The force causes a torque and angular acceleration.

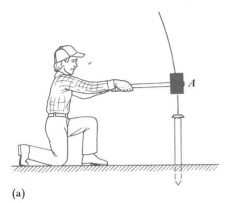

(a)

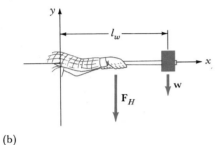

(b)

FIG. 7.4. (a) A person exerts a torque while swinging a sledgehammer. (b) A force diagram for the sledgehammer.

7.4a. (b) If the hammerhead's tangential speed at A is 2.0 m/s, calculate its speed 0.10 s later, assuming the acceleration remains constant. Ignore the mass of the hammer's handle.

SOLUTION (a) Two forces cause torques (see Fig. 7.4b). The person's hands exert a torque $\tau_H = 50$ N·m, and the hammerhead's weight causes a torque given by the expression

$$\tau_w = l_w w = l_w(mg)$$
$$= (0.80 \text{ m})(7.0 \text{ kg} \times 9.8 \text{ m/s}^2) = 55 \text{ N·m}.$$

Rearranging Eq. (7.11), we can determine the hammerhead's angular acceleration caused by these torques:

$$\alpha = \frac{\tau_H + \tau_w}{mr^2} = \frac{50 \text{ N·m} + 55 \text{ N·m}}{(7.0 \text{ kg})(0.80 \text{ m})^2}$$
$$= 23.4 \text{ N/kg·m}.$$

Using the fact that $1 \text{ N} = 1 \text{ kg·m/s}^2$, we find that the angular acceleration is

$$\alpha = 23.4 \text{ rad/s}^2.$$

The radian unit is included because α is an angular quantity.

(b) The rest of the problem involves kinematics. At time zero, the hammerhead's speed is 2.0 m/s. Later, at $t = 0.10$ s, the speed has increased to $v = v_0 + a_t t$, where

$$a_t = r\alpha = (0.80 \text{ m})(23.4 \text{ rad/s}^2).$$

Thus,

$$v = (2.0 \text{ m/s}) + (0.80 \text{ m})(23.4 \text{ rad/s}^2)(0.10 \text{ s})$$
$$= 3.9 \text{ m/s}. \qquad \blacksquare$$

7.5 Moment of Inertia

Equation (7.11), $\Sigma \tau = (mr^2)\alpha$, is the rotational equivalent of Newton's second law. There is a strong analogy between each of the three quantities in this equation and the three quantities in Newton's second law:

	Rotational Motion	Linear Motion
Cause of acceleration	$\Sigma \tau$	$\Sigma \mathbf{F}$
Acceleration	α	$\mathbf{a}$
Inertia of accelerated object	mr^2	m

For linear motion, mass is a measure of an object's inertia—that is, how much force is needed to cause the object to experience a certain acceleration. For rotational motion, the mass times the square of the distance of the mass from the axis of rotation (mr^2) is a measure of the object's rotational inertia. The quantity mr^2 is called the object's **moment of inertia** and is given the symbol I. The moment of inertia measures the difficulty in starting or stopping rotation. Equation (7.11) can now be written as

$$\Sigma \tau = \tau_1 + \tau_2 + \cdots = I\alpha. \qquad (7.12)$$

Notice that the moment of inertia depends not only on the mass m of the object but also on the distance r of the mass from the point about which it rotates. The farther the mass is from the point about which it rotates, the harder it is to change its rotational motion. A heavy mallet is harder to swing when it is held at the end of the handle than when it is held in the middle of the handle, nearer the head. This dependence of rotational inertia on the distance r is part of our everyday experience. If a baseball player has trouble getting the bat around in time to hit the ball, he or she is told to "choke up"on the bat. The result is a reduction of the moment of inertia of the bat, since its mass is brought closer to the hands. With the same torque, then, the batter can increase the angular acceleration of the bat because its moment of inertia has been reduced.

It is simple to calculate the moment of inertia of a single mass. Most objects, however, have a continuous distribution of mass in a certain shape. To get a sense of how we might calculate the moment of inertia of these objects, let us look first at several simple objects with more than one mass. Consider first the barbell shown in Fig. 7.5. Two masses are at the ends of a rod of length L. The rod is assumed to have a negligible weight compared to the weights on the end. Each mass is a distance $L/2$ from the axis of rotation and thus contributes an amount $mr^2 = m(L/2)^2$ to the total moment of inertia. The moment of inertia of the barbell about an axis through its center is therefore

$$I = m\left(\frac{L}{2}\right)^2 + m\left(\frac{L}{2}\right)^2 = 0.5mL^2.$$

In Fig. 7.6 we see a different rod that pivots about one end. Four equal masses are attached to the rod at quarter-length intervals along the rod. The rod itself has negligible mass. The moment of inertia of this rod is

$$I = m\left(\frac{1}{4}L\right)^2 + m\left(\frac{2}{4}L\right)^2 + m\left(\frac{3}{4}L\right)^2 + m(L)^2$$
$$= 1.88mL^2.$$

Once again, each mass contributes to the moment of inertia of the rod. However, masses farther from the axis of rotation contribute much more than those near the axis. The mass at the right-hand end of the rod in Fig. 7.6 contributes more to the rod's moment of inertia than the other three masses combined. This one mass makes the rod harder to rotate than the other three masses do.

The rotational inertia of an extended body is calculated in a similar fashion. An **extended body** is one with mass distributed continuously in a certain shape. The disc-shaped flywheel on a car is an extended body. Your leg is an extended body, as is a pencil or a bat.

To calculate the moment of inertia of an extended body, we assume that it is made of many small masses (Fig. 7.7). Each small mass contributes to the moment of inertia in proportion to its mass and the square of its distance from the axis about which we calculate the moment. For example, mass 7 in Fig. 7.7 contributes an amount $m_7(r_7)^2$ to the moment of inertia of a person's leg. The

FIG. 7.5. A barbell.

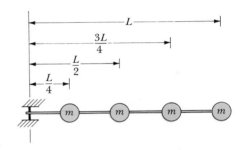
FIG. 7.6. Each mass contributes differently to the moment of inertia of this rod about an axis through its end.

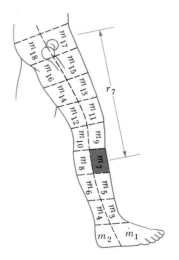
FIG. 7.7. To calculate the moment of inertia of a leg, we add the moments caused by all the small masses that are part of the leg.

moment of inertia of the whole leg is then the sum of all the small terms:

$$I = m_1 r_1^2 + m_2 r_2^2 + m_3 r_3^2 + \cdots. \tag{7.13}$$

Often this summation process is done using integral calculus; sometimes I is determined experimentally. In most of our examples and problems we either will be given the moment of inertia of an extended body or will use the values given in Table 7.1 for objects with common shapes.

TABLE 7.1 Moments of Inertia of Various Bodies

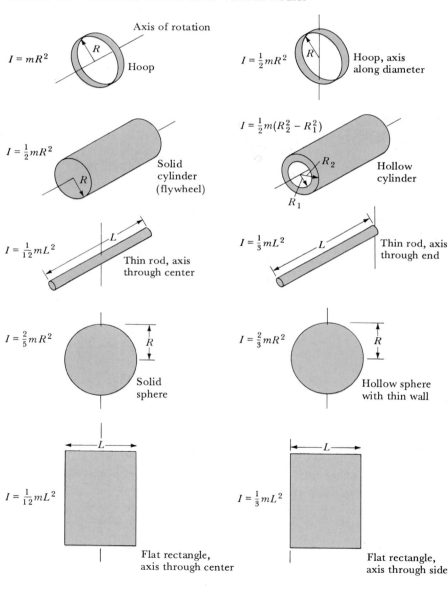

7.6 Rotational Motion Problem Solving and Examples

The technique for solving problems involving rotational motion is very similar to that used in dynamics and statics. Examples 7.5 and 7.6 illustrate the following technique.

1. Draw a picture of the whole situation described in the problem.

2. Choose one particular object and construct a force diagram showing all the forces acting on the object. It is important to place force arrows in the force diagram at the position where they exert their force on the object. This position affects the magnitude of the torque that the force exerts.

3. Superimpose a coordinate system on the force diagram. This coordinate system may be placed anywhere; however, all subsequent calculations must be consistent with this placement. Usually the origin is placed at the center of rotation.

4. Write the torque–angular acceleration equation ($\Sigma \tau = I\alpha$) for the object and substitute all the known information. The moment of inertia must be calculated using the same axis as that used to calculate the torques.

5. Solve for the unknown. Kinematic equations may be needed to calculate the angular acceleration or some other angular property of the object.

EXAMPLE 7.5 A boy chased by a bull runs through a gate with a uniform mass of 153 kg. The gate is 2.8 m wide and 1.5 m high. The bull will escape if the gate is not closed in 3.0 s. The boy pushes the gate with a 100-N force directed perpendicular to the gate at its edge 2.8 m from the hinges. Calculate (a) the moment of inertia of the gate (considered a flat rectangle), (b) the angular acceleration of the gate, and (c) the time needed to close the gate if it starts at rest and rotates through 50° before closing.

SOLUTION (a) The situation is shown in Fig. 7.8a. The gate is considered to be a flat rectangle with its axis at one side (the situation in the lower right of Table 7.1). We see from the table that the moment of inertia I of the gate equals $(1/3)mL^2$, where L is the gate's width (2.8 m):

$$I = \frac{1}{3}mL^2 = \frac{1}{3}(153 \text{ kg})(2.8 \text{ m})^2 = \underline{400 \text{ kg} \cdot \text{m}^2}.$$

(b) A force diagram for the gate is shown in Fig. 7.8b. The origin of coordinates is located along a line at the gate's hinges. The torque caused by the boy's force is

$$\tau = lF = (2.8 \text{ m})(100 \text{ N}) = 280 \text{ N} \cdot \text{m}.$$

Substituting this and the gate's moment of inertia ($I = 400 \text{ kg} \cdot \text{m}^2$) into Eq. (7.12) and rearranging to solve for α, we find that

$$\alpha = \frac{\Sigma \tau}{I} = \frac{280 \text{ N} \cdot \text{m}}{400 \text{ kg} \cdot \text{m}^2} = \underline{0.70 \text{ rad/s}^2}.$$

By now you should be able to make the preceding unit conversion. Try it.

A motor (not shown) exerts a torque that causes the potter's table to rotate. The potter's hands exert an opposing torque. If the torques balance, the table rotates at constant angular velocity. If the torques do not balance, the table experiences an angular acceleration—its angular velocity changes. (Paul S. Conklin, Photographer)

(a)

(b)

FIG. 7.8. (a) A boy being chased by a bull exerts a torque on a gate to close it before the bull escapes. (b) A force diagram for the gate.

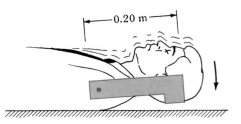

FIG. 7.9. A person's head rotates toward the floor after a faint. The rotating block serves as a model of the rotating head as it is stopped by the floor.

Remember that radians are unitless and can be placed in an expression or removed at will. The radian is included here because α, the angular acceleration, is an angular quantity.

(c) To determine the time needed to close the gate, we use Eq. (7.9) from kinematics:

$$\theta - \theta_0 = \omega_0 t + \frac{1}{2}\alpha t^2,$$

where, for this problem, $\theta - \theta_0 = 50°$, $\omega_0 = 0$, and $\alpha = 0.70 \text{ rad/s}^2$. Substituting these values and rearranging, we find that

$$t^2 = \frac{2(\theta - \theta_0)}{\alpha} = \frac{2\left(50° \times \dfrac{2\pi \text{ rad}}{360°}\right)}{0.70 \text{ rad/s}^2} = 2.49 \text{ s}^2,$$

or

$$t = \underline{1.6 \text{ s}}.$$

As expected, our young hero has time to close the gate and escape. ∎

EXAMPLE 7.6 A person faints and falls to the floor. Just before the person's head hits the floor, it is rotating about a pivot point at the shoulders with an angular velocity $\omega_0 = 8.0 \text{ rad/s}$ (see Fig. 7.9). The head is stopped by the floor in 0.0050 s as it hits 0.20 m from the pivot point. The moment of inertia of the rotating part of the body is 0.25 kg·m². Calculate (a) the average torque of the floor on the head and (b) the magnitude of the average force of the floor on the head. Ignore the torque caused by the head's weight.

SOLUTION (a) The torque caused by the force of the floor on the head is given by Eq. (7.12), $\tau = I\alpha$, where $I = 0.25 \text{ kg·m}^2$. We can determine α, the head's angular acceleration, using Eq. (7.7) from kinematics:

$$\alpha = \frac{\omega - \omega_0}{t} = \frac{0 - 8.0 \text{ rad/s}}{0.0050 \text{ s}} = -1.6 \times 10^3 \text{ rad/s}^2.$$

The negative sign implies that the angular velocity is decreasing. Substituting for α and I in Eq. (7.12), we find that the average torque of the floor on the head is

$$\tau = I\alpha = (0.25 \text{ kg·m}^2)(-1.6 \times 10^3 \text{ rad/s}^2) = \underline{-400 \text{ N·m}}.$$

The negative sign implies that the torque opposes the initial motion.

(b) The moment arm l of the floor's force on the head is 0.20 m. Since the torque caused by the force is given by the expression $\tau = lF$ and equals -400 N·m, we find that the average force is

$$F = \frac{\tau}{l} = \frac{-400 \text{ N·m}}{0.20 \text{ m}} = \underline{-2000 \text{ N}}.$$

This force, equal to 450 lb, could cause a concussion. ∎

7.7 Angular Momentum and Its Conservation

One of the important conserved quantities in science is **angular momentum,** a measure of the tendency of a rotating object to continue its rotational motion without change. A spinning Frisbee has small angular momentum—you can stop it with your hand. On the other hand, the earth while rotating about its axis has a very large angular momentum—an astronomical intervention would be needed to alter the earth's rotation.

We can write an expression for angular momentum based on the analogous expression for linear momentum. Linear momentum p is the product of an object's mass and its linear velocity:

$$\text{Linear momentum} = \mathbf{p} = m\mathbf{v}.$$

The rotational quantities analogous to mass and linear velocity are moment of inertia I and angular velocity ω, respectively. Angular momentum L can be defined in analogy to linear momentum as follows:

The **angular momentum** L of an object is the product of its moment of inertia I and its angular velocity ω:

$$L = I\omega. \tag{7.14}$$

The units of angular momentum are $\text{kg} \cdot \text{m}^2/\text{s}$.

To develop the principle of conservation of angular momentum, let us start with the rotational equivalent of Newton's second law:

$$\Sigma \, \tau = I\alpha.$$

Using Eq. (7.7) from kinematics to substitute for α in the preceding equation, we find that

$$\Sigma \, \tau = I \frac{\omega - \omega_0}{t},$$

or

$$(\Sigma \, \tau)t = I\omega - I\omega_0. \tag{7.15}$$

The right side of Eq. (7.15) is the change in angular momentum of an object with moment of inertia I in time t. The left side is the cause of this change, called the **rotational impulse,** which is defined as the sum of all torques acting on the object multiplied by the time t that the torques act to cause the change in angular momentum.

We found in Chapter 4 that if two (or more) objects are involved in a collision and if no external forces act on those objects during the collision, or if the external forces add to zero, then the linear momentum of the objects is conserved. The linear momentum remains the same after the collision as it was before. If one object gains linear momentum, the other object loses the same amount. A similar principle, the **conservation of angular momentum,** applies to rotational motion:

If no torques of external origin act on a system or if the torques acting on a

(a)

(b)

FIG. 7.10. A figure skater spinning into a pirouette.

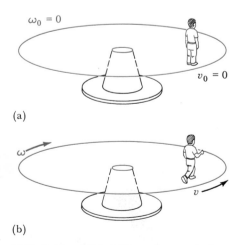

$\omega_0 = 0$

$v_0 = 0$

(a)

ω

v

(b)

FIG. 7.11. (a) A child stands on a platform at rest. (b) When the child runs counterclockwise, the platform must rotate clockwise to conserve angular momentum.

system add to zero, then the angular momentum of the objects in the system is constant.

We can see from Eq. (7.15) that this principle is true for a single object. If no torques act on the object, then

$$\Sigma \tau = 0,$$

and the right side of Eq. (7.15) must also be zero; or

$$0 = I\omega - I_0\omega_0.$$

Thus,

$$I\omega = I_0\omega_0, \tag{7.16}$$

where I_0 and I are the object's initial and final moments of inertia, respectively, and ω_0 and ω are its initial and final angular velocities, respectively.

Let us apply this principle to three examples. In each, no torques of external origin act on a single object. When the object's moment of inertia changes, its angular velocity must also change so that $I\omega$ remains constant.

EXAMPLE 7.7 A figure skater with her arms and a leg extended, as shown in Fig. 7.10a, spins with an angular velocity of 1.0 rev/s. Her moment of inertia is 4.0 kg·m². As she draws her arms and leg in close to her axis of rotation, as shown in Fig. 7.10b, her moment of inertia decreases to 0.80 kg·m².* At what angular velocity does she now rotate? Ignore any friction forces acting on her skates.

SOLUTION We are given that $I_0 = 4.0$ kg·m², $\omega_0 = 1.0$ rev/s, and the final moment of inertia $I = 0.80$ kg·m². Since no external torques act on her body, the skater's angular momentum must remain constant. Her final angular velocity can be calculated using Eq. (7.16):

$$\omega = \frac{I_0\omega_0}{I} = \frac{(4.0 \text{ kg·m}^2)(1.0 \text{ rev/s})}{0.80 \text{ kg·m}^2} = \underline{5.0 \text{ rev/s}}. \quad \blacksquare$$

EXAMPLE 7.8 A 40-kg child stands at the edge of a circular platform that can rotate freely in a horizontal plane (Fig. 7.11a). The child and platform are initially at rest. The platform has a radius of 2.0 m and a moment of inertia of 500 kg·m². The child starts running around the edge of the platform in a counterclockwise direction and attains a speed of 3.0 m/s relative to the ground (Fig. 7.11b). What is the angular velocity of the platform?

SOLUTION The child's counterclockwise angular momentum will be equal and opposite to the clockwise angular momentum of the platform:

$$\underbrace{L_{\text{child}} - L_{\text{platform}}}_{\substack{\text{Final angular} \\ \text{momentum}}} = \underbrace{0.}_{\substack{\text{Initial} \\ \text{angular} \\ \text{momentum}}} \tag{7.17}$$

---

*Remember that moment of inertia depends on mr^2. As the skater's arms are brought close to her body, the value of r for the hands and arms decreases, as does the moment of inertia. Because moment of inertia depends on r^2, a change in r can have a dramatic effect on the value of I.

The final angular momentum of the child will be

$$L_{\text{child}} = I\omega = (mr^2)\left(\frac{v}{r}\right) = mrv,$$

where r is the distance of the child from the center about which he is running, m is the child's mass, and v is his speed in the circular path. The final angular momentum of the platform will be

$$L_{\text{platform}} = I_p\omega_p,$$

where I_p is the moment of inertia of the platform and ω_p is its unknown angular velocity. Substituting these expressions for L into Eq. (7.17) and rearranging, we find that

$$\omega_p = \frac{mrv}{I_p} = \frac{(40 \text{ kg})(2.0 \text{ m})(3.0 \text{ m/s})}{500 \text{ kg} \cdot \text{m}^2}$$

$$= 0.48 \text{ rad/s.} \qquad \blacksquare$$

We should point out that the expression for the child's angular momentum derived in the last example is a general expression for the angular momentum of a mass m when moving at speed v in a circle of radius r. This expression for angular momentum,

$$L = mrv, \qquad (7.18)$$

is very important in atomic physics when considering the motion of electrons in orbit about a nucleus. It is also useful in the study of the orbital motion of planets, comets, and the like.

Most of the objects in our solar system, such as the planets and comets, move in elliptical orbits, with the sun at one of the two foci of the ellipse, as illustrated in Fig. 7.12. In 1705, the British astronomer Edmund Halley suggested that the comets that crossed the sky in 1531, 1607, and 1682 were really one comet making repeated trips around its orbit. Halley predicted the comet's return in 1758, and it did appear on Christmas night of that year, 16 years after his death. This comet, which completes its orbit about the sun each 74 to 79 years, has now been named in Halley's honor.

Comets consist of a bright head and a diffuse, long tail that stretches out from the head away from the sun (Fig. 7.13). The head has at its center a small nucleus consisting of frozen molecules such as water, carbon dioxide, ammonia, and a little dust. The nucleus is surrounded by a much larger coma made of gas and dust, partly liberated from the nucleus. The tail (sometimes there are two) consists of dust and ionized gas.

Planets and comets that move in elliptical orbits have constant angular momentum because the torque caused by the gravitational pull of the sun is zero, as illustrated for the orbiting comet shown in Fig. 7.12. The force points from the comet toward the focus, that is, the sun. Thus, the line of action of that force passes through the sun, which is also the origin of the coordinate system. Hence, the moment arm of the force about the origin is zero, as is the torque. But, according to Eq. (7.15), if the net torque acting on a rotating object is zero, then the object's angular momentum does not change. Let us apply this idea to Halley's comet.

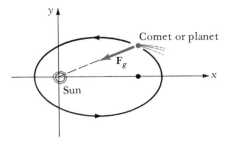

FIG. 7.12. A comet (or planet) moves in an elliptical orbit around the sun. The gravitational force of the sun on the comet has a zero moment arm and causes a zero torque on the comet. Consequently, its angular momentum is constant.

FIG. 7.13. Halley's comet as seen May 6, 1910. (Jay Pasachoff)

EXAMPLE 7.9 Halley's comet moves in an elliptical orbit. On February 9, 1986, the comet has its nearest approach to the sun—0.587 AU (1 AU, astronomical unit, is the average distance from the earth to the sun, equal to 1.50×10^{11} m). At its nearest approach, the comet's speed is 54.6 km/s. What will be the comet's speed 38 years later, when it is farthest from the sun, at a distance of 35.3 AU? (It will have gone beyond Neptune.)

SOLUTION The comet's angular momentum when nearest the sun is given by Eq. (7.18). The angular momentum is the same when farthest away since the comet's angular momentum does not change:

$$mr_{\text{near}}v_{\text{near}} = mr_{\text{far}}v_{\text{far}},$$

or

$$v_{\text{far}} = \left(\frac{r_{\text{near}}}{r_{\text{far}}}\right)v_{\text{near}} = \left(\frac{0.587 \text{ AU}}{35.3 \text{ AU}}\right)54.6 \text{ km/s}$$

$$= \underline{0.908 \text{ km/s}},$$

a little less than three times the speed of sound (0.34 km/s). Notice that as the comet plods around the most distant part of its orbit, where it is 60 times its distance of nearest approach, its speed is reduced to 1/60 of its speed when shooting by the sun at its distance of nearest approach. The product mrv is the same at both positions, but if r is small, v is large, and vice versa. ■

Summary and Additional Readings

By now you may feel that a bewildering variety of equations must be used to work problems involving rotational acceleration. Almost everything we have done in this chapter is analogous to what we did when we considered linear motion in Chapters 2, 3, and 4. You should find that this correlation between linear motion and rotational motion is helpful in remembering and applying rotational equations. The important equations and concepts used in this chapter and the equivalent equations used to describe linear motion are summarized in the accompanying table.

In addition to these equations, we have also used several kinematic equations. Four of these are presented on p. 138; they relate the angular quantities θ, ω, and α at different times. Four other equations on the same page relate the quantities s, v, and a_t—quantities that describe motion along the arc of the circular path followed by the rotating object. These tangential quantities are related to their angular analogue by multiplying by r; $a_t = r\alpha$, $v = r\omega$, and $s = r\theta$.

David Clark, Garry Hunt, and William McCrea, "Celestial Chaos and Terrestrial Catastrophes," *New Scientist* **14,** December (1978). Summarizes the possible consequences of celestial events such as passing through a comet's tail.

R. L. Page, "The Mechanics of Swimming and Diving," *The Physics Teacher* **14,** 72 (1976).

Peter Brancazio, "Physics of Basketball," *American Journal of Physics* **49,** 356 (1981).

Quantity or Relation	Linear Motion	Angular Motion	
Velocity	$v = \Delta x/\Delta t$	$\omega = \Delta\theta/\Delta t$	**(7.1)**
Acceleration	$a = \Delta v/\Delta t$	$\alpha = \Delta\omega/\Delta t$	
Newton's second law	$\Sigma F = ma$	$\Sigma\tau = I\alpha$	**(7.12)**
Cause of acceleration	F	τ The torque of a force is the product of the force and its moment arm ($\tau = lF$).	
Inertia of an object	m	I The rotational moment of inertia of a single object is the product of its mass and the square of its distance from the axis of rotation ($I = mr^2$). For an extended body, we divide the body into many small masses and add the moments due to each of these masses ($I = m_1r_1^2 + m_2r_2^2 + \cdots$). The larger the value of I, the harder it is to change its state of rotation.	
Momentum of an object	$p = mv$	$L = I\omega$	**(7.14)**
Conservation of linear and angular momenta for an isolated system	$p_{\text{initial}} = p_{\text{final}}$	$L_{\text{initial}} = L_{\text{final}}$	

Questions

1. Why is the force needed to open a heavy door much greater when the door is pushed near the hinges than when pushed at the side opposite the hinges?

2. A child is given a push on a swing. Why doesn't the child continue in a vertical loop over the top of the swing. Explain in terms of torques.

3. In terms of the torque needed to rotate your leg as you run, would it be better to have a large calf and small thigh, or vice versa? Explain.

4. Suppose that two bicycles have equal overall mass, but one has thin, lightweight tires while the other has heavier tires. Why is the bicycle with thin tires easier to accelerate?

5. When riding a ten-speed bicycle up a hill, a cyclist shifts the chain to a larger-diameter gear attached to the back wheel. Why is this gear preferred to a smaller gear?

6. A meter stick is supported horizontally at each end by your fingers. A mass rests on one end of the stick. If you remove your fingers under the end with the mass, that end of the meter stick falls faster than the mass. Why?

7. If all the people on the earth took elevators to the tops of high buildings in their communities, how would the length of the day be affected? Explain.

8. The Mississippi River carries sediment from higher latitudes toward the equator. How does this affect the length of the day? Explain.

9. A spinning raw egg, if stopped momentarily and then released by the fingers, will resume spinning. Explain. Will this happen with a hard-boiled egg? Explain.

10. When we start an automobile engine, the fan and many other parts of the car start to rotate. If angular momentum is conserved, where does this angular momentum originate?

11. If you could accelerate from zero to 8 mph anywhere on the earth's surface, where and in which direction would you run to increase the length of the day most?

12. Compare the magnitude of the earth's angular momentum about its axis to that of the moon about the earth. The tides exert a torque on the earth and moon so that eventually they will both rotate with the same period. The object with the greatest angular momentum will experience the smallest percent change in the period of rotation. Will the earth's solar day increase more than the moon's period of rotation decreases? Explain.

Problems

7.1, 7.2, and 7.3 Angular Acceleration, Tangential Acceleration, and Kinematics

1. (a) When a record player is stopped, the turntable, initially rotating at 33.3 rpm, slows down and stops in 60 s. Calculate the angular acceleration of the turntable. (b) A grinding wheel, when turned on, accelerates with an angular acceleration of 190 rad/s². How much time is required to attain a final angular velocity of 1800 rpm?

2. (a) A centrifuge accelerates from rest with an angular acceleration of 30 rad/s². What is the angular velocity of the centrifuge after 30 s? (b) The angular velocity of an automobile engine accelerates from 800 rpm to 3000 rpm in 15 s. Calculate the engine's angular acceleration.

3. During a tennis serve, the top of the racquet, which is 1.5 m from its pivot point, accelerates from rest to a speed of 20 m/s in a time of 0.10 s. Calculate the average tangential acceleration of the top of the racquet and also its angular acceleration.

4. A record player is switched from 33 rpm to 45 rpm. The turntable changes speed in 2.0 s. Calculate its average angular acceleration and the average tangential acceleration of a point on the turntable that is 15 cm from the axis of rotation.

■ 5. When a person punts a football, the foot accelerates from a speed of zero to a final speed of 11 m/s just before it hits the ball. This acceleration occurs in a time of 0.30 s. (a) Calculate the average tangential acceleration of the foot. (b) If the foot rotates at the end of a leg that is 1.0 m long, what is the angular velocity of the leg just before hitting the ball? (c) What is the average angular acceleration of the leg during the 0.30 s?

■ 6. An ant clings to the outside edge of the tire of an exercise bicycle. The ant's speed increases from zero to 10 m/s in 2.5 s. (a) Calculate the average tangential acceleration of the ant. (b) If the wheel's angular acceleration is 13 rad/s², how far is the ant from the axis of rotation and (c) what distance does the ant travel in that time?

■ 7. According to the speedometer on an exercise bicycle, a point on the rim of its 0.30-m-radius wheel travels in a circle a distance of 200 m while its speed increases from 0 to 10 m/s. Calculate (a) the average tangential acceleration of the rim, (b) the wheel's angular acceleration, and (c) the time needed for the change in speed.

■ 8. The angular velocity of the wheel of an exercise bicycle changes from 5.0 rad/s to 8.0 rad/s in 6.0 s. Calculate (a) the wheel's average angular acceleration, (b) the angle through which it turns during the 6.0 s, and (c) the distance along the circular path traveled by a point 0.60 m from the axle.

■ 9. A moth sits on a potter's wheel 0.30 m from its axle. The wheel's angular velocity decreases from 4.0 rad/s to 2.0 rad/s in 10 s. Calculate (a) the wheel's average angular acceleration, (b) the angle through which it turns during the 10 s, and (c) the distance traveled by the moth.

■ 10. A Ferris wheel starts from rest and acquires an angular velocity of 0.50 rad/s after completing one revolution. Calculate (a) the wheel's angular acceleration (assumed constant), (b) the time required for the first revolution, and (c) the distance traveled in this time by a person seated 20 m from the axis of rotation.

■ 11. Calculate what the earth's angular acceleration would

be in rad/s² if the length of a day increased from 24 h to 48 h during the next 100 years.

■ 12. A father pushes the edge of a disc-shaped platform of 2.0-m radius on which his children sit. The platform starts at rest and experiences an angular acceleration of 0.30 rad/s². Calculate the distance the father must run while pushing the platform to increase its speed at the edge to 7.0 m/s.

■ 13. *Estimate* the average angular acceleration of a car tire as you leave an intersection after a light has turned green. Discuss the choice of numbers used in your estimate.

7.4 Torque and Angular Acceleration

14. A 1.2-kg ball is attached at one end of a 0.80-m stick of negligible mass. The other end of the stick has a pin through it that allows the stick to rotate freely in a vertical circle. (a) Calculate the gravitational torque on the ball when the stick is held in a horizontal position. (b) Calculate the stick's angular acceleration just as the ball is released.

15. A 0.30-kg ball moves in a horizontal circle of radius 0.90 m at the end of a stick of negligible mass. Because of air resistance, a torque of 0.036 N·m is needed to keep the stick and ball rotating at constant angular velocity. Calculate the magnitude of the air-resistance force opposing the ball's motion. The air resistance force on the stick can be ignored.

■ 16. A machine used to test golf balls accelerates a golf club with a 0.24-kg head at the end of a 0.50-m handle of negligible mass. The head swings in an arc, starting from rest and attaining a final speed of 40 m/s in a time of 0.20 s. Calculate (a) the average angular acceleration of the golf club and (b) the net torque required to produce this acceleration.

■ 17. A 2.0-kg block attached to the end of a 0.80-m-long stick of negligible mass can rotate in a circular path on a horizontal frictionless surface. The mass starts at rest and is pushed by a 4.0-N force in a direction tangent to the circle. Calculate (a) the angular acceleration of the block and (b) the time needed for it to acquire a speed of 3.0 m/s.

■ 18. A 1.6-kg block attached to the end of a 0.60-m-long stick of negligible mass rotates in a horizontal circular path about a bearing at the other end of the stick. The mass initially moves with an angular velocity of 5.4 rad/s. Calculate the angular deceleration of the block and the time needed to stop its motion if it is opposed by a 1.8-N friction force.

7.5 Moment of Inertia

19. Calculate the moment of inertia of the four balls shown in Fig. 7.14 about an axis perpendicular to the paper and passing

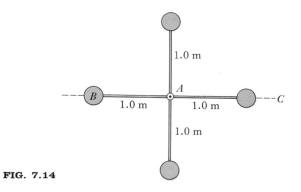

FIG. 7.14

through point *A*. The mass of each ball is 2.0 kg. Ignore the mass of the rods to which the balls are attached.

20. Repeat Problem 19 for an axis perpendicular to the paper through point *B*.

21. Repeat Problem 19 for axis *BC*, which passes through two of the balls.

■ 22. *Estimate* the moment of inertia of a 200-g meter stick about an axis at its end and perpendicular to the meter stick. To do this, divide the meter stick into five equal sections and add the contributions of each section to the moment of inertia.

■ 23. A 1.0-m stick can rotate in a horizontal circle about a pin in one end of the stick. A 1.0-kg mass sits on top of the pin and another at the end of the stick, 1.0 m from the pivot point. If you put both masses together, where on the stick should they be placed so that the moment of inertia is the same as in the original situation?

■ 24. *Estimate* the moment of inertia of a baseball bat about an axis passing perpendicular to and through its handle.

7.6 Rotational Motion Problem Solving

25. A Ferris wheel with a moment of inertia of 2.0 × 10⁵ kg·m² is to accelerate from rest to an angular velocity of 0.20 rad/s in 20 s. What is the minimum torque that its motor must provide to cause this acceleration?

26. A turntable is rotating with an angular velocity of 33.3 rpm when it is turned off. The turntable, whose moment of inertia is 1.0 × 10⁻² kg·m², stops in 50 s. Calculate the resistive torque that slows the turntable.

■ 27. Determine the average torque needed to accelerate the turbine of a jet engine from rest to an angular velocity of 160 rad/s in 25 s. The turbine's rotating parts have a 32-kg·m² moment of inertia.

■ 28. A centrifuge with a 0.40-kg·m² moment of inertia is to experience an angular acceleration of 100 rad/s² when the power is turned on. (a) Calculate the minimum torque that the motor must supply. (b) How much time is needed for the centrifuge's angular velocity to increase from zero to 5000 rad/s?

■ 29. A merry-go-round, considered as a uniform disc of radius 5.0 m and mass 25,000 kg, must have a new motor that will accelerate it from rest to 1.5 rad/s in 8.0 s. Calculate the torque that the motor must provide to the merry-go-round.

■ 30. Calculate the average force needed to cause a 1.0 m × 1.0 m gate of mass 30 kg and initially at rest to swing open 90° in 1.8 s. The force is applied 1.0 m from the hinges and perpendicular to the door.

■ 31. Suppose that a force **F** is applied at the earth's equator tangent to its surface for a time of one year. How large must the force be to stop the earth's rotation? Assume that the earth's mass is distributed uniformly.

■ 32. As the tides slide across the earth, they exert a friction force that opposes the earth's rotation. Because of this, the time required for one earth rotation increases by 0.0016 s every 100 years. (a) Calculate the angular deceleration of the earth. (b) With the assumption that the earth is a uniform solid sphere, what torque do the tides exert on the earth? (c) If the friction force caused by the tides were all concentrated at the equator, how large would this force be?

■■ 33. A rope wrapped around a flywheel is attached to a 3.0-kg pail hanging at the end. The flywheel has a radius of 0.10 m and a mass of 8.0 kg. (a) Make a force diagram for the pail and another for the flywheel. (b) Use Newton's second law and the pail force diagram to obtain an expression relating the rope tension and the acceleration of the pail. (c) Use Eq. (7.12) and the flywheel force diagram to obtain an expression relating the rope tension to the angular acceleration of the flywheel. (d) The pail acceleration a and flywheel angular acceleration α are related by the equation $a = r\alpha$, where r is the flywheel radius. Combine this with your equations in parts (b) and (c) to obtain the value of the rope tension and pail acceleration.

■■ 34. You are to buy a motor that will be used to lift a 46-kg vat of plaster to the top of a building being constructed. To save time, the cable between the motor and the vat must provide enough tension to accelerate the vat upward at 1.5 m/s². The pulley on the motor has a radius of 12 cm. Calculate the tension in the cable and the torque that the motor and pulley must be able to exert on the cable to provide this acceleration.

■■ 35. *Estimate* the average force of a car door when closing on a person's fingers. Indicate all numbers you use (the moment of inertia of the door, the angular velocity of the door just as it reaches the fingers, the stopping time, and so on).

7.7 Angular Momentum and Its Conservation

36. (a) Calculate the angular momentum of a 1500-kg race car traveling with a speed of 40 m/s around a circular racetrack whose radius is 800 m. (b) An electron rotating about a proton in a hydrogen atom has an angular momentum of 1.05×10^{-34} kg·m²/s. The electron's mass is 9.1×10^{-31} kg, and it moves in a circular path of 0.53×10^{-10} m radius. Calculate the average speed of the electron.

37. A ballerina spins with an initial angular velocity of 1.5 rev/s when her arms and a leg are extended. As she draws her arms and leg in toward her body, her moment of inertia becomes 1.0 kg·m² and her angular velocity is 4.0 rev/s. Calculate her initial moment of inertia.

■ 38. (a) Calculate the angular momentum of a 10-kg disc-shaped flywheel of radius 9.0 cm when rotating with an angular velocity of 320 rad/s. (b) With what angular velocity must a 10-kg solid sphere of 9.0-cm radius rotate to have the same angular momentum as the flywheel?

■ 39. A 0.20-kg block moves at the end of a 0.50-m string in a circular path on a frictionless table. The block's initial angular velocity is 2.0 rad/s. As the block rotates, the string wraps around the stick at the axis of rotation. Calculate the final angular velocity and tangential speed of the block when the string is 0.20 m from the axis.

■ 40. A student sitting on a chair on a circular platform of negligible mass rotates freely on an air table with initial angular velocity 2.0 rad/s. The student's arms are initially extended with a 6.0-kg mass in each hand. As the student's arms are pulled in toward his body, the masses move from a distance of 0.80 m to 0.10 m from the axis of rotation. Calculate the student's final angular velocity. The initial moment of inertia of his body (not including the masses) with arms extended is 6.0 kg·m², and the final moment of inertia with his arms drawn in is 5.0 kg·m².

■ 41. A turntable whose moment of inertia is 1.0×10^{-3} kg·m² rotates on a frictionless air cushion with an angular velocity of 2.0 rev/s. A 1.0-g beetle falls to the center of the turntable and then walks 0.15 m to its edge. Calculate the angular velocity of the turntable with the beetle on the edge.

■ 42. A large disc-shaped platform can rotate about a center axle. A 60-kg woman stands at rest at the edge of the platform 4.0 m from its center. The platform is also at rest. She starts running around the edge of the platform and attains a speed relative to the ground of 2.0 m/s. Calculate the angular velocity of the platform. Its mass is 120 kg.

■■ 43. At present, the motion of people on the earth is fairly random; the number moving east equals the number moving west. Assume that we could get all the earth's inhabitants lined up on the land at the equator. If they all ran to the west at 3.0 m/s, *estimate* the change in angular velocity of the earth and the change in the length of the earth's day. [*Note:* You have not been given all the information you need. Justify any other numbers used in your calculations.]

■■ 44. An extremely dense neutron star with mass equal to that of the sun has a radius of about 10 km. (A teaspoonful of neutron star would weigh nearly a billion tons!) These stars are thought to rotate once about their axis every 0.03 to 4 s, depending on their size and mass, and are thought to be the pulsating sources of radiation called *pulsars*. Suppose that the neutron star described in the first sentence rotated once every 0.04 s. If its volume then expanded to occupy a uniform sphere of radius 1.4×10^8 m (most of the sun's mass is in a sphere of this size) with no change in mass or angular momentum, what time would be required for one rotation? By comparison, the sun rotates once about its axis each month.

Energy and Its Transformations

In Part I of this book our approach to problem solving was based on Newton's second law of motion and kinematics. By adding all forces acting on an object, we could use the second law to determine the object's acceleration. Kinematic equations could then be used to determine the object's future position and velocity. A similar approach was used in dealing with torques and the changing rotational motion of an object.

In Part II we develop a new technique for solving problems, a technique based on the conservation of energy. We will find that as an object or group of objects move about, some forms of their energy decrease while others increase. However, the net change of energy is zero. The fact that energy is conserved allows us to calculate some unknown property of the objects as they move from an initial situation to a final situation.

In Chapter 8 we will be primarily interested in developing the energy concept as a problem-solving tool. As we progress further into the subject, we will consider many interesting, everyday examples that depend on energy transformations. These include the energy changes in our bodies, in the earth and its climate, and in all aspects of nature.

CHAPTER 8

Work and Energy

Concepts related to energy are perhaps the most important in all of science. Energy appears in many different forms: the energy an object has because of its motion or its elevation above the earth, the energy of a compressed or stretched spring, the energy of positive and negative electric charges when separated, the thermal energy of hot substances, the mass energy of nuclei on the sun, the chemical energy of a glucose molecule, and so forth. Every living and dynamic process in nature involves the conversion of energy from one form to another. Energy transformations are important for powering the machines built and used by humans. Energy transformations cause ocean currents, rain, wind, temperature variations, and all other phenomena related to the earth's climate. The ability to capture, transform, and store various forms of energy is the essential feature of living organisms. Energy has been called the *vis viva*—the force of life.

In this chapter we will learn how to calculate the changes of several different forms of energy. We will also consider one way in which energy is transferred between different objects—that is, by doing work. But the most important topic of this chapter is the development of the principle of conservation of energy and its use as the foundation for a new way to solve problems. We start by analyzing one way in which energy is transferred between different objects.

8.1 Work

Our development of energy principles in this chapter evolves from the definition of a quantity called work. We will find that this definition of work differs somewhat from everyday expressions involving the word *work*—"I'm going to work" or "I have to work a problem."

In physics, work is done when a force acts on an object as the object moves from one place to another. Because of the work, energy is transferred from the object that causes the force to the object on which the force acts, or vice versa. Consider a bowler who exerts a force on a bowling ball, causing it to accelerate forward. The ball gains energy of motion as its speed increases. The bowler, in turn, loses a little chemical energy in his or her muscles while pushing the ball. The energy transfer from bowler to ball has occurred because of the work done by the bowler on the ball.

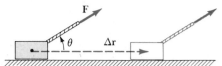

FIG. 8.1. The work done by the rope pulling the block is $W = F \Delta r \cos \theta$, where θ is the angle between **F** and **Δr**.

In physics, work is defined in terms of the force exerted on an object as it undergoes a displacement:

If a constant force **F** acts on an object while the object undergoes a displacement **Δr** (Fig. 8.1), the **work** done by the force on the object during the displacement is

$$W = F \Delta r \cos \theta, \qquad (8.1)$$

where θ is the angle (180° or less) between the direction of **F** and the direction of **Δr**.*

The F and Δr that appear in Eq. (8.1) are both scalar quantities—the magnitudes of the force and displacement vectors, respectively. Since work is defined as the product of three scalars (F, Δr, and $\cos \theta$), work is also a scalar quantity.

We also see from our definition of work that an object must experience a displacement before any work is done on it. According to the physics definition of work, holding an 80-lb bag of cement all day is no work, since Δr, the cement's displacement while being held, is zero.

Note also that the work done on an object by a force depends very much on the direction of the force relative to the object's displacement. The angle θ appearing in the definition of work (see Fig. 8.1) is the *angle between the vectors* **F** *and* **Δr** and is *not* the angle that either **F** or **Δr** makes with some axis of a set of coordinates. Consider carefully the angle θ used in the next two examples.

EXAMPLE 8.1 A carpenter pulls a cart load of lumber 10 m across a floor (Fig. 8.2a). The tension in the rope is 200 N and is directed 37° above the horizontal. Calculate the work done by the carpenter in pulling the lumber.

SOLUTION The tension force **T** and displacement **Δr** make an angle of 37° with respect to each other (Fig. 8.2b). Hence, the work done by the rope tension force in moving the load of lumber is

$$W = T \Delta r \cos 37° = (200 \text{ N})(10 \text{ m})(0.8) = \underline{1600 \text{ N·m}}. \qquad \blacksquare$$

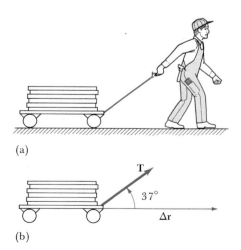

(a)

(b)

FIG. 8.2. (a) A carpenter pulls a load of lumber. (b) The rope tension force makes an angle of 37° with the displacement. $W = T \Delta r \cos 37°$.

EXAMPLE 8.2 A woman jumps from a high wall and lands in soft sand (Fig. 8.3a). The sand exerts an average upward force of 50,000 N while stopping her. The woman stops after sinking 5 cm into the sand. How much work is done by the force of the sand on her?

SOLUTION The force of the sand on the woman points upward, whereas the woman's displacement while she is stopping is 5 cm downward (Fig. 8.3b); hence there is a 180° angle between the direction of the force and the direction of the displacement. The work done by the force of the sand on the woman is

$$W = F \Delta r \cos 180° = (50,000 \text{ N})(0.05 \text{ m})(-1) = \underline{-2500 \text{ N·m}}. \qquad \blacksquare$$

*This definition of W is sometimes written in an abbreviated form as $W = \mathbf{F} \cdot \mathbf{\Delta r}$ and is called the "dot or scalar product" of vectors **F** and **Δr**. In general, the scalar product of any two vectors **A** and **B** is

$$\mathbf{A} \cdot \mathbf{B} \equiv AB \cos \theta,$$

where θ is the angle (180° or less) between the directions of **A** and **B**.

If negative work is done on an object, as in Example 8.2, the object's energy decreases. For example, the woman in Example 8.2 lost a form of energy called *kinetic energy* (energy of motion) because the sand's force opposed her motion. The sand did a negative amount of work on the woman and caused her to stop moving.

Units of Work

We see from the previous examples that the units of work are the product of the units of force (newtons or pounds) and the units of displacement (meters or feet). In fact, these are the units of all forms of energy. Thus, work and energy units are newton-meters (N·m) or pound-feet (lb·ft).

Because work and energy are such important quantities, their unit has been given a special name in the SI metric system; it is called the **joule** (J) in honor of the English physicist James Joule (1818–1889):

$$1 \text{ joule} \equiv 1 \text{ newton-meter,}$$

or

$$1 \text{ J} = 1 \text{ N·m.} \tag{8.2}$$

Joule, a physicist whose scientific interests were financially supported by a family brewery, spent his honeymoon near a waterfall. During his honeymoon, Joule measured with great accuracy the temperature of the water at the top of the falls and at the bottom after the water had dashed onto the rocks below. This and other experiments eventually led to the very important conservation-of-energy theorem discussed later in the chapter. Such devotion to physics surely deserves a unit named in Joule's honor.

Work Done by a Variable Force

The work done by a constant force can be calculated nicely using Eq. (8.1). But how do we calculate the work done by a force whose magnitude varies, such as the varying force of a golf club on a golf ball, as plotted in Fig. 8.4a? In this figure, the magnitude of the force is given on the vertical axis and the ball's position on the horizontal axis. The ball starts at position $x_0 = 1.0$ cm with no force acting on it. As the club strikes the ball, it moves forward (x increases) and the magnitude of the force also increases. When the ball reaches position $x = 2.0$ cm, the force is a maximum, 10,000 N. As the ball moves farther forward, the force decreases. Evidently, the ball is starting to leave the club head. Finally, when the ball reaches position $x = 3.0$ cm, the club no longer touches the ball and the force is zero.

If we knew the average force $\bar{F}$ on the ball as it moved from $x_0 = 1.0$ cm to $x = 3.0$ cm, then the work done by the club head on the ball would be

$$W = \bar{F}(x - x_0) \cos 0° = \bar{F}(x - x_0).$$

We can perform a similar calculation for the variable force by breaking the distance along the x axis into many small, equal displacements Δx, as shown in Fig. 8.4b. The work done during the first small displacement (the darkly shaded rectangle on the left) is

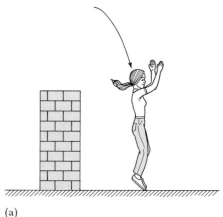

(a)

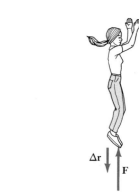

(b)

FIG. 8.3. (a) A woman jumps from a wall and is stopped while landing in sand. (b) The force of the sand on the woman is opposite her displacement as she sinks in the sand. Thus, the work done on her by the sand is negative.

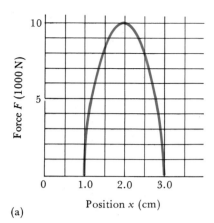

(a)

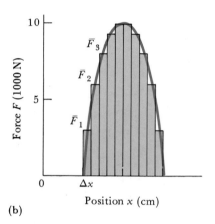

(b)

FIG. 8.4. (a) The varying force of a golf club on a golf ball is plotted for various positions during contact. (b) By dividing the force-versus-displacement graph into small displacements Δx, we can determine the work done by the club on the ball during each small displacement ($\Delta W = F\,\Delta x$).

$$\Delta W_1 = F_1\,\Delta x,$$

where F_1 is the average force on the ball as it moves the first short distance Δx. Similarly, the work done during the second short displacement is $\Delta W_2 = F_2\,\Delta x$, and so forth. The total work done as the ball moves from x_0 to x is the sum of the work done during each of the short displacements between those positions:

$$W = \Delta W_1 + \Delta W_2 + \cdots = F_1\Delta x + F_2\Delta x + \cdots.*$$

The $F\Delta x$ work done during each short displacement is the area of the narrow rectangle above the displacement since area equals height (F) times width (Δx). Thus, the right side of the preceding equation is just the sum of the areas of all the rectangles between x_0 and x. If we make the rectangles very narrow, the area of the rectangles between x_0 and x equals the area under the F-versus-x curve between those positions.

> The work done by a variable force acting on an object as it moves from an initial position x_0 to a final position x is the area under the force-versus-displacement curve between those positions.

8.2 Energy

Whenever work is done on an object, it gains energy. If 100 J of work is done, the object gains 100 J of energy. The energy may be stored by the object in a variety of forms:

Kinetic energy—the energy an object has because it is moving. For example, a moving hammerhead has kinetic energy while one at rest has none.

Gravitational potential energy—the energy an object has because of its vertical separation from the earth. For instance, a pile driver high above a post has more energy to drive a post into the ground than one at a lower elevation.

Elastic potential energy—the energy stored in a stretched or compressed elastic material such as a spring. For example, the spring on the handle of a pinball machine has more energy when compressed than when in a relaxed position.

Internal energy—the atomic and molecular energy of matter consisting of (1) the kinetic energy of the atoms and molecules due to their random motion (called **thermal energy**) and (2) the energy atoms and molecules have as a result of their bonds and interactions with each other.

These forms of energy are introduced in this chapter. Other forms of energy, including electrical energy and nuclear energy, are discussed in later chapters. In this chapter we derive equations that can be used to calculate changes in these forms of energy under a variety of conditions. These equations are, for the most part, very general and will be used extensively throughout the book. The derivations rely on the fact that work done on an object causes its energy to increase (positive work) or decrease (negative work).

---

*In calculus, the summation on the right is called the integral of F with respect to x and is represented in symbols as $W = \int_{x_0}^{x} F\,dx$.

8.3 Kinetic Energy (*KE*)

Translational Kinetic Energy

Kinetic energy is the energy an object has because it is moving. A moving car has more kinetic energy than one sitting at rest. To derive an equation for calculating the change in kinetic energy of an object whose speed changes, we combine the equation for work with Newton's second law and kinematics.

Consider a beam for a space station being constructed in outer space (Fig. 8.5). If an astronaut applies a force to the beam, the beam accelerates, increasing its kinetic energy. The work done by the force *F* that acts on the beam for a distance $x - x_0$ is

$$W = F \, \Delta r \cos \theta = F(x - x_0) \cos 0° = F(x - x_0). \tag{8.3}$$

The beam's acceleration is related by Newton's second law to the force pushing it:

$$F = ma, \tag{4.1}$$

where *m* is the beam's mass. The displacement is related to the acceleration and change of speed by Eq. (2.12) from kinematics:

$$x - x_0 = \frac{v^2 - v_0^2}{2a}, \tag{2.12}$$

where v_0 is the initial speed of the beam and *v* is its final speed. Substituting Eqs. (4.1) and (2.12) into Eq. (8.3), we find that

$$W = F(x - x_0) = (ma)\left(\frac{v^2 - v_0^2}{2a}\right) = \frac{1}{2}mv^2 - \frac{1}{2}mv_0^2.$$

In this situation, all the work done by the force has been converted to kinetic energy of the beam. Thus, the change in its kinetic energy must equal $1/2mv^2 - 1/2mv_0^2$. This is a general expression for calculating the change in an object's kinetic energy when its speed changes.

The **kinetic energy** *KE* of a mass *m* moving with a speed *v* is $KE = \frac{1}{2}mv^2$. The change in kinetic energy when the speed of the mass changes from an initial value v_0 to a final value *v* is

$$\Delta KE = KE - KE_0 = \frac{1}{2}mv^2 - \frac{1}{2}mv_0^2. \tag{8.4}$$

EXAMPLE 8.3 A 0.10-kg stone is thrown from the edge of an ocean cliff with an initial speed of 20 m/s (Fig. 8.6). When it strikes the water below, it is traveling at 45 m/s. What is the change in kinetic energy of the stone?

SOLUTION We are given that $m = 0.10$ kg, $v_0 = 20$ m/s, and $v = 45$ m/s. Thus,

$$\Delta KE = \frac{1}{2}mv^2 - \frac{1}{2}mv_0^2 = \frac{1}{2}(0.10 \text{ kg})(45 \text{ m/s})^2 - \frac{1}{2}(0.10 \text{ kg})(20 \text{ m/s})^2$$

$$= 100 \text{ kg}\frac{m^2}{s^2} - 20 \text{ kg}\frac{m^2}{s^2} = 80 \text{ kg}\frac{m^2}{s^2} = 80\left(\text{kg}\frac{m}{s^2}\right)m$$

$$= 80 \text{ N·m} = \underline{80 \text{ J}}.$$

FIG. 8.5. A force **F** acting on an object floating in outer space causes it to accelerate.

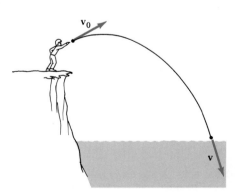

FIG. 8.6. The change in kinetic energy of the rock depends on the initial and the final speeds of the rock but not on its direction.

Notice that the change in kinetic energy depends only on the object's speed and not on its direction of motion.

Rotational Kinetic Energy

The expression just derived for kinetic energy applies primarily to objects moving from one place to another—that is, to translational motion. An analogous equation can be used to calculate the kinetic energy of a rotating object such as a yo-yo or a potter's wheel.

Recall from Chapter 7 that the rotational equivalent of mass is moment of inertia I and that the rotational equivalent of speed is angular velocity ω. If we replace m in the equation for translational kinetic energy ($KE = \frac{1}{2}mv^2$) by I, and v by ω, we have an expression for an object's rotational kinetic energy: $KE_r = \frac{1}{2}I\omega^2$:

If the angular velocity of a rotating object with moment of inertia I changes from an initial value ω_0 to a final value ω, then its change in **rotational kinetic energy** is

$$\Delta KE_r = KE_r - KE_{r_0} = \frac{1}{2}I\omega^2 - \frac{1}{2}I\omega_0^2. \tag{8.5}$$

A flywheel is a device built to store energy by its rotational motion. Usually it consists of a heavy, disc-shaped mass that spins about an axis through its center. By spinning faster, a flywheel stores more energy; when spinning slower, it stores less energy. In the past, flywheels have been used primarily to reduce variations in energy provided by intermittent sources. For example, the engine in a conventional car runs on energy provided by a series of small explosions in its cylinders. A heavy flywheel rotating on the main shaft smooths out the bursts of energy released by the explosions. A potter's wheel serves the same purpose; the flywheel stores and smooths the irregular pedaling energy supplied by the potter.

8.4 Gravitational Potential Energy (PE_g)

Gravitational potential energy is a form of energy an object has because of its vertical separation from the earth. Consider the pile driver shown in Fig. 8.7. When high above the earth, as in Fig. 8.7b, the pile driver has the potential to drive a post into the ground. At a lower elevation, as in Fig. 8.7a, the pile driver has less potential for driving a post into the ground. The only difference in the two situations is the vertical separation of the pile driver from the earth. When separated farther, it has more gravitational potential energy.

We can calculate the extra gravitational potential energy of the pile driver in Fig. 8.7b compared to that in Fig. 8.7a by calculating the work required to lift the pile driver from position y_0 to position y. We lift it in such a way that there are no changes in any other form of energy. To do this, we lift the pile driver with a force **F** that just balances the downward force of gravity (Fig. 8.8). This

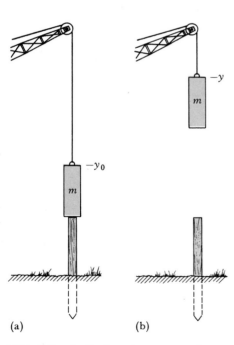

(a) (b)

FIG. 8.7. A pile driver has more gravitational potential energy when higher above the earth as in (b) than when closer to the earth as in (a).

will ensure that the pile driver's kinetic energy does not increase.* The work done by the cable while lifting the pile driver is

$$W = F(y - y_0) \cos 0°.$$

Since the average force of the cable on the pile driver is equal in magnitude to its weight—that is, $F = w = mg$, the work done by F is

$$W = F(y - y_0) \cos 0° = mg(y - y_0).$$

Because all the work has been used to increase the pile driver's gravitational potential energy, we find that $W = \Delta PE_g = mg(y - y_0)$. This equation can be used to calculate the change in gravitational potential energy of any mass whose vertical position relative to the earth changes from an initial position y_0 to a final position y.

If the vertical separation of a mass m from the earth is changed from an initial position y_0 to a final position y, then there is a change in **gravitational potential energy** ΔPE_g given by

$$\Delta PE_g = PE_g - PE_{g_0} = mgy - mgy_0 = wy - wy_0 \tag{8.6}$$

Three points about Eq. (8.6) need emphasizing: (1) Gravitational potential energy depends only on the vertical separation (the height) of a mass from the earth, not on any horizontal change in position. If a mass m is pulled a distance d up a plane inclined at an angle θ, as in Fig. 8.9, the change in the block's gravitational potential energy is $mg(d \sin \theta)$, where $d \sin \theta$ is the additional vertical separation of the mass and the earth. The change in gravitational potential energy is not mgd. (2) Changes in gravitational potential energy depend only on the initial and final heights of the mass, not on the path taken to change heights. In Fig. 8.10, the net change in gravitational potential energy is $mg(y - y_0)$. The change does not depend on the fact that the roller-coaster car went up and down several times between its initial and final positions. (3) When using Eq. (8.6), the origin of the y axis may be placed anywhere along the axis. The change in gravitational energy depends on the *change* in vertical position ($y - y_0$), which is independent of the origin.

EXAMPLE 8.4 A 400-kg roller-coaster car starts 48 m above the ground and is 3 m above the ground at the end of the ride (Fig. 8.10). Calculate the change in gravitational potential energy of the car.

SOLUTION For this problem, $y_0 = 48$ m and $y = 3$ m. Hence, the change in gravitational potential energy is

$$\begin{aligned}
\Delta PE_g &= mg(y - y_0) \\
&= (400 \text{ kg})(9.8 \text{ m/s}^2)[(3 \text{ m}) - (48 \text{ m})] = (400 \text{ kg})(9.8 \text{ m/s}^2)(-45 \text{ m}) \\
&= -180{,}000 \text{ (kg·m/s}^2) \text{ m} = -180{,}000 \text{ N·m} = \underline{-180{,}000 \text{ J}}.
\end{aligned}$$

At the end of the ride the roller-coaster car is closer to the earth, and so it has less

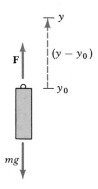

FIG. 8.8. The work done in lifting the pile driver from y_0 to y by the force **F** is $W = F(y - y_0)$.

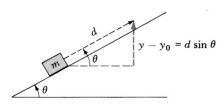

FIG. 8.9. The change in gravitational potential energy when a mass is moved a distance d up the incline is $\Delta PE_g = mgd \sin \theta$, not mgd.

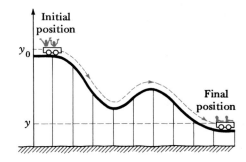

FIG. 8.10. The change in gravitational potential energy depends on the initial and final vertical positions of an object and not on the path followed in moving between those positions.

*To get the pile driver started moving up, we must make F just a little larger than the pile driver's weight. When we get close to position y, we can make F slightly less than its weight so that the pile driver stops. On the average, the magnitude of F just equals the pile driver's weight.

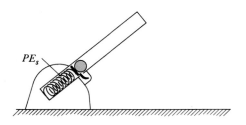

FIG. 8.11. A compressed spring has elastic potential energy.

gravitational potential energy. This energy change is not affected by the fact that the roller-coaster car went up and down several times on its trip. It depends only on its initial and final positions and on its mass. ◼

8.5   Elastic Potential Energy (PE_s)

Gravitational potential energy is caused by the relative position of two masses, the mass of the earth and the mass of some object near its surface. Potential energy can also be stored in an object because of its shape. In Fig. 8.11 the energy in a compressed spring can be used to shoot a ball from a cannon. A relaxed spring does not have this potential. The elastic potential energy of a spring depends on its shape—how much it is compressed or stretched.

To find the energy stored in a compressed spring, we calculate the work required to compress it. In Fig. 8.12 a force **F** slowly pushes on a spring to compress it from its **equilibrium position** at $x = 0$ to some other position x. (The equilibrium position is the place where the end of the spring rests when not compressed or stretched.) The force needed to compress a spring increases as the spring becomes more compressed. In fact, this force is directly proportional to the amount the spring is already compressed, an experimental observation made by the English scientist Robert Hooke three hundred years ago. This variation of force with the distance the spring is already compressed is illustrated in Fig. 8.12. When the end of the spring is at $x = 0$, as in Fig. 8.12a, no force is needed to hold the spring. When the spring is compressed to position x', as in Fig. 8.12b, the force needed to hold it there is $F' = kx'$. If compressed farther to position x, as in Fig. 8.12c, the force needed to hold the spring in place is $F = kx$. The proportionality constant k (called the **force constant**) depends on how stiff the spring is. A stiff spring has a large value of k. A spring that is easily stretched or compressed has a small value of k.

Since the compression force is zero when $x = 0$ and is $+kx$ with a displacement from equilibrium of x, the average force F_{av} needed to compress the spring from position zero to position x is

$$F_{av} = \frac{k0 + kx}{2} = \frac{1}{2}kx.$$

The displacement of the spring is in the same direction as F_{av} and has a magnitude

$$\Delta x = x - 0.$$

The work done in compressing the spring is

$$W = F_{av}\Delta x = \left(\frac{1}{2}kx\right)(x - 0) = \frac{1}{2}kx^2.$$

This work causes the elastic potential energy of the spring to increase. Equation (8.7), which follows, summarizes the calculation of elastic potential energy change for springs or other elastic objects.

The **elastic potential energy** of a spring with force constant k that is stretched or compressed a distance x from its equilibrium position is $PE_s = \frac{1}{2}kx^2$. The change in elastic energy when the spring is stretched or compressed from an initial position x_0 to a final position x is

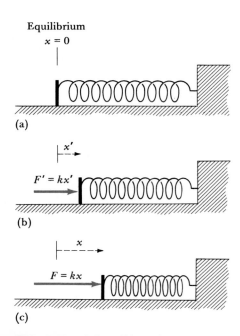

Equilibrium
$x = 0$

(a)

x'

$F' = kx'$

(b)

x

$F = kx$

(c)

FIG. 8.12. A force **F** is used to compress a spring. (a) No force is required to hold the spring at its equilibrium position. (b) and (c) Progressively larger forces are needed to hold the spring or to slowly compress it farther. The magnitude of the force F varies with the compression distance x according to the equation $F = +kx$, where k is the force constant of the spring.

$$\Delta PE_s = PE_s - PE_{s_0} = \frac{1}{2}kx^2 - \frac{1}{2}kx_0^2. \tag{8.7}$$

The equilibrium position of the spring is assumed to be at $x = 0$.

Equation (8.7) applies equally to springs that are compressed and to springs that are stretched from their equilibrium position. The equation can be used to estimate the change in elastic energy of other elastic materials such as bones, metal rods, wire, string, and the membranes of blood vessels. The subscript s is used in the symbol for elastic potential energy because of the similarity in behavior of elastic materials and springs.

EXAMPLE 8.5   The spring in a BB gun has a force constant of 1.8×10^4 N/m. When loaded, the spring is compressed a distance of 1.2 cm. (a) Calculate the change in elastic potential energy of the spring when it is released and returns to its equilibrium position. (b) If all the elastic potential energy is converted to the kinetic energy of a 0.36-g BB fired from the gun, what is the final speed of the BB?

SOLUTION   (a) The initial and final displacements of the spring are $x_0 = 0.012$ m and $x = 0.000$ m (the spring ends at its equilibrium position). The change in elastic potential energy is calculated using Eq. (8.7):

$$\Delta PE_s = \frac{1}{2}kx^2 - \frac{1}{2}kx_0^2$$

$$= \frac{1}{2}(1.8 \times 10^4 \text{ N/m})(0.000 \text{ m})^2 - \frac{1}{2}(1.8 \times 10^4 \text{ N/m})(0.012 \text{ m})^2$$

$$= 0 - 1.3 \text{ N} \cdot \text{m} = \underline{-1.3 \text{ J}}.$$

The negative sign implies that the spring lost elastic energy as it went from its initial compressed position to its final relaxed position at $x = 0$.

(b) If the BB gains in kinetic energy all of the elastic energy lost by the spring, then the BB's kinetic energy as it leaves the gun is $+1.3$ J. Thus, according to Eq. (8.4), we have

$$\Delta KE = 1.3 \text{ J} = \frac{1}{2}mv^2 - \frac{1}{2}mv_0^2.$$

But since the ball's original speed v_0 is zero, we find that

$$v^2 = \frac{2(\Delta KE)}{m}$$

$$= \frac{2(1.3 \text{ J})}{0.36 \times 10^{-3} \text{ kg}} = 7.2 \times 10^3 \text{ m}^2/\text{s}^2,$$

or

$$v = \underline{85 \text{ m/s}}. \qquad \blacksquare$$

Example 8.5 is representative of the type of calculations we will use in the rest of the chapter to determine some unknown property of a system involved in energy change. In this example we were able to determine the increase in kinetic energy of the BB and its final speed by calculating the decrease in elastic potential energy of the spring. This procedure is based on a very important principle called the conservation-of-energy principle.

8.6 Conservation-of-Energy Principle

One of the most important and useful aspects of energy is the fact that energy is a conserved quantity. Supposedly, the amount of energy in the universe is the same now as it was millions of years ago. How can we make such a universal statement when we cannot even measure the amount of coal and oil in the earth's crust? For over one hundred years scientists have performed a variety of controlled experiments monitoring the amount of energy involved in various processes. In these experiments the form of the energy changes, but the total amount always remains constant. Since energy is conserved under a great variety of conditions on the earth, we assume that energy is conserved in the whole universe.

Occasionally experiments have challenged this faith in the energy-conservation principle. In the early 1930s, for example, some experiments were performed involving nuclear reactions. During these nuclear processes, energy seemed to disappear. A physicist, Wolfgang Pauli, proposed that a new, unseen particle carried away the missing energy. This particle, called a *neutrino,* was discovered twenty-three years later, in 1956, and it accounted exactly for the missing energy. Pauli's faith in the energy-conservation principle had been justified.

The basic philosophy underlying conservation principles was outlined in Section 4.3 and is reviewed here with special emphasis on energy conservation. Consider Fig. 8.13, which represents the whole universe. We choose one small part, called the **system,** in which we have a special interest. The system can be a single atom, a solution of chemicals, a human body, a desert, a lake, a supernova, or whatever. Everything outside the system is called the **environment.** The **conservation-of-energy principle** says that the total energy in the system, E_{system}, plus that in the environment, $E_{\text{environment}}$, must be a constant:

$$E_{\text{system}} + E_{\text{environment}} = \text{Constant.}$$

Energy can, however, be transferred between the system and the environment and can also be converted between different forms while in the system.

Energy can be transferred between the system and the environment in several different ways: as heat, Q; by the process called work, W, defined earlier in the chapter; as energy carried by mass that crosses the boundary of the system; and as electromagnetic energy, EM, such as sunlight, that crosses the boundary of the system. These energy-transfer processes are represented in Fig. 8.13 by arrows. If energy is transferred from the environment to the system as heat Q, by doing work W, or in any other way, the change in the energy of the system ΔE_{system} equals its final energy E minus its initial energy E_0:

$$\begin{aligned} Q + W &= E - E_0 \\ &= \Delta E_{\text{system}}. \end{aligned} \tag{8.8}$$

The environment loses just as much energy as the system gains. The total energy change of the universe is zero.

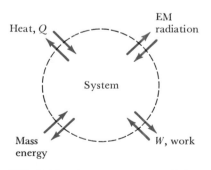

FIG. 8.13. A system, any part of the universe, and its environment have a total energy that is constant. However, energy can be transferred between the system and the environment by work, by heat transfer, as electromagnetic (EM) radiation, and by mass energy transfer.

There can also be changes in the type of energy already in a system. These changes appear on the right side of Eq. (8.8), which is the sum of the changes of all the different types of energy already in a system—such as its kinetic energy, gravitational potential energy, elastic potential energy, and internal energy. Equation (8.8) then becomes

$$Q + W = (KE - KE_0) + (PE_g - PE_{g_0}) + (PE_s - PE_{s_0}) + (E_{int} - E_{int_0}) \tag{8.9}$$

$$\underbrace{}_{\substack{\text{Energy} \\ \text{added to} \\ \text{or removed} \\ \text{from system}}} = \underbrace{\Delta KE \quad + \quad \Delta PE_g \quad + \quad \Delta PE_s \quad + \quad \Delta E_{int}.}_{\substack{E - E_0, \text{ the algebraic sum of the} \\ \text{changes of different forms of} \\ \text{energy in system}}}$$

Equations (8.8) and (8.9) may be easier to understand conceptually if all of the initial energy terms are moved to the left-hand side of the equations. We have then

$$E_0 + (Q + W) = E, \tag{8.10}$$

$$\underbrace{(KE_0 + PE_{g_0} + PE_{s_0} + E_{int_0})}_{\substack{\text{Initial energy} \\ \text{in system}}} + \underbrace{(Q + W)}_{\substack{\text{+ Energy added} \\ \text{to or removed} \\ \text{from system}}} = \underbrace{(KE + PE_g + PE_s + E_{int}).}_{\substack{\text{Final energy} \ \textbf{(8.11)} \\ \text{in system}}}$$

The sum of the initial energy of all types in a system plus the energy added to or removed from the system during some time period equals the final energy of the system.

Equation (8.11) is similar to one an accountant might use to monitor your wealth. Your initial wealth may consist of art works, stocks and bonds, money in savings accounts, bicycles, cars, and so forth [the E_0 part of Eqs. (8.10) and (8.11)]. If you gain wealth because of pay for work done or lose wealth because of money spent, such as on heat for your apartment, then your wealth increases or decreases [the $Q + W$ term of Eqs. (8.10) and (8.11)]. Thus, after some period of time, the final wealth on the right-hand side is simply the sum of what you had originally and what you added or removed.

Often when dealing with energy, we can choose a system so that no energy is added to or removed from the system ($Q = 0$ and $W = 0$). The system, as far as energy is concerned, is said to be **isolated.** Its initial energy equals its final energy:

$$\text{Isolated system:} \quad E_0 = E. \tag{8.12}$$

However, even though the system is isolated, the form of the energy often changes. For example, a skier initially at the top of a hill has considerable gravitational potential energy. But as the skier descends, the gravitational energy decreases and the skier's kinetic energy increases (he or she is moving faster). Also, there is an increase in internal energy as the friction forces between the skis and the snow cause the snow to warm or melt slightly. The net energy change is zero; the skier has converted gravitational potential energy into an equal amount of kinetic energy plus internal energy. It is as though you took money from your savings account to purchase a bicycle. Your net wealth is the same—it's just in the form of a bicycle rather than of money.

The work-energy problems in this chapter can be calculated using either Eq. (8.9) or Eq. (8.11). They are the same and produce the same results. Techniques for using these equations are illustrated in the next section. In this chapter, we use a restricted form of Eq. (8.11) in which the system's energy is changed only by the process of work W. All other mechanisms for transferring energy across the system's boundary, as illustrated in Fig. 8.13, are zero except for work. With $Q = 0$, Eq. (8.11) becomes a modified form of what is sometimes called the **work-energy equation:**

$$\underbrace{(KE_0 + PE_{g_0} + PE_{s_0} + E_{\text{int}})}_{\substack{\text{Initial energy } E_0 \text{ in} \\ \text{system}}} + \underbrace{W}_{\substack{\text{Energy en-} \\ \text{tering or} \\ \text{leaving} \\ \text{system}}} = \underbrace{(KE + PE_g + PE_s + E_{\text{int}})}_{\substack{\text{Final energy} \\ E \text{ in system}}}. \tag{8.13}$$

8.7 Work-Energy Calculations

Earlier we used the definition of work to derive mathematical equations for calculating changes in different types of energy. Now we use these equations and the work-energy equation to solve for some unknown quantity involved in an energy change. A procedure for solving such problems is outlined shortly. You should be warned that physicists use a variety of techniques to solve energy problems, and your instructor may use a different technique than is outlined here. Each technique is correct and has advantages and disadvantages. In fact, in the first edition of this book, we solved work-energy problems using a method based on Eq. (8.9). That method has the advantage that the basic equation, Eq. (8.9), is similar to the energy-conservation equation used in thermodynamics in Chapters 10 and 11—called the first law of thermodynamics. However, this time we will try a method based on Eq. (8.13). Some people feel that this way of expressing energy conservation is easier to understand.

Solving Problems with the Work-Energy Equation

Each step of the energy problem-solving technique based on Eq. (8.13) is described in general and illustrated for the following problem: A ball with initial speed zero is dropped from a roof. Determine the ball's speed as it reaches the ground 15 m below.

1. Choose a system and draw a picture or diagram indicating its initial and final situations. The diagram often needs a coordinate axis or several axes. A vertical axis, if needed, must be oriented with the positive direction pointing up because Eq. (8.6), used to calculate gravitational potential energy, requires this choice. Include in your diagram the known information and identify in symbols the quantity you wish to determine.

   The system is anything you wish; everything outside the system is the environment. Sometimes it is convenient to choose a large system so that forces from outside the system that act on objects in the system are small and can be ignored.

For the problem illustrated here, we choose a system consisting of the ball and the earth. With this choice, no forces from outside the system affect the ball's fall.

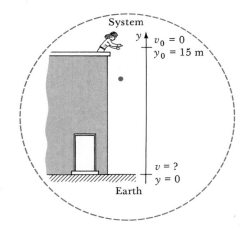

2. Identify the external forces acting on objects in the system. External forces are forces that originate from outside the system. Then calculate the work done on the system by these external forces as the system changes from its initial to its final state. Remember that work is a way to transfer energy between objects that exert forces on each other. When calculating work, we are interested only in work-type energy transfer across the boundary of the system. We do not need to calculate the work done by the force of one object in the system or another object in the system. We calculate only the work done on the system by forces originating in the environment. The system gains energy if the work is positive and loses energy if the work is negative.

For the problem illustrated here, no forces from outside the system affect the ball's fall. Thus, $W = 0$.

3. Write expressions for the initial and final energies of the system. You will need to include in these expressions only the types of energy that are changing.

In our example, both the gravitational potential energy and the kinetic energy of the falling ball change, so they are included in the energy expressions. However, since there are no springs involved in this problem, no change occurs in elastic energy, and we do not need to include it in the energy expressions.

$$\text{Initial energy} = E_0 = \frac{1}{2}mv_0^2 + mgy_0$$

$$= 0 + mgy_0,$$

$$\text{since } v_0 = 0$$

$$\text{Final energy} = E = \frac{1}{2}mv^2 + mgy$$

$$= \frac{1}{2}mv^2 + 0,$$

$$\text{since } y = 0$$

4. Substitute the expressions for work, initial, and final energies, determined in Steps 2 and 3, into Eq. (8.13). By rearranging the equation and substituting for the known information, you should be able to solve for the unknown.

$$E_0 + W = E, \quad \text{or}$$

$$(0 + mgy_0) + 0 = \left(\frac{1}{2}mv^2 + 0\right)$$

After rearranging, we find that

$$v^2 = 2gy_0, \quad \text{or}$$

$$v = \sqrt{2gy_0} = \sqrt{2(9.8 \text{ m/s}^2)15 \text{ m}} = \underline{17 \text{ m/s}}.$$

The form of the water's energy changes as it goes over the waterfall. The gravitational potential energy that the water has at the top converts to kinetic energy as the water falls. After hitting the bottom, the kinetic energy converts to thermal energy—the water is slightly warmer. (Susan Van Etten, Photographer)

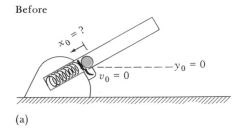

(a)

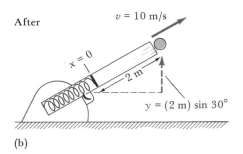

(b)

FIG. 8.14. (a) Before the spring gun has fired, all the energy is elastic potential energy of the compressed spring. (b) The kinetic and gravitational energies increase after firing.

Equations used to calculate the different types of energy and work are summarized as follows:

Work and Types of Energy	Initial	Final
Work (W)	$F \, \Delta r \cos \theta$	
Translational kinetic energy (KE)	$\frac{1}{2} mv_0^2$	$\frac{1}{2} mv^2$
Rotational kinetic energy (KE_r)	$\frac{1}{2} I\omega_0^2$	$\frac{1}{2} I\omega^2$
Gravitational potential energy (PE_g)	mgy_0	mgy
Elastic potential energy (PE_s)	$\frac{1}{2} kx_0^2$	$\frac{1}{2} kx^2$

These expressions for energy and work are used in Eq. (8.13):

$$(KE_0 + PE_{g_0} + PE_{s_0} + \cdots) + W = (KE + PE_g + PE_s + \cdots). \qquad \textbf{(8.13)}$$

EXAMPLE 8.6 A spring gun, shown in Fig. 8.14, is loaded by compressing a spring that has a force constant $k = 1.2 \times 10^4$ N/m. A 1.0-kg ball rests against the spring. When the spring is released, the ball is shot out of the end of a barrel 2.0 m long. If the ball leaves the end of the barrel with a speed of 10 m/s, how far was the spring compressed before the gun was fired? Ignore friction.

SOLUTION If we choose the ball, gun, and earth as the system, then there are no objects outside the system that exert forces on objects in the system; all energy changes relating to this problem occur in the system. Since no external forces affect the problem, the energy transferred to or from the system by doing work is zero:

$$W = 0.$$

For the situation pictured in Fig. 8.14, the ball's initial kinetic and gravitational potential energies are zero ($v_0 = 0$ and $y_0 = 0$). However, the spring is initially compressed an unknown distance x_0 and, consequently, the spring's initial elastic energy is $(\frac{1}{2})kx_0^2$. Thus, the initial energy is

$$E_0 = 0 + 0 + \frac{1}{2} kx_0^2.$$

In the final situation, as the ball leaves the barrel, its kinetic energy is $(\frac{1}{2})mv^2$ and its gravitational energy has increased to mgy, where its final vertical elevation $y = (2.0 \text{ m}) \sin 30°$. However, the spring has returned to its equilibrium position ($x = 0$), so the final elastic energy is zero. Thus, the final energy is

$$E = \frac{1}{2} mv^2 + mgy + 0.$$

Substituting for E_0, W, and E in Eq. (8.13), we find that

$$\frac{1}{2} kx_0^2 = \frac{1}{2} mv^2 + mgy.$$

The initial elastic energy has been converted to kinetic energy and to gravitational potential energy. To solve for the spring's initial compression x_0, we rearrange the preceding equation in symbols before substituting the values of known quantities:

$$x_0 = \left(\frac{mv^2 + 2\,mgy}{k}\right)^{1/2}$$

$$= \left[\frac{(1.0\ \text{kg})(10\ \text{m/s})^2 + 2(1.0\ \text{kg})(9.8\ \text{m/s}^2)(2\ \text{m}) \sin 30°}{1.2 \times 10^4\ \text{N/m}}\right]^{1/2}$$

$$= \underline{0.10\ \text{m}.}$$ ∎

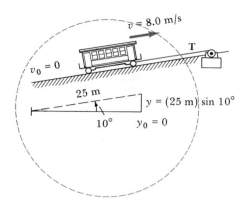

FIG. 8.15. The work done by the cable causes the car's kinetic and gravitational potential energies to increase.

EXAMPLE 8.7 A cable attached to a motor pulls a 1.2×10^4-kg cable car up a hill in San Francisco (Fig. 8.15). If the car starts at rest and reaches a speed of 8.0 m/s in 25 m, calculate the average tension in the cable. The street is inclined at an angle of 10° with the horizontal.

SOLUTION   In this example, we choose as the system the cable car, the earth, and part of the cable pulling the car. The motor is excluded from the system because the energy transformations occurring in the motor are complicated and are not needed to solve the problem. The motor's effect on the car can be determined easily by calculating the work done by the cable that connects the car to the motor.

We see from Fig. 8.15 that the portion of the cable outside the system does work on the system as it pulls the car up the hill. The work done by the cable while pulling the car 25 m is

$$W = T\,\Delta r \cos 0° = T(25\ \text{m})(1),$$

where T is the unknown tension in the cable.

The cable car's initial kinetic and gravitational energies are zero because its initial speed is zero ($v_0 = 0$) and its initial vertical position is zero ($y_0 = 0$). Thus,

$$E_0 = 0 + 0.$$

The car's final energy consists of kinetic energy and gravitational potential energy:

$$E = \frac{1}{2}mv^2 + mgy.$$

Substituting for E_0, W, and E in Eq. (8.13), we find that

$$0 + T(25\ \text{m}) = \frac{1}{2}mv^2 + mgy.$$

The work done by the cable has increased the car's kinetic and gravitational energies. We can now solve for the tension in the cable:

$$T = \frac{\frac{1}{2}mv^2 + mgy}{25\ \text{m}}$$

$$= \frac{\frac{1}{2}(1.2 \times 10^4\ \text{kg})(8.0\ \text{m/s})^2 + (1.2 \times 10^4\ \text{kg})(9.8\ \text{m/s}^2)(25\ \text{m}) \sin 10°}{25\ \text{m}}$$

$$= 3.6 \times 10^4\ \text{kg·m/s}^2 = \underline{3.6 \times 10^4\ \text{N}.}$$ ∎

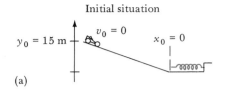

Initial situation

$v_0 = 0$

$y_0 = 15$ m

$x_0 = 0$

(a)

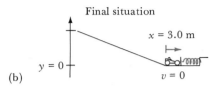

Final situation

$x = 3.0$ m

$y = 0$

$v = 0$

(b)

FIG. 8.16. The initial gravitational potential energy of a soapbox derby race car is converted into elastic energy as it compresses a spring at the end of its run.

You might try working this last example using Newton's second law and kinematics. Your answer should be the same. When using Newton's second law, however, you must use vector quantities, and the effort required is somewhat greater. The energy approach to problem solving is usually much easier. The next example would be very difficult to solve using Newton's second law because it involves the variable force of a spring. The energy approach is fairly simple.

EXAMPLE 8.8 A soapbox derby race car starts at rest at the top of a track that has a vertical drop of 15 m. The car is to be stopped at the end of the track by colliding with and compressing a spring. To avoid injury to the driver, the spring must be compressed 3.0 m while stopping the car. What should be the force constant of the spring? The car and driver together have a mass of 113 kg. Ignore friction forces.

SOLUTION We choose the car, the spring, and the earth as the system (Fig. 8.16). There are no external forces acting on the system, and hence $W = 0$.

The system's initial energy consists of the gravitational energy of the car:

$$E_0 = 0 + mgy_0 + 0.$$

Note that the car's initial speed and kinetic energy are zero. Also, the spring's initial compression distance is zero ($x_0 = 0$); hence its initial elastic energy is zero.

At the end of the ride, the car has again stopped ($v = 0$) at a vertical position zero ($y = 0$). But now the spring is compressed 3.0 m ($x = 3.0$ m). Thus, the system's final energy is

$$E = 0 + 0 + \frac{1}{2}kx^2.$$

Substituting for E_0, W, and E in Eq. (8.13), we find that

$$(0 + mgy_0 + 0) + 0 = 0 + 0 + \frac{1}{2}kx^2.$$

The initial gravitational energy was converted to elastic energy as the spring was compressed. Solving for the force constant k, we find that

$$k = \frac{2mgy_0}{x^2} = \frac{2(113 \text{ kg})(9.8 \text{ m/s}^2)(15 \text{ m})}{(3.0 \text{ m})^2}$$

$$= 3700 \text{ (kg·m/s}^2)/\text{m} = \underline{3700 \text{ N/m}}. \quad ■$$

EXAMPLE 8.9 Several reports have been made of parachutists whose parachutes did not open and who survived their impact with the earth by landing in a snowbank. One such event occurred in 1955 during a large airborne parachute jump exercise in Alaska. A paratrooper dropped 370 m from a C-119 airplane and landed in a snowbank. "The impact looked like a mortar round exploding in the snow. When the airmen reached the spot, they found a young paratrooper flat on his back at the bottom of a $3\frac{1}{2}$ foot crater in the snow. . . . He could talk and did not appear injured." * He did suffer several minor injuries, including a fractured clavicle and bruises, but soon rejoined his unit. Calculate the average

*R. G. Snyder, "Terminal Velocity Impacts into Snow," *Military Medicine* **131**, 1290–1298 (1966).

force of the snow in stopping his fall. We will assume that the paratrooper's mass was 90 kg and that he was falling at a speed of 54 m/s, or 120 mi/h, as he entered the snow.

SOLUTION We choose as the system the falling paratrooper and the earth's mass. We exclude the snow from the system. Its upward force **F** on the paratrooper is then an external force that does work on him as he sinks into the snow 1.07 m (see Fig. 8.17):

$$W = Fd\,(\cos 180°) = F(1.07\text{ m})(-1) = -(1.07\text{ m})F.$$

The other forms of energy that change are the paratrooper's kinetic and gravitational energies. With the origin of the y axis at the paratrooper's final position, his initial position as he reaches the snow is $y_0 = 1.07$ m and his initial speed at that point is $v_0 = 54$ m/s. His initial energy is

$$E_0 = \frac{1}{2}mv_0^2 + mgy_0.$$

His final position is $y = 0$, and his final speed is $v = 0$. Thus, his final energy is zero:

$$E = 0 + 0.$$

Substituting for E_0, the work W done by the snow, and E in Eq. (8.13), we find that:

$$\left(\frac{1}{2}mv_0^2 + mgy_0\right) - (1.07\text{ m})F = 0 + 0,$$

or

$$F = \frac{\frac{1}{2}mv_0^2 + mgy_0}{1.07\text{ m}} = \frac{\frac{1}{2}(90\text{ kg})(54\text{ m/s})^2 + (90\text{ kg})(9.8\text{ m/s}^2)(1.07\text{ m})}{1.07\text{ m}}$$

$$= \underline{1.2 \times 10^5\text{ N}},$$

an average force of about 12 tons! In this example, the paratrooper's initial kinetic and gravitational energies were reduced to zero by the negative work done by the opposing force of the snow. ■

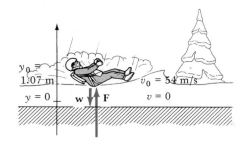

FIG. 8.17. The forces acting on a paratrooper as he falls into a snowbank after his parachute fails to open.

8.8 Friction Forces and Internal Energy

Friction plays an important role in most dynamic processes. In this section we introduce two ways to include the effects of friction in energy calculations.

The first method is illustrated in Fig. 8.18a. A car moving with initial speed v_0 skids to a halt after its brakes are jammed to the floor. In the situation shown in Fig. 8.18a, we have chosen the car alone as the system. Since the road is not in the system, the friction force $\mathbf{F}_k$ of the road on the car is an external force, and its effect is included in the work-energy equation by calculating the work done by friction. If the car travels a distance Δs while being stopped, the work done by friction is

$$W_{\text{friction}} = F_k\,\Delta s \cos 180°$$
$$= -F_k\,\Delta s.$$

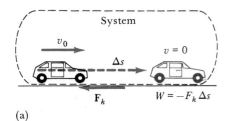

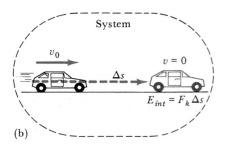

(a)

(b)

FIG. 8.18. The effect of friction can be included in the work-energy equation (a) as the work done by friction or (b) as the change in the system's internal energy.

Note that the friction force opposes the motion and does negative work on the car. For this whole process, the car's initial energy is $E_0 = \frac{1}{2}mv_0^2$, and its final energy after stopping is zero, that is, $E = 0$. Substituting E_0, W, and E in Eq. (8.13), we find that

$$\frac{1}{2}mv_0^2 + (-F_k\,\Delta s) = 0.$$

The car lost its initial kinetic energy because of the work done by the opposing friction force.

The effect of friction can be included in the work-energy equation by calculating the **work done by the friction force**:

$$W_{\text{friction}} = -F_k\,\Delta s, \tag{8.14}$$

where F_k is the magnitude of the average friction force acting on an object as it moves a distance Δs on the surface causing the friction. The surface is not considered part of the system.

We can use a second method to account for the effect of friction. In this method, the two surfaces in contact with each other are both included in the system. As one surface rubs across the other, the internal energy of the objects in contact increases. The internal energy consists of (1) thermal energy, that is, the random kinetic energy of the atoms and molecules of the objects in the system (they become warmer because of friction) and (2) the energy due to the bonds between atoms and molecules of these objects (friction may cause some bonds to be broken). Internal energy is produced in many different ways including the friction force between objects that rub against each other (we occasionally warm our hands this way) and the drag force of the earth's atmosphere on a meteorite, which gets red hot as it falls through the air.

This second method for including the effect of friction in the work-energy equation is illustrated in Fig. 8.18b. In this case the car and road are both considered as part of the system. The friction force is no longer an external force because it does not originate outside the system. Thus, for the situation pictured in Fig. 8.18b, no work is done on the system. The car's initial energy is still $E_0 = \frac{1}{2}mv_0^2$ and its final *kinetic* energy is still zero. But the system's final energy is not zero because it has gained some internal energy due to the friction between the car and the road. In this case the gain in internal energy due to friction is $+F_k\,\Delta s$. Substituting $E_0 = \frac{1}{2}mv_0^2$, $W = 0$, and $E = E_{\text{int}} = +F_k\,\Delta s$ into Eq. (8.13), we have

$$KE_0 + 0 = E_{\text{int}},$$

or

$$\frac{1}{2}mv_0^2 + 0 = F_k\,\Delta s.$$

The car's initial kinetic energy is converted to internal energy of the system. Thus, if the sliding object and the surface on which it slides are both part of the system, then the effect of friction is included in the work-energy equation by calculating the system's change in internal energy, as summarized next.

The **internal energy** of an object consists of (1) the random kinetic energy of the atoms and molecules of which the object is made (called its **thermal energy**) and (2) the chemical energy stored in the bonds between the atoms and molecules of the object. The internal energy added to a system because of friction when an object moves a distance Δs across another surface in the system is given by

$$\Delta E_{\text{int}} = E_{\text{int}} - E_{\text{int}_0} = F_k \, \Delta s, \qquad \textbf{(8.15)}$$

where F_k is the magnitude of the kinetic friction force between the object and the surface on which it slides. The internal energy is shared by the moving object and the surface.

We see by comparing Eqs. (8.14) and (8.15) that the effect of friction can be included in the work-energy equation by calculating either the work done by friction or the increase in the system's internal energy caused by friction. Either approach produces the same result because putting $W_f = -F_k \, \Delta s$ on the left-hand (work) side of Eq. (8.13) has the same effect as putting $\Delta E_{\text{int}} = F_k \, \Delta s$ on the right-hand side.

EXAMPLE 8.10   A 1100-kg car traveling at 18 m/s skids to a stop on a level road in a distance of 24 m. Calculate the average kinetic friction force acting on the car as it stops.

SOLUTION   If we choose the car, the road, and the earth as the system, no external forces act on the system. Thus

$$W = 0.$$

The only energy changes that occur are a decrease in the car's kinetic energy and a corresponding increase in internal energy of the tires and the road. The system's initial energy consists of the car's initial kinetic energy:

$$E_0 = \frac{1}{2} m v_0^2.$$

The system's final energy has an extra amount of internal energy caused by the friction of the road on the car:

$$E = F_k \, \Delta s.$$

Substituting E_0, W, and E in Eq. (8.13), we find that

$$\left(\frac{1}{2} m v_0^2 \right) + 0 = F_k \, \Delta s,$$

or

$$F_k = \frac{\frac{1}{2} m v_0^2}{\Delta s} = \frac{\frac{1}{2}(1100 \text{ kg})(18 \text{ m/s})^2}{24 \text{ m}}$$

$$= 7.4 \times 10^3 \text{ kg} \cdot \text{m/s}^2 = \underline{7.4 \times 10^3 \text{ N}}. \qquad \blacksquare$$

EXAMPLE 8.11   You wish to build a ski lift for the small hill next to your house. The cable is to pull a 100-kg skier up a hill inclined 20° to the horizontal. The skier's speed is zero at the bottom and is 6.0 m/s at the top. The hill is 50 m

FIG. 8.19. The work done by the cable causes the skier's kinetic and gravitational energies to increase. The system's internal energy also increases because of friction.

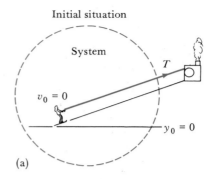

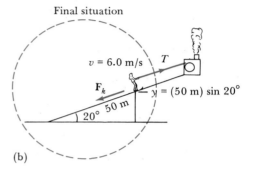

FIG. 8.19. The work done by the cable causes the skier's kinetic and gravitational energies to increase. The system's internal energy also increases because of friction.

long, and a 150-N friction force opposes the skier's motion. Calculate the average tension needed in the cable pulling the skier.

SOLUTION We choose the skier, the hill, and the earth as the system (Fig. 8.19). However, the motor and the cable are outside the system and do work on the skier. The work done by the cable is

$$W = T \Delta r \cos 0° = T(50 \text{ m})(1),$$

where T is the unknown cable tension.

Three forms of energy change occur as the skier is pulled up the hill: The skier's kinetic energy increases, the skier's gravitational potential energy increases, and the internal energy of the skis and snow increases because of the friction. The initial energy of the system is zero ($v_0 = 0$ and $y_0 = 0$):

$$E_0 = 0.$$

The final energy is

$$E = \frac{1}{2}mv^2 + mgy + F_k \Delta s,$$

where $y = (50 \text{ m}) \sin 20°$ and $\Delta s = 50$ m, the actual distance traveled along the hill. Substituting for E_0, W, and E in Eq. (8.13), we have

$$0 + T(50 \text{ m}) = \frac{1}{2}mv^2 + mgy + F_k \Delta s.$$

Notice that the work done by the cable causes the increases in the three forms of energy shown on the right. Rearranging to solve for T, we find that

$$T = \frac{\frac{1}{2}mv^2 + mgy + F_k\,\Delta s}{50\ m}$$

$$= \frac{\frac{1}{2}(100\ \text{kg})(6.0\ \text{m/s})^2 + (100\ \text{kg})(9.8\ \text{m/s}^2)(50\ \text{m})\sin 20° + (150\ \text{N})(50\ \text{m})}{50\ \text{m}}$$

$$= \underline{520\ \text{N}}.$$ ∎

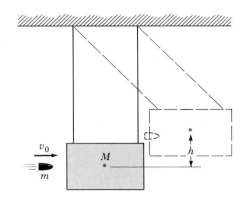

FIG. 8.20.   A ballistic pendulum.

8.9   Inelastic and Elastic Collisions

Energy and momentum are both conserved when two objects collide. However, the form of the energy almost always changes, some of the original kinetic energy being converted during the collision to other forms such as thermal energy. This happens when a basketball bounces on the floor. Each bounce produces a little thermal energy as the ball's leather or rubber skin flexes and bends. The ball leaves the floor with less speed and less kinetic energy than it had when it hit the floor. This event is called an **inelastic collision**—one in which kinetic energy is converted to other forms of energy. A car collision is inelastic because a considerable portion of the car's initial kinetic energy is lost as its shape is changed during the collision. Even though kinetic energy is lost during inelastic collisions, momentum is conserved.

EXAMPLE 8.12   A ballistic pendulum is used to measure the speed of a fast-moving projectile. The projectile hits and embeds in a wooden block hanging from strings, as shown in Fig. 8.20. Derive an expression for the initial speed v_0 of the projectile in terms of the height h that the block and projectile swing up into the air after the collision and in terms of their masses.

SOLUTION   Consider the whole process shown in Fig. 8.20 as consisting of two separate steps. First, the projectile, moving at speed v_0, hits and embeds itself in the hanging block. The collision is inelastic because some of the projectile's kinetic energy is used to split the wood fibers in the block. However, even during this collision, momentum is conserved:

Initial momentum = Final momentum,

or

$$mv_0 = (m + M)v, \qquad (8.16)$$

where v_0 is the initial speed of the projectile of mass m, and v is the speed of the block of mass M and projectile moving together just after the collision has occurred.

In the second part of the process, the block and projectile, moving together with an initial speed v just after the collision, swing in an arc and end at rest a vertical distance h above where they started. In this part of the process, we can account for all the energy changes that occur, and so we use the work-energy equation. The initial energy, just after the collision, consists of the kinetic energy of the block and projectile:

$$E_0 = \frac{1}{2}(M + m)v^2.$$

The final energy at the top of the arc consists of the gravitational energy of the block and projectile:

$$E = (M + m)gh.$$

No work is done on the system during the upward swing. Substituting in Eq. (8.13), we have

$$\frac{1}{2}(M + m)v^2 = (M + m)gh.$$

Rearranging the preceding equation and substituting from Eq. (8.16) for v, we find that

$$v_0 = \sqrt{2gh}\left(\frac{m + M}{m}\right).$$　∎

Often, when atoms and molecules collide with each other, the total kinetic energy of the particles before the collision equals the sum of their kinetic energy afterward. If one atom or molecule gains kinetic energy, the other loses an equal amount. These events are called **elastic collisions**—ones in which both kinetic energy and momentum are conserved. Most collisions involving macroscopic objects are inelastic, but occasionally they can be approximated as elastic, as in Example 8.13.

EXAMPLE 8.13 You may be familiar with a toy that was developed years ago for demonstrations in physics classes and is now sold in stores. The device consists of several identical steel balls, each hanging from two strings, as shown in Fig. 8.21. Consider an experiment in which one ball swings down and hits a second ball hanging at rest. Confirm that the first ball stops when it hits the second and that the second moves off at about the same speed that the first ball had before the collision. We assume that the thermal energy created by the collision is small and that the collision is elastic.

SOLUTION Since the collision is assumed elastic, both kinetic energy and momentum are conserved:

Kinetic energy:　　$\frac{1}{2}mv_{1_0}^2 + \frac{1}{2}mv_{2_0}^2 = \frac{1}{2}mv_1^2 + \frac{1}{2}mv_2^2,$　　**(8.17)**

Momentum:　　　　$mv_{1_0} + mv_{2_0} = mv_1 + mv_2,$　　**(8.18)**

where m is the mass of each ball, v_{1_0} and v_{2_0} are the initial velocities of balls 1 and 2, and v_1 and v_2 are their final velocities. In this problem, $v_{2_0} = 0$. Substituting for v_{2_0} in Eq. (8.18), cancelling the m's, and rearranging, we find that

$$v_1 = v_{1_0} - v_2.$$　　**(8.19)**

Substituting Eq. (8.19) into Eq. (8.17) leads to the equation

$$\frac{1}{2}mv_{1_0}^2 + \frac{1}{2}m0^2 = \frac{1}{2}m(v_{1_0} - v_2)^2 + \frac{1}{2}mv_2^2.$$

Cancelling the m's and rearranging to solve for v_2, we find that

$$v_2 = v_{1_0}.$$

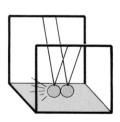

FIG. 8.21. A ball swings down and hits a second ball, initially hanging at rest.

After the collision, ball 2 has the same velocity as ball 1 had before the collision. Substituting this result into Eq. (8.19), we see that $v_1 = 0$. The first ball is stopped by the collision.　■

8.10　Power

We have been considering a variety of processes involving energy changes but have not discussed the time required for these changes. Power is an important quantity used to describe the rate of doing work or the rate of the transformation of one form of energy into another.

Power is defined as the ratio of the work ΔW done and the time Δt required to do that work:

$$\text{Power} = P = \frac{\Delta W}{\Delta t}. \tag{8.20}$$

Power can also be defined as the ratio of the change in a certain form of energy ΔE, such as electrical or chemical energy, and the time Δt required for that energy change:

$$\text{Power} = P = \frac{\Delta E}{\Delta t}. \tag{8.21}$$

The SI metric unit of power is the watt, W, where 1 watt = 1 J/s.

A 60-W lightbulb converts 60 J of electrical energy to other forms of energy (heat and light) each second. A cyclist while pedaling a bicycle at moderate speed typically converts chemical energy in the body to other uses at a rate of about 400–500 W, or 400–500 J/s.

　　Another power unit commonly used to describe the power of engines is the **horsepower** (hp):

$$1 \text{ hp} = 746 \text{ W}.$$

A $\frac{1}{2}$-hp engine converts electrical energy to other forms of energy at a rate of $0.5 \times 746 \text{ W} = 373 \text{ W}$, or 373 J/s.

EXAMPLE 8.14　A crane lifts a 300-kg load at constant speed a vertical distance of 30 m in 10 s. Calculate the rate at which the crane is doing work on the load.

SOLUTION　The work done by the crane per unit time is

$$P = \frac{\Delta W}{\Delta t} = \frac{F(\Delta y)\cos 0°}{\Delta t} = \frac{mg\,(\Delta y)(1)}{\Delta t}$$

$$= \frac{(300 \text{ kg} \times 9.8 \text{ m/s}^2)(30 \text{ m})(1)}{10 \text{ s}} = \frac{8800 \dfrac{\text{kg} \cdot \text{m}}{\text{s}^2}\,\text{m}}{\text{s}}$$

$$= 8800 \text{ N} \cdot \text{m/s} = 8800 \text{ J/s} = \underline{8800 \text{ W}} = \underline{8.8 \text{ kW}}.　■$$

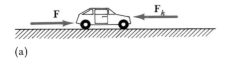

(a)

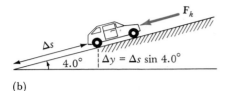

(b)

FIG. 8.22. (a) A road exerts a force **F** on the car to balance the opposing friction force. (b) When the car goes up a hill, the road must exert a greater force **F′** to balance both the friction force and the component of the car's weight parallel to the road.

EXAMPLE 8.15 A 1400-kg car traveling at 27 m/s is opposed by air and rolling friction forces of magnitude 680 N. (a) Calculate the power that the car, traveling on a level road, must expend at its wheels to overcome this friction. (b) If the car drives up a 4.0° incline at the same speed, what power must now be expended at the wheels to both overcome friction and propel the car up the hill?

SOLUTION (a) The car moves forward by pushing back on the road. The road in turn exerts a forward reaction force, **F**, on the car (Fig. 8.22a). This reaction force must have magnitude 680 N to balance the opposing friction forces that act on the car. The power expended by the car at its wheels is the work done to move the car a distance Δx ($W = F\,\Delta x$) divided by the time Δt needed to do this work:

$$P = \frac{\Delta W}{\Delta t} = \frac{F\,\Delta x}{\Delta t} = \frac{F_k\,\Delta x}{\Delta t} = F_k v, \qquad (8.22)$$

where $v = \Delta x/\Delta t$ is the car's speed. Thus,

$$P = (680\ \text{N})(27\ \text{m/s}) = 1.8 \times 10^4\ \text{N·m/s} = 1.8 \times 10^4\ \text{J/s}$$
$$= \underline{1.8 \times 10^4\ \text{W}}\ (25\ \text{hp}).$$

(b) Besides doing work to counteract the friction forces, the car must also increase its gravitational potential energy. If the car moves a distance Δs up the hill, its vertical elevation changes by $\Delta y = \Delta s \sin \theta = \Delta s \sin 4.0°$ (see Fig. 8.22b). The change in gravitational energy is $mg\,\Delta y = mg\,\Delta s \sin 4.0°$. The car must now expend power at its wheels as follows:

$$P' = \frac{\Delta W + \Delta PE_g}{\Delta t} = \frac{F_k\,\Delta s + mg \sin 4.0°\,\Delta s}{\Delta t}$$

$$= (F_k + mg \sin 4.0°)\frac{\Delta s}{\Delta t} = (F_k + mg \sin 4.0°)v$$

$$= [680\ \text{N} + (1400\ \text{kg})(9.8\ \text{m/s}^2)(0.070)](27\ \text{m/s})$$

$$= (680\ \text{N} + 960\ \text{N})(27\ \text{m/s}) = \underline{4.4 \times 10^4\ \text{W}}\ (59\ \text{hp}).$$

The car uses more energy to climb the hill than to oppose friction. ∎

Metabolic Rate

The energy for bodily activity is provided by the chemical energy of the foods we eat. The body breaks the larger molecules of these foods into smaller molecules having less chemical energy. The absolute value of the rate of conversion of this chemical energy $|\Delta E/\Delta t|$ into other forms of energy by our bodies is called our **metabolic rate.** The metabolic rate depends on many factors, including a person's weight, physical activity, and efficiency of bodily processes. Table 8.1 lists the metabolic rates of people under several different conditions and in several different units of measure. The kilocalorie, kcal, is a unit of energy related to the joule as follows:

$$1\ \text{kcal} = 1000\ \text{calories} = 4186\ \text{J}.$$

Dieticians call a kcal simple a Cal. A piece of bread provides about 70 kcal of metabolic energy.

TABLE 8.1   Energy Usage in Various Activities

System	$|\Delta E/\Delta t|$, the metabolic rate		
	Watts (W)	kcal/h	kcal/day
100-lb person at rest	80	70	1,600
150-lb person at rest	100	90	2,100
200-lb person at rest	120	110	2,600
Person walking at 3 mi/h	280	240	5,800
Moderate exercise	470	400	10,000
Heavy exercise	700	600	14,000

As we can see from Table 8.1, a 150-lb person when at rest uses energy at about the same rate as a 100-W bulb. During a day of rest, the 150-lb person would use about 2100 kcal of food. To keep his or her weight constant, the person must eat roughly 2100 kcal of food. A 200-lb person would require about 2600 kcal per day. Exercise or other activity increases the caloric requirement. A person loses about 1 lb of body weight for each 3500 kcal of food omitted from the diet. A person can also lose a pound of weight by increasing the metabolic rate by exercising so that 3500 extra kcal of energy is used.

Summary and Additional Readings

1.   **Work:** Work is one way of transferring energy to or from an object. If a constant force **F** acts on an object as it undergoes a displacement $\Delta \mathbf{r}$, the work done by the force on the object during the displacement is

$$W = F \, \Delta r \cos \theta, \qquad (8.1)$$

where θ is the angle (180° or less) between the direction of **F** and the direction of $\Delta \mathbf{r}$. The object gains energy if the work done on it is positive and loses energy if the work is negative.

2.   **Energy changes:** The objects in a system can have different forms of energy. Four different types of energy are described as follows, along with equations used to calculate the change in each form.

Kinetic energy (KE): Kinetic energy is energy an object has because of its motion. The change in kinetic energy of a mass m whose speed changes from v_0 to v is

$$\Delta KE = KE - KE_0 = \frac{1}{2}mv^2 - \frac{1}{2}mv_0^2. \qquad (8.4)$$

The change in an object's **rotational kinetic energy** is

$$\Delta KE_r = KE_r - KE_{r_0} = \frac{1}{2}I\omega^2 - \frac{1}{2}I\omega_0^2, \qquad (8.5)$$

where I is its moment of inertia and ω_0 and ω are its initial and final angular velocities.

Gravitational potential energy (PE_g): Gravitational potential energy is energy an object has because of its vertical separation from the earth's surface. The change in gravitational potential energy when a mass m moves from a vertical position y_0 to a vertical position y is

$$\Delta PE_g = PE_g - PE_{g_0} = mg(y - y_0). \qquad (8.6)$$

Elastic potential energy (PE_s): Elastic potential energy is energy an object such as a spring has because it is compressed or stretched from its equilibrium position. The change in elastic potential energy when the spring is stretched or compressed from position x_0 to position x is

$$\Delta PE_s = PE_s - PE_{s_0} = \frac{1}{2}kx^2 - \frac{1}{2}kx_0^2, \qquad (8.7)$$

where k is the force constant of the spring.

Internal energy (E_{int}): The internal energy of an object consists of (1) the random kinetic energy of the atoms and molecules of which the object is made (called its **thermal energy E_{th}**) and (2) the chemical energy due to the bonds and interactions between atoms and molecules in the object. The increase in internal energy when one object slides a distance Δs across another surface while being opposed by a kinetic friction force F_k is

$$\Delta E_{int} = E_{int} - E_{int_0} = F_k \, \Delta s. \qquad (8.15)$$

3.   **Work-energy equation:** Energy is a conserved quantity. If a small part of the universe called the system gains or loses energy from its environment because of work done on the system by forces originating in the environment, then the change in the system's energy is

$$W = E - E_0 = \Delta E_{system}. \qquad (8.8)$$

Rearranging and substituting for the different types of energy, we get the equation

$$\underbrace{(KE_0 + PE_{g_0} + PE_{s_0} + E_{int_0})}_{\substack{\text{Initial energy in} \\ \text{system}}} + \underbrace{W}_{\substack{\text{Energy} \\ \text{added to or} \\ \text{removed} \\ \text{from system}}} = \underbrace{(KE + PE_g + PE_s + E_{int})}_{\substack{\text{Final} \\ \text{energy} \\ \text{in system}}}.$$ (8.13)

This equation is a modified form of the **work-energy equation.**

4. **Power:** Power is defined as the rate of doing work or the rate of energy conversion from one form to another:

$$P = \frac{\Delta W}{\Delta t}, \quad \text{or} \quad \frac{\Delta E}{\Delta t}. \qquad \textbf{(8.20) or (8.21)}$$

P. W. Bridgman, *The Nature of Thermodynamics*, Harper and Brothers, New York (1961). An excellent, qualitative discussion of energy concepts by a Nobel laureate.

V. V. Raman, "Where Credit Is Due—The Energy Conservation Principle," *The Physics Teacher* **13**, 80 (1975). An interesting history of the subject.

W. R. Magie, *Source Book in Physics*, Harvard University Press, Cambridge (1963). Excerpts of papers by Joule, Mayer, and other scientists who developed ideas concerning energy.

Questions

1. Several activities are described in this question. For each, show the force exerted by the person and the displacement of the object on which the force acts. Using the definition of work, explain whether the force does positive, negative, or zero work. (a) A person slowly lifts a box from the floor to a table. (b) A person slowly lowers a box from a table to the floor. (c) A person carries a bag horizontally from one location to another. No lifting or lowering takes place. (d) A Dutch boy holds his thumb in a hole in a dike, preventing water from rushing through the hole and flooding Holland.

2. A person rides a bicycle along a horizontal road. Consider the following forces and decide, by showing the direction of the force and the direction of the displacement, whether the work done is positive, negative, or zero. (a) The force of the pedal on the person's foot as the pedal and foot move down. (b) The force of the person's foot on the pedal as it moves down. (c) The gravitational force of the earth *on* the person as he or she moves forward a horizontal distance of 100 m.

3. Give three examples of activities that are often called work in everyday life but that, according to the physics definition, are not work. Explain why, in each case, no physics work is done.

4. *Estimate* the change in gravitational potential energy when you rise from bed to a standing position each morning (or afternoon).

5. A pendulum is displaced from the vertical and released. (a) As it swings down toward the vertical axis, its kinetic energy increases. From where does this energy come? (b) Eventually the pendulum will stop swinging and will hang at rest, straight down. Where has the energy gone?

6. The heart does about 1 J of work each heartbeat. (a) *Estimate* the total work done by the heart in an average lifetime. (b) Mt. Everest is 9400 m high. Calculate the mass that would have to be raised from sea level to the top of Mt. Everest to cause an increase in gravitational potential energy equal to the work done by the heart in a lifetime.

7. A jogger who bounces up and down while running is said to waste more energy than a jogger with a smooth stride whose center of gravity remains at about a constant elevation. Why?

8. Describe a process in which gravitational potential energy is converted mostly into (a) kinetic energy, (b) elastic potential energy, and (c) thermal energy caused by friction.

9. *Estimate* the maximum horsepower you can exert when raising your body mass as fast as possible up a flight of ten stairs. Justify any numbers you use in your estimate. The only energy change you should consider is the change in gravitational potential energy—that is, $P = \Delta PE_g / \Delta t = [mg(y - y_0)] / \Delta t$.

10. Calculate the lifting power exerted by a person during each of the following activities: (a) the person lifts a 10-kg mass 1 m higher in 4 s, (b) the person lifts the 10-kg mass 1 m higher in 2 s, (c) the person lifts the 10-kg mass 2 m higher in 4 s, and (d) the person lifts a 20-kg mass 1 m higher in 4 s.

11. *Estimate* the horsepower exerted by a weight lifter who lifts 150 kg from the ground to his chest. Justify any numbers you use in your estimate.

Problems

8.1 Work

1. (a) A boy pulls a wagon 30 m along a sidewalk with a rope tension force of 80 N. The force is directed 25° above the horizontal. How much work is done by the boy? (b) How much work is done if the same tension force is directed parallel to the ground?

2. Calculate the work done by a woman (a) when slowly lifting a 15-kg suitcase 0.80 m upward, (b) when lowering it 0.80 m, and (c) when holding it at rest.

3. (a) An 80-N rope tension force is used to slowly pull a wagon 50 m up a hill inclined at an angle of 20°. Calculate the work done by the rope if it pulls parallel to the hill. (b) Repeat part (a), but assume that the rope slowly lowers the wagon 50 m down the hill.

4. A rope attached to a truck pulls a water skier along an irrigation canal. The tension in the rope is 400 N, and the rope makes an angle of 15° with the canal. Calculate the work done by the rope in pulling the skier 300 m.

5. A bellhop lifts a 20-kg suitcase 0.40 m, then slowly carries it 30 m along a hall, and finally sets it back down on the floor. Calculate the work done by the bellhop on the suitcase during each part of the trip.

■ **6.** A car is stopped in 0.80 m as it runs into a pile of sand. The work done on the car by the sand is -6.0×10^5 J. (a) Calculate the average force of the sand on the car. (b) Why was the work done by the sand negative? (c) How is the average force altered if the stopping distance is doubled? Justify your answer.

■ **7.** A man does 5800 J of work while pulling a rope attached to a cart loaded with bricks. The tension force in the rope is 320 N, and the cart moves 20 m along level ground. Draw a picture showing the orientation of the rope relative to the ground. Justify with appropriate calculations the orientation shown in the picture.

■ **8.** A lever 1.1 m long is used to raise the back bumper of a car. The fulcrum of the lever is 0.1 m from the bumper. (a) How much work is done by a man pushing with a 500-N force on the long end of the lever while it is displaced downward a distance of 0.5 m? (b) How far up is the bumper raised? (c) Calculate the work done by the lever on the car.

■ **9.** A person slowly pulls the free end of the rope shown in Fig. 8.23 with a force of 90 N. The rope moves a distance of 1.0 m. (a) Calculate the work done by the person on the rope. (b) Calculate the work done by the hanging pulley on the 180-N weight. Justify carefully the latter calculation.

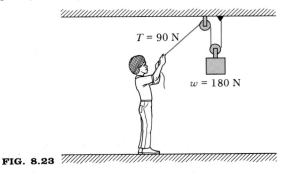

$T = 90$ N

$w = 180$ N

FIG. 8.23

■ **10.** (a) *Estimate* the work done by the golf club on the golf ball as the ball moves from 1.0 cm to 3.0 cm for the force-versus-displacement curve shown in Fig. 8.4. (b) What would be the magnitude of an average force that would produce the same work?

8.3 Kinetic Energy

11. Calculate the change in kinetic energy when a 1000-kg car accelerates (a) from 0 to 10 m/s and (b) from 10 m/s to 20 m/s.

12. The kinetic energy of a 1500-kg car increases by 2.6×10^5 J. Its final speed is 20 m/s. Calculate its initial speed.

13. A 5.0-kg rabbit and a 12-kg Irish setter have the same kinetic energy. If the setter is running at a speed of 4.0 m/s, how fast is the rabbit running?

14. *Estimate* your average kinetic energy when walking to physics class.

15. A 950-kg car traveling at 22 m/s loses 1.4×10^5 J of kinetic energy while slowing for a 56-km/h (35 mi/h) speed zone. Is the car's speed reduced enough?

■ **16.** A rope exerts a 120-N average force while pulling a 12-kg mass for 16 m on a horizontal, frictionless surface. The rope's force is exerted parallel to the surface. If the mass starts with speed 4.0 m/s and all of the work done by the rope is converted to kinetic energy, what is the final speed of the mass?

■ **17.** The kinetic energy of a car is 1.0×10^5 J. Calculate its kinetic energy if the car's speed is (a) increased by 20 percent; (b) decreased by 20 percent.

■ **18.** Block A with 10-kg mass and block B with 20-kg mass have equal momentum. Calculate the ratio of their kinetic energies, KE_A/KE_B.

■ **19.** A 1500-kg car travels 15 km along a city road. Instead of driving at a steady speed of 15 m/s and synchronizing his motion with the stoplights, the driver accelerates the car from rest to 30 m/s and then stops for the next stoplight. The lights are 1 km apart. (a) Calculate the kinetic energy the car uses each time it accelerates from zero to 30 m/s. (b) Calculate the total energy it uses for these accelerations during the 15-km trip. (c) If the car is only 5 percent efficient, it uses 20 times more gasoline energy than the kinetic energy generated by its accelerations. How much gasoline energy does the car use because of the accelerations? (d) If one piece of bread releases 2.5×10^6 J of energy, how many pieces of bread, in terms of energy, does this start-and-stop driver waste?

■ **20.** Huge amounts of air circulate across the earth in regular patterns. A tropical air current called the Hadley cell carries about 2×10^{11} kg of air per second past a cross section of the earth's atmosphere while moving toward the equator. The average air speed is about 1.5 m/s. (a) Calculate the kinetic energy of the air that passes the cross section each second. (b) The United States in 1980 used about 7×10^{16} J of energy. Calculate the ratio of the kinetic energy of the air that passes toward the equator each second and the energy consumed in the United States each second.

21. Calculate the change in rotational kinetic energy when the angular velocity of the turntable of a stereo system increases from 0 to 33 rpm. Its moment of inertia is 6×10^{-3} kg·m^2.

22. A grinding wheel with a (4.0×10^{-3})-kg·m^2 moment of inertia gains 1.2×10^3 J of rotational kinetic energy after starting from rest. Calculate the flywheel's final angular velocity.

■ **23.** The U.S. Department of Energy at one time had plans for a 1500-kg automobile to be powered completely by the rotational kinetic energy of a flywheel. (a) If the 300-kg flywheel had a 6.0-kg·m^2 moment of inertia and could turn at a maximum angular velocity of 3600 rad/s, calculate the energy stored in the flywheel. (b) How many accelerations from a speed of zero to 15 m/s could the car make before the flywheel's energy was dissipated, assuming 100 percent energy transfer and no flywheel regeneration during braking?

■ **24.** The angular velocity of a flywheel increases by 40 percent. By what percent does its rotational kinetic energy increase?

8.4 Gravitational Potential Energy

25. A 5200-kg cable car in San Francisco is pulled a distance of 360 m up a hill inclined 12° from the horizontal. Calculate the change in gravitational potential energy of the car.

■ **26.** *Estimate* the change in gravitational potential energy of a person who successfully climbs Mt. Everest from sea level.

■ **27.** A ski slope drops at an angle of 24° to the horizontal and is 500 m long. (a) Calculate the change in gravitational potential energy of a 60-kg skier who goes down this slope. (b) If 20 percent of the gravitational potential energy lost is converted into kinetic energy, how fast is the skier traveling at the bottom of the slope? The skier's initial velocity is zero.

■ 28. (a) Calculate the change in gravitational energy of a 50-kg mountain climber after moving 30 m up a vertical slope. (b) If the muscles in the body can convert chemical energy into gravitational potential energy with an efficiency of no more than 5 percent, calculate the least muscle chemical energy used to climb the slope.

■ 29. A rope pulls a 10-kg crate 6.0 m along a frictionless plane inclined at an angle of 30°. The tension in the rope is 79 N, and its force is directed parallel to the plane. The crate's speed increases from zero to 6.0 m/s. Calculate (a) the work done by the rope on the crate, (b) the change in kinetic energy, and (c) the change in gravitational potential energy of the crate.

■ 30. The racetrack for a soapbox derby is 300 m long and has a vertical drop of 28 m. (a) Calculate the change in gravitational potential energy of a race car and its driver (a total mass of 113 kg) from start to finish of a race. (b) If all this energy is converted to kinetic energy, how fast is the car moving at the end of the race?

■ 31. (a) Calculate the net change in gravitational potential energy of the barrel of bricks and the bricklayer described in Example 3.8 on page 68. (b) If this gravitational potential energy change is converted into the kinetic energy of the bricks and the bricklayer, how fast are they moving just before the bricklayer hits the beam?

8.5 Elastic Potential Energy

32. A spring-loaded car bumper is built to be compressed 0.30 m when it hits a wall while moving at a speed of 9.0 m/s. The energy the spring must absorb is 80,000 J. What should be the force constant of the bumper's spring?

33. A spring with a (1.2×10^4)-N/m force constant is stretched 6.0 cm from its equilibrium position. (a) Calculate the change in elastic potential energy of the spring. (b) Calculate its change in elastic potential energy as it returns from the 6.0-cm position to a position 3.0 cm from equilibrium. (c) Calculate its elastic potential energy change as it moves from the 3.0-cm position back to its equilibrium position.

34. An elevator cable breaks. The elevator's fall is stopped by compression of a spring at the bottom of the shaft. The spring, whose force constant is 2.0×10^5 N/m, must absorb 3.0×10^4 J of energy when being compressed by the elevator. How far is the spring compressed?

■ 35. A spring with a 1200-N/m force constant is compressed 0.20 m, and a 2.0-kg block that rests on a horizontal surface is placed against the spring. When the spring is released, the block shoots forward. (a) Calculate the change in elastic potential energy of the spring. (b) If the elastic potential energy change of the spring is balanced by an increase in kinetic energy of the block, how fast does the block move as it leaves the spring?

■ 36. A partially compressed spring is compressed farther, causing its elastic potential energy to increase by 50 percent. Calculate the percent increase in the spring's compression distance.

■ 37. The force required to stretch a slingshot by different amounts is graphed in Fig. 8.24. (a) Calculate the force constant k of the slingshot. (b) How much energy is required to stretch the sling 15 cm from equilibrium? [*Hint:* For part (a), use the equation $F = kx$ for one or two known points on the force-versus-displacement curve to calculate k.]

8.7 Work-Energy Calculations

In solving Problems 38–50, be sure to use the work-energy equation.

■ 38. You throw a 1.0-kg rock straight up at a Frisbee resting in a tree. If the rock's speed as it reaches the Frisbee is 4.0 m/s, what was its speed as it left your hand 2.8 m below the Frisbee?

■ 39. A car traveling on a level surface at 12 m/s drives off an embankment and lands in an arroyo 6.0 m lower. Calculate the car's speed as it lands.

■ 40. A person on a bicycle traveling at 10 m/s on a horizontal surface stops pedaling as she starts up a hill inclined at an angle of 3.0°. How far along the incline will the person travel before stopping? Ignore friction forces.

■ 41. A 72-kg daredevil driver wishes to buy a spring for the ejector seat of his car (see Fig. 2.14). When loaded, the spring is to be compressed 0.30 m and, when released, should launch the driver to a maximum height of 7.0 m above the top of the loaded spring. What should be the force constant of the spring the driver buys?

■ 42. A 16-m-long rope attached to the limb of a tall tree is pulled to the side at a 53° angle from the vertical and held by a 62-kg student standing on a balcony. If the student swings down on the rope, calculate his or her speed when the rope is vertical. [*Hint:* What is the student's change in elevation and in gravitational energy?]

■ 43. A 1200-kg elevator must be lifted by a cable that causes the elevator's speed to increase from zero to 4.0 m/s in a vertical distance of 6.0 m. Calculate the cable tension needed.

■ 44. A vehicle built to test the effects of high accelerations on people is moving at a speed of 20 m/s. The vehicle runs into a piston that compresses the air in a cylinder and stops the car in 0.30 m. A 70-kg person is also stopped in the same distance by the force of shoulder straps and seat belts on her body. Calculate the average force of these restraints on the person during the stop.

■ 45. A 55-kg woman jumps from the window of a burning building. (a) Calculate her speed just as she reaches a fire fighter's net 8.0 m below. Ignore air resistance. (b) She is stopped by the net after falling 1.2 m into it. Calculate the average force of the net on her body.

■ 46. A 0.20-kg egg is dropped from a ladder a vertical distance of 4.0 m. (a) Calculate the speed of the egg at the bottom of

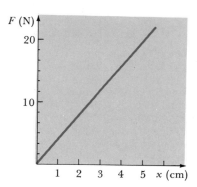

FIG. 8.24. A graph of the force F needed to compress a spring versus the compression x of the spring.

its fall. (b) The egg will break if subjected to an impulsive force greater than 80 N. Over what minimum distance must a constant stopping force be exerted to avoid breaking the egg?

■ 47.   An object of mass m moving at speed v_0 on a horizontal, frictionless surface is stopped in a distance Δx by an opposing force of average magnitude F. (a) Using the work-energy method, show that $F = (mv_0^2)/(2\,\Delta x)$. (b) If the stopping distance Δx increases by 50 percent, by what percent does the average force needed to stop the object change, assuming that m and v_0 are unchanged? Justify your result carefully.

■ 48.   A rope exerts a 254-N force while pulling a 60-kg skier upward along a hill inclined at 25°. (The rope's force is parallel to the surface of the hill.) If the skier starts at rest, calculate his speed after moving 100 m up the slope. Ignore friction forces.

■ 49.   A rope exerts a 40-N force while lowering a 20-kg mass down a frictionless plane inclined at an angle of 15° (the force acts parallel to the plane). If the speed of the mass is initially 2.0 m/s, calculate its final speed after moving 10 m down the plane.

■ 50.   A spring with an 8540-N/m force constant rests on a frictionless plane inclined at a 37° angle. The spring is compressed 0.050 m, and a 0.60-kg block is placed against the spring. When the spring is released, the block shoots up the plane. Calculate the speed with which it leaves the end of the plane 1.0 m from where it started.

8.8   Friction Forces and Internal Energy

Use the work-energy technique to solve Problems 51–62.

■ 51.   A 1500-kg car traveling at 20 m/s is brought to a stop by the opposing forces of friction. If these opposing forces have an average magnitude of 800 N, how far will the car travel before stopping?

■ 52.   A 1200-kg car traveling at 24 m/s drives 18 m through some wet mud in which the net force on the car from all causes (mostly the resistive force of the mud) is 1.7×10^4 N opposite the direction of motion. Calculate the car's speed as it leaves the mud.

■ 53.   An 80-N friction force opposes the motion of a bicycle rider. (a) How much internal energy is created because of friction by the cyclist in traveling 1.0 mi? (b) How far must the cyclist travel to produce 3×10^5 J of internal energy? By comparison, this is the energy released to the body after a person eats a slice of bread.

■ 54.   A 0.057-kg tennis ball after falling 18 m has a speed of 12 m/s (the ball's initial speed is zero). Calculate the average value of the resistive force opposing the ball's motion.

■ 55.   A water slide is 42 m long and has a vertical drop of 12 m. If a 60-kg person starts down the slide with a speed of 3.0 m/s, calculate his or her speed at the bottom. A 120-N average friction force opposes the motion.

■ 56.   A spring with a (1.2×10^4)-N/m force constant is compressed 6.0 cm, and a 10-kg block is placed against it. When the spring is released, the block shoots forward along a horizontal surface whose coefficient of friction is 0.20. How far will the block travel before coming to a stop?

■ 57.   A 900-kg car initially at rest rolls down a hill inclined at an angle of 5.0°. A 400-N friction force opposes its motion. Calculate the car's speed after moving 50 m.

■ 58.   A 1500-kg car is initially coasting at 20 m/s down a hill inclined at 2.0° with the horizontal. If an 800-N friction force

opposes the car, how far will it travel down the hill before stopping?

■ 59.   A 6.0-kg block slides from position A down a frictionless curve to position B (Fig. 8.25). At B, a friction force opposes the motion of the block so that it comes to a stop 2.5 m from B. Calculate the coefficient of kinetic friction between the block and the surface after position B.

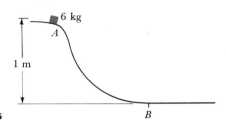

FIG. 8.25

■ 60.   A spring with force constant 10,000 N/m is compressed 0.20 m, and a 2.0-kg block rests against the spring. The block is shot up a plane 3.0 m long, inclined at 37°. A 10-N friction force opposes the block as it slides up the plane. Calculate the speed with which the block leaves the plane.

■ 61.   A rope exerts a 274-N force while pulling an 80-kg skier upward along a hill inclined at 12°. (The rope pulls parallel to the hill.) A 110-N friction force opposes the skier's motion. If the skier starts at rest, calculate his speed after moving 100 m up the slope.

■ 62.   A rope exerts an 18-N force while lowering a 20-kg crate down a plane inclined at an angle of 20° (the force acts parallel to the plane and opposite to the crate's motion). A 24-N friction force also opposes the motion. If the crate starts at rest, calculate its speed after moving 10 m down the plane.

8.9   Inelastic and Elastic Collisions

■ 63.   An 80-g arrow moving at 80 m/s hits and embeds in a 10-kg block resting on ice. (a) Use the conservation-of-momentum principle to determine the speed of the block and arrow just after the collision. (b) Use the work-energy principle to determine how far the block slides on the ice following the collision if it is opposed by a 9.2-N friction force.

■ 64.   An 1800-kg car moving west at 10 m/s hits and becomes locked to a 1000-kg stationary car. (a) Calculate the velocity of the cars just after the collision. (b) After the collision, the cars' motion is opposed by a (1.2×10^4)-N friction force. How far do they skid before coming to a stop?

■ 65.   A 32-kg girl sitting at rest on a 2.4-m-long swing catches an 8.0-kg medicine ball moving at 4.0 m/s. (a) Use momentum conservation to calculate the speed of the girl and the ball just after she catches the ball. (b) Use energy conservation to calculate the vertical distance the girl and the ball rise on the swing.

■■ 66.   A billiard ball at rest is hit head-on by a second billiard ball moving 1.5 m/s toward the east. If the collision is elastic and we ignore rotational motion, calculate the final speed of each ball.

■■ 67.   A block of mass 4.0 kg moving at 2.0 m/s toward the west on a frictionless surface has an elastic head-on collision with a second, stationary block of mass 1.0 kg. Calculate the final velocity of each block.

■■ 68.   A block of mass 1.0 kg moving north at 2.0 m/s has an elastic, head-on collision with a second, stationary block of mass 4.0 kg. Calculate the final velocity of each block.

8.10 Power

69. Calculate the power needed from the motor of an elevator for it to lift a (1.0×10^4)-kg mass a distance of 20 m in 5.0 s.

70. How much time is needed for a 1-hp motor to lift a student's 72-kg body 16 m to the window of his dormitory room or apartment?

71. Calculate the output power needed for the pump of a fire engine that must lift 30 kg of water a vertical distance of 20 m each second.

■ **72.** A large tree can evaporate 500 kg of water a day. (a) If the water is raised 8.0 m, how much energy must be provided by the tree to lift the water this distance? (b) If the evaporation occurs over a 12-hr period, what is the average power in watts needed to provide this increase in gravitational energy?

■ **73.** An 82-kg hiker climbs to the summit of Mount Mitchell in western North Carolina. During one 2.0-hr period, the climber's vertical elevation increases 540 m. (a) Calculate the climber's change in gravitational potential energy and (b) the power generated to increase the gravitational energy. (c) If only 10 percent of the climber's total energy expenditure goes into this gravitational energy change, what is the climber's average metabolic rate during the hike?

■ **74.** A rope exerts a force F while displacing an object a distance Δx in a time Δt. Show that the power exerted by the force is $P = Fv$, where v is the object's speed.

■ **75.** (a) Calculate the power output of the motor of a 1500-kg car traveling on a level road at a speed of 90 km/h if the car is opposed by a total friction force of 750 N. (b) If the car travels at the same speed and is opposed by the same friction force while climbing a 3.2° hill, what power must the engine generate?

■ **76.** A 76-kg cyclist has a maximum power output (the rate at which the cyclist can do work against friction and gravity) of 680 W. Calculate the cyclist's maximum speed while moving up a hill inclined at 5.0° with an opposing friction force from all causes of 20 N.

■ **77.** In the past, salmon would swim more than 700 miles to spawn at the headwaters of the Salmon River in central Idaho. The trip took about 22 days, and the fish consumed energy at a rate of 2.0 W for each kilogram of body weight. (a) What is the total energy used by a 3-kg salmon while making this 22-day trip? (b) About 80 percent of this energy is released by burning fat and the other 20 percent by burning protein. How many grams of fat are burned? One gram of fat releases 3.8×10^4 J of energy. (c) If the salmon is about 15 percent fat by weight at the beginning of the trip, how many grams of fat does it have left at the end of the trip?

■ **78.** A 150-lb person wishes to lose 10 lb of body weight in 30 days. Estimate the number of kilocalories the person should have in his or her diet to achieve this goal (see Table 8.1). Remember that omission of 3500 kcal from the diet results in a loss of 1 lb of weight.

■ **79.** A 150-lb person walks at 3 mi/h one hour a day for one year. (a) Use the numbers in Table 8.1 to calculate the extra number of kilocalories of energy used because of the walking. (b) How much less should the person expect to weigh at the end of the year because of the walking?

CHAPTER 9

Thermal Energy, Temperature, and Heat

In Chapter 8 we used the work-energy equation to analyze energy transformations involving macroscopic objects, objects easily seen with our eyes. In this and the next three chapters we look at energy transformations involving the atoms and molecules of which matter is made. This atomic and molecular energy of a system, called *internal energy*, consists of (1) thermal energy, the random kinetic energy of the atoms and molecules, and (2) the potential energy of these tiny particles resulting from their bonds and interactions with each other.

Temperature is an important indicator of thermal energy, and we will define it along with different temperature scales. We will also define a quantity called heat—a mechanism, like work, for transferring energy from one object to another. Several consequences of heat transfer are discussed, including temperature changes; changes between the solid, liquid, and gaseous states of matter; and changes in the dimensions of objects as they become warmer or cooler.

Processes involving thermal energy and heat are very important, affecting our metabolic rates, the climate of the earth, and every sensation of comfort and discomfort we feel that is related to temperature.

9.1 Thermal Energy

The **thermal energy** in a substance is the kinetic energy associated with the random motion of all its atoms and molecules. Consider a cup of water. The water molecules in the cup experience three different types of random motion, as depicted in Fig. 9.1. They move, or translate, from one position to another (Fig. 9.1a). Collisions with other water molecules frequently interrupt this motion, causing a molecule to change direction and speed—hence the use of the word *random* to describe the motion. The molecules also vibrate (Fig. 9.1b). The hydrogen atoms attached to the oxygen atom (H–O–H) vibrate in and out and bend back and forth. The molecules also spin or rotate as they move through space (Fig. 9.1c). Because all three types of motion—translational, vibrational, and rotational—occur simultaneously, there is much variation in the motion of

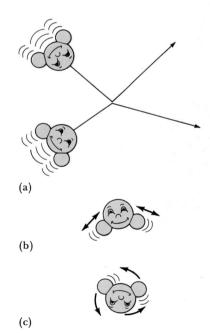

(a)

(b)

(c)

FIG. 9.1 Types of motion of a water molecule: (a) translational, interrupted by a collision with another water molecule; (b) vibrational; and (c) rotational.

different molecules at any instant in time. Some may be rotating more and translating less, while some move up and others down.

Each molecule's motion changes almost one hundred million million (10^{14}) times per second because of collisions with other molecules. A molecule moving slowly before a collision may move fast afterward. However, when averaged for one second or more, the average motion of every molecule of water in the cup is about the same—even though there is great variation in the motion of the molecules at each instant of time.

If the water becomes warmer, the molecules translate from one place to another faster, become more distorted in shape as they vibrate, and rotate more rapidly. The average energy associated with each type of motion increases. If the water becomes cooler, the translational, vibrational, and rotational motions decrease, as does the average energy associated with each type of motion. We see that thermal energy is related closely to the idea of temperature.

9.2 Temperature

Temperature is an indicator of the average thermal energy of the atoms or molecules in a substance. As the temperature increases, so does the average thermal energy of its atoms or molecules. Thermal energy and temperature, then, are closely related quantities—but there is an important distinction. Temperature depends on the average thermal energy *per* molecule or atom, whereas thermal energy is the total random kinetic energy of *all* the atoms or molecules in a substance. If, for example, the ocean and a glass of water are at the same temperature, then the average thermal energy of one water molecule in the ocean is the same as that of one in the glass. However, the ocean contains much more thermal energy because there are so many more water molecules in the ocean than in the glass.

Temperature is measured using a variety of substances that change in a predictable way as their temperatures change. For example, the length of a metal rod increases as it warms. This variation of length with temperature is called a **thermometric property**—a property of matter that varies in a predictable manner with temperature. Other thermometric properties are the increase in the pressure of an enclosed gas as it warms, the increase in the volume of space occupied by a liquid such as mercury as it warms, the change in color of light emitted by a flame as the flame becomes hotter, and the increase in electrical resistance of a metal wire as its temperature increases. Every thermometer uses one of these or some other thermometric property as the basis for measuring temperature.

One of the most familiar thermometers is the liquid mercury thermometer (Fig. 9.2). A narrow tube is attached to a bulb containing liquid mercury. As the bulb becomes warm, the mercury expands farther into the tube. Marks are placed on the outside of the tube at points to which the mercury rises when the bulb is placed in a temperature bath whose temperature is fixed. For example, one mark can be placed at the point to which the mercury rises when the bulb is placed in a solution of freezing water. Another mark can be inscribed at the point to which the mercury rises when the bulb is in boiling water at atmospheric pressure.

Temperature scales are constructed by assigning numbers to these fixed temperatures. On the Celsius temperature scale, the number zero is assigned to

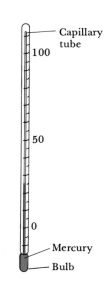

FIG. 9.2. A liquid mercury thermometer. As mercury in the bulb is warmed, it expands farther up into the capillary tube.

the freezing point of water and the number 100 to its boiling point. The space between these two marks is divided into 100 equal divisions or intervals. A temperature is expressed as a **point** on the scale in units of degrees Celsius (°C). For example, the average temperature of a person's body is 37.0°C. The difference of two temperatures is called an **interval** whose unit is Celsius degrees (C°). For instance, if the outdoor temperature increases from 20°C to 26°C, the change in temperature, an interval, equals 6 C°.*

Another temperature scale commonly used in the United States is the Fahrenheit scale. This scale, developed by Gabriel Fahrenheit (1686–1736), a German who emigrated to the Netherlands, was originally standardized on the low end by assigning a temperature of zero to an ice, water, and salt solution that was colder than the freezing temperature of water because of the presence of the salt. The higher end of the scale was standardized by assigning the value 96 to the normal temperature of the human body. Using his new scale, Fahrenheit was the first person to realize that water always freezes at the same temperature, 32° on his scale. Over time, Fahrenheit's scale has been modified slightly so that today, the sea-level boiling point of water at 212°F is the standard at the upper end, the freezing temperature of water at 32°F is the standard at the low end, and the normal temperature of the body falls at 98.6°F.

On the Fahrenheit scale the boiling and freezing points of water are separated by 180 F° (that is, 212°F − 32°F), while on the Celsius scale these same two points are separated by only 100 C°. The ratio of these two numbers, 180 F°/100 C° = 9 F°/5 C°, can be used to relate temperature changes on one scale to those on the other. To calculate a Fahrenheit temperature T_F from a Celsius temperature T_C we must first multiply the Celsius temperature by $\frac{9}{5}$. We then add 32 to account for the fact that the freezing temperature of water on the Fahrenheit scale is 32°F and not zero. Thus,

$$T_F = \frac{9}{5}T_C + 32. \qquad (9.1)$$

To calculate an equivalent Celsius temperature from a Fahrenheit temperature, Eq. (9.1) can be rearranged to give

$$T_C = \frac{5}{9}(T_F - 32). \qquad (9.2)$$

Let us consider two examples of temperature conversions.

EXAMPLE 9.1 If the temperature of a room is 20°C, what is its temperature in degrees Fahrenheit?

SOLUTION Using Eq. (9.1), we find that

$$T_F = \frac{9}{5}(20) + 32 = 36 + 32 = \underline{68°F}. \qquad \blacksquare$$

EXAMPLE 9.2 The normal core temperature of the body is 98.6°F. What is the core temperature in degrees Celsius?

*This scale is named in honor of its inventor, the Swedish astronomer Anders Celsius (1701–1744). It was formerly called the centigrade scale, the Latin prefix *centi* meaning one-hundredth.

FIG. 9.3. Several common temperatures on the Kelvin, Celsius, and Fahrenheit temperature scales.

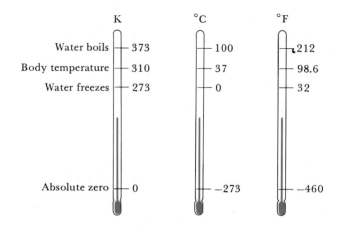

SOLUTION Using Eq. (9.2), we find that

$$T_C = \frac{5}{9}(98.6 - 32) = \frac{5}{9}(66.6) = \underline{37.0°C}.$$

Note that we first subtracted 32°F from 98.6°F to get 66.6°F. This is the number of Fahrenheit degrees that the body is above the freezing point of water. Since each Fahrenheit degree is 5/9 of a Celsius degree, we find that the normal core temperature of the body is $(\frac{5}{9})66.6°F = 37.0°C.$ ∎

Another very important temperature scale in physics is the **Kelvin** or **absolute temperature scale.** In this scale, the lowest possible temperature attainable by an object is assigned the value zero, that is, 0 K.* At this temperature the random motion of the atoms and molecules in a substance is at a minimum, and the temperature cannot be lowered further by decreasing this motion. The lowest possible temperature is called **absolute zero.** Absolute zero is equivalent to −273.15°C or −459.67°F. Temperature intervals on the Kelvin scale are the same as on the Celsius scale. The freezing point of water on the Kelvin scale is 273.15 K, and the boiling point is 100 degrees higher—373.15 K. Figure 9.3 shows the relation among the three scales.

9.3 Heat and the Conservation-of-Energy Principle

The thermal energy of a system can change in many ways. The engine of an idling car becomes warm because much of the energy released when gasoline burns in the engine is converted into thermal energy. If you rub your hands together vigorously, they become warm because work done by the muscles in your arms is converted into thermal energy caused by friction as your hands rub together.

Thermal energy in a system also can be changed by a process called heat transfer. **Heat transfer** occurs when two objects at different temperatures are

*The Kelvin degrees are abbreviated K, not °K, according to a convention established by the General Conference of Weights and Measures, an international standards organization.

brought into contact with each other. Thermal energy in the hotter object is transferred to the cooler one. Consider the effect of the flame of a Bunsen burner on a glass of water (Fig. 9.4). The gas molecules in the flame have a temperature of about 900°C. Since they are so hot, the random kinetic energy of a molecule in the flame is much greater than the random kinetic energy of an atom in the glass. When a fast-moving gas molecule in the flame collides with the glass, an atom in the glass gains kinetic energy, and the gas molecule of the flame loses kinetic energy. The atom in the glass is warmed. There has been an energy transfer from the flame to the glass; we call it a heat transfer.

Heat is energy transferred to or from an object because of a difference in its temperature and that of some other object it contacts in its environment. The cooler object gains energy and the warmer body loses energy. Heat is given the symbol Q and has units of energy (such as joules, calories, and Btu).

The terms *heat* and *thermal energy* are often used interchangeably. They are, however, different quantities. The thermal energy in a system is the amount of random kinetic energy of the atoms and molecules of objects in the system. Heat, on the other hand, is energy transferred between a system and its environment because of a difference in their temperatures (see Fig. 9.5). If the system gains energy, Q is positive; if the system loses energy, Q is negative.

Typically, when scientists discuss energy transformations that involve heat, the energy-conservation idea is expressed in the following form (you might review Section 8.6, in which energy conservation was presented in two different ways):

$$W + Q = \Delta E_{\text{system}}$$
$$= \Delta KE + \Delta PE_g + \Delta E_{\text{th}} + \Delta E_{\text{chem}} + \cdots. \quad (9.3)$$

Work W and heat Q represent the energy added to or removed from a system during a certain time period or during some process of interest. This energy added to or removed from the system equals the net change in all the different forms of energy within the system—the terms on the right-hand side of Eq. (9.3). Since thermal energy and the chemical energy stored in molecular bonds are two forms of energy in a system, expressions for these forms of energy change appear on the right-hand side of the equation. This equation, introduced in Section 8.6, is called the **first law of thermodynamics** when used to describe processes involving thermal and chemical energy changes.

9.4 Specific Heat Capacity

To calculate the heat energy added to or removed from an object, we need to introduce a new quantity. The **specific heat capacity** c of a substance is the amount of heat that must be added to a unit mass of that substance to raise its temperature by 1 C°. The units of specific heat capacity are energy per mass per degree temperature—for example, J/kg·C°, cal/g·C°, Btu/slug·F°. Table 9.1 lists the values of the specific heat capacity of some common substances. The specific heat capacity of water is 4180 J/kg·C°. This means that 4180 J of energy must be added to 1 kg of water to raise its temperature 1 C°. If we know the specific heat capacity of a substance, we can find the amount of heat required to raise the temperature of a mass m of that substance by an amount ΔT.

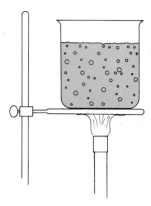

FIG. 9.4. When the "hot" molecules in the flame hit the glass, kinetic energy is transferred to the atoms in the glass. The glass and its contents become warmer. This is a form of heat transfer.

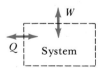

Environment

FIG. 9.5. Energy can enter or leave a system by heat transfer Q or by work W done on or by the system. The change in the energy of the system ΔE_{system} is given by the equation $Q + W = \Delta E_{\text{system}}$.

TABLE 9.1 Specific Heat Capacities of Various Substances

Substance	c (J/kg·C°)	Liquid Substances	c(J/kg·C°)	Gaseous Substances*	c (J/kg·C°)
Iron or steel	450	Water	4180	Steam	1970
Copper	390	Methanol	2550	Oxygen	910
Aluminum	900	Ethanol	2480	Nitrogen	1040
Lead	130	Ethylene glycol	2380	Dry air	~1000
Glass	840	Benzene	1720		
Sodium chloride	880				
Ice	2090				
Wood	1680				
Sand	820				
Brick	840				
Concrete	880				
Human body	3470				

*At constant pressure.

EXAMPLE 9.3 You wish to take a bath and will need to warm 160 kg of water by 14 C°. How much heat is required?

SOLUTION Note first that to warm 1 kg of water by 1 C° requires 4180 J. To warm 160 kg by 1 C° would require 160×4180 J $= 670,000$ J. However, we wish to warm the water 14 C° insted of 1 C°. We will need 14 times as much energy, or $14 \times 160 \times 4180$ J $= \underline{9.4 \times 10^6 \text{ J}}$.* ∎

We see from this example that the heat Q needed to raise the temperature of a mass m of substance with specific heat capacity c by a temperature ΔT is the product of m, c, and ΔT:

$$Q = mc\Delta T = mc(T - T_0). \tag{9.4}$$

The ΔT in Eq. (9.4) is the final temperature T of the object minus its initial temperature T_0 before the heat was added or removed. When the object's temperature increases, ΔT and Q are positive numbers (energy is added to the object). When the object's temperature decreases, ΔT and Q are negative numbers (the object loses energy).

EXAMPLE 9.4 A 30-kg child has a temperature of 39.0°C (102.2°F). How much heat must be removed from the child's body to lower his temperature to 37.0°C (98.6°F)? (See Fig. 9.6.)

SOLUTION From Table 9.1 we see that the specific heat capacity of the human body is $c = 3470$ J/kg·C°. This means that 3470 J of energy must be removed from 1 kg of a person's body to lower his or her temperature by 1 C°. The child's mass is $m = 30$ kg. The desired temperature change is $\Delta T = T - T_0 = (37.0°C - 39.0°C) = -2.0$ C°. Using Eq. (9.4), we find that

FIG. 9.6. Heat transfer occurs from a hot child to a cool tub of water.

*The average cost of electricity in the United States in 1984 was about 1 cent for 5×10^5 J (or 8 cents per kilowatt-hour). The bath in this example cost about 20 cents.

the amount of heat that must be removed to lower the child's temperature is

$$Q = \left(3470 \, \frac{J}{kg \cdot C^\circ}\right)(30 \, kg)(-2.0 \, C^\circ) = \underline{-2.1 \times 10^5 \, J}.$$

The negative sign reminds us that heat was transferred out of the child's body. ∎

As we see in Table 9.1, the specific heat capacity for different materials varies considerably. This variation affects our lives in many ways. Notice, for example, that the specific heat capacity of water is about five times greater than that of sand. This is one reason why the temperatures of cities near large bodies of water remain more stable on a daily and seasonal basis than the temperatures of cities inland. During the daylight hours, absorption of sunlight causes a temperature increase roughly five times greater in sandy terrain than in a large body of water. At night, as heat is lost, the temperature of the sandy terrain drops more rapidly than that of land near a body of water.

Specific heat capacity is also important to people interested in solar heating and cooling for houses. They must carefully consider the specific heat capacity of materials used to store solar energy. During the day, energy gained from sunlight is used to warm the storage material. At night the energy is released to warm the house. It is important to have a storage material with a high specific heat capacity, one that will store large amounts of energy per unit mass.

In passive solar heating, sunlight coming through a window is absorbed by some surface in the house (Fig. 9.7). If the surface has a low specific heat capacity and a small mass, it will warm quickly, making the daytime temperature uncomfortably high. At night it will cool quickly, resulting in uncomfortably low temperatures. It is better to have the sunlight absorbed by a massive wall with a high heat capacity. The resulting temperature change $\Delta T = Q/mc$ will be small because both m and c are large.

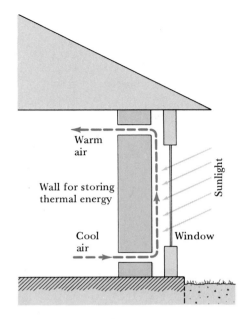

FIG. 9.7. A massive wall is heated by sunlight. Air that circulates along the wall is warmed.

9.5 Calorimetry

Calorimetry is an experimental technique used to measure the specific heat capacity of a substance. Calorimetry is also used to measure the energy required to melt or boil a substance, to measure the metabolic rates of animals, and to calculate the chemical energy released during chemical reactions and even during explosions.

Calorimetry measurements are made using a device called a *calorimeter*. Such devices vary considerably in construction, depending on the type of measurement to be made. A calorimeter used to measure the specific heat capacity of a solid material is shown in Fig. 9.8. It consists of an insulated can containing a known amount of water. The insulation around the can (for example, Styrofoam or a vacuum jacket) prohibits the transfer of heat into or out of the can.

The can and water are set at some known initial temperature, and the object whose specific heat capacity is to be measured is warmed or cooled to a temperature different from that of the water and can. The object is then placed in the water. The can, water, and object eventually reach the same final temperature. If the object starts at a higher temperature, it cools by transferring heat to the water and can. The water and can, in turn, are warmed. The net heat change is zero since heat gained by the water and can equals that lost by the object, or vice versa.

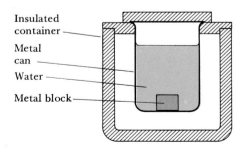

FIG. 9.8. A schematic representation of a calorimeter.

We are able to write this heat-transfer process as follows:

$$Q_{can} + Q_{water} + Q_{object} = 0, \tag{9.5}$$

where Q_{can}, Q_{water}, and Q_{object} are the heat gained or lost by the can, water and object, respectively. If heat is gained, Q is positive; if heat is lost, Q is negative.

We can use Eq. (9.4) ($Q = cm \, \Delta T$) to calculate the heat change for each term in Eq. (9.5). We find that

$$c_{can} m_{can} \Delta T_{can} + c_{water} m_{water} \Delta T_{water} + c_{object} m_{object} \Delta T_{object} = 0.$$

The only unknown in the above equation is c_{object}. Thus, we are able to solve for the unknown specific heat capacity.

EXAMPLE 9.5 An insulated aluminum can whose mass is 0.20 kg contains 0.30 kg of water at 25°C. A 0.10-kg metal block whose temperature is 80°C is lowered into the water. The final temperature of the water, can, and block is 30°C. What is the specific heat capacity of the metal block?

SOLUTION In this example the metal block gives up heat as its temperature is lowered from 80°C to 30°C:

$$\begin{aligned} Q_{block} &= (c_{block})(m_{block})(T - T_0)_{block} \\ &= (c_{block})(0.10 \text{ kg})(30°C - 80°C) \\ &= (c_{block})(0.10 \text{ kg})(-50 \text{ C}°) = -(c_{block})(5 \text{ kg} \cdot \text{C}°). \end{aligned}$$

The water and can both gain energy as their temperatures rise from 25°C to 30°C:

$$\begin{aligned} Q_{water} &= (c_{water})(m_{water})(T - T_0)_{water} \\ &= \left(4180 \, \frac{\text{J}}{\text{kg} \cdot \text{C}°} \right)(0.30 \text{ kg})(30°C - 25°C) = +6270 \text{ J}, \\ Q_{can} &= (c_{can})(m_{can})(T - T_0)_{can} \\ &= \left(900 \, \frac{\text{J}}{\text{kg} \cdot \text{C}°} \right)(0.20 \text{ kg})(30°C - 25°C) = +900 \text{ J}. \end{aligned}$$

Note that we used the specific heat capacity of aluminum when calculating Q_{can}.

Since no heat has left the calorimeter, the sum of the three heat terms must be zero [Eq. (9.5)]:

$$900 \text{ J} + 6270 \text{ J} - (c_{block})(5 \text{ kg} \cdot \text{C}°) = 0.$$

After rearranging the equation, we find that

$$c_{block} = \frac{(900 + 6270) \text{ J}}{5 \text{ kg} \cdot \text{C}°} = \underline{1430 \text{ J/kg} \cdot \text{C}°}. \qquad \blacksquare$$

9.6 Melting and Freezing

When heat is added to or removed from matter, other changes besides a temperature change can occur. For example, the heat can cause a change in state of the matter. Most forms of matter can exist in three different states—solid, liquid, and gas. Water, for example, can exist as ice, as liquid water, or as gaseous water vapor. In this section we consider the energy involved in melting a solid or in freezing a liquid. The subject of Section 9.7 is boiling and condensation—transitions between the liquid and gaseous states of matter.

Melting Solids

The atoms and molecules in a solid are usually bonded to each other in a well-defined structure. They can vibrate about an equilibrium position but are unable to rotate or move to new positions in the solid. The organization of water molecules in ice is depicted in Fig. 9.9a. The oxygen atom of a water molecule shares hydrogen atoms with each of four neighboring water molecules.

If heat is added to a solid such as ice, the temperature of the solid increases. The atoms and molecules vibrate with greater amplitude. As more and more heat is added, eventually the vibrations become so violent that the atoms or molecules break away from their neighbors and move about more freely. The matter has melted; it has changed from the solid state to the liquid state. When water melts, neighboring water molecules no longer share hydrogen atoms (Fig. 9.9b). Each water molecule has its own two hydrogen atoms. This change of state usually occurs at a well-defined temperature, called the **melting temperature,** which varies from substance to substance—the melting temperature of water is 0°C, whereas that of copper is 1083°C.

The amount of energy required to break the crystalline structure of a solid to form a liquid also differs for different substances. A quantity called the heat of fusion (or, more formally, the latent heat of fusion) is a measure of the energy required to melt a solid. The **heat of fusion** L_f of a solid substance is defined as the energy needed to melt 1 kg of the solid into the liquid state at the melting temperature of the substance. The units of L_f are J/kg. For example, 3.35×10^5 J of energy is required to melt 1 kg of ice at 0°C; thus, the heat of fusion of ice is 3.35×10^5 J/kg. A list of melting temperatures and heats of fusion of different substances appears in Table 9.2.

The amount of energy needed to melt larger or smaller amounts of matter is directly proportional to the mass m of the matter involved. For example, twice as much energy is required to melt 2 kg of ice as to melt 1 kg of ice.

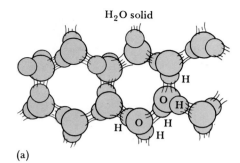

H$_2$O solid

(a)

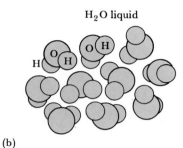

H$_2$O liquid

(b)

FIG. 9.9. (a) The crystalline structure of the water molecules in ice. (b) The more random arrangement of the molecules in liquid water.

TABLE 9.2 Heats of Fusion and Vaporization

Substance	Melting		Boiling	
	Melting Temperature (°C)	L_f, Heat of Fusion (J/kg)	Boiling Temperature (°C)	L_v, Heat of Vaporization (J/kg)
Water	0.00	3.35×10^5	100.00	2.256×10^6
Ethanol	−114	10.42×10^5	78	0.854×10^6
Hydrogen (H$_2$)	−259.31	0.586×10^5	−252.89	0.452×10^6
Oxygen (O$_2$)	−218.83	0.138×10^5	−182.97	0.213×10^6
Nitrogen (N$_2$)	−210.10	0.257×10^5	−195.81	0.201×10^6
Aluminum	658.7	3.984×10^5	2300	10.5×10^6
Copper	1083	1.34×10^5	2600	3.1×10^6
Iron	1530	2.721×10^5	2500	6.364×10^6

Why has snow melted on the roof above the inside of the house but not on the roof above the unheated garage and eaves? What is the source of heat that causes the melting? (Photo by A. A. Bartlett, Boulder, Colo.)

In general, the heat Q needed to **melt** a mass m of solid is

$$Q = mL_f. \tag{9.6}$$

This is always a positive number because energy must be added to the material to melt it.

Freezing Liquids

Freezing is the opposite of melting. If energy is removed from a liquid, its temperature decreases; the vibrational, rotational, and translational motions of the atoms and molecules in the liquid are reduced. When the liquid reaches what is called its freezing temperature, the random motion of the atoms and molecules is slow enough that they start to fuse or bond with each other; the solid state begins to form. This transition from the liquid to the solid state is called freezing and occurs at the same temperature at which the solid melts. Thus, the freezing and melting temperatures of a substance are the same.

As the atoms or molecules fuse together into the structure of the solid state, energy is released. The amount of energy released equals that needed to break the bonds when the solid melts.

The energy Q released when a mass m of liquid **freezes** to form the solid state is

$$Q = -mL_f. \tag{9.7}$$

The negative sign reminds us that the liquid loses or releases energy when it freezes.

It is important to note that a substance does not change temperature while it is melting or freezing. For example, when heat is added to ice at $0\,°C$, no temperature change occurs in the ice-water mixture until all of the ice melts. Similarly, if heat is removed from water at $0\,°C$, it does not change temperature until all of the water has been converted to ice.

EXAMPLE 9.6 The $1.0 \times 10^7\,kg$ of ice in a small pond has an average temperature of $-5.0\,°C$ during the middle of winter. How much heat must be added to the ice to convert it to water at $27.0\,°C$ for summer swimming?

SOLUTION The conversion of the ice to water takes place in three steps.

1. The ice is warmed from $-5.0°C$ to $0.0°C$, its melting temperature. We can calculate the heat required for this process using Eq. (9.4) and the specific heat capacity of ice from Table 9.1:

$$Q(\text{to warm ice}) = mc\Delta T = (1.0 \times 10^7 \text{ kg})(2090 \text{ J/kg·C°})[(0°C) - (-5°C)]$$
$$= 1.0 \times 10^{11} \text{ J}.$$

2. Next, the ice at $0°C$ is melted to form an equal mass of water at $0°C$. We use Eq. (9.6) with the heat of fusion for water taken from Table 9.2:

$$Q(\text{to melt ice}) = mL_f = (1.0 \times 10^7 \text{ kg})(3.35 \times 10^5 \text{ J/kg})$$
$$= 33.5 \times 10^{11} \text{ J}.$$

3. Finally, the ice having been converted to water at $0°C$, it must now be warmed to $27.0°C$. We again use Eq. (9.4) to calculate the heat needed to cause this temperature change, this time using the specific heat capacity of water from Table 9.1:

$$Q(\text{to warm water}) = mc\Delta T$$
$$= (1.0 \times 10^7 \text{ kg})(4180 \text{ J/kg·C°})(27.0°C - 0)$$
$$= 11.3 \times 10^{11} \text{ J}.$$

The total heat needed is the sum of the three terms:

$$Q_{\text{total}} = [(1.0 \times 10^{11}) + (33.5 \times 10^{11}) + (11.3 \times 10^{11})] \text{ J}$$
$$= \underline{46 \times 10^{11} \text{ J}}.$$

Our final answer is rounded off to two significant digits. Notice that most of the heat is needed to melt the ice.   ∎

EXAMPLE 9.7   Ten grams of ice at $0°C$ is placed in a Styrofoam cup containing 300 g of coffee initially at $90°C$. Calculate the final temperature T of the coffee after all the ice has melted. Assume that no heat leaves the Styrofoam cup. The specific heat capacity of coffee is 4180 J/kg·C°, the same as that of water.

SOLUTION   Since no heat leaves the cup, the coffee must lose an amount of heat equal to that gained by the ice, and the net heat change is zero:

$$Q_{\text{ice}} + Q_{\text{coffee}} = 0. \tag{9.8}$$

The heat transferred to the ice has two parts: (1) that needed to melt it and (2) that needed to raise the temperature of the melted ice from $0°C$ to its final temperature T. Thus,

$$Q_{\text{ice}} = m_{\text{ice}}L_f + m_{\text{ice}}c_{\text{water}}(T - 0°C).$$

The heat transferred from the coffee is given by

$$Q_{\text{coffee}} = m_{\text{coffee}}c_{\text{coffee}}(T - 90°C).$$

Notice that the final temperature T is less than $90°C$; hence Q_{coffee} is negative. Substituting these two expressions into Eq. (9.8), we find that

$$\overbrace{\underbrace{m_{\text{ice}}L_f}_{\substack{\text{To melt} \\ \text{ice}}} + \underbrace{m_{\text{ice}}c_{\text{water}}(T - 0°C)}_{\substack{\text{To raise } T \text{ of} \\ \text{melted ice}}}}^{\text{Heat gained by ice}} + \overbrace{m_{\text{coffee}}c_{\text{coffee}}(T - 90°C)}^{\text{Heat lost by coffee}} = 0.$$

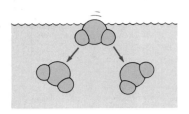

FIG. 9.10. A water molecule at the surface of the water is prevented from leaving the water by its attraction to neighboring molecules. Only the most energetic molecules can escape.

Solving for T, we find that

$$T = \frac{m_{\text{coffee}}c_{\text{coffee}}(90\,^\circ\text{C}) - m_{\text{ice}}L_f}{m_{\text{ice}}c_{\text{water}} + m_{\text{coffee}}c_{\text{coffee}}}$$

$$= \frac{(0.3\text{ kg})(4180\text{ J/kg}\cdot\text{C}^\circ)(90\,^\circ\text{C}) - (0.01\text{ kg})(3.35 \times 10^5\text{ J/kg})}{(0.3\text{ kg})(4180\text{ J/kg}\cdot\text{C}^\circ) + (0.01\text{ kg})(4180\text{ J/kg}\cdot\text{C}^\circ)} = \underline{85\,^\circ\text{C}}.\ \blacksquare$$

9.7 Boiling and Condensation

Boiling Liquids

Even in the liquid state, forces of attraction exist between neighboring atoms and molecules. For a molecule to leave the surface of a liquid, it must have enough energy to break away from the attracting forces of its neighbors (Fig. 9.10). When heat is added to the liquid, its temperature increases. The atoms and molecules move about faster and bump into each other harder. Eventually, some molecules in the liquid have enough energy to break away from their neighbors and leave the surface to become free molecules in the gaseous atmosphere above the liquid. This is called **evaporation.** As the liquid's temperature increases, the rate of evaporation increases because more molecules have enough energy to break away from their neighbors.

If enough heat is added to a liquid, eventually so many molecules will be leaving the liquid that all the extra energy added to the liquid is used by the escaping molecules. No energy is left to increase the temperature of the molecules remaining behind, and the liquid is said to be **boiling.**

The **boiling temperature** of a liquid is the temperature at which any heat added to the liquid causes the liquid to evaporate but does not cause an increase in temperature. Like freezing temperature, boiling temperature varies from substance to substance; at atmospheric pressure water boils at 100 °C, and copper boils at 2300 °C.

The amount of energy needed to convert a liquid to a gas depends on the type of liquid and is indicated by a quantity called the liquid's heat of vaporization (or the latent heat of vaporization). The **heat of vaporization** L_v of a substance is the energy required to convert 1 kg of the substance from the liquid to the gaseous state at the liquid's boiling temperature. The units of L_v are J/kg. The heat of vaporization of water is 2.256×10^6 J/kg. Table 9.2 lists the boiling temperatures and heats of vaporization for a variety of substances.

The energy required to boil a liquid also depends on its mass—the greater the mass the more energy that is needed to boil it.

The amount of heat Q required to **boil** a mass m of substance whose latent heat of vaporization is L_v is

$$Q = mL_v. \tag{9.9}$$

EXAMPLE 9.8 How much heat is required to evaporate 20 g of ethanol?

SOLUTION From Table 9.2, the latent heat of vaporization of ethanol is $L_v = 8.5 \times 10^5$ J/kg. Thus, the heat required to evaporate 20 g is

$$Q = (0.020\text{ kg})(8.5 \times 10^5\text{ J/kg}) = \underline{1.7 \times 10^4\text{ J}}.$$

Ethanol (rubbing alcohol) is often applied to a person's skin to cool the body. Heat is transferred from the body and skin to the ethanol, causing it to evaporate. ■

EXAMPLE 9.9 A woman carrying a 15-kg pack on a hike in the desert sun evaporates perspiration from her skin at a rate of about 0.28 g/s. Estimate the rate at which heat must be transferred out of her body to the liquid so that it can evaporate at this rate. At the surface temperature of the skin, the heat of vaporization of perspiration is approximately 2.4×10^6 J/kg.

SOLUTION We divide each side of Eq. (9.9) by the same time interval t:

$$\frac{Q}{t} = \frac{m}{t} L_v.$$

Q is heat that must be added to the liquid on the skin from the body in time t to cause a mass m of the liquid to evaporate in that time. For this problem, $m/t = 0.28$ g/s $= 0.28 \times 10^{-3}$ kg/s, $L_v = 2.4 \times 10^6$ J/kg, and we wish to determine the heat transfer rate, Q/t:

$$\frac{Q}{t} = \frac{m}{t} L_v = (0.28 \times 10^{-3} \text{ kg/s})(2.4 \times 10^6 \text{ J/kg}) = 670 \text{ J/s} = \underline{670 \text{ W}}.$$

The body loses heat by evaporation at a rate of 670 W, a very important factor in cooling the body. ■

Condensation

Condensation, the reverse of evaporation, occurs when gas molecules are cooled to their boiling temperature. The further removal of energy causes the gas molecules to come together (that is, condense) to form a liquid. The **condensation temperature** is the same as the boiling temperature.*

The heat given up when 1 kg of gas condenses to form 1 kg of liquid is called the **heat of condensation** and equals the heat of vaporization.

The heat energy Q transferred from a mass m of gas that **condenses** to the liquid state is

$$Q = -mL_v. \quad \text{(9.10)}$$

The negative sign reminds us that the gas releases energy when it liquefies.

The temperature at which a liquid boils or a gas condenses depends on the pressure of the gas above the liquid. Table 9.3 shows the variation of the boiling temperature of water as a function of pressure. As the air pressure above the water decreases, so does its boiling temperature. This variation of boiling temperature with pressure has a significant effect on cooking times. In Denver, Colorado, the air pressure is about 610 millimeters of mercury (mm Hg) compared to a pressure of 760 mm Hg at sea level. We see from Table 9.3 that water

*Some materials pass directly from the solid state to the gaseous state without becoming a liquid. When this occurs, the solid is said to **sublime.** The heat required to make this transition is called the **heat of sublimation.** Dry ice, the solid form of carbon dioxide, is such a substance. It sublimes from the solid, dry-ice state to the gaseous state at a temperature of about −79°C.

TABLE 9.3 Boiling Temperature of Water as a Function of Pressure

Pressure (mm Hg)	Boiling T(°C)
1500	120.3
1000	108.0
760 (atmospheric pressure at sea level)	100.0
700	97.8
600	93.6
400	83.0
100	51.6
50	38.1
20	22.2
10	11.3
4.6	0.0
2.0	−10.8

FIG. 9.11. A schematic representation of a fossil fuel electric power plant.

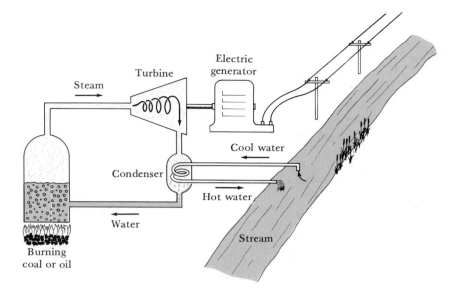

boils at about 94°C at Denver. It takes longer to boil an egg in Denver because the cooking temperature is lower.

The moon has no atmosphere, and its air pressure is nearly zero. On the moon, water boils at a very low temperature—below 0°C. An astronaut standing on the moon must wear a pressurized suit to prevent his or her blood from boiling.

EXAMPLE 9.10 (Electric power plants) An electric power plant is illustrated in Fig. 9.11. Water is heated by burning coal, oil, or gas (a fossil fuel plant) or by energy released by nuclear fusion (a nuclear boiling-water reactor). The water is heated under very high pressure from about 80°C to 260°C; it does not boil until it reaches about 250°C. High-pressure steam produced above the water pushes through a steam turbine, causing its blades to rotate. An electric generator attached to the turning blades produces electricity. After leaving the turbine, the steam is cooled in a condenser and converts back to liquid water, which returns to the boiler for another cycle.

We will make three calculations: (a) Estimate the energy transfer rate $Q/\Delta t$ to the water in the boiler needed to convert 8 kg/s of water to steam. (b) If 35 percent of the energy transferred to the water results in production of electrical energy, what is the rate of electric power production? (c) If the other 65 percent of the energy is carried away by cooling water at the condenser, what is the rate of heat transfer to the environment?

SOLUTION (a) We wish to convert 8 kg/s of water at 80°C into steam at 260°C. Because of the high pressure in the boiler, water does not boil until about 250°C. The rate of heat transfer will be

$$\frac{Q}{\Delta t} = \underbrace{\frac{\Delta m}{\Delta t} c_{\text{water}}(250°\text{C} - 80°\text{C})}_{\substack{\text{To raise water} \\ \text{to } 250°\text{C}}} + \underbrace{\frac{\Delta m}{\Delta t} L_v}_{\substack{\text{To boil} \\ \text{water}}} + \underbrace{\frac{\Delta m}{\Delta t} c_{\text{steam}}(260°\text{C} - 250°\text{C})}_{\substack{\text{To raise } T \text{ of steam} \\ \text{from } 250°\text{C to } 260°\text{C}}}$$

$$= \left(8\,\frac{kg}{s}\right)\left(4180\,\frac{J}{kg\cdot C^\circ}\right)(170\;C^\circ) + \left(8\,\frac{kg}{s}\right)\left(2.3\times10^6\,\frac{J}{kg}\right)$$
$$+ \left(8\,\frac{kg}{s}\right)\left(2000\,\frac{J}{kg\cdot C^\circ}\right)(10\;C^\circ)$$
$$= 24\times10^6\,J/s = 24\times10^6\,W = \underline{24\;MW}.$$

(b) The electric power produced is 0.35×24 MW $= \underline{8.4\;MW}$.

(c) The remaining 65 percent of the power, or 0.65×24 MW $= \underline{15.6\;MW}$ is transferred to the environment as thermal pollution. This huge amount of heat is one of the environmental problems of most types of electric power plants.

■

9.8 Thermal Expansion

In addition to causing changes in the temperature and the state of a substance, the transfer of heat can also cause a change in the volume of a substance. As its temperature increases, the atoms and molecules of a substance vibrate with greater amplitude. If free to translate, they move faster and have more energetic collisions. Because of their increased motion, the atoms and molecules occupy slightly more space than when moving less vigorously. They are, on the average, spread farther apart when the substance is hot than when cool. Therefore, almost all forms of matter expand when heated and contract when cooled.

Architects and engineers must consider thermal expansion and contraction of metal beams when designing tall buildings and bridges. Expansion joints are usually placed in sidewalks, bridges, and highways to prevent their buckling during summer heat. The road shown in Fig. 9.12 did not have an expansion joint, and as its length increased on a hot summer day, the road buckled. Dentists must use filling materials that expand and contract with temperature change at the same rate as a tooth; otherwise, fillings will become loose.

The change in length—**linear expansion**—ΔL of an object that is warmed or cooled depends on its change in temperature ΔT, on its original length L, and

FIG. 9.12. A road that buckled because of expansion during the summer heat. (Courtesy of the New York Department of Transportation.)

TABLE 9.4 Coefficients of Thermal Expansion at 20°C

Substance	Linear α (C$^{\circ -1}$)	Volume γ (C$^{\circ -1}$)
Aluminum	25×10^{-6}	72×10^{-6}
Steel and iron	12×10^{-6}	36×10^{-6}
Glass (Pyrex)	3×10^{-6}	9×10^{-6}
Glass (jar)	$\sim 10 \times 10^{-6}$	$\sim 30 \times 10^{-6}$
Brick and concrete	$\sim 10 \times 10^{-6}$	$\sim 30 \times 10^{-6}$
Rubber	$\sim 80 \times 10^{-6}$	$\sim 240 \times 10^{-6}$
Ethanol	250×10^{-6}	750×10^{-6}
Methanol	400×10^{-6}	1200×10^{-6}
Gasoline	$\sim 300 \times 10^{-6}$	$\sim 900 \times 10^{-6}$
Air		3670×10^{-6}

on the type of material of which it is made. These quantities are related by the equation

$$\Delta L = \alpha L \, \Delta T. \qquad (9.11)$$

The quantity α is called the **coefficient of thermal expansion.** From Eq. (9.11) we see that $\alpha = \Delta L/(L \, \Delta T)$ and has units of inverse degrees (C$^{\circ -1}$) because the length units cancel. Table 9.4 lists the values of α for various materials. The coefficient α is not a constant for a particular material but changes slightly with temperature. However, because the change is usually small, we ignore it in our problems and examples.

EXAMPLE 9.11 A steel beam in a bridge extends 25 m across a small stream. What is its change in length from the winter, when its temperature is $-20°C$, to the summer, when it is $38°C$?

SOLUTION The change in temperature of the bridge is

$$\Delta T = 38°C - (-20°C)$$
$$= 58 \, C°.$$

Using Eq. (9.11), we find that the bridge's change in length is

$$\Delta L = \alpha L \, \Delta T = (12 \times 10^{-6} \, C^{\circ -1})(25 \text{ m})(58 \, C°)$$
$$= 0.017 \text{ m} = \underline{1.7 \text{ cm}}. \qquad \blacksquare$$

EXAMPLE 9.12 The aluminum lid of a jar of dill pickles is stuck to the jar. To loosen the lid so it can be opened, hot water is poured over the lid, causing it to expand. If the temperature increase of the lid and glass is 40 C°, calculate the change in circumference of the lid and of the glass on which it is screwed. The diameter of the lid before heating is 22 cm.

SOLUTION The lid and glass expand uniformly in all directions. Thus, as each warms, its diameter d increases by an amount Δd equal to

$$\Delta d = \alpha d \, \Delta T,$$

where α is the coefficient of linear expansion of the aluminum lid ($25 \times 10^{-6} \, C^{\circ -1}$) or glass rim ($10 \times 10^{-6} \, C^{\circ -1}$) and ΔT is the temperature change. The circumference changes from πd to $\pi(d + \Delta d)$ or by an amount $\pi \, \Delta d = \pi \alpha d \, \Delta T$. For the aluminum lid, the circumference change is

$$\pi \alpha_{\text{alum}} \, d \, \Delta T = \pi(25 \times 10^{-6} \, C^{\circ -1})(22 \text{ cm})(40 \, C°)$$
$$= 0.069 \text{ cm} = \underline{0.69 \text{ mm}}.$$

The circumference of the glass changes by

$$\pi \alpha_{\text{glass}} d \, \Delta T = \pi(10 \times 10^{-6} \, C^{\circ -1})(22 \text{ cm})(40 \, C°)$$
$$= 0.028 \text{ cm} = \underline{0.28 \text{ mm}}.$$

Since the aluminum expands more than the glass, we should now be able to open the jar of pickles. $\blacksquare$

Because a material usually expands uniformly in all directions as it becomes warmer, its volume increases. The change in volume ΔV of a substance depends on its change in temperature, its original volume V, and the material of which it is made:

$$\Delta V = \gamma V \Delta T. \tag{9.12}$$

The quantity γ (the Greek letter gamma) is called the **coefficient of volume expansion;** its value depends on the substance that is expanding or contracting. The units of γ are $C^{\circ -1}$, the same as those of the coefficient of linear expansion. A list of the values of γ for a variety of substances appears in Table 9.4.

The coefficient of volume expansion γ of most substances is about three times the coefficient of linear expansion α of the same substance. This seems reasonable since the volume of an object is the product of three linear dimensions, such as length, height, and width, and each expands as it is heated. Although the mathematics needed to prove this statement are a little more difficult than this simple argument implies, we can see that the statement is true by comparing the values of α and γ in Table 9.4.

EXAMPLE 9.13 During a summer night when the temperature is 20°C, your house contains 453 m^3 of air. What volume of air leaves the house through an open window if the air warms to 40°C on a very hot summer day? Assume that the dimensions of the house experience negligible change and that all other conditions are constant.

SOLUTION As the air warms, its volume increases by an amount $\Delta V = \gamma V \Delta T$ where, according to Table 9.4, the coefficient of volume expansion of air is $\gamma = 3670 \times 10^{-6} \, C^{\circ -1}$. Thus,

$$\Delta V = (3670 \times 10^{-6} \, C^{\circ -1})(453 \, m^3)(40°C - 20°C) = \underline{33 \, m^3}.$$

Since the house stays about the same size, the air must find extra space outside. Thus, about 33 m^3 of air will leave the house, approximately the volume of air in a typical bedroom. ■

The change in volume of air as it warms and cools has significant effects on our atmosphere. As a gas such as air expands, it becomes less dense; its molecules on the average are spaced farther apart than in a cool gas. For this reason a hot (less dense) gas will rise and float on a cool (more dense) gas. This property of gases is very important in cleansing the atmosphere surrounding large cities. Air warmed at the earth's surface rises and carries with it air contaminants. Cool, clean air sinks down from above to replace the warm air.

Occasionally, an inversion occurs. For some reason, the air at the earth's surface remains cool and is covered by warmer air above. The temperature is said to be inverted. The cool, dense air at the surface does not rise, and pollution accumulates until the inversion breaks.

Another interesting environmental consequence of thermal expansion and contraction is related to the freezing of water at the surface of a lake or pond. Why doesn't the 0°C water that is ready to freeze sink to the bottom, allowing slightly warmer water to rise to the surface? This would happen with most substances that expand as they warm and contract as they cool. Water is a

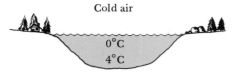

Cold air

0°C
4°C

FIG. 9.13. Since water at 4°C is denser than water at 0°C, the warmer water sinks. The cold water freezes at the top of a lake rather than at the bottom.

notable and important exception. When water cools from 4°C to 0°C, it expands and becomes less dense than the slightly warmer water; the 0°C water has less weight per unit volume than the 4°C water. Thus, in a lake or pond, the less dense 0°C water floats at the top while the more dense, slightly warmer water sinks to the bottom, as depicted in Fig. 9.13. Consequently, a lake freezes from the top down rather than from the bottom up. During the winter, fish remain near the bottom of the lake in the slightly warmer water.

Summary and Additional Readings

1. **Temperature scales:** An object's temperature depends on the average random kinetic energy of its atoms and molecules. Three temperature scales are used to measure temperature: the Celsius scale, the Fahrenheit scale, and the Kelvin scale. Temperatures in the Celsius and Fahrenheit scales are related by the equations $T_F = \frac{9}{5}T_C + 32$ and $T_C = \frac{5}{9}(T_F - 32)$. The Kelvin temperature is calculated by adding 273 degrees to the Celsius temperature.

2. **Thermal energy** E_{th} is the random kinetic energy of all of the atoms and molecules in a substance. **Heat** Q is the energy transferred to or from a substance because of a difference in its temperature and that of its environment. When calculating the energy changes of a system using the conservation-of-energy principle,

$$Q + W = \Delta E_{system}$$
$$= \Delta KE + \Delta PE_g + \Delta E_{th} + \cdots, \quad (9.3)$$

heat appears on the left-hand side because it is energy added to or removed from the system. Thermal energy is on the right-hand side because it is a form of energy in the system.

3. **Specific heat capacity:** The amount of heat Q that must be added to a mass m of substance to change its temperature by ΔT is calculated using the equation

$$Q = mc \, \Delta T, \quad (9.4)$$

where c *is* the specific heat capacity of the substance.

4. The amount of energy consumed or released by a mass m of substance when it **changes state** can be calculated using the following equations:

To melt a solid: $\quad Q = mL_f, \quad$ **(9.6)**

To freeze a liquid: $\quad Q = -mL_f, \quad$ **(9.7)**

To vaporize a liquid: $\quad Q = mL_v, \quad$ **(9.9)**

To condense a gas: $\quad Q = -mL_v, \quad$ **(9.10)**

where L_f and L_v are the latent **heats of fusion** and **vaporization,** respectively.

5. The **linear expansion** ΔL of an object with length L caused by a temperature change ΔT is calculated using the equation

$$\Delta L = \alpha L \, \Delta T, \quad (9.11)$$

where α is the **coefficient of thermal expansion.**

The **change in volume** ΔV of a substance caused by a temperature change ΔT is given by

$$\Delta V = \gamma V \, \Delta T, \quad (9.12)$$

where V is the original volume of the substance and γ is the **coefficient of volume expansion.**

Edwin R. Jones, Jr., "Fahrenheit and Celsius: A History," *The Physics Teacher,* pp. 594–595, November (1980).

W. R. Magie, *Source Book in Physics,* Harvard University Press, Cambridge (1963). Includes excerpts from original articles concerning the development of the conservation-of-energy principle and the second law of thermodynamics.

P. W. Bridgman, *The Nature of Thermodynamics,* Harper and Brothers, New York (1961). Contains interesting and illuminating discussion of the first and second laws of thermodynamics.

Questions

1. We have said that a water molecule in a glass of water has 10^{14} collisions with other molecules in the glass each second. *Estimate* the number of years a college football player would have to play (24 hours a day) to have the same number of collisions. Explain all the assumptions used in your estimate.

2. Deserts often have much greater daily temperature variations than vegetated land at the same latitude. Explain.

3. A person has a solar collector that requires 2000 kg of water to store the solar energy. Approximately what mass of rocks would store the same amount of energy (assuming the same temperature change)?

4. Dublin, Ireland, and Edmonton, Alberta, Canada, are at the same latitude (53.5°). The January and July average temperatures in Dublin are 40°F and 60°F, respectively, while in Edmonton they are 10°F and 70°F, respectively. Why is the temperature variation so much greater in Edmonton than in Dublin?

5. A farmer's fruit storage cellar is unheated. To prevent the fruit from freezing, the farmer places a barrel of water in the cellar. Why does this help prevent the fruit from freezing?

6. Why does an egg take the same time to cook in water that is just barely boiling as in water that boils vigorously? *Moral:* Don't waste energy by boiling eggs vigorously.

7. Why does food cook faster in a pressure cooker than in an open kettle?

8. Your property must be surveyed, and measurements will be made with a metal tape. Your tax rate will be based on the amount of land you have. For tax purposes, is it better to have the survey done in the summer or in the winter? Explain.

9. Do you get more gasoline for your money if the gasoline is cold or hot when purchased? Explain.

10. Suppose that water contracted instead of expanded as its temperature decreased from 4°C to 0°C. Explain two ways in which this might affect us or our environment.

Problems

9.2 Temperature

1. (a) The body temperature of a chicken is about 107°F. Calculate its temperature in degrees Celsius. (b) The body temperature of a pig is 39.4°C. Calculate its temperature in degrees Fahrenheit.

2. The moon's temperature on its bright side is about 100°C and on its dark side, about −173°C. Calculate both these temperatures in degrees Fahrenheit and in degrees Kelvin.

3. (a) The highest temperature recorded on earth was 136°F on September 13, 1922, in Azizia, Libya. The U.S. record is 56.7°C on July 10, 1913, in Death Valley, California. By how much in Fahrenheit degrees did the United States record miss the earth's record? (b) The earth's record cold temperature was −127°F on August 24, 1960, at Vostok Station, Antarctica. Calculate this temperature in degrees Celsius and Kelvin.

4. A frog can survive for long periods with a body temperature anywhere between −1.2°C and 30°C. Calculate the difference in these temperature extremes in degrees Fahrenheit.

5. Nitrogen changes from a liquid to a gas at −196°C. Calculate this temperature in degrees Fahrenheit and Kelvin.

■ 6. Suppose that a person's body temperature, initially 37.0°C, increased by 1.00 percent on the Kelvin scale. What is the new temperature (a) on the Celsius scale and (b) on the Fahrenheit scale?

9.3 and 9.4 Heat, the Conservation-of-Energy Principle, and Specific Heat Capacity

7. A person will die if his or her body core cools from 37°C to about 31°C. Calculate the heat that must be removed from the body of a 70-kg person to cause this temperature change.

8. A 60-kg person consumes about 2000 kcal of food in one day. If 10 percent of this food energy is converted to thermal energy and cannot leave the body, calculate the temperature change of the person. [*Note:* 1 kcal = 4180 J.]

9. Calculate the amount of heat required to raise the temperature (a) of 0.5 kg of water by 10 C°, (b) of 0.5 kg of ethanol by 10 C°, (c) of 0.5 kg of iron by 10 C°.

10. Calculate the change in temperature of the 90 kg of air in a house to which 2.7×10^5 J of heat is added.

■ 11. Solar energy entering the windows of your house is absorbed and stored by a concrete wall of mass m. The wall's temperature increases by 10 C° during the sunlight hours. What mass of water, in terms of m, would have the same temperature increase if it absorbed an equal amount of energy?

■ 12. A tree leaf has a mass of 0.8 g and a specific heat capacity of 3700 J/kg·C°. The energy of the sunlight striking the leaf per second is 2.8 J/s. If all the solar energy striking the leaf is absorbed, calculate the energy absorbed by the leaf (a) in 1 second, (b) in 1 minute. (c) If this energy were not removed from the leaf, calculate the temperature change of the leaf in 1 minute. [*Note:* Do not be surprised if your answer is large. A leaf clearly needs ways to remove heat absorbed from the sun. These mechanisms are discussed in Chapter 10.]

■ 13. Calculate the time required for a 200-kg cast-iron car engine to warm from 30°C to 1000°C (approximately the melting temperature of iron) if the engine while idling absorbs 8000 J/s.

■ 14. *Estimate* the time needed for an 850-W coffee maker to prepare 10 cups of coffee. State clearly all numbers used in your estimate.

■ 15. The Superdome in New Orleans holds 76,791 people for a sporting event. The volume of air in the dome is about $3 \times 10^6 \text{m}^3$. If all the seats in the dome are filled and each person transfers his or her metabolic thermal energy to the air in the dome at a rate of 100 W (100 J/s), calculate the temperature change of the air in 2 h. Assume that no heat leaves the air through the walls, floor, or ceiling of the dome. The density of air is 1.3 kg/m³ and its specific heat capacity is about 1000 J/kg·C°.

■ 16. It is recommended for purposes of ventilation that the inside air in a home be replaced with outside air once every 2 h. This air infiltration occurs naturally by leakage through tiny cracks around doors and windows, even in well-caulked and weather-stripped homes. The home's volume is 450 m³. The density of air is 1.3 kg/m³, and its specific heat capacity at constant pressure is about 1000 J/kg·C°. (a) Calculate the mass of air lost every 2 h. (b) Calculate the mass of air lost per second. (c) Calculate the heat lost per second by this air if its temperature changes by 38 C°, the difference in the temperature of the inside and outside air. (This figure is the energy per second needed to warm outside air leaking into the house.)

■ 17. A 5.0-g lead bullet traveling at 300 m/s penetrates a wooden block and stops. If 50 percent of the initial kinetic energy of the bullet is converted into thermal energy in the bullet, by how much does its temperature increase?

■ 18. During a study break, a 64-kg student repeatedly lifts a 42-kg barbell 0.90 m from her chest to an extended position above her head. If her body retains 10 J of thermal energy for each joule of work done while lifting, how many times must she lift the barbell to warm her body 0.50 C°?

■ 19. (a) The great Arizona crater was created by the impact of a meteorite whose mass was estimated to be 5×10^8 kg. The meteorite's speed before impact was about 10,000 m/s. Calculate the kinetic energy of the meteorite before the collision. (b) If 20 percent of this energy was converted to thermal energy in the meteorite, whose specific heat capacity was 900 J/kg·C°, by how much did its temperature increase? [*Note:* Large amounts of rock found near the crater had melted on impact and then resolidified, indicating that the temperature of the rock during impact reached at least 1700°C.]

9.5 Calorimetry

■ 20. A 150-g insulated aluminum calorimeter containing 250 g of water is initially at 20°C. A 200-g metal block at 60°C is added to the water, resulting in a final temperature of 22.8°C. Calculate the specific heat capacity of the block.

■ 21. A 300-g insulated aluminum calorimeter holds 150 g of water. The water and aluminum are initially at 30°C; 200 g of ethanol at 55°C is added to the water. Calculate the final temperature of the mixture.

■ 22. A 500-g aluminum calorimeter contains 300 g of water. The water and aluminum are initially at 40°C. A 200-g iron block with a temperature of 0°C is added to the water. Calculate the final temperature of the mixture.

■ 23. To 200 g of coffee (essentially water) at a temperature of 70°C in an insulated cup, 25 g of milk at 10°C is added. If the specific heat capacity of milk is 3800 J/kg·C°, by how much is the coffee temperature lowered when the milk is added? Ignore the specific heat capacity of the cup.

■ 24. Bath water of 150-kg mass is at a temperature of 44°C. How much cool water at 20°C must be added to lower the temperature to 38°C? Ignore heat lost to the tub and air.

■ 25. An insulated bowl containing 0.20 kg of soup at 40°C is cooled by adding 20°C water. Calculate the mass of water that must be added so that the mixture has a final temperature of 34°C. Ignore the heat transfer to the bowl. The heat capacity of the soup is 3800 J/kg·C°.

■ 26. A small lake initially at 16°C and containing 5.0×10^9 kg of water absorbs 1.0×10^4 kg/s of 18°C water from a stream used to cool an electric power plant. Calculate the time needed to warm the lake 0.5 C°, assuming no heat is lost to the ground or air.

9.6 Melting and Freezing

27. Calculate the amount of energy needed to change a 0.50-kg block of ice at 0°C into water at 20°C.

■ 28. When 1.4×10^5 J of energy is removed from 0.60 kg of water initially at 20°C, will all the water freeze? If not, how much remains unfrozen?

■ 29. An electric heater warms a large 0°C block of ice at a rate of 200 J/s. Calculate the mass of ice that melts in 10 min.

■ 30. A 300-g insulated aluminum calorimeter containing 200 g of water has 100 g of ice at 0°C placed in it. The water and calorimeter are initially at 40°C. Calculate the final temperature of the calorimeter and its contents.

■ 31. Calculate the number of grams of ice at 0°C that must be added to an insulated cup with 250 g of tea at 40°C to cool the tea to 35°C. Ignore heat transfer to the cup.

■ 32. An ice-making machine removes heat from 0°C water at a rate of 280 J/s. Calculate the time needed to form 2.0 kg of ice at 0°C.

■ 33. A tub containing 50 kg of water is placed in a farmer's canning cellar, initially at 10°C. On a cold evening the cellar loses heat through the walls at a rate of 1200 J/s. Without the tub of water, the fruit would freeze in 4 h (the fruit freezes at −1°C because the sugar in the fruit lowers the freezing temperature). By how much time will the presence of the water delay the freeze? [*Hint:* Calculate the energy released in cooling the water to 0°C and then freezing it. This energy is made available to the room. The time delay is the extra time required to lose this energy through the walls.]

■■ 34. A Dow Chemical product called TESC-81 (mainly calcium chloride hexahydrate) is used as an energy-storage material for solar applications. Energy from the sun raises the temperature of the solid material, causing it to melt at 27°C (81°F). At night the energy is released as the salt cools and returns to the solid state. (a) Calculate the energy required to raise the temperature of 1.0 kg of solid TESC-81 from 20°C to the liquid state at 27°C. (b) How warm would 1.0 kg of water become if it started at 20°C and absorbed the same energy? (c) Discuss the desirability of TESC-81 as a heat-storage material compared to water. For TESC-81, c(solid) = 1900 J/kg·C° and $L_f = 1.7 \times 10^5$ J/kg.

9.7 Boiling and Condensation

35. How much energy is required to convert (a) 0.10 kg of water at 100°C to steam at 100°C? (b) 0.10 kg of liquid ethanol at 78°C to ethanol vapor at 78°C?

36. During a back rub, 80 g of ethanol (rubbing alcohol) is converted from a liquid to a gas. Calculate the heat removed from a person's body for this conversion. Ignore the ethanol's change in temperature before it changes state.

■ 37. An enclosed, insulated container holds 50 g of ethanol. To this, 50,000 J of heat is transferred. Calculate the final state and temperature of the ethanol if it is initially at 28°C. The heat capacity of ethanol in the gaseous state is 1800 J/kg·C°.

■ 38. A lightning flash releases about 10^{10} J of electrical energy. If all this energy is added to 50 kg of water (the amount of water in a 165-lb person) at 37°C, what is the final state and temperature of the water?

■ 39. A kettle containing 0.75 kg of boiling water absorbs heat from a gas stove at a rate of 600 J/s. How much time is required for the water to boil away, leaving a charred kettle?

■ 40. A nuclear power plant generates 1000 MW of waste heat. If this heat is disposed of in an evaporative cooling tower, how much water must be evaporated (a) per second; (b) per day?

■ 41. While jogging, a man must lose 320 J/s of thermal energy by transferring it from his body to moisture on his skin, thus causing the moisture to evaporate. What mass of perspiration must evaporate each second? The heat of vaporization of water at the temperature of the skin is approximately 2.4×10^6 J/kg.

■■ 42. Moisture in the air in the gaseous state condenses to form a cloud. A rainstorm follows, resulting in 2 cm of rain over an area 2 km × 2 km. *Estimate* the energy released as the water condenses and drops to the earth. Note that the mass of 1 m³ of water is 1000 kg.

■■ 43. A vacuum jar contains 100 g of water initially at 30°C. A vacuum pump lowers the pressure in the jar. (a) Using Table 9.3, *estimate* the pressure at which the water will start to boil. (b) If the latent heat of vaporization of water at this temperature is about 2.7×10^6 J/kg, how much heat is removed from the water when 1 g of the water boils away? (c) Calculate the temperature change in the remaining 99 g of water when this heat is removed.

9.8 Thermal Expansion

44. If one steel beam extended the entire height of the Sears Tower in Chicago (434 m), by how much would the beam contract from its summer high temperature of 37°C to its winter low of −20°C?

45. The main steel span of the Golden Gate Bridge is about 1350 m long. By how much will it expand if its temperature is increased from 10°F to 100°F?

46. An interstate highway is made of concrete slabs 25 m long placed end to end. (a) What expansion gap must be left between the slabs to prevent buckling when the temperature changes from 20°C to 50°C? (b) Starting with the gap calculated in part (a), find the gap when the temperature decreases to −20°C.

■ 47. A 100-m-long steel tape is used to measure the length of your property. When measured in the winter at a temperature of 10°C, your property is 85.000 m long. What is the measured length of the property using the same tape in the summer when the temperature is 30°C? Assume that the property does not change dimensions.

■ 48. (a) The Celsius temperature of a 100-m-long aluminum wire originally at 15°C doubles. By how much does its length increase? (b) If the wire is cut in half, how much does the length of each half increase for the same temperature change as in part (a). (c) A 100-m-long aluminum wire with twice the radius as that in part (a) experiences the same temperature change. How much does its length change?

■■ 49. Two concrete roadway slabs 25 m long are accidentally laid without expansion gaps (Fig. 9.14). How high will the slabs buckle up if their temperature increases from 10°C to 50°C?

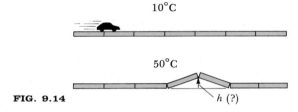

FIG. 9.14

■ 50. The volume of an iron ball of radius r is given by $\frac{4}{3}\pi r^3$. A particular ball has a diameter of 10.00 cm and is 0.01 cm too large to fit through a hole in a metal plate. What temperature change of the ball will allow it to fit through the hole?

■ 51. A 50-gal (1.9×10^5 cm³) steel drum at 5°C is filled to the brim with gasoline. The drum and its contents are warmed to 40°C. How much gasoline is lost through the open cap of the drum?

■ 52. The radiator of a car is filled to the overflow level with 5.4 liters of water. The water is warmed from 25°C to 95°C. How much water spills out of the overflow tube? The coefficient of volume expansion of water in this temperature range is $550 \times 10^{-6}\,\mathrm{C°^{-1}}$.

■■ 53. (a) Calculate the change in volume of the water in the earth's oceans if it warms from an average temperature of 15°C to 18°C. The present volume is approximately 1.4×10^{18} m³. (b) Approximately how high will the level of the ocean rise? The coefficient of volume expansion of water at this temperature range is approximately $180 \times 10^{-6}\,\mathrm{C°^{-1}}$. [*Note:* You need one other important quantity whose value is probably given in an encyclopedia.]

■■ 54. A rectangular lake that is 2000 m wide, 4000 m long, and 15 m deep contains 1.2×10^{11} kg of water. The lake is warmed by the addition of 2.0×10^{15} J of heat. Calculate its change in volume. For the water temperature in this problem, $\gamma = 210 \times 10^{-6}\,\mathrm{C°^{-1}}$.

■■ 55. Prove that the coefficient of volume expansion for a solid cube equals approximately three times the coefficient of linear expansion for one dimension of the cube.

CHAPTER 10

Heat Transfer

In Chapter 9 we learned that addition or removal of heat causes matter to change temperature or state and to expand or contract. In this chapter we consider the actual mechanisms by which heat is transferred to or from an object. These heat-transfer mechanisms include (1) conduction, (2) convection, and (3) radiation. We will also discuss a fourth mechanism, evaporation, which is really not a form of heat transfer but has the effect of cooling a body. These four mechanisms, either singly or in combination, are involved in the temperature control of a wide variety of systems such as our bodies, the buildings and homes in which we live, and the earth. These subjects are considered later in the chapter.

10.1 Heat Transfer by Conduction

If we place a metal pan over the flame of a gas stove, the handle soon becomes too hot to hold (Fig. 10.1). Heat entering the pan from the flame is transferred from the pan to the handle by a process called *conduction*. The atoms in the hot part of the pan have more thermal energy than those in the cooler handle. When these hot atoms vibrate and collide with their slower-moving neighbors, kinetic energy is transferred. The atoms that have just gained energy collide with their neighbors, and more thermal energy passes along the pan toward the handle. Eventually, atoms at the outer end of the handle are warmed by this bumping-type motion that passes from atom to atom along the pan. Heat transfer by **conduction** is a transfer of thermal energy from atom to atom through a material and is caused by the collision of hotter atoms with cooler, slower-moving atoms. The atoms themselves do not move from one end of the material to the other; only their random kinetic energy is transferred.

In most of the examples in this chapter we will consider the rate at which heat is transferred. The **heat-transfer rate** H is the amount of heat ΔQ that passes from one region of an object to another region divided by the time Δt needed for that heat transfer:

$$\text{Heat-transfer rate} = H = \frac{\Delta Q}{\Delta t}. \tag{10.1}$$

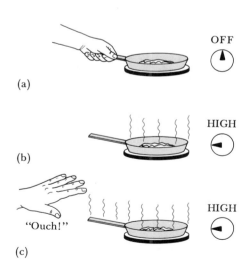

(a)

(b)

(c)

"Ouch!"

OFF

HIGH

HIGH

FIG. 10.1. (a) A pan is placed on a stove. (b) The stove, on high, warms the pan. (c) Eventually the handle, heated by conduction, becomes too hot to hold.

The units of H are watts, since 1 W is the ratio of a unit of heat, the joule, and a unit of time, the second (1 W = 1 J/s).

The heat-conduction rate across an object, such as shown in Fig. 10.2, can be determined using the following equation:

$$\left(\frac{\Delta Q}{\Delta t}\right)_{\text{conduction}} = H_{\text{cd}} = \frac{KA(T_2 - T_1)}{L}, \tag{10.2}$$

where H_{cd} is the **conductive heat-transfer rate** from region 1 of an object at temperature T_1 to another part of the object, region 2, at temperature T_2; L is the distance the heat must travel from region 1 to region 2; and A is the cross-sectional area across which the heat travels. The quantity K, **thermal conductivity,** indicates the ease or difficulty of heat transfer by conduction through different types of materials. A list of the thermal conductivities of several materials appears in Table 10.1. Metals, which are good heat conductors, have large values of K; poor heat conductors such as air, wood, and Styrofoam have small values of K. These latter materials are often called heat **insulators** and are used to prevent the escape of heat from ice chests, thermos bottles, and houses. In the SI metric system of units, K is in units of W/m·C°; A is in m²; L is in m; T_1 and T_2 are in °C; and H_{cd} is in W.

In our heat-transfer calculations in this chapter we will be interested in the heat transfer to or from one particular region of some object. For example, we may want to know the rate of heat lost from our skin to the surrounding air or the rate at which heat passes from inside a room to the outside air. Equation (10.2) and *all of our heat-transfer equations are written so that we can calculate the rate at which heat is entering or leaving region 1 at temperature T_1.* The signs in Eq. (10.2) and in equations we develop later are such that, if $\Delta Q/\Delta t$ is positive, heat is entering region 1; if $\Delta Q/\Delta t$ is negative, heat is leaving region 1.

EXAMPLE 10.1   One end of a cylindrical copper rod 1.2 m long with a 1.0-cm radius is in boiling water at 100°C. The other end is in ice at 0°C. (a) Calculate the conductive heat-transfer rate to the ice. (b) How much time is required to melt 10 g of ice?

SOLUTION   (a) We are given the following information:

$$L = 1.2 \text{ m},$$
$$A = \pi r^2 = \pi(0.010 \text{ m})^2 = 3.14 \times 10^{-4} \text{ m}^2,$$
$$T_2 - T_1 = T_{\text{water}} - T_{\text{ice}} = 100°C - 0°C = 100 \text{ C}°.$$

We have chosen to call region 1 the ice side of the rod and region 2 the boiling-water side. Our calculations will then tell us the heat-transfer rate to the ice, region 1.

The conduction coefficient for copper is $K_{\text{copper}} = 385$ W/m·C° (see Table 10.1). Substituting in Eq. (10.2), we find that

$$H_{\text{cd}} = \left(385 \, \frac{\text{W}}{\text{m·C°}}\right)(3.14 \times 10^{-4} \text{ m}^2)\left(\frac{100 \text{ C}°}{1.2 \text{ m}}\right) = \underline{10 \text{ W}}.$$

(b) We now know the heat-transfer rate $H_{\text{cd}} = \Delta Q/\Delta t$ and wish to determine the

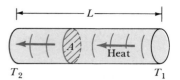

FIG. 10.2. Heat transfer by conduction occurs along a metal rod of cross-sectional area A and length L. The rate of heat transfer from region 1 at temperature T_1 to region 2 at temperature T_2 is given by Eq. (10.2).

TABLE 10.1   Thermal Conductivity K of Different Materials

Material	K(W/m·C°)
Metals	
Aluminum	205
Brass	109
Copper	385
Steel	50.2
Other solids	
Dry soil	0.2
Red brick	0.6
Concrete	0.84
Glass	0.8
Ice	1.6
Styrofoam insulation	0.016
Wood	0.04
Fat	0.021
Muscle	0.042
Bone	0.042
Other materials	
Air	0.024
Water	0.59

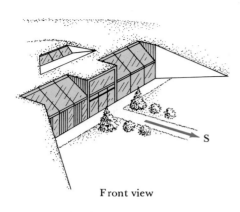

Front view

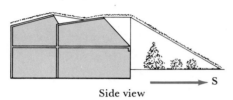

Side view

FIG. 10.3. An earth house. The roof and three walls are covered with earth. A fourth wall of windows provides a view and collects solar energy in the winter.

time needed to transfer enough heat to melt 10 g of ice. After rearranging the heat-transfer equation to find the time, substituting Eq. (9.7) for the heat needed to melt the ice, and taking the heat of fusion of ice L_f from Table 9.2, we find that

$$\Delta t = \frac{\Delta Q}{H_{cd}} = \frac{mL_f}{H_{cd}} = \frac{(0.010 \text{ kg})(335,000 \text{ J/kg})}{10 \text{ J/s}}$$
$$= \underline{340 \text{ s}}. \qquad \blacksquare$$

EXAMPLE 10.2 (a) A mobile home has walls that are 0.030 m thick and have a thermal conductivity of 0.10 W/m·C°. The total area of the walls is 45 m². Calculate the heat-transfer rate from the inside of the home to the outside if the inside temperature is 20°C and the outside temperature is −5°C. (b) Compare the conductive heat-transfer rate of the mobile home walls to that of the roof of a large 45-m² room in an earth house covered with 0.50 m of dry soil (Fig. 10.3).

SOLUTION (a) In this problem the heat travels a short distance $L = 0.03$ m across a large area $A = 45$ m². The small value of L and large value of A contribute to a high heat transfer by conduction. The temperature difference across which the heat travels is

$$T_2 - T_1 = T_{\text{outside}} - T_{\text{inside}} = (-5°C) - (20°C) = -25 \text{ C}°.$$

Substituting these values into Eq. (10.2), we find that

$$H_{cd} = \left(0.10 \frac{\text{W}}{\text{m·C}°}\right)(45 \text{ m}^2)\left(\frac{-25 \text{ C}°}{0.030 \text{ m}}\right) = \underline{-3750 \text{ W}}.$$

The negative sign indicates that the heat flows away from region 1, the region inside the home. The home is losing energy to the outside by heat conduction at a rate of 3750 W.

(b) Dry soil, according to Table 10.1, has a thermal conductivity $K_{\text{soil}} = 0.2$ W/m·C°. Substituting $L = 0.50$ m, $A = 45$ m², $T_2 - T_1 = -25$ C°, and the value of K_{soil} into Eq. (10.2), we find that the conduction heat-transfer rate from the roof of the earth house up through the earth is

$$H_{cd} = \left(0.2 \frac{\text{W}}{\text{m·C}°}\right)(45 \text{ m}^2)\frac{(-25 \text{ C}°)}{0.50 \text{ m}} = \underline{-450 \text{ W}}.$$

The extra thickness of the soil provides a barrier that greatly reduces the rate of heat loss from the house compared to the −3750-W heat-loss rate through the thin walls of the mobile home. $\blacksquare$

10.2 Heat Transfer by Convection

Remember that conductive heat transfer means the transfer of kinetic energy from atom to atom: The atoms do not move from one place to another. **Convective heat transfer** is the transfer of thermal energy from one place to another by mass that moves between the places and carries the energy with it. The heat transfer occurs from the warmer to the cooler material. An example of forced-convective heat transfer is the forced-air heating system used to warm houses in the winter. A furnace warms air, the air is blown or forced by a fan through ducts to different parts of the house, and then cold-air ducts return cool air to the furnace for reheating. The thermal energy is carried by the air from the furnace to other parts of the house; hence, this is a form of convective heat transfer.

The engine in an automobile is cooled by forced convection (Fig. 10.4). Water cooled in the radiator is pumped through the engine, where it absorbs heat generated by the engine. This warmed water then returns to the radiator for cooling. Heat is carried away from the engine by the flowing water.

A bicycle rider can lose energy as cool air moves across his skin (Fig. 10.5). The air carries heat away from the skin. Since the air mass is moving relative to the skin, we call this convective heat transfer. The examples just discussed are called **forced convection** because the fluid's motion past the object is caused by a mechanical force (a fan in the furnace pushes the air, a radiator pump pushes the water, and a bicycle rider pedals the bicycle through the air).

Natural convection occurs when a fluid such as air or water moves because of its change in temperature and density caused by absorption of heat from another object. During the day, for example, air near the earth's surface is warmed by its contact with the earth. As it warms, its volume increases. The air becomes less dense and rises, allowing cooler, more dense air from above to come down and replace the warm air. The air movement is a natural convection current. In a room in a house, air that is warmed near a radiator rises, and cool air near the ceiling falls. This natural convection current causes the room to become uniformly heated.

The calculation of the rate of convective heat transfer is a complex subject and depends on many factors. We can approximate the convective heat-transfer rate using the following equation:

$$\left(\frac{\Delta Q}{\Delta t}\right)_{\text{convection}} = H_{cv} = hA(T_2 - T_1) = hA(T_{\text{fluid}} - T_{\text{surface}}), \quad \textbf{(10.3)}$$

where ΔQ is the heat transferred convectively in a time Δt from or to a surface of area A across which a fluid such as air or water moves. T_{surface} (that is, T_1) is the temperature of the surface to or from which heat is being transferred, and T_{fluid} (that is, T_2) is the temperature of the fluid such as air or water that moves across the surface. The quantity h in Eq. (10.3) is called the **convection coefficient** and accounts for many of the other variables on which the convective heat-transfer rate depends, such as the type of fluid that crosses the surface, the speed of the fluid, and the shape and texture of the surface across which the fluid moves. The convection coefficient must be determined for each unique situation. Tables and empirical equations are often used to estimate the value of h for different situations. Figure 10.6 gives the value of h for dry air crossing a person's bare skin. We see that the value of h varies considerably with air speed. For example, if the air speed is 1 m/s, we see from Fig. 10.6 that $h \cong 15\ \text{W/m}^2 \cdot \text{C}°$. If the air speed is 3 m/s, then $h \cong 26\ \text{W/m}^2 \cdot \text{C}°$.

It is important when using Eq. (10.3) to consider the sign of the convective heat-transfer rate. If the fluid is warmer than the surface across which it moves, then $T_2 - T_1$ is greater than zero, and H_{cv} is positive. Heat is transferred from the warm, moving fluid to the cooler surface. For example, hot air from a hair dryer warms the skin as the air blows across it. If, on the other hand, the moving fluid is cooler than the surface across which it moves, then $T_2 - T_1$ is less than zero, and H_{cv} is negative. The cool fluid removes heat from the warmer surface.

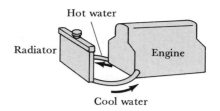

FIG. 10.4. A car engine is cooled by convection. Moving water carries thermal energy away from the engine to the radiator.

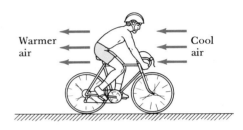

FIG. 10.5. The body of a cyclist is cooled convectively by air moving across his skin.

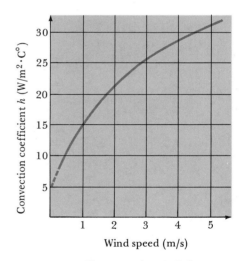

FIG. 10.6. The convection coefficient h as a function of air speed for dry air moving across bare skin. [Adapted from G. Brengelmann, *Physiology and Biophysics III*, 20th ed., T. C. Ruch and H. D. Patton, editors, p. 108, W. B. Saunders, Philadelphia (1973).]

EXAMPLE 10.3 A woman rides a bicycle at a speed of 4 m/s in still air with a temperature of 30°C. Her skin temperature is 35°C and her surface area is 1.4 m². If one-half her skin area is exposed to the air, what is the rate of convective heat loss from her skin?

SOLUTION From Fig. 10.6 we see that the convection coefficient h when $v = 4$ m/s is about 29 W/m²·C°. Although the rider's surface area is 1.4 m², only half that area is exposed. For the half that is covered, the air speed across the skin is zero, and no convection occurs. Thus, in our calculations we use only one-half the rider's area or 0.7 m². Substituting into Eq. (10.3), we find that

$$H_{cv} = \left(29 \frac{W}{m^2 \cdot C°}\right)(0.7 \text{ m}^2)(30°C - 35°C) = \underline{-100 \text{ W}}.$$

The negative sign means that the rider is losing energy from her skin at a rate of 100 W. Under these conditions, convection would account for about 30 percent of her heat loss. ∎

Wind-Chill Temperature

Our skin temperature is normally about 33°C, but it can vary from about 30°C to 35°C depending on the conditions of the environment. In the winter when the air temperature is low, there is convective heat loss from our bodies to the cool air that moves across our skin. As we see from Eq. (10.3) and Fig. 10.6, the rate of this heat loss increases as the speed of the cool air increases. We feel colder on a cold, windy day than on a cold, still day.

A quantity called the wind-chill temperature accounts for this effect of air speed on how cold we feel. The **wind-chill temperature** is the temperature of still, cold air that would cause the same heat-loss rate from our bodies as occurs when cool air is moving at a particular speed. Table 10.2 gives the wind-chill temperatures for different temperatures and wind speeds. For example, if the air temperature is 0°C and the wind speed is 10 m/s, the body loses heat at the same rate as if the temperature were −15°C with no wind speed. One of the functions of clothing is to produce a layer of still air around the skin, which decreases air movement and heat losses by convection.

10.3 Heat Transfer by Radiation

A third method of heat transfer involves a form of energy known as **electromagnetic radiation**, which is produced by vibrating and accelerating electric charges. When electrons vibrate back and forth in the antenna of a radio station, radio waves, a form of electromagnetic radiation, leave the antenna much as water waves move away from a beach ball pushed up and down in a pond. These radio waves travel at a speed of 3.0×10^8 m/s (186,000 mi/s) in a vacuum and in air, and they travel slower in other media. A small fraction of the waves leaving the radio station can be absorbed by the antenna of a car radio as electrons in the antenna are forced to move up and down on the passing wave. An energy transfer has thus occurred from the antenna of the radio station to that of the car. We do not call this particular process a heat transfer, but since energy moves from one place to another with the waves, it is an important energy-transfer mechanism.

TABLE 10.2 Wind-Chill Temperatures*

Air temperature (°C)	Wind speed, m/s (mi/h)				
	2 (4.5)	5 (11.2)	10 (22.4)	15 (33.6)	20 (44.7)
2	1.0	−6.6	−12	−16	−18
0	−1.3	−8.4	−15	−18	−20
−5	−7.0	−15	−22	−26	−29
−10	−12	−21	−29	−34	−36
−20	−23	−34	−44	−50	−52

*The numbers in color indicate the temperature of still air at which the heat-loss rate is the same as when warmer air whose temperature is indicated on the left moves past the body at the speeds indicated.

Another process, commonly termed **heat transfer by radiation,** is analogous to energy transfer by radio waves. Consider a log burning in a fire. Because of their high temperature, many of the molecules present in and near the fire vibrate somewhat more violently than usual. These large-amplitude molecular vibrations produce a form of radiation called *infrared waves*. The waves travel from the fire in all directions at the same speed as radio waves. Infrared waves falling on your skin are absorbed and cause the molecules in your skin to vibrate with greater amplitude. Your skin feels warmer. This process is called heat transfer by radiation.

There are many different forms of electromagnetic radiation, the most familiar being light—the visible radiation detected by our eyes. Radio waves, microwaves, infrared and ultraviolet radiation, x-rays, and gamma rays are all forms of electromagnetic radiation. All these types of radiation travel at the same speed (3.0×10^8 m/s in air or a vacuum), are produced by electric charges that vibrate or experience some other type of accelerated motion, and are absorbed by matter that lies in their path.

The sun emits all these forms of radiation, although most of the radiative energy leaving the sun consists of light and infrared radiation. The earth is bathed by approximately 1.7×10^{17} J of this radiative energy *each second*—more than the total energy consumed in the United States *each year*. If the earth had absorbed this energy for five million years (about the age of humans) without losing an equal amount of energy in some other manner, the temperature of the earth would now be well over 1000°C. However, our planet has managed to remain at a constant temperature (within a few degrees) by emitting into space each second an amount of electromagnetic radiation equal to that absorbed from the sun (Fig. 10.7). The earth is not special in this ability to emit radiation. Every object with a temperature above absolute zero emits electromagnetic radiation. A fire, a wall heater, a table, our bodies, and even ice cubes emit radiative energy.

The rate at which radiation is emitted from an object depends on its temperature T_1, its surface area A, and the type of surface it has. The surface is characterized by a quantity called emissivity e. **Emissivity** is a unitless number ranging from 0 to 1. A good emitter of radiation, such as a dark or black surface, has an emissivity close to 1. For a poor emitter, such as a white or shiny surface (for example, a mirror), e is close to zero. For the earth, the average value of e is about 0.95. For the human body it is 0.98.

The rate of emission of electromagnetic radiation from an object depends, then, on these three quantities—temperature T_1, area A, and emissivity e—according to the following equation:

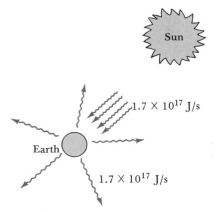

FIG. 10.7. The earth absorbs 1.7×10^{17} J/s of radiative energy from the sun and emits energy at the same rate into outer space.

$$\left(\frac{\Delta Q}{\Delta T}\right)_{\text{emission}} = R_e = e\sigma A T_1^4, \qquad (10.4)$$

where $\sigma = 5.67 \times 10^{-8}\,\text{W/m}^2 \cdot \text{K}^4$ is a constant called the Stefan-Boltzmann constant. Equation (10.4) is called Stefan's law. Note that the rate of emission of electromagnetic radiation from an object depends on the *fourth power of its temperature*. When this law is used, the *temperature must be in degrees Kelvin* (K)!

A macroscopic object at uniform temperature emits electromagnetic radiation at a rate proportional to the fourth power of its temperature. For example, a human body of surface area 1.7 m² emits radiation at a rate approximately equal to

$$R_e \cong (0.98) \left(5.67 \times 10^{-8}\,\frac{\text{W}}{\text{m}^2 \cdot \text{K}^4}\right) (1.7\,\text{m}^2) [\overbrace{(273 + 35)\,\text{K}}^{\text{Skin } T \text{ in K}}]^4 = 850\,\text{W}.$$
$$\underbrace{}_{\substack{\text{Skin } T \\ \text{in } °C}}$$

This is about eight times our resting metabolic rate. We would certainly cool to very low temperatures if we did not gain energy from some other external energy source. Our bodies gain considerable energy by absorbing electromagnetic energy emitted by objects in our environment. When an object, such as your body, is in an environment that has a uniform temperature T_2, the environment emits radiation in proportion to T_2^4. Some of this radiation is absorbed by the object. The rate of absorption by the object is given by

$$\left(\frac{\Delta Q}{\Delta t}\right)_{\text{absorption}} = R_a = e\sigma A T_2^4, \qquad (10.5)$$

where A is the object's area and e its emissivity. If the object is a good emitter with an emissivity near 1, it is also a good absorber.

For the object, then, the **net radiative heat-transfer rate** is the radiation absorbed minus the radiation emitted:

$$H_r = R_a - R_e,$$

or

$$H_r = e\sigma A T_2^4 - e\sigma A T_1^4 = e\sigma A (T_2^4 - T_1^4). \qquad (10.6)$$

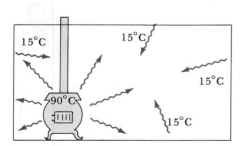

FIG. 10.8. A stove emits more radiation per unit area to the room than it absorbs because its surface temperature is higher than that of the room.

EXAMPLE 10.4 A black potbellied stove has a surface area of 1.5 m² and a surface temperature of 90°C (Fig. 10.8). Calculate the net radiative heat-transfer rate of the stove in a room whose wall temperature is 15°C. The emissivity of the black surface of the stove is about 1.0.

SOLUTION First we must convert the temperatures to degrees Kelvin by adding 273 to the temperature in degrees Celsius:

$$T_1 = T_{\text{stove wall}} = 273 + 90 = 363\,\text{K},$$
$$T_2 = T_{\text{room wall}} = 273 + 15 = 288\,\text{K}.$$

The net radiative heat-transfer rate from the stove, then, is

$$H_r = R_a - R_e = e\sigma A(T_2^4 - T_1^4)$$

$$= (1.0)\left(5.67 \times 10^{-8}\,\frac{W}{m^2 \cdot K^4}\right)(1.5\,m^2)[(288\,K)^4 - (363\,K)^4]$$

$$= \underline{-890\,W}.$$

The negative sign indicates that the stove is losing energy to its surroundings.

■

Thermography

The fact that objects emit radiation in proportion to the fourth power of temperature provides a useful technique for locating warmer objects surrounded by slightly cooler objects. Suppose, for example, that an object is 10 percent warmer than the temperature of its surroundings. If neighboring objects or material has a temperature T, then the warmer object has a temperature $1.1T$. The rate of radiative emission from the cool objects is proportional to T^4, whereas the rate of emission from the hotter object is proportional to $(1.1T)^4$. But $(1.1T)^4 = 1.1^4 T^4 = 1.46 T^4$. We see that the warmer object emits 1.46 times as much radiative energy per unit surface area (that is, 46 percent more) than neighboring objects, even though its temperature is only 10 percent hotter. Infrared cameras can easily detect the difference in radiative emission and in temperature from slightly warmer objects in cool surroundings.

Figure 10.9 shows an infrared picture called a **thermogram.** The change in temperature of a person's hand is apparent as he starts smoking a cigarette. Smoking causes the circulation to decrease and, hence, lowers the skin temperature. Less infrared radiation is emitted. Thermography is also used to detect cancerous growths near the skin's surface. These growths are usually warmer than their surroundings because of increased chemical activity in them.

This type of thermal photography is also used for other purposes. Figure 10.10 shows an infrared film of a house. Hot spots, such as the chimney and windows, appear white, indicating where the most energy is being lost from the house.

10.4 Heat Transfer by Evaporation

The molecules in a liquid such as water continually move about and collide with each other. At any instant there is much variation in their speed—some move fast and others slow. Slow-moving, less energetic molecules may reach the water surface, but they are usually prevented from escaping because of their attraction to neighboring molecules at the surface. However, energetic, fast-moving molecules that reach the surface can escape (Fig. 10.11). The continual escape of these energetic molecules from a liquid into the gas above, called **evaporation,** is a process that occurs at all temperatures.

Evaporation tends to cool a liquid because only the energetic, "hot" molecules escape. Those that remain in the liquid are less energetic and cooler. Therefore, the average thermal energy and temperature of molecules that remain in the liquid are less because of evaporation.

Often, energy from some other source is added to the liquid to compensate for the decrease in thermal energy caused by evaporation. When you perspire, a

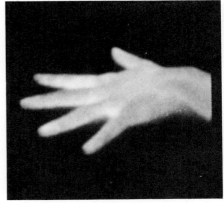

0 min

(b)   3 min

FIG. 10.9. The general reduction in heat emission from the hand as a person smokes a cigarette. Commercial thermographic systems, consisting of an infrared detector, camera, and display unit, can measure temperatures in the range 15–45°C with a minimum detectable temperature difference of 0.1 C° at 30°C object temperature. (Courtesy AGA Corporation, Pine Brook, N.J.)

layer of moisture covers your skin, and as the moisture evaporates, it becomes cooler. However, thermal energy created by metabolic processes inside your body is carried to the skin by your blood. The layer of moisture, then, is cooled by evaporation and warmed by heat removed from inside your body. Sweating is one of the important mechanisms for maintaining a constant body temperature when air temperatures are high or when we exercise.

The rate of heat loss by evaporation depends on the rate of evaporation—that is, the ratio of the mass Δm of liquid evaporated and the time Δt required for this evaporation:

$$\text{Evaporation rate} = \frac{\Delta m}{\Delta t}. \tag{10.7}$$

The evaporation rate is affected by many factors: (1) The evaporation rate of water depends on the relative humidity of the air. If the humidity is high, the air contains many water molecules, which may hit the liquid surface and join the liquid. This process, *condensation*, cancels the effect of evaporation. This is why it is more difficult to remain cool on a hot, humid day than on a hot, dry day. On a humid day, water condenses on our skin almost as fast as it evaporates, and there is no net cooling effect. (2) The rate of evaporation depends on whether the air near the body is moving. When water is evaporated from a lake on a still

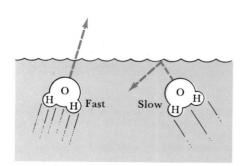

FIG. 10.11.   A fast-moving water molecule can escape from a liquid when it hits the surface. A slow-moving one cannot escape.

day, the water vapor remains above the water's surface (Fig. 10.12a). Many molecules return to the lake from this vapor. This condensation cancels the effect of evaporation. If a breeze blows, the gaseous water vapor moves away, and fewer molecules return to the lake's surface (Fig. 10.12b). Thus, the net evaporation rate is greater when a breeze blows. (3) The evaporation rate depends on the temperature of the liquid. As the liquid is warmed, more molecules gain the energy needed to leave the liquid surface. (4) The rate of evaporation depends on the surface area of the liquid. If the liquid is spread over a thin layer, as when the body sweats, evaporation is faster than when the same amount of liquid is held in a compact container.

With all of these variables to consider, it is difficult to write a general equation for calculating the rate of evaporation. We can say, however, that the heat ΔQ lost by a liquid when a mass Δm of the liquid evaporates is given by

$$\Delta Q \cong -L_e \, \Delta m,$$

where L_e is the heat of vaporization of the liquid at the temperature and pressure of the liquid that is evaporating. The heat of vaporization changes a little as the temperature and pressure change. For example, the heat of vaporization of water from a person's skin when at 35°C is approximately 2.4×10^6 J/kg, whereas the boiling-temperature heat of vaporization is 2.256×10^6 J/kg. We will use this approximate skin temperature value in our examples and problems. Thus, the heat lost by water when a mass Δm of the water evaporates is approximately

$$\Delta Q = -(2.4 \times 10^6 \, \text{J/kg}) \, \Delta m \qquad \text{(for water)}.$$

The negative sign indicates that the liquid remaining after evaporation has lost energy because of evaporation.

The evaporative heat-transfer rate—the heat lost per unit time—is calculated using the following equation:

$$\left(\frac{\Delta Q}{\Delta t} \right)_{\text{evaporation}} = H_e = -L_e \frac{\Delta m}{\Delta t}, \qquad \textbf{(10.8)}$$

or

$$\left(\frac{\Delta Q}{\Delta T} \right)_{\text{evaporation}} = H_e = -(2.4 \times 10^6 \, \text{J/kg}) \frac{\Delta m}{\Delta t} \qquad \text{(for water)}, \quad \textbf{(10.9)}$$

where $\Delta m / \Delta t$ is the evaporation rate. Several representative values of the evaporation rate for a person in various desert conditions are listed in Fig. 10.13. For example, with the outside temperature at 37.8°C, an average person sitting nude in the sun evaporates about 0.2 g of water each second.

EXAMPLE 10.5 Calculate the heat-loss rate by evaporation for an individual walking clothed in the desert sun at 37.8°C (100°F).

SOLUTION We can use the evaporation-rate numbers given in Fig. 10.13. The person's evaporation rate ($\Delta m / \Delta t$) when walking clothed in a desert is about 0.24 g/s. From Eq. (10.9) we find that

$$H_e = -\left(\frac{2.4 \times 10^6 \, \text{J}}{\text{kg}} \right) \left(\frac{0.24 \, \text{g}}{\text{s}} \times \frac{10^{-3} \, \text{kg}}{\text{g}} \right) = -580 \, \text{J/s}$$

$$= \underline{-580 \, \text{W}}. \qquad ■$$

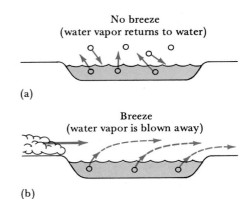

No breeze
(water vapor returns to water)

(a)

Breeze
(water vapor is blown away)

(b)

FIG. 10.12. (a) When there is no breeze, evaporated water remains above the lake, and some water vapor returns to the lake. This reduces the net evaporation rate. (b) When a breeze blows, evaporated water is removed from the lake and does not return. The net evaporation rate is greater with a breeze.

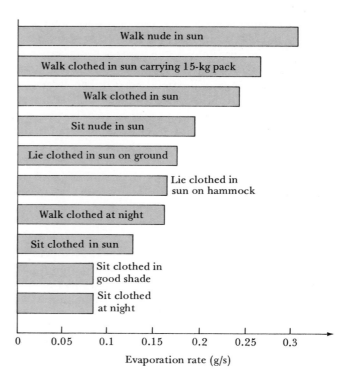

Evaporation is an extremely important heat-loss mechanism in the desert (and elsewhere). There is a net gain of radiative energy when walking in the sun of almost 500 W. This energy, along with thermal energy produced by metabolism (about 200 J/s when walking), must be removed from the body. Evaporation is responsible for most of the heat removal. We would be unable to live outdoors on warm days if we could not sweat.

EXAMPLE 10.6 The sun irradiates the earth's outer atmosphere with about 1.7×10^{17} J of energy each second. Of this energy, 47 percent is actually absorbed by the earth's surface, and the other 53 percent is absorbed or reflected by the atmosphere. Of the 47 percent absorbed, almost half is used to evaporate water. The water that evaporates condenses in the atmosphere to form clouds. How much water is removed from the earth each second by this evaporation?

SOLUTION We have been told that the amount of energy used for evaporation per second is half of 47 percent of 1.7×10^{17} J. The evaporative heat-loss rate is

$$H_e = \frac{(0.5)(0.47)(1.7 \times 10^{17}\,\text{J})}{1\,\text{s}}.$$

In general, however, the evaporative heat-loss rate is

$$H_e = -(2.4 \times 10^6\,\text{J/kg})\frac{\Delta m}{\Delta t}.$$

Substituting the value of H_e for the earth and rearranging, we find the earth's evaporation rate:

$$\frac{\Delta m}{\Delta t} = -\frac{(0.5)(0.47)(1.7 \times 10^{17}\,\text{J/s})}{(2.4 \times 10^6\,\text{J/kg})} = \underline{-1.7 \times 10^{10}\,\text{kg/s}}.$$

This is equivalent to the evaporation *each second* of all the water in a lake one-half mile in diameter and over 30 m deep!   ■

10.5   Temperature Control

The conservation-of-energy principle $(Q + W = \Delta E_{\text{system}})$ introduced in Section 8.6 and discussed again in Section 9.3 provides a useful framework for analyzing systems whose temperature is controlled by various heat-transfer and energy-conversion mechanisms. In this section we use the principle to study temperature control in our bodies, in our houses, and on the earth. We are interested in the heat ΔQ added to or removed from a system, the work ΔW done on the system, and the subsequent energy change of the system ΔE_{system} in a time Δt. Dividing each term in the conservation-of-energy equation by Δt results in the equation

$$\frac{\Delta Q}{\Delta t} + \frac{\Delta W}{\Delta t} = \frac{\Delta E_{\text{system}}}{\Delta t}. \qquad (10.10)$$

We consider briefly each of these terms before we use them in our analysis.

Rate of Change of a System's Energy: $\Delta E_{\text{system}}/\Delta t$

A system's energy can change with time in many ways. In this and the next chapter we consider primarily changes in the system's internal energy, in particular (1) its change in thermal energy with time $(\Delta E_{\text{th}}/\Delta t)$ and (2) its change in chemical energy with time $(\Delta E_{\text{chem}}/\Delta t)$. This latter form of energy, E_{chem}, is a form of electrical potential energy that holds atoms and molecules together. Examples of chemical energy change are the melting of ice, the conversion of natural gas and oxygen to combustion products, and the conversion by our bodies of sugar and oxygen to carbon dioxide and water. Whenever bonds between atoms in a molecule or bonds between different molecules are broken or formed, chemical energy changes.

For now, we consider only the changes in these two forms of internal energy and not the changes in mechanical energy of a system (for example, in its gravitational potential energy or its kinetic energy). Thus,

$$\frac{\Delta E_{\text{system}}}{\Delta t} = \frac{\Delta E_{\text{th}}}{\Delta t} + \frac{\Delta E_{\text{chem}}}{\Delta t}, \qquad \text{(when there are no changes} \quad (10.11)$$
$$\text{in mechanical energy).}$$

The energy conservation equation $Q + W = \Delta E_{\text{system}}$ when applied only to changes in a system's internal atomic and molecular energy is called the **first law of thermodynamics.**

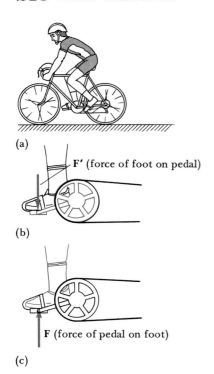

(a)

F′ (force of foot on pedal)

(b)

F (force of pedal on foot)

(c)

FIG. 10.14. (a) A rider pedaling a bike. (b) The rider pushes down on the pedal with a force **F′**. (c) The pedal pushes up on the rider with a force **F**.

Power: $\Delta W / \Delta t$

The work done on a system per unit time is called *power*. When calculating work, we consider only forces originating outside the system that act on the system. The work done by one of these external forces **F** as the system or some part of the system undergoes a displacement $\Delta \mathbf{r}$ is $\Delta W = F \Delta r \cos \theta$, where θ is the angle between the directions of **F** and $\Delta \mathbf{r}$.

Consider the work done on the bicycle rider pictured in Fig. 10.14a. We choose the rider as the system. The rider pushes down on the pedal with a force **F′** (Fig. 10.14b), the force of the rider on his environment. To calculate the work done on the system, we must determine the work done by the force **F** of the pedal on the rider's foot (Fig. 10.14c). This force points upward, whereas the displacement of the foot is down. Since **F** and $\Delta \mathbf{r}$ point in opposite directions, we find that the work done by the pedal on the system is a negative number:

$$\Delta W = F \Delta r \cos \theta = F \Delta r \cos 180° = -F \Delta r.$$

A negative value for work means simply that the system loses energy by doing work. The rider has less energy because of the work that was done. Often, when we choose an engine or a person as our system, the values of W and $\Delta W / \Delta t$ will be negative. The engine or person does work and hence has less energy.*

Heat-Transfer Rate: $\Delta Q / \Delta t$

In this chapter we have discussed four ways to transfer heat into or out of a system: conduction, convection, radiation, and evaporation. The net heat-transfer rate $\Delta Q / \Delta t$ is the sum of the four heat-transfer rates:

$$\frac{\Delta Q}{\Delta t} = \left(\frac{\Delta Q}{\Delta t}\right)_{\text{cond}} + \left(\frac{\Delta Q}{\Delta t}\right)_{\text{conv}} + \left(\frac{\Delta Q}{\Delta t}\right)_{\text{rad}} + \left(\frac{\Delta Q}{\Delta t}\right)_{\text{ev}}$$
$$= H_{cd} + H_{cv} + H_r + H_e. \tag{10.12}$$

Heat transfer is positive if energy is added to the system, negative if energy is lost by the system.

First Law of Thermodynamics

Substituting Eqs. (10.11) and (10.12) into Eq. (10.10) provides us with a general equation for describing the rate of a system's internal energy change:

$$\underbrace{H_{cd} + H_{cv} + H_r + H_e}_{\dfrac{\Delta Q}{\Delta t}} + \underbrace{\frac{\Delta W}{\Delta t}}_{+\dfrac{\Delta W}{\Delta t}} = \underbrace{\frac{\Delta E_{\text{th}}}{\Delta t} + \frac{\Delta E_{\text{chem}}}{\Delta t}}_{\dfrac{\Delta E_{\text{system}}}{\Delta t}} + \cdots, \tag{10.13}$$

---

*The first law of thermodynamics is sometimes written as $Q - W' = \Delta E_{\text{system}}$. When written this way, W' is defined as the work done by the system on the environment. For the bicycle rider example, $W' = F' \Delta r \cos \theta = F' \Delta r \cos 0° = +F' \Delta r$. The rider did positive work on the environment. The negative sign in front of W' in the first law of thermodynamics indicates that the system's energy decreased by $-W' = -F' \Delta r$ because of this work. Since the magnitudes of **F** and **F′** are equal, both sign conventions produce the same result.

We will apply Eq. (10.13), the time rate of change of quantities in the first law of thermodynamics, to three special situations involving temperature control: (1) body temperature control; (2) home temperature control; and (3) temperature control of the earth.

Body Temperature Control

The temperature of the central core of the body is maintained at about 37.0°C (98.6°F), even though the conditions outside and inside the body vary dramatically. Our metabolic rates, for instance, can increase by a factor of twenty during vigorous exercise. The extra thermal energy generated by the exercise must be removed from our body. If not removed, the body's core temperature would increase by 6°C in about 15 min, and brain damage and convulsions would occur.

Not only must we be able to remove great excesses of energy, we must be able to do it in a variety of external conditions. We live and work in outdoor environments that range in temperature from well below freezing to over 40°C (104°F). The ability of the body to adapt to different conditions is illustrated in Example 10.7.

EXAMPLE 10.7 A runner with a metabolic rate of 800 W (eight times the resting rate) runs on an indoor track at a speed of 3 m/s. The runner's surface area is 1.7 m² and her skin temperature is 35°C. Air resistance and friction forces oppose the runner and do negative work on her at a rate of −100 W (the runner loses energy by doing work against friction and air resistance). The runner's core temperature remains constant. The building in which the track is located has an air and wall temperature of 28°C. (a) Calculate the rate of heat transfer by convection and radiation (we ignore conductive heat transfer because it is small compared to the other forms of heat transfer). (b) Using Eq. (10.13), estimate the heat-loss rate by evaporation. (c) Use the result of part (b) to calculate the runner's evaporation rate, $\Delta m/\Delta t$.

SOLUTION (a) First we use Eq. (10.3) to calculate the convective heat-transfer rate. According to Fig. 10.6, the convection coefficient for a person moving at 3 m/s is $h = 25$ watts/m²·C°. Thus,

$$H_{cv} = hA(T_{air} - T_{skin}) = \left(25 \frac{W}{m^2 \cdot C^\circ}\right)(1.7 \text{ m}^2)(28°C - 35°C)$$

$$= \underline{-300 \text{ W}}.$$

The radiative heat-transfer rate is determined using Eq. (10.6):

$$H_r = e\sigma A(T_{walls}^4 - T_{skin}^4)$$

$$\cong (0.98)\left(5.67 \times 10^{-8}\frac{W}{m^2 \cdot K^4}\right)(1.7 \text{ m}^2)[(301 \text{ K})^4 - (308 \text{ K})^4]$$

$$= \underline{-75 \text{ W}}.$$

Notice that we used an emissivity for the runner's body of 0.98 and that we converted temperatures to degrees Kelvin before substituting in Eq. (10.6). (b) We are now ready to substitute numbers into Eq. (10.13). We must remember that the change in chemical energy of the runner ($\Delta E_{chem}/\Delta t$) is just the negative of her metabolic rate. The runner loses chemical energy at a rate of -800 W due to the "burning" of sugar-type molecules with oxygen to form carbon dioxide and water. Also, since the runner's temperature is constant, $\Delta E_{th}/\Delta t$ is zero. We find, then, that Eq. (10.13),

$$H_{cd} + H_{cv} + H_r + H_e + \frac{\Delta W}{\Delta t} = \frac{\Delta E_{th}}{\Delta t} + \frac{\Delta E_{chem}}{\Delta t},$$

becomes

$$0 - 300 \text{ W} - 75 \text{ W} + H_e - 100 \text{ W} \cong 0 - 800 \text{ W},$$

or

$$H_e = -325 \text{ W}.$$

(c) The evaporative heat-loss rate, as expressed by Eq. (10.9), equals -325 W:

$$H_e \cong -(2.4 \times 10^6 \text{ J/kg}) \frac{\Delta m}{\Delta t} = -325 \text{ W},$$

or

$$\frac{\Delta m}{\Delta t} = \frac{325 \text{ W}}{2.4 \times 10^6 \text{ J/kg}} = 0.14 \times 10^{-3} \text{ kg/s} = 0.14 \text{ g/s} = \underline{490 \text{ g/h}}.$$

The runner will lose about one-half liter of water in one hour. A normal individual can easily sweat over one liter per hour, so the runner will be able to maintain a constant temperature by moderate sweating. Notice that 13 percent (100 W) of the runner's energy was used to do work, 38 percent (300 W) was lost by convection, 10 percent (75 W) was lost by radiation, and 40 percent (325 W) was lost by evaporation—that is, sweating. ■

Home Temperature Control

Equation (10.13) provides a satisfactory basis for examining home temperature control. We can, for example, estimate the rate at which energy must be supplied to a home by a furnace, wood stove, or other heating system to maintain the home's temperature. The changes in energy for two homes are summarized in Table 10.3, along with anticipated heating costs for three different heating systems.

Heat losses from the homes are primarily by conduction through the ceilings, walls, windows, and floors and by convection leakage of warm air out of and cold air into the house. The energy required to evaporate water for humidification purposes is also a source of heat loss. This humidified air leaks out of the house through door openings, cracks around windows, and other small openings.

In home 1, with a minimum of insulation, 22,280 W of power must be provided to compensate for conductive, convective, and evaporative heat losses. Of the total energy needed, 800 W are supplied radiatively by sunlight entering windows; 2000 W are provided by energy generated by the home's occupants (about 100 W per person) and by thermal energy made available by the use of electric and gas appliances; the remaining 19,480 W must be supplied by the heating system.*

---

*The $\Delta E_{chem}/\Delta t$ term is chemical in origin. For example, gas that is burned in a furnace has less chemical energy after burning than before; hence the negative sign for this quantity.

TABLE 10.3 Heat-Loss Rates and Heating Costs for Homes of About 1500-ft^2 Area*

		Home 1 (Minimum insulation)	Home 2 (Adequate insulation and storm windows)
Conduction	Walls	−4,580	−1,330
	Ceiling	−2,240	−1,440
	Floor	−3,670	−1,220
	Windows	−7,750	−3,720
+		+	+
Convection	Ventilation (one air change every two hours)	−2,940	−2,940
+		+	+
Radiation	Solar through windows	+800	+800
+		+	+
Evaporation	Humidification	−1,100	−1,100
=		=	=
$\dfrac{\Delta E_{chem}}{\Delta t}$	Energy-loss rate by people and appliances	−2,000	−2,000
	Energy-loss rate from heating system	−19,480	−8,950

	Yearly heating costs		
1. Gas furnace		$1000	$500
2. Electric radiative		$3000	$1500
3. Heat pump		$1500	$800

*An inside-to-outside temperature difference of 39°C is assumed. Home 1 has 4-in insulation in the ceiling and none in the walls. Home 2 has 6-in insulation in the ceiling, 4-in insulation in the walls, a foil with air gap under the floor, and storm windows. All units are in watts except the heating costs. The yearly heating costs are calculated for locations with an annual degree-day heating requirement of about 5700, including these cities: Boise; Boston; Columbus, Ohio; Indianapolis; Lincoln, Nebraska; Pittsburgh, and Providence. The costs are based on a charge of $0.08 per kilowatt hour for electricity and $0.44 per therm for natural gas.

Three types of heating systems are commonly used: (1) forced-air gas furnaces; (2) electric radiative heating systems; and (3) heat pumps (discussed in Chapter 11). Table 10.3 gives estimates of the heating costs per year for each of these systems. Of course, the cost of energy varies substantially in different parts of the country, and these costs increase each year. Costs also depend on the efficiency with which the heating system operates. The numbers in Table 10.3 are representative of energy costs in Ohio for the year 1985.

Home 2 is better insulated than home 1 and has storm windows. Notice that heat losses through the walls have been reduced by 70 percent for home 2 compared to home 1. The storm windows have reduced heat loss through windows by 50 percent. Overall, one-half as much energy is needed to heat home 2 as home 1.

We should look a little more closely at the convective heat losses caused by warm air leaking out of the house and by cool air leaking through cracks around windows and doors and through other small openings. These leaks into and out of the house, called **infiltration,** can be reduced by placing weather stripping around doors and windows and by caulking small cracks in the walls. Some

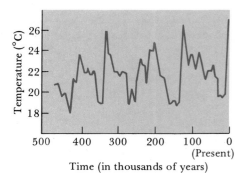

FIG. 10.15. The temperature of the earth during the last 500,000 years. [Adapted from J. D. Hays et al., "Variations in Earth's Orbit: Pace Maker of the Ice Ages," *Science* **194**, 1121–1132, December 10 (1976). Copyright 1976 by the American Association for the Advancement of Science.]

infiltration is desirable for ventilation. It is recommended that there be one complete air change in a house every one to two hours. The houses in our example are assumed to make one air change every two hours—very "tight" houses.

Temperature Stability of the Earth

The earth's temperature has fluctuated between 19°C and 27°C for the last five hundred thousand years (Fig. 10.15), during which there have been about five major changes in climate. When the earth's temperature was low, continental ice sheets 260 m thick extended into the middle latitudes. The most recent period of glaciation ended about six thousand years ago, when the Wisconsin glacier melted.

In recent years concern has grown that increasing amounts of carbon dioxide in the atmosphere will absorb radiation normally leaving the earth and cause a small increase in the earth's temperature. A rise of several Celsius degrees could supposedly melt the polar ice caps and cause major changes in the earth's climate.

Let us apply the first law of thermodynamics to the system pictured in Fig. 10.16 to see why this change might occur. The system inside the dashed lines consists of the earth and its atmosphere. The system gains radiant energy from the sun, R_{sun}, and loses radiant energy emitted by the earth and its atmosphere to outer space, R_{earth}. Thus, for this system, Eq. (10.13) becomes

$$R_{sun} - R_{earth} = \frac{\Delta E_{th}}{\Delta t} + \frac{\Delta E_{chem}}{\Delta t}.$$

If $R_{sun} - R_{earth} = 0$, then the internal energy and temperature of the system remain constant. If $R_{sun} - R_{earth} > 0$, the system's internal energy increases, and if $R_{sun} - R_{earth} < 0$, its internal energy decreases.

The radiant energy entering the system from the sun consists primarily of visible and short-wavelength infrared radiation. Energy emitted by the earth and its atmosphere is at much longer wavelengths—long-wavelength infrared radiation. Because carbon dioxide is a strong absorber in this long-wavelength region, increased carbon dioxide in the atmosphere would absorb radiation normally leaving the earth, R_{earth} would be reduced, and $R_{sun} - R_{earth}$ would be greater than zero. The internal energy of the earth would increase, as would its temperature. As the earth's temperature increased, it would radiate energy at a faster rate. Eventually its radiation rate would again balance energy coming from the sun, and the earth's temperature would stabilize at a new, higher temperature.

It has been estimated that doubling the CO_2 concentration in the atmosphere would cause about a 3-C° increase in the earth's temperature. If present trends continue, the CO_2 concentration will double before the middle of the next century. The CO_2 concentration is increasing for two reasons: (1) when fossil fuels such as oil, coal, and gas are burned, CO_2 is emitted into the atmosphere; and (2) forests normally absorb CO_2 as part of their photosynthetic growth process, but many forests on the earth are being cut for uses such as paper and building materials and are not being replenished.

It is difficult to calculate the changes expected because of the decreased

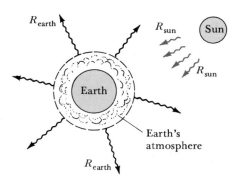

FIG. 10.16. The earth and its atmosphere constitute a system that gains energy by radiation from the sun (R_{sun}) and loses it by radiation to outer space (R_{earth}).

radiation leaving the earth. We can see, however, that if $R_{sun} - R_{earth} > 0$, then $\Delta E_{th}/\Delta t$ and $\Delta E_{chem}/\Delta t$ for the earth system will be greater than zero. A positive value of $\Delta E_{th}/\Delta t$ would cause the earth's temperature to increase. A positive value for $\Delta E_{chem}/\Delta t$ would cause ice to melt. Remember that chemical energy is needed to break the bonds of water molecules in ice when converting ice to liquid water. Thus, our simple model predicts a temperature increase and some melting of ice.

A two-year study by the United States Nuclear Regulatory Commission published in 1983 (see Summary and Additional Readings at the end of the chapter) anticipates average temperature increases of 1.5–4.5 C° in the next century and the melting of enough ice to increase the sea level by 70 cm or more. Climatic changes will also occur. The warming trend will increase the growing season in the northern United States by 10 days. The southern United States will receive less precipitation, which could reduce the agricultural output of the Texas Gulf coast, the Rio Grande valley, the upper and lower Colorado River regions, and California and other western areas.

Summary and Additional Readings

Heat is transferred between objects or between different parts of an object by four mechanisms: conduction, convection, radiation, and evaporation.

1. **Conduction:** Heat is transferred from the hot to the cold part of an object by the transfer of thermal energy from atom to atom through the object. The atoms do not leave their equilibrium positions. Only thermal energy moves through the object. The rate of heat transfer is calculated using the equation

$$H_{cd} = \frac{KA(T_2 - T_1)}{L}. \qquad (10.2)$$

2. **Convection:** Heat is transferred from one place to another by mass that moves between the places. The mass carries thermal energy with it. The rate of heat transfer by convection when a fluid (such as air) crosses the body is calculated using the equation

$$H_{cv} = hA(T_2 - T_1). \qquad (10.3)$$

3. **Radiation:** Heat is transferred by electromagnetic radiation, such as light and infrared waves. This radiation is emitted by any object with a temperature above absolute zero. The net rate of radiative heat transfer from an object is the difference of its rate of absorption and its rate of emission of radiation:

$$H_r = R_a - R_e = e\sigma A T_2^4 - e\sigma A T_1^4, \qquad (10.6)$$

where T_2 is the temperature of the surroundings in degrees Kelvin and T_1 is the object's temperature in degrees Kelvin.

4. **Evaporation:** When a liquid evaporates, the liquid that remains behind is cooled. The rate of heat transfer by evaporation is

$$H_e = -L_e \frac{\Delta m}{\Delta t}, \qquad (10.8)$$

where L_e is the latent heat of evaporation of the liquid and $\Delta m/\Delta t$ is the rate of evaporation (the mass of liquid evaporated per unit time). For water near 35°C,

$$H_e = -(2.4 \times 10^6 \text{ J/kg}) \frac{\Delta m}{\Delta t}. \qquad (10.9)$$

5. **Temperature control and the first law of thermodynamics:** When these heat-transfer mechanisms are incorporated into the first law of thermodynamics, we have an equation that is useful in the analysis of a wide variety of heat-transfer systems:

$$\underbrace{\frac{\Delta Q}{\Delta t} + \frac{\Delta W}{\Delta t}}_{} = \underbrace{\frac{\Delta E_{system}}{\Delta t}}_{}$$

or

$$H_{cd} + H_{cv} + H_r + H_e + \frac{\Delta W}{\Delta t} = \frac{\Delta E_{th}}{\Delta t} + \frac{\Delta E_{chem}}{\Delta t} + \cdots . \qquad (10.13)$$

This equation can be used to calculate an unknown heat-transfer rate, work rate, or internal energy change if all other terms in the equation are known.

Paul R. Ryan, "High Sea Levels and Temperatures Seen Next Century," *Oceanus*, pp. 63–67, Winter (1983/84). Assessment by Nuclear Regulatory Commission on effects of increased CO_2 concentration in the atmosphere.

Jennifer Adams, "Comparing Collectors," *Solar Age*, pp. 36–41, April (1982).

Joe Kohler and Dan Lewis, "Glass and Mass," *Solar Age*, pp. 32–35, February (1982). Data-rich article on passive solar principles.

Gaylon S. Campbell, *An Introduction to Environmental Biophysics*, Springer-Verlag, New York (1977). Contains many environmental applications of heat-transfer mechanisms.

Questions

1.   A potato into which several nails have been pushed bakes faster than a similar potato with no nails. Explain.

2.   Explain carefully why storm windows are effective at reducing heat losses in a home.

3.   The water in a paper cup can be boiled by placing the cup directly over a flame (such as from a candle or Bunsen burner). Explain why this is possible without burning a hole in the cup. [*Note:* The flame may even "touch" the cup!]

4.   Explain two reasons why blowing across hot soup or coffee helps lower its temperature.

5.   Placing a moistened finger in the wind can help identify the wind's direction. Explain why.

6.   Joggers often accumulate large amounts of water on their skin when running with the wind. When running against the wind, their skin may seem almost dry. Give two explanations.

7.   Why does covering a keg of beer with wet towels on a warm day help keep the beer cool?

8.   Explain why dogs can cool themselves by panting.

9.   Why is it more difficult to work vigorously on a hot, humid day than on a hot, dry day?

10.   The water in an outdoor swimming pool is heated faster and remains warmer if covered by a transparent, plastic cover than if not covered. Give two reasons.

11.   If you place one bare foot on a hot concrete swimming pool deck and the other bare foot on an adjacent wooden deck at the same temperature as the concrete, the concrete feels hotter. Why?

12.   A burning candle is placed between two cans containing equal amounts of water. One can is black; the other is silver. In which can will the water be heated most rapidly? Explain.

13.   A woman has a cup of hot coffee and a small container of milk. The milk is at room temperature and will be added to the coffee before she drinks it. If the woman must wait ten minutes before drinking the coffee, and she wants it to be as hot as possible at that time, should she add the milk and wait ten minutes or wait ten minutes and add the milk? Explain.

14.   Look carefully at your surroundings for one or more days. Make four recommendations of ways in which energy lost by heat transfer might be reduced. For one of these recommendations, make a rough *estimate* of the percentage of energy now lost that you think would be saved if your recommendation were accepted. Show your calculations.

15.   When the body is exposed to very cold temperatures, thermal energy must be conserved so that the body core temperature is not lowered. To do this, blood circulation to the skin and extremities is reduced, resulting in lower skin and extremity temperatures. Explain why this helps conserve body energy. [*Note:* One possible consequence of this strategy is frostbite of the extremities. The body sacrifices the extremities rather than the core!]

Problems

10.1   Heat Transfer by Conduction

1.   The wall of an adobe home in Mexico is 0.51 m thick and has an area of 30 m^2 and a thermal conductivity of 0.40 W/m·C°. (a) If the inside temperature is 15°C and the outside temperature is −5°C, what is the rate of energy loss through the wall? (b) How much energy is lost in 24 h?

2.   A goose down sleeping bag has an exposed area of 2.5 m^2 and is insulated with 6.0 cm of goose down whose thermal conductivity is 2.5×10^{-2} W/m·C°. (a) If a person in the bag is at 37°C and the outside temperature is −10°C, will the person lose heat through the bag faster than it is produced by metabolic processes (about 100 W)? Explain. (b) What if the bag is wet and has the thermal conductivity of water?

3.   A windowpane is 1.0 m × 1.3 m and is 3.0 mm thick. Its inner surface is at 10.0°C and its outer surface at 9.5°C. Calculate the conductive heat-transfer rate through the glass.

4.   A surfer wearing a 0.62-cm-thick wet suit of thermal conductivity 0.080 W/m·C° and area 1.4 m^2 bobs in the 15°C Pacific Ocean water. Calculate the conductive heat-transfer rate from the surfer's 35°C skin through the suit into the water.

■ 5.   A solid brass rod of radius 0.030 m sticks 1.0 m into the ground, where the bottom of the rod is at a temperature of 12°C. The other end is attached to the bottom of the metal water reservoir for a farmhouse. At what rate is heat transferred along the rod from the earth below to the reservoir whose temperature on a freezing night is 0°C? Ignore heat transfer from the sides of the rod.

■ 6.   Insulation rated as R-12 will conduct $\frac{1}{12}$ Btu/h of heat through each square foot of surface if there is a 1-F° temperature difference across the material. We say that R-12 insulation has a thermal resistance R of 12 h·ft^2 · F°/Btu. The heat-conduction rate through a material of area A, across which there is a temperature difference $T_2 - T_1$, is

$$\left(\frac{\Delta Q}{\Delta T}\right)_{cond} = \frac{1}{R} A(T_2 - T_1).$$

Calculate the conductive heat-flow rate across (a) an 8.0 ft × 16.0 ft R-15 wall, (b) a 3.0 ft × 7.0 ft R-4 door, (c) a 3.0 ft × 4.0 ft R-1.5 window if the inside temperature is 68°F and the outside temperature is 20°F. (d) Convert each answer from Btu/h to J/s = W.

■ 7.   One end of a steel bar with cross-sectional area 0.060 m^2 and length 0.50 m is in the flame of a Bunsen burner, and the other end is on a block of ice. The hot end is at 250°C. (a) Calculate the heat transferred from the hot end to the ice in 30 min. (b) Calculate the number of kilograms of ice that will melt.

■ 8.   You are to design an ice chest of dimensions 0.80 m × 0.60 m × 0.60 m that will store a 1.0-kg chunk of ice initially at 0°C for 5.0 h before the ice converts entirely to water at 0°C. If all of the heat loss is due to conduction through the 0.035 W/m·C° walls when their outside temperature is 28°C, how thick should the walls be?

■ 9. An igloo shaped like a hemisphere of radius 1.5 m has walls 0.36 m thick. Calculate the conductive heat-transfer rate through the walls if the inside temperature is 10°C and the outside temperature is −10°C.

■■ 10. Wood of thermal conductivity K_1, thickness L_1, and cross-sectional area A contacts a second insulating material of equal area, conductivity K_2, and thickness L_2. If the total temperature difference across the two materials is $T_2 − T_1$, show that the conductive heat transfer rate is given by

$$H_{cd} = \frac{A(T_2 − T_1)}{(L_1/K_1 + L_2/K_2)}.$$

[*Hint:* The conduction rate is the same across both materials. Use this fact to find the temperature between the two materials and then the conduction rate across one of the materials.]

10.2 Heat Transfer by Convection

11. A person in shorts stands in front of a fan that blows 25°C air at a speed of 2.5 m/s. If the person's surface area is 1.6 m² and skin temperature is 35°C, estimate the convective heat-loss rate from the person's body.

12. The temperature of a leaf is 5 C° above the air temperature. The surface area of the leaf is 2 cm² (1 cm² per side). When the air is moving at 5 mi/h, the convection coefficient is 24 W/m²·C°. What is the convective heat-loss rate from the leaf? [*Note:* The leaf gains heat from the sunlight at a rate of 0.14 W. Thus, the leaf must have other heat-loss mechanisms besides convection.]

■ 13. A woman runs with a speed of 3 m/s at the beach. Her skin temperature is 36°C, and her surface area is 1.4 m². The air is at 30°C and is blowing at 2 m/s (the air moves across the woman's body at a relative speed of 3 + 2 = 5 m/s). (a) *Estimate* her heat loss by convection. (b) What is the heat-loss rate by convection if she runs with the wind? Explain.

■ 14. A person blows 35°C air across the top of a bowl of soup with a temperature of 90°C. The air moves at a speed of 3 m/s, and the top surface area of the soup is 80 cm². (a) *Estimate* the convective heat-loss rate of the soup. (b) How much heat is lost in 10 min? (c) If the soup has a mass of 160 g and a specific heat capacity of 4000 J/kg·C°, what is its temperature change due to convection in 10 min?

■ 15. A person loses heat by convection at a rate of 250 W when standing in air whose speed past him is 1.0 m/s. Estimate the convective heat-loss rate if the air speed increases to 3.0 m/s. The skin-air temperature difference remains the same. Justify your answer clearly so that a classmate could understand.

10.3 Heat Transfer by Radiation

16. (a) Calculate the radiative heat-transfer rate of a woman with a surface area of 1.5 m² and emissivity 1.0 while standing in a bikini in a room at 25°C. Her skin temperature is 35°C. (b) Repeat the calculations for a woman in a sauna where the wall and air temperature is 45°C (113°F).

17. A man in shorts has a surface area of 1.7 m² and emissivity 0.95; he rides a bicycle on a cloudy day at a speed of 5 m/s. The air temperature is 30°C, and his skin temperature is 36°C. Calculate the man's net radiative heat-transfer rate.

■ 18. A woman sits in a room whose temperature is 27°C. A radiation-detection device indicates that each square meter of her skin has a radiative heat transfer rate of −57 W. Calculate her skin temperature, assuming that the emissivity of her skin is 0.98.

■ 19. The radiant energy emitted by the earth's surface per unit of area, R_e/A, is about 420 W/m², and its emissivity is 0.95. (a) *Estimate* the temperature of the earth. (b) The moon's average emittance is about 320 W/m². *Estimate* its temperature. [*Note:* Without our atmosphere, the earth's emittance rate and temperature would be the same as the moon's. Our atmosphere absorbs and reflects back to the earth much of the radiant energy that would otherwise leave. Our atmosphere is a "blanket" that helps keep the earth warm.]

■ 20. A leaf with an area of 1 cm² on each side and emissivity 1.0 absorbs radiation on each side at a rate of about 700 W/m². What should the leaf's temperature be if it must lose radiative energy at the same rate?

■ 21. The average sunlight incident on a photoelectric solar collector on a factory roof for the 8-h working day is 700 W/m². What must the area of the collector be to provide 3000 W of electricity? Assume that the radiant energy is converted to electricity with an efficiency of 15 percent and that the collector's emissivity is 1.0.

■ 22. A carbon rod, when heated to 1000 K, emits radiant energy at a rate of 20 W. Calculate the radiant energy emission rate if the rod is (a) at 1200 K and (b) at 1500 K.

■ 23. Calculate the ratio of the radiant energy *emitted* from a surface at 37°C and an adjacent surface of equal area and emissivity at 35°C.

■■ 24. The walls, floor, and ceiling of a room are at 20°C. By how many degrees Celsius must the ceiling temperature be increased to provide 1.2 kW of extra radiant energy to the room compared to what it would provide if at 20°C? The ceiling has an area of 50 m² and an emissivity of 0.95.

■■ 25. Show that if $T_2 − T_1$ is small (that is, $T_2 \cong T_1$), then an object's net radiative heat-transfer rate as given by Eq. (10.6) can be calculated approximately using the equation

$$H_r \cong e\sigma A 4 T_2^3 (T_2 − T_1).$$

[*Hint:* Recall that $x^4 − y^4 = (x^2 + y^2)(x^2 − y^2)$.]

10.4 Heat Transfer by Evaporation

26. A marathon runner loses heat by evaporation at a rate of 380 W. Calculate the moisture lost while she runs for 3 h.

27. A bowl of hot soup loses 0.40 g of water by evaporation in 1 min. Calculate the average evaporative heat-transfer rate of the soup during that minute, assuming the soup is primarily water.

■ 28. (a) *Estimate* the evaporative heat-transfer rate of a person walking clothed in the desert sun while carrying a 15-kg pack. (b) How much heat does the person lose in 30 min of walking?

■ 29. A keg of beer gains heat from its surroundings at a rate of 20 W. (a) At what rate in grams per second must water evaporate from a towel placed over the keg to cool the keg at the same rate that heat is being absorbed? (b) How much water in grams is lost in 2 h?

■ 30. A canteen is covered with wet canvas. If 15 g of water evaporates from the canvas and if 50 percent of the heat used to evaporate the water is supplied by the 400 g of water in the canteen, calculate the temperature change of the water in the canteen.

■■ 31. A man sitting clothed in the desert sun has a body temperature of 37.8°C. (a) *Estimate* his heat loss rate by evaporation. (b) If the moisture evaporates uniformly from the man's 1.5-m² body surface, what thickness of sweat evaporates in 30 min? The volume of 1.0 kg of water is 1.0×10^{-3} m^3.

■ ■ **32.** Each year a layer of water of average depth 0.8 m evaporates from each square meter of the earth's surface. *Estimate* the average energy-transfer rate in watts needed to continue this process. Note that 1 m^3 of water has a mass of 1000 kg.

10.5 Temperature Control

■ **33.** The water in a fishbowl has an evaporation rate of 0.050 g/s. The heat-transfer rate to the bowl by conduction, convection, and radiation is +36 W. At what rate must the water be heated by an electric heater to keep its temperature constant?

■ **34.** A marathon runner does 150 J of work each second and uses 1000 J of chemical energy each second. (a) How much heat must be removed from the runner's body per second to keep his temperature constant? (b) If 50 percent of this heat loss is caused by evaporation, how much water does the runner lose per second? (c) How much water does the runner lose in 3 h?

■ **35.** A man wearing shorts stands in a 23°C room. His skin temperature is 35°C and he does no work. Air moves slowly past his skin, resulting in a convection coefficient of 5.0 W/m^2·C°. The emissivity of his skin is 0.98 and its area is 1.6 m^2. (a) Calculate the rates of heat transfer by convection and radiation. We ignore conductive heat transfer and we assume that no evaporative heat loss occurs, since the man needs to conserve thermal energy. (b) Using the results of part (a) and Eq. (10.13), estimate the man's metabolic rate ($-\Delta E_{chem}/\Delta t$) if his temperature is to remain constant. [*Note:* The man can double or triple his resting metabolic rate if he shivers. Does this man need to shiver to maintain constant temperature?]

■ **36.** A cyclist with an exposed surface area of 1.5 m^2 and emissivity 0.95 rides on an indoor track in a building with walls and air at 28°C. The cyclist's metabolic rate is 600 W, and her skin temperature is 35°C. The bicycle and rider travel at a speed of 4 m/s. The cyclist expends work energy at a rate of 100 W (that is, $\Delta W/\Delta t = -100$ W). (a) *Estimate* the rate of heat transfer by convection and radiation. (Ignore the small rate of conductive heat transfer.) (b) Use the time rate of change of the first law of thermodynamics to estimate the heat-loss rate by evaporation. (c) Calculate the expected rate of evaporation.

■ **37.** A hot tub in an insulated container that prevents heat loss from the bottom and sides contains water at 37°C. The tub's top surface area is 2.0 m^2, and it sits in a room at 27°C. The emissivity of the water is 0.90 and the convection coefficient is 7.0 W/m^2·C°. Calculate (a) the convective heat-transfer rate and (b) the radiative heat-transfer rate. (c) If the water evaporation rate is 0.12 g/s, at what rate must a heater provide thermal energy to the water to keep its temperature constant? Ignore conductive heat transfer.

■ **38.** The water in a fishbowl must remain 10 C° above room temperature (22°C). It has the following heat-transfer rate losses: conduction (−20 W), convection (−40 W), radiation (−30 W), and evaporation (−30 W). (a) Calculate the surface area that is exposed to radiation if the emissivity is 0.90. (b) Calculate the amount of water lost per second by evaporation. (c) At what rate must an electric heater provide energy to keep its temperature constant?

■ **39.** A well-insulated bowl contains 250 g of soup. The ex-

posed surface of the soup has an area of 0.03 m^2 and an emissivity of 1.0. The soup is initially at 50°C, and the air temperature is 20°C. The evaporation rate $\Delta m/\Delta t = 4.0 \times 10^{-3}$ g/s. Air blows across the soup at a speed of 1.0 m/s. (a) Calculate the sum of the evaporative, convective, and radiative heat-loss rates from the soup (ignore the conductive loss). (b) How much heat will the soup lose in 5 min? (c) What will the soup's temperature be after 5 min? The specific heat capacity of the soup is 4000 J/kg·C°.

40.* On an average winter day (38°F or 3°C average temperature) a "standard" house requires heat because the energy already in the house is lost at the following rates: (i) 2.1 kW is lost through partially insulated walls and roof by conductance, (ii) 0.3 kW is lost through the floor by conductance, and (iii) 1.9 kW is lost through the windows (there are no storm windows) by conductance. Additional heat is also needed at the following rates: (iv) 2.3 kW to heat the air entering the house through cracks, flues, and other openings (infiltration losses) and (v) an additional 1.1 kW to humidify the incoming air (because warm air must contain more water vapor than cold air for people to be comfortable). What is the total rate at which energy is lost from this house?

41. On the same average winter day as in Problem 40, some heat energy is supplied to the same standard house in the following amounts (on a day with average cloudiness): (i) sunlight through windows, 0.5 kW; (ii) people's body heat, 0.2 kW; and (iii) heat from appliances, 1.2 kW. How many kilowatts must be supplied to the standard house in Problem 40 by the heating system of the house to keep it at constant temperature?

42. Suppose that the following design changes are made in the house in Problem 40: (i) added insulation of walls, roof, and floors, cutting the losses incurred there by 60 percent; (ii) tightly fitting double-glazed windows with selective coatings to reduce the passage of infrared light or with special shutters, cutting conductance losses by 70 percent; (iii) elimination of cracks, closing of flues, and so on, cutting infiltration losses by 70 percent. What is the total rate at which energy is lost from this house?

43. The "conservation" house of Problem 42 should include other design changes, such as shifting windows away from the north to the south sides and improvements in appliances (which might involve a microwave oven, venting of dryer heat internally in winter, substitution of new types of fluorescent bulbs for incandescent bulbs, and a well-insulated hot water tank). Heat energy is supplied to a house with these improvements at the following rates: (i) sunlight through windows, 1.0 kW; (ii) people's warmth, 0.2 kW; and (iii) appliance warmth, 0.8 kW. How many kilowatts must be supplied by the heating system of this house to keep it at constant temperature?

■ ■ **44.** Assume that because of increasing CO$_2$ concentration in the atmosphere, the net radiation heat-transfer rate for the earth and its atmosphere is $+0.02 \times 1350$ W/m^2, corresponding to a 2 percent decrease in radiation leaving the earth. (a) Calculate the extra heat added to the earth and its atmosphere in 10 yr. [*Note:* The radiation falls on an area equal to πr_{earth}^2, where $r_{earth} = 6.4 \times 10^6$ m.] (b) If 30 percent of this energy is used to melt the polar ice caps, how many kilograms of ice will be melted in 10 yr? (c) How many cubic meters of ice will be melted? [*Note:* The density of water is 10^3 kg/m^3.] (d) By how much will the level of the ocean rise in 10 yr? [*Note:* The ocean's surface area is 4.0×10^{14} m^2.]

*Problems 40–43 are adapted from Joan Ross and Marc Ross, *The Physics Teacher* **16,** 272 (May 1978). Copyright 1978 by The American Association of Physics Teachers.

CHAPTER 11

Entropy and Thermodynamics

Thermodynamics, the study of processes involving energy transformations and heat, originated in the early part of the nineteenth century, about sixty years after the start of the industrial revolution. During the industrial revolution, there was considerable interest in developing machines to do mechanical work. The steam engine, for instance, was built to pump water from mine shafts in England. Early models of the steam engine were extremely inefficient, but as they were gradually improved and their efficiency increased, scientists began to wonder if these machines had an ultimate level of efficiency. This wondering led to the formulation of thermodynamic principles that allowed scientists to calculate the maximum efficiency of a machine such as a steam engine.

Thermodynamics has remained an important tool for analyzing the efficiency or lack of efficiency of all types of mechanical devices such as steam engines, electric power plants, and automobiles. In recent years the laws of thermodynamics have also been used to analyze energy transformations in biological and geological systems. Thermodynamic laws can be used to study energy transformations occurring in living cells, energy transformations of global concern, and even philosophical ideas about the expected distribution of energy billions of years from now.

We saw in Chapter 10 how the first law of thermodynamics, based on the idea that energy is neither created nor destroyed, helps us understand a variety of processes such as the temperature control of our bodies, our homes, and the earth. In this chapter we focus on the second law of thermodynamics, which deals with the degradation or conversion of energy from useful to less useful forms.

In thermodynamics the usefulness of energy is rated according to the amount of work that can be done with it. The chemical energy stored in the bonds of octane molecules can do more work and is more useful than an equal amount of thermal energy in hot pavement or in seawater. The usefulness of the energy is rated using a quantity called **entropy.** Entropy, like golf, has an inverted scale: the lower the score, the more useful is the energy. We will find that entropy is related closely to the idea of order and disorder. Organized, useful energy has low entropy. The organized bonds between atoms in an octane or sugar molecule are an ordered way of storing energy. Less useful energy has high entropy and is more disordered. The thermal energy of a water molecule caused by its random motion is a disordered and less useful form of energy.

According to the second law of thermodynamics, the entropy of the universe increases during any process in nature. For example, the ordered form of the energy in octane molecules is converted to thermal energy as octane molecules combine with oxygen while burning. The burning of these molecules causes our entropy score to go up. Your body increases the entropy of the universe as you read this book; the chemical energy of sugar molecules in your body is being converted to thermal energy. The universe is not losing energy—the energy is just being converted to less useful forms.

The consequences of the second law of thermodynamics extend from the practical to the profound. The law affects the efficiency with which machines can operate and also influences the costs of maintaining a comfortable environment in which to live. If it were not for nature's observance of this law, it seems likely that we could find a fountain of youth to prevent our aging. Even philosophical ideas about predestination and the destiny of the universe have been viewed in terms of entropy and the second law of thermodynamics.

To begin our discussion, we look at reasons why the first law does not account for all aspects of energy transformation, which leads to an introduction of entropy and the second law of thermodynamics.

11.1 Irreversible Processes

Our discussions of energy until now have been based on the conservation-of-energy principle and the first law of thermodynamics. If a part of the universe called a system gains energy because of work done on it or by heat transfer into it, then the environment surrounding the system loses an equal amount of energy, or vice versa. If a system's kinetic energy increases, there must be an equal decrease of some other form or forms of energy.

While this principle was being developed, scientists observed that most processes in nature move in only one direction—people grow only older, not younger; a log can burn with oxygen to produce thermal energy, carbon dioxide, and ashes, but ashes in a fireplace do not spontaneously combine with thermal energy and carbon dioxide to produce a log and oxygen; a rock will roll unaided down a hill, but it will not spontaneously roll up a hill. None of these reverse processes is prohibited by the conservation-of-energy principle.

Consider a rock. As a rock rolls down a hill, some of its kinetic energy is converted by friction into thermal energy in the rock and in the ground. When the rock stops at the bottom of the hill, all its kinetic energy has been converted into thermal energy—the rock and ground are slightly warmer. Why is the rock unable to extract thermal energy from the ground when at the bottom of the hill and convert it to an equal amount of kinetic energy to roll up the hill? Conservation of energy does not prohibit this. It just says that the amount of thermal energy lost must equal the kinetic energy gained. Yet even though this process of rolling up a hill is allowed by energy conservation, we never see it happen (except in cartoons, such as Fig. 11.1).

Our experiences in life indicate that some processes are irreversible—they move only in one direction. There must be some explanation to indicate why these processes are irreversible. The explanation is embodied in the second law of thermodynamics and is related to the idea of entropy. One way to understand entropy is based on a statistical approach.

FIG. 11.1. This could happen to you, according to the first law of thermodynamics.

TABLE 11.1 Different Ways in Which Four Coins Can Be Arranged

Macrostate Label, i	Different Arrangements of Four Coins ● = Tails ○ = Heads	Count, W_i	Entropy $S_i = k \ln W_i$
0	○ ○ ○ ○	1	0
1	● ○ ○ ○ ○ ● ○ ○ ○ ○ ● ○ ○ ○ ○ ●	4	1.4k
2	● ● ○ ○ ● ○ ● ○ ● ○ ○ ● ○ ● ● ○ ○ ● ○ ● ○ ○ ● ●	6	1.8k
3	● ● ● ○ ● ● ○ ● ● ○ ● ● ○ ● ● ●	4	1.4k
4	● ● ● ●	1	0

11.2 Statistical Definition of Entropy

To arrive at a statistical definition of entropy, imagine a tray on which four identical coins rest. We will calculate the number of ways that a particular distribution of heads and tails can occur following a shake of the tray. We can obtain five different distributions by shaking four coins. These different distributions are distinguished by the symbol i, which can assume the integer values from 0 through 4. We see in Table 11.1 that the $i = 0$ state has zero tails and four heads, the $i = 1$ state has one tail and three heads, and so forth. Each of these five states is called a **macrostate** of the system of four coins. Each macrostate has a different number of tails n_t and of heads n_h. In thermodynamics we call n_t and n_h **state variables** because they are used to distinguish one particular macrostate of a system from other macrostates.

Next we want to determine the number of ways in which the ith macrostate can occur. We will call this the **count** W_i of the ith state. Macrostate 0 with zero tails and four heads can occur in only one way—if all coins lie with heads up. For macrostate 0, $W_0 = 1$. Macrostate 1, with one tail and three heads, can be attained in four different ways, since there are four coins, and we have four choices for the coin that will be tails. For macrostate 1, $W_1 = 4$. Column 3 in Table 11.1 lists the number of ways of getting the different macrostates. Each of these ways is called a **microstate** of that particular macrostate. For example, macrostate 1 has four microstates. Macrostate 2 has six; that is, there are six unique ways in which the four coins can be arranged to get two tails and two heads.

In general, for any number of coins, we can calculate the number W_i of microstates of the ith macrostate by using the equation

$$W_i = \frac{n!}{n_t! n_h!}, \tag{11.1}$$

where n is the total number of coins and n_t and n_h are the number of tails and heads for the ith state.* Using Eq. (11.1), we find that the number of microstates of macrostate 2 is

$$\mathcal{W}_2 = \frac{4!}{2!2!} = \frac{4 \cdot 3 \cdot 2 \cdot 1}{(2 \cdot 1)(2 \cdot 1)} = 6.$$

If we add all the numbers in the third column of Table 11.1, we find that there are 16 microstates. This means that there are 16 unique ways in which the four coins can be arranged, and each is equally likely to occur if the tray is given a good shake. The **probability** of a particular macrostate occurring after each shake is the number of microstates for that macrostate (the count) divided by the total number of microstates. For example, the probability of getting all heads after a shake is 1 in 16, the probability of getting three heads and one tail is 4 in 16, and the probability of getting two heads and two tails is 6 in 16. Obviously, the distribution with two heads and two tails is most likely to occur, and the distributions with all heads or all tails are least likely to occur. These latter states are said to be more **ordered**—each coin is the same in the all-heads or all-tails macrostates. These ordered states have the least probability of occurring, and we will find that they also have the least entropy. The two-heads, two-tails macrostate is more **disordered**—there is more variation from coin to coin. Also, this macrostate can be formed in six different ways. When the tray is being shaken, a particular coin might jump from heads to tails and another from tails to heads. The distribution of coins is still in the $i = 2$ macrostate if it started in that state. This probable state allows more variation and randomness. In the statistical approach to thermodynamics, we measure the randomness or disorder of a state in terms of its probability of occurring—*the greater the probability of the state, the greater is its disorder or randomness.*

Entropy also increases with disorder. The entropy S_i of the ith macrostate can be defined in terms of the count $\mathcal{W}_i$ of the state:

$$\text{Entropy of }i\text{th macrostate} = S_i = k \ln \mathcal{W}_i, \qquad \textbf{(11.2)}$$

where $\ln \mathcal{W}_i$ is the abbreviation for the natural logarithm of the number $\mathcal{W}_i$ (the use of logarithms is reviewed briefly in Appendix A); k is a proportionality constant called **Boltzmann's constant** and has a value

$$k = 1.3805 \times 10^{-23}\,\text{J/K.}†$$

The fourth column of Table 11.1 lists the values of entropy S_i for the different macrostates. The values are listed in units of k.

The *second law of thermodynamics* says that the universe, as it changes, moves toward states with the greatest probability and count, and therefore toward states of maximum entropy. Consequently, if we start with a tray of four coins showing all heads—entropy zero—and shake the tray, supposedly each shake will produce a distribution with greater entropy. Eventually, the entropy

Ludwig Boltzmann's equation for calculating entropy ($S = k \ln W$) is faintly visible at the top of his gravestone. (Courtesy of D. Flamm)

*The exclamation marks (!) in Eq. (11.1) mean that we multiply the number before the exclamation mark by all the positive integers smaller than the number. Thus, $8! = 8 \cdot 7 \cdot 6 \cdot 5 \cdot 4 \cdot 3 \cdot 2 \cdot 1$, and $2! = 2 \cdot 1$. This mathematical shorthand is called "factorial". 8! is said to be eight factorial, and $n!$ is said to be n factorial. Zero factorial is defined as one: $0! = 1$.

†Ludwig Boltzmann (1844–1906) made many significant contribuⁿons to the statistical development of thermodynamics. His gravestone carries this important defining equation for entropy: $S = k \ln W$.

should reach a maximum possible value, $1.8k$, the entropy of the two-heads, two-tails distribution.

This law does not work well for distributions with small numbers of objects. For the tray with four coins, there is an excellent chance that after one shake the system will not be in a state with greater entropy. There is even a 1-in-16 chance of getting all heads or all tails, states with zero entropy. With a large number of coins, the situation is quite different. With 100 coins, for instance, the state with zero tails and 100 heads is about 10^{29} less likely to occur than the state with 50 heads and 50 tails. We would have to shake the tray once per second for about 3×10^{22} years (about seven trillion times longer than the age of the earth) to get one shake resulting in zero tails and 100 heads.

If we increase the number of coins to one million, states with equal numbers of heads and tails are overwhelmingly more probable than states with an unequal number of heads and tails. There is, for example, almost a 10^{87} better chance of getting an equal number of heads and tails than of getting 51 percent heads and 49 percent tails after one shake of one million coins. If we started with one million coins having all heads up, we would eventually, after many shakes, expect to have a 50–50 distribution with perhaps imperceptible fluctuations about this equal distribution. This 50–50 distribution is the state with highest probability and highest entropy.

In general, for systems with large numbers of objects, the move toward high-probability, high-entropy states is inevitable. Once the system is in the state of maximum entropy, it cannot leave that state. The system can perhaps bump into neighboring states of high probability. But usually these states are so close to the maximum-probability state that we do not know the system has changed.

This final state of maximum entropy is often called the **equilibrium state.** If the system is in the most probable state, it does not change. The system has reached equilibrium. A chemical reaction moves toward the equilibrium state as time progresses. Once the state is reached, the reaction stops and we see no more changes in the chemical properties of the system.

11.3 Favored Forms of Energy

The statistical definition of the second law of thermodynamics says that during any process of nature, the universe moves into more probable states—states that are favored by statistics. These favored states have greater entropy and disorder than those from which they evolve. The move toward statistically favored, high-entropy states causes energy to be converted from ordered to disordered forms. The sequence shown in Fig. 11.2 depicts the progression of nature toward more probable states. Unfortunately, this progression results in the conversion of energy to less useful forms. Organized energy is more useful for running engines or for use as energy for our bodies. Random thermal energy with hot and cold reservoirs is next most useful—it is used in power plants to produce electricity. Random thermal energy at uniform temperature is least useful—heat does not flow, and work cannot be done with this energy.

It used to be thought that the evolution of living animals, such as humans, was perhaps an exception to the second law of thermodynamics (the molecules in living animals are very organized—see Fig. 11.3). It is now understood that the growth of order in these living beings must be considered relative to the

FIG. 11.2. The transformation of energy into forms favored by statistics. Unfortunately, when in statistically favored forms, the energy is least useful. (Charles Moore/Black Star.)

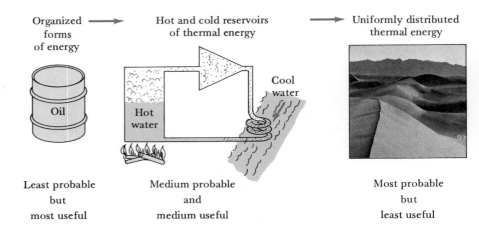

Organized forms of energy → Hot and cold reservoirs of thermal energy → Uniformly distributed thermal energy

Oil

Hot water

Cool water

Least probable but most useful

Medium probable and medium useful

Most probable but least useful

great disorder produced in their environment to allow for their growth. A person consumes and converts complicated molecules with low entropy, such as carbohydrates, fats, and proteins, into much smaller molecules (CO_2, H_2O). A healthy adult consumes about 500 kg, or a half-ton, of these foods each year. The degradation of this food results in the transfer to the environment of two billion joules of thermal energy. The human may remain ordered, but our environment is greatly disordered. A UNESCO document on environmental pollution gives these figures and concludes:

> Maintenance of life is an expensive process in terms of generation of disorder, and no one can understand the full implications of human ecology and environmental pollution without understanding that first.

The figures just mentioned represent only the human entropy increase due to food consumption. In industrialized countries the entropy increase is almost 100 times greater per person, owing to the degradation of high-quality fuels that are used to maintain our mechanized society.

The scientific basis of the energy crisis rests with the second law of thermodynamics and ideas of entropy.* The problem is the rate of degradation of energy to a less useful form rather than the actual loss of energy.

11.4 Thermodynamic Definition of Entropy

The statistical approach to entropy makes sense—it seems reasonable that nature moves toward more probable distributions. For most real-life problems, though, it is not practical to determine statistically the probability that a system will be in different possible states. Even small systems are made of huge numbers of atoms and molecules. A simple bacterial cell consists of roughly 10^{11} atoms.

Fortunately, there is another, more practical approach to entropy and the second law of thermodynamics, which was developed in the 1800s, before the atomic nature of matter was understood and years before the statistical definition of entropy was determined. Unfortunately, entropy as defined thermodynamically is difficult to perceive. It does not affect our senses the way temperature or force does. In this book we will not fully develop the thermodynamic

FIG. 11.3. Is this organized being a mistake—a contradiction of the second law of thermodynamics?

*There are also many social and political ramifications of the energy crisis.

approach to entropy; we will, however, define entropy and use this definition to analyze some representative energy transformations that occur in machines and in other parts of nature.

In thermodynamics, the **change in entropy** ΔS of a system as it moves from an initial state to some arbitrary final state is defined as

$$\Delta S = \frac{\Delta Q}{T}, \qquad\qquad (11.3)$$

where ΔQ is the heat added to or removed from the system as it moves between the two states and T is the temperature of the system *in degrees Kelvin* during the heat transfer. This definition is valid only during reversible changes in the system—changes involving no degradation of energy. No friction forces can be acting on a system or on its environment during the reversible process. All entropy changes are due only to heat transfer. We will not pursue further this subject of reversibility; it receives considerable attention in more comprehensive treatments of thermodynamics. We will instead illustrate the use of the thermodynamic definition of entropy change, Eq. (11.3), in several examples in this and later sections. These examples involve energy transformations of macroscopic objects—objects large enough to be seen easily with our unaided eyes.

EXAMPLE 11.1 Calculate the change in entropy of a block of ice to which 10^4 J of heat is added, causing part of the ice to melt. The temperature of the ice is 0°C.

SOLUTION When using Eq. (11.3) to calculate changes in entropy, temperatures must be in degrees Kelvin. The temperature of the block of ice is 0°C, or 273 K. Thus, the change in entropy is

$$\Delta S = \frac{\Delta Q}{T} = \frac{10^4 \text{ J}}{273 \text{ K}} = \underline{37 \text{ J/K}}.^* \qquad\blacksquare$$

EXAMPLE 11.2 Calculate the change in entropy of a pot of boiling water to which 10^4 J of energy is added. The water's temperature is 100°C.

SOLUTION The temperature of the boiling water in degrees Kelvin is $100 + 273 = 373$ K. Thus, the change in entropy of the water is

$$\Delta S = \frac{\Delta Q}{T} = \frac{10^4 \text{ J}}{373 \text{ K}} = \underline{27 \text{ J/K}}. \qquad\blacksquare$$

The water discussed in Example 11.2 received the same heat as the ice discussed in Example 11.1, yet their changes in entropy were different. A hot object is already quite disordered (water is less ordered than ice). Hence, when heat is added to a hot object, its change in disorder is more difficult to notice and is less than the entropy change of a cold object to which heat is added. This dependence of entropy change on temperature is very important when we consider the overall entropy change of the universe during processes involving heat transfer, the topic of the next section.

*A J/K is sometimes called an entropy unit (eu). Thus, 1 J/K = 1 eu.

11.5 Second Law of Thermodynamics

The **second law of thermodynamics** can now be stated as follows:

During any process of nature, the entropy change of the universe—the entropy change of a system plus that of its environment—must be greater than or equal to zero. This law can be written in mathematical symbols as

$$\Delta S_{\text{system}} + \Delta S_{\text{environment}} \geq 0.^* \qquad (11.4)$$

If a system is isolated from its environment, there is no interaction between the two. The change in entropy of the isolated system during some process, then, is independent of changes occurring in the environment. In such a case, a system's entropy must increase or, at best, remain constant. Thus,

$$\Delta S_{\text{system}} \geq 0 \qquad \text{(isolated system)}. \qquad (11.5)$$

If Eq. (11.4) [or Eq. (11.5) for an isolated system] is not satisfied for some process, then the process being described cannot occur.

EXAMPLE 11.3 A 10-g cube of ice at 0°C lies on ground whose temperature is 20°C. Calculate the change in entropy that results from heat transfer from the ground to the ice, causing the ice to melt and become 10 g of water at 0°C. The ground is considered a very large reservoir whose temperature remains constant at 20°C.

SOLUTION We will calculate the entropy changes of the ice and ground separately, then add them to find the net entropy change.

The heat needed to melt 10 g of ice is given by Eq. (9.6):

$$\Delta Q = mL_f = (0.010 \text{ kg})(3.35 \times 10^5 \text{ J/kg}) = +3350 \text{ J}.$$

The change in entropy of the ice when it melts is

$$\Delta S_{\text{ice}} = \frac{\Delta Q}{T_{\text{ice}}} = \frac{+3350 \text{ J}}{273 \text{ K}} = +12.3 \text{ J/K}.$$

The ground loses just as much heat as the ice gained. Thus, the entropy change of the ground is

$$\Delta S_{\text{ground}} = \frac{\Delta Q}{T_{\text{ground}}} = \frac{-3350 \text{ J}}{(273 + 20) \text{ K}} = -11.4 \text{ J/K}.$$

The net change in entropy is

$$\Delta S = \Delta S_{\text{ice}} + \Delta S_{\text{ground}} = +12.3 \text{ J/K} - 11.4 \text{ J/K} = \underline{+0.9 \text{ J/K}}.$$

Since the entropy change is positive, the process is allowed by the second law of thermodynamics. ∎

EXAMPLE 11.4 Calculate the entropy change when a container holding 0.50 kg of water at 30°C contacts a second container holding 0.50 kg of water at

*The symbol $\geq$ means "greater than or equal to"; that is, the left-hand side of the equation must be greater than or equal to the right-hand side.

70°C. Heat is transferred from the hot to the cool water until both containers are at 50°C.

SOLUTION To calculate the change in entropy, we consider separately the entropy change of each container of water, then add these changes to find the net change. The heat added to the cooler water as it warms from 30°C to 50°C is given by Eq. (9.4):

$$\Delta Q = mc\,\Delta T = (0.50\text{ kg})(4180\text{ J/kg}\cdot\text{C}°)(50°\text{C} - 30°\text{C}) = 41,800\text{ J}.$$

The warmer water loses an equal amount of energy when cooling from 70°C to 50°C.

We now have a problem. To calculate the entropy change using the equation $\Delta S = \Delta Q/T$, we must associate some temperature with the material gaining or losing the heat. But the temperature of each container changes as heat flows into or out of it. To do the calculation correctly, we must transfer the heat in very small increments, each of which causes a small change in temperature. The total change in entropy is the sum of these small entropy changes.

As a rough approximation, we will pretend that all the heat was transferred from the warm water at its average temperature (half of 70°C + 50°C, or 60°C) to the cold water at its average temperature (40°C). The total change in entropy is, then,

$$\Delta S = \Delta S_{\text{cool}} + \Delta S_{\text{hot}} \cong \frac{\Delta Q}{\overline{T}_{\text{cool}}} - \frac{\Delta Q}{\overline{T}_{\text{hot}}}$$

$$= \frac{41,800\text{ J}}{(273 + 40)\text{ K}} - \frac{41,800\text{ J}}{(273 + 60)\text{ K}}$$

$$= +134\text{ J/K} - 126\text{ J/K} = \underline{+8\text{ J/K}}.$$

The negative sign appears in front of the entropy change for the hot container because it lost heat. An exact calculation using integral calculus indicates that the entropy change was 8.04 J/K. Our approximate method using average temperatures provides satisfactory results when temperature changes are not too great. Since the net change in entropy is positive, the process is allowed by the second law of thermodynamics. ■

11.6 Heat Engines

Most of the world's electricity is produced by electric power plants that are examples of thermodynamic heat engines. These engines generally operate at about 35 percent efficiency, which means that only 35 percent of the energy input to the engine results in the production of useful electrical energy. The other 65 percent is lost as waste heat to the environment. The first and second laws of thermodynamics are very useful for determining the best possible efficiency of different energy-conversion devices. We should be able to decide if we can generate electricity at 100 percent efficiency using a heat engine, or if the 35 percent now achieved is the best we can expect.

A heat engine used to generate electricity is represented schematically in Figs. 11.4a and 11.4b. Heat enters the engine on the left side, where water is continually warmed. The heat is produced by burning gas, oil, or coal or by nuclear fission. The temperature of the water on the hot side is typically 560°C.

FIG. 11.4. (a) A heat engine to produce electricity. (b) A thermodynamic representation of the engine.

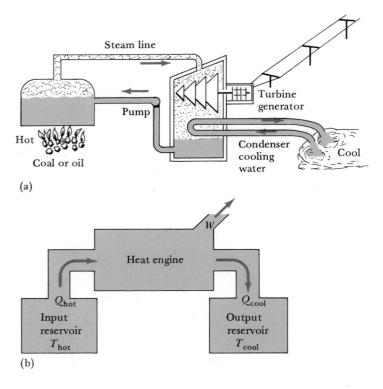

(a)

(b)

In Fig. 11.4b this heat source is represented by a box on the left labeled "Input reservoir," whose temperature is T_{hot}. An amount of heat Q_{hot} flows from this reservoir into the heat engine and represents the input energy used to operate the engine.

In the engine the water, under high pressure, boils and is converted to gaseous water vapor (Fig. 11.4a), which causes the blades of a turbine to rotate. As the turbine rotates, an electric generator turns, and electricity is produced. The electricity can then be used to do work. In our block diagram (Fig. 11.4b) the electrical energy leaving the heat engine, symbolized by W for work, represents the useful output of the generator.

Finally, after leaving the turbine (Fig. 11.4a), the steam is condensed back to the liquid state and returns to the boiler for another trip around the loop. When the steam is condensed, heat is released. This heat is carried away by evaporating water in evaporative cooling towers or by passage of the steam through coils immersed in the cool water of a lake or stream. In Fig. 11.4b the temperature at the condenser coils is T_{cool}, and an amount of heat Q_{cool} leaves the engine at the condenser. The condenser temperature is typically about 100°C.

Let us now use the first and second laws of thermodynamics to determine the efficiency of this engine and its heat reservoirs. The **efficiency** e of a device is defined as the ratio of its useful output divided by its input:

$$\text{Efficiency} = e = \frac{\text{Useful output}}{\text{Input}}. \tag{11.6}$$

For the heat engine and its reservoirs, the useful output is the work done, and the input is the heat entering the engine from the hot reservoir:

$$\text{Efficiency of heat engine} = e = \frac{W}{Q_{\text{hot}}}. \tag{11.7}$$

Once operating, a heat engine remains in a stable or steady state, and its internal energy does not change. Thus, according to the first law of thermodynamics, the energy entering the engine must equal that leaving the engine, or

$$Q_{\text{hot}} - Q_{\text{cool}} - W = 0.$$

This equation has been written so that Q_{hot}, Q_{cool}, and W are all positive numbers. W is the work done *by* the engine.

Solving the above equation for W and then substituting in Eq. (11.7), we find that

$$e = \frac{Q_{\text{hot}} - Q_{\text{cool}}}{Q_{\text{hot}}} = 1 - \frac{Q_{\text{cool}}}{Q_{\text{hot}}}. \tag{11.8}$$

If no heat is emitted from the engine—if $Q_{\text{cool}} = 0$—then the efficiency is 1. This means that all the input heat is converted to work, and the engine is 100 percent efficient. We will find, however, that this is not possible.

The maximum possible efficiency of this process can be determined using the second law of thermodynamics. We start by evaluating the entropy changes of the three parts of the system pictured in Fig. 11.4b: the hot reservoir that supplies heat to the engine, the engine, and the cool reservoir that takes waste heat from the engine. The entropy evaluation will take place over some time period (or over a cycle of operation for engines that go through repetitive cycles, such as automobile engines). The entropy change of the hot reservoir due to heat Q_{hot} leaving it and entering the engine is

$$\Delta S_{\text{hot}} = \frac{-Q_{\text{hot}}}{T_{\text{hot}}}.$$

The engine itself is in a steady state that does not change; its hot side stays hot and its cool side stays cool. Thus, its net entropy change over the operating period is zero. Finally, the entropy change of the cool reservoir as it gains heat Q_{cool} from the engine is

$$\Delta S_{\text{cool}} = \frac{Q_{\text{cool}}}{T_{\text{cool}}}.$$

The total entropy change is the sum of these three entropy changes, which according to the second law of thermodynamics, must be greater than or equal to zero:

$$\Delta S_{\text{total}} = \Delta S_{\text{hot}} + \Delta S_{\text{engine}} + \Delta S_{\text{cool}} = -\frac{Q_{\text{hot}}}{T_{\text{hot}}} + 0 + \frac{Q_{\text{cool}}}{T_{\text{cool}}} \geq 0.$$

This last inequality can be rewritten as

$$\frac{Q_{\text{cool}}}{Q_{\text{hot}}} \geq \frac{T_{\text{cool}}}{T_{\text{hot}}}. \tag{11.9}$$

If we now rearrange Eq. (11.8), substituting Eq. (11.9), we find that

$$1 - e = \frac{Q_{\text{cool}}}{Q_{\text{hot}}} \geq \frac{T_{\text{cool}}}{T_{\text{hot}}}.$$

After some further rearrangement, we find the **efficiency e of a heat engine** when operating between reservoirs at temperatures T_{hot} and T_{cool}:

$$e = \frac{W}{Q_{hot}} \leq 1 - \frac{T_{cool}}{T_{hot}}. \qquad (11.10)$$

The **maximum efficiency of a heat engine** equals $1 - (T_{cool}/T_{hot})$ and depends only on the temperatures of the input and output reservoirs. For example, if $T_{hot} = 560°C = 833$ K and if $T_{cool} = 100°C = 373$ K, then the efficiency can be no more than

$$e \leq 1 - \frac{373 \text{ K}}{833 \text{ K}} = 0.55.$$

The engine can be at best 55 percent efficient. The efficiency in practice of a modern coal-fired power plant is less than 40 percent.

Equation (11.10) was first derived in 1824 by a French engineer named Sadi Carnot. His achievement is truly remarkable in that neither the first nor the second laws of thermodynamics had been developed at the time of his discovery. Carnot constructed a theoretical model of a heat engine and determined the maximum possible work it could do by adding heat from a hot reservoir and removing it to a cooler reservoir. His theoretical heat engine was shown to have the maximum efficiency possible when operating between reservoirs at temperatures T_{hot} and T_{cool}.

EXAMPLE 11.5 An inventor claims to have developed a heat engine that can supplement a car's power by attaching the heat engine to the car's exhaust system. The engine operates on heat emitted from the exhaust ($T = 150°C$). The output side of the engine is at air temperature ($10°C$). (a) Calculate the maximum efficiency of the engine. (b) If the car exhaust system deposits 2.0×10^5 J/s of heat into this heat engine, what is the maximum power of the engine? (Power $= P = W/t$.)

SOLUTION (a) The engine operates between reservoirs at temperatures $T_{hot} = 150 + 273 = 423$ K and $T_{cool} = 10 + 273 = 283$ K and has a maximum efficiency given by Eq. (11.10):

$$e_{max} = 1 - \frac{T_{cool}}{T_{hot}} = 1 - \frac{283 \text{ K}}{423 \text{ K}} = 0.33.$$

The engine is at best 33 percent efficient.

(b) If we divide the top and bottom of the right-hand side of Eq. (11.7) by time t and rearrange, we find that $(W/t) = e(Q_{hot}/t)$. But Q_{hot}/t is just the rate at which heat enters the engine from the hot reservoir (the exhaust); it equals 2.0×10^5 J/s. W/t is the work done per unit time (the power of the supplementary heat engine), and its maximum value is

$$P_{max} = \left(\frac{W}{t}\right)_{max} = e_{max}\left(\frac{Q_{hot}}{t}\right)$$
$$= 0.33(2.0 \times 10^5 \text{ J/s})$$
$$= 6.6 \times 10^4 \text{ J/s} = \underline{6.6 \times 10^4 \text{ W}},$$

enough energy to operate five or six automobile air conditioners. Unfortunately, the actual efficiency of most engines is usually somewhat less than the maximum possible efficiency, but we should be able to operate at least one air conditioner. ■

11.7 Heat Pumps and Refrigerators

Heat flows naturally from hot objects to cool objects. A heat pump is a device that reverses the direction of flow; heat is transferred from a cold reservoir to a warmer one. A heat pump can be used to warm your house in the winter by extracting thermal energy from cool air outside and transferring it into the house to keep the house warm. A refrigerator is another example of a heat pump; heat is transferred from cool air inside the refrigerator to warmer air outside the refrigerator.

Heat Pumps

A heat pump used to warm a house in the winter is depicted in Fig. 11.5a. Heat, removed from cold air at temperature T_{cool} outside the house, is absorbed by a fluid circulating through the heat pump. This input heat is labeled Q_{cool} in Fig. 11.5b. The heat pump cannot operate without some outside source of energy that does work W circulating and compressing the fluid in the heat pump. The input work is usually done by an electric motor and is represented by the symbol W in Fig. 11.5b. Finally, heat Q_{hot} is transferred from the heat pump into the house, whose temperature is T_{hot}.

The useful output of a heat pump is the heat Q_{hot} transferred into the house. The energy needed to do this is the work W done by its motor. To be efficient, a heat pump should transfer a large amount of heat while doing little work. The heat pump is rated using an efficiency quantity called the **coefficient of performance,** which is the ratio of the heat transferred into the hot reservoir and the work needed to transfer that heat:

$$\text{Coefficient of performance of heat pump} = \eta = \frac{Q_{hot}}{W}. \quad \textbf{(11.11)}$$

The larger the value of η, the better is the heat pump. Typical heat pumps used for home heating have coefficients of performance of 2 or 3; in other words, two or three times more heat is transferred into the house than the work that is done transferring the heat. We can use the first and second laws of thermodynamics to see if this is the best efficiency we can expect.

The heat pump, while operating, remains in a stable situation. Its internal energy does not change. According to the first law of thermodynamics, the energy entering the heat pump must equal that leaving it:

$$Q_{cool} + W - Q_{hot} = 0.$$

Once again, all quantities are positive. Solving the preceding equation for W and substituting into Eq. (11.11), we find that

$$\eta = \frac{Q_{hot}}{Q_{hot} - Q_{cool}}. \quad \textbf{(11.12)}$$

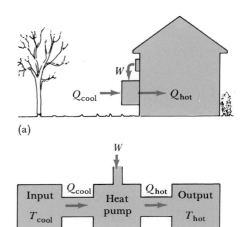

(a)

(b)

FIG. 11.5. (a) A heat pump to warm a house in the winter. (b) A schematic representation of a heat pump.

This equation can be rewritten in terms of temperatures, using the second law of thermodynamics. The entropy change of the input reservoir is

$$\Delta S_{cool} = \frac{-Q_{cool}}{T_{cool}},$$

and that of the output reservoir is

$$\Delta S_{hot} = \frac{+Q_{hot}}{T_{hot}}.$$

The heat pump itself is in steady state and has no entropy change. According to the second law of thermodynamics, the net entropy change of this heat-pump system, including the two reservoirs, is

$$-\frac{Q_{cool}}{T_{cool}} + \frac{Q_{hot}}{T_{hot}} \geq 0. \tag{11.13}$$

With a little manipulation of Eqs. (11.12) and (11.13), we find that the maximum possible value for the **coefficient of performance of a heat pump** is limited by the following equation:

Coefficient of performance (heat pump) $= \eta$

$$= \frac{Q_{hot}}{W} \leq \frac{T_{hot}}{T_{hot} - T_{cold}}. \tag{11.14}$$

EXAMPLE 11.6 A heat pump deposits 14,000 J of heat into a home. The temperature of the outside air is $-10°C$ (263 K), and the air in the house is at $20°C$ (293 K). (a) Calculate the maximum possible coefficient of performance of the heat pump. (b) Calculate the amount of electrical work done in operating the heat pump at this performance level.

SOLUTION (a) We are given that $T_{cool} = 263$ K and $T_{hot} = 293$ K. According to Eq. (11.14), the coefficient of performance of a heat pump operating between these two temperatures is limited to the following value:

$$\eta \leq \frac{T_{hot}}{T_{hot} - T_{cool}} = \frac{293}{293 - 263} = \underline{9.8}.$$

A heat pump could in principle deposit almost ten times more heat into the house than the work used to extract the heat.

(b) Rearranging our definition of coefficient of performance, Eq. (11.11), we find that

$$W = \frac{Q_{hot}}{\eta} = \frac{14,000 \text{ J}}{9.8} = \underline{1430 \text{ J}}.$$

Only 1430 J of work is needed to transfer 14,000 J of heat. Unfortunately, most real heat pumps operate with coefficients of performance of only 2 or 3. ∎

Refrigerators

A refrigerator works much like a heat pump, except that heat is withdrawn from inside the refrigerator and deposited in the warmer air outside, as depicted in

Fig. 11.6. The useful function achieved by a refrigerator is the removal of heat Q_{cool} from inside the refrigerator, and an amount of work W is expended to achieve this removal. The efficiency quantity used to rate refrigerators, also called a coefficient of performance, is

$$\text{Coefficient of performance of refrigerator} = \eta = \frac{Q_{cool}}{W}. \qquad (11.15)$$

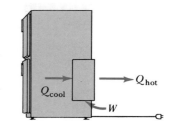

(a)

A calculation similar to that performed for a heat pump shows that the **coefficient of performance of a refrigerator** is limited by the equation

$$\text{Coefficient of performance (refrigerator)} = \eta$$
$$\qquad (11.16)$$
$$= \frac{Q_{cool}}{W} \leq \frac{T_{cool}}{T_{hot} - T_{cool}}.$$

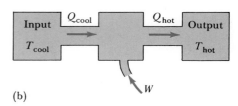

(b)

FIG. 11.6. (a) A refrigerator is a heat pump. (b) A schematic representation of the energy transformations in a refrigerator.

EXAMPLE 11.7 (a) How much electrical energy is required to remove 700,000 J of heat from the freezer of a refrigerator operating at its maximum coefficient of performance? The freezer temperature is $-5°C$, and the room temperature is $20°C$. (b) How much heat is deposited in the room by the refrigerator?

SOLUTION (a) We are given the following information: $T_{cool} = -5°C = 268$ K; $T_{hot} = 20°C = 293$ K; $Q_{cool} = 700,000$ J. We wish to find the work done by the electrical compressor of the refrigerator. Using Eq. (11.16), we find that

$$Q_{cool} \frac{T_{hot} - T_{cool}}{T_{cool}} \leq W,$$

or

$$700,000 \text{ J} \left(\frac{293 \text{ K} - 268 \text{ K}}{268 \text{ K}} \right) = 65,000 \text{ J} \leq W.$$

Thus, at least $\underline{65,000 \text{ J}}$ of electrical work is needed to remove the heat. Most refrigerators use much more than the optimum electrical work needed.

(b) Using the first law of thermodynamics for the refrigerator shown in Fig. 11.6, we see that $Q_{cool} + W - Q_{hot} = 0$, or $Q_{hot} = Q_{cool} + W = 700,000 \text{ J} + 65,000 \text{ J} = \underline{765,000 \text{ J}}$. Refrigerators actually deposit considerable heat in a kitchen. ∎

11.8 Recurrence Paradox

In this chapter we have considered entropy and the second law of thermodynamics from two points of view: (1) the statistical view and (2) a view in which we consider the conversion of high-quality energy sources (gas and oil) into thermal energy to do work. In the statistical view, the universe moves toward states that are most probable. In terms of energy, states in which the energy is dispersed randomly in the form of thermal energy are most probable. Thus, it seems that the two views are consistent.

Yet several interesting controversies have evolved from a supposed conflict between these two views of the second law of thermodynamics. One of these, the

recurrence paradox, is based on the fact that, statistically, if the universe consists of a finite number of particles, then there is a finite although extremely small chance for every arrangement of these particles to occur. If we wait long enough, the particles should at some time in the future be arranged again just as they are now. It is like shaking a tray full of coins. If we do it long enough, at some time in the future we will certainly get a distribution of all heads or of any other arrangement of heads and tails. The German philosopher Friedrich Nietzsche stated this idea as follows:

> If the universe may be conceived as a definite quantity of energy, as a definite number of centres of energy—and every other concept remains indefinite and therefore useless—it follows therefrom that the universe must go through calculable number of combinations in the great game of chance which constitutes its existence. In infinity [of time], at some moment or other, every possible combination must once have been realized; not only this, but it must have been realized an infinite number of times.

On the basis only of the ideas of statistical probability and random distributions, we would expect that at some future time all the atoms and molecules of the universe will be back in their present arrangement. It seems reasonable, then, that history should move forward in about the same way it does now—the drama of life will be repeated.

This idea seems to conflict with the approach to the second law based on heat. In this approach energy continues to be dispersed, and there seems to be no way in which it can be shaken back into order.

Recent theories of cosmology have been addressing this conflict not so much in terms of entropy but by using ideas of mechanics and the origin of the universe. The universe supposedly started its evolution about 20 billion years ago as an extremely dense, high-temperature quantity of energy that erupted in an explosive expansion called the **big bang.** The mass of the universe was thrown into motion away from this explosion. It is now generally accepted that the universe is expanding—that all parts are moving apart—as if in consequence of a big bang. Cosmologists are considering whether this expansion will continue forever or will eventually stop. By analogy, consider a rocket leaving the earth. If its velocity is great enough, it will be able to escape the earth's gravitational pull; if its velocity is too small, the earth's gravitation will eventually pull it back.

Cosmologists are trying to calculate the density of mass in the universe to decide whether it is great enough to eventually halt this expansion. If it is, then the halt will occur, and thereafter the mass of the universe will be drawn back toward its center, eventually resulting in the **big crunch** and possibly a recurrence of the big bang. Some theories predict that such a crunch will occur in about 50 billion years. If the universe does oscillate out and in, then perhaps Nietzsche's idea about the recurrence of events in history might be valid. There would, however, be an extremely long waiting time between similar events.

Some cosmologists say that the measurements of the mass density of the universe indicate that it is too small to halt the expansion. The expansion will continue indefinitely into parts of space incredibly far from us. We will be left with dead stars and possibly some interstellar gas. Although the high-quality forms of energy needed to support life will decrease, we will have enough to maintain life for a long time. This fate is called the *heat death* of the universe (Fig. 11.7).

FIG. 11.7. Flammarion's conception of the continued expansion of the universe, sometimes called its "heat death." From the novel *Le Fin du Monde* (1894).

Summary and Additional Readings

1. **Irreversible processes:** Many processes of nature are observed to move in only one direction and are said to be irreversible. Examples are aging, the burning of a log, and the rolling of a rock down a hill. The reverse of these processes is allowed by the first law of thermodynamics, yet we do not see the reverse processes occur. The concept of entropy has been developed to help us understand why these reverse processes do not occur. **Entropy** is a measure of disorder and the randomness of the distribution of a system's energy. The greater the disorder, the greater the entropy.

2. **Statistical development of entropy and the second law of thermodynamics:** Entropy can be defined quantitatively by using either statistics or a thermodynamic definition involving heat and temperature. In statistics, the entropy S_i of the ith macrostate of a system is defined as

$$S_i = k \ln \mathcal{W}_i, \qquad (11.2)$$

where $\mathcal{W}_i$ is the **count** of the ith state, the number of different ways the objects in the system can be arranged to form that macrostate. Each of these different ways is called a microstate. The second law of thermodynamics says that the universe continually moves toward states of greater entropy. Since entropy increases as the count and probability of a state increase, we see that the universe moves toward more probable states. This movement toward probable states causes thermal energy to be favored over more organized forms of energy such as the chemical energy of molecular bonds. Most processes in nature move toward increasing thermal energy, while the amount of organized, useful energy decreases.

3. **Thermodynamic development of entropy and the second law of thermodynamics:** In thermodynamics, the **change in entropy** ΔS of a system or object at temperature T (in degrees Kelvin) to which heat ΔQ is added or removed is

$$\Delta S = \frac{\Delta Q}{T}. \qquad (11.3)$$

The **second law of thermodynamics** says that during any process, the entropy change of a system plus that of its environment must be greater than zero:

$$\Delta S_{\text{system}} + \Delta S_{\text{environment}} \geq 0. \qquad (11.4)$$

The first and second laws of thermodynamics are especially useful for analyzing the efficiency of heat engines, heat pumps, and other systems involved in the transfer of heat for useful purposes.

Wolfgang Sassin, "Energy," *Scientific American*, pp. 119–132, September (1980). Discusses the growth of energy-based technology and the energy crisis.

P. W. Bridgman, *The Nature of Thermodynamics*, Harper and Brothers, New York (1961).

Jay M. Pasachoff, "The Future of the Universe," *The Physics Teacher*, pp. 291–298, May (1979).

James S. Trefil, "How the Universe Will End," *Smithsonian*, pp. 72–83, June (1983).

"Efficient Use of Energy," *Physics Today* **28**, 23 (1975).

S. W. Angrist, "Perpetual Motion Machines," *Scientific American* **218**, 115 (1968).

Questions

1. Describe five everyday examples of processes that involve increases in entropy. Be sure to state all the parts of nature involved in these entropy increases.

2. A cup of hot coffee sits on your desk. Assume that some of the thermal energy in the coffee can be converted to gravitational potential energy—enough so that the coffee rises up above the rim of the cup. *Estimate* the drop in temperature needed to accomplish this energy conversion. Why does this conversion not occur?

3. In terms of the statistical definition of entropy, why is there a good chance that the entropy of five coins being shaken on a tray will decrease, whereas this chance for 10^6 coins is negligible?

4. The entropy of the molecules that form leaves on a tree decreases in the spring of each year. Is this a contradiction of the second law of thermodynamics? Explain.

5. Give three examples of processes, other than those described in the book, in which potentially useful forms of energy are converted to less useful forms of energy.

6. The efficiency of nuclear power plants used to produce electricity is about 40 percent. Under what condition could this conversion efficiency be increased to 100 percent? What difficulty might be encountered in doing this?

7. There is a huge amount of thermal energy in the oceans of the earth. Indicate two problems that limit the usefulness of this energy as a source of power for a heat engine.

8. An air conditioner transfers heat from a cold reservoir to a hot reservoir. How can this heat transfer be reconciled with the second law of thermodynamics?

9. Would it be possible to use a refrigerator with its door open as a device to cool a house in the summer? Explain.

10. Would it be possible, in terms of the first and second laws of thermodynamics, for a refrigerator to remove as much heat from inside the refrigerator as it gave off into the air outside? Explain.

11. An air conditioner and a refrigerator both use work (electrical energy) to remove heat from a cold reservoir and deposit it in a hot reservoir. Why does an air conditioner cool a house and a refrigerator warm the house (even with its door open)?

12. List five ways in which the universe would be different if it had reached its final equilibrium state.

Problems

11.2 Statistical Definition of Entropy

■ 1. (a) Identify, in terms of the numbers of heads and tails, all the different macrostates of a system of five coins. (b) Calculate the number of microstates of each macrostate. (c) Calculate the entropy of each state.

■ 2. Repeat Problem 1 for a system with six coins.

■ 3. Calculate the ratio of the number of microstates (count) of a system of eight coins when in a macrostate with 4 heads and 4 tails and when in a macrostate with 7 heads and 1 tail. Which state has the greatest entropy?

■ 4. The probability that a lost person wandering about an island will be on the north part is one-half and on the south part is also one-half. (a) Calculate the probability that three lost people wandering about independently will all be on the south half. (b) Repeat part (a) for the probability of six lost people all being on the south half of the island.

■ 5. Parachutists have an equal chance of landing on the south half of a small island as on the north half. If eight parachutists jump at one time, what is the ratio of the probability that six land on the north half and two on the south half to the probability that four land on each half?

■ 6. (a) Calculate the ratio of the counts of a system of twenty coins when in a macrostate with 10 heads and 10 tails and when in a macrostate with 18 heads and 2 tails. (b) Do the same for a system with ten coins and for the states with 5 heads and 5 tails, compared to 9 heads and 1 tail. (c) When you compare your answers to parts (a) and (b), what do you infer about a similar ratio for a system with 10,000 coins?

■ 7. Nine numbered balls are dropped randomly into three boxes. The numbers of balls falling into each box are labeled n_1, n_2, and n_3 (the numbers must add to nine). (a) Identify five possible arrangements or macrostates of the balls (there are many more than five). (b) Calculate the ratio of the count of the equal distribution ($n_1 = 3$, $n_2 = 3$, and $n_3 = 3$) and of the 0, 0, 9 distribution. (c) Calculate the ratio of the count of the equal distribution and of the 2, 3, 4 distribution. [*Note:* The count is given by $W = n!/(n_1!n_2!n_3!)$, where n is the total number of balls.]

■ 8. Two dice are rolled. Macrostates of these dice are distinguished by the total number for each roll (that is, 2, 3, 4, . . . , 12). (a) Calculate the number of microstates for each macrostate. For example, there are three microstates for macrostate 4: 2, 2; 3, 1; and 1, 3. (b) What is the macrostate with greatest entropy? (c) What is the macrostate with least entropy? Explain.

■ 9. Three dice are rolled. (a) Identify the different possible macrostates. (b) Indicate the possible microstates for the macrostate that shows 17 on the dice; (c) for the macrostate that shows 16. (d) How many more times probable is the 16 state than the 17 state?

11.4 Thermodynamic Definition of Entropy

10. (a) Calculate the change in entropy when 2.0 kg of ice at 0°C melts. (b) Calculate the change in entropy when 2.0 kg of steam at 100°C condenses to water at 100°C.

11. (a) Calculate the change in entropy when 0.50 kg of water freezes at 0°C. (b) Calculate the change in entropy when 0.50 kg of water boils at 100°C.

■ 12. *Estimate* the entropy changes of (a) 200 g of water that is warmed from 20°C to 50°C, (b) 200 g of ethanol that is warmed from 20°C to 50°C, and (c) 200 g of copper that is warmed from 20°C to 50°C.

■ 13. Heat is added to 1.0 kg of ice at −10°C. *Estimate* the entropy change of the ice, (a) as it warms to 0°C, (b) as it melts at 0°C, (c) as it warms from 0°C to 100°C, (d) as it boils at 100°C, and (e) as the steam warms from 100°C to 120°C. (f) What is the total entropy change?

■ 14. A towel at 30°C, while drying on a line, absorbs 1000 J of energy from the sun, thus causing water to evaporate. Calculate the entropy change of the evaporated water. The towel does not change temperature, and the latent heat of vaporization at this temperature is 2.4×10^6 J/kg.

■ 15. *Estimate* the entropy change of 1.5 kg of water vapor at 100°C that is converted to 1.5 kg of ice at 0°C.

11.5 Second Law of Thermodynamics

■ 16. *Estimate* the total change in entropy of two containers of water. One container holds 0.10 kg of water at 70°C and is warmed to 90°C by heat emitted from the other container. This other container, also holding 0.10 kg of water, cools from 30°C to 10°C. Is this heat-transfer process allowed by the first law of thermodynamics? By the second? Explain.

■ 17. (a) 0.10 kg of ice at 0°C is added to 0.30 kg of tea at 30°C. Calculate the final temperature of the mixture after reaching equilibrium. The heat capacity of ice tea is the same as that of water. (b) Estimate the entropy change of ice tea of the universe during this process. Is it allowed by the second law of thermodynamics?

■ 18. A house at 20°C loses 1.0×10^5 J of heat to outside air at a temperature of −15°C. Calculate the entropy change of the universe. Is this process allowed by the second law of thermodynamics? Explain.

■ 19. A vat containing 200 kg of water is kept in a fruit cellar during the winter. On a cold night the water freezes, releasing heat to the room. The heat passes from the cellar to the outside air, whose temperature is −20°C. Calculate the entropy change for this process if the cellar remains at 0°C.

■ 20. (a) Calculate the final temperature when 0.10 kg of water at 10°C is added to 0.30 kg of soup at 40°C (the soup has the same heat capacity as water). (b) *Estimate* the entropy change of the universe during this process. Is the process allowed by the second law of thermodynamics?

■ 21. A 250-g aluminum calorimeter cup containing 200 g of water is at a temperature of 10°C. A 300-g block of copper at 60°C is placed in the cup. (a) Calculate the final equilibrium temperature. (b) Estimate the change in entropy of the cup, water, and copper block during this process.

■■ 22. A 5.0-kg block sliding at 8.0 m/s on a level surface at 20°C is stopped by friction after sliding 10 m. Calculate the change in entropy of the universe for this process. [*Hint:* Use the heat definition of entropy. The heat that would be added to the system is the same as the internal energy gained by friction.]

■■ 23. A 5.0-kg block slides from an initial speed of 8.0 m/s to a final speed of zero. It travels 12 m down a plane inclined at 15° with the horizontal before being stopped by friction. Calculate the change in entropy of the universe for this process if the block and plane are at a temperature of 27°C. (See the hint with Problem 22.)

11.6 Heat Engines

24. A heat engine operates between a heat reservoir at temperature 700 K and another at 300 K. Calculate (a) the maximum possible efficiency of the engine and (b) the maximum work the engine can do if it receives 10,000 J of heat from the hot reservoir.

■ 25. Near Bermuda, ocean water at the surface has an average temperature of about 24°C. At a depth of 800 m, the water temperature is about 10°C. (a) Calculate the maximum efficiency of a heat engine operating between these two heat reservoirs. (b) How much heat per second would have to be transferred down from the surface water to the engine and then exhausted to the water below to produce 10,000 kW of electrical power?

■ 26. A thermal hot springs 400 m deep at a temperature of 147°C provides heat for a heat engine. After leaving the engine, heat is exhausted into the air at a temperature of 27°C. Calculate the minimum amount of heat that must be removed from the hot springs for the engine to do 1.0×10^5 J of work.

■ 27. If the maximum efficiency of an automobile engine is 0.56 and the exhaust temperature is 120°C, what is the temperature of the burning gas in the engine? How much work could this engine produce when burning 1 gal of gasoline? (One gallon releases 1.4×10^8 J of heat.) [*Note:* There are many inefficiencies other than those of the engine that reduce the overall efficiency of the car.]

■ 28. A 60-kg woman walking on level ground at 1 m/s metabolizes energy at a rate of 230 W. When she walks up a 5° incline at the same speed, her metabolic rate increases to 370 W. Calculate her efficiency at converting chemical energy into gravitational potential energy.

■ 29. Gasoline, when burned in a car, provides about 3×10^7 J of thermal energy per liter of gasoline. If the car's mileage is 12 km/liter when traveling at a speed of 96 km/h, and it does work at a rate of 25 hp, what is the car's efficiency?

■ 30. A heat engine has a maximum efficiency of 0.15 when receiving heat from an 89°C thermal reservoir. What must the reservoir's temperature be to produce a maximum efficiency of 0.25?

■■ 31. A nuclear power plant operates between a high-temperature heat reservoir at 560°C and a low-temperature stream at 20°C. (a) Calculate the maximum possible efficiency of this heat engine. (b) How much heat per second must be put into the engine to produce 1000 MW of electrical power (work/time)?

(c) How much heat per second is dumped into the environment? (d) If this heat is absorbed by the Niagara River with a flow rate of 2×10^6 kg/s, by how much will the river's temperature change?

11.7 Heat Pumps and Refrigerators

■ 32. (a) Calculate the maximum possible coefficient of performance of a heat pump that operates between a room at 20°C and metal plates 10 m in the ground at a temperature of 15°C and one that operates between a room at 20°C and outside air at −10°C. (b) If heat is pumped into the room at a rate of 14,000 W, what is the least work that must be done by each pump during 1 s? Assume that the heat pump operates at the maximum coefficient of performance.

■ 33. A heat pump collects heat from outside air at 5°C and delivers it into a house at a duct at 40°C (at this temperature the air is slightly warmer than body temperature and feels warm if moving past your body). (a) Calculate the maximum possible coefficient of performance of the heat pump. (b) If the heat pump's motor does 1000 J of electrical work during a certain time, how much heat is delivered into the house, assuming the heat pump works at maximum coefficient of performance? (c) Repeat part (c) for a heat pump whose coefficient of performance is 2.0 (the efficiency of a real-life heat pump).

■ 34. A refrigerator transfers heat from inside a refrigerator at 5°C to room air at 20°C. Calculate (a) the maximum possible coefficient of performance of the refrigerator and (b) the minimum work needed to cool 5.0 kg of water from 20°C to 5°C when in the refrigerator.

■ 35. The freezer compartment of a refrigerator operates between a temperature of −15°C and room temperature (22°C). (a) Calculate the maximum coefficient of performance of the freezer. (b) How much energy is required to freeze 0.20 kg of water (an ice-cube tray) at 0°C and then lower its temperature to −15°C? (c) How much work must the freezer do to extract this much heat if operating at maximum efficiency? (d) How much heat is deposited into the room?

■ 36. An ice-making machine is to convert 2.0 kg of water at 0°C into 2.0 kg of ice at 0°C. The room temperature surrounding the ice machine is 20°C. (a) How much heat must be removed from the water? (b) Calculate the minimum work in the form of electrical energy needed to extract this heat by the ice-making machine. (c) How much heat is deposited in the room?

■■ 37. Using the first and second laws of thermodynamics and the definition of the coefficient of performance of a refrigerator, Eq. (11.15), derive Eq. (11.16).

PART III

Gases, Liquids, and Solids

In Parts I and II we have concentrated on two of the basic theories of science—Newton's laws of motion and the conservation-of-energy principle. In Part III we will apply these theories to the study of the physical properties of gases, liquids, and solids. A knowledge of the properties of these three forms of matter will allow us to better understand the operation of mechanical devices such as car engines, the physiology of our bodies, physical factors that affect the climate, the evolving structure of the earth, and other interesting properties of nature.

CHAPTER 12

Gases

A gas is a form of matter in which the atoms and molecules are relatively far apart. The atoms and molecules are in constant motion and collide frequently and violently with each other. Collisions also occur when gas molecules hit solid and liquid surfaces in their environment. The collisions cause a force and pressure to be exerted on these surfaces.

Gases are described in terms of quantities such as pressure, temperature, the volume of space occupied, and the amount of gas in that space. These quantities are called *state variables* because they describe the state or status of a gas—how hot it is, the pressure it exerts on a surface, and so forth. State variables are related to each other by equations called *equations of state*. Scientists use these equations of state in conjunction with the laws of thermodynamics to study phenomena such as climate changes in the atmosphere, the operation of gas-driven heat engines, and the change in pressure in a person's skull due to clogged sinuses.

12.1 Description of a Gas

Suppose for a moment that you have been reduced to the size of a molecule—about 10^{-10} times your normal size. What would life be like if you, in your reduced state, became a molecule in the air? Other molecules would be constantly moving past you in all directions at speeds of about 1000 mi/h (see Fig. 12.1); some would be moving faster, others slower.

On the average, you would be separated from the nearest molecule by a distance of about ten times your own size, a situation comparable to having twenty people spread over the area of a football field. At this separation, you feel little effect of the presence of other air molecules. Nevertheless, your isolation is interrupted by frequent collisions. The collisions are violent, since the molecules move at speeds of about 1000 mi/h. Each time a molecule crashes into you, you are knocked randomly in a new direction. After a collision you move only about 10^{-7} m before being hit by another molecule. During the course of a minute, you experience about 300 billion collisions, each causing your direction of motion to change. Because of the random nature of these collisions, you find after one minute that you have moved only about 4 cm from your starting position, even though your speed during that time has averaged about 1000 mi/h. Dur-

FIG. 12.1. Life as you would see it if you were an air molecule.

FIG. 12.2. This dumbbell-shaped object is about 20 million times the size of an oxygen or nitrogen molecule. On the same scale, the nearest air molecule would be near the opposite edge of this page.

ing this time, you have had no meaningful relations with any particular air molecule, just a succession of abrupt collisions.

This picture of air is characteristic of all gases. The space between atoms and molecules in any gas is large compared to their size (see Fig. 12.2). The molecules move fast, but because frequent collisions interrupt and change their direction of motion, they undergo little net displacement during any short period.

Besides colliding with each other, molecules and atoms in a gas also bounce off solid objects and liquid surfaces in their path. If a molecule of air hits your skin, it rebounds. Although we cannot sense these individual collisions, the molecule does exert a small force on the skin during the collision. This force is the cause of what we call *pressure.*

12.2 Pressure

When molecules of air or of any gas collide with a solid or liquid surface, they exert force against that surface. The force is of an impulsive nature, much like the force of a tennis ball hitting a practice board. Our skin cannot sense the force of individual collisions, but we can sense the average effect of a large number of the collisions. Each second about 10^{23} molecules of air strike each square centimeter of our skin. The net effect of these collisions is a force that seems constant in nature. Later in this chapter we discuss the magnitude and atomic nature of the force caused by these abrupt collisions. For now, the pressure caused by the collisions of the atoms and molecules in a gas with a solid or liquid surface is our main interest.

Pressure P is defined as the force F exerted perpendicular to a surface divided by the area A over which the force is exerted:

$$P = \frac{F}{A}. \tag{12.1}$$

At sea level, collisions of air molecules with a solid surface cause an average force of 1.01×10^5 N to be exerted on each square meter of surface. Thus, atmospheric pressure at sea level is 1.01×10^5 N/m^2. In terms of the English system of units, the pressure of air at sea level is 14.7 lb/in^2 (or 14.7 psi—pounds per square inch).

Units of Pressure

Since pressure is defined as the ratio of force and area, the unit for pressure is the N/m^2 (lb/ft^2 in the English system). A N/m^2 is now called a **pascal** (Pa), the official SI metric unit of pressure: 1 Pa = 1 N/m^2.

Other hybrid units of pressure based on particular measuring techniques are in common use today. The pressure units millimeter of mercury (mm Hg), also called **torr,** and inch of mercury (in Hg) evolved because of the use of manometers to measure pressure. A manometer consists of a U-shaped tube containing mercury or water (Fig. 12.3). If the tube is open on both ends, air pushes down on each column with equal pressure, and the columns on each side of the tube remain at the same elevation (Fig. 12.3a). However, if one end of the tube is closed and the air removed from it, creating a vacuum, the air pressure forces the mercury down on the open side and up into the evacuated side, as shown in Fig. 12.3b. The height h of the mercury column in the evacuated side above that in the side open to the air is a measure of the pressure of the air. At sea level the mercury column will be about 760 mm or 29.9 in higher on the left side than on the right side of the tube. Thus, we say that the pressure of the earth's atmosphere at sea level is 760 **millimeters of mercury** (760 mm Hg) or 29.9 **inches of mercury** (in Hg). Other pressure units occasionally used include the **atmosphere** (the pressure of air at sea level equal to 1.01×10^5 N/m^2) and the **bar** (a unit slightly less than one atmosphere). Factors needed to convert between these and other units of pressure are listed in Table 12.1.

12.3 Gauge Pressure

The pressure we have been discussing is called *absolute pressure,* or simply *pressure.* It is the pressure we would measure with a manometer having one side of the tube evacuated (Fig. 12.3b). This device compares the pressure of the gas pushing down on the mercury in the open side of the tube to the zero pressure in the evacuated side of the tube.

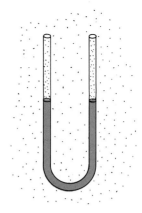

(a)

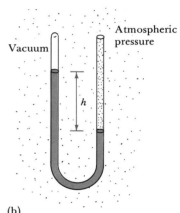

(b)

FIG. 12.3. (a) When both tubes of the manometer are open to the atmosphere, the mercury on each side is at the same elevation. (b) If one side of the manometer is closed and evacuated, the atmospheric pressure on the open side pushes the mercury down and up into the evacuated side. The height of the mercury column in the evacuated side above the column in the open side is a measure of the atmospheric pressure.

TABLE 12.1 Conversion Factors for Common Units of Pressure.*

	N/m^2 (pascal)	lb/in^2	atm	bar	mm Hg (torr)	cm H$_2$O
N/m^2 or Pa†	1	1.45×10^{-4}	9.87×10^{-6}	10^{-5}	7.50×10^{-3}	1.02×10^{-2}
lb/in^2	6.89×10^3	1	6.80×10^{-2}	6.89×10^{-2}	51.7	70.3
atm	1.01×10^5	14.7	1	1.01	760	1.03×10^3
bar	10^5	14.5	0.987	1	750	1.02×10^3
mm Hg (torr)	133	1.93×10^{-2}	1.32×10^{-3}	1.33×10^{-3}	1	1.36
cm H$_2$O	98.1	1.42×10^{-2}	9.68×10^{-4}	9.81×10^{-4}	0.736	1

*Each number gives the value of 1 unit of the pressure named at the left in terms of the unit named at the top; for example, 1 lb/in^2 equals 51.7 mm Hg.
†A N/m^2 is also called a pascal (Pa).

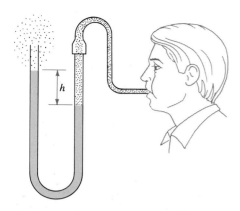

FIG. 12.4. When a person blows into the right side of an open-tube manometer, the extra pressure on this side forces water down and up into the left side.

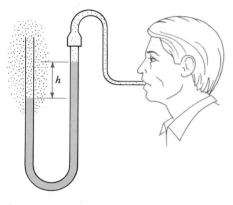

FIG. 12.5. When a person sucks on the right side of a manometer, the reduced pressure causes the water to rise on this side.

Many actual pressure-measuring devices do not have an evacuated side but are open on both sides. An open-tube manometer compares an unknown pressure of gas above one side of the tube to the atmospheric pressure above the other side. If both sides are exposed to air as in Fig. 12.3a, the pressure is the same on each side, and the mercury or water in the tube will be at the same level on each side.

If, however, one side of the manometer is attached to a container of gas at a pressure different from atmospheric pressure, the column of mercury or water on that side will be displaced either upward or downward, depending on whether the pressure is less or greater than atmospheric pressure. For example, if you blow into a rubber hose attached to the right side of a manometer containing water, as in Fig. 12.4, the water column will be displaced down on the right, owing to the extra pressure above it.* The difference in the height h of the water column on the left compared to the right indicates the pressure difference. This pressure difference is called **gauge pressure.** The gauge—that is, the measuring device—compares the unknown air pressure on the right to atmospheric pressure on the left. The absolute pressure P on the right is just the gauge pressure plus atmospheric pressure P_{atm}:

$$P = P_{atm} + P_{gauge}. \tag{12.2}$$

For the situation shown in Fig. 12.4, the gauge pressure is positive.

If a person sucks on a manometer as shown in Fig. 12.5, the gauge pressure is negative—that is, less than atmospheric pressure. The column of water is forced up into the partially evacuated right side by the atmospheric pressure on the left side.

In this and later chapters we will usually need to use absolute pressures in our equations. For simplicity we will use the word *pressure* to mean "absolute pressure." When necessary, Eq. (12.2) can be used to convert gauge pressure to (absolute) pressure and vice versa.

EXAMPLE 12.1 The gauge pressure inside a straw when you begin to suck on it while drinking a soda is −15 mm Hg. What is the (absolute) pressure in the straw?

SOLUTION

$$P = P_{atm} + P_{gauge} = 760 \text{ mm Hg} - 15 \text{ mm Hg}$$
$$= \underline{745 \text{ mm Hg.}}$$

∎

12.4 Equations of State

Pressure is one of several measurable characteristics of gases. Others are temperature, the volume of space occupied by the gas, and the amount of gas in that space. These quantities, called **state variables,** are used to describe the condition or state of the gas: Is the gas hot or cold? Compressed under high pressure into a small space or spread over a large volume under low pressure?

In the case of gases, scientists have found that state variables are related to each other in the form of equations called **equations of state.** They help us

---

*You would not want to blow or suck on the air above a mercury manometer; mercury is a highly toxic substance!

determine how changes in one state variable will affect other state variables. For instance, we know that if we close the nozzle of a bicycle pump so that air cannot leave it and then try to compress the gas by pushing down on the handle, the force with which we must push increases as the gas becomes more compressed. The pressure of the gas increases as its volume decreases. An equation of state is a mathematical equation that can be used to calculate this expected change in pressure when the volume of gas changes.

Robert Boyle (1627–1691), a contemporary of Newton, developed the equation of state that relates the pressure of a gas to its volume. Boyle used a manometer closed on one side (Fig. 12.6). A fixed amount of air was trapped in the closed side by a column of mercury, and the pressure on the air could be increased by adding mercury to the open column. Boyle found that if the temperature and amount of air in the tube remained constant, the volume of space occupied by the air decreased in inverse proportion to the pressure acting on the air; that is, $V \propto 1/P$. If the pressure doubled, the volume decreased to one-half its original volume. Boyle described this relation between pressure and volume with the following equation, now called **Boyle's law:**

$$PV = \text{Constant} \quad \text{(for a fixed amount of gas at constant temperature)} \quad \textbf{(12.3)}$$

For the product of P and V to remain constant, one of the quantities must decrease if the other increases, and vice versa. Absolute pressures must be used in this equation.

We often use Boyle's law by comparing the product of the pressure and volume in a gas before some process occurs and after it is over. If the amount of gas and the temperature of the gas remain constant, then the product of the final values of pressure and volume must equal the product of the initial values, since both products equal the same constant:

$$PV = P_0 V_0 = \text{Constant.} \quad \textbf{(12.4)}$$

Once again, absolute pressures must be used in this equation.

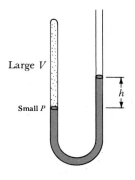

(a)

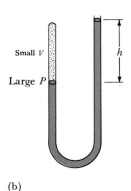

(b)

FIG. 12.6. (a) A small column of mercury creates little pressure, and the gas occupies a large volume. (b) A larger pressure is created by a high column of mercury; the gas occupies less volume.

EXAMPLE 12.2 Suppose that a person's skull is filled with air at a gauge pressure of +8 mm Hg. A skull fracture reduces the skull volume and, in turn, increases the gauge pressure in the skull to +16 mm Hg. Assuming that the air temperature remains constant, calculate the ratio of the final skull volume to the initial skull volume.

SOLUTION By converting from gauge to absolute pressures and then solving Eq. (12.4) for V/V_0, we get the desired ratio:

$$\frac{V}{V_0} = \frac{P_0}{P} = \frac{(8 \text{ mm Hg} + 760 \text{ mm Hg})}{(16 \text{ mm Hg} + 760 \text{ mm Hg})} = \frac{768}{776} = \underline{0.99}. \quad \blacksquare$$

The observed pressure change in Example 12.2 is caused by only a 1 percent decrease in volume. In terms of pressure, a person's skull behaves much like a skull filled with air. The gauge pressure is normally about 8 mm Hg. Pressures exceeding 16 to 20 mm Hg are cause for concern.

The calculation in Example 12.2 shows that pressure measurements easily reveal changes in skull volume and can therefore help in the detection of a number of head problems including brain disorders, tumors, hemorrhage, infections, swelling of the brain, and injuries to the skull. A sinus headache, for instance, occurs when sinus fluids occupy space in the skull, reducing the volume of space available for its usual contents. The reduction in volume increases the pressure on the brain, causing a headache.

Boyle's law is just one of many empirical equations developed to calculate expected changes in one state variable caused by changes in others. For example, with Charles' law (P/T = Constant) we can calculate expected changes in the pressure of a constant volume of gas when the temperature of the gas changes. Perhaps the most famous and most used of these equations of state is the ideal gas law.

12.5   Ideal Gas Law

The ideal gas law is a particularly simple example of an equation of state relating the variables pressure, temperature, volume, and amount of gas. For many gases at temperatures significantly above their condensation temperature, these variables are related approximately by the equation

$$PV = NkT, \tag{12.5}$$

where P is the gas pressure, N is the number of molecules of gas in a volume of space V, T is the gas temperature *in degrees Kelvin,* and k is a constant, called **Boltzmann's constant,** that equals 1.38×10^{-23} J/K.

FIG. 12.7. A typical house contains about 100 kg of air.

EXAMPLE 12.3   A house whose volume equals 80 m^3 contains air at a pressure of 1 atmosphere and at a temperature of $20°C$ (Fig. 12.7). How many molecules of air are in the house? If the average mass of one molecule is 4.8×10^{-26} kg, what is the mass of all the air in the house?

SOLUTION   When using the ideal gas law, we must convert temperatures to degrees Kelvin [$T = 20°C = (273 + 20)$ K = 293 K]. Also, pressure must be in units consistent with other quantities in the equation. In this example we convert from atm to N/m²; $P = 1$ atm = 1.01×10^5 N/m². Substituting in Eq. (12.5), we find that the number of molecules in the house is

$$N = \frac{PV}{kT} = \frac{(1.01 \times 10^5 \text{ N/m}^2)(80 \text{ m}^3)}{(1.38 \times 10^{-23} \text{ J/K})(293 \text{ K})} = \underline{2.0 \times 10^{27}}.*$$

Since the average mass of each molecule equals 4.8×10^{-26} kg, the total mass of the air in the house is

$$m = (2.0 \times 10^{27} \text{ molecules})(4.8 \times 10^{-26} \text{ kg/molecule}) = \underline{96 \text{ kg}}.$$

The air weighs about 200 lb.   ■

EXAMPLE 12.4   A bubble of water rises from the bottom of a lake, where the pressure is 1.8 atm, to the lake's surface, where the pressure is 1.0 atm. If the bubble's volume at the bottom of the lake is 1.0 cm^3, what is its volume just before it reaches the surface? The water temperature changes from $7°C$ at the bottom to $22°C$ at the surface.

*Remember that $1 \text{ N·m} = 1 \text{ J}$.

SOLUTION The problem can be worked using a proportionality technique with Eq. (12.5): $PV/T = Nk$. Since, the number of air molecules in the bubble does not change as it rises, the right-hand side of this equation is unchanged. Thus, the final value of PV/T as the bubble reaches the surface equals its initial value at the bottom:

$$\frac{PV}{T} = \frac{P_0 V_0}{T_0} = Nk.$$

The preceding equation can be rearranged to determine the final volume:

$$V = \left(\frac{P_0}{P}\right)\left(\frac{T}{T_0}\right)V_0. \qquad (12.6)$$

The ratio $P_0/P = 1.8\,\text{atm}/1.0\,\text{atm} = 1.8$. Since the units cancel, any pressure unit can be used. The pressures must, however, be absolute pressures, not gauge pressures. The ratio of the temperatures is $T/T_0 = (22 + 273)\,\text{K}/(7 + 273)\,\text{K} = 295/280 = 1.054$. Note that *temperatures must be in degrees Kelvin when using the ideal gas law*. We now substitute these ratios and $V_0 = 1.0\,\text{cm}^3$ in Eq. (12.6) to find the bubble's final volume: $V = (1.8)(1.054)(1.0\,\text{cm}^3) = \underline{1.9\,\text{cm}^3}$. ∎

Avogadro's Number, Moles, and the Ideal Gas Law

Most containers of gas, when near atmospheric pressure, hold huge numbers of atoms or molecules of gas. A cubic centimeter of air contains about 3×10^{19} molecules. This is such a large number that it has little meaning.

Sir James Jeans (1877–1946), the British astrophysicist, offered an interesting example of the enormous number of molecules involved in normal activities. If George Washington's last breath, which contained more than 10^{22} molecules, is now distributed uniformly in the air of the earth's atmosphere, then there are about five molecules from his last breath in each $2000\,\text{cm}^3$ of the atmosphere. Since your lungs hold about $2000\,\text{cm}^3$ of air, it is likely that five of the molecules now in your lungs were also part of George Washington's last breath.

We can avoid using large numbers in the ideal gas law by defining a quantity, the mole, and a related number, called Avogadro's number. One **mole** of a substance (solid, liquid, or gas) consists of 6.023×10^{23} atoms or molecules of that substance. The number 6.023×10^{23} is called **Avogadro's number** and is given the symbol N_A. One mole of any form of matter contains a mass in grams equal to the atomic or molecular mass of that form of matter. For example, one mole of atomic hydrogen (H) has a mass of 1 g, one mole of molecular hydrogen (H_2) has a mass of 2 g, one mole of helium (He) has a mass of 4 g, one mole of molecular oxygen (O_2) has a mass of 32 g, and so on.

Using the preceding definition, we can now define the number of molecules of a substance in terms of Avogadro's number and the number of moles of that substance. If we have 0.4 mole of oxygen in a container, the number of molecules of oxygen will be 0.4 times Avogadro's number, or $0.4(6.023 \times 10^{23}) = 2.4 \times 10^{23}$ molecules of oxygen. In general, the number N of atoms or molecules of a particular substance is related to the number n of moles of that substance and to Avogadro's number N_A by the equation

$$N = nN_A. \qquad (12.7)$$

If we substitute $N = nN_A$ into Eq. (12.5), we find that

$$\frac{PV}{T} = Nk = nN_A k = nR,$$

where $R \equiv N_A k = 8.314 \text{ J/mole}\cdot\text{K}$, the **universal gas constant.** Using this constant, we can rewrite the **ideal gas law** in terms of the number of moles of gas, a much more manageable number than the number of gas molecules:

$$PV = nRT. \tag{12.8}$$

Equation (12.8) provides a reasonable relation between the state variables for many real gases of low density at temperatures well above the temperature at which the gas condenses to a liquid. As before, the temperature in this equation must be in degrees Kelvin, and absolute pressures must be used.

EXAMPLE 12.5   A large cylinder of compressed helium gas contains 970 g of helium. How many balloons can be filled with the gas if the volume of each balloon is 0.015 m^3 (about 30 cm in diameter), the pressure inside each balloon is $1.2 \times 10^5 \text{ N/m}^2$, and the temperature of the helium is 27°C.

SOLUTION   We first use the ideal gas equation, Eq. (12.8), to determine the number of moles of helium needed for each balloon:

$$n = \frac{PV}{RT} = \frac{(1.2 \times 10^5 \text{ N/m}^2)(0.015 \text{ m}^3)}{(8.314 \text{ J/mole}\cdot\text{K})(27 + 273)\text{ K}}$$
$$= 0.72 \text{ mole}.$$

Since the atomic mass of helium is 4.0 g/1 mole, the mass of helium needed for one balloon is

$$0.72 \text{ mole} \left(\frac{4.0 \text{ g}}{1 \text{ mole}}\right) = 2.9 \text{ g}.$$

The cylinder contains 970 g of helium, so we should be able to fill 970 g/2.9 g = <u>330 balloons</u>.   ∎

EXAMPLE 12.6   A scuba diver breathes 600 cm^3 of air per breath and takes 30 breaths/min. (a) What volume of air does the diver breathe in 20 min? (b) The diver is under the water at a depth where the air is inhaled at a pressure of 2.5 atm. The tank from which the air is released is at 240 atm. What must be the volume of the tank so that the air will last 20 min? (c) Calculate the mass of the air in the tank at a temperature of 290 K. One mole of air has a mass of 29 g.

SOLUTION   (a) The volume of air inhaled in 20 min is

$$V = (600 \text{ cm}^3/\text{breath})(30 \text{ breaths/min})(20 \text{ min})$$
$$= 3.6 \times 10^5 \text{ cm}^3 = (3.6 \times 10^5 \text{ cm}^3)\left(\frac{10^{-6} \text{ m}^3}{1 \text{ cm}^3}\right) = \underline{0.36 \text{ m}^3}.$$

(b) This volume V of air at a pressure of 2.5 atm is to be compressed at a pressure of 240 atm so that its volume in the tank is V_t. We can find V_t by rearranging Boyle's law ($P_t V_t = PV$):

$$V_t = \frac{P}{P_t}V = \left(\frac{2.5 \text{ atm}}{240 \text{ atm}}\right)(0.36 \text{ m}^3) = \underline{3.8 \times 10^{-3} \text{ m}^3}.$$

(c) By Eq. (12.8), the number of moles of air in the tank is

$$n = \frac{P_t V_t}{RT},$$

where, for the tank, $V_t = 3.8 \times 10^{-3} \text{ m}^3$. The pressure must be converted from atm to N/m²:

$$P_t = (240 \text{ atm})\left(\frac{1.01 \times 10^5 \text{ N/m}^2}{1 \text{ atm}}\right) = 2.4 \times 10^7 \text{ N/m}^2.$$

We find, then, that

$$n = \frac{(2.4 \times 10^7 \text{ N/m}^2)(3.8 \times 10^{-3} \text{ m}^3)}{(8.314 \text{ J/mole} \cdot \text{K})(290 \text{ K})} = 38 \text{ moles}.$$

Since each mole has a mass of 29 g, the mass of the air is

$$m = (38 \text{ moles})(29 \text{ g/mol}) = \underline{1100 \text{ g}.} \qquad \blacksquare$$

A Note About Units

Often we can work a problem using ratios. For example, in part (b) of Example 12.6, the volume of the air in the tank depended on the ratio of the pressure at the diver's lungs to the pressure in the tank. In that situation any absolute pressure unit can be used, since the units cancel. In part (c), however, where we calculated the number of moles of gas, the units do not cancel. It is essential when making this type of calculation to *use a common system of units for all quantities*. The SI metric system is highly recommended and is used almost exclusively in this text.

12.6 Kinetic Theory of Gases

We have seen that quantities such as pressure, temperature, volume, and number of molecules are related by equations such as Boyle's law and the ideal gas law. We can substitute numbers into these equations and find the value of some unknown quantity. Doing this gives us little extra insight into the reason gases behave as they do. Why, for example, does the pressure of a gas double when its volume is reduced to one-half its original volume? Boyle's law describes this behavior but does not help us understand why this happens.

This variation of gas pressure with volume is really very surprising. Remember that a gas is made of molecules separated by distances much greater than their own size (Fig. 12.2). Almost 99.9 percent of the volume of a container of air is empty. It should be easy to "crowd" these air molecules into a smaller space. If we reduce the volume by half, the volume occupied by the gas is still about 99.8 percent empty. The molecules are still not rubbing elbows with each other. Yet when we do reduce the volume by a factor of one-half, the molecules exert twice as much pressure on the walls of their container. How can this be?

To answer such questions, scientists have proposed models concerning the nature and behavior of gases. Boyle proposed that a gas might be made of tiny

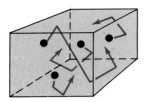

FIG. 12.8. Molecules of gas in constant motion cause a pressure because of their collisions with the walls of their container.

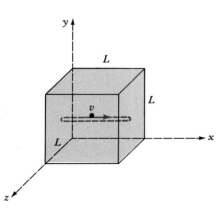

FIG. 12.9. A molecule that hits the wall, moving with speed v, rebounds and moves with speed v in the opposite direction. The impulse of the force is $f \Delta t = 2mv$.

particles that were springy—like little balls of curly wool piled together. The "wool," if coarse and springy enough, could resist compression much as air does. Isaac Newton, on the other hand, proposed that air was made of tiny particles somewhat farther apart. His idea was that particles resisted compression because of a repulsive force between the particles that increased as their separation decreased.

Although both these models seem to explain the relation between volume and pressure, neither explains the dramatic dependence of pressure on temperature. If the temperature of a constant volume of gas is doubled, its pressure doubles. In Newton's model the pressure depends only on the separation of the molecules, which does not explain the dependence of pressure on temperature. Similarly, Boyle's model does not account for variations in gas pressure with temperature.

A successful model for gases was not developed until the nineteenth century, about 150 years after Boyle discovered his law. This theory, the kinetic theory of gases, is used to derive an equation of state for gases, which, when compared to the ideal gas law, results in new insights into the meaning of temperature and thermal energy. The **kinetic theory of gases** states that gases consist of atoms and molecules in constant and rapid motion. The pressure caused by these atoms and molecules is a result of collisions of the atoms and molecules with the walls of their container (Fig. 12.8).

This model of gases and Newton's second law are used now to derive an expression for the pressure caused by a gas against the walls of its container. To complete this derivation, we will make several assumptions that simplify our calculations without affecting the general nature of the results. These assumptions are needed because of the huge number of atoms and molecules in a gas and because of the complicated motion of the individual atoms and molecules. We assume that individual atoms and molecules are point masses having no size, that they all move at the same speed v, and that one-third of the molecules move parallel to the x axis, one-third move parallel to the y axis, and one-third move parallel to the z axis. Also, we assume that the container holding the gas has length L on each side and that the x, y, and z axes are oriented parallel to the walls of the container, as shown in Fig. 12.9.

Our assumption of zero size allows us to ignore the internal structure of a molecule—the fact that it is made of nuclei surrounded by electrons. We can also ignore the vibrational and rotational motion of the molecules and the collisions of one molecule with another; they are too small to collide. The only collisions that occur are with the walls of the container. These collisions are assumed to be elastic; that is, there is no loss of kinetic energy during a collision.

Let us now consider the pressure of the gas on a wall lying in the y-z plane. When a molecule hits this wall, it rebounds in the opposite direction with the same speed (Fig. 12.9) because the collision is assumed to be elastic. The wall must therefore cause a force f' on the molecule, given by Eq. (4.3),

$$f' \Delta t = mv_x - mv_{x_0}, \qquad \textbf{(4.3)}$$

where Δt is the time of duration of the collision, f' is the average force of the wall on the molecule during that time, m is the molecule's mass, v_{x_0} is its velocity in the x direction before the collision, and v_x is its velocity in the x direction after the collision. For a molecule initially moving with a speed v in the positive x direction, its x components of velocity before and after a collision with the wall are $v_{x_0} = v$ and $v_x = -v$. Thus,

$$f' = \frac{m(-v) - mv}{\Delta t} = \frac{-2mv}{\Delta t}.$$

The force f of the molecule on the wall is the reaction to the force f' of the wall on the molecule. Since action-reaction forces have equal magnitudes but opposite directions, we find that

$$f = \text{Force of molecule on wall} = +\frac{2mv}{\Delta t}. \qquad (12.9)$$

After colliding with the wall, the molecule rebounds and flies across the box to hit the other wall. Rebounding from this wall, it returns again for another collision with the original wall. Each collision causes an impulse $f\Delta t$, the result of an abrupt force acting for a very short time. Let us multiply and divide this impulse by τ, the time needed for one round-trip across the box and back. Since the box has length L, the distance traveled is $2L$. This transit time for one round-trip at speed v is

$$\tau = \frac{2L}{v}. \qquad (12.10)$$

The impulse can now be written as

$$f\Delta t = f\Delta t\frac{\tau}{\tau} = \left(f\frac{\Delta t}{\tau}\right)\tau = (\overline{f})\tau.$$

We can think of $\overline{f}$ as an average force, smaller than f by a factor $\Delta t/\tau$, that acts during the whole time of transit τ. The impulse caused by this small average force $\overline{f}$ spread over the longer time τ has the same effect—that is, impulse—as the larger, abrupt force f acting for a short time Δt. According to Eq. (12.9), $f\Delta t = 2mv$. But since $f\Delta t = \overline{f}\tau$, then $\overline{f}\tau$ must also equal $2mv$. Rearranging the above equation and substituting for τ from Eq. (12.10) and for f from Eq. (12.9), we find that

$$\overline{f} = \frac{f\Delta t}{\tau} = \frac{2mv}{\tau} = \frac{2mv}{2L/v} = \frac{mv^2}{L}.$$

If there are N molecules in the box and one-third of them are hitting the wall (the other two-thirds move parallel to the wall and do not hit it), then the average force due to all of these molecules is

$$F = \frac{N}{3}\overline{f} = \frac{N}{3}\left(\frac{mv^2}{L}\right).$$

The pressure exerted on the wall by this force equals the force divided by the area of the wall. For a cubic box, the area is L^2. Thus,

$$P = \frac{F}{A} = \frac{1}{L^2}\left(\frac{N}{3}\right)\left(\frac{mv^2}{L}\right) = \frac{1}{3}\left(\frac{Nmv^2}{L^3}\right).$$

The volume of a cube of length L on each side is L^3. After substituting V for L^3 and rearranging, we find that

$$PV = \frac{1}{3}Nmv^2.$$

If we multiply and divide the right-hand side by 2, we can see the explicit dependence of the product PV on the kinetic energy of a molecule:

$$PV = \frac{2}{3}N\left(\frac{1}{2}mv^2\right).\qquad(12.11)$$

Now let us compare the ideal gas law ($PV = NkT$) to Eq. (12.11). Both equations have PV on the left. Thus, for both equations to be true simultaneously, we find that

$$\frac{2}{3}N\left(\frac{mv^2}{2}\right) = NkT,\quad\text{or}$$

$$\frac{1}{2}mv^2 = \frac{3}{2}kT.$$

The kinetic energy of a molecule in the gas equals a constant times the temperature of the gas.

Although our derivation was for molecules moving at constant speed v, the result is the same for molecules that move with a variety of speeds. For this real gas, we replace v^2 by the average of the square of a molecule's speed ($\overline{v^2}$). Thus,

$$\frac{1}{2}m\overline{v^2} = \frac{3}{2}kT.$$

If a container holds N molecules at temperature T, the average kinetic energy due to the random motion of all molecules is $N(3/2)kT$. This energy due to the random motion of molecules is called thermal energy.

The **average thermal energy** per molecule in a gas at temperature T is

$$E_{\text{th}}/\text{molecule} = \frac{1}{2}m\overline{v^2} = \frac{3}{2}kT.\qquad(12.12)$$

The **thermal energy of N molecules** is

$$E_{\text{th}} = N\left(\frac{3}{2}\right)kT.\qquad(12.13)$$

Temperature must be in degrees Kelvin when we use these equations.

We see that temperature is a measure of thermal energy!

The preceding expressions for thermal energy include the random kinetic energy due to translational motion only. We have ignored a molecule's thermal energy because of its rotational and vibrational motions. In our examples and problems in this chapter we assume that Eq. (12.13) includes all the thermal energy of a gas of N atoms. In a more rigorous treatment of gases consisting of diatomic molecules at room temperature, the 3/2 in front of kT should be replaced by 5/2 to include the molecule's rotational thermal energy.

EXAMPLE 12.7 A cup contains about 5.0×10^{21} gaseous helium atoms at a temperature of $27°C$. Calculate the thermal energy of these atoms.

SOLUTION The thermal energy of these atoms is

$$E_{\text{th}} = N\left(\frac{3}{2}kT\right) = (5.0 \times 10^{21})\frac{3}{2}\left(1.38 \times 10^{-23}\frac{\text{J}}{\text{K}}\right)(300\text{ K}) = \underline{31\text{ J}},$$

approximately equal to the work done by a person's heart in 30 beats. ∎

Root-Mean-Square Speed

Equation (12.12) allows us to estimate the speeds of different types of atoms or molecules in a gas. If we solve the equation for $\overline{v^2}$, we find that

$$\overline{v^2} = \frac{3kT}{m}.$$

FIG. 12.10. Particles of gas have the same average kinetic energy. If their mass is large, their speed is small; if their mass is small, their speed is large.

This is the average of the square of the speed of a molecule in the gas. The square root of $\overline{v^2}$ is called the **root-mean-square speed**:

$$v_{\text{rms}} = \text{Root-mean-square speed} = \sqrt{\overline{v^2}} = \sqrt{\frac{3kT}{m}}, \qquad \textbf{(12.14)}$$

where T is the temperature of the gas and m is the mass of the atom or molecule whose root-mean-square speed is being calculated.

We should remember that at any instant there is great variation in the speeds of the different atoms and molecules. The equation allows us to estimate roughly the average speed of an atom or molecule in the gas.

Equation (12.14) applies not only to single atoms and molecules in a gas but also to small particles consisting of many atoms or molecules, such as specks of dust, grains of pollen, and particles of smog (see Fig. 12.10).

EXAMPLE 12.8 Calculate the root-mean-square speed of an air molecule at a temperature of 27°C. One mole of air has a mass of 29 g. How does v_{rms} of air molecules compare to the speed of sound in air (340 m/s)?

SOLUTION Since one mole (6.023×10^{23} molecules) of air has a mass of 29 g, the average mass of a molecule is

$$m = \frac{29 \text{ g}}{6.023 \times 10^{23}} = 4.8 \times 10^{-23} \text{ g} = 4.8 \times 10^{-26} \text{ kg}.$$

We find, then, that

$$v_{\text{rms}} = \sqrt{\frac{3kT}{m}} = \sqrt{\frac{3(1.38 \times 10^{-23} \text{ J/K})(27 + 273) \text{ K}}{4.8 \times 10^{-26} \text{ kg}}} = \underline{510 \text{ m/s}}.$$

Sound waves are caused by air molecules oscillating in and out of regions of varying air density. Since the molecules cannot move into or out of these regions faster than their random speed, the speed with which sound waves are formed and travel is limited by this random speed. The speed of sound in a gas is about two-thirds the root-mean-square speed of its molecules. ∎

12.7 Work Done by a Gas

The technique used to calculate the thermal energy in a gas can be used with the laws of thermodynamics to analyze energy-transformation processes involving gases, such as those occurring in the cylinders of automobile engines or in the earth's atmosphere. These processes involve work, and our equation for calculating work must be modified slightly when we think of the work done on or by a gas as its volume changes.

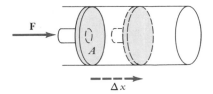

FIG. 12.11. The work done on the gas by the force of the piston when displaced by a distance Δx is $W = F \Delta x = P \Delta V$, where ΔV is the volume change of the gas and P is its pressure.

Recall from Chapter 8 that work is defined as

$$W = F \Delta r \cos \theta, \tag{8.1}$$

where F is the magnitude of a force acting on an object as it undergoes a displacement of magnitude Δr and θ is the angle between the direction of **F** and $\Delta \mathbf{r}$. Let us apply this definition to determine the work needed to change the volume of a cylinder containing gas. In Fig. 12.11 a force is shown pushing against a movable piston that forms one wall of the cylinder. The work done by the force as the piston undergoes a small displacement Δx is $W = F \Delta x \cos 0° = F \Delta x$. If we multiply and divide this expression for work by A, the cross-sectional area of the piston and cylinder, we find that

$$W = F \Delta x = F \Delta x \frac{A}{A} = \frac{F}{A}(\Delta x A) = P \Delta V,$$

where F/A is the pressure exerted by the gas against the piston and ΔV is the change in volume of the gas (the cross-sectional area of the cylinder times its change in length). This is a general expression for the work done on a gas during any change in volume. But if we use the convention that ΔV is negative when the volume decreases, then a negative sign should be placed in front of the $P \Delta V$ so that the work done on the gas is positive when its volume decreases.

The **work** W done **on a gas** when at pressure P as its volume changes by an amount ΔV is

$$W = -P \Delta V. \tag{12.15}$$

If the volume decreases, ΔV is negative and W is positive. If the volume increases, ΔV is positive and W is negative.

In our earlier use of the first law of thermodynamics and conservation-of-energy principle ($Q + W = \Delta E$), W was defined as the work done *on* an object in the system. If the gas in a cylinder is the system, then W is the work done *on the gas* and is defined as $W = -P \Delta V$. In some treatments of thermodynamics, the first law is written as $Q - W' = \Delta E$, where W' is defined as the work done *by* the system on its environment. For a gas, W', the work done *by the gas,* is defined as $W' = +P \Delta V$. Both conventions produce the same result, since a negative sign appears in front of W' in the first law when we use the latter convention.

EXAMPLE 12.9   A cylinder of gas has a movable piston of cross-sectional area 0.12 m^2 (Fig. 12.12). The piston can move freely up or down and exerts a constant pressure of $2.40 \times 10^5 \text{ N/m}^2$ on the gas. If the gas is heated, it expands and the piston moves 4.0 cm upward. Calculate the work done by the piston on the gas.

SOLUTION   The volume of the gas increases by $\Delta V = A \Delta x = +(0.12 \text{ m}^2)(0.04 \text{ m}) = +4.8 \times 10^{-3} \text{ m}^3$. The work done by the piston on the gas is

$$W = -P \Delta V = -(2.40 \times 10^5 \text{ N/m}^2)(4.8 \times 10^{-3} \text{ m}^3) = -1.15 \times 10^3 \text{ J}.$$

The gas in turn does work $W' = +1.15 \times 10^3 \text{ J}$ on the piston while lifting it. ∎

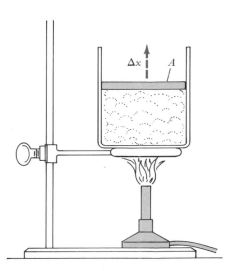

FIG. 12.12. The weight of a piston causes a constant pressure to be exerted on the gas.

Temperature Change
When the Volume of a Gas Changes

A change in volume of a gas can affect its temperature. Suppose that the volume of a gas decreases by an amount $-|\Delta V|$ and that no heat is added to or removed from the gas. The work done on the gas is $W = -P(-|\Delta V|) = P|\Delta V|$ and is a positive number. Substituting into the first law of thermodynamics ($Q + W = \Delta E$) for the case when $Q = 0$, we find that

$$0 + P|\Delta V| = \Delta E.$$

The internal energy ΔE of the gas must increase. For an ideal gas, most of this increase results in increased thermal energy. Since thermal energy is proportional to temperature [Eq. (12.13)], the temperature of the gas increases as its volume decreases. By similar reasoning we can show that the temperature of a gas decreases if its volume increases—that is, if no heat is added to or removed from the gas.

This change in temperature with volume change has many interesting consequences. For instance, when air moves up into the atmosphere from the earth below, it expands because of the decrease in pressure. If the change in elevation and expansion is rapid, the gas exchanges little heat with surrounding air, and its temperature is lowered because of the expansion. If the temperature is lowered below the dew point, water condenses and rain falls.

Another example in which the change in volume of air affects our climate occurs in the fall in California. Warm air known as the Santa Ana wind blows west from the desert toward the ocean. As the wind moves down off the mountains at the west coast and piles up against dense ocean air, its volume contracts, and its temperature increases. The air, already warmed by its passage across the desert, becomes even warmer. The change in temperature can be significant, as is apparent from the next example.

EXAMPLE 12.10 Estimate the change in temperature of air if its volume V contracts by 1 percent with no addition or removal of heat. The original number of molecules per unit volume (N/V) in the air is approximately 2.4×10^{25} molecules/m^3 and its average pressure is one atmosphere (1.01×10^5 N/m^2).

SOLUTION The volume decreases by 1 percent, or $\Delta V = -0.01V$, where V is the original volume of the gas. The work done in compressing the gas is

$$W = -P\,\Delta V = -P(-0.01V) = 0.01\,PV.$$

Heat is neither added to nor removed from the gas, so

$$Q = 0.$$

The change in internal thermal energy of the gas is

$$\Delta E_{\text{th}} = N\left(\frac{3}{2}\right)k(T - T_0) = N\left(\frac{3}{2}\right)k\,\Delta T,$$

where ΔT is the temperature change of the gas and N is the number of gas molecules.

Substituting for W, Q, and ΔE into the first law of thermodynamics, we find that

$$0 + 0.01PV = N\left(\frac{3}{2}\right)k\,\Delta T,$$

or

$$\Delta T = \frac{2(0.01)PV}{3Nk} = \frac{0.02}{3}\left(\frac{V}{N}\right)\frac{P}{k}$$

$$= \frac{0.02}{3}\left(\frac{1}{2.4 \times 10^{25}\ \text{molecules/m}^3}\right)\frac{(1.01 \times 10^5\ \text{N/m}^2)}{(1.38 \times 10^{-23}\ \text{J/K})} = \underline{2.0\ \text{K}}.$$

The temperature will increase by 2 K (2 C° or 3.8 F°) with only a 1 percent volume decrease. A 5 percent volume decrease would cause a 10-C° temperature increase.* ∎

12.8 P-V Diagrams and Thermodynamics of Gases

P-V Diagrams

When the pressure in a gas remains constant as it expands or contracts, the work calculation is quite easy; we multiply the pressure times the change in volume. However, if the pressure changes as the volume changes, the calculation is more difficult. In this case, a technique using P-V diagrams is helpful for making the work calculation.

A P-V diagram consists of a graph in which the pressure of a gas is plotted on the vertical axis for different values of volume plotted on the horizontal axis. When a gas expands under constant pressure, the P-V diagram consists of a straight, horizontal line as shown in Fig. 12.13a. The magnitude of work done on the gas during this expansion is $P(V - V_0)$ and equals the shaded area under the curve, since the area equals the product of the height P of the rectangle and its width, $V - V_0$. The sign of the work is negative if $V > V_0$ and positive if $V < V_0$.

If pressure changes while a gas expands or contracts, as represented by the P-V diagram shown in Fig. 12.13b, *the magnitude of the work done on the gas still equals the shaded area under the curve.* The expansion $(V - V_0)$ can be divided into many small expansions, each of magnitude ΔV. The work done during each small expansion is $P\,\Delta V$, where P is the average pressure during the small expansion and ΔV is the small change in volume. The area of each narrow rectangle in Fig. 12.13b equals the product of P and ΔV and therefore equals the work done during each small expansion. As the pressure decreases, the area of the rectangles becomes smaller, as does the work done during that small expansion. By adding the area of all of these rectangles, we have the total area under the curve and the total work done during the volume change.

Gases and the First Law of Thermodynamics

We are now in a position to use the first law of thermodynamics ($Q + W = \Delta E_{\text{system}}$) to analyze a variety of processes, such as the cyclical expansion and

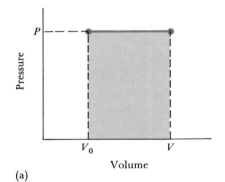

(a)

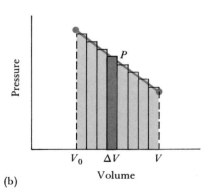

(b)

FIG. 12.13. (a) A change in volume at constant pressure. (b) A change in volume as the pressure changes. The area under the curve equals the magnitude of the work done on the gas during the volume change.

*Since air consists mostly of diatomic molecules whose thermal energy at room temperature equals $\frac{5}{2}kT$ rather than $\frac{3}{2}kT$ (the extra kT is present because of the molecules' rotational motion), our estimates are a little high. A 5 percent volume decrease should lead to a 6-C° temperature increase.

contraction of gases in an engine or the change in state of air as it rises on a hot summer day. The first law allows us to determine one unknown quantity (Q, W, or ΔE_{system}) if the other two are known. The quantities used in this equation are either given or determined as follows:

Q is the known or unknown heat transferred to or from the system.

W is the work done on the system and equals the negative area under the curve in a *P-V* diagram.

ΔE_{system} is calculated using an expression for the total internal energy stored in the system, such as Eq. (12.13) for an ideal gas.

If the only energy change of a monatomic, ideal gas is its thermal energy change, then $\Delta E_{system} = (\frac{3}{2}) Nk \, \Delta T$. This equation is often used with the ideal gas law (or with some other gas law). For example, we might use the ideal gas law to calculate the temperature change of a gas:

$$\Delta T = T - T_0 = \frac{PV}{nR} - \frac{P_0 V_0}{nR},$$

where P and V are the final pressure and volume of the gas and P_0 and V_0 are the initial values. We could then use Eq. (12.13) to calculate the change in thermal energy of the gas.

EXAMPLE 12.11 As $0.90 \, \text{m}^3$ of air at temperature $37°C$ rises on a hot summer day, the pressure of surrounding air decreases from $1.00 \, \text{atm}$ to $0.70 \, \text{atm}$, and the volume of the air increases to $1.15 \, \text{m}^3$ (represented by the *P-V* diagram in Fig. 12.14). (a) Calculate the work done on the air. (b) If the air gains $0.70 \times 10^4 \, \text{J}$ of heat, by how much does its internal energy change? (c) Calculate the final temperature of the gas.

SOLUTION (a) The work done on the gas during the expansion can be calculated by counting the squares under the curve representing this expansion in the *P-V* diagram shown in Fig. 12.14. Each square of height $0.1 \times 10^5 \, \text{N/m}^2$ and width $0.05 \, \text{m}^3$ corresponds to work of $(0.1 \times 10^5 \, \text{N/m}^2)(0.05 \, \text{m}^3) = 500 \, \text{J}$. Since there are 43 squares under the curve, the total work done on the gas during the expansion is

$$W = -(43 \text{ squares})(500 \text{ J/square}) = \underline{-2.2 \times 10^4 \, \text{J}}.$$

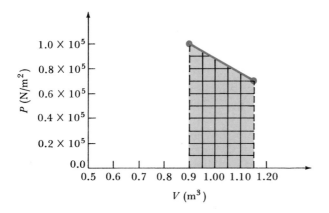

FIG. 12.14. The curve represents a process in which a gas expands from $0.9 \, \text{m}^3$ at pressure $1.0 \times 10^5 \, \text{N/m}^2$ to $1.3 \, \text{m}^3$ at pressure $0.7 \times 10^5 \, \text{N/m}^2$.

The work is negative, since $W = -P\,\Delta V$ is negative if ΔV is positive.

(b) To determine the air's change in internal energy, we substitute $W = -2.2 \times 10^4$ J and $Q = 0.70 \times 10^4$ J into the first law of thermodynamics ($Q + W = \Delta E$). Thus,

$$\Delta E = Q + W = 0.70 \times 10^4 \, \text{J} - 2.2 \times 10^4 \, \text{J} = \underline{-1.5 \times 10^4 \, \text{J}}.$$

(c) To determine the temperature change of the gas, we can first determine the number of moles of gas that have expanded by using the ideal gas law and the initial state of the gas:

$$n = \frac{P_0 V_0}{R T_0} = \frac{(1.01 \times 10^5 \, \text{N/m}^2)(0.90 \, \text{m}^3)}{(8.314 \, \text{J/mole·K})(310 \, \text{K})} = 35.3 \text{ mole.}$$

We can then calculate the final temperature of the gas using the ideal gas law and the final state of the gas:

$$T = \frac{PV}{nR} = \frac{(0.7 \times 1.01 \times 10^5 \, \text{N/m}^2)(1.15 \, \text{m}^3)}{(35.3 \, \text{mole})(8.314 \, \text{J/mole·K})} = 277 \, \text{K} = \underline{4°C}.$$

We could also have calculated the temperature using the calculated change in internal energy (-1.5×10^4 J) and Eq. (12.13) with the substitution $N = n N_A = (35.3 \text{ mole})(6.023 \times 10^{23} \text{ molecules/mole}) = 2.13 \times 10^{25}$ molecules:

$$\Delta T = T - T_0 = \frac{2}{3} \frac{\Delta E_{\text{th}}}{Nk} = \frac{2(-1.5 \times 10^4 \, \text{J})}{3(2.13 \times 10^{25})(1.38 \times 10^{-23} \, \text{J/K})} = -33 \, \text{K},$$

or a change in temperature of -33 C°. Since $T_0 = 37°C$, $T = \underline{4°C}$. It is interesting to note that the gas gains energy by heat transfer, but its temperature still decreases because of energy lost due to the work done by the gas as it expands. This cooling as the gas rises and expands could lead to condensation and perhaps rain. ∎

12.9 Diffusion

Earlier we pictured the molecules in a gas as moving at high speed and having frequent collisions. A molecule follows a zigzag path, its course changing abruptly at each collision. If we followed the motion of one molecule of air, it would be displaced only several centimeters in one minute even though it travels with an average speed of about 500 m/s. During the minute, it would experience more than 10^{11} collisions, each collison causing its speed and direction of motion to change.

If a group of special molecules, such as those from a bottle of perfume, enter the air at one location, they also experience this random motion. As time progresses, the molecules slowly move or diffuse away from their point of origin into the surrounding space (Fig. 12.15). **Diffusion** is the slow spreading due to random collisions of atoms and molecules away from areas in which they are concentrated into areas in which they are less concentrated.

Diffusion plays an important role in biological processes. Oxygen released in capillaries diffuses from the blood to cells near the capillaries, and muscle fibers receive oxygen that diffuses from surrounding material into the fibers. Since a fiber needs oxygen to twitch, the action of muscles may be limited by the rate of diffusion of oxygen into the fiber.

$t = 0$

Later

FIG. 12.15. As time passes, molecules of perfume diffuse farther from their point of origin.

12.10 Osmosis

Osmosis is a form of diffusion that occurs across a membrane. The membrane allows small molecules to pass but prohibits the passing of larger molecules. The skin of a wrinkled prune will allow water, but not the sugar molecules dissolved inside the skin, to pass. When the prune is placed in water, it absorbs water through its skin until it becomes swollen. It does not, however, lose sugar or any of its nutritious ingredients through the skin. A red blood cell, if placed in water, swells from the absorption of extra water through its membrane, but the cell does not lose hemoglobin or other large molecules from inside.

During osmosis, *there is a net fluid flow across the membrane into the region with the greatest concentration of dissolved molecules or ions.* This is why water flows into a prune toward the region of high sugar concentration and why water flows into a red blood cell toward the region of high hemoglobin concentration. If a person soaks an infected sore in epsom salts, fluid flows out of the sore into the water with high salt concentration.

What causes this flow of fluid through a membrane toward a region of higher concentration of dissolved molecules? Consider the two solutions pictured in Fig. 12.16a. The solution on the left of the membrane is a mixture of water and sugar molecules; that on the right is pure water. The solutions are separated by a semipermeable membrane that allows only water, not the sugar molecules, to pass. The two types of molecules (water and sugar) act much like molecules of gas. When the water molecules hit the membrane, they can pass through, but the sugar molecules cannot; they bounce off the membrane much as gas molecules bounce off the walls of a container. These sugar molecules create a pressure on the membrane that happens to be the same as the pressure P we would expect from a group of atoms of gas hitting a wall:

$$P = \frac{NkT}{V} = \frac{nRT}{V},$$

where N is the number of sugar molecules in the left chamber of volume V at temperature T, and n is the number of moles of sugar in this chamber.

Next consider the pressure caused by the water molecules. More water molecules per unit volume are on the right side of the membrane than are on the left because some of the space that could have been occupied by water in the left chamber is occupied by sugar molecules. Thus, more water molecules will be hitting the membrane per unit time from the right side than from the left. Since water can pass through the membrane, we find that a net movement of water occurs from the right side to the left. As this continues, the water level on the left

A wrinkled prune (top), when placed in water, absorbs water by osmosis and becomes swollen (bottom). Water flows toward the high sugar concentration in the prune. (Paul Conklin)

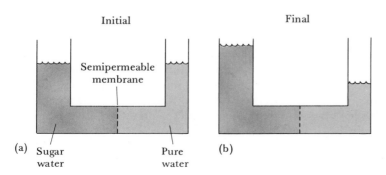

(a) Sugar water · Pure water · Semipermeable membrane · Initial · Final · (b)

FIG. 12.16. (a) A semipermeable membrane allows water but not sugar molecules to pass. More water molecules pass through the membrane from the right side, where they are most concentrated, to the left, where sugar molecules occupy the space normally occupied by water. (b) Eventually the water level rises until this osmosis stops. The extra pressure of the solution on the left compared to that on the right is called an osmotic pressure.

will rise and the level on the right will fall (Fig. 12.16b). This flow continues until the added pressure on the left causes the frequency of collisions of water molecules from the left to balance that from the right.

Because the diffusion stops when the number of water molecules hitting the membrane is the same on each side, we can see that the net water pressure on the membrane is zero. There is, however, the extra pressure on the membrane from the left caused by the dissolved sugar molecules. This pressure, called the **osmotic pressure,** equals the difference in pressure across the membrane, the higher pressure existing on the side of the membrane containing the dissolved molecules. The value of the osmotic pressure is

$$\text{Osmotic pressure} = P_{os} = \frac{nRT}{V}, \tag{12.18}$$

where n is the number of moles of dissolved molecules of any type in the chamber that is at the higher pressure.*

When working with solutions, it is common to define the **molar concentration** c of the solution—the number of moles of solute n (such as sugar) divided by the volume V of the solution in which it is dissolved:

$$\text{Molar concentration} = c = \frac{n}{V}. \tag{12.19}$$

The osmotic pressure across the membrane is, then,

$$P_{os} = cRT. \tag{12.20}$$

EXAMPLE 12.12 A balloon-shaped membrane containing sugar water of concentration 0.20 mole/1000 cm^3 and at a temperature of 27°C is placed in a container of pure water. If the membrane is permeable to water but not to sugar, what is the pressure difference between the inside and outside of the balloon when osmosis stops?

SOLUTION Water will move toward the solution with the highest concentration of solute—in this case, into the membrane. When osmosis has stopped, the water pressure on each side of the membrane is equalized. However, the sugar exerts a net outward pressure equal to

$$P_{os} = cRT = \left(\frac{0.2 \text{ mole}}{1000 \text{ cm}^3}\right)\left(8.314 \frac{J}{\text{mole} \cdot K}\right)(300 \text{ K})$$
$$= 0.50 \text{ J/cm}^3 = 5.0 \times 10^5 \text{ J/m}^3 = \underline{5.0 \times 10^5 \text{ N/m}^2}.$$

This is a pressure difference of about 5 atm and could easily cause the balloon to burst. A similar fate is suffered by red blood cells stored in water. ∎

Osmosis is important in the transport of water up the stalk of a plant. If the water in the plant stalk contains dissolved molecules, such as sugar in sap, then groundwater flows across the membranes into plant roots, forcing water higher in the plant stalk. However, if the groundwater contains too much salt concentration, osmosis causes water to flow out of the plant into the groundwater. This

*The situation in which molecules are dissolved on both sides of the membrane is considered in Example 12.13.

is one reason for watering plants with distilled water; salt does not accumulate in the soil. Salt can be leached from the soil of houseplants by running water through the soil for several minutes.

EXAMPLE 12.13 A 1.5-m-tall Norfolk pine is planted in soil that contains 0.100 mole/liter of salt ions (Na^+, K^+, Cl^-, and so forth). Calculate the concentration of dissolved molecules in the tree fluid that causes water to move by osmosis to the top of the tree. The plant grows at a temperature of 27°C, or 300 K.

SOLUTION From Table 12.1, we see that a pressure difference of 1 atm causes water to rise 1030 cm, or 10.3 m, higher on the evacuated side of a manometer than on the high-pressure side. In this example, the water is to rise 1.5 m up the tree. Thus, the osmotic pressure must be

$$P_{os} = \frac{1.5 \text{ m}}{10.3 \text{ m}}(1 \text{ atm}) = 0.15 \text{ atm} = 0.15 \times 10^5 \text{ N/m}^2.$$

The salt in the soil draws water out of the plant, whereas molecules dissolved in the tree sap draw water into the plant. These dissolved molecules must be greater in concentration than the salt. The c in Eq. (12.20) is the difference in concentrations of dissolved molecules or ions across the membrane of the plant roots. The net osmotic pressure in the desired direction is calculated as follows:

$$P_{net} = P_{in} - P_{out} = c_{sap}RT - c_{salt}RT = (c_{sap} - c_{salt})RT,$$

or

$$c_{sap} - c_{salt} = \frac{P_{net}}{RT} = \frac{0.15 \times 10^5 \text{ N/m}^2}{(8.314 \text{ J/mole} \cdot \text{K})(300 \text{ K})} = 6 \frac{\text{moles}}{\text{m}^3}$$

$$= 6 \frac{\text{mole}}{\text{m}^3}\left(\frac{1 \text{ m}^3}{10^6 \text{ cm}^3}\right)\left(\frac{10^3 \text{ cm}^3}{1 \text{ liter}}\right) = 6 \times 10^{-3} \frac{\text{mole}}{\text{liter}}.$$

Therefore

$$c_{sap} = 0.006 \text{ mole/liter} + c_{salt}$$

$$= (0.006 + 0.1) \text{ mole/liter}$$

$$= 0.106 \text{ mole/liter}. \quad \blacksquare$$

Summary and Additional Readings

1. **Gases** consist of atoms and molecules in constant motion that are separated far from each other except during frequent collisions. When the atoms and molecules collide with a solid or liquid surface, a small impulsive force is exerted on the surface. As a result of the many collisions per unit time, fairly uniform pressure is exerted by the gas on the surface.

2. **Pressure** is defined as the force that is exerted by a fluid such as air perpendicular to a surface divided by the area of the surface:

$$P = \frac{F}{A}. \quad (12.1)$$

3. **Equations of state:** In addition to pressure, gases are described using other quantities such as temperature T, the volume of space occupied by the gas V, the number of atoms or molecules N in that space, and the number of moles n in the space. These quantities are related to each other by equations of state such as **Boyle's law:**

$$PV = \text{Constant (if } T \text{ and } N \text{ are constant)}, \quad (12.3)$$

and the **ideal gas law:**

$$PV = NkT \quad (12.5)$$

or

$$PV = nRT, \tag{12.8}$$

where

$$n = \text{Number of moles of gas} = \frac{N}{N_A}. \tag{12.7}$$

$N_A = 6.023 \times 10^{23}$ is called **Avogadro's number;** $k = 1.38 \times 10^{-23}$ J/K is called **Boltzmann's constant;** and $R \equiv N_A k = 8.314$ J/mole·K is the **universal gas constant.**

4. **Thermal energy:** For a gas at temperature T (in degrees Kelvin) the average thermal energy due to translational motion of one of its molecules is

$$E_{th}/\text{molecule} = \frac{1}{2}m\overline{v^2} = \frac{3}{2}kT. \tag{12.12}$$

The **root-mean-square speed** of a particle of mass m in the gas is

$$v_{rms} = \sqrt{\overline{v^2}} = \sqrt{\frac{3kT}{m}}. \tag{12.14}$$

The **thermal energy** due to translational motion of N molecules is

$$E_{th} = N\left(\frac{3}{2}\right)kT. \tag{12.13}$$

5. **Work done on a gas:** The work W done *on a gas* as its volume changes is

$$W = -P\,\Delta V, \tag{12.15}$$

where ΔV is positive for a volume increase and negative for a decrease. This expression for work can be used with the first law of thermodynamics ($Q + W = \Delta E$) to analyze energy-transformation processes involving gases. Sometimes it is more convenient to consider the work done *by a gas,*

$$W' = +P\,\Delta V,$$

in which case the first law of thermodynamics is written as $Q - W' = \Delta E$.

6. **Osmosis** is the flow of a liquid such as water across a semipermeable membrane toward a region that contains the highest concentration of dissolved ions and molecules. The **osmotic pressure** caused by these dissolved ions and molecules is

$$P_{os} = \frac{n}{V}RT = cRT, \quad \textbf{(12.18) and (12.20)}$$

where $c = n/V$ is the **molar concentration** of the dissolved ions and molecules.

Russell K. Hobbie, *Intermediate Physics for Medicine and Biology*, John Wiley & Sons, New York (1978). Chapters 3–5 provide an interesting and detailed analysis of biological transport processes.

L. Peusner, *Concepts in Bioenergetics,* Prentice-Hall, Englewood Cliffs, N.J. (1974). Presents biological occurrences of diffusion and osmosis.

Questions

1. A skin diver when 20 m under the water is exposed to a pressure from the water of about 3×10^5 N/m². *Estimate* the net force of the water pushing in on his chest (equal to the force pushing in on his back). Why doesn't the diver's chest collapse under this large force?

2. *Estimate* the force of air pushing down on your head and shoulders in units of newtons and pounds. Does this downward force of the air on your body affect the reading of a bathroom scale when you weigh yourself? Explain.

3. One gram of air contains about 2×10^{22} molecules. If these molecules were distributed uniformly on the earth's surface, estimate the number that would be under your feet right now. The radius of the earth is 6.38×10^6 m.

4. The weight of the earth's atmosphere on each 1 m² of surface is about 10^5 N. *Estimate* the thickness of the atmosphere if it condensed to a liquid with each 1000 kg occupying 1 m³ of space.

5. Why do air bubbles increase in size as they rise?

6. A paper towel is crumpled and stuffed in the bottom of a tall glass. The glass is pushed, open end down, into some water until it is completely immersed. Will the towel remain dry? Explain.

7. Place a flat piece of cardboard over the top of a glass *filled* with water. Hold the glass over a sink and turn it upside down while holding the cardboard in place. If you remove your hand from the cardboard, it should remain in place and prevent water from falling out of the glass. Explain why this happens.

8. One mole of chicken feathers is spread uniformly over the surface of the earth. *Estimate* the thickness of this layer of feathers. Justify any assumptions made in your calculations. [*Note:* The radius of the earth is about 6×10^6 m.]

9. The temperature of the air in a room is doubled from 15°C to 30°C. By what fraction does the thermal energy of the air molecules increase?

10. A room is at a temperature of 20°C, and the speed of sound is 340 m/s. To about what temperature must the room air be increased to double the speed of sound? Explain.

11. Ten grams of nitrogen molecules have 1400 J of thermal energy, as do 10 g of oxygen molecules (held in a separate container). Are both containers at the same temperature? If not, which is hotter? Explain.

12. Why is air as it leaves a cylinder in which it has been compressed cooler than when in the cylinder?

13. Ammonia molecules move with a speed of about 600 m/s. Why do we not smell ammonia throughout a room the instant it is spilled on a table?

14. Why don't air molecules diffuse from near the earth's surface (where they are concentrated) into outer space (where they are not concentrated)?

Problems

12.1 Description of a Gas

■ 1.   There are about 3×10^{25} molecules of air in each cubic meter of the atmosphere. If each molecule had its own cube of space, what is the length of one side of the cube? How many times greater is the length of a side of the cube than the length of a molecule that is about 3×10^{-10} m long?

■ 2.   What is the length of the side of a cube that holds 4×10^9 molecules of air (equal in number to the people on the earth)? There are 3×10^{25} molecules in each cubic meter.

■ 3.   An air molecule traveling at 500 m/s has about 10^{10} collisions per second with other air molecules. (a) What distance does the molecule travel between collisions? (b) If the average distance between the nearest molecules is 3×10^{-9} m, how many other molecules does a molecule pass on the average between collisions?

12.2 Pressure

4.   Calculate the force of the earth's atmosphere on the cover of your physics book.

5.   Calculate the force of the earth's atmosphere (a) on one side of an 8×11-inch piece of paper, (b) on one side of a 3×5-inch index card.

6.   The air pressure drops by 0.30 bar when you climb from sea level to an elevation of 3000 m. Convert this pressure change to units of (a) mm Hg, (b) lb/in^2, and (c) N/m^2.

■ 7.   The pressure of the atmosphere at sea level is 1.0×10^5 N/m^2. Calculate the area over which the force of the atmosphere equals the weight of a 1000-kg car.

■ 8.   Air is withdrawn from inside a cube. The air outside at a pressure of 1 atm exerts a force of 1.0×10^4 N on each wall of the cube. What is the volume of the cube?

■ 9.   A cylindrical iron plunger is held against the ceiling, and the air is pumped from inside it. A 72-kg person hangs by a rope from the plunger (Fig. 12.17). What minimum radius must the plunger have so that atmospheric air pressure keeps it forced against the ceiling?

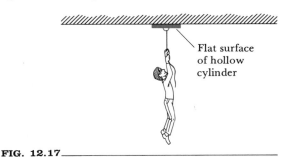

Flat surface of hollow cylinder

FIG. 12.17

12.3 Gauge Pressure

10.   (a) The pressure in a cylinder of compressed air is 52.0 lb/in^2. Calculate its gauge pressure. (b) An open-tube manometer with one side attached to a spinal tap has mercury that is 11 mm higher on the side open to the atmosphere. What is the gauge pressure in the spine? The absolute pressure?

■ 11.   Air is sucked from a straw, reducing the absolute pressure in the straw by 10 percent below atmospheric pressure. What is the absolute pressure in units of N/m^2 in the straw after sucking? The gauge pressure?

■ 12.   The gauge pressure of blood in the aorta increases from 80 to 120 mm Hg when blood is pumped into the aorta from the heart. Calculate the percent increase in the absolute blood pressure in the aorta.

■ 13.   During a hurricane, the gauge pressure in a house is zero and that above the roof is -1060 Pa (N/m^2). If the area of the roof is 200 m^2, what is the net force of the air on the roof—that is, the difference of the force caused by the air from below and that exerted from above?

12.4 Equations of State

14.   The gauge pressure of gas in a balloon decreases from 0.40×10^5 N/m^2 to 0.30×10^5 N/m^2 while the temperature and amount of gas remain constant. Calculate the ratio of the final volume to the initial volume of the gas.

■ 15.   The middle ear has a volume of 0.60 cm^3 when at atmospheric pressure at sea level. Calculate the volume of that same air when the air pressure is 0.83×10^5 Pa at an elevation of 1500 m above sea level. The air temperature remains constant. (If the volume of the middle ear remains constant, some air will have to leave as the elevation increases. The ears are said to "pop.")

■ 16.   The inside of a student's skull has a volume of 1130 cm^3, and the gauge pressure in the skull is $+6$ mm Hg. While cramming for an exam, the volume of the student's brain increases, causing an increase in gauge pressure to $+25$ mm Hg. Estimate the increased volume of the brain. [*Note:* This increase equals the decrease in space available inside the skull.]

■ 17.   The pressure of the air in a skin diver's lungs when she is 15 m under the water surface is 2.5 atm, and the air occupies a volume of 4.8 liters (4.8×10^{-3} m^3). Calculate the volume of the air in the diver's lungs when she quickly moves to the surface, where the pressure is 1.0 atm. Why is this not a good move?

■ 18.   When surrounded by air at a pressure of 1 atm, a basketball has a radius of 0.12 m. (a) Calculate its volume. (b) Calculate its volume when held 15 m under the surface of water where the absolute pressure is 2.5 atm. (c) Calculate its radius when at this pressure. Assume that the air temperature remains constant.

■ 19.   A cork gun is fired by compressing a piston until the pressure in a cylinder with the cork at the end increases to 5.0×10^5 Pa. If the initial pressure is 1.0×10^5 Pa with the piston 0.40 m from the cork, how far is the piston from the cork when it is dislodged from the gun? Assume that the air temperature does not change.

■■ 20.   A spherical air bubble in water has a radius of 0.060 cm when at a certain depth. As the bubble rises, the pressure decreases by 50 percent. Now, what is the bubble's radius, assuming no temperature change?

■■ 21.   A cylindrical diving bell, open at the bottom and closed at the top, is 4 m tall. When at the ocean's surface, the bell is filled with air at atmospheric pressure. The pressure increases by 1 atm for each 10 m the bell is lowered below the ocean surface. If the bell is lowered 36 m below the ocean surface, how many meters of air space are left at the top of the inside of the bell?

12.5 Ideal Gas Law

22. (a) What are the approximate molecular masses (in grams) of lithium (Li), argon (Ar), and water (H_2O)? Approximately how many moles is (b) 33.6 g of water (H_2O)? (c) 33.6 g of oxygen (O_2)? (d) 33.6 g of helium (He)?

23. (a) Calculate the number of molecules per unit volume (N/V) in the atmosphere at the top of Mt. Everest. The pressure is 0.31 atm and the temperature is $-30°C$. (b) Calculate the number of molecules per unit volume at sea level, where the pressure is 1.0 atm and the temperature is $20°C$.

24. A high-altitude weather balloon filled with helium atoms has a volume of 3.8 m³ when the pressure and temperature are $0.64 \times 10^5 \, N/m^2$ and $17°C$, respectively. Calculate (a) the number of helium atoms and (b) the number of moles of helium in the balloon.

■ 25. During the compression stroke of the cylinder of a diesel engine, the air pressure in the cylinder increases from 1 atm to 50 atm and the temperature increases from $26°C$ to $517°C$. Calculate the ratio of the initial volume to the final volume of the cylinder.

■ 26. Homemade beer is capped into a bottle at a temperature of $27°C$ and a pressure of 1.2 atm. The cap will pop off if the pressure inside the bottle exceeds 1.5 atm. What temperature must the gas inside the bottle reach to pop the cap? The gas is assumed to be ideal.

■ 27. A cork gun is shot by heating the gas in a cylinder with both ends closed (one end is closed by the cork). The gas in the cylinder starts at a temperature of $20°C$ with a pressure of 1.2 atm, and the cork explodes from the cylinder when the gas temperature reaches $200°C$. What was the pressure in the cylinder when the cork left?

■ 28. An ideal gas at $0°C$ is held in a cylinder with a movable piston in one end. Calculate the percent change in the temperature of the gas if (a) the pressure increases by 10 percent while its volume and the amount of gas remain constant, (b) the volume increases by 10 percent while the pressure and amount of gas remains constant, and (c) the amount of gas increases by 10 percent while its pressure and volume remain constant.

■ 29. A 600-m³ house contains air at $17°C$ and 1 atm. If the air pressure remains at 1 atm, how many moles of air are removed from the house as the air warms to $27°C$?

■ 30. The air in the middle ear has a volume of 0.6 cm³, a pressure of 1 atm, and a temperature of $37°C$ when at sea level. After a person climbs to an altitude of 3000 m, the volume is still 0.6 cm³, the pressure is now 0.66 atm, and the temperature is $35°C$. How many moles of air have been removed from the middle ear?

■ 31. A car tire initially contains 5.1 moles of air at a gauge pressure of $2.1 \times 10^5 \, N/m^2$ and a temperature of $27°C$. The temperature increases to $37°C$, the volume decreases to 0.80 times its original volume, and the gauge pressure decreases to $1.6 \times 10^5 \, N/m^2$. Calculate the number of moles of gas that leaked out of the tire.

12.6 Kinetic Theory of Gases

32. The hydrogen atoms on the sun's surface are at a temperature of 6000 K. Each atom has a mass of 1.66×10^{-27} kg. Calculate the root-mean-square speed of these atoms. Would the speed of sound be faster or slower on the sun than on the earth? Explain.

33. A nitrogen molecule (4.7×10^{-26} kg), a virus (1.6×10^{-20} kg), and a golf ball (0.045 kg) are all in a room at a temperature of 300 K. Calculate the root-mean-square speed of each object.

■ 34. The root-mean-square speed of a gas of hydrogen molecules (H_2) is 1700 m/s. Calculate the molecular mass of a smoke particle suspended in the gas if its root-mean-square speed is 0.14 m/s ($m_{H_2} = 3.35 \times 10^{-27}$ kg).

■ 35. The speed that a projectile must have to escape from the earth is 1.12×10^4 m/s. (a) At what temperature would an average molecule of oxygen in the earth's atmosphere be able to escape from the earth ($m_{O_2} = 5.31 \times 10^{-26}$ kg)? (b) Repeat part (a) for hydrogen ($m_{H_2} = 3.35 \times 10^{-27}$ kg). Comment on the earth's ability to retain hydrogen and oxygen.

■ 36. (a) If the total mass of a car is 1500 kg and an average atom in the car has a mass of 1.0×10^{-25} kg, how many atoms does the car contain? (b) If each atom has an average thermal energy of $3/2 \, kT$ and the temperature of the car is $30°C$, what is the total thermal energy of the car? (c) At what speed must the car drive so that its nonrandom kinetic energy along a highway ($1/2 \, mv^2$) equals its thermal energy?

■ 37. *Estimate* the thermal energy of all the atoms in your body. [*Note:* The mass of a single carbon atom is about 2×10^{-26} kg.]

■ 38. (a) Show that the root-mean-square speed of gas molecules at temperature T_1 is related to that at temperature T_2 by the equation

$$\frac{v_1}{v_2} = \sqrt{\frac{T_1}{T_2}}.$$

(b) The speed of sound is roughly proportional to the root-mean-square speed of the gas molecules. If the speed of sound is 331 m/s at $0°C$, what is its speed at $20°C$?

■ 39. The root-mean-square speed of atom A is 1.5 times that of atom B when the atoms are at the same temperature. Calculate the ratio of the mass of atom A to that of atom B.

■ 40. If the speed of sound is proportional to the root-mean-square speed of a molecule, show that the speed of sound in helium (He) is approximately 2.6 times the speed in nitrogen (N_2).

■ 41. A box is divided in the middle by a wall. Hydrogen molecules (H_2) moving with a root-mean-square speed of 300 m/s are placed on the left side, and nitrogen molecules (N_2) moving at 100 m/s are on the right. Calculate the ratio of the temperature of the hydrogen gas and the temperature of the nitrogen gas.

■■ 42. One insulated container of gas holds a certain number of hydrogen molecules, and a second insulated container of equal volume holds 0.80 times as many oxygen molecules. If the pressures are the same in each cylinder, calculate the ratio of (a) the temperature of the hydrogen gas and that of the oxygen gas and (b) the ratio of the root-mean-square speed of a hydrogen molecule and that of an oxygen molecule. $m_{O_2}/m_{H_2} = 16$.

■■ 43. A 0.058-kg tennis ball, traveling at 25 m/s, hits a wall, rebounds with the same speed in the opposite direction, and is hit again by a tennis player, causing the ball to return to the wall at the same speed. The ball returns to the wall once each 0.60 s. (a) Calculate the force of the ball on the wall averaged over the time between collisions. (b) If 10 people are practicing against the same wall, whose area is 30 m², calculate the average pressure of the 10 tennis balls against the wall.

■■ 44. Several students throw snowballs at the wall of a barn with dimensions 3.0 m × 6.0 m. The snowballs have mass 0.10 kg and hit the wall with an average speed of 6.0 m/s. They do not rebound. Calculate the average pressure of the snowballs on the wall if 40 snowballs hit the wall each second.

12.7 and 12.8 Work Done by a Gas, P-V Diagrams, and Thermodynamics

45. A gas expands from a volume of 2.0×10^3 cm^3 to a volume of 3.4×10^3 cm^3 while at a pressure of 1.2×10^5 N/m^2. Calculate (a) the work done on the gas during the expansion and (b) the work done by the gas.

■ 46. One cubic meter of air containing 2.4×10^{25} molecules expands to 1.2 times its original volume as it moves up in the atmosphere. Suppose that the average pressure acting on the gas is 0.9 atm (the pressure actually decreases as the gas expands). (a) Calculate the work done on the gas during its expansion. (b) If no heat enters or leaves the gas, calculate its change in temperature using the first law of thermodynamics and the expression for the thermal energy of a gas.

■ 47. In a certain process involving a gas, -1000 J of work is done on the gas as it absorbs 3000 J of heat. Calculate its change in internal energy. Does the gas expand or contract? Does its temperature increase or decrease? Explain.

■ 48. An ideal gas at a constant pressure of 1.10×10^5 N/m^2 contracts from an initial volume of 0.300 m^3 to 0.280 m^3. As it contracts, the internal thermal energy of the gas changes by $+1200$ J. Calculate (a) the work done on the gas and (b) the heat added to or removed from the gas. (c) Does its temperature increase or decrease? Explain.

■ 49. An ideal gas is compressed slowly at constant temperature to 0.70 times its original volume. While the gas is being compressed, 5000 J of heat is removed from it. (a) Calculate the change in thermal energy of the gas. (b) Calculate the work done on the gas during compression.

■■ 50. Ten moles of an ideal gas has an initial volume of 0.30 m^3 at 1 atm pressure. Heat is added to the gas, causing its temperature to increase by 200 C° while its pressure remains constant at 1 atm. Calculate (a) the increase in internal energy of the gas, (b) the initial and final temperatures of the gas, (c) its final volume, (d) the work done on the gas, and (e) the heat added to the gas.

■ 51. A 0.035-m^3 cylinder containing 1.5 moles of an ideal gas at 40°C is closed at both ends, one end by a piston. The piston is compressed to a final volume of 0.0031 m^3. The compression occurs rapidly so that no heat is lost during compression (called an *adiabatic compression*). The final temperature of the gas is 540°C. How much work is done during compression?

■ 52. For the P-V diagram shown in Fig. 12.18, calculate the

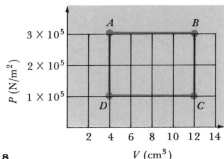

FIG. 12.18

work done on a gas as it moves (a) from A to B, (b) from B to C, (c) from C to D, and (d) from D to A.

53. In Fig. 12.19, estimate the work done on a gas during the expansion (a) from A to B, (b) from B to C, and (c) from C to A.

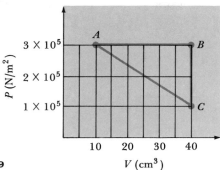

FIG. 12.19

■■ 54. The changes of internal thermal energy during each portion of the cycle shown in Fig. 12.18 are $A \rightarrow B$ (5.6 J); $B \rightarrow C$ (-8.2 J); $C \rightarrow D$ (-4.8 J); and $D \rightarrow A$ ($+7.4$ J). Calculate the work done on the gas and the heat added to or removed from the gas during each step of the cycle.

■■ 55. The changes in internal thermal energy of an ideal gas moving along each step of the cycle shown in Fig. 12.19 are $A \rightarrow B$ (10 J); $B \rightarrow C$ (-9 J); $C \rightarrow A$ (-1 J). Calculate the work done on the gas during each step of the cycle and the heat added to or removed from the gas during each step.

12.10 Osmosis

56. A semipermeable membrane separates distilled water on one side from 2.0×10^{-3} m^3 of water on the other side in which 0.030 moles of sugar is dissolved. Calculate (a) the molar concentration of sugar molecules and (b) the osmotic pressure across the membrane. The water is at 27°C.

57. Seawater with 400 mole/m^3 of salt ions dissolved in it and distilled water at 27°C are placed on opposite sides of a semipermeable membrane. Calculate the osmotic pressure across the membrane. In which direction is the net flow of water across the membrane?

■ 58. A rigid balloon with semipermeable walls contains 27°C water of volume 5 cm^3. Dissolved in the water is 0.20 g of glucose with molecular weight 180. (a) What is the molar concentration of glucose in the balloon? (b) The balloon is placed in a glass of water containing salt with dissolved ions at a molar concentration of 1.6×10^2 mole/m^3. Calculate the osmotic pressure due to the glucose and due to the salt. (c) Calculate the net osmotic pressure, the difference of the pressures calculated in part (b). Will the balloon expand or shrink? Explain.

■ 59. The osmotic pressure at the base of a houseplant kept at 27°C is 1.6×10^4 N/m^2. (a) Calculate the molar concentration of dissolved molecules in the plant's veins if there are no dissolved molecules in the soil water. (b) Suppose that the soil water contains 0.010 mole/liter of dissolved molecules and ions and that the concentration of dissolved molecules in the plant is that calculated in part (a). What is the osmotic pressure across the plant roots?

■ 60. A typical red blood cell has a volume of 95 μm^3 (95 × 10^{-18} m^3) and holds about 3.4×10^8 molecules of hemoglobin that are impermeable to the cell walls. Most other molecules in the blood can cross the cell walls. If a red blood cell is placed in pure water, what osmotic pressure is developed in the cell relative to the water? Actually, the cell stretches and its walls become permeable to hemoglobin before this pressure is attained. Assume that the temperature is 17°C.

■■ 61. The osmotic pressure in a prune when placed in water at 300 K is about 15 atm. (a) Calculate the molar concentration of sugar in the prune when at room temperature. (b) *Estimate* the volume of a prune, and then calculate the number of moles of sugar in the prune. (c) If the molecular weight of sugar is 180, about how many grams of sugar are in the prune?

■■ 62. A *Ficus benjamina* has 1.28 mole/m³ of dissolved molecules in the fluid (sap) in the plant. Calculate the osmotic pressure between this fluid and water in the soil if the water has a concentration of dissolved molecules and ions equal to 1.00 mole/m³. Does the osmotic pressure cause fluid to flow into the plant or out of the plant? If fluid flows into the plant, how high will the water rise because of the osmotic pressure? The temperature is 17°C.

CHAPTER 13

Fluid Statics

Any form of matter that flows, such as a liquid or a gas, is called a **fluid.** In this chapter we use the ideas of statics to describe static fluids—fluids at rest. In the next chapter we consider fluid dynamics—fluids in motion.

Fluids are of paramount importance to living organisms. The human body is two-thirds water, and liquids in the body not only transport nutrients to its 10^{14} living cells but also carry waste products from those cells. The oxygen we need for combustion of foods flows in and out of the respiratory system along with other molecules in the air.

Fluids are also responsible for most of the climatic conditions on the earth. Air and ocean currents moderate the climates of different parts of the globe, while convection currents in the atmosphere carry pollutants away from large cities. Even the drift of the continents on the earth's mantle can be considered a form of flow.

Because fluids are not rigid, they can flow and change shape when exposed to external forces. If you walk into a wall, its rigid structure opposes your motion. If you walk into air, it simply flows past your body. Because fluids often lack definite size and shape, it is convenient to use the quantities pressure and density rather than force and mass when studying fluid statics and dynamics. We will use these quantities with Newton's laws to develop some of the important principles concerning static fluids.

13.1 Pressure

When we study fluids, the quantity pressure assumes a role of primary importance. What force is to solids in the study of statics and dynamics, pressure is to fluids.

Pressure P was defined in Chapter 12 as the ratio of the magnitude of a force applied perpendicular to a surface and the area over which the force is exerted:

$$P = \frac{F}{A}. \tag{12.1}$$

Pressure is a scalar quantity and has no direction. However, the force caused by the pressure of a fluid has a magnitude and a direction. The magnitude of the

The pressure in a fluid, such as blood, depends upon elevation. If a giraffe did not have special valves in its circulatory system, the pressure increase as it lowered its head could cause brain damage, and the pressure decrease as it raised its head could cause it to faint. (Tom Fix/Peter Arnold)

FIG. 13.1. The pressure of the earth's atmosphere causes a force on the skin that is always perpendicular to the skin.

force on a surface of area A is

$$F = PA.$$

The direction of the force caused by the pressure is perpendicular to and toward the surface. For example, the molecules of air colliding with a person's head cause a pressure of about 10 N/cm^2, so the magnitude of the force on a 1-cm^2 area is 10 N. The direction of the force depends on the orientation of the skin, as depicted in Fig. 13.1. The air hitting a 1-cm^2 area at the top of the man's head causes a 10-N force directed downward, whereas a 1-cm^2 area under his chin experiences a 10-N force pointing up. A horizontal force acts on the sides of the face.

A static fluid at pressure P exerts a force $\mathbf{F}$ perpendicular to and toward a surface. The magnitude of the force against a surface of area A is

$$F = PA. \tag{12.1}$$

EXAMPLE 13.1 Estimate the force of the earth's atmosphere on the skin under your chin when your head is held upright.

SOLUTION The area of the skin under the chin is about $5 \text{ cm} \times 8 \text{ cm} = 40 \text{ cm}^2 = 40 \times 10^{-4} \text{ m}^2$. The pressure of the earth's atmosphere at sea level is $1.0 \times 10^5 \text{ N/m}^2$. Thus, the magnitude of the air's force on the skin is

$$F = PA \cong (1.0 \times 10^5 \text{ N/m}^2)(40 \times 10^{-4} \text{ m}^2) = \underline{400 \text{ N}}$$

(about 100 lb)! The direction of the force is perpendicular to and toward the surface. For this example the force points up (approximately). ∎

13.2 Density

Just as pressure is a more convenient quantity than force when we discuss fluids, density is more useful than mass to describe the inertial property of fluids. The mass density—that is, **density**—of a substance is given the Greek letter ρ (rho) and is defined as the ratio of the mass m of the substance and its volume V:

$$\text{Density} = \rho = \frac{m}{V}. \tag{13.1}$$

The units of density are kg/m^3, g/cm^3, slug/ft^3, and so on. The densities of various substances are listed in Table 13.1 in the SI unit kg/m^3. Notice that water has a density of 1000 kg/m^3. Hence, 1000 kg of water occupies 1 m^3. Platinum has a density of $21{,}450 \text{ kg/m}^3$, whereas the density of air is about 1.3 kg/m^3.

Densities are often quoted in units of g/cm^3 rather than kg/m^3. A density of 1 g/cm^3 equals a density of 1000 kg/m^3, as we can see by the following conversion:

$$1 \frac{\text{g}}{\text{cm}^3} = 1 \frac{\text{g}}{\text{cm}^3} \left(\frac{1 \text{ kg}}{10^3 \text{ g}} \right) \left(\frac{10^6 \text{ cm}^3}{1 \text{ m}^3} \right) = 10^3 \frac{\text{kg}}{\text{m}^3}.$$

Equation (13.1) can be used to calculate the unknown mass of a known volume of a fluid whose density is known.

TABLE 13.1   Densities of Various Substances

Solids		Liquids		Gases	
Substance	Density (kg/m³)	Substance	Density (kg/m³)	Substance	Density (kg/m³)*
Aluminum	2,700	Acetone	792	Dry air, 0°C	1.29
Copper	8,920	Ethyl alcohol	791	10°C	1.25
Gold	19,300	Methyl alcohol	810	20°C	1.21
Iron	7,860	Gasoline	660–690	30°C	1.16
Lead	11,300	Mercury	13,600	Helium	0.178
Platinum	21,450	Milk	1,028–1,035	Hydrogen	0.090
Silver	10,500	Seawater	1,025	Oxygen	1.43
Bone	1,700–2,000	Water, 0°C	999.87		
Brick	1,400–2,200	3.98°C	1,000.00		
Cement	2,700–3,000	20°C	1,001.80		
Clay	1,800–2,600	Blood Plasma	1,030		
Glass	2,400–2,800	Blood, whole	1,050		
Ice	917				
Balsa wood	120				
Oak	600–900				
Pine	500				
Planet earth	5,252				
Moon	3,340				
Sun	1,410				
Universe (average)	10^{-27}				
Pulsar	10^{8}–10^{11}				

*At 1-atm pressure and 0°C temperature.

EXAMPLE 13.2   We see from Table 13.1 that the density of a pulsar, thought to be an extremely dense rotating star made of neutrons, is about 10^{11} kg/m³. Calculate the mass of a pulsar contained in a volume the size of your fist (about 200 cm³).

SOLUTION   Using Eq. (13.1), we find that the mass of the material is

$$m = \rho V = (10^{11} \text{ kg/m}^3)(200 \text{ cm}^3)\left(\frac{1 \text{ m}}{100 \text{ cm}}\right)^3$$

$$= \underline{2 \times 10^7 \text{ kg}}.$$

This is about one-tenth the mass of the Empire State Building!   ■

EXAMPLE 13.3   The radius of a collapsing star destined to eventually become a pulsar decreases by 10 percent while at the same time 12 percent of its mass escapes. Calculate the percent change in its density.

SOLUTION   The volume of a sphere is $(\frac{4}{3})\pi r^3$. Thus, the star's density is $\rho = m/V = m/(\frac{4}{3}\pi r^3) = (3\,m)/(4\,\pi r^3)$. The star's mass m after decreasing by 12 percent is related to its initial mass m_0 by the equation $m = 0.88m_0$. Similarly, since the star's radius decreases by 10 percent, $r = 0.90r_0$. We can now determine the ratio of the star's density ρ after these changes and its initial density ρ_0:

$$\frac{\rho}{\rho_0} = \frac{(3m)/(4\pi r^3)}{(3m_0)/(4\pi r_0{}^3)} = \left(\frac{m}{m_0}\right)\left(\frac{r_0}{r}\right)^3$$

$$= \left(\frac{0.88m_0}{m_0}\right)\left(\frac{r_0}{0.90r_0}\right)^3 = 0.88(1.11)^3 = 1.21$$

The density increases by 21 percent. ∎

Mass densities are rarely used in the English system of units. It is more common to use a quantity called **weight density** D, the ratio of an object's weight w and its volume V:

$$\text{Weight density} = D = \frac{w}{V}.$$

Since $w = mg$, we find that mass density and weight density are related by the equation $D = w/V = mg/V = (m/V)g = \rho g$, or

$$\rho = \frac{D}{g}.$$

To convert a weight density to a mass density, we simply divide by g, the acceleration due to gravity.

Many tables in handbooks list what is called the **relative density** or **specific gravity** of a substance instead of its density. The relative density of a substance is defined as the ratio of its density to the density of water at $3.98\,°C$ (1000 kg/m^3). For example, since the relative density of lead is 11.34, its density is 11.34 times that of water, or $11,340 \text{ kg/m}^3$. Relative density is a unitless quantity since it is the ratio of two quantities having the same units.

13.3 Pressure Variation with Depth

The pressure exerted by a fluid varies with depth; the deeper an object is in a fluid, the greater the pressure acting on the object. You may have felt the increase in water pressure on your ears while swimming under water or the decrease in air pressure when driving or hiking from lower to higher elevations in the mountains, when taking off in an airplane, or while riding up in a skyscraper's elevator. When our ears "pop," air is released from the middle ear. This release of air causes a reduction in air pressure inside the middle ear that compensates at least partially for the reduced pressure outside the ear when at a higher elevation.

The pressure at a particular depth in a fluid is due to the weight of the fluid above. For instance, the blood pressure in the veins of a person's feet exceeds the pressure in the head because the blood in the feet supports the weight of the blood above it. This increased pressure may cause one's feet to swell, especially if one must stand while working.

The variation of pressure with depth is somewhat like stacking books on a table (Fig. 13.2). Imagine that each book is a layer of air. The only pressure on the top book is that due to the air pushing down from above. However, the second book from the top must support the weight of the top book plus the atmospheric air above it. The bottom book in the stack must support the weight of the five books above it plus the atmospheric air, and so forth.

FIG. 13.2. The pressure is greatest on books deepest in the stack because they must support the weight of all the books above them.

We can derive an equation that allows us to calculate the difference in pressure at different depths in a fluid. Consider the shaded portion of the fluid shown in Fig. 13.3. The bottom surface is at vertical elevation y_1 and the top surface is at elevation y_2. The difference in pressure between y_1 and y_2 should depend on the extra weight of fluid that must be supported at elevation y_1 compared to that at elevation y_2.

The vertical forces acting on the shaded volume are shown in Fig. 13.4. The fluid above the shaded volume pushes down with a force $\mathbf{F}_2$. If the pressure at elevation y_2 is P_2 and the cross-sectional area of the cylinder is A, then the magnitude of the force from above is

$$F_2 = P_2 A.$$

Similarly, fluid from below the shaded section of fluid pushes up with a force $\mathbf{F}_1$ equal in magnitude to

$$F_1 = P_1 A.$$

The third force acting on the shaded volume is its weight $\mathbf{w}$, a downward-directed force.

Since the fluid is not moving and therefore is in equilibrium, these three vertically directed forces must add to zero. With the y axis pointing upward, this equilibrium condition is written as

$$\Sigma F_y = F_1 - F_2 - w = 0,$$

or

$$P_1 A - P_2 A - w = 0. \tag{13.2}$$

The weight of the shaded section of fluid depends on its density and volume. The volume is

$$V = A(y_2 - y_1),$$

where $(y_2 - y_1)$ is the height of the shaded section of fluid. The weight of this part of the fluid (assuming constant density) is

$$w = mg = (\rho V)g = \rho A(y_2 - y_1)g.$$

Substituting for w in Eq. (13.2), we find that

$$P_1 A - P_2 A - \rho g A(y_2 - y_1) = 0.$$

After dividing the equation by A, we have the desired result:

$$P_1 - P_2 = \rho g(y_2 - y_1)$$

or

$$P_1 = P_2 + \rho g(y_2 - y_1). \tag{13.3}$$

This equation allows us to calculate the pressure in a fluid at one elevation in terms of the pressure at a second elevation. The equation has been derived with the assumption that the y axis points upward and that the density of the fluid does not change with elevation.

EXAMPLE 13.4 Calculate the pressure on a skin diver who is 10 m below the surface of the water. The density of seawater is 1025 kg/m^3, and the air pressure at the water's surface is 1.01×10^5 N/m^2.

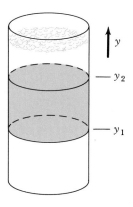

FIG. 13.3. A cylinder of fluid.

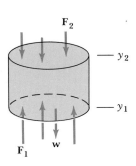

FIG. 13.4. The forces acting on the shaded section of fluid shown in Fig. 13.3.

SOLUTION Let $y_2 = 0$ m at the water's surface. Then $P_2 = 1.01 \times 10^5$ N/m^2. The elevation of the skin diver is then $y_1 = -10$ m, and the pressure of the water on the skin diver at that elevation is

$$\begin{aligned} P_1 &= P_2 + \rho g(y_2 - y_1) \\ &= 1.01 \times 10^5 \,\text{N/m}^2 + (1025 \,\text{kg/m}^3)(9.8 \,\text{m/s}^2)[0 - (-10 \,\text{m})] \\ &= 1.01 \times 10^5 \,\text{N/m}^2 + 1.01 \times 10^5 \,\text{N/m}^2 = \underline{2.02 \times 10^5 \,\text{N/m}^2}. \end{aligned}$$

This is equivalent to a pressure of 2 atm. The pressure increases by 1 atm for each 10 m under the water's surface. For example, if the pressure at the surface is 1 atm, the pressure 30 m below the surface is 1 atm + 3 atm = 4 atm. ∎

EXAMPLE 13.5 Estimate the net force on your eardrum due to the air inside and out after you drive from Denver at an elevation of 1609 m to the top of Pikes Peak at 4301 m. Assume that the area of the eardrum is 0.20 cm^2, that the pressure of the outside air and the air in the middle ear are balanced at Denver, and that no air leaves the middle ear during the trip. The average density of the outside air is about 0.80 g/m^3.

SOLUTION The net force of air on the ear is the force of air in the middle ear pushing out minus the force of the atmosphere pushing in, where each force, according to Eq. (12.1), equals the pressure of the air times the area of the eardrum ($F = PA$).

  The air inside the ear does not change, so the force of this air pushing out remains the same. However, as the person's elevation increases, the pressure of the atmosphere decreases, as does its force on the eardrum. According to Eq. (13.3),

$$\begin{aligned} P_{\text{atm Pikes Peak}} - P_{\text{atm Denver}} &= \rho g(y_{\text{Denver}} - y_{\text{Pikes Peak}}) \\ &= (0.80 \,\text{kg/m}^3)(9.8 \,\text{m/s}^2)(1609 \,\text{m} - 4301 \,\text{m}) \\ &= -2.1 \times 10^4 \,\text{N/m}^2. \end{aligned}$$

Thus, the force of the atmosphere on the eardrum decreases by

$$\begin{aligned} \Delta F_{\text{atm}} &= (P_{\text{atm Pikes Peak}} - P_{\text{atm Denver}})A \\ &= (-2.1 \times 10^4 \,\text{N/m}^2)(0.20 \,\text{cm}^2 \times 10^{-4} \,\text{m}^2/\text{cm}^2) \\ &= -0.42 \,\text{N}. \end{aligned}$$

Since the air in the middle ear pushes out with an unchanging force and the force of the atmosphere decreases by 0.42 N, the net force on the ear at the top of Pikes Peak is $\underline{0.42 \,\text{N pushing out}}$, about the weight of half of a small apple. ∎

13.4 Externally Applied Pressure

We have seen that fluid pressure varies with depth. Another important property of fluids is the uniform increase in pressure throughout an enclosed volume of fluid that is caused by increased pressure at one location in the fluid. This consequence is predicted by Eq. (13.3):

$$P_1 = P_2 + \rho g(y_2 - y_1).$$

If the pressure P_2 at elevation y_2 increases, there must be an equal increase in pressure P_1 at elevation y_1. This property was first recognized in the seventeenth century by the French mathematician and philosopher Blaise Pascal (1623–1662). The principle, named in his honor, is stated as follows:

Pascal's principle: An external pressure exerted on a static, enclosed fluid is transmitted uniformly throughout the volume of the fluid.

If the pressure at the pumping station of a city water system is increased by 10 pounds per square inch (psi), then the pressure at all the homes connected by water lines to the pumping station also increases by 10 psi. The pressure varies in different homes, depending on their elevation relative to the pumping station, but an increase in pressure at the station causes an equal increase throughout the water system.

Pascals' principle has important consequences in biology and medicine. For example, glaucoma, an eye disease, involves Pascal's principle. A clear fluid called aqueous humor fills two chambers in the front of the eye (Fig. 13.5). In the normal eye, new fluid is continually secreted into these chambers while old fluid drains from the eye through small sinus canals. When a person has glaucoma, these drainage canals close, fluid accumulates in the front of the eye, and the pressure inside the eye increases. This pressure increase is transmitted to the retina at the back of the eye and causes degenerative changes in the retina that can eventually lead to blindness.

A practical application of Pascal's principle is the hydraulic press, a form of simple machine that converts small forces into larger forces, or vice versa. Hydraulic presses are used by automobile mechanics to lift cars and by dentists and barbers to raise and lower the chairs on which their clients sit. The hydraulic brakes of an automobile are also a form of hydraulic press. Most of these devices work on the simple principle illustrated in Fig. 13.6, although the actual devices are usually more complicated in construction.

A force F_1 is applied to a piston that has a cross-sectional area A_1. The pressure in the fluid just under the piston is

$$P = \frac{F_1}{A_1}.$$

Because the pressure is transmitted uniformly throughout the fluid, the pressure of the fluid under piston 2 is also $P = F_1/A_1$.* But since the area of piston 2 is greater than that of piston 1, we find that the upward force of the fluid on piston 2 is greater than the force causing the pressure at piston 1. The upward force of the fluid on piston 2 is

$$F_2 = PA_2 = \left(\frac{F_1}{A_1}\right)A_2 = \left(\frac{A_2}{A_1}\right)F_1. \tag{13.4}$$

Since A_2 is greater than A_1, F_2 is also greater than F_1.

EXAMPLE 13.6 How large a force is needed on a small piston of area 2 cm² to support a 1000-N weight resting on a piston of area 20 cm²?

*We are assuming that piston 2 is at the same elevation as piston 1. If not, Eq. (13.3) will tell us the pressure differences under the two pistons because of their differences in elevation.

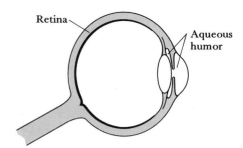

FIG. 13.5. When extra fluid cannot drain out of the eye's cornea, pressure increases throughout the eye, possibly causing blindness.

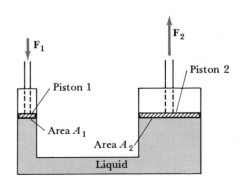

FIG. 13.6. A hydraulic press. A force on piston 1 causes an increase in pressure that is transmitted through the fluid. This increased pressure causes a large force on piston 2 because of its larger area.

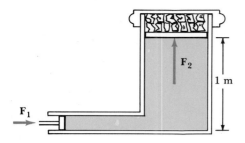

FIG. 13.7. A hydraulic can crusher.

SOLUTION We are given that $A_1 = 2$ cm^2, $A_2 = 20$ cm^2, $F_2 = 1000$ N. We are asked to find F_1. After rearranging Eq. (13.4), we find that

$$F_1 = \left(\frac{A_1}{A_2}\right)F_2 = \left(\frac{2 \text{ cm}^2}{20 \text{ cm}^2}\right)1000 \text{ N} = \underline{100 \text{ N}}.$$

A 100-N force can support a 1000-N weight—like balancing a heavy weight on the short arm of a lever by a small force applied to the long arm. ■

EXAMPLE 13.7 A hydraulic can crusher is shown in Fig. 13.7. The large piston has an area of 8 m^2 and exerts a force $\mathbf{F}_2$ of magnitude 2×10^6 N on the cans. Calculate the magnitude of the force $\mathbf{F}_1$ exerted by the small piston (area 10 cm^2) on the fluid. Do not ignore the fact that the large piston is 1 m higher than the small piston.

SOLUTION The pressure P_2 in the fluid at the top of the crusher is

$$P_2 = \frac{F_2}{A_2} = \frac{2.0 \times 10^6 \text{ N}}{8.0 \text{ m}^2} = 2.5 \times 10^5 \text{ N/m}^2.$$

Then, according to Eq. (13.3), the pressure on the fluid at the small piston is

$$\begin{aligned}
P_1 &= P_2 + \rho g(y_2 - y_1) \\
&= 2.5 \times 10^5 \text{ N/m}^2 + (1000 \text{ kg/m}^2)(9.8 \text{ m/s}^2)(1.0 \text{ m} - 0) \\
&= 2.5 \times 10^5 \text{ N/m}^2 + 0.098 \times 10^5 \text{ N/m}^2 \cong 2.6 \times 10^5 \text{ N/m}^2.
\end{aligned}$$

Since the small piston has an area of 10 cm^2 (10^{-3} m^2), the magnitude of the force $\mathbf{F}_1$ on this piston is

$$F_1 = P_1 A_1 = (2.6 \times 10^5 \text{ N/m}^2)(10^{-3} \text{ m}^2) = \underline{260 \text{ N}}. \qquad ■$$

13.5 Archimedes' Principle

A fluid exerts an upward buoyant force on objects in the fluid. Such buoyant forces either completely or partially support the object's weight. For example, the buoyant force of water can support the weight of a swimmer or a million-pound iceberg. The atmosphere exerts a buoyant force that helps support the weight of a balloon.

To derive an equation for calculating the buoyant force of a fluid on an object, consider a metal block immersed in a liquid, as shown in Fig. 13.8. The fluid, a liquid in this case, exerts forces on all six surfaces of the block. The forces exerted on opposite sides of the block cancel each other. However, the upward force $\mathbf{F}_1$ due to the fluid pressure from below the block and the downward force $\mathbf{F}_2$ due to the pressure from above the block do not cancel, since the pressure of the liquid on the bottom is greater than that on the top. The net force of the fluid on the block $\mathbf{F}_1 - \mathbf{F}_2$, points upward, since $\mathbf{F}_1$ is greater in magnitude than $\mathbf{F}_2$. This net fluid force is called the **buoyant force B** of the fluid, and its magnitude is

$$B = F_1 - F_2 = P_1 A - P_2 A = (P_1 - P_2)A, \qquad (13.5)$$

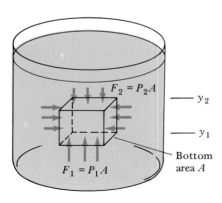

FIG. 13.8. The forces acting on a metal block immersed in a liquid.

where A is the area of the bottom and top surfaces of the block, and P_1 and P_2 are the fluid pressures at the bottom and top of the block, respectively.

The difference in pressure at two different elevations in the fluid is given by Eq. (13.3):

$$P_1 - P_2 = \rho g(y_2 - y_1).$$

Substituting in Eq. (13.5), we find that the buoyant force is

$$B = \rho g(y_2 - y_1)A.$$

The product $(y_2 - y_1)A$ is the volume V of the block—its height times its area—and also equals the space taken up by the block in the fluid. This space is called the **volume V of fluid displaced** by the object in the fluid. Substituting V in place of $(y_2 - y_1)A$ results in the following expression for the buoyant force of the fluid:

$$B = \rho g V.$$

It is important to note that the density ρ in this equation is the density of the fluid, not of the object in the fluid. According to Eq. (13.1), the product ρV is the mass of a volume V of fluid having density ρ. When we multiply mass times g, we convert mass to weight. Thus, $\rho g V$ is the weight of fluid whose volume equals V. From now on we will place subscript f on ρ and on V to remind us that these quantities refer to "fluid" density and "fluid" volume.

The **buoyant force B** exerted by a fluid on an object in the fluid is equal in magnitude to the weight of fluid displaced by the object. If the displaced fluid has density ρ_f and volume V_f, then the buoyant force is

$$B = \rho_f g V_f. \tag{13.6}$$

If an object is completely immersed in a fluid, the volume of the displaced fluid equals the volume of the object. If the object floats, then the volume of the displaced fluid equals the volume of space taken up by the object below the fluid's surface.

The derivation of Eq. (13.6) was for a solid cube, but the results apply to objects of any shape. The pressure on the lower sides of irregular-shaped objects exceeds that on the upper sides, and the resultant buoyant force is given by Eq. (13.6).

The principle just derived, called **Archimedes' principle** was supposedly discovered by the Greek mathematician Archimedes while sitting in his bath. He had been asked by the king to determine whether a crown was made of pure gold or of less valuable metals made to look like gold. In the bath Archimedes thought of a technique he could use to measure the density of the crown: he would weigh the crown first in air and then when immersed in water. Using these two measurements and a small amount of calculation, he would be able to tell whether the crown's density was that of gold or of a less dense and less valuable metal. Archimedes was so excited by his flash of brilliance that he ran from his house shouting, "Eureka!"—the Greek for "I have found [it]!" Whether he was naked or wore a towel is still unknown.

Problems involving Archimedes' principle are solved in the same way that we solved problems in statics and dynamics. We start by drawing a force diagram of an object. The buoyant force is included with other forces in this diagram. If the object is at rest, then the sum of the vertical components of all forces acting on the object must be zero. If the object is accelerating, then the sum of

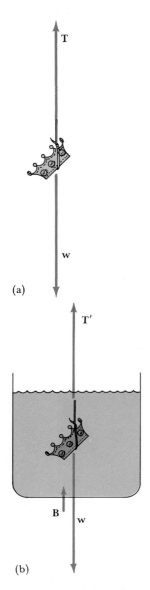

(a)

(b)

FIG. 13.9. A force diagram for a crown hanging in (a) air and (b) water.

The weight of the extra water displaced by the *President Lincoln* in the photograph on the far right equals the weight of the cargo added to the ship. (Craig Dawson)

the vertical components of all forces acting on the object equals its mass times its vertical component of acceleration (Newton's second law). In addition to these fundamental equations, we often use the equation that relates mass and weight ($w = mg$) and the defining equation for density ($\rho = m/V$).

In summary, when solving problems that involve buoyant forces, we draw a force diagram and then use one or more of the following equations to solve the problem depicted in the force diagram:

$$\Sigma F_y = 0 \qquad \text{if } a_y = 0 \qquad\qquad\text{(6.2b)}$$

or

$$\Sigma F_y = ma_y \qquad \text{if } a_y \neq 0, \qquad\qquad\text{(3.6)}$$

$$B = \text{magnitude of the upward-pointing buoyant force}$$

$$= \rho_f g V_f, \qquad\qquad\text{(13.6)}$$

$$\rho = m/V, \qquad\qquad\text{(13.1)}$$

$$w = mg. \qquad\qquad\text{(3.4)}$$

EXAMPLE 13.8 The tension in a string supporting a crown is found to be 25.0 N when the crown is suspended in air and 22.6 N when suspended in water. Calculate the density of the crown to see if it is made of gold, as claimed by its donor. The density of water is 1000 kg/m³.

SOLUTION When the crown is supported in air (Fig. 13.9a), the 25.0-N tension force and the weight balance. Thus, the crown's weight is 25.0 N. The weight of the crown is related to its density ρ and its volume V by Eqs. (3.4) and (13.1):

$$w = mg = \rho V g. \qquad\qquad\text{(13.7)}$$

We know the crown's weight (25.0 N) but not its volume; hence we cannot

determine its density from Eq. (13.7). We learn more by considering the forces shown in Fig. 13.9b that act on the crown when in water: **B** is the upward buoyant force of the water on the crown, **w** is the crown's weight (25.0 N), and **T'** is the upward tension force of the string (22.6 N) that is attached to the crown. Since the crown hangs at rest, these three forces must add to zero:

$$\Sigma F_y = T' + B - w = 0. \tag{13.8}$$

We substitute into Eq. (13.8) for the buoyant force: $B = \rho_f g V$, where V is the volume of displaced water, which equals the volume of the crown since it is completely immersed, and ρ_f is the density of the water. After substituting for B and rearranging, we find that

$$V = \frac{w - T'}{\rho_f g}.$$

Rearranging Eq. (13.7) to solve for the crown's density ρ and substituting for the crown's volume from above, we find that

$$\rho = \frac{w}{gV} = \frac{w}{g} \frac{\rho_f g}{(w - T')} = \frac{w}{(w - T')} \rho_f$$

$$= \frac{25.0 \text{ N}}{(25.0 \text{ N} - 22.6 \text{ N})} 1000 \frac{\text{kg}}{\text{m}^3} = \underline{10,400 \text{ kg/m}^3}.$$

We see from Table 13.1 that $10,400 \text{ kg/m}^3$ is less than the density of gold. Thus, the crown is not made of pure gold. ∎

EXAMPLE 13.9 A life raft of cross-sectional area 2.0 m × 3.0 m has its top edge 0.36 m above the waterline when unloaded. How many 75-kg passengers can the raft hold before water starts to leak over the edges? The raft is in seawater of density 1025 kg/m³.

SOLUTION As people enter the raft, it sinks deeper into the water. The maximum buoyant force available for supporting the passengers is the weight of water displaced by the raft as it sinks 0.36 m deeper into the water (Fig. 13.10). This increased buoyant force balances the weight of an unknown number N of passengers, each of average weight $w = mg$, where $m = 75$ kg. The first condition of equilibrium becomes

$$\Sigma F_y = B_{\text{max}} - Nw = 0, \qquad \text{or}$$

$$N = \frac{B_{\text{max}}}{w} = \frac{\rho_f g V}{mg}$$

$$= \frac{(1025 \text{ kg/m}^3)(0.36 \text{ m} \times 2.0 \text{ m} \times 3.0 \text{ m})}{75 \text{ kg}}$$

$$= 29.5.$$

The raft can precariously hold 29 passengers. ∎

Volume of water displaced as people enter raft

0.36 m

FIG. 13.10. As people enter a raft, it sinks deeper and displaces more water.

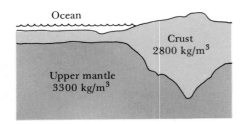

Ocean

Crust
2800 kg/m³

Upper mantle
3300 kg/m³

FIG. 13.11. A continent "floats," partially submerged, on the earth's mantle.

EXAMPLE 13.10 It is believed that the earth's crust is composed of about thirty continent-sized pieces of rock that "float" on a semifluid "plastic" mantle lying below the earth's surface (Fig. 13.11). If the average density of the continental mass is 2800 kg/m³ and that of the mantle is 3300 kg/m³, what fraction of the continental crust is floating above the mantle (like an iceberg floating above water)?

SOLUTION Since the crust is floating, its weight w must be balanced by the upward buoyant force of the mantle:

$$\Sigma F_y = B - w = 0, \quad \text{or} \quad B = w. \tag{13.10}$$

The weight of the continental crust is related to its density ρ and volume V by Eqs. (3.4) and (13.1):

$$w = mg = \rho_c g V_c.$$

The buoyant force of the mantle (considered as a fluid) is

$$B = \rho_m g V_m,$$

where ρ_m is the density of the mantle and V_m is the volume of space in the mantle that is occupied by the floating crust. For example, if 60 percent of the continent is in the mantle, then $V_m = 0.6 V_c$.

Substituting for B and w in Eq. (13.10), we find that

$$\rho_m g V_m = \rho_c g V_c,$$

or

$$\frac{V_m}{V_c} = \frac{\rho_c}{\rho_m} = \frac{2800 \text{ kg/m}^3}{3300 \text{ kg/m}^3} = 0.85.$$

We see that 85 percent of the continent is submerged in the mantle. Only 15 percent is above the mantle, and much of that is covered with water. ∎

13.6 Surface Tension

Most of us would be surprised to see a "pile" of water shaped like a giant drop sitting on a road (Fig. 13.12) or an unsupported wall of water around a home. Whenever we see a large amount of water, a container is always holding it in place, such as the concrete walls of a swimming pool, the plastic mattress of a water bed, or a bathtub.

Yet it is possible to form small, unsupported walls and piles of water. We have all seen small drops of water on an oil-covered driveway, and children routinely form small, thin walls of water when blowing soap bubbles. What supports these tiny walls and piles of water?

A bubble of soap water is not supported by a membrane at its surfaces. The water itself forms a very thin network of molecules at its surface that holds the bubble together. This surface network is held in place by attractive forces between adjacent molecules. These forces are caused by the nonuniform distribution of positive and negative electric charges on the molecules. The molecules arrange themselves so that the part of one molecule with a small positive charge is, on the average, nearer the part of a neighboring molecule with a small negative charge, resulting in a small attractive force between the molecules.

Water?

FIG. 13.12. Could water remain "piled" on a road without any support from the sides?

The net force on a molecule inside a liquid is zero, since the forces of other molecules pull it from all directions (see Fig. 13.13). However, molecules at the surface do not have attractive forces pulling on them from outside the water. They are attracted only toward molecules at the side and to those deeper in the solution (Fig. 13.13). Consequently, a small layer of molecules on the liquid surface is held together and pulled back into the liquid. This surface layer acts much like a plastic membrane that helps to contain the water.

The strength of this liquid surface layer is measured in terms of a quantity called surface tension. To define surface tension, consider a canvas bag with a slash of length l across its surface (Fig. 13.14a). When we sew the cut together and fill the bag with water, we find that a force pulls on the stitches from each side of the cut (Fig. 13.14b). If we represent the total force of the stitches in one direction by the symbol F, then the surface tension γ is defined as the ratio of F and l:

$$\text{Surface tension} = \gamma = \frac{F}{l}. \qquad (13.11)$$

Surface tension is the force per unit length needed to hold the surface together. The force is tangent to the liquid surface and is caused by the mutual attraction of the surface molecules to each other and to other molecules below the surface. The force has the same effect as a thin membrane holding the liquid in place (see Fig. 13.14b). Table 13.2 lists the surface tension of several liquids in units of newtons per meter.

EXAMPLE 13.11 An inverted U-shaped wire bounded on the bottom by a slide wire that can be moved up or down surrounds a thin film of soap water (Fig. 13.15a). If the surface tension of the soap water is 0.025 N/m and the slide wire weighs 1.0×10^{-3} N and is 0.030 m long, how much weight w must be hung from the slide wire to keep it in equilibrium?

SOLUTION Let us choose the slide wire, hanging weight, and a small section of soap film as the object for a force diagram (Fig. 13.15b). Since this group of objects is at rest, the upward forces must be equal in magnitude to the downward forces. The downward forces are the weight of the slide wire w' and the weight w of the hanging mass.

The upward force F is caused by the surface tension on each side of the soap film (Fig. 13.15c). Since both sides of the film help support the weights, the total surface tension force is twice the tension force on one surface of the film:

$$F = 2\gamma l.$$

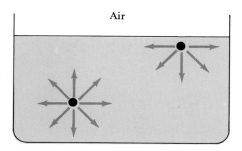

FIG. 13.13. A molecule under the surface of a liquid is attracted from all sides by other molecules. A molecule at the surface is attracted toward molecules at its sides and to others deeper in the liquid. These forces create a tension in the liquid's surface, which has the same effect as a membrane containing the liquid.

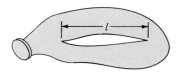

(a)

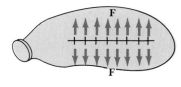

(b)

FIG. 13.14. (a) A water bag with a cut of length l. (b) When sewed together, a net surface force **F** pulls in each direction on the stitches.

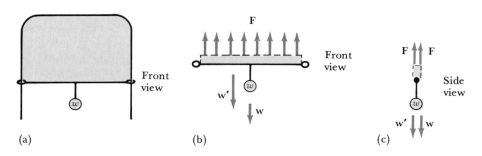

(a)   (b)   (c)

FIG. 13.15. (a) A soap film supported in a wire frame. (b) The forces acting on the slide wire. **F** is the force caused by the tension in the soap film. (c) Side view of the film supporting the weights.

Since the sum of the vertical forces is zero, we find that

$$2\gamma l - w' - w = 0,$$

or

$$w = 2\gamma l - w' = 2(0.025 \text{ N/m})(0.03 \text{ m}) - 1.0 \times 10^{-3} \text{ N} = \underline{0.5 \times 10^{-3} \text{ N}}.$$

∎

Liquid surface tension is important in a variety of situations. For example, bugs such as water striders are supported by the surface tension of water. The water sags under their weight, but the surface does not "tear." The striders float even though they are more dense than the water.

Surface tension is of importance in the fabric and cleansing industries. The spaces between threads in tent and umbrella material are much larger than the size of water molecules, yet these materials are waterproof because the surface tension in water is so great that individual water molecules are unable to leave the water to pass through the material. The water molecules hang together and bridge the gaps in the fabric. Cleansing agents such as soaps are forms of matter that, when added to water, reduce its surface tension, allowing the water to break apart and penetrate more easily into the fabric of a material, where it dissolves dirt.

Disinfectants often are solutions of low surface tension. The solution, instead of forming drops, spreads easily over the surface of a wound and over the surface of bacteria cells in the wound. Insect sprays also have low surface tension. The spray will spread easily into the insect trachea, causing it to become clogged.

13.7 Capillary Action

The force of attraction between similar type molecules is called **cohesion.** Cohesive forces are responsible for the surface tension in liquids. There can also be an attractive force between different types of molecules; this force is called **adhesion.** Adhesive tape sticks to the surface of the skin because of the adhesion between the molecules of the tape and those of the skin. The meniscus that you observe when water partially fills a glass cylinder is caused by the adhesive force between water molecules and the atoms and molecules in the glass (Fig. 13.16). The water molecules crawl up the side of the glass because of this adhesion.

Capillary action is a result of both adhesion and cohesion. A capillary is a hole or bore of small radius in a solid material. If the entrance to the capillary is in contact with a liquid that adheres to the solid capillary walls, the liquid will be drawn up into the capillary. The rise of liquid into a capillary is called **capillary action.** Examples are the rise of water in the bore of a glass capillary tube, the rise of water up the small capillaries in plants, the absorption of water into the small pores of a paper towel, and the movement of kerosene up the wick of a kerosene lamp.

The height to which a liquid will rise into a capillary is limited not by the adhesion of the water to the sides of the capillary but by the ability of the thin film of liquid at the sides of the meniscus to support the weight of liquid that has been lifted. Consider the rise of water in the glass capillary pictured in Fig. 13.17a. The thin film of water at the sides of the meniscus pulls like suspenders on the water below (Fig. 13.17b). The surface tension in this meniscus supports the weight of fluid that has been lifted up the capillary tube. The upward

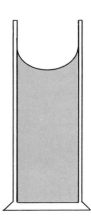

FIG. 13.16. Water molecules attracted to atoms in glass cause a meniscus at the top edge of a glass cylinder.

surface tension force exerted by the meniscus on the liquid below is given by Eq. (13.11):

$$\text{Up:} \quad F = \gamma 2\pi r,$$

where $2\pi r$ is the circumference of the capillary tube—that is, the length around the edge of the meniscus that is supporting the liquid below.*

The downward weight force on the liquid that has been lifted is

$$\text{Down:} \quad w = mg = (\rho V)g = \rho(\pi r^2 h)g,$$

where h is the height of the liquid that has risen in the capillary and V is the volume of liquid that has risen in the capillary.

Since the raised portion of liquid shown in Fig. 13.17b is in equilibrium, the sum of the vertical components of the forces acting on it must be zero:

$$\Sigma F_y = \gamma 2\pi r - \rho(\pi r^2 h)g = 0.$$

After rearranging, we find the height to which the fluid rises because of capillary action:

$$h = \frac{2\gamma}{\rho g r}$$

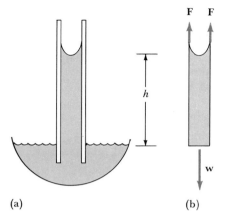

FIG. 13.17. (a) Capillary action causes water to rise in the bore of a capillary tube. (b) A force diagram of the raised water.

EXAMPLE 13.12 The small capillary tubes in a plant have a radius of about 0.0050 mm. To what height will 20°C water be raised in these tubes by capillary action?

SOLUTION According to Tables 13.1 and 13.2, the density of water is 1000 kg/m^3, and its surface tension is 0.073 N/m. The height to which the water will rise by capillary action is

$$h = \frac{2\gamma}{\rho g r} = \frac{2(0.073 \text{ N/m})}{(1000 \text{ kg/m}^3)(9.8 \text{ m/s}^2)(5.0 \times 10^{-6} \text{ m})} = \underline{3.0 \text{ m}}. \quad \blacksquare$$

*We have assumed that the liquid at the top of the meniscus is parallel to the surface of the capillary. This is usually the case when water contacts glass. However, often the liquid at the top of the capillary makes an angle θ with the side of the capillary wall (see Fig. 13.18). The upward component of the surface tension force is then

$$F = \gamma 2\pi r \cos\theta.$$

TABLE 13.2 Surface Tensions of Several Liquids

Surface	In Contact with	Temperature (°C)	Surface Tension γ (N/m)
Acetone	Air	20	0.0237
Ammonia	Air	34	0.0181
Benzene	Air	20	0.0289
Ethyl alcohol	Air	20	0.0228
Glycerol	Air	20	0.0634
Methyl alcohol	Air	20	0.0226
Water	Air	0	0.0756
		20	0.0728
		60	0.0662
		100	0.0589
Soap water	Air	20	0.025
Mercury	Air	20	0.465

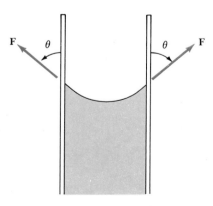

FIG. 13.18

Summary and Additional Readings

1. A **fluid** is a form of matter, such as a liquid or gas, that can flow. A static fluid—a fluid that is not moving—exerts a force perpendicular to and toward a surface it contacts. The magnitude of the force is

$$F = PA, \qquad (12.1)$$

where P is the pressure in the fluid contacting the surface and A is the area of the surface.

2. The **density** ρ of a substance is defined as the ratio of its mass m and volume V:

$$\rho = \frac{m}{V}. \qquad (13.1)$$

3. The pressure in a fluid increases with depth. The pressure P_1 at a vertical elevation y_1 is related to the pressure P_2 at vertical elevation y_2 by the equation

$$P_1 = P_2 + \rho g(y_2 - y_1), \qquad (13.3)$$

where ρ is the density of the fluid. The equation applies when the positive y axis is directed upward. This variation of pressure with depth occurs because the fluid at a certain depth must support the weight of fluid above it. At greater depths, more weight must be supported.

4. **Pascal's principle** says that if external pressure is exerted on an enclosed static fluid, the pressure increase will be transmitted uniformly through the whole volume of the fluid. Equation (13.3) predicts this pressure increase, since if P_2 is increased by an external pressure, then there must be an equal increase in the pressure P_1.

5. A fluid exerts an upward **buoyant force** on an object that is partially or totally submerged in the fluid. According to **Archimedes' principle,** the magnitude of this buoyant force equals the weight of fluid displaced by the object. If the density of the fluid is ρ_f and the volume of fluid displaced by the object is V_f, then the buoyant force B is

$$B = \rho_f g V_f. \qquad (13.6)$$

6. **Surface tension:** The mutual attraction of molecules at the surface of a liquid to each other and to molecules deeper in a fluid gives rise to a surface tension that has the effect of a membrane containing the fluid. The surface tension γ is the force per unit length needed to hold these surface molecules together:

$$\gamma = \frac{F}{l}. \qquad (13.11)$$

The force is directed tangent to the surface.

7. **Capillary action** occurs when liquids are drawn up a small capillary tube because of the adhesion of the liquid to the surface of the tube and because of cohesion of molecules of liquid to each other. The height to which the liquid is drawn is given by the equation

$$h = \frac{2\gamma}{\rho g r}, \qquad (13.12)$$

where r is the radius of the capillary, γ is the surface tension of the liquid, and ρ is its density.

Jearl Walker, *The Flying Circus of Physics,* John Wiley and Sons, New York (1975). Chapter 4 contains interesting, provocative questions and real-world experiments involving fluids.

R. M. Alexander, *Animal Mechanics,* University of Washington Press, Seattle (1968). Presents some applications of static fluids in biology.

Alan C. Burton, *Physiology and Biophysics of the Circulation: An Introductory Text,* 2nd ed., Year Book Medical Publishing, Chicago (1972). Interesting use of introductory physics to explain the circulatory system.

Questions

1. Give two examples in which a fluid exerts an upward force on an object and two examples in which it exerts a downward force.

2. A milk carton is filled mostly with milk but also with some air at atmospheric pressure. What would happen to the carton (a) if it were immersed about 50 m under the surface of the water? (b) if it were taken many thousands of meters above the earth's surface?

3. Water extends 400 m behind a dam that is 100 m wide and 10 m deep. Another dam, also 100 m wide, is only 5 m deep, but water extends 1600 m behind this dam. On which dam does the water exert the greatest force? Explain.

4. How was it possible for the Dutch boy to hold back the Atlantic Ocean by placing his thumb in a hole in a dike in Holland?

5. When placed in a lake, an object either floats on the surface or sinks. It does not float at some intermediate location between the surface and the bottom of the lake. However, a weather balloon floats at some intermediate elevation between the earth's surface and the top of its atmosphere. Explain.

6. A metal rowboat floats in a pool. Does the level of water in the pool rise, fall, or remain the same if the boat sinks? Explain.

7. A rubber raft floating in a swimming pool holds a heavy rock. Does the level of water in the pool rise or fall when the rock is thrown out of the raft and into the water? Explain.

8. Will a boat float higher or lower in saltwater or in freshwater? Explain.

9. A hollow sphere hanging from one arm of an equal-arm balance is balanced exactly by a solid weight of the same material on the other arm. The air is then removed from the room holding the balance. Will the arms remain balanced? If not, which arm moves lower? Explain.

10. A glass with one ice cube in it is filled to the *rim* with water. Will the water level remain unchanged, decrease, or overflow when the ice melts? Explain. [*Hint:* You must consider the density of the ice and the volume of space displaced by the ice before melting, compared to the volume of the melted ice.]

11. A helium balloon floats in a car. In which direction will the balloon move when the car accelerates from a stop sign? Explain.

12. Your skin's surface area in square meters can be estimated using the equation

$$A = 0.20m^{0.425}h^{0.725},$$

where m is your mass in kilograms and h is your height in meters. (a) Use this equation to *estimate* your own surface area. (b) If an atom on the surface of your skin is about 1.2×10^{-10} m in radius, roughly how many atoms are in the outside layer of skin on your surface?

13. Hot water leaks more easily from small holes in faucets than cold water at the same pressure. Explain.

14. Sprinkle pepper over the surface of a dish of water. What do you think will happen to the pepper when you gently touch a bar of soap to the water surface? Try it and describe the results? Why do you think this happened? [*Hint:* Think of two membranes fastened together, one stretched somewhat more tightly than the other.)

15. How can capillary action be used to estimate the radii of glass capillary tubes?

Problems

13.1 Pressure

1. You hold your physics book in the air so that its back cover faces the ground. Approximately how large a force in newtons and in pounds does the atmosphere exert on the cover? Why does this force not support the book's weight?

2. Estimate the downward force of the earth's atmosphere on the state where you attend college.

3. Suppose that all the air inside your bedroom were removed (you would have to be in a pressurized space suit). *Estimate* the force in units of newtons, pounds, and tons with which you would have to push on the door to get out.

■ 4. The atmosphere at a pressure of 1.0 atm exerts a downward force of 0.80×10^5 N on a circular table. Calculate the radius of the table.

■ 5. Calculate the average pressure of the needle of a turntable on a record if the needle's radius is 10 μm and it supports a mass of 1.5 g. Compare this to the pressure of your shoes on the ground when standing.

■ 6. The tires of a 980-kg car have absolute pressure of 3.0×10^5 N/m^2. Determine the average area of contact of each tire with the road.

13.2 Density

7. A single-level home has a floor area of 200 m^2 with ceilings that are 2.6 m high. Determine the mass of the air in the house when at 20°C.

8. About two-thirds of your body mass consists of water. Calculate the volume of water in a 70-kg person.

9. Calculate the average mass density of the earth. The earth's mass is 5.98×10^{24} kg, and its radius is 6.38×10^6 m.

■ 10. A 75.0-kg person's density is changed from 970 kg/m^3 to 990 kg/m^3. Calculate the person's percent change in volume.

■ 11. If the atmosphere had a uniform density of 1.3 kg/m^3 from the earth's surface up to the "top" of the atmosphere, how high would the atmosphere be? The pressure at the earth's surface is 1.0×10^5 N/m^2. Explain why there is air even at an elevation of 2.5×10^4 m above the earth's surface.

■■ 12. A person's mass while dieting decreases by 5 percent. Exercise creates muscle and reduces fat, causing the person's density to increase by 2 percent. Calculate the percent change in the person's volume.

13.3 Pressure Variation with Depth

13. The pressure at the top of the water in a city's gravity-fed water reservoir is 1.0×10^5 N/m^2. Calculate the pressure at the faucet of a home 42 m below the reservoir.

14. *Estimate* the gauge pressure of blood in your brain and in your feet when standing, relative to the average 1.3×10^4 N/m^2 gauge pressure (100 mm Hg) in your heart.

15. A glucose solution of density 1050 kg/m^3 is transferred from a bottle exposed to the atmosphere through a tube and syringe into the vein of a person's arm. The pressure in the arm exceeds atmospheric pressure by1400 N/m^2. How high above the arm must the top of the liquid in the bottle be so that the pressure in the glucose solution at the needle exceeds the pressure of the blood in the arm? Ignore the pressure drop across the needle and tubing due to viscous forces.

16. Determine the change in air pressure as you climb from an elevation of 1650 m at the timberline of Mount Rainier to its 4392-m summit, assuming an average air density of 0.82 kg/m^3.

■ 17. A person climbing a vertical distance of 2000 m up a mountain carries a half-filled jar of water. At the start of the hike the pressure inside and outside the jar is 1.0×10^5 N/m^2. Calculate the force on the lid (magnitude and direction) of area 75 cm^2 at the end of the hike. Assume that the air density is constant at 1.0 kg/m^3.

■ 18. The window of a deep-sea diving vessel has an area of 0.36 m^2. The window can withstand a net force of 1.4×10^6 N before breaking. Calculate the maximum depth in the ocean that the vessel can go. The air pressure inside is the same as the air pressure at the ocean's surface.

■ 19. Your car slides off an embankment into a pond. *Estimate* the force needed to open the door if the top of the door is 0.50 m below the water's surface. How might you escape without opening the door?

■ 20. You can develop a gauge pressure of -30 mm Hg in your lungs while sucking on a straw. Up what length straw can you suck a fruit drink with a density of 1200 kg/m^3?

■ 21. A device known as Hare's apparatus (Fig. 13.19) is used to measure the relative density of a fluid. Suction in a common, center tube causes water to rise in the tube on the right and a different type of fluid of unknown density to rise in the tube on the left. When the valve in the center tube is closed, the water has risen to a level of 10.5 cm and the fluid on the left has risen to a

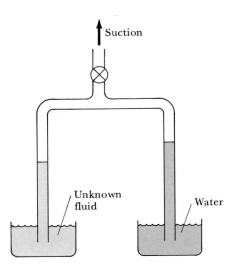

FIG. 13.19. Hare's apparatus for determining relative density in liquids.

level of 9.2 cm. Calculate the density of the fluid on the left. The density of water is 1000 kg/m^3.

22. Olive oil is poured into a U-tube containing mercury with the result shown in Fig. 13.20. Both sides of the tube are open to the atmosphere. Calculate the density of the olive oil.

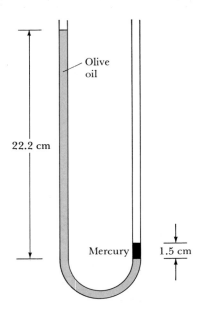

FIG. 13.20

13.4 Externally Applied Pressure

23. A hydraulic lift is used to raise a 900-kg car. Compressed air with a pressure of 4.0×10^5 N/m^2 is applied to a small cylinder. Calculate the area of a large piston supporting the car's weight.

24. A 10-kg mass is placed on top of a piston of a hydraulic lift; the piston's area is 3.0 cm^2. A 100-kg woman wrestler sits on a

piston of 60-cm^2 area. Will she cause the 10-kg mass to rise? Explain.

25. A person is poked in the eye with a finger that exerts a 60-N force over an area of 0.3 cm^2 at the front of the eye. The increased pressure in the eye is transmitted to the back of the eye. (a) Calculate the force on the 0.05-cm^2 area from which the optic nerve leaves the eye. (b) Calculate the force at the back of the eye on a single rod whose diameter is about 10^{-6} m.

26. As shown in Fig. 13.21, a 100-N force pulling up on a lever causes an increase in pressure under a small piston of 8.0-cm^2 area. (a) Calculate the pressure created under the small piston. (b) Calculate the upward force **F** on the large, 160-cm^2 piston of the hydraulic press.

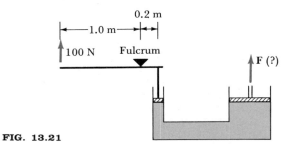

FIG. 13.21

27. The lever shown in Fig. 13.21 is applied to piston 1 of the hydraulic can crusher in Fig. 13.7. With what force must the end of the lever's handle be pulled to exert a (2.4×10^6)-N force F_2 on the cans? $A_1 = 24$ cm^2 and $A_2 = 2400$ cm^2.

13.5 Archimedes' Principle

Construct force diagrams when solving problems 28–42.

28. A person with a density of 980 kg/m^3 floats in seawater of density 1025 kg/m^3. What fraction of the person is submerged?

29. Calculate the buoyant force on a rock of mass 32 kg and density 2800 kg/m^3 when fully submerged in water of density 1000 kg/m^3.

30. Calculate the density of a whale having 5.0 percent of its mass above seawater of density 1025 kg/m^3. Be sure to justify your calculation technique.

31. A gold crown weighs 40 N when in air. What will it weigh when suspended (a) in water and (b) in ethyl alcohol?

32. A 60-kg woman having a density of 980 kg/m^3 stands on a bathroom scale. Calculate the reduction of the reading on the scale due to the buoyant force of air of density 1.2 kg/m^3.

33. A 30-kg rock of density 2400 kg/m^3 is immersed in a lake and must be moved from near a boat dock. Calculate the force needed to lift the rock while it is under the water.

34. While analyzing a sample of ore, a geologist finds that it weighs 2.00 N in air and 1.13 N when immersed in water. Calculate the density of the ore.

35. A helium balloon of volume 0.12 m^3 has a total mass (the helium plus the balloon) of 0.12 kg. Calculate (a) the buoyant force on the balloon when in air of density 1.3 kg/m^3 and (b) the initial acceleration of the balloon when released.

36. A protein molecule of mass 1.1×10^{-22} kg and density 1.3×10^3 kg/m^3 is placed in a vertical tube of water of density 1000 kg/m^3. Calculate the initial acceleration of the protein.

37. Logs of density 600 kg/m^3 are used to build a raft. What is the weight of the maximum load that can be supported by a raft built from 300 kg of logs?

■ 38. How much deeper into the water does a barge of cross section 12 m × 28 m sink after receiving a (1.6 × 10⁵)-kg load. The water's density is 1020 kg/m³.

■■ 39. A life preserver is manufactured to support a 70-kg person with 20 percent of his volume out of the water. The person's density and that of the water are both about 1000 kg/m³. If the density of the life preserver is 100 kg/m³ and it is completely submerged, what must its volume be?

■■ 40. (a) Calculate the tension in a vertical string supporting an 0.80-kg rock of density 3300 kg/m³ when fully submerged in water of density of 1000 kg/m³. (b) If the tension needed to support the rock increases by 12 percent when the rock is submerged in a different fluid, calculate the fluid's density.

■■ 41. Derive an equation for determining the unknown density of a liquid by measuring the tension force T needed to support an object of known mass m and density ρ while submerged in the liquid.

■■ 42. A container of water resting on a balance has a measured mass of 1.48 kg. A mass weighing 12.0 N is attached to a string and lowered into the water without touching the bottom or sides of the container. The tension in the string is now 7.2 N. What does the scale supporting the container of water now read? Justify your answer clearly by using force diagrams for the mass and the container with water.

13.6 Surface Tension

■ 43. If the apparatus with a slide wire 0.050 m long shown in Fig. 13.15 supports a total weight of 4.2 × 10⁻³ N, what is the surface tension of the liquid?

■ 44. A horizontal platinum wire having a 0.015-g mass and 5.0-cm length is dipped in ethyl alcohol. Calculate the force needed to pull the wire slowly from the alcohol.

■ 45. The tension in the skin of the left ventricle of the heart is 290 N/m. A 4-cm cut is made in the left ventricle during open-heart surgery. How many stitches must be used to repair the cut? Each stitch will safely support a tension of 0.40 N.

■■ 46. The surface tension of a liquid can be determined by measuring the force F needed to lift a horizontal, circular platinum ring of radius r and weight w from the liquid's surface. (a) Derive an equation for γ in terms of F, r, and w. (b) If $F = 12.2 \times 10^{-3}$ N, $r = 2.0$ cm, and $w = 5.4 \times 10^{-3}$ N, calculate the value of γ for a particular fluid.

■■ 47. A thin wire twisted in a circular shape of 1.12-cm radius is supported horizontally in a liquid and withdrawn slowly by an upward force of 5.53 × 10⁻³ N. If the surface tension of the liquid is 0.032 N/m, what is the mass of the wire? Note that the ring is pulled down by two surfaces, one inside the ring and one outside.

■■ 48. One leg of a water strider standing on 20°C water makes a depression of radius 2.0 mm. The water surface tension force is exerted at an angle of 53° with the horizontal (Fig. 13.22). (a) Calculate the weight supported by surface tension at the sides of this depression. (b) Calculate the weight and mass of the water strider if each of its six legs causes a similar depression.

■■ 49. A hollow, cylindrical rod with an outer radius of 0.12 cm is closed at the bottom (Fig. 13.23). Weights dropped into the rod cause it to float vertically in 20°C water with the weighted end down. The rod and weights together have a mass of 0.18 g. (a) If the water contacts the rod parallel to its surface, calculate the water surface tension force **F** that supports the rod. (b) If the

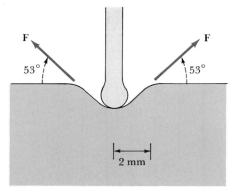

FIG. 13.22. A water strider's leg supported by surface tension.

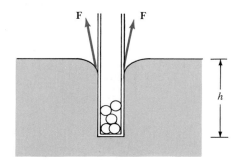

FIG. 13.23. A cylinder supported by surface tension and a buoyant force.

bottom of the rod is a distance h below the surface, calculate the buoyant force on the rod in terms of h. (c) Combine the forces in parts (a) and (b) and the known weight of the rod and its contents to calculate the distance h.

13.7 Capillary Action

50. Calculate the heights to which water at 20°C will rise in glass capillary tubes of radii 0.2 mm, 0.5 mm, 1.0 mm, and 2.0 mm.

51. Water rises to a height of 15.6 cm in a glass capillary tube. Estimate the radius of the tube. To what height would ethyl alcohol rise in the same tube?

52. If capillary action were responsible for lifting water to the tops of redwood trees 100 m tall, what would be the maximum radii of capillaries in these trees?

53. It has been calculated that capillary action will lift water to the following heights in various types of soils: (a) 0.1 m in fine gravel, (b) 0.5 m in coarse sand, (c) 2 m in fine sand, (d) 10 m in silt, and (e) 50 m in clay. Estimate the radii of pores in each type of soil.

■ 54. Water is raised 14 cm in the bore of a capillary tube. How high will soap water be raised in the same tube? Explain.

■ 55. Two large parallel glass plates separated by a distance of 0.5 mm are dipped into water. To what height will water rise between the plates?

■ 56. Two capillary tubes of different radii are placed in water. The water rises 12 cm in the first tube and 45 cm in the second. Calculate the ratio of the area of the first tube and the area of the second.

CHAPTER 14

Fluid Dynamics

In the last chapter we studied fluid statics—fluids at rest. In this chapter we will learn about fluid dynamics—moving fluids. The study of fluid dynamics will permit us to answer a variety of questions concerning moving fluids: What causes the roof to be lifted off a house during a strong wind? Why does air moving through the bronchial passages cause us to snore? How does the speed of the flow of blood affect the chance of dislodging a plaque from the wall of an artery? These and countless other questions can be answered using one basic equation developed in 1738 by the Swiss physicist Daniel Bernoulli (1700–1782). Bernoulli's equation allows us to calculate variations in fluid pressure caused by a fluid's motion.

Friction in fluids, a subject that is ignored in Bernoulli's equation, becomes important when fluids move through narrow vessels. This friction in moving fluids and the equally important friction, or drag force, of a fluid on a moving object are considered toward the end of the chapter.

14.1 Types of Fluid Flow

When the pressure is lower on one side of a fluid than on the other, the fluid will flow toward the low-pressure region. For instance, large masses of air in the earth's atmosphere move from regions of high pressure into regions of low pressure. Blood flows from the arterial side of the circulatory system at a pressure of about 100 mm Hg to the venous side at a pressure of about 5 mm Hg.

Fluid flow is characterized by two main types: streamline flow and turbulent flow. In **streamline flow,** sometimes called laminar flow, every particle of fluid that passes a particular point follows the same path as particles that passed the point before it. The particles move in an unchanging line called a *streamline*. The streamlines of water molecules passing through a constriction in a pipe are depicted in Fig. 14.1a. The streamlines of air moving past the wing of an airplane are shown in Fig. 14.1b.

Turbulent flow is characterized by agitated, disorderly motion. Instead of following a given path, molecules of a fluid swirl and form whirlpool patterns called *eddies*. The turbulent flow of water moving through a narrow constriction in a pipe is illustrated in Fig. 14.2a. The turbulent motion of air behind an inclined airplane wing is shown in Fig. 14.2b. Usually, turbulent flow occurs in

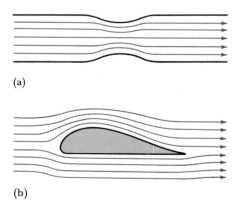

FIG. 14.1. Laminar flow (a) of water through a constriction in a pipe and (b) of air over an airplane wing.

fluids moving at high speed. Friction forces are normally much greater during turbulent flow than during streamline flow. In this chapter we concentrate mostly on situations involving streamline flow.

14.2 Flow Rate and Equation of Continuity

We are often interested in the volume V of fluid that flows past some cross section of a vessel in a time t. The ratio of this volume and the time required for it to flow is called the **flow rate** of the fluid and is given the symbol Q:

$$Q \equiv \frac{V}{t}. \tag{14.1}$$

The units of flow rate are expressed in m³/s, ft³/s, ft³/min, and so forth.

Often we cannot directly measure the actual volume of fluid that passes along a vessel per unit time. For example, we cannot directly measure the volume of blood pumped through an artery in a certain time without somehow collecting and measuring that volume. Doubtless, a person would object to having his blood collected for a routine flow-rate measurement, but there are indirect techniques for measuring the flow rate. One of these methods, discussed in Chapter 17, utilizes a noninvasive measurement of the blood's speed and the cross-sectional area of the vessel carrying the blood.

Consider the fluid moving along the pipe pictured in Fig. 14.3a. We are interested in the volume of fluid that passes a cross section of area A at some position along the pipe. For the pipe shown in Fig. 14.3a, the darkened portion of the fluid moves forward and passes the cross section A in a time t. Thus, after a time t, the back part of this fluid has moved forward to the position shown in Fig. 14.3b.

The volume V of fluid in this darkened portion of the cylinder is the product of its length l and the cross-sectional area of the pipe:

$$V = lA.$$

We find, then, that

$$Q = \frac{V}{t} = \frac{lA}{t} = \frac{l}{t}A.$$

However, l is also the distance the fluid moves in a time t. Thus, l/t is the average speed $\bar{v}$ of the fluid. Substituting $\bar{v} = l/t$ into the above equation, we find that

$$Q = \bar{v}A. \tag{14.2}$$

This is a general equation that allows us to calculate the **flow rate** of fluid past any cross-sectional area along a pipe or vessel.

EXAMPLE 14.1 The heart pumps blood at a flow rate of 80 cm³/s into the aorta, the diameter of which is 1.5 cm. Calculate the average speed of blood in the aorta.

SOLUTION The cross-sectional area of the aorta is

$$A = \pi r^2 = \pi (1.5 \text{ cm}/2)^2 = 1.8 \text{ cm}^2.$$

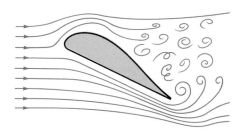

(a)

(b)

FIG. 14.2. Turbulent flow (a) of water in a pipe and (b) of air behind a tilted airplane wing.

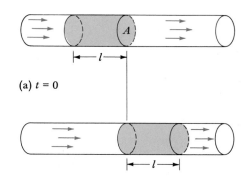

(a) $t = 0$

(b) Time, t

FIG. 14.3. The darkened section of fluid requires a time t to pass the cross section A. The volume of fluid that flows in this time is $V = lA$.

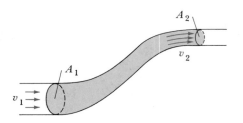

FIG. 14.4. A fluid flows through a vessel with changing cross-sectional area. For incompressible fluids, the flow rate across A_1 must equal that across A_2.

The average speed of blood in the aorta is determined by rearranging Eq. (14.2):

$$\bar{v} = \frac{Q}{A} = \frac{80 \text{ cm}^3/\text{s}}{1.8 \text{ cm}^2} = \underline{44 \text{ cm/s}}.$$ ∎

Often, the cross-sectional area of a vessel through which fluid flows varies from one part of the vessel to another. If the fluid is incompressible—if its volume undergoes negligible change when the pressure acting on the fluid increases—then the volume of fluid that passes one cross section of the vessel equals that passing another cross section. During any period of time, the fluid entering the region between the cross sections (for example, A_1 and A_2 in Fig. 14.4) must equal that leaving the region. The flow rate in must equal the flow rate out. In terms of the cross sections A_1 and A_2 in Fig. 14.4, we can say that

$$Q = \bar{v}_1 A_1 = \bar{v}_2 A_2, \tag{14.3}$$

where $\bar{v}_1$ is the average speed of fluid passing cross section A_1, and $\bar{v}_2$ is the average speed of fluid passing A_2. In a narrow section of pipe where A is small, the speed of the fluid will have to be great in order for the flow rate to remain constant.

Equation (14.3) is called the **continuity equation** and is used to relate the cross-sectional areas and speeds of fluid flow in different parts of a vessel carrying the fluid. It will be an especially useful equation when we discuss applications of Bernoulli's equation later in this chapter.

EXAMPLE 14.2 Blood normally flows with an average speed of about 10 cm/s in the large arteries whose radii are about 0.3 cm. Suppose that a small section of artery is reduced in radius by one-half because of thickening of its walls (called atherosclerosis). Calculate the blood speed past the constriction.

SOLUTION The blood speed $\bar{v}_2$ past the constriction is related to the speed $\bar{v}_1$ in the clear artery by the equation

$$\bar{v}_2 = \frac{A_1}{A_2}\bar{v}_1 = \left(\frac{\pi r_1^2}{\pi r_2^2}\right)\bar{v}_1$$

$$= \left(\frac{r_1}{r_2}\right)^2 \bar{v}_1 = \left(\frac{r_1}{0.5 r_1}\right)^2 \bar{v}_1 = 4\bar{v}_1.$$

Thus, $\bar{v}_2$ will be $\underline{40 \text{ cm/s}}$ if $\bar{v}_1$ is 10 cm/s. ∎

14.3 Bernoulli's Equation

In 1738, Daniel Bernoulli formulated an equation that could be used to determine the variation of pressure in fluids as a function of the speed of the fluid and the elevation of the vessel through which the fluid passed. This equation can be used to explain why airplane wings have lift, how to avoid having the roof of your house blown off during a tornado or hurricane, why plaque in arteries becomes dislodged (possibly causing a stroke), and many other phenomena involving moving fluids.

If Bernoulli had developed this equation in the late 1800s or in this century, it is unlikely that the equation would have been named for him because it is just the work-energy equation (Eq. 8.9) applied to a special type of fluid flow. Bernoulli, however, devised the equation long before thermodynamics and ideas about energy had been introduced to the world of science. Since these topics are now well understood, we can use the work-energy equation to derive Bernoulli's equation.

A pipe through which fluid flows is shown in Fig. 14.5a. We assume (1) that the fluid undergoes streamline flow, (2) that is incompressible, and (3) that no friction forces act on the fluid. The work-energy equation is used to describe the behavior of the fluid as it moves a short distance along the vessel. We choose as our system the darkened section of fluid pictured in Figs. 14.5a and 14.5b.

As the system flows to the right, fluid behind the system at position 1 pushes it to the right with a pressure P_1. The fluid ahead of the system at position 2 pushes back on the system with a pressure P_2. The speed of the fluid varies from position to position according to the equation of continuity ($A_1v_1 = A_2v_2$). As a result, the fluid moves faster at position 2 than at position 1 since the vessel is narrower at 2 than at 1.

The net effect of the movement of the fluid a short distance to the right is summarized in Fig. 14.5c. A volume of fluid initially at position 1 moving at speed v_1 has now been transferred, in effect, to position 2, where it moves at

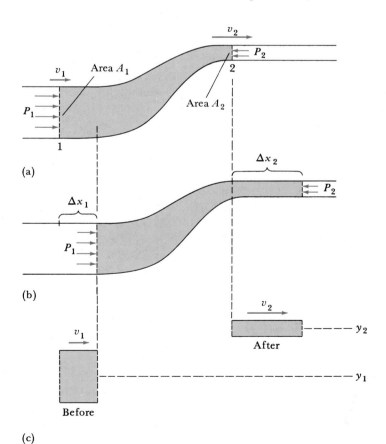

(a)

(b)

(c)

FIG. 14.5. (a) A vessel carrying fluid. The darkened part of the fluid is the system—the part to which we apply the work-energy equation. (b) The system a short time later after it has moved along the vessel. (c) The net effect of this movement is that part of the system has been transferred from the *before* region to the *after* region.

speed v_2. Both the kinetic and gravitational potential energies of the fluid in the system have changed. These changes have been caused by the work done on the fluid by the forward and backward pressures at positions 1 and 2, respectively. The work-energy equation, in the form of Eq. (8.9), relates nicely these energy changes of the system to the work done to cause the energy change:

$$W = \Delta KE + \Delta PE_g.$$

Let us now calculate each of these terms in the above equation.

Work: Work is done on the system by the force of the fluid pushing from the left with pressure P_1 and of the fluid on the right pushing back on the system with pressure P_2. If the fluid moves a distance Δx_1 on the left side (Fig. 14.5b), it will have to move a farther distance Δx_2 on the right side since the vessel is narrower on the right. Using Eq. (8.1), we find the net work done on the fluid to be

$$W = F_1 \Delta x_1 \cos 0° + F_2 \Delta x_2 \cos 180°$$
$$= F_1 \Delta x_1 - F_2 \Delta x_2 = P_1 A_1 \Delta x_1 - P_2 A_2 \Delta x_2.$$

The volume of fluid ΔV that has moved from the left to the right is $\Delta V = A_1(\Delta x_1) = A_2(\Delta x_2)$. The preceding expression for work becomes

$$W = (P_1 - P_2) \Delta V.$$

Change in kinetic energy: The mass of fluid that has moved is related to its density and volume by the equation

$$m = \rho \, \Delta V.$$

As the system moves, a small amount of fluid originally moving with speed v_1 at position 1 has in effect been transferred to position 2, where it moves with speed v_2. Thus, the kinetic energy change of the mass m of fluid shown in Fig. 14.5c is

$$\Delta KE = \frac{1}{2} m v_2^2 - \frac{1}{2} m v_1^2$$
$$= \frac{1}{2}(\rho \, \Delta V) v_2^2 - \frac{1}{2}(\rho \, \Delta V) v_1^2.$$

Change in gravitational potential energy: The gravitational potential energy of the system has also changed because a mass m of fluid has effectively moved from elevation y_1 to elevation y_2. The change in gravitational potential energy is

$$\Delta PE_g = mg(y_2 - y_1) = (\rho \, \Delta V)g(y_2 - y_1).$$

If we put these three energy changes into Eq. (8.9) and cancel the common ΔV in each term, we find that

$$P_1 - P_2 = \frac{1}{2}\rho(v_2^2 - v_1^2) + \rho g(y_2 - y_1). \tag{14.4}$$

This is one form of **Bernoulli's equation.** It relates the pressures, speeds, and elevations of a fluid at any two points in the fluid (our positions 1 and 2 in Fig. 14.5 were arbitrary). Equation 14.4 applies to the nonturbulent flow of a frictionless, incompressible fluid. The term on the left side of Eq. (14.4) is a consequence of the work done because of the difference in fluid pressure pushing forward and that pushing back. The two terms on the right result from the changes in kinetic and gravitational potential energies of the fluid.

Bernoulli's equation is often written in a form that is easier to use and perhaps easier to remember. A rearrangement of Eq. (14.4) yields

$$P_1 + \frac{1}{2}\rho v_1^2 + \rho g y_1 = P_2 + \frac{1}{2}\rho v_2^2 + \rho g y_2. \qquad \textbf{(14.5)}$$

We see that the sum of the three terms on the left for position 1 in the fluid must equal the sum of the same three terms for position 2. Since the choice of positions is arbitrary, the sum of these three terms must be constant at all points in the fluid:

$$P + \frac{1}{2}\rho v^2 + \rho g y = \text{Constant}. \qquad \textbf{(14.6)}$$

These three terms must add to the same value at any point along a streamline.

EXAMPLE 14.3 Water is pumped at a rate of 24 cm³/s through a 0.50-cm-radius pipe on the main floor of a house to a 0.35-cm-radius pipe in a solar hot water collector 4.0 m higher on the roof. If the pressure in the pipe on the roof is $1.20 \times 10^5 \, \text{N/m}^2$, what is the pressure in the larger pipe on the main floor?

SOLUTION We use Bernoulli's equation to determine the pressure. But we must first determine the speed of the water in each pipe. Using Eq. (14.6) with the area of a pipe equal to πr^2, we find that the speed of the water in the pipe on the main floor (position 1 in Fig. 14.6) is

$$v_1 = \frac{Q}{\pi r_1^2} = \frac{24 \, \text{cm}^3/\text{s}}{\pi(0.50 \, \text{cm})^2} = 31 \, \text{cm/s} = 0.31 \, \text{m/s},$$

and the speed at position 2 in the pipe on the roof is

$$v_2 = \frac{Q}{\pi r_2^2} = \frac{24 \, \text{cm}^3/\text{s}}{\pi(0.35 \, \text{cm})^2} = 62 \, \text{cm/s} = 0.62 \, \text{m/s}.$$

The pressure in the pipe on the main floor can now be determined using Bernoulli's equation:

$$P_1 = P_2 + \frac{1}{2}\rho(v_2^2 - v_1^2) + \rho g(y_2 - y_1)$$

$$= (1.20 \times 10^5 \, \text{N/m}) + \frac{1}{2}(1000 \, \text{kg/m}^3)[(0.62 \, \text{m/s})^2 - (0.31 \, \text{m/s})^2]$$

$$+ (1000 \, \text{kg/m}^3)(9.8 \, \text{m/s})(4.0 \, \text{m} - 0)$$

$$= 1.20 \times 10^5 \, \text{N/m}^2 + 140 \, \text{N/m}^2 + 39{,}300 \, \text{N/m}^2$$

$$= \underline{1.60 \times 10^5 \, \text{N/m}^2}. \qquad \blacksquare$$

When using Bernoulli's equation, we learn some surprising consequences of the effect of motion on fluid pressure. Consider the air above the roof of a house during a windstorm. If the air's kinetic energy increases, some other term in Eq. (14.6), such as the pressure, must decrease so that the sum of the three terms

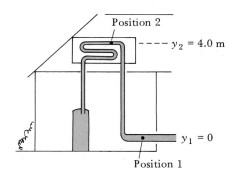

FIG. 14.6. A water line for a solar collector.

FIG. 14.7. Two pages held apart are pulled together because of the reduced pressure of air blown between them.

remains constant. We find that the pressure above the roof of a house during a windstorm is less than normal. If the windows are closed, the normal pressure inside the house may blow the roof off.

A simple experiment can help convince you that the pressure of air on a surface *decreases* because the air moves. Hang two sheets of paper straight down as shown in Fig. 14.7. When you blow air between the sheets, the pressure is reduced and the sheets move together. After you stop blowing, the sheets separate.

The reed of an oboe closes in a similar fashion. When air is blown across the reed, the pressure is reduced compared to pressure outside the reed where the air is not moving. The reed closes just as the papers in Fig. 14.7 come together. When the reed is closed, air flow stops, the pressures inside and outside the reed equalize, and the reed reopens, causing air flow to start. The pressure is again lowered inside the reed due to the moving air, and it again closes. The rhythmic opening and closing of the reed initiates the sound we hear coming from the oboe.

Snoring is caused in a similar way when the bronchial passage alternately opens and closes. As air flows through the passage, air pressure inside is reduced and air pressure outside causes the passage to close. When air flow stops, the pressures equalize and the passage reopens. The rhythmic opening and closing of the passage resembles the vibration of an oboe reed and creates the snoring sound.

14.4 Applications of Bernoulli's Equation

When using Bernoulli's equation, we choose two different positions in a fluid. One position is located at a place where we want to find an unknown property of the fluid—for example, the pressure. The second position is chosen at a place where the pressure, density, speed, and elevation of the fluid are known. When all of this known information is substituted into Bernoulli's equation, we can solve for the unknown.

EXAMPLE 14.4 (a) Calculate the speed with which water flows from a hole in the dam of a large irrigation canal. The hole is 0.80 m below the surface of the water (Fig. 14.8). (b) If the hole has a radius of 2.0 cm, what is the flow rate of water from the hole?

SOLUTION (a) We choose position 1 as the place where the water leaves the hole (see Fig. 14.8). Position 2 can be chosen at a place where the pressure, elevation, and speed of the water are known. A convenient place is at the top of the water behind the dam. There is so much water behind the dam that its level moves downward very slowly. Thus, the water speed at position 2 is nearly zero ($v_2 \simeq 0$). Since position 2 is 0.80 m higher than position 1, $y_2 - y_1 = 0.80$ m. The water pressure at position 2 is atmospheric pressure. At position 1, where the water flows through the hole into the atmosphere, the pressure is also atmospheric. Thus, $P_1 = P_2 = P_{atm}$.

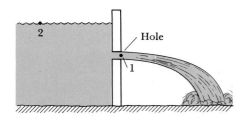

FIG. 14.8

We next substitute all of this information into Eq. (14.4):

$$P_{atm} - P_{atm} = \frac{1}{2}\rho(0^2 - v_1^2) + \rho g(y_2 - y_1).$$

Solving for v_1, we find that

$$v_1 = \sqrt{2g(y_2 - y_1)}.$$

This is called **Torricelli's theorem.** Substituting for g and $(y_2 - y_1)$, we find that

$$v_1 = \sqrt{2(9.8 \text{ m/s}^2)(0.80)} = \underline{4.0 \text{ m/s}}.$$

(b) The flow rate of the water from the hole is determined using Eq. (14.2), $Q = A_1 v_1$, where A_1 is the area of the hole: $A_1 = \pi r_1^2 = \pi(0.020 \text{ m})^2 = 1.3 \times 10^{-3} \text{ m}^2$. The flow rate is, then,

$$Q = A_1 v_1 = (1.3 \times 10^{-3} \text{ m}^2)(4.0 \text{ m/s}) = \underline{5.2 \times 10^{-3} \text{ m}^3/\text{s}}. \quad \blacksquare$$

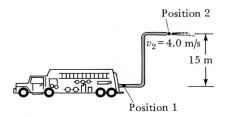

FIG. 14.9. What is the pressure in the hose at position 1 in order for an adequate flow of water from the nozzle at position 2?

EXAMPLE 14.5 Calculate the pressure needed for the pump of a 3.0-cm-radius firehose to pump water through its nozzle of radius 2.0 cm at an average speed of 4.0 m/s when the nozzle is 15 m above the fire truck.

SOLUTION We wish to determine the water pressure at position 1 in Fig. 14.9 by comparing the flow at this point to the flow at the exit of the nozzle, position 2 in Fig. 14.9. The average speed of flow at position 1 is determined using the continuity equation, Eq. (14.3):

$$v_1 = \frac{A_2}{A_1}v_2 = \frac{\pi(2.0 \text{ cm})^2}{\pi(3.0 \text{ cm})^2}(4.0 \text{ m/s}) = 1.78 \text{ m/s}.$$

We now know the following information: $y_1 = 0$, $v_1 = 1.78$ m/s, $y_2 = 15$ m, $v_2 = 4.0$ m/s, $P_2 = 1.0 \times 10^5$ N/m^2 (the water is exposed to the atmosphere as it leaves the nozzle), and $\rho(\text{water}) = 1000$ kg/m^3. The pressure P_1 in the hose by the pump can be determined by rearranging and substituting in Bernoulli's equation:

$$P_1 = P_2 + \frac{1}{2}\rho(v_2^2 - v_1^2) + \rho g(y_2 - y_1)$$

$$= (1.0 \times 10^5 \text{ N/m}^2) + \frac{1}{2}(1000 \text{ kg/m}^3)[(4.0 \text{ m/s})^2 - (1.78 \text{ m/s})^2]$$

$$+ (1000 \text{ kg/m}^3)(9.8 \text{ m/s}^2)(15 \text{ m} - 0)$$

$$= 1.0 \times 10^5 \text{ N/m}^2 + 0.064 \times 10^5 \text{ N/m}^2 + 1.47 \times 10^5 \text{ N/m}^2$$

$$= \underline{2.5 \times 10^5 \text{ N/m}^2}.$$

The pump needs to exert a pressure of about 2.5 atm. $\quad \blacksquare$

EXAMPLE 14.6 A 45-m/s (100-mph) wind blows across the top of a flat-roofed house (Fig. 14.10). Calculate the net force on the roof due to the pressure differential from inside the house to the outside. Assume that the pressure in the house is atmospheric pressure and that the roof has an area of 200 m^2. The density of air equals 1.3 kg/m^3.

FIG. 14.10. Air blowing across a roof causes the pressure to be reduced. If the wind blows fast enough, the internal air pressure can blow the roof off.

SOLUTION Air above the house pushes down with a force $F_1 = P_1 A$, where P_1 is the air pressure above the house and A is the area of the roof. The air inside the house pushes up with a force $F = P_{atm}A$, where P_{atm} is the atmospheric pressure of the air inside the house. The net upward force is

$$F_{net} = F - F_1 = P_{atm}A - P_1 A = (P_{atm} - P_1)A. \qquad (14.7)$$

We can use Bernoulli's equation to find pressure P_1. To do this, we choose a second point in the air that is at the same elevation as air above the roof, but where the air is not blowing. This point may be miles away. We must be able to move freely from position 1 to position 2 without passing across or through any artificial barrier. If the house is completely closed, there is an artificial barrier between position 1 and the inside of the house, so we cannot choose position 2 inside the house.

If we go far from the house, we will find another place where the wind is not blowing and which is at the same elevation as position 1. At this place, if we are at sea level, the pressure is also atmospheric pressure. Thus,

$$P_2 = \text{Atmospheric pressure} = P_{atm}.$$

The air is at rest at this distant position ($v_2 = 0$), and its elevation is the same as at position 1 ($y_2 = y_1$). From Eq. (14.5),

$$P_1 + \frac{1}{2}\rho v_1^2 + \rho g y_1 = P_{atm} + \frac{1}{2}\rho(0)^2 + \rho g y_1.$$

Solving for P_1, we find that

$$P_1 = P_{atm} - \frac{1}{2}\rho v_1^2.$$

Substituting for P_1 into Eq. (14.7), we find that

$$F_{net} = (P_{atm} - P_1)A = [P_{atm} - (P_{atm} - \frac{1}{2}\rho v_1^2)]A = \frac{1}{2}\rho v_1^2 A$$

$$= \frac{1}{2}(1.3 \text{ kg/m}^3)(45 \text{ m/s})^2(200 \text{ m}^2)$$

$$= \underline{2.6 \times 10^5 \text{ N}},$$

or an upward force of about 30 tons. This huge upward force is caused by the reduced pressure over the roof compared with the atmospheric pressure in the house. ■

According to scientists at the Disaster Research Institute at Texas Tech University, your best course of action in a very high wind such as occurs in a tornado is to move to the basement of a home or to take cover in the center of a home on its lowest floor. If you live in a mobile home, you should evacuate it when a tornado warning is issued. In public buildings, avoid large, open rooms such as gymnasiums. Look instead for shelter in an interior hallway on the lowest floor. If traveling in an auto, leave it and take refuge in a ditch. (But be alert for flash flooding.) The technique of opening a window in a building to equalize the air pressure inside and out is no longer recommended—it does not seem to make much difference.

Bernoulli's principle can play a villainous role in causing heart attacks. A person's arteries may become constricted by the growth of an atherosclerotic plaque such as shown in Fig. 14.11. If dislodged, the plaque moves downstream and may block a smaller vessel so that no blood flows through it. Should this happen to one of the arteries supplying blood to the heart muscles, a heart attack may occur. Bernoulli's principle helps us understand how a plaque becomes dislodged, as described in the simplified model of a plaque in the next example.

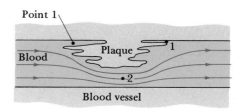

FIG. 14.11. Blood must accelerate as it moves past a plaque. The reduced blood pressure of this fast-moving blood may cause the plaque to become dislodged.

EXAMPLE 14.7 Blood of density $1030 \, \text{kg/m}^3$ flows through the unobstructed part of the blood vessel shown in Fig. 14.12 at a speed of 0.50 m/s. The cross-sectional area through which blood flows past the plaque is one-sixth the normal area of the vessel. The bottom surface of the plaque facing down toward position 2 has an area of $0.60 \, \text{cm}^2$, as do the top surfaces in the channels facing position 1. Calculate the net downward force that tends to dislodge the plaque.

SOLUTION The blood pressure at position 2 is less than at position 1 because the blood must flow at high speed through the constricted artery, whereas it sits at rest in the channels at position 1 ($v_1 = 0$). We determine the blood speed at position 2, using the continuity equation, by comparing the speed of the blood and area of the blood vessel in the unconstricted artery (v and A) with those in the constricted artery (v_2 and A_2):

$$v_2 = \frac{Av}{A_2} = \left(\frac{A}{A_2}\right)v = (6)(0.50 \text{ m/s}) = 3.0 \text{ m/s}.$$

We next calculate the difference in pressure from position 1 to position 2 in Fig. 14.12. We also assume that positions 1 and 2 are at approximately the same elevation; that is, $y_1 \simeq y_2$. According to Eq. (14.4),

$$P_1 - P_2 = \frac{1}{2}\rho v_2^2 - \frac{1}{2}\rho v_1^2 + \rho g(y_2 - y_1)$$

$$= \frac{1}{2}(1030 \text{ kg/m}^3)(3.0 \text{ m/s})^2 - 0 + 0 = 4640 \text{ N/m}^2.$$

Since the pressure is greater from above the plaque at position 1 than from below at position 2, a net downward force acts on the plaque. If the top and bottom areas against which the fluids press are $0.60 \, \text{cm}^2 = 6.0 \times 10^{-5} \, \text{m}^2$, then the net force trying to dislodge the plaque is

$$F_{\text{net}} = F_1 - F_2 = (P_1 - P_2)(\text{Area})$$
$$= (4640 \text{ N/m}^2)(6.0 \times 10^{-5} \text{ m}^2) = \underline{0.28 \text{ N}}.$$

This is about the weight of one-fourth of a medium-sized apple pulling on the plaque.

In addition to the force calculated using Bernoulli's principle that tends to suck the plaque off the wall, there is also an "impact" pressure and force caused by blood hitting the plaque's upstream side. ∎

14.5 Viscous Fluid Flow

In our previous discussion and examples we have ignored the role of friction forces. This is not appropriate in many practical cases. In fluids, just as in solids, friction is often responsible for the generation of thermal energy, but the way in

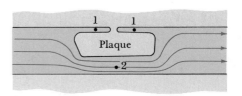

FIG. 14.12. A simplified version of a plaque.

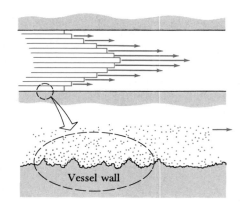

FIG. 14.13. The fluid moving down the center of a vessel moves fastest. Fluid at the edge does not move at all. The molecules of fluid in this surface layer stick to the molecules in the ridges and trenches of the solid surface of the vessel.

TABLE 14.1 Viscosities of Liquids and Gases

Substance	Viscosity ($N \cdot s/m^2$)
Gases	
Air (30°C)	1.9×10^{-5}
Water vapor (100°C)	1.25×10^{-5}
Liquids	
Water (0°C)	1.8×10^{-3}
(20°C)	1.0×10^{-3}
(40°C)	0.66×10^{-3}
(60°C)	0.47×10^{-3}
(80°C)	0.36×10^{-3}
(100°C)	0.28×10^{-3}
Alcohol (20°C)	1.2×10^{-3}
Blood plasma (37°C)	1.5×10^{-3}
Blood, whole (37°C)	$\sim 4 \times 10^{-3}$
Oil, light machine (37°C)	0.113
Oil, SAE No. 10	0.20
Glycerin	0.629

which this thermal energy is created differs for liquids compared to solids. When one solid slides across another, the microscopic ridges and trenches on one surface hook and bond to the trenches and ridges on the other surface, and the thermal energy is created at the surface between the solids.

When a fluid moves across a solid, thermal energy is not generated at the interface. A thin layer of fluid, perhaps just a few molecules thick, does not move at all. The molecules in this layer continually encounter the microscopic ridges and trenches of the solid surface. However, layers of fluid above this interface do slide across each other, much as pictured in Fig. 14.13. This sliding of one layer of fluid across another causes the friction force in fluids. Consequently, the friction in fluids is generated within the fluid rather than at the interface of the fluid and a solid.

The net effect of this friction is known as the **viscosity** of the fluid. The more tightly one layer of fluid is held to an adjacent layer, the greater is the viscosity of the fluid, and the greater is the amount of thermal energy generated as the fluid moves. Molasses and honey are very viscous fluids, whereas water is less viscous.

If a fluid is pushed through a cylindrical pipe, a viscous retarding force F_v opposes its motion. The magnitude of this **viscous force** is related to the speed v of the fluid at the center of the pipe by the equation

$$F_v = 4\pi\eta l v,$$

where l is the length of the pipe and the Greek letter eta η is the **coefficient of viscosity** of the fluid at a particular temperature. Its magnitude indicates the difficulty with which a fluid flows—or how viscous the fluid is. The coefficient of viscosity for honey is larger than that for water. A list of the coefficients of viscosity (or, simply, viscosities) of several fluids appears in Table 14.1. The SI unit of viscosity is $N \cdot s/m^2$. Another commonly used viscosity unit is the **poise** (P), where $1 \, N \cdot s/m^2 = 10 \, P$.

For a fluid to move through a pipe at constant speed, a driving force must balance the opposing viscous retarding force. If the pressure at the entrance of a pipe is P_1, then the force pushing the fluid forward is $F_1 = P_1 A$, where A is the cross-sectional area of the pipe. The pressure force opposing the motion at the exit of the pipe at pressure P_2 is $F_2 = P_2 A$. The net driving force is, then,

$$F_1 - F_2 = (P_1 - P_2)A.$$

If the fluid moves at constant speed, this driving force must be equal in magnitude to the viscous retarding force. Thus,

$$(P_1 - P_2)A = 4\pi\eta l v.$$

If the pipe has a radius r, then its cross-sectional area is $A = \pi r^2$. Substituting for A and rearranging, we find that

$$P_1 - P_2 = \left(\frac{4\eta l}{r^2}\right)v. \tag{14.8}$$

This equation relates the entrance and exit pressure difference to properties of the fluid and to properties of the vessel carrying the fluid. Remember that v is the speed of the fluid along the center of the pipe.

EXAMPLE 14.8 Blood flows with a speed of 0.5 m/s down the center of the aorta. If the length of the aorta is 0.4 m and its radius is 0.8 cm, what is the drop in pressure needed to force the blood through the aorta?

SOLUTION   We have been given the following information: $l = 0.4$ m, $r = 0.8$ cm $= 8 \times 10^{-3}$ m, and $v = 0.5$ m/s. The coefficient of viscosity of blood, according to Table 14.1, is $\eta = 4 \times 10^{-3}$ N $\cdot$ s/m^2. Substituting these numbers into Eq. (14.8), we find that

$$P_1 - P_2 = \frac{4(4 \times 10^{-3} \text{ N} \cdot \text{s/m}^2)(0.4 \text{ m})}{(8 \times 10^{-3} \text{ m})^2}(0.5 \text{ m/s}) = \underline{50 \text{ N/m}^2}.$$

Since 133 N/m^2 = 1 mm Hg, the pressure drop along the aorta due to friction, in units of mm Hg, is

$$P_1 - P_2 = (50 \text{ N/m}^2)\left(\frac{1 \text{ mm Hg}}{133 \text{ N/m}^2}\right) = 0.4 \text{ mm Hg}.$$

This is a very small part of the 100–mm Hg drop in blood pressure from the exit of the heart through the circulatory system and back again to the entrance of the heart. Most of the blood pressure drop in the circulatory system occurs across the capillaries and the small vessels, called arterioles, that feed the capillaries. These vessels have small radii, and, according to Eq. (14.8), the smaller the value of r, the greater the pressure drop.   ■

We are often interested in the pressure needed to maintain a certain fluid flow rate. The flow rate, defined as the volume of fluid that passes through a vessel per unit time, is related to the average speed $\bar{v}$ and cross-sectional area A of the vessel by the equation

$$Q = \frac{V}{t} = \bar{v}A.$$

For a cylindrical pipe, the fluid flows fastest at the center of the pipe and not at all at the edge of the pipe (the fluid is stuck on the microscopic ridges and trenches of this surface). Overall, the average speed $\bar{v}$ of the fluid is half its speed v at the center of the pipe ($\bar{v} = v/2$). For a pipe of cross-sectional area $A = \pi r^2$,

$$Q = \bar{v}A = \frac{v}{2}(\pi r^2). \tag{14.9}$$

Solving Eq. (14.8) for v, we find that

$$v = \left(\frac{r^2}{4\eta l}\right)(P_1 - P_2).$$

Substituting the above expression for v into Eq. (14.9) produces the equation

$$Q = \left(\frac{\pi r^4}{8\eta l}\right)(P_1 - P_2). \tag{14.10}$$

This is called **Poiseuille's law,** named in honor of the French physician Jean Louis Marie Poiseuille (1799–1869). Poiseuille established this law empirically with remarkable accuracy.

Notice that the flow rate is proportional to the fourth power of the radius of the vessel carrying the fluid. If the radius of a vessel carrying fluid is reduced by a factor of 2, the flow rate will be reduced by a factor of 16—that is, 2^4.

EXAMPLE 14.9   Suppose that over a period of years, the radius of an artery in a person's heart decreases by 40 percent. Calculate the ratio of the present

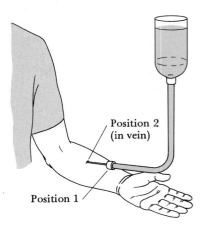

FIG. 14.14. The flow rate of blood into a vein depends on the radius of the needle.

flow rate and the original flow rate if the pressure across the artery, its length, and the viscosity of blood are unchanged.

SOLUTION The radius r of the vessel at the present time is related to the radius r_0 years ago by the equation $r = 0.60r_0$ (a 40 percent reduction in radius). The ratio of the flow rates, given by Eq. (14.10), is

$$\frac{Q}{Q_0} = \frac{\left(\frac{\pi r^4}{8\eta l}\right)(P_1 - P_2)}{\left(\frac{\pi r_0^4}{8\eta l}\right)(P_1 - P_2)} = \left(\frac{r}{r_0}\right)^4 = \left(\frac{0.60r_0}{r_0}\right)^4 = 0.60^4 = \underline{0.13}.$$

The flow rate is only 13 percent of its value years earlier! The heart must increase the pressure (high blood pressure) to increase the flow rate. ∎

EXAMPLE 14.10 Blood is to be transferred from the reservoir shown in Fig. 14.14 into a person's vein. The flow rate is regulated by the length and radius of the needle since all other vessels carrying the blood have large radii and cause little resistance to the flow. The flow rate is to be $1.0 \times 10^{-6}\,\mathrm{m^3/s}$ (about 1 liter each 15 minutes). The pressure at the entrance of the needle (position 1) is $9000\,\mathrm{N/m^2}$ greater than that in the vein at position 2. The viscosity of blood is $4.0 \times 10^{-3}\,\mathrm{N \cdot s/m^2}$. If the length of the needle is 0.030 m, calculate the radius of the needle needed to get the desired flow rate.

SOLUTION When Poiseuille's law is solved for the radius of the needle, we find that

$$r^4 = \frac{8\eta l Q}{\pi(P_1 - P_2)} = \frac{8(4.0 \times 10^{-3}\,\mathrm{N \cdot s/m^2})(0.030\,\mathrm{m})(1.0 \times 10^{-6}\,\mathrm{m^3/s})}{\pi(9000\,\mathrm{N/m^2})}$$

$$= 3.73 \times 10^{-14}\,\mathrm{m^4},$$

$$r = (3.73 \times 10^{-14}\,\mathrm{m^4})^{1/4} = 4.4 \times 10^{-4}\,\mathrm{m} = \underline{0.44\,\mathrm{mm}}. \qquad ∎$$

14.6 Reynolds Number and Turbulence

Poiseuille's law applies only to laminar, or streamline, flow. As the speed of laminar flow increases, an abrupt transition to turbulent flow may occur. Once turbulent flow is started, greater pressure is needed to maintain the flow because the viscous friction force increases during turbulent flow.

The abrupt onset of turbulence is characterized by a number called the **Reynolds number** (Re):

$$\mathrm{Re} = \frac{2\bar{v}r\rho}{\eta}, \qquad (14.11)$$

where $\bar{v}$ is the average speed of the fluid, ρ is its density, η is the fluid's coefficient of viscosity, and r is the radius of the vessel that carries the fluid. The Reynolds number is dimensionless since all of its units cancel. It has been determined experimentally that if Re is less than 2000, laminar flow occurs; if Re is greater than 3000, turbulent flow occurs; and if Re is between 2000 and 3000, the flow is unstable and may be either laminar or turbulent.

EXAMPLE 14.11 Estimate the maximum rate of laminar flow of water in a garden hose whose radius is 1.0 cm. The water temperature is 20°C.

SOLUTION When the Reynolds number exceeds 2000, flow may become turbulent. Thus, the maximum average speed for laminar flow of water having density $\rho = 1000 \text{ kg/m}^3$ and viscosity $\eta = 1.0 \times 10^{-3} \text{ N} \cdot \text{s/m}^2$ (see Table 14.1) is, according to Eq. (14.11),

$$\bar{v} = \frac{\eta \cdot \text{Re}}{2r\rho} \simeq \frac{(1.0 \times 10^{-3} \text{ N} \cdot \text{s/m}^2)(2000)}{2(1.0 \times 10^{-2} \text{ m})(1000 \text{ kg/m}^3)} = 0.10 \text{ m/s}.$$

The flow rate is

$$Q = \bar{v}A = \bar{v}(\pi r^2) = (0.10 \text{ m/s})\pi(1.0 \times 10^{-2} \text{ m})^2 = \underline{3.1 \times 10^{-5} \text{ m}^3/\text{s}},$$

about 2 liter/min. ∎

14.7 Drag Force

In the last two sections we analyzed the effect of viscosity on a fluid flowing through a tube. In this section we discuss the viscous friction force that opposes the motion of an object relative to a fluid. Examples include the resistive force of water on a swimmer and the resistive force of air on a car, bicycle, or skydiver. These viscous friction forces are often called **drag forces.**

A Ping-Pong ball dropped from a building is illustrated in Fig. 14.15. If the ball falls slowly, laminar flow occurs. Streamlines of air pass the ball as it falls. As the ball's speed increases, the flow of air past the ball becomes turbulent. The onset of turbulence is predicted by a different form of the Reynolds number:

$$\text{Re} = \frac{vL\rho}{\eta}, \tag{14.12}$$

where v is the object's speed, L is the object's length, ρ is the density of the fluid, and η is the fluid's viscosity. This number is used for an entirely different purpose than the Reynolds number introduced in the last section, and the two should not be confused. When Re given by Eq. (14.12) is less than about 1, laminar flow occurs past an object moving through a fluid, and when Re is greater than 1, the flow past the object is turbulent. The previous form of the Reynolds number is used to predict the type of flow of fluid through a vessel.

The drag force acting on an object moving relative to a fluid depends on whether the flow past the object is laminar or turbulent. If laminar, the drag force increases approximately in proportion to the object's speed relative to the fluid:

$$F_D = Dv,$$

where D is a constant. For the situation where a spherical object of radius r falls at speed v through a liquid with viscosity η, the drag force for laminar flow is

$$F_D = 6\pi\eta rv. \tag{14.13}$$

This equation is called **Stokes' law.**

If fluid flow past the moving object is turbulent, the drag force increases

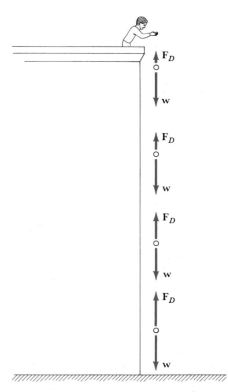

FIG. 14.15. As a ball falls faster, the magnitude of the drag force increases. Eventually, $\mathbf{F}_D$ and $\mathbf{w}$ are equal in magnitude, and the ball falls at a constant terminal speed, v_T.

approximately in proportion to the square of the object's speed relative to the fluid:

$$F_D = kv^2.$$

For instance, for objects moving through air with turbulent flow,

$$F_D \simeq \frac{1}{2} C_D \rho A v^2, \tag{14.14}$$

where ρ is the density of air, C_D is a unitless number called the *drag coefficient*, and A is the cross-sectional area of the object as seen along its line of motion.

EXAMPLE 14.12 Estimate the air drag force on a 1130-kg compact car when it is moving at a speed of 27 m/s (60 mph). The cross-sectional area of the car is roughly 2.0 m², and the drag coefficient C_D is approximately 0.5 for a well-designed car. The density of air is 1.3 kg/m³.

SOLUTION From Eq. (4.14), we find that

$$F_D = \frac{1}{2}(0.5)(1.3 \text{ kg/m}^3)(2.0 \text{ m}^2)(27 \text{ m/s})^2 = 470 \text{ kg m/s}^2 = \underline{470 \text{ N}},$$

a force of about 100 lb. ∎

Terminal Speed

An object falling through a fluid eventually reaches a speed where the force pulling it in one direction (for example, the weight force pulling it down) is balanced by the opposing drag force. The net force on the object is zero; it no longer accelerates, and it continues to move at a constant speed, called its **terminal speed** v_T. The terminal speed of a skydiver is approximately 54 m/s (120 mph), whereas that of a falling raindrop is about 12 m/s. A red blood cell falls in a beaker of water with a terminal speed of about 10^{-7} m/s. The next example shows how an object's terminal speed can be calculated.

EXAMPLE 14.13 Estimate the terminal speed of a Ping-Pong ball weighing 2.2×10^{-2} N whose radius is 1.9×10^{-2} m. The ball experiences a turbulent drag force with a drag coefficient of 0.60.

SOLUTION The ball reaches terminal speed when its weight is balanced by the upward drag force:

$$\Sigma F_y = F_D - w = 0.$$

Substituting Eq. (14.14) for F_D and rearranging, we find that

$$v_T = \sqrt{\frac{2w}{C_D \rho A}} = \sqrt{\frac{2w}{C_D \rho \pi r^2}} = \sqrt{\frac{2(2.2 \times 10^{-2} \text{ N})}{(0.60)(1.3 \text{ kg/m}^3)\pi(1.9 \times 10^{-2} \text{ m})^2}}$$
$$= \underline{7.1 \text{ m/s}}. \qquad ∎$$

Summary and Additional Readings

1. **Fluid flow:** When the pressure is greater in one region of a fluid than in another, the fluid flows from the high-pressure to the low-pressure region. If the fluid flows smoothly in streamlines (each molecule of fluid following the one in front of it), then the flow is said to be **laminar.** If the fluid swirls and forms eddies, the flow is said to be **turbulent.**

2. **Flow rate:** The flow rate Q is defined as the volume of fluid V that flows past a cross section of the vessel carrying the fluid divided by the time t for the fluid to pass:

$$Q \equiv \frac{V}{t}. \tag{14.1}$$

When fluid flows through a pipe of cross-sectional area A at an average speed $\bar{v}$, then the flow rate is

$$Q = \bar{v}A. \tag{14.2}$$

If the fluid is incompressible, then the flow rate past one cross section in a pipe equals the flow rate past another cross section. Thus,

$$Q = v_1 A_1 = v_2 A_2, \tag{14.3}$$

where v_1 is the speed of the fluid past a cross section of area A_1, and v_2 is its speed past a cross section of area A_2. Equation (14.3) is called the **continuity equation.**

3. **Bernoulli's equation** results from applying the work-energy equation to a section of incompressible, frictionless fluid flowing along a pipe. We find that the pressure, speed, and elevation of the fluid at one position are related to those same quantities at a second position in the fluid by the equation

$$P_1 + \frac{1}{2}\rho v_1^2 + \rho g y_1 = P_2 + \frac{1}{2}\rho v_2^2 + \rho g y_2, \tag{14.5}$$

where ρ is the density of the fluid.

4. **Friction in fluids:** When fluids flow through vessels, the layer of fluid nearest the vessel does not move. But layers of fluid away from the surface do move and slide across layers nearer the vessel. This sliding of one layer across another causes a friction force in fluids that opposes their motion. This friction is called **viscosity.** A coefficient of viscosity η characterizes the degree of viscosity in fluids.

Viscous forces oppose the motion of all fluids. We can often ignore these viscous forces when fluids move through vessels with large cross-sectional areas and when the flow is laminar. However, the flow rate can be greatly affected by viscous forces when the fluid flows through vessels of small cross-sectional area. For such vessels the flow rate Q is given by **Poiseuille's equation:**

$$Q = \left(\frac{\pi r^4}{8\eta l}\right)(P_1 - P_2), \tag{14.10}$$

where $P_1 - P_2$ is the pressure change across a length l of a vessel whose radius is r.

5. **Drag force:** A viscous drag force opposes the motion of an object through a fluid. If turbulence is caused in the fluid as the object passes, the drag force is proportional to the square of the object's speed. If the object does not create turbulence, the drag force is usually proportional to the first power of the object's speed. We can calculate the **Reynolds number** [Eq. (14.12)] to decide whether turbulence does or does not occur.

Alan C. Burton, *Physiology and Biophysics of the Circulation: An Introductory Text,* 2nd ed., Year Book Medical Publishing, Chicago (1972). Contains excellent examples of the use of introductory physics to understand the circulatory system.

Jearl Walker, *The Flying Circus of Physics,* John Wiley & Sons, New York, (1975). Chapter 4 has many interesting questions and simple experiments dealing with fluids.

J. V. Warren, "The Physiology of the Giraffe," *Scientific American* **321,** 96 (Nov. 1974).

Questions

1. The stream of water falling from a faucet often becomes more narrow as it falls. Why might this be?

2. Explain why a wind blowing past the window of a building causes the window to be blown out rather than in.

3. Why does the top of a convertible bulge when the car is riding along a highway?

4. How does Bernoulli's principle help explain the fact that air is drawn up the chimney of a house?

5. A person can open an envelope by cutting the end and blowing air past the cut end, thus causing the envelope to bulge open. Explain, using Bernoulli's principle.

6. Hold the handle of a spoon lightly between two fingers and let water from a faucet run across the back of the spoon bowl as it hangs from your fingers. Now slowly move the spoon's handle away from the stream. The bowl remains in the stream as though glued to it! Try the experiment and then explain why you think this happens.

7. An air gun blowing air over a Ping-Pong ball will support it, as shown in Fig. 14.16. Construct a force diagram for the ball. Explain in terms of forces how the ball can remain in equilibrium.

8. If the friction in a moving fluid occurs in the fluid rather than at the boundary of the fluid, why doesn't blood get very hot? Explain in detail.

9. A ball of mass m and another of the same material but with mass $2m$ are dropped simultaneously from the Leaning Tower of Pisa. Will the balls hit the ground at the same time? If not, which will hit first? Explain. Do not ignore air resistance.

FIG. 14.16. Air moving across the top of a Ping-Pong ball reduces the pressure. The air below the ball has less speed and greater pressure and helps support the ball's weight.

Problems

Unless stated otherwise, assume in these problems that atmospheric pressure is 1.01×10^5 N/m^2 and that the densities of water and air are 1000 kg/m^3 and 1.3 kg/m^3, respectively.

14.2 Flow Rate and Equation of Continuity

1. (a) Calculate the flow rate of water moving at an average speed of 32 cm/s through a garden hose of radius 1.2 cm. (b) Calculate the speed of the water in a second hose of radius 1.0 cm that is connected to the first hose.

2. (a) An irrigation canal has a rectangular cross section of 5.0-m width and 1.2-m depth. If water flows at a speed of 0.80 m/s, what is its flow rate? (b) If the width of the stream is reduced to 3.0 m and the depth to 1.0 m as the water passes a flow-control gate, what is the speed of the water past the gate?

3. Each second, 0.070 m^3 of water flows through a firehose. If the water is to leave the nozzle at a speed of 25 m/s, what should its diameter be?

4. A main waterline for a housing project must deliver water at a maximum flow rate of 0.010 m^3/s. If the speed of the water at this flow rate is 0.30 m/s, calculate the diameter of the pipe carrying the water.

■ 5. The flow rate of blood in the aorta is 80 cm^3/s. Beyond the aorta, this blood travels through about 6×10^9 capillaries. If the radii of these capillaries is 8×10^{-4} cm, what is the speed of the flow of blood through the capillaries?

■ 6. A farmer's 4.0-cm-diameter pipe from an irrigation canal takes 20 h to flood a small field. How much time is required using a 6.0-cm pipe, assuming the water flows at the same average speed in both pipes?

14.3 and 14.4 Bernoulli's Equation and Applications

■ 7. The pressure of water flowing through a pipe of radius 0.060 m at a speed of 1.8 m/s is 2.2×10^5 N/m^2. Calculate (a) the flow rate of the water and (b) the pressure in the water after it goes up a 5.0-m-high hill and flows in a pipe of radius 0.050 m.

■ 8. A garden hose of 0.80-cm radius is connected to one of 1.0-cm radius. The smaller hose is held on the roof of a house 4.0 m above the larger hose. Water leaves the smaller hose at a speed of 6.0 m/s and at atmospheric pressure. Calculate the pressure in the larger hose on the ground.

■ 9. The large pipe of a waterline has radius 0.060 m and feeds ten smaller pipes of radius 0.020 m that carry water to homes. The flow rate of water in each of the smaller pipes is to be 6.0×10^{-3} m^3/s, and the pressure is 4.00×10^5 N/m^2. The homes are 10.0 m above the main pipes. Calculate the average speed of the water in (a) a smaller pipe and (b) in the main pipe. (c) Calculate the pressure in the main pipe.

■ 10. Blood flows at an average speed of 0.40 m/s in a horizontal artery of radius 1.0 cm. The average gauge pressure is 1.4×10^4 N/m^2. Calculate (a) the average speed of the blood past a constriction where the radius of the opening is 0.30 cm and (b) the gauge pressure of the blood as it moves past the constriction.

■ 11. Air of density 1.3 kg/m^3 blows past a 1.2 m × 2.2-m window of a skyscraper at a speed of 25 m/s. The air inside the building is at atmospheric pressure. (a) Calculate the difference in pressure between the inside of the window and the outside. (b) Calculate the net force of the air on the window (magnitude and direction).

■ 12. Calculate the lift caused by the Bernoulli effect on an airplane wing of area 30 m^2. Air of density 1.1 kg/m^3 moves across the top surface at a speed of 180 m/s and across the bottom at 165 m/s.

■ 13. At what speed must air of density 1.2 kg/m^3 move across the flat roof of a house of area 160 m^2 to be able to lift the weight of the 2.1×10^4-kg roof. Air inside the house is at atmospheric pressure and at rest.

■ 14. A straw extends 3.0 cm out of a glass of water. How fast must air blow across the top of the straw to draw water to the top of the straw?

■ 15. A U-shaped tube, open at both ends, contains water. If air blows across the top of one end at a speed of 10 m/s and not across the other, how much higher will the water be in the side below the moving air compared with that below the air at rest?

■ 16. A large, open barrel is filled with water to a height of 1.0 m above a cork. Calculate the initial flow rate of water from the 1.2-cm-radius hole when the cork is removed.

■ 17. A large, closed barrel of water has a 0.01-m-diameter opening at the bottom through which water leaves at a speed of 2.0 m/s. Calculate the pressure at the top of the water level in the barrel 0.50 m above the hole.

■ 18. Water sits at rest behind an irrigation dam. The water is 1.2 m above the bottom of a gate that, when lifted, allows water to flow under the gate. To what height h from the bottom of the dam should the gate be lifted to allow a flow rate of 1.0×10^{-2} m^3/s? The gate is 0.50 m wide.

■ ■ 19. The gauge pressure inside a 1.0-cm radius vessel carrying blood at a speed of 0.50 m/s is 5200 N/m^2, and the gauge pressure outside the vessel is 3200 N/m^2. To what radius must the vessel be reduced by a constriction so that the outside pressure is greater than that inside, thus causing the vessel to close at the constriction? (The vessel will flutter open and closed like a vibrating reed.)

14.5 Viscous Fluid Flow

20. A 5.0-cm-diameter firehose 60 m long carries 20°C water at a speed of 12 m/s. Calculate the drop in pressure along the hose due to viscous friction.

■ 21. A horizontal waterline of 5.0-cm radius is 500 m long. Water at 20°C flows at a rate of 1.0×10^{-2} m^3/s. (a) Calculate the pressure difference from the beginning to the end of the line. (b) If the pressure difference is kept constant and the flow rate is to be doubled, what must the radius of the pipe be?

■ 22. The pump for a firehose can develop a maximum gauge pressure of 5.0×10^5 N/m^2. The hose is to carry water of viscosity 1.0×10^{-3} N·s/m^2 at a flow rate of 1.0 m^3/s along a horizontal hose that is 50 m long. What is the minimum radius for the hose?

■ 23. Water flows in a solar collector through a copper tube of 0.60-cm radius that is 20 m long. The average temperature of the water is 60°C and the flow rate is 200 cm^3/s. Calculate the pressure drop along the tube, assuming the water does not change elevation.

■ 24. Your heart pumps blood at a flow rate of about 80 cm³/s. This blood flows through about 9×10^9 capillaries, each of radius 4×10^{-4} cm and 0.1 cm long. Calculate the pressure drop across a capillary, assuming a blood viscosity of 4×10^{-3} N/s · m².

■ 25. The radius of a person's arterioles, small vessels carrying blood to the capillaries, decreases by 5 percent when the person smokes. (a) Calculate the percent change in flow rate if the pressure across the arterioles remains constant. (b) Calculate the percent change in pressure if the flow rate remains constant.

■ 26. Calculate the ratio of the flow rate through capillary tubes A and B (that is, Q_A/Q_B). The length of A is twice that of B, and the radius of A is one-half that of B. The pressure across both tubes is the same.

■ 27. A piston pushes 20°C water through a horizontal tube of 0.20-cm radius and 3.0-m length. One end of the tube is open and at atmospheric pressure. (a) Calculate the force needed to push the piston so that the flow rate is 100 cm³/s. (b) Repeat the problem if the piston pushes SAE 10 oil instead of water.

■■ 28. A glucose solution of viscosity 2.2×10^{-3} N · s/m² and of density 1030 kg/m³ flows from an elevated bag into a vein. The needle into the vein has a radius of 0.30 mm and is 2.0 cm long. All other tubes leading to the needle have much larger radii, and viscous forces in them can be ignored. The pressure in the vein is 1000 N/m² above atmospheric pressure. (a) Calculate the gauge pressure needed to maintain a flow rate of 0.10 cm³/s. (b) To what elevation should the bag containing the glucose be raised to maintain this pressure at the needle?

■ 29. We can include the effect of friction in Bernoulli's equation by adding a term for the thermal energy generated by the viscous retarding force acting on the fluid. Show that the term to be added to Eq. (14.4) for flow in a vessel of uniform cross-sectional area A is

$$\frac{\Delta E_{th}}{\Delta V} = \frac{4\pi\eta lv}{A},$$

where v is the speed of the fluid of viscosity η along the center of a pipe whose length is l.

■■ 30. (a) Show that the work done per unit time by viscous friction in a fluid with a flow rate Q across which there is a pressure drop P is

$$\frac{W}{t} = PQ = Q^2 R = \frac{P^2}{R},$$

where $R = 8\eta l/\pi r^4$ is called the **flow resistance** of the fluid moving through a vessel of radius r. (b) By what percentage must the work per unit time increase if the radius of a vessel decreases by 10 percent and all other quantities including the flow rate remain constant (the pressure does not remain constant)?

14.6 Reynolds Number and Turbulence

31. Blood flows at an average speed of 0.40 m/s in an artery whose radius is 0.50 cm. Based on a calculation of the Reynolds number, determine whether the flow is laminar or turbulent.

■ 32. The radius of a vessel carrying blood at a flow rate of 20 cm³/s is slowly reduced at a constriction until the flow, initially laminar, becomes turbulent. At the constriction, the Reynolds number now equals 2500. Calculate the radius of the vessel when turbulence starts. [*Note:* The blood speed changes as the radius changes.]

■ 33. Air flows from a furnace through a circular duct having a diameter of 0.30 m. Do you think the flow is laminar or turbulent? Explain.

■ 34. Water at a temperature of 40°C flows in a copper tube whose radius is 0.30 cm. Calculate the maximum flow rate before turbulence starts.

14.7 Drag Force

35. A 2300-kg "luxury" car has a 0.6 drag coefficient and an effective frontal area of 2.8 m². Calculate the air drag force on the car (a) at 55 mi/h (24 m/s) and (b) at 70 mi/h (31 m/s). The density of air is 1.3 kg/m³.

36. *Estimate* the drag force opposing your motion when you ride a bicycle at 5 m/s.

37. Calculate the drag force on a red blood cell having a radius of 1.0×10^{-5} m and moving through 20°C water at a speed of 1.0×10^{-5} m/s. (Assume that laminar flow occurs.)

■ 38. A protein of radius 3.0×10^{-9} m falls through a tube of water whose viscosity η is 1.0×10^{-3} N · s/m². A constant downward force of 3×10^{-22} N acts on the protein. (a) Use Stokes' law and the other information provided to *estimate* the terminal speed of the protein. (b) How many hours would be required for the protein to fall 0.1 m?

■ 39. Colloidal clay particles each have a constant downward force of 3×10^{-17} N acting on them. The particles settle 10 cm in water in 10^2 minutes. Calculate the radii of these particles. The viscosity of water is 1.0×10^{-3} N · s/m².

■ 40. A sphere falls through a fluid. A constant downward force of 0.50 N acts on the sphere. The drag force opposing the sphere's motion is given by

$$F_D = 2v,$$

where F_D is in N if v is in m/s. What is the terminal speed of the sphere?

41. A balloon floats down through the air, pulled by a constant downward force of 0.050 N. A drag force opposing the balloon's motion is given by the equation

$$F_D = 0.03v^2,$$

where F_D is in N if v is in m/s. Calculate the terminal speed of the balloon.

42. A parachutist, whose mass including that of the parachute is 80 kg, falls toward the earth with parachute open at a constant terminal speed of 8.5 m/s. If the drag force on the parachutist is given by Eq. (14.14) and the drag coefficient $C_D = 0.5$, what is the area of the parachute?

43. A balloon, whose radius when inflated is 0.20 m, falls slowly through air with a terminal speed of 0.40 m/s. Its drag coefficient C_D is approximately 0.5. (a) What is the drag force on the balloon? (b) What is the balloon's weight (remember that it is falling at terminal speed)? (c) What is the balloon's mass?

■■ 44. A round grain of sand with radius 0.15 mm and density 2300 kg/m³ is placed in a 20°C lake. Calculate the terminal velocity of the sand as it sinks into the lake. Do not forget to include the buoyant force acting on the sand.

■■ 45. Show that the terminal velocity for a skier opposed by a turbulent-air drag force and by a kinetic friction force is

$$v_T = \left[\frac{2mg(\sin\theta - \mu\cos\theta)}{C_D \rho A}\right]^{1/2},$$

where θ is the angle of the ski slope, μ is the coefficient of kinetic friction between the skis and the snow, m is the skier's mass, ρ is the density of air, A is the skier's frontal area, and C_D is the drag coefficient. [See Angelo Armenti, Jr., "How Can a Downhill Skier Move Faster Than a Skydiver," *The Physics Teacher,* p 109, February (1984).]

■ ■ 46. On June 30, 1908, a monstrous cometary fragment of mass greater than 10^9 kg is thought to have devastated a 2000-km^2 area of remote Siberia (called the Tunguska event). *Estimate* the terminal velocity of such a comet in air of density 0.7 kg/m^3. State clearly all of your assumptions.

CHAPTER **15**

Elastic Properties
of Solids

In the last three chapters we have been studying fluids—gases and liquids. Now we come to solids, a form of matter in which bonds between neighboring atoms and molecules are strong. Whereas both liquids and gases flow and take on the shape of their container, solids tend to have definite shapes that they retain unless acted on by external forces. When opposing external forces are applied, the atomic and molecular bonds may be stretched, compressed, bent, or changed in other ways. When the forces are removed, however, the bonds between the atoms often return to their original orientation and length, and the solid returns to its original shape. When this happens, we say that the solid is **elastic**—it returns to its original shape after the external force deforming it has been removed.

Not all solids have this elastic property. Clay, when deformed by a force, remains deformed. This permanent alteration of the shape of clay is actually another form of flow, such as we studied in the last chapter. Here we will be concerned primarily with elastic deformations of solids, such as the compression of a bone, the stretching of a blood vessel, and the shearing of a continental plate as it moves on the earth's surface. We will develop and use equations that relate the forces causing these deformations to the actual magnitude of the deformations. We start by considering a simple and familiar example—stretching a spring.

15.1 Elastic Deformations

Stretching and compressing a spring illustrates many of the more complex types of elastic deformation we will encounter later. When no force acts on the spring and mass shown in Fig. 15.1, the end of the spring rests as its **equilibrium position.** As the force pulling the mass increases, the spring stretches and moves farther from its equilibrium position. We find by experiment that the spring's displacement x from equilibrium is proportional to the applied force F:

$$F \propto x.$$

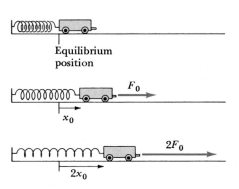

FIG. 15.1. As the force pulling on a spring increases, its displacement from equilibrium increases.

313

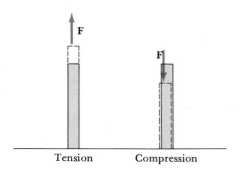

Tension   Compression

(a)

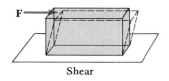

Shear

(b)

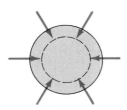

Change in volume

(c)

FIG. 15.2. The three basic types of deformation of solids: (a) A tensile or compressive force causes a change in an object's length. (b) A shear force causes a change in an object's shape. (c) A pressure change causes a change in an object's volume.

For example, if a force of magnitude F_0 causes a displacement x_0, then a force $2F_0$ will cause a displacement $2x_0$. If the force is removed from the spring, it returns to its equilibrium position ($x = 0$).

The proportionality between the magnitude of the applied force and the displacement from equilibrium can be written as an equation, called **Hooke's law:**

$$F = kx. \qquad (15.1)$$

The proportionality constant k, called the **force constant,** depends on how stiff the spring is: the value of k is large for a stiff spring and small for an easily stretched spring.

All types of elastic deformations can be described in an analogous manner. The description has three elements: (1) a cause of the deformation such as the force pulling the spring in Fig. 15.1; (2) a measure of the deformation such as the x coordinate used to indicate a spring's displacement from equilibrium; and (3) a proportionality constant that relates the deformation and its cause in an equation such as Eq. (15.1). For deformations of elastic solids, the cause of the deformation is called a **stress,** the resulting deformation is called a **strain,** and the proportionality constant relating stress and strain is called a **modulus of elasticity.** The definition of each of these three quantities depends on the particular type of deformation being described.

15.2 Types of Solid Deformations

Although a solid can be deformed in many ways, such as compression, stretching, twisting, bending, folding, and volume changes, there are only three basic types of deformations. All others are combinations of these three types. As illustrated in Fig. 15.2, the three basic deformations and the forces causing them are (a) a **change in length** caused by a tension or compression, (b) an **angular change** in the orientation of the surfaces of a solid caused by a shearing force, and (c) a **volume change** caused by a uniform change in pressure on all of the surfaces of a solid.

Other types of deformation are either a special example of one of these deformations or a combination of two or three of them. For example, bending is a combination of a stretch and a compression. The fibers in the top half of the board shown in Fig. 15.3 are being stretched; those in the bottom half are being compressed. The fibers along the center of the board (the dashed line in Fig. 15.3) are unchanged in length.

Stress and strain are two quantities used to describe the nature of the applied force and the resulting deformation, respectively. Our next task is to define these terms.

15.3 Stress and Strain

When a force acts on a solid to produce a change in its dimensions or shape, the cause of the deformation is called a stress.

FIG. 15.3. Bending is a combination of tension in the upper portion of the board and compression in the bottom portion.

Stress is defined as the force acting on a solid divided by the area over which the force acts:

$$\text{Stress} = \frac{\text{Force}}{\text{Area}} = \frac{F}{A}. \qquad \textbf{(15.2)}$$

The units of stress are N/m^2, lb/in^2, and so forth.

When defined as force per unit area, stress is proportional to the deformation it causes. If defined simply as force, stress would not be proportional to the deformation. Consider the groups of identical rubber bands pictured in Fig. 15.4. Group A consists of one rubber band elongated a distance ΔL by a force F. Group B consists of two rubber bands that are also elongated the distance ΔL, but a force $2F$ is needed to produce this elongation since there are two rubber bands instead of one. A force $4F$ is needed to elongate the four rubber bands in group C by a distance ΔL.

Although the elongation is the same for each group of rubber bands, the force causing the elongation is different. Thus, the force is not directly proportional to the deformation. If we define the stress as the force per cross-sectional area of the rubber bands, however, then the stress is the same for each group and is proportional to the deformation. For example, if the cross-sectional area of one rubber band is A, the ratio of force and area for each group is the same:

$$\frac{F}{A} = \frac{2F}{2A} = \frac{4F}{4A}.$$

Each group of rubber bands was stretched the same amount because the force per area (the stress) acting on each group was the same.

The quantity **strain** is a measure of the deformation caused by a stress. The specific definition of strain depends on which of the three types of deformations shown in Fig. 15.2 are being described. In most cases we can think of the strain as being the actual change in an object divided by the original value of the quantity that has changed:

$$\text{Strain} = \frac{\text{Change in dimension}}{\text{Original dimension}}. \qquad \textbf{(15.3)}$$

For instance, if a bone is 100 mm long when unstressed and 99 mm long when under stress, the strain is the change in length divided by the original length, or

$$\text{Strain} = \frac{1 \text{ mm}}{100 \text{ mm}} = 0.01.$$

If a uniform external pressure causes the volume of a rock to change from 10.00 cm^3 to 9.98 cm^3, then the volume strain is the change in volume divided by the original volume, or

$$\text{Strain} = \frac{0.02 \text{ cm}^3}{10.00 \text{ cm}^3} = 0.002.$$

Shearing strain is a little more complicated and will be defined later in the chapter.

For elastic materials, the strain increases in proportion to the stress placed on the material. The proportionality constant relating the stress and strain is called a **modulus of elasticity:**

$$\text{Stress} = (\text{Modulus of elasticity})(\text{Strain}). \qquad \textbf{(15.4)}$$

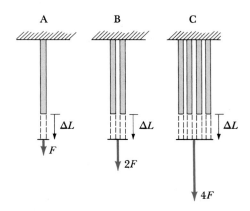

FIG. 15.4. Three groups of rubber bands are each displaced the same distance. If stress is defined as Force/Area, then in each case the stress is the same as is the elongation caused by the stress.

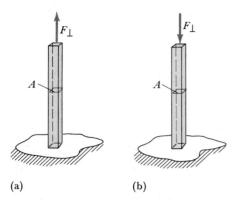

(a)   (b)

FIG. 15.5. (a) A tensile stress and (b) a compressive stress.

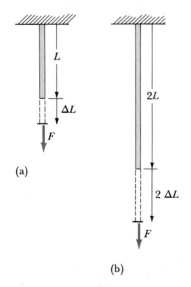

FIG. 15.6. The distance an elastic object is stretched by a force is proportional to the object's length.

The modulus of elasticity plays the same role as the force constant k in the equation relating the force F pulling a spring and its displacement from equilibrium x:

$$F = kx.$$

In the next three sections the stress, strain, and modulus of elasticity are defined carefully for each of the three basic types of solid deformations.

15.4 Length Deformation

Forces that stretch a solid produce what is called **tensile stress** since the force causes tension in the stretched object. Its atoms and molecules are pulled farther apart. Pulling a rope and stretching a tendon are examples of tensile stresses (Fig. 15.5a).

A **compressive stress** is caused by a force pushing against one side of an object while it is supported by an equal force on the other side (Fig. 15.5b). Compressing a bone and pounding a nail are examples of compressive stresses.

To define stress caused by tension and compression forces, we choose a cross-sectional area through some plane of the object (see Fig. 15.5). The stress across that plane is the component of the applied force that is perpendicular or normal to that plane $F_\perp$ divided by the area A of the plane:

$$\text{Tensile or compressive stress} = \frac{F_\perp}{A}.$$

The strain caused by a tensile or compressive stress is defined as the change in length ΔL of the object divided by its original length L:

$$\text{Tensile or compressive strain} = \frac{\Delta L}{L}.$$

We have included the object's original length in our definition because strain, when defined this way, depends only on molecular properties of the material being stretched or compressed and not on its dimensions. Consider the rubber bands shown in Fig. 15.6. If an equal force is applied to each, the rubber band of length $2L$ will be displaced twice as far as the one of length L because twice as many molecules are in the longer rubber band, each being stretched in the same direction with the same tension. We see, however, that the strain produced is the same for each rubber band:

$$\text{Strain} = \frac{\Delta L}{L} = \frac{2\,\Delta L}{2L}.$$

Strain is independent of the dimensions of the object; it depends only on its molecular structure.

Elastic materials are those for which the stress and strain are proportional to each other. If the stress is doubled, the strain is doubled. The proportionality constant that relates the compressive or tensile stress and the strain of a particular type of material is called the **Young's modulus** of the material and is given the symbol Y. Thus, for compressions and tensions of elastic materials,

Compressive or tensile stress

$$= (\text{Young's modulus})(\text{Compressive or tensile strain}),$$

or

$$\left(\frac{F_\perp}{A}\right) = Y\left(\frac{\Delta L}{L}\right). \qquad (15.5)$$

Some representative values of Young's modulus are listed in Table 15.1. Notice that for some materials, Y is different for tension than for compression. This is because the attractive force between atoms when stretched does not equal the repulsive force when compressed. We see that Young's modulus for steel is approximately 10^5 times greater than that for rubber; that is, steel is 10^5 times harder to stretch than rubber. Bone is about 10^4 times more difficult to stretch than rubber, and a tendon about 10^3 times more difficult.

It is common when testing materials to make a stress-versus-strain graph of the material. A compressive stress-versus-strain graph for the bone in the upper arm (the humerus) is shown in Fig. 15.7. The slope of the straight-line portion of this graph from the origin to point a (the elastic region) is Young's modulus for bone. Beyond point a, the material may still try to return to its original shape, but nonelastic structural changes have occurred. At point b, the material breaks or fractures. The maximum stress that the material can experience before fracture occurs is called the **ultimate tensile** or **compressive strength.** Some representative values of ultimate strength of different materials appear in Table 15.1.

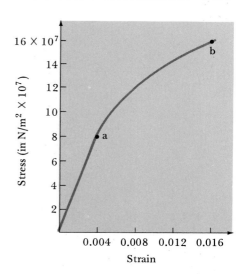

FIG. 15.7. The stress-strain graph of the humerus. The values of stress plotted on the vertical axis should be multiplied by 10^7.

EXAMPLE 15.1 A parachutist falling at 5.8 m/s becomes rigid just before landing and is stopped in an upright position as his heels sink a distance $d = 3.2$ cm into the ground. (a) Calculate the change, while landing, in the length of the femur, the bone in the thigh, whose original length is 0.36 m and whose cross-sectional area is 9.4 cm^2. The mass of the upper body supported by the femur of the two legs is 60 kg. (b) Will the bone shatter from the compression during impact?

TABLE 15.1 Young's Modulus and Ultimate Strengths of Various Materials

| Material | Young's Modulus (N/m^2) | | Ultimate Strength (N/m^2) | |
	Tension	Compression	Tension	Compression
Aluminum	7.1×10^{10}	7×10^{10}	2.0×10^8	3.5×10^8
Steel	20×10^{10}	20×10^{10}	5.2×10^8	
Brick	2×10^{10}	2×10^{10}	0.4×10^8	
Glass	7×10^{10}	7×10^{10}	0.7×10^8	3.5×10^8
Bone	1.6×10^{10}	0.9×10^{10}	1.2×10^8	1.7×10^8
Oak	1.4×10^{10}	10^{10}		10^8
Pine	0.9×10^{10}			
Silk	6×10^9			
Spider's thread	3×10^9			
Catgut	3×10^9			
Hair	10^{10}		10^8	
Tendon	10^9			
Rubber	10^6			
Tooth (human crown)				1.5×10^8

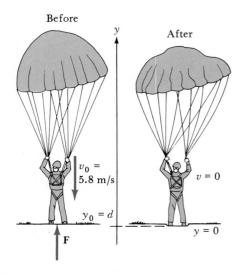

FIG. 15.8. The force of the ground on the parachutist's feet causes a stress in the bones of the legs.

SOLUTION (a) We must first estimate the average force exerted by the upper body on the femur while landing. We can use Newton's second law (used to work Problems 28 and 29 in Chapter 3), the impulse-momentum technique (see Example 4.4), or the work-energy approach (see Example 8.9). Using the latter technique for the situation shown in Fig. 15.8, we find that

$$\frac{1}{2}mv_0^2 + mgd + Fd\cos 180° = \frac{1}{2}m0^2 + mg0,$$

or

$$F = \frac{\frac{1}{2}mv_0^2 + mgd}{d}$$

$$= \frac{\frac{1}{2}(60\text{ kg})(5.8\text{ m/s})^2 + (60\text{ kg})(9.8\text{ m/s}^2)(0.032\text{ m})}{0.032\text{ m}}$$

$$= 32{,}000\text{ N}.$$

The average force on each femur while the parachutist stops is one-half this value, or 16,000 N.

We can now rearrange Eq. (15.5) to determine the change in length ΔL of the bone when subjected to this force. We are given that $L = 0.36$ m, $A = 9.4$ cm^2 $= 9.4 \times 10^{-4}$ m^2, and $Y = 0.9 \times 10^{10}$ N/m^2 (from Table 15.1). Thus,

$$\Delta L = \frac{FL}{AY} = \frac{(16{,}000\text{ N})(0.36\text{ m})}{(9.4 \times 10^{-4}\text{ m}^2)(0.9 \times 10^{10}\text{ N/m}^2)} = 7 \times 10^{-4}\text{ m} = \underline{0.7\text{ mm}}.$$

The bone's length changes by about 0.7 mm while the parachutist is being stopped.

(b) We can decide if the bone is likely to shatter due to compression by determining if the stress on the bone exceeds the ultimate compressional stress given in Table 15.1. For the parachutist in this example, the stress on each femur while landing is

$$\text{Stress} = \frac{F}{A} = \frac{16{,}000\text{ N}}{9.4 \times 10^{-4}\text{ m}^2} = 1.7 \times 10^7\text{ N/m}^2.$$

According to Table 15.1, a bone will break if the compressional stress exceeds 1.7×10^8 N/m^2 (the bone's ultimate strength). Our parachutist will probably not break a femur due to its compression. There is a chance, however, that the knob at the top of the femur that is part of the hip joint might break due to a shear stress, discussed next.

Parachutists are taught to land on their toes with their knees flexed and to fall to the side. By spreading the velocity change over a longer distance, the force and stress on bones are reduced considerably. ∎

15.5 Shear Deformation

A second type of solid deformation is caused by a **shear force.** If you place your hand on the top cover of a book and gently push it to the side, your hand generates a shear stress (Fig. 15.9). The resulting distortion of the book as the pages slide across each other is a shear strain. The **shear stress** is defined as

$$\text{Shear stress} = \frac{F_\parallel}{A},$$

where $F_\parallel$ is the force exerted parallel to a given cross-sectional area. In this example, the area is that of the book's top surface.

This choice of area may seem strange, since the force of your hand is not spread over the whole top cover of the book. The cover, however, is held together as a unit by the forces between its atoms, and the amount of displacement caused by the force depends on this area. We may think of adjacent pages in the book as being bonded loosely to each other by frictional and molecular forces. The strength of the bonding between adjacent pages increases as the area increases. It is harder to make adjacent layers of a solid "slide" across each other when the layers have large cross-sectional areas. Our definition of stress, then, accounts not only for the force but also for this effect of area on the shearing of the object exposed to the force.

The result of the stress is a **shear strain,** defined as the distance Δs that the top surface is displaced in the direction of the force divided by the thickness L of the book or whatever is undergoing a shear strain (see Fig. 15.9c):

$$\text{Shear strain} = \frac{\Delta s}{L}.$$

Note carefully the symbols and corresponding geometrical dimensions used in this definition compared with those used to define tensile and compressive strain.

Shear strain is independent of the dimensions of the object undergoing the strain. Imagine two books equal in every way, except that one is twice as thick as the other. If an equal shear stress is applied to each book (Fig. 15.10), the thicker book will be displaced twice as far as the thinner book. But since shear strain is defined as $\Delta s/L$, we find that the strain is the same for both books ($\Delta s/L = 2\,\Delta s/2L$). Shear strain, then, is an indicator of the displacement per unit thickness.

If the object being sheared is elastic, there is a proportionality between the shear strain and the shear stress. The proportionality constant that relates the shear stress and strain is called the **shear modulus** and is given the symbol S. Thus, for shearing forces of elastic materials,

$$\text{Shear stress} = (\text{Shear modulus})(\text{Shear strain}),$$

or

$$\frac{F_\parallel}{A} = S\left(\frac{\Delta s}{L}\right). \tag{15.6}$$

The shear modulus is roughly one-third the value of Young's modulus for similar materials. Some representative values of shear modulus are listed in Table 15.2.

EXAMPLE 15.2 According to the geological theory of plate tectonics, the surface of the earth consists of 13 to 15 rigid plates that move slowly past each other on the mantle of the earth. Earthquakes can occur when the edges of the plates, originally stuck together, slip, causing a cleavage in the earth's surface. Let us estimate the size of the shear force that results in an earthquake.

(a)

(b)

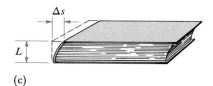

(c)

FIG. 15.9. (a) A book with area A. (b) A force parallel to the cover causes a stress $F_\parallel/A$. (c) The strain caused by this stress is the ratio of Δs and L. Note carefully the meaning of the symbols!

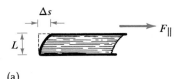

(a)

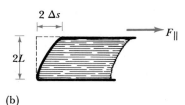

(b)

FIG. 15.10. The shear strain is the same for both books since $\Delta s/L = 2\,\Delta s/2L$.

TABLE 15.2 The Shear Modulus of Various Materials*

Material	Shear Modulus (N/m^2)
Aluminum	3×10^{10}
Steel	8×10^{10}
Glass	$(0.8–2.3) \times 10^{10}$
Bone	$\sim 10^{10}$
Rubber (soft, vulcanized)	$(0.8–1.6) \times 10^6$
Quartz	$(0.8–3) \times 10^{10}$
Wood	$(0.8–1) \times 10^{10}$

*The value of the shear modulus can vary somewhat between different samples of the same material.

SOLUTION Two plates, as seen from above, are pictured in Fig. 15.11a. They move slowly past each other in opposite directions. Friction forces prevent the sides of the plates that are touching from moving. As each plate moves forward (Fig. 15.11b), a shearing strain is caused by the force of each plate holding the other's edge in place.

Let us estimate roughly the magnitude of the shear force of one plate on the other. We use the following numbers, which represent, at best, the order of magnitude of quantities appearing in Eq. (15.6).

Area: A fault may be several hundred kilometers long (we will use a distance of 100 km) and extends perhaps 10 km into the ground. The area of contact between the two places is roughly

$$A \cong (1 \times 10^5 \text{ m})(1 \times 10^4 \text{ m}) = 10^9 \text{ m}^2.$$

Length: The plate is sheared for perhaps 10 km on each side of the edge between the two plates: $L \cong 10^4$ m.

Displacement Δs: The plates move at a speed of about 4 cm/yr. Since major earthquakes occur roughly every 50 to 100 years, we estimate that the unimpeded part of each plate away from the fault has moved a distance of about (75 years)(0.04 m/yr) $\cong$ 3 m.

Shear modulus: A representative value of the shear modulus for rock is 3×10^{10} N/m².

Solving Eq. (15.6) for shear force $F_\parallel$, we find that

$$F_\parallel = \frac{AS\,\Delta s}{L} = \frac{(10^9 \text{ m}^2)(3 \times 10^{10} \text{ N/m}^2)(3 \text{ m})}{10^4 \text{ m}} = \underline{0.9 \times 10^{16} \text{ N.}}*$$

When this force becomes too great, the rocks will slip at their boundary, causing an earthquake. After the slip, the plates start shearing again, as depicted in Fig.

*This is about equal to the weight of the top 3 m of land in the whole state of New York.

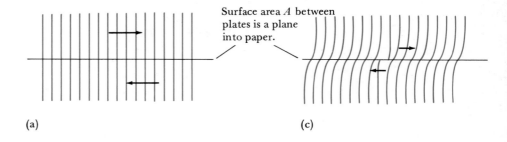

(a)

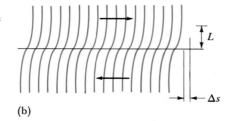

(b)

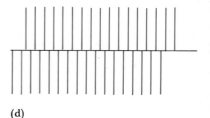

(c)

(d)

FIG. 15.11. (a) Two plates on the earth's crust moving in opposite directions are held together at a surface of contact (the center line). (b) Strain develops as the parts of the plates away from this contact area move in opposite directions. (c) and (d) When the strain becomes too great, the plates slip, causing a cleavage in the earth at the interface between the two plates.

15.11b. Another earthquake can be expected at some future time when the force again becomes too great to restrain the rocks.

The strain that occurs in a plate stores a great amount of elastic potential energy—like the energy of a stretched or compressed spring. This energy is released when the plates slip. A large earthquake can release about 10,000 times more energy than that released by the first atomic bomb! Such great releases of energy are avoided if the faults slip more frequently. Much less energy is released per earthquake in a larger number of small earthquakes. ■

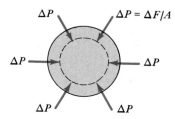

FIG. 15.12. The change in volume of a solid is caused by a uniform change in pressure ΔP (equal to the change in force ΔF per area A).

15.6 Volume Deformation

A volume deformation is caused by a uniform change in the pressure exerted perpendicular to all surfaces of a solid. The **volume stress** is defined as

$$\text{Volume stress} = \frac{\Delta F}{A} = \Delta P,$$

where ΔF is the magnitude of the increased or decreased force acting perpendicular to a portion of the surface of area A (Fig. 15.12) and ΔP is the corresponding pressure change.

The **volume strain** is defined as the change in volume ΔV of the solid divided by the original volume:

$$\text{Volume strain} = \frac{\Delta V}{V}.$$

If a solid is elastic, then the volume stress increases in proportion to the volume strain. The proportionality constant relating volume stress and strain is called the **bulk modulus** B. For elastic materials, the volume stress and strain are related by the equation

$$\text{Volume stress} = -(\text{Bulk modulus})(\text{Volume strain})$$

or

$$\Delta P = -B\left(\frac{\Delta V}{V}\right). \tag{15.7}$$

The negative sign is included in the definition because an increase in the pressure acting on the solid causes a decrease in volume; that is, ΔV is negative. With the use of this negative sign convention, the bulk modulus is then a positive number.

The inverse of the bulk modulus is called **compressibility**:

$$\text{Compressibility} = \frac{1}{B} = -\frac{1}{V}\frac{\Delta V}{\Delta P} = -\frac{1}{\Delta P}\frac{\Delta V}{V}. \tag{15.8}$$

The compressibility is the fractional change in volume $\Delta V/V$ caused by a pressure change, ΔP. The easier it is to compress a material, the greater is its compressibility. Representative values of the bulk modulus and compressibility of various solids and liquids are listed in Table 15.3.

EXAMPLE 15.3 Calculate the density of water near the bottom of the ocean where the pressure is about 500 atm.

TABLE 15.3 The Bulk Modulus and Compressibility of Various Materials

Material	Bulk Modulus (N/m^2)	Compressibility (m^2/N)
Aluminum	7.7×10^{10}	0.13×10^{-10}
Copper	14×10^{10}	0.07×10^{-10}
Steel	16×10^{10}	0.06×10^{-10}
Marble	7×10^{10}	0.14×10^{-10}
Granite	4.7×10^{10}	0.21×10^{-10}
Mercury	27×10^{9}	0.37×10^{-10}
Water	2.0×10^{9}	4.9×10^{-10}
Ethyl alcohol	0.9×10^{9}	11×10^{-10}

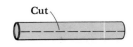

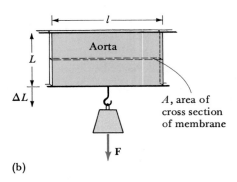

FIG. 15.13. (a) A cylindrical vessel is cut and (b) hung from a support. As weights are added, the membrane stretches.

SOLUTION The density of water at the ocean's surface, where the pressure is 1 atm, is about 1025 kg/m³ (see Table 13.1). Thus, 1 m³ of water at the surface has a mass of 1025 kg. Let us calculate the change in volume ΔV of this water when it is at a pressure of 500 atm. According to Eq. (15.7),

$$\Delta V = -\frac{(\Delta P)V}{B}.$$

The pressure change is (500 atm − 1 atm) = 499 atm, which is equivalent to 499×10^5 N/m². Thus the volume change is

$$\Delta V = -\frac{(499 \times 10^5 \text{ N/m}^2)(1 \text{ m}^3)}{2.0 \times 10^9 \text{ N/m}^2} = -0.025 \text{ m}^3.$$

The density of the water is then

$$\rho = \frac{m}{V + \Delta V} = \frac{1025 \text{ kg}}{1.000 \text{ m}^3 - 0.025 \text{ m}^3} = \frac{1025 \text{ kg}}{0.975 \text{ m}^3} = \underline{1051 \text{ kg/m}^3}.$$

The water is compressed only 2 to 3 percent at the ocean bottom where the pressure is several hundred atmospheres. Liquids and solids are not easily compressed. ∎

15.7 Elastic Membranes

One important example of tensile stress and strain is the stretching that occurs in the vessels of the human circulatory system. The volume of the aorta and large arteries increases as blood is pumped into these vessels once each heartbeat. The tension in the walls of the vessels also increases as they are stretched. Let us consider the stress-versus-strain curve for one of these blood vessels.

A vessel of length l is pictured in Fig. 15.13a. In Fig. 15.13b, it has been cut lengthwise along the side and hung from a support. If we hang weights from the bottom, the vessel stretches much as a sheet of rubber would stretch. As more weight is added, the vessel stretches farther.

The stress on the vessel is the force F hanging from its end divided by its cross-sectional area A. The strain caused by this stress is the change in length ΔL of the vessel divided by its original length L. The solid curve in Fig. 15.14 is a

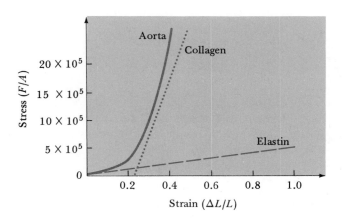

FIG. 15.14. The stress-versus-strain curve of a human aorta and its individual elastic constituents.

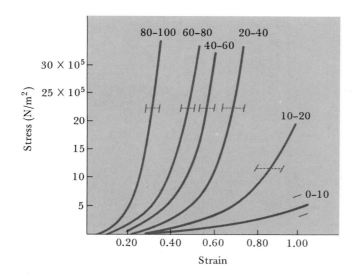

FIG. 15.15. The stress graphs of the human aorta for different age groups. As a person ages, the aorta contains more collagen. The collagen also loses its slack. Both effects cause a greater slope in the curve (a greater Young's modulus). Reproduced by permission of the National Research Council of Canada from *Canadian Journal of Biochemistry and Physiology* **37**, 557–569 (1959).

typical stress-versus-strain curve for a human artery. The varying slope of this curve equals the varying Young's modulus for the vessel.

Two different types of elastic fibers in human arteries are responsible for the elasticity of these vessels. One fiber, elastin, is abundant in arteries of the circulatory system. The elastin is easily stretched—about six times more easily than rubber. The stress-versus-strain curve of an elastin fiber is shown by the dashed line in Fig. 15.14.

Collagen is a second type of fiber found in arteries. Because it is much stiffer than elastin, a small number of collagen fibers can give an artery a high degree of resistance to distension. Usually, the collagen in the walls of an artery is somewhat slack. It does not start stretching until the vessel has already been stretched somewhat beyond its normal size. A stress-versus-strain curve for collagen is shown by the dotted line in Fig. 15.14.

An actual artery is a combination of elastin and collagen. Its stress-versus-strain curve is a composite of the two. The initial portion of the curve is quite flat, indicating an easily extended membrane. As the membrane is stretched farther, more and more collagen fibers lose their slack and contribute to the total membrane tension. The final steep slope is representative of a membrane whose resistance to extension is limited primarily by collagen fibers.

As a person ages, the blood vessels acquire a greater amount of collagen. Also, these collagen fibers lose their slack. Representative stress-versus-strain curves of blood vessels of different age groups are shown in Fig. 15.15. As we see, the vessels become somewhat stiffer with age.

Summary and Additional Readings

1. An **elastic deformation** is one in which an object returns to its original shape when the force causing the deformation is removed. The force causing the deformation is often proportional to the deformation itself. For example, for a spring, the stretching force F is proportional to the spring's displacement x from equilibrium ($F = kx$). The proportionality constant k, called the **spring's force constant,** depends on the stiffness of the spring.

The forces causing deformations of solids are characterized in terms of a quantity called stress. **Stress** is defined as the ratio of the force acting on the solid and the area over which the force is exerted:

$$\text{Stress} = \frac{\text{Force}}{\text{Area}}. \tag{15.2}$$

The deformation of a solid caused by a stress is characterized in terms of a quantity called strain. **Strain** is defined as the change in some dimension of a solid (its length, orientation, or volume) divided by an original dimension of the object:

$$\text{Strain} = \frac{\text{Change in dimension}}{\text{Original dimension}}. \quad (15.3)$$

For an elastic solid, the stress is proportional to the strain. The proportionality constant is called a **modulus of elasticity:**

$$\text{Stress} = (\text{Modulus of elasticity})(\text{Strain}). \quad (15.4)$$

All possible elastic deformations of solids can be analyzed in terms of a combination of three basic types (length, shear, and volume deformations).

2. **Length deformation,** as in Fig. 15.16, caused by compression or tensile forces:

$$\frac{F_\perp}{A} = Y\left(\frac{\Delta L}{L}\right), \quad (15.5)$$

where Y is the Young's modulus.

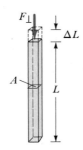

FIG. 15.16

3. **Shear deformation,** as in Fig. 15.17, caused by a shear force:

$$\frac{F_\parallel}{A} = S\left(\frac{\Delta s}{L}\right), \quad (15.6)$$

where S is the shear modulus.

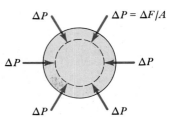

FIG. 15.17

4. **Volume deformation,** as in Fig. 15.18, caused by a change in pressure:

$$\frac{\Delta F}{A} = \Delta P = -B\left(\frac{\Delta V}{V}\right), \quad (15.7)$$

where B is the bulk modulus.

FIG. 15.18

D. M. Boore, "Motion of the Ground in Earthquakes," *Scientific American* **237**, 68 (1977).

Gilbert Wilson, *Introduction to Small-Scale Geological Structures*, George Allen & Unwin, Boston (1982). A qualitative discussion of geological stresses and strains.

Alan C. Burton, *Physiology and Biophysics of the Circulation: An Introductory Text*, 2nd Ed., Year Book Medical Publishing, Chicago (1972). Discusses the stress and strain of elastic membranes in the circulatory system.

Questions

1. Two springs, each of force constant k, hang next to each other from the ceiling. At the bottom, the springs are attached to a handle. Determine the effective force constant of this parallel combination of two springs.

2. A spring with force constant k is cut in half. What is the force constant of one of the halves? Explain.

3. Does the "old-fashioned" rotating telephone dial seem to obey a Hooke's law–type equation in which the force needed to rotate the dial increases as its angular displacement increases? Perform a careful experiment to justify your answer.

4. Why do mountain climbers use ropes that have Young's moduli less than those of most other ropes?

5. Based on your own experimental measurements, *estimate* the Young's modulus of a single strand of a rubber band. Justify any numbers used and explain your procedure carefully.

6. How does the shear stress change (a) when the force acting on a cube of Jell-O parallel to its top surface is doubled, (b) when the force remains constant but each dimension of the Jell-O is doubled? (c) How does the strain change for these two situations?

7. Partially fill a rubber balloon with sand and water so there is more than enough water to cover the sand but not enough to fill the entire balloon. Then tie the top and try squeezing the balloon. Pretty easy at first, isn't it? As you continue to compress the balloon, you'll suddenly find a point where the balloon just refuses to bulge even though you squeeze for all you're worth. What causes this sudden and determined resistance to further squeezing? [See Jearl Walker, *The Flying Circus of Physics*, p 172, John Wiley and Sons, New York (1975).]

Problems

15.1 Elastic Deformations

1. A spring having a force constant of 1.5×10^3 N/m is stretched 0.12 m from its equilibrium position. (a) Calculate the force pulling on the end of the spring. (b) If the force calculated in part (a) is increased by 60 N, how far is the spring now displaced from equilibrium?

2. A force of 20 N causes a spring to be displaced from equilibrium by 2.5 cm. (a) Calculate the force constant of the spring. (b) What is its displacement when a force of 45 N pulls on the spring?

■ 3. A 1.5-kg mass hangs at the end of a spring. When 0.5 kg more is added, the spring is displaced by an extra 7.0 cm. How far was the spring displaced when only 1.5 kg hung at its end?

15.3 Stress and Strain

4. (a) A brick wall 2 m tall, 0.1 m thick, and 20 m long has a total weight of 8.0×10^4 N. Calculate the stress on the foundation under the bottom row of bricks. (b) Oil in a barrel of volume 0.20 m^3 expands by 10 cm^3 when heated. What is the volume strain of the oil?

5. (a) A table with four identical legs, each of cross-sectional area 1.0×10^{-3} m^2, supports a weight of 600 N. Calculate the stress in each leg. (b) A rope 2.4 m long, used to pull a car from a mud hole, has a strain of 0.050. How much is it stretched when pulling the car?

6. (a) A rope 10 m long is stretched 3.0 cm when a person hangs at its end. Calculate the strain in the rope. (b) A rope of radius 0.50 cm supports a weight of 400 N. Calculate the stress in the rope.

7. (a) A violin string is stressed to produce a strain of 2.0×10^{-4}. If its length is increased by 1.4×10^{-2} cm, what is the unstretched length of the string? (b) A wire supporting a 50-N metal ball has a stress of 6.4×10^7 N/m^2. Calculate the cross-sectional area and radius of the wire.

■ 8. Calculate the stress and strain for each of the following. (a) A 400-N weight sits on a brick whose area is 10 cm $\times$ 20 cm and whose thickness is 5 cm. The weight causes the brick's thickness to change by 2.0×10^{-6} m. (b) As a helium balloon rises, the force of the atmosphere on each 1.0 cm^2 of surface decreases by 0.40 N, and the radius of the balloon increases from 30 cm to 32 cm.

15.4 Length Deformation

9. A 2.5-m-long upright beam supports a load of 3000 N. The beam is compressed 0.20 cm by the load. Young's modulus of the beam is 1.0×10^{10} N/m^2. Calculate its cross-sectional area.

10. A tendon originally 7.5 cm long is stretched 0.30 cm. (a) Calculate the strain in the tendon. (b) Calculate the stress causing this strain. (c) If the cross-sectional area of the tendon is 0.30 cm^2, how great is the force causing the stress?

11. Calculate the change in length of a rope used to pull a car as its tension increases from zero to 8000 N. The radius of the rope is 1.0 cm, its original length is 3.0 m, and its Young's modulus is 3.0×10^9 N/m^2.

12. A nylon rope used in mountain climbing stretches 1.2 m when supporting a 75-kg climber. The rope is 40 m long and has a diameter of 0.80 cm. (a) Calculate Young's modulus of the rope. (b) How far will it stretch when supporting a 90-kg climber?

13. When in tune, a steel banjo string has a tension of 1300 N. If the ultimate tensile strength of the string is 6.0×10^8 N/m^2, what must be the area of the string so that it does not break? Allow a factor of five [that is, it should be able to support a tension of $5 \times (1300$ N)].

14. Is it safe for a 52-kg ballerina to support her weight on a single toe of cross-sectional area 2.0 cm^2 if the ultimate strength of the bone is 1.7×10^8 N/m^2? Explain.

■ 15. The ultimate compressive strength of a tooth is about 1.5×10^8 N/m^2. *Estimate* the force a molar can support before it fractures. Justify any numbers used in your calculation.

■ 16. (a) Calculate the stress, strain, and Young's modulus of a wire of cross-sectional area 3.0×10^{-7} m^2 and length 0.60 m that is stretched a distance 1.2×10^{-3} m by a 120-N force. (b) By what factor do the stress and strain change if the same force is applied to a wire of equal Young's modulus and length but whose radius is twice the radius of the wire in part (a)?

■ 17. A wire of length L and cross-sectional area A with Young's modulus Y is stretched a distance ΔL by a force F. (a) If the wire is cut in half, how far is one of the halves stretched by the same force? (b) What force relative to F must be used to stretch one of the halves a distance ΔL? (c) If the same force F pulls the two cut wires placed side by side, how far are they stretched? Be sure to justify your answer carefully for each part.

■ 18. Each leg of a table is made of a different material: aluminum, steel, oak, and pine. If the cross-sectional area of the steel leg is 1.2 cm^2, calculate the area needed for each of the other legs if they are to have equal strains when a large mass is placed at the center of the tabletop. The legs start with equal lengths.

■ ■ 19. A 6400-kg elevator has a maximum acceleration of 2.5 m/s^2. Calculate the minimum-diameter steel cable needed to support the elevator. Allow a safety factor of five, that is, the cable should be able to provide five times the force needed to cause this acceleration.

■ ■ 20. Three steel cables each 0.50 m long and of cross-sectional area 1.2×10^{-6} m^2 support a 240-kg mass in an arrangement such as shown in Fig. 6.18. If $\theta_2 = 53°$ and $\theta_3 = 0°$, calculate the change in length of cable 2 caused by the load.

■ ■ 21. A 60-kg person falling at 7.0 m/s is stopped stiff-legged by sinking 0.010 m on one heel into the ground. (a) Calculate the average force of the ground on the person's heel. (b) If the person's thighbone has an ultimate compressive strength of 1.4×10^8 N/m^2 and a cross-sectional area of 2.4×10^{-4} m^2, will the bone break? Explain.

■ ■ 22. An 80-kg person hangs from a 10-m-long weightless crossbar that is supported by steel cables (Fig. 15.19). The cable on the left has a radius of 1.0 mm and that on the right a radius of 2.0 mm. Calculate the distance x from the left cable that a person should hang so that both cables stretch the same amount.

■ ■ 23. In Chapter 8 we learned that the elastic potential energy of a stretched spring is $PE_s = \frac{1}{2}kx^2$, where x is the displacement of the spring from equilibrium and k is the force constant. Use a derivation similar to that in Section 8.4 to show that the elastic potential energy of a stretched wire is

$$PE_s = \frac{1}{2}\left(\frac{AY}{L}\right)\Delta L^2,$$

where A is the cross-sectional area of the wire, L is its length, Y is

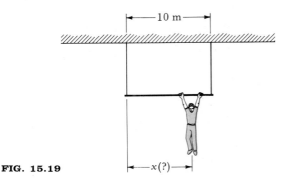

FIG. 15.19

the Young's modulus of the wire, and ΔL is the distance it is stretched.

■■ 24. A 75-kg mountain climber falls at a speed of 10 m/s when the rope attached to her begins to become taut. The climber falls an extra 1.2 m before the rope stops her. (a) Calculate the change in kinetic energy of the climber. (b) Calculate the change in her gravitational potential energy while being stopped by the rope. (c) Based on your answers to parts (a) and (b), what is the change in elastic potential energy of the rope? (d) If the Young's modulus of the rope is 4.5×10^7 N/m^2 and its length is 10 m, what is the cross-sectional area of the rope (see the equation derived in Problem 23)?

15.5 Shear Deformation

25. A wooden block 0.20 m wide, 0.40 m long, and 0.50 m high and with a shear modulus 9.0×10^9 N/m^2 is glued to a floor. A shearing force of 2.5×10^5 N is applied to the top surface. Calculate the distance that the top edge of the block is displaced by the shear force.

26. A block of gelatin of area 4.0 cm × 6.0 cm and 3.0 cm high has a 0.80-N force applied parallel to its top surface. (a) Calculate the displacement of the surface if the shear modulus of the gelatin is 1.0×10^3 N/m^2. (b) Calculate the displacement of a larger block of gelatin whose dimensions are each twice as large as those of the block in part (a). The same force pushes the top surface.

27. A plate of the earth's crust is sheared because one edge cannot move while the rest of the plate moves. Energy released during an earthquake indicates that the force retarding the motion at the edge is 10^{16} N. Following the earthquake, the plate edge slips 1.5 m. The length of the plate perpendicular to the surface that bends as the plate moves is 15 km. *Estimate* the area of the side of the plate that is prevented from moving.

28. The elastic potential energy change of a solid object that is sheared a distance Δs is given by

$$\Delta PE_{\text{el}} = \frac{1}{2}\left(\frac{AS}{L}\right)\Delta s^2$$

(see Sec. 15.5 for the meaning of the symbols). *Estimate* the elastic energy that will be released when the plate discussed in Example 15.2 "slips" during an earthquake.

■ 29. A horizontal cylindrical beam made of wood and having a radius of 0.070 m and a shear modulus of 1.0×10^{10} N/m^2 projects out 0.10 m from the wall of an adobe house in which it is embedded. A 100-kg mass hangs from its end. Estimate the downward shear displacement of the end of the beam.

■ 30. A steel block of bottom area 1.0 m^2 is attached to the floor, and a force is applied parallel to its top surface. Determine the area of an aluminum block of the same height that will experience an equal displacement when the same force is applied parallel to its top surface.

15.6 Volume Deformation

31. Calculate the volume changes of 0.500 m^3 of each of the following types of materials when exposed to a pressure increase of 10 atm: (a) steel, (b) aluminum, (c) water, (d) ethyl alcohol.

32. The pressure on a container of water in the cargo area of an airplane decreases from 1.00×10^5 N/m^2 to 0.85×10^5 N/m^2 as the plane ascends. Calculate the change in volume of the water. The original volume was 1.6 m^3.

33. The oil in a hydraulic press occupies a volume of 0.500 m^3 when the pressure is 2 atm. Calculate the change in volume when the pressure is increased to 12 atm. The bulk modulus of oil is about 1.5×10^{10} N/m^2.

■ 34. A head-shrinker decides to mechanize his operation by applying a high-pressure gas to the heads. Estimate the pressure needed if each head is to be decreased in volume by 30 percent. Assume that the bulk modulus of the head is one-tenth that of water.

■ 35. The pressure on a granite rock 100 km below the earth's surface is approximately 3.7×10^4 atm. Estimate the percent increase in the density of granite at this depth compared to its density at the earth's surface.

■ 36. At what depth below the surface of water will the density of the water increase by 0.1 percent because of the increased pressure? [*Note:* The pressure increases by 1 atm for each 10 m of depth below the surface.]

■ 37. The pressure on two cubes of copper increases from 1.0×10^5 N/m^2 to 20.0×10^5 N/m^2. Each dimension of one cube is twice the respective dimension of the other. Calculate the ratio of the volume change of the larger cube and the smaller cube.

15.7 Elastic Membranes

■ 38. (a) Estimate the Young's modulus of elastin and of collagen in their elastic regions using the stress-versus-strain curves shown in Fig. 15.14. (b) Which type of material is more difficult to stretch and by what factor?

■ 39. Use the equation derived in Problem 23 to estimate the elastic potential energy stored in a section of aorta that is 50 cm long when its radius increases from 1.00 cm to 1.18 cm. The thickness of the aorta wall is about 0.2 cm, and its Young's modulus in the elastic region is about 8×10^5 N/m^2. [*Hint:* The initial length of the membrane being stretched is the circumference of the aorta.]

INTERLUDE III Scaling

When we watch ants pursuing their daily activities, the environment in which they move seems similar to ours, except smaller. For an insect like an ant, a patch of grass must look like a forest, a puddle of water like a lake, and a pebble like a boulder. If our bodies could be reduced to the size of an ant's, would life go on as usual, or would we be unable to adapt to this small world because of some basic physiological problem caused by our new size?

Scaling is used in science to determine the effect of size on function. Can we live normal, active lives if reduced to the size of an ant or if increased to the size of King Kong? Scaling is also used to construct models of larger or smaller objects. For example, a geologist can build a model whose changes in a short time represent structural changes in the earth that take millions of years to occur. Engineers and architects use scaling to make models for testing designs of buildings, airplane wings, power plants, and so forth.

To start our discussion of scaling, look at the two groups of blocks shown in Fig. III.1. The first group consists of two cubes, one sitting on the other. Each cube has sides of length 1 (we omit the units). The two blocks together have dimensions $1 \times 1 \times 2$.

The second group of blocks has the same shape as the first but is three times longer in each direction. Its dimensions are $3 \times 3 \times 6$. The scaling factor of the second group of blocks compared to the first is 3 since each dimension of the second group is three times longer than the similar dimension of the first group.

The **scaling factor** λ is the ratio of a dimension d' of one object and the similar dimension d of another object whose shape is the same as the first:

$$\text{Scaling factor} = \lambda = \frac{d'}{d}. \qquad \textbf{(III.1)}$$

If the objects have similar shapes, any dimension can be used to calculate the scaling factor, just as long as the same dimension is used for each object.

How does the area of similarly shaped objects depend on the scaling factor? Consider the top surface area of the blocks shown in Fig. III.1. For the small set of blocks, the top area A is $1 \times 1 = 1$, whereas for the larger set of blocks the area A' is $3 \times 3 = 9$. The ratio of the two areas, $A'/A = 9/1 = 9$ is the square of the scaling factor. The ratio of the areas of the side surfaces of the blocks also equals the scaling factor squared since, for the side surfaces, $A'/A = (3 \times 6)/(1 \times 2) = 18/2 = 9$.

Areas scale as the square of the scaling factor:

$$\frac{A'}{A} = \lambda^2. \qquad \textbf{(III.2)}$$

This seems reasonable. Each dimension has increased by λ, and area is the product of two dimensions.

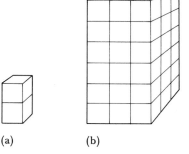

(a)   (b)

FIG. III.1. Each dimension in the large group of blocks is three times longer than the corresponding dimension in the small group.

327

How do the volumes of these two similarly shaped groups of blocks compare? The volume of the small group of blocks is $1 \times 1 \times 2 = 2$, whereas that of the larger groups is $3 \times 3 \times 6 = 54$. The ratio of their volumes, $V'/V = 54/2 = 27$, is the cube of the scaling factor ($3^3 = 27$). This also seems reasonable. Each dimension has increased by λ, and the volume is the product of three dimensions. Thus, the volume of one object should be λ^3 times that of the other object.

Volumes scale as the cube of the scaling factor:

$$\frac{V'}{V} = \lambda^3. \tag{III.3}$$

If two objects have equal and uniform density, then their masses and weights also scale in proportion to the cube of the scaling factor because mass and weight are proportional to volume [recall Eq. (13.1): $m = \rho V$]. If the volume of one object is λ^3 bigger than that of another, its mass is also λ^3 bigger, assuming their densities are equal.

EXAMPLE III.1 During a one-year period, a young girl grows so that each dimension in her body increases by 5 percent. By what percentage does her weight increase, assuming that her density remains unchanged?

SOLUTION One of the dimensions of her body, such as her height, increases in one year from a value d to a value d', where, for a 5 percent increase, $d' = 1.05d$; the scaling factor is 1.05. Her weight increases in proportion to the scaling factor cubed, or

$$\frac{w'}{w} = \lambda^3 = 1.05^3 = 1.16.$$

Thus, her weight increases by <u>16 percent.</u> ∎

EXAMPLE III.2 (King Kong's Achilles tendon) A giant such as the popular movie figure King Kong would have many structural problems if shaped like a man or woman. Suppose the giant is 60 ft tall and stands on the balls of his feet with his heels slightly elevated. Will his Achilles tendon tear? The tension in the Achilles tendon of a 6-ft tall, 180-lb man of similar shape and stance is 310 lb. The man's tendon tears if the tension exceeds 1500 lb.

SOLUTION The giant is 10 times taller than the man; thus, $\lambda = 10$. The giant's mass and weight are $\lambda^3 = 10^3 = 1000$ times greater than that of the man. However, the strength of the giant's muscles and tendons is only λ^2, or 100 times greater than that of the man, since strength depends on the cross-sectional area of muscles, tendons, and bones. Thus, the giant can exert 100 times more force with his muscles and tendons than can the man and can support 100 times more force on his bones. But, unfortunately, he is 1000 times heavier.

In Example 6.7, we found that the tension in the Achilles tendon of a 180-lb man when his heels are slightly elevated is

$$T = \frac{(w/2)(0.14 \text{ m})}{0.04 \text{ m}} = \frac{[(180 \text{ lb})/2](0.14 \text{ m})}{0.04 \text{ m}} = 310 \text{ lb}.$$

For the giant, the tension in the Achilles tendon will be

$$T' = \frac{[(\lambda^3 w)/2](\lambda \, 0.14 \text{ m})}{\lambda \, 0.04 \text{ m}} = \lambda^3(310 \text{ lb}) = 310,000 \text{ lb}.$$

A man's Achilles tendon will rip when the tension exceeds about 1500 lb. Since the giant's tendon is λ^2, or 100 times larger in cross section, it will rip if the tension exceeds $\lambda^2(1500 \text{ lb})$, or 150,000 lb. But to stand with his heel slightly off the ground requires a tension of 310,000 lb. *Merely standing on the ball of his foot will rip the tendon.* If the giant runs or jumps, the tension is even greater. The compression force on his bones and joints could cause fractures when standing. A 60-ft giant shaped like a man would be unable to stand, walk, or run. ■

EXAMPLE III.3 Parachutists have survived impacts in snowbanks even though their parachutes did not open. Their speed at impact was about 50 m/s (120 mi/hr), the terminal speed of a falling skydiver. Will a giant skydiver 10 times the length of an average-sized woman ($\lambda = 10$) or an elf one-tenth the length of an average-sized woman ($\lambda = 0.1$) have a greater terminal speed as they reach the snowbank?

SOLUTION At terminal speed, the weight of the falling person is balanced by the opposing drag force of the air $\mathbf{F}_D$ (see Fig. III.2):

$$\Sigma F_y = w - F_D = 0.$$

Using Eq. (14.14) to substitute for F_D, we find that

$$w - \frac{1}{2}C_D \rho A v_T^2 = 0,$$

or

$$v_T = \sqrt{\frac{2w}{C_D \rho A}},$$

where C_D is the drag coefficient (the same roughly for all three divers), ρ is the density of the air, A is the cross-sectional area of the diver, and v_T is the diver's terminal speed.

The ratio of the terminal speed of a giant or elf and that of a woman with a similar shape is

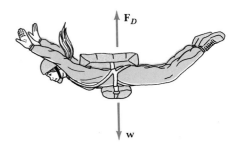

FIG. III.2. Forces on a falling skydiver.

329

$$\frac{v'_T}{v_T} = \frac{\sqrt{2w'/C_D\rho A'}}{\sqrt{2w/C_D\rho A}} = \sqrt{\frac{2w'}{C_D\rho A'}\frac{C_D\rho A}{2w}} = \sqrt{\frac{w'}{w}\frac{A}{A'}} = \sqrt{(\lambda^3)\frac{1}{(\lambda^2)}} = \sqrt{\lambda}.$$

Note that weight scales as the third power of the scaling factor, and area scales as the second power. We see that the ratio of the terminal speeds of similarly shaped falling objects scales as the square root of the scaling factor. Using $v_{T_{\text{woman}}} = 50$ m/s, we find that

$$v'_{T_{\text{giant}}} = \sqrt{10} \times 50 \text{ m/s} = \underline{160 \text{ m/s}} \ (380 \text{ mi/h}),$$

whereas

$$v'_{T_{\text{elf}}} = \sqrt{0.1} \times 50 \text{ m/s} = \underline{16 \text{ m/s}} \ (38 \text{ mi/h}).$$

The elf moves at a much lower speed when arriving at the snowbank. An extension of calculations such as this indicates that the force per unit area needed to stop the giant is over 100 times greater than that needed to stop the elf, if they stop in the same distance. ∎

Gravity is a less important factor for small animals than for large ones. An insect can hang easily from a ceiling, but a giant such as King Kong will break bones and rip tendons while walking because of the huge mass that must be supported by his legs. It appears that gravity limits the size of animals living on the earth. Had life evolved on a less massive planet, animals might have been larger.

Geologists have used scaling to construct models of geological formations. The materials for these models are chosen so that their change with time represents similar structural changes expected in the earth. The time scale can be reduced so that changes in the model occurring in several hours represent similar changes in the earth that may occur in millions of years. To build these models, the scaling of distance, time, and mass must be considered carefully, as must the effect of these choices of scale on other quantities such as density, force, pressure, and flow rate. The following example illustrates the effect of scale on a derived quantity such as flow rate.

EXAMPLE III.4 A reservoir fills with water at a rate of 1000 m³/min. A model of the reservoir is built so that each dimension in the model is 10^{-3} times the similar dimension of the reservoir, and the filling of water in the model takes 10^{-2} less time than filling the actual reservoir. At what rate should water flow into the model?

SOLUTION The scaling factors for distance and for time are:

$$\text{Distance scaling factor} = \lambda = \frac{d_{\text{model}}}{d_{\text{reservoir}}} = 10^{-3},$$

$$\text{Time scaling factor} = \tau = \frac{t_{\text{model}}}{t_{\text{reservoir}}} = 10^{-2},$$

where t_{model} represents the time between events happening in the model (such as the start and completion of filling the model with water) and $t_{\text{reservoir}}$ represents the time for the same events in the actual reservoir.

The flow rate of water into the model Q_{model} is the volume V_{model} of water that flows into the model divided by the time t_{model} required for this flow:

$$Q_{\text{model}} = \frac{V_{\text{model}}}{t_{\text{model}}}.$$

Since volume scales as the cube of the distance scaling factor, we find that

$$Q_{\text{model}} = \frac{V_{\text{model}}}{t_{\text{model}}} = \frac{\lambda^3 V_{\text{reservoir}}}{\tau t_{\text{reservoir}}} = \frac{\lambda^3}{\tau} Q_{\text{reservoir}}$$

$$= \frac{(10^{-3})^3}{(10^{-2})}(1000 \text{ m}^3/\text{min}) = 10^{-4} \text{ m}^3/\text{min} = \underline{100 \text{ cm}^3/\text{min}}. \quad \blacksquare$$

Problems

1. A 55-kg boy is 1.5 m tall. What will be his mass when his height increases 0.2 m, assuming that he retains the same body shape and density?

■ 2. A watermelon 30 cm long costs \$3, and a similarly shaped watermelon 24 cm long costs \$1.80. Which watermelon is the better buy based on price per unit volume? Explain.

■ 3. A 36-kg girl who is 1.40 m tall and her 1.80-m-tall mother have exactly the same shape and density. Each lies on a board of nails for a neighborhood carnival. The boards have an equal number of nails per unit area. (a) If the girl is supported by 70 nails, how many nails support the mother? (b) What is the mother's weight? (c) If the pain experienced depends on the average force per nail, which person experiences the greater pain? Explain.

■ 4. A 70-kg man and an 80-kg man have similar builds. (a) Calculate the scale factor of the heavier man compared with the lighter man. [*Note:* It is not 1.14.] (b) If the smaller man is 1.60 m tall, how tall is the larger man? (c) A person's strength depends primarily on the cross-sectional area of his muscles. If the smaller man can lift a 300-N weight with his forearm, how much weight can the larger man lift with his forearm?

■ 5. A person's strength is proportional to the cross-sectional area of his or her muscles. Why is it reasonable that an ant can lift many times its own weight, while a man or woman cannot?

■ 6. A girl and her mother (heavier than her daughter) each use identical parachutes for a 1-mi jump. Who will reach the ground in the shortest time? Explain.

■ 7. In Example 6.7 we found that a 310-lb tension in the Achilles tendon was needed to lift the heels of a 6-ft, 180-lb person slightly off the ground. This tension is directly proportional to a person's weight. The cross-sectional area of a 180-lb person's Achilles tendon is approximately 0.25 in². (a) What is the tension per unit area in the person's tendon? (b) An Achilles tendon will tear if the tension per area exceeds 6000 lb/in². What is the height of the tallest giant who can lift his heels off the ground (stand on his toes) without tearing a tendon?

■ 8. A model of a dam is built. The model is 0.5×10^{-3} times smaller than the actual dam in every dimension. A time-scale factor of 10^{-2} is chosen for the model. (a) If the water flow rate into the actual dam is 20 m³/s, what should it be for the model? (b) If the model fills in six hours, what time is needed to fill the actual dam?

■ ■ 9. You are to make a geological model of a mountain. The modeling material must be chosen such that its compression strength and other properties are related by scaling as closely as possible to the properties of the actual mountain material (limestone). The differential compression strength of limestone (the unbalanced weight that it will support per unit area without breaking) is approximately 40,000 N/cm². If our model is to be 10^{-3} times the size of the mountain, what should the compression strength of the model material be? Explain. [*Note:* Weak, wet clay is often used.]

PART IV

Vibrations and Waves

Much of our study of physics has involved moving objects: a golf ball flying through the air, a car skidding to a stop, blood rushing through our arteries. In this section we analyze a different type of motion called *wave motion*. A wave consists of a disturbance that moves from one place to another. Examples are the disturbance in the level of water caused by a water wave, the disturbance of the pressure and density of air as a sound wave passes, and the disturbance of the compression and shear strain in the earth as an earthquake passes.

Waves are of utmost importance for the transmission of information. Our senses of sight and hearing depend on light and sound waves. Our ability to think depends on information transferred by nerve impulses, a form of wave motion. X-rays and ultrasound are used in medicine for diagnostic purposes. Geologists use radio waves and sound waves to explore the earth's crust. Microscopes bend light waves so that we see tiny objects such as body cells. Telescopes gather light and other waves from distant galaxies.

Waves are also responsible for energy transfer. Most of the earth's energy comes from the sun via light and infrared waves. A surfer derives his energy of motion from a water wave.

Our analysis of wave motion begins in Chapter 16 with a look at the physics of vibration. This is a natural starting place since waves are produced and detected by vibrating objects. Following this introduction, we take up in Chapter 17 the general properties of all waves, with particular reference to mechanical-type waves such as sound waves and water waves. A special kind of wave important in music (called a standing wave) is considered in Chapter 18.

CHAPTER 16

Vibration

Vibration, also called oscillation, is used to describe the motion of an object that moves back and forth in a repetitive fashion. The world is full of objects that vibrate: banjo strings, electrons in a television antenna, our eardrums, the atoms in molecules, and ducks on a rough sea. Vibrations are not always desirable. Tall buildings must be constructed so that they do not sway excessively, and shock absorbers in your car dampen its up-and-down vibrations while traveling on a bumpy road. At other times much effort is made to produce vibrations. The hills and valleys on a record groove initiate mechanical and electrical vibrations in a stereo system, and when amplified, the electrical vibrations cause the diaphragm of a speaker to vibrate. The sound waves produced by the speaker cause our eardrums to vibrate, and we hear sound. Sound, like most forms of communication, is created and detected by objects that vibrate.

The study of vibrations involves no new laws or principles of physics. Instead, we use Newton's second law, kinematics, and energy principles to analyze vibratory motion. The analysis starts with the study of a single mass vibrating at the end of a spring. The mathematical equations developed for this simple system can be easily extended to describe the motion of other vibrators such as a pendulum, a swinging leg, or a string on a musical instrument. We conclude our study by considering the conditions under which one vibrator can transfer energy to another.

16.1 Vibratory Motion

Objects that vibrate have an **equilibrium position** and are subject to a **restoring force** that tries to return the object to its equilibrium position. Consider a cart that is attached to the end of a spring (Fig. 16.1). The spring shown in part (a) is neither compressed nor extended, and the cart sits at rest at its equilibrium position. If the cart is pulled to the right, as in part (b), the spring exerts a restoring force on the cart toward the left. If the cart is pushed to the left as in part (c), the spring exerts a restoring force toward the right. The spring always exerts a force on the cart toward the equilibrium position.

The restoring force F_r of the spring on the cart is an example of a **Hooke's law-type force** that can be written as

$$F_r = -kx, \qquad (16.1)$$

The energy of a mass swinging at the end of a string fluctuates continuously between gravitational potential energy and kinetic energy. At what positions is the gravitational energy large and the kinetic energy small? At what position is the kinetic energy large and gravitational energy small? (Courtesy of Harold Edgerton, M.I.T.)

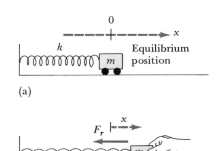

FIG. 16.1. (a) An object is at its equilibrium position. (b) If displaced to the right, the spring exerts a force to the left. (c) If displaced to the left, the spring exerts a restoring force to the right.

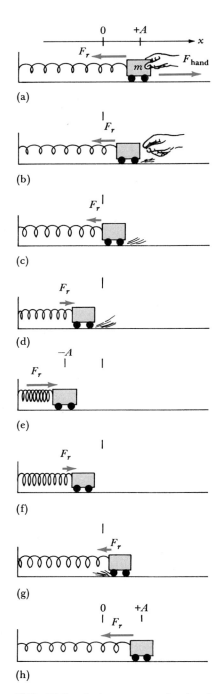

(a)

(b)

(c)

(d)

(e)

(f)

(g)

(h)

FIG. 16.2. A time sequence showing the effect of the restoring force on an oscillator during one cycle of motion. The speed of the cart is at maximum as it passes through the equilibrium position.

where x is the displacement of the cart from equilibrium, and k is the force constant for the spring. The negative sign means that the force points opposite the direction of the displacement. If the displacement is toward the right in the positive x direction, the force is toward the left and has a negative sign, and vice versa.* The **force constant** k is a measure of a spring's stiffness. If k is large, the spring is stiff—hard to compress or stretch. If k is small, the spring is easier to compress and stretch.

Why does a restoring force cause an object to vibrate? Another cart is shown in Fig. 16.2. If we pull the cart to the right as in part (a) and release it as in part (b), the spring's force on the cart causes it to accelerate toward the left. When the cart reaches the equilibrium position, it has momentum and over-shoots. As the cart passes the equilibrium position, the spring's restoring force starts pushing in the opposite direction. The cart decelerates and is finally stopped for an instant on the left (part e) and then begins to move back to the right. It once again overshoots the equilibrium position and finally stops for an instant at its starting position. We say that the cart has completed one **cycle of motion**; it has made one **vibration.** If the friction forces in the system are small, the cart will continue to vibrate back and forth many times. The vibratory motion is called **simple harmonic motion** if the restoring force is proportional to the object's displacement from equilibrium.

This vibrational motion is described using several quantities, defined below.

Amplitude A: The amplitude of a vibration is the magnitude of the maximum displacement of the object from equilibrium. For the cart shown in Fig. 16.2, the amplitude of vibration is the distance that the cart is displaced to the right at the start of its vibration.

Period τ (Greek letter tau): The period of a vibration is the time required for one vibration. For the cart shown in Fig. 16.2, the period is the time needed for the cart to move from position A at the right to position $-A$ at the left and back again to position A.

Frequency f: The frequency of vibration is the number of vibrations (also called cycles) that occur in a given time period. The most commonly used frequency unit is called the **hertz** (Hz), where 1 hertz equals 1 vibration per second:

$$1 \text{ Hz} = 1 \frac{\text{vib}}{\text{s}} = 1 \frac{\text{cycle}}{\text{s}}.$$

The words *vibration* and *cycle* indicate what is being counted but are not really units; these words can be removed from the units of equations involving frequency. Thus, an appropriate unit for frequency is $1/\text{s}$ or s^{-1}.

If an object's frequency of vibration is known, its period can be calculated, and vice versa. Suppose, for example, that the mass shown in Fig. 16.2 has a frequency of 5 vib/s. To complete 5 vibrations in 1 second, each vibration must take one-fifth of a second. Thus, the period is $\frac{1}{5}$ s. If an object's frequency is 10 vib/s, the time needed for each vibration (the period) is $\frac{1}{10}$ s. As we see, the

---

*In Chapters 8 and 15, when discussing springs, we used an equation $F = kx$. The force in that case was the external force needed to displace the cart and spring from equilibrium. In this section we are talking about the restoring force of the spring on the cart. This force opposes an external force and hence has the opposite sign.

frequency and period are inversely related:

$$f = \frac{1}{\tau}. \tag{16.2}$$

EXAMPLE 16.1 The Empire State Building sways back and forth at a vibration frequency of 0.125 Hz. Calculate its period of vibration.

SOLUTION Rearranging Eq. (16.2) to solve for τ, we find that

$$\tau = \frac{1}{f} = \frac{1}{0.125 \text{ Hz}} = \frac{1}{0.125 \text{ vib/s}} = \underline{8 \text{ s/vib.}}$$

Each vibration requires 8 s. You might be interested to know that the amplitude of vibration of the top of the Empire State Building in a high wind is less than 4 cm. ∎

16.2 Energy of Simple Harmonic Motion

When a system vibrates, its energy continually interchanges between kinetic energy and some form of potential energy. Consider the system shown in Fig. 16.3. It includes a vibrating cart attached to a spring that alternately becomes stretched and compressed. Two forms of energy are involved in the vibration of this system: (1) the kinetic energy, $\frac{1}{2}mv^2$, of the moving cart (see Section 8.3) and (2) the elastic potential energy, $\frac{1}{2}kx^2$, of the stretched or compressed spring (see Section 8.5).

Suppose that we start this system vibrating by first stretching the spring so that the cart is a distance $x = A$ from its equilibrium position. We then release the cart (Fig. 16.3a). At the instant of release, the cart is not yet moving; its kinetic energy is zero. However, the spring is stretched. Thus, the initial energy of the system is all in the form of the elastic potential energy of the stretched spring:

$$\text{At } x = A: \quad E_0 = \frac{1}{2}m0^2 + \frac{1}{2}kA^2 = \frac{1}{2}kA^2.$$

As the cart moves left (Fig. 16.3b), its displacement from equilibrium x *decreases,* as does its elastic potential energy. However, the cart's speed and kinetic energy *increase* as the spring pulls it back toward the equilibrium position. The energy of the system is now part kinetic and part elastic:

$$\text{At any } x: \quad E = \frac{1}{2}mv^2 + \frac{1}{2}kx^2. \tag{16.3}$$

As the cart passes the equilibrium position (Fig. 16.3c), its displacement from equilibrium is zero ($x = 0$), as is the elastic potential energy of the spring. But now the cart is moving at its maximum speed v_{max}; the spring has been pulling the cart toward the left since its release. At this point, the system's energy is all in the form of kinetic energy:

$$\text{At } x = 0: \quad E = \frac{1}{2}mv_{max}^2 + \frac{1}{2}k0^2 = \frac{1}{2}mv_{max}^2.$$

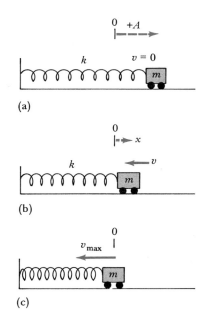

FIG. 16.3. (a) When the cart is displaced a maximum distance from equilibrium, the energy is all elastic energy of the spring. (b) The total energy of a harmonic oscillator is the sum of its kinetic energy and its elastic energy. (c) When the cart is passing through the equilibrium position, the energy is all kinetic energy of the cart.

We now have three expressions for the energy of the system, each appropriate for different locations of the cart. How are these energies related? Consider the work-energy equation, Eq. (8.13):

$$E_0 + W = E.$$

According to this principle, the initial energy of a system plus the energy added to or removed from the system because of work done by forces from outside the system equals the final energy of the system. If we can ignore friction and air-drag forces, then no work is done on the system. Consequently, $E_0 = E$; the system's energy does not change. The energy remains the same in each position. However, the form of the energy shifts in a continuous manner between kinetic energy and elastic potential energy.

If we include friction and air-drag forces, the elastic and kinetic energy will eventually be converted to thermal energy as the amplitude of the system's vibration decreases until it eventually stops.

The energy of most systems that vibrate can be analyzed analogously. There will always be a shifting back and forth of the energy between some form of kinetic energy and some form of potential energy.

EXAMPLE 16.2   A spring with a 1.6×10^4 N/m force constant and a 0.10-kg mass at its end has a total vibrational energy of 3.2 J. (a) Calculate its amplitude of vibration. (b) Calculate its maximum speed. (c) Calculate its speed when displaced 0.010 m from equilibrium. (d) What is its amplitude of vibration if its energy is doubled? Ignore friction forces.

SOLUTION   (a) When at its maximum displacement from equilibrium, the speed of the mass is zero (it has stopped for an instant to change directions), and its displacement from equilibrium is either $+A$ or $-A$. The energy of the system is

$$E = \frac{1}{2}m0^2 + \frac{1}{2}kA^2.$$

Solving for A, we find that

$$A = \sqrt{\frac{2E}{k}} = \sqrt{\frac{2(3.2\ \text{J})}{1.6 \times 10^4\ \text{N/m}}} = \underline{0.020\ \text{m}.}$$

(b) The maximum speed occurs when the mass is passing its equilibrium position (when $x = 0$):

$$E = \frac{1}{2}mv_{\text{max}}^2 + \frac{1}{2}k0^2,$$

or

$$v_{\text{max}} = \sqrt{\frac{2E}{m}} = \sqrt{\frac{2(3.2\ \text{J})}{0.10\ \text{kg}}} = \underline{8.0\ \text{m/s}.}$$

(c) The system's energy when displaced 0.010 m from equilibrium is a combination of kinetic energy and elastic potential energy:

$$E = \frac{1}{2}mv^2 + \frac{1}{2}kx^2,$$

or

$$v^2 = \frac{2E}{m} - \frac{k}{m}x^2 = \frac{2(3.2\,\text{J})}{0.10\,\text{kg}} - \frac{1.6 \times 10^4\,\text{N/m}}{0.10\,\text{kg}}(0.010\,\text{m})^2 = 48\,\text{m}^2/\text{s}^2.$$

Thus,

$$v = \underline{6.9\,\text{m/s}}.$$

(d) When the energy is doubled so that $E' = 6.4\,\text{J}$, the vibrational amplitude of the mass is

$$A' = \sqrt{\frac{2E'}{k}} = \sqrt{\frac{2(6.4\,\text{J})}{1.6 \times 10^4\,\text{N/m}}} = \underline{0.028\,\text{m}}.$$

(Notice that its amplitude of vibration did not double when its energy doubled since E is proportional to A^2, not to A.) ∎

16.3 Frequency of Simple Harmonic Motion

The period of vibration of a simple harmonic oscillator can be estimated using a little reasoning along with the ideas of energy developed in the last section. If the period is known, the frequency can be calculated, since $f = 1/\tau$.

Recall the motion of the cart pictured in Fig. 16.2. It moves a total distance $4A$ during one vibration: $2A$ to the left and $2A$ back to the right.

If the average speed $\bar{v}$ of the mass is known, the time τ needed to move the distance $4A$ is $\tau = 4A/\bar{v}$.* Estimating the average speed of the oscillator to be about half its maximum speed, then

$$\tau = \frac{4A}{\bar{v}} \sim \frac{4A}{(v_{max}/2)} = 8\frac{A}{v_{max}}. \tag{16.4}$$

The expression is approximate, since $\bar{v}$ is not exactly equal to $v_{max}/2$.

We can use our knowledge of the oscillator's energy to rewrite this equation for τ in terms of such quantities as k and m—quantities that are intrinsic properties of the oscillator. Recall that the energy of the oscillator when displaced a maximum distance A from its equilibrium position is

$$E = \frac{1}{2}kA^2.$$

This also equals the kinetic energy of the mass as it passes through its equilibrium position at $x = 0$:

$$E = \frac{1}{2}mv_{max}^2.$$

Since the energy is the same in each position, these two expressions for energy are equal:

$$\frac{1}{2}kA^2 = \frac{1}{2}mv_{max}^2.$$

*Remember that Speed = Distance/Time. Therefore, Time = Distance/Speed, or $\tau = 4A/\bar{v}$.

Rearranging, we find that

$$\frac{A}{v_{\text{max}}} = \sqrt{\frac{m}{k}}.$$

The above can be substituted into Eq. (16.4):

$$\tau \simeq 8\sqrt{\frac{m}{k}}.$$

This is a surprisingly good estimate. An exact calculation using calculus shows that the period of vibration of a mass at the end of a spring is

$$\tau = 2\pi\sqrt{\frac{m}{k}}. \qquad (16.5)$$

Since the frequency of vibration is inversely proportional to the period ($f = 1/\tau$), the **frequency of vibration of a simple harmonic oscillator** is

$$f = \frac{1}{2\pi}\sqrt{\frac{k}{m}}. \qquad (16.6)$$

Notice that the frequency is greater for objects fastened to stiff springs (large values of k) and less for objects with large mass.

EXAMPLE 16.3 A 0.20-kg mass hangs at the end of a rubber band that vibrates at a frequency of 1.2 Hz. Calculate the force constant of the rubber band.

SOLUTION Equation (16.6) can be rearranged to solve for k:

$$k = 4\pi^2 f^2 m = 4\pi^2 (1.2\ \text{s}^{-1})^2 (0.20\ \text{kg}) = 11.4\ \text{kg/s}^2 = \underline{11.4\ \text{N/m}}.* \qquad ■$$

A Mass Hanging at the End of a Spring

The vibrational motion of a mass hanging at the end of a spring is slightly more complicated than its motion when vibrating horizontally on a frictionless surface. We must now include the force of gravity in our analysis.

When no mass hangs at the end of a spring, as in Fig. 16.4a, it has a length L; when a mass is added to the spring, its length increases by ΔL. The equilibrium position of the mass is now a distance $L + \Delta L$ from the spring's support (part b). The forces acting on the mass, shown in part (c), are the upward force of the spring, equal in magnitude to $k\,\Delta L$, and the downward force of gravity, mg. Since the mass is in equilibrium, the sum of the forces acting on the mass add to zero (we call the vertical direction the x axis to conform with our previous equations for springs):

$$\Sigma F_x = k\,\Delta L - mg = 0.$$

*Note that $1\,\dfrac{\text{N}}{\text{m}} = \left(1\ \text{kg}\,\dfrac{\text{m}}{\text{s}^2}\right)\dfrac{1}{\text{m}} = 1\,\dfrac{\text{kg}}{\text{s}^2}$.

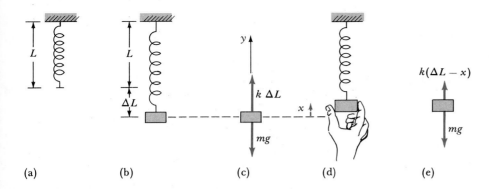

FIG. 16.4. The forces acting on a mass at the end of a spring are shown with the mass at different positions.

(a)   (b)   (c)   (d)   (e)

What happens if we displace the mass a small distance above this new equilibrium position and then release it? Let us choose the x axis pointing up, with $x = 0$ at the equilibrium position indicated by the horizontal dashed line in parts (b) and (c). If we displace the mass a distance x above this equilibrium position (part d) and then let go (part e), the mass will accelerate downward because the upward force of the spring on the mass has been reduced from $k\,\Delta L$ to $k(\Delta L - x)$, whereas the downward gravitational force remains unchanged. Newton's second law for motion along the x direction becomes

$$\Sigma F_x = k(\Delta L - x) - mg = ma_x.$$

But since $k\,\Delta L - mg = 0$, we find that

$$-kx = ma_x,$$

which is exactly the same force and acceleration equation we get for a horizontally vibrating mass when displaced a distance x from its equilibrium position. The frequency of vibration, then, must be the same for a hanging mass as for one vibrating on a horizontal surface at the end of the spring:

$$f = \frac{1}{2\pi}\sqrt{\frac{k}{m}}.$$

EXAMPLE 16.4 A simple harmonic oscillator consists of a mass hanging at the end of a spring. At the end of one vibration, when $x = A$, 25 percent of the mass falls off. Calculate the ratios of (a) the energy, (b) the amplitude, and (c) the frequency of the oscillator after the mass falls off compared to before it falls off.

SOLUTION (a) When the vibrating mass is at its maximum displacement from equilibrium ($x = A$), its energy is given by Eq. (16.3), with $v = 0$: $E = 1/2kA^2$. This expression depends not on the mass but only on the force constant of the spring and on the amount by which the spring is stretched. Since neither of these quantities changes when the mass falls off at position $x = A$, the energy is unchanged:

$$E' = E, \quad \text{or} \quad E'/E = \underline{1.0},$$

where E is the original energy and E' is the energy of the oscillating system with less mass.

(b) The amplitude also does not change. Both masses start their vibration from the same maximum displacement. Thus,

$$A' = A, \quad \text{or} \quad A'/A = \underline{1.0}.$$

(c) Since the mass decreases by 25 percent, $m' = 0.75m$. The ratio of the frequency f' of the system with mass m' and the frequency f with the full mass m is determined using Eq. (16.6):

$$\frac{f'}{f} = \frac{\left(\dfrac{1}{2\pi}\right)\sqrt{\dfrac{k}{m'}}}{\left(\dfrac{1}{2\pi}\right)\sqrt{\dfrac{k}{m}}} = \sqrt{\frac{m}{m'}}$$

$$= \sqrt{\frac{m}{0.75m}} = \sqrt{1.33} = \underline{1.15}.$$

With the reduced mass, the frequency is 15 percent greater. ∎

16.4   The Simple Pendulum

A **simple pendulum** consists of a mass hanging at the end of a cord of negligible weight. Pulled to the side and released, the mass swings back and forth in approximate simple harmonic motion.

A pendulum, like all simple harmonic oscillators, has a surprising property. The same time is taken for one swing of small amplitude as for one of larger amplitude. This property was first observed by Galileo, who used his heartbeat to time the period of a pendulum. He realized that a pendulum could be used as a standard of time for a clock. Pendulum clocks are still sold today, more than three hundred years after their invention.

An expression for the frequency at which a pendulum swings can be determined by considering the forces acting on the mass at its end (Fig. 16.5a). The cord attached to the mass exerts a tension force **T** in the direction of the cord and perpendicular to the arc along which the mass moves. The downward weight force mg has two components (Fig. 16.5b). One component of magnitude $mg \cos \theta$ is perpendicular to the arc of motion and is balanced by **T**, the tension force. The other component is tangent to the arc and restores the pendulum toward its equilibrium position. This component serves as a restoring force and is given by the equation

$$F_r = -mg \sin \theta.$$

(The negative sign indicates that the restoring force points opposite the direction of the displacement.)

For small angles, the value of θ, when using radian units, is approximately equal to $\sin \theta$ (see Table 16.1). Notice, for example, that when $\theta = 0.3491$ radians (equal to $20°$), $\sin \theta = 0.3420$. They differ by only 2 percent, even for this moderately large angle.

Because $\theta \simeq \sin \theta$ for small angles, the restoring force can be written as

$$F_r \simeq -mg\theta.$$

The angle θ is related to x, the displacement of the mass from its equilib-

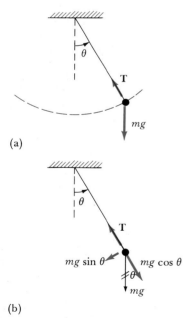

FIG. 16.5. (a) Two forces act on the pendulum ball: the rope-tension force **T** and the ball's weight mg. (b) The component of the weight force tangent to the arc along which the ball travels serves as the restoring force for the ball.

rium position (the heavy line on the arc in Fig. 16.6), by a variation of Eq. (5.1):

$$x = L\theta,$$

where L is the length of the pendulum. Thus, we find that the restoring force is

$$F_r = -mg\theta = -mg\frac{x}{L} = -\frac{mg}{L}x.$$

The preceding equation for restoring force is very similar to that for a mass at the end of a spring ($F_r = -kx$), the only difference being that the force constant k for a spring is replaced by mg/L for a pendulum. The quantity mg/L is called the effective force constant of the pendulum and can be substituted in place of k for all the equations derived to describe a mass vibrating at the end of a spring. Consequently, the frequency of a pendulum is given by Eq. (16.6) with mg/L replacing k:

$$f = \frac{1}{2\pi}\sqrt{\frac{k}{m}} = \frac{1}{2\pi}\sqrt{\frac{mg/L}{m}},$$

or

$$f = \frac{1}{2\pi}\sqrt{\frac{g}{L}}. \qquad (16.7)$$

We see that the frequency of a pendulum depends only on its length and on the acceleration of gravity, not on the mass of the object hanging at its end or on the amplitude of vibration.

Before 1950, Eq. (16.7) was the basis of a technique used by geologists to locate salt domes, which are likely spots for oil. Since the density of these geological structures is less than that of surrounding parts of the earth's crust, the value of g is less above a salt dome than above surrounding parts of the earth, so a pendulum should swing slower when hanging above a salt dome.

EXAMPLE 16.5 Use the pendulum-frequency equation to estimate the number of steps that is "natural" for a leg to take per unit time while walking—in other words, the natural frequency of vibration of the leg.*

SOLUTION The center of mass of the leg is approximately at the knee, and the length L that is used in Eq. (16.7) is about half the actual length of the leg. If a leg is 1 m long, then $L = 0.5$ m and its natural swinging frequency is approximately

$$f = \frac{1}{2\pi}\sqrt{\frac{g}{L}} \simeq \frac{1}{2\pi}\sqrt{\frac{\overset{.3.2}{9.8 \text{ m/s}^2}}{0.5 \text{ m}}} = \underline{0.7 \text{ s}^{-1}}.$$

The period or time for one step is the inverse of f, or

$$\tau = \frac{1}{f} \simeq 1.4 \text{ s.} \qquad 1.5\chi \qquad \blacksquare$$

*The leg is not a simple pendulum but is an example of what is called a **physical pendulum**—an extended body that swings back and forth about a pivot point. The swinging frequency of a physical pendulum is $f = (\frac{1}{2}\pi)\sqrt{mgL/I}$, where L is the distance from the pivot point to its center of mass, m is its mass, and I is the moment of inertia of the pendulum about its pivot point. Since our calculations in this example are approximate, we will treat the leg as a simple pendulum.

TABLE 16.1 Values of θ and $\sin\theta$ for Small Angles

θ, radians	(degrees)	$\sin\theta$
0	(0)	0
0.0872	(5)	0.0872
0.1745	(10)	0.1736
0.3491	(20)	0.3420
0.5236	(30)	0.5000

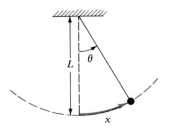

FIG. 16.6. The linear displacement x of a pendulum from its equilibrium position is related to its angular displacement θ by the equation $x = L\theta$, where θ must be in units of radians.

FIG. 16.7. The natural frequency of vibration of your leg is about the same as a comfortable swinging frequency for your leg as you walk.

According to our equation, then, it is natural to take one step with your right leg (or your left leg) each 1.4 s or so. To test this estimate, you can stand on the toes of one foot and swing the free leg back and forth at what seems like a natural frequency (Fig. 16.7). Count the number of swings in 60 s. When you divide this number into 60 s, you will have the period for one swing. It should be about 1.4 s.

If your legs are short, the value of L is less than 0.5 m, and the period is shorter than 1.4 s. Similarly, longer legs have longer periods of vibration. Giraffes take relatively few steps per unit time because of their long legs. A Chihuahua dog, on the other hand, has very short legs and a high "walking frequency."

16.5 Kinematics
of Simple Harmonic Motion

We now turn to the kinematics of simple harmonic motion. How do the position, velocity, and acceleration of a simple harmonic oscillator vary with time? The variation of the position with time can be illustrated with the aid of the experiment shown in Fig. 16.8. A wheel with a peg fastened to its edge rotates next to a mass vibrating at the end of a spring. Light from a projector shines on both objects, and the shadows from the vibrating mass and the rotating peg move up and down on a screen. The mass is caused to vibrate so that the amplitude of its vertical motion equals the radius of the rotating circle. If the rotational motion of the circle is adjusted to make one complete rotation each time the mass completes one vibration, we find that their shadows move up and down on the screen in exactly the same pattern of motion. The vertical position of the peg on the edge of the rotating wheel changes with time in the same fashion as a mass vibrating at the end of a spring.

If we use the coordinate x to represent the vertical position of the peg, we see from Fig. 16.9 that

$$x = R \sin \theta, \tag{16.8}$$

where R is the radius of the circle and θ is the angle between a horizontal line and a radial line drawn from the center of the circle to the peg. If we start our clock running when the peg is at the side of the circle—that is, $\theta = 0$ when $t = 0$—then at a later time t the angular position θ of the peg is $\theta = \omega t$, where ω is the angular velocity of the rotating wheel. [Remember from Eq. (5.4) that $\omega = \Delta\theta/\Delta t = (\theta - \theta_0)/(t - t_0) = (\theta - 0)/(t - 0) = \theta/t$, or $\theta = \omega t$]. Since the wheel completes one rotation through 2π radians in a time of one period τ, we

FIG. 16.8. The shadow of the peg on a screen moves up and down in the same motion as a mass vibrating at the end of a spring.

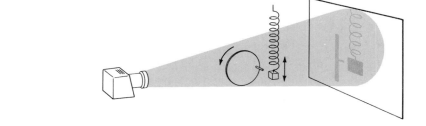

see that $\omega = 2\pi/\tau$. Thus,

$$\theta = \omega t = \frac{2\pi}{\tau}t = 2\pi ft,$$

where we have substituted f for $1/\tau$ in the last part of this equation. The frequency f represents the number of times the peg rotates in a circle each second and also represents the vibrational frequency of the mass attached to the spring. Substituting for θ in Eq. (16.8), we find that the position of the vibrating mass changes with time according to the equation

$$x = A \sin 2\pi \frac{t}{\tau} = A \sin 2\pi ft. \qquad (16.9)$$

The radius of the circle has been replaced by the amplitude of vibration. The position x of the vibrating mass is plotted as a function of time t in the graph shown in Fig. 16.10a. The general pattern is called a sinusoidal-type curve, and an object whose motion can be described using Eq. (16.9) is said to undergo **sinusoidal motion**.

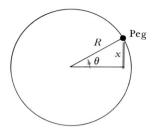

FIG. 16.9. The distance of the peg above the horizontal is $x = R \sin \theta$.

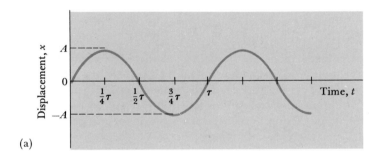

(a)

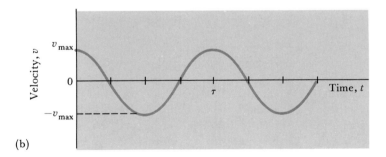

(b)

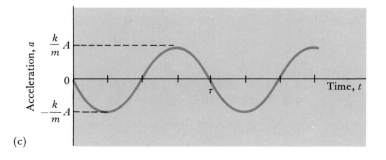

(c)

FIG. 16.10. Graphs showing the (a) displacement, (b) velocity, and (c) acceleration of a simple harmonic oscillator as a function of time.

The velocity of the object at a particular time can be determined by taking the slope of the position-versus-time graph at that time (see Section 2.10). Notice that the slope of the graph in Fig. 16.10a is steepest at $t = 0$ when the object is speeding past the equilibrium position in the positive x direction. At $t = (\frac{1}{4})\tau$, when $x = A$, the slope of the graph is zero; the object has stopped for an instant to change direction at its farthest point from equilibrium. At $t = (\frac{1}{2})\tau$, the slope is again steep, but this time in the negative direction as the object speeds past the equilibrium position in the negative x direction. If we continue this analysis, we find that the velocity is given by a sinusoidal-looking graph such as shown in Fig. 16.10b.

The acceleration is plotted as a function of time in Fig. 16.10c. Note that the acceleration graph looks like the position-versus-time graph but with the opposite sign. According to Newton's second law, $F = -kx = ma$, or $a = -(k/m)x$. The acceleration does in fact have the opposite sign of the displacement. If the mass is displaced in the positive x direction, the spring is pulling back (causing it to accelerate) in the negative direction toward the equilibrium position.

16.6 Damping

In the preceding sections, we have ignored the effect of friction on vibrating objects. Yet friction is very important in real-world vibrations. If friction were not present in the elastic suspension system of a car, for instance, the car would continue vibrating up and down for miles after crossing a bump on a road. Tall buildings rely on friction to reduce their vibrational energy when they sway in the wind.

Friction is present to varying degrees in every object that vibrates or oscillates. The vibrations are said to be **damped** by the friction. A **weakly damped oscillator** has little friction and continues to vibrate for many periods, but the amplitude of each successive vibration is slightly reduced because of friction (Fig. 16.11a).

If the friction force opposing the vibration is increased, the vibrational energy is transferred more quickly to thermal energy. The amplitude of vibration decreases more rapidly, as shown in Fig. 16.11b. In an **overdamped oscillator,** the damping is so great that all of the elastic energy of the stretched oscillator is converted to thermal energy as the oscillator makes its first move toward its equilibrium position (Fig. 16.11c). When overdamped, the oscillator takes a very long time to return to the equilibrium position.

A **critically damped oscillator** is one in which the oscillator returns toward equilibrium in the shortest possible time but does not overshoot the equilibrium position (Fig. 16.11d). The springs and suspension system of a car should be critically damped to avoid continued vibrations when traveling on a bumpy road.

16.7 Energy Transfer Between Oscillators

Much of our contact with the world around us is a result of the transfer of energy between different oscillators. Vibrational energy created in one person's throat is transmitted as sound waves to another person's ears. The emission of

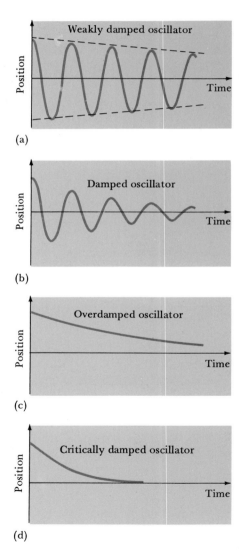

(a)

(b)

(c)

(d)

FIG. 16.11. The oscillators represented in the graphs have varying degrees of damping. With damping, the oscillator energy is converted into thermal energy, thus reducing or even stopping the vibrations.

(a)

(b)

FIG. 16.12. The Tacoma Narrows Bridge is an unusual example of energy transfer between oscillators. A strong wind induced complicated vibrations in the bridge's supporting cables and in the bridge itself. The bridge oscillated for one hour before collapsing. (Wide World)

light and its absorption in our eyes is a result of energy transfer between atomic and molecular oscillators.

A dramatic example of the effects of energy transfer between different types of oscillators is shown in Fig. 16.12. The Tacoma Narrows Bridge near Tacoma, Washington, was opened for traffic on July 1, 1940. It was constructed in a way that allowed several vibrations to occur with little damping. A complicated interaction of the wind with the bridge caused large-amplitude vibrations of the bridge. On November 7, 1940, a strong wind caused a twisting vibration whose amplitude at the edge of the road was over 3 m (Fig. 16.12a). The bridge withstood these twisting, up-and-down vibrations for about one hour before it collapsed (Fig. 16.12b). The Tacoma Narrows Bridge, a vibrator, was excited by absorbing energy from the air blowing past the bridge and its supporting cables.

The ability of one vibrator to transfer energy to another depends on two factors: (1) the frequency of one vibrator relative to the frequency of the other vibrator and (2) the contact or coupling between the vibrators. Consider first how the frequencies of two vibrators affect their ability to exchange energy. For efficient energy transfer from one to the other, they should vibrate at nearly the same frequency. Imagine a person pushing a child on a swing. To push the child higher, the person must push so that positive work is done on the child—that is, the force must be exerted in the direction of the child's displacement, as shown in Fig. 16.13. To transfer additional energy to the child, the next push must occur one period later, when the child is again swinging forward.

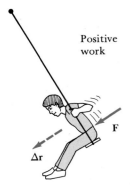

Positive work

F

Δr

FIG. 16.13. The swinger gains energy if positive work is done by the force—that is, if **F** and **Δr** are in the same directions.

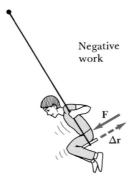

Negative work

F

Δr

FIG. 16.14. The swinger loses energy if negative work is done by the force—that is, if **F** and **Δr** are in opposite directions.

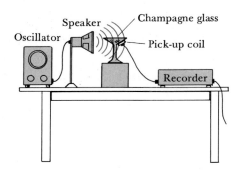

FIG. 16.15. A system to record the transfer of vibrational energy from a speaker to a champagne glass at different frequencies.

If the pushes are at a frequency different from that at which the child swings, the next push may occur as the child swings back, in which case the push opposes the child's motion (Fig. 16.14). The push will reduce the amplitude of vibration, and the child will have less energy. Maximum energy transfer occurs when the pushing is at the same frequency as the swinging.

The importance of having the vibrator frequencies equal for efficient energy transfer to occur is illustrated by another experiment involving a sound system and a champagne glass (Fig. 16.15). Suppose we make the speaker of the sound system vibrate at different frequencies by connecting it to an electrical oscillator. The sound energy leaving the speaker is kept constant as the sound frequency is varied. The relative amplitude of vibration of the glass at each frequency is detected by using a small pick-up coil attached to the glass.

As frequency varies, the amplitude of vibration of the glass also varies, as shown in Fig. 16.16. The graph of vibration amplitude versus frequency is called the **response function** of the glass and indicates the ability of the glass to vibrate at different frequencies. The peak of the curve at frequency f_0 is called the **resonant frequency** of the glass—that is, the frequency at which the glass vibrates with greatest amplitude. For the vibration of the speaker to transmit energy to the glass, the speaker's frequency must overlap with the resonant curve of the glass; the greater the overlap, the more energy is transferred from one to the other.

For example, if the speaker is vibrating at a frequency of 250 Hz and the resonant frequency of the glass is 700 Hz, there is no overlap and no transfer of energy from the speaker to the glass (Fig. 16.17a). As the speaker's frequency is moved closer to that of the glass, the speaker frequency begins to overlap with the glass resonant curve, and energy is transferred to the glass. Maximum energy is transferred when both frequencies are the same, as shown in Fig. 16.17d. Singers have supposedly fractured glassware by inducing large-amplitude vibrations at the resonant frequency of the glass.

The second condition needed to transfer energy from one vibrating object to another is a coupling mechanism through which energy can travel from one oscillator to the other. The stronger the coupling, the greater the amount of energy that is transferred in a certain time. For the speaker and glass, the medium coupling the speaker vibrations to the glass is the air between the speaker and the glass. Pressure variations caused by the speaker travel as sound waves through the air to the glass. The coupling can be reduced by moving the glass farther from the speaker so that the sound from the speaker is diffused more and is less intense at the glass. If the experiment is done in a vacuum, no energy transfer occurs at any frequency because the coupling medium (the air) has been removed.

These conditions for energy transfer affect our ability to communicate information by sound and by light. They also affect a variety of daily occurrences that may or may not have been part of your experience. For example, your car may vibrate when traveling at a certain speed. The car probably has a wheel that is slightly imbalanced. When rotating at a frequency equal to the frequency at which some other part of the car can vibrate, energy is transferred from the imbalanced wheel to the other vibrator. As the car moves slower or faster, the wheel's rotational frequency differs from that of the car part, and little energy transfer occurs. Electric motors on devices such as refrigerators can cause some other part of the refrigerator to vibrate if the armature of the motor rotates at the same frequency at which the other part of the refrigerator vibrates. Your

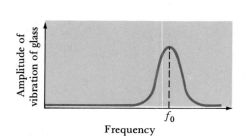

FIG. 16.16. The resonant vibration of a glass occurs at frequency f_0.

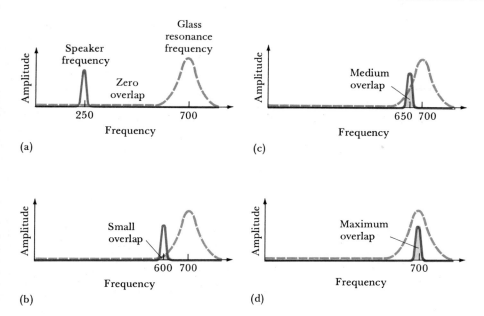

FIG. 16.17. The amount of energy transfer between vibrators depends on the overlap of their frequencies.

arm swings when you walk if the "walking frequency" of your legs equals the "swinging frequency" of your arms. Raising or lowering the arms changes their resonant frequency and may cause them to stop swinging.

Summary and Additional Readings

1. **Restoring force:** Objects that vibrate are acted on by a restoring force that tries to return them to their equilibrium position if they become displaced from it. For simple harmonic motion, the restoring force $\mathbf{F}_r$ is proportional to the object's displacement x from equilibrium. For a mass at the end of a spring,

$$F_r = -kx, \qquad (16.1)$$

where k is the **force constant** of the spring.

2. **Vibration:** The vibrational motion of an object is characterized in terms of three quantities:

The **amplitude** A of the vibration is the magnitude of the maximum displacement from equilibrium of the vibrating object.

The **period** τ is the time required for one complete vibration.

The **frequency** f is the number of vibrations per unit time. Frequency and period are related by the equation

$$f = \frac{1}{\tau}. \qquad (16.2)$$

Frequency is measured most often in units of hertz (Hz):

$$1\ \text{Hz} = 1\ \text{vib/s} = 1\ \text{s}^{-1}.$$

3. The **energy** of a freely vibrating, simple harmonic oscillator that is acted on by negligible friction force is

$$E = \frac{1}{2}mv^2 + \frac{1}{2}kx^2. \qquad (16.3)$$

4. The **frequency of vibration** of different devices depends on intrinsic properties of the devices. For a mass m at the end of a spring with force constant k,

$$f = \frac{1}{2\pi}\sqrt{\frac{k}{m}}. \qquad (16.6)$$

For a pendulum with a mass at the end of a cord of length L, the vibration frequency is

$$f = \frac{1}{2\pi}\sqrt{\frac{g}{L}}, \qquad (16.7)$$

where g is the acceleration of gravity ($9.8\ \text{m/s}^2$).

5. **Damping** occurs when friction converts the vibrational energy of an oscillator to thermal energy, causing the amplitude of vibration to decrease.

6. **Energy transfer** occurs between two vibrators if (1) they can vibrate at approximately the same frequency and (2) there is a coupling medium or mechanism through which energy can transfer from one oscillator to the other.

Peter F. Hinrichsen, "Practical Applications of the Compound Pendulum," *The Physics Teacher,* pp 286–292, May (1981).

R. D. Edge, "Coupled and Forced Oscillations," *The Physics Teacher,* pp 485–488, October (1981).

Questions

1. A banjo string, a pendulum, and water sloshing in a glass all undergo vibratory motion. Draw a force diagram indicating the cause of the restoring force for each type of motion.

2. At what position(s) during simple harmonic motion is an oscillator's speed zero? Where is its acceleration zero? Explain.

3. A known mass placed at the end of a spring causes it to stretch a distance ΔL. Explain how you can calculate the frequency of vibration of this mass if you do not know the value of the spring's force constant k.

4. Describe two different ways to determine the force constant of a rubber band using only a 100-g mass, a stopwatch, and a meter stick.

5. A pendulum clock is running too fast. How should the length be adjusted? Explain.

6. The length of the metal rod on a pendulum clock that sits in an outdoor area is carefully adjusted to give the correct time in the summer. Will it run fast or slow in the winter when the temperature is colder? Explain.

7. A pendulum clock is moved from the Mississippi Delta region to the Rocky Mountains. Will the clock run faster or slower in the Rockies? Why? How can it be adjusted to compensate for this change?

8. Oil is often found in a geological structure called a salt dome. The rock in the salt dome is less dense than surrounding rock and is pushed up so that it "floats" on surrounding rock. Explain how a sensitive pendulum can detect the location of a salt dome.

9. By experiment and by calculation, *estimate* the natural period and frequency of vibration of your arm. Explain carefully the numbers used in each estimate.

10. A pendulum and a mass at the end of a spring are both carefully adjusted to make one vibration each second when on the earth's surface. Will the period of vibration of the pendulum or of the mass be affected if they are placed on the moon? Explain.

11. Will the frequency of vibration of a swing be greater or less when you stand on the swing than when you sit? Explain.

12. *Estimate* the time required for one step with one leg of a Chihuahua and of an elephant. Assume that they walk at their natural frequency.

13. The amplitude of vibration of a swing slowly decreases to zero if you do not pump your legs while swinging. Explain carefully what happens to the vibrational energy.

14. If you walk with your arms hanging down, they often oscillate forward and back at the same frequency as your stepping frequency. However, if your arms are bent at the elbow so that the forearms are horizontal, they often do not oscillate. Why?

Problems

16.1 Vibratory Motion

1. (a) Musicians in an orchestra tune their instruments to what is called "concert A," a frequency of 440 Hz. Calculate the period for one vibration. (b) The atoms in an oxygen molecule complete one vibration in a time of 2.11×10^{-14} s. Calculate the frequency of vibration of O_2.

2. A mass at the end of a spring makes one vibration in 0.20 s. (a) Calculate the frequency of vibration. (b) If the period is doubled by increasing the mass, what is the frequency?

3. (a) The period of the lowest-frequency sound the normal ear can hear is 0.050 s. Calculate its vibrational frequency. (b) The highest-frequency sound heard by a normal ear is about 20,000 Hz. Calculate the time for one vibration.

4. An FM radio station broadcasts at a frequency of 90 MHz. What distance does a radio wave leaving the antenna travel in one vibration if the wave travels at a speed of 3.0×10^8 m/s?

5. A spring stretches 5.0 cm when pulled by a 100-N force. Calculate the force constant of the spring.

■ 6. A spring is stretched by a 100-N force. The displacement of the end of the spring increases 5.0 cm beyond its original stretched position when the force increases 20 percent to 120 N. Calculate the force constant of the spring and its original displacement.

■ 7. *Estimate* the effective force constant of the suspension system of a car. Describe carefully your technique.

■■ 8. Determine the effective force constant of the two-spring system shown in Fig. 16.18 in terms of the force constant for each spring. [*Hint:* The restoring force $F_r = -k_{\text{effective}}x$, where x is the displacement of the mass from equilibrium. Show the relation between $k_{\text{effective}}$ and k.]

16.2 Energy of Simple Harmonic Motion

9. A spring with a force constant of 1200 N/m has a 55-g mass at its end. (a) If the vibrator energy is 6.0 J, what is the amplitude of vibration? (b) What is the maximum speed of the mass? (c) What is the speed when the mass is at a position $x = A/2$?

10. A 15-N force applied to a mass attached to a spring causes the mass to move 0.060 m from its equilibrium position. When the force is removed, the mass undergoes simple harmonic motion. (a) Calculate the force constant of the spring. (b) Calculate the energy of the oscillator.

11. A spring with force constant 2.5×10^4 N/m has a 1.4-kg mass at its end. (a) If its amplitude of vibration is 0.030 m,

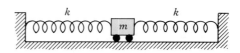

FIG. 16.18

what is the total energy of the oscillator? (b) What is the maximum speed of the mass? (c) If the energy is tripled, what is the new amplitude? (d) What is the maximum speed of the oscillator?

12. When exposed to the weakest sound it can hear, the ear's energy of vibration is about 10^{-19} J. If the force constant of the ear is 20 N/m, what is the amplitude of vibration of the eardrum? How does this compare to the size of an atom ($\sim 5 \times 10^{-10}$ m)?

■ 13. By what factor must we increase the amplitude of vibration of a mass at the end of a spring in order to double its maximum speed during a vibration? Explain.

■ 14. A 0.10-kg mass at the end of a spring of force constant 120 N/m undergoes horizontal vibrations of amplitude 0.080 m. As the mass reaches its maximum amplitude of vibration, a 0.04-kg piece of clay is added to the mass. Calculate the ratio of the maximum speed of the oscillator before the mass is added and that of the oscillator after the mass is added.

■ 15. A simple harmonic oscillator vibrates with an amplitude A. (a) What fraction of the total energy of the oscillator is elastic potential energy and what fraction is kinetic energy when the oscillator is at position $x = A/2$? (b) At what position is the oscillator when its kinetic energy equals its elastic potential energy?

■ 16. A 0.12-kg calculator is to be tested for its ability to withstand vibration and acceleration. It is strapped to a cart of negligible mass at the end of a spring. The vibration amplitude is to be 5.0 cm. (a) Calculate the force constant of the spring if the cart is to experience an initial acceleration of $10g$. (b) Calculate the energy of vibration.

■■ 17. A 5.0-g bullet traveling horizontally at an unknown speed hits and embeds itself in a 0.195-kg block resting on a frictionless table. The block slides into and compresses a spring of force constant 180 N/m a distance of 0.10 m before stopping the block and bullet. Calculate the initial speed of the bullet. [*Hint:* See Section 8.9 concerning inelastic collisions.]

16.3 Frequency of Simple Harmonic Motion

18. A 1.6-kg mass attached to a spring undergoes simple harmonic motion of amplitude 0.12 m and has an energy of 72 J. Calculate the force constant of the spring and the frequency of vibration.

19. The vibrations of a 30-g mass attached to a spring are to serve as the time standard of a clock. Calculate the desired force constant of the spring if the mass makes one vibration each second.

20. A 70-kg gymnast sitting on a trampoline causes it to sag 0.10 m at the center. (a) Calculate an effective force constant for the trampoline. (b) The trampoline is pulled downward an extra 0.05 m by a strap sewed under the center of the trampoline. When the strap is released, what is the energy and frequency of vibration of the trampoline? Ignore the mass of the material that is stretched.

■ 21. A 1.2-kg block sliding at 6.0 m/s on a frictionless surface runs into and sticks to a spring. The spring is compressed 0.10 m before stopping the block and starting its motion back in the opposite direction. Calculate (a) the energy of vibration, (b) the force constant of the spring, and (c) the frequency of vibration.

■ 22. If you double the amplitude of vibration of a mass at the end of a spring, how does this affect the values of τ, k, E, and v_{max}?

■ 23. A simple harmonic oscillator consisting of a spring with a mass at its end vibrates at frequency 6.0 Hz. (a) Calculate the period of vibration. (b) Calculate the frequency if the mass is doubled while the spring's force constant remains unchanged and (c) the frequency if the spring's force constant is doubled while the mass remains the same as in part (a).

■ 24. A 0.10-kg mass vibrating at the end of a spring has an extra mass added to it when its displacement is $x = A$. How large a mass should be added to reduce the frequency to half its initial value?

■ 25. A hydrogen atom of mass 1.67×10^{-27} kg is attached to a very large protein by a bond that behaves much like a spring. (a) If the vibrational frequency of the hydrogen is 1.0×10^{14} Hz, what is the "effective" force constant of this spring-type bond? (b) If the total vibrational energy is kT (k is Boltzmann's constant), approximately what is the classical amplitude of vibration at room temperature? By comparison, the diameter of a hydrogen atom is about 10^{-10} m.

■ 26. A 1.00-kg chair hangs from a spring that is attached at the other end to the ceiling. When a 4.00-kg mass is placed in the chair, the system vibrates with a period of 3.00 s. Determine the mass of a child that vibrates with a period of 4.20 s when placed in the same chair.

■ 27. The Sears Tower in Chicago has a mass of about 5×10^8 kg and sways back and forth at a frequency of about 0.1 Hz. (a) Estimate the force constant for this swaying motion. (b) A gust of wind hitting the building exerts a force of about 4×10^6 N. By approximately how much is the top of the building displaced by a steady force equal in magnitude to that of the wind?

■ 28. A hole is drilled through the center of the earth. The gravitational force on a mass in that hole is mgr/R, where r is the distance of the mass from the earth's center and R is the radius of the earth (6.4×10^6 m). A mass dropped into the hole executes simple harmonic motion. Find the period of the motion. How does one-half this time compare with the time needed to fly in an airplane halfway around the earth?

■■ 29. A sound wave traveling through water must vibrate approximately 1000 times more mass than one traveling through air. Calculate the ratio of the vibration amplitude of sound in air and that in water, assuming equal frequency and energy of vibration.

■■ 30. Low-frequency vibrations (less than 5 Hz) are annoying to humans if the product of the amplitude and the frequency squared (Af^2) equals 0.5×10^{-2} m $\cdot$ s^{-2} (or more). Determine the frequency and amplitude of a carnival vibrator that produces these annoying vibrations if the device vibrates a total mass of 120 kg and has a vibrational energy of 12 J.

■■ 31. Calculate the frequency of vibration of the mass shown in Fig. 16.19.

■■ 32. Derive an expression for the frequency of oscillation of the mass shown in Fig. 16.18.

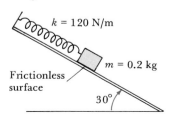

FIG. 16.19

■■ 33. A 0.50-kg wooden cart sits on a horizontal, frictionless surface at the end of a spring that is attached to the wall at the other end. When a 0.050-kg arrow flying through the air penetrates and sticks into the cart, the cart and arrow together compress the spring and start the system vibrating at a frequency of 2.0 Hz with an amplitude of 0.20 m. Calculate the initial speed of the arrow. [*Hint:* See Section 8.9 concerning inelastic collisions.]

16.4 The Pendulum

34. In 1851, the French physicist Jean Foucault hung a large iron ball on a wire about 61 m (200 ft) long to show that the earth rotates. The pendulum appears to swing in different directions as the earth turns under it. Calculate the swinging frequency of this pendulum.

35. You wish to build a clock whose pendulum makes one swing back and forth per second. (a) Calculate the desired length of the rod (assumed to have negligible mass) holding the weight at its end. (b) Will the rod need to be shorter or longer if you include the mass of the rod in your calculations? Explain.

36. Show that the expression for the frequency of a pendulum is dimensionally correct.

■ 37. A geologist finds that the frequency of a pendulum is 0.3204 Hz when at a location where the acceleration of gravity is 9.800 m/s². What is the value of g at a location where the pendulum's frequency is 0.3196 Hz?

■ 38. The frequency of a pendulum is 39 percent less when on the surface of Mars than when on the earth's surface. Use this fact to calculate the acceleration of gravity on Mars.

■ 39. You are designing a pendulum clock whose period can be adjusted by 10 percent by changing the length of the pendulum. By what percent must you be able to change the length to provide this flexibility of period? Explain.

■■ 40. (a) Show that the gravitational potential energy of a simple pendulum of mass m and length L when displaced at an angle θ from the vertical is $PE_g = mgL(1 - \cos\theta)$. (b) Using an expansion of $\cos\theta$ in terms of powers of θ, show that this expression for gravitational potential energy becomes, for small angles, $PE_g \simeq (\frac{1}{2})(mgL)\,\theta^2$.

■■ 41. A 500-kg ball at the end of a 30-m cable suspended from a crane is used to demolish a building. (a) If the ball has an initial angular displacement of 15° from the vertical, what is its gravitational potential energy compared with its gravitational potential energy when hanging straight down? (b) If the ball is released from this position, calculate its speed at the bottom of the arc.

■■ 42. Calculate the maximum speed of a child on a swing 3.0 m long when the amplitude of vibration is 1.2 m.

■■ 43. (a) Determine an expression for the change $\Delta\tau$ in the period of a pendulum caused by a small change Δg in the acceleration of gravity. [*Hint:* The period will be related to the acceleration of gravity by the equation

$$\tau + \Delta\tau = \frac{1}{2\pi}\sqrt{\frac{g + \Delta g}{L}}.$$

To find $\Delta\tau$, expand $(g + \Delta g)^{1/2}$ using the binomial theorem:

$$(g + \Delta g)^{1/2} \simeq g^{1/2} + \frac{1}{2}g^{-1/2}\,\Delta g + \cdots.$$

If Δg is small, you can ignore higher-power terms of Δg.] (b) Use your results to find the fractional change in period $\Delta\tau/\tau$ in terms of $\Delta g/g$. (c) What is $\Delta\tau/\tau$ for a 0.2 percent variation in g that occurs when moving across the surface of a salt dome?

16.5 Kinematics of Simple Harmonic Motion

■ 44. A graph of position versus time for an object undergoing simple harmonic motion is shown in Fig. 16.20. Estimate from the graph the amplitude and period of the motion, and calculate the object's frequency.

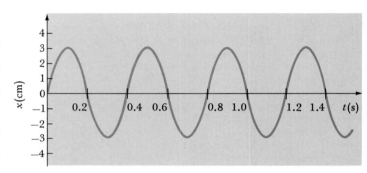

FIG. 16.20

■ 45. A mass at the end of a spring undergoes simple harmonic motion of amplitude 10 cm and frequency 5.0 Hz. (a) Calculate the period of vibration. (b) Write an expression for the spring's position at different times assuming it is at $x = 0$ at $t = 0$. (c) Determine its position at 0.050 s and (d) at 0.100 s.

■ 46. Draw a graph showing the position-versus-time curve of a simple harmonic oscillator (a) with twice the frequency of that shown in Fig. 16.20 and (b) with the same frequency but twice the amplitude as shown in Fig. 16.20.

■ 47. Suppose that at time zero the cart attached to the spring shown in Fig. 16.2 is released from rest at position $x = +A$ and that its period of vibration is 8 s. Draw nine pictures showing approximately the cart's position after successive 1-s time intervals starting at 0 s and ending at 8 s. Draw and label arrows indicating the relative velocity and acceleration of the cart at each position.

CHAPTER 17

Traveling Waves

Vibrations, such as discussed in the last chapter, are essential for the creation, transmission, and detection of waves; and without waves the earth would be a quiet, dull place—no sound, no light, and no nerve impulses transferring information between your brain and other body parts. Nor would there be communication by radio, television, or telephone. Even the ocean would be a less enchanting place.

All these different types of waves have common features, the most important of which is their ability to transfer energy and information. In this chapter we will analyze general principles that apply to all waves and will use these principles to study mechanical waves such as sound waves, waves on stretched strings, shock waves that pass through the earth following an earthquake, and water waves.

17.1 Properties of Waves

The word *wave* brings to mind a picture of water waves moving across the ocean. A water wave consists of repeated disturbances in the level of the water, and these disturbances move one after the other toward the shore. A sound wave is a moving disturbance of the equilibrium pressure and density of matter. If you clap your hands, for instance, air molecules move rapidly from between your hands, causing a compression of nearby air. Following the compression, the molecules return to their original positions, and the density of the air at the source of the disturbance returns to its equilibrium value. However, the compression spreads rapidly at a speed of about 340 m/s to more distant parts of the air. In general, a **wave** can be defined as a disturbance that moves from one place to another.

The sound wave produced by clapping your hands is an example of an **impulsive wave,** as is the shock wave caused by an earthquake. An impulsive water wave occurs when you throw a rock into a pond, as shown in Fig. 17.1. A single disturbance with a few trailing ripples moves at constant speed away from the source of the disturbance.

A **periodic wave** is caused by a repeated disturbance. For instance, the diaphragm of a high-fidelity speaker vibrates back and forth, and the air pressure in front of the diaphragm alternately increases and decreases. This repeated

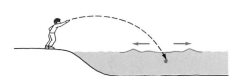

FIG. 17.1. A rock thrown into a pond causes a single impulsive wave to spread in all directions from the splash.

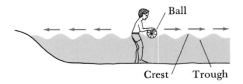

FIG. 17.2. A periodic wave is produced by a repetitive disturbance—a vibration.

or periodic disturbance of air pressure moves away from the speaker in the form of a sound wave. A periodic water wave can be created in a pond by pushing a large beach ball up and down in the water (Fig. 17.2). The disturbances created by each up-and-down motion of the ball move one after the other, at constant speed, away from the source. The peaks of the waves are called **crests,** and the depressed regions between crests are called **troughs** (see Fig. 17.2).

All waves have several features in common. One important property is that *a wave does not transfer matter from one place to another.* A wave is a moving disturbance, but no net displacement of matter occurs. Consider again water waves on a pond. The molecules of water at the surface undergo a circular, vibratory motion about an equilibrium position at the center of the circle (Fig. 17.3). However, the water as a whole does not move from one side of a pond to the other. There is no transfer of matter across the pond.

A second property of waves is that they *transfer energy.* A wave in water passing under a boat is momentarily able to lift the boat—increase its gravitational potential energy. The boat also gains a little upward kinetic energy as the wave passes. The wave could not momentarily increase the boat's energy if it did not carry energy. Waves traveling on the earth's surface after an earthquake can carry enough energy to level a building miles from the quake. Such a wave does indeed transfer energy.

A third important property of waves is their ability to *transmit information.* Two blindfolded children on opposite sides of a swimming pool could communicate by creating and detecting impulsive-type waves with beach balls—like sending messages by Morse code. Much of the information in the human nervous system is transmitted in a similar way—that is, by coded electrical impulses. Periodic waves can also carry information by the frequency and amplitude of the waves. Our ability to recognize different sounds depends on the frequency and amplitude of sound waves.

We conclude, therefore, that waves are traveling disturbances that can transfer energy and information but not matter.

17.2 Types of Waves

Waves are classified in a variety of ways. In the last section we distinguished impulsive waves from periodic waves. Waves can also be categorized according to the type of disturbance that moves through space. A sound wave consists of a fluctuation in the pressure and density of matter. A water wave consists of a variation in the level of water at its surface. A radio wave consists of a fluctuating electric and magnetic disturbance that moves through space, causing electric charges in the path of the wave to vibrate.

Waves are also classified according to the direction in which objects in the path of the wave vibrate relative to the direction the wave travels. A **transverse**

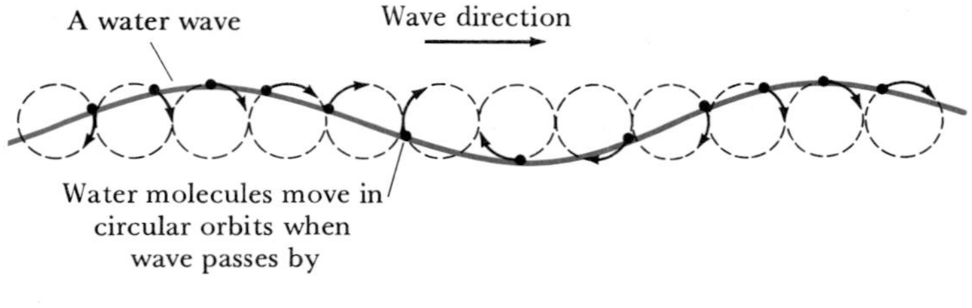

FIG. 17.3. As water waves pass, water molecules at the surface of the water move in a circular path. The water does *not* as a whole move in the direction of travel of the wave.

A water wave   Wave direction

Water molecules move in circular orbits when wave passes by

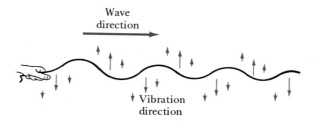

wave causes an object to vibrate perpendicular to the direction the wave travels, whereas a **longitudinal wave** causes the object to vibrate parallel to the direction the wave travels. Consider the transverse wave shown in Fig. 17.4. The fibers of a rope vibrate up and down perpendicular to the direction of travel of the wave. Water waves, light waves, and radio waves are other examples of transverse waves.

For longitudinal waves, the particles in the medium vibrate back and forth in the same direction that the wave travels. A longitudinal wave created on a Slinky is shown in Fig. 17.5. As the person's hand pushes forward and pulls back on one end of the Slinky, the coils in front of the hand are alternately compressed, or crowded together, and then expanded, or made less crowded. This latter situation is called a **rarefaction**—a decrease in the density or crowding of a part of a medium. Each region of compression and rarefaction moves to the right as new compressions and rarefactions are produced by the hand's vibration. Individual coils do not move from one end of the Slinky to the other; only the disturbance of coil spacing moves. The individual coils do, however, vibrate back and forth about their equilibrium position. Since their vibrational motion is parallel to the direction in which the disturbance travels, the wave is called a longitudinal wave. Sound waves are also longitudinal waves.

Longitudinal waves are often pictured by plotting their density or pressure at one instant of time at different positions in space. Two examples are shown in Fig. 17.6. In part (a) the number of Slinky coils per unit length (linear density) at one instant of time is plotted on the vertical axis versus position on the horizontal axis. In part (b) the density of air as a sound wave passes is plotted on the vertical axis as a function of position on the horizontal axis. Notice that the crests of the waves in parts (a) and (b) are regions of greater than normal density while the troughs are regions of reduced density. The graphs in Fig. 17.6 are called **waveforms** and represent the disturbance caused by the wave at different locations at one instant of time. Since the wave moves, an object in the path of the wave feels the changing pressure and density fluctuations as the wave passes. The variation of pressure and density at one position in space at different times caused by the passage of a wave has the same shape as the waveform and is itself called a waveform.

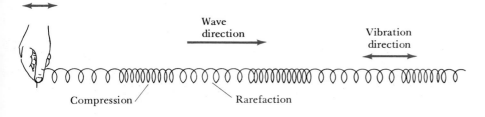

FIG. 17.5. A longitudinal wave is created on a Slinky by vibrating the hand back and forth. The coils vibrate in the same direction that the wave travels.

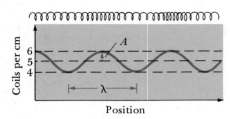

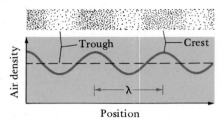

(a)

(b)

FIG. 17.6. (a) The concentration of coils on a Slinky is plotted versus coil position. (b) The density of air as a sound wave passes is plotted versus position in space.

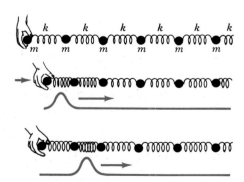

FIG. 17.7. The speed at which a pulse travels along this chain of spheres and springs depends on the force constant of the springs and the mass of the spheres.

17.3 Amplitude, Speed, Frequency, and Wavelength

Periodic waves of all types are described using quantities such as amplitude, speed, frequency, and wavelength.

Amplitude

The **amplitude** of a wave is the magnitude of the maximum disturbance caused by the wave. Consider the longitudinal wave moving along the Slinky shown in Fig. 17.6a. The number of coils in each centimeter length of Slinky is plotted for different positions along the Slinky. When no wave is present, there are 5 coils/cm. This is the equilibrium number of coils per centimeter. As the wave passes, the number fluctuates between 4 coils/cm and 6 coils/cm. The amplitude of the wave equals the maximum disturbance from equilibrium, or 1 coil/cm.

Speed

The **speed** v of a wave is the distance that a disturbance travels per unit time. The speed of a particular type of wave depends on properties of the medium in which the wave travels. Consider an impulse sent along the row of springs and masses shown in Fig. 17.7. The speed of the impulse depends on how stiff the springs are. If they are very stiff, an impulse on one side of a spring is transmitted quickly to the other side. We expect that an impulse should travel faster along a row of masses connected together by stiff springs than along a row of masses connected by easily flexed springs. In general, the speed of a wave in a medium is proportional to the square root of some elastic property of the medium, a property that measures the "stiffness" of the bonds or interactions between neighboring atoms and molecules in the medium:

$$v \propto \sqrt{\text{Elastic property}}.$$

The mass of the spheres shown in Fig. 17.7 also affects the speed of the wave. When the first spring on the left pushes the second mass toward the right, its acceleration depends on how large its mass is. The greater its mass, the less its acceleration and the longer the time needed to compress the next spring. The impulse travels slowly if the mass of the spheres is large and vice versa. In general, the speed of a wave is inversely proportional to the square root of a quantity that depends on the density of mass in a medium—that is, on an inertial property:

$$v \propto \frac{1}{\sqrt{\text{Inertial property}}}.$$

When these two factors that determine a wave's speed are combined, we see that speed depends on an elastic property and on an inertial property of a medium in the following way:

$$v \propto \sqrt{\frac{\text{Elastic property}}{\text{Inertial property}}}. \qquad \textbf{(17.1)}$$

TABLE 17.1   Equations for Calculating the Speed of Different Types of Waves

Type of Wave	Speed of Wave*	Sample Values		
Sound in a gas or liquid	$\sqrt{\dfrac{B}{\rho}}$	Air Water	340 m/s 1500 m/s	**(17.2)**
Compressional or longitudinal wave in solids (P-wave)	$\sqrt{\dfrac{B + \frac{4}{3}S}{\rho}}$	Clay Sandstone Limestone Granite Salt	1000 m/s 2000 m/s 4000 m/s 5000 m/s 6000 m/s	**(17.3)**
Shear or transverse wave in solids (S-wave)	$\sqrt{\dfrac{S}{\rho}}$	About 0.6 times the speed of compressional waves given above		**(17.4)**
Compressional wave in thin rod or bone	$\sqrt{\dfrac{Y}{\rho}}$	Bone	3000 m/s	**(17.5)**
Transverse wave on a string	$\sqrt{\dfrac{T}{\mu}}$	Violin A-string Violin G-string	288 m/s 128 m/s	**(17.6)**

* ρ = density, μ = mass per unit length, B = bulk modulus, S = shear modulus (also called modulus of rigidity), Y = Young's modulus, and T = tension in the string. These different elastic moduli are defined in Chapter 15.

We will not derive the exact equations for the speed of different types of waves. Several equations are listed in Table 17.1 along with representative values of the speed of these waves in different media. Notice, for example, that the speed of sound in a gas or liquid is given by

$$v = \sqrt{\frac{B}{\rho}},$$

where B is the bulk modulus of the gas or liquid (an elastic property) and ρ is its density (an inertial property).

EXAMPLE 17.1   An explosion is detonated at the surface of the ocean over the Japanese trench, the deepest part of the ocean. The impulse travels to the bottom of the trench, is reflected, and returns to a ship at the surface in 14.6 s. Calculate the depth of the ocean trench.

SOLUTION   We first calculate the speed of sound in seawater using Eq. (17.2) in Table 17.1. The density of seawater is about 1030 kg/m³. The bulk modulus of water from Table 15.3 is 2.0×10^9 N/m². Thus, the speed of sound in seawater is

$$v = \sqrt{\frac{B}{\rho}} = \sqrt{\frac{2.0 \times 10^9\,\text{N/m}^2}{1030\,\text{kg/m}^3}} = 1.4 \times 10^3\,\text{m/s}.$$

The sound travels a distance 2 times the ocean depth D (down and back). This depth is related to the speed and time of travel by Eq. (2.2):

$$2D = vt = (1.4 \times 10^3\,\text{m/s})(14.6\,\text{s}) = 2.0 \times 10^4\,\text{m}.$$

Thus the ocean depth at this trench is 1.0×10^4 m, or <u>10 km</u>. ∎

Frequency

The **frequency** f of a wave is the number of back-and-forth or up-and-down vibrations of a part of the medium through which the wave travels. For example, for the wave traveling along the Slinky shown in Fig. 17.6a, the frequency of the wave equals the number of back-and-forth vibrations per unit time of a coil in the Slinky. Usually, the frequency of a periodic wave equals the frequency of vibration of its source.*

Wavelength

The **wavelength** λ (lambda) of a periodic wave is the distance between adjacent troughs, between adjacent crests, or between any two identical parts of adjacent disturbances (see Fig. 17.6). The wavelength depends on the wave's speed and frequency and is not an independent quantity. If the hand pushes the Slinky shown in Fig. 17.5 back and forth at a low frequency, for instance, then a crest created by one compression will travel a relatively long distance before the next compression is created. Thus, the crests of the wave will be separated by large distances (long wavelength). If, on the other hand, the hand vibrates in and out at a high frequency, there will be little distance between adjacent crests. A previous crest can travel only a short distance before another crest is created. The separation of crests (wavelength) is short.

We can, in fact, derive an equation that allows us to calculate the wavelength of a wave if we know its speed and frequency. Recall that the time required for one vibration is called the period of vibration and is given the symbol τ. Note that if we produce one crest by pushing in on a coiled spring, the next crest will be created one vibration or one period τ later. The original crest has traveled a distance of one wavelength. Its speed is, then,

$$v = \frac{\text{Distance}}{\text{Time}} = \frac{\lambda}{\tau}. \qquad (17.7)$$

But the period τ and frequency f are related by Eq. (16.2), $f = 1/\tau$. Substituting for $1/\tau$ in Eq. (17.7), we find that

$$v = f\lambda \qquad (17.8)$$

EXAMPLE 17.2 An oboe creates a sound wave of frequency 440 Hz (concert A) that travels at a speed of 340 m/s. Calculate the wavelength of the sound.

SOLUTION We rearrange Eq. (17.8) to find wavelength:

$$\lambda = v/f = (340 \text{ m/s})/(440 \text{ s}^{-1}) = \underline{0.77 \text{ m}}.$$

The adjacent crests (high-pressure regions) of the wave are separated by 0.77 m as the wave travels from its source. Notice that we substituted the frequency unit s^{-1} in place of the equivalent unit Hz. ∎

---

*We will learn later that the frequency is altered if the source of the waves is moving or if the observer is moving. We ignore these effects of motion until Section 17.8.

17.4   Reflection and Transmission

Until now, we have been considering waves moving through a homogeneous medium, a medium with the same properties everywhere. What happens to a wave when it reaches the end of a medium or when there is an abrupt change from one medium to another? At these boundaries between different media, a wave separates—part is **transmitted** to the new medium and part is **reflected** back to the old. For example, a sound wave traveling in air is partially reflected at a wall and partially transmitted into the wall. Light waves traveling in air when striking the surface of glass are partly reflected and partly transmitted into the glass.*

The amount of reflection at these interfaces and the nature of the reflected wave can provide considerable information about differences of the two media at their interface. Geologists, for instance, have determined that the antarctic ice sheet rests on earth rather than water because radio waves reflected from the bottom of the ice sheet are characteristic of an ice-earth interface rather than an ice-water interface.

To understand why waves are partially reflected at these interfaces between media, consider two ropes woven together, as shown in Fig. 17.8. The rope on the left is thin and has a small mass per unit length, while the rope on the right has a large mass per unit length. By flicking the hand quickly up and down, a wave pulse is started in the left rope and travels toward the interface.

When the pulse reaches the interface, only part of it is transmitted to the large rope. Because of its large mass per unit length, the large rope is harder to accelerate upward than the light rope. Only a small-amplitude upward pulse is initiated in the large rope (Fig. 17.8g). This is called the **transmitted** pulse.

Because the large rope does not "give," it exerts a downward force on the small rope (Fig. 17.8f). The small rope is "jerked" below its equilibrium position by the downward force, starting an inverted, reflected pulse back toward the left (Fig. 17.8g). We say that there has been a 180° **change in phase** of the reflected wave. An upward pulse is converted to a downward pulse.

A change of phase does not always occur at an interface. A pulse traveling along a heavy rope toward an interface with a light rope is depicted in Fig. 17.9. When the pulse reaches the interface, the small rope restrains the large rope less than usual. The large rope overshoots and initiates a large transmitted pulse in the small rope. A small, upright, reflected pulse is also created by this overshoot. In this situation the incident and reflected pulses are both upright. There is no phase change.

The degree to which waves are reflected and transmitted at the boundary between different media can be analyzed using a quantity called impedance. The **impedance** Z of a medium is a measure of the difficulty a wave has in distorting the medium and depends on the elasticity and density of the medium. Impedance can be defined as follows:

$$\text{Impedance} = Z = \sqrt{(\text{Elastic property})(\text{Inertial property})}. \quad (17.9)$$

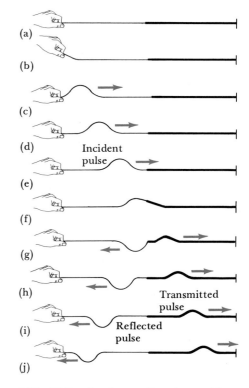

(a)

(b)

(c)

(d)   Incident
pulse

(e)

(f)

(g)

(h)   Transmitted
pulse

(i)   Reflected
pulse

(j)

FIG. 17.8.   A pulse traveling along a thin rope is partially reflected and partially transmitted at the interface with a large rope. The reflected pulse is inverted—there is a 180° change in phase.

*Some absorption of a wave's energy also takes place when it reaches the boundary between two media. In fact, **absorption** can occur at any part of a medium through which the wave travels. The coordinated vibrations of the atoms in the medium are turned into random kinetic energy—that is, thermal energy. The rate of this conversion varies greatly. The slower the conversion, the longer the wave lives.

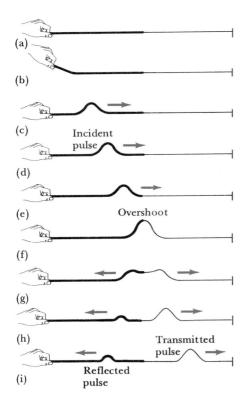

FIG. 17.9. A pulse traveling along a heavy rope is partially reflected and partially transmitted at the interface with a thin rope. The reflected pulse in this case is upright, as was the incident pulse. There is no change in phase.

TABLE 17.2 Impedances to Sound of Different Media

Medium	Impedance (kg/m²·s)
Air	450
Water	1.4×10^6
Clay	2×10^6
Sandstone	5×10^6
Limestone	10×10^6
Salt	12×10^6
Granite	13×10^6
Bone	6×10^6

The preceding expression for impedance seems reasonable. The greater a medium's elasticity, such as having stiff springs, the harder it is to distort the medium. Thus, impedance should be proportional to some power of an elastic constant. Also, the greater the mass or inertia of the medium, the less easily it is distorted and the greater is its impedance.

The impedance of a medium for a particular type of wave is calculated by substituting the same elastic and inertial constants into Eq. (17.9) as were used to calculate the wave's speed in the equations listed in Table 17.1. For example, according to Table 17.1, the speed of sound in water is $v = \sqrt{B/\rho}$, where B is the bulk modulus of the water and ρ is its density. The same two constants are used to calculate the impedance of water for a sound wave ($Z = \sqrt{B\rho}$). The speed of a transverse wave on a string is $v = \sqrt{T/\mu}$ where T is the tension in the string and μ is its mass per unit length. The impedance of the string is $Z = \sqrt{T\mu}$. The approximate values of the impedances of several different media to sound waves are listed in Table 17.2.

Whenever a wave passes between two media with different impedances, part of the wave is transmitted and part of the wave is reflected. If we know the impedances of the two media, it is possible to calculate the amplitudes of the reflected and transmitted waves relative to the amplitude of the incident wave. For example, when a wave traveling in a medium with impedance Z_1 hits an interface with a medium of impedance Z_2, the amplitudes of the reflected wave A_r and transmitted wave A_t compared with that of the incident wave A_i are calculated using the following equations (which we will not derive):

$$A_r = \frac{(Z_1 - Z_2)}{(Z_1 + Z_2)} A_i = R A_i \qquad \textbf{(17.10)}$$

and

$$A_t = \frac{2\sqrt{Z_1 Z_2}}{(Z_1 + Z_2)} A_i = T A_i. \qquad \textbf{(17.11)}$$

The ratio of the reflected and incident wave amplitudes (A_r/A_i) is called the **reflection coefficient** R, and the ratio of the transmitted and incident amplitudes is called the **transmission coefficient** T.

Notice that when $Z_1 = Z_2$, the reflection coefficient is zero. There is no reflected wave. All of the wave is transmitted. We say that there is an **impedance match** at the interface between the two media. Whenever it is important to transfer information and energy by waves, considerable care must be taken in passing the waves between media with similar impedances. If not, much of the energy and information is reflected at their interface.

Notice also that when $Z_2 > Z_1$, the reflection coefficient is a negative number. This corresponds to the situation pictured in Fig. 17.8. The reflected wave in this case is inverted—there is a 180° phase change. A knowledge of the reflection coefficient can be used to indicate the kind of medium reflecting sound waves.

EXAMPLE 17.3 A geologist finds that the reflection coefficient at the interface between the sandstone at the earth's surface and an unknown material below the sandstone is −0.45. The negative sign means that the reflected wave is inverted. What is the impedance of this unknown material below the sandstone?

SOLUTION The impedance Z_1 of sandstone is, according to Table 17.2, about 5×10^6 kg/m²·s. After rearranging Eq. (17.10) to find Z_2, the unknown impedance of the material below the sandstone, we find that

$$Z_2 = \left(\frac{1-R}{1+R}\right)Z_1 = \left(\frac{1-(-0.45)}{1+(-0.45)}\right)5 \times 10^6 \;\; \text{kg/m}^2\cdot\text{s} \qquad (17.12)$$

$$= \underline{13 \times 10^6 \,\text{kg/m}^2\cdot\text{s}}.$$

According to Table 17.2, 13×10^6 kg/m^2·s is approximately the impedance of granite or possibly salt, but not water or the other materials listed in the table. Thus, the limestone probably rests on a layer of granite or possibly on a salt dome, a likely place for gas and oil. We should note that the actual experimental determination of the reflection coefficient is a complex process that we will not discuss here. ■

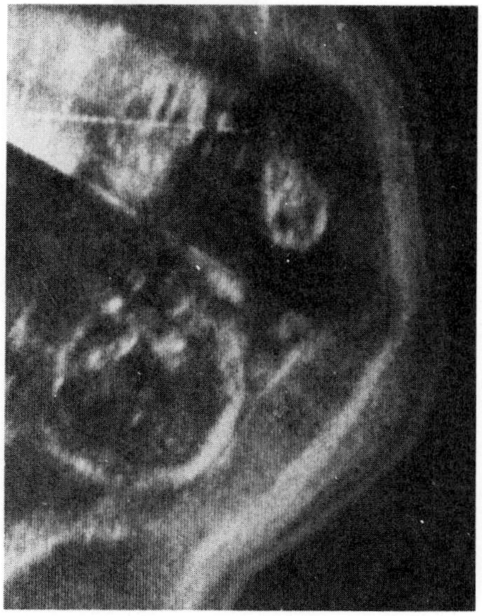

FIG. 17.10. Ultrasound photographs of a 28-week fetus. (Martin M. Rotker/Taurus Photos).

Sound waves are used routinely in medicine to probe different organs of the body. In most instances the reflected waves are used only to determine the location of some interface between two media and not to determine the impedance of the reflecting medium. The sound used for medical imaging is of high frequency, 1–10 MHz,* and is called ultrasound.

Ultrasound scanners can produce a two-dimensional image of internal interfaces in the body by using an array of receiving elements to detect high-frequency sound reflected from these surfaces. The scanners are ideal for detecting and diagnosing tumors, cysts, foreign bodies, and the altered location of interfaces (see Fig. 17.10).

17.5 Principle of Superposition

In the past sections, we have been studying the behavior of a single wave or pulse. A wave characterized by a single frequency and wavelength is often called a **sinusoidal wave.** Such waves are initiated by simple harmonic oscillators. Most periodic or repetitive disturbances of a medium are not sinusoidal waves but are combinations of two or more waves of different frequencies traveling through the same medium at the same time. The sound coming from a violin, for instance, may be a combination of almost 20 sinusoidal waves, each of different frequency and wavelength.

In this and later chapters on waves, we are often concerned with what happens in a medium through which two or more waves simultaneously pass. The waves or pulses are added according to the principle of superposition:

Principle of superposition: When two or more waves pass through the same medium at the same time, the net displacement of any point in the medium is the sum of the displacements that would be caused by each wave if alone in the medium at that time.

An example will enable us to understand this principle more clearly. In Fig. 17.11a, two pulse waves are shown moving toward each other along a string. When the pulses reach the same region of space (Fig. 17.11d), the resultant disturbance is the sum of the disturbances caused by all the waves. Since the pulse moving left causes a negative displacement of the string, it cancels the

*1 MHz = 10^6 Hz, approximately 50 times higher than the highest-frequency sound detectable by our ears (20,000 Hz).

FIG. 17.11. Two wave pulses, one upright, the other inverted, pass at the center of a rope. The two pulses interfere destructively to form a resultant wave smaller than either of the component pulses.

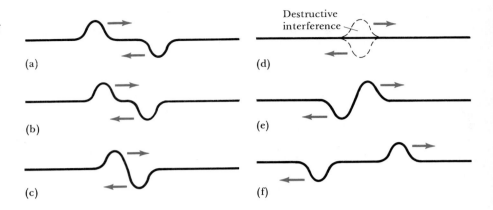

equal-amplitude positive displacement caused by the pulse moving toward the right. The net disturbance is zero (the solid line in Fig. 17.11d). The resultant pulse shown in part (d) is an example of **destructive interference**—the resultant displacement is smaller than the displacement caused by either wave if moving alone through the medium. After passing through the common region in the center of the string, the pulses continue moving in the same direction with the same amplitude as though they had never met (Fig. 17.11f).

If two pulses that cause displacements in the same direction (as shown in Fig. 17.12) pass each other on a string, the resultant displacement when they reach the same part of the string (Fig. 17.12d) is greater than either pulse. This is called **constructive interference**—the resultant displacement has greater amplitude than its components.

17.6 Beat Frequencies

An interesting application of the principle of superposition is the formation of beats. Beats occurs when two sinusoidal waves of slightly different frequency, such as waves 1 and 2 in Fig. 17.13, are combined. The amplitude of the resultant wave alternately increases and decreases, as shown in Fig. 17.13. Each segment of rising and falling amplitude is called a **beat**.

To see why a beat pattern is formed, consider the varying pressure of sound produced at the ears of a person who is equidistant from two different sound

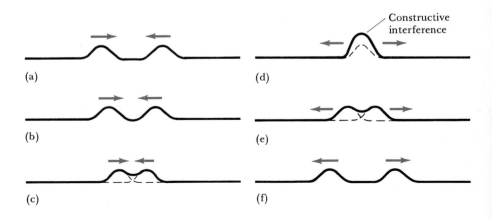

FIG. 17.12. Two upright pulses pass at the center of a rope. The resultant pulse is larger than either component pulse. They are said to interfere constructively.

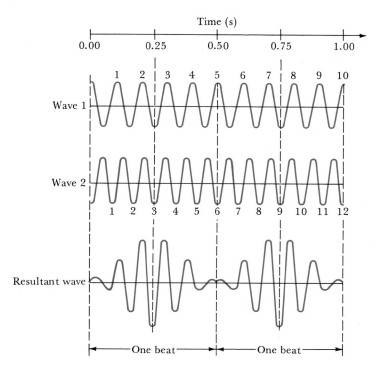

Time (s)

1 2 3 4 5 6 7 8 9 10

Wave 1

Wave 2

1 2 3 4 5 6 7 8 9 10 11 12

Resultant wave

←————One beat————→←————One beat————→

FIG. 17.13. The fluctuating pressure as a function of time for (a) a 10-Hz sound wave and (b) a 12-Hz sound wave. (c) If both waves are present at the same time, the resultant complex wave has a fluctuating amplitude. Each region of rising and falling amplitude is called a beat.

sources, one of which vibrates at frequency 10 Hz and the other at 12 Hz. The fluctuating pressure at each ear as a function of time for each wave and for their resultant is shown in Fig. 17.13. At time $t = 0$, wave 1 causes high air pressure whereas wave 2 causes low pressure. The two pressure variations cancel, and the complex wave has zero amplitude at this time. At $t = 0.25$ s, wave 1 has completed 2.5 vibrations and now causes reduced pressure at the ear. Wave 2, vibrating at higher frequency, has completed 3.0 vibrations and again causes reduced pressure. At this time, the two waves are said to be *in phase,* and the resultant complex wave has a large negative amplitude. At $t = 0.50$ s, the two waves are again out of phase—that is, wave 1 causes increased pressure that is canceled by the decreased pressure of wave 2. The beat pattern, the rise and fall of the amplitude of the resultant wave, is now complete. During the time from $t = 0.50$ s to $t = 1.00$ s, another beat occurs.

One beat pattern occurs every 0.50 s, and the beat frequency is 1 beat/0.50 s or 2 beats/s. This equals the frequency difference of the waves used to construct the complex wave (12 Hz − 10 Hz). In general, the number of beats produced per unit time, called the **beat frequency** f_{beat}, equals the magnitude of the difference of the frequencies of the two waves used to construct the beat pattern:

$$f_{\text{beat}} = |f_1 - f_2|. \tag{17.13}$$

Beats are useful in precise frequency measurements. For example, a piano tuner can easily set the frequency of middle C on a piano to 262 Hz by observing the beat frequency produced when the piano and a 262-Hz tuning fork are sounded simultaneously. If a beat frequency of 2 Hz is observed, then the piano must be vibrating either at 260 Hz or at 264 Hz. The string is then tightened or loosened until the beat frequency is reduced to zero.

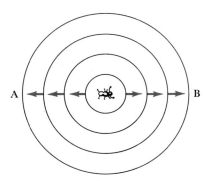

FIG. 17.14. The circular lines represent wave crests on the water caused by a beetle bobbing up and down at constant frequency in the same place.

17.7 Doppler Effect

Until now, we have been studying situations in which the source of a wave and the observer are both stationary. In this case, the frequency of a wave as it leaves its source is the same as the frequency of the wave when detected by an observer. If either the source or observer is moving, though, a change in frequency occurs. Suppose you are standing near a racetrack as fast-moving race cars pass. While a car approaches, a high-pitched, whirring sound is heard, but as the car passes, the frequency or pitch of the sound suddenly and noticeably drops. A similar phenomenon occurs when you listen to the sound from the horn of a passing car or the whistle of a passing train. The change from a higher pitch as the sound source approaches to a lower pitch as it moves away is an example of the **Doppler effect**—any change in the observed frequency of a wave that is caused by the motion of the wave source or the observer.

We can see why the observed frequency changes by examining the waves created by a water beetle bobbing up and down in the water at a constant frequency. If the beetle bobs up and down in the same place, the pattern of wave crests appear as shown in Fig. 17.14. The crests move away from the source at a constant speed. The large circle represents a crest produced by the beetle some time ago, and the small circle is a crest produced more recently. The distance between adjacent crests is one wavelength and is the same toward observer A as toward observer B. The frequency of waves reaching both observers is the same.

Now suppose the bobbing beetle moves to the right at a slower speed than the speed at which the waves move. Each new wave produced by the beetle originates from a point farther to the right (Fig. 17.15). Wave crest 1, the first to be created, is a large circle centered at position 1 where the beetle started bobbing. Wave crest 2 is a smaller circle centered a step to the right at position 2, and so forth.

Wave 1, which started an equal distance from observers A and B, reaches them at the same time. However, wave 4 reaches B somewhat sooner than it reaches A because the wave originated to the right of center, closer to B than to A. Thus, B observes four waves in a shorter time period than does A. In general, an observer detects a higher frequency than that emitted by the source if the source moves toward the observer and a lower frequency if the source moves away from the observer.

To derive an equation for calculating this shift, we first calculate the separation of crests in front of the beetle. Take a look at the separation of wave crest 4 and the wave initiated by the beetle's fifth step (wave crest 5 shown in Fig. 17.16). If the beetle bobs up and down at frequency f, the time between the start of wave 4 and the start of wave 5 is $\tau = 1/f$. During this time, the beetle moves a distance $v_s\tau$ to the right, where v_s is the speed of the "source" of the waves, in this case the beetle. Wave 4 has moved a distance $v\tau$ during that same time, where v is the speed of the wave through the water. We see in Fig. 17.16 that the crests of waves 4 and 5 are separated by a distance λ' (a wavelength) given by

$$\lambda' = v\tau - v_s\tau = (v - v_s)\tau.$$

These waves are incident on observer B at a frequency

$$f' = \frac{\text{Wave speed}}{\text{Wavelength}} = \frac{v}{\lambda'} = \frac{v}{(v - v_s)\tau}.$$

The time τ between beetle bobs is related to the beetle's bobbing frequency by

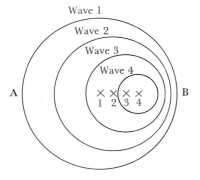

FIG. 17.15. If a beetle moves to the right as it creates new waves, the crests are crowded closer together in front of the beetle and are spread farther apart behind the beetle.

Eq. (16.2): $f = 1/\tau$. Substituting for $1/\tau$ in the preceding equation, we find that

$$f' = f\frac{v}{(v - v_s)}. \tag{17.14}$$

This equation can be used to calculate the frequency f' of the waves detected by the observer when emitted from a moving source at frequency f. The speed of the source v_s is positive if the source moves toward the observer and negative if the source moves away from the observer.

A change in the observed frequency also occurs if the observer moves and the source is stationary. Assume, for instance, that a friend remains stationary in the water of a swimming pool and creates waves by pushing a beach ball up and down. If you swim toward the source, you will encounter the waves more frequently than if you remain stationary. This is the Doppler effect due to a moving observer. If the observer moves toward the source at speed v_o, then the waves appear to move past the observer at a speed of $v + v_o$. Since the wavelength λ of the waves from a stationary source is uniform in all directions, the observer will encounter waves at a frequency

$$f' = \frac{\text{Speed of observer relative to waves}}{\text{Wavelength}} = \frac{v + v_o}{\lambda}.$$

But for the source, $\lambda = v/f$. Thus,

$$f' = f\left(\frac{v + v_o}{v}\right). \tag{17.15}$$

This equation can be used to calculate the frequency f' detected by an observer moving at speed v_o toward (v_o positive) or away from (v_o negative) a stationary source emitting waves at frequency f.

The general equation for the **Doppler effect** for sound waves combines Eqs. (17.14) and (17.15). The frequency f' of waves detected by the observer is related to the frequency f of waves generated by the source by the following equation:

$$f' = f\left(\frac{v + v_o}{v - v_s}\right). \tag{17.16}$$

In this equation, v is the speed of the wave through the medium in which it travels; v_o is the speed of the observer relative to the medium and is positive ($+$) if the observer moves toward the source and negative ($-$) if the observer moves away from the source; v_s is the speed of the source relative to the medium and is positive ($+$) if the source moves toward the observer and negative ($-$) if the source moves away from the observer. *

EXAMPLE 17.4 A train traveling at 40 m/s has a horn that vibrates at a frequency of 200 Hz. Calculate the frequency of the horn's sound heard by a bicycle rider traveling at 10 m/s in the same direction as the train when the cyclist is (a) ahead of the train and (b) behind the train after it has passed.

*This equation applies only to motion of the source and observer along the direction of a line joining them.

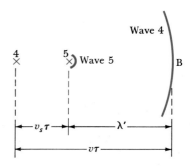

FIG. 17.16. A source emits a wave from position 4. During the period of time τ before the next wave is emitted, the source moves a distance $v_s\tau$ to the right. Wave 4 has moved a distance $v\tau$ from position 4. The separation λ' of wave 4 and the new wave 5 is $\lambda' = v\tau - v_s\tau$.

(a)

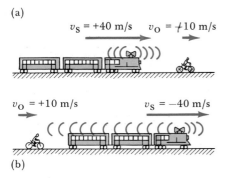

(b)

FIG. 17.17. (a) The train's velocity is positive because it is moving toward the observer, whereas the observer's velocity is negative because it moves away from the source. (b) The signs are reversed after the train passes the bike. The train now moves away from the observer and has a negative velocity. The bike rider has a positive velocity while moving toward the source.

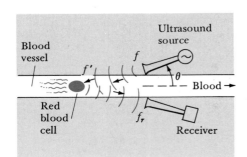

FIG. 17.18. Sound waves leave a source, are reflected from a moving red blood cell, and are detected by a receiver. The frequency shift of the detected waves indicates the speed of the blood.

SOLUTION (a) The observer's velocity $v_o = -10$ m/s (v_o is negative because the observer is moving *away* from the train—see Fig. 17.17a). The train's velocity $v_s = +40$ m/s (v_s is positive because the train is moving *toward* the observer). Thus, the frequency heard by the observer, the cyclist, is

$$f' = f\left(\frac{v + v_o}{v - v_s}\right) = 200 \text{ Hz}\left[\frac{340 \text{ m/s} + (+10 \text{ m/s})}{340 \text{ m/s} - 40 \text{ m/s}}\right] = \underline{220 \text{ Hz}}.$$

(b) After the train passes, the bicycle rider's velocity $v_o = +10$ m/s and the train's velocity $v_s = -40$ m/s (see Fig. 17.17b). Thus,

$$f' = 200 \text{ Hz}\left[\frac{340 \text{ m/s} + 10 \text{ m/s}}{340 \text{ m/s} - (-40 \text{ m/s})}\right] = \underline{184 \text{ Hz}}.$$

The frequency of sound heard by the cyclist decreases from 220 Hz to 184 Hz as the train passes. ∎

The Doppler effect is the basis of a technique used to measure the speed of flow of blood. High-frequency sound waves called ultrasound are directed into an artery, as shown in Fig. 17.18. The waves are reflected by red blood cells back to a receiver. The frequency detected at the receiver f_r relative to that emitted by the source f indicates the cell's speed and the speed of the blood. (A similar arrangement is used to measure the speed of cars, but microwaves are used instead of ultrasound).

To derive an equation for the frequency shift, we consider the process in two parts: (1) The waves leave the source at frequency f and strike the cell at frequency f', and (2) the waves reflected from the cell at frequency f' return to the receiver where they are detected at frequency f_r. In all of our analysis, the velocities we use must be velocities *relative to the blood*, the medium through which the waves travel.

If the angle θ is small, the frequency of sound f' striking a red blood cell is related to the frequency f emitted from the source by the equation

$$f' = f\left(\frac{v + 0}{v - v_b}\right), \tag{17.17}$$

where v_b is the speed of the blood (also the speed at which the source moves *relative to* the blood) and v is the speed of sound in blood.

The cells reflect this sound to a receiver, which detects the sound. The cells act as a source at frequency f'. The frequency detected at the receiver f_r is related to f' by the equation

$$f_r = f'\left(\frac{v + v_b}{v - 0}\right).$$

Substituting for f' from Eq. (17.17), we find the frequency detected at the receiver f_r relative to the frequency emitted by the sound source f:

$$f_r = f\left(\frac{v + 0}{v - v_b}\right)\left(\frac{v + v_b}{v - 0}\right) = f\left(\frac{v + v_b}{v - v_b}\right) \tag{17.18}$$

EXAMPLE 17.5 A source of ultrasound emits waves at a frequency of 100,000 Hz. The waves are reflected by red blood cells moving toward the source at a speed of 0.30 m/s. Calculate the frequency of sound detected at a receiver next to the source. The speed of sound in the blood is 1500 m/s.

SOLUTION   According to Eq. (17.18), the frequency of sound detected at the receiver is

$$f_r = (100{,}000 \text{ Hz}) \left(\frac{1500 \text{ m/s} + 0.3 \text{ m/s}}{1500 \text{ m/s} - 0.3 \text{ m/s}} \right) = \underline{100{,}040 \text{ Hz}.}$$

There is a 40-Hz shift in frequency. The shift can be detected easily by combining the reflected sound at frequency 100,040 Hz with some of the sound leaving the source at frequency 100,000 Hz. The combined waves produce a beat frequency of 40 Hz, which is easily measured and provides an accurate indication of the speed of the flow of the blood.   ∎

The equation used to calculate the Doppler frequency shift for electromagnetic waves such as microwaves, radio waves, and light waves is different from that used for sound waves. The observed frequency f' of an electromagnetic wave is related to the source frequency f by the equation

$$f' = f \left(\frac{1 + v/c}{1 - v/c} \right)^{1/2}. \tag{17.19}$$

The speed of the electromagnetic wave, c, equals 3.0×10^8 m/s. The speed of the observer relative to the source v is positive if observer and source are approaching each other and negative if they are moving apart.

Some of the most interesting applications of the Doppler effect are in astronomy. Astronomers have found that, like the frequency of sound from a train moving away from us, the frequency of light coming from distant galaxies is shifted to lower frequency than light from similar sources in our own galaxy. Based on calculations using the Doppler effect, it appears that nearby galaxies are moving away from us at speeds of about 250,000 m/s. Distant galaxies are moving away at speeds up to 90 percent the speed of light. The universe is moving apart and expanding in all directions.

EXAMPLE 17.6   Hydrogen atoms in a laboratory discharge lamp on the earth emit a band of light at wavelength 656 nm (1 nm = 10^{-9} m), called the H_α line of hydrogen. The same band of light coming from hydrogen atoms in a galaxy sighted in the Corona Borealis constellation is observed by astronomers on the earth to have a wavelength of 704 nm (the light is shifted toward the red part of the light spectrum). Calculate the speed of this galaxy relative to the earth.

SOLUTION   The light when emitted from hydrogen atoms in the galaxy has the same frequency as light emitted by hydrogen atoms on the earth:

$$f = \frac{c}{\lambda} = \frac{3.0 \times 10^8 \text{ m/s}}{656 \times 10^{-9} \text{ m}} = 4.57 \times 10^{14} \text{ Hz}.$$

However, because we are moving away from the Corona Borealis galaxy, the frequency f' of waves coming from its hydrogen atoms is observed by us to be less than the frequency at the source. The frequency we observe is

$$f' = \frac{c}{\lambda'} = \frac{3.0 \times 10^8 \text{ m/s}}{704 \times 10^{-9} \text{ m}} = 4.26 \times 10^{14} \text{ Hz}.$$

Equation (17.19) can be rearranged to determine the relative velocity v of the observer and source. We find that

$$v = \frac{[(f'/f)^2 - 1]}{[(f'/f)^2 + 1]}c = \left[\frac{(4.26/4.57)^2 - 1}{(4.26/4.57)^2 + 1}\right](3.0 \times 10^8 \text{ m/s}) \quad \textbf{(17.20)}$$
$$= -2.1 \times 10^7 \text{ m/s}.$$

(The 10^{14} exponents in f' and f cancel.) The negative sign means that the Corona Borealis (the source) and the astronomers on the earth (the observers) are moving apart at a relative speed of 2.1×10^7 m/s, approximately one-tenth the speed of light! ∎

Summary and Additional Readings

1. **Waves:** A **wave** is a disturbance that travels from one place to another. A single disturbance moving away from a source is called an **impulse.** A repetitive disturbance is called a **periodic wave.** A **transverse wave** is one in which the disturbance causes objects in its path to vibrate perpendicular to the direction in which the wave travels. A **longitudinal wave** is one in which the disturbance causes vibration along the same direction in which the wave travels.

The motion of waves is characterized in terms of the following quantities:

Amplitude A is the maximum displacement from equilibrium caused by the wave motion.

Speed v is the distance that some part of the wave travels per unit time and is an intrinsic property of the medium in which the wave travels. The speed is the square root of the ratio of an elastic property of the medium and an inertial property. For example, for sound waves in air or liquid, $v = \sqrt{B/\rho}$, where B is the bulk modulus of the medium (an elastic constant) and ρ is its density (an inertial constant).

Frequency f is the number of vibrations per unit time of some point in the medium along which the wave travels. It is usually equal to the frequency of the source causing the wave.

Wavelength λ is the distance between two identical points of adjacent waves (for example, the distance between two crests). The wavelength depends on the frequency and speed according to the following equation:

$$v = \lambda f. \quad \textbf{(17.8)}$$

2. **Reflection and transmission:** Whenever a wave passes the boundary between two different media, part of the incident wave is reflected and part is transmitted. The amplitude of the reflected wave depends on the difference in impedance of the two media, where impedance is a measure of the difficulty a wave has in causing the medium to vibrate.

3. **Principle of superposition:** When two or more waves pass through the same medium at the same time, the net disturbance of any point in the medium is the sum of the disturbances that would be caused by each wave if alone in the medium at that time.

4. **Beats:** When the frequencies of two waves are slightly different, their addition results in a beat pattern. The beat frequency is related to the frequencies of the two waves by the equation

$$f_{\text{beat}} = |f_1 - f_2|. \quad \textbf{(17.13)}$$

5. **Doppler effect:** The frequency at which we observe a wave usually equals the frequency of vibration of the wave's source. However, if either the source or the observer is moving, the observed frequency f' will differ from the source frequency f. The Doppler frequency shift for sound waves is given by the equation

$$f' = f\left(\frac{v + v_o}{v - v_s}\right). \quad \textbf{(17.16)}$$

Here v is the speed of the wave; v_o, the observer speed, is positive $(+)$ if the observer moves toward the source and negative $(-)$ if the observer moves away from the source; and v_s, the source speed, is positive $(+)$ if the source moves toward the observer and negative $(-)$ if the source moves away from the observer.

For electromagnetic waves, the observed frequency f' is related to the source frequency f by the equation

$$f' = f\left(\frac{1 + v/c}{1 - v/c}\right)^{1/2}, \quad \textbf{(17.19)}$$

where c is the speed of light $(3.0 \times 10^8$ m/s$)$ and v is the relative speed of the source and observer (positive if toward each other and negative if away from each other).

David Taylor and Joan Whamond, Editors, *Non-Invasive Clinical Measurement,* University Park Press, Baltimore (1977). Discusses the use of ultrasound in medicine.

D. G. Aubrey, "Our Dynamic Coastline," *Oceanus,* pp 4–13 Winter (1980/81). Presents startling effects of waves and geological factors on coastlines.

R. A. Peterson and W. C. Walter, *Through the Kaleidoscope: A Doodlebugger in Wonderland,* United Geophysical Corporation, Pasadena, April (1974). Includes many interesting (in subject and in presentation) applications of waves in geophysics.

John N. Shive and Robert L. Weber, *Similarities in Physics,* John Wiley & Sons, New York (1982). Chapters 6–13 describe vibration and wave phenomena applied to many different subjects.

Questions

1. Describe three phenomena that show that waves transfer energy from one place to another.

2. When a bell is rung, what happens to the vibrational energy initially present in the bell?

3. A television antenna has horizontal rods whose length is approximately half the wavelength of the waves carrying the television signal. Use this fact to *estimate* the frequency of the waves. Their speed is 3×10^8 m/s.

4. Invent and describe an experiment to *estimate* the speed of sound in air. You can use only everyday items found in a drugstore or general store.

5. When an earthquake occurs, you feel the tremor before hearing the sound. Explain. If you are 20 km from the earthquake, *estimate* by calculation the time delay between the tremor and the sound.

6. Explain why the speed of sound is greater in humid air than in dry air at the same pressure and temperature.

7. How can you use the time delay between a lightning flash and the sound of thunder to estimate the distance to the place where the flash occurred? Justify any assumptions you make by using an example calculation.

8. Describe two useful types of information a geologist can obtain by detecting sound reflected from different materials under the earth's surface.

9. Two speakers hang from racks placed in an open field. When sound of the same frequency comes from both speakers, there are places on the field where no sound is heard and other places at about the same distance from the speakers where the sound is loud. Explain.

10. Two identical sound waves are sent down a long hall. Does the resultant wave necessarily have an amplitude twice that of the constituent waves? Explain. Under what conditions would we hear no sound?

11. Sound waves of all frequencies in the audio-frequency range (20–20,000 Hz) travel at the same speed in air. Explain how some common experience supports this statement.

12. How can we use the phenomena of beats to tune a piano string to exactly the same frequency as an electronic oscillator?

13. A child traveling on a merry-go-round blows a whistle. Describe the variation in sound frequency heard by a stationary observer some distance from the merry-go-round. Be sure to indicate what direction the child is moving relative to the observer at different times in this frequency variation. How is this similar to the variation of the frequency of light we detect coming from a spinning galaxy?

Problems

The speed of sound in air is assumed to be 340 m/s for all of these problems unless stated otherwise.

17.3 Amplitude, Speed, Frequency, and Wavelength

1. People can hear sounds ranging in frequency from about 20 Hz to 20,000 Hz. Calculate the wavelengths for each of these sounds.

2. A dolphin has a sonar system that emits sounds having a frequency of 2.0×10^5 Hz. Calculate the wavelength of the sound. Remember that the dolphin is not in the air but that the sounds are emitted and travel in water.

3. (a) The speed of sound in water, which has a density of 1000 kg/m³, is 1450 m/s. Calculate the bulk modulus of water. (b) Calculate the speed of sound in a long, thin bone whose Young's modulus Y is 1.6×10^{10} N/m² and whose density is 1800 kg/m³.

4. Radio waves travel at a speed of 1.7×10^8 m/s through ice. A radio-wave pulse sent into the antarctic ice reflects off the earth at the bottom and returns to the surface in 32.9×10^{-6} s. How deep is the ice?

5. A flash of lightning is seen, and 2.4 s later thunder is heard coming from the same location. (a) Why is there this long delay? (b) How far away did the lightning flash occur? Justify any assumptions you make in your calculations.

■ 6. In Table 17.1 the speed of waves on an A-string and a G-string of a violin is given. Use this information to calculate the ratio of the mass per unit length of the strings (μ_A/μ_G) assuming they are under equal tension.

■ 7. An earthquake produces a longitudinal pressure wave called a P-wave* that travels through the body of the earth at speed $v = \sqrt{(B + \frac{4}{3}S)/\rho} = 7.8$ km/s. The quake also produces a transverse wave called an S-wave that travels at speed $v = \sqrt{S/\rho} = 4.5$ km/s. If the earth's density ρ equals 2400 kg/m³, use the preceding information to calculate the bulk modulus B and shear modulus S of the earth.

■ 8. A telephone lineman is told to stretch the wire between poles to a tension of about 800 N. Not having a tension scale, the lineman decides to measure the speed of a pulse created in the wire when he hits it with a wrench. The pulse travels from one pole 60 m to another pole and back again in 2.6 s. The 60-m wire has a mass of 15 kg. Should the wire be tightened or loosened? Explain.

9. A basketball player's teammate shouts at her to catch a ball. *Estimate* the time required for the basketball player to respond—for a sound to travel a distance of 10 m, be detected by the ear, travel as a nerve impulse to the brain, be processed, then travel as a nerve signal for muscle action back to an arm, and finally cause a muscle contraction. Nerve impulses travel at a speed of about 120 m/s in humans. You will have to make reasonable guesses for any numbers not known.

■ 10. A pulse travels at speed v on a stretched rope. By what factor must the tension be increased to cause the speed to increase by a factor of 1.30?

*The P stands for primary. These waves travel faster than transverse waves and are the first ones we feel following an earthquake. Slower-traveling S-waves (secondary) are felt second.

■ 11. Two ropes have equal length and are stretched to the same tension. The speed of a pulse on rope 1 is 1.4 times the speed on rope 2. Calculate the ratio of the mass of rope 1 and of rope 2.

12. *Estimate* the separation of adjacent crests or hills needed in the groove of a long-playing record to produce a sound of frequency 500 Hz.

17.4 Reflection and Transmission

13. A bat receives a reflected sound wave from a fly. If the reflected wave is returned to the bat 0.042 s after it is sent, how far is the fly from the bat? Would you expect the reflected wave pulse to be in or out of phase with the incident wave pulse? Explain.

14. A sound wave created by an explosion at the earth's surface is reflected by a discontinuity of some type under the earth's surface (Fig. 17.19). Calculate the distance from the surface to the discontinuity. Does it have a greater or lesser impedance to sound than the earth above it? Explain. Sound travels at about 3000 m/s through the top layer of the earth.

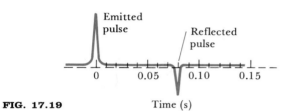

FIG. 17.19

15. Rearrange Eq. (17.10) to get Eq. (17.12).

■ 16. Show that the impedance of a gas or liquid for the transmission of sound is equal to the product of the speed of sound in the medium times its density, that is, $Z = \rho v$.

■ 17. A 5-kg rope that is 20 m long is woven to an 8-kg rope that is 16 m long. The ropes are pulled taut, and a pulse initiated in one is reflected at their interface. (a) Calculate the ratio of the reflected and incident wave amplitudes for an incident wave started in the 5-kg rope. (b) Calculate the same ratio as in part (a) but for an incident wave started in the 8-kg rope.

■ 18. A sound wave whose pressure amplitude in air is 0.0020 N/m² is incident on a marble wall. Calculate the pressure amplitude of the reflected sound wave. The impedance to sound in marble is about 16×10^6 kg/m²·s. Is the reflected wave upright or inverted?

■ 19. An ultrasound source emits waves through a very thin layer of air to a bone in the skull. (a) Calculate the transmission coefficient from the air to the skull. (b) If the sound travels from the emitter through a layer of water and then into the skull, calculate the transmission coefficient from the water to the skull. (c) Based on your answers, explain why water is used to "match" the impedance between an ultrasound emitter and a body surface.

■ 20. Sound waves incident on your forehead and transmitted into the skull can reach the inner ear by transmission through the bones and tissue in the skull. One reason why this is an inefficient sound detection mechanism is that most of the sound is reflected at the forehead. Calculate the ratio of the reflected and incident wave amplitudes and the ratio of the transmitted and incident wave amplitudes at the air-forehead interface (air to bone).

■ 21. The energy of a sound wave is proportional to its amplitude squared. Calculate the fraction of the energy of a sound wave that is transmitted from air into water at their interface. As you see, you will have a hard time hearing poolside sounds when swimming underwater.

22. The total energy of a wave pulse is proportional to its amplitude squared. Show that the sum of the amplitude squared of a reflected pulse and of a transmitted pulse equals the amplitude squared of an incident pulse ($A_r^2 + A_t^2 = A_i^2$)—wave energy is conserved at a boundary between media.

17.5 Principle of Superposition

■ 23. The pulses shown in Fig. 17.20 (time zero) are moving toward each other at speeds of 10 m/s. Draw graphs showing the resultant pulse at times of 0.1, 0.2, and 0.3 s.

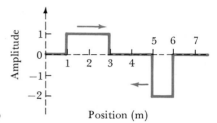

FIG. 17.20   Position (m)

■ 24. Repeat Problem 23 for the case where the pulse on the right in Fig. 17.20 is upright rather than inverted.

■ 25. Two waves shown in Fig. 17.21 at time zero move toward each other at a speed of 10 m/s. Draw graphs of the resultant wave at times of 0.1, 0.2, and 0.3 s.

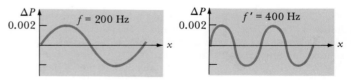

FIG. 17.21   Position (m)

■ 26. Carefully add the waves shown in Fig. 17.22 to get the resultant complex waveform.

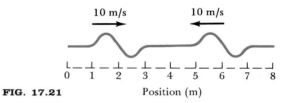

FIG. 17.22

■■ 27. Two sound speakers separated by 100 m face each other and vibrate in unison at a frequency of 85 Hz. Determine three places between the speakers where the sound is almost zero.

17.6 Beat Frequencies

28. A 30-Hz sound is annoying or dissonant to the ear. If one clarinet plays a sound at a frequency of 360 Hz, at what frequencies (more than one) might another clarinet simultaneously play so that dissonant 30-Hz beats are created? Explain.

29. A sound of wavelength 1.2 m and another of wavelength 1.0 m are played simultaneously. What beat frequency is heard?

■ 30. A tuning fork vibrates at a frequency of 440 Hz. When it and another tuning fork vibrate together, a beat frequency of 2 Hz is produced. (a) What are the possible frequencies at which the second fork is vibrating? (b) Will a small piece of wax added to one of the prongs of the second fork cause its frequency to increase or decrease? Explain. (c) When the wax is added and both forks are sounded, the beat frequency decreases. Which of the frequencies in part (a) was the actual frequency of vibration of the second tuning fork? Explain.

■ 31. A tuning fork of frequency 440 Hz produces a 5-Hz beat frequency when played simultaneously with a violin A-string. When the tension in the string is decreased slightly, the beat frequency becomes 2 Hz. What was the violin's original frequency? Explain.

■ 32. A hyperactive being from the planet Krypton is found to have cycles of activity triggered by hormones. One cycle has a period of eight days, and the other cycle a period of ten days. (a) Calculate the beat frequency—that is, the difference in frequency of these two cycles. (b) Calculate the beat period, the inverse of the beat frequency. This is the time from one large maximum of hyperactivity (both hormones working at their maximum) to the next.

17.8 Doppler Effect

33. A car horn vibrates at a frequency of 250 Hz. (a) Calculate the frequency a stationary observer hears as the car approaches at a speed of 20 m/s and (b) departs at 20 m/s. If the car is stationary, what frequency is heard (c) by an observer approaching the car at 20 m/s and (d) by an observer departing from the car at 20 m/s?

34. A car drives at a speed of 25 m/s along a road parallel to a railroad track. A train traveling at 15 m/s sounds a horn that vibrates at 300 Hz. (a) If the train and car are moving toward each other, what frequency of sound is heard by a person in the car? (b) If the train and car are moving away from each other, what frequency of sound is heard in the car?

35. A whistle with frequency 400 Hz moves at speed 20 m/s in a horizontal circle at the end of a rotating stick. Calculate the highest and lowest frequencies heard by a person riding a bicycle at speed 10 m/s toward the whistle. (See note at the end of Problem 39.)

36. A red blood cell travels at a speed of 0.40 m/s in a large artery. A sound with a frequency of 100,000 Hz enters the blood opposite the direction of flow. (a) Calculate the frequency of sound reflected from the cell and detected by a receiver. (b) If the emitted and received sounds are combined in the receiver, what beat frequency is measured?

■ 37. A bat emits short pulses of sound at a frequency of 3.90×10^5 Hz. As the bat swoops toward a flat wall at a speed of 30 m/s, this sound is reflected from the wall back to the bat. (a) What is the frequency of sound incident on the wall? (b) Consider the wall as a sound source at the frequency calcu-

lated in part (a). What frequency of sound does the bat hear coming from the wall?

■ 38. A hungry student working in a cafeteria decides to eat plates of food that pass on a conveyor belt. The plates are separated by 3 m, and the belt moves at a speed of 9 m/min. (a) How many plates of food does the student eat per minute? (b) As the student's hunger is appeased, he moves with the belt at a speed of 6 m/min. Calculate the number of plates of food that now reach the student each minute. (This shift is similar to the Doppler shift for a moving listener.)

■ 39. A whistle is placed at the end of a stick that rotates in a horizontal plane. A stationary observer hears a sound coming from the whistle that varies from a minimum value of 395 Hz (when the whistle is rotating away from the observer) to a maximum of 405 Hz one-half turn later when the whistle is rotating toward the observer. When at rest, the whistle vibrates at 400 Hz. Calculate the speed of the whistle (a) when moving away from the observer and (b) when moving toward the observer. (*Note:* Astronomers use a similar technique to determine the angular velocity of galaxies. The speed v due to rotational motion, the angular velocity ω, and the radius of the galaxy are related by the equation $v = r\omega$. They use the Doppler shift of light to calculate v.)

■ ■ 40. A Doppler speed meter operating at exactly 1.02×10^5 Hz emits sound waves and detects the same waves after they are reflected from a baseball thrown by the pitcher Nolan Ryan. The receiver "mixes" the reflected wave with a small amount of the emitted wave and measures a beat frequency of 0.30×10^5 Hz. How fast is the ball moving?

■ ■ 41. You are one of a committee of angels asked to design a Doppler "sight" system for an evolving species on a planet of the star Polaris. You, in particular, are to choose the frequency of sound for this system. There are two requirements: (a) The wavelength should be such that you can distinguish the features of a fly. Thus, the wavelength must be smaller than a fly. (b) The Doppler part of the detection system must be able to produce a beat frequency of 1 Hz for each 0.5 m/s of motion of an object from which the sound is deflected. *Estimate* the minimum frequency you can use to satisfy each condition. The speed of sound on this planet is 800 m/s.

■ 42. The wavelength of the H_α line observed coming from 3C 273, the nearest quasar (distant objects that emit enormous amounts of energy) is 761 nm. Calculate the velocity of 3C 273 relative to the earth.

■ 43. A galaxy sighted in the Ursa Major constellation is moving away from us at a speed of 6.1×10^7 m/s. Calculate the wavelength of the H_α hydrogen line coming to us from this galaxy.

■ 44. At what velocity must you move relative to a stoplight so that red light of wavelength 650 nm emitted by the stoplight appears to you to be green with wavelength 510 nm?

■ 45. Rearrange Eq. (17.19) to get Eq. (17.20).

■ ■ 46. Using the binomial theorem for the situation in which the relative velocity v is much less than the speed of light, show that Eq. (17.19) reduces to $v \simeq (f' - f)c/f$.

INTERLUDE IV Dimensional Analysis

You may recall from Chapter 1 that a small number of physical quantities, seven to be exact, are called *basic quantities*. Since beginning our study of physics, we have used five of these quantities—length, time, mass, temperature, and quantity of matter. In the SI metric system the units of these quantities are, respectively, the meter, second, kilogram, kelvin, and mole. Other derived quantities such as speed, acceleration, and force can be defined in terms of these basic quantities. The **dimensions** of a derived quantity are the basic quantities used in its definition. For example, the dimensions of speed are length divided by time (l/t or lt^{-1}). The dimensions of acceleration are lt^{-2} and those of force are mlt^{-2}.

Dimensional analysis involves the use of the dimensions of quantities to derive mathematical equations relating the quantities. A person unfamiliar with dimensional analysis will be surprised at the simplicity of deriving these equations. To illustrate, we will introduce the technique with an example.

EXAMPLE IV.1 Derive an equation for the speed of a wave on a stretched string.

SOLUTION Dimensional analysis relies to a certain extent on our physical intuition. We must make an educated guess at what quantities affect the wave's speed. Fortunately, if we guess wrong, the error is often made apparent by our subsequent calculations. For this example, let us assume that the speed of a wave on a stretched string depends on its tension T, on its mass per unit length μ, and on the amplitude of the wave A. Actually, the speed does not depend on A, but we may not realize that before starting our derivation. In this example our error is corrected for us by the techniques of dimensional analysis.

We wish now to determine the type of equation relating these quantities to speed. Is the equation of the form $v = T/\mu A$, $T^2 A/\mu$, T/A, $\sqrt{T/\mu}$, or of some other form? We can write a general equation that encompasses all of the above expressions as follows:

$$v = CT^x \mu^y A^z, \tag{IV.1}$$

where x, y, and z are unknown exponents of the quantities T, μ, and A, respectively, and C is an unknown constant that has no dimensions (such as π, 8, or some other number). If the speed is given by the equation $v = T^{1/2}\mu^{-1/2}$, then $x = \frac{1}{2}$, $y = -\frac{1}{2}$, $z = 0$, and $C = 1$. Dimensional analysis can be used to determine the values of x, y, and z but not the value of C, which must be determined by experiment or by a more detailed derivation using basic principles.

Next we consider the dimensions of all quantities appearing in Eq. (IV.1):

Physical Quantity	Symbol	Dimensions
Speed of wave	v	lt^{-1}
Tension in string (a force)	T	mlt^{-2}
Mass per unit length	μ	ml^{-1}
Amplitude of wave	A	l

Substituting the dimensions for each quantity in Eq. (IV.1), we find that

$$lt^{-1} = (mlt^{-2})^x(ml^{-1})^y(l)^z$$
$$= (m^x l^x t^{-2x})(m^y l^{-y})(l^z) \qquad \text{(IV.2)}$$
$$= m^{x+y} l^{x-y+z} t^{-2x}.$$

Remember that C has no dimensions. For the last equation to be dimensionally correct, each dimension must appear to the same power on the left side and on the right side of the equation. Length, for instance, appears to power 1 on the left side of the equation and to power $x - y + z$ on the right side. For the equation to be dimensionally correct, we require that

$$1 = x - y + z.$$

For time to have the same dimensions on each side of the equation, we require that

$$-1 = -2x.$$

Mass does not appear on the left side of Eq. (IV.2). We say that mass appears to the power zero, since $m^0 = 1$. For mass to have the same dimensions on each side of Eq. (IV.2), we require that

$$0 = x + y.$$

We now have three equations relating x, y, and z:

$$1 = x - y + z,$$
$$-1 = -2x,$$
$$0 = x + y.$$

Solving these equations, we find that $x = 1/2, y = -1/2$, and $z = 0$. We have determined that the speed of a wave on a string is calculated using an equation of the form

$$v = CT^{1/2}\mu^{-1/2}A^0 = \underline{CT^{1/2}\mu^{-1/2}}.$$

The constant C cannot be determined by dimensional analysis. A derivation of this same equation using Newton's second law indicates that $C = 1$ and that

$$v = T^{1/2}\mu^{-1/2} = \sqrt{T/\mu},$$

which is Eq. (17.6) given in Table 17.1. ∎

We see that deriving equations through dimensional analysis involves the following steps: (1) Choose a set of quantities on which a quantity such as wave speed depends. If you choose wrong, dimensional analysis may indicate your error by showing that the quantity should be to the power zero (it does not belong in the expression), or you may find that there is no satisfactory set of exponents for the set of quantities you have chosen. In the latter case, you will have to try again with a different set of quantities. (2) Having selected the quantities, construct an equation of the form $v = CT^x\mu^y A^z$ and substitute the dimensions for each quantity in this equation. (3) For the equation to be dimensionally correct, the exponent of each dimension must be the same on each side of the equation. This requirement allows us to write several equations relating x, y, and z. (4) Solve these equations for x, y, and z.

EXAMPLE IV.2   Use dimensional analysis to derive an equation for the magnitude of the centripetal acceleration of a mass m moving in a circle of radius r at constant speed v.

SOLUTION   We will assume that the acceleration depends on all three of these quantities: m, r, and v. The equation relating centripetal acceleration to these quantities is written in the general form

$$a_c = Cm^x r^y v^z.$$

Substituting the dimensions into the above equation (remember that C is dimensionless), we find that

$$lt^{-2} = (m)^x (l)^y (lt^{-1})^z$$
$$= m^x l^{y+z} t^{-z}.$$

For the equation to be dimensionally consistent, we find that

$$0 = x,$$
$$1 = y + z,$$
$$-2 = -z.$$

These equations are solved to determine the values of the exponents: $x = 0$, $y = -1$, and $z = 2$. Thus,

$$a_c = Cm^0 r^{-1} v^2 = \underline{Cv^2/r}.$$

A derivation such as done in Chapter 6 indicates that $C = 1$ and that $a_c = v^2/r$. ∎

EXAMPLE IV.3   Suppose that a hole is drilled through the earth (Fig. IV.1). Use dimensional analysis to determine an equation for the time τ needed for a person to fall down the hole from one side of the earth to the other and then back again. Ignore the drag force of air on the falling person.

SOLUTION   The person's acceleration as he starts to fall is the acceleration due to gravity g, but as he approaches the center of the earth, the acceleration decreases to zero. Even though the acceleration changes, the time for the trip must depend on his initial acceleration g and also on the distance of the trip (four times the earth's radius R). Our previous experience with falling masses leads us to believe that all masses should fall with the same acceleration if we ignore drag forces. So we will not include the person's mass as a quantity affecting the time for the trip. The effect of the earth's mass M is included in the value

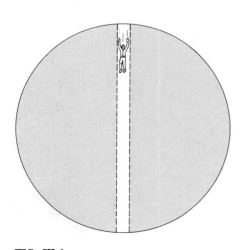

FIG. IV.1

374

of g. To be sure that the earth's mass does not affect the time for the trip in some other way, we will include it in our dimensional analysis.

The equation for the time τ for the round trip is written in terms of the acceleration of gravity g, the radius of the earth R, and the earth's mass M as follows:

$$\tau = CR^x g^y M^z.$$

Substituting the dimensions for each quantity, we find that

$$t = (l)^x (lt^{-2})^y (m)^z$$
$$= l^{x+y} t^{-2y} m^z.$$

To make this equation dimensionally correct, we require that

$$0 = x + y,$$
$$1 = -2y,$$
$$0 = z.$$

Solving for x, y, and z, we find that $z = 0$, $y = -1/2$, and $x = 1/2$. The equation for the time of one trip through the earth and back again is of the form

$$\tau = CR^{1/2} g^{-1/2} M^0 = C\sqrt{R/g}.$$

A calculation using Newton's second law and techniques developed in Chapter 16 leads to the equation $\tau = 2\pi\sqrt{R/g}$. Substituting $R = 6.4 \times 10^6$ m and $g = 9.8$ m/s^2, we find that $\tau = 2\pi\sqrt{(6.4 \times 10^6 \text{ m})/(9.8 \text{ m/s}^2)} = 5.1 \times 10^3$ s $= 1.4$ h. Actually, the drag force of air would greatly increase the time needed for the trip. ∎

The examples of dimensional analysis presented here and in the problems that follow are fairly simple. In the hands of an expert, dimensional analysis provides guidance in solving much more complex problems. Lord Rayleigh, a brilliant and versatile physicist of the early twentieth century, combined a keen insight into the essentials of a physics problem with the techniques of dimensional analysis to provide neat and simple solutions to a wide variety of problems. He and others have found dimensional analysis to be an important technique for guiding and checking their work in physics.*

Problems

■ 1. Use dimensional analysis to determine the swinging frequency of a simple pendulum consisting of a mass m hanging in the earth's gravitational field g at the end of a string of length L.

■ 2. Use dimensional analysis to determine the dependence of the speed of sound v in air on the density ρ and bulk modulus B of the air.

■ 3. A string of length L and mass m when stretched with tension T vibrates at frequency f. Use dimensional analysis to determine the functional dependence of f on these other quantities.

■ 4. An object falls from rest a distance h in a time t. Use dimensional analysis to determine the functional form of an equation for calculating h at different times.

*A particularly good dimensional analysis reference for the beginning student is H. E. Huntley, *Dimensional Analysis*, Holt, Rinehart and Winston, New York (1951).

CHAPTER 18

Standing Waves and Sound

The sounds of an old-time fiddler or a concert violinist start as the vibration of a string fixed at both ends. The sound is different from the whine of truck tires on a highway or the groan of a furnace. It is called music.

Much of the uniqueness of musical sounds depends on what physicists call *standing-wave vibrations*. For a violinist, a standing-wave vibration is the unique vibration of the string that occurs when it is excited by a bow (Fig. 18.1). The string's vibration is the result of waves moving back and forth in opposite directions along the string. These waves interfere with each other and produce a vibration of the string as a whole that looks like a wave that is not moving; hence the name *standing wave*. The standing-wave vibration of a string affects the pitch, quality, and loudness of the violin's sound. The study of standing waves is a natural place to start this chapter, much of which is devoted to musical types of sound.

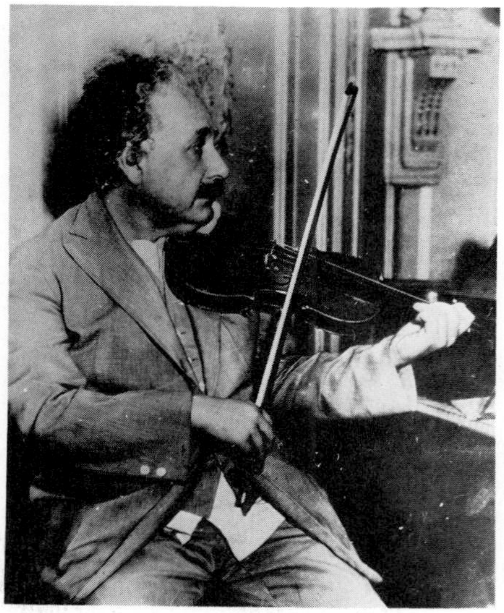

18.1 Standing Waves on Strings

To begin our study of standing waves on a string, let us consider the rope shown in Fig. 18.2. One end of the rope is attached to a fixed support while the other end is held in a woman's hand. The woman wishes to shake the rope up and down at a frequency that will produce large-amplitude vibrations on the rope.

To start, she sends a pulse down the rope (parts b–d). When it reaches the fixed end, the upward pulse is reflected and inverted to a downward pulse traveling back toward her hand (parts e–g). When it reaches her hand, it is once again reflected and inverted to an upward pulse (part h).

To make this pulse bigger, she can give her hand another upward shake, making sure to time the shake so that it adds to the previous upward pulse that has just been reflected (part h). If her upward shake is too soon, the new pulse she creates will interfere destructively with the inverted pulse that is returning to her hand. To make each new pulse interfere constructively with a previous pulse, then, she must shake her hand upward each time a previous pulse returns to her hand. If she continues to synchronize her shaking frequency with the start of reflected pulses from her hand, the amplitude of the disturbance traveling along the rope grows and looks like the vibration shown in part (1).

FIG. 18.1. Albert Einstein producing standing-wave vibrations on his violin strings. (Brown Brothers)

The frequency at which she should vibrate her hand to cause the rope vibration shown in part (1) depends on the length L of the rope and the speed v of the pulse along the rope. The time for one complete round-trip of a pulse down the rope and back is

$$\tau = \frac{2L}{v}.$$

Thus, the frequency $f_1{}^*$ of the upward shakes should be one shake per time τ, or

$$f_1 = \frac{1}{\tau} = \frac{1}{2L/v} = \frac{v}{2L}. \qquad \textbf{(18.1)}$$

If the frequency is less than this, each new pulse will interfere destructively with previous pulses, and the rope will vibrate little. Should the woman raise and lower her hand at a frequency $f_1 = v/2L$, each new pulse will add to the last and create large-amplitude vibrations.

The vibration shown in part (1) is called the **fundamental frequency vibration** or the **first harmonic vibration** of the string or rope with fixed ends. It is the lowest frequency of the allowed large-amplitude vibrations on the rope. If you pluck a guitar string of length L, the sound you hear coming from the guitar is at the fundamental frequency given by Eq. (18.1). *The speed that appears in this equation is the speed of a pulse on the string and not the speed of sound in air.*

EXAMPLE 18.1 Calculate the fundamental frequency of vibration of the A-string on a violin. Its mass is 0.30 g, it is 33 cm long, and it is stretched with a tension of 77 N.

SOLUTION The speed of a wave on a string is given by $v = \sqrt{T/(m/L)}$ (Eq. 17.6), where m is the mass of the vibrating portion of the string and L is its length. For the violin A-string

$$v = \sqrt{\frac{77 \text{ N}}{0.30 \times 10^{-3} \text{ kg}/0.33 \text{ m}}} = 290 \text{ m/s}.$$

The fundamental frequency of vibration is

$$f_1 = \frac{v}{2L} = \frac{290 \text{ m/s}}{2(0.33 \text{ m})} = 440 \text{ s}^{-1} = \underline{440 \text{ Hz}}. \qquad \blacksquare$$

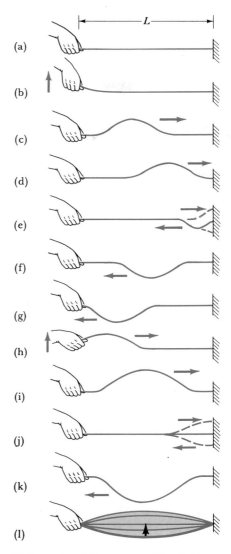

FIG. 18.2. Pulses generated by shaking a rope up and down cause the vibration pattern shown in (l).

A string will vibrate with large amplitude at other frequencies besides the fundamental. Consider again the woman shaking a rope. If she shakes the rope at a frequency twice the fundamental frequency, two pulses travel along the rope, as shown in Fig. 18.3a. Each pulse that returns to her hand after reflection from the opposite end is reinforced by another upward shake of her hand.

However, as a downward, reflected pulse passes the middle of the rope, it interferes destructively with an upward pulse that has just left the hand (part c), and the middle of the rope never becomes displaced. As the inverted, reflected pulse approaches the hand and the upright pulse approaches the fixed end, as shown in part (d), the rope's shape appears inverted compared with its shape in

---

*The subscript "1" on f_1 indicates the first or lowest frequency at which the woman can shake the rope to get large-amplitude vibrations. There are higher frequencies at which large-amplitude vibrations also occur, as we will soon learn.

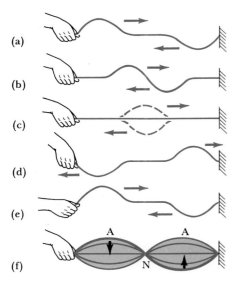

(a)

(b)

(c)

(d)

(e)

(f)

FIG. 18.3. If two pulses are generated on a stretched rope in the time required for one round-trip of a pulse, the second harmonic vibration pattern shown in (f) is produced.

part (a). After the pulses are reflected from the hand and from the fixed end of the rope, as in part (e), the rope returns to its original shape (part a).

The vibration pattern of the rope when shaken twice during the time it takes a pulse to travel up the rope and back again is shown in Fig. 18.3f. In this case the rope's vibration frequency is twice the frequency of the fundamental:

$$f_2 = 2f_1 = 2\left(\frac{v}{2L}\right).$$

This vibration is called the **second harmonic vibration** or the **first overtone vibration.** The place where the rope undergoes no vibration is called a **node** and is labeled by an N in part (f). The places where the vibration amplitude is greatest are called **antinodes** (A in part f).

If the rope is shaken at three times the fundamental frequency,

$$f_3 = 3f_1 = 3\left(\frac{v}{2L}\right),$$

three pulses travel along it. Each pulse is reinforced by an upward motion of the hand at just the right time, and large-amplitude vibrations are again produced. The vibration pattern that results is shown in Fig. 18.4d and is called the **third harmonic vibration** or the **second overtone.**

In general, a rope or string will vibrate with large amplitude at any frequency for which a pulse produced by the wave source (for example, the vibrating hand) is reinforced by a new pulse after completing a round-trip up the string and back again. This condition is met when the string's *standing-wave vibration frequency* is

$$f_n = n\left(\frac{v}{2L}\right), \qquad \text{where } n = 1, 2, 3, 4, \ldots . \qquad \textbf{(18.2)}$$

Thus, there are many frequencies at which we can shake the rope or string and get large-amplitude vibrations. Each frequency corresponds to a different value of the positive integer n. The wave patterns that are produced on the rope for these different values of n do not appear to travel or move. They appear to be stationary vibrations, called **standing waves,** and are caused by the interference of waves moving in opposite directions. The vibration patterns for several standing waves are shown again in Fig. 18.5. Notice that they are all integral multiples of the fundamental frequency.

The wavelengths of these standing-wave vibrations are calculated using Eq. (17.8), $\lambda = v/f$. For example, the wavelength of the fundamental vibration is

$$\lambda_1 = \frac{v}{f_1} = \frac{v}{(v/2L)} = 2L.$$

For the second harmonic, the wavelength is

$$\lambda_2 = \frac{v}{f_2} = \frac{v}{2(v/2L)} = L.$$

In general, the wavelengths of the string's harmonic vibrations are expressed by the equation

$$\lambda_n = \frac{v}{f_n} = \frac{v}{n(v/2L)} = \frac{2L}{n}, \qquad \text{where } n = 1, 2, 3, 4, \ldots . \qquad \textbf{(18.3)}$$

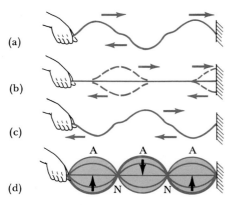

(a)

(b)

(c)

(d)

FIG. 18.4. If we initiate three pulses in the time needed for a pulse to make one round-trip along the rope, we get the third harmonic vibration pattern as shown in (d).

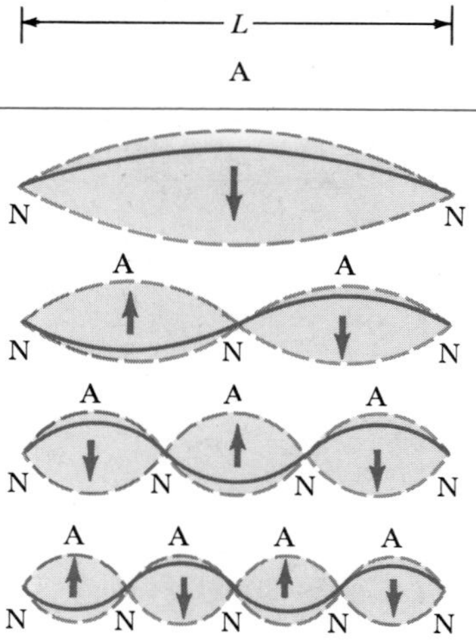

Frequency	Wavelength	A	Name
$f_1 = \dfrac{v}{2L}$	$\lambda_1 = 2L$		First harmonic (fundamental)
$f_2 = 2\left(\dfrac{v}{2L}\right)$	$\lambda_2 = \dfrac{2L}{2}$		Second harmonic (first overtone)
$f_3 = 3\left(\dfrac{v}{2L}\right)$	$\lambda_3 = \dfrac{2L}{3}$		Third harmonic (second overtone)
$f_4 = 4\left(\dfrac{v}{2L}\right)$	$\lambda_4 = \dfrac{2L}{4}$		Fourth harmonic (third overtone)

FIG. 18.5. The first four standing-wave vibration patterns on a string fixed at both ends.

Remember that these wavelengths are for waves traveling on the string and not for the sound waves in the air caused by the string's vibrations.

EXAMPLE 18.2 If tuned correctly, the fundamental frequency of the A-string of a violin is 440 Hz. List the frequencies of several other standing-wave vibrations of the A-string.

SOLUTION The string, when bowed or plucked, vibrates simultaneously at many different frequencies. These frequencies are integral multiples of the fundamental and include 880 Hz, 1320 Hz, 1760 Hz, and so forth. ■

EXAMPLE 18.3 A banjo D-string is 0.69 m long and has a fundamental frequency of 587 Hz. The string vibrates at a higher fundamental frequency if its length is reduced by holding the string down over a fret. Where should a fret be placed to play the note F-sharp, which has a fundamental frequency of 734 Hz?

SOLUTION The fundamental frequency of vibration f_1 is related to the string length L by the equation $f_1 = v/2L$, or $L = v/2f_1$. We cannot solve immediately for the unknown length when the string is vibrating at 734 Hz since we do not know the speed v of the wave on the string. But we can solve for v using the fact that the string vibrates at 587 Hz when at its full 0.69-m length:

$$v = f_1(2L) = (587 \text{ s}^{-1})(2 \times 0.69 \text{ m}) = 810 \text{ m/s}.$$

Since the speed is constant as the length is changed, we can now solve for the unknown length L' when the string is vibrating at the higher frequency f_1' equal to 734 Hz:

$$L' = \frac{v}{2f_1'} = \frac{810 \text{ m/s}}{2 \times 734 \text{ s}^{-1}} = 0.55 \text{ m}.$$

The fret must be located a distance 0.69 m − 0.55 m = 0.14 m from the end of the string.

This problem can be solved even more easily by taking the ratio of the two different lengths, as given by rearranging Eq. (18.1). The common velocity of the wave along the string cancels out.

$$\frac{L'}{L} = \frac{(v/2f_1')}{(v/2f_1)} = \frac{f_1}{f_1'},$$

or

$$L' = \frac{f_1}{f_1'}L = \frac{(587\text{ Hz})}{(734\text{ Hz})}0.69\text{ m} = 0.55\text{ m}. \qquad \blacksquare$$

18.2 Standing Waves in Air Columns

Standing-wave vibrations occur in many objects besides strings. They are manifested in the swaying of tall buildings, the twisting vibrations of bridges, the motion of a drumhead, and the vibration of the gas in an air column. This last type of vibration is important for the generation of sound by brass and wood-wind musical instruments.

The formation of standing-wave vibrations in air columns (pipes) is very similar in concept to the formation of such vibrations on strings. A pressure pulse is initiated at one end of the pipe. This might be done by blowing air across an open-hole tube, such as in a flute. Or it might be done by blowing air through a vibrating reed, such as in a clarinet. Little swirls or puffs of air enter the air column in a repetitive fashion.

Consider the fate of one of these pressure pulses while moving in an air column with both ends open, called an **open pipe.** The pressure pulse moves along the pipe as shown in Fig. 18.6a–c. At the open end, part of the pulse is transmitted out of the pipe and part is reflected. It may surprise you that sound is reflected at the open end of a pipe. The pipe confines the movement of air molecules and causes the air inside the pipe to have a different impedance to sound waves than the air outside the pipe. Whenever there is an abrupt change of impedance, such as occurs at the open end, waves are partially reflected.

The reflected pressure pulse has an inverted phase. If the incident pulse causes increased pressure, the reflected pulse causes decreased pressure.* This pulse of decreased pressure moves back toward the end of the pipe where it started (parts d–f) and is reflected at the open end on the left (part g). After reflections from both open ends, the pulse again has increased pressure and is moving toward the right. If a new puff of air is initiated at this time, it interferes constructively with the reflected pulse, and its amplitude is increased. A large-amplitude standing-wave vibration is beginning to be formed in the open pipe. If, however, the new puff of air is too early or too late, it interferes destructively with the reflected pulse, and the amplitude is reduced.

The frequency of standing-wave vibrations in pipes depends on the time needed for a pulse to travel down the pipe and back again. If the sound pulse travels at a speed v along a pipe of length L, then the time τ needed to travel a distance $2L$ is $\tau = 2L/v$. To reinforce the vibration of air in the pipe, puffs

*The pressure pulse changes phase at the end of an open tube. A displacement pulse, which indicates the motion of air molecules, does not change phase at the boundary of an open tube. We are discussing pressure pulses, not displacement pulses.

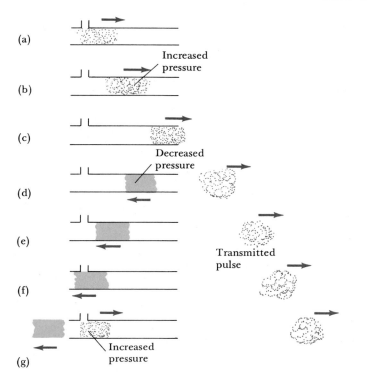

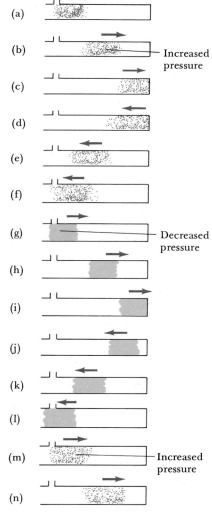

FIG. 18.6. A pulse of increased pressure traveling along a pipe (a–c) is partially transmitted and partially reflected at the end (d). After being reflected from the end where it started, a pulse of increased pressure again moves to the right (g).

should occur once each time period τ or at a frequency $f_1 = 1/\tau = v/2L$. This equation is the same as that for the fundamental frequency of vibration of a string. However, in this case, v is the speed of a pulse through the air inside the pipe (about 340 m/s).

Just as a string can be excited to vibrate at multiples of the fundamental frequency, the pipe also vibrates with large amplitude if excited by puffs of air at integral multiples of the fundamental frequency. In general, an *open pipe* can vibrate with large amplitude at the following *standing-wave frequencies:*

$$f_n = n\left(\frac{v}{2L}\right), \qquad \text{for } n = 1, 2, 3, \ldots, \qquad \textbf{(18.2)}$$

where v is the speed of air in the pipe and L is its length. Musical instruments that function as open pipes include the flute, recorder, oboe, and bassoon.

Next let us turn to a pipe called a **closed pipe,** which is closed at one end and open at the other end. Musical instruments that behave somewhat like closed pipes include the clarinet, trumpet, trombone, and human voice. The closed end of these instruments is at the reed, mouthpiece, and vocal cords, respectively. The determination of the standing-wave vibration frequencies of closed pipes is slightly more complicated than for open pipes.

Suppose that a pulse of increased pressure is initiated near the open end of the pipe (Fig. 18.7a). When the pulse is reflected at the closed end, its phase is unchanged (part d). When the reflected pulse travels back to the open end, it is once again reflected. However, this time there is a change in phase; it becomes a pulse of decreased pressure (part g). If a new pulse of increased pressure is initiated, it cancels the effect of the reflected pressure pulse. To add to the re-

FIG. 18.7. A pulse of increased pressure must make two trips up the pipe and back before it again becomes a pulse of increased pressure.

flected pulse in a constructive way, we must wait for it to make an additional trip down the pipe and back again. During each round-trip, its phase is inverted once. After two round-trips it is again a pulse of increased pressure (part m), and a new pulse adds constructively to increase its amplitude. For a large-amplitude standing wave to be produced, then, the time τ between pulses must be $4L/v$. The frequency of the fundamental is $f_1 = 1/\tau = v/4L$.

The closed pipe is unique in that overtone vibrations do not occur at all frequencies that are integral multiples of the fundamental frequency. For example, at twice the frequency of the fundamental, a pulse makes only one trip up and down the pipe before another pulse is started. The reflected pulse, which is inverted, interferes destructively with the new pulse. The pipe cannot vibrate at twice the fundamental frequency. A closed pipe can, however, vibrate at *odd* integral multiples of the fundamental frequency. The *standing-wave vibration frequencies of a closed pipe* are

$$f_n = n\left(\frac{v}{4L}\right), \qquad \text{where } n = 1, 3, 5, 7, \dots . \tag{18.4}$$

EXAMPLE 18.4 Assume that the auditory canal of the outer ear is a 3.0-cm-long pipe closed at one end by the eardrum (look ahead to Fig. 18.12). Calculate the fundamental frequency of this pipe when (a) in air and (b) in water. How might this resonance of the auditory canal affect our ability to hear sounds at these frequencies?

SOLUTION (a) The air in a closed pipe, such as our auditory canal, vibrates most easily at the standing-wave frequencies given by Eq. (18.4). The fundamental frequency ($n = 1$) for a closed pipe of length $L = 3.0\,\text{cm} = 0.030\,\text{m}$ filled with air through which sound travels at a speed $v = 340\,\text{m/s}$ is

$$f_1 = n\left(\frac{v}{4L}\right) = 1\left(\frac{340\,\text{m/s}}{4(0.030\,\text{m})}\right) = \underline{2800\,\text{Hz}}.$$

The ear is, in fact, most sensitive to sound in this frequency range.
(b) If the auditory canal is filled with water, the sound travels at speed $v = 1500\,\text{m/s}$ (see Table 17.1). The fundamental frequency of a closed pipe filled with water is

$$f_1 = 1\left(\frac{1500\,\text{m/s}}{3(0.030\,\text{m})}\right) = \underline{17{,}000\,\text{Hz}}.$$

We should be able to hear very-high-frequency sounds better while underwater. ∎

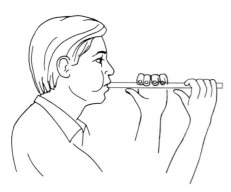

FIG. 18.8. A flute made by cutting holes in a straw. The length of the air column is the distance from the mouth to the nearest open hole.

EXAMPLE 18.5 A flute made from a straw behaves like an open pipe and has six holes cut in it (Fig. 18.8). The length of the flute's air column (called the *bore length*) is the distance from the mouth to the nearest open hole. Opening a hole is like cutting the end off the straw and decreasing its length.

With all holes closed, the bore length is 16 cm, and the flute vibrates with a fundamental frequency f_1. With the third hole open, the fundamental frequency f_1' equals $\frac{4}{3}f_1$. What is the flute's bore length with the third hole open?

SOLUTION The ratio of f_1' and f_1 equals $\frac{4}{3}$. Substituting Eq. (18.2) for the fundamental frequencies and canceling the speed of sound, which is the same for both vibrations, we find that the fundamental frequencies and bore lengths are related by the equation

$$\frac{4}{3} = \frac{f_1'}{f_1} = \frac{v/2L'}{v/2L} = \frac{L}{L'},$$

or

$$L' = \frac{3}{4}L = \frac{3}{4}(16\ \text{cm}) = \underline{12\ \text{cm}}. \qquad \blacksquare$$

18.3 Complex Waves

We have seen that a stretched string, an open pipe, and a closed pipe can vibrate at only certain discrete frequencies (the standing-wave vibration frequencies discussed in the previous two sections). Suppose that we excite one of these strings by plucking or bowing it or excite a wind instrument by blowing across its reed. At what standing-wave frequency or frequencies will the string or pipe vibrate? Usually, but not always, the fundamental frequency is excited. In addition, other standing-wave vibrations (higher harmonics) are often simultaneously excited. The resultant sound wave leaving a musical instrument consists of waves of several (sometimes many) different frequencies combined in what is called a **complex wave**—a wave consisting of two or more sinusoidal waves each of different frequency and wavelength.

Before looking at examples of complex waves from musical instruments, let us see how these waves are added together to form the complex wave. What does its waveform look like? As a simple example, consider two sound waves sent simultaneously down a hallway (Fig. 18.9a). The variation of air pressure caused by the first wave at one instant of time at different positions along the hall is shown in part (b). The graph is like a photograph of the change in air pressure at each position along the hall if only the first wave is present. The variation of air pressure caused by the second wave at the same instant is shown in part (c). Notice that sound speaker 2 produces sound of twice the frequency (one-half the wavelength) and twice the amplitude of sound from speaker 1.

If both waves are present in the hall at that same instant of time, we do not see either pattern. Instead, we see a complex wave pattern that results from the combination of the pressure variations of both waves. This complex wave is calculated using the principle of superposition and can be constructed on a graph by adding the pressure variations caused by each wave at each point in space. For example, at a position 10 m down the hall, wave 1 in Fig. 18.9 causes a pressure increase of $0.001\ \text{N/m}^2$—that is, $\Delta P_1 = +0.001\ \text{N/m}^2$. Wave 2 causes a pressure decrease of $0.002\ \text{N/m}^2$: $\Delta P_2 = -0.002\ \text{N/m}^2$. Thus, the actual pressure change at this point due to both waves is $\Delta P_1 + \Delta P_2 = (0.001 - 0.002)\ \text{N/m}^2$, or $-0.001\ \text{N/m}^2$. By adding the pressures at each point along the hall, we are able to construct the complex wave shown in part (d). Since both sinusoidal waves move to the right at the speed of sound, the complex wave also moves to the right at the speed of sound. If a microphone were placed in the hall, it would detect a fluctuating pressure such as shown in part (d), a pattern called the **waveform** of the complex wave.

FIG. 18.9. (a) Two sound waves emitted by different speakers are sent down a hall. (b) and (c) The variation in pressure of the air caused by the respective sound waves. (d) The complex pressure wave is produced by adding these two waves.

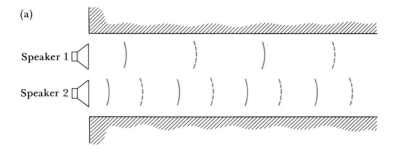

(a)

Speaker 1

Speaker 2

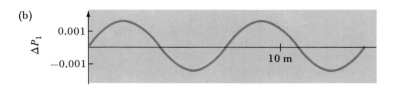

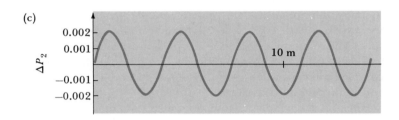

(b)

ΔP_1

0.001

−0.001

10 m

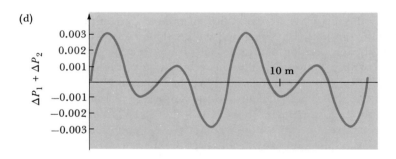

(c)

ΔP_2

0.002
0.001

−0.001
−0.002

10 m

(d)

$\Delta P_1 + \Delta P_2$

0.003
0.002
0.001

−0.001
−0.002
−0.003

10 m

Practical reasons arise for wanting to construct a complex wave. Suppose that you wish to build an electric organ that can produce sounds like those of a clarinet or violin. The organ must be able to produce complex waves that look similar to those from the actual instruments. This can be done by combining sinusoidal waves to obtain the appropriate complex wave. To do this, we must first determine what frequency waves are part of a complex wave coming from a clarinet or violin when playing a particular pitch sound.

A variety of techniques are used to separate a complex wave into its sinusoidal components. For example, a sound wave detected by a microphone can be fed into a device called a spectrum analyzer. This device electronically analyzes the wave pattern and plots on a screen the amplitude of waves at different frequencies that are present in the complex wave. This amplitude-versus-frequency graph is called a **frequency spectrum** of the complex wave. A frequency spectrum of the complex wave we constructed in Fig. 18.9 is shown in Fig. 18.10. The height of the line at each frequency is proportional to the amplitude of the wave at that frequency. Since the amplitude of wave 2 is twice that of wave 1, its vertical line is twice the height of the vertical line for wave 1.

Examples of actual sound-wave pressure variations and the corresponding frequency spectra are shown in Fig. 18.11. The complex wave shown in part (a) is that of a piano when playing concert A at 440 Hz. The wave pattern is repetitive and represents a complex wave made mostly of a sinusoidal wave of frequency 440 Hz. The complex wave also consists of components of smaller amplitude at 880 Hz, 1320 Hz, and higher frequency, as is indicated by the frequency spectrum of the wave shown in part (b). The complex wave from a

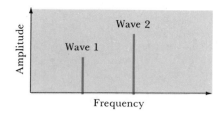

FIG. 18.10. A frequency spectrum for the complex wave shown in Fig. 18.9d.

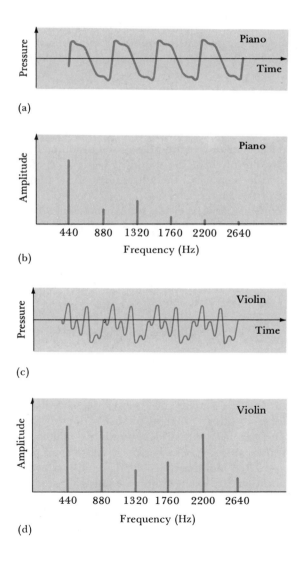

FIG. 18.11. (a) The waveform of sound coming from a piano playing concert A (440 Hz) and (b) the frequency spectrum of the waveform. (c) and (d) The waveforms and frequency spectrum, respectively, of a violin when playing concert A.

violin (part c) when playing the same pitch sound has many more peaks and valleys than that from the piano. The frequency spectrum of the violin complex wave is shown in part (d). Notice that many more sinusoidal waves have been combined to form the violin complex wave. A component at 2200 Hz is especially prominent in the violin and is responsible for much of the bumpy appearance of the complex wave shown in part (c).

The presence or absence of various sinusoidal waves in complex waves can be detected by our ears. This is one reason a violin sounds different from a piano even though they play the same pitch sound. Our ears in conjunction with our brains have the remarkable ability to serve as a spectrum analyzer for sound waves. How this is done is still a matter of active research.

18.4   Pitch

The word **pitch** is used to indicate a person's subjective impression about the frequency of a sound. A high-pitch sound like that from a piccolo has a high vibration frequency, while a low-pitch sound like that from a foghorn has a low vibration frequency. A normal ear is sensitive to sounds in the audio-frequency range from about 20 Hz to 20,000 Hz. As a person grows older, the limits of this hearing range shrink.

Most sounds we hear are complex waves consisting of sinusoidal waves of many different frequencies. The sound we call noise includes waves of a broad range of frequencies, and we seldom associate a pitch with the noise. However, if the complex sound consists of a fundamental frequency vibration and higher harmonics whose frequencies are multiples of the fundamental frequency, then we usually identify the pitch of the sound as being that of the fundamental frequency. This is true even when the amplitudes of the higher harmonics are greater than that of the fundamental.

Several physiological aspects of our hearing system can cause variations in the pitch we hear even though the frequency of the sound is unchanged. For example, a loud, high-frequency sound may seem to have a higher pitch than the same frequency sound played softly. Under some conditions these variations may be considerable. Fortunately, these conditions do not often arise in music, and the pitch change can usually be ignored.

18.5   Quality of Sound

A violin playing concert A at 440 Hz does not sound like an oboe or French horn playing the same note. Why do musical instruments sound differently even when they play equally loudly at the same pitch? We say that the sounds of the instruments have different **quality** or timbre, the characteristic that distinguishes one sound from others of equal pitch and loudness.

Differences in the quality of various sounds are related to their frequency spectra. When a musical instrument is played, more than one of its standing-wave vibrations is excited. For example, when a violin string is bowed, 10 or 20 standing-wave vibrations are excited simultaneously. The pitch of the sound is that of the lowest standing-wave frequency present in the complex wave. *The **quality** of the sound depends on the number and relative amplitudes of the other standing-wave frequencies that make up the complex sound wave and on the way that this spectrum of tones changes with time.*

18.6 Intensity and Intensity Level

We have distinguished musical and spoken sounds by two properties: pitch and quality. Sounds are also distinguished by a third property, loudness. The loudness of a sound depends partly on the amplitude of the vibrations the sound induces in our eardrums, which in turn depends on the magnitude of the pressure variation caused by the sound wave. The pressure fluctuation of a barely audible sound is called the **threshold of audibility.** For the normal ear of a young person, the threshold pressure is about 2×10^{-5} N/m^2, which is less than one-billionth atmospheric pressure. A sound that is painful or harmful to the ear is about 10^6 times greater in pressure, or about 20 N/m^2. The pressures just mentioned are fluctuations about the normal pressure of the atmosphere (1.0×10^5 N/m^2 at sea level).

Intensity

Instruments built to measure the loudness of a sound do not measure the sound's pressure amplitude but instead measure two other quantities called intensity and intensity level. The **intensity** I of a wave of any type is defined as the wave energy that crosses an area A perpendicular to the wave's direction in a time t divided by the area and time:

$$I = \frac{E}{tA} = \frac{P}{A}. \qquad (18.5)$$

Intensity can also be defined in terms of power P since power is the energy that flows per unit time. The units of intensity are the units of energy (joule) divided by the product of the units of area (m^2) and time (s), or J/m$^2 \cdot$s. The unit J/s is called a watt; thus, the units of intensity are often given as watt/m^2, or W/m^2.

EXAMPLE 18.6 The intensity of sound at a construction site is 0.1 W/m^2. Assuming that the area of the eardrum is 0.2 cm^2, how much sound energy is absorbed by one ear in an 8-h work day?

SOLUTION Rearranging Eq. (18.5) to solve for energy, we find that

$$E = IAt = \left(0.1\,\frac{\text{J}}{\text{s}\cdot\text{m}^2}\right)(0.2 \times 10^{-4}\,\text{m}^2)\left(8\,\text{h} \times \frac{3600\,\text{s}}{1\,\text{h}}\right) = \underline{0.06\,\text{J}}. \qquad \blacksquare$$

The intensities of some familiar sounds are listed in Table 18.1. Our environment includes sounds that differ in intensity from about 10^{-12} W/m^2 (barely audible) to nearly 1 W/m^2 (a painful sound). The noise in a classroom is about 10^{-7} W/m^2.

Intensity Level

Because of the wide variation in the range of sound intensities, a different quantity for measuring sound intensities, called *intensity level,* has been developed. Intensity level is not a measure of a sound's intensity but is a comparison of the intensity of one sound and the intensity of a reference sound. **Intensity level β** is defined on a logarithmic scale as follows:

$$\beta = 10 \log \frac{I}{I_0}, \tag{18.6}$$

where I_0 is a reference intensity to which other intensities I are compared (for sound, $I_0 = 10^{-12} \text{ W/m}^2$). The log in Eq. (18.6) represents the logarithm to the base ten of the ratio I/I_0.* Intensity level is a unitless quantity because the units of I and I_0 in Eq. (18.6) cancel. Nevertheless, intensity level has a dimensionless unit called the **decibel** (dB). The unit serves as a reminder of what we are calculating, much like the radian is a dimensionless reminder of one way of specifying angles. The intensity levels β of several familiar sounds are listed in Table 18.1.

EXAMPLE 18.7 The sound in an average classroom has an intensity of 10^{-7} W/m^2. (a) Calculate the intensity level of that sound. (b) If the sound intensity is doubled, what is the new sound intensity level?

SOLUTION (a) The ratio of the sound intensity I and the reference intensity I_0 is

$$\frac{I}{I_0} = \frac{10^{-7} \text{ W/m}^2}{10^{-12} \text{ W/m}^2} = 10^5.$$

*The use of logarithms is reviewed in Appendix A.

TABLE 18.1 Intensities and Intensity Levels of Common Sounds

Source of Sound	Intensity (W/m²)	Intensity Level (dB)	Description
Large rocket engine (nearby)	10^6	180	
Jet takeoff (nearby)	10^3	150	
Pneumatic riveter; machine gun	10	130	
Rock concert with amplifiers (2 m); jet takeoff (60 m)	1	120	Pain threshold
Construction noise (3 m)	10^{-1}	110	
Subway train	10^{-2}	100	
Heavy truck (15 m); Niagara Falls	10^{-3}	90	Constant exposure endangers hearing
Noisy office with machines; average factory	10^{-4}	80	
Busy traffic	10^{-5}	70	
Normal conversation (1 m)	10^{-6}	60	
Quiet office	10^{-7}	50	Quiet
Library	10^{-8}	40	
Soft whisper (5 m)	10^{-9}	30	
Rustling leaves	10^{-10}	20	
Normal breathing	10^{-11}	10	Barely audible
	10^{-12}	0	Hearing threshold

Since log 10^5 is 5, the sound intensity level is

$$\beta = 10 \log 10^5 = 10(5) = \underline{50 \text{ dB}}.$$

(b) If I is doubled, the ratio of I and I_0 becomes

$$\frac{I}{I_0} = \frac{2 \times 10^{-7} \text{ W/m}^2}{10^{-12} \text{ W/m}^2} = 2 \times 10^5.$$

The sound intensity level is, then,

$$L = 10 \log (2 \times 10^5) = 10(5.3) = \underline{53 \text{ dB}}. \qquad \blacksquare$$

The intensity levels listed in Table 18.1 can be calculated easily by considering the power of 10 by which the sound's intensity exceeds the reference intensity. For example, the sound intensity in a noisy office is about 10^{-4} W/m², which is 10^8 times more intense than the reference intensity of 10^{-12} W/m². Since the noisy office is 8 powers of 10 greater than I_0, the intensity level is $10(8) = 80$ dB.

Determining Intensity from Intensity Level

Sometimes in our problem solving we need to calculate the intensity of a sound from its known intensity level. To do this, we use another general property of logarithms [see Eq. (A.4) in Appendix A.3]. If

$$\frac{\beta}{10} = \log (I/I_0), \quad \text{then} \quad I/I_0 = 10^{\beta/10}$$

or

$$I = 10^{\beta/10} I_0. \qquad (18.7)$$

For example, suppose that a shout produces an intensity level of 75 dB. The intensity of the sound is

$$I = 10^{75/10} I_0 = 10^{7.5} I_0 = (10^{0.5})(10^7) I_0$$
$$= 3.2 \times 10^7 I_0 = 3.2 \times 10^7 (10^{-12} \text{ W/m}^2) = 3.2 \times 10^{-5} \text{ W/m}^2.*$$

EXAMPLE 18.8 The noise from a vacuum cleaner is 85 dB while that from a television set is 78 dB. What is the intensity level of the two sounds together?

SOLUTION We first calculate the intensities of the two sounds and then add them to find the total sound intensity. We then convert this total intensity to intensity level in the usual fashion. The intensity of the sound from the vacuum cleaner is

$$I = 10^{85/10} I_0 = (3.2 \times 10^8)(10^{-12} \text{ W/m}^2) = 3.2 \times 10^{-4} \text{ W/m}^2.$$

The intensity of the sound from the television is

$$I' = 10^{78/10} I_0 = (6.3 \times 10^7)(10^{-12} \text{ W/m}^2) = 0.6 \times 10^{-4} \text{ W/m}^2.$$

*Remember that $10^{a+b} = (10^a)(10^b)$. Thus, $10^{7.5} = 10^{0.5+7.0} = (10^{0.5})(10^{7.0}) = 3.2 \times 10^7$.

The total intensity of both sounds is

$$I_{total} = 3.2 \times 10^{-4}\,\text{W/m}^2 + 0.6 \times 10^{-4}\,\text{W/m}^2 = 3.8 \times 10^{-4}\,\text{W/m}^2.$$

The intensity level of both sounds together is

$$\beta_{total} = 10\log\frac{I_{total}}{I_0} = 10\log\frac{3.8 \times 10^{-4}\,\text{W/m}^2}{10^{-12}\,\text{W/m}^2}$$
$$= 10\log(3.8 \times 10^8) = 10(8.6) = \underline{86\,\text{dB}},$$

only 1 dB more than the intensity level from the vacuum cleaner alone. ∎

When two sounds of different intensity levels are heard simultaneously, the intensity level of the two sounds is not much more than that of the louder sound. The loud sound can sometimes mask or hide the less-loud sound.

Comparing Intensity Levels

Scientists often compare the intensity levels of two sounds by taking their difference. For example, to compare the intensity level β_1 of the sound at an intersection during normal hours with the intensity level β_2 during rush hours, we subtract β_1 from β_2. Using general properties of logarithms, this difference $\beta_2 - \beta_1$ can be calculated easily from the ratio of the intensities of the two sounds (I_2/I_1), as we see next:*

$$\beta_2 - \beta_1 = 10\log(I_2/I_0) - 10\log(I_1/I_0)$$
$$= (10\log I_2 - 10\log I_0) - (10\log I_1 - 10\log I_0)$$
$$= 10\log I_2 - 10\log I_1,$$

or

$$\beta_2 - \beta_1 = 10\log(I_2/I_1). \tag{18.8}$$

Relative Loudness of a Sound

Suppose that intensity measurements at an intersection and calculations using Eq. (18.8) indicate that the rush hour intensity level is 10 dB greater than the intensity level caused by normal traffic ($\beta_2 - \beta_1 = 10$ dB). How much louder does the sound seem to our ears?

Our ears do not sense that the loudness of a sound increases in direct proportion to either the intensity or the intensity level of the sound. If the intensity of a sound doubles, we do not think that the sound is twice as loud. Only when the sound intensity is 10 times greater do we sense that the sound is twice as loud. If a sound is 10^2 times more intense than another sound, it seems 2^2 times louder to our ears; if 10^3 times more intense, it seems 2^3 times louder.

Since the intensity level of a sound increases by 10 dB each time the intensity increases by a factor of 10, the loudness of a sound seems to double each time the intensity level increases by 10 dB. A sound that is 30 dB greater than another sound seems 2^3 times louder.

*Note that $\log(ab) = \log a + \log b$ and that $\log(a/b) = \log a - \log b$.

EXAMPLE 18.9 A chicken farmer decides to increase the size of her flock from 500 chickens to 2000 chickens. She must relieve her husband's anxiety about the increased noise (he says that 2000 chickens will be four times louder than 500 chickens). (a) Calculate the change in intensity level and estimate how much louder the sound will be. (b) How big can her flock become before it seems four times as loud?

SOLUTION (a) We will represent the sound intensity of 500 chickens by the symbol I. When the number of chickens is quadrupled, the intensity increases to $4I$. The difference in the intensity levels is calculated using Eq. (18.8):

$$\beta_{2000} - \beta_{500} = 10 \log (I_{2000}/I_{500}) = 10 \log (4I/I) = 10 \log 4 = \underline{6 \text{ dB}}.$$

The number of chickens belonging to the farmer increased by a factor of 4, as did the intensity of the sound coming from the chickens. However, the sound would seem less than two times louder to our ears because the intensity level increased by less than 10 dB.

(b) If the farmer increased the flock from 500 to 5000 chickens, the intensity would increase by a factor of 10, and the intensity level would be 10 dB greater. But the loudness as perceived by our ears would be only two times greater! If she increased her flock from 500 to 50,000 chickens, the intensity would increase by $10^2 = 100$, and the intensity level by 20 dB. The chickens would now seem $2^2 = 4$ times louder to our ears. ■

18.7 The Human Ear

A simplified sketch of the human ear is shown in Fig. 18.12. The part of the ear at the side of our head, called the *pinna*, gathers and guides sound energy into the auditory canal. Because of the pinna, the pressure variation at the eardrum due to a sound wave is about *two times* the pressure variation that would occur without the pinna and auditory canal.

The pressure variation at the eardrum causes it to vibrate. A large fraction of the vibrational energy is transmitted through three small bones in the middle ear known collectively as the ossicles and individually as the hammer, anvil, and stirrup. These three bones constitute a lever system that increases the pressure that can be exerted on a small membrane of the inner ear known as the oval window. The pressure increase is possible for two reasons: (1) The hammer feels

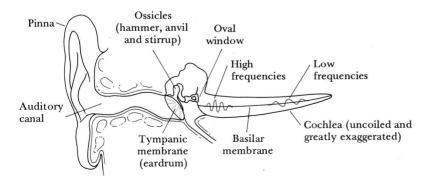

FIG. 18.12. A simplified depiction of the ear.

a small pressure variation over a large area of its contact with the eardrum, whereas the stirrup exerts a large pressure variation over the small area of its contact with the oval window. This difference in areas causes the pressure to be increased by a factor of about 15 to 30. (2) The three bones act as a lever that causes increased force of the stirrup against the oval window, although the displacement caused by the force is less than if the bones were not present. The pressure variation caused by sound can be increased up to 3 times by this lever action. Because of all these effects, pressure fluctuation of the stirrup against the oval window can be *180 times* greater than the pressure fluctuation of a sound wave in air before it reaches the ear (a factor of 2 due to the pinna and auditory canal, a factor of up to 30 because of the area change from the eardrum to the oval window, and a factor of 3 from the lever action of the ossicles).

The increased pressure fluctuation against the oval window causes a fluctuating pressure in the fluid inside the cochlea of the inner ear. The fluctuating pressure is sensed by nerve cells along the basilar membrane. Nerves nearest the oval window respond to high-frequency sounds whereas nerves farther from the window respond to low-frequency sounds. Thus, our ability to distinguish the frequency of a sound depends on the variation in sensitivity of different cells along the basilar membrane to different frequencies. This ability is remarkable in that the whole membrane is only about one centimeter long, and yet a normal ear can distinguish sounds that differ in frequency by about 0.3 percent (a frequency difference of 3 Hz can be distinguished for two sounds near 1000 Hz). This ability to distinguish sounds of slightly different frequency spans the human hearing range from about 20 Hz to 20,000 Hz.

Summary and Additional Readings

1. **Standing waves** are produced by the interference of waves reflecting back and forth on or in an object of length L. For a **string** and for an **open pipe,** the frequencies of the standing-wave vibrations are

$$f_n = n(v/2L), \qquad \text{where } n = 1, 2, 3, 4, \ldots . \quad \textbf{(18.2)}$$

For strings, v is the speed of the wave on the string and equals $\sqrt{T/(m/L)}$. For air columns or pipes, v is the speed of sound in the pipe. For a **closed pipe,** the *resonant frequencies* are

$$f_n = n(v/4L), \qquad \text{where } n = 1, 3, 5, 7, \ldots . \quad \textbf{(18.4)}$$

2. The actual sound waves that we hear are usually **complex waves** produced by adding two or more sinusoidal waves, each of different frequency and wavelength.

3. The **pitch** of a sound is our subjective impression of how high or low its frequency is. The pitch of a musical sound is very nearly the same as the lowest standing-wave frequency of the sound's source.

4. The **quality** of a sound is the characteristic of a sound that distinguishes it from other sounds of the same pitch and loudness. Quality depends on the frequency spectra of the complex waves coming from the sound source.

5. The **loudness** of a sound depends on the energy in a sound wave and the fluctuations in air pressure it causes against our ear.

The relation between loudness and sound energy is quantified using the quantities intensity and intensity level.

Sound **intensity** I is defined as the wave energy E that passes a cross section of area A in a time t, divided by that area and time:

$$I = E/At = P/A. \quad \textbf{(18.5)}$$

Since E/t is defined as power, intensity is also the sound power P passing a cross section of area A.

The **intensity level** β of a sound of intensity I is defined as

$$\beta = 10 \log (I/I_0) \quad \textbf{(18.6)}$$

in units of dB, where I_0 is the least intense sound heard by the normal ear (10^{-12} W/m^2) and log is the logarithm to the base 10 of the ratio I/I_0. A 10-dB change in intensity level corresponds roughly to a doubling of the loudness of the sound.

Readings from Scientific American: The Physics of Music, W. H. Freeman, San Francisco (1978).

J. Backus. *Acoustical Foundations of Music,* W. W. Norton, New York (1969).

Thomas D. Rossing, *The Science of Sound,* Addison-Wesley, Reading, Mass. (1982). This book covers everything from instruments to room acoustics to electronic music.

Questions

1. The low-frequency strings on a piano are about the same length as those used for the middle frequencies. The low-frequency strings, however, are more massive. Why?

2. How are the fundamental frequencies of the strings of a stringed instrument affected by increasing their tension? Explain.

3. The wavelength of the wave on a string of fixed length is unchanged when the string is tightened. However, the sound wave coming from the string does have a different frequency and wavelength. Explain.

4. Standing-wave vibrations analogous to those on a vibrating string can occur on a bridge. What would the second harmonic bridge vibration look like? How might it be excited?

5. A mechanical device produces vibrations at one end of a string of fixed length. (a) Show qualitatively on a graph the expected vibration amplitude-versus-frequency curve that you expect to observe if the frequency of the exciting device is increased from zero to some arbitrary high frequency. (b) If you lightly rest your finger one-fourth of the way from one end of the string, how is the curve in part (a) modified? Explain and show the expected curve.

6. If you look at the third harmonic of a vibrating string, you will see two nodes between the ends. The string vibrates in three sections. How does kinetic energy get across nodes from one antinode to another?

7. After inhaling helium, the frequency at which you speak is increased so that you sound like Donald Duck. Why does this occur?

8. A hot air duct blows warm air on the pipes of an organ. Explain how this affects their vibration frequency.

9. Assume that the members of an orchestra did not warm their instruments before a concert. Compare the change in pitch of the stringed instruments to that of the woodwinds as the instruments and air around them warm. Be sure to justify your conclusions.

10. *Estimate* the sound frequencies that resonate between the ceiling and floor of your physics classroom or lecture hall.

11. You fill a pop bottle with water. Describe and explain the changing sound you hear as the water nears the top.

12. If you touch the middle of a violin string very lightly while it is being bowed, you hear a sound twice the normal frequency. What has your finger done to cause the change in pitch of the sound?

13. The pitch of notes played by instruments in an orchestra ranges from about 50 Hz to 4000 Hz. Why do high-fidelity systems amplify sounds up to 20,000 Hz? How would the sound of an orchestra be altered if the system reproduced sounds to only 4000 Hz?

14. Why is the quality of the human voice affected by a head cold?

15. Sound can travel from inside your mouth to your ear via bones in your head. How might this affect your perception of the sound of your voice compared to the sound you hear from a tape recorder? [*Note:* Bones transmit low-frequency sounds better than high.]

16. When a drummer crashes the cymbals, what happens to the sound energy created by the crash?

17. A sound intensity is increased by 10^4. (a) About how much louder does the sound seem? (b) By how much has the sound intensity level increased?

18. The intensity level of a sound is decreased by 20 dB. (a) About how much more quiet does it seem? (b) By what factor has the sound intensity changed?

Problems

For all problems, unless stated otherwise, assume that the speed of sound in air is 340 m/s.

18.1 Standing Waves on Strings

1. A banjo G-string is 0.69 m long and has a fundamental frequency of 392 Hz. (a) Calculate the speed of a wave or pulse on the string. (b) Identify three other frequencies at which the string can vibrate.

2. A pulse travels at speed 16 m/s on a 24-m-long electric power line. Calculate three standing-wave frequencies at which the line will vibrate.

■ 3. A 0.33-m-long violin string has a mass of 1.42 g. Calculate its fundamental frequency of vibration when pulled by a tension of 72 N.

■ 4. How far from the end of the banjo string discussed in Problem 1 must a fret be placed so that its fundamental frequency is increased to 490 Hz when the string is held down at the fret?

■ 5. A violin A-string 0.33 m long vibrates at a fundamental frequency of 440 Hz. Where should your finger press the string against the finger board so that its decreased length causes it to vibrate at (a) five-fourths times the original frequency and (b) four-thirds times the original frequency?

■ 6. A 5.0-m-long rope of mass 0.40 kg is secured at one end and stretched with a 120-N tension at the other end. The rope is vibrating in three segments with nodes separating each segment. Calculate its vibration frequency.

■ 7. A canary sits 10 m from the edge of a clothesline 30 m long, and a grackle sits 5 m from the other end. At what frequency can the line be vibrated so that the grackle is dislodged by the vibration while the canary sits undisturbed? The tension in the line is 200 N and the mass per unit length is 0.10 kg/m.

■ 8. Perform an *estimation* calculation to determine approximately the fundamental frequency of vibration of a telephone line between adjacent poles near where you live.

9. Two wires on a piano are the same length and are pulled by the same tension. Wire 1 vibrates at a frequency 1.5 times that of wire 2. Calculate the ratio of their masses (m_2/m_1).

10. By what percent does the frequency of a piano string change if its tension is increased by 10 percent?

11. The beat frequencies produced by several like instruments that are slightly out of tune with each other cause a pleasing sensation called the "chorus effect." By what factor must the tension of the A-string of one violin initially vibrating at 440 Hz be increased to produce 2-Hz beats with another violin with which it was initially in tune?

18.2 Standing Waves in Air Columns

12. A clarinet (considered a closed pipe) has an effective bore length of 0.60 m. Calculate the first three standing-wave frequencies of the clarinet when all the finger holes are closed.

13. (a) Calculate the first three standing-wave frequencies of an open pipe 40 cm long. (b) Do the same for a closed pipe 40 cm long.

14. The 2779-m Brooklyn-Battery Tunnel, connecting Brooklyn and Manhattan, is one of the world's longest underwater car tunnels. (a) Calculate its fundamental frequency of vibration. (b) What harmonic must be excited so that it resonates in the audio region at 20 Hz or greater?

15. A wooden flute, open at both ends, is 0.48 m long. (a) Calculate its fundamental vibration frequency. (b) How far from one end should a finger hole be placed to produce a sound whose frequency is four-thirds that calculated in part (a)? Be sure to justify how you arrive at your answer.

16. The lowest three standing-wave vibration frequencies of an organ pipe are 120 Hz, 360 Hz, and 600 Hz. (a) Is the pipe open or closed and what is its length? (b) Calculate the frequencies of the first two harmonic vibrations on a pipe of the same length but of the other type than that described in part (a).

17. A person hums in a well and finds strong resonances at frequencies of 60 Hz, 100 Hz, and 140 Hz. (a) What is the fundamental frequency of the well? Explain. (b) How deep is the well?

18. (a) Use the dimensions of a 12-oz pop bottle to *estimate* its fundamental resonant frequency when empty. (b) Calculate the depth of water that must be added to increase its frequency by a factor of 4/3.

19. Assume that a friend has a hole in his head extending from one ear out the other. *Estimate* the frequency at which air in this hole would vibrate when excited by moving air as the friend jogs.

20. The speed of sound can be measured using an apparatus such as the one shown in Fig. 18.13. A 440-Hz tuning fork vibrating above a tube partially filled with water initiates sound waves in the tube. The air inside the tube vibrates in sympathy with the tuning fork when the water in the tube is lowered 0.20 m and 0.60 m from the top of the tube. Use this information to calculate the speed of sound in air.

21. Use dimensional analysis to determine the functional dependence of the standing-wave vibration frequency of an open pipe on the following quantities: the pipe's length L, and the density ρ and bulk modulus B of the air in the pipe.

22. The fundamental frequency of a closed pipe, such as your vocal tract, is 240 Hz when filled with air whose molecules have an average molecular mass of 29. Using ideas from the ki-

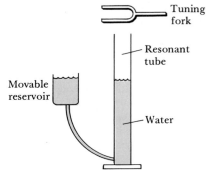

FIG. 18.13

netic theory of gases, estimate the vibration frequency when the pipe is filled with helium atoms of atomic mass 4. Be sure to show clearly the reasoning behind your answer.

18.3, 18.4, and 18.5 Complex Waves, Pitch, and Quality of Sound

23. (a) Carefully add the waves shown in Fig. 18.14 to get the resultant complex waveform. (b) Construct the frequency spectrum for the complex wave.

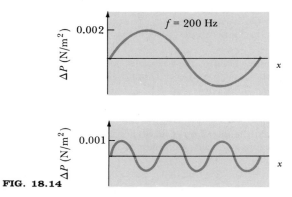

FIG. 18.14

24. Construct the complex wave whose frequency spectrum is shown in Fig. 18.15. Include two full wavelengths of the lowest-frequency wave. Start both waves with zero amplitude on the left and rising.

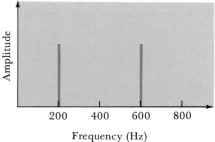

FIG. 18.15

25. A complex wave is made of two sinusoidal waves of equal amplitude. One wave has a frequency of 175 Hz and the other a frequency of 525 Hz. (a) Make a frequency spectrum of the complex wave. (b) Construct the waveform of the complex wave.

It should include two full wavelengths of the lowest-frequency wave. Start both waves with zero amplitude on the left and rising.

■ **26.** Repeat Problem 25 for a complex wave made of three sinusoidal waves: wave 1 at 175 Hz and amplitude 0.002 N/m^2, wave 2 at 350 Hz and amplitude 0.001 N/m^2, and wave 3 at 525 Hz and amplitude 0.001 N/m^2.

■ **27.** Construct a frequency spectrum and waveform (as in Problem 25) for a clarinet playing at a fundamental frequency of 400 Hz. The sound consists of three harmonics of equal amplitude. [*Note:* A clarinet behaves like a closed pipe.]

■ **28.** The frequency spectrum of an alto saxophone when playing a note is shown in Fig. 18.16. (a) What pitch sound do you hear? (b) Construct a waveform for the sound.

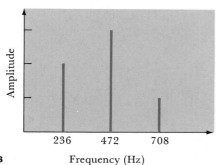

FIG. 18.16

18.6 Intensity and Intensity Level

29. The sound intensity at a gasoline station next to a turnpike averages 10^{-3} W/m^2. The owner decides to collect this energy and convert it to thermal energy for heating his building. If his conversion is 100 percent efficient, what is the length of the side of a square sound collector that is needed to provide thermal energy at a rate of 500 W? Is this practical? Explain.

30. The sound intensity 3 mi from the takeoff point of a Concord supersonic jet is 0.60 W/m^2. What area sound collector would absorb enough energy to run a 40-W lightbulb?

31. In music a very soft sound called "triple piano" (ppp) has an intensity of about 10^{-8} W/m^2. A very loud sound called "triple forte" (fff) has an intensity of about 10^{-2} W/m^2. Convert these intensities to intensity levels (units of dB).

32. Convert the following intensities to intensity levels: (a) $3.0 \times 10^{-8} \text{ W/m}^2$ in a quiet living room; (b) $2.0 \times 10^{-5} \text{ W/m}^2$ inside a car; (c) $6.0 \times 10^{-3} \text{ W/m}^2$ near a passing truck; and (d) $5.0 \times 10^{-1} \text{ W/m}^2$ at a construction site where an air hammer is operating.

33. Calculate the change in intensity level when a sound intensity is increased by (a) a factor of 8 ($I_2 = 8I_1$); (b) a factor of 80; and (c) a factor of 800.

■ **34.** One violin creates a sound whose intensity level is 60 dB. Calculate the intensity level of 16 violins, each playing at this intensity.

■ **35.** A swarm of 10,000 locusts causes a sound of intensity level 75 dB. Calculate the intensity level caused by one locust.

■ **36.** A person has a 34-dB hearing loss. By what factor must a hearing aid amplify the sound intensity to restore normal hearing?

■ **37.** A high-fidelity speaker is said to have a 3-dB flat response to sounds from 100 Hz to 10,000 Hz. This means that if electrical signals of equal power are sent to the speaker at any two frequencies in this range, the sound intensity levels coming from the speaker at these frequencies will differ by 3 dB or less. What is the maximum variation of the sound intensities produced by the speakers?

■ **38.** The average intensity level of a person's voice is 58 dB. A microphone and amplification system increases the intensity level to 70 dB. By what factor did the amplification system increase the sound intensity?

■ **39.** Two separate sounds are incident on the ear. The intensity level of one sound is 80 dB and of the other 85 dB. Calculate the intensity level of the two sounds together.

■ **40.** The average intensity level of sound near a busy intersection is 86 dB. Calculate the sound energy incident on a person's eardrum of area 0.20 cm^2 in 8 h.

■ **41.** The average sound intensity level from an interstate highway with 1 car per 100 m is 78 dB. Calculate the average intensity and average intensity level of the sound if there are 14 cars per 100 m.

■■ **42.** While camping, you record a thunderclap whose intensity is 10^{-2} W/m^2. The clap reaches you 3 s after a flash of lightning that caused the thunder. Calculate the total acoustical power generated by the lightning. [*Hint:* Assume that the sound intensity is the same over all parts of the surface of a sphere centered at the place where the lightning flash occurred.]

■■ **43.** A window whose area is 1.5 m^2 opens onto a street. Sound at an intensity level of 75 dB enters the open window. How long would it take to collect enough sound energy to warm a cup of tea (0.20 kg of water) from 20°C to 40°C?

18.7 The Human Ear

■ **44.** Inhabitants of a planet in the constellation Ursa Minor have hearing systems that consist of cylindrical holes in their chests. Each hole is a different length and has a fixed membrane of nerves at the bottom for detecting sound. (a) If the holes vary in length from 5 cm to 50 cm and the speed of sound on the planet is 900 m/s, what range of frequencies can these people detect? (b) If there is a total of 51 tubes that vary in length by ±1 cm from the neighboring tubes, what is the ratio of the frequency of the longest tube to that of the next shorter tube? (c) How does this hearing system compare in range and discrimination (ability to distinguish sounds of different frequency) to our own? The human ear can distinguish two sounds that differ in frequency by about 0.3 percent.

PART V

Light and Optics

In Part IV we examined several general properties of waves and their applications in the study of mechanical waves such as sound waves. Now we introduce several new ideas about waves. These ideas and those developed earlier will be used in the study of light. In Chapter 19 we learn about the reflection of light at the boundary between two media and about the refraction (bending) of the light transmitted into a new medium. This information is important in Chapter 20, where we study the bending of light that passes through lenses and the reflection of light from mirrors. We discuss how lenses and mirrors change the visual appearance of objects viewed with cameras, projectors, eyeglasses, microscopes, and telescopes. In Chapter 21 we encounter some of the interesting phenomena that depend on the interference of light waves coming from different sources. We learn how the wavelength of light is measured in spectrographs, about the diffraction or spreading of light as it passes through narrow openings, and about phenomena related to the polarization of light waves.

CHAPTER 19

Reflection and Refraction

In our previous discussions of waves, we learned that a wave traveling in one medium is partly reflected and partly transmitted at its boundary with a second medium. We limited our discussion to waves that strike a boundary perpendicular to its surface. In this chapter we examine the more general situation in which the directions of the incident, reflected, and transmitted waves are not perpendicular to the boundary between media. We will find that if the incident wave does not strike perpendicular to the surface, then the transmitted wave moves in a different direction from the incident wave; the transmitted wave is said to be *refracted*.

Refraction is a general property of all waves (sound, water, light, and so forth). The refraction of light waves is especially important because it affects the ability of glass lenses in microscopes, telescopes, and eyeglasses to focus and change the direction of light. Most of our examples in this chapter involve the reflection and the refraction of light, although the principles apply to all types of waves. Before turning to the reflection and refraction of waves, we briefly discuss the nature of light.

19.1 The Nature of Light

The study of the nature of light has extended over thousands of years. Four hundred years before Christ, during the Golden Age of Enlightenment in Greece, men like Socrates and Plato speculated that light rays left our eyes, were reflected from objects, and then returned, thereby allowing us to see. Euclid's book *Optics*, published around 300 B.C., contained 58 theorems about the nature of light, especially its geometrical behavior. According to legend, Archimedes, a scientist who studied Euclid's work, built an optical device to defend his hometown in Sicily against the Romans. The device, a huge concave mirror, would reflect sunlight on enemy ships and cause them to catch fire and burn. According to a popular myth, the device supposedly burned several of the 60 ships beseiging the town during one battle. The other ships fled in panic.

In more recent history, our understanding of the nature of light has centered on two conflicting theories. During the 1700s, a **corpuscular theory** was

On a hot summer day, the distant highway often appears wet. What we are seeing is skylight that traveled toward the highway but bent or refracted upward into our eyes before striking the pavement. We are seeing the sky. (Photo by Roderick Beebe/Photo Researchers, Inc.)

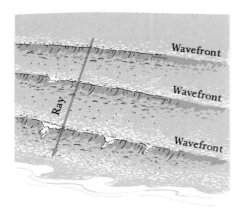

FIG. 19.1. A series of water waves moves toward a shore. The crests are examples of wavefronts. A ray is an arrow drawn perpendicular to the wavefronts and pointing in the direction the wave travels.

championed by Newton and widely accepted by other scientists. According to this theory, light streaming from objects consisted of huge numbers of tiny particles or corpuscles of light. The corpuscular theory was supported by the fact that light seemed to move in straight lines instead of spreading out as waves do. Because Newton also observed that sunlight passing through a prism was broken into a rainbow of different colors, he felt that white light consisted of many kinds of tiny particles, each a different color. When the different light corpuscles were brought back together, white light was reformed.

Not everyone believed the corpuscular theory. A conflicting theory of light was proposed by the Dutch scientist Christian Huygens (1629–1695), a contemporary of Newton. According to Huygens, light was a **wave** that traveled through an invisible substance called ether (not to be confused with the liquid used as an anesthetic). As light passed, the ether vibrated. In 1801, an English scientist, Thomas Young (1773–1829), showed that light beams from two sources interfered with each other in the same way that two sound waves interfere. This constructive and destructive interference of light was inconsistent with the corpuscular theory. It was, however, consistent with the wave theory, and so, during the 1800s, light was regarded as a wave phenomenon. The nature of light waves was eventually described in detail in the 1860s by the British physicist James Clerk Maxwell (1831–1879). Maxwell's four equations embraced all of the previously known behavior of electrical and magnetic phenomena. His equations predicted the existence of light waves (and other types of electromagnetic waves, to be discussed in Chapter 31) and the speed at which they traveled. Maxwell's equations left little doubt about the wave nature of light.

The reign of the wave theory was short. New experiments performed in the latter part of the nineteenth century and early in this century produced results inconsistent with the wave theory. In 1905 Albert Einstein published a theory concerning what was called the *photoelectric effect*. According to this theory, light absorbed by atoms consisted of individual, massless bundles of energy that are now called **photons.** The light absorption was inconsistent with the wave theory of light.

Since the early 1900s, scientists have viewed light as a phenomenon with a dual nature, part particle and part wave. For now, all phenomena we discuss can be explained in terms of the wave nature of light. In Chapter 31 we will examine in greater detail the particle nature of light.

19.2 Coherence, Wavefronts, and Rays

Like all waves, light can be viewed as a series of crests and troughs moving away from a source. If several sources produce waves simultaneously, the surrounding space is disturbed by the interference of the waves from each source. If the different sources synchronize their vibrations, the resultant wave disturbance can have an orderly apperance—like a marching band on parade. This is called a **coherent wave.** A wave created by random disturbances (such as by rocks thrown randomly into a small pond) has a disorganized appearance that changes unpredictably. This is called an **incoherent wave.** In this chapter and Chapter 20, we consider the direction of travel of all types of waves (sound waves, water waves, light waves, and so on). Our attention is restricted to coherent waves because they are easiest to picture and analyze.

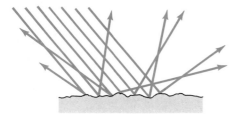

FIG. 19.2. Diffuse reflection from an uneven surface.

Coherent waves can be drawn as one crest following another at a distance of one wavelength. These crests are called **wavefronts.** The disturbance caused by the wave is the same at every point on the wavefront. Wavefronts moving across a large body of water are pictured in Fig. 19.1.

A ray is an arrow drawn perpendicular to a wavefront that points in the direction of the wave's motion (Fig. 19.1). Wavefronts and rays are convenient geometrical devices to help visualize the changing direction of a wave as it is reflected and transmitted at the boundary between two media.

19.3   Reflection

As we learned in Chapter 17, when a wave is incident on a boundary between two media that have different properties, part of the wave is reflected. For example, when light traveling in air strikes glass, a fraction of the light is reflected and the remaining portion is transmitted into the glass. The fraction that is reflected depends on the type of glass and on the angle at which the light meets the surface of the glass. If the light strikes perpendicular to the glass, about 4 percent is reflected. If it strikes at other angles, a larger percent is reflected.

The direction of travel of a reflected light beam depends on the roughness of the reflecting surface relative to the wavelength of the reflected wave. If the wave is incident on a rough surface, the reflected wave travels in many directions (Fig. 19.2). This is called **diffuse reflection** and occurs if the irregularities in the surface are about the same size as or larger than the wavelength of the wave striking the surface. The walls of concert halls are often made a "fluted" shape (Fig. 19.3a) so that sound waves striking the walls are reflected in different directions. A person in the audience receives a small amount of reflected sound from many parts of the wall (Fig. 19.3a) rather than a larger amount of reflected sound from one part of the wall (Fig. 19.3b). Light is being reflected diffusely from the pages of this book because the irregularities in its surface are larger than the wavelength of the light waves hitting the page.

If, however, a surface is flat, the reflected wave travels in only one direction after reflection. For example, if the wall of a concert hall is flat, as in Fig. 19.3b, sound from the stage will be reflected from only one part of each wall onto your ears. It may sound as if the performing group is actually located at these parts of

Does light experience diffuse or specular reflection from this ordinary sheet of paper (as seen with an electron microscope)? The irregularities in the paper are larger than the wavelength of light. (Photo by David Scharf/Peter Arnold, Inc.)

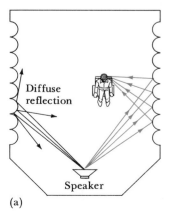

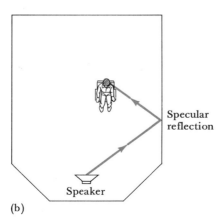

(a)   (b)

FIG. 19.3. (a) With fluted walls, sound is deflected from each small section of a wall to a listener. (b) If the walls are flat, an intense reflected sound comes from only one part of the wall.

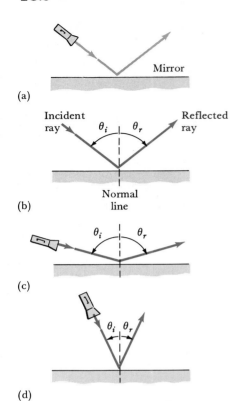

(a)

(b)

(c)

(d)

FIG. 19.4. (a) A narrow beam of light from a flashlight is reflected from a mirror. (b) The directions of the incident and reflected rays are indicated by the angles θ_i and θ_r that they make with a normal line. (c) and (d) As the angle of incidence increases, so does the angle of reflection.

the walls. The reflection of waves from surfaces whose irregularities are smaller than the wavelength of the wave is called **specular reflection.** Examples are the reflection of sound from flat walls and the reflection of light from a mirror.

The directions of the incident and reflected wave for specular reflection are shown by drawing rays. In Fig. 19.4a, a narrow beam of light is reflected from a smooth surface such as a mirror. The incident ray shown in Fig. 19.4b makes an angle θ_i between itself and a dashed line drawn perpendicular to the surface (called the normal line). The reflected ray makes an angle θ_r with respect to the normal line. According to the **law of reflection,** the angle of incidence θ_i equals the angle of reflection θ_r:

$$\theta_i = \theta_r. \tag{19.1}$$

The incident and reflected rays are on opposite sides of the normal line and in the same plane with it.

The law of reflection is easily confirmed by shining a flashlight or laser beam on a mirror. As the incident angle is changed, the reflected angle also changes (Fig. 19.4b–d).

19.4 Index of Refraction

When light or any type of wave strikes the interface between two media at an acute angle, the transmitted wave usually travels in a different direction from that of the incident wave. The transmitted wave is said to be **refracted**—that is, its direction changes. Refraction is caused by a change in speed of the wave as it passes from one medium to another. Geophysicists usually analyze the refraction of sound waves at the boundaries between different media in the earth in terms of the speeds of sound in these media. In the case of light waves, another quantity called the index of refraction is used.

The **index of refraction** n of a medium is the ratio of the speed of light in a vacuum (given the symbol c) and the speed of light v in the medium:

$$n = \frac{c}{v}. \tag{19.2}$$

The index of refraction has no units because it is the ratio of two quantities with identical units. A list of the speeds of light in several different media and of their indices of refraction appears in Table 19.1. Notice that in a vacuum, light travels at a speed

$$c = 2.997925 \times 10^8 \text{ m/s} \simeq 3.00 \times 10^8 \text{ m/s}.$$

Unless great accuracy is desired, the rounded-off value of 3.00×10^8 m/s will be used in our examples and problems. The speed of light c in a vacuum is always greater than its speed v in other media. Since c is greater than v for all media, the index of refraction $n = c/v$ will always be greater than 1. However, the speed of light in air is so close to its speed in a vacuum that the index of refraction of air for practical purposes is 1.000.

TABLE 19.1 Speed and Index of Refraction of Yellow Light in Various Substances

Material	Speed ($\times 10^8$ m/s)	Index of refraction (c/v)
Vacuum	$c = 2.997925 \times 10^8$	1.00000
Air	$v = 2.99706$	1.00029
Carbon dioxide	2.99658	1.00045
Helium	2.99782	1.000034
Water (20°C)	2.2490	1.3330
Ethyl alcohol	2.2016	1.3617
Methyl alcohol	2.2555	1.3292
Benzene	1.9968	1.5014
Carbon disulfide	1.8415	1.6279
Sugar solution, 25%	2.1846	1.3723
Sugar solution, 50%	2.1112	1.4200
Sugar solution, 75%	2.0292	1.4774
Glass, light crown	1.976	1.517
Glass, dense crown	1.888	1.588
Glass, light flint	1.899	1.579
Glass, heavy flint	1.820	1.647
Canada balsam	1.959	1.530
Fluorite	2.091	1.434
Diamond	1.240	2.417

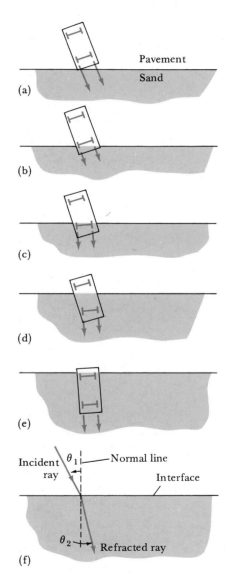

FIG. 19.5. (a) A car drives from a concrete road onto a sandy desert. (b) As the right front wheel hits the sand, the jeep veers right. (e) When it has moved completely onto the sand, its direction has changed. (f) A "ray diagram" summarizing this direction change.

EXAMPLE 19.1 Calculate the thickness of the arctic ice sheet using the fact that a radio-wave pulse travels from the top of the ice to the bottom and back again to the top in 2.1×10^{-5} s. The index of refraction of ice is 1.305.

SOLUTION The radio wave's speed v in the ice can be calculated using Eq. (19.2):

$$v = \frac{c}{n} = \frac{3.00 \times 10^8 \text{ m/s}}{1.305} = 2.30 \times 10^8 \text{ m/s}.$$

We can now rearrange Eq. (2.4) to find the depth D of the ice sheet. Note that it takes the radio wave a time $t = 2.1 \times 10^{-5}$ s to travel a distance $2D$ at a speed $v = 2.30 \times 10^8$ m/s. Thus, $v = 2D/t$, or

$$D = \frac{vt}{2} = \frac{(2.30 \times 10^8 \text{ m/s})(2.1 \times 10^{-5} \text{ s})}{2} = \underline{2400 \text{ m}}. \quad \blacksquare$$

19.5 Refraction

As we said in the last section, the direction of a wave usually changes as it passes from one medium to another; the transmitted wave is said to be refracted. Refraction is caused by the change in speed of a wave as it passes from one medium to another. This phenomenon is illustrated crudely by considering an automobile that moves from pavement onto a field of soft sand or mud, as shown in Fig. 19.5. The two front tires and their axle represent a wave crest, as do the two back tires.

As the right front tire leaves the pavement, it is slowed by the sand (Fig. 19.5b). The front left wheel, still on the pavement, continues at its normal speed. The car veers to the right as it is held back on the right side by the sand (Fig. 19.5c). When all four wheels finally enter the sand, the car straightens but is now moving in a different direction (Fig. 19.5e). If a ray diagram is drawn for the car, it looks as shown in Fig. 19.5f.

The incident angle is usually labeled θ_1. The direction of the refracted ray in the second medium is indicated by the angle θ_2 in Fig. 19.5f. Note that both angles are relative to the normal line and not to the interface! We see that the car's direction bends toward the normal line as it moves from pavement to sand.

If a car were to move out of the sand onto pavement, the front tire that reached the pavement first would move forward more easily than the front tire still in the sand. The axle would again turn. This time, however, the car's direction of travel would bend away from the normal line. (You might try to draw your own diagram for this situation.) The change in direction is similar for waves. **If a wave enters a medium in which its speed decreases, its direction bends toward the normal; if it enters a medium in which its speed increases, its direction bends away from the normal.**

To derive an equation for calculating the expected change in direction, consider the wavefront of the wave shown in Fig. 19.6a as it approaches the boundary between two media. The wavefront could represent the crest of a sound wave moving from air into a wall or the wavefront of light moving from air into glass or water. Point B on the wavefront requires a time t to move forward to the boundary between the two media. Since the wave's speed in medium 1 is v_1, point B moves a distance $v_1 t$ during that time (Fig. 19.6b). Point A, which has just entered medium 2, moves a distance $v_2 t$ in the second medium during that same time. If v_2 is less than v_1, then point A moves a shorter distance than point B, and the wave refracts toward the normal line. However, if v_2 is greater than v_1, point A moves a longer distance than point B, and the wave refracts away from the normal line.

The change in direction can be calculated by comparing the triangles shown in Fig. 19.6c. They each have the same hypotenuse—the length L along the boundary between the media. But the angles θ_1 and θ_2 differ because the sides of the triangles opposite θ_1 and θ_2 are different lengths. Consider the sines of these angles:

$$\sin \theta_1 = \frac{v_1 t}{L} \quad \text{and} \quad \sin \theta_2 = \frac{v_2 t}{L}.$$

If we take the ratio of these two sines, we find that

$$\frac{\sin \theta_1}{\sin \theta_2} = \frac{v_1 t/L}{v_2 t/L} = \frac{v_1}{v_2},$$

or

$$\frac{\sin \theta_1}{v_1} = \frac{\sin \theta_2}{v_2}. \tag{19.3}$$

This equation, known as **Snell's law,** was determined experimentally in a different form in 1621 by the Dutch scientist Willebrord Snell (1591–1626).

We have derived the equation using the angles that the wavefronts make with the boundary between the media. However, it is customary when consider-

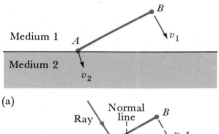

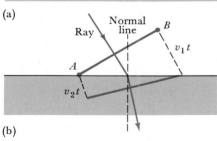

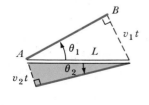

FIG. 19.6. (a) A wavefront moves from medium 1 toward medium 2. (b) During a time t, point B travels a distance $v_1 t$, while point A travels a distance $v_2 t$. (c) By comparing these triangles, we can derive Snell's law.

ing refraction to use the angle that a ray moving in the direction of travel of the wave makes the normal line. We see from Fig. 19.7 that the angle θ_1 between the ray and the normal line equals the angle θ_1 between the wavefront and the boundary. The same applies to angle θ_2. Thus, Eq. (19.3) can be used to calculate the change in direction of wave rays relative to normal lines. In the future **we will always refer to angle θ_1 as the direction of an incident ray relative to the normal line and θ_2 as the direction of the transmitted or refracted ray relative to the normal line.** Snell's law is then written as

$$\frac{\sin \theta_1}{v_1} = \frac{\sin \theta_2}{v_2}, \tag{19.3}$$

where v_1 is the wave's speed in the medium of the incident wave and v_2 is its speed in the medium of the refracted wave. Equation (19.3) is used by geophysicists when analyzing the refraction of sound waves passing between different layers of the earth.

A different form of Snell's law is used to analyze the refraction of light. In this form, Eq. (19.2) is used to replace the wave's speed by the indices of refraction. Substituting $v_1 = c/n_1$ and $v_2 = c/n_2$ into Eq. (19.3), we rewrite Snell's law:

$$\frac{\sin \theta_1}{c/n_1} = \frac{\sin \theta_2}{c/n_2},$$

or

$$n_1 \sin \theta_1 = n_2 \sin \theta_2. \tag{19.4}$$

This form of Snell's law will be used in all of our calculations involving light. (See Fig. 19.8 for the meaning of the symbols.)

EXAMPLE 19.2 A ray of light passes from air ($n = 1.00$) into glass ($n = 1.52$) and then into Jell-O (see Fig. 19.9). The incident ray makes a 58.0° angle with the normal as it enters the glass and a 36.4° angle with the normal in the Jell-O. (a) Calculate the angle of the refracted ray in the glass. (b) Calculate the index of refraction of the Jell-O.

SOLUTION (a) The ray bends toward the normal as it is slowed while entering the glass. For the first interface, the following information is given: $\theta_1 = 58.0°$, $n_1 = 1.00$ (air), and $n_2 = 1.52$ (glass). The angle θ_2 of the refracted ray in the glass is determined using Eq. (19.4):

$$\sin \theta_2 = \frac{n_1 \sin \theta_1}{n_2} = \frac{1.00 \sin 58.0°}{1.52} = 0.558,$$

or

$$\theta_2 = \underline{33.9°}.$$

(b) The angle of incidence for the second interface (θ_3 in Fig. 19.9) equals the refracted angle θ_2 at the first interface. Can you explain why? Thus, for the second interface, the known information is $\theta_3 = 33.9°$, $\theta_4 = 36.4°$ (this was given), and $n_3 = 1.52$ (glass). The index of refraction n_4 of the Jell-O can be determined using Eq. (19.4):

$$n_4 = \frac{n_3 \sin \theta_3}{\sin \theta_4} = \frac{1.52 \sin 33.9°}{\sin 36.4°} = \underline{1.43}.$$

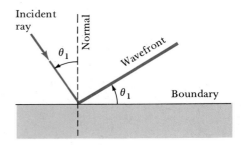

FIG. 19.7. A ray makes the same angle with the normal that the wavefront makes with the interface.

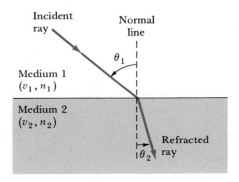

FIG. 19.8. Light is refracted at an interface between two media.

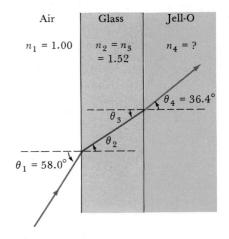

FIG. 19.9. Light passes from air through glass into Jell-O.

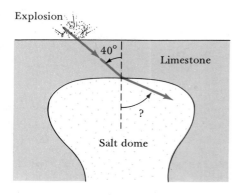

FIG. 19.10. A ray of sound is refracted at the interface between limestone and the top of a salt dome.

Notice that the Jell-O's index is less than that of the glass (the wave moves faster in the Jell-O than in the glass). Because the wave moves faster, it bends away from the normal ($\theta_4 > \theta_3$ in Fig. 19.9). ■

EXAMPLE 19.3 Sound caused by an explosion travels at 4000 m/s in limestone and strikes a boundary at the top of a salt dome (Fig. 19.10). If part of the sound strikes the boundary at an angle of incidence of 40°, calculate the angle of refraction. The speed of sound in salt is about 6000 m/s.

SOLUTION In this problem, speeds rather than indices of refraction are known. Thus, Eq. (19.3) will be used rather than Eq. (19.4). The known information is $\theta_1 = 40°$, $v_1 = 4000$ m/s, and $v_2 = 6000$ m/s. The sine of the refracted wave is

$$\sin \theta_2 = v_2 \frac{\sin \theta_1}{v_1} = (6000 \text{ m/s}) \frac{\sin 40°}{4000 \text{ m/s}} = 0.964 ,$$

or

$$\theta_2 = \underline{74.6°}.$$

Because the wave traveled from a slower to a faster medium, the wave bent away from the normal as it entered the salt (Fig. 19.10). ■

Mirages

One of the interesting consequences of the refraction of light waves is the formation of mirages. On a hot day there may be a layer of very hot air just above the ground. This hot air is less dense than cooler air farther above the surface. Light travels slightly faster through the less dense, hot air than through the cooler air. Wavefronts of light leaving a distant object are bent as shown in Fig. 19.11a because the lower part of a wavefront travels faster than the upper part. Some of the light originally slanting downward is refracted up through the hot air and into our eyes; the light never hits the ground. The refracted light from a distant tree (Fig. 19.11b) appears to come from the surface in front of the tree, like the reflection of light off a pond of water. A mirage is caused not by a loss of mental facility on a hot, dry desert, but by refracted light that appears to be reflected light from the smooth surface of a pond in front of an object.

Occasionally, a motorist observes what appears to be a wet spot on a distant part of a hot highway. This is just light that originates in the sky and travels toward the highway. Before reaching the pavement, it is bent or refracted up into your eye by the hot air above the road. You are seeing light from the sky that curved down toward the highway but missed.

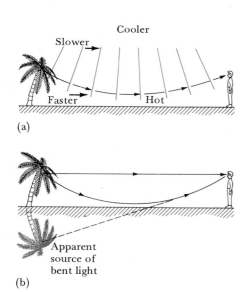

FIG. 19.11. (a) Wavefronts are bent because the light travels faster in the hot air than in cooler air. (b) Rays enter a person's eye as though coming from an inverted object in the ground. The mirage looks like the reflection of the tree off a pond.

19.6 Total Internal Reflection

When the speed of light increases as it passes from one medium to another, the light bends away from the normal as shown in Fig. 19.12a. As the angle of the incident ray increases, the refracted ray bends farther from the normal. For a particular incident angle called the **critical angle** θ_c (Fig. 19.12c), the refracted

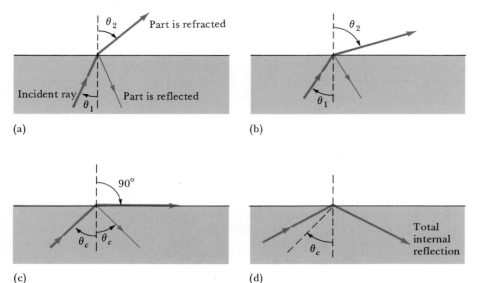

(a)

(b)

(c)

(d)

FIG. 19.12.   (a) and (b) If the speed of a wave increases as it moves from one medium to another, its direction bends away from the normal. (c) At an incident angle called the critical angle θ_c, the refracted wave moves parallel to the boundary between the media ($\theta_2 = 90°$). (d) If θ_1 is greater than θ_c, total internal reflection occurs and there is no refracted wave.

ray makes an angle of 90° with the normal and skims along the boundary between the two media. The critical angle is determined by setting $\theta_1 = \theta_c$ and $\theta_2 = 90°$ in Snell's law:

$$\sin \theta_c = \frac{n_2}{n_1} \sin 90° = \frac{n_2}{n_1}. \qquad (19.5)$$

If a ray strikes a surface at an angle less than θ_c, part of the wave is refracted into the second medium (Fig. 19.12b). However, for incident angles greater than θ_c (Fig. 19.12d), we find using Snell's law that the sine of the refracted angle ($\sin \theta_2$) is greater than 1. The sine of an angle can never be greater than 1. This impossible result is avoided because there is no refracted wave for incident angles greater than θ_c; the wave is *totally reflected* (100 percent) at the boundary between the media (Fig. 19.12d). This phenomenon, called **total internal reflection,** occurs when a wave travels at an incident angle greater than the critical angle while trying to move into a second medium in which its speed increases— that is, into a medium with a smaller index of refraction.

EXAMPLE 19.4   Suppose that medium 1 in Fig. 19.12 is an antifreeze solution and that medium 2 is air. Calculate the index of refraction of the antifreeze solution if light is totally reflected at the antifreeze-air interface for angles of 45.3° and greater.

SOLUTION   The known information is $n_2 = 1.00$ (air), $\theta_1 = 46.4°$ (the critical angle), and $\theta_2 = 90°$. (The refracted angle is 90° when the incident angle equals the critical angle.) Now, according to Snell's law, the index of refraction of the solution (medium 1) is

$$n_1 = \frac{n_2 \sin \theta_2}{\sin \theta_1} = \frac{1.00 \sin 90°}{\sin 45.3°} = \underline{1.41}. \qquad ∎$$

Measurements of the critical angle, as in Example 19.4, can be used to indicate the concentration of a solution—such as the concentration of antifreeze

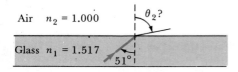

FIG. 19.13. A ray of light in glass strikes an interface with air. Is the light totally reflected?

in a car's radiator. This is because the concentration affects the index of refraction of the solution. For example, a 25:75 ratio of water and ethylene glycol produces an index of refraction of 1.41 and a critical angle of 45.3°. On the other hand, a 75:25 water–ethylene glycol solution produces an index of refraction of 1.35 and a critical angle of 47.8°. Thus, an accurate measurement of the critical angle indirectly indicates the ethylene glycol concentration in the water.

EXAMPLE 19.5 Light travels inside a block of light crown glass (Fig. 19.13). If the beam strikes a surface between the glass and air at an incident angle of 51°, will part of the beam be refracted into the air?

SOLUTION In Fig. 19.13, we call medium 1 the glass and medium 2 the air. We see from Table 19.1 that the index of refraction of light crown glass is 1.517 and that of air is 1.000. Thus, according to Snell's law, the sine of the refracted angle is

$$\sin \theta_2 = \frac{n_1 \sin \theta_1}{n_2} = \frac{1.517 \sin 51°}{1.000} = 1.18.$$

If, when using Snell's law, we find that the sine of the refracted angle is greater than 1.00, we interpret it as meaning that no refraction occurs. The ray experiences total internal reflection.

Our result can be checked by calculating the critical angle for light striking an interface between crown glass and air:

$$\sin \theta_c = \frac{n_2 \sin 90°}{n_1} = \frac{1.000(1.00)}{1.517} = 0.659,$$

or

$$\theta_c = 41.2°.$$

Since 51° is greater than θ_c, the light is totally reflected. ∎

Fiber Optics

Suppose that the glass discussed in Example 19.5 was a long, thin fiber of glass. If light moving through the glass strikes the top edge of the glass at an angle of 51° (or at any angle greater than θ_c), the light is totally reflected. The reflected light moves toward the bottom surface of the glass and strikes it at an incident angle of 51° (see Fig. 19.14). The light is once again totally reflected. The light continues to move along the fiber bouncing from one side to the other, each bounce resulting in total reflection. The fiber, which is called a **light pipe,** carries light along the fiber much the way water flows along a pipe. No light is lost from the sides.

These fibers that carry light have become an important new tool in technology. The communications industry is developing laser communications systems in which telephone conversations and other forms of information are transmitted by light waves traveling along flexible glass fibers. These glass fibers are also used to make ornamental lamps consisting of a large number of flexible fibers that carry light upward along the fibers from a light source at the base.

Fiber optic devices have been built to view places that are normally inaccessible to our eyes, such as the inside of car radiators or the inside of a person's

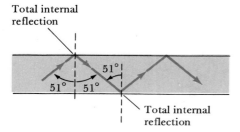

FIG. 19.14. After being totally reflected at the top of the glass, light is again totally reflected at the bottom. This glass is a "light pipe."

(a)

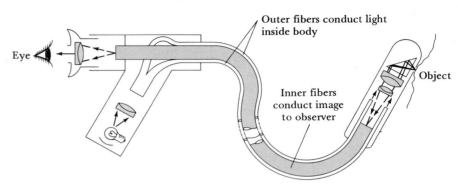

(b)

FIG. 19.15. (a) A photograph of cartilage in a knee joint. A thin metal suction device is cleaning damaged areas. (b) A schematic representation of the fiber optic device used to view the joint. (Photograph courtesy of Dr. William H. Dunbar.)

stomach. In these devices, a bundle of tiny glass fibers transmits a view of an inaccessible region at one end of the bundle to a person's eye looking at the other end of the bundle. If a lens is placed at the entrance end of the bundle, a larger field of view can be transmitted along the rods. A picture taken inside the knee joint is shown in Fig. 19.15a. An outer group of fibers carries light into the joint to illuminate it; an inner group carries an image of the illuminated surfaces of the joint to the viewer.

Prisms

Since the critical angle for total internal reflection from a glass-air interface is less than 45°, glass prisms with 45° and 90° angles are used to reflect light through 90° or 180° angles in many optical instruments such as telescopes and binoculars. Examples of the reflecting ability of prisms are shown in Fig. 19.16.

The reflective ability of prisms is preferred over mirrors for several reasons. First, prisms reflect almost 100 percent of the light incident on the prism, whereas mirrors reflect somewhat less than 100 percent. Second, mirrors tarnish and lose their reflective ability with age, whereas prisms retain their reflective ability. Finally, prisms can invert an image—that is, make it appear upside down (Fig. 19.16c). This inversion may seem like a disadvantage. However, the lenses in optical devices such as telescopes and binoculars cause an object viewed

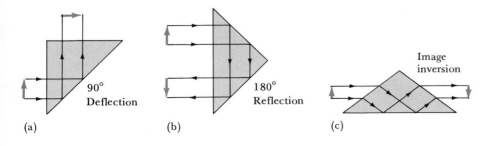

(a)

(b)

(c)

FIG. 19.16. A prism is used in three different ways to change either (a) and (b) the direction of light from an object or (c) the orientation of light from an object.

TABLE 19.2 Speed and Index of Refraction for Different Wavelengths of Light in Silicate Flint Glass

Color	Wavelength (nm)	v(m/s)	n
Red	650	1.860×10^8	1.613
Yellow	580	1.851×10^8	1.621
Green	530	1.843×10^8	1.628
Blue	470	1.834×10^8	1.636
Violet	400	1.806×10^8	1.661

through the telescope to appear inverted. In binoculars, a form of telescope, prisms are placed in the light path so that the inverted image of the telescope lenses is reinverted and appears right side up.

19.7 Dispersion

Light waves vary in wavelength from about 700 nm (red light) to 400 nm (violet light). In a vacuum the different wavelengths of light all travel at the same speed, but in media such as glass and water the speed of the different colors and wavelengths of light vary. For example, red light having a wavelength of 650 nm travels at 1.860×10^8 m/s in silicate flint glass, whereas violet light of 400-nm wavelength travels at 1.806×10^8 m/s (Table 19.2). Any medium that causes waves of different wavelength to move at different speed is said to be a **dispersive medium.** The medium can cause white light that consists of all different wavelengths of light to be separated or dispersed into its different wavelength components. A prism such as that shown in Fig. 19.17 causes the dispersion of white light.

To understand why the dispersion shown in Fig. 19.17 occurs, consider a wavefront and ray of light entering the prism shown in Fig. 19.18. A wavefront speeds through the air until its right edge strikes the glass (Fig. 19.18a). This edge slows down and bends as it enters the glass, while the left edge that is still in the air continues speeding forward (Fig. 19.18b and c). The slowing of the wave causes it to bend and change direction. As the wavefront reaches the second surface (Fig. 19.18e), the left edge of the wavefront moves into the air first. It speeds forward while the right edge is still shuffling along through the glass at slower speed. As the wavefront emerges from the prism, it is deflected even farther downward.

All colors of light undergo a similar change in direction, but the change is greatest for those wavelengths (colors) that are slowed most in the glass. Hence, violet is deflected farthest downward and red the least (see Fig. 19.17).

This nonuniform deflection of light composed of many wavelengths is useful for analyzing the wavelengths of light coming from a source. Light from the sun, for example, is made of all different wavelengths. Thus, deflection of sunlight through a prism produces a rainbow of all the different colors (Fig. 19.17). On the other hand, a hot gas of hydrogen atoms emits light at only certain discrete wavelengths: red light at about 656 nm and blue light at 486 nm, 434 nm, and 410 nm. If the light given off by hydrogen atoms is passed through the prism and then projected on a screen, separate lines for each discrete wavelength are seen. The 410-nm blue light is deflected most; the red light is de-

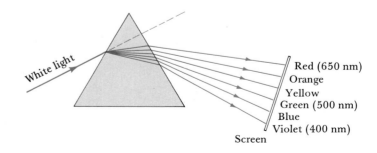

FIG. 19.17. White light is separated into its many wavelength components by refraction through a prism.

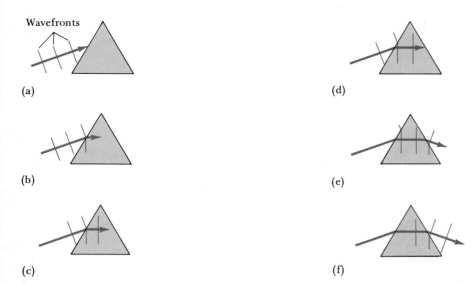

Wavefronts

(a)

(b)

(c)

(d)

(e)

(f)

FIG. 19.18. Wavefronts of light are bent at the surfaces of a prism.

flected least. If light from a star has lines at these four hydrogen wavelengths, then an astronomer knows that hydrogen is one type of atom on the star. Other atoms can be identified in a similar manner.

Rainbows

Rainbows are a dramatic example of the effect of dispersion. Water, like glass, causes the dispersion of light. When white light from the sun enters a spherical raindrop as shown in Fig. 19.19, the light is refracted or bent. The different

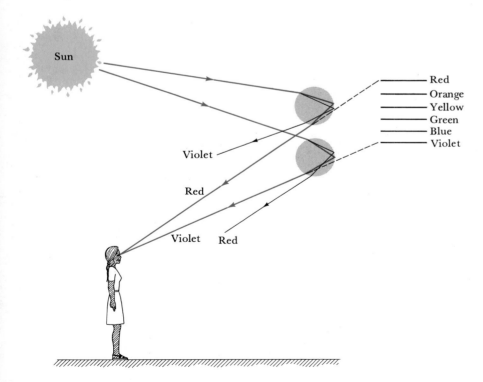

FIG. 19.19. Red light is refracted and reflected from a raindrop at a different angle from violet light. As a result, we see different colors of light coming from different parts of the sky. A rainbow is formed.

colors of light are bent by different amounts. After reflection off the back surface of the drop, the different colors of light are refracted again as they leave the front surface.

Each drop disperses a full spectrum of colors. Yet an observer on the ground facing away from the sun sees at most only one color of light coming from a particular drop. If the person sees red light from a drop, the violet light is deflected above her head. Or if the person looks at drops lower in the sky, she may see violet light coming from the drops. The red light from these drops is deflected below her eyes onto the ground. She sees red light when her line of view makes an angle of 42° with the beam of sunlight and violet light when the angle is 40°. Other colors of light are seen at intermediate angles.

Summary and Additional Readings

1. **Coherence: A coherent wave** is caused by the interference of waves coming from several sources whose vibrations are synchronized with each other. A coherent wave usually has an orderly appearance, like a marching band on parade. An **incoherent wave** is caused by the interference of waves coming from different sources whose disturbances are random and unsynchronized. The incoherent wave has a disorganized appearance that changes unpredictably.

2. **Wavefronts and rays:** A coherent wave can be drawn as one crest following another. These moving crests are called **wavefronts,** and the disturbance caused by the wave is the same at every point on the wavefront. A **ray** is an arrow drawn perpendicular to the wavefront that points in the direction of travel of the wave.

3. **Reflection: Diffuse reflection** results in a wave being reflected in many directions and occurs when a wave strikes a surface whose irregularities are larger than the wavelength of the wave. **Specular reflection** results in a reflected wave that moves in only one direction and occurs when the wave strikes a smooth surface. The angle that the reflected wave makes with a normal line drawn perpendicular to the reflecting surface equals the angle that the incident ray makes with that line (see Fig. 19.20).

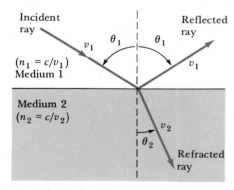

FIG. 19.20. The angles used to describe the directions of the incident, reflected, and refracted waves.

4. **Index of refraction:** The index of refraction n of a medium is defined as the speed of light in a vacuum c (3.00×10^8 m/s) divided by the speed of light in the medium v:

$$n = \frac{c}{v}. \tag{19.2}$$

5. **Refraction and Snell's law:** When a wave strikes the interface between two media, the transmitted wave usually travels in a different direction from the incident wave. The transmitted wave is said to be **refracted**—that is, to change direction—and is called the refracted wave. The direction of the refracted ray θ_2 relative to the normal line (see Fig. 19.20) is determined using Snell's law in one of two forms:

$$\frac{\sin \theta_1}{v_1} = \frac{\sin \theta_2}{v_2} \tag{19.3}$$

or

$$n_1 \sin \theta_1 = n_2 \sin \theta_2, \tag{19.4}$$

where θ_1 is the angle of the incident ray relative to the normal line, v_1 and v_2 are the speeds of the wave in the first and second media, respectively, and n_1 and n_2 are the indices of refraction of these media.

6. **Total internal reflection:** When the speed of a wave increases as it moves from one medium to another, the refracted ray bends farther away from the normal line. If the incident ray makes an angle θ_c, called the **critical angle,** with the normal line, the refracted ray is at 90° and skims along the surface between the media. If the incident angle is greater than θ_c, the wave is **totally reflected**—there is no refracted wave. The critical angle is determined by substituting $\theta_1 = \theta_c$ and $\theta_2 = 90°$ into Snell's law.

7. **Dispersion:** Any medium that causes waves of different wavelengths to move at different speeds is said to be a **dispersive medium.** A glass prism causes different colors and wavelengths of light to travel at different speeds and is able to separate white light into a rainbow of colors by the varying angular refraction of the different colors of light.

A. B. Fraser and W. H. March, "Mirages," *Scientific American* **234**(1), 102 (1976).

F. Wolf Helmut, Editor, *Handbook of Fiber Optics: Theory and Applications*, Garland STPM Press, New York (1976). See especially Chapter 10 (Endoscopy) and 11 (Optical Communications Activities).

John N. Shive and Robert L. Weber, *Similarities in Physics*, pp 170–181, John Wiley & Sons, New York (1982). Contains a useful discussion about dispersion.

Questions

1. (a) *Estimate* the thickness of a page of this book. With a little ingenuity you should be able to do this to within about 10 percent. (b) Calculate the number of 500-nm wavelengths of light that equal this thickness.

2. The Cathedral of Girgenti in Sicily is in the shape of half an ellipse that has been rotated about its long axis. The confessional was unknowingly placed at one focus of the ellipse. A person standing at the other focus can easily overhear a sinner speaking in private to a priest. Explain. [*Hint:* Construct several rays that leave one focus and then reflect from the walls one or more times. The sound will be very intense where the rays intersect.]

3. A wave passes from one medium to another. Why does the wave's frequency not change, whereas its wavelength does?

4. A person sits in the balcony of a concert hall. Does more sound reach the person when the air is warm near the ground floor and cool near the ceiling, or vice versa? Explain.

5. Lumps of sugar are added to a container of water (Fig. 19.21). The water is not stirred while the sugar dissolves slowly at the bottom. A narrow laser beam enters the container, bends downward, is reflected from the bottom, and then bends again as it leaves the container. Explain the bending of the laser beam. [Adapted from W. M. Strouse, *American Journal of Physics* **40**, 913 (1972).]

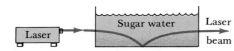

FIG. 19.21

6. Why does the air shimmer above a hot highway?

7. A rod partly immersed in water is shown in Fig. 19.22. Use ray diagrams to explain the apparent break in the rod at the place where it enters the water.

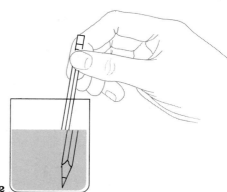

FIG. 19.22

8. Place a penny in the bottom of an empty cup whose walls are not transparent to light (a styrofoam cup works well). Now look down into the cup at an angle that puts the penny just out of sight. When you slowly add water to the cup, the penny comes into view even though you have not moved your eyes or the penny. Explain why this occurs.

9. A fish looks upward along a line making a 45° angle with the water-air interface. What does it see? Explain.

Problems

19.3 Reflection

■ 1. Construct on a piece of paper a half-circle representing a mirror whose inside surface is silvered (see Fig. 19.23). Suppose you place a small lightbulb on the axis halfway between point R in Fig. 19.23 and the surface of the mirror. Now carefully draw three different rays that leave the lightbulb, reflect from the mirror, and move away. Would this mirror and light be useful for a headlight on your car? Explain. (Ignore the colored ray in Fig. 19.23.)

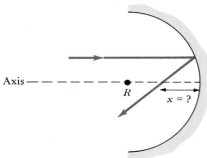

FIG. 19.23

■ 2. Two mirrors are oriented at right angles. A ray strikes the horizontal mirror at an incident angle of 65°, reflects from it, and then hits the vertical mirror. (a) Calculate the angle of incidence at the vertical mirror. (b) Show that the ray leaves the vertical mirror parallel to its original direction.

■■ 3. A spherical mirror is made by silver-coating half of the inside of a sphere of radius $R = 20$ cm. A light ray moves horizontally 5.0 cm above the axis of the mirror as in Fig. 19.23. When reflected from the mirror, how far from the mirror does the ray cross the axis? You will need to use some trigonometry and geometry to solve this problem.

19.4 Index of Refraction

4. (a) Calculate the index of refraction of a sugar solution through which light travels at 2.06×10^8 m/s. (b) Calculate the speed of light in blood whose refractive index is 1.35.

5. (a) Light travels 0.700 times as fast in Jell-O as it does in a vacuum. Calculate the index of refraction of Jell-O. (b) If it travels 1.10 times faster in bouillon soup than in Jell-O, what is the soup's index of refraction? Be sure to justify your calculation technique.

■ 6. The index of refraction for radio waves passing through ice is 1.305. If the ice is 1860 m thick, how much time is needed for a radio-wave pulse to travel from the surface of the ice sheet to the bottom and back to the surface again?

■ 7. Two light pulses travel side by side, one in a vacuum of refractive index 1.00000 and the other in air of refractive index 1.00029. Which pulse travels fastest? After moving 1 mi, how far ahead of the slower pulse is the fast pulse?

■ 8. A water molecule vibrating at a frequency of 9.66×10^{13} Hz in a glass of water emits infrared radiation. (a) Calculate the wavelength of the wave when in the water. (b) Calculate the wavelength after it leaves the water and is in the air.

■ 9. When a wave passes from one medium to another, its frequency remains the same but its wavelength changes. Derive an expression for the ratio of the wavelength in medium 1 to that in medium 2 as a function of the indices of refraction of the two media ($\lambda_1/\lambda_2 = ?$).

19.5 Refraction

10. (a) Light passes from air into a 25 percent sugar solution at an incident angle of 35°. Calculate the angle of the refracted wave. (b) Light passes from ethyl alcohol to air at an incident angle of 12°. Calculate the angle of the refracted wave in the air. (c) Draw pictures of each situation showing the interface between the media, the normal line, the incident, reflected, and refracted rays, and the angles of these rays relative to the normal line.

11. (a) Light passes from glass with index of refraction 1.58 into water with index of refraction 1.33. The angle of the refracted ray in water is 58.0°. Calculate the angle of the incident ray at the glass-water interface. (b) A ray passing from air to cyclohexane is incident at 48° and has an angle of refraction of 31°. Calculate the index of refraction of the cyclohexane. (c) Draw pictures, such as requested in Problem 10(c), for each situation.

12. A sound wave traveling in air ($v = 340$ m/s) is incident on a surface of water at an angle of 5.0°. (a) Calculate the direction of the refracted sound wave in the water ($v_{water} = 1400$ m/s). (b) Draw a picture such as requested in Problem 10(c).

13. Sound passes from limestone, where it moves at a speed of 4000 m/s, into another unknown material. The angle of incidence at the interface is 24° and the angle of refraction in the unknown material is 38°. (a) Calculate the speed of sound in this material. (b) Draw a picture such as requested in Problem 10(c).

14. A bagpipe marching band moving at a speed of 0.90 m/s on a hard surface enters a mud field and slows to 0.50 m/s. If the front row enters the mud at an incident angle of 25°, calculate its direction in the mud.

■ 15. A light ray traveling in air ($n = 1.00$) is incident on an interface with another medium at an angle of 43°. The reflected ray and the refracted ray make an angle of 108° with respect to each other. Calculate the index of refraction of the second medium.

■ 16. A light ray passes from air ($n = 1.00$) through a glass plate ($n = 1.56$) into water ($n = 1.33$). The angle of the refracted ray in the water is 42.0°. Calculate the angle of the incident ray at the air-glass interface.

■ 17. Sound traveling in limestone enters the top surface of a cubical salt dome at an angle of incidence of 35°. After passing through the salt dome, the sound leaves the right side and reenters the limestone. Calculate its final direction in the limestone and show in a figure the directions of the ray as it enters, passes through, and leaves the salt dome. ($v_{limestone} = 4000$ m/s and $v_{salt} = 6000$ m/s.)

■ 18. Light moving up and toward the right in air enters the side of a cube of gelatin ($n = 1.30$) at an incident angle of 80°. Calculate the angle at which the light leaves the top surface of the cube.

■■ 19. The eyes of a person standing at the edge of a 1.2-m-deep swimming pool are 1.6 m above the surface of the water. Light coming from a silver dollar at the bottom of the pool enters the person's eyes at an angle of 37° below the horizontal. Draw a picture of a light ray that leaves the dollar and enters the person's eyes and calculate the horizontal distance from the person to the dollar.

20. When a light ray passes through a prism with two equal-length sides (see Fig. 19.24), the angle of deviation δ is a minimum when the ray's path inside the prism is parallel to the base of the prism. Show that the minimum deviation angle δ is related to the index of refraction n of the prism and the angle ϕ of the top corner of the prism by the equation

$$n = \frac{\sin \frac{1}{2}(\phi + \delta)}{\sin (\phi/2)}.$$

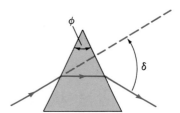

FIG. 19.24

19.6 Total Internal Refraction

21. Calculate the critical angle for light inside a diamond incident on an interface with air.

22. A scuba diver swimming underwater at night shines a beam of light so that it hits the water-air interface at an incident angle of 52°. Will the light be seen above the water? Explain.

23. Calculate the index of refraction of a sugar solution for which the critical angle for light traveling in the solution incident on an interface with air is 42.5°.

■ 24. (a) Calculate the critical angle of reflection for sound moving in air at 340 m/s that strikes a surface of water where its speed is 1500 m/s. (b) Is there much chance that a swimmer whose head is underwater will hear sounds from a lifeguard on the edge of a pool? Explain.

■ 25. Light is incident on the boundary between two media at an angle of 32°. If the refracted light makes an angle of 42°, what is the critical angle for light incident on the same boundary?

■ 26. (a) Rays of light are incident on a glass-air interface (Fig. 19.25a). Calculate the critical angle for total internal reflection ($n_{glass} = 1.58$). (b) If there is a thin, horizontal layer of water ($n = 1.33$) on the glass, will a ray incident on the glass-water interface at the critical angle calculated in part (a) be able to leave the water (Fig. 19.25b)? Justify your answer.

■ 27. What must be the minimum value of the index of refraction of the prism shown in Fig. 19.16a in order that light be totally reflected where indicated? Is the index of refraction of most types of glass greater than this value?

■ 28. The prism shown in Fig. 19.16a is immersed in water of refractive index 1.33. Calculate the minimum value of the index

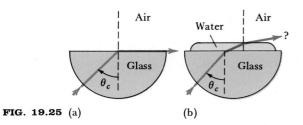

FIG. 19.25 (a)   (b)

of refraction for the prism in order that the light is totally reflected where shown.

■ 29.   Rays of light enter the end of a light pipe from air at an angle θ_1 (Fig. 19.26). The index of refraction of the pipe is 1.64. Calculate the greatest angle θ_1 for which the ray is totally reflected at the top surface of the glass-air interface inside the pipe.

19.7 Dispersion

■ 30.   Red light of wavelength 650 nm strikes the side of the prism ($n_{red} = 1.613$) shown in Fig. 19.17 at an incident angle of

53.8°. (a) Calculate the refracted angle relative to a line perpendicular to the first air-glass interface. (b) If the prism has three 60° angles, calculate the incident angle at the second glass-air interface. (c) Calculate the refracted angle relative to a line perpendicular to this surface. (d) Calculate the net angular change in direction of the light.

■ 31.   Repeat Problem 30 for blue light for which $n_{blue} = 1.636$.

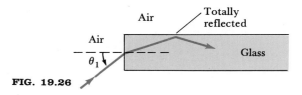

FIG. 19.26

CHAPTER 20

Optical Instruments

In 1590 Zacharias Janssen of Holland made the first microscope, and 18 years later another Dutchman, Hans Lippershey, constructed the first telescope. Science was embarking on a new era that would eventually allow us to see the tiny details of living organisms and the grand formations in the sky.

Microscopes and telescopes are only two of many optical instruments used to alter the way we see things. Other instruments that will be considered in this chapter, besides the human eye, include the camera, slide projector, and eyeglasses. These devices depend on glass lenses that gather and focus light from an object to form an image seen by our eyes. We will learn how these images are formed to make things look bigger or smaller, closer or farther away, upright or inverted. We will also find out how curved mirrors can be used to produce images that are different (for example, bigger) than the object being viewed.

20.1 Lenses

Our eyes perceive an object as being located at a point from which rays of light leave the object. A **lens** is an optical device that changes the apparent location of an object by altering the path of light rays coming from the object.

The effect of a lens on light is easily visualized by constructing a lens from two prisms, as shown in Fig. 20.1a. When parallel rays of light pass through the prism lens, they bend toward an axis through the center of the lens. Because the rays converge or move toward each other, this lens is called a **converging lens** (also called a convex lens). The rays of light, after passing through the prism lens, intersect at different places. If, instead of using prisms, we make a converging lens whose surfaces are properly curved, all the rays will pass through one point (Fig. 20.1b). This point where *parallel* rays meet after passing through a converging lens is called the **focal point** of the lens. (The word *parallel* is emphasized because parallel rays are the only ones that converge at the focal point. Rays not parallel to the axis cross it at other points.) The **focal length** f is the distance of this focal point from the lens. For a converging lens, the focal length is always a positive number.

Another type of lens commonly used in optical instruments is a concave or **diverging lens.** Instead of bringing rays of light together, a diverging lens causes them to diverge or move apart. This effect can be illustrated using two prisms, as

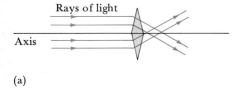

(a)

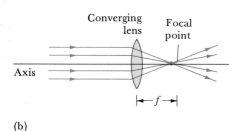

(b)

FIG. 20.1. (a) A converging lens built from two prisms causes rays of light to cross at different points after passing through the lens. (b) A lens with properly curved surfaces causes parallel rays to converge at the focal point.

416

shown in Fig. 20.2a. If parallel rays of light strike one side of the prism, the refracted rays diverge when leaving the other side. If the surfaces of a lens are properly curved, all parallel rays will appear to diverge from the same point on the left side of the lens (Fig. 20.2b). The point from which parallel rays appear to diverge after leaving a diverging lens is called the **focal point** of the lens. The **focal length** is the distance from this focal point to the lens. For diverging lenses, the focal length is a negative number.

20.2 Formation of Images

The use of lenses in optical instruments, such as cameras, movie projectors, microscopes, and telescopes, can be illustrated using **ray diagrams** that show the path of light along several carefully chosen rays. The intersection of these rays after passing through a lens indicates the effect of the lens on our perception of an object—does it appear bigger, smaller, closer to our eye, farther away, inverted, or upright?

To construct a ray diagram, draw a lens with a line passing through the center of the lens and perpendicular to its surface (Fig. 20.3). The line is called the principal axis or the **axis** of the lens. Place dots on the axis on each side of the lens to indicate the focal points of the lens. The distance from the lens to the focal points represents the focal length of the lens, a distance that is usually given for a particular lens.

Next draw an arrow to the left side of the lens perpendicular to and resting on the axis, as shown in Fig. 20.3a. The arrow represents an object that we wish to view in one way or another. For example, the object might be a crater on the moon viewed through a telescope, an amoeba viewed through a microscope, or a slide whose picture we wish to project on a screen. The distance in the drawing between the arrow (the object of our interest) and the lens represents the separa-

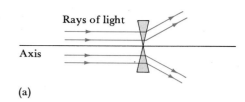

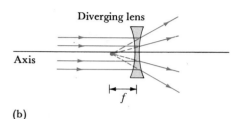

FIG. 20.2. (a) Two prisms join to form a diverging lens. Rays of light move apart, or diverge, after passing through the prisms. (b) A diverging lens with properly curved surfaces causes parallel rays to appear to diverge from a single point called the focal point.

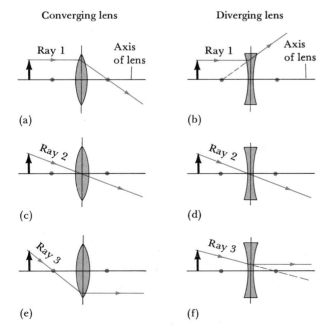

FIG. 20.3. Three special rays passing through converging and diverging lenses. Two or three of these rays can be used to locate the image of any object.

tion of the actual object from the lens. The convention of always placing the object on the left side of the lens simplifies our later mathematical calculations. Finally, draw several rays of light that pass from the object through the lens. These rays represent light reflected from the object. We do not, however, use all rays of light leaving the object, but only two or three special rays described below and pictured in Fig. 20.3. These rays are chosen because we can in general predict their direction after they pass through a lens. Knowing their direction allows us to determine the altered size and location of the object as seen with the lens.

The rays as they pass through the lens are refracted at its front and back surfaces. However, to simplify our analysis, we will assume that the rays bend only at a vertical plane through the lens, represented by the vertical line through the lenses in Fig. 20.3.

Ray 1 leaves the object and moves parallel to the axis. As the ray passes through a converging lens, it is refracted toward the axis and passes through the focal point on the right (Fig. 20.3a). If it passes through a diverging lens, ray 1 is refracted away from the axis and appears to come from the focal point on the left (the dashed line in Fig. 20.3b indicates the ray's apparent origin).

Ray 2 (Fig. 20.3c and d) leaves the object and passes directly through the middle of the lens. Although the ray may be refracted slightly at the front and back surfaces of the lenses, the two refractions cancel each other. If the lens is thin, the ray will move straight through its center.

For converging lenses, Ray 3 leaves the object and passes through the focal point on the left (Fig. 20.3e). The lens refracts the ray so that it moves parallel to the axis on the right. For a diverging lens, ray 3 leaves the object and moves toward the focal point on the right (Fig. 20.3f). Before reaching the focal point, the ray is refracted parallel to the axis on the right. The dashed line indicates the ray's path if the lens were not present.

If two or three of these rays are drawn on the same diagram, they may converge on the right side of the lens at a point called the *real image* of the object. If the rays diverge on the right side, the point on the left from which they seem to diverge is called the *virtual image* of the object.

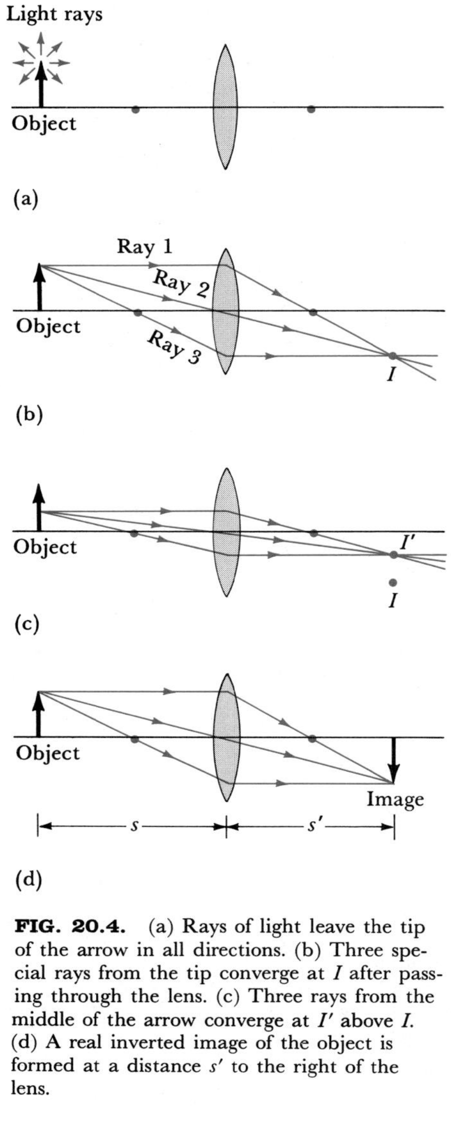

Formation of a Real Image

To illustrate the formation of real images, we will locate an object to the left side of a converging lens at a distance of about two times the focal length of the lens (Fig. 20.4a). Light rays leave the object in all directions, but only some of these rays pass through the lens. The paths of our special three rays from the tip of the object are shown in Fig. 20.4b. We find that the rays converge on the right side of the lens at point I.

Rays of light are also coming from other points on the object. For example, three rays leaving the middle of the arrow are brought together at point I', a little above point I (Fig. 20.4c). When a ray diagram is constructed for rays leaving each point on the object, we find that an inverted image is formed at a distance s' on the right of the lens (Fig. 20.4d). If a screen is placed at a distance s' to the right of the lens, we will see a real, inverted picture of the object on the screen. This is called a **real image** since the rays actually come together to form an image of the object. The distance s' is called the **image distance** and the distance s from the object to the lens is called the **object distance.**

FIG. 20.4. (a) Rays of light leave the tip of the arrow in all directions. (b) Three special rays from the tip converge at I after passing through the lens. (c) Three rays from the middle of the arrow converge at I' above I. (d) A real inverted image of the object is formed at a distance s' to the right of the lens.

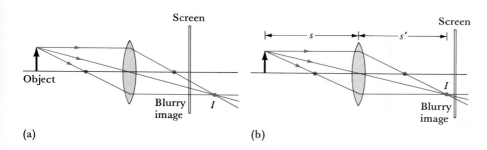

(a)

(b)

FIG. 20.5. If a screen is placed at any distance from the lens other than s', a blurry image of the object is formed.

A sharp, clear image of an object appears in only one place. If a screen is placed closer to the lens than distance s' (Fig. 20.5a), where rays leaving the tip of the arrow have not yet converged to a point, the image of the screen will be blurry and out of focus. If the screen is placed to the right of s' (Fig. 20.5b), where rays from the tip are diverging after having converged at I, a blurry image is again the result. Only at distance s' from the lens is a sharp, clear, focused image seen.

The distance s' of the focused image from the lens can be determined by experiment, by drawing a ray diagram, or by calculation using a thin-lens equation (to be introduced later). The ray diagram technique is the least accurate of these methods and is used mainly as a check on calculations using the thin-lens equation. The image position depends on where the object is located and on the focal length of the lens. Notice in Fig. 20.6a and b that the image distance increases as the object distance decreases. However, if an object is placed between the focal point and the lens, a real image is no longer formed, as we will see next.

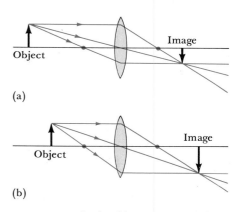

(a)

(b)

FIG. 20.6. As the object moves toward the focal point from the left, the image moves farther toward the right.

Virtual Image

When the rays from an object diverge after passing through a lens (Fig. 20.7a), a real image is not formed. We cannot place a screen on the right side of the lens and see an image of the object. But if we look with our eyes at light passing through the lens, the rays appear to diverge from a position behind the lens (Fig. 20.7b). At this position we see the **virtual image** of the object.

If an object is placed between a converging lens and its focal point, a virtual image is formed. To locate the image, consider the rays of light shown in Fig. 20.7a. Ray 3 does not pass through the lens and cannot be used to locate the

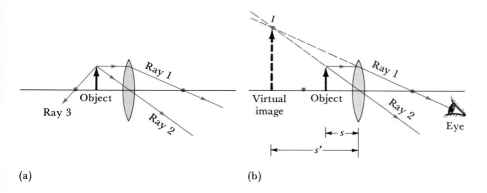

(a)

(b)

FIG. 20.7. (a) The three special rays leaving an object placed inside the focal point of a converging lens. (b) The rays, after passing through the lens, appear to come from point I. This is a virtual image since light does not actually come from this point.

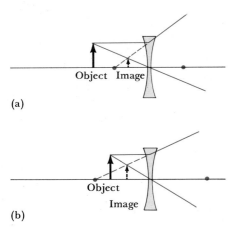

(a)

(b)

FIG. 20.8. Virtual images formed by diverging lenses.

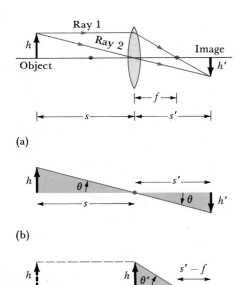

(a)

(b)

(c)

FIG. 20.9. (a) A ray diagram for a converging lens that forms a real image of an object. (b) and (c) Two pairs of similar triangles are used to derive the thin-lens equation.

image. However, rays 1 and 2 do pass through the lens and diverge on the right side. These rays from the tip of the arrow appear to our eyes to diverge from point I on the left side of the lens (where the dashed lines in Fig. 20.7b cross). Since the directions of these rays imply that light is actually coming from I, we think the tip of the arrow is located at I. This apparent location of the arrow is said to be a virtual image of the arrow.

A virtual image is usually drawn using a dashed arrow rather than a solid arrow. The image is always located on the same side of the lens as the object. Examples of virtual images of objects seen through diverging lenses are shown in Fig. 20.8.

20.3 Thin-Lens Equation

As we saw in the last section, a ray diagram can be used to locate the image of light rays passing through a lens from an object. The ray diagram technique provides a means of determining the type of image (real or virtual, inverted or upright) and its approximation location. The image location can be determined more accurately using an equation called the **thin-lens equation:**

$$\frac{1}{f} = \frac{1}{s} + \frac{1}{s'}, \tag{20.1}$$

where f is the focal length of the lens, s is the object distance, and s' is the image distance. This equation can be solved for any one of the three quantities (f, s, or s') if the other two quantities are known. The equation is used in the design of optical instruments such as telescopes, microscopes, slide projectors, and cameras. The equation is also the basis for choosing lenses for eyeglasses.

Several **sign conventions** are important when using the thin-lens equation. These conventions apply for light rays moving from left to right.

1. The **focal length** f is positive for converging lenses and negative for diverging lenses.

2. The **object distance** s is positive if the object is to the left of the lens and negative if the object is to the right of the lens.*

3. The **image distance** s' is positive for real images formed to the right of the lens and negative for virtual images formed to the left of the lens.

The thin-lens equation is derived geometrically by considering the triangles formed by the special rays shown in Fig. 20.9a. Look at the two triangles formed by ray 2 and the axis of the lens (Fig. 20.9b). The angle θ is the same for each triangle, as is the tangent of each angle. Thus,

$$\tan \theta = \frac{h}{s} = \frac{h'}{s'}.$$

The right side of ray 1 and the axis of the lens also form two triangles (Fig.

*This rule is useful for systems containing more than one lens. The object of the second lens is the image of the first. Occasionally, the object of the second lens may actually be to the right of the lens. We will not need to use this rule until Section 20.7. For now, the object distance is always positive.

20.9c). The angle θ' is the same for each of these triangles, as is the tangent of each angle. Thus,

$$\tan \theta' = \frac{h}{f} = \frac{h'}{s' - f}.$$

The last two equations can be rearranged so that the quantity h'/h appears on one side of each equation. We find then that

$$\frac{s'}{s} = \frac{h'}{h} \quad \text{and} \quad \frac{s' - f}{f} = \frac{h'}{h}.$$

Since the right sides of these two equations are identical, their left sides must be equal:

$$\frac{s'}{s} = \frac{s' - f}{f}.$$

Dividing each term in this equation by s' yields

$$\frac{1}{s} = \frac{1}{f} - \frac{1}{s'}, \quad \text{or} \quad \frac{1}{s} + \frac{1}{s'} = \frac{1}{f}.$$

The thin-lens equation can also be derived for diverging lenses and for converging lenses in which virtual images are formed.

EXAMPLE 20.1 An object is placed 20 cm to the left of a diverging lens whose focal length is -10 cm (the negative sign is used for a diverging lens). Where is the image of this object located and is the image real or virtual?

SOLUTION A ray diagram for this system is shown in Fig. 20.10. On the right side of the lens, light appears to come from a virtual image between the lens and the left focal point—about 7 cm left of the lens. Thus, we expect that $s' \simeq -7$ cm. The negative sign indicates that the image is virtual and to the left of the lens. Let us see if the same result is obtained using the thin-lens equation.

Rearranging Eq. (20.1) to solve for s', we find that

$$\frac{1}{s'} = \frac{1}{f} - \frac{1}{s} = \frac{1}{-10 \text{ cm}} - \frac{1}{20 \text{ cm}}$$
$$= -0.100 \text{ cm}^{-1} - 0.050 \text{ cm}^{-1} = -0.150 \text{ cm}^{-1},$$

or

$$s' = \frac{1}{-0.150 \text{ cm}^{-1}} = \underline{-6.67 \text{ cm}}.$$

Remember that to find s', you must invert the value of $1/s'$ obtained from Eq. (20.1). *Forgetting to invert is a common student error in optics problems!* Our thin-lens calculation ($s' = -6.67$ cm) checks well with the estimated solution ($s' \simeq -7$ cm) determined using a ray diagram.

If we looked through a diverging lens of focal length -10 cm at a picture 20 cm behind the lens, we would see light coming from a picture of reduced size that seems to be about 7 cm behind the lens. ∎

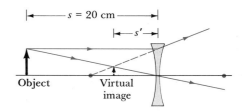

FIG. 20.10. A ray diagram for a diverging lens.

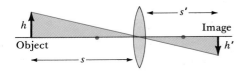

FIG. 20.11. A ray passing through the center of the lens forms two similar triangles that are used to calculate the linear magnification of the lens.

Linear Magnification

For some optical devices, the height of the image compared to the height of the object must be known. You might, for instance, want to know whether the image of a person 2 m tall will fit on a film that is 35 mm high. The ratio of the height of the image h' and the height of the object h is called the **linear magnification** m:

$$m = \frac{h'}{h}.$$

The linear magnification is a unitless number that simply tells how much taller the image is than the object; for example, if $m = 1$, the image and object are the same height. By convention, the height is a positive number for an upright object or an upright image and a negative number for an inverted object or image. For example, in Fig. 20.11, h' has a negative value because the image is inverted. If m is a negative number, the image is inverted relative to the object.

The linear magnification of an optical system can be calculated using the object and image distances, s and s', rather than their heights. In Fig. 20.11, the two shaded triangles are similar. Thus,

$$\frac{(-h')}{s'} = \frac{h}{s}.$$

The negative sign in the above equation is needed so that both ratios have the same sign. Recall that h' shown in Fig. 20.11 is a negative number. Thus, $-h'$ is a positive number, as is the ratio $-h'/s'$. This ratio equals the positive ratio h/s. We can rearrange the preceding equation as follows: $h'/h = -s'/s$. Therefore, the linear magnification m of an optical system can be calculated using either ratio:

$$m = \frac{h'}{h} = -\frac{s'}{s}. \tag{20.2}$$

Equation (20.2) applies to any type of lens and to real and virtual images. If m is negative, the image is inverted relative to the object.

Cameras

A camera, shown in Fig. 20.12, has a lens of fixed focal length. To focus sharply on objects at different distances, the film moves forward and back relative to the lens. Actually, the lens is moved relative to the film, but the effect is the same. As the object approaches the lens (decreasing object distance s), the film must be moved farther from the lens (increasing image distance s'). For a camera, then, the focal length f is a constant, and the image distance s' is varied by adjusting the camera's distance setting to accommodate different object distances s.

Besides adjusting the distance setting, on many cameras you must also adjust the shutter speed and the f-stop. The shutter speed regulates the length of time that the shutter is open and that light is exposing the film. The "speed" can typically be varied from about one or more seconds to 1/500 s. Unless a camera is securely mounted on a tripod or made secure in some other way, speeds of 1/60 s or faster are recommended to avoid a blurry picture due to camera movement.

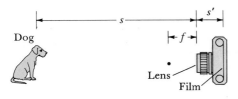

FIG. 20.12. A real image of a dog is formed by a lens on the film in a camera.

The f-stop setting on a camera regulates the opening size of an iris diaphragm in front of the shutter. On a bright day the opening should be small to prevent overexposure of the film. In darker conditions, a larger opening is needed to allow enough light to enter the camera to properly expose the film.

The **f-stop** is a number that specifies the opening diameter:

$$\text{f-stop} = \frac{f}{D},$$

where f is the focal length of the lens and D is the diameter of the opening for the iris diaphragm. Notice that as D gets smaller, the f-stop number gets larger. Typical f-stop settings are 1.0, 1.4, 2.0, 2.8, 4.0, 5.6, 8, 11, 16, and 22. Why these choices? Suppose that your camera has a lens of focal length 5.0 cm. With an f-stop setting of 1.0, the diameter of the opening is $D = f/(\text{f-stop}) = (5.0 \text{ cm})/1.0 = 5.0 \text{ cm}$. The area of the diaphragm when open is $A = \pi(D/2)^2 = \pi(2.5 \text{ cm})^2 = 20 \text{ cm}^2$. Using the next f-stop setting of 1.4, $D = (5.0 \text{ cm})/1.4 = 3.6 \text{ cm}$ and $A = \pi(1.8 \text{ cm})^2 = 10 \text{ cm}^2$. By moving the f-stop from 1.0 to 1.4, we have reduced the area of the opening by one-half. If you repeat this procedure for the other f-stop settings listed earlier, you will find that *adjacent f-stop settings have areas that differ by approximately a factor of* 2. For example, if you change the f-stop from 8 to 11, you have reduced the size of the opening by one-half. Your film will be exposed by one-half as much light.

Since adjacent shutter-speed settings and adjacent f-stop settings on a camera are related by a factor of 2 in terms of light exposure, you can easily adjust your camera for the proper film exposure under a variety of conditions. For example, shutter-speed and f-stop settings of $\frac{1}{120}$ s and 5.6 cause the same exposure of a film as settings of $\frac{1}{60}$ s and 8 or as $\frac{1}{30}$ s and 11. The smaller shutter-speed settings are preferred for action shots of moving objects. The larger shutter-speed settings should be reserved for still shots with the camera in a stable position. Large f-stops (small openings) allow light only through the center of the lens, and the depth of field is improved; near and distant objects are simultaneously in better focus.

EXAMPLE 20.2 A camera has a converging lens with a focal length of 5.00 cm. A dog 1.0 m tall standing 3.0 m from the lens is photographed. (a) Where must the film be located relative to the lens so that a sharp image is produced on the film? (b) How large is the image? (c) The film will receive the correct exposure with the shutter speed of $\frac{1}{60}$ s and an f-stop setting of 11. What would be a better setting to photograph a running dog?

SOLUTION The optical system appears as shown in Fig. 20.12. The known information is $f = 5.00 \text{ cm}$, $s = 300 \text{ cm}$, and $h = 1.0 \text{ m}$.

(a) To find where the film should be placed relative to the camera lens, we must solve Eq. (20.1) for the image distance s':

$$\frac{1}{s'} = \frac{1}{f} - \frac{1}{s} = \frac{1}{5.00 \text{ cm}} - \frac{1}{300 \text{ cm}}$$
$$= 0.2000 \text{ cm}^{-1} - 0.0033 \text{ cm}^{-1} = 0.1967 \text{ cm}^{-1},$$

or

$$s' = \frac{1}{0.1967 \text{ cm}^{-1}} = \underline{5.08 \text{ cm}}.$$

(b) To calculate the height h' of the image, we must first solve for the linear magnification of the lens using Eq. (20.2):

$$m = -\frac{s'}{s} = -\frac{5.08 \text{ cm}}{300 \text{ cm}} = -0.0169.$$

The negative sign means that the image is inverted. Again, using Eq. (20.2), we can find the image height:

$$h' = mh = (-0.0169)(1.0 \text{ m}) = -0.0169 \text{ m} = \underline{-16.9 \text{ mm}}.$$

The inverted image of the dog should occupy about one-half the height of a 35-mm film.

(c) To prevent a blurry picture, the shutter speed should be increased to $\frac{1}{120}$ s (or to $\frac{1}{250}$ s), in which case the f-stop should be reduced to $\underline{8}$ (or 5.6). ∎

20.4 The Human Eye

The human eye resembles in many ways an expensive camera. The eye is equipped with a built-in cleaning and lubricating system, an exposure meter, an automatic field finder, and a continuous supply of film. Light from an object enters the cornea (Fig. 20.13), a transparent covering over the surface of the eye, and passes through a transparent lens held in place by ciliary muscles. An iris in front of the lens opens or closes like the shutter on a camera to regulate the amount of light entering the eye. The cornea and lens together act as a lens of variable focal length that focuses light from an object to form a real image on the back surface of the eye, called the retina.

The retina acts like the film of a camera. It contains about 130 million light-sensitive cells called *cones* and *rods*. Light absorbed by these cells initiates photochemical reactions that cause electrical impulses in nerves attached to the cones and rods. The signals from individual cones and rods are combined in a complicated network of nerve cells and transferred from the eye to the brain via the optic nerve. What we see depends on which cones and rods are excited by absorbing light and on the way in which the electrical signals from different cones and rods are combined and interpreted by the brain.

The cones are concentrated in one part of the retina called the *fovea*. The fovea is about 0.3 mm in diameter and contains 10,000 cones and no rods. Each cone in this region has a separate nerve fiber that leads to the brain along the optic nerve. Because of the large number of nerves coming from this small area, the fovea is the best part of the retina for resolving the fine details of a bright object. Besides providing a region of high visual acuity, the cones in the fovea and in other parts of the retina are specialized for detecting different colors of light.

The concentration of cones decreases outside the fovea. In these peripheral regions, the rods predominate. Their density in the retina (about 150,000/m^2) is about the same as that of the cones in the fovea region. However, the light signals from perhaps 100 adjacent rods are brought together into a single nerve cell that leads to the brain. This combining of the rod signals reduces our ability to see the fine details of an object but helps us see dimly lit objects since many small signals are combined to produce a larger signal.

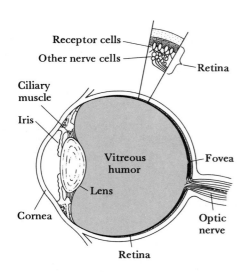

FIG. 20.13. The human eye.

Optics of the Eye

A normal eye can focus on objects located anywhere from about 25 cm to hundreds of miles away. This ability to focus on objects at different distances is called **accommodation**. Unlike the camera, which uses a fixed focal length lens and a variable image distance to accommodate different object distances, the eye has a fixed image distance of about 2.1 cm (roughly the distance from the cornea and lens to the retina) and a variable focal length lens system.

The changing focal length of the eye's lens is illustrated in Fig. 20.14. When the eye looks at distant objects, the ciliary muscle attached to the lens of the eye relaxes, and the lens becomes less curved (Fig. 20.14a). When less curved, the focal length increases and an image is formed at the retina. If the lens remains flattened and the object moves closer to the lens, the image will then move back behind the retina, causing a blurred pattern of light on the retina. To avoid this, the ciliary muscles contract and cause an increase in the curvature of the lens, reducing its focal length (Fig. 20.14b). With reduced focal length, the image moves forward and again forms a sharp, focused image on the retina. If your eyes become tired after reading for many hours, it is because the ciliary muscles have been tensed to keep the lenses of your eyes curved.

The **far point** of the eye is the greatest distance to an object on which the relaxed eye can focus. The **near point** of the eye is the closest distance of an object on which the tensed eye can focus. For the normal eye, the far point is effectively infinity (we can focus on the moon and on distant stars) and the near point is about 25 to 50 cm. We turn next to the changes in the eye that cause nearsightedness and farsightedness.

Nearsightedness

A nearsighted person's eyes can focus clearly on objects that are close to the eye but not on those that are distant. The far point of a nearsighted person may be only a few meters rather than infinity.

Nearsightedness occurs if a person's eyeball is larger than the usual diameter (variations in diameter of 1 mm can easily cause a person to be nearsighted). In such cases, the image of a distant object is formed in front of the retina (Fig. 20.15a), even when the eye lens is relaxed. Tensing the ciliary muscles increases the curvature of the lens and causes the image to move even farther in front of the retina. Nearsightedness also occurs if the cornea and relaxed lens are too curved—the image of a distant object is again formed in front of the retina.

To correct nearsighted vision, a diverging eyeglass lens is placed in front of the cornea. The rays from an object diverge slightly while passing through the eyeglass lens so that when they pass through the lens of the eye, an image is formed farther back in the eye (Fig. 20.15b). You can think of the eyeglass lens as performing a trick on the eye. If an object is very distant (for example, $s \simeq \infty$), then the focal length of the eyeglass lens is chosen so that its virtual image is formed at the far point of the eye (Fig. 20.15c). Light passing through the eyeglass lens appears to come from the image at the far point, not from the more distant object. To calculate the desired focal length of the eyeglass lens, we set $s = \infty$ and $s' = -$(Far point). The negative sign accounts for the fact that the image of the object as seen through the lens of the eyeglass is virtual and to the left of the lens. The value of f is then calculated using the thin-lens equation.

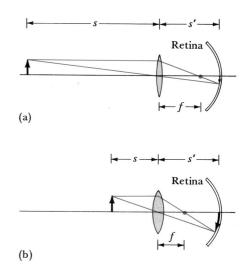

(a)

(b)

FIG. 20.14. The eye has a fixed image distance s' and a variable focal length f to accommodate different object distances s.

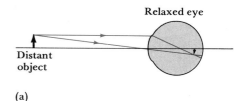

(a)

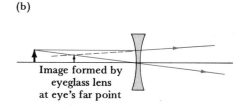

(b)

(c)

FIG. 20.15. (a) When a nearsighted person views a distant object, the image is formed in front of the retina. (b) A diverging eyeglass lens causes rays to diverge so that a final image is formed on the retina. (c) The corrective lens forms a virtual image at the eye's far point.

EXAMPLE 20.3 The far point for a nearsighted person is 2.0 m from the eye. What focal length lens should the person use in order to focus on a very distant object ($s = \infty$)?

SOLUTION We are given that $s = \infty$ and that the image of this distant object should be at the far point of the eye—that is, at $s' = -2.0$ m. Inserting these values into Eq. (20.1), we can calculate the focal length of the lens for the eyeglasses:

$$\frac{1}{f} = \frac{1}{s} + \frac{1}{s'} = \frac{1}{\infty} + \frac{1}{-2.0 \text{ m}} = 0 + \frac{1}{-2.0 \text{ m}} = -\frac{1}{2.0 \text{ m}},$$

or

$$f = -2.0 \text{ m}.$$

The focal length of the lens for a nearsighted person equals the negative value of the person's far point. ∎

Farsightedness

Unlike nearsighted people, farsighted persons are able to see distant objects but cannot focus on those nearby. Whereas the normal eye has a near point of 25–50 cm, a farsighted person may have a near point several meters from the eye. If the person's vision is uncorrected by glasses, a farsighted person will have to hold a book or newspaper at the eye's near point, several meters from the eye, in order to focus on the print. If the book is held closer to the eye, the image of the book is formed behind the retina, as shown in Fig. 20.16a. Light striking the retina is blurred and out of focus. Farsightedness may occur if the diameter of a person's eyeball is smaller than usual or if the lens is unable to curve enough when the ciliary muscles contract.

Converging eyeglass lenses are used to correct farsighted vision. The eyeglass lens bends light rays from an object toward the axis of the lens. The rays converge even farther as they enter the eye's cornea and lens, and the image is formed on the retina (Fig. 20.16b). Without the bending caused by the eyeglass lens, the image is formed behind the retina.

The converging lens used to correct farsightedness must have a focal length that is greater than the near point of a normal eye, which by convention is said to be 25 cm. When an object such as a book is held at the normal near point and viewed through the eyeglass lens, a virtual image of the object is formed behind the object by the glasses (Fig. 20.16c). The eye looking through the eyeglass lens perceives light to be coming from the virtual image. An eyeglass lens of the correct focal length will cause the virtual image to be formed at the actual near point of the farsighted person's eye. To determine the focal length of the eyeglass lens, we need to solve the thin-lens equation for f, given that $s = 25$ cm (the actual distance of the book or nearby object from the eyeglass lens) and $s' = -(\text{Near point})$ (the position of the virtual image produced by the eyeglass lens). Light entering the eye appears to come from s'. The negative sign means that the virtual image is on the same side of the lens as the object.

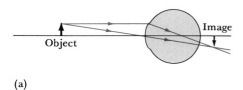

(a)

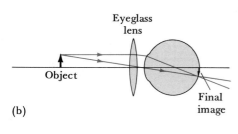

Eyeglass lens

Object

Final image

(b)

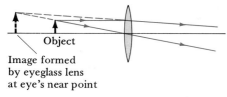

Object

Image formed by eyeglass lens at eye's near point

(c)

FIG. 20.16. (a) The image of a near object is formed behind the retina for a farsighted person. (b) A converging eyeglass lens will cause the rays to converge on the retina. (c) The corrective lens forms a virtual image at the eye's near point.

EXAMPLE 20.4 A farsighted woman has a near point of 1.5 m. Calculate the focal length of lenses needed for her eyeglasses so that she can read a book held 25 cm from her eyes.

SOLUTION We know that $s = 25$ cm (the position at which the book is held) and $s' = -150$ cm (the position from which we want her eye to perceive the light as coming). Substituting into Eq. (20.1) and solving for f, we find that

$$\frac{1}{f} = \frac{1}{s} + \frac{1}{s'} = \frac{1}{25 \text{ cm}} + \frac{1}{-150 \text{ cm}} = 0.0333 \text{ cm}^{-1},$$

or

$$f = +30 \text{ cm}. \qquad \blacksquare$$

Diopter

An optometrist prescribes lenses for eyeglasses in terms of the power P of the lens in diopters rather than in terms of the focal length. The **power** of a lens is defined as the inverse of its focal length in meters:

$$P \text{ (diopters)} = \frac{1}{f \text{ (meters)}}. \qquad (20.3)$$

For example, the power of a diverging lens of focal length $f = -50$ cm is $P = 1/(-0.50 \text{ m}) = -2$ diopters. A lens of high power has a short focal length and causes rays to converge rapidly after passing through the lens.

20.5 Angular Magnification and Magnifying Glasses

The linear magnification of an optical system compares only image and object heights, but the apparent size of an object as judged by the eye depends not only on its height but also on its location relative to the eye. For example, a pencil held 25 cm from your eye appears longer than one held 100 cm away (Fig. 20.17). In fact, the pencil may appear longer than a 100-story building that is several miles away, even though the pencil is much smaller in size.

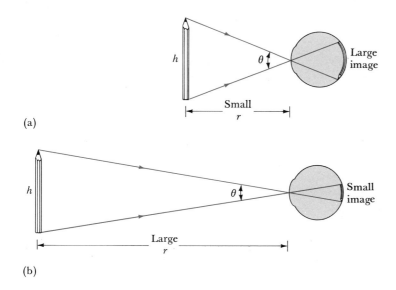

(a)

(b)

FIG. 20.17. The size of the image of a pencil on the retina depends on its distance from the retina.

A person's impression of an object's size depends on the angle θ shown in Fig. 20.17. This angle, called the **angular size,** is related to an object's height h and its distance r from the eye by the equation

$$\theta = \frac{h}{r}. \tag{20.4}$$

Remember that an angle θ, in radians, is the ratio of an arc length (h in Fig. 20.17) and the radius of a circle (r in Fig. 20.17).*

If a person looks at an object through a system of lenses, he or she sees light that appears to come from the final image of the system of lenses. If the person's eye is close to the last lens, then the angular size θ' as seen through the lenses is

$$\theta' = \frac{h'}{s'}, \tag{20.5}$$

where h' is the height of the final image and s' is the distance of the final image from the last lens. The **angular magnification** M of an optical system is defined as the ratio of θ' and θ:

$$M = \frac{\theta'}{\theta} = \frac{\text{Angular size of final image of optical system}}{\text{Angular size of object as seen by unaided eye}}. \tag{20.6}$$

A magnifying glass is the simplest optical device that provides angular magnification. The device consists of a single converging lens, as shown in Fig. 20.18. If the object is held between the focal point and the lens, a magnified virtual image is formed behind the object. When looking through the magnifying glass, the viewer thinks the light is coming from this enlarged image. By looking carefully at Fig. 20.18, you will see that both the object and the image make the same angle θ' with the optic axis ($\theta' = h'/s' = h/s$). If we look directly at the object when it is located a distance s from our eye or if we look through the lens at the image a distance s' from our eye, both produce equally large images on our retina. Their angular size is the same. So why use the magnifying glass?

Unfortunately, we cannot focus on objects brought closer to our eyes than our near point. A magnifying glass, however, does allow us to bring the object nearer. However, instead of looking directly at the object, we look at the enlarged image produced by the magnifying glass. And we can focus on the image because it is at or beyond our eyes' near point.

To calculate the angular magnification of the magnifying glass, we compare the angular size θ' of the image seen through the magnifying glass and the angular size θ of the object seen with the unaided eye. The angular size of the image viewed through the magnifying glass is $\theta' = h'/s'$. By considering ray 2 in Fig. 20.18, we see that $h'/s' = h/s$. Thus,

$$\theta' = \frac{h}{s},$$

where h is the actual height of the object and s is its distance from the magnifying glass.

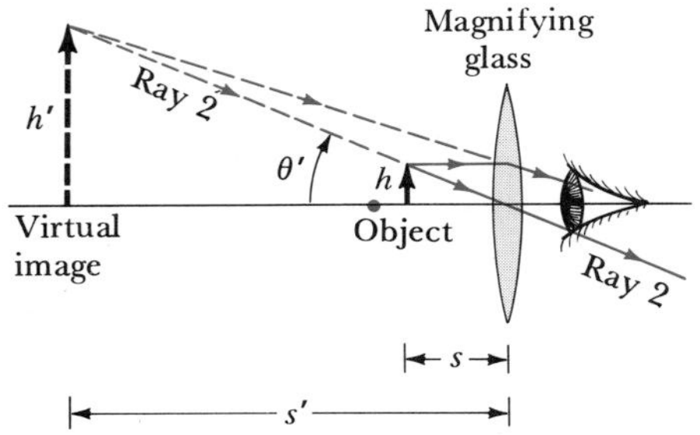

FIG. 20.18. A magnifying glass produces an enlarged virtual image of an object held near the glass.

*This expression for θ is exact only if h is the curved length on the arc of a circle ($s\,\bigcirc$) and not the straight line between the corners of the triangle ($h\,\triangleright$). For small angles, $h \simeq s$, and Eq. (20.4) is a good approximation.

Next we consider the angular size of the object as seen with the unaided eye. The angular size θ of an object of height h held a distance r from the eye is $\theta = h/r$ (Fig. 20.19a). As the object is brought closer to the eye (Fig. 20.19b and c), its angular size increases. Should the object be brought closer than the near point, a blurred image is formed on the retina (Fig. 20.19d). The maximum angular size θ_{max} of an object of height h, when viewed by the naked eye and when focused on the retina, is

$$\theta_{max} = \frac{h}{\text{Near point}}.$$ (20.7)

The **angular magnification of the magnifying glass** (called **magnifying power**) is the ratio of the angular size θ' as seen through the magnifying glass and the maximum angular size θ_{max} as seen with the unaided eye. Thus,

$$M = \frac{\theta'}{\theta_{max}} = \frac{h/s}{h/(\text{Near point})} = \frac{\text{Near point}}{s}.$$ (20.8)

EXAMPLE 20.5 A stamp is held 4.2 cm from a magnifying glass whose focal length is $+5.0$ cm. Locate the virtual image of the stamp and calculate the angular magnification of the magnifying glass. The near point of the eye is 25 cm.

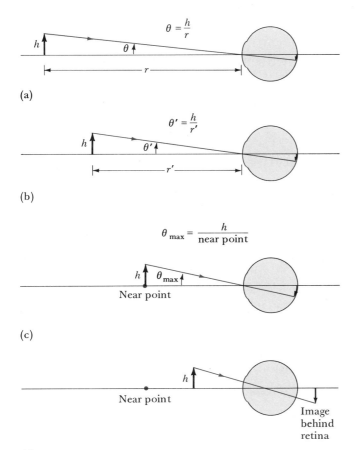

(a)

(b)

(c)

(d)

FIG. 20.19. The angular size θ of an object increases as it is brought closer to the eye. If brought too close, as in (d), an image is formed behind the eye, causing a blurred image on the retina.

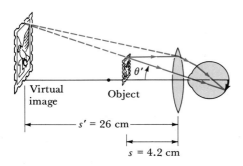

Virtual image

Object

θ'

$s' = 26$ cm

$s = 4.2$ cm

FIG. 20.20. A magnifying glass forms a virtual image of a stamp held near the glass.

SOLUTION The image distance s' is calculated using the thin-lens equation:

$$\frac{1}{s'} = \frac{1}{f} - \frac{1}{s} = \frac{1}{5.0 \text{ cm}} - \frac{1}{4.2 \text{ cm}} = -0.0381 \text{ cm}^{-1},$$

or

$$s' = -26 \text{ cm}.$$

The image is virtual and is located 26 cm behind the lens, as shown in Fig. 20.20. Since 26 cm is greater than the near point of the eye, a clear image is formed on the retina. The eye perceives rays of light as coming from this image.

The angular magnification of the magnifying glass is

$$M = \frac{\text{Near point}}{s} = \frac{25 \text{ cm}}{4.2 \text{ cm}} = \underline{6.0}. \qquad \blacksquare$$

20.6 Combinations of Lenses

Many optical instruments that have more than one lens (such as microscopes and telescopes) form an image of an object that is magnified more than is possible using only a single lens. The technique for locating the final image and calculating its magnification is not difficult but requires careful attention to several details.

Consider the two-lens system shown in Fig. 20.21a. The lenses are separated by a distance d and have focal lengths f_1 and f_2. The object is a distance s_1 from lens 1. The location of the final image of the two-lens system can be determined as follows.

Solving Multiple-Lens Problems

1. Using Eq. (20.1), determine the location s_1' of the image formed by the first lens (Fig. 20.21b). If s_1' is positive, the image is to the right of lens 1. If s_1' is negative, the image is to the left of lens 1. In our example, s_1' is positive.

2. The image of lens 1 is now the object for lens 2. The object distance s_2 (Fig. 20.21c) is

$$s_2 = d - s_1', \qquad (20.9)$$

where d, the separation of lenses, is a positive number and s_1' is either positive or negative, depending on the value calculated in step 1. If s_1' is a larger positive number than d, the image of lens 1 is formed to the right of lens 2; s_2 will then be a negative number (be sure to retain this negative sign in subsequent calculations).

3. Use the thin-lens equation to calculate the image distance s_2' of lens 2 (Fig. 20.21d):

$$\frac{1}{s_2'} = \frac{1}{f_2} - \frac{1}{s_2}.$$

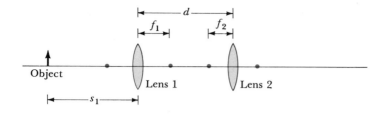

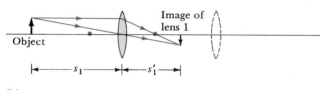

(a)

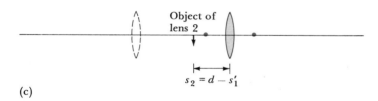

(b)

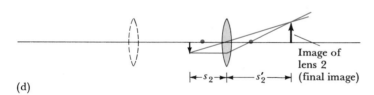

(c)

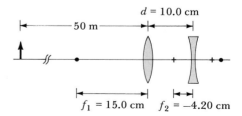

(d)

FIG. 20.21. A step-by-step scheme for locating the final image of the two-lens system shown in (a). (b) First locate the image of the first lens. (c) Calculate the object distance for the second lens. (d) Locate the final image formed by the second lens.

This is the location of the final image relative to the second lens. If s'_2 is positive, the final image is real and located to the right of lens 2. If s'_2 is negative, the final image is virtual and to the left of lens 2. In our example, the final image is real and to the right of lens 2.

4. The total linear magnification m equals the product of the linear magnifications of each lens: $m = m_1 m_2$. If m is positive, the final image has the same orientation as the object. If m is negative, the image is inverted relative to the object. Techniques for calculating the total angular magnification M are illustrated in the discussions and examples that follow.

The order of the steps used in the problem-solving procedure is often altered, depending on the known and desired information. You should find the procedure simple to use; just be sure to follow the appropriate sign conventions.

EXAMPLE 20.6 An object is located 50 m from a +15.0-cm focal length converging lens (Fig. 20.22). A diverging lens with a −4.20-cm focal length is placed 10.0 cm to the right of the first lens. Locate the final image of the two-lens system.

FIG. 20.22. A Galilean two-lens system that is similar to an opera glass.

SOLUTION The following information is given: $f_1 = +15.0$ cm, $f_2 = -4.20$ cm, $d = 10.0$ cm, and $s_1 = 5000$ cm. First locate the image of the first lens using Eq. (20.1):

$$\frac{1}{s_1'} = \frac{1}{f_1} - \frac{1}{s_1} = \frac{1}{+15.0 \text{ cm}} - \frac{1}{5000 \text{ cm}} = 0.06647 \text{ cm}^{-1},$$

or

$$s_1' = \frac{1}{0.06647 \text{ cm}^{-1}} = +15.05 \text{ cm}.$$

Next locate the object of the second lens using Eq. (20.9):

$$s_2 = d - s_1' = 10.0 \text{ cm} - 15.05 \text{ cm} = -5.05 \text{ cm}.$$

The negative sign indicates that the object of the second lens is to its right. Finally, using Eq. (20.1), locate the image of the second lens:

$$\frac{1}{s_2'} = \frac{1}{f_2} - \frac{1}{s_2} = \frac{1}{-4.20 \text{ cm}} - \frac{1}{-5.05 \text{ cm}} = -0.0401 \text{ cm}^{-1},$$

or

$$s_2' = \frac{1}{-0.0401 \text{ cm}^{-1}} = \underline{-25.0 \text{ cm}}.$$

The negative sign indicates that the final image is 25.0 cm to the left of the second lens. Since it is to the left, the final image is virtual.

Telescopes

A telescope (Fig. 20.23a) is a two-lens optical system that helps us see distant objects. The most common telescope consists of two converging lenses separated by a distance approximately equal to the sum of their focal lengths. When observing a distant object, the first lens produces a real image just beyond its own focal length, as shown in part (b). The second lens is located so that this image of the first lens is just inside the focal point of the second lens. The image of the second lens is a magnified virtual image, as shown in part (c). Our eye perceives that light is coming from this enlarged virtual image. The magnification of the telescope is the product of the magnification of each lens.

EXAMPLE 20.7 A lion 1.2 m tall stands 50 m from the first lens of the telescope shown in Fig. 20.23a. (a) Locate the final image of the lion. (b) Calculate the linear magnification of the telescope and the height of the final image. (c) Calculate the angular magnification of the telescope.

SOLUTION (a) We will leave the details of calculating the final image position to you. The results of these calculations are as follows and are shown in Fig. 20.23b and c:

$$s_1 = 50 \text{ m} = 5000 \text{ cm},$$
$$s_1' = 20.08 \text{ cm},$$
$$s_2 = 4.42 \text{ cm},$$
$$s_2' = \underline{-38.1 \text{ cm}}.$$

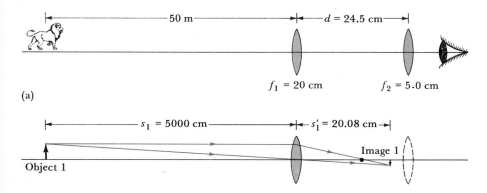

(a)

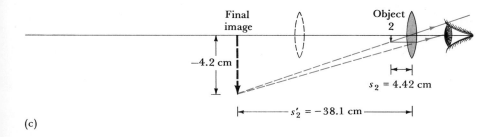

(b)

(c)

FIG. 20.23. (a) A lion is viewed through a telescope. (b) The first lens produces a real inverted image 20.08 cm from the lens. (c) This image is the object for the second lens. The final image is a virtual inverted image that is 38.1 cm to the left of the second lens. The distance to and height of the lion are not drawn to scale. The final image is actually much smaller than the lion.

The final virtual image is 38.1 cm to the left of the second lens.

(b) The total linear magnification is the product of the linear magnifications of both lenses [Eq. (20.2)]:

$$m = m_1 m_2 = \left(-\frac{s_1'}{s_1}\right)\left(-\frac{s_2'}{s_2}\right) = \frac{(s_1')(s_2')}{(s_1)(s_2)} = \frac{(20.08 \text{ cm})(-38.1 \text{ cm})}{(5000 \text{ cm})(4.42 \text{ cm})} = \underline{-0.035}.$$

The negative sign indicates that the image is inverted. The height of the final image is

$$h' = mh = (-0.035)(1.2 \text{ m}) = -0.042 \text{ m} = \underline{-4.2 \text{ cm}}.$$

(c) To calculate the angular magnification, we compare the angular size of the lion as seen through the optical system and its angular size as seen with the unaided eye. Through the optical system, the lion seems to be a distance of 38.1 cm from the second lens, and its inverted height is 4.2 cm. Thus, according to Eq. (20.5),

$$\theta' = \frac{h'}{s'} = \frac{-4.2 \text{ cm}}{38.1 \text{ cm}} = -0.110 \text{ rad}.$$

To the unaided eye, the lion is 50.245 m away and 1.2 m tall. Its angular size to the unaided eye is

$$\theta = \frac{h}{r} = \frac{1.2 \text{ m}}{50.245 \text{ m}} = 0.024 \text{ rad}.$$

The angular magnification is given by Eq. (20.6):

$$M = \frac{\theta'}{\theta} = \frac{-0.110 \text{ rad}}{0.024 \text{ rad}} = \underline{-4.6}.$$

The lion appears 4.6 times taller when viewed through the telescope. The negative sign indicates that the image is inverted. Prisms, if placed between the two lenses, will reinvert the image (see Fig. 19.16) so that the final image is upright; binoculars use such prisms. ∎

The Compound Microscope

The compound microscope is another common example of a two-lens system. Microscopes have been used for more than three hundred years to examine the details of small objects. In 1665 Robert Hooke published a book, *Micrographia*, that contained detailed drawings of insects, snowflakes, and other small objects that had been examined with microscopes whose angular magnifications were 100 or less. Today, the very best light microscopes provide angular magnifications of about 1000. With these instruments, scientists can distinguish objects separated by 2×10^{-7} m (about the length of 700 atoms lined in a row).

Although modern microscopes may contain as many as ten or more lenses, they work in principle much like the two-lens microscope used by Robert Hooke (depicted schematically in Fig. 20.24a). In this microscope both lenses are converging. The lens nearest the object is called the *objective lens*, and that nearest the eye of the observer is called the *eyepiece* or *ocular lens*. Both lenses have relatively short focal lengths (on the order of 2 cm or less) and are typically separated by 10 to 20 cm. When the object is placed a little to the left of the focal point of the objective lens (Fig. 20.24b), a real, magnified image of this object is formed in the microscope tube. If a screen were placed in the tube at the image distance s'_1, a real image of the object would be formed on the screen. This image of the objective lens has a linear magnification given by Eq. (20.2):

$$m_1 = -\frac{s'_1}{s_1}.$$

The image of the first lens is the object for the second. Since this object is located just inside the focal point of the second lens, a final, virtual image is formed outside the microscope (Fig. 20.24c). The second lens of the microscope is actually a magnifying glass used to view the real image of the first lens. The angular magnification of this magnifying glass is, according to Eq. (20.8),

$$M_2 = \frac{\text{Near point}}{s_2}.$$

Since the original object has already been magnified by an amount $m_1 = -(s'_1/s'_1)$, the total angular magnification M of the microscope is the product of the two magnifications:

$$M = m_1 M_2 = \left(-\frac{s'_1}{s_1}\right)\left(\frac{\text{Near point}}{s_2}\right). \tag{20.10}$$

This expression for angular magnification can be rewritten in terms of the focal lengths of the lenses and their separation d. Consider Fig. 20.25. Notice that $s_1 \simeq f_1$; $s'_1 \simeq d - f_2$; and $s_2 \simeq f_2$. By substituting these values in Eq. (20.10), the angular magnification becomes

$$M \simeq \left(-\frac{d - f_2}{f_1}\right)\left(\frac{\text{Near point}}{f_2}\right). \tag{20.11}$$

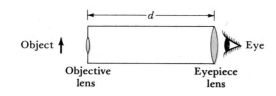

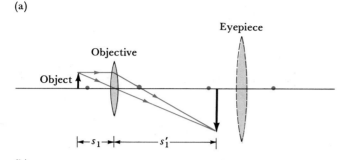

(a)

(b)

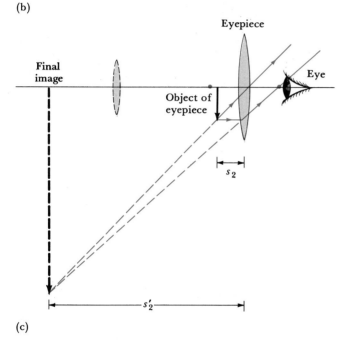

(c)

FIG. 20.24. (a) A two-lens microscope. (b) A magnified, real image of an object is formed by the objective lens. (c) This image is the object for the eyepiece, which serves as a magnifying glass for examining the object.

The negative sign indicates that the final image is inverted. This expression for magnification is easier to use than Eq. (20.10) because we need not locate the image and object positions. It depends only on the focal lengths of the lenses, on their separation, and on the near point of the eye.

EXAMPLE 20.8 A compound microscope has an objective lens of focal length 0.80 cm and an eyepiece of focal length 1.25 cm. The lenses are separated by 18.0 cm. If a red blood cell is located 0.84 cm in front of the objective lens, (a) where is the final image of the cell located, and (b) what is its angular magnification? The viewer's near point is 25 cm.

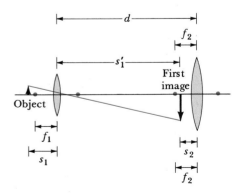

FIG. 20.25

SOLUTION (a) We are given the following information: $f_1 = +0.80$ cm, $f_2 = +1.25$ cm, $d = 18.00$ cm, and $s_1 = +0.84$ cm. The image of the first lens is found using the thin-lens equation:

$$\frac{1}{s_1'} = \frac{1}{f_1} - \frac{1}{s_1} = \frac{1}{0.80 \text{ cm}} - \frac{1}{0.84 \text{ cm}} = 0.0595 \text{ cm}^{-1},$$

or

$$s_1' = \frac{1}{0.0595 \text{ cm}^{-1}} = +16.8 \text{ cm}.$$

The image of the first lens is the object of the second. The object distance is, according to Eq. (20.9), $s_2 = d - s_1' = 18.0$ cm $- 16.8$ cm $= 1.2$ cm. The image of the second lens, which is the final image, is then found from the thin-lens equation:

$$\frac{1}{s_2'} = \frac{1}{f_2} - \frac{1}{s_2} = \frac{1}{1.25 \text{ cm}} - \frac{1}{1.20 \text{ cm}} = -0.0333 \text{ cm}^{-1},$$

or

$$s_2' = \frac{1}{-0.0333 \text{ cm}^{-1}} = \underline{-30 \text{ cm}}.$$

(b) The linear magnification of the first lens is, according to Eq. (20.2), $m_1 = -s_1'/s_1 = -(16.8 \text{ cm})/(0.84 \text{ cm}) = -20$. The angular magnification of the second lens is $M_2 = (\text{Near point})/s_2 = 25 \text{ cm}/1.2 \text{ cm} = 21$. Thus, the total angular magnification is

$$M = m_1 M_2 = (-20)(21) = \underline{-420}.$$

The negative sign indicates that the image is inverted.

The angular magnification can be calculated approximately using Eq. (20.11):

$$M \simeq -\left(\frac{d - f_2}{f_1}\right)\left(\frac{\text{Near point}}{f_2}\right) = -\left(\frac{18.0 \text{ cm} - 1.25 \text{ cm}}{0.80 \text{ cm}}\right)\left(\frac{25 \text{ cm}}{1.25 \text{ cm}}\right) = -420.$$

As we see, Eq. (20.11) produces the same result in this example as the more exact calculation using expressions for the magnification of each lens. ∎

Scientists using microscopes have found that the eye feels the least strain when the final image is about 100 cm from the eyepiece. This is called the position of minimum eyestrain. The final image can be shifted to this position through a small shift in the location of the eyepiece; the change in angular magnification is usually minor.

20.7 Aberrations

We have been using the thin-lens equation to locate the image produced by lenses that gather and focus light from an object. Most lenses form images that are slightly out of focus, even at the location where a sharp image is supposed to be formed. The lens defects causing this lack of focus are called **aberrations.**

Some aberrations result from improper lens construction and flaws in the material used to construct the lens. Others, such as spherical aberrations, chromatic aberrations, and astigmatisms, result from inherent refractive properties of light passing through the lens.

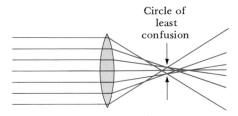

FIG. 20.26. Light rays moving parallel to the axis of a spherical lens do not all cross exactly at the focal point. The image of the rays is smallest at the circle of least confusion.

Spherical Aberration

Lenses are easiest to make if their surface has a spherical shape. Unfortunately, light rays moving parallel to the principal axis of a spherical lens do not all cross the axis at the same point (Fig. 20.26). Rays passing through the lens near its edge are refracted more than desired and cross the principal axis before rays that pass through the center region of the lens. Thus, there is no single focal point for all parallel rays passing through the lens. There is, however, one position along the axis where the transmitted light forms the smallest circle, called the **circle of least confusion** (Fig. 20.26). Parallel rays of light seem to focus best at this position on the axis.

Spherical aberration can be minimized by collecting light from a smaller region or aperture near the center of the lens (using a high f-stop on a camera) or by combining several lenses that provide the same net focal length as a single lens but whose spherical aberrations cancel each other.

Chromatic Aberration

Chromatic aberration is caused by the difference in the degree to which different colors of light are refracted. Blue light, for example, is refracted more than red light at the surfaces of a converging lens (Fig. 20.27a), so the blue light focuses closer to the lens than the red light. This variation in focal length for different colors of light can be corrected by making a compound lens of two different materials such as shown in Fig. 20.27b. If properly constructed, this compound lens, called an achromatic lens (from the Greek *achrōmatos*, "without color"), can focus all colors of light at the same point.

Astigmatism

Astigmatism is a common defect of the eye caused by a lack of symmetry in the eye's cornea. For instance, if the cornea is flatter along a vertical axis parallel to the corneal surface than along a horizontal axis parallel to the surface, light passing through each axis is brought to focus at different distances from the cornea. Astigmatism is detected by viewing a set of lines such as shown in

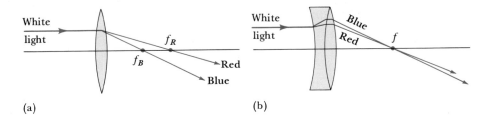

(a) (b)

FIG. 20.27. (a) Different colors of light are focused at different distances from a lens. (b) An achromatic compound lens brings different color rays to focus at the same point.

FIG. 20.28. If you have astigmatism, some lines will appear fainter and less focused than others.

Fig. 20.28. The lines should appear equally intense if you do not have astigmatism. If you do have astigmatism, lines in one direction may appear fainter and less focused than lines perpendicular to that direction. Astigmatism of the eye is corrected by eyeglass lenses that are less curved in one direction than in another.

Another form of astigmatism, called *off-axis astigmatism,* has a noticeable effect on glass lenses used in optical devices such as cameras. The image of a point O (Fig. 20.29) that is above the principal axis of a converging lens does not form a point image but instead forms a horizontal line image and a vertical line image at slightly different distances from the lens. Once again, a circle of least confusion occurs between these lines where the point is focused best. Off-axis astigmatism is reduced in photography by placing the important part of a photographed scene in the center of the field of view. Less-important background that is farther from the principal axis will suffer more from off-axis astigmatism. Expensive cameras reduce off-axis astigmatism by combining several lenses whose off-axis astigmatism effects cancel each other.

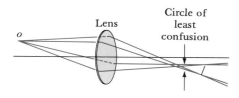

FIG. 20.29. Off-axis astigmatism results in a slightly confused image.

20.8 Mirrors

We have been discussing optical devices that focus and change the direction of light by refracting it as the light passes through glass lenses. We now consider devices such as mirrors that focus light by reflecting it. Mirrors can be used in much the same way as lenses. For example, most modern research telescopes use a mirror to focus light from distant stars and galaxies onto a film or into our eyes. The mirror of the large telescope at Mt. Palomar is 5.1 m (17 ft) in diameter. A converging lens of this size would sag under its own weight if supported at its edges.

A more commonly used mirror, such as that found on a wall, is called a **plane mirror** (the word *plane* indicates that it is flat). If an object is held in front of the mirror, as in Fig. 20.30, an upright image is formed an equal distance behind the mirror. The image is the place from which reflected rays of light seem to come—the place from which the dashed lines in Fig. 20.30 diverge. The height h' of the image is the same as the height h of the object. Thus, the linear magnification [Eq. (20.2)] of a plane mirror is $m = h'/h = 1$. The image shown in Fig. 20.30 is virtual since the light does not actually pass through the image. It just appears to come from the image.

Although plane mirrors only change the direction of light, curved mirrors can focus light to produce expanded or contracted images. Most of these curved mirrors are portions of either the inside or outside surface of a sphere. If the mirrored surface is on the inside of a spherical surface, it is said to be a **concave** or **converging mirror** (Fig. 20.31a). If the mirrored surface is on the outside of the sphere, it is said to be a **convex** or **diverging mirror** (Fig. 20.31b). Two techniques are used to find the size and location of the image produced by these mirrors: the ray technique and the mirror equation. Both techniques are similar to image location techniques used with lenses.

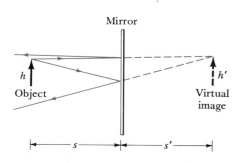

FIG. 20.30. An object and its virtual image, as formed by a plane mirror. For a plane mirror, $s = s'$.

Ray Techniques

An image of an object formed by a mirror can be located by drawing two special rays that leave the object (Fig. 20.32). Ray 1 moves parallel to the axis of the

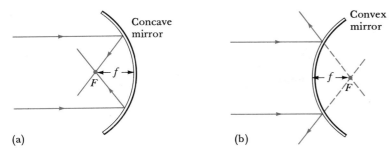

FIG. 20.31. (a) Parallel rays cross at the focal point after reflection from a converging (concave) mirror. (b) Parallel rays appear to leave the focal point after reflection from a diverging (convex) mirror.

mirror. For a converging mirror, the reflected ray passes through the focal point on the axis in front of the mirror (Fig. 20.32a). For a diverging mirror, the reflected ray appears to diverge from the focal point behind the mirror (the dashed-line path shown in Fig. 20.32c). Ray 2 reflects from the center of each type of mirror at the place where the principal axis crosses the mirror (Fig. 20.32b and d). The angle of the reflected ray equals the angle of the incident ray.

The image of an object is located at the place where these rays intersect after reflecting from the mirror (a real image). If the rays diverge after leaving the mirror, the image is located at the position from which the rays appear to diverge from behind the mirror (a virtual image). Several examples of the ray technique of locating mirror images are shown in Fig. 20.33.

Mirror Equation

The location of the image can be calculated using a **mirror equation** [the same as the thin-lens equation—Eq. (20.1)]:

$$\frac{1}{f} = \frac{1}{s} + \frac{1}{s'}.$$

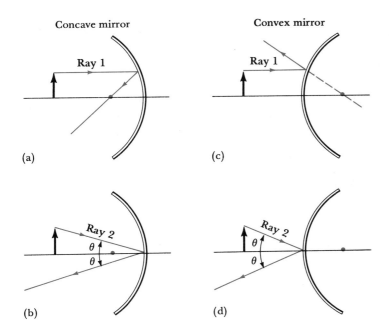

FIG. 20.32. The paths followed by two special rays after reflection from mirrors. The rays are used to construct mirror images.

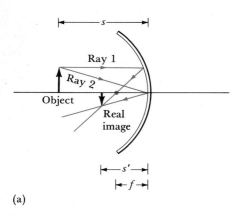

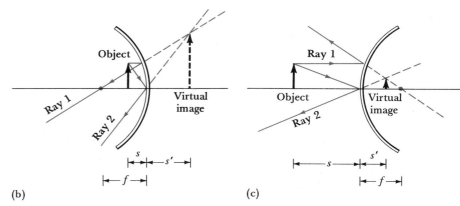

(a)   (b)   (c)

FIG. 20.33. Examples of image location for mirrors.

The linear magnification of the image is calculated using the same equation that was used for lenses [Eq. (20.2)]:

$$m = \frac{h'}{h} = -\frac{s'}{s}.$$

The symbols in these equations are illustrated in Fig. 20.33 and described as follows:

1. The **object distance** s is the distance of the object from the mirror and is a positive number.

2. The **image distance** s' is the distance of the image from the mirror and is positive for real images in front of the mirror (to the left in Fig. 20.33a) and negative for virtual images behind the mirror (Fig. 20.33b and c).

3. The **focal length** f is the distance of the focal point from the mirror and is positive for converging mirrors (focal point in front of the mirror) and negative for diverging mirrors (focal point behind the mirror).

4. The **object height** h and **image height** h' are positive if upright and negative if inverted.

EXAMPLE 20.9 A dentist examines the cavity in a person's tooth using a concave mirror of focal length $+2.0$ cm (Fig. 20.34). If the cavity is 1.5 cm from the mirror's surface, (a) where is the image located, (b) is it real or virtual, and (c) what is its linear magnification?

SOLUTION (a) The following information is given: $f_1 = +2.0$ cm (a concave mirror has a positive focal length); $s = 1.5$ cm. The mirror equation [Eq. (20.1)] is used to find the image position:

$$\frac{1}{s'} = \frac{1}{f} - \frac{1}{s} = \frac{1}{2.0 \text{ cm}} - \frac{1}{1.5 \text{ cm}} = -0.167 \text{ cm}^{-1},$$

or

$$s' = \frac{1}{-0.167 \text{ cm}^{-1}} = \underline{-6.0 \text{ cm}}.$$

FIG. 20.34

(b) Since s' is negative, the image is <u>virtual</u> and appears to be 6.0 cm behind the mirror.

(c) Using Eq. (20.2), we find that the image magnification is

$$m = -\frac{s'}{s} = -\frac{-6.0 \text{ cm}}{1.5 \text{ cm}} = \underline{+4.0}.$$

The cavity will appear to be four times its actual size. Since the magnification is positive, the image has the same orientation as the cavity. ■

EXAMPLE 20.10 The observation mirror in a store, shown in Fig. 20.35, is convex and has a focal length of -1.8 m. Calculate (a) the image location and (b) the image height of a 1.50-m-tall boy standing 4.0 m from the mirror.

SOLUTION (a) The following information is given: $s = 4.0$ m, $f = -1.8$ m, and $h = 1.50$ m. The image seen in Fig. 20.35 appears to be behind the mirror. Let us use the mirror equation [Eq. (20.1)] to see if our impression agrees with formal optics. Solving for s', the image distance, we find that

$$\frac{1}{s'} = \frac{1}{f} - \frac{1}{s} = \frac{1}{-1.8 \text{ m}} - \frac{1}{4.0 \text{ m}} = -0.56 \text{ m}^{-1} - 0.25 \text{ m}^{-1}$$

$$= -0.81 \text{ m}^{-1}, \quad \text{or}$$

$$s' = \frac{1}{(-0.81 \text{ m}^{-1})} = \underline{-1.2 \text{ m}}.$$

The boy does produce a virtual image behind the mirror.

(b) The boy is magnified by an amount

$$m = -\frac{s'}{s} = -\frac{(-1.2 \text{ m})}{4.0 \text{ m}} = 0.30.$$

His image height will be

$$h' = mh = 0.30(1.50 \text{ m}) = \underline{0.45 \text{ m}}.$$

He appears to be 0.45 m tall and is upright. ■

FIG. 20.35. A diverging mirror. (Photograph by Douglas Rachele.)

Summary and Additional Readings

1. **A lens** is an optical device that changes the apparent location of an object by altering the path of light coming from the object. The image—that is, the apparent location of the object as seen with the lens—can be located by using ray diagrams or by using the thin-lens equation.

2. **Ray diagrams:** A **real image** can be located by determining the place where two or three rays from the object (Fig. 20.3) cross after passing through the lens. If the rays diverge after passing through the lens, we say that a **virtual image** is formed at the point from which the rays seem to originate on the same side of the lens as the object.

3. **Thin-lens equation:** The image position can be calculated using the thin-lens equation:

$$\frac{1}{f} = \frac{1}{s} + \frac{1}{s'}, \qquad (20.1)$$

where f is the focal length of the lens, s is the distance of the object from the lens, and s' is the distance of the image from the lens.

If we arrange the lens so that light moves through it from left to right, then the sign conventions for the thin-lens equation are as follows. (1) The **focal length** f is positive for converging (convex) lenses and negative for diverging (concave) lenses. (2) The **object distance** s is positive if the object is to the left of the lens and negative if to the right. (This rule is needed only for systems with two or more lenses.) (3) The **image distance** s' is positive for real images formed to the right of the lens and negative for virtual images formed to the left of the lens.

For a two-lens optical system, the image of the first lens is the object of the second. The object distance s_2 for the second lens is

$$s_2 = d - s_1', \qquad (20.9)$$

where d is the separation of the lenses and s_1' is the image distance of the first lens.

4. **Linear magnification:** The linear magnification m of a single lens is the ratio of the image height h' and object height h. The height is positive if the image or object is upright and negative if it is inverted. The linear magnification can also be calculated using the image and object distances:

$$m = \frac{h'}{h} = -\frac{s'}{s}. \qquad (20.2)$$

For a two-lens system, the total linear magnification is just the product of the individual linear magnifications of each lens.

5. The **eye's optical system** consists of a cornea and lens that together act as a converging lens to produce a real image of objects on the eye's retina. By varying the focal length of the lens, the normal eye can focus on objects from a **far point** of infinity to a **near point** of 25 to 50 cm.

6. **Angular magnification:** The **angular size** θ of an object as seen by the unaided eye is $\theta = h/r$, where h is its height and r is its distance from the eye. The maximum angular size $\theta_{\max}$ of an object is $\theta_{\max} = h/(\text{Near point})$, where the near point is the closest an object can be held to the eye while remaining in focus. When viewed through an optical device, the angular size is $\theta' = h'/s'$, where h' is the height of the final image of the optical system and s' is the distance from the lens nearest the eye to the final image. The angular magnification M of distant objects is

$$M = \frac{\theta'}{\theta}. \qquad (20.6)$$

The angular magnification of nearby objects when seen through a **magnifying glass** is

$$M = \frac{\theta'}{\theta_{\max}} = \frac{\text{Near point}}{s}. \qquad (20.8)$$

7. **Mirrors:** The images of objects viewed in a mirror are located by the ray technique using the rays shown in Fig. 20.32 or by using the mirror equation [the same as Eq. (20.1)]. The sign conventions for the mirror equations are as follows. (1) The **focal length** f is positive for a converging (concave) mirror and negative for a diverging (convex) mirror. (2) The **object distance** s is positive. (3) The **image distance** s' is positive for real images in front of the mirror and negative for virtual images behind the mirror.

Michael J. Ruiz, "Camera Optics," *The Physics Teacher*, pp. 372–380, September (1982).

Reuben E. Alley, Jr., "The Camera Obscura in Science and Art," *The Physics Teacher*, pp 632–638, December (1980). Explains the pinhole camera's role in science and art from Aristotle to recent times.

F. D. Smith, "How Images Are Formed," *Scientific American* **219**(3), 97 (1968).

Questions

1. A bubble of air is suspended underwater. Draw a ray diagram showing the approximate path followed by light rays passing from the water through the upper half of the bubble, through its center, and through its lower half. Does the bubble act like a converging or diverging lens? Explain.

2. A person underwater cannot focus clearly on any object. Yet if the person wears goggles, he or she can see objects clearly. Explain.

3. The lens of a slide projector is adjusted so that a clear image of the slide is formed on a screen. If the screen is moved farther away, must the lens be moved toward or away from the slide? Explain.

4. The retina has a blind spot at the place where the optic nerve leaves the retina. Design and describe a simple experiment that allows you to perceive the presence of this blind spot. Why doesn't the blind spot affect your normal vision?

5. Borrow the glasses of a person who is known to be farsighted. Invent and describe a simple experiment to measure the approximate focal lengths of the lenses of his or her glasses.

6. You run toward a building with walls of a metallic, reflective material. Your speed is 1.5 m/s. How fast does your image appear to move toward you? Explain.

7. Explain why your image in a plane mirror has its left and right sides reversed, while the image of your head and legs are not inverted.

8. You stand in the Hall of Mirrors at Versailles. You are 4 m from one wall of mirrors. What is the desired distance setting on your camera when photographing your image in the mirrors? Explain.

9. The image of three children standing in front of a "trick mirror" is shown in Fig. 20.36. Why is the face of the child on the right elongated whereas his hands seem shortened? Justify your answer in terms of the magnification produced by different types of mirrors.

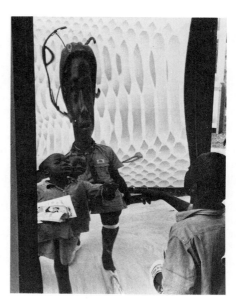

FIG. 20.36. (Marilyn Silverstone © 1964 Magnum Photos.)

Problems

20.2 Formation of Images

■ 1. Using a ruler, carefully draw ray diagrams to locate the images of the following objects: (a) an object that is 30 cm from a converging lens of 10-cm focal length; (b) an object that is 14 cm from the same lens; (c) an object that is 5 cm from the same lens. (Choose a scale so that your drawing fills a significant portion of the width of a paper.) Measure the image locations on your drawings and indicate if they are real or virtual, upright or inverted.

■ 2. Repeat the procedure described in Problem 1 for the following lenses and objects: (a) an object that is 30 cm from a diverging lens of −10-cm focal length; (b) an object that is 14 cm from the same lens; (c) an object that is 5 cm from the same lens.

■ 3. Repeat the procedure described in Problem 1 for the following lenses and objects: (a) an object that is 7 cm from a converging lens of 10-cm focal length; (b) an object that is 7 cm from a diverging lens of −10-cm focal length.

■ 4. Repeat the procedure described in Problem 1 for the following lenses and objects: (a) an object that is 20 cm from a converging lens of 10-cm focal length; (b) an object that is 5 cm from the same lens; (c) an object that is 20 cm from a diverging lens of −10-cm focal length; (d) an object that is 5 cm from the lens in part (c).

20.3 Thin-Lens Equation

5. Use ray diagrams to locate the images of the following objects: (a) an object that is 10 cm from a converging lens of 15-cm focal length, (b) an object that is 10 cm from a diverging lens of −15-cm focal length. (c) and (d) Calculate the image locations for parts (a) and (b) using the thin-lens equation.

6. Use ray diagrams to locate the images of the following objects: (a) an object that is 6 cm from a converging lens of +4-cm focal length, (b) an object that is 8 cm from a diverging lens of −4-cm focal length. (c) and (d) Calculate the image locations for parts (a) and (b) using the thin-lens equation.

7. Light passes through a narrow slit, then through a lens, and onto a screen. The slit is 20 cm from the lens. The screen, when adjusted for a sharp image of the slit, is 15 cm from the lens. What is the focal length of the lens?

8. A camera with a lens of focal length 6.00 cm is used to photograph a painting located 3.00 m from the camera lens. Where should the film be located?

9. A photographer sets her camera at 1/120 s and f-5.6 for a picture at an afternoon football game. The next day under the same lighting conditions she takes a picture in a meadow using an f-stop of 16 (for better depth of focus). What shutter speed should she use on the second day? Explain.

■ 10. A camera with 5.00-cm focal length lens has an iris diaphragm opening of diameter 1.25 cm when the shutter speed is 1/240 s for particular lighting conditions. What shutter speed should be used if the lighting conditions are the same and the camera's f-stop is set at 8?

■ 11. A camera with an 8.00-cm focal length lens is used to photograph a person that is 2.00 m tall. The height of the image on the film must be no greater than 3.50 cm. (a) Calculate the closest distance the person can stand to the lens. (b) For this object distance, where should the film be located from the lens?

■ 12. A slide is located 12.6 cm from the 12.0-cm focal length lens of a projector. Calculate (a) the distance the screen should be placed from the lens and (b) the height of the image of a person on the screen that is 2.0 cm tall on the slide.

■ 13. A slide projector produces an inverted image on a screen that is 30 times larger than the view on the slide. The lens has a focal length of 12.0 cm. Calculate (a) the distance of the slide from the lens and (b) the distance of the screen from the lens.

■ 14. A carpenter ant being photographed is 18 cm from the film. At what places can the 4.0-cm converging camera lens be located so that a sharp image of the ant is produced on the film?

■ 15. A camera with a 5.0-cm focal length lens is used to photograph a secret document whose height is 10 cm. At what distance from the lens must the document be held so that an image 2.5 cm high is produced on the film? [*Note:* The real image is inverted.]

■ 16. An enlarging camera has a lens of +12.5-cm focal length. The object is a negative that is 35 mm tall. If the image to be produced on the film is 150 mm tall, how far from the lens should the object be placed?

■ 17. An aerial camera with a lens of 0.40-m focal length is used to photograph a landscape 2.0 km wide from a height of 5.0 km. Calculate the width of the image on the film.

■ 18. When viewed through a diverging lens, a scene 12 m from the lens appears to be reduced in size by one-fifth. Calculate the focal length of the lens. [*Note:* In the past, movie directors used such a lens to reduce a scene to the size it would appear on a screen.]

■ ■ 19. Make a rough graph of linear magnification versus object distance for a converging lens of 20-cm focal length as the object distance is varied from ∞ to 0. Indicate in which regions the image is real and in which regions it is virtual.

■ ■ 20. Repeat Problem 17 for a diverging lens of −20-cm focal length.

20.4 The Human Eye

21. The image distance for the lens of a person's eye is 2.10 cm. Calculate the focal length of the eye's lens system for an object (a) at infinity, (b) 500 cm from the eye, and (c) 25 cm from the eye.

22. Assume that the eye accommodates to objects at different distances by altering the distance from the lens system to the retina. If the lens system has a focal length of 2.10 cm, what is the lens-retina distance needed to view objects at (a) infinity, (b) 300 cm, and (c) 25 cm?

■ 23. (a) A woman can focus only on objects that are from 150 cm to 25 cm from her eyes. Indicate the type of vision problem she has and calculate the focal length of eyeglass lenses that will correct her problem. (b) Repeat part (a) for a man who can focus only on objects that are 3.0 m or more from his eyes. He would like to be able to read a book held 30 cm from his eyes.

■ 24. A man who can focus only on objects that lie from 80 cm to 240 cm from his eyes needs bifocal lenses. (a) Calculate the desired focal length of the upper half of the glasses used to see distant objects. (b) Calculate the focal length of the lower half used to read a paper held 25 cm from his eyes.

■ 25. A woman who can focus only on objects that lie from 100 to 300 cm from her eyes needs bifocal lenses. (a) Calculate the desired power of the upper half of the glasses used to see distant objects. (b) Calculate the power of the lower half used to read a book held 30 cm from her eyes.

■ 26. (a) A woman while reading wears glasses of +50-cm focal length. What eye defect is being corrected and what approximately are the near and far points of her unaided eye? (b) Repeat part (a) for a man whose glasses have a −350-cm focal length. He wears the glasses while driving a car.

■ 27. A 35-year-old patent clerk needs glasses of 50-cm focal length to read patent applications held 25 cm from his eyes. Five years later, while wearing the same glasses, he must hold the patent applications 40 cm from his eyes to see them clearly. What should be the focal length of new glasses so that he can read again at 25 cm?

20.5 Angular Magnification and Magnifying Glasses

■ 28. An aphid on a plant leaf is examined with a magnifying glass of +6.0-cm focal length. The glass is held so that the final virtual image is 40 cm from the lens. Assuming a near point of 30 cm, calculate the angular magnification.

■ 29. The fine print in a legal contract is examined with a magnifying glass of focal length 5.0 cm. (a) How far from the lens should the print be located to form a final virtual image 30 cm from the lens (at the eye's near point)? (b) Calculate the angular magnification of the magnifying glass.

■ 30. A person has a near point of 150 cm. (a) What is the nearest distance that a magnifying glass of 5.0-cm focal length can be held from print on a page and still form an image beyond the person's near point? (b) Calculate the angular magnification for an image at the near point.

■ 31. A stamp is viewed through a magnifying glass of 5.0-cm focal length. Calculate the object distance for virtual images formed at (a) negative infinity, (b) −200 cm, and (c) −25 cm. (d) Calculate the angular magnification in each case.

■ 32. A magnifying glass of focal length 8.0 cm is held 6.9 cm from a page of fine print. (a) Where is the image located? (b) How large is the image of a letter that is 1.6 mm high? (c) What is the angular size of the image as seen through the magnifying glass? (d) What is the maximum angular size of the letter when viewed directly by a person with a near point of 25 cm? (e) Calculate the angular magnification of the magnifying glass when used as described.

20.6 Combinations of Lenses

■ 33. A converging lens of focal length 20 cm is placed 30 cm in front of another converging lens of focal length 4.0 cm. An object is placed 100 cm in front of the first lens. Determine (a) the location of the final image, (b) its orientation, and (c) whether it is real or virtual.

■ 34. A converging lens with a 25-cm focal length is placed 50 cm in front of a diverging lens having a focal length of −40 cm. An object 2 cm tall is placed 30 cm in front of the converging lens. Determine (a) the location of the final image, (b) its orientation, and (c) whether it is real or virtual.

■ 35. An object is placed 10 cm in front of a converging lens of focal length 4 cm. A second converging lens, also of focal length 4 cm, is located 12 cm from the first lens. (a) By carefully constructing a ray diagram, locate the final image. Using measurements on your ray diagram, *estimate* (b) the linear magnification of the object. Be sure to show clearly your rays and/or estimation technique for each step. Do not use equations!

■ 36. Repeat Problem 35 for an object located 6 cm from a converging lens of focal length 3 cm separated by 11 cm from another converging lens of focal length 1 cm.

■ 37. A telephoto lens system consists of a converging lens of

focal length 10.0 cm followed by a diverging lens of focal length −5.0 cm. The lenses are separated by 8.0 cm. (a) Where should the film be placed to view an object that is 100 m from the first lens? (b) If the object is 1.5 m tall, calculate its height on the film.

■ 38. A Galilean telescope such as shown in Fig. 20.22 produces an upright image, whereas an astronomical telescope produces an inverted image. A Galilean telescope is used to view a deer 2 m tall that is 50 m from the objective lens ($f = +100$ cm). The eyepiece of the telescope has a focal length of −5 cm. (a) Locate the image of the objective lens. (b) What must be the object distance s_2 for the eyepiece in order that the final image be at the eye's near point (that is, $s_2' = −25$ cm)? (c) What is the separation of the lenses? (d) What is the height of the final image? (e) What is the total angular magnification?

■ 39. A Galilean telescope, sometimes called an opera glass, such as shown in Fig. 20.22, consists of a +10.0-cm objective lens and a −2.0-cm eyepiece. The lenses are separated by 8.0 cm. A Victoria butterfly 20 m from the objective lens is viewed. (a) Locate the final image. (b) Calculate the total linear magnification. (c) Calculate the total angular magnification.

■ 40. During a laboratory experiment, the focal length of a diverging lens is measured by first forming a real image of a light source using a converging lens. The image is formed on a screen 20 cm from the lens. The diverging lens is then placed halfway between the converging lens and screen. To obtain a focused image, the screen must now be moved 15 cm farther to the right (away from the lenses). Calculate the focal length of the diverging lens.

■ 41. Confirm that the effective focal length f of two thin lenses of focal lengths f_1 and f_2, when in contact, is

$$\frac{1}{f} = \frac{1}{f_1} + \frac{1}{f_2}.$$

■ 42. Three lenses each of +10-cm focal length are placed in a line and separated from each other by 15 cm. If an object is placed 20 cm to the left of the first lens, where is the final image?

Telescopes

■ 43. A telescope consists of a +4.0-cm objective lens and a +0.80-cm eyepiece that are separated by 4.78 cm. Calculate (a) the location and (b) the height of the final image for an object that is 1.0 m tall and is 100 m from the objective lens.

■ 44. The world's largest telescope made only from lenses (that is, it has no mirrors) is located at the Yerkes Observatory near Chicago. Its objective lens is 1 m in diameter and has a focal length f_1 of +18.9 m. The eyepiece has a focal length f_2 of +7.5 cm. The objective lens and eyepiece are separated by 18.970 m. (a) What is the location of the final image of a moon crater 3.8×10^5 km from the earth? (b) If the crater has a diameter of 2 km, what is the size of its final image? (c) Calculate the angular magnification of the telescope by comparing the angular size of the image as seen through the telescope and the object as seen by the unaided eye. [*Note:* The angular magnification should be approximately f_1/f_2, as it is for most telescopes of this type.]

■ 45. A telescope consisting of a +3.0-cm objective lens and a +0.6-cm eyepiece is used to view an object that is 20 m from the objective lens. (a) What must the distance between the objective lens and eyepiece be to produce a final virtual image 100 cm to the left of the eyepiece? (b) What is the total angular magnification?

■ 46. You are marooned on a tropical island. Design a telescope from a cardboard map tube and the lenses of your eyeglasses. One lens has a +1.0-m focal length and the other has a +0.3-m focal length. The telescope should allow you to view an animal 100 m from the objective with the final image being formed 1.0 m from the eyepiece. Indicate the location of the lenses and the expected angular magnification.

Microscopes

■ **47.** A microscope has a +0.5-cm objective lens and a +3.0 cm eyepiece that is 20 cm from the objective lens. (a) Where should the object be located to form a final virtual image 100 cm to the left of the eyepiece? (b) What is the total angular magnification of the microscope assuming a near point of 25 cm?

■ **48.** A microscope has a +0.4-cm objective and a +2.5-cm eyepiece. (a) Where must the object for the eyepiece be located to form a final virtual image 100 cm in front of the eyepiece? (b) If the lenses are separated by 20 cm, what is the distance between the object being viewed and the objective lens? (c) Assuming a near point of 25 cm, calculate the total angular magnification.

■ **49.** A dissecting microscope is designed with a larger than normal distance between the object and the objective lens. One of these microscopes has an objective lens of 5-cm focal length and an eyepiece of 2-cm focal length. The lenses are separated by 15 cm. The final virtual image is located 100 cm to the left of the eyepiece. (a) Calculate the distance of the object from the objective lens. (b) Calculate the total angular magnification.

■ **50.** A microscope has an objective lens of focal length 0.8 cm and an eyepiece of focal length 2.0 cm. An object is placed 0.9 cm in front of the objective lens. The final virtual image is 100 cm from the eyepiece at the position of minimum eyestrain. (a) Calculate the separation of the lenses. (b) Calculate the total angular magnification.

■■ **51.** Calculate the lens separation and object location for a microscope made from an objective lens of focal length 1.0 cm and an eyepiece of focal length 4.0 cm. Arrange the lenses so that a final virtual image is formed 100 cm to the left of the eyepiece and so that the angular magnification is −260 for a person with a near point of 25 cm.

20.8 Mirrors

■ **52.** Use ray diagrams and the mirror equation to locate the position, orientation, and type of image formed by an object held in front of a concave mirror of focal length 20 cm. The object distance is (a) 200 cm, (b) 40 cm, and (c) 10 cm.

■ **53.** Repeat Problem 52 for a convex mirror of focal length −20 cm.

■ **54.** Use ray diagrams and the mirror equation, to locate the images of the following objects: (a) an object that is held 10 cm from a concave mirror of focal length 7 cm; (b) an object that is held 10 cm from a convex mirror of focal length −7 cm.

■ **55.** Calculate the focal length of the mirror Archimedes might have used to burn ships that were 150 m away. Justify your answer.

■ **56.** A fortune teller looks into her silvered-surface crystal ball whose radius is 10 cm and whose focal length is −5 cm. (a) If her eye is 30 cm from the ball, where is the image of her eye? (b) Estimate the size of that image.

■ **57.** You view yourself in a large convex mirror of −1.2-m focal length from a distance of 3 m. (a) Locate your image. (b) If you are 1.7 m tall, what is your image height?

■ **58.** The moon's diameter is 3.5×10^3 km, and its distance from the earth is 3.8×10^5 km. Calculate the position and size of the image formed by the Hale Observatory reflecting mirror, whose focal length is 16.5 m.

■ **59.** You wish to order a mirror from a scientific supply company. You want to use the mirror for shaving or makeup. The mirror should produce an image that is upright and magnified by a factor of 2 when held 15 cm from your face. What type and focal length mirror should you order?

■ **60.** You view your face in a converging mirror of focal length 20 cm. How far from the mirror should your face be located to form an image that is magnified by a factor of 1.5?

■ **61.** A clothing store has a cylindrical mirror that causes the customer's width to have a linear magnification of +0.85 when the customer stands 2.5 m from the mirror. (a) Calculate the focal length of the mirror. (b) Calculate the linear magnification if the person stands 5.0 m from the mirror.

■■ **62.** A large concave mirror of focal length 3 m stands 20 m in front of you. Describe the changing appearance of your image as you move from 20 m to 1 m from the mirror. Indicate distances from the mirror where the change in appearance is dramatic.

■■ **63.** Two mirrors (one plane and one diverging) at the side of a truck are shown in Fig. 20.37. *Estimate* the focal length of the curved mirror. Carefully explain the reasoning behind your estimate.

FIG. 20.37. (© 1979 by Ted Cowell/Black Star.)

CHAPTER 21

Wave Interference and Polarization

In the last chapter, when we analyzed optical instruments using ray diagrams and the thin-lens equation, we had little need to consider the wave nature of light. However, a variety of phenomena involving light and all forms of waves must be explained in terms of the wave nature of light. These phenomena depend on the constructive and destructive interference of waves originating from different points in space.

We learned about wave interference in Chapter 17. Our interest then was primarily with sound waves moving in one direction. If the crests (or the troughs) of two sound waves overlapped at the same place at the same time, constructive interference occurred; the two waves combined to form a new wave of increased amplitude. However, if a crest of one wave overlapped a trough of another wave, destructive interference occurred; the resultant wave had a smaller amplitude than either of the two waves.

In this chapter we will analyze the interference of waves originating from different points in space. Our analysis will allow us to understand the patterns formed when light passes through a single small opening such as a narrow slit or through several closely spaced slits. We will also learn about the interference of light reflected from thin films such as soap bubbles and the transparent films that often cover the surface of a lens. Finally, we will learn about the polarization of waves. As with all phenomena discussed in this part of the book, the interference effects to be described apply to all types of waves, but we will restrict most of our attention to light waves, for which the effects are easily demonstrated.

21.1 Light Waves

In Chapter 17 we defined a wave as a disturbance that moves from one place to another. Water waves consist of moving disturbances of the level of water. Sound waves consist of moving disturbances of the pressure and density of a medium such as air. What is the disturbance caused by a light wave, and through what does the disturbance travel?

The interference fringes produced by the overlapping teeth of two combs are similar to fringes formed by the overlapping crests and troughs of light waves that originate from two or more sources, the subject of this chapter. (Courtesy of Alan Van Heuvelen)

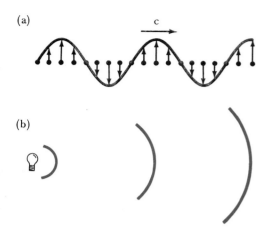

(a)

(b)

FIG. 21.1. (a) A light wave travels to the right at speed c. The vertical arrows indicate the direction that the wave exerts a force on a positive electric charge at different positions along its path. (b) A top view of a wave showing the positions of crests (wavefronts) that have left the light source.

Unlike other waves, light waves need no medium in which to travel. Light can travel through a vacuum. Light waves are produced by the oscillation of electrons in atoms as the electrons jump between different orbits in the atoms. If an electric charge lies in the path of a light wave, the light exerts a force on the charge that causes it to vibrate back and forth or up and down. Thus, light waves are produced by oscillating electric charges and disturb electric charges in their path, but need no medium through which to travel.

We will postpone until Chapter 31 (after our study of electricity and magnetism) a more precise definition of the nature of a light wave. For now we picture a light wave as shown in Fig. 21.1a. The arrows indicate the direction of the wave's force on a positive electric charge. The whole pattern shown in Fig. 21.1a moves in a vacuum at a speed c equal to 3.0×10^8 m/s and slower in other media. We will sometimes draw a "top" view of the wave by showing wave crests (wavefronts) that emanate from a light source (Fig. 21.1b). The changing shape of a wavefront as it moves can be determined using a principle developed over 300 years ago by Christian Huygens.

21.2 Huygens' Principle

From the time of Alexander the Great (350 B.C.) to that of the kings of France, the rulers of countries often hired scientists to bring prestige and fame to their courts. One of the top recruits in the seventeenth century was a Dutch mathematician, Christian Huygens, who was selected by King Louis XIV to direct all scientific research in France. During his tenure with the king, Huygens developed the best telescopes of the time and invented a pendulum clock so accurate that it easily detected small differences in the acceleration of gravity at different parts of the earth.

Huygens' most famous contribution to science was his wave theory of light, proposed in *Treatise on Light* (1673). This treatise contained many ideas about the nature of light that have since been confirmed. His technique for determining the future position of a wave based on its present location is the subject of this section. His technique, called *Huygens' principle*, is illustrated in Fig. 21.2a. A

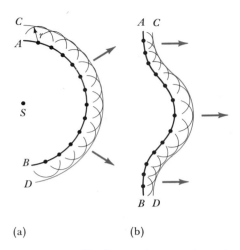

(a)   (b)

FIG. 21.2. Circular wavelets move forward from wavefront AB to form a new wavefront CD. (a) The wavelets leaving each part of AB move at the same speed. (b) The wavelets move faster at the center of the wavefront than at the sides.

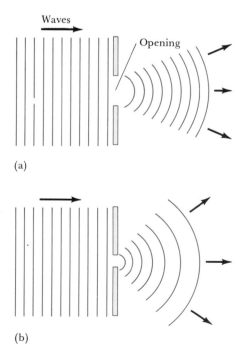

(a)

(b)

FIG. 21.3. Waves spread or diffract more after passing through a (b) small opening than through (a) a larger opening.

wavefront represented by line AB is shown moving at speed v away from a point source S. The location of the wavefront a short time t after it is at position AB is determined by drawing a large number of circular arcs, each centered on a different point of wavefront AB. Each arc represents a small wave, called a **wavelet,** that moves away from a point on the original wavefront. The radius of a wavelet r depends on the speed v of the wavefront at the point where the wavelet originates and on the time t that the wavelet has traveled away from that point. The three quantities are related by the equation $r = vt$. The interference of these wavelets produces a new wavefront whose position CD is a line drawn tangent to the edges of these wavelets, as shown in Fig. 21.2a and b. **Huygens' principle** can be summarized as follows:

Every point on a wavefront can be considered as the source of a circular-shaped wavelet that moves in the forward direction at the same speed at which the wave moves. The new wavefront is a line drawn tangent to the edges of these wavelets.

Huygens' principle is useful for determining the shape of wavefronts that pass through narrow openings or that move past obstacles. For example, Huygens' principle could be used to decide whether ocean waves during a gale will bend around a small island to strike a harbor on the lee side of the island.

The spreading of waves that pass through a narrow opening or move past an obstacle is called **diffraction** and is illustrated in Fig. 21.3. In this example, the opening is about the same size as the wavelength of the wave passing through the opening. The Huygens' wavelets form a new wavefront that moves forward and spreads to the sides as shown in Fig. 21.3a. As the opening is made smaller, the wave spreads more after passing through the opening (Fig. 21.3b)—like a single arc-shaped wavelet spreading from a point. In general, the spreading of the wave increases as the size of the opening decreases.

21.3 Young's Double-Slit Experiment

Although Christian Huygens proposed his wave theory of light in 1673, it received little support during his lifetime. None of the experiments with light at that time required an explanation based on waves, and the corpuscular theory of light was most popular. This changed, however, more than a hundred years later because of the work of the brilliant English physician and physicist Thomas Young (1773–1829). At age two, Young read fluently; by four, he had read the Bible twice; by age fourteen he knew eight languages. During his adult life, Young made many contributions to the understanding of fluids, concepts of work and energy, and the elastic properties of materials. He also made contributions to deciphering Egyptian hieroglyphics.

Young's most important contributions to science were in the field of light and optics. He discovered that the eye's lens could change shape, allowing it to focus on objects at different distances. He also developed a successful theory explaining color vision. Young received his greatest acclaim among physicists for his double-slit interference experiment (reported in 1803), which became the cornerstone of the wave theory of light. With this experiment Young laid to rest the corpuscular theory of light that had been popular during the 1700s.

The **double-slit experiment** is illustrated in Fig. 21.4a. Light first passes through a small opening, like a narrow slit; in Young's experiment sunlight passed through a small hole in his window shutter. Wavelets produced in the single slit form wavefronts that spread and fall on two closely spaced slits. If light is made of individual little bullet-like bundles (corpuscles), then these bundles should move straight through the slits and produce a sharp image of each slit on a screen placed behind the slits (Fig. 21.4b). However, the pattern Young actually observed on the screen consisted of alternating bright and dark lines (called fringes), as shown in Fig. 21.4c.

Young explained that the pattern of fringes was caused by the interference of waves coming from each slit. Bright bands of light occurred where the light waves interfered constructively; dark bands occurred where the waves interfered destructively. This phenomenon is illustrated in Fig. 21.5. Circular wavefronts spread from each narrow slit. In some directions (indicated by the arrows in Fig. 21.5) a crest of one wave overlaps a crest of another (the dots in Fig. 21.5)—that is, they constructively interfere—and the wave amplitude is doubled. In other directions the waves interfere destructively—a crest of one wave overlaps a trough of another, and the resultant amplitude of the light is zero. If you observe the waves striking a screen, you will see large-amplitude waves (bright light) at the positions of the arrows and no disturbance (darkness) in between.

We can predict the directions in which large-amplitude waves are formed as follows (see Fig. 21.6). First, note that the same wave enters each slit from the left; the disturbances at the slits are synchronized or in phase with each other. A crest (or a trough) leaves slit 1 in Fig. 21.6b at the same time that a crest (or a trough) leaves slit 2. The waves spread in all directions, but we consider only the waves moving toward the numbered positions of bright light (large-amplitude waves) on the screen shown in Fig. 21.6a. At the center of the interference pattern is a band of bright light called the **zeroth-order interference maximum.** At this point, light has traveled an equal distance from slit 1 (S_1) and from slit 2 (S_2), as shown in Fig. 21.6b. Since position 0 is equidistant from the slits, crests (and troughs) that simultaneously leave the slits arrive simultaneously at position 0, producing constructive interference and large-amplitude waves.

Constructive interference also occurs at position 1, which is one wavelength λ farther from slit 1 than from slit 2 (Fig. 21.6c). At position 2, the difference in distance is two wavelengths, and again constructive interference occurs (Fig. 21.6d). Between the bright bands, a crest from one slit arrives at the same time as a trough from the other slit. Destructive interference occurs and creates a dark band.

At the nth bright band to the right or left of the center band, the difference in distance from the two slits is n wavelengths of light ($n\lambda$ in Fig. 21.7). This bright band is called the **nth-order interference maximum.** The angle θ of rays moving from the slits to the nth-order interference maximum is calculated with

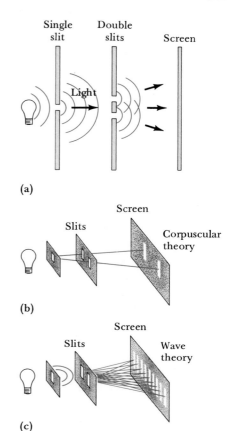

(a)

(b)

(c)

FIG. 21.4. (a) A schematic representation of Young's double-slit experiment. (b) The expected pattern of light on a screen if light is made of corpuscles. (c) The actual pattern observed by Thomas Young.

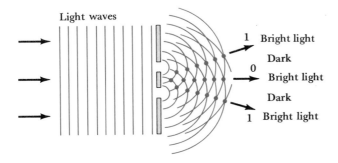

FIG. 21.5. Circular Huygens' wavelets originate at each slit. Interference of crests from each wave (the dots) produce large-amplitude waves (bright light). In between, relative calm exists because of the destructive interference of a crest from one wave with a trough from the other (darkness).

FIG. 21.6. (a) Light passes through a single slit. Coherent waves then pass through a pair of slits and produce a pattern of alternating bright and dark bands on a screen. (b) Constructive interference occurs at the center band because the waves from each slit travel an equal distance to the band. (c) The distance from the slits to band 1 differs by one wavelength and (d) to band 2 by two wavelengths.

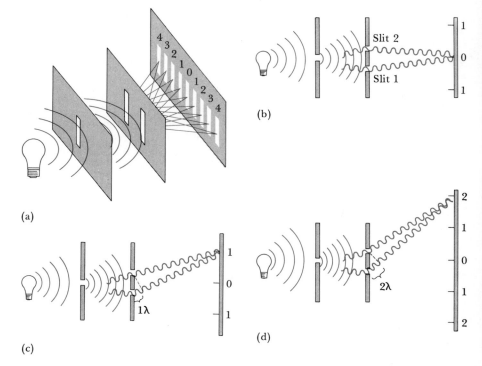

(a)

(b)

(c)

(d)

reference to Fig. 21.7. The small, shaded triangle shown next to the slits has a hypotenuse d equal to the slit separation. The side of the triangle opposite the angle θ has a length $n\lambda$. Thus, we see that $\sin \theta = n\lambda/d$. The angle θ in the small triangle equals the angular deflection θ of light moving from the closely spaced slits toward the nth bright band and is calculated using the same equation:

$$\sin \theta = \frac{n\lambda}{d}, \qquad (21.1)$$

where λ is the wavelength of the light, d is the slit separation, and n is an integer (0 for the center maximum, 1 for the first-order bright bands on each side, and so on).

EXAMPLE 21.1 When red light of wavelength 650 nm passes through a pair of slits, a series of bands appears on a screen. (a) The fifth band to the left of the center maximum has an angular deflection of 8.5°. Calculate the slit separation d. (b) Calculate the angular deflections θ of the zeroth-, first-, and second-order interference maximum.

SOLUTION (a) The following information is given: $\lambda = 650 \times 10^{-9}$ m, $n = 5$, and $\theta = 8.5°$. Equation (21.1) is solved for the slit separation d:

$$d = \frac{n\lambda}{\sin \theta} = \frac{5(650 \times 10^{-9} \text{ m})}{\sin 8.5°} = \underline{2.2 \times 10^{-5} \text{ m}}.$$

(b) The sines of the angular deflections of the other bands are

$$\sin \theta = \frac{n\lambda}{d} = \frac{n(650 \times 10^{-9} \text{ m})}{2.2 \times 10^{-5} \text{ m}} = n(0.030).$$

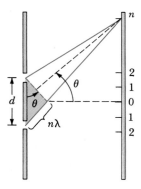

FIG. 21.7. Rays are drawn from the slits to the nth bright band above the center maximum.

Thus, for $n = 0$, $\sin\theta = 0$, and $\theta = \underline{0°}$; for $n = 1$, $\sin\theta = 0.030$, and $\theta = \underline{1.7°}$; and for $n = 2$, $\sin\theta = 0.060$, and $\theta = \underline{3.4°}$. ∎

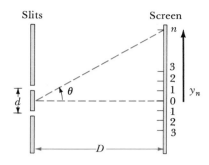

Slits   Screen

FIG. 21.8

EXAMPLE 21.2 Blue light, rather than red light, is incident on the pair of slits discussed in Example 21.1. The angular deflection of the fifth band to the left of the center is 6.3°. Calculate the wavelength of this light.

SOLUTION We are given that $n = 5$, $\theta = 6.3°$, and $d = 2.2 \times 10^{-5}$ m (calculated in the previous example). We solve Eq. (21.1) for the unknown wavelength:

$$\lambda = \frac{d\sin\theta}{n} = \frac{(2.2 \times 10^{-5}\text{ m})(\sin 6.3°)}{5} = 480 \times 10^{-9}\text{ m} = \underline{480\text{ nm}}. \quad ∎$$

In the last example, known values of d, θ, and n were used to calculate the unknown wavelength of light passing through two slits. The angular deflection θ is usually small and difficult to measure accurately. It can, however, be determined indirectly, as illustrated in Fig. 21.8. The distance y_n of the nth bright band from the center (zeroth-order) maximum is measured, along with the distance D of the slits from the screen. The value of θ is determined from the equation

$$\tan\theta = \frac{y_n}{D}. \tag{21.2}$$

Because $\tan\theta \simeq \sin\theta$ for small angles (10° or less) such as occur in most of these experiments, we can set the right sides of Eqs. (21.1) and (21.2) equal to each other. Thus,

$$\sin\theta = \frac{n\lambda}{d} \simeq \frac{y_n}{D} \qquad \text{(for small } \theta). \tag{21.3}$$

If the slit separation d is known, the wavelength of light can be calculated using Eq. (21.3) after measuring y_n and D.

Young's double-slit experiment is historically very important because of its confirmation of the wave nature of light, but the analysis of double-slit interference patterns has little practical use. A modern descendant of a double slit, called a *grating*, has great practical use in the analysis of the wavelengths of light emitted and absorbed by different types of atoms and molecules.

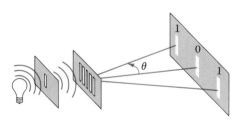

(a)

21.4 Gratings

A **grating** consists of many closely spaced slits (for example, 10,000 in one centimeter) and can be used to separate light of different wavelengths and to measure those wavelengths.

The operation of a grating with only five slits is illustrated in Fig. 21.9. Light that strikes one of these slits spreads after passing through the slit. The waves from each of the slits interfere to produce beams of very intense light in several directions and little light elsewhere.

The first intense band to the left of the center maximum strikes the screen at position 1 (Fig. 21.9a). The extra distance of this position from adjacent slits is

(b)

FIG. 21.9. (a) Circular wavelets of light spread from each slit in a grating. A bright band occurs on a screen at point 1, where the distance from adjacent slits differs by one wavelength as shown in (b).

one wavelength (Fig. 21.9b). Rays from all slits interfere constructively to form a bright spot. The angle of deflection θ and the positions y_n of the **intensity maxima of light from a grating** are calculated using the same equations as were used for a double slit:

$$\sin \theta = \frac{n\lambda}{d}, \qquad \text{where } n = 0, 1, 2, \ldots, \tag{21.1}$$

and

$$\tan \theta = \frac{y_n}{D}, \tag{21.2}$$

where d is the separation of adjacent slits, λ is the wavelength of the light, D is the distance from the grating to the screen, and y_n is the distance on the screen from the center bright spot to the nth bright spot. As with double slits, the positions of bright spots are called **orders** (zeroth order for $n = 0$, first order for $n = 1$, and so forth).

The bright lines or spots are somewhat more intense and better defined than those from double slits because many more slits contribute to the interference pattern. Instead of finding alternating light and dark regions, we find a large region of darkness followed by a narrow, bright band or spot of light.

Spectrograph

Wavelength measurements using gratings are made routinely in many fields of science. Astronomers, for instance, determine what types of atoms are present in distant stars by identifying the wavelengths of light coming from atoms in the stars. This is done using a **spectrograph** (Fig. 21.10). Light, such as that from a star, passes through a narrow slit S_1 and then through a lens L_1 (Fig. 21.10a) that bends and collimates the rays of light into a parallel beam. The beam of light then passes through a grating that causes the light to separate into different-order beams. The zeroth-order beam is undeflected. A first-order beam is deflected both left and right. The angle of the deflection depends on the wavelength of the light.

Another lens L_2 is placed beyond the grating (Fig. 21.10b) to focus the deflected beams of light through another narrow slit onto a film. The lens L_2 and slit S_2 are wavelength selectors since the wavelength of light that exposes the film depends on the orientation of the lens and slit. As this collimating system is rotated, the film is exposed at different wavelengths.

Different types of atoms emit different wavelengths of light. For example, mercury emits a purple doublet line at 408 nm, a blue line at 436 nm, a greenish-yellow line at 546 nm, a yellowish doublet at 579 nm, and two weak red lines near 620 nm. The 546-nm line is being viewed by the spectrograph pictured in Fig. 21.10b. As the collimating system is rotated, an exposed film, such as shown in Fig. 21.10c, is produced. The wavelengths at which the film is exposed indicate the types of atoms producing the light. If a star contains mercury atoms, then the film should be exposed at those wavelengths shown in part (c). It will also be exposed at the wavelengths of light emitted by other atomic constituents of the stars. The amount by which the film is exposed at different wavelengths indicates the abundance of the different types of atoms on the star. (The spectra

VISIBLE-REGION SPECTRA FOR SELECTED ELEMENTS

Courtesy of Bausch & Lomb

Tungsten Lamp

Iron Arc

Molecular Hydrogen

Atomic Hydrogen

Neon

Barium

Fraunhofer Lines

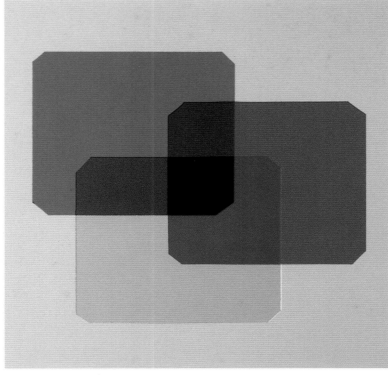

Subtractive method of producing colors Dyes or pigments, such as those in inks, paints, and photographic materials like the three transparencies shown here, absorb and effectively subtract some wavelengths of light and pass only a part of the spectrum. The *subtractive primary colors* are yellow, cyan, and magenta. When white light passes through overlapping sheets of these colors, all wavelengths are blocked (subtracted) and we have black. Where only yellow and cyan overlap, we see green; where cyan and magenta overlap, we see blue; and where yellow and magenta overlap, all the wavelengths are subtracted but red. Various proportions of subtracted wavelengths by yellow, cyan, and magenta dyes will produce nearly any color in the spectrum. The color of this and other printed pages, for example, is accomplished with only yellow, cyan, and magenta ink. Three separate printing plates print the three colors of inks in tiny dots of varying intensity. A fourth plate in black ink provides shading and depth.

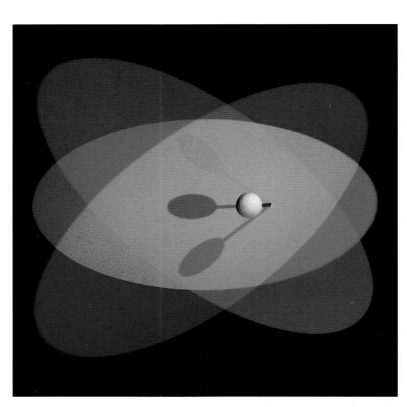

Additive method of producing colors When light from all parts of the visible spectrum overlap, the additive mixture of colors appears white. But the eye doesn't require a mixture of all the colors of the spectrum to see white. Light from the lower-frequency end of the spectrum (red) combined with light from the middle of the spectrum (green) combined with light from the higher-frequency end (blue) will also appear white. Red, green, and blue are the *additive primary colors.* Various mixtures of these primary colors will produce nearly any color in the spectrum. This view shows the overlapping of three spotlights: red, green, and blue. Note the different-colored shadows of the golf ball. The middle shadow is cast by the green spotlight and is not dark because it is illuminated by the red and blue lights, which overlap and add to produce magenta. Similarly, the shadow cast by the blue light appears yellow because it is illuminated by red and green light. Can you see why the shadow cast by the red light appears cyan?

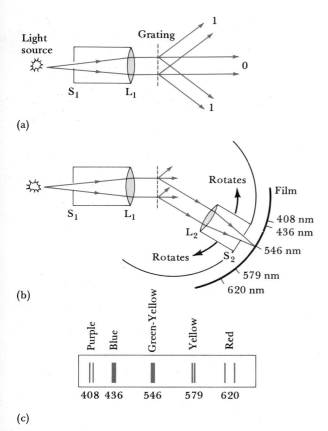

(a)

(b)

(c)

FIG. 21.10. (a) The input side of a spectrograph. (b) A spectrograph with the collimating system. (c) An exposed film from mercury gas.

of light emitted by several different types of atoms are shown on the color insert on the facing page.)

EXAMPLE 21.3 A spectrograph has a grating with 4000 lines/cm. Light from a hot gas of atomic hydrogen passes through the grating and strikes a film that is 0.500 m from the grating. The developed film appears as shown in Fig. 21.11 and also as shown in the center band of the color insert on the facing page. Use the information given here and in Fig. 21.11 to determine the wavelengths of the hydrogen line in the red part of the visible spectrum and the line in the blue-green region.

SOLUTION We see from Fig. 21.11 that the distances on the film of the first order ($n = 1$) red and blue-green lines from the central maximum are

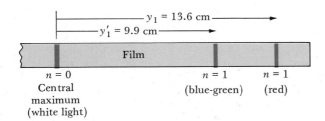

FIG. 21.11. Film produced by the spectrograph described in Example 21.3.

$y_1 = 13.6$ cm and $y_1' = 9.9$ cm, respectively. Since $D = 0.500$ m (the distance of the film from the grating), we find using Eq. (21.2) that the angular deflection of these lines after leaving the grating is

Red: $\tan \theta = \dfrac{y_1}{D} = \dfrac{0.136 \text{ m}}{0.500 \text{ m}} = 0.272$, or $\theta = 15.2°$;

Blue-green: $\tan \theta' = \dfrac{y_1'}{D} = \dfrac{0.099 \text{ m}}{0.500 \text{ m}} = 0.198$, or $\theta' = 11.2°$.

We can now use Eq. (21.1) to find the wavelengths of each line. First, however, we must calculate the separation d of adjacent slits on the grating:

$$d = \frac{1}{4000 \text{ lines/cm}} = 2.5 \times 10^{-4} \text{ cm} = 2.5 \times 10^{-6} \text{ m}.$$

We find then that the wavelength of the red line is

$$\lambda = d \sin \theta = (2.5 \times 10^{-6} \text{ m}) \sin 15.2°$$
$$= 6.55 \times 10^{-7} \text{ m} = 655 \times 10^{-9} \text{ m} = \underline{655 \text{ nm}},$$

and the wavelength of the blue-green line is

$$\lambda' = d \sin \theta' = (2.5 \times 10^{-6} \text{ m}) \sin 11.2°$$
$$= 4.86 \times 10^{-7} \text{ m} = 486 \times 10^{-9} \text{ m} = \underline{486 \text{ nm}}. \quad \blacksquare$$

21.5 Thin-Film Interference

Another example of wave interference of light is the rainbow of colors reflected from soap bubbles and from thin films of oil floating on water. This effect is caused by the interference of light reflected from different surfaces of the film. In Fig. 21.12a, part of an incident ray of light reflects from the top surface of a film at A and the rest of the light is transmitted into the film. Part of the transmitted ray reflects from the bottom surface at B and moves back toward the top. As a portion of this ray leaves the top surface at C and moves toward our eye, it interferes with the ray reflected at A. The interference may be either constructive (bright light) or destructive (reduced light intensity), depending on the wavelength of the light and on the extra distance ABC traveled by the ray passing through the film. If multicolored light of many wavelengths strikes the surface, the light leaving the surface consists primarily of those wavelengths that undergo constructive interference.

For constructive interference to occur, the light leaving the surface at C must be in phase with light reflected at A (Fig. 21.12b); the crests and troughs of the two waves must coincide. Two factors affect the relative phase of the waves: (1) The phase of a wave can change when it is reflected from the surface of a medium with greater impedance, and (2) a phase difference is caused by the extra distance ABC traveled by the wave passing through the film.

Consider first the phase change caused by reflection. Remember from Section 17.4 that a wave changes phase when reflected from the boundary of a medium with greater impedance (an upright incident pulse on a string is converted to a downward reflected pulse). If the pulse or wave reflects from a boundary with less impedance, however, no phase change occurs.

The same type of phenomenon occurs for light. The impedance of a medium through which light travels is proportional to its index of refraction n. If

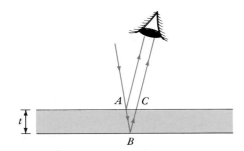

(a)

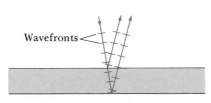

(b)

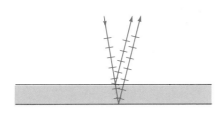

(c)

FIG. 21.12. (a) Light rays reflecting from the top and bottom surface of a thin film. (b) The crests of rays leaving the surface (indicated by the horizontal lines) are in phase, and constructive interference occurs. (c) The crests are out of phase, resulting in destructive interference.

light moving in air ($n = 1.00$) reflects from a soap bubble ($n \simeq 1.3$), the light changes phase. However, if light moving in a soap bubble reflects from a boundary of a medium with lesser index of refraction, as occurs at point B in Fig. 21.12a, no phase change takes place. The two rays shown leaving the surface in Fig. 21.12 differ in phase by $1/2\lambda$ because of the change in phase caused by reflection at A but not at B.

The phase also differs because of the extra distance ABC traveled by the wave passing through the film. If the rays move perpendicular to the surface and the thickness of the film is t, then the extra distance traveled by the ray passing through the film is $2t$. If $2t = 1/2\lambda$, then this half-wavelength phase change is balanced by the $\lambda/2$ phase change caused by reflection at A. The rays shown in Fig. 21.12b leave the surface in phase and interfere constructively. Constructive interference also occurs if $2t = 3/2\lambda, 5/2\lambda, \ldots$ because when the $1/2\lambda$ phase change caused by reflection is added to the $3/2\lambda, 5/2\lambda, \ldots$ change caused by the extra distance traveled, the waves differ by an *integral multiple* of wavelengths. Since a crest of one wave still coincides with a crest of the other, the waves are in phase and interfere constructively. Destructive interference occurs when the crest of one reflected ray overlaps the trough of the other ray, as illustrated in Fig. 21.12c.

We have ignored one important factor that affects the interference. When a light wave travels in a substance having an index of refraction n greater than 1.0, its speed decreases according to Eq. (19.2): $v = c/n$. When the wave moves at slower speed, its wavelength decreases according to Eq. (17.8): $\lambda = f/v$. We can combine these two equations to determine the wavelength of light in air λ_{air} that will undergo constructive interference in a soap bubble or oil film with index of refraction n:

$$\lambda = \frac{v}{f} = \frac{1}{f}\left(\frac{c}{n}\right) = \left(\frac{c}{f}\right)\frac{1}{n} = \frac{\lambda_{air}}{n}.$$

Thus, the *wavelengths of light moving in air that interfere constructively with a soap bubble or oil film* satisfy the condition that

$$2t = \frac{1}{2}\left(\frac{\lambda_{air}}{n}\right), \ \frac{3}{2}\left(\frac{\lambda_{air}}{n}\right), \ \frac{5}{2}\left(\frac{\lambda_{air}}{n}\right), \ \ldots \tag{21.4}$$

where t is the thickness of the film, λ_{air} is the wavelength of the light in the air, and n is the index of refraction of the thin film.

EXAMPLE 21.4 A soap bubble appears blue when viewed perpendicular to its surface. Determine several possible values for the thickness of the soap film whose index of refraction is 1.38.

SOLUTION Blue light has a wavelength of about 450 nm. Rearranging Eq. (21.4), we find that the thickness could be one of several possible values:

$$t = \frac{\lambda_{air}}{4n} = \frac{450 \text{ nm}}{4(1.38)} = \underline{82 \text{ nm}},$$

$$t = \frac{3\lambda_{air}}{4n} = \underline{245 \text{ nm}},$$

$$t = \frac{5\lambda_{air}}{4n} = \underline{408 \text{ nm}}. \qquad \blacksquare$$

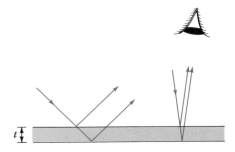

FIG. 21.13. The path length through a thin film is shorter when one is looking directly at light coming from the nearest surface than when one is looking at light from regions at the side.

If we look at the surface of a soap bubble that is nearest our eyes, the light that we see travels perpendicular to the surface of the film. The extra distance traveled by light passing through the film and back is $2t$. However, light reaching our eyes from regions to the side of this nearest surface travels through the film along an angular path that is longer than $2t$ (Fig. 21.13). Thus, constructive interference occurs for longer-wavelength light coming from these regions. This is why we see a rainbow of colors coming from different parts of the thin film.

The phenomenon of thin-film interference is used to produce nonreflective coatings for glass lenses and prisms. The reduction of reflection is especially important for expensive optical devices such as microscopes that use many lenses to produce an image without aberrations. For example, if four percent of the light striking a glass-air interface is reflected from each surface, then for the twenty surfaces of an optical system with ten lenses, only $(0.96)^{20} = 0.44$ or 44 percent of the light will be transmitted through all the lenses.

To reduce the amount of light reflected, a thin layer of hard transparent material with an index of refraction less than that of glass is coated on the glass. In this case we want light reflected from the top surface of the coating at A (Fig. 21.12a) to interfere destructively with light passing through the film and back out at C. Both rays are reflected from media with greater impedance causing a $\lambda/2$ phase change at both A and B. Thus, the relative phase of the two rays is unchanged by reflection.

For **destructive interference** to occur, the path difference must cause a one-half wavelength phase change. The *thickness of the nonreflective film on a glass lens* must satisfy the following equation:

$$2t = \frac{1}{2}\left(\frac{\lambda_{\text{air}}}{n}\right), \ \frac{3}{2}\left(\frac{\lambda_{\text{air}}}{n}\right), \ \frac{5}{2}\left(\frac{\lambda_{\text{air}}}{n}\right), \ \dots \qquad \textbf{(21.5)}$$

The coating on glass lenses is usually made the correct thickness to reduce reflection at a wavelength of 550 nm in the center of the visible spectrum of light. The reflection of this yellow-green light is reduced by the coating from about 4 percent to about 1 percent. However, the coating is less effective at reducing the reflection at extreme wavelengths of light (red at 700 nm and violet at 400 nm). An expensive lens with a thin-film coating has a purple hue because the red and violet light are reflected more and produce a purple color.

EXAMPLE 21.5 Magnesium fluoride, MgF, with an index of refraction of 1.38 is used to coat a glass lens of index of refraction 1.50. What should the thickness of the coating be to reduce the reflection of 550-nm light?

SOLUTION According to Eq. (21.5),

$$t = \frac{\lambda_{\text{air}}}{4n} = \frac{550 \text{ nm}}{4(1.38)} = \underline{100 \text{ nm}}.$$

The film can also have a thickness of 300 nm, 500 nm, or other odd multiples of 100 nm. ∎

21.6 Single-Slit Diffraction

The spreading and bending of waves behind obstacles that lie in their path is called **diffraction.** The diffraction of light waves passing through a single slit is illustrated in Fig. 21.14. In Fig. 21.14a, the slit width is large compared to the

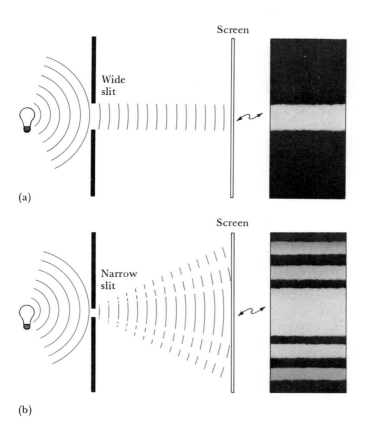

(a)

(b)

FIG. 21.14. (a) Light casts a sharp image of a slit on a screen whose width is much wider than the wavelength of light. (b) If the slit width is about the same as the wavelength of light, the diffraction or spreading of the light is easily observed.

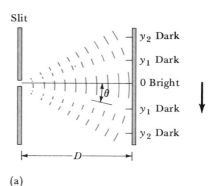

(a)

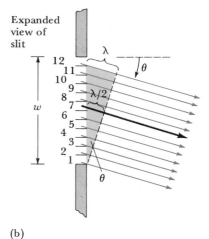

(b)

FIG. 21.15. (a) Light incident on a narrow slit produces alternating bands of bright light and darkness on a screen. (b) The first dark band is caused by the destructive interference of light coming from different parts of the slit; region 1 cancels 7, 2 cancels 8, and so on.

wavelength of the light, and the pattern of light cast on the screen after passing through the slit has a shape almost exactly like that of the slit. A small amount of light does bend behind the slit but is not apparent when compared with the intense light at the center of the pattern in part (a). In Fig. 21.14b, the slit width has been reduced to almost the same size as the wavelength of the light. The light spreads after passing the slit and forms a broad central band of light on the center of the screen and several weaker bands or fringes of light at the sides. The diffraction or spreading of light (or of any type wave) *increases* as the width of the opening through which the light passes *decreases*.

The angular spread of the light beam after passing through a slit can be calculated with reference to Fig. 21.15. The slit opening is divided into a number of smaller regions (twelve for the slit shown in Fig. 21.15b). Every region in the slit opening vibrates in synchronization with other regions as waves strike the slit from the left. If the slit is excited by light of only one wavelength (monochromatic light), then in certain directions the light from some regions interferes destructively with light from other regions, producing a dark band of light on a screen. If light from different regions interferes constructively, a bright band is produced.

Look at the formation on the screen of the first dark band at position y_1 shown in Fig. 21.15a. Light from region 7 shown in Fig. 21.15b travels a half-wavelength farther to y_1 than does light from region 1. Therefore, the waves from these two regions interfere destructively and produce no net light at position y_1. The same is true for regions 2 and 8, 3 and 9, 4 and 10, and so on. Since the light from each region is cancelled by that from another, there is no net light intensity at position y_1.

The angle θ between a line from the slit to the center of the diffraction pattern on the screen and a line from the slit to position y_1 can be determined by considering the shaded triangle shown in Fig. 21.15b. The hypotenuse of this triangle is the slit width w, and the length of the side of the triangle opposite the angle θ equals the wavelength λ. We see, then, that the light from the slit produces the **first dark band** on the screen when the angular deflection θ from the slit to the dark band satisfies the equation $\sin \theta = \lambda/w$.

If the angle θ is slightly larger so that waves from region 12 travel $3\lambda/2$ farther to the screen than waves from region 1, we no longer have total destructive interference. Light from regions 1 and 5 are $\lambda/2$ out of phase and cancel each other, as does light from regions 2 and 6, 3 and 7, and 4 and 8. However, this leaves the light from regions 9, 10, 11, and 12 uncancelled, and we find a moderately intense beam of light when $\sin \theta = (3\lambda/2)w$. The intensity at these first bright fringes on the sides is less than one-third the intensity at the center maximum.

At even larger angular deflection, the light from region 12 travels 2λ farther to the screen than the light from region 1. Light from region 4 travels $\lambda/2$ farther to the screen than light from region 1; the light from the two regions interferes destructively to cancel each other as does light from regions 2 and 5, 3 and 6, 7 and 10, 8 and 11, and 9 and 12. All light is cancelled by light from another region, and another dark band or fringe is formed on the screen. This band occurs when the angular deflection is calculated from the equation $\sin \theta = 2\lambda/w$.

In general, **dark bands in single-slit diffraction patterns** occur when the angular deflection θ to the *dark* band is calculated using the following equation:

$$\sin \theta = \frac{n\lambda}{w} \qquad \text{for } n = 1, 2, 3, 4, \dots,$$

where w is the slit width and λ is the wavelength of the light. Bright bands are formed between the dark bands. As with double-slit interference fringes, the angular deflection can also be determined using the distance y_n on a screen of the nth dark band from the central maximum for a screen that is separated from the slit by a distance D (see Fig. 21.15a). We find that

$$\sin \theta = \frac{n\lambda}{w} \simeq \frac{y_n}{D} \qquad \text{for } n = 1, 2, 3, 4, \dots. \tag{21.6}$$

EXAMPLE 21.6 Light of wavelength 630 nm passes through a single slit. The angular deflection to the tenth dark band on the side of the center maximum is 3.6°. (a) Calculate the slit width. (b) Calculate the angular deflection θ to the tenth dark band using light of 450-nm wavelength.

SOLUTION (a) The following information is given: $\lambda = 630$ nm, $n = 10$ (tenth dark band), and $\theta = 3.6°$. We rearrange Eq. (21.6) to find the slit width:

$$w = \frac{n\lambda}{\sin \theta} = \frac{(10)(630 \times 10^{-9}\,\text{m})}{(\sin 3.6°)} = 1.0 \times 10^{-4}\,\text{m} = \underline{0.10\text{ mm}}.$$

(b) When 450-nm light passes through the same 0.10-mm-wide slit, the sine of the angular deflection is

$$\sin \theta = \frac{n\lambda}{d} = \frac{(10)(450 \times 10^{-9} \text{ m})}{(1.0 \times 10^{-4} \text{ m})} = 0.045,$$

$$\theta = \underline{2.6°}. \qquad \blacksquare$$

EXAMPLE 21.7 Sound of frequency 400 Hz and light of wavelength 550 nm are incident on a small 1.0-cm-wide opening in a doorway. Calculate the angular deflection of the central diffraction maximum for each type of wave.

SOLUTION Sound travels at a speed of about 340 m/s. The wavelength of the sound is given by Eq. (17.8):

$$\lambda = \frac{v}{f} = \frac{340 \text{ m/s}}{400 \text{ Hz}} = \frac{340 \text{ m/s}}{400 \text{ s}^{-1}} = 0.85 \text{ m}.$$

The angular deflection of the central maximum of the sound is given by Eq. (21.6):

$$\sin \theta = \frac{n\lambda}{w} = \frac{(1)0.85 \text{ m}}{0.010 \text{ m}} = 85.$$

Remember that the sine of an angle can never exceed 1.0. Our impossible result ($\sin \theta = 85$) means that the central "bright" region of sound is spread over all angles in the forward direction—we never see a dark band (no sound) at the side. This phenomenon occurs whenever the wavelength of the wave is greater than the width of the opening, that is, when λ/w in Eq. (21.6) is greater than 1.0. Notice that the wavelength of sound in this example (0.85 m) greatly exceeds the width of the opening (0.01 m), and the sound is diffracted uniformly throughout the room.

In contrast, the wavelength of light (550 nm) is much less than the width of the opening (0.010 m). The angular deflection to the first dark band of the diffraction pattern at the side of the transmitted light is very small:

$$\sin \theta = \frac{(1)(550 \times 10^{-9} \text{ m})}{(0.010 \text{ m})} = 5.5 \times 10^{-5},$$

$$\theta = \underline{0.0032°}.$$

The diffraction is so small that it will hardly be noticed. The light will appear to form a sharp image of the opening with no observable diffraction. ■

21.7 Polarization

Polarization is a property of transverse waves related to the direction or directions in which vibrations occur. The vibrations of waves traveling along a rope provide a concrete example of polarization. The wave shown in Fig. 21.16a is said to be **plane polarized**—the rope vibrates in a single plane, the vertical plane in this case. The wave shown in Fig. 21.16b is also plane polarized since it vibrates in a single plane, the horizontal plane. The wave in Fig. 21.16c is not plane polarized—it vibrates simultaneously in more than one plane.

The wave shown in Fig. 21.16c can be made polarized by placing a slit in its path, as in Fig. 21.16d. With the first slit oriented vertically, only the vertical component of vibration passes through the slit. Thus, on the other side of the

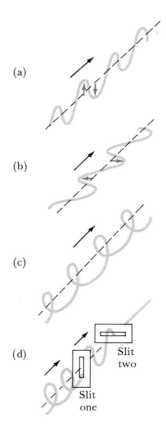

(a)

(b)

(c)

(d)

Slit two

Slit one

FIG. 21.16. (a) A vertically plane polarized wave on a rope, (b) a horizontally plane polarized wave, and (c) a wave that is not polarized in a plane. (d) The wave in (c) becomes plane polarized after passing through slit one. The wave's amplitude is reduced to zero by a second slit oriented perpendicular to the first slit.

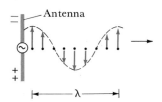

FIG. 21.17. A plane polarized radiowave leaves an antenna.

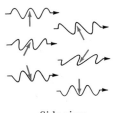

(a)   Side view

(b)   Front view

FIG. 21.18. (a) A side view of unpolarized light that causes vibrations in many different planes. (b) If we looked into a narrow beam of this light, the vibrations would oscillate in many different directions.

first slit, the wave is polarized vertically. If a second horizontal slit is placed in the wave's path, both components are blocked and the wave's amplitude of vibration is reduced to zero after passing the second slit.

A radio antenna in which charge vibrates up and down along a single axis produces a plane polarized radiowave, represented schematically in Fig. 21.17. The wave vibrates in the plane of the paper, which is also the plane of the wave's source (the antenna).

The production of light waves by the vibration of electrons in atoms and molecules is often quite different. Many atoms or molecules independently contribute to the total intensity of the radiation. Each atom or molecule that emits light acts for a short time (typically, about 10^{-10} s) like an antenna whose electric charge vibrates along an axis that is oriented differently from the axes for other atoms. Thus, the resultant light consists of many different waves originating at random times, that is, **incoherently,** from atoms and molecules that are oriented randomly relatively to each other (Fig. 21.18a). The emitted light is said to be **unpolarized**—it oscillates in randomly oriented directions perpendicular to the direction in which the waves travel. Light from the filament of a light bulb and from the surface of the sun are examples of unpolarized light. If we could observe the many separate waves in a beam of unpolarized light moving directly toward our eyes, the vibrations of these waves would look like a porcupine (Fig. 21.18b). Vibrations would occur along many axes.

This unpolarized light beam can be polarized in much the same way that the unpolarized wave traveling along the rope shown in Fig. 21.16d was polarized by a slit. Instead of using a single slit, the light passes through a material that has the effect of many narrow slits, all oriented parallel to each other.

The most common polarizing material is sold commercially under the name Polaroid. A Polaroid sheet consists of long molecules with their axes parallel to each other. Each molecule can absorb light that vibrates perpendicular to its axis, but does not absorb light that vibrates parallel to the axis. Thus, light waves that vibrate parallel to the axes pass through the material with little absorption. However, if the wave vibrates perpendicular to the molecular axes, the wave is absorbed by the Polaroid sheet. An unpolarized light beam becomes polarized by the first Polaroid sheet in Fig. 21.19 and its intensity is reduced by one-half.

If a second Polaroid sheet is placed after the first and the molecule axes of the second sheet are oriented at an angle θ relative to the first (Fig. 21.19), the amplitude of the light is reduced by the second Polaroid by a factor of $\cos \theta$. The

FIG. 21.19. Unpolarized light becomes polarized after passing through Polaroid one. The plane of polarization is changed and the intensity is reduced by $\cos^2 \theta$ as the light passes through a second Polaroid oriented at angle θ relative to the first.

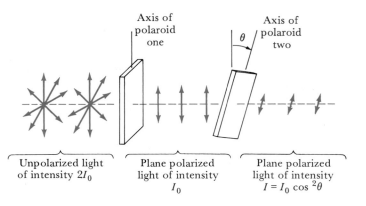

Axis of polaroid one

Axis of polaroid two

θ

Unpolarized light of intensity $2I_0$

Plane polarized light of intensity I_0

Plane polarized light of intensity $I = I_0 \cos^2 \theta$

intensity of a wave (the energy it transmits per unit time and area) is proportional to the amplitude squared. Thus, the second Polaroid reduces the intensity by a factor $\cos^2 \theta$:

$$I = I_0 \cos^2 \theta \qquad (21.7)$$

where I_0 is the intensity of the polarized light entering the second Polaroid, I is the intensity of light leaving the second Polaroid, and θ is the angle between the direction of polarization and the axis of the second Polaroid. For $\theta = 90°$, $\cos 90° = 0$, and light leaving the second Polaroid is zero. The "crossed" Polaroids together absorb all the light incident on the first Polaroid.

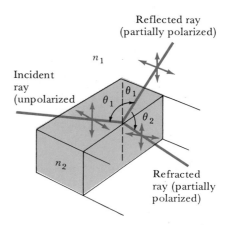

FIG. 21.20. When unpolarized light is reflected from a nonmetallic surface, the reflected wave is partially polarized. The horizontal electric field component has larger amplitude than the vertical component.

Polarization by Reflection

When a ray of unpolarized light strikes a nonmetallic surface at an incident angle greater than zero,* the reflected ray is partially polarized. The component of the light beam that vibrates perpendicular to the surface is reflected less than the component vibrating parallel to the surface (Fig. 21.20). If you look at the the morning sun reflected from a lake, for example, the light waves that vibrate parallel to the lake's surface have larger amplitude than the reflected waves that vibrate up and down. To reduce the glare of this reflected light, we can wear Polaroid sunglasses whose axes are oriented vertically. The glasses then absorb the more intense horizontally polarized light. Photographers sometimes use Polaroid sheets in front of a lens to reduce the glare of light reflected from a surface. Polaroid sunglasses absorb light reflected from smooth surfaces, thus reducing the glare.

The degree to which reflection causes an unpolarized wave to become partially polarized depends on the angle of incidence. For example, if an unpolarized ray is incident perpendicular to a surface (an incident angle of zero), the reflected wave remains completely unpolarized. However, at an angle of incidence known as the **polarizing angle** or **Brewster angle** θ_B, the reflected ray is completely polarized—the electric field only vibrates parallel to the reflecting surface. The Brewster angle depends on the index of refraction n_1 of the medium through which the wave travels and the index of refraction n_2 of the reflecting medium according to the equation

$$\tan \theta_B = \frac{n_2}{n_1}. \qquad (21.8)$$

EXAMPLE 21.8 Suppose that you are wearing Polaroid sunglasses with the axes of the molecules oriented vertically. (a) At what angle can you look at a sheet of glass of refractive index 1.60 and observe no reflected light? (b) What fraction of the light reflected at the Brewster angle will pass through your sunglasses if you tilt your head and glasses 30° to the side?

SOLUTION (a) At the Brewster angle the reflected light is completely polarized in the horizontal direction. Your Polaroid glasses with their axes oriented vertically will absorb all of this horizontally polarized light. The Brewster angle

---

*Recall from Chapter 19 that the incident angle is the angle between the ray and a line drawn perpendicular to the surface.

is determined using Eq. (21.8):

$$\theta_B = \text{invtan}\,\frac{n_2}{n_1} = \text{invtan}\,\frac{1.60}{1.00} = \underline{58.0°}.$$

No reflected light will be seen if you look down at an angle of 58.0° relative to a line perpendicular to the surface (or at an angle 32.0° above the horizontal).

(b) Light reflected at the Brewster angle is horizontally polarized, and the axes of your sunglasses when tilted 30° to the side now make a 60° angle with the horizontal. We can use Eq. (21.7) to calculate the ratio of the intensity I of light passing through the lens and of the polarized light I_0 incident on the lens:

$$\frac{I}{I_0} = \cos^2 \theta = \cos^2 60° = \underline{0.25}.$$

Twenty-five percent of the light reflected at the Brewster angle passes through the Polaroid glasses. ■

Summary and Additional Readings

1. **Huygens' principle:** Every point on the wavefront of a wave can be considered as the source of a small, circular **wavelet** that moves in the forward direction at the same speed as the wave. The new wavefront a short time later is a line drawn tangent to the front edges of these wavelets.

2. **Young's double-slit interference:** If light waves strike two narrow, closely spaced slits, the circular wavelets leaving each slit on the other side interfere constructively at some places to form bright bands of light and destructively at other places to form no light. The angular deflection θ of rays moving in the direction of **bright bands** is determined using the equation

$$d \sin \theta = n\lambda \qquad \text{for } n = 0, 1, 2, \dots, \qquad (21.1)$$

where d is the slit separation, λ is the wavelength of the light, and n is the order of the bright band (zero for the central maximum, 1 for the first bands on each side, and so on). If the waves are projected on a screen, the nth bright band is separated on the screen from the central maximum by a distance y_n given by

$$\tan \theta = \frac{y_n}{D}, \qquad (21.2)$$

where D is the distance from the slits to the screen.

3. **Gratings:** A **grating** consists of many closely spaced slits. Equations (21.1) and (21.2) can be used to locate the angular deflection to bright spots on a screen for waves passing through a grating. When using Eq. (21.1) for gratings, d is the separation of adjacent slits. Gratings are used to separate a beam of light into its different colors and wavelengths.

4. **Thin-film interference:** The bright colors coming from a soap bubble or thin film of oil floating on water are caused by the constructive and destructive interference of rays reflected from the top surface of the film with rays reflected from the bottom surface of the film. For constructive interference, the two rays must differ in phase by an integral number of wavelengths; for destructive interference, their phase must differ by $1/2\lambda$, $3/2\lambda$, and so on.

Phase differences occur because of phase change caused by reflection and because of differences in the path lengths of the rays.

5. **Diffraction:** When a wave passes the edge of an obstacle or through a small opening, it bends or spreads as it leaves the other side, a property called **diffraction.** Diffraction is caused by the interference of light passing through different regions in the slit or at the side of the obstacle. As light passes through a narrow slit of width w, it forms a band of light in the center and alternating bands of darkness and light on each side. The angular deflection θ to **dark bands** is determined using the equation

$$\sin \theta = \frac{n\lambda}{\omega} \simeq \frac{y_n}{D} \qquad \text{for } n = 1, 2, 3, \dots, \qquad (21.6)$$

where λ is the wavelength and n is a positive integer that labels the dark bands (1 for the first-order dark band on the sides of the central maximum, 2 for the second-order dark band on the sides, and so on). If the diffraction pattern is projected on a screen separated by a distance D from the slit, the distance y_n of the nth dark band from the central maximum is related to the angular deflection θ by Eq. (21.6).

6. **Polarization:** A transverse wave is **plane polarized** if its vibrations are in a single plane. If plane polarized light of intensity I_0 strikes a Polaroid sheet whose axis is oriented at an angle θ relative to the plane of polarization, the intensity of the wave passing through the Polaroid is reduced to I where

$$I = I_0 \cos^2 \theta. \qquad (21.7)$$

Light reflected from a smooth, nonmetallic surface is completely polarized parallel to the surface when the angles of incidence and reflection are at the polarizing or **Brewster angle** θ_B given by

$$\tan \theta_B = \frac{n_2}{n_1}, \qquad (21.8)$$

where n_1 and n_2 are the indices of refraction of the incident and reflecting media, respectively.

E. B. Sparberg, "Misinterpretation of Theories of Light," *American Journal of Physics* **34**, 377 (1966).

Y. P. Hwu, "Two Good Demonstrations," *The Physics Teacher,* p 258 April (1979). Contains demonstrations to show the "beat" of gratings and the diffraction pattern of blood corpuscles.

Questions

1. Describe the appearance of the double-slit interference pattern produced by white light.

2. Devise and carefully describe a double-slit-type interference experiment for sound waves. You can use one speaker, excited by an electronic oscillator, as your source of sound waves.

3. An onion skin is held in front of a light. As you move your head sideways relative to the onion skin, a rainbow spectrum of colors appears. Explain.

4. Suppose that a soap bubble has a uniform thickness. If you see green light of 520-nm wavelength when looking at the nearest part of the bubble, why will you see longer-wavelength light to the sides of this green region?

5. One edge of a flat piece of glass rests on a mirror while the other edge rests on a thin wire that separates the glass and mirror. When looking down from above at the glass, which is illuminated with light of only one wavelength, we see alternating bright and dark fringes. Explain.

6. Monochromatic light illuminates a single slit. After passing through the slit, the light falls on a screen. Describe the way in which the light pattern on the screen changes as the width of the slit is slowly decreased.

7. Sound passes through an open doorway into a field. As the sound frequency increases, the angular spread of the central intensity maximum changes. Describe this change qualitatively. Support your answer with calculations at two different frequencies of your choice.

8. A foghorn has an opening shaped like an elongated rectangle. To spread sound over a broad horizontal angle, should the long side of the horn be oriented up and down or sideways? Explain.

9. Why do Polaroid sunglasses reduce the glare of light reflected from lakes better than sunglasses that are not Polaroid?

10. While wearing Polaroid sunglasses, you first look at light reflected by a pond of water or a sheet of glass with your head upright and your eyes horizontal. You then rotate your head 90° to the side so that your eyes are vertical, one above the other. In which case is the glare greatest? Explain.

Problems

21.2 Huygens' Principle

■ 1. Draw on a piece of paper a horizontal straight wavefront that is about 6 cm long. Suppose that the left half travels downward at 2 cm/s and the right half travels downward at 1 cm/s. Use Huygens' principle to construct the wavefront 1/4 s and 1/2 s after it has the original straight shape.

■ 2. Draw a wave crest that extends about 15 cm across a paper. Wavelets that emanate from the central third of the crest travel at a speed of 6 cm/s. On the sides, the wavelets travel at 3 cm/s. (a) Using Huygens' wavelets, construct the shape of the crest 1/4 s in the future. (b) Using the new crest formed in part (a), construct the crest's shape as seen 1/4 s later (that is, 1/2 s after the initial crest).

■ 3. A wedge-shaped wave crest, such as shown in Fig. 21.21, spreads outward. The top edge travels at a speed of 6 cm/s and the bottom edge at 3 cm/s. Using Huygens' wavelets, construct the wave's shape 1/4 and 1/2 s in the future. Pay special attention to the "point" of the wedge.

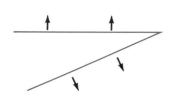

FIG. 21.21

■ 4. The wave crest shown in Fig. 21.22 has been traveling in medium 1 at a speed of 6 cm/s. Use Huygens' wavelets to construct the shape of the crest 1/4 and 1/2 s after it reaches medium 2 where its speed is 3 cm/s.

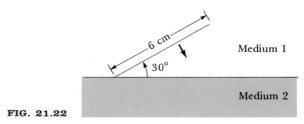

FIG. 21.22

21.3 Young's Double-Slit Experiment

5. Green light of wavelength 540 nm is incident on two slits that are separated by 0.50 mm. (a) Calculate the angular deflection of light to the zeroth-, first-, and second-order bright bands. (b) Repeat the calculations for infrared waves of wavelength 1080 nm.

■ 6. Blue light of wavelength 440 nm is incident on two slits separated by 0.30 mm. Calculate (a) the angular deflection of the third-order bright band and (b) its spatial separation from the zeroth-order band when projected on a screen located 3.0 m from the slits. (c) Draw a picture (not to scale) that schematically represents the situation described and label all known distances and angles.

■ 7. Red light of wavelength 630 nm passes through two slits and then onto a screen that is 1.2 m from the slits. The third-order bright band on the screen is separated from the central maximum by 0.80 cm. (a) Calculate the separation of the slits. (b) Draw and label a picture such as requested in Problem 6(c).

■ 8. Sound of frequency 680 Hz is synchronized as it leaves two speakers that are separated on an open field by 0.80 m. Calculate the angular deflection from a line straight in front of the speakers, where sound is intense, to the next position at the side where the sound is also intense. The speed of sound is 340 m/s.

■ 9. Two stereo speakers separated by a distance $d = 2.0$ m play the same musical note of frequency 1000 Hz. A listener starts from position 0 (Fig. 21.23) and walks along a line parallel to the speakers. (a) Is the sound easily heard at position zero? (b) Calculate the distance from position 0 to positions 1 and 2 where intense sound is also heard. The speed of sound is 340 m/s.

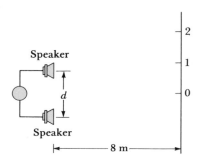

FIG. 21.23

■ 10. Monochromatic light passes through two slits and then strikes a screen. The distance separating the central maximum and the first bright fringe at the side is 2.0 cm. (a) Calculate the fringe separation if the slit separation is doubled while everything else remains constant. (b) Calculate the fringe separation if the wavelength is doubled while everything else remains unchanged. (c) Calculate the fringe separation if the screen distance is doubled while everything else remains unchanged.

■ 11. Draw a picture of two narrow slits separated by 4.0 cm. Then construct the crests of six waves of wavelength 1.0 cm that have left each slit. Next draw lines from a point halfway between the slits in the directions of the zeroth- and first-order intensity maxima. Finally, draw a screen 7.0 cm from the slits and *estimate*, based on your drawing, the distance between the central maximum on the screen and a first-order bright spot to the side. Check your results using equations from the text (20 percent accuracy is fine).

■■ 12. A mercury lamp emits light of wavelength 436 nm. After passing through a narrow single slit, the light strikes two slits separated by 0.20 mm and located 0.50 m from the single slit. After passing through the pair of slits, the light strikes a screen 9.0 m away. (a) Make a sketch on paper of the interference pattern that you expect to see on the screen (use the correct dimensions). (b) Below the interference pattern, sketch the pattern expected if corpuscles of light cause a geometric image of the slits to be formed on the screen.

21.4 Gratings

13. Light of wavelength 520 nm passes through a grating with 4000 lines/cm. (a) Calculate the angular deflection of the

second bright band. (b) The light falls on a screen located 1.6 m from the grating. Calculate the separation of the second-order bright spot from the central maximum. (c) Draw a picture (not to scale) that schematically represents the situation described and label all known distances and angles.

14. Light of wavelength 656 nm and 486 nm emitted from a hot gas of hydrogen atoms strikes a grating with 4800 lines/cm. Calculate the angular deflection of both wavelengths in the first and second orders.

15. You are to purchase a grating that will cause the deflection of the second-order bright band of 1000-nm infrared radiation to be 42°. How many lines per centimeter should your grating have?

■ 16. Most light emitted by hot sodium atoms has a wavelength of 589.3 nm. When the light passes through a grating, its first-order angular deflection is 16.1°. (a) Calculate the separation of adjacent lines in the grating and the number of lines per centimeter. (b) Calculate the angular deflection of the third-order line. (c) Calculate the distance on a film located 2.0 m from the grating of the first-order line from the central maximum.

■ 17. Light of wavelength 630 nm passes through a grating and then onto a screen located several meters from the grating. The first-order bright band is located 0.28 m from the central maximum. Light from a second source produces a band 0.20 m from the central maximum. Calculate the wavelength of the second source. Be sure to show your calculation procedure clearly. (Hint: If the angular deflection is small, $\tan\theta \simeq \sin\theta$.)

■ 18. Hydrogen atoms emit purple bands of light of wavelength 389 nm and 397 nm. The light passes through a grating and then falls on a film located 0.50 m from the grating. *Estimate* the number of lines per meter for the grating so that the first-order bands of the two wavelengths of light are separated by 1.0 mm on the film. (See the hint for Problem 17.)

■ 19. A fence consists of alternating slats and openings, the openings being separated from each other by 0.40 m. Parallel wavefronts of a single-frequency sound wave irradiate the fence from one side. A person 20 m from the fence walks parallel to it. Intense sound is heard directly in front of the fence and in another region 15 m farther along a line parallel to the fence. Calculate the wavelength and frequency of the sound ($v_{sound} = 340$ m/s).

■ 20. A reflection grating reflects light from adjacent lines in the grating instead of allowing the light to pass through slits as in a transmission grating. Interference between the reflected light waves produces different-order reflection maxima. The angular deflection of bright bands, assuming perpendicular incidence, is calculated using Eq. (22.1). White light is incident on the wing of a Morpho butterfly (whose wings act as a reflection grating). Red light of wavelength 660 nm is deflected in first order at an angle of 1.2°. (a) Calculate the angular deflection in first order of blue light (460 nm). (b) Calculate the angular deflection in third order of yellow light (560 nm).

■ 21. Two different wavelengths of light shine on the same grating. The third-order line of wavelength A (λ_A) has the same angular deflection as the second-order line of wavelength $B(\lambda_B)$. Calculate the ratio λ_B/λ_A. Be sure to show the reasoning leading to your solution.

■ 22. Monochromatic light from a laser is used to illuminate two different gratings. The angular deflection of the second-order band of light leaving grating A equals the angular deflection of the third-order band from grating B. Calculate the ratio of the number of lines per centimeter for grating A and for grating B.

■ ■ 23. An astronomer has a grating spectrograph with 5000 lines/cm. A film 0.500 m from the grating records bands of light passing through the grating. (a) Calculate the wavelength and frequency of the H_α line of hot hydrogen gas in a laboratory discharge tube that produces a first-order band separated on the film by 17.4 cm from the central maximum. (b) Calculate the wavelength and frequency of the H_α line coming from the galaxy Hydra. The first-order band of the H_α line of this light is 28.6 cm from the central maximum. (c) Use the Doppler equation for electromagnetic radiation to estimate the relative speed of Hydra and the earth.

21.5 Thin-Film Interference

24. A soap bubble of index of refraction 1.40 appears blue ($\lambda = 440$ nm) when viewed perpendicular to its surface. Does the light change phase when reflected from (a) the outside surface of the bubble and (b) the inside surface? (c) Calculate the wavelength of the light when passing through the bubble. (d) Calculate the thickness of the thinnest bubble for which the 440-nm blue light reflected from the outside surface of the bubble interferes constructively with light reflected from the inside surface.

25. Light of 690-nm wavelength interferes constructively when reflected from a soap bubble having index of refraction 1.33. Determine two possible thicknesses of the soap bubble.

26. A film of oil with index of refraction 1.50 is spread on water whose index of refraction is 1.33. Calculate the least thickness of the film for which green light of wavelength 520 nm is reflected constructively.

27. A lens coated with a thin layer of material having an index of refraction 1.25 reflects the least amount of light at a wavelength of 590 nm. Calculate the minimum thickness of the coating.

■ 28. A film of transparent material 120 nm thick and having index of refraction 1.25 is placed on a glass sheet having index of refraction 1.50. Calculate (a) the longest wavelength of light that interferes destructively when reflected from the film and (b) the longest wavelength that interferes constructively.

■ 29. Two flat glass surfaces are separated by a 160-nm gap of air. (a) Explain why 640-nm-wavelength light illuminating the air gap is reflected brightly. (b) What wavelength of radiation is not reflected from the air gap?

■ ■ 30. A wedge of glass of refractive index 1.64 has a silver coating on the bottom, as shown in Fig. 21.24. Calculate the least distance x to a position where 500 nm light reflected from the top surface of the glass interferes constructively with light reflected from the silver coating on the bottom. The light changes phase when reflected at the silver coating.

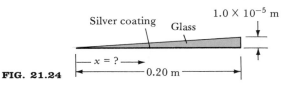

FIG. 21.24

21.6 Single-Slit Diffraction

31. Light of wavelength 630 nm is incident on a long, narrow slit. Calculate the angular deflection of the first diffraction minimum if the slit width is (a) 0.020 mm, (b) 0.20 mm, and (c) 2.0 mm.

32. Light of wavelength 520 nm is incident on a long, narrow slit of width 0.050 mm. Calculate the angular deflection of the fifth-order diffraction minimum.

33. The wavelength of water waves entering a harbor is 14 m. The angular deflection of the first-order diffraction minimum of the waves in the water beyond the harbor is 38°. Calculate the width of the opening into the harbor.

■ 34. Sound of frequency 442 Hz passes through an opening that is 2.10 m wide. Calculate the angular deflection to the second diffraction minimum. ($v_\text{sound} = 340$ m/s.)

■ 35. Light of wavelength 624 nm passes through a single slit and then strikes a screen that is 1.2 m from the slit. The third dark band is 0.60 cm from the central bright band. Calculate the slit width.

■ 36. Babinet's principle states that the diffraction pattern of complementary objects is the same. For example, a slit in a screen (☐) produces the same diffraction pattern as a screen the same size as the slit (■); a hair should produce the same diffraction pattern as a slit of the same width. Calculate the width of a hair that, when irradiated with laser light of wavelength 630 nm, produces a diffraction pattern on a screen with the first minimum 2.5 cm on the side of the central maximum. The screen is 2.0 m from the hair.

■ 37. The opening of a stereo speaker is shaped like a slit. Calculate the maximum width such that the first diffraction minimum of sound is at least 45° on each side of the direction in which the speaker points. Perform the calculations for sound waves of frequency (a) 200 Hz, (b) 1000 Hz, and (c) 10,000 Hz. The speed of sound is 340 m/s.

■ 38. The angular deflection of the first-order bright band of light passing through double slits separated by 0.20 mm is 0.15°. Calculate the angular deflection of the third-order diffraction minimum when the same light passes through a single slit of width 0.30 mm.

■ ■ 39. Assume that the earth, its structures, and its inhabitants are all decreased in size by the same factor. *Estimate* the decrease required in order that the first-order diffraction dark band of 500-nm light entering a typical room window is at 90° (the central bright band would light most of the room). Explain all aspects of your calculations.

21.7 Polarization

40. At what angle of incidence (and reflection) does light reflected from a smooth pond become completely polarized parallel to the pond's surface?

41. At what angle is the light that is reflected from a water-glass interface at the bottom of a cake pan holding water completely polarized parallel to the surface of the glass for glass of refractive index 1.65?

■ 42. An unpolarized beam of light passes through two Polaroid sheets that are aligned initially so that the transmitted beam is a maximum. By what angle should the second Polaroid be rotated to reduce the transmitted intensity to (a) one-half and (b) one-tenth the intensity that was transmitted through both Polaroid sheets when aligned.

■ 43. Two Polaroid sheets are oriented at an angle of 60° relative to each other: (a) Calculate the factor by which the intensity of an unpolarized light beam is reduced after passing through both sheets. (b) Calculate the factor by which the intensity of a polarized beam oriented at 30° relative to each Polaroid is reduced after passing through both Polaroids. (Hint: The light is polarized parallel to the first Polaroid after passing through it.)

Most of classical physics provides what is called a **deterministic** description of nature; if we know the present status of an object, then we can determine its future status with absolute certainty. Newtonian mechanics is deterministic; if the forces acting on an object and its initial position and velocity are known, Newton's laws and kinematics can be used to determine with great accuracy the future position and velocity of the object. Using Newtonian mechanics, for example, we can in theory predict the future motions of the planets with an accuracy limited only by our ingenuity, skill, and patience in measuring their present positions and velocities and in calculating their future motions.

In theory, the future course of events in the life of a human is equally predictable. An extremely elaborate machine would have to measure and record the position and velocity of every particle in your body and of other particles in your environment that exert forces on your body. This information would then be transferred to a giant computer that would calculate the whole course of your future life.

The task of calculating your future is impossible for a variety of reasons. For one thing the body consists of over 10^{28} particles (electrons, protons, and neutrons), whereas big computers store only about 10^{20} bits of information. But the idea that a person's future is determined does not depend on whether instruments can actually measure and calculate that future. It depends on the supposition that a particle's position and velocity and the forces acting on it at each instant of time are precisely defined. If this is true, then the particle's future motion is also precisely defined, even though we do not know what that future will be. If the future of each particle in a person is determined, then the person's future must also be determined. Based on this idea, a philosopher might argue that free will—our ability to make choices that affect the future—is an illusion.*

The idea of determinism is hard to accept, given the large number of decisions we seem to make each day and the number of seemingly random events that affect our lives. The development early in this century of a new theory of the atom, called *quantum mechanics,* appeared to free electrons and other small particles from a future that was completely determined. According to the theory of quantum mechanics, a particle's position and velocity cannot be measured simultaneously with unlimited accuracy. If we measure the particle's position accurately, its velocity is uncertain. If we measure its velocity accurately, its position is uncertain. In classical physics we need to know both the particle's initial position and velocity to determine its future motion. Thus, according to quantum mechanics, we lack half the information needed to predict the future motion of the electrons and other small particles in our body. If the motion of these small particles is uncertain, then the future of a person must also be uncertain. Quantum mechanics seemed to restore support for the idea of free will.

We will have to postpone until Chapter 31 a discussion of the reason for the indeterminacy in the position and velocity of small particles. We can, however,

---

*Philosophy is a complex and sophisticated subject, and the comments of a physicist as an amateur philosopher should be taken with a grain of salt.

think about the problems encountered in measuring accurately the positions of small particles. These position-measurement ideas will be useful when we develop the ideas of quantum mechanics. The ideas are also of great practical importance for persons using optical instruments such as cameras, microscopes, and telescopes. Can the fine structure of a cell be seen with a microscope? Can a reconnaissance aircraft take photographs of a missile site with sufficient detail? We will find that diffraction (studied in Chapter 21) limits our ability to see the fine structure of small objects or the details of distant objects.

Consider the light passing through the small hole shown in Fig. V.1a. Light leaving the hole produces a diffraction pattern on a screen behind the hole. The pattern resembles that formed by light passing through a narrow slit, except that alternating rings of bright light and darkness are formed rather than bands of light and darkness. The angle θ between a line drawn from the hole to the center of the pattern and a line drawn toward the first dark ring is calculated using the equation

$$\sin \theta = 1.22 \frac{\lambda}{d}, \qquad \textbf{(V.1)}$$

where λ is the wavelength of the wave passing through the hole and d is the diameter of the tiny hole. Equation (V.1) is similar to Eq. (21.6), which was used to calculate the angular deflection to dark bands in a single-slit diffraction pattern. The factor 1.22 that appears in Eq. (V.1) and not in Eq. (21.6) is a result of the circular geometry of the opening.

The diffraction pattern produced by a circular hole is exactly like the pattern produced by a circular disc of the same size (Fig. V.1b). The only difference is that the light regions in the diffraction pattern of the hole are dark in the diffraction pattern of the disc, and vice versa.

The angular deflection θ to the first ring in either diffraction pattern shown in Fig. V.1a or b depends on the diameter of the hole or disc and on the wavelength of the radiation used to form the pattern. According to Eq. (V.1), the sine of the angle θ (and θ itself) increases as the diameter of the opening decreases ($\sin \theta \propto 1/d$). The rings are spread farther apart for small obstacles than for large obstacles. This latter variation of angular spread with diameter provides a convenient way to measure the size of small objects. The actual diffraction patterns of laser light that diffracted around two spheres of different size is shown in Fig. V.2a and b. The spheres are approximately the size of human body cells. The smaller sphere produces the more spread-out diffraction pattern shown in Fig. V.2a. The diffraction pattern shown in Fig. V.2c is that of a red cell of a chicken. The irregularities in this pattern provide information about distortions in the spherical geometry of the cell and also indicate the presence of the cell nucleus, which has different light-transmission properties than the rest of the cell.

Scientists are now developing instruments that can measure the size of a large number of cells in a short time. The cells pass one by one in a narrow

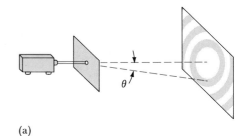

(a)

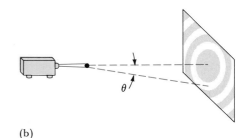

(b)

FIG. V.1 (a) The diffraction pattern of light passing through a small hole has the same spread and shape as (b) the diffraction pattern of light passing a disc with the same diameter as the hole.

467

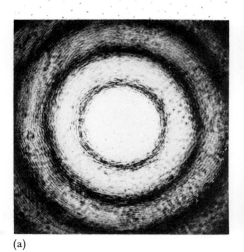
(a)

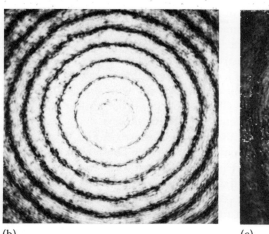
(b)

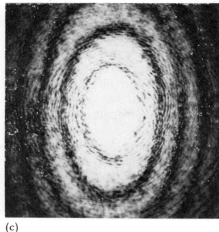
(c)

FIG. V.2 The diffraction patterns produced by 633-nm light from a He-Ne laser passing (a) a spherical bead 6 μm in diameter, (b) a spherical bead 16 μm in diameter, and (c) a chicken's elliptical red blood cell 13 μm long by 7 μm wide. (Courtesy of Dr. Douglas Burger, Yale University School of Medicine and the Los Alamos Scientific Laboratory.)

stream of water through a laser beam. The diffraction pattern caused by a cell as it passes through the beam is recorded by an array of light detectors behind the cell. The information from the detectors is transmitted to a computer that calculates the size of the cell. The cell size indicates the age of the cell. In some human diseases, such as cervical cancer, the cells in a sample of tissue have an abnormal size and age distribution. At the present time technicians must measure the size of a large number of cells by examining them one by one with a microscope. An automated system based on diffraction measurements should be more accurate, faster, and less expensive.

We see that diffraction provides a convenient method for measuring the size of a cell. But we find also that diffraction limits our ability to see objects such as electrons that are much smaller than cells, and this has an important impact on our earlier discussion of determinism. Consider again Eq. (V.1):

$$\sin \theta = 1.22 \frac{\lambda}{d}.$$

Remember that θ is the angle between a line from the particle toward the center of the diffraction pattern and another line from the particle toward the first ring to the side of the central region of the pattern. If $1.22\lambda/d$ is 1, then $\sin \theta = 1$ and $\theta = 90°$. In this case, the central region of the diffraction spreads over the whole space beyond the particle. We cannot tell where the object causing the pattern is located because all we see on a screen beyond the object is a uniform pattern of light (no rings). If $1.22\lambda/d$ is greater than 1, then according to Eq. (V.1), $\sin \theta > 1$. This impossible result means that the diffraction pattern is spread even farther, and we see only the center of the central portion of the diffraction pattern. It has no rings, and we cannot determine the location of the object causing the pattern.

If a diffraction pattern with rings is to be formed to locate an object and determine its size, the quantity on the right of Eq. (V.1) must be less than 1:

$$\sin \theta = 1.22 \frac{\lambda}{d} < 1,$$

or

$$\lambda < \frac{d}{1.22}. \qquad \qquad \textbf{(V.2)}$$

It is common to ignore the 1.22 in Eq. (V.2). In that case, we can say that *a particle's position and size can be determined using waves only if the wavelength of the radiation is less than the diameter of the particle.* The diameters of protons and neutrons are about 10^{-15} m, and the diameter of an electron is not known but is much less than that of a proton. The wavelength of light is more than 10^8 times longer than the diameter of a proton and cannot be used to locate these particles. Even x-rays and gamma rays have wavelengths that are longer than the size of a proton or neutron and much longer than the size of an electron. If we do manage to produce waves whose wavelengths are shorter than these particles, the waves will disrupt the future motion of the particle in an unpredictable way, as we will see in Chapter 31. Either we cannot locate the particle because the wavelength is too long, or we disrupt its future motion in an uncertain way because the wavelength is too short. Thus, we cannot determine simultaneously an object's position and velocity with complete accuracy. The future of a small particle is uncertain.

EXAMPLE V.1 Suppose that you are reduced to the size of an ant. One of your ant friends holds a tiny flashlight in front of your eyes. Over what angular region does light leaving the flashlight spread after passing through the pupil of your eye on its way toward the retina?

SOLUTION If all of your body parts are reduced proportionately, the pupil of your eye becomes very small when your body is reduced to the size of an ant. Light passing through the small opening into your eye is diffracted over a broad angular region. To calculate the angular spread, we first estimate the diameter of the eye's pupil if your body is decreased to the size of an ant. An ant is approximately 2 mm long and a person is approximately 2 m long. Thus, when shrunk, each dimension in the person's body decreases by a factor of about $2 \text{ mm}/2 \text{ m} = 2 \times 10^{-3} \text{ m}/2 \text{ m} = 10^{-3}$. The pupil in a person's eye has a diameter of about 1.5 mm. After shrinking, the diameter is reduced by 10^{-3}:

$$d = (10^{-3})1.5 \text{ mm} = 1.5 \times 10^{-3} \text{ mm} = 1.5 \times 10^{-6} \text{ m}.$$

Next consider green light of 500-nm wavelength entering this small, circular opening. The light forms a diffraction pattern on the retina with a bright region in the center and alternating dark and bright rings circling the central bright region. The angular deflection from a line drawn toward the center of the pattern to a line drawn toward the first dark ring is determined using Eq. (V.1):

$$\sin \theta = 1.22 \frac{\lambda}{d} = 1.22 \left(\frac{500 \times 10^{-9} \, \text{m}}{1.5 \times 10^{-6} \, \text{m}} \right) = 0.41,$$

$$\theta = 24°.$$

Two lines moving toward opposite sides of the first dark ring are separated by twice this angle, or by 48°, as shown in Fig. V.1a. Thus, the bright region in the center of the pattern spreads over an angular region that is a little less than 48° wide. Most of our retina would be covered by light from this one small light source. It would be difficult for an ant-sized person to see the details of his or her environment since light reflected from two closely spaced objects would excite the same group of cones and rods over a large region of the retina. We would see a flood of light but could not determine from where it was coming. ∎

EXAMPLE V.2   A biologist builds a device to detect and measure the size of insects. The device emits sound waves. If an insect passes through the beam of sound waves, it produces a diffraction pattern on an array of sound detectors behind the insect. What is the lowest-frequency sound that can be used to detect a fly that is about 3 mm in diameter? The speed of sound is 340 m/s.

SOLUTION   The wavelength of a sound wave is related to its frequency f by Eq. (17.8), $f = v/\lambda$, where v is the speed of sound. To produce a diffraction pattern that is spread over a reasonable angular region, the biologist builds the device so that the angular deflection to the first ring of the diffraction pattern is approximately 30° or less. By combining Eqs. (17.8) and (V.1), we find that the frequency of the sound wave needed to produce a diffraction minimum at 30° when passing an insect of size $d = 3 \, \text{mm} = 3 \times 10^{-3} \, \text{m}$ is

$$f = \frac{v}{\lambda} = v \left(\frac{1.22}{d \sin \theta} \right) = (340 \, \text{m/s}) \left[\frac{1.22}{(3 \times 10^{-3} \, \text{m})(\sin 30°)} \right]$$

$$= 2.8 \times 10^5 \, \text{s}^{-1} = \underline{2.8 \times 10^5 \, \text{Hz}}. \quad ∎$$

Problems

1. Laser light of wavelength 630 nm passes through a tiny hole. The angular deflection of the transmitted light to the first dark ring is 26°. Calculate the diameter of the hole.

■ 2. Infrared radiation of wavelength 1020 nm passes a dark, round glass bead and produces a circular diffraction pattern 0.80 m beyond. The diameter of the first bright circular ring on the screen is 6.4 cm. Calculate the bead's diameter.

■ 3. Sound leaves your mouth. (a) *Estimate* the diameter of your mouth when open wide. (b) Calculate the angular deflection of sound of frequency 200 Hz and of frequency 15,000 Hz. If, during your calculations, you find that sin $\theta > 1$, explain the meaning of this result. The speed of sound is 340 m/s.

■ 4. Light of 626-nm wavelength from a helium-neon laser passes two different-sized body cells. The angular deflection of the first ring in the diffraction patterns formed as the light passes one cell is 0.060 rad and 0.085 rad as it passes the other cell. (a) Calculate the size of each cell. (b) Determine the radius of the first ring of each pattern on a screen 1.2 m from the cells.

■ 5. A sound of frequency 1000 Hz passes a basketball. *Estimate* the angular deflection from the basketball to the first ring in the diffraction pattern. The speed of sound is 340 m/s.

■ 6. What is the diameter of the smallest object that forms a diffraction pattern when irradiated by the (8×10^4)-Hz sound from a bat?

PART VI

Electricity and Magnetism

Much of our past study of physics has used Newton's second law and energy principles to analyze the motion of objects. In this part of the book these same techniques will be applied to the study of electric and magnetic phemonena.

Electric and magnetic forces, first discovered by the ancient Greeks, were enshrouded in mystery for centuries. The practical use of electricity began only about a hundred years ago. Today, a relatively short time later, electric and magnetic devices provide instant communication by television, telephone, and microwave networks that carry international news, worldwide weather reports, stock quotations, and other, exotic forms of information. Electrical energy is transmitted from centralized power plants for operating household appliances, lighting office buildings and homes, and providing power for myriad other business and domestic uses. Modern computer systems store information and perform calculations that help to control space flights, the operations of businesses, and charge accounts.

We cannot examine the details of all the interesting and important applications of modern electricity and magnetism. We can, however, develop a basic understanding of the underlying principles that made these applications possible. Our analysis begins with a discussion of the basic force between two charged particles. As we continue through Part VI, our interest will move toward the electronic applications of electricity and magnetism.

CHAPTER 22

Electric Force

Your body contains more than 10^{28} protons and nearly an equal number of electrons. The protons have a positive electric charge; the electrons have a negative charge. The electrical force between these charges is one of the most important forces in nature.

Suppose for a moment that you could borrow all the electrons from a friend's body and put them into your pocket. The mass of the electrons would be about 20 grams. With no electrons, your friend would have a huge positive charge. You, on the other hand, would have a huge negative charge in your pocket. If you stood 10 m from your friend, the attractive force between the two of you would be about 10^{23} tons—more than 100,000 times greater than the gravitational force between the earth and the sun. Obviously, your pocket would be ripped off long before you had gathered even a tiny fraction of your friend's electrons.

We never find 10^{28} protons or 10^{28} electrons gathered in one place on the earth. Equal numbers of positive and negative charges almost always combine to form atoms—tiny particles that have no net charge. So the huge electrical force just described does not occur.

There are, however, many instances in which smaller charge imbalances occur. In this chapter we will learn how to calculate the magnitude and direction of the forces between such charges. Our concern is mostly with forces between stationary charges—that is, with **static electricity.**

22.1 Electric Charge

Many words and concepts must be defined in terms of the effects they produce. *Love*, for example, can be defined as a strong feeling that causes a person to delight in and desire the presence of another. Love, then, is defined in terms of the effects it produces—delight and desire.

Electric charge is also defined by the effect it produces. No one has ever seen electric charge; it has no known weight, color, length, or width. Yet the effects of electric charge are among the most important in nature. Almost every dynamic process in your body and in its environment involves a rearrangement of particles with electric charge, and these rearrangements cause dramatic changes in the forces acting on the particles.

Atoms, the basic units of which all matter is built, consist of three types of particles: electrons, protons, and neutrons. Two of these particles—electrons and protons—have electric charge, and the neutron has no charge. The magnitude of the electric charge of an electron equals that of a proton (even though the electron has only 1/2000 the mass of a proton). The proton is said to have a positive electric charge; the magnitude of the charge is given the symbol e. An electron has a negative charge whose value is represented by the symbol $-e$. An atom has no net charge since it contains equal numbers of protons and electrons.

More than one hundred different types of atoms have been discovered. The atoms differ in the number of protons, neutrons, and electrons of which they are made. The central core of an atom, called the **nucleus,** consists of protons and neutrons. Electrons circle the nucleus in a variety of different orbits. Hydrogen, for example, has a nucleus of one proton surrounded by one orbiting electron. Carbon, on the other hand, has six protons and six neutrons in its nucleus, and six electrons move in orbits around the nucleus (Fig. 22.1).

As you can see in Fig. 22.1, the electrons in an atom occupy different orbits. Electrons in orbits nearest the nucleus are tightly bound to it by their electrical attraction. The electrons farthest from the nucleus (called the outer electrons) are loosely bound. It is common for an outer electron to actually leave one atom and move into an outer orbit of another type of atom. When such a charge transfer occurs, the transferred electron is usually bound more tightly to the new atom than it was to the old atom. An atom that has lost an electron has a net charge of $+e$ (it has one more proton than electron) and is called a **positive ion.** An atom that gains an extra electron has a net charge of $-e$ and is called a **negative ion.**

This type of charge transfer often occurs when materials made of different types of atoms are brought into contact with each other. The small electrical shock you receive when you touch another person after shuffling across a synthetic rug is the result of removal of electrons from the rug through your shoes and onto your body. Some of these electrons leave your body quickly (the shock) when you touch the other person.

The existence of electric charge and of charge transfer can be demonstrated using a balloon, a wool scarf, and a thin plastic clothes bag (Fig. 22.2). The balloon is suspended by a string. Rubbing the balloon with a plastic bag (Fig. 22.2a) causes the balloon to accumulate excess electrons from the bag and to become negatively charged. The plastic is left with an equal number of positive charges (positive ions). We find that the plastic bag and balloon attract each other (part b). *Opposite charges attract.*

If the plastic bag is then rubbed against a wool scarf (part c), electrons will be transferred from the scarf to the bag because electrons are more attracted to the atoms in plastic than to those in wool. The plastic bag, which started with a deficiency of electrons, gains so many electrons from the scarf that it becomes negatively charged. The plastic bag and balloon will now repel each other (part d). *Like charges repel.* The scarf, which loses electrons when rubbed with plastic, becomes positively charged and will attract the balloon, since opposite charges attract (part e).

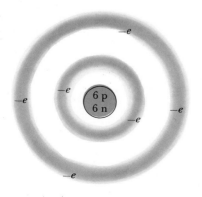

FIG. 22.1. A carbon atom consists of a nucleus with six protons (charge $+6$) and six neutrons that are surrounded by six electrons in different orbits.

Unit of Charge

The SI metric system unit of charge is called the coulomb (C) after the French engineer Charles Augustin de Coulomb (1736–1806). One **coulomb** equals the

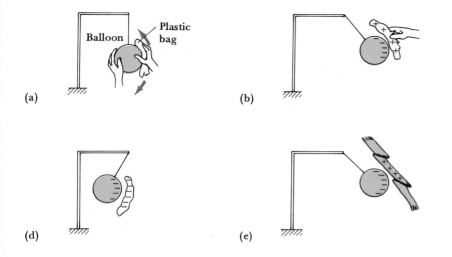

(a)

(b)

(c) Wool scarf

(d)

(e)

FIG. 22.2. (a) A plastic bag is rubbed against a rubber balloon. (b) This transfers negative charge from the bag to the balloon, resulting in their attraction. (c) The plastic bag is rubbed against a wool scarf. (d) The bag is now charged negatively and repels the balloon. (e) The positively charged scarf attracts the balloon.

charge of an aggregate of 6.25×10^{18} protons. The charge e of a single proton, in units of coulombs, is

$$e = 1.60 \times 10^{-19} \text{ C.}$$

An electron has a charge $-e$ equal to -1.60×10^{-19} C. Balloons rubbed with plastic, as illustrated in Fig. 22.2, typically accumulate about 10^{12} excess electrons, each with a charge of -1.60×10^{-19} C. The total charge q on the balloon is then

$$q = 10^{12}(-e) = 10^{12}(-1.60 \times 10^{-19} \text{ C}) = -1.60 \times 10^{-7} \text{ C.}$$

22.2   Coulomb's Law

The electrical force exerted by one charged object on another depends on the amount of charge on each object, the sign of the charges (positive or negative), and the distance between them. The magnitude of the electrical force can be calculated using an equation determined experimentally in 1788 by Charles Coulomb. The equation is called **Coulomb's law:**

$$F = \frac{kq_1q_2}{r^2}. \tag{22.1}$$

The amount of electric charge on the two objects in units of coulombs is represented by the symbols q_1 and q_2 and r represents the distance in meters separating the charges.* The quantity k is a proportionality constant whose value in the SI metric system is

$$k = 8.98742 \times 10^9 \text{ N} \cdot \text{m}^2/\text{C}^2 \simeq 9.0 \times 10^9 \text{ N} \cdot \text{m}^2/\text{C}^2.$$

The value 9.0×10^9 N·m/C² will be used for k in all of our calculations.

The force between charges also depends on the medium in which the charges are embedded. This medium dependence is considered in Section 22.4.

---

*Coulomb's law applies only to situations in which the size of the objects with charges q_1 and q_2 is much less than their separation r. We will occasionally use the law for making estimates when this condition is not satisfied.

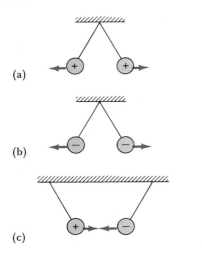

(a)

(b)

(c)

FIG. 22.3. (a) and (b) Like charges repel. (c) Unlike charges attract.

The direction of the force that one charge exerts on another depends on the relative signs of the two charges. If the charges have like signs, then they exert repulsive forces on each other, as shown in Fig. 22.3a and b. If the charges have opposite signs, they exert attractive forces on each other, as shown in Fig. 22.3c. *Coulomb's law is used to determine only the magnitude of the force; we will ignore the signs of the charges when substituting in Coulomb's law. The direction of the force of one charge on another must be determined by considering the signs of their charges.*

EXAMPLE 22.1 A young woman accumulates a charge q_1 of $+2.0 \times 10^{-5}$ C while sliding out of the front seat of a car. Her boyfriend, who has been waiting in the wind, has gained electrons and now has a charge q_2 of -8.0×10^{-5} C. (a) Estimate the magnitude of the electrical force that each person exerts on the other when separated by a distance of 6.0 m. Is the force attractive or repulsive? (b) Suppose the two people move toward each other. Calculate the magnitude of the electrical force of one on the other when their separation is reduced by a factor of 0.50.

SOLUTION (a) The magnitude of the force exerted by one charged person on another is determined using Eq. (22.1):

$$F = \frac{kq_1q_2}{r^2} = \frac{(9.0 \times 10^9 \text{ N} \cdot \text{m}^2/\text{C}^2)(2.0 \times 10^{-5} \text{ C})(8.0 \times 10^{-5} \text{ C})}{(6.0 \text{ m})^2} = \underline{0.40 \text{ N}.}$$

When using Eq. (22.1), remember that only the magnitude of the charges q_1 and q_2 are used to determine the magnitude of the force. Thus, the negative sign is not included when substituting the value of q_1. Because the charges have opposite signs, the woman and man are attracted to each other. The man exerts a force toward the left on the woman ($\mathbf{F}_{2 \text{ on } 1}$ in Fig. 22.4) and the woman exerts an equal magnitude force toward the right on the man ($\mathbf{F}_{1 \text{ on } 2}$ in Fig. 22.4).

(b) As the people move closer, the magnitude of the forces increases. Their new separation r' is related to the original separation by the equation $r' = 0.50r = 0.50(6.0 \text{ m}) = 3.0 \text{ m}$.

The new force can be calculated in two ways. First, by direct substitution, we find that

$$F' = \frac{kq_1q_2}{(r')^2} = \frac{(9.0 = 10^9)(2.0 \times 10^{-5})(8.0 \times 10^{-5})}{(3.0)^2} = \underline{1.6 \text{ N}.}$$

The force can also be calculated by proportionality-type reasoning, as follows:

$$F' = \frac{kq_1q_2}{(r')^2} = \frac{kq_1q_2}{(0.50r)^2} = \frac{kq_1q_2}{0.50^2r^2} = \frac{F}{0.50^2}.$$

But we learned in part (a) that $F = 0.40$ N. Thus,

$$F' = \frac{0.40 \text{ N}}{0.50^2} = \underline{1.6 \text{ N}.}$$

q_2 q_1

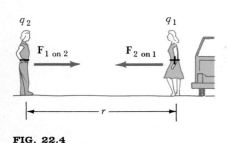

$\mathbf{F}_{1 \text{ on } 2}$ $\mathbf{F}_{2 \text{ on } 1}$

r

FIG. 22.4

You should stop for a moment and think about this result. Does it make sense? The separation of the charges was reduced by 0.50. Since the charges are closer, the force must be larger. Now what about the change in magnitude of the force? If the force had a $1/r$ dependence, then reducing r by one-half would double the force ($1/0.50 = 2.0$). But the force depends on $1/r^2$. Thus, a reduction in r by 0.50 causes the force to increase by a factor of $1/0.50^2 = 1/0.25 = 4.0$.

To test yourself in this type of reasoning, use the proportionality method to confirm that a decrease in the distance by a factor of $1/4$ causes the force to increase by a factor of 16 to 6.4 N; if the distance increases by a factor of 2, the force decreases by $1/4$ to 0.10 N. ∎

22.3 Static Electrical Forces

Most electrical systems consist of many charged particles, each of which exerts an electrical force on all the other charges, and vice versa. To calculate the net force acting on one charge due to the other charges, we must add each electrical force acting on the charge using the techniques of vector addition developed in Chapters 1 and 3. You may wish to review the summaries of these chapters to freshen your memory about vector problem-solving methods. The technique is illustrated qualitatively in Fig. 22.5a.

Three charges are shown. The upper positive and negative charges reside on electrodes placed by a geologist at the earth's surface, and the lower charge represents a sodium ion (Na^+) dissolved in the groundwater under the earth. In this situation we are interested in determining the direction in which the sodium ion will be pushed by the charges on the electrodes.

Whenever considering the effect of forces on an object, *we always start our problems by choosing an object of interest and then drawing a force diagram showing all forces acting on that object.* A picture such as a force diagram is often an essential first step in problem solving. Since the sodium ion is the object of interest in the situation shown in Fig. 22.5a, we look only at the forces acting on the sodium ion. (We are not concerned with the effects of the sodium ion on other charges or with the effects of the other charges on each other. You might review the comments about force diagrams in Section 3.6).

The force diagram for the sodium ion (Fig. 22.5b) shows two electrical forces acting on it: (1) a repulsive force F_+ caused by the charge of the positive

FIG. 22.5. (a) Positive and negative charges on electrodes near the earth's surface exert forces (b) on a sodium ion under the ground. (c) The resultant electrical force acting on the sodium.

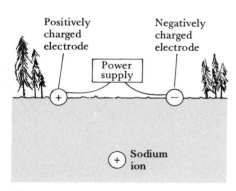

(a)

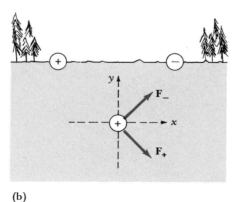

(b)

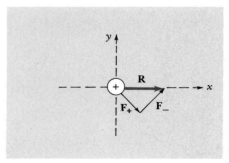

(c)

electrode and (2) an attractive force $\mathbf{F}_-$ caused by the charge of the negative electrode. We ignore for now drag forces that oppose the motion of the ion through the groundwater. The net electrical force on the ion is the vector sum of the forces. They can be added graphically as in Fig. 22.5c or by using the component technique developed in Chapter 1. Examples using both techniques will be presented shortly.

In Fig. 22.5c we see that the net force on the sodium ion is toward the right. Therefore, if only the net electrical force acts on the ion, it accelerates to the right. However, when it starts moving, a drag force will oppose its motion. The drag force represents the average effect of retarding collisions between the sodium ion and other atoms, ions, and molecules in its path. The effect of this drag force will be addressed in Chapter 24. For now, our primary concern is learning how to calculate the net electrical force on a particular charge near other charges.

EXAMPLE 22.2 From the information shown in Fig. 22.6a, calculate the net electrical force of charges q_2 (-1.0×10^{-5} C) and q_3 ($+3.0 \times 10^{-5}$ C) on q_1 ($+2.0 \times 10^{-5}$ C).

SOLUTION The object of interest is charge q_1. A force diagram for that charge is shown in Fig. 22.6b. $\mathbf{F}_2$ is the attractive force of q_2 on q_1, and $\mathbf{F}_3$ is the repulsive force of q_3 on q_1. The graphical addition of the forces, illustrated in Fig. 22.6c, indicates that the resultant force points below the negative x-axis into the third quadrant.

To calculate the magnitude and direction of the resultant force, we add the vectors by the component technique. Before doing this, we first calculate the magnitude of each force by using Coulomb's law, Eq. (22.1):

$$F_2 = \frac{kq_1q_2}{r^2} = \frac{(9.0 \times 10^9 \text{ N} \cdot \text{m}^2/\text{C}^2)(2.0 \times 10^{-5} \text{ C})(1.0 \times 10^{-5} \text{ C})}{(1.0 \text{ m})^2}$$
$$= 1.8 \text{ N},$$

and

$$F_3 = \frac{kq_1q_3}{r^2} = \frac{(9.0 \times 10^9 \text{ N} \cdot \text{m}^2/\text{C}^2)(2.0 \times 10^{-5} \text{ C})(3.0 \times 10^{-5} \text{ C})}{(2.0 \text{ m})^2}$$
$$= 1.4 \text{ N}.$$

FIG. 22.6. (a) A system of three charges. (b) A force diagram showing the forces on charge q_1 due to q_2 and q_3. (c) The graphical addition of these two forces to find the resultant force.

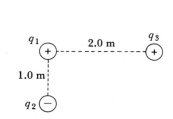

(a)

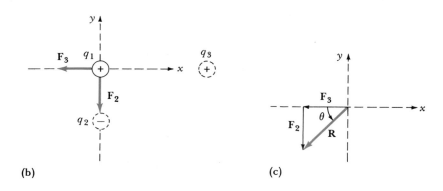

(b)

(c)

The x and y components of the resultant are then calculated using Eqs. (1.3)–(1.6):

$$R_x = F_{2x} + F_{3x} = F_2 \cos 90° - F_3 \cos 0° = 0 - F_3 = -1.4 \text{ N},$$

and

$$R_y = F_{2y} + F_{3y} = -F_2 \sin 90° + F_3 \sin 0° = -F_2 + 0 = -1.8 \text{ N}.$$

The magnitude of the resultant force is found using Eq. (1.7):

$$R = \sqrt{R_x^2 + R_y^2} = \sqrt{(-1.4 \text{ N})^2 + (-1.8 \text{ N})^2}$$
$$= \underline{2.3 \text{ N}}.$$

To completely specify a vector quantity, we must also know its direction. The angle θ that the vector makes with the negative x axis is determined using Eq. (1.8):

$$\tan \theta = \left| \frac{R_y}{R_x} \right| = \frac{1.8 \text{ N}}{1.4 \text{ N}} = 1.29,$$
$$\theta = 52°.$$

Thus, the resultant points 52° below the negative x axis. The fact that it points below the negative x axis is apparent from the signs of the components and from the vector addition shown in Fig. 22.6c. ∎

EXAMPLE 22.3 Two balloons hang at the ends of strings that are 0.80 m long and have negligible weight. Each balloon has a mass of 1.5 g. The balloons have equal but unknown electrical charges that cause the balloons to repel each other. If the balloons are separated from each other by 0.80 m, as shown in Fig. 22.7a, what is the magnitude of the charge on each balloon?

SOLUTION To solve the problem we must first draw a force diagram for one of the balloons and use the first condition of equilibrium to determine the electrical force acting on that balloon. Although either balloon can be chosen as the object of interest for the force diagram, we will choose the balloon on the right. A force diagram for this balloon is shown in Fig. 22.7b. Three forces act on the balloon: (1) a string-tension force **T**, (2) a downward weight force **w** of magnitude mg, and (3) an electrical force **F** caused by the repulsion between the like charges on the balloons. The forces are analyzed using the coordinate system shown in Fig. 22.7b.

Since the balloon is at rest, the sum of the x and y components of the forces must each be zero (the first condition of equilibrium). Thus, using Eqs. (1.3) and (1.5) for the x components, we find that

$$\Sigma F_x = T_x + w_x + F_x = -T \cos 60° + mg \cos 90°^{\,0} + F \cos 0° = 0.$$

Solving this last equation for F, we find that $F = 0.5T$. Next, using Eqs. (1.4) and (1.6) for the y components, we find that

$$\Sigma F_y = T_y + w_y + F_y = T \sin 60° - mg \sin 90° + F \sin 0°^{\,0} = 0.$$

Substituting for the sines and rearranging, we find that $T = mg/0.87$. Substituting this result into our earlier result using the x-component equation

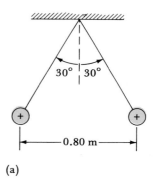

(a)

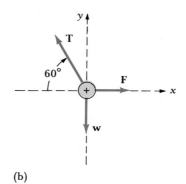

(b)

FIG. 22.7. (a) Two positively charged balloons repel each other. (b) A force diagram for the balloon on the right.

$(F = 0.5T)$, we find that

$$F = 0.5T = 0.5\left(\frac{mg}{0.87}\right) = 0.58mg.$$

The magnitude of the electrical force **F** is given by Coulomb's law, Eq. (22.1):

$$F = 0.58mg = \frac{kqq}{r^2},$$

where q is the unknown charge on each balloon and g is the acceleration of gravity (9.8 m/s²). Solving for q, we find that

$$q = \sqrt{\frac{0.58mgr^2}{k}} = \sqrt{\frac{0.58(1.5 \times 10^{-3}\,\text{kg})(9.8\,\text{m}\cdot\text{s}^{-2})(0.80\,\text{m})^2}{9.0 \times 10^{-9}\,\text{N}\cdot\text{m}^2/\text{C}^2}}$$
$$= 0.79 \times 10^{-6}\,\text{C} = \underline{0.79\,\mu\text{C}}. \qquad \blacksquare$$

In the next example the forces acting on the charge of interest lie along a line, in which case the vector addition of the forces is greatly simplified.

EXAMPLE 22.4 Suppose that at one instant, the electric charge distribution on a person's heart can be represented by charges q_1 and q_2 as shown in Fig. 22.8a. * Calculate the force on the sodium ion (charge q_3) in tissue to the right of the heart due to the positive and negative charges on the heart.

SOLUTION The object of interest is the sodium ion. We are concerned only with forces acting on it and not with forces it exerts on other charges. To indicate the directions of the forces on the sodium ion, we first construct a force diagram (shown in Fig. 22.8b). $\mathbf{F}_1$ is the attractive force of the negative charge q_1 on the positively charged sodium ion q_3, and $\mathbf{F}_2$ is the repulsive force of q_2 on q_3. Since q_1 and q_2 have equal magnitudes and since the separation between q_2 and q_3 is less than that between q_1 and q_3, force $\mathbf{F}_2$ has a larger magnitude than $\mathbf{F}_1$. Thus, when the two forces are added, the resultant force points toward the right in the direction of $\mathbf{F}_2$ (see the graphical addition in Fig. 22.8c.)

Coulomb's law, Eq. (22.1), is used to calculate the magnitude of each force:

$$F_1 = \frac{kq_1q_3}{r^2}$$
$$= \frac{(9.0 \times 10^9\,\text{N}\cdot\text{m}^2/\text{C}^2)(6.0 \times 10^{-6}\,\text{C})(1.6 \times 10^{-19}\,\text{C})}{(0.20\,\text{m})^2}$$
$$= 2.16 \times 10^{-13}\,\text{N},$$

and

$$F_2 = \frac{kq_2q_3}{r^2}$$
$$= \frac{(9.0 \times 10^9\,\text{N}\cdot\text{m}^2/\text{C}^2)(6.0 \times 10^{-6}\,\text{C})(1.6 \times 10^{-19}\,\text{C})}{(0.10\,\text{m})^2}$$
$$= 8.64 \times 10^{-13}\,\text{N}.$$

*The cause of these charges is discussed at the end of the example.

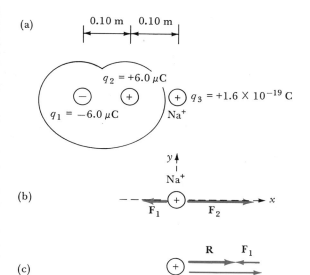

(a)

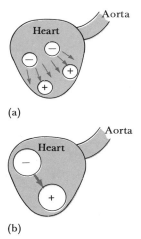

Since the forces point in opposite directions along the x axis, the resultant force equals the difference of their magnitudes,

$$R = F_2 - F_1 = 8.64 \times 10^{-13}\,\text{N} - 2.16 \times 10^{-13}\,\text{N}$$
$$= 6.5 \times 10^{-13}\,\text{N},$$

and points toward the right along the positive x axis. The force on a negative ion, such as chlorine Cl^-, would be in the opposite direction. ∎

Electrocardiography

An important clinical application of these ideas concerning electric forces is **electrocardiography,** a technique for analyzing the condition of the heart. As each muscle cell in the heart contracts, positive and negative charges are separated as shown in Fig. 22.9a. The effect of the charges separated across all of the contracting muscle cells at one instant of time can be represented by a single positive and a single negative charge, as in Fig. 22.9b. The magnitude and relative location of these charges depends on the number and orientation of muscle cells that are contracting at any one time. Since the heart repeats a cycle of contraction about one each second, it produces charge distributions that change magnitude and location in a repetitive fashion—once each second.

At any instant of time, the electric charges on the heart exert forces on ions in the body's surface tissue that cause the ions to move. For example, the charges on the heart shown in Fig. 22.10 force sodium ions in the tissue toward the left and chlorine ions toward the right. Electrodes placed on the arms absorb charges of different sign. The heart acts much like a battery pushing opposite-sign charges in opposite directions in the body tissue. An electrocardiogram (ECG or EKG) is a recording of the charge separation on the body's surface caused by the electric charge of the heart. Abnormalities in the size or timing sequence of different phases of the heartbeat cycle are easily detected by these electrodes.

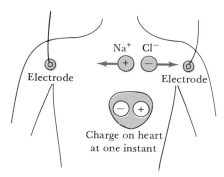

FIG. 22.9. (a) Positive and negative charges are separated as heart muscle cells contract. (b) The net charge distribution on the heart at one instant of time depends on the number and orientation of the cells that are contracting at that time.

FIG. 22.10. The heart's charge distribution causes positive and negative charges on the body surface to move in opposite directions. This charge separation is detected by electrodes of an electrocardiograph.

22.4 Dielectric Constant

In the previous sections of this chapter, we have ignored the effect of a medium in which two charges reside on the force exerted by the charges on each other. This effect can be significant. For example, the force between a sodium ion and a chlorine ion when in water is only 1/80 the force between the same ions when in a vacuum or in air. The effects of the medium are particularly important in biochemical processes, many of which involve the rearrangement of electric charges in molecules. In living organisms these rearrangements are facilitated by water, a solvent that reduces the electrical forces holding the molecules together.

To account for the effect of a medium on the magnitude of the force between two charges, a quantity called the dielectric constant is included in Coulomb's law. The **dielectric constant** K of a medium is the ratio of the electrical force between two charges when in a vacuum (called F_{vacuum}) and the force between the same two charges when kept at the same separation in the medium (called F_{medium}):

$$K = \text{Dielectric constant of a medium} = \frac{F_{vacuum}}{F_{medium}}.$$

Coulomb's law as written in Eq. (22.1) applies only to charges in a vacuum (or in air): $F_{vacuum} = kq_1q_1/r^2$. Substituting this expression for F_{vacuum} into the above equation and rearranging, we find that the force F (we drop the subscript "medium" from now on) between the charges when separated by a distance r in a medium of dielectric constant K is

$$F = \frac{F_{vacuum}}{K} = \frac{kq_1q_2}{Kr^2}. \tag{22.2}$$

The dielectric constants of a number of media are listed in Table 22.1. Notice that, according to Eq. (22.2), the larger the value of K, the smaller is the force between the charges. *

TABLE 22.1 Dielectric Constants of Different Substances

Substance	Dielectric Constant (K)*
Vacuum	1.0000
Dry air	1.0006
Wax	2.25
Mica	2–7
Glass	4–7
Benzene	2.28
Paper	3.5
Axon membrane	8
Body tissue	8
Ethanol	26
Methanol	31
Water	80

*At 20°C.

EXAMPLE 22.5 Calculate the ratio of the electrical force between a potassium ion (K^+) and a chlorine ion (Cl^-) when separated by a distance r in air and when separated by the same distance in water.

SOLUTION The ratio of the forces is

$$\frac{F_{air}}{F_{water}} = \frac{\dfrac{kq_1q_2}{(K_{air})r^2}}{\dfrac{kq_1q_2}{(K_{water})r^2}} = \frac{K_{water}}{K_{air}} = \frac{80}{1.0} = \underline{80}. \qquad \blacksquare$$

*The constant k is often written as $k = 1/4\pi\varepsilon_0$, where ε_0 is called the permittivity of free space ($\varepsilon_0 = 8.85 \times 10^{-12}\ \text{C}^2/\text{m}^2\cdot\text{N}$). Coulomb's law is then written as

$$F = \frac{k}{K}\frac{q_1q_2}{r^2} = \frac{1}{K4\pi\varepsilon_0}\frac{q_1q_2}{r^2} = \frac{1}{4\pi\varepsilon}\frac{q_1q_2}{r^2}.$$

The product $K\varepsilon_0$ is called the permittivity of a medium with dielectric constant K. This unusual way of writing Coulomb's law allows the proportionality constant ε_0 to be related more naturally to other electromagnetic phenomena to be discussed in later chapters.

We learned in the last example that the force between two ions in air is about 80 times more than the force between the same ions when in water. For this reason, potassium chloride and any salt should separate into ions (dissolve) more easily in water than in air. In the human body, most charged particles (ionized salts, charged macromolecules, cell membranes, and so forth) reside in water. The effect of water on the electrical interactions of these charges is great.

The reason the electrical force between charges is reduced greatly in water and in certain other media is related to the dipole nature of these media, as we learn next.

22.5 Electric Dipoles

We have seen that the electrical force between charges is less when the charges are in a medium such as water than when in air or a vacuum. This reduction in the electrical force is caused by the **polar** nature of water molecules. Although a water molecule has a net charge of zero, the charge is not distributed uniformly throughout the molecule. Excess positive charge resides on one side of the molecule, and an equal excess of negative charge is on the other side.

Many different types of molecules have this polar electric charge distribution. These molecules are characterized in terms of their electric dipole moments. Consider the equal-magnitude but opposite-sign charges shown in Fig. 22.11a. The **electric dipole moment p** of this charge distribution is a vector, shown in Fig. 22.11b, that points from the negative charge toward the positive charge. The magnitude of the dipole moment is the product of the positive charge q and the distance a separating the charges:

$$\text{Electric dipole moment } p = qa. \qquad (22.3)$$

Let us see how the electric dipole moment of a polar molecule such as water (H_2O in Fig. 22.12a) is formed. The molecule (Fig. 22.12b) has three nuclei: an oxygen nucleus with a charge $+8e$ and two hydrogen nuclei (protons) each with charge $+e$. The ten electrons in the molecule are arranged somewhat as shown in Fig. 22.12b. Two electrons circle close to the oxygen nucleus while the other eight electrons are located in four orbits, often called lobes. Two of the lobes have a proton and two do not. This distribution of electrons and protons results in a charge separation with two electron pairs pointing in one direction and two electron pairs with protons pointing in the other direction. This charge separation produces an electric dipole moment **p** whose magnitude 6.1×10^{-30} C·m is the same as would be produced by a charge e separated from a charge $-e$ by a distance a equal to 0.38×10^{-10} m, or about one-fifth the size of a water molecule.

We can now understand why the electrical force between ions when in water is less than when in air. If a positive ion or any positively charged particle is placed in water, the water molecules nearest the ion are no longer randomly oriented. The negative part of the water molecules tend to rotate so that they point toward the positive ion (the dashed circle on the left in Fig. 22.13). With a negative ion in the water, the positive sides of the water molecules tend to point toward the negative ion (the right circle in Fig. 22.13). As far as the negative ion is concerned, the positive ion appears to have less charge than usual because its positive charge is reduced by the negative charges hovering near it. The charge of the negative ion is reduced in a similar way by the presence of the positive ends of the water molecules. Consequently, the force between the ions is reduced.

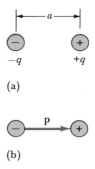

(a)

(b)

FIG. 22.11. (a) An electric dipole consists of a positive charge q and an equal-magnitude negative charge $-q$ separated by a distance a. (b) The electric dipole moment **p** of these charges is a vector from $-q$ toward q whose magnitude $p = qa$.

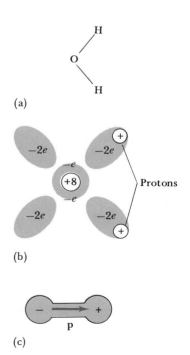

(a)

(b)

(c)

FIG. 22.12. (a) A water molecule. (b) The distribution of nuclei and electrons in a water molecule results in (c) an effective dipole charge distribution and dipole moment.

FIG. 22.13. Dipolar water molecules reduce the effective charge of ions dissolved in the water.

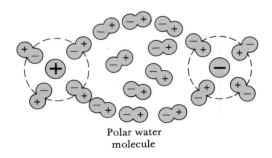

Polar water
molecule

It should be emphasized that the water molecules are continually bumping each other. The orientation of one molecule may change 10^{14} times each second. The reduction in the effective charge of an ion in water is a statistical phenomenon that depends on the average orientation of the water molecules.

An interesting consequence of the polar nature of water molecules is related to the ability of microwave ovens to cook foods. The microwave radiation produced in these ovens interacts with the dipole-charge distribution of water molecules and exerts a torque on the dipole, causing the molecule to rotate. The heating in a microwave oven results from the extra rotational energy imparted to the water molecules by the microwaves. This rotational energy is converted to thermal energy as the molecules collide with each other while rotating. A dry paper plate placed by itself in a microwave oven will remain cool because its molecules do not have electric dipole moments and are thus unaffected by microwaves.

22.6 Electric Fields

In the previous sections we used Coulomb's law and vector addition to calculate the resultant electrical force acting on a charge due to the presence of other charges. What if we do not know the distribution of electric charges in a certain region of space? How might we probe the surrounding space to learn about that distribution? The answer to this question leads to the idea of electric fields and to an alternative method for describing electric forces.

To understand the concept of electric fields, suppose that your instructor asks you to determine the charges on two small metal spheres that have been placed under a box (Fig. 22.14). Your assignment is to devise a technique to determine the sign, magnitude, and location of the charges without removing the box.

One possible technique might involve building a device to measure the electric force on a small test charge held at different places outside the box. You could attach a tiny metal ball with a small positive electric charge (q') to a spring balance that measures the magnitude and direction of the electrical force acting on q'. Knowing the force on q' will help you determine the charge distribution that causes the force. For example, if the spheres under the box are charged as shown in Fig. 22.14, the resultant force on test charge q' is downward when q' is above and to the left side of the box, toward the left when q' is over the center of the box, and upward when q' is above and to the right side of the box. This information indicates that a dipole charge is under the box and that the

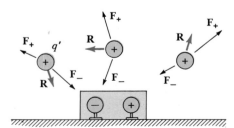

FIG. 22.14. The force on a charge q' varies from place to place, depending on the positions and types of charges under the box.

positive charge is on the right. If the force on test charge q' is measured at every point in the space around the box, we can determine the exact charge distribution under the box.

Now suppose that another person who does not know about the charges under the box is given the device that measures the force on the test charge q'. The person discovers that the force acting on q' changes as q' is moved from one place to another. Knowing nothing of the charges under the box, the person moving the device might say that the space outside the box varies in some way. There seem to be invisible "elastic bands" pulling on the test charge. These invisible bands constitute an electric field that exists throughout the space around the charges. A charge placed in the electric field will feel a force that depends on the magnitude and direction of the field and on the magnitude and sign of the charge.

An electric field is analogous to a gravitational field. The mass of the earth is said to create a gravitational field just as the charges under the box in Fig. 22.14 create an electric field. If another mass, such as the mass of a skydiver, is placed in the earth's gravitational field, it experiences a gravitational force. If another charge, such as q' in Fig. 22.14, is placed in an electric field, it experiences an electric force.

An electric field is defined in terms of the force acting on a charge placed in the field rather than in terms of the charges causing the field:

An **electric field** $\mathbf{E}$ at a point in space is the ratio of the net electric force $\mathbf{F}$ acting on a small, positive test charge q' placed at that point, divided by the value of the test charge:

$$\mathbf{E} = \frac{\mathbf{F}}{q'}. \tag{22.4}$$

The units of electric field are newtons/coulomb, N/C.

The reason for dividing the force by q' is to define the electric field in a way that is independent of the size of the test charge. The electric field is a property of the space and not of the charge testing the field. For example, if the test charge is doubled, the electric force acting on it would also double. However, the ratio $2F/2q'$ remains the same. The electric field depends only on the charges causing the field and not on the test charge used to measure it. Although the field is defined to be independent of the charge q' that measures the field, q' can affect the field if its value is too large. The value of q' must be so small that it does not cause the charges that produce the field to move.

Equation (22.4) is often used to calculate the force on a charge q that is placed in a known electric field $\mathbf{E}$. In this case, Eq. (22.4) becomes

$$\mathbf{F} = q\mathbf{E}.$$

This equation can be used to determine the force on both positive and negative charges. If q is positive, then $\mathbf{F}$ and $\mathbf{E}$ are in the same direction; if q is negative, $\mathbf{F}$ is in the opposite direction of $\mathbf{E}$.

Two techniques are commonly used to calculate the electric field at a point in space. One technique involves a principle developed by the German mathematician and physicist Karl Friedrich Gauss (1777–1855). This principle, called

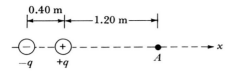

FIG. 22.15

Gauss' law, is beyond the scope of this book, although we will occasionally use the results of a Gauss' law determination of **E** in an example or problem.

A second method for determining the electric field at a point is illustrated below.

Calculating the Electric Field at a Point in Space

We wish to determine the electric field at point *A* in Fig. 22.15 caused by the dipole charges *q* and −*q*, each of magnitude 0.50×10^{-6} C.

(Step 1)

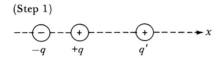

1. Place a very small, positive, "imaginary" test charge *q'* at the point where you wish to calculate the field. Notice the positive charge at point *A* in the accompanying figure. We need not specify the magnitude of *q'* as its value cancels out of our later calculations.

2. Construct a force diagram for the "imaginary" test charge. Include in the force diagram all electric forces acting on *q'*. The charges that cause the forces on *q'* are those producing the electric field at the point where *q'* is located. For the problem illustrated in the accompanying figure, both *q* and −*q* exert electric forces on *q'*.

(Step 2)

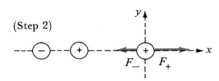

3. Use Coulomb's law to determine the magnitude of each force. The forces can be expressed in terms of *q'*, which is not specified.

For the problem illustrated here,

$$F_+ = \frac{kqq'}{r^2} = \frac{(9.0 \times 10^9 \, \text{N} \cdot \text{m}^2/\text{C}^2)(0.50 \times 10^{-6} \, \text{C})q'}{(1.20 \, \text{m})^2}$$
$$= (3130q') \, \text{N/C},$$

$$F_- = \frac{kqq'}{r^2} = \frac{(9.0 \times 10^9 \, \text{N} \cdot \text{m}^2/\text{C}^2)(0.50 \times 10^{-6} \, \text{C})q'}{(1.60 \, \text{m})^2}$$
$$= (1760q') \, \text{N/C}$$

4. Use vector addition to determine the resultant electric force acting on *q'*. The resultant force will also be expressed in terms of *q'*.

For forces that lie along a line, their magnitudes can simply be added if they point in the same direction or subracted if they point in opposite directions (as in the example illustrated above.) For forces that do not all point along the same line, vector addition by components must be used (see the chapter summary on page 15.)

Since **F**₊ points right in the positive *x* direction and **F**₋ points left in the negative *x* direction, the resultant also points in the *x* direction and equals

$$R = F_+ - F_-$$
$$= (3130q' - 1760q') \, \text{N/C}$$
$$= 1370q' \, \text{N/C}.$$

The resultant points right since F_+ has greater magnitude than F_-.

5. The electric field **E** at the point equals the resultant electric force acting on the positive test charge *q'* at that point divided by *q'*. Because we divide by

q', the electric field depends not on the test charge but only on the other charges that set up the field. The test charge was just a device to help us calculate the field at that point. Notice that q' cancelled out of the expression for the electric field calculated here. Since electric field is a vector quantity, we must specify both its magnitude and direction.

$$E = \frac{R}{q'} = \frac{1370\,q'\,\text{N/C}}{q'} = \underline{1370\,\text{N/C}}$$

and points in the same direction as the resultant force on the positive charge q', that is, <u>toward the right</u>.

As we mentioned earlier, electric fields are often determined in more advanced treatments by using Gauss' law. One result of the use of Gauss' law is illustrated in Fig. 22.16. Two large parallel plates separated by a distance d hold opposite charges, $+q$ distributed uniformly on the top plate and $-q$ on the bottom. If the area of each plate is A, then the charge per unit area on the positively charged plate is

$$\sigma = \text{Charge per unit area} = \frac{q}{A}.$$

Using Gauss' law, we would find that the electric field between the plates is given by

$$E = 4\pi k\sigma = 4\pi k\left(\frac{q}{A}\right) \tag{22.5}$$

and points down from the positively charged plate toward the negatively charged plate.

Charged plates such as shown in the figure are used in oscilloscopes and television sets to deflect electron beams. The next example illustrates this process.

EXAMPLE 22.6 An electron moves toward the screen of an oscilloscope (Fig. 22.16). On its way, it passes between two parallel plates. The upper plate has a positive charge of 1.2×10^{-10} C, and the lower plate has an equal-magnitude negative charge. The plates are 0.10 m wide and 0.20 m long. (a) Calculate the electric field between the plates, (b) the force on an electron while moving between the plates, and (c) the vertical component of the electron's velocity as it leaves the plates. Its initial vertical component of velocity is zero, and it spends 2.1×10^{-8} s while traveling horizontally between the plates.

SOLUTION (a) The area of each plate is $(0.10\,\text{m})(0.20\,\text{m}) = 2.0 \times 10^{-2}\,\text{m}^2$. The <u>electric field points down</u> from the positive toward the negative plate and has magnitude given by Eq. (22.5):

$$\begin{aligned}
E &= 4\pi k\sigma = 4\pi k(q/A) \\
&= 4\pi(9.0 \times 10^9\,\text{N}\cdot\text{m}^2/\text{C}^2)(1.2 \times 10^{-10}\,\text{C})/(2.0 \times 10^{-2}\,\text{m}^2) \\
&= \underline{680\,\text{N/C}.}
\end{aligned}$$

(b) The force on an electron between the plates is determined using Eq. (22.4). Because the electron has a negative charge of magnitude 1.6×10^{-19} C, the force on the electron is opposite the direction of the electric field (the <u>force points</u>

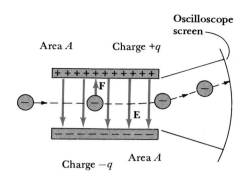

FIG. 22.16. A downward electric field created by opposite-sign charges on two parallel plates causes an upward force on an electron.

up). The magnitude of the force is

$$F = qE = (1.6 \times 10^{-19}\,\text{C})(680\,\text{N/C}) = \underline{1.1 \times 10^{-16}\,\text{N}}.$$

(c) The electron's motion through the field is much like that of a projectile (see Section 2.9). No forces act on the electron in the horizontal direction. Consequently, its horizontal acceleration is zero, and it moves at constant horizontal speed between the plates. However, the upward force due to the electric field causes the electron to accelerate upward (much like a bale of hay accelerates downward when dropped from a plane in the earth's gravitational field). The electron also feels a gravitational force, but the gravitational force on an electron is usually very small compared to the electric force and will be ignored.

According to Newton's second law, the acceleration in the vertical direction is

$$a_y = \frac{\Sigma\,F_y}{m} = \frac{(qE)}{m},$$

where m = mass of electron = 9.11×10^{-31} kg (see the Table of Constants on the front flyleaf). Using Eq. (2.9y) from kinematics with $v_{0_y} = 0$, we can now determine the vertical speed of the electron as it leaves the region between the plates 2.1×10^{-8} s after entering the region:

$$v_y = v_{y_0} + a_y t = 0 + \frac{(qE)}{m}t$$

$$= \frac{(1.1 \times 10^{-16}\,\text{N})}{(9.11 \times 10^{-31}\,\text{kg})}(2.1 \times 10^{-8}\,\text{s})$$

$$= \underline{2.5 \times 10^{6}\,\text{m/s}}.$$

Now, besides the horizontal velocity that the electron maintains while crossing the plates, it also has a vertical component of velocity as it leaves the plates. An electron beam can easily be deflected up or down or to the sides if relatively small charges of opposite signs are placed on parallel plates, as illustrated in Fig. 22.16. ∎

22.7 Lines of Force

Electric fields can be represented by drawing what are called lines of force. Lines of force were first used by an English scientist, Michael Faraday (1771–1867). Faraday, who had to leave school at thirteen to help support his destitute family, was self-educated and had no formal training in mathematics. Because of his lack of mathematical expertise, Faraday developed the notion of lines of force to help himself visualize the effects of electric and magnetic fields.

A **line of force** is a line drawn such that its direction at any point is in the direction of the electric field at that point. The separation of neighboring lines indicates the magnitude of the field in that region. If the lines are close to each other, the electric field in that region is relatively strong; if the lines are far apart, the field is weak. Lines of force always originate on a positive charge and end on a negative charge.

Several examples of lines of force are shown in Fig. 22.17. In Fig. 22.17a, the lines leaving a single positive charge are shown. A small positive charge

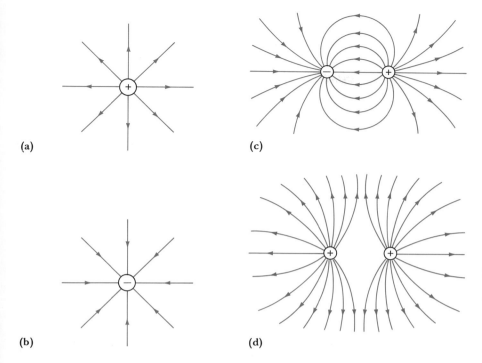

(a)

(b)

(c)

(d)

FIG. 22.17. Lines of force.

placed on one of the lines would feel an electric force in the direction of the line. A negative charge would feel a force in the opposite direction of the line.

Lines moving in toward a single negative charge are shown in part (b). In part (c), the lines of force caused by a dipole charge move from the positive to the negative charge. In part (d), the lines of force leave two positive charges. The negative charges on which the lines end are far away and are not shown in the drawing.

The young woman shown in Fig. 22.18 has a large negative charge on her body and hair. The electric field (and lines of force caused by this charge) points in toward her head as in Fig. 22.17b. The force on the excess negative charge on her hair is opposite the direction of the electric field and causes her hair to point out away from her head.

FIG. 22.18. Lines of force along Jodi's hair. (Photograph by John Womack)

Summary and Additional Readings

1. **Coulomb's law:** A proton has a positive electric charge e equal to 1.6×10^{-19} C; an electron has a negative charge $-e$. The magnitude of the force between a charge q_1 and a charge q_2 when separated by a distance r is given by Coulomb's law:

$$F = \frac{k q_1 q_2}{K r^2}, \qquad (22.2)$$

where $k = 9.0 \times 10^9$ N·m²/C² and K is the dielectric constant of the material in which the charges are immersed. If the charges are in a vacuum or in dry air, $K = 1$; if in water, $K = 80$. The direction of the electrical force between two charges depends on the signs of the charges; like charges repel and unlike charges attract.

2. **Multiple charges:** When several electric charges exert forces on one particular charge, the resultant electrical force is the vector sum of all the forces:

$$\mathbf{R} = \mathbf{F}_1 + \mathbf{F}_2 + \mathbf{F}_3 + \cdots.$$

The forces can be added graphically or by using the component technique (both techniques are explained in detail in Chapter 1):

$$R_x = F_{1x} + F_{2x} + F_{3x} + \cdots, \qquad (1.5)$$
$$R_y = F_{1y} + F_{2y} + F_{3y} + \cdots, \qquad (1.6)$$
$$R = \sqrt{R_x^2 + R_y^2}, \qquad (1.7)$$
$$\tan \theta = |R_y / R_x|. \qquad (1.8)$$

3. **Electric dipole:** An electric dipole is a positive charge q and a negative charge $-q$ of equal magnitude that are separated by a distance a. The **electric dipole moment p** points from $-q$ to q and has a magnitude

$$p = qa. \qquad (22.3)$$

4. **Electric field:** The electric field **E** at a point equals the electric force **F** acting on a small, positive test charge q' placed at that point, divided by the value of the charge:

$$\mathbf{E} = \frac{\mathbf{F}}{q'}. \qquad (22.4)$$

If a charge q is placed in a known electric field **E**, the force on the charge is determined by rearranging Eq. (22.4):

$$\mathbf{F} = q\mathbf{E}.$$

Daniel J. Kevles, "Robert A. Millikan," *Scientific American*, pp 142–151, January (1979). A fascinating story about the triumphs and failures of the 1923 Nobel laureate who first measured the charge of an electron.

J. L. Heibron, "Franklin's Physics," *Physics Today* **29**, 32 (1976).

A. D. Moore, "Electrostatics," *Scientific American*, p 47, March (1972). Describes the modern applications of static electricity.

Questions

1. "In all processes that occur in nature, electric charge is conserved." Explain what is meant by this statement and describe one particular example.

2. After a snowstorm or windstorm, a wire fence in the Plains states may occasionally give an intense shock to a person moving near it. Why might this happen?

3. An object acquires a positive electric charge. Does its mass increase, decrease, or remain the same? Explain. How is its mass affected if it acquires a negative electric charge?

4. At one time it was thought that electric charge was a weightless fluid. An excess of this fluid resulted in a positive charge; a deficiency resulted in a negative charge. Describe an experiment for which this hypothesis provides a satisfactory explanation. Describe an experiment that is not explained by this hypothesis.

5. The electrical force of one electric charge on another is proportional to the products of their charges—that is, $F_q \propto q_1 q_2$. Invent an experiment to show that the force is not proportional to the sum of their charges ($q_1 + q_2$).

6. An electric dipole such as a water molecule is in a constant electric field. Will the force of the field cause the dipole to have a linear acceleration along a line in the direction of **E**? Explain. Will the field exert a torque on the dipole? Explain.

7. A dipole charge, such as pictured in Fig. 22.19, exists on your heart. A circular wire is placed on your chest. Where on the wire will positive charge accumulate and where will negative charge accumulate because of the force exerted on the charge by the heart's dipole? Explain.

FIG. 22.19

8. A very small, positive charge when placed at one point in space has an electric force of magnitude zero acting on it. (a) What must the value of the electric field be at that point? (b) Does this mean there are no other electric charges nearby? Explain. (c) Think of at least one charge distribution (two or more other charges) that would produce a zero electric field at the point.

9. A metal sphere has no charge on it. A positive charge is brought near, but does not touch, the sphere. Show that the positive charge can exert a force on the sphere even though the sphere has no net charge. [*Hint:* Electrons move about freely on the surface of the metal.]

Problems

In some of these problems you may need to know that the mass of an electron is 9.11×10^{-31} kg.

22.1 Electric Charge

1. (a) The earth has an excess of 6.0×10^5 electrons on each square centimeter of surface. Calculate the electric charge in coulombs on each square centimeter of surface. (b) As you walk across a rug, 9.0×10^{-23} kg of electrons transfer to your body. Calculate the number of electrons and the total charge in coulombs on your body.

■ 2. (a) A drop of rain accumulates a charge of -3.2×10^{-18} C. By how much does its mass increase? (b) A person standing in the wind loses 1.0×10^{-21} kg of electrons. Calculate the charge on the person after the electrons are lost.

■ 3. Sodium chloride (table salt) consists of sodium ions (charge e) arranged in a crystal lattice with an equal number of chlorine ions (charge $-e$). The mass of each sodium ion is 3.82×10^{-26} kg and of each chlorine ion 5.89×10^{-26} kg. Suppose that the sodium ions could be separated into one pile and the chlorine ions in another. What mass of salt would be needed to get 1.00 C of charge in the sodium ion pile and -1.00 C in the chlorine ion pile?

22.2 Coulomb's Law

4. The membrane of a body cell has a positive ion of charge $+e$ on the outside wall and a negative ion of charge $-e$ on the inside wall. Calculate the magnitude of the electrical force between these ions if the membrane thickness is 0.80×10^{-9} m.

5. Calculate the electrical force between two protons in the nucleus of a helium atom when separated by 2.0×10^{-15} m.

6. A cloud with a net charge of -12.0 C exerts an attractive electrical force of 7.7×10^{-3} N on a person 1500 m below. Calculate the magnitude and sign of the charge on the person.

■ 7. (a) A man and a woman each with no net charge are separated by 10 m. What negative charge must be transferred from the woman to the man to cause an attractive force of 200 N? (b) Calculate the mass of the transferred charge.

■ 8. A man and a woman each contain 4.0×10^{28} electrons and an equal number of protons. Calculate the electrical force between them if 1 percent of the woman's electrons are transferred to the man, who is 100 m away.

■ 9. Two metal spheres are held 1.6 m apart by a board that fractures when compressed by a force greater than 3.0×10^4 N. Each sphere starts with 10^{27} electrons and 10^{27} protons. Calculate the fraction of electrons that must be moved from one sphere to the other to fracture the board.

■ 10. Two charges exert a 4.0-N force on each other when separated by 1.0 m. Indicate the magnitude of the force when (a) the separation is doubled, (b) the separation is reduced by one-half, (c) the magnitude of one charge is reduced by one-half, and (d) both charges are doubled.

■ 11. Changes q_1 and q_2 exert 20-N repulsive forces on each other. Calculate the repulsive force of one charge on the other when their separation is increased by a factor of 1.4 while the magnitude of each charge is decreased by a factor of 0.80.

■ 12. Calculate the number of electrons that must be transferred from the earth to the moon so that the electrical attraction between them is equal in magnitude to their present gravitational attraction. What is the mass of this number of electrons?

■ 13. Suppose that the electron in a hydrogen atom moves in a circular orbit of radius 0.53×10^{-10} m about the proton nucleus. (a) Calculate the magnitude of the electrical force between the charges. (b) Calculate the electron's centripetal acceleration and speed.

■■ 14. Two identical metal spheres are separated by 1.00×10^5 m. The spheres exert an 18.9-N repulsive force on each other. A wire is then connected between the spheres so that they become equally charged. The spheres now repel each other with a 22.5-N force. Calculate the original charges on the spheres.

■■ 15. Suppose that the earth and moon initially have zero charge. Then 1000 kg of electrons are transferred from the earth to the moon. Calculate the radius of a stable moon orbit when both the electrical and gravitational forces of attraction are acting on the moon and it completes one rotation about the earth in 29.5 days.

22.3 Static Electrical Forces

■ 16. Three 1.0-C charges are equally spaced on a straight line. The separation of each charge from its neighbor is 100 m. Find the force on the center charge if (a) all charges are positive, (b) all charges are negative, and (c) the rightmost charge is negative and the other two positive.

■ 17. Three charges are placed in a row: a -3.0-C charge on the left, a $+2.0$-C charge in the middle, and a $+6.0$-C charge on the right. The distance separating each charge is 1.0×10^4 m. Calculate the electrical force (magnitude and direction) on the $+2.0$-C charge due to the other two charges.

■ 18. Four ions (Na$^+$, Cl$^-$, Na$^+$, and Cl$^-$) each separated from its neighbors by 3.0×10^{-10} m are in a row. The charge of a sodium ion is $+e$ and that of a chlorine ion is $-e$. Calculate the force on the chlorine ion at the end of the row due to the other three ions.

■ 19. A 1.0-C charge and a 2.0-C charge are separated by 100 m. Where should a charge of $-(1.0 \times 10^{-3})$ C be located on a line between the positive charges so that the net electrical force on the negative charge is zero?

■■ 20. Two charges q and $4q$ are separated by 1.0 m. Determine the sign, magnitude, and position of a third charge that causes all three charges to remain in equilibrium.

■ 21. Calculate the magnitude and direction of the resultant force caused by charges q_1 and q_3 on charge q_2 (see Fig. 22.20).

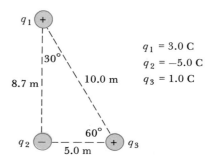

$q_1 = 3.0$ C
$q_2 = -5.0$ C
$q_3 = 1.0$ C

FIG. 22.20

■ 22. Calculate the magnitude and direction of the resultant force caused by charges q_2 and q_3 on charge q_1 (see Fig. 22.20).

■ 23. Three $+2.0$-C charges are placed in a vertical column each separated by a distance of 3.0×10^4 m from their neighbor. Calculate the electrical force caused by these three positive charges on a -1.0-C charge placed 4.0×10^4 m in the horizontal direction to the right of the center positive charge.

■ 24. A triangle with equal sides of length 1.0×10^3 m has a -2.0-C charge at each corner. Calculate the electrical force (magnitude and direction) acting on the charge at the top corner due to the two charges at the base of the triangle.

■■ 25. Four charges of 1.0×10^{-4} C are located at the corners of a square whose sides are 2.0 m long. Calculate the net electrical force on the charge at the lower left corner due to the other three charges.

22.4 Dielectric Constant

26. Calculate the force between a sodium ion (charge e) and a chlorine ion (charge $-e$) when separated by 2.8×10^{-10} m in air and when separated by the same distance in water.

27. Repeat Problem 4 using the fact that the dielectric constant of a cell membrane is 8.

■ 28. An African fish called the *Gymnarchus niloticus* has a positive charge q of magnitude 1.0×10^{-7} C at its head and a negative charge $-q$ at its tail (see Fig. 22.21). Calculate the magnitude and direction of the electrical force on a hydroxide ion (charge $-e$) at position A. Remember that the fish and ion are in water.

■ 29. Two small metal spheres are attached to the ends of a thin, 1.6-m-long rod that fractures if compressed by a force greater

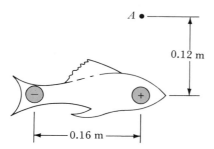

FIG. 22.21

0.12 m

0.16 m

than 640 N. If the spheres and rod are immersed in benzene, what is the maximum charge that can be transferred from one sphere to the other before the rod fractures?

■ 30. A positive charge and a negative charge are separated by 0.10 m in water. Use a ratio technique to calculate the distance that the charges must be separated in ethanol so that the force between them is the same as when they were in water.

22.5 Electric Dipoles

■ 31. The heart has a dipole charge distribution with a charge of $+1.0 \times 10^{-7}$ C 6.0 cm above a charge of -1.0×10^{-7} C. Calculate the electrical force (magnitude and direction) caused by the heart's dipole on a sodium ion (charge 1.6×10^{-19} C) that is in the body tissue of dielectric constant 7 and lies along a line with the heart's charges a distance of 8.0 cm above the heart's positive charge.

■ 32. An electric dipole with $q = 2.0$ C and $a = 80$ m is oriented horizontally with the positive charge on the right in a medium of dielectric constant 10. Calculate the force (magnitude and direction) on a charge of -4.0×10^{-4} C (also in the dielectric medium) that is 60 m to the left of the negative dipole charge.

■ 33. Calculate the force on a sodium ion (charge e) at position A in Fig. 22.22 due to the dipole charges produced by a geologist's electrodes. The dielectric constant of the soil is 8.0 and $q = 4.0$ C.

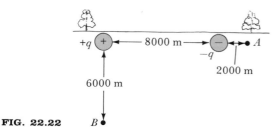

$+q$ 8000 m $-q$ A

6000 m

2000 m

FIG. 22.22 B

■ 34. Repeat Problem 33 but for a sodium ion at position B in Fig. 22.22.

■■ 35. A shuttle train moves from one station to another. It is powered by equal-magnitude, opposite-sign charges on top of each station (see Fig. 22.23). If a 7000-N friction force opposes the

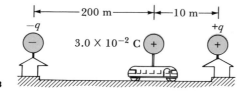

200 m   10 m

$-q$ $+q$

3.0×10^{-2} C

FIG. 22.23

train's initial motion, how great must the dipole charge be to induce an initial acceleration of 1.0 m/s²? The train's mass is 2.0×10^4 kg, and it has a positive charge of 3.0×10^{-2} C on its roof.

22.6 Electric Fields

36. Calculate the magnitude and direction of the electric field (a) 0.40 m from a charge of $+(1.0 \times 10^{-3})$ C; (b) 0.80 m from a charge of $+(1.0 \times 10^{-3})$ C; (c) 0.40 m from a charge of $-(1.0 \times 10^{-3})$ C; and (d) 0.80 m from a charge of $-(1.0 \times 10^{-3})$ C.

37. A uranium nucleus has 92 protons. (a) Calculate the magnitude of the electric field at a distance of 0.58×10^{-12} m from the nucleus (about the radius of the innermost electron orbiting the nucleus). (b) What is the magnitude of the force on an electron due to this electric field?

■ 38. Two horizontal parallel plates, one above the other, each have an area of 2.0 m² and opposite charges of magnitude 4.0×10^{-4} C. (a) Calculate the magnitude and direction of the electric field between the plates if the bottom plate is positively charged. (b) What is the force (magnitude and direction) on an electron between the plates?

■ 39. A +2.0-C charge is 1.0×10^4 m to the right of a −4.0-C charge. Calculate the electric field (magnitude and direction) at a point 2.0×10^4 m to the right of the positive charge and along a line passing through the two charges.

■ 40. A +4.0-C charge is 400 m along a horizontal line toward the right of a −3.0-C charge. Calculate the electric field at a point 300 m directly above the negative charge.

■ 41. Calculate the electric field (magnitude and direction) needed to support the weight of a metal sphere of mass 0.10 kg that has a charge of -1.2×10^{-5} C.

■ 42. (a) Calculate the magnitude of an electric field that will accelerate an electron at 4.5×10^{12} m/s². (b) Calculate the electron's speed after traveling 100 m (it starts at rest).

■■ 43. Four 1.0-C charges are placed at the corners of a square whose sides are 50 m long. (a) Calculate the electric field at the center of the square (if you wish, you may use a verbal symmetry argument to justify your answer). (b) Repeat the calculations in part (a), but for a square with two negative 1.0-C charges at the top and two positive 1.0-C charges at the bottom.

■■ 44. An electron passes between two parallel plates as shown in Fig. 22.16. A 0.20-N/C electric field points down. The electron moves a horizontal distance of 4.0 cm in the field and has an initial velocity in the horizontal direction of 5.0×10^4 m/s. Calculate the change in vertical position of the electron as it leaves the 4.0-cm horizontal region.

■■ 45. Calculate the electrical torque on a water molecule about a point at its center when the molecule is oriented perpendicular to a 500-N/C electric field. Consider the molecule as a charge $+e$ and another charge $-e$ separated by a distance a equal to 0.38×10^{-10} m.

■■ 46. An electron is projected with a horizontal velocity of magnitude v_0 into a constant downward electric field E between two parallel plates. Show that the equation of its trajectory while in the field is

$$y = \left(\frac{eE}{2mv_0^2}\right) x^2,$$

where m is the electron's mass and y is its vertical displacement when its horizontal displacement in the field is x.

■ ■ **47.** An atom normally has no electric dipole moment. However, when it is placed in an electric field, the nucleus is displaced in the direction of the field and the electrons are displaced opposite the direction of the field. The induced electric dipole moment is **p** = α**E**, where α is called the **polarizability** of the atom. An atom with polarizability of 1.7×10^{-41} C$^2 \cdot$m/N is placed 6.0×10^{-8} m from a charge $+e$. Calculate the induced electric dipole moment.

22.7 Lines of Force

■ **48.** (a) Indicate approximately the direction of the net electrical force on a small, positive charge placed at positions A, B, and C in Fig. 22.24. (b) Continue along the directions of these vector arrows and draw lines of force in the directions they are expected to move beyond points A, B, and C.

■ **49.** Repeat the procedure outlined in Problem 48 for the charges shown in Fig. 22.25.

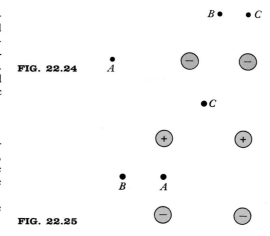

FIG. 22.24

FIG. 22.25

CHAPTER 23

Electrical Energy

In the last chapter we looked at the force that one or more electric charges exert on another charge. These electrical forces, along with other types of forces, can be used in Newton's second law to calculate the acceleration of an object $[\mathbf{a} = (\Sigma\,\mathbf{F})/m]$. The force-acceleration approach to solving problems was introduced and used in Part I of the book but is seldom used in problems dealing with electricity. Instead, the energy approach introduced in Chapter 8 is preferred.

The energy problem-solving technique is based on the conservation-of-energy principle, $Q + W = E - E_0 = \Delta E_{\text{system}}$, where Q and W represent energy added to or removed from a system by heat transfer and work, respectively. The expression ΔE_{system} represents the sum of the changes in the many different forms of energy (kinetic, gravitational potential, elastic potential, and internal) that a system of particles can have. In this chapter we introduce another form of energy to include with ΔE_{system}—the change in electrical potential energy that occurs when charged particles move relative to each other. After learning how to calculate the change in electrical potential energy of a system, we will use the energy approach to solve problems involving electric charges.

23.1 Electrical Potential Energy (PE_q)

In Chapter 8, we found that the amount of potential energy in a system always depends on the relative position of the parts of that system. For example, a mass m and the earth have more gravitational potential energy (PE_g) when separated by several meters than when resting against each other, and a spring has more elastic potential energy (PE_s) when its coils are stretched or compressed than when relaxed. Similarly, two or more electric charges possess electrical potential energy (PE_q), and the amount depends on their positions relative to each other.

Consider the cannon in Fig. 23.1a. Two positive charges are held near each other in the barrel of the cannon. When the trigger is released, the positively charged cannonball is repelled from the other positive charge, causing it to fly out the end of the barrel (Fig. 23.1b). The initial electrical potential energy (PE_q in Fig. 23.1a) is converted to kinetic energy (KE) and gravitational potential energy (PE_g) as the charges move apart. The situation shown in part (a) is

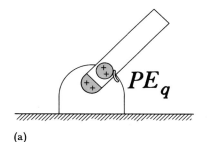

(a)

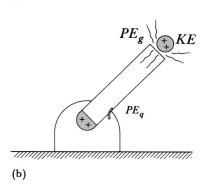

(b)

FIG. 23.1. (a) Two positive charges have considerable electrical potential energy when held near each other—like a compressed spring. (b) The electrical energy is converted to other energy forms as the charges move apart.

similar to that of a compressed spring; instead of the coils of the spring being squeezed together, two like charges are squeezed or forced together. The *electrical potential energy of like charges increases as they are forced nearer each other and decreases as they move apart.*

Next, imagine the electrical potential energy of the two opposite-sign charges used as a nutcracker in Fig. 23.2. When separated by considerable distance, as in Fig. 23.2a, the opposite charges have more electrical potential energy. When released, the negatively charged block rushes toward the positive charge (Fig. 23.2b). The kinetic energy of the block increases, and the electrical potential energy of the charge decreases. We see that *the electrical potential energy of unlike charges increases as they are pulled apart and decreases as they move together.*

To derive an expression for the change in electrical potential energy of two electric charges whose separation changes, let us return to the conservation-of-energy principle:

$$Q + W = \Delta E_{\text{system}}, \tag{8.8}$$

where Q is the heat flow into or out of the system and W is the work done on objects in the system by forces that act from outside the system. The system's energy can change in many ways:

$$Q + W = \overbrace{\Delta KE + \Delta PE_g + \Delta PE_s + \Delta PE_q + \Delta E_{\text{int}} + \cdots}^{\Delta E_{\text{system}}}, \tag{8.9}$$

where ΔKE is the system's change in kinetic energy; ΔPE_g, ΔPE_s, and ΔPE_q are its changes in gravitational, elastic, and electrical potential energies, respectively; and ΔE_{int} is its change in internal energy.

To derive an expression for calculating ΔPE_q, the change in electrical potential energy, we will choose a hypothetical system so that when work is done on this system only its electrical potential energy changes. The system shown in Fig. 23.3 consists only of charges q_1 and q_2, initially separated by a distance r_0. The like charges repel each other with an electrical force whose magnitude is given by Eq. (22.1): $F = kq_1q_2/r_0^2$. A person can prevent charge q_1 from moving

(a)

(b)

FIG. 23.2. (a) Opposite charges have considerable electrical potential energy when held far apart—like a stretched spring. (b) The electrical energy is converted to other forms as the charges move closer.

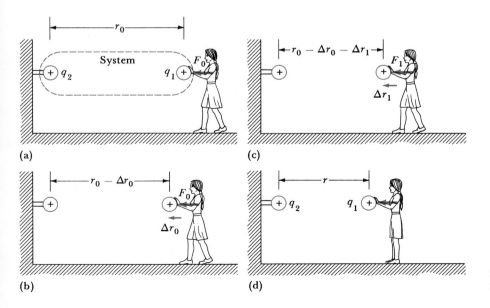

(a)

(b)

(c)

(d)

FIG. 23.3. A person does work in pushing two like charges closer together.

away from q_2 by pushing q_1 with an equal-magnitude force $\mathbf{F}_0$ toward q_2 (Fig. 23.3a).

If the person pushes q_1 just a tiny bit harder, the charge can be displaced a small distance Δr_0 toward q_2 (Fig. 23.3b). The work ΔW_0 done by the pushing force during this small displacement can be calculated using Eq. (8.1):

$$\Delta W_0 \simeq F_0 \Delta r_0 = \frac{kq_1 q_2}{r_0^2} \Delta r_0.$$

The calculation is approximate because the force needed to push q_1 changes as its separation from q_2 changes. If Δr_0 is small, the equation is a good approximation for the work done during that small displacement. If we wish to push the charge farther, however, more force must be exerted to move q_1 because it is now closer to q_2. The work done during the next small displacement Δr_1, shown in Fig. 23.3c, is approximately

$$\Delta W_1 \simeq F_1 \Delta r_1 = \frac{kq_1 q_2}{(r_0 - \Delta r_0)^2} \Delta r_1.$$

For each step closer, the amount of work needed to move q_1 closer to q_2 increases.

The total work done in moving the charge from an initial separation r_0 to a final separation r (Fig. 23.3d) equals the sum of the work for each small step. This type of infinitesimal addition is done easily using calculus. The results of such a calculation are

$$W = \Delta W_1 + \Delta W_2 + \Delta W_3 + \cdots = \frac{kq_1 q_2}{r} - \frac{kq_1 q_2}{r_0} = kq_1 q_2 \left(\frac{1}{r} - \frac{1}{r_0} \right).$$

At each step of the process represented in Fig. 23.3, there is no acceleration and, therefore, no change in kinetic energy; nor are there changes in gravitational, elastic, or internal energies. Also no heat flows into or out of the system. Only the electrical potential energy changes because the charges are separated by less distance. By substituting the above expression for W into Eq. (8.9), and zeros for all energy changes except the electrical potential energy change, we find

$$0 + kq_1 q_2 \left(\frac{1}{r} - \frac{1}{r_0} \right) = 0 + 0 + 0 + \Delta PE_q + 0 + \cdots.$$

This equation applies to any situation in which two charges in a system move closer to each other or farther apart.

The change in **electrical potential energy** ΔPE_q when the separation of two charges q_1 and q_2 is changed from an initial separation r_0 to a final separation r is given by the equation

$$\Delta PE_q = PE_q - PE_{q_0} = kq_1 q_2 \left(\frac{1}{r} - \frac{1}{r_0} \right). \tag{23.1}$$

This equation applies for charges in a vacuum. If the charges are in a medium with dielectric constant K, the force between the charges is reduced by a factor $1/K$. The change in electrical potential energy is then

$$\Delta PE_q = PE_q - PE_{q_0} = \frac{kq_1 q_2}{K} \left(\frac{1}{r} - \frac{1}{r_0} \right). \tag{23.2}$$

Electrical potential energy is measured in units of joules.

Equations (23.1) and (23.2) can be used for both positive and negative charges; *the sign of the charges must be included in calculations involving these two equations.* Notice that electrical potential energy is proportional to $1/r$ and not to $1/r^2$ as in Coulomb's force law.

EXAMPLE 23.1 Suppose the electric nutcracker shown in Fig. 23.2a has a stationary positive charge of $+5.0 \times 10^{-5}$ C initially separated 0.40 m from a negatively charged block with charge -2.0×10^{-5} C. When the block is released, it accelerates toward the positive charge and the nut. (a) Calculate the change in electrical potential energy when the negatively charged block moves to 0.10 m from the positive block. (b) If all of the lost electrical potential energy from part (a) is converted to kinetic energy of the 0.50-kg block, how fast is it moving when 0.10 m from the positive charge, assuming that the block starts at rest?

SOLUTION (a) If we call the positive charge q_1 and the negative charge q_2, then $q_1 = +5.0 \times 10^{-5}$ C; $q_2 = -2.0 \times 10^{-5}$ C; r_0 equals the initial separation of the charges, 0.40 m; and r equals their final separation, 0.10 m. The dielectric constant K of the air surrounding the charges is 1.0. Using Eq. (23.2), we find that the change in electrical potential energy is

$$\Delta PE_q = \frac{kq_1q_2}{K}\left(\frac{1}{r} - \frac{1}{r_0}\right)$$

$$= \frac{(9.0 \times 10^9 \text{ N}\cdot\text{m}^2/\text{C}^2)(5.0 \times 10^{-5}\text{ C})(-2.0 \times 10^{-5}\text{ C})}{1.0}$$

$$\times \left(\frac{1}{0.10 \text{ m}} - \frac{1}{0.40 \text{ m}}\right)$$

$$= -68 \text{ N}\cdot\text{m} = \underline{-68 \text{ J}}.$$

Notice that the signs of the charges must be included in these calculations.

(b) The unlike charges lose 68 J of electrical potential energy as they move closer together—like the loss of elastic energy as a stretched spring relaxes. The lost electrical energy is converted into 68 J of kinetic energy. Using the expression for kinetic energy [Eq. (8.4)] with initial speed zero, we find that the block's final kinetic energy is

$$\frac{1}{2}mv^2 = 68 \text{ J}$$

and the block's final speed is

$$v = \sqrt{\frac{2(68 \text{ J})}{m}} = \sqrt{\frac{2(68 \text{ J})}{0.50 \text{ kg}}} = \sqrt{272 \text{ m}^2/\text{s}^2}$$

$$= \underline{16 \text{ m/s}}. *$$ ∎

EXAMPLE 23.2 A sodium ion (charge e) and chlorine ion (charge $-e$) are initially separated by a distance r_0 equal to 3.5×10^{-10} m. When they are dissolved in water, their separation r increases to 100×10^{-10} m. Compare the

*Note that 1 J/kg = 1 N·m/kg = 1 (kg·m/s^2)m/kg = 1 m^2/s^2.

changes in electrical potential energy required for this change in separation when (a) in air ($K = 1.0$) and (b) in water ($K = 80$).

SOLUTION (a) Using Eq. (23.2), we find that the change in electrical potential energy when in air is

$$\Delta PE_q = \frac{kq_1q_2}{K}\left(\frac{1}{r} - \frac{1}{r_0}\right)$$

$$= \frac{(9.0 \times 10^9\,\text{N}\cdot\text{m}^2/\text{C}^2)(1.6 \times 10^{-19}\,\text{C})(-1.6 \times 10^{-19}\,\text{C})}{1.0}$$

$$\times \left(\frac{1}{100 \times 10^{-10}\,\text{m}} - \frac{1}{3.5 \times 10^{-10}\,\text{m}}\right)$$

$$= \underline{640 \times 10^{-21}\,\text{J}}.$$

For those of you with some background in chemistry, 640×10^{-21} J is equivalent to 100 kcal/mole—a large amount of energy on the atomic and molecular level. Sodium and chlorine ions are not easily separated when in air.

(b) To remove a sodium ion from a chlorine ion over the same distance when the ions are in water, only 1/80 as much energy is needed. (The dielectric constant of water is 80 whereas that of air is 1.0). Thus, the electrical potential energy change when in water is

$$\Delta PE_q = \frac{1}{80}(640 \times 10^{-21}\,\text{J}) = \underline{8.0 \times 10^{-21}\,\text{J}},$$

or about 1 kcal/mole—a small energy change on the atomic and molecular level. Table salt (NaCl) dissolves (separates into ions) easily when in water.

◼

23.2 Electrical Potential Energy Involving Multiple Charges

If one particular electric charge q' moves relative to several other stationary charges, the total change in electrical potential energy is the sum of the changes caused by each charge. For example, charge q', pictured in Fig. 23.4a, is initially separated from three other charges, q_1, q_2, and q_3, by distances r_{1_0}, r_{2_0}, and r_{3_0}, respectively. When q' is moved to a new location (Fig. 23.4b), its separations from these charges are now r_1, r_2, and r_3. The total change in electrical potential energy is the sum of the changes in energy caused by each charge:

$$\Delta PE_q = PE_q - PE_{q_0} = \frac{kq_1q'}{K}\left(\frac{1}{r_1} - \frac{1}{r_{1_0}}\right)$$

$$+ \frac{kq_2q'}{K}\left(\frac{1}{r_2} - \frac{1}{r_{2_0}}\right) + \frac{kq_3q'}{K}\left(\frac{1}{r_3} - \frac{1}{r_{3_0}}\right).$$

The preceding equation is usually written in an abbreviated form as

$$\Delta PE_q = PE_q - PE_{q_0} = \sum_{i=1}^{3} \frac{kq_iq'}{K}\left(\frac{1}{r_i} - \frac{1}{r_{i_0}}\right), \tag{23.3}$$

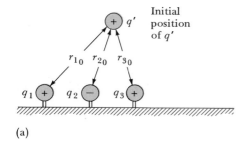

(a)

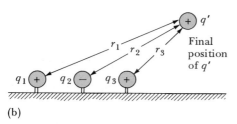

(b)

FIG. 23.4. A charge q' moves from (a) an initial position to (b) a final position.

where the symbol Σ (sigma) stands for the sum of the expression following it for each possible value of i. For the example illustrated in Fig. 23.4, i can assume three values ($i = 1$, 2, and 3); we add the changes in electrical potential energy that q' has because of its change in position relative to q_1, q_2, and q_3. If q' moves relative to five other charges, then the summation would include five terms.

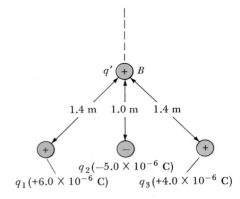

EXAMPLE 23.3 Calculate the change in electrical potential energy when a 10-μC charge (q') is brought from infinity to position B, shown in Fig. 23.5.

SOLUTION The 10-μC charge moves relative to three fixed charges. The known information is shown in Fig. 23.5. Because q' starts infinitely far from all three charges, $r_{1_0} = r_{2_0} = r_{3_0} = \infty$. The change in electrical potential energy is calculated using Eq. (23.3):

$$\Delta PE_q = kq_1q'\left(\frac{1}{r_1} - \frac{1}{\infty}\right) + kq_2q'\left(\frac{1}{r_2} - \frac{1}{\infty}\right) + kq_3q'\left(\frac{1}{r_3} - \frac{1}{\infty}\right)$$

$$= \frac{9.0 \times 10^9(6.0 \times 10^{-6})(10 \times 10^{-6})}{1.4}$$

$$+ \frac{9.0 \times 10^9(-5.0 \times 10^{-6})(10 \times 10^{-6})}{1.0}$$

$$+ \frac{9.0 \times 10^9(4.0 \times 10^{-6})(10 \times 10^{-6})}{1.4}$$

$$= 0.39 - 0.45 + 0.26 = \underline{0.20 \text{ J}}.$$

FIG. 23.5

Notice that when the positive charge q' moved closer to the negative charge q_2, a -0.45-J energy change occurred; whenever opposite charges move closer, the electrical potential energy decreases. This decrease was more than balanced by the positive increases in PE_q as q' moved closer to q_1 and q_3 (PE_q increases as like charges move closer). Since the net electrical potential energy increased, we can assume that some external agent, not pictured, did work to push q' to position B. ∎

EXAMPLE 23.4 A sodium ion (charge q' equal to $+e$) moves through body tissue from position A to position B, as shown in Fig. 23.6. Calculate the change in electrical potential energy of the sodium ion due to its change in position relative to the dipole charge on the heart ($q = 4.0 \times 10^{-6}$ C). The dielectric constant of the body is about 8.

SOLUTION In this example, Eq. (23.3) has two terms, one for the change in energy of the sodium ion relative to charge $+q$ and another for its change relative to $-q$. The sodium ion starts 0.15 m from charge $+q$ and 0.10 m from charge $-q$ and ends 0.10 m from $+q$ and 0.05 m from $-q$. The change in electrical potential energy is

$$\Delta PE_q = \frac{k(+q)q'}{K}\left(\frac{1}{r_+} - \frac{1}{r_{+_0}}\right) + \frac{k(-q)q'}{K}\left(\frac{1}{r_-} - \frac{1}{r_{-_0}}\right)$$

$$= \frac{(9.0 \times 10^9)(+4.0 \times 10^{-6})(1.6 \times 10^{-19})}{8}\left(\frac{1}{0.10} - \frac{1}{0.15}\right)$$

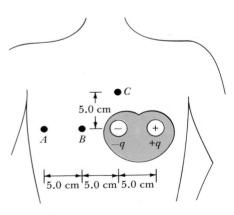

FIG. 23.6. A sodium ion moves through the body tissue.

$$+ \frac{(9.0 \times 10^9)(-4.0 \times 10^{-6})(1.6 \times 10^{-19})}{8} \left(\frac{1}{0.05} - \frac{1}{0.10} \right)$$

$$= (2.4 \times 10^{-15}\,\text{J}) - (7.2 \times 10^{-15}\,\text{J}) = \underline{-4.8 \times 10^{-15}\,\text{J}}.$$

In Example 23.4 we found that a sodium ion would have less electrical potential energy at position B than at A. In moving from A to B, the sodium ion resembles a rock rolling down a hill; the rock's gravitational potential energy decreases, as does the electrical potential energy of the sodium ion.

It is natural for systems to move toward states in which their potential energy is minimized. A rock rolls down a hill, not up. A sodium ion moves from A to B—that is, in the direction of decreasing electrical potential energy—rather than from B to A. This tendency can be explained in two ways: (1) Forces responsible for potential energy push objects in a direction that minimizes their potential energy. For example, the net electrical force on a sodium ion if at A in Fig. 23.6 is toward B and in the direction of decreasing potential energy. (2) Our analysis of entropy in Chapter 11 indicated that nature moves toward the most probable states. In Example 23.4, as the sodium ion moves, it collides with atoms and molecules along its path, causing thermal energy to be produced. The situation in which a large number of molecules have a small amount of thermal energy is more probable than one in which the sodium ion has all the energy in the form of electrical potential energy.

23.3 Conservation-of-Energy Calculations

Now that we have learned how to calculate changes in electrical potential energy, we will use the work-energy technique developed in Chapter 8 to solve some problems. We will ignore problems involving heat flow; thus, $Q = 0$. The conservation-of-energy equation with electrical potential energy included becomes

$$W = \Delta KE + \Delta PE_g + \Delta PE_s + \Delta PE_q + \Delta E_{\text{int}} + \cdots, \tag{8.9}$$

where

$$W = F\,\Delta r \cos\theta, \tag{8.1}$$

$$\Delta KE = \frac{1}{2}mv^2 - \frac{1}{2}mv_0^2, \tag{8.4}$$

$$\Delta PE_g = mg(y - y_0), \tag{8.6}$$

$$\Delta PE_s = \frac{1}{2}kx^2 - \frac{1}{2}kx_0^2, \tag{8.7}$$

$$\Delta PE_q = \sum_{i=1}^{N} kq_i q' \left(\frac{1}{r_i} - \frac{1}{r_{i_0}} \right), \tag{23.3}$$

$$\Delta E_{\text{int}} = F_k\,\Delta x \quad \text{(due to friction)}. \tag{8.15}$$

Review Chapter 8 for the notation or meaning of these different expressions. To use Eq. (8.9), we first calculate the work done on the system and the known energy changes as the system moves from an initial to a final state; then we substitute these known quantities into the conservation-of-energy equation to calculate one unknown quantity.*

*When dealing with electrical energy, it is customary and convenient to solve problems using the conservation-of-energy principle in the form of Eq. (8.9) rather than in the form of Eq. (8.11).

EXAMPLE 23.5 An electron (charge $-e$) starts at rest at an infinite distance r_0 from a stationary proton (charge e). Calculate the speed of the electron after it moves unimpeded to a distance r of 10^{-9} m from the proton. The mass of an electron is 9.11×10^{-31} kg.

SOLUTION In the problem there are no springs ($\Delta PE_s = 0$), no friction forces ($\Delta E_{int} = 0$), and no work is done ($W = 0$). The change in gravitational potential energy of the electron with respect to the proton is very small and can be ignored ($\Delta PE_g = 0$). We must account for only the following energy changes:

$$\Delta KE = \frac{1}{2}mv^2 - \frac{1}{2}mv_0^2 = \frac{1}{2}(9.11 \times 10^{-31} \text{ kg})v^2 - 0 \text{ (note that } v_0 = 0),$$

$$\Delta PE_q = kq_1q_2\left(\frac{1}{r} - \frac{1}{r_0}\right) = (9 \times 10^9 \text{ N·m}^2/\text{C}^2)(-1.6 \times 10^{-19} \text{ C})$$

$$\times (1.6 \times 10^{-19} \text{ C})\left(\frac{1}{10^{-9} \text{ m}} - \frac{1}{\infty}\right) = -2.3 \times 10^{-19} \text{ N·m}.$$

Substituting into the conservation-of-energy equation, we find that

$$0 = \frac{1}{2}(9.11 \times 10^{-31} \text{ kg})v^2 + 0 + 0 - 2.3 \times 10^{-19} \text{ N·m} + 0.$$

The electron and proton lose electrical potential energy as their opposite charges move closer together. This loss is made up by an equal increase in the electron's kinetic energy. Rearranging the above equation to solve for v, we find that

$$v^2 = \frac{2(2.3 \times 10^{-19} \text{ N·m})}{(9.11 \times 10^{-31} \text{ kg})} = 5.05 \times 10^{11} \left(\frac{m^2}{s^2}\right)$$

or

$$v = \underline{7.1 \times 10^5 \text{ m/s}}. \qquad \blacksquare$$

EXAMPLE 23.6 A soapbox derby race car, shown in Fig. 23.7, has a charge q_1 of $-(2.3 \times 10^{-4})$ C under its hood. This charge is attracted by a 0.30-C positive charge q_2 placed beyond the end of the racetrack. Calculate the speed v of the race car and racer (total mass 113 kg) at the end of the race. A 40-N friction force opposes the motion of the car. The dielectric constant of air is 1.0.

SOLUTION Our system consists of everything in Fig. 23.7. No work is done on the system by forces outside the system, and there are no springs, so $\Delta PE_s = 0$. The changes in the other forms of energy are as follows:

$$\Delta KE = \frac{1}{2}mv^2 - \frac{1}{2}mv_0^2 = \frac{1}{2}(113 \text{ kg})v^2,$$

$$\Delta PE_g = mg(y - y_0) = (113 \text{ kg})(9.8 \text{ m/s}^2)(0 - 28 \text{ m}) = -3.1 \times 10^4 \text{ J},$$

$$\Delta PE_q = \frac{kq_1q_2}{K}\left(\frac{1}{r} - \frac{1}{r_0}\right)$$

$$= \frac{(9.0 \times 10^9 \text{ N·m}^2/\text{C}^2)(-2.3 \times 10^{-4} \text{ C})(+0.30 \text{ C})}{1.0}\left(\frac{1}{30 \text{ m}} - \frac{1}{330 \text{ m}}\right)$$

$$= -1.9 \times 10^4 \text{ J},$$

$$\Delta E_{int} = F_k \Delta x = (40 \text{ N})(300 \text{ m}) = 1.2 \times 10^4 \text{ J}.$$

FIG. 23.7. The soapbox derby race car gains kinetic energy and loses gravitational and electrical potential energies as it rolls down the hill.

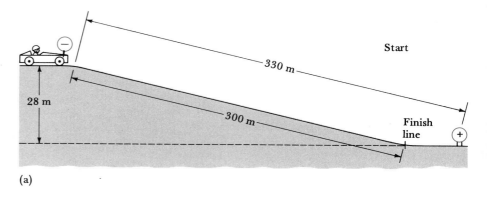

(a)

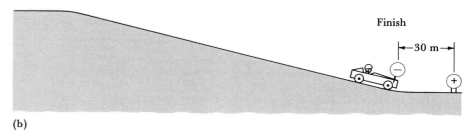

(b)

When these values are substituted in the conservation-of-energy equation to solve for v,

$$0 = \frac{1}{2}(113\,\text{kg})v^2 - (3.1 \times 10^4\,\text{J}) + 0 - (1.9 \times 10^4\,\text{J}) + (1.2 \times 10^4\,\text{J}),$$

we find that

$$v^2 = \frac{2[(3.1 \times 10^4\,\text{J}) + (1.9 \times 10^4\,\text{J}) - (1.2 \times 10^4\,\text{J})]}{113\,\text{kg}} = 673\,\text{m}^2/\text{s}^2,$$

$$v = \underline{26\,\text{m/s}}.$$ ∎

23.4 Electrical Potential Difference and Potential

We learned in Sections 23.1 and 23.2 that the electrical potential energy of a system changes if one electric charge q' moves relative to other stationary charges. Equation (23.3) is used to calculate the change in electrical potential energy caused by the altered position of q' relative to the stationary charges.

This situation can be treated in another way using the concept of electrical potential difference. Instead of considering the motion of q' relative to other charges, we think of q' as moving in a space that has electrical hills and valleys that are created by the other charges. *Electrical potential difference* is a measure of the change in elevation of these electrical hills or valleys. If q' is allowed to move freely in this space, it moves toward a region where its electrical potential energy decreases (like a boulder rolling down a hill toward a region of lower gravitational potential energy).

The quantity *electrical potential difference* is defined to reflect the change in electrical energy a particle undergoes when moving between different points in space. The definition is made in terms of the space rather than in terms of the stationary charges causing the space to be modified.

The **electrical potential difference** V_{AB} from point A to point B is defined as the change in electrical potential energy $\Delta PE_{q_{AB}}$ that occurs when a small, positive charge q' moves from A to B, divided by the value of q':

$$V_{AB} = \frac{PE_{q_B} - PE_{q_A}}{q'} = \frac{\Delta PE_{q_{AB}}}{q'}. \tag{23.4}$$

The unit of electrical potential difference is the **volt,** where one volt is the change in electrical potential difference that occurs when a one-coulomb charge experiences a change in electrical potential energy of one joule. We see that

$$1 \text{ volt} = 1\frac{\text{joule}}{\text{coulomb}}, \quad \text{or} \quad 1 \text{ V} = 1\frac{\text{J}}{\text{C}}.$$

We often use the less formal expression *voltage change* or simply *voltage* when referring to the potential difference between two points. For example, the potential difference or voltage across the terminals of a car battery is 12 V. If a positive charge sits at the negative terminal of the car battery, the positive terminal looks like an electrical hill that is 12 V high.

Electronic instruments called voltmeters are used to measure the voltage between two different points. For example, voltmeters of a special type can measure the potential differences between different parts of your body caused by the dipole change of your heart; geologists use voltage measurements to probe the material in the earth's crust; and technicians make voltage measurements between different points in electronic circuits of television sets, radios, and other electronic instruments.

Since voltage is measured so routinely, Eq. (23.4) is often used to calculate the change in electrical potential energy ΔPE_q that occurs when a charge q moves through a known change in voltage V_{AB}:

$$\Delta PE_{q_{AB}} = qV_{AB}.$$

This equation is usually written in a more abbreviated form as

$$\Delta PE_q = qV, \tag{23.5}$$

where V is the potential difference between the final and initial positions for which the energy change is calculated. The equation can be used for positive or negative charges. The sign of the charge must be included when using the equation.

In summary, we should think of a voltage change—a potential difference—as an indicator of the difference in the electrical elevation of two points as seen by a positive charge; the higher the voltage, the higher the electrical hill. A positive charge tends to roll down an electrical hill toward regions of lower voltage. However, a negative charge tends to move in the opposite direction toward regions of higher voltage where its electrical potential energy, according to Eq. (23.5), becomes less.

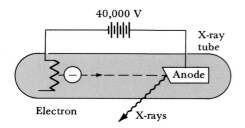

40,000 V

X-ray tube

Anode

Electron   X-rays

FIG. 23.8. An x-ray tube. When electrons are accelerated through a 40,000-V potential difference and crash into the anode, x-rays are produced.

EXAMPLE 23.7 The potential difference V_{AB} from an electron source (called a cathode) of an x-ray tube to the positively charged anode is $+40,000$ V (Fig. 23.8). If an electron begins at rest ($v_0 = 0$) and moves toward the anode, calculate its speed v just before it reaches the anode. The mass m of an electron is 9.11×10^{-31} kg and its charge q is -1.6×10^{-19} C.

SOLUTION In the system shown in Fig. 23.8, no external forces of significance act on the system. Therefore, $W = 0$. The only energy changes are the electrical potential energy change due to the electron moving across the tube and the corresponding change in the electron's kinetic energy. The energy changes are calculated using Eqs. (23.5) and (8.4):

$$\Delta PE_q = qV = (-1.6 \times 10^{-19}\,\text{C})(40,000\,\text{V})$$
$$= -6.4 \times 10^{-15}\,\text{C·V} = -6.4 \times 10^{-15}\,\text{J},$$

$$\Delta KE = \frac{1}{2}mv^2 - \frac{1}{2}mv_0^2 = \frac{1}{2}(9.11 \times 10^{-31}\,\text{kg})v^2.$$

These quantities are substituted into the work-energy equation [Eq. (8.9)], which is then solved for the final speed v:

$$0 = \frac{1}{2}(9.11 \times 10^{-31}\,\text{kg})v^2 + 0 + 0 - (6.4 \times 10^{-15}\,\text{J}) + 0,$$

$$v^2 = \frac{2(6.4 \times 10^{-15}\,\text{J})}{(9.11 \times 10^{-31}\,\text{kg})} = 1.4 \times 10^{16}\,\text{m}^2/\text{s}^2,$$

$$v = \underline{1.2 \times 10^8\,\text{m/s}}.$$

The electron is traveling at almost 0.4 times the speed of light! Whenever masses move at speeds of the order of 0.1 or more times the speed of light, theories of relativity should be used to calculate their speed. Our calculation is only an approximation; nevertheless, we can see from our calculations that the electron (a negative charge) loses electrical potential energy as it moves toward higher voltage. The loss of electrical energy is balanced by an increase in kinetic energy. ∎

EXAMPLE 23.8 During 30 s, -90 C of charge flows through the 1/4-hp electric motor of a Weedwacker from its 0-V terminal to the $+120$-V terminal. Calculate the efficiency of the motor, defined as the ratio of the work W it does during the 30 s and the magnitude of the electrical potential energy used to do that work.

SOLUTION The electrical potential energy (ΔPE_q) expended by the motor is given by Eq. (23.5), where q is the charge (-90 C) that traverses the $+120$-V potential difference across the motor:

$$\Delta PE_q = qV = (-90\,\text{C})(+120\,\text{V}) = -1.08 \times 10^4\,\text{J}.$$

The motor converts 1.08×10^4 J of electrical potential energy into work and other forms of energy.

Next we calculate the work done by the motor during the 30 s. The power of a motor (1/4-hp in our example) is defined by Eq. (8.20) as the ratio of the work W done in a time t divided by that time ($P = W/t$). Rearranging, we can calculate the work done by the motor in 30 s:

$$W = Pt = (0.25 \text{ hp})(30 \text{ s})$$

$$= \left(0.25 \text{ hp} \times \frac{746 \text{ W}}{1 \text{ hp}}\right)(30 \text{ s})$$

$$= (187 \text{ J/s})(30 \text{ s})$$

$$= 0.56 \times 10^4 \text{ J}.$$

We have used the unit conversions 1 hp = 746 W and 1 W = 1 J/s.

The motor's efficiency is the ratio of the useful work done and the electrical energy expended to produce that work:

$$\text{Efficiency} = \frac{W}{\Delta PE_q} = \frac{0.56 \times 10^4 \text{ J}}{1.08 \times 10^4 \text{ J}}$$

$$= \underline{0.52}.$$

We see that 52 percent of the energy used by the motor results in useful work. For most motors, the electrical energy that does not produce work (48 percent in our example) ends up being wasted by the production of thermal energy—the motor and air surrounding the motor get hot. ∎

Absolute Potential

Potential difference is an indicator of the difference in electrical elevation of two points. We are usually interested in the electrical potential difference between two points in space, but in some situations it is convenient to ascribe an absolute value to the electrical potential at each point in space. For example, one point in an electric circuit is often called the **ground point.** The voltage at that point is defined as zero. **Absolute potential** is the value of the voltage at other points in the electrical circuit relative to this reference point of zero potential.

The word *ground* is used because in many electrical instruments and appliances the ground point actually connects via a wire to a pipe that passes under the earth. The voltage at the ground point is exactly the same as that of the earth outside a house. Since pipes in the ground maintain a constant voltage all over a city, it is a good reference level for electrical potential difference measurements.

The reference point used for potential difference measurements and calculations is completely arbitrary. For instance, when describing the electrical potential at different points surrounding the nucleus of an atom, it is common to designate a point infinitely far from the nucleus as a point of zero absolute potential or zero voltage. The absolute potential or voltage at other points relative to this reference point is determined using Eqs. (23.1) and (23.4) in the following manner. If another point is separated by a distance r from a nucleus with charge q, its potential relative to the zero-voltage reference point at infinity is

$$V = \frac{PE_{q_r} - PE_{q_\infty}}{q'} = \frac{\dfrac{kqq'}{r} - \dfrac{kqq'}{\infty}}{q'} = \frac{\dfrac{kqq'}{r} - 0}{q'} = \frac{kq}{r}. \tag{23.6}$$

In this equation, q' is a small, positive test charge whose electrical potential energy changes by $PE_{q_r} - PE_{q_\infty}$ as its separation from q changes from infinity

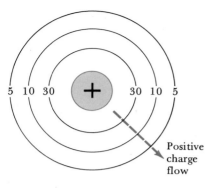

FIG. 23.9. Equipotential lines surround a positive charge.

to r. The voltage at r is the ratio of this energy change divided by the imaginary charge q' that moved.

Notice that the absolute potential at a point a distance r from q depends only on q and r and not on the test charge q' used to calculate that potential. Our charge q sets up a potential field surrounding itself; it creates electrical hills or valleys that can be represented by an electrical contour map. The test charge q' is just a mental device to help us determine an expression for that potential, but it does not affect its value.

EXAMPLE 23.9 Calculate the absolute potential at a point 0.53×10^{-10} m from a proton. This is the average distance of an electron from the proton nucleus of a hydrogen atom. Assume that the potential is zero at a point infinitely far from the proton.

SOLUTION The following information is known: $q = 1.6 \times 10^{-19}$ C (charge of proton) and $r = 0.53 \times 10^{-10}$ m. The absolute potential at a distance r from the proton is calculated using Eq. (23.6):

$$V = \frac{kq}{r} = \frac{(9.0 \times 10^9 \text{ N} \cdot \text{m}^2/\text{C}^2)(1.6 \times 10^{-19} \text{ C})}{(0.53 \times 10^{-10} \text{ m})}$$

$$= 27 \frac{\text{N} \cdot \text{m}}{\text{C}} = 27 \frac{\text{J}}{\text{C}} = \underline{27 \text{ V}}. \qquad \blacksquare$$

Equipotential Surfaces

Contour maps can be drawn to represent electrical potential. In Fig. 23.9, a large, positive charge q resides at one point in space, and the circular lines surrounding the charge represent lines of constant electrical potential. A small test charge placed at one point on a line has the same electrical potential energy as at any other point on that line. The line is part of a three-dimensional surface called an **equipotential surface;** every point on the surface is at the same electrical potential. The lines drawn on Fig. 23.9 are analogous to lines drawn on a contour map that represent regions of constant vertical elevation. Just as water tends to flow from regions of higher to regions of lower elevation, positive electrical charge tends to flow perpendicular to equipotential lines toward regions of lower potential (see the dashed line in Fig. 23.9). As the voltage decreases, so does the electrical potential energy of the positive charge ($\Delta PE_q = qV$). Negative charge also tends to move toward regions where its electrical potential energy is minimized. ΔPE_q will be negative for a negative charge if V is positive. Thus, negative charges tend to move in the opposite direction of positive charges toward regions of higher electrical potential.

The dipole charge on a person's heart creates equipotential contours on a person's body surface. An example of these contours at one instant of time is shown in Fig. 23.10 (the dipole charges on the heart causing the contours are not shown). At any point on one of these lines, a charge in the skin will have the same electrical potential energy as when located at another point on the same line.

A positive charge tends to move toward regions of lower electrical potential. The -1.0-mV circle in Fig. 23.10 is like a low-elevation bog in which water

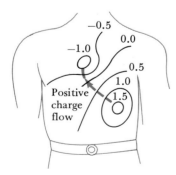

FIG. 23.10. Equipotential lines on the surface of a person's body at one time during a heartbeat cycle. The lines are caused by the dipole charge on the heart.

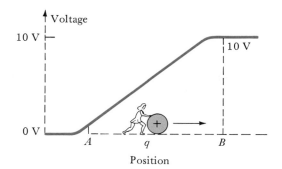

FIG. 23.11. The electrical potential energy of charge q increases as it is pushed toward the region of higher voltage at B.

accumulates after a rain. On the body, positive charge accumulates there. Negative charge accumulates at high "electrical elevations"—near the $+1.5$-mV peak.

23.5 Potential Difference and Electric Field

As we learned earlier, potential difference (change in voltage) is a measure of the change in the electrical elevation that an electric charge experiences as it moves between different positions in space. If, for example, the change in voltage from point A to point B in Fig. 23.11 is $+10$ V, then a positive electric charge will have to be pushed up this electrical hill.

Such a hill could be caused by the group of charges shown in Fig. 23.12—work must be done to move q toward B, the location of a row of positive charges.

These charges shown in Fig. 23.12 produce an electric field that points from B to A. The electric field exerts a force $\mathbf{F}_q$ on a charge q placed in the field. The force is given by Eq. (22.4):

$$\mathbf{F}_q = q\mathbf{E}.$$

For a positive charge q to be moved in a direction opposite the field, it must be pushed with an opposing force $\mathbf{F}$ (represented by the push of the person in Fig. 23.12) whose magnitude is at least as large as $\mathbf{F}_q$:

$$\mathbf{F} = -q\mathbf{E}.$$

The negative sign means that the person's force $\mathbf{F}$ is opposite $\mathbf{E}$, or toward the right.

The electric field pointing from B to A in Fig. 23.12 and the voltage hill from A to B in Fig. 23.11 are just two different ways of representing the electrical effect of the row of charges at A and B. These two quantities, E and V, are not independent but are intimately related, as we learn next.

To move the charge q a small distance Δx to the right requires work. If the charge moves at a constant speed, then all of the work is converted into electrical potential energy:

$$W = \Delta PE_q.$$

The work done by the pushing force F shown in Fig. 23.12 is

$$W = F\,\Delta x.$$

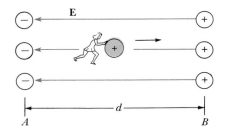

FIG. 23.12. The electrical hill shown in Fig. 23.11 could be caused by the charge distribution here. These charges at A and B produce an electric field that points to the left and opposes the motion of q.

(a)

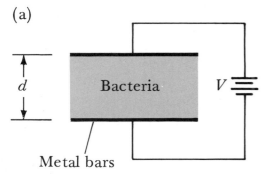

Metal bars

(b)

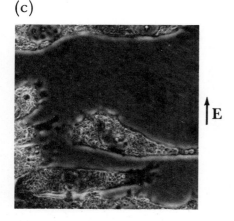

(c)

E

FIG. 23.13. (a) The schematic representation of an apparatus used to orient bacteria and animal cells. Randomly oriented axolotl neural crest cells (b) before and (c) 1 h after application of an electric field of 1180 V/m. Besides becoming oriented by the field, the cells move at an average speed of $3.6\ \mu m/min$. (Courtesy of M. S. Cooper and R. E. Keller, Biophysics and Medical Physics Group and Department of Zoology, University of California, Berkeley.)

Since the charge does not accelerate, $\mathbf{F}$ exactly balances the electrical force caused by the electric field: $\mathbf{F} = -\mathbf{F}_q = -q\mathbf{E}$. Thus,

$$W = -qE\,\Delta x.$$

But according to Eq. (23.5), the change in electrical potential energy when the charge q passes through a change in voltage ΔV that occurs when it moves the distance Δx is

$$\Delta PE_q = q\,\Delta V.$$

When these values of W and ΔPE_q are substituted into the work-energy equation ($W = \Delta PE_q$), we find that

$$-qE\,\Delta x = q\,\Delta V,$$

or

$$E = -\frac{\Delta V}{\Delta x}. \tag{23.7}$$

The electric field equals the negative change in voltage ΔV divided by the distance Δx over which that voltage change occurs. If the voltage changes by amount V in distance d, then the average electric field in that region is

$$E_{av} = -\frac{V}{d}. \tag{23.8}$$

Notice that if V is positive while moving from point A to point B, then $\mathbf{E}$ points in the opposite direction—from B to A. A positive change in voltage produces an electric field pointing in the opposite direction—hence the negative sign in Eqs. (23.7) and (23.8).

If the voltage change across a region is known, Eqs. (23.7) and (23.8) can be used to calculate the electric field in that region. The equations can also be used to calculate a change in voltage when moving in a known electric field.

EXAMPLE 23.10   A colony of bacteria is placed between two long, parallel metal bars separated by 0.20 m (see Fig. 23.13a). A potential difference of 100 V is placed across the metal rods. (a) Calculate the average electric field between the bars in the region where the bacteria are located. (b) Calculate the average force on a calcium ion (Ca^{2+}) in the solution containing the cells.

SOLUTION   (a) The average electric field between the bars, which are separated by a distance $d = 0.20$ m and have a potential difference $V = 100$ V connected across them, is given by Eq. (23.8):

$$E_{av} = -\frac{V}{d} = -\frac{(100\ \text{V})}{(0.20\ \text{m})} = \underline{-500\ \text{V/m}}.$$

The negative sign implies that the field points away from the higher-voltage terminal.

(b) The force on a charge $q = 2e = 3.2 \times 10^{-19}$ C when in an electric field is given by Eq. (22.4):

$$F = qE = (3.2 \times 10^{-19} \text{ C})(-500 \text{ V/m})$$
$$= \underline{-1.6 \times 10^{-16} \text{ N}}.$$

The force on a positive charge points from the positive to the negative terminal. The following unit conversions have been used:

$$1 \text{ V} = 1 \text{ J/C} \quad \text{and} \quad \text{J} = 1 \text{ N} \cdot \text{m}. \qquad \blacksquare$$

Scientists have found that bacteria orient themselves and sometimes move in the direction of fairly small electric fields (Figs. 23.13b and c). As the direction of the field changes, the bacteria reorient themselves to remain aligned with the field. Electric fields can also induce the growth of biological tissue along the direction of the field. For example, recalcitrant leg fractures that fail to heal under normal treatment are successfully healed by a so-called pulsing electromagnetic field (PEMF) treatment [*Science News* **121**, 119, Feb. 20 (1982)].

Summary and Additional Readings

1. **Electrical potential energy:** The change in electrical potential energy when the separation of charges q_1 and q_2 is changed from r_0 to r is

$$\Delta PE_q = \frac{kq_1q_2}{K}\left(\frac{1}{r} - \frac{1}{r_0}\right), \qquad (23.2)$$

where K is the dielectric constant of the material in which the charges move. If one charge q moves relative to several other fixed charges q_1, q_2, q_3, . . . , the change in electrical potential energy is the sum of the energy changes for q relative to each fixed charge:

$$\Delta PE_q = \sum_{i=1} \frac{kq_iq}{K}\left(\frac{1}{r_i} - \frac{1}{r_{i_0}}\right), \qquad (23.3)$$

2. **Work-energy calculations:** Electrical potential energy is one of the many forms of energy that can change when a system moves from an initial to a final state. The conservation-of-energy principle can be used to determine an unknown change when other energy changes are known:

$$W = \frac{\text{Energy changes in system}}{\Delta KE + \Delta PE_g + \Delta PE_s + \Delta PE_q + \Delta E_{\text{int}} + \cdots}. \qquad (8.9)$$

The different expressions in this equation are discussed in Chapter 8 (see the Chapter 8 summary).

3. **Electrical potential difference (change of voltage):** Electrical potential difference is an indicator of the difference in electrical

"elevation" of two points in space. When a charge q moves up or down one of these electrical hills or valleys from point A to point B, its electrical potential energy changes by an amount $PE_{q_B} - PE_{q_A} = qV_{AB}$, which is usually written in the more abbreviated form

$$\Delta PE_q = qV, \qquad (23.5)$$

where V is the potential difference (change in voltage) from A to B. The unit of potential difference called the **volt** is defined as one joule per coulomb (1 V = 1 J/C).

4. **Potential difference and electric field:** The average electric field E_{av} between two points separated by a distance d is

$$E_{\text{av}} = -\frac{V}{d}, \qquad (23.8)$$

where V is the difference in the voltage from the first to the second point. The negative sign reminds us that the electric field points from the higher toward the lower voltage.

H. W. Lissmann, "Electric Location by Fishes," *Scientific American*, p 50, March (1963).

Russell K. Hobbie, "The Electrocardiogram as an Example of Electrostatics," *American Journal of Physics* **41**, 824 (1973).

Questions

1. A person's heart has a dipolar charge on it, with the positive charge on the bottom and the negative charge on the top of the heart. Will the electrical potential energy of a sodium ion (charge $+e$) on the person's left side increase or decrease if the sodium ion moves up? What will happen to the potential energy of a chlorine ion (charge $-e$) if it moves up? Justify each answer.

2. The metal sphere on the top of a Van de Graaff generator has a large, negative charge. In which direction will a negatively charged ion in the air move relative to the sphere in order to experience a decrease in electrical potential energy? Justify your answer.

3. To get more paint from spray guns on metal car parts, the paint droplets are charged negatively as they leave the gun, and a voltage source is attached to the metal car part. Should the car part be at a higher or lower voltage than the spray gun to attract the negatively charged paint drops? Justify your answer.

4. A bird perches on a 100,000-V power line. Why is the bird not killed by electric charge passing through its body due to the high voltage?

5. Your left hand touches a metal rack at 0 V, and your right hand accidentally touches the 100-V power source in a piece of electronic equipment. Indicate the direction in which sodium ions in the tissue of your right arm flow. In which direction do chlorine ions flow? Justify your answer. (*Do not try this experiment!*)

6. If the voltage in a region is constant, is the electric field necessarily zero? Explain your answer.

7. Can electric charges be arranged so that the electric field that they produce and the absolute potential are both zero at the same point? Show such an arrangement of charges and a point or points that satisfy this condition.

8. Show a charge arrangement and a point in space where the absolute potential produced by the charges is zero but the electric field is not zero.

9. Show a charge arrangement and a point in space where the electric field produced by the charges is zero but the absolute potential is not zero.

Problems

23.1 Electrical Potential Energy

1. Calculate the change in electrical potential energy when a sodium ion (charge e) and chlorine ion (charge $-e$) are moved from an initial separation of 2.8×10^{-9} m to 1.0×10^{-7} m when (a) in air and (b) in water.

2. Calculate the change in electrical potential energy when a hydrogen atom is ionized. Its electron moves from an orbit that is 0.53×10^{-10} m from the proton nucleus to a relatively large separation (effectively infinity).

3. Calculate the change in electrical potential energy when a -1.5-C charge and a -4.0-C charge are moved from an initial separation of 500 km to a final separation of 100 km. The charges are in air.

■ 4. Two Ping-Pong balls suspended in air from strings have electric charges of $+1.0 \times 10^{-6}$ C and $+3.0 \times 10^{-6}$ C. The balls are initially separated by 20 cm. (a) Calculate their final separation if their electrical potential energy increases by $+0.045$ J. (b) Calculate their final separation if the charges, initially separated by 20 cm in water, gain $+0.045$ J of electrical potential energy.

■ 5. Two protons each of mass 1.67×10^{-27} kg and charge e are initially at rest and separated by 1.0×10^{-14} m (approximately the radius of a nucleus). When released, the protons fly apart because of their electrical repulsion. (a) Calculate the change in their electrical potential energy when they are 1.0×10^{-10} m apart (approximately the radius of an atom). (b) If the electrical potential energy lost is converted entirely into the kinetic energy of the protons (shared equally), what is the speed of one proton when 1.0×10^{-10} m from the other proton?

■■ 6. Two protons, initially separated by a very large distance ($r_0 \simeq \infty$), move directly toward each other with the same initial speed v_0. Calculate their initial speed if the distance of closest approach when their speeds are reduced to zero is 4.0×10^{-14} m.

■ 7. An electron is 0.10 cm from an electric charge of $+3.0 \times 10^{-3}$ C. (a) Calculate the electrical force F_0 on the electron. (b) The electron is pulled slowly by a force so that it moves to a distance of 0.11 cm from the charge. Calculate the electrical force F on the electron when at this distance. (c) Estimate the work done by the average force pulling the electron $[(F + F_0)/2] \Delta x$. (d) Compare this number to the change in electrical potential energy of the electron as it moves away from the charge. Why should the numbers be approximately equal?

23.2 Electrical Potential Energy Involving Multiple Charges

■ 8. A -0.10-C stationary charge is 1000 m to the left of a $+0.20$-C stationary charge. Calculate the change in electrical potential energy when a $+0.050$-C charge initially 500 m left of the negative charge moves 1500 m farther to the left.

■ 9. The magnitude q of charges on electrodes placed by a geophysicist in the ground (Fig. 23.14) is 2.0×10^{-3} C. (a) Calculate the change in electrical potential energy of a positive ion of charge e that is moved from position A to B. (b) Repeat the calculations for a negative ion of charge $-e$. (c) On the basis of these energy changes, which ion would prefer position A and which would prefer position B? Explain. Assume that the earth's dielectric constant is 7.0.

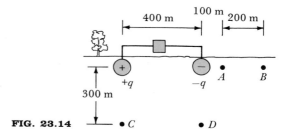

FIG. 23.14

■ 10. Calculate the change in electrical potential energy when a -0.050-C charge moves from position C to position D as shown in Fig. 23.14. The magnitude of the stationary charges q is 0.100 C and the dielectric constant of the earth surrounding the charges is 6.0.

■ 11. Calculate the change in electrical potential energy when a $+5.0$-C charge moves to position C (Fig. 23.15) from infinity: $q_1 = q_2 = q_3 = +10.0$ C.

■ 12. Repeat the calculation described in Problem 11, but for $q_1 = q_3 = -10.0$ C and $q_2 = +10.0$ C.

■ 13. Calculate the change in electrical potential energy when a $+2.0$-C charge moves from position C to position D (Fig. 23.15): $q_1 = q_3 = +10.0$ C and $q_2 = -20.0$ C.

■ 14. A chlorine ion (charge $-e$) moves from position B to position C as shown in Fig. 23.6. Calculate its change in electrical potential energy. The magnitude of the dipole charges on the

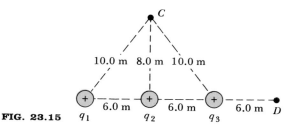

FIG. 23.15

heart ($+q$ and $-q$) is 4.0×10^{-6} C and the charges are in tissue having an average dielectric constant of 7.0.

■ 15. Four charges ($-e$, $+e$, $-e$, $+e$) in a row are each separated from neighboring charges by 3.0×10^{-10} m. Calculate the change in electrical potential energy needed to remove the positive ion on the right side to infinity.

23.3 Conservation-of-Energy Calculations

■ 16. Two electrons are initially separated from each other by 0.50×10^{-9} m. When released, the electrons fly apart. Calculate the speed of each electron when separated by a large distance ($r \simeq \infty$). [*Hint:* Each electron gains kinetic energy equal to one-half the total electrical potential energy lost.]

■ 17. An alpha particle (two protons and two neutrons together in one nucleus) with a mass of 6.64×10^{-27} kg and charge $+2e$ flies head-on from a large distance toward a stationary gold nucleus (charge $+79e$) at a speed of 3.0×10^7 m/s. Calculate the distance of the alpha particle from the gold nucleus when it stops.

■ 18. A stationary block has a charge of $+6.0 \times 10^{-4}$ C. A 0.80-kg cart with a charge of $+4.0 \times 10^{-4}$ C is initially at rest and separated by 4.0 m from the block. Calculate the cart's speed after it is released and moves along a frictionless surface to a distance of 10.0 m from the block.

■ 19. Consider 10^{12} chlorine ions (each with mass 5.9×10^{-26} kg and charge $-e$) moving through the body tissue from position B (Fig. 23.6) to position A. If the ions start at rest and their speed at A is 0.10 m/s, calculate the internal energy added to the body. The average dielectric constant of the body tissue is 7.0 and $q = 4.0 \times 10^{-6}$ C.

■ 20. A 5.0-kg cart, shown in Fig. 23.16, has on its roof a small metal sphere of charge -2.0×10^{-4} C. The sphere moves toward another sphere of charge $+1.5 \times 10^{-3}$ C located on the wall. As it moves, the cart pulls a cable that passes over a pulley and lifts a 10-kg mass. If the cart starts at rest 5.0 m from the wall charge, calculate its speed when 2.0 m from the wall charge. Ignore friction.

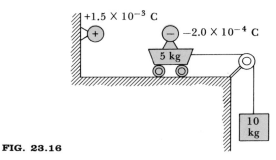

FIG. 23.16

■■ 21. Repeat the calculation for Problem 20, but assume that a 40-N friction force opposes the motion of the cart.

■■ 22. An ion cannon, such as shown in Fig. 23.1, consists of a positively charged ($+2.0 \times 10^{-4}$)-C metal ball compressed into a plastic barrel so that it is 0.10 m from an equal charge at the closed end of the barrel. When the ball is released, it is shot out the open end of the barrel by the repulsive force between the charges. Calculate the speed of the 10-kg ball as it leaves the end of the barrel 10.0 m from where it started. Assume that no friction force exists in the barrel. The barrel is oriented at 37° with respect to the horizontal.

■■ 23. Repeat Problem 22, assuming that a 90-N friction and drag force opposes the motion of the ball up the barrel.

23.4 Electrical Potential Difference and Potential

24. During a lightning flash, -15 C of charge moves through a potential difference of 8.0×10^7 V. How much electrical potential energy is released?

25. A spark of electricity jumps from a person's finger to your nose and through your body. While passing through the air, the spark travels across 2.0×10^4 V and releases 3.0×10^{-7} J of electrical potential energy. How much charge in coulombs and how many electrons flow?

■ 26. The potential difference across the tungsten heating element of a toaster is 120 V. Electrons flowing along the wire experience no change in kinetic energy (all of their electrical potential energy loss is converted to thermal energy). Calculate the number of electrons that must flow across the 120-V potential difference each second to produce 500 J of thermal energy.

■ 27. A battery charger forces 5.0 C of positive charge to move across a potential difference of 12 V in a battery each second. Calculate the increase in electrical potential energy of the battery in 1 h of charging.

■ 28. A 1/2-hp motor produces useful work at a rate of 373 J/s. (a) If the motor's efficiency is 0.60 (60 percent), at what rate in J/s is electrical energy used to produce this work? (b) If the motor is connected across a 120-V potential difference, how much charge passes through the motor each second? Explain.

■ 29. The potential difference from the cathode to the screen of a television set is $+22,000$ V. If an electron of mass 9.11×10^{-31} kg leaves the cathode with an initial speed of zero, what is its speed just before hitting the screen?

■ 30. Across what potential difference must an electron, initially at rest, move to attain a speed of 1.0×10^7 m/s?

■ 31. An electron starting at rest accelerates across a potential difference V_0. If the potential difference is doubled, by what factor does the electron's final speed increase?

■ 32. Show that the speed of a mass m with positive charge q when accelerated from rest across a potential difference $-V$ is given by the expression

$$v = \sqrt{\frac{2qV}{m}}.$$

■ 33. A 10,000-kg shuttle bus carries a $+15$-C charge on a sphere on its top. Through what potential difference must the bus travel to acquire a speed of 10 m/s, assuming no friction force?

■ 34. A nerve cell is shaped like a cylinder. The membrane wall of the cylinder has a $+0.090$-V potential difference from the inside to the outside of the wall. To help maintain this potential difference, sodium ions are pumped from inside the cell to the outside. For a typical cell, 10^9 ions are pumped each second. (a) Calculate the change in chemical energy each second required to produce this increase in electrical potential energy. (b) If there are roughly 7×10^{11} of these cells in the body, how much chemical energy is used in pumping sodium ions each second. (c) What fraction of an average person's metabolic rate is this energy that is used to pump ions (see Table 8.1)?

■■ 35. Protons are accelerated from the end of an ion gun used to propel a rocket ship. (a) Calculate their speed if they start at rest and pass through a potential difference of 50,000 V. The mass of a proton is 1.67×10^{-27} kg. (b) Calculate the momentum given to 3.0 kg of these protons. (c) Calculate the final speed of a 4000-kg rocket ship after emitting the protons. Assume that the ship starts at rest while floating in space and that all the protons are emitted in one burst from the resting ship.

■■ 36. A lightning flash occurs when -40 C of charge jumps from a cloud to the earth through a potential difference of 4.0×10^8 V. Calculate the change in temperature and state of 50 kg of water (the mass of water in a 150-lb person) if all the electrical potential energy is converted to thermal energy in the water. The water starts at 37°C.

■■ 37. In a hot water heater, water is warmed by converting electrical potential energy into thermal energy. (a) Calculate the energy needed to warm 180 kg of water by 10 C°. (b) If -10.0 C of electric charge passes through a $+120$-V potential difference in the heating coils each second, calculate the time needed to warm the water 10 C°.

■ 38. Two charges of -3.0×10^{-6} C each are separated by 0.20 m. Calculate the absolute potential (assuming zero volts at infinity) at a point (a) halfway between the charges and (b) 0.20 m to the side of one of the charges along a line joining them.

■ 39. (a) Calculate the absolute potential at a point halfway between electric dipole charges that are separated by 0.10 m and

for which $q = 5.0 \times 10^{-6}$. (b) Calculate the absolute potential at a distance 0.20 m to the side of the positive charge along a line joining the two charges.

23.5 Potential Difference and Electric Field

40. An average electric field of magnitude 200 V/m exists between two plates separated by 5.0 cm. How large a potential difference exists from one plate to the other?

41. If the electric field between your finger and some other object exceeds 3.0×10^6 V/m, a spark will jump. If you shuffle across a rug, a spark will jump when your finger is 0.20 cm from another person's nose. Calculate the difference in voltage between your body and the other person's body just before the spark jumps.

■ 42. The average electric field between two large metal plates separated by 0.10 cm is 3.0×10^5 V/m. (a) Calculate the potential difference across the plates. (b) What is the change in electrical potential energy when an electron moves across this potential difference?

■ 43. A parallel-plate capacitor is an electronic device that consists of two large parallel plates separated by a small distance. (a) Calculate the average electric field between the plates if a 120-V potential difference exists from one plate to the other and if their separation is 0.50 cm. (b) A spark will jump if the electric field exceeds 3.0×10^6 V/m. What is the closest the plates can be placed to each other to prevent sparking?

■ 44. (a) Calculate the average electric field across a body cell membrane. A 0.085-V potential difference exists from one side to the other, and the membrane is 7.0×10^{-9} m thick. (b) Calculate the magnitude of the electrical force on a sodium ion (charge $+e$) in the membrane.

CHAPTER 24

Electric Current and Ohm's Law

In the last two chapters we learned how to incorporate electrical force and energy into the general principles of physics that were introduced early in the book. With this background we now begin our study of electric charge that moves or flows from one place to another. All the electrical appliances we use, such as radios, electric stoves, and flashlights, depend on the flow of electric charge.

In this chapter we analyze some very simple electric circuits that allow electric charge to flow in a single loop, such as shown in Fig. 24.1. An electric circuit of any type must include (1) a voltage source, such as a battery, that forces electric charge to move through the rest of the circuit from one terminal of the voltage source back to the other and (2) an unbroken path through the rest of the circuit along which the charge can move. The path may consist of wires connected to lightbulbs, the heating elements of toasters, or whatever. The charge that moves through this circuit is called an electric **current.**

We will learn how to determine the amount of current that flows through a circuit and the rate at which electrical energy is used by an element in the circuit, such as a lightbulb. The circuits considered in this chapter consist of a voltage source connected by wires to one other element. In the next chapter, more complicated circuits are presented.

24.1 Electric Power Sources

The first step in producing an electric current is to separate positive and negative charges in an electrical power source such as a battery. Energy is needed for this charge separation. For example, hydroelectric power generation depends on the conversion of the kinetic and gravitational potential energy of water into the electrical potential energy needed to separate charges. The chemical energy released by the burning of coal, oil, and natural gas provides the electrical potential energy used by many electric power plants.

One of the most common electrical power sources, the battery, also derives its electrical energy from the energy released by chemical reactions. These chemical reactions can be complex, yet many of the basic properties of all batteries

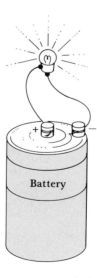

FIG. 24.1. A simple electric circuit.

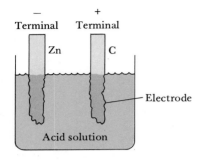

FIG. 24.2. A battery.

can be understood with reference to the simple battery shown in Fig. 24.2. Two dissimilar metal rods or plates, called **electrodes,** are immersed in a dilute acidic solution called the **electrolyte.** In a typical flashlight battery, one electrode is made of carbon and the other (the outer casing of the battery) is made of zinc; the electrolyte is absorbed in a powdery, sawdustlike material that is placed between the electrodes.

The metal electrodes dissolve in the electrolyte. When zinc dissolves, positively charged zinc ions (Zn^{2+}) enter the electrolyte and leave two electrons behind on the zinc electrode. Eventually the electrode accumulates such an excess of electrons that dissolved zinc ions are attracted back to the electrode at the same rate as zinc ions dissolve and leave the electrode; the net charge on the electrode remains constant. The other metal electrode may dissolve at a slower rate and accumulate a smaller number of electrons than the first electrode. Thus, although both electrodes become negatively charged, one electrode has less negative charge and is at a higher electrical potential than the other. In a flashlight battery, the potential difference between the electrodes is 1.5 V, the carbon electrode being at a higher potential than the zinc electrode.

The parts of the electrodes that stick out of the electrolyte are called the battery **terminals.** If the battery terminals are connected by wires to a flashlight bulb, electrons leave the negative terminal and move through the wire, through the filament of the flashlight bulb, and then back along the other wire to the positive terminal. As this electric charge flows, more of the electrode dissolves to provide a steady supply of electric charge. The battery becomes "dead" when one or the other of the electrodes becomes dissolved to the extent that it can no longer supply electric charge.

Electric power sources such as batteries are represented in electric circuits by the symbol shown in Fig. 24.3. By convention, the positive terminal is on the side with the longer vertical line; the negative terminal is on the side with the shorter line.

Any device such as a battery that separates positive and negative electric charge and therefore produces a voltage is called a **source of electromotive force (emf).** The emf of such a device (represented by the symbol $\mathcal{E}$) is the steady potential difference across its terminals when no net charge is moving onto or off the terminals. The emf $\mathcal{E}$ of a flashlight battery is 1.5 V, of a car battery 12 V. It would have been better to call these devices voltage sources since they provide a potential difference or voltage across their terminals rather than force. Years of common usage have left us with this unfortunate terminology.

24.2 Electric Current (I)

In the previous section we learned how a source of emf separates positive and negative electric charges, causing a voltage to exist between the terminals of the source. If a wire is attached from one terminal to the other, electric charge flows through the wire. As the source of emf continues to separate charges, a steady flow of charge continues through the wire (Fig. 24.4). The flow of charge is called an **electric current** and is given the symbol I.

In the case of a wire, the electric current is caused by a stream of electrons moving from the negative voltage terminal toward the positive terminal. If we could look into the wire, we might see approximately 10^{20} electrons pass our line of sight each second.

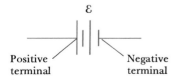

FIG. 24.3. The symbol $\mathcal{E}$ represents the voltage across the terminals of a source of emf when no current is flowing.

Current is not always caused by electron flow. For example, if you touch the positive terminal of a flashlight battery with your index finger and the negative terminal with your thumb, negative ions in your hand such as Cl⁻ flow toward the positive terminal, and positive ions such as Na⁺ and K⁺ flow toward the negative terminal. This is called an **ion current.**

In general, electric current I is defined as the number of coulombs of electric charge Δq that pass a cross section of a material divided by the time Δt required for the charge to pass. Thus,

$$I = \frac{\Delta q}{\Delta t}. \qquad (24.1)$$

The unit of current called the **ampere,** or **amp** (A), is defined as the flow of one coulomb of charge each second:

$$1 \text{ A} = 1 \text{ C/s}. \qquad (24.2)$$

The direction of the current is the direction in which positive electric charges flow and opposite the direction in which negative charges flow.

The convention for describing the direction of an electric current is occasionally confusing because most electric currents are caused by moving electrons (negative charges), whereas the current is by definition in the opposite direction to that in which the electrons move. The confusion can be traced back to Benjamin Franklin. When Franklin rubbed a glass rod with a piece of silk, electrons moved from the rod to the silk, and the rod was left with a deficiency of electrons. Franklin chose arbitrarily to call the charge on the rod positive and the electron charge on the silk negative. The opposite choice would have been more convenient. Electric current would then have been defined as flowing in the same direction as that in which "positive" electrons move through a wire. We cannot, however, blame Benjamin Franklin for his choice. The electron causing the negative charge was not discovered until a hundred years later. Franklin could not have known that these were the charges that moved when a current flowed. In any case, since the electron was assigned a negative charge, current is now defined as flowing in the opposite direction to that in which electrons move.

EXAMPLE 24.1 Each second, 10^{17} electrons flow from right to left across a cross section of wire attached to the two terminals of a battery. Calculate the magnitude and direction of the electric current in the wire.

SOLUTION The magnitude of the charge e of an electron is 1.6×10^{-19} C. Since 10^{17} electrons pass a cross section in the wire each second, the total charge flowing each second is

$$\Delta q = 10^{17} e = 10^{17}(1.6 \times 10^{-19} \text{ C}) = 1.6 \times 10^{-2} \text{ C}.$$

Substituting Δq and Δt in Eq. (24.1), we find that the magnitude of the electric current is

$$I = \frac{\Delta q}{\Delta t} = \frac{1.6 \times 10^{-2} \text{ C}}{1 \text{ s}} = 1.6 \times 10^{-2} \text{ C/s} = \underline{1.6 \times 10^{-2} \text{ A}}.$$

The direction of the current is <u>from left to right</u>, opposite the direction of electron flow. ∎

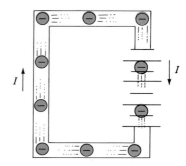

FIG. 24.4. Electrons moving through the wire and battery constitute an electric current. The current is defined such that it points in the opposite direction of the electron motion.

An electric current produced by a battery is an example of a **direct current.** When a direct current flows, charge always moves in the same direction—electrons move away from the negative terminal of the emf source toward the positive terminal.

In many emf sources, one terminal will have first an excess of positive charge and then, a short time later, an excess of negative charge. Charges in a wire attached to the terminals of this type of electric power supply are first pushed in one direction, then back in the opposite direction, creating an **alternating current** (ac). The voltage across the wall socket in a home changes polarity at a frequency of 60 Hz (60 times each second). Charges in lightbulbs and in other electronic devices that are connected to the terminals of a wall socket oscillate back and forth at 60 Hz. In this chapter we will consider only direct currents.

24.3 Ohm's Law and Resistance

We have now defined electric current and have seen how a battery develops a potential difference (voltage) across its terminals. The next step is to see what magnitude current flows through an object when a potential difference is placed across its ends. Suppose, for example, that we connect the bulb of a flashlight across a 3.0-V potential difference. Enough current flows through the bulb's filament to cause it to become so hot that the filament emits light. On the other hand, if you place the thumb and forefinger of your hand across the same 3.0-V potential difference, nothing seems to happen—certainly, your hand does not "light up." Your skin offers considerable resistance to the flow of electric charge, causing the current to be very small. In this section one of our primary interests will be in a quantity called **electrical resistance,** which is given the symbol R and is a measure of the opposition that an object offers to the flow of electric current. (A more precise definition of resistance follows shortly.)

A German physicist, Georg Simon Ohm, in the early 1800s found experimentally that for many types of materials, the three quantities current I, potential difference V, and resistance R were related by the equation

$$V = IR. \tag{24.3}$$

The equation, called **Ohm's law,** is not one of the fundamental principles of physics, as are Newton's laws of motion and the conservation-of-energy principle. Yet Ohm's law has been considered important enough to be included in the physics curriculum of students for more than one hundred years. When Ohm presented his idea, it was received with scorn.

Ohm, at the age of 38, had served for ten years as a poorly paid mathematics and science teacher at the Jesuit College of Cologne. To qualify for a university position, he was required to produce some kind of scientific masterpiece, the value of which would bring recognition and university job offers. After many years of experimenting with electricity, during which time he published numerous short papers, Ohm produced a 250-page manuscript entitled *Mathematical Measurements of Electrical Currents.*

The paper was ignored by most of Ohm's German colleagues. One critic who did not ignore it said, "A physicist who professed such heresies was unworthy to teach science." Ohm, unfortunately, had presented his work at a time and place where experiment was disdained as a means of acquiring knowledge. He did not receive a university position and even had to resign from the Jesuit College. After six dismal years, King Ludwig I of Bavaria helped Ohm obtain a professorship at the Polytechnic School of Nuremberg. Ohm eventually received recognition for his work. In 1841 he was awarded the Copley Medal by the Royal Society of London, and in 1842 he was honored as the Society's most distinguished foreign member.*

Rather than derive Ohm's law from theoretical considerations, let us instead perform an experiment, much like Ohm himself might perform if he could join us today in the laboratory. Our apparatus is illustrated in Fig. 24.5a, and an electric circuit representing the apparatus appears in Fig. 24.5b.

In place of a battery, we will use what is called a **direct current (dc) power supply,** represented in Fig. 24.5b by the symbol ⊣⊦⊦. The arrow indicates that the potential difference produced by this power supply is variable. We next take a long, thin loop of wire and attach one end to the positive terminal of the power supply. The wire, if long and thin, offers significant resistance to the flow of electric current. When drawing an electric circuit, we represent an element in the circuit that has electrical resistance by the symbol ⌇. The loop of wire and all other connecting wires are collectively represented by this symbol in Fig. 24.5b. Next we connect the other end of the loop of wire to one terminal of an **ammeter,** a device that measures the electric current I flowing through the circuit. The other ammeter terminal is connected to the negative terminal of the power supply. When all the connections have been made, current now flows from the positive terminal of the power supply through the loop of wire to the ammeter, where the current is measured, then to the negative terminal of the power supply, and finally back to its positive terminal for another trip around the circuit. The flow of this electric charge around the circuit is represented by the letter I in Fig. 24.5b and by an arrow that indicates the current's direction. The resistance to this current is caused primarily by the loop of wire since ammeters are constructed with very low resistance.

Suppose now that we adjust the potential difference V across the ends of the loop of wire so that a particular current flows, say 0.10 A. We will record this potential difference and current. Then we change the potential difference to get a new current, say 0.20 A. We can construct a table giving the potential difference needed to produce a variety of currents (see Fig. 24.6a). When we plot these data in a graph with potential difference on the vertical axis and current on the horizontal axis, as in Fig. 24.6b, a straight line passes nicely through the data points.

A straight-line graph on linear graph paper implies that the two variables being plotted are related by a linear equation of the form $y = ax + b$, where y is the variable plotted on the vertical axis, x is the variable on the horizontal axis, a is the slope of the graph, and b is the vertical intercept. For the graph in Fig. 24.6b, the vertical intercept is zero and the variables V and I are related by a linear equation

$$V = RI, \tag{24.3}$$

where R is the slope of the graph. Equation (24.3) is **Ohm's law,** and the slope R of the graph is called the **electrical resistance** of the object through which the current flows.

*Jay E. Greene, ed., *100 Great Scientists,* pp. 214–217, Washington Square Press, New York (1964).

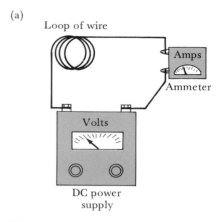

(a)
Loop of wire

Amps
Ammeter

Volts

DC power supply

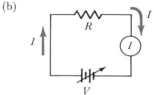

(b)

R

I

I

I

V

FIG. 24.5. (a) An apparatus for measuring the electrical resistance of a loop of wire. (b) The electric circuit for the apparatus in part (a).

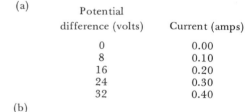

(a)

Potential difference (volts)	Current (amps)
0	0.00
8	0.10
16	0.20
24	0.30
32	0.40

(b)

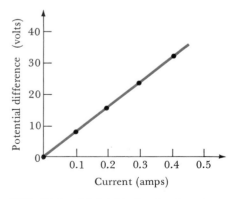

FIG. 24.6. (a) A table showing the potential difference across an object needed to produce different currents through the object. (b) A graph of the data listed in part (a). The slope of the graph equals the object's resistance.

It makes sense to call this quantity resistance. Suppose, for example, that we apply an equal potential difference V across two different objects, a piece of wire and a wooden dowel of the same length. We would find that a large current flows through the wire because its resistance to the flow of electric charge is small:

$$I_{wire} = \frac{V}{R_{wire}}.$$

On the other hand, the wooden dowel has very large electrical resistance that causes the current to be small:

$$I_{wood} = \frac{V}{R_{wood}}.$$

Resistance can now be defined more precisely.

The **electrical resistance** of an object is a measure of its opposition to the flow of current through the object. It can be determined by taking the slope of a V-versus-I graph for the object. For *ohmic* devices that produce a linear V-versus-I graph, the resistance is the ratio of the potential difference and current for any current ($R = V/I$).

The unit of electrical resistance is called the **ohm** and is given the symbol Ω (omega). One ohm is defined as the electrical resistance of an object that will allow a one-ampere (1-A) electrical current to flow when a 1-V potential difference is placed across the object. Rearranging Ohm's law to solve for resistance ($R = V/I$), we see that

$$1 \text{ ohm} = 1 \, \Omega = \frac{1 \text{ V}}{1 \text{ A}}. \qquad (24.4)$$

Resistive devices that produce straight-line V-versus-I graphs are said to be **ohmic**. The current through the device increases in direct proportion to the increase in the potential difference across the device. Examples of ohmic devices include loops of wire, many parts of the human body, some geological formations, and a body of saltwater.

There are, however, other types of devices and materials that are **nonohmic**. For example, the current passing through a transistor does not increase in direct proportion to an increase in voltage. The graph shown in Fig. 24.7a is for another nonohmic device called a **diode**. A small potential difference placed across this device with the positive terminal in the "forward" direction (Fig. 24.7b) causes a large current. However, if the potential difference is reversed, very little current flows in the opposite direction (the negative direction). The diode offers little resistance to current in the forward direction but very large resistance to current in the opposite direction.

Just as the resistance of an ohmic device is the slope of the V-versus-I graph, so also is the resistance of a nonohmic device at a particular current the slope of the V-versus-I graph for that device at that current. Notice that the slope of the V-versus-I graph in Fig. 24.7a is small for forward currents and very large for currents in the opposite direction. In most of our examples and problems, we will assume that the resistive objects we are considering have ohmic behavior.

(a)

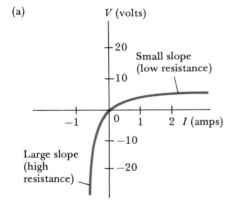

(b)

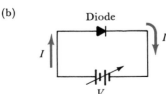

FIG. 24.7. (a) A V-versus-I graph for a nonohmic device called a diode. (b) With the positive voltage placed on the forward side of the diode, large current flows.

24.4 Resistance, Resistivity, and Applications

Ohm's law provides a nice working definition for measuring the resistance of an ohmic device. The definition does not, however, give us a good sense of what causes some objects to have higher resistances than others. An object's resistance depends on properties of the object including among other things its dimensions and the type of material of which it is made. For example, scientists have determined that the electrical resistance of cylindrical objects is directly proportional to their length L and inversely proportional to their cross-sectional area A:

$$R = \rho \frac{L}{A}. \qquad (24.5)$$

The proportionality constant ρ (rho) is called the object's **resistivity**; it depends on the type of material of which it is made and on its temperature (discussed in Section 24.5). Equation (24.5) seems reasonable. The longer an object is, the more difficult it is for charge to cross it (like water that must travel a long distance along a slope that has a very small inclination). Also, resistance increases as an object's cross-sectional area decreases because few electric charges are available to pass the small cross section.

The resistivity ρ of a material depends on intrinsic properties of that type of material, such as the number of electrons per unit volume that are able to move and the hindrance these electrons experience as they travel through the material. Copper has a low resistivity because of its large concentration of free electrons and the relative lack of hindrance experienced by the electrons while moving through the copper. On the other hand, the resistivity of glass is about 10^{20} times greater than that of copper because it contains so few free electrons. Copper is an example of a material called an electrical **conductor** (low resistivity) whereas glass is an example of an electrical **insulator** (high resistivity that prohibits the flow of electricity). The resistivities of a variety of substances are listed in Table 24.1. Substances are sometimes characterized in terms of another quantity called conductivity. The **conductivity** σ of a substance is the inverse of its resistivity ($\sigma = 1/\rho$). Thus, a material with low resistivity has high conductivity and is a good conductor of electricity, and vice versa.

Applications

We are now ready to solve some problems. Our analysis will involve the equation that defines electric current ($I = \Delta q/\Delta t$), the equation for the resistance of a cylindrical object ($R = \rho L/A$), and Ohm's law ($V = IR$).

EXAMPLE 24.2 (a) If an automobile starter has a resistance R of 0.10 ohm, how much current flows through the starter when it is connected to a 12-V battery? (b) How much charge passes through the starter if it runs for 10 s?

SOLUTION (a) Using Eq. (24.4), we find that

$$I = \frac{V}{R} = \frac{12\,\text{V}}{0.10\,\Omega} = 120\,\frac{\text{V}}{\Omega} = \underline{120\,\text{A}}.$$

TABLE 24.1 Resistivity (ρ) of a Variety of Materials

Material	Resistivity (ohm · m at 20°C)
Metals	
Silver	1.6×10^{-8}
Copper	1.7×10^{-8}
Aluminum	2.8×10^{-8}
Electrical insulators	
Ordinary glass	9×10^{11}
Hard rubber	1×10^{16}
Shellac	1×10^{14}
Wood	$10^8 – 10^{12}$
Body parts	
Blood	1.5
Lung	20
Fat	25
Human trunk	5
Membrane	5×10^7
Geological materials	
Igneous rocks	$10^2 – 10^7$
Sedimentary rocks	$1 – 10^5$
Groundwater	~ 10

The conversion of units from V/Ω to A (amperes) is made using Eq. (24.7).

(b) The total electric charge that flows is determined by rearranging Eq. (24.1):

$$\Delta q = I(\Delta t) = (120 \text{ A})(10 \text{ s}) = 1200 \text{ A} \cdot \text{s} = \underline{1200 \text{ C}}.$$

The unit conversion $A \cdot s$ to C is possible because $1 \text{ A} = 1 \text{ C/s}$. Thus, $1 \text{ A} \cdot \text{s} = 1 \text{ (C/s)s} = 1 \text{ C}$. ■

EXAMPLE 24.3 A device called a *fluid-filled rubber-tube displacement transducer* can be used to convert mechanical motion into a varying electric current. When the tube's length increases, its electrical resistance also increases and the current through the fluid in the tube decreases. (a) Calculate the electrical resistance of the fluid in a fluid-filled rubber-tube displacement transducer that is 0.75 m long with a radius of 0.30 cm. The resistivity of the fluid is $0.80 \, \Omega \cdot \text{m}$. (b) Calculate the resistance of the fluid if the tube is stretched so that its length increases by a factor of 1.20, that is, by 20 percent. The volume of the fluid remains unchanged. (c) Calculate the current through the fluid for both situations if a 9.0-V potential difference is placed across its ends.

SOLUTION (a) The following information is used in our calculations:

$$\text{Resistivity of fluid } \rho = 0.80 \, \Omega \cdot \text{m},$$
$$\text{Length of fluid } L = 0.75 \text{ m},$$
$$\text{Area of fluid } A = \pi r^2 = (3.14)(0.30 \times 10^{-2} \text{ m})^2$$
$$= 2.83 \times 10^{-5} \text{ m}^2.$$

The electrical resistance R of the fluid is found using Eq. (24.6):

$$R = \rho \frac{L}{A} = (0.80 \, \Omega \cdot \text{m})\left(\frac{0.75 \text{ m}}{2.83 \times 10^{-5} \text{ m}^2}\right)$$
$$= \underline{2.1 \times 10^4 \, \Omega}.$$

(b) When the tube is stretched, its length increases and its cross-sectional area decreases (see Fig. 24.8). However, the volume V of fluid remains the same because no fluid has been added to or removed from the tube. This fact allows us to calculate the area A' of the fluid in the stretched tube of length L'. For the unstretched tube $V = LA$, and for the stretched tube of equal volume $V = L'A'$. Using $L' = 1.20L$ and $V = L'A' = LA$, we find that

$$A' = \frac{L}{L'}A = \frac{L}{1.20L}A = \frac{A}{1.20}.$$

The resistance R' of the fluid in the stretched tube is

$$R' = \rho \frac{L'}{A'} = \rho \frac{(1.20L)}{(A/1.20)}$$
$$= \rho \frac{L}{A} 1.20^2 = R \, 1.44$$
$$= (2.1 \times 10^4 \, \Omega) 1.44 = \underline{3.0 \times 10^4 \, \Omega}.$$

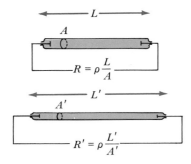

FIG. 24.8. As a flexible tube containing salt is stretched, its length increases and its cross-sectional area decreases. Both changes lead to a high electrical resistance.

You might like to check this result by actually substituting numbers for ρ, L', and A'. Notice that the resistivity ρ does not change because it depends only on the type of material and not on its dimensions.

(c) The electric current in each case is given by Ohm's law:

Unstretched tube: $I = \dfrac{V}{R} = \dfrac{9.0\,\text{V}}{2.1 \times 10^4\,\Omega} = 4.3 \times 10^{-4}\dfrac{V}{\Omega} = \underline{4.3 \times 10^{-4}\,\text{A}}.$

Stretched tube: $I = \dfrac{V}{R'} = \dfrac{9.0\,\text{V}}{3.0 \times 10^4\,\Omega} = \underline{3.0 \times 10^{-4}\,\text{A}}.$ ∎

As we see, the transducer described in Example 24.3 can be used to convert mechanical motion into a varying electric current. For example, if a flexible tube filled with saltwater is placed around your chest with a battery connected across the fluid at the ends of the tube, this transducer would produce a varying electric current each time you breathed, that is, each time the circumference of your chest changed. The tube might be used as part of an alarm system to alert a nurse if a person in a hospital stops breathing.

EXAMPLE 24.4   In the human body, most electric current is caused by the movement of sodium (Na^+), potassium (K^+), chlorine (Cl^-), and magnesium (Mg^{2+}) ions. The resistivity ρ of the tissue beneath your skin depends on the concentration of these ions and is approximately $5\,\Omega \cdot \text{m}$. (The resistivity from the outside surface of the skin to the tissue inside is much greater and prevents large electric currents from flowing through our bodies.) (a) Estimate the electrical resistance of the tissue under the skin of the lower leg of a human. (b) Calculate the current through the leg if connected directly to a 120-V potential difference, the average voltage across a wall socket. (You should never place 120 V across parts of your body. Dangerously high currents may flow as we see next.)

SOLUTION   (a) Resistance depends on the resistivity and the dimensions of the object. Let us assume that the lower leg has a length $L = 0.3\,\text{m}$ and a radius of about $5\,\text{cm} = 0.05\,\text{m}$. Its cross-sectional area is, then,

$$A = \pi r^2 = \pi(0.05\,\text{m})^2 = 8 \times 10^{-3}\,\text{m}^2.$$

Thus,

$$R = \rho \frac{L}{A} = (5\,\Omega \cdot \text{m})\frac{0.3\,\text{m}}{8 \times 10^{-3}\,\text{m}^2} = \underline{200\,\Omega}.$$

(b) The current flowing through the leg when connected to 120 V is determined using Ohm's law:

$$I = \frac{V}{R} = \frac{120\,\text{V}}{200\,\Omega} = \underline{0.6\,\text{A}}.$$

This is about six times as great as the minimum current through the heart that will cause death to a human being (see Fig. 24.9). ∎

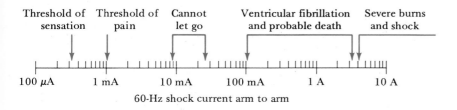

FIG. 24.9. Physiological effects of electric current. [From Peter Strong, *Biophysical Measurements*, p. 257, Tektronix, Inc., Beaverton, Ore. (1970). Reprinted by permission.]

The resistance calculated in the last example is the resistance of the internal body tissue, which is an excellent conductor of electricity. Many more deaths caused by electric shock would occur if it were not for our skin, which is a poor electrical conductor. When skin is dry, the electrical resistance across it ranges from several thousand ohms to hundreds of thousands of ohms; if the skin is moist, the electrical resistance across it is lowered. Consequently, electrical appliances, such as radios and light switches, should not be touched while you are in the bathtub. The resistance across the skin at your hands and feet may be low enough for a fatal current to pass through your body if you accidentally touch the 110-V line entering the appliance from the wall socket. Normally, this line is well insulated and not exposed. Occasionally, the wire becomes cracked with age and your wet skin can make electrical contact through the crack with the high-voltage wire inside the insulation. Sometimes, a frayed wire contacts the outside case of an appliance. If the case conducts electricity, you can receive a potentially dangerous shock by touching the case, especially when your body is wet. The appliance does not have to be on for these hazards to be present. To avoid a dangerous shock, do not touch switches and appliances when your body is wet.

Another hazard is appliances sitting on shelves near bathtubs. Approximately 20 children under age five are killed each year when hair dryers and other appliances plugged into wall sockets are accidentally bumped into the water of a bathtub. The water completes the circuit, with a child in the water being part of the circuit.

The currents at which various physiological effects are produced on the human body are listed in Fig. 24.9.

24.5 Resistance and Temperature

We have seen that the electrical resistance of an object depends on the resistivity of the material of which it is made and on its dimensions. The resistivity of a material is not constant; it changes as the material's temperature changes. For example, the resistance of the tungsten filament in a lightbulb increases by more than a factor of 10 as the filament warms after the light is turned on. At the higher temperature, the atoms and ions in the tungsten vibrate with greater amplitude. Electrons have a more difficult time finding a path through the hot material.

The resistivity of metals and their alloys varies with temperature roughly according to the equation

$$\rho_T = \rho_0(1 + \alpha \, \Delta T), \tag{24.6}$$

where ρ_0 is the resistivity at some reference temperature (usually 0 °C), ρ_T is the resistivity at a temperature interval ΔT above or below the reference temperature, and α is called the **temperature coefficient of resistance.** Representative values for α are given in Table 24.2. The units for α are C$°^{-1}$. The resistance R of a metal wire of length L and cross-sectional area A also varies in a similar fashion:

$$R_T = R_0(1 + \alpha \, \Delta T). \tag{24.7}$$

The variation of an object's electrical resistance with temperature provides a convenient method for measuring temperature. For example, the resistance of

TABLE 24.2 Temperature Coefficients of Resistance

Material	α (C$°^{-1}$)
Aluminum	0.0039
Constantan (60% Cu, 40% Ni)	0.000002
Copper (annealed)	0.00393
Iron	0.0050
Nichrome	0.0004
Platinum	0.00392
Silver	0.0038
Tungsten	0.0045
Carbon	−0.0005

a coil of very thin wire is first measured at a reference temperature. The coil is then moved to a region of unknown temperature and its resistance is again measured. Equation (24.7) is used to calculate the difference in the unknown temperature and the reference temperature. Platinum is a common metal used for these measurements. It has a high melting temperature and is free from corrosion and oxidation.

EXAMPLE 24.5 A platinum resistance thermometer has a resistance of 100.0 Ω when placed in an ice bath (0°C). A geologist lowers the thermometer into a hot springs, and the resistance increases to 123.5 Ω. What is the temperature of the hot springs?

SOLUTION The temperature coefficient of resistance of platinum is 0.00392 C°⁻¹ (see Table 24.2). Equation (24.9) is rearranged to solve for the difference in temperature ΔT of the hot springs and the reference temperature:

$$\Delta T = \frac{R_T - R_0}{\alpha R_0} = \frac{123.5\ \Omega - 100.0\ \Omega}{(0.00392\ \mathrm{C°^{-1}})(100.0\ \Omega)} = 60.0\ \mathrm{C°}.$$

Since the reference temperature is 0°C, the hot springs is at a temperature of 60°C. ■

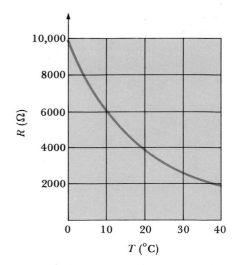

FIG. 24.10. The variation of the resistance of a thermistor with temperature.

The resistance of some materials, known as semiconductors, does not vary as just described. **Thermistors,** for instance, are semiconductor devices whose resistance decreases dramatically as their temperature increases. The resistance of a thermistor may decrease by several percent for each 1-C° rise in temperature (see Fig. 24.10). Thermistors provide an accurate and rapid indication of the temperature of objects in the temperature range from −100°C to 300°C.

24.6 Electrical Power

Most electrical appliances are rated in terms of **power,** the rate at which they convert electrical potential energy into other forms of energy (such as thermal energy):

$$\text{Power} = P = \frac{\Delta PE_q}{\Delta t}. \tag{24.8}$$

The unit of power is the watt, where

$$1\ \mathrm{W} = 1\ \mathrm{J/s}.$$

For example, a 60-W lightbulb converts 60 J of electrical energy into heat and light each second.

To calculate the electric power used by some electrical devices such as electric heaters, we rewrite Eq. (24.10) in a form that depends on the current I passing through the device and on the potential difference V across the device. Recall from Eq. (23.5) that the change in electrical potential energy when a charge Δq flows through a potential difference V is $\Delta PE_q = (\Delta q)V$. Substituting this expression for ΔPE_q in Eq. (24.10), we find that

$$P = \frac{\Delta PE_q}{\Delta t} = \frac{(\Delta q)V}{\Delta t} = \left(\frac{\Delta q}{\Delta t}\right)V = IV,$$

where $I = \Delta q/\Delta t$ is the current flowing through the device across which a potential difference V exists. If the object's resistance R is known, the preceding expression can be rewritten in different forms by using Ohm's law ($V = IR$) to substitute for either V or I:

$$\begin{aligned} P &= IV \\ &= I(IR) = I^2R \\ &= \left(\frac{V}{R}\right)V = V^2/R. \end{aligned} \tag{24.9}$$

In summary, we see that the electrical power expended by a device of resistance R through which a current I flows when a potential difference V is placed across the device is given by any one of the three forms of Eq. (24.9).

EXAMPLE 24.6 A 60-W lightbulb connected to a 120-V potential difference runs for 4.0 h. (a) Calculate the current through the bulb's filament, (b) the electrical resistance of the bulb, and (c) the total electrical energy used during the 4.0 h.

SOLUTION (a) To calculate the current, we use the first form of Eq. 24.9 for the bulb's power ($P = 60$ W) when connected across a potential difference $V = 120$ V. Rearranging, we find that

$$I = \frac{P}{V} = \frac{60\text{ W}}{120\text{ V}} = \underline{0.50\text{ A}}.$$

(b) Ohm's law can now be used to calculate the bulb's resistance:

$$R = \frac{V}{I} = \frac{120\text{ V}}{0.50\text{ A}} = \underline{240\ \Omega}.$$

(c) Rearranging the definition of power [Eq. (24.8)], we find that the electrical energy used in a time Δt is

$$\begin{aligned} \Delta PE_q &= P\Delta t = (60\text{ W})(4.0\text{ h}) \\ &= \left(60\frac{\text{J}}{\text{s}}\right)\left(4.0\text{ h} \times \frac{3600\text{ s}}{1\text{ h}}\right) \\ &= \underline{8.6 \times 10^5\text{ J}}. \end{aligned}$$

We will find shortly that the electric power plant will charge about two cents for this energy. ∎

EXAMPLE 24.7 The resistivity of nichrome wires used in electric heaters is $1.0 \times 10^{-6}\ \Omega \cdot \text{m}$. Calculate the length of wire needed for a 1200-W heater that is connected across a 120-V potential difference. The wire's radius is 0.40 mm.

SOLUTION We are given the following information:

$$\text{Resistivity of wire } \rho = 1.0 \times 10^{-6}\,\Omega \cdot m,$$

$$\text{Heater's power } P = 1200\,W,$$

$$\text{Potential difference across heater } V = 120\,V,$$

$$\text{Radius of wire } r = 0.40\,mm.$$

The problem involves two steps. First, to find the wire's resistance, we use the third form of Eq. (24.9) since both P and V are given. Rearranging that equation, we calculate the heater's resistance:

$$R = \frac{V^2}{P} = \frac{(120\,V)^2}{1200\,W} = 12\,\Omega.$$

Next, we rearrange Eq. (24.5) to find the length of wire that produces that resistance:

$$L = \frac{RA}{\rho} = \frac{R(\pi r^2)}{\rho}$$

$$= \frac{(12\,\Omega)[\pi(0.40 \times 10^{-3}\,m)^2]}{1.0 \times 10^{-6}\,\Omega \cdot m}$$

$$= \underline{6.0\,m}. \qquad \blacksquare$$

Electric power companies charge their customers according to the amount of electrical potential energy the customer uses each month, but instead of charging for each joule of energy, the companies charge by an energy unit called the kilowatt-hour. A **kilowatt-hour** (kW $\cdot$ h) is the energy expended when a device converts electrical potential energy to other forms at a rate of 1000 W (1 kilowatt or 1 kW) for a time of one hour. The conversion of kilowatt-hours to joules is shown below:

$$1\,kW \cdot h = 1\,kW \times 1\,h\left(\frac{1000\,W}{1\,kW}\right)\left(\frac{3600\,s}{1\,h}\right)$$

$$= 3.6 \times 10^6\,W \cdot s = 3.6 \times 10^6\,J.$$

In 1986 electric utility companies charged about 9 cents for 1 kW $\cdot$ h of electrical energy. By comparison, the chemical energy released by 14 slices of bread is about 1 kW $\cdot$ h and in 1986 cost approximately 70 cents.

EXAMPLE 24.8 An electric hot water heater has two resistive heating elements that together produce thermal energy at a rate of 7000 W. You are the last in a group of people to take a shower and find that there is no more hot water available. (a) Calculate the energy needed to warm the 120 kg of water in the heater from 15°C to 60°C. (b) How long will you have to wait before the water is warm? (c) How much did warming the water cost at 9 cents per kW $\cdot$ h?

SOLUTION (a) Recall from Chapter 9 that the heat Q that must be added to a mass m of water of specific heat capacity $c = 4180\,J/kg \cdot C°$ (see Table 9.1) to warm it from an initial temperature T_0 to a final temperature T is given by Eq. (9.4):

$$Q = mc(T - T_0)$$
$$= (120 \text{ kg})(4180 \text{ J/kg} \cdot \text{C}°)(60°\text{C} - 15°\text{C}) = 2.26 \times 10^7 \text{ J}.$$

The electric resistive wires will have to convert $2.26 \times 10^7 \text{ J}$ of electrical energy into thermal energy.

(b) We are given that the rate of electrical energy conversion, that is, the power $P = 7000 \text{ W} = 7000 \text{ J/s}$. Rearranging Eq. (24.8), we find the time needed to supply the needed energy at this power:

$$\Delta t = \frac{\Delta PE_q}{P} = \frac{2.26 \times 10^7 \text{ J}}{7000 \text{ J/s}}$$
$$= 3.2 \times 10^3 \text{ s} = \underline{0.90 \text{ h}}.$$

(c) The energy used to warm the water in units of kW · h is:

$$\Delta PE_q = 2.26 \times 10^7 \text{ J}\left(\frac{1 \text{ kW} \cdot \text{h}}{3.6 \times 10^6 \text{ J}}\right) = 6.3 \text{ kW} \cdot \text{h}.$$

The cost of warming the water will be

$$\text{Cost} = 6.3 \text{ kW} \cdot \text{h}\left(\frac{9 \text{ cents}}{1 \text{ kW} \cdot \text{h}}\right) = \underline{57 \text{ cents}}. \qquad \blacksquare$$

Summary and Additional Readings

1. A **source of electromotive force** (emf), such as a battery, uses various forms of energy to separate positive and negative electric charges. The **emf ℰ** of the source is the steady voltage across its terminals when no current flows out of or into the terminals.

2. When a conducting object is attached to the terminals of a source of emf, electric charge flows through the object. The **current** I is defined as the charge Δq that flows past a cross section in the object divided by the time Δt needed for the charge to pass:

$$I = \frac{\Delta q}{\Delta t}. \qquad (24.1)$$

The current is in units of **amperes** (A), where 1 A = 1 C/s. The direction of the current is opposite the direction in which negatively charged electrons flow.

3. The magnitude of the current depends on the voltage V across the object and on its electrical resistance R. The **resistance** of the object is

$$R = \rho\frac{L}{A}, \qquad (24.5)$$

where ρ is its **resistivity** (see Table 24.1), L is the length of the object, and A is its cross-sectional area. The unit of resistance is the **ohm** (Ω), where 1 ohm = 1 V/A.

4. For many objects, the current I passing through an object of resistance R when a voltage V is placed across the object is determined using **Ohm's law:**

$$V = IR. \qquad (24.3)$$

5. **Power** is the rate at which electrical energy is converted to other forms of energy. When an electrical current I passes through an object of resistance R across which a voltage V exists, the power usage can be determined using any of the following expressions:

$$\text{Power} = P = \frac{\Delta PE_q}{\Delta t} \qquad (24.8)$$
$$= IV = I^2R = V^2/R. \qquad (24.9)$$

W. F. Magie, *Source Book in Physics,* Harvard University Press, Cambridge (1963). Contains excerpts of papers by Ohm, Volta, and others.

T. H. Gebolle and J. K. Hulm, "Superconductors in Electric-Power Technology," *Scientific American,* pp. 138–172, November (1980).

G. M. Shepher, "Microcircuits in the Nervous System," *Scientific American* p. 92, February (1978).

Questions

1. If electrons in wires that connect the terminals of a battery to a lightbulb "drift" through the wire at an average speed of about 0.3 mm/s, why does the light turn on the instant the wires are connected, even though the wires leading to the bulb may be several meters long?

2. When you open the valve to a water faucet, water rushes out. Explain why electric charge does not get pushed out the end of a wire attached to the terminal of a battery.

3. Think of five uses (practical or impractical) for the fluid-filled rubber-tube displacement transducer described in Example 24.3.

4. At one time aluminum rather than copper wires were used to transmit electricity through homes. Which wire must have the larger radius if they are the same length and have the same electrical resistance? Explain.

5. Why can birds perch on a 100,000-V power line with no adverse effects?

6. When repairing electronic equipment in which high voltages are present, you should keep one hand in your back pocket and work only with the other hand. Explain carefully why this procedure is recommended.

7. Does a 60-W lightbulb have more or less resistance than a 100-W bulb? Explain.

8. When electric current flows through a resistor, the voltage drop across the resistor is $V = IR$ and the thermal energy lost by the resistor per unit time (power lost) is $P = VI$. A skydiver's fall at terminal velocity is analogous to this electrical situation. Describe verbally the analogous physical quantities that might be used to describe the skydiver's fall.

9. Explain the difference between a kilowatt and a kilowatt-hour in terms of the quantities for which they are units.

Problems

24.2 Electric Current

1. A 60-W lightbulb typically has a current of 0.50 A flowing through it. Calculate (a) the number of electrons passing a cross section of the bulb's filament each second and (b) the number that pass the cross section in 1 h.

2. A long wire is connected to the terminals of a battery. In 8 s, 9.6×10^{20} electrons pass a cross section along the wire. Calculate the current in the wire. If the electrons flow from left to right, in which direction is the current?

3. A typical flashlight battery will produce a 0.5-A current for about 3 h before losing its charge. Calculate the total number of electrons that flow during that time past a cross section of wire connecting the battery and lightbulb.

■ 4. Suppose that all the people on the earth moved at the same speed around a circular track. Approximately how many times per second would each person have to pass the starting line so that the "people current" would be the same as the number of electrons passing a cross section in a wire when a 1-A current flows?

■ 5. A 5.0-A current caused by moving electrons flows through a wire. (a) Calculate the number of electrons that flow past a cross section each second. (b) Suppose that the same number of water molecules flow along a tiny stream. Calculate the volume of water that flows past a cross section of the stream each second. [*Hint:* The density of water is 1000 kg/m^3, and one water molecule has a mass of 3.0×10^{-26} kg.]

24.3 and 24.4 Ohm's Law, Resistance, and Resistivity

6. (a) Calculate the current through a 2.5-Ω flashlight filament when connected across two 1.5-V batteries (a potential difference of 3.0 V). (b) Calculate the resistance of the heating elements in a toaster if a 10-A current flows when connected across a 120-V potential difference.

7. A person accidentally touches a 120-V power line with one hand while touching a 0-V ground wire with the other hand. Calculate the current through the body when the hands are dry (100,000-Ω resistance) and when wet (5000-Ω resistance). Are either or both currents dangerous? Explain.

8. (a) The light in an automobile draws a current of 1.0 A when connected to a 12-V battery. What is the light's resistance?

(b) What potential difference is needed to produce a current of 5.0 mA through a 2.0-MΩ resistor?

9. A long wire connected to the terminals of a 12-V battery has 6.4×10^{19} electrons passing a cross section of the wire each second. Calculate its resistance.

10. A human nerve cell can be considered as a long, thin cylinder of radius 5×10^{-6} m and length 0.3 m. The resistivity of the fluid inside the cell is 0.5 Ω · m. Calculate the resistance of this fluid.

11. A 100-m-long copper wire of radius 0.12 mm is connected across a 1.5-V battery. Calculate (a) the resistance of the wire and (b) the current passing through it.

12. The BMT subway line in New York City stretches roughly 30 km from the Bronx to Brooklyn. Calculate the electrical resistance of its rail that carries electric current. The rail has a cross section of about 40 cm² and is made of steel whose resistivity is 10×10^{-8} ohm · m.

■ 13. A platinum resistance thermometer consists of a coil of 0.10-mm-diameter platinum wire wrapped in a coil. Calculate the length of wire needed so that the coil's resistance at 20°C is 25 Ω. The resistivity of platinum at this temperature is 1.0×10^{-7} Ω · m.

■ 14. As the potential difference in volts across a thin platinum wire increases, the current in amperes changes as follows: $(V, I) = (0, 0)$, $(1.0, 0.112)$, $(3.0, 0.337)$, and $(6.1, 0.675)$. Plot a graph of potential difference as a function of current and indicate whether the platinum wire satisfies Ohm's law. If so, what is the resistance of the wire?

■ 15. Saltwater in a cylindrical rubber hose has a resistivity of 5.0 Ω · m. Calculate the factor by which the resistance of the water changes when the hose is stretched so that its length increases by a factor of 1.4. The water volume remains constant.

■ 16. A wire whose resistance is 2.0 Ω is stretched so that its length is tripled while its volume remains unchanged. Calculate the resistance of the stretched wire.

■ 17. Calculate the ratio of the resistances of two wires that are identical except that (a) wire A is twice as long as wire B, (b) wire A has twice the radius of wire B, (c) wire A is made of copper and wire B is made of aluminum. Be sure to show clearly how you arrive at each answer.

■ 18. The wires in a home carry 15 A of current. The voltage drop across the wires because of their resistance should be 1.0 V or less. If the wires are made of aluminum and have a radius of 1.2 mm, what is their maximum length?

■ 19. You fall from a cliff and grab a high-voltage power line to break your fall. Your hands are separated by 0.60 m. The wire carries 300 A of current through a copper wire with 3.0×10^{-4} m^2 cross-sectional area. Calculate the voltage between your hands. Would you be electrocuted by this voltage? Explain. You will be in serious trouble if you grab another line with one hand while holding the high-voltage line with the other hand. Why?

24.5 Resistance and Temperature

20. If the copper wire of an electromagnet has a resistance of 0.300 Ω when at room temperature (20°C) with no current flowing, what is the temperature of the wire when the current is turned on and the resistance increases to 0.347 Ω?

21. Calculate the resistance of a carbon rod when at 60°C if its resistance at 0°C is 0.0180 Ω.

■ 22. (a) A nichrome wire has a resistance of 12.000 at 0°C. Calculate its resistance when placed in a hot springs at 55°C. (b) What is the temperature of bath water in which the wire's resistance is 12.146 Ω?

■ 23. (a) A thermistor, whose resistance varies with temperature as shown in Fig. 24.10, is placed in a stream of air, causing it to have a resistance of 2420 Ω. What is the temperature of the air? (b) What is the approximate resistance of the thermistor if the temperature of the air decreases by 10°C from the answer in part (a)?

■ 24. The tungsten filament of a lightbulb has a resistance of 8.00 Ω when no current flows, and its temperature is 20°C. Estimate the filament's temperature when a 1.00-A current flows after a 120-V potential difference is placed across the filament.

■ 25. The resistance of a copper wire increases by a factor of 1.32 when current flows through the wire compared to its resistance when no current flows and its temperature is 0°C. What is its temperature when the current flows?

■ 26. (a) Calculate the resistance of a 0°C copper wire of length 50 m and radius 1.0 mm. (b) What is the wire's resistance at 120°C?

24.6 Electrical Power

27. A lightbulb when connected to a 110-V potential difference uses electrical energy at a rate of 60 W. Calculate the resistance of the bulb and the current flowing through the bulb.

28. (a) A pocket calculator draws 0.40 A of current when connected to a 6.0-V battery. At what rate does the calculator use electrical energy? (b) An electric heater is to provide 1.4 kW of power when connected to a 120-V potential difference. What should the resistance of the heater be when turned on?

■ 29. A mercury cell has a 1.5 A · h capacity (the product of the current it supplies and the time the current is supplied before the battery loses its charge). If the potential difference across its terminals is 1.35 V, how long can the cell provide the 0.10-mW power needed for a heart pacemaker before the cell stops operating?

30. The 12-V battery of a car produces a 0.20-A current in the car radio. Calculate (a) the power used by the radio and (b) the electrical energy used in 30 min of operation.

31. During a lightning flash, −6 C of charge is transferred across a potential difference of about 10^7 V. The flash lasts approximately 1 ms. Calculate the average power during the flash—that is, the average rate of electrical energy decrease.

■ 32. A refrigerator connected to a 120-V line runs for 180 h during one month. While on, it draws 3.0 A of current. (a) Calculate the electrical energy used by the refrigerator during the month in kilowatt-hours and in joules. (b) Calculate the cost of the electricity at a rate of 9 cents per kilowatt-hour.

■ 33. A 1/2-hp motor draws a current of 4.0 A from a 120-V line. (a) Calculate the electric power used by the motor. (b) Calculate the motor's efficiency—that is, the work it does per unit time (convert 1/2 hp to W) divided by the electrical power it uses.

■ 34. An x-ray tube accelerates electrons across a 50,000-V potential difference. The current of the electron beam is 5.0 mA. Calculate (a) the number of electrons passing a cross section of the electron beam per second and (b) the electrical energy dissipated in the 0.30 s the beam is on.

■ 35. A wire of resistivity 2.8×10^{-8} Ω · m and length 20 m must dissipate electrical energy at a rate less than 3.0 W when a 15-A current flows. Calculate the minimum cross-sectional area of the wire.

■ 36. A wire is connected to the terminals of a constant-voltage source of emf. By what factor does (a) the current through the wire change and (b) the power dissipated by the wire change when its length is increased by a factor of 2.0? The wire's volume remains constant.

■ 37. Two long wires are connected across the same voltage. Calculate the ratio of the power used in wire A to that used in B if the wires are identical except that (a) wire A is twice as long as B, (b) the radius of wire A is twice the radius of wire B, (c) wire A is made of copper and wire B of aluminum.

■ 38. *Estimate* the cost of electricity used to complete the homework assignment that includes this problem. The only restriction is that the assignment is worked at night (for estimation purposes only). Justify any numbers used in your estimate.

■ 39. The United States expects to switch more to coal-fired electric power plants as our gas and oil become depleted. (a) *Estimate* the mass of coal that will be burned each year if 100 percent of our electricity is provided by burning coal. (b) *Estimate* the thermal energy transferred to the environment by this coal burning. The following numbers may be helpful. One kilogram of coal provides 2.9×10^7 J of thermal energy. About 35 percent of the coal's thermal energy is converted to electrical energy, and the other 65 percent is transferred to the environment. The average person in the United States uses 31 kW · h of electricity per day.

■ ■ 40. An electric grill made of iron has a specific heat of 460 J/kg · C° and a mass of 2.8 kg. To cook French toast, the grill is warmed from 20°C to 350°C by resistive heating wires that produce thermal energy at a rate of 1500 W when connected to a 115-V potential difference. Fifty percent of the thermal energy is radiated into the room as the grill warms. How many minutes are required to warm the grill?

■ ■ 41. An airplane deicer melts 0.10 kg of ice from the wings of an airplane each minute. The deicer consists of resistive heating wires connected to a 24-V battery. Calculate the current through the heating wires and their resistance. Assume that the deicer transfers 100 percent of its energy to the ice?

■ ■ 42. The wires in a large electromagnet carry 150 A of current across a voltage of 40 V. The wires are cooled by water flowing near the coiled wire of the electromagnet. If water flows at 2.0×10^4 cm^3/min, what is the difference in temperature of water leaving the magnet compared to water entering the magnet?

CHAPTER 25

Direct-Current Electric Circuits

Until now, we have discussed only the electric current flowing from a voltage source through a single electronic device that has resistance. Most electric circuits consist of a combination of many resistive elements, one or more voltage sources, switches, and a variety of other electronic devices. Your home, for example, can be represented crudely by the electronic circuit shown in Fig. 25.1.

In this chapter we will develop techniques to determine the electric current flowing through each element in one of these more complicated circuits and the electric power used by each element. Our attention is restricted to circuits powered by direct-current sources of emf that have a variety of different resistive elements.

25.1 Kirchhoff's Rules

All of the circuits we will describe in this chapter have one or more sources of electromotive forces, which in our case can be considered as batteries. Inside the batteries, positive and negative electric charge are separated and made available at the battery terminals to an external circuit. When connected to the wires and resistive elements in an external circuit, the potential difference across the battery terminals causes electric charge to move through the circuit—electric current is produced. For the more complicated circuits of this chapter, the electric charge moves along several paths, like water running down different arroyos during a rainstorm. We have two main goals in this chapter: (1) to determine the current flowing along each part of a circuit and (2) to determine the potential difference (voltage change) across different elements in the circuit. Two important rules are used for this purpose: Kirchhoff's loop rule and Kirchhoff's junction rule.

Kirchhoff's Loop Rule

The loop rule is based on the conservation-of-energy principle and states that the algebraic sum of the voltage changes around any closed path or loop in a

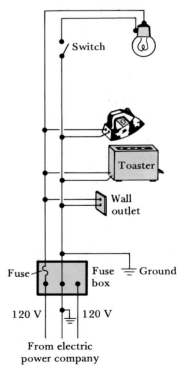

FIG. 25.1. An electric circuit for a home wiring system.

FIG. 25.2. (a) A lightbulb is attached to a battery by wires. (b) An electronic circuit representing the equipment shown in part (a).

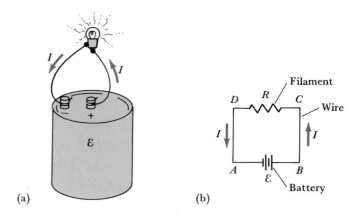

(a)

(b)

circuit is zero. The loop rule can be understood by examining the flow of electric charge from a battery through wires connected to the filament of a lightbulb (Fig. 25.2a). An electronic circuit representing this arrangement of battery, wires, and filament is shown in Fig. 25.2b.

Let us follow the motion of a positive charge as it makes one complete counterclockwise trip around the circuit (the charge moves in the same direction as the current). When using the loop rule, we will always pretend that moving positive charges cause the current. It is much easier to keep track of the signs of voltage changes caused by a positive charge moving in the direction of the current than to keep track of the signs of voltage changes caused by negative charges moving opposite the direction of the current. The result is the same in either case.

Consider first the change in voltage as a charge q moves from position A through the battery. Chemical energy must be expended to pull the charge through the battery away from the negative terminal and toward the positive terminal. For a positive charge, the trip across the battery is like being pulled up a hill whose height in units of volts is $\mathcal{E}$, the emf from the negative to positive terminal of the battery.

During the rest of the trip, the charge moves through resistors. The potential difference across a resistor of resistance R through which a current I flows is given by Ohm's law: $V = IR$. If the charge is moving in the direction of the current, the potential decreases as the charge moves through the resistor. The motion of the charge is analogous to the fall of a skydiver in the earth's gravitational field. The skydiver's elevation and gravitational potential energy decrease as the skydiver is pulled downward in the gravitational field. The electrical potential and potential energy of a positive electric charge also decreases as it is pulled thorugh a resistive material by an electric field.

In the case of the circuit shown in Fig. 25.2b, there is no change in potential along the wires since they have almost zero resistance, and, according to Ohm's law, $V = IR = I(0) = 0$. However, the tungsten filament of the lightbulb does have resistance, and the change in voltage across this resistance is $V_{CD} = -IR$, the negative sign being used to indicate a lower voltage at D than at C.

The charge then moves along the second wire from position D back to position A where it started. The net voltage change for the whole trip is zero. The charge is at the same position and electrical elevation as it was when it started the trip. Thus, the sum of the changes in voltage for each step along the trip must add to zero:

Voltage changes				
Through battery	Along wire	Through resistor	Along wire	
ε +	0 −	IR +	0 =	0.

Kirchhoff's loop rule: The algebraic sum of the voltage changes around any closed loop in an electric circuit is zero.

The rule applies to circuits in which a steady current flows. If abrupt changes in current occur, such as when a switch is closed, the rule may not be applicable.

When using Kirchhoff's loop rule, we must be careful to use the corrent signs for the voltage change across each element. These voltage-change conventions are summarized in Fig. 25.3. The symbol V_{AB} represents the voltage change across an element from position A to position B. Note that the voltage change across a battery is $+\varepsilon$ if we move from the negative to the positive terminal and $-\varepsilon$ if we move from the positive to the negative terminal.*

For resistors, the voltage change is $-IR$ if we move across the resistor in the direction of the current and $+IR$ if we move across the resistor opposite the direction of current flow. The voltage is higher on the side at which current enters the resistor than on the side at which it leaves.

EXAMPLE 25.1 (a) Write a loop-rule equation for the circuit shown in Fig. 25.4a. (b) Solve the equation for the current that is flowing. (c) Determine the voltage change from point A to point B, that is, V_{AB}.

SOLUTION (a) To start the problem, we first draw an arrow indicating the direction that we expect the current will flow (Fig. 25.4b). For this circuit, the current flows counterclockwise because the 10-V battery "pushes harder" than the 4-V battery, which opposes the current as drawn. We will discover later that if the direction of the current is incorrectly chosen, the equations will indicate our error with no extra effort on our part.

Starting at A in Fig. 25.4b and moving counterclockwise, we find the following voltage changes:

$$+10\text{ V} - I(2\,\Omega) - 4\text{ V} - I(6\,\Omega) - I(4\,\Omega) = 0.$$

Notice that the voltage change across the 4-V battery was negative; we moved from the positive to the negative terminal.

(b) We now solve the preceding equation for the current:

$$I = \frac{10\text{ V} - 4\text{ V}}{2\,\Omega + 6\,\Omega + 4\,\Omega} = \frac{6\text{ V}}{12\,\Omega} = \underline{0.5\text{ A}}.$$

(c) To determine the voltage change from A to B, we follow the circuit along any continuous path from A to B and add the voltage changes across each element in the circuit along that path. For this problem, we can travel from A to B by two different routes (either can be used). We first move counterclockwise from A through the 10-V battery and 2-Ω resistor:

$$V_{AB} = +10\text{ V} - (0.5\text{ A})(2\,\Omega) = \underline{+9\text{ V}}.$$

*We are ignoring the internal resistance of the battery.

Device	Symbol	Voltage Change
Battery	$A \bullet\!\!-\!\!|\!|\!|\!-\!\!\bullet B$ ε	$\begin{cases} V_{AB} = +\varepsilon \\ V_{BA} = -\varepsilon \end{cases}$
Resistor	$A \bullet\!\!-\!\!\text{\Large ww}\!\!-\!\!\bullet B$ R, I	$\begin{cases} V_{AB} = -IR \\ V_{BA} = +IR \end{cases}$

FIG. 25.3. Potential difference across batteries and resistors.

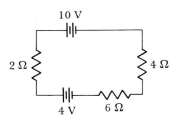

(a)

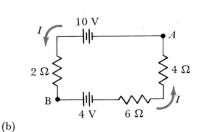

(b)

FIG. 25.4. (a) An electric circuit and (b) the same circuit with the current identified (including its direction).

The voltage is 9V higher at B than at A. If we placed the terminals of a voltmeter from A to B (the black terminal at A and the red at B), the voltmeter would read $+9$ V.

We can also calculate the voltage change from A to B by moving clockwise through the 4-Ω and 6-Ω resistors (opposite the current flow) and through the 4-V battery:

$$V_{AB} = +(0.5\ \text{A})(4\ \Omega) + (0.5\ \text{A})(6\ \Omega) + 4\ \text{V}$$
$$= 2\ \text{V} + 3\ \text{V} + 4\ \text{V} = \underline{+9\ \text{V}}.$$

The voltage change is the same no matter which route we take from A to B.

■

Kirchhoff's Junction Rule

We have used the loop rule to analyze a circuit with one loop. However, many electric circuits contain several loops. The current moving through a wire in one loop may come to a junction where it divides to travel in two or more wires, or the currents from several wires may join at a junction to move through a single wire. The flow of electric charge into and out of these *junctions* is much like the flow of water in a pipe. If the pipe divides into several pipes at a junction, the water flowing in the original pipe must equal the water leaving the junction in the other pipes.

For electric charge, the current flowing along one or more wires into a junction must equal the current flowing out of the junction along other wires. Consider the circuit shown in Fig. 25.5a. The battery at the top causes current to flow out of the positive terminal on the left in a counterclockwise sense (I in Fig. 25.5b). When this current reaches point A, two paths are available. The current I actually splits at this junction, the part labeled I_1 going through resistor R_1 and the remainder labeled I_2 going through resistor R_2. For point A, the current I entering that junction equals the sum of the currents I_1 and I_2 leaving that junction:

$$I = I_1 + I_2.$$

At point B, the two currents recombine to form a single, larger current I that leaves junction B. For B, the current entering equals that leaving:

$$I_1 + I_2 = I.$$

For the circuit in Fig. 25.5b, the current I moving up the right side away from B equals the current I moving down the left side toward A. This is true because there are no junctions along this upper part of the circuit in which the current can separate.

We can summarize this analysis of the splitting of currents at a junction as follows.

Kirchhoff's junction rule: The algebraic sum of all currents flowing into the junction of a circuit equals the algebraic sum of the currents flowing out:

Sum of currents into junction = Sum of currents out.

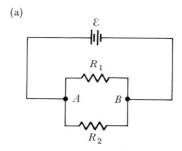

(a)

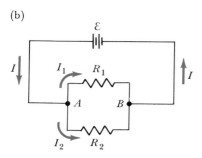

(b)

FIG. 25.5. The current divides into two branches at point A.

Solving Problems with Kirchhoff's Rules

A procedure for using Kirchhoff's rules to determine the currents through and voltages across various parts of a circuit is described and illustrated as follows. Our goal in the illustration problem is to determine the current in each branch of the circuit shown in Fig. 25.6.

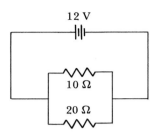

1. Redraw the original circuit. Then label each separate current with a symbol and an arrow that points in the direction of flow of that current. A current label and arrow should be made for each branch of the circuit. A branch is a section of the circuit along which the current does not change. If you mistakenly assign the wrong direction to a current, do not worry. When your equations are solved, you will find that the current has a negative sign. This indicates that the current actually moves in the opposite direction to the one you assigned.

FIG. 25.6

The circuit in the accompanying figure has three branches, and the currents for these branches have been labeled I, I_1, and I_2 and assigned arrows indicating the anticipated directions of the currents.

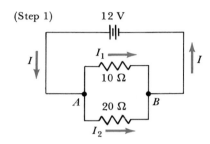

2. Apply Kirchhoff's junction rule to enough junctions so that each current appears in at least one equation.

For point A in the figure, the junction rule is

$$I = I_1 + I_2.$$

Since all three currents appear in this equation, we do not need to use the junction rule again.

3. Apply Kirchhoff's loop rule to enough closed loops so that each current appears in at least one loop equation. Remember to use the sign conventions listed in Fig. 25.3.

We will use the loop rule for two different loops. In the first loop, we start at B and move counterclockwise through the battery and the 10-Ω resistor back to B again:

$$+12 \text{ V} - (10 \text{ }\Omega)I_1 = 0.$$

For the second loop, we move counterclockwise from B through the battery and the 20-Ω resistor and then back to B:

$$+12 \text{ V} - (20 \text{ }\Omega)I_2 = 0.$$

The current I does not appear in these loop equations because there are no resistors in the branch of the circuit with current I.

4. Solve the equations for the desired unknowns.

We now have three equations and three unknowns (I, I_1, and I_2). The latter two equations can be solved for I_1 and I_2:

$$I_1 = \frac{12 \text{ V}}{10 \text{ }\Omega} = \underline{1.2 \text{ A}},$$

$$I_2 = \frac{12 \text{ V}}{20 \text{ }\Omega} = \underline{0.6 \text{ A}}.$$

Substituting these results in our first junction-rule equation, we find the current I:

FIG. 25.7

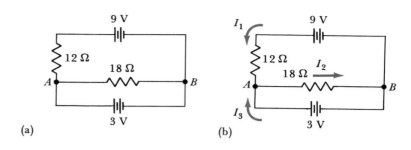

(a) (b)

$$I = I_1 + I_2 = 1.2\,\text{A} + 0.6\,\text{A}$$
$$= \underline{1.8\,\text{A}}.$$

EXAMPLE 25.2 (a) Calculate the currents flowing in each branch of the circuit shown in Fig. 25.7a. (b) Calculate the potential difference across the 18-Ω resistor from A to B.

SOLUTION (a) The circuit is redrawn in Fig. 25.7b with the current symbols and directions identified. We have intentionally drawn the direction of current I_3 incorrectly to show that the proper application of Kirchhoff's rules will indicate our error. The junction rule applied to point A results in the equation

$$I_1 + I_3 = I_2. \tag{25.1}$$

The loop rule is applied for the two different loops shown in Fig. 25.8. First, starting at B in Fig. 25.8a and moving clockwise through the 3-V battery and the 18-Ω resistor, we find that

$$+3\text{V} - (18\,\Omega)I_2 = 0,$$

or

$$I_2 = \frac{3\,\text{V}}{18\,\Omega} = \underline{\frac{1}{6}\text{A}}.$$

For the loop in Fig. 25.8b, we move counterclockwise from B through the 9-V battery, the 12-Ω resistor, and the 3-V battery back again to B:

$$+9\text{ V} - (12\,\Omega)I_1 - 3\text{ V} = 0,$$
$$I_1 = \frac{9\text{ V} - 3\text{ V}}{12\,\Omega} = \underline{\frac{1}{2}\text{A}}.$$

Then, using Eq. (25.1), we find that

$$I_3 = I_2 - I_1 = \frac{1}{6}\text{A} - \frac{1}{2}\text{A} = \underline{-\frac{1}{3}\text{A}}.$$

The negative sign indicates that the direction of I_3 was chosen incorrectly. The 9-V battery actually forces charge through the 3-V battery from left to right.

(b) The voltage from A to B, that is V_{AB}, can be calculated by adding the voltage changes across each element in the circuit as we move from A to B. Since there are three unique paths between A and B in the circuit shown in Fig. 25.7b, we can calculate the voltage change in three independent ways:

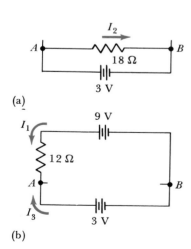

(a)

(b)

FIG. 25.8. Two loops of the circuit shown in Fig. 25.7.

A to *B* through 18-Ω resistor: $V_{AB} = -I_2 R$

$$= -\left(\frac{1}{6}\text{A}\right)(18\,\Omega) = \underline{-3\text{ V}}.$$

A to *B* through 12-Ω resistor and 9-V battery: $V_{AB} = +I_1(12\,\Omega) - 9\text{ V}$

$$= +\left(\frac{1}{2}\text{A}\right)(12\,\Omega) - 9\text{ V}$$

$$= -3\text{ V}.$$

A to *B* through 3-V battery: $V_{AB} = \underline{-3\text{ V}}.$

We see that the voltage between points *A* and *B* is independent of which route we choose. The voltage is 3 V lower at *B* than at *A* ∎

EXAMPLE 25.3 A 1380-W electric heater, a 180-W heating pad, a 120-W lightbulb, and a 1200-W iron are connected in parallel across the terminals of a single wall outlet (Fig. 25.9). If the current through the wire from the fuse box to the wall outlet exceeds 20 A, a fuse is blown. Will the fuse be blown when all these appliances are in operation at the same time? The voltage across each appliance is 120 V.

SOLUTION The wires connecting the different appliances have negligible resistance. Consequently, there is no $V = IR$ voltage drop along these wires. It is as though they were all connected together as shown in Fig. 25.9b. Thus, the potential difference across each appliance is 120 V and the current through each appliance can be calculated using Eq. (24.9):

Heater: $\quad I_1 = \dfrac{P}{V} = \dfrac{1380\text{ W}}{120\text{ V}} = 11.5\text{ A},$

Heating pad: $\quad I_2 = \dfrac{180\text{ W}}{120\text{ V}} = 1.5\text{ A},$

Light: $\quad I_3 = \dfrac{120\text{ W}}{120\text{ V}} = 1.0\text{ A},$

Iron: $\quad I_4 = \dfrac{1200\text{ W}}{120\text{ V}} = 10\text{ A}.$

Using Kirchhoff's junction rule for point *A*, we find the current *I*:

$$I = I_1 + I_2 + I_3 + I_4 = (11.5 + 1.5 + 1.0 + 10)\text{ A}$$

$$= \underline{24\text{ A}}.$$

Since the current *I* exceeds 20 A, the fuse, a thin piece of copper wire, becomes hot and melts. Thus, when all the appliances are operating, the fuse melts or blows and the current stops. If a fuse were not included in the circuit and the current exceeded 20 A, the main wires that carry the current through the walls of the house would become very warm and could possibly start a fire. ∎

EXAMPLE 25.4 Solve for the currents in all branches of the circuit shown in Fig. 25.10. $\mathcal{E}_1 = 12\text{ V}$, $\mathcal{E}_2 = 3\text{ V}$, $R_1 = 50\,\Omega$, $R_2 = 200\,\Omega$, and $R_3 = 100\,\Omega$.

SOLUTION The currents are identified in Fig. 25.10. The junction rule for point *B* produces the equation

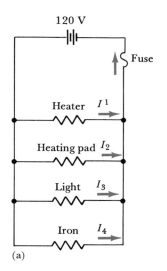

(a)

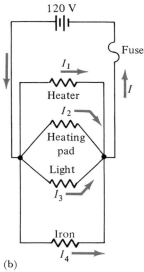

(b)

FIG. 25.9. (a) The circuit for one outlet of a home wiring system. (b) The same circuit redrawn in a different but equivalent manner.

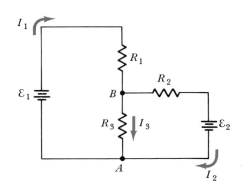

FIG. 25.10

$$I_1 = I_2 + I_3. \tag{25.2}$$

Since there are three unknown currents, we will have to apply the loop rule twice to get three independent equations (including the junction-rule equation above) to solve for the three unknowns. Moving clockwise around the outermost loop starting at A, we find, omitting units, that

$$\mathcal{E}_1 - R_1 I_1 - R_2 I_2 - \mathcal{E}_2 = 0, \quad \text{or}$$
$$12 - 50 I_1 - 200 I_2 - 3 = 0. \tag{25.3}$$

Moving clockwise around the left loop starting at A, we find that

$$\mathcal{E}_1 - R_1 I_1 - R_3 I_3 = 0, \quad \text{or}$$
$$12 - 50 I_1 - 100 I_3 = 0. \tag{25.4}$$

Equations (25.2)–(25.4) must now be solved for the three unknown currents. We can first use Eq. (25.2) to substitute for I_1 in Eqs. (25.3) and (25.4):

$$12 - 50(I_2 + I_3) - 200 I_2 - 3 = 0,$$
$$12 - 50(I_2 + I_3) - 100 I_3 = 0.$$

Combining terms with common factors, the preceding equations become:

$$9 - 250 I_2 - 50 I_3 = 0,$$
$$12 - 50 I_2 - 150 I_3 = 0.$$

If we multiply the first equation by -3 and then add it and the second equation, the I_3 drops out:

$$-27 + 750 I_2 + 150 I_3 = 0$$
$$\underline{\quad 12 - 50 I_2 - 150 I_3 = 0 \quad}$$
$$-15 + 700 I_2 \qquad\quad = 0,$$

or
$$I_2 = \frac{15}{700} = \underline{0.021 \text{ A}}.$$

Substituting this result into the first equation above, we find that

$$-27 + 750(0.21) + 150 I_3 = 0, \quad \text{or}$$
$$I_3 = \frac{10.93}{150} = \underline{0.073 \text{ A}}.$$

We can now solve for I_1 using Eq. (25.2):

$$I_1 = I_2 + I_3 = 0.021 \text{ A} + 0.073 \text{ A}$$
$$= \underline{0.094 \text{ A}}.$$

25.2   Equivalent Resistance

When analyzing an electric circuit, it is often convenient to replace several of its resistors with a single equivalent resistor that produces the same effect as the resistors it replaces. These replacements can be made for several resistors in series and for several resistors in parallel.

Resistors in Series

A circuit containing three resistors in series is shown in Fig. 25.11a. Resistors are in series if the same current flows through each with no branching between resistors. If a source of voltage $\mathcal{E}$ is placed across the resistors, a current I flows through them. We wish to replace the three resistors in series by a single equivalent resistor R_{eq} that will cause the same current to flow when attached to an identical voltage source $\mathcal{E}$ (Fig. 25.11b).

The loop rule applied to the circuit shown in Fig. 25.11a results in the equation

$$\mathcal{E} - IR_1 - IR_2 - IR_3 = 0, \quad \text{or}$$

$$I = \frac{\mathcal{E}}{R_1 + R_2 + R_3}.$$

When applied to the circuit shown in Fig. 25.11b, the loop rule results in the equation

$$\mathcal{E} - IR_{eq} = 0, \quad \text{or}$$

$$I = \frac{\mathcal{E}}{R_{eq}},$$

where R_{eq} is the resistor that is equivalent to R_1, R_2, and R_3. Since the current is the same in both circuits,

$$I = \frac{\mathcal{E}}{R_1 + R_2 + R_3} = \frac{\mathcal{E}}{R_{eq}}.$$

This will be true if

$$R_{eq} = R_1 + R_2 + R_3.$$

In general, any group of **series resistors** can be replaced by a single equivalent resistor whose resistance is

$$R_{eq} = R_1 + R_2 + R_3 + \cdots . \tag{25.5}$$

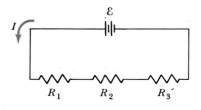

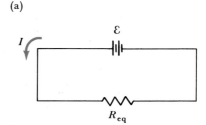

FIG. 25.11. (a) A circuit with three series resistors and (b) an equivalent circuit.

Resistors in Parallel

Three parallel resistors are shown in Fig. 25.12a. Resistors are in parallel if one side of each resistor is connected to a common junction and the other side is connected to another common junction. If a source of voltage $\mathcal{E}$ is connected across these resistors, the current I splits when it reaches the first junction (point A in Fig. 25.12a). The voltage across each resistor in Fig. 25.12a is the same and has a value $\mathcal{E}$. Thus, the current through the resistors is given by

$$I_1 = \frac{\mathcal{E}}{R_1}, \quad I_2 = \frac{\mathcal{E}}{R_2}, \quad I_3 = \frac{\mathcal{E}}{R_3}.$$

When the junction rule is applied to point A, the total current I is seen to be

$$I = I_1 + I_2 + I_3.$$

Substituting for I_1, I_2, and I_3, we find that

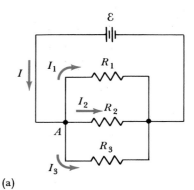

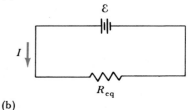

FIG. 25.12. (a) A circuit with three parallel resistors and (b) an equivalent circuit.

$$I = \frac{\mathcal{E}}{R_1} + \frac{\mathcal{E}}{R_2} + \frac{\mathcal{E}}{R_3} = \mathcal{E}\left(\frac{1}{R_1} + \frac{1}{R_2} + \frac{1}{R_3}\right).$$

If the three parallel resistors are replaced by a single equivalent resistor (Fig. 25.12b), the current passing through it is

$$I = \frac{\mathcal{E}}{R_{eq}} = \mathcal{E}\left(\frac{1}{R_{eq}}\right).$$

For the current I to be the same for each circuit, the equivalent resistance must be given by the equation:

$$\frac{1}{R_{eq}} = \frac{1}{R_1} + \frac{1}{R_2} + \frac{1}{R_3}.$$

In general, any group of **parallel resistors**, $R_1, R_2, R_3, \ldots$, can be replaced by an equivalent resistor whose resistance is calculated using the equation:

$$\frac{1}{R_{eq}} = \frac{1}{R_1} + \frac{1}{R_2} + \frac{1}{R_3} + \cdots . \qquad (25.6)$$

These two rules for replacing series and parallel resisters by an equivalent resistor can now be applied to replacing, step by step, a group of many resistors by one or more equivalent resistors that produce the same effect as all the resistors. Each step involves replacing either several series resistors or several parallel resistors by an equivalent resistance. We cannot, however, simultaneously replace combinations of series and parallel resistors by a single equivalent resistor. Consider the next examples.

EXAMPLE 25.5 (a) Calculate the equivalent resistance of the arrangement of resistors shown in Fig. 25.13a. (b) Calculate the total current I flowing through the battery.

SOLUTION (a) It is simpler to calculate the current I by using equivalent resistors rather than by using Kirchhoff's rules. The procedure is done in steps. The first step involves replacing the 3-Ω and 6-Ω parallel resistors by an equivalent resistor:

$$\frac{1}{R_{eq}} = \frac{1}{3\,\Omega} + \frac{1}{6\,\Omega} = \frac{2}{6\,\Omega} + \frac{1}{6\,\Omega} = \frac{3}{6\,\Omega}, \quad \text{or}$$

$$R_{eq} = \frac{6\,\Omega}{3} = 2\,\Omega.$$

The new circuit with the parallel resistors replaced by a single 2-Ω resistor is shown in Fig. 25.13b.

We are now left with three resistors in series. The equivalent resistance of these series resistors is:

$$R_{eq} = 5\,\Omega + 2\,\Omega + 8\,\Omega = \underline{15\,\Omega}.$$

(b) The new circuit, shown in Fig. 25.13c, has only one 15-Ω resistor attached to the 12-V power source. The current I through this resistor and power source is given by

$$\mathcal{E} - IR = 0.$$

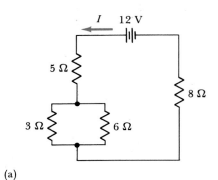

(a)

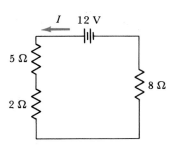

(b)

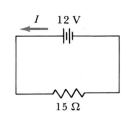

(c)

FIG. 25.13. (a) A circuit with four resistors is reduced in steps to one with a single equivalent resistor (c).

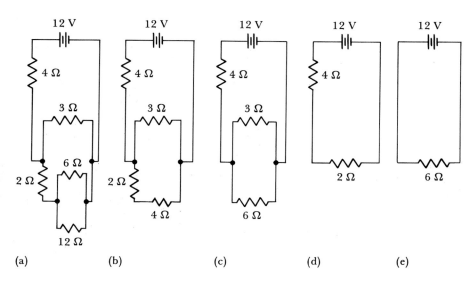

Thus,

$$I = \frac{\mathcal{E}}{R} = \frac{12 \text{ V}}{15 \text{ } \Omega} = \underline{0.80 \text{ A}}.$$ ■

EXAMPLE 25.6 A more complicated step-by-step reduction of a group of resistors is shown in Fig. 25.14. You should follow each step to see how a single resistor replaces two or more resistors that were part of the previous circuit.

■

25.3 Measuring Current and Voltage

Kirchhoff's laws and the ideas of equivalent resistance are used to design electrical circuits. Once built, the circuits can be tested using meters or other electrical instruments that measure the voltage at various parts of the circuit and the current flowing through different elements in the circuit.

Care must be taken when making any type of measurement that the measuring device does not alter the quantity being measured. A ruler used to measure a person's height does not change the height. A meter used to measure current or voltage should not change the current or voltage that is being measured. The effect of a meter on a circuit depends on what is called its **internal resistance**—the net electrical resistance of all the parts inside the meter. We will find that meters used to measure current must have a very low internal resistance and meters used to measure voltage must have a large internal resistance.

Measuring Current

To measure the current flowing in one part of a circuit, one must open the circuit (Fig. 25.15) and insert into it a current-measuring meter called an **ammeter.** The current being measured must now pass through the ammeter, which indicates the value of the current.

FIG. 25.15. (a) To measure the current in a circuit, (b) the circuit should first be opened. (c) The current is measured by inserting an ammeter into the circuit.

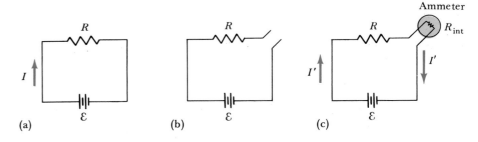

So the meter does not alter the current being measured, the meter's internal resistance must be much less than the resistance of other resistors in the circuit through which the current flows. For example, in the circuit shown in Fig. 25.15a, the current is $I = \mathcal{E}/R$, where R is the resistance of the circuit before the meter is added. After the meter is placed in the circuit, the total resistance of the circuit increases to $R + R_{int}$, where R_{int} is the internal resistance of the meter. With the meter in the circuit, the current is reduced to $I' = \mathcal{E}/(R + R_{int})$. If $R_{int} \ll R$, then we can ignore R_{int} in the expression for I', and the measured current I' is for practical purposes equal to I, the current when the meter is not present. The internal resistance of ammeters is typically 1 ohm or less.

The fact that the internal resistance of an ammeter is small can result in damage to the meter if it is used incorrectly. For instance, if the wires (called **leads**) that are used to connect the meter with the circuit being tested are accidentally placed across a resistor or across a battery, the voltage across the resistor or battery may be large enough to produce a very large current through the meter. The current is likely to damage the ammeter. *An ammeter should always be inserted into a circuit in a way that allows the normal current of the circuit to pass through it without producing a significant voltage across the leads.*

Measuring Voltage

The potential difference or voltage between two different points in a circuit, such as from one side of a resistor or battery to the other side, is measured by connecting the leads from a **voltmeter** to these points (Fig. 25.16). Unlike an ammeter, a voltmeter should have a much higher internal resistance than the resistance of the section of the circuit across which the voltage is being measured. Suppose, for example, that you wish to measure the voltage across a resistor of resistance R that has a current I flowing through it. When the voltmeter is not connected (Fig. 25.16a), the voltage across the resistor is, according to Ohm's law, $V = IR$. If a voltmeter having internal resistance R_{int} less than R is connected in parallel to R (Fig. 25.16b), then most of the current flows through the voltmeter rather than through the resistor R. If the current through the resistor is reduced, the voltage drop (IR) across the resistor is also reduced, and an incorrect voltage reading will be made. The internal resistance of the meter should be roughly 100 times or more greater than the resistance of the elements in the circuit across which the voltage is being measured.

The internal resistance of a voltmeter is usually stated on the voltmeter. For example, if the maximum or full-scale reading of the voltmeter is 30 V and its internal resistance is 20,000 Ω per volt full scale, then its internal resistance is $30 \times 20,000 \ \Omega = 6 \times 10^5 \ \Omega$.

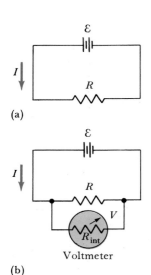

FIG. 25.16. Voltage is measured by placing the leads from a voltmeter across the section of the circuit for which you wish to measure the voltage change.

Summary and Additional Readings

1. **Kirchhoff's loop rule:** The algebraic sum of the changes in voltage around any closed loop in an electric circuit is zero. The positive terminal of a battery is at $+\mathcal{E}$ higher potential than the negative terminal (we ignore the internal resistance of the battery). The voltage drop across a resistor in the direction of a current is $-IR$.

2. **Kirchhoff's junction rule:** The electric currents flowing into any junction in an electric circuit equal the currents flowing out of the junction.

3. **Equivalent resistors:** A group of **series resistors,** R_1, R_2, R_3, . . . , can be replaced by an equivalent resistor R_{eq} of resistance

$$R_{eq} = R_1 + R_2 + R_3 + \cdot\cdot\cdot\cdot \qquad (25.5)$$

A group of **parallel resistors,** R_1, R_2, R_3, . . . , can be replaced by an equivalent resistor whose resistance R_{eq} is determined using the equation

$$\frac{1}{R_{eq}} = \frac{1}{R_1} + \frac{1}{R_2} + \frac{1}{R_3} + \cdot\cdot\cdot\cdot \qquad (25.6)$$

4. **Meters:** An **ammeter** is a device used to measure the current flowing through a part of an electric circuit. Ammeters have low internal resistance and are inserted in the circuit when measuring current. A **voltmeter** is a device used to measure the voltage, or potential difference, between two different points of a circuit. A voltmeter has high internal resistance and is connected in parallel with the part of the circuit across which the voltage is being measured.

———————

P. H. Baker, "The Nerve Axon," *Scientific American*, p 74, March (1966).

Questions

1. Construct an electric circuit that is analogous to your circulatory system. Indicate the corresponding parts of the two systems.

2. You wish to be able to switch a light in the center of a hallway on and off from either end of the hall. Draw an electric circuit that will allow you to do this.

3. It is possible to make Christmas tree lights that work when connected in parallel. If lights having a different resistance are used, they can also be connected in series. (a) Describe one advantage of lights connected in parallel. (b) Should the parallel lights or the series lights have a greater resistance so that each light will use the same electrical power in each situation? Explain.

4. Three 9-Ω resistors are to be connected to a 12-V battery. To produce the most thermal energy in the resistors, should they be connected in series or in parallel? Justify your answer.

5. Two lightbulbs, one 30 W and the other 120 W, are built to be connected across a potential difference of 120 V. Which bulb uses the most electrical power when the bulbs are connected in series to a 240-V potential difference? Explain.

6. Describe one difference between an ammeter and a voltmeter.

Problems

25.1 Kirchhoff's Rules

■ 1. (a) Write one junction-rule equation and two loop-rule equations for the circuit shown in Fig. 25.17. (b) Use these equations to calculate the current in each branch of the circuit for the case where $R_1 = 0\,\Omega$, $R_2 = 18\,\Omega$, $R_3 = 9\,\Omega$, and $\mathcal{E} = 6.0$ V.

■ 2. Repeat Problem 1 for the case where $R_1 = 50\,\Omega$, $R_2 = 30\,\Omega$, $R_3 = 15\,\Omega$, and $\mathcal{E} = 120$ V.

■ 3. The current through resistor R_1 in Fig. 25.17 is 2.0 A. Calculate the currents through resistors R_2 and R_3 and the emf of the battery. $R_1 = 4\,\Omega$, $R_2 = 10\,\Omega$, and $R_3 = 40\,\Omega$.

■ 4. Use Kirchhoff's rules to prove in general that the currents I_2 and I_3 through resistors R_2 and R_3, respectively, in Fig. 25.17 satisfy the relation $I_2/I_3 = R_3/R_2$.

■ 5. (a) Write Kirchhoff's loop rule for the circuit shown in Fig. 25.18 for the case where $\mathcal{E}_1 = 20$ V, $\mathcal{E}_2 = 8$ V, $R_1 = 30\,\Omega$, $R_2 = 20\,\Omega$, and $R_3 = 10\,\Omega$. (b) Calculate the current flowing in

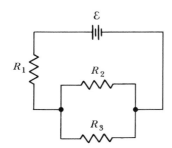

FIG. 25.17

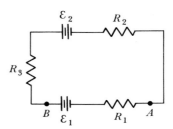

FIG. 25.18

the circuit. (c) Using this value of current, start at position A and move clockwise around the circuit calculating the voltage change across each element in the circuit (be sure to indicate the sign of each voltage change). (d) Add these voltage changes around the whole circuit.

■ 6. Repeat parts (a) and (b) of Problem 5 for the case $\mathcal{E}_1 = 12$ V, $\mathcal{E}_2 = 3$ V, $R_1 = R_2 = 1\Omega$, and $R_3 = 16\,\Omega$. (c) Calculate the voltage change from A to B.

■ 7. (a) Calculate the value of $\mathcal{E}_1$ so that a current of 1 A flows clockwise in the circuit shown in Fig. 25.18, where $\mathcal{E}_2 = 12$ V, $R_1 = 2\,\Omega$, $R_2 = 1\,\Omega$, and $R_3 = 12\,\Omega$. (b) Calculate the voltage change from B to A.

■■ 8. (a) Write the loop rule for two different loops in the circuit shown in Fig. 25.19 and the junction rule for point A. Solve the equations to find the current in each loop when $\mathcal{E}_1 = 3$ V, $\mathcal{E}_2 = 6$ V, $R_1 = 10\,\Omega$, $R_2 = 20\,\Omega$, and $R_3 = 30\,\Omega$. (b) Calculate the voltage change from A to B. Check your answer by taking a different path from A to B.

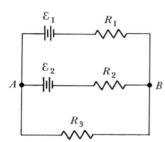

FIG. 25.19

■ 9. Determine the value of R_2, shown in Fig. 25.19, so that the current through R_3 equals twice that through R_2. The values of the other circuit elements are $\mathcal{E}_1 = 12$ V, $\mathcal{E}_2 = 15$ V, $R_1 = 15\,\Omega$, and $R_3 = 30\,\Omega$.

■ 10. A simplified electrical circuit for a home is shown in Fig. 25.20. Calculate the currents through the fuse, lightbulb, electric crock, and toaster.

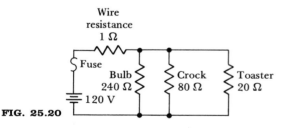

FIG. 25.20

■ 11. Determine the current in each branch of the circuit shown in Fig. 25.20 and the electric power used by each element in the circuit.

■ 12. (a) Use Kirchhoff's rules to calculate the current in each branch of the circuit shown in Fig. 25.17 for the case where $\mathcal{E} = 90$ V, $R_1 = 10\,\Omega$, $R_2 = 30\,\Omega$, and $R_3 = 60\,\Omega$. (b) Repeat the current calculations if R_1 increases to $20\,\Omega$ while $\mathcal{E}$, R_2, and R_3 remain unchanged. (c) Discuss briefly the analogy of this electrical system to the circulatory system of a person and the effect of a decrease in the size of a main artery on the flow of blood to other parts of the body.

■ 13. (a) Write Kirchhoff's rules for two loops and one junction for the circuit shown in Fig. 25.21. (b) Solve the equations for the current in each branch of the circuit when $\mathcal{E}_1 = 10$ V, $\mathcal{E}_2 = 2$ V, $R_1 = 50\,\Omega$, $R_2 = 200\,\Omega$, and $R_3 = 20\,\Omega$. (c) Calculate the voltage across resistor R_3 from point B to point A.

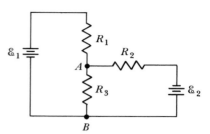

FIG. 25.21

■ 14. The circuit shown in Fig. 25.22 is an example of a voltage divider. The potential difference $\mathcal{E}$ is divided by R_1 and R_2, and only a fraction of the voltage appears across the load resistor, R_L. Calculate the value of R_2 so that the voltage across R_L is 4.0 V when $\mathcal{E} = 20$ V, $R_1 = 80\,\Omega$, and $R_L = 600\,\Omega$.

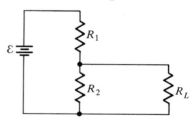

FIG. 25.22. A voltage divider.

■■ 15. The load resistor R_L in Fig. 25.22 has resistance $1000\,\Omega$ and uses 10 W of power. Calculate (a) the current through R_L and (b) the resistance of R_1 for the case where $R_2 = 100\,\Omega$ and $\mathcal{E} = 155$ V.

■■ 16. A Wheatstone bridge circuit used to measure resistance is shown in Fig. 25.23. Resistances R_1 and R_2 equal $20\,\Omega$, and R_3 is adjusted so that no current flows through the *galvanometer* at the center. The current I_g is zero when R_3 is $67.5\,\Omega$. Calculate the value of the unknown resistance R_x. Be sure to show clearly your calculation technique.

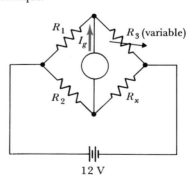

FIG. 25.23. A Wheatstone bridge.

■■ 17. A battery has a small **internal resistance** r to charge flowing through it. The battery can be represented by a circuit

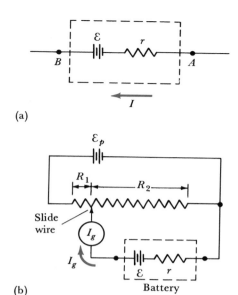

(a)

(b)

FIG. 25.24. (a) A battery with internal resistance. (b) A potentiometer is used to measure the emf of the battery.

such as shown in Fig. 25.24a, where points A and B represent the battery terminals. If a current flows through the battery, the voltage across the terminals is

$$V_{AB} = \varepsilon - Ir.$$

For most purposes, the small Ir drop can be ignored. A potentiometer is a device that can measure the value of the battery's emf ε when no current flows. The slide wire of the potentiometer circuit shown in Fig. 25.24b is moved along the resistor until the current I_g through the battery stops. Derive an equation showing how the unknown emf ε of the battery is related to ε_p, R_1, and R_2 when I_g is zero.

■ **18.** One 120-V electrical line in a home is connected to a 60-W lightbulb, a 1050-W toaster, a 600-W heater, and the 2000-W burner of a stove. The devices, connected in parallel in a circuit similar to that shown in Fig. 25.9, cause the circuit's fuse to blow when all are on at the same time. However, if all but the light are on, the fuse does not blow. Roughly what current causes the fuse to blow?

■ ■ **19.** Three resistors are connected as shown in Fig. 25.25. Each resistor can dissipate no more than 0.50 W of electrical power. Calculate the maximum current that can flow through the circuit.

■ ■ **20.** A battery has an emf of 12 V and an internal resistance of 3 Ω (see the discussion with Problem 17). (a) Calculate the power delivered to a resistor R connected to the battery terminals for values of R equal to 1, 2, 3, 4, 5, and 6 Ω. (b) Plot on a graph the calculated values of P versus the different values of R. Connect the points by a smooth curve. You should confirm that the maximum power is delivered when R has the same resistance as the internal resistance of the power source (3 Ω in this example).

25.2 Equivalent Resistance

21. Calculate the equivalent resistance of the resistors shown in Fig. 25.25.

22. Calculate (a) the equivalent resistance of resistors R_1, R_2, and R_3 in Fig. 25.17 for $R_1 = 28\ \Omega$, $R_2 = 30\ \Omega$, and $R_3 = 20\ \Omega$ and (b) the current through the battery if $\varepsilon = 10$ V.

■ **23.** (a) Calculate the equivalent resistance of resistors R_1, R_2, and R_3 in Fig. 25.17 for $R_1 = 50\ \Omega$, $R_2 = 30\ \Omega$, and $R_3 = 15\ \Omega$. (b) Determine the current through R_1 if $\varepsilon = 120$ V. (c) Use Kirchhoff's loop rule and your result from part (b) to determine the current through R_2 and through R_3.

■ **24.** Calculate the equivalent resistance of the resistors shown in Fig. 25.26 if $R_1 = 60\ \Omega$, $R_2 = 30\ \Omega$, $R_3 = 20\ \Omega$, $R_4 = 20\ \Omega$, $R_5 = 60\ \Omega$, $R_6 = 20\ \Omega$, and $R_7 = 10\ \Omega$.

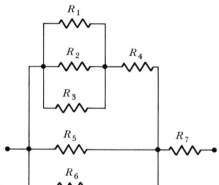

FIG. 25.26

■ **25.** Calculate the equivalent resistance of the resistors in Fig. 25.26 if $R_1 = R_2 = 20\ \Omega$, $R_3 = 10\ \Omega$, $R_4 = 25\ \Omega$, $R_5 = 30\ \Omega$, $R_6 = 10\ \Omega$, and $R_7 = 50\ \Omega$.

■ **26.** Calculate (a) the equivalent resistance of the resistors in the circuit in Fig. 25.27 and (b) the current through the battery.

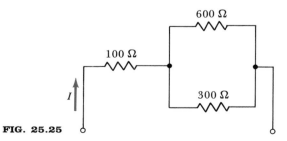

600 Ω

100 Ω

I

300 Ω

FIG. 25.25

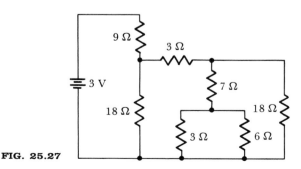

9 Ω

3 Ω

3 V

7 Ω

18 Ω

18 Ω

3 Ω

6 Ω

FIG. 25.27

■ 27. Calculate the equivalent resistance of the group of resistors shown in Fig. 25.28.

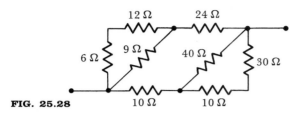

FIG. 25.28

■ 28. A fuse for one line in your home's electrical system melts if more than 30 A of current flows. Determine if the fuse will melt if the following appliances are all connected in parallel to that line: a 13-Ω toaster, an 18-Ω dishwasher, a 24-Ω refrigerator, and a 15-Ω heater. The potential difference across the devices is 120 V.

■ 29. Calculate (a) the equivalent resistance, (b) the current through the battery, and (c) the power supplied by the battery for the circuit shown in Fig. 25.29. (d) Use Kirchhoff's loop rule to calculate the currents through the 60-Ω, 40-Ω, and 30-Ω resistors.

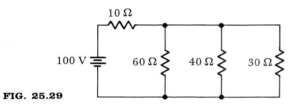

FIG. 25.29

■ 30. Calculate the total power that must be supplied by a 12-V car battery connected in parallel to two lightbulbs each of resistance 6.0 Ω and a cigar lighter of resistance 4.0 Ω.

■ 31. A battery produces a 2.0-A current when connected to an unknown resistance R. When a 10-Ω resistor is connected in series with R, the current drops to 1.2 A. Calculate the emf ℰ of the battery and the resistance R.

■ 32. Nine Christmas tree lights are connected in parallel across a 120-V potential difference. The cord to the wall socket carries a current of 0.36 A. (a) Calculate the resistance of one of the bulbs. (b) What would the current be if the bulbs were connected in series?

■ 33. Two lightbulbs use 30 W and 60 W, respectively, when connected in parallel to a 120-V power source. How much power does each bulb use when connected in series across the 120-V power source, assuming that their resistances remain the same?

■ 34. A bus has twenty 15-W lightbulbs connected in parallel to a 12-V battery. The battery is rated at 120 A·hr—that is, it will provide 1 A of current for 120 hours, 2 A of current for 60 hours, and so on. How long will the battery last if all of the lights are accidentally left on while the bus is parked?

■ 35. Three equal resistors, when connected in series, together consume electrical energy at a rate of 15 W. Calculate the power consumed by the resistors when connected in parallel to the same potential difference.

■■ 36. Your high-fidelity amplifier has one output for a speaker whose resistance is 8 Ω. How can you arrange two 8-Ω speakers, one 4-Ω speaker, and one 12-Ω speaker so that all are powered by

the amplifier and their equivalent resistance when connected together in this way is 8 Ω.

25.3 Measuring Current and Voltage

■ 37. (a) Calculate the current in the circuit shown in Fig. 25.30 if ℰ = 9.0 V, R_1 = 6.0 Ω, and R_2 = 12.0 Ω. (b) Calculate the current if an ammeter of internal resistance 2.0 Ω is used. Does the measuring device (the ammeter) affect the quantity being measured? Explain.

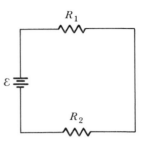

FIG. 25.30

■ 38. (a) and (b) Repeat Problem 37 for ℰ = 9.0 V, R_1 = 60 Ω, and R_2 = 120 Ω using the same ammeter. (c) In Problem 37, inserting the ammeter into the circuit reduced the current by 10 percent (from 0.50 A to 0.45 A). By what percent is the current reduced in this problem? Why is the percent reduction less even though the same ammeter is used?

■ 39. (a) Calculate the voltage across R_2 in Fig. 25.30 if ℰ = 120 V, R_1 = 2000 Ω, and R_2 = 4000 Ω. (b) Calculate the measured voltage across R_2 using a voltmeter with internal resistance 10,000 Ω. Does the measuring device affect the quantity being measured? Explain.

■ 40. (a) and (b) Repeat Problem 39 for ℰ = 120 V, R_1 = 20 Ω, and R_2 = 40 Ω using the same voltmeter. (c) In Problem 39, placing the voltmeter across R_2 reduced the voltage across it by 12 percent (from 80 V to 70.6 V). By what percent is the voltage reduced in this problem? Why is the percent reduction less even though the same voltmeter is used?

■ 41. A voltmeter measures a voltage V when placed across both an ammeter of internal resistance R_A and an unknown resistance R in series with the ammeter. A current I flows through the ammeter and through R. Show that

$$R = \frac{V}{I} - R_A.$$

■ 42. Show that for the meter arrangement shown in Fig. 25.31,

$$R = \left(\frac{V}{IR_V - V}\right)R_V,$$

where R_V is the voltmeter resistance.

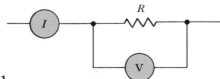

FIG. 25.31

CHAPTER 26

Capacitors

In the last two chapters we studied electric circuits and some of the elements commonly found in them, such as direct-current power supplies and resistors. Another device frequently used in electronic circuits is a capacitor, a device that stores electric charge on its conducting surfaces. Capacitors are used in radio and television sets, in automobile ignition systems, in electronic flash attachments for cameras, in the circuits of heart pacemakers, in the starting circuits for electric motors, as electronic filters, and as parts of radiation detection devices.

In this chapter, we will consider the structure of capacitors, how they store electric charge and electrical energy, and how they are used in electric circuits with resistors, batteries, and switches.

26.1 Capacitors

A **capacitor** consists of two electrical conductors separated by a nonconducting material. For example, a parallel-plate capacitor has two parallel metal plates separated by air, rubber, paper, or some other material that does not conduct electric charge (Fig. 26.1a). When the conducting plates are connected to the terminals of a voltage source, as in Fig. 26.1b, negative charge is forced onto one plate, leaving the other plate with a net positive charge.

While some capacitors are made of parallel metal plates, most (especially those used in electronic equipment) consist of strips of metal foil separated by a thin piece of paper or some other nonconducting material. The strips are then rolled in a cylinder so that the unit occupies little volume (Fig. 26.2).

Not all capacitors are manufactured. Nerve cells and other cells in the body have properties of a capacitor (Fig. 26.3), the conductors of which are the fluids on the inside and the outside of the cell. The cell membrane serves as a moderately nonconductive medium that separates the two conducting fluids. In the membrane, chemical processes, which are only partly understood, cause ions to be "pumped" across the membrane. The membrane's inner surface accumulates a small excess of negative charge, and the outer surface accumulates a small excess of positive charge. Approximately 10 percent of our metabolic energy is used to pump electric charge across the membranes of body cells. In nerve cells the energy is released when a nerve impulse travels along the cell. For other body cells the energy stored by these separated charges becomes involved in a

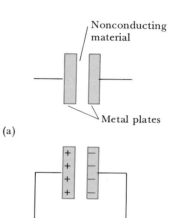

FIG. 26.1. (a) A parallel-plate capacitor. (b) When the capacitor is attached to a voltage source, opposite-sign charges move onto the two plates.

FIG. 26.2. A capacitor used in electronic equipment.

547

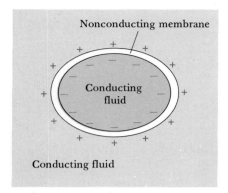

FIG. 26.3. A body cell has properties of a capacitor. The inside and outside fluids serve as conducting plates separated by the non-conducting membrane.

complex way with the chemical reactions that provide energy for our body.

All capacitors, when drawn in an electric circuit, are represented by the following symbol: ⊣⊢. Occasionally one of the lines is drawn slightly curved: ⊣⊢.

26.2 Capacitance

In earlier chapters we learned that the current flowing through a resistor depends on the voltage across the resistor and on its resistance. In the case of capacitors we are usually interested in the positive and negative electric charge that accumulates on its plates when a voltage is placed across them. The positive charge that accumulates on one plate and the negative charge on the other plate depends on the voltage across the plates and on another quantity, called the capacitance of the capacitor.

Capacitance is an indicator of the amount of charge a particular capacitor will store per volt of potential difference across its plates; the larger the capacitance, the more the charge that is stored. The capacitance of a capacitor is analogous to the resistance of a resistor in that capacitance depends on the materials of which the capacitor is made and on its dimensions. Specifically, the capacitance depends on the size of its conducting plates or surfaces, on the separation of these plates, and on the type of material separating them.

To understand how these factors affect capacitance, look at the parallel-plate capacitors shown in Fig. 26.4. First, capacitance is greater for capacitors with plates of large area A; the more space available on which to store charge,

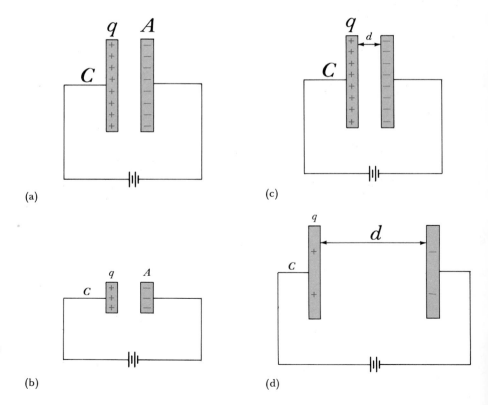

(a)

(c)

(b)

(d)

FIG. 26.4. (a) A capacitor with plates of large area can store more charge than (b) one with plates of small area. More charge is stored if (c) the plates are close to each other rather than (d) far apart.

the greater the capacitance (Fig. 26.4a). Similarly, the smaller the plate area, the less the ability to store charge and the less the capacitance (Fig. 26.4b).

Capacitance also depends on the separation d of the plates (Figs. 26.4c and d). When the plate separation is small, as in part (c), more charge is drawn onto the plates because of the greater mutual attraction between the opposite-sign charges. When the plates are separated by a large distance, as in part (d), little charge is attracted to them. Negative charge on the right plate feels little attraction for positive charge on the left one. The ability of capacitor plates to attract and store charge is inversely related to the separation d of the plates; as d increases, the capacitance decreases, and vice versa.

The third quantity on which the capacity to store electric charge depends is the dielectric constant K of the insulating material between the plates. Recall from Section 22.4 that materials with high dielectric constant are often made of polar molecules (molecules with an excess positive charge on one side and an equal excess negative charge on the other side). When capacitor plates become charged, polar molecules between them become oriented so that their negative side points toward the positively charged plate and the positive side points toward the negatively charged plate. The polar molecules near the plate tend to cancel the effect of some of the charge on the plate and make it easier for more charge to move onto the plates. Thus, more charge moves onto the capacitor plates for capacitors whose plates are separated by a material of high dielectric constant. The capacitance increases in proportion to the dielectric constant of the material. We will have more to say about this in Section 26.3.

The capacitance of a parallel-plate capacitor is calculated using the equation

$$C = \frac{K}{4\pi k}\left(\frac{A}{d}\right), \tag{26.1}$$

where k is the constant in Coulomb's law ($k = 9.0 \times 10^9\,\text{N}\cdot\text{m}^2/\text{C}^2$). For other types of capacitors (cylindrical ones or those shaped like a body cell), different equations must be used. In most of our work, capacitance will be a given quantity. Keep in mind, though, that capacitance depends on intrinsic geometrical and material properties of the capacitor.

Charge on a Capacitor

If a capacitor of capacitance C is attached to a voltage source with voltage V across its terminals, the magnitude of charge q on each plate (positive on one plate and negative on the other) is

$$q = CV. \tag{26.2}$$

Equation (26.2) is often used to define capacitance:

$$C = \frac{q}{V}.$$

However, the capacitance C of a capacitor is an intrinsic property of its dimensions and the material separating the conducting plates. Equation (26.2) simply relates the three quantities C, q, and V and can be used to calculate the value of any one of the three quantities if the other two are known.

Unit of Capacitance

We see by rearranging Eq. (26.2) that $C = q/V$. Thus, the unit of capacitance must equal the units of the quantity q/V—that is, coulomb/volt. The unit of capacitance called the **farad** (named after Michael Faraday) is defined as one coulomb/volt:

$$1 \text{ farad} = 1\frac{\text{coulomb}}{\text{volt}}, \quad \text{or} \quad 1 \text{ F} = 1 \text{ C/V}.$$

A 1-F capacitor attached to a 1-V power supply would have a charge of $+1$ C on one plate and a charge of -1 C on the other plate. Most capacitors store much less than 1 C of charge when attached to a 1-V power supply. Capacitors found in electronic equipment usually have capacitances between 5000 μF and 1 pF, where $1 \mu\text{F} = 1$ microfarad $= 10^{-6}$ F and $1 \text{ pF} = 1$ picofarad $= 10^{-12}$ F.

EXAMPLE 26.1 (a) Estimate the capacitance of your physics book (its capacity to store electric charge) if it had metal covers. The dielectric constant of paper pages is approximately 6. (b) What voltage must be placed across its covers to get 10^{-6} C of charge on each cover (positive on one cover and negative on the other)?

SOLUTION (a) The covers of the book have an area of about 0.20 m $\times$ 0.25 m $= 0.050$ m^2. The covers are separated by about 3.6 cm $= 3.6 \times 10^{-2}$ m. The material between the covers is paper, which has a dielectric constant of about 6. The capacitance of the book with metal covers is calculated using Eq. (26.1):

$$C = \frac{6(0.050 \text{ m}^2)}{(4\pi 9.0 \times 10^9 \text{ N}\cdot\text{m}^2/\text{C}^2)(3.6 \times 10^{-2} \text{ m})} = 74 \times 10^{-12} \text{ F} = \underline{74 \text{ pF}}.^*$$

(b) The answer to part (a) means that the book capacitor will have 74×10^{-12} C of charge on each cover (positive on one cover and negative on the other) for each volt of potential difference across the plates of the capacitor. Using Eq. (26.2), we find the voltage needed to get 10^{-6} C of charge on each cover:

$$V = \frac{q}{C} = \frac{10^{-6} \text{ C}}{74 \times 10^{-12} \text{ F}} = 14 \times 10^3 \frac{\text{C}}{\text{F}} = \underline{14 \times 10^3 \text{ V}}.$$

Even with this large voltage, we still have only 10^{-6} coulombs of charge on the metal book covers. ∎

EXAMPLE 26.2 Calculate (a) the total capacitance C and (b) the charge q on all the membranes of a person's 10^{13} body cells if each cell has a surface area of 1.8×10^{-9} m^2 and if the membranes are 8.0×10^{-9} m (8.0 nm) thick and have a 0.090-V potential difference across them and a dielectric constant of 8.0.

SOLUTION The cell diameters are much greater than the thickness of their membranes. Thus, each cell can be considered as a parallel-plate capacitor. The plates are the inside and outside cell fluids. The membrane is a relatively non-conducting material separating the fluids.

*Note that the unit C^2/N$\cdot$m $=$ C^2/J $=$ C/V $=$ F and that 1 pF $= 10^{-12}$ F.

(a) The total membrane area is the number of cells multiplied by the area of each cell:

$$A = (10^{13}\,\text{cells})(1.8 \times 10^{-9}\,\text{m}^2/\text{cell}) = 1.8 \times 10^4\,\text{m}^2.$$

This is the area of about three football fields. Using Eq. (26.1), we find that the total capacitance of these cells is

$$C = \frac{K}{4\pi k}\left(\frac{A}{d}\right) = \frac{8(1.8 \times 10^4\,\text{m}^2)}{(4\pi 9.0 \times 10^9\,\text{N}\cdot\text{m}^2/\text{C}^2)(8.0 \times 10^{-9}\,\text{m})} = \underline{160\,\text{F}}.$$

(b) The total charge separated by the membranes of the cells when a 0.090-V potential difference exists across the membranes is calculated using Eq. (26.2):

$$q = CV = (160\,\text{F})(0.090\,\text{V}) = \underline{14\,\text{C}}.$$

Although these calculations are approximate, it is clear that the separation of electric charge across cell membranes must be an important part of our metabolic processes. The 14-C charge stored on our membrane capacitors is about the same as the charge on a cloud before a lightning flash. Discharging our cell capacitors in one surge would be an event of no small significance. ∎

26.3 Dielectrics

In Chapter 22, we found that materials with large dielectric constant can greatly reduce the force between charges immersed in them. These materials are often made of polar molecules. If placed between the plates of a capacitor, a material with large dielectric constant can greatly increase the charge on the plates of a capacitor—assuming the voltage across the plates remains constant. This can be understood easily by referring to Fig. 26.5.

Two parallel plates separated by air are attached to a voltage source. The charges on the plates create an electric field between the plates (Fig. 26.5a). If a material made of polar molecules replaces the air, as in Fig. 26.5b, the molecules become oriented by the electric field. The negative ends of the polar molecules point toward the positive plate and effectively neutralize some of its positive charge. Similarly, the positive ends of the molecules point toward the negative plate and neutralize some of its negative charge. Because part of the electric charge on each plate is neutralized by the polar molecules, more charge is pushed by the battery onto the plate. The total charge on the plates increases by a factor K, where K is the dielectric constant of the material between the plates.

The capacitance of any capacitor increases by a factor K when a material of dielectric constant K is placed between the conductors. If its capacitance is C_0 when the conductors are separated by air, its capacitance is $C = KC_0$ when they are separated by the dielectric material. When a voltage V is placed across the plates of the capacitor, the magnitude of the charge on the plates increases by a factor K:

$$q = CV = KC_0V.$$

EXAMPLE 26.3 A capacitor can be used to detect motion as the dielectric material between its plates moves. Consider a 5.0-μF capacitor with air initially separating the plates, which have been charged by a 100-V power supply. The capacitor is disconnected from the supply, and a sheet of Plexiglas with dielec-

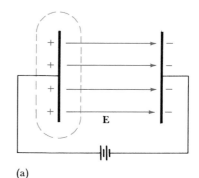

(a)

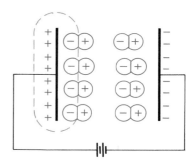

(b)

FIG. 26.5. (a) An electric field exists between the plates of a charged capacitor. (b) Dipole molecules between the plates are oriented by the field. The charge inside the circle in part (b) is the same as that inside the circle in part (a), but the capacitor plate in part (b) holds more charge because the polar molecules are present.

tric constant 3.40 is placed between the plates so that it fills the space. Calculate the voltage across the plates.

SOLUTION When the capacitor is removed from the power supply, charge cannot leave the plates. Thus, the charge on the plates remains constant as the Plexiglas is inserted between them.

Because Plexiglas has a greater dielectric constant than air, the capacitor's capacitance increases as the Plexiglas is inserted between its plates. But since q is constant, we see from Eq. (26.2) that V must decrease as the capacitance increases. Thus,

$$q = C_0 V_0 = CV = (KC_0)V,$$

where C_0 is the initial capacitance with air between the plates, $V_0 = 100$ V is the initial voltage, C is the capacitance when the Plexiglas of dielectric constant $K = 3.40$ is inserted in the air gap, and V is the voltage after the Plexiglas is inserted. We find that

$$V = \frac{C_0 V_0}{C} = \frac{C_0 V_0}{(KC_0)} = \frac{V_0}{K} = \frac{100 \text{ V}}{3.40} = \underline{29 \text{ V}}.$$

A device such as just described could be used to detect motion. A burglar alarm, for example, could be constructed so that a Plexiglas sheet was inserted between the plates of a capacitor when a door or window was opened. The resulting decrease in voltage would activate an alarm. ■

26.4 Dielectric Breakdown

The magnitude of the electric field between the plates of a parallel-plate capacitor is given by Eq. (23.8):

$$E = \frac{V}{d},$$

where V is the voltage across the plates and d is the distance between them.

Normally, the material between the plates does not conduct electric charge, but if the electric field in the material is very large, the atoms and molecules in this material may become ionized. The electrons are then pulled toward the positive capacitor plate while the positive ions left behind attract electrons from the negatively charged plate. As a result, a very intense current flows for a brief time as the charge on the capacitor plates flows or discharges through the material between them.

This phenomenon, called **dielectric breakdown,** occurs when a nonconducting material is subjected to a very large electric field that causes it to become ionized and to conduct electricity. In such situations, the onset of conduction occurs suddenly and often produces a spark. The **dielectric strength** of a material is defined as the value of the electric field for which breakdown of the material occurs. Nonconducting materials, such as plastic, rubber, paraffin, and transformer oils, have dielectric strengths of 8×10^6 to 20×10^6 V/m.

If the dielectric strength of a material is 8×10^6 V/m, then the voltage across a 1-m-thick slab of the material would have to be 8×10^6 V or greater for breakdown to occur. If the slab were 0.1 m thick, then a voltage of 8×10^5 V or greater would cause breakdown. [The electric field in the material is still

8×10^6 V/m or greater since, according to Eq. (23.8), $E = V/d = 8 \times 10^5$ V/0.1 m $= 8 \times 10^6$ V/m.]

EXAMPLE 26.4 The dielectric strength of air is 3×10^6 V/m. As you walk across a synthetic rug, your body accumulates electric charge, causing a potential difference of 6000 V between it and a doorknob. Estimate the separation of your finger from the doorknob when dielectric breakdown of air occurs causing a spark of electric charge to jump between your body and the knob.

SOLUTION In this problem the two conducting surfaces are the doorknob and your body. Air is the nonconducting material separating them. Breakdown occurs when the electric field between your finger and the door exceeds 3×10^6 V/m. Rearranging Eq. (23.8) to determine the separation d of your finger and the knob when the field exceeds this amount, we find that

$$d = \frac{V}{E} = \frac{6000 \text{ V}}{3 \times 10^6 \text{ V/m}} = 2 \times 10^{-3} \text{ m} = \underline{2 \text{ mm}}. \qquad \blacksquare$$

Explosions in grain elevators have been ignited by sparks caused by dielectric breakdown. Similar explosions of anesthetics could occur in operating rooms. To prevent such explosions, physicians and nurses wear shoes with soles that conduct electricity. This prevents the accumulation of charge on their bodies and hence prevents sparks.

EXAMPLE 26.5 Calculate the electric field across the membrane of a body cell if the membrane's thickness is 7.0 nm and the potential difference across it is 90 mV.

SOLUTION The magnitude of the electric field in the membrane is determined using Eq. (23.8):

$$E = \frac{V}{d} = \frac{0.090 \text{ V}}{7.0 \times 10^{-9} \text{ m}} = \underline{13 \times 10^6 \text{ V/m}}.$$

If our cell membranes were made of a material with dielectric strength less than 13×10^6 V/m, dielectric breakdown would occur and the 14 C of charge separated by these membranes (calculated in Example 26.2) would be discharged. Fortunately, the dielectric strength of our cell membranes is about twice the electric field in the membranes. $\qquad \blacksquare$

Lightning

Lightning is a complex phenomenon that depends on dielectric breakdown. Perhaps the most common type of lightning (there are many types) is cloud-to-ground lightning. As shown in Fig. 26.6a, a cloud becomes positively charged at the top (the P region) and negatively charged at the bottom (the N region). A small portion of the cloud at the bottom also has a positive charge (the p region). The reasons for this charge distribution are complex and only partially understood.

(Michael Weisbrot/Stock Boston)

FIG. 26.6. (a) Charge distribution in a cloud just before cloud-to-ground lightning. (b) A discharge between the N and p regions of the cloud starts the lightning. (c) The stepped leader moves toward the ground. (d) Ionization of air occurs near the ground, causing light and an intense downward flow of electrons. (e) Electric charge from farther up flows down.

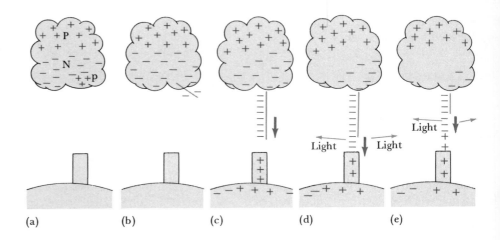

(a)   (b)   (c)   (d)   (e)

If the electric field caused by this charge distribution is large enough, dielectric breakdown occurs, and electrons leap from the N region toward the p region. The freed electrons continue to move down and away from the rest of the negative charge in the cloud (Fig. 26.6b). The electrons make successive 50-m jumps toward the earth at intervals of one-millionth of a second. This downward trip of the electrons is called a **stepped leader** (Fig. 26.6c).

As the electrons approach the earth, positive charge on the earth's surface is attracted toward the leader, while negative charge is repelled (Fig. 26.6c). The electric field in the air just above the surface becomes so intense that it ionizes atoms and molecules in the air. The air experiences dielectric breakdown (Fig. 26.6d), and an intense flash of light is produced. Breakdown often occurs at elevated parts of the earth or at the tops of buildings or trees, since the electric field is strongest at places that are nearest the approaching leader (recall that $E = V/d$; thus E is greatest for small d).

After breakdown occurs, negative charge in the stepped leader farther above the earth can now rush down through the region of ionized air. (Fig. 26.6e). This intense flow of electrons originating farther above the earth causes a flash of light that, when photographed, appears to move upward because electron flow starts from successively higher and higher positions. The flash, which appears to move up, and the large downward electron flow make up what is called the **return stroke.** Eventually, a large portion of electrons originally on the cloud discharge from the cloud to the earth. About five coulombs of electrons move down during the stepped leader, and another 20 to 30 coulombs of electrons move down during the return stroke.

In the United States, about 100 to 200 deaths are caused each year by lightning. In eighteenth-century Europe, church bells were often rung to protect against lightning. Some persons thought, erroneously, that the sound of the bells disturbed the path of the lightning and protected anyone near the bells. Unfortunately, the metal bells were usually at high elevations and thus were likely spots for lightning to strike. In one 33-year period during the eighteenth century, 386 bell towers were struck, and 103 bell ringers were killed at their ropes.

What should you do during a lightning storm? You should not make a lightning rod of yourself or stand beneath one (such as a tree). If you are in an open space, find a ravine, valley, or depression in the ground. Crouch so that you do not project above the surrounding landscape. Lightning usually strikes objects farthest above the earth's surface. Do not lie on the ground. A large voltage may develop between your head and feet, causing an undesirable flow of electric charge through your body. When crouched, keep your feet close together.

26.5 Energy in a Charged Capacitor

As we learned in Chapter 23, whenever positive and negative electric charges move apart, electrical potential energy increases. In Fig. 26.7a, two parallel plates with no net charge are shown. We will charge the capacitor by moving small increments of charge Δq from one plate to the other; one plate gains a charge $+\Delta q$, leaving the other plate with a charge deficiency $-\Delta q$.

Little energy is used to move the first Δq since there is no charge on the left plate to oppose its motion (Fig. 26.7b). The next small charge Δq is harder to move (Fig. 26.7c) since there is now a $+\Delta q$ charge on the left plate that repels its motion while a $-2\Delta q$ charge on the right plate pulls it back. The more charge that is transferred, the harder it becomes to move another increment Δq. Eventually, we will have moved a total charge q from one plate to the other.

Some external energy source, such as a battery, usually provides the increased electrical potential energy needed to charge a capacitor. The energy needed can be calculated using Eq. (23.5):

$$\Delta PE_q = qV_{av}$$

where V_{av} is the average potential difference between the capacitor plates while being charged. Since the initial voltage across the plates is zero and the final voltage is V, the average potential difference between the plates is

$$V_{av} = \frac{0 + V}{2} = \frac{1}{2}V.$$

Substituting this expression for V_{av} in Eq. (23.5), we find that

$$\Delta PE_q = qV_{av} = q\left(\frac{1}{2}V\right).$$

The above equation for the *change in electrical potential energy* that occurs when a capacitor receives a charge q from a potential difference V can be rewritten in three different forms using Eq. (26.2), $q = CV$:

$$\Delta PE_q = \frac{1}{2}qV = \frac{1}{2}CV^2 = \frac{1}{2}q^2/C. \qquad (26.3)$$

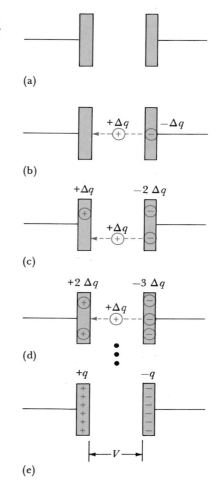

(a)

(b)

(c)

(d)

(e)

FIG. 26.7. The charging of a capacitor starts with the transfer of a small charge Δq from the right to the left plate. This transfer is fairly easy. However, as more and more charge is transferred, additional transfer is more difficult.

EXAMPLE 26.6 Estimate the increase in electrical potential energy needed to charge all the membrane capacitors of a person's body. In Example 26.2 we estimated that the total capacitance of the membranes of 10^{13} cells was 160 F, the total charge on these cells was 14 C, and the voltage across the membranes was 0.090 V.

SOLUTION The increased electrical potential energy when the membrane capacitors of the body are charged can be calculated using any of the three expressions in Eq. (26.3):

$$\Delta PE_q = \frac{1}{2}qV = \frac{1}{2}(14 \text{ C})(0.090 \text{ V}) = \underline{0.6 \text{ J}},$$

$$\Delta PE_q = \frac{1}{2}CV^2 = \frac{1}{2}(160 \text{ F})(0.090 \text{ V})^2 = \underline{0.6 \text{ J}},$$

$$\Delta PE_q = \frac{1}{2}q^2/C = \frac{1}{2}(14 \text{ C})^2/(160 \text{ F}) = \underline{0.6 \text{ J}}. \qquad \blacksquare$$

EXAMPLE 26.7 As the pressure against one plate of a capacitor with an air gap increases, the plate separation decreases. Suppose that such a 5.0-μF capacitor is connected to a 12-V battery. When added pressure is applied to the plates, their separation decreases by 20 percent. By what percentage do the capacitor's (a) capacitance, (b) stored charge, and (c) stored electrical potential energy change? The voltage across the capacitor remains constant.

SOLUTION (a) Let us first calculate the ratio of the final capacitance C and the initial capacitance C_0. Using Eq. (26.1) and the fact that $d = 0.80d_0$ (a 20 percent decrease in plate separation), we find that

$$\frac{C}{C_0} = \frac{(KA/4\pi kd)}{(KA/4\pi kd_0)} = \frac{d_0}{d} = \frac{d_0}{0.80 \, d_0} = 1.25, \quad \text{or}$$
$$C = 1.25C_0.$$

Thus, the capacitance increases by $\underline{25 \text{ percent}}$. (Note that the dielectric constant K and plate area A did not change).

(b) Using the same method and Eq. (26.2), we find that

$$\frac{q}{q_0} = \frac{CV}{C_0V} = \frac{C}{C_0} = 1.25.$$

The charge on the plates also increases by $\underline{25 \text{ percent}}$.

(c) Each of the three forms of Eq. (26.3) produces the same result:

$$\frac{\Delta PE_q}{\Delta PE_{q_0}} = \frac{qV/2}{q_0V/2} = \frac{q}{q_0} = 1.25,$$

$$\frac{\Delta PE_q}{\Delta PE_{q_0}} = \frac{CV/2}{C_0V/2} = \frac{C}{C_0} = 1.25,$$

$$\frac{\Delta PE_q}{\Delta PE_{q_0}} = \frac{q^2/2C}{q_0^2/2C_0} = \left(\frac{q}{q_0}\right)^2\left(\frac{C_0}{C}\right) = \left(\frac{1.25}{1}\right)^2\left(\frac{1}{1.25}\right) = 1.25.$$

The stored electrical energy also increases by $\underline{25 \text{ percent}}$.

As we see, a capacitor is a convenient device for producing an electrical signal when there is a change in pressure. $\qquad \blacksquare$

26.6 Charging and Discharging a Capacitor Through a Resistor

A charged capacitor stores electrical potential energy by separating positive charge on one plate from negative charge on the other. This energy is released if the negative charge on one plate returns to the positively charged plate by flowing through a circuit attached to the plates. The electrical potential energy lost by the capacitor is transferred to elements in the circuit through which the charge flows. For example, the light from a camera's flashlamp is produced by discharging a capacitor through a xenon lamp. The energy for pulsed lasers is supplied by discharging large capacitors. Even the electrical energy that a heart pacemaker uses to initiate a heartbeat can be supplied by a repetitive charging and discharging of a capacitor.

For many of these applications, the time needed to charge or discharge the capacitor is important. To cause a bright flash of light in your camera's flashlamp, its capacitor must discharge almost completely in less than one millisecond (10^{-3} s). The discharge time depends on both the capacitance of the capacitor and the resistance of the circuit through which the charge flows.

Charging a Capacitor

Figure 26.8a shows the charging of a capacitor through a circuit consisting of a battery attached to a resistor R in series with a capacitor C (called an **RC series circuit**). An open switch ($-\diagdown-$) initially prevents the battery from charging the capacitor.

When the switch is closed (Fig. 26.8b), the battery forces current around the circuit in a clockwise direction. However, the charge cannot flow between the capacitor plates. Instead, opposite-sign charges start to accumulate on the plates. This charge makes it difficult for more charge to flow onto the plates, and the current decreases (Fig. 26.8c). When the capacitor has become fully charged with a final charge q_f, the current stops (Fig. 26.8d).

We can better understand the time variation of the current i through the resistor and the charge q on the capacitor by analyzing the voltage across the three parts of the circuit shown in Fig. 26.8. (We will use lowercase letters to denote quantities that change with time.) The voltage across the battery is constant and equals V. The voltage across the resistor varies with time and is $v_R = iR$ at the instant of time that a current i flows through the resistor. The voltage across the capacitor also varies with time and depends on the instantaneous charge on its plates according to Eq. (26.2): $v_C = q/C$. The sum of the voltages across the resistor and capacitor equals the constant voltage across the battery:

$$V = v_R + v_C = iR + \frac{q}{C}.$$

Rearranging, we find that

$$i = \frac{V}{R} - \frac{q}{RC}.$$

At the instant the switch is closed, $q = 0$ and the initial current is $i_0 = V/R$. As charge accumulates on the capacitor, q/RC increases, and the current decreases. Eventually, when the capacitor is fully charged, $i = 0$ and

$$0 = \frac{V}{R} - \frac{q_f}{RC}, \quad \text{or} \quad q_f = VC.$$

The final charge q_f on the capacitor is just the charge we expect using Eq. (26.2) when a voltage V is across the capacitor.

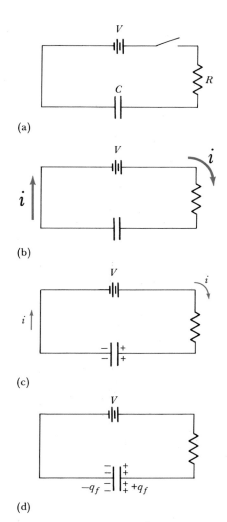

(a)

(b)

(c)

(d)

FIG. 26.8. (a) An RC series circuit. (b) When the switch is closed, a large current starts to flow and charge begins to accumulate on the capacitor plates. (c) As more charge accumulates, the current is reduced because of its repulsion from the charge already on the plates. (d) Finally, the plates become completely charged and no more current flows.

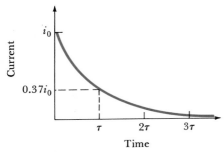

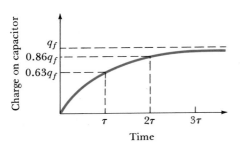

(a)

(b)

FIG. 26.9. The variation of (a) current and (b) charge with time as a capacitor is charged through a resistor.

The variation with time of the current through the resistor and of the charge on the capacitor are shown in Fig. 26.9. The current starts abruptly with an initial value i_0 and decreases with time. The charge on the capacitor increases from zero toward a final value q_f. Equations for the variation of i and q with time can be derived using techniques involving calculus. We present the results without derivation:

$$i = i_0 e^{-t/RC}, \tag{26.4}$$

$$q = q_f(1 - e^{-t/RC}). \tag{26.5}$$

We see that the current decreases exponentially with time and that the charge also varies in a way that depends on an exponential function of time.

One very important feature of Eqs. (26.4) and (26.5) is the appearance of the quantity RC, called the **time constant** τ of the circuit:

$$\tau = RC. \tag{26.6}$$

At a time $t = RC$, the current has decreased to e^{-1} of its initial value. Since $e = 2.72$, we find that at $t = RC$, $i = i_0 e^{-1} = i_0 0.37$ and that the charge on the capacitor has increased to $q = q_f(1 - e^{-1}) = 0.63q_f$. At twice this time $(t = 2RC)$, $i = i_0 e^{-2} = i_0 0.14$ and $q = q_f(1 - e^{-2}) = q_f 0.86$. At $t = 3RC$, $i = i_0 e^{-3} = i_0 0.05$ and $q = q_f(1 - e^{-3}) = q_f 0.95$. We see that the time needed to charge a capacitor through a resistor depends on the values of R and C. At a time $t = RC$, the capacitor is 63 percent charged; at a time $t = 2RC$, the capacitor is 86 percent charged; at a time $t = 3RC$, the capacitor is 95 percent charged, and so forth.

Discharging a Capacitor

The discharge of a fully charged capacitor through a resistor is represented by the series of steps shown in Fig. 26.10. Initially, when the switch is open (Fig. 26.10a), the capacitor has a charge q_0. When the switch is closed, charge leaves the capacitor and flows through the resistor. The charge q on the plate

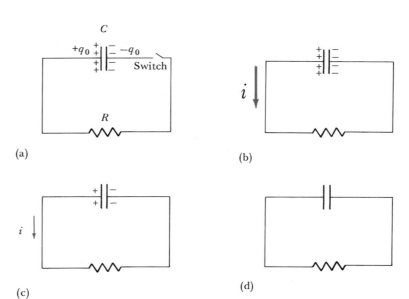

FIG. 26.10. (a) A charged capacitor. (b) When the switch is closed, current flows through the resistor, and (c) and (d) the capacitor eventually becomes discharged.

decreases exponentially with time according to the equation

$$q = q_0 e^{-t/RC}, \tag{26.7}$$

where q_0 is the initial charge on the plates (Fig. 26.11).

Once again, the time constant $\tau = RC$ appears in the equation for the discharging of a capacitor through a resistor. At a time $t = RC$, $q = q_0 e^{-1} = q_0 0.37$; at $t = 2RC$, $q = q_0 e^{-2} = q_0 0.14$; at $t = 3RC$, $q = q_0 e^{-3} = q_0 0.05$; and so forth. Thus, the capacitor is 95 percent discharged at a time $3RC$.

Although it is not immediately obvious, the quantity RC that has units resistance times capacitance is also a unit of time, as the following manipulations confirm:

$$R = \frac{V}{I}, \quad \text{or} \quad \text{ohm} = \frac{\text{volt}}{\text{ampere}},$$

and

$$I = \frac{\Delta q}{\Delta t}, \quad \text{or} \quad \text{ampere} = \frac{\text{coulomb}}{\text{second}}.$$

Thus,

$$\text{ohm} = \frac{\text{volt}}{\text{ampere}} = \text{volt}\left(\frac{\text{second}}{\text{coulomb}}\right).$$

Also note that

$$C = \frac{q}{V}, \quad \text{or} \quad \text{farad} = \frac{\text{coulomb}}{\text{volt}}.$$

The unit of the time constant τ is the same as the unit of RC. Thus,

$$\text{ohm} \cdot \text{farad} = \left[\text{volt}\left(\frac{\text{second}}{\text{coulomb}}\right)\right]\left(\frac{\text{coulomb}}{\text{volt}}\right) = \text{second}.$$

EXAMPLE 26.8 If you take a picture using a flash attachment, there is a delay of 5 to 10 s before you can take the next. The delay is caused by the time needed to charge a capacitor through a resistor (see Fig. 26.12).* Once charged, the capacitor is ready to release its charge through a xenon lamp, producing light. Estimate the series resistance R of the charging circuit if the capacitor has a capacitance of 450 μF,

SOLUTION Since you must wait 5 to 10 s between pictures, let us estimate that the time τ needed to get the capacitor 63 percent charged is 5 s.

Then, according to Eq. (26.6),

$$R = \frac{\tau}{C} = \frac{5 \text{ s}}{450 \times 10^{-6} \text{ F}} = 1.1 \times 10^4 \ \Omega = \underline{11 \text{ k}\Omega}. \quad \blacksquare$$

EXAMPLE 26.9 Determine the time needed for a 10-μF capacitor in series with a (5.0×10^5)-Ω resistor to become 80 percent charged after a switch connecting them to a battery is closed.

*The actual circuit in a flash attachment is somewhat more complicated than this.

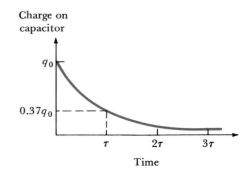

Charge on capacitor

FIG. 26.11. The variation of charge with time as a capacitor is discharged through a resistor.

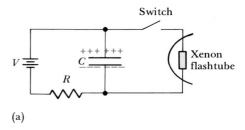

(a)

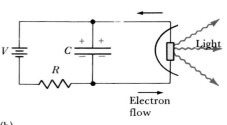

(b)

FIG. 26.12. (a) A circuit for a camera flashtube. When not in use, a capacitor is charged by a battery through resistor R. (b) When the switch is closed, the capacitor discharges quickly through a xenon lamp, thus producing light.

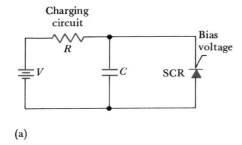

(a)

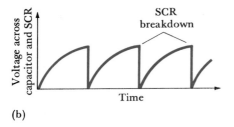

(b)

FIG. 26.13. (a) A circuit such as this could be used for a heart pacemaker. (b) The variation of voltage across the capacitor as a function of time.

SOLUTION We wish to know the time t when the charge q equals $0.80\, q_f$, that is, 80 percent of its final charge. Rearranging Eq. (26.5), we find that

$$e^{-t/RC} = 1 - \frac{q}{q_f} = 1 - \frac{0.80 q_f}{q_f} = 0.20.$$

Taking the natural logarithm of each side of the preceding equation, we have

$$-t/RC = \ln 0.20, \quad \text{or}$$
$$t = -RC \ln 0.20$$
$$= -(5.0 \times 10^5\ \Omega)(10 \times 10^{-6}\ \text{F})(-1.61)$$
$$= 8.0\ \Omega \cdot \text{F} = \underline{8.0\ \text{s}}.\qquad\blacksquare$$

The charging and discharging of capacitors is also used in the electrical circuits of some heart pacemakers. A simplified circuit for a pacemaker is shown in Fig. 26.13a. A battery charges the capacitor C through a resistor R. The values of R and C are chosen so that the capacitor becomes almost fully charged in about 1 s.

Another device, called a silicon-controlled rectifier (SCR), is placed in parallel with the capacitor. Little current flows through the rectifier until the voltage across it reaches some predetermined value where breakdown occurs. At this breakdown voltage, the rectifier becomes a conductor with very low resistance, and the charge on the capacitor is discharged almost instantaneously through the rectifier.

In summary, then, the voltage across the capacitor builds up slowly in about 1 s as it is charged by the battery through the resistor R. When the rectifier breaks down, the capacitor discharges quickly, and the process restarts. The variation of voltage with time is shown in Fig. 26.13b.

When a pacemaker is connected to the heart, the voltage produced stimulates the heart to beat at a fixed frequency. A new heartbeat is initiated by each increase in voltage.

EXAMPLE 26.10 A 0.40-μF capacitor is used in the circuit for a heart pacemaker (Fig. 26.13a). If the time constant for the RC-charging circuit is to be 0.70 s, what should the resistance of the resistor be?

SOLUTION According to Eq. (26.6),

$$R = \frac{\tau}{C} = \frac{0.70\ \text{s}}{0.40 \times 10^{-6}\ \text{F}} = 1.8 \times 10^6\ \Omega = \underline{1.8\ \text{M}\Omega}.\qquad\blacksquare$$

26.7 Capacitors in Parallel and in Series

We have been considering electric circuits that contain only one capacitor. Often, it is convenient or necessary to use several capacitors in these circuits. For example, the power for a pulse laser is provided by discharging a large capacitor through a flashlamp, thereby exciting the laser. The flash can be made greater by simultaneously discharging several capacitors through the lamp. In this section we consider the total capacitance when several capacitors are connected in parallel or in series.

Capacitors in Parallel

Several capacitors connected in parallel are shown in Fig. 26.14. The potential difference V across each capacitor is the same, so the charge on any one capacitor depends on its capacitance: $q_i = C_i V$, where i identifies a particular capacitor. The sum of the charges on the left side of all of the capacitors is

$$
\begin{aligned}
q &= q_1 + q_2 + q_3 + \cdots \\
&= C_1 V + C_2 V + C_3 V + \cdots, \quad \text{or} \\
q &= (C_1 + C_2 + C_3 + \cdots)V.
\end{aligned}
$$

A single, equivalent capacitor, built to replace the parallel capacitors, should hold the same amount of charge when a voltage V is placed across its plates (Fig. 26.14b). Thus,

$$
q = C_{eq} V.
$$

By comparing the last two equations, we see that the *equivalent capacitance of one capacitor that replaces several parallel capacitors* is

$$
C_{eq} = C_1 + C_2 + C_3 + \cdots. \tag{26.8}
$$

Capacitors in Series

Now consider capacitors attached in series, as shown in Fig. 26.15a. The battery will push a charge $-q$ to the right plate of capacitor C_3. An equal charge will be repelled from its left plate onto the right plate of capacitor C_2. The net charge in the region between capacitors C_2 and C_3 is zero, as it was before the capacitors were charged (see the region inside the dashed lines in Fig. 26.15a). A similar phenomenon occurs between capacitors C_1 and C_2. Thus, the left plate of each capacitor has a charge $+q$, and the right plate of each capacitor has a charge $-q$. The potential difference V across all the capacitors is just the sum of the potential differences across each capacitor:

$$
V = V_1 + V_2 + V_3 + \cdots.
$$

For each capacitor, $V_i = q/C_i$ [Eq. (26.2)]; thus,

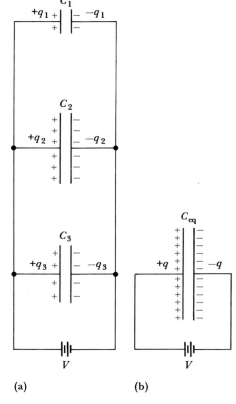

FIG. 26.14. (a) Three capacitors in parallel and (b) an equivalent capacitor.

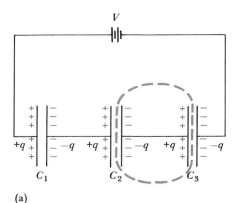

(a)

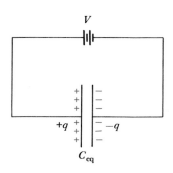

(b)

FIG. 26.15. (a) Three capacitors in series and (b) an equivalent capacitor.

$$V = \frac{q}{C_1} + \frac{q}{C_2} + \frac{q}{C_3} + \cdots$$

$$= q\left(\frac{1}{C_1} + \frac{1}{C_2} + \frac{1}{C_3} + \cdots\right).$$

If a single, equivalent capacitor replaces the series capacitors, its capacitance should produce the same charge on each plate when the same voltage is applied across its plates (Fig. 26.15b). Thus,

$$V = q\left(\frac{1}{C_{eq}}\right).$$

For V and q to be the same in the last two equations, the *equivalent capacitance of a group of series capacitors* must satisfy the following equation:

$$\frac{1}{C_{eq}} = \frac{1}{C_1} + \frac{1}{C_2} + \frac{1}{C_3} + \cdots. \qquad (26.9)$$

Notice that the expressions for equivalent capacitance are just the opposite of those for equivalent resistances—series resistors are added linearly, whereas series capacitors are added by first taking their inverses.

EXAMPLE 26.11 Calculate the total energy stored in three 10-μF capacitors when attached in parallel and when attached in series. A 100-V power supply is used to charge the capacitors.

SOLUTION First calculate the equivalent capacitance for each situation using Eqs. (26.8) and (26.9):

Parallel: $\quad C_{eq} = C_1 + C_2 + C_3 = 3(10\ \mu\text{F}) = 30\ \mu\text{F},$

Series: $\quad \dfrac{1}{C_{eq}} = \dfrac{1}{C_1} + \dfrac{1}{C_2} + \dfrac{1}{C_3} = \dfrac{1}{10\ \mu\text{F}} + \dfrac{1}{10\ \mu\text{F}} + \dfrac{1}{10\ \mu\text{F}} = \dfrac{3}{10\ \mu\text{F}},$

or

$$C_{eq} = \frac{10\ \mu\text{F}}{3} = 3.3\ \mu\text{F}.$$

The energy stored in the capacitors is calculated using Eq. (26.3):

Parallel: $\quad \Delta PE_q = \frac{1}{2} C_{eq} V^2 = \frac{1}{2}(30 \times 10^{-6}\ \text{F})(100\ \text{V})^2 = \underline{0.15\ \text{J}},$

Series: $\quad \Delta PE_q = \frac{1}{2} C_{eq} V^2 = \frac{1}{2}(3.3 \times 10^{-6}\ \text{F})(100\ \text{V})^2 = \underline{0.033\ \text{J}}.$

As you can see, much more energy can be stored in parallel capacitors. Consequently, if we wished to release a large amount of electrical potential energy by simultaneously discharging several capacitors through the flashlamp of a laser, we should combine the capacitors in parallel, not in series. ∎

Summary and Additional Readings

1. A **capacitor** is a device that stores electric charge. It consists of two electrical conductors separated by a nonconducting material. A capacitor is given the symbol —||— in diagrams of electric circuits.

2. The **capacitance** C of a capacitor depends on the area A of its conducting surfaces, the separation d of these surfaces, and the dielectric constant K of the nonconducting material between the conductors. For a parallel-plate capacitor

$$C = \frac{KA}{(4\pi k)d}, \qquad (26.1)$$

where k is a constant equal to $9.0 \times 10^9 \text{ N·m}^2/\text{C}^2$.

3. **Charging a capacitor:** When a capacitor is attached to a power source of potential difference V, a charge $+q$ accumulates on one plate and $-q$ on the other. The magnitude of the charge on each plate is

$$q = CV. \qquad (26.2)$$

4. The **energy** needed to separate the charges onto the plates of a capacitor is given by any of the following expressions:

$$\Delta PE_q = \frac{1}{2}qV = \frac{1}{2}CV^2 = \frac{1}{2}q^2/C. \qquad (26.3)$$

5. **RC series circuit:** If a capacitor C is charged through a series resistor R, the charge on the capacitor increases with time according to the equation

$$q = q_f(1 - e^{-t/RC}), \qquad (26.5)$$

where q_f is the final charge on the capacitor and RC is called the **time constant** τ of the circuit. If a capacitor is discharged through a resistor, the variation of the charge with time is given by the equation

$$q = q_0 e^{-t/RC}, \qquad (26.7)$$

where q_0 is the initial charge on the capacitor.

6. The **equivalent capacitance** C_{eq} of a group of capacitors connected in **parallel** is

$$C_{eq} = C_1 + C_2 + C_3 + \cdots . \qquad (26.8)$$

The equivalent capacitance of a group of capacitors connected in **series** is determined by solving Eq. (26.9) for C_{eq}:

$$\frac{1}{C_{eq}} = \frac{1}{C_1} + \frac{1}{C_2} + \frac{1}{C_3} + \cdots . \qquad (26.9)$$

Henry Lansford, "The Frightening Mystery of the Electrical Storm," *Smithsonian,* pp 74–81, August (1979).
Devere Logan, "The ABCs of Lightning Protection," *Ham Radio Horizons,* pp 12–19, September (1980).

Questions

1. A dielectric material vibrates in and out of the space between two parallel plates. The plates are connected to a small resistor and a battery, as shown in Fig. 26.16. What happens to the charge on the plates as the dielectric material is pushed between them? As the material is pulled out, what happens to the charge? Explain why a voltage is measured across the resistor as the dielectric material moves in and out between the plates. Devices such as this are used to measure vibrations.

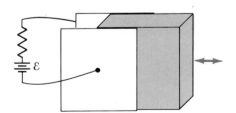

FIG. 26.16

2. Would you expect the dielectric constant of a material, such as water, to increase, decrease, or remain the same as the temperature of the material increases? Explain. [*Hint:* When water molecules are placed in an electric field, more molecules (a small fraction) point in the direction of the field than in any other direction. Consider how temperature would affect that fraction.]

3. A 10-μF capacitor and a 0.01-μF capacitor are both charged by a 100-V potential difference. From which capacitor will you get the bigger shock if it is discharged through your body? Explain how you arrive at your answer.

4. The plates of a capacitor with an air gap are charged by a battery. A sheet of mica with dielectric constant 5.0 is then inserted between the plates while the battery remains connected. Explain how each of the following quantities changes: the capacitance, the charge on the plates, the voltage across the plates, and the energy stored by the capacitor. Justify each answer.

5. Two plates of a capacitor are charged by a battery. The plates are then disconnected from the battery and pulled apart so that their separation is doubled. Explain how each of the following quantities changes: the capacitance of the capacitor, the charge on the capacitor plates, the voltage across the plates, and the energy stored by the capacitor. Justify each answer.

6. A capacitor is connected in series to a resistor and a battery. (a) Does the time constant for charging the capacitor increase, decrease, or remain the same if a second resistor is connected in parallel with the first resistor? Explain. (b) Explain what happens if a second resistor is added to the circuit in series.

7. Two capacitors of capacitance C_1 and C_2, where $C_1 > C_2$, are connected in parallel to a battery. Which capacitor has the largest charge on its plates, voltage across the plates, and stored electrical energy (the quantities may be the same in some cases)? Explain carefully.

8. Repeat Question 7, but for the same two capacitors connected in series.

Problems

26.2 and 26.3
Capacitance and Dielectrics

1. Two strips of aluminum foil, separated by a strip of paper 0.070 mm thick and with dielectric constant 6.0, are wrapped together in a cylinder. Calculate the area of the aluminum sheets needed to make (a) a 5.0-μF capacitor and (b) a 500-pF capacitor.

2. Two round metal plates 0.40 m in diameter are separated by a sheet of mica 0.30 mm thick, whose dielectric constant is 6.3. Calculate the capacitance of the plates.

3. A parallel-plate "trimmer" capacitor has a screw adjustment that can be used to vary the separation of its plates. The plates have an area of 8.0 cm^2 and are separated by a 0.100-mm air gap. (a) Calculate the capacitor's capacitance. (b) To what separation should the plates be moved to change the capacitance to 100 pF?

4. A cylindrical capacitor consists of two cylindrical conducting materials, one inside the other. The capacitance of a cylindrical capacitor is calculated using the equation:

$$C = \frac{KL}{2k \ln (R_2/R_1)}, \qquad (26.10)$$

where $k = 9.0 \times 10^9$ N·m^{2}C^{-2}, K is the dielectric constant of the material between the two cylinders, L is the length of the capacitor, R_1 is the radius of the smaller cylinder, and R_2 is the radius of the larger cylinder. Calculate the capacitance of a coaxial cable with an air dielectric, a 1.0-mm inner radius, a 2.0-mm outer radius, and a 1.5-m length.

5. A 4.2-μF capacitor has a charge of $+7.6 \times 10^{-4}$ C on one plate and a charge of -7.6×10^{-4} C on the other. Charge slowly "leaks" through the dielectric material separating the plates so that after 1 h the magnitude of charge on the plates is 5.4×10^{-4} C. Calculate the change in voltage across the plates during that hour.

■ 6. Consider a cloud to be one plate of a large capacitor and the earth below to be the other plate. *Estimate* the voltage between the cloud and earth if the cloud has a charge of -50 C (the earth below has an equal positive charge), is 2 km above the earth, has an area of 5 km $\times$ 5 km, and is separated from the earth by dry air.

■ 7. Two square plates 1.0 m long on each side are separated by 1.0 cm of air. (a) Calculate the capacitance of the plates. (b) If the plates are cut into fourths to make four capacitors, what must the plate separation be so that each of the four new capacitors will have the same capacitance as the original? Explain. (c) Transformer oil with dielectric constant 2.1 is placed between the plates of one of the small capacitors. To what separation must its plates now be adjusted so that its capacitance will be the same as in part (b)?

■ 8. A capacitor with square metal plates separated by air has a capacitance of 4.0 μF. (a) What does the capacitance become if the separation of the plates is doubled? (b) What is the capacitance using plates with the same separation as the original plates but whose sides are each two times longer than the original plates? (c) What does the capacitance become if a dielectric material of dielectric constant 1.5 is placed between the original plates?

■ 9. Aluminum foil is glued on the top and bottom surfaces of a desk in your physics lecture room. Roughly, how much voltage is needed to separate $+10^{-7}$ C of charge onto one sheet of foil and -10^{-7} C of charge onto the other? Justify any numbers used in your estimation.

■ 10. In Example 26.2 we estimated the capacitance of all of the cells in a human body. Show that our estimate of the surface area for a single cell is reasonable if the person has a mass of 70 kg and density of 980 kg/m^3 and consists entirely of 10^{13} spherical cells.

■ 11. Two square, parallel metal plates that are 20 cm long on each side are separated by a 3.0-mm air gap. The plates are charged by a 100-V power supply and then disconnected from it. A sheet of rubber of dielectric constant 3.0 is placed between the plates. (a) Does the charge on the plates change? Explain. If your answer is yes, consider where the new charge originates. (b) Does the voltage change? If it does, calculate the change. (c) Does the electric field between the plates change? If it does, calculate the change.

■ 12. Two plates of an area 100 cm^2 and separated by a 1.0-mm air gap have opposite charges of magnitude 8.0×10^{-8} C on the plates. (a) What is the voltage across the plates? (b) If the plates are disconnected from the voltage source and then connected across another identical pair of plates that are originally uncharged, what now is the magnitude of the charge on one plate and the voltage across the plates?

■ 13. A 2.0-μF parallel-plate capacitor with an air gap separating the plates is connected to a 12-V battery. The air gap is then filled with polyethylene, which has a dielectric constant of 2.25. Calculate the magnitude of the charge added to or removed from a plate when the polyethylene is placed between them. The capacitor remains connected to the battery.

■ 14. A parallel-plate capacitor with paper of dielectric constant 5.0 in the gap has a capacitance of 0.20 μF. The capacitor is connected to a 12-V battery. (a) Calculate the charge on its plates. (b) While the voltage source remains attached, the paper is withdrawn from between the plates, leaving behind an air gap. Now what is the capacitance of the capacitor and the charge on the plates?

■ 15. A 2.0-μF capacitor and a 4.0-μF capacitor are both charged by a 100-V power supply. The capacitors are then disconnected from the supply. Each plate of each capacitor is connected by a wire to a plate of the opposite charge on the other capacitor. (a) Calculate the total charge on two plates connected together. (b) Calculate the magnitude of the charge on each capacitor. [*Hint:* The voltage is the same across each capacitor.] (c) What is the voltage across the capacitors?

■ 16. Two plates of 200-cm^2 area and separated by air are charged by a battery with equal but opposite charges of magnitude 3.6×10^{-7} C. When the battery is disconnected and the space between the plates is filled with a dielectric material, the electric field between the plates has a magnitude of 4.2×10^5 V/m. Calculate the dielectric constant of the dielectric material.

26.4 Dielectric Breakdown

17. The potential difference from the metal sphere on the top of a Van de Graaff generator to your body is 1.0×10^6 V. Calculate the farthest distance of a surface of your body from the sphere when dielectric breakdown of the air occurs.

18. Your body becomes charged as you shuffle across a carpet. A spark leaps 0.15 cm from your finger to another person's nose. Estimate the potential difference between your body and the other person's body just before the spark jumped.

19. Electrons in the stepped-leader portion of a lightning flash jump a distance of 50 m. Estimate the potential difference between the two parts of the cloud across which the jump occurs.

■ 20. The potential difference between the cloud described in Problem 6 and the earth is 4.5×10^8 V. (a) Estimate the electric field between the cloud and earth. (b) Does dielectric breakdown occur if dry air of dielectric strength 3.0×10^6 V/m separates the cloud and earth? Explain.

■ 21. Calculate the voltage at which dielectric breakdown occurs between the plates described in Problem 2. The dielectric strength of mica is 2.0×10^8 V/m.

■ 22. A parallel-plate capacitor consists of plates of area 100 cm^2 that are separated by a 0.20-mm-thick material with dielectric constant 4.0 and dielectric strength 12×10^6 V/m. Calculate (a) the maximum potential difference that can be put across the plates without causing dielectric breakdown and (b) the charge on the plates with that potential difference.

26.5 Energy in a Charged Capacitor

23. The flashlamp for a laser must discharge 8.0 J of energy to provide an intense flash to excite the laser. A capacitor, charged by a 10,000-V power supply, releases this much energy when discharged. (a) Calculate the capacitance of the capacitor. (b) How much charge does the capacitor store?

24. A 120-μF capacitor, when charged, stores 2.4 J of energy. Calculate the voltage across the capacitor and the magnitude of charge on its plates.

25. A defibrillation unit used to electrically stimulate the heart muscles following a heart attack consists of a 20-μF capacitor charged by a 6000-V potential difference. (a) Calculate the energy released when the defibrillator is discharged through the body. (b) How much charge is released?

■ 26. A 0.20-μF parallel-plate capacitor with an air gap is connected to a 40-V battery. (a) Calculate the charge on one of its plates and the electrical potential energy stored. (b) With the battery connected, a material with dielectric constant 4.0 is inserted between the plates. Now what is the capacitance of the capacitor, the charge on its plates, and the electrical potential energy stored.

■ 27. Suppose a large capacitor charged at the beginning of your physics lecture could provide all of the metabolic energy needed by the students during the class. *Estimate* the capacitance of a capacitor that, when charged by 100,000 V, would provide the needed energy. Justify any numbers used in your calculations.

■ 28. A 1.5-μF parallel-plate capacitor must store 0.60 J of electrical potential energy when fully charged. (a) What voltage must be placed across the plates? (b) The gap between the plates is 0.80 mm and is filled with material of dielectric constant 4.0 and dielectric strength 4.0×10^8 V/m. Does dielectric breakdown occur before the capacitor becomes fully charged? Explain.

■ 29. A 1.6-μF capacitor will break down (spark) if the voltage across it exceeds 200 V. (a) Calculate the maximum electrical energy the capacitor can store. (b) If the capacitor plates are separated by polystyrene insulation whose dielectric strength is 2.0×10^7 V/m, calculate the separation of the plates.

■■ 30. A cloud has -25 C excess charge. The potential difference from the cloud to the earth is 6.0×10^6 V. (a) Calculate the capacitance of this capacitor (the cloud is considered as one plate and the earth as the other). (b) How much electrical potential energy is released during a lightning flash that discharges the capacitor? (c) Is the energy enough to boil all of the water in a home's hot water heater? Explain.

■■ 31. A parallel-plate capacitor with air between its plates has a capacitance of 0.060 μF. When paper is inserted between the plates, the capacitance increases to 0.300 μF. (a) Calculate the dielectric constant of the paper. (b) The capacitor with a paper gap receives a charge of 1.5×10^{-8} C. Calculate the stored energy. (c) The voltage source is removed and the paper is pulled out of the gap. The charge on the plates remains the same. Calculate the stored energy. (d) Account for any gain or loss of energy.

■■ 32. A 0.100-μF parallel-plate capacitor with a variable air gap is charged by a 100-V potential difference. The voltage source is removed from the capacitor, leaving it with a fixed charge. (a) Calculate the energy stored in the capacitor. (b) The capacitor plates are forced apart, reducing its capacitance to 0.080 μF while leaving it with the same charge. Calculate the voltage across the plates. (c) Calculate the stored energy. (d) Using the work-energy equation, account for the change in electrical potential energy.

26.6 Charging and Discharging a Capacitor Through a Resistor

33. (a) A 0.20-μF capacitor is to be charged through a resistor so that it becomes 63 percent charged in 0.10 s. What should the resistance of the resistor be? (b) What is the time constant if the capacitor is charged through a 20-MΩ resistor?

34. A 10-μF capacitor is connected in series with a 2.0-MΩ resistor and a 100-V power source. (a) Calculate the time constant of this circuit. (b) What is the final charge on the capacitor? How many seconds are required to get the capacitor (c) 63 percent charged and (d) 95 percent charged?

35. A 1.5-μF capacitor in a television set is charged by a 22,000-V potential difference. To prevent harm to the television repair technician, a large "bleeder" resistor is connected across the capacitor's ends. When the television set is turned off, the capacitor discharges through the bleeder resistor rather than through the technician, who might accidentally touch its ends. (a) Calculate the time needed to discharge 95 percent of the charge on the capacitor through a 1.0-MΩ resistor. (b) What is the total charge that must be discharged?

■ 36. A defibrillation unit used to resuscitate a heart attack victim was discussed in Problem 25. Calculate the time needed for 86 percent of the charge on the defibrillator's capacitor plates to discharge through the body if the resistance across the body is 10,000 Ω.

■ 37. A heart pacemaker has a capacitor that is charged by a 6-V potential difference before it discharges. When discharged, the capacitor releases 8.0×10^{-3} J of energy. Calculate the resistance of the resistor in the charging circuit if the pacemaker time constant is 0.8 s.

■ 38. The circuit for a flashing light is shown in Fig. 26.17. A 100-V battery charges a capacitor through a resistor. When the voltage across the capacitor exceeds 0.86 times the voltage that is present when the capacitor is fully charged, the neon lamp breaks down and becomes a conductor of electricity. The capacitor discharges through the neon lamp, causing a flash of light. The recharging of the capacitor starts again. (a) Calculate three different pairs of values for the resistance and capacitance to make a circuit

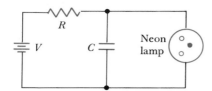

FIG. 26.17

that will flash 5 times a second. (b) Calculate the energy released during a flash for each of your answers to part (a).

■ 39. You are to construct a circuit similar to that described in Problem 38. Your circuit should produce one flash of light each second to keep time with the ticks of a clock. The neon lamp breaks down (offers zero resistance) when the voltage across it exceeds 86 V. The capacitor discharge must provide 0.20 J of energy. Calculate suitable values for V, C, and R.

26.7 Capacitors in Parallel and in Series

40. Calculate the net capacitance of three capacitors of capacitances 2.0 μF, 4.0 μF, and 6.0 μF when connected (a) in series and (b) in parallel.

41. (a) Three capacitors of capacitances 10 μF, 12 μF, and 16 μF are connected in parallel across a 120-V potential difference. Calculate the equivalent capacitance of a single capacitance that can replace the three capacitors, and calculate the magnitude of the charge on its plates. (b) Repeat the problem for the same three capacitors arranged in series.

42. Using techniques similar to those used when calculating equivalent resistance, calculate the equivalent capacitance of the capacitors shown in Fig. 26.18.

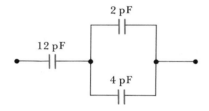

FIG. 26.18

■ 43. Calculate the energy stored by the capacitors in the circuit shown in Fig. 26.19.

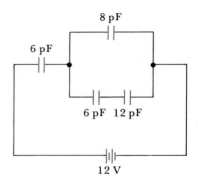

FIG. 26.19   12 V

■ 44. (a) A 10-μF capacitor and a 6-μF capacitor are connected in parallel. Find the energy stored by the capacitors when charged by a 1000-V potential difference. (b) Repeat the problem for a circuit in which the capacitors are connected in series.

■ 45. A 3.0-μF and a 6.0-μF capacitor when connected in parallel together store 4.0 J of electrical potential energy. Calculate the energy stored if capacitors are connected in series across the same voltage.

■ 46. A 0.080-μF capacitor is given a charge of 9.6 μC. The capacitor is removed from the voltage source and connected in parallel to an uncharged capacitor. The potential difference across the capacitors is now 80 V. Calculate the capacitance of the second capacitor. [*Hint:* A single, equivalent capacitor has a charge of 9.6 μC when charged by the 80-V potential difference.]

CHAPTER 27

Magnetic Forces and Fields

Our study of magnetism in this chapter focuses on two basic ideas: (1) An electric charge experiences a magnetic force when moving through a magnetic field, and (2) a moving electric charge produces a magnetic field. With these two ideas we can understand many of the fascinating phenomena that depend on magnetic forces and fields, such as the operation of a mass spectrometer and of electric motors, the use of magnetic fields to contain extremely hot gases, the way in which the earth's magnetic field shields the earth from high-energy particles flying toward us from the sun, the cause of the northern lights, and the way in which a permanent magnet works. We hope you will find that magnetism is a surprising and interesting subject.

27.1 Magnetic Fields

We have all seen a compass needle rotate in the earth's magnetic field. Perhaps you have seen iron filings oriented along the magnetic field lines that emanate from the north pole of a bar magnet and terminate at the south pole (Fig. 27.1).

At this point we will not examine the cause of these magnetic-field lines—their existence is implied by the simple experiments just mentioned. However, before we begin our study of magnetism, we must define magnetic fields. The fields produced by three different magnets are represented by the lines shown in Fig. 27.2. The symbol **B** indicates the direction and the strength (magnitude) of the field at a point and can be determined using a compass needle. The direction of the field is the direction in which a compass needle points when in the field. The magnitude of the magnetic field might be determined by measuring the torque exerted on the compass needle by the field when the needle is held perpendicular to the field direction. The greater the torque, the greater is the magnitude of the field. However, other techniques are usually used to measure the magnitude of the field, as we learn later.

The magnitude of the field is given in a unit called the **tesla** (T). A magnetic field with a magnitude of 1 tesla is quite large—by comparison, the earth's magnetic field is about 5×10^{-5} T. Representative values of the magnetic fields produced by different magnets are listed in Fig. 27.2. Magnetic fields can also be

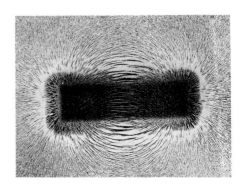

FIG. 27.1. Iron filings line up along the magnetic field lines of a bar magnet. (Fundamental Photographs, New York.)

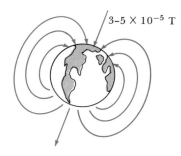

3-5 × 10⁻⁵ T

0.01-0.1 T

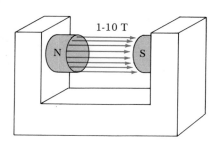

1-10 T

FIG. 27.2. Representative values of magnetic fields.

listed in terms of weber/m² and gauss. The weber/m² is an older name for a tesla. The three units are related to each other as follows:

$$1 \text{ gauss} = 10^{-4} \text{ T} = 10^{-4} \text{ weber/m}^2,$$
$$1 \text{ T} = 1 \text{ weber/m}^2 = 10^4 \text{ gauss}$$

27.2 Magnetic Force on a Moving Charge

The force caused by a magnetic field is different in a number of ways from the forces caused by gravitational and electric fields. (1) While an electric field exerts a force on an electric charge, and a gravitational field exerts a force on a mass (a gravitational "charge"), there seems to be no magnetic charge on which a magnetic field exerts a force. Instead, a magnetic field exerts a force on an electric charge. (2) Although electric and gravitational forces are exerted on both stationary and moving objects, a magnetic force is exerted on an electric charge only when it moves. No magnetic force is exerted on a charge at rest. (3) Gravitational and electric forces do not depend on the direction in which objects move, whereas a magnetic force is exerted only if a component of the motion of the charge is perpendicular to the field. If the charge moves parallel to the field, no magnetic force is exerted. (4) Finally, while gravitational and electric fields always exert force in the direction of the field, the magnetic force does not point in the direction of the field, nor does it point in the direction of the charge's velocity. The magnetic force is perpendicular to both the magnetic field and the charge's velocity.

The unusual nature of the magnetic force might be contrasted to driving a car along a highway across which a wind blows (Fig. 27.3a). If the "wind field" acted like a magnetic field, the car would be lifted rather than blown off the highway (Fig. 27.3b). The lifting force would be perpendicular to both the car's velocity and the direction of the wind. If the car were at rest in the field, no force would be exerted on it. If the car drove into the wind or with the wind, it would feel no force from the wind because no force is exerted on an object moving parallel to the field.

The nature of the magnetic force produces many interesting phenomena and also provides special opportunities for useful applications. Before discussing these phenomena and applications, we define the magnetic force.

FIG. 27.3. If a car entered a region where air blew across a highway and if the blowing air (a wind field) acted like a magnetic field, an upward force would be exerted by the air on the car.

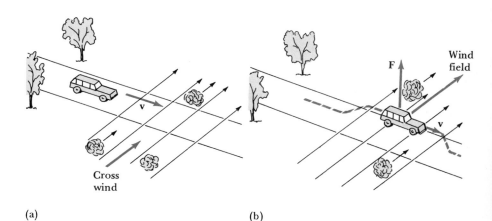

(a)

(b)

Magnetic force on a moving charge: If an electric charge q moves with a velocity $\mathbf{v}$ through a magnetic field $\mathbf{B}$, a magnetic force $\mathbf{F}_m$ is exerted on the charge. The magnitude of the force is

$$F_m = qvB \sin \theta, \qquad (27.1)$$

where θ is the angle ($180°$ or less) between the directions of $\mathbf{v}$ and $\mathbf{B}$. The force is exerted perpendicularly to both $\mathbf{v}$ and $\mathbf{B}$, as discussed shortly.

The vectors $\mathbf{v}$ and $\mathbf{B}$ lie in a plane (see Fig. 27.4a). The magnetic force $\mathbf{F}_m$ on a moving charge is perpendicular to that plane. From what we have said, the force could be up, as in Fig. 27.4a, or down, since both directions are perpendicular to the plane of $\mathbf{B}$ and $\mathbf{v}$. The right-hand rule illustrated in Fig. 27.4b is used to determine which is the correct choice for the direction of the force.

Right-hand rule: Point the fingers of your open right hand in the direction of the magnetic field (Fig. 27.4b). Orient your hand so that the thumb points along $\mathbf{v}$, as though you were "thumbing" a ride (hitchhiking) in the direction of motion. The direction of the magnetic force on a positive charge is the direction in which your open palm would push. The force on a negative charge is in the opposite direction.

When Eq. (27.1) is used, the unit of $\mathbf{B}$ is the tesla, the unit of $\mathbf{v}$ is the meter per second, the unit of q is the coulomb, and the unit of force is the newton. We see that

$$1\,\text{N} = 1\,\text{C(m/s)\,T} \quad \text{or} \quad 1\,\text{T} = 1\frac{\text{N} \cdot \text{s}}{\text{C} \cdot \text{m}}. \qquad (27.2)$$

EXAMPLE 27.1 Each of the lettered dots shown in Fig. 27.5 represents an electric charge of $+2.0 \times 10^{-6}\,\text{C}$ moving at speed $3.0 \times 10^7\,\text{m/s}$ in the directions shown. Determine the magnetic force (magnitude and direction) acting on each charge due to the 0.10-T magnetic field that points in the positive y direction.

SOLUTION (a) The charge at position a moves at a $90°$ angle relative to the magnetic field. Thus, the magnitude of the magnetic force on charge a is

$$F_m = qvB \sin 90° = (2.0 \times 10^{-6}\,\text{C})(3.0 \times 10^7\,\text{m/s})(0.10\,\text{T})1.0$$
$$= \underline{6.0\,\text{N}}.$$

To determine the direction of the force, we point the fingers of our right hand toward the top of the page in the direction of $\mathbf{B}$. Our hand must be oriented such that our thumb points to the left in the direction of $\mathbf{v}$. With this hand orientation, our palm points into the paper, the direction of the magnetic force.

(b) Charge b moves in a direction opposite to $\mathbf{B}$. Thus, $\theta = 180°$. Since $\sin 180° = 0$, the magnetic force on charge b is zero.

(c) Charge c moves up out of the paper perpendicular to $\mathbf{B}$. The magnitude of the force on charge c is $\underline{6.0\,\text{N}}$, the same as for charge a. To determine the direction of the force, our right hand is oriented so that the fingers point toward the top of the page and our thumb points up out of the paper. Our palm then faces left—the magnetic force on charge c points in the negative x direction.

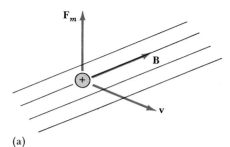

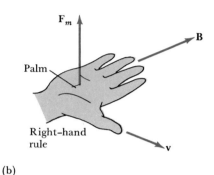

(b)

FIG. 27.4. (a) The magnetic force on a moving charge is perpendicular to $\mathbf{B}$ and $\mathbf{v}$. (b) If your fingers point in the direction of $\mathbf{B}$ and your thumb toward $\mathbf{v}$, your palm faces the direction of $\mathbf{F}_m$.

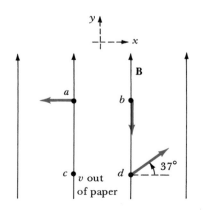

FIG. 27.5

(d) The angle between the velocity of charge d and the magnetic field is $53°$ (not $37°$). Thus,

$$F_m = qvB \sin 53° = (2.0 \times 10^{-6}\text{ C})(3.0 \times 10^7\text{ m/s})(0.10\text{ T})0.80$$
$$= \underline{4.8\text{ N}}.$$

With our right hand lying with the palm up, our fingers point toward the top of the page in the direction of the magnetic field and our thumb points off toward the right in the direction of **v**. Thus, the magnetic force on charge d points up out of the paper, in the direction that our palm faces. ∎

Ions Moving Through Crossed Magnetic and Electric Fields

The situation depicted in Fig. 27.6 is the basis of a variety of technological uses of magnetic fields, including the magnetohydrodynamic generation of electric power and the measurement of the speed at which blood flows. A magnetic field is produced that points perpendicular to the direction in which positive and negative electric charges move (into the paper in Fig. 27.6a). As positive charges move through the magnetic field, they experience a magnetic force perpendicular to both the field and to their direction of motion (toward the right in Fig. 27.6b). These positive charges are forced to the side where they collect on the wall of a vessel through which they move or on a metal plate.

The magnetic force on negative charges points in the opposite direction (to the left in Fig. 27.6c). Consequently, negative charges move to the side opposite the positive charges. The magnetic field has separated positive and negative charges on opposite walls of a vessel or on metal plates at the sides of the moving charges (Fig. 27.6d).

As the vessel walls or plates become charged, they produce an electric field that opposes further accumulation of charge on the walls. A trailing ion that enters the magnetic field now feels two opposing forces: (1) the magnetic force $\mathbf{F}_m$ and (2) an electric force $\mathbf{F}_q$ caused by the electric field produced by positive and negative charges that have already accumulated on the walls at the sides (Fig. 27.6d). The accumulation stops when the electric field becomes large enough so that the net force on a trailing charge entering the field is zero (Fig. 27.6e):

$$F_m - F_q = 0, \tag{27.3}$$

These trailing charges move straight ahead rather than toward a plate. This condition allows us to calculate the voltage across the plates when they stop accumulating charge.

The electric force on the moving charges is given by Eq. (22.4):

$$F_q = qE,$$

where E is the electric field between the plates. The electric field depends on the voltage V across the plates and on the plate separation d according to Eq. (23.8):

$$E = \frac{V}{d}.$$

Therefore,

$$F_q = q\frac{V}{d}.$$

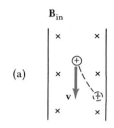

(a)

$\mathbf{B}_{in}$

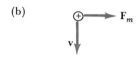

(b)

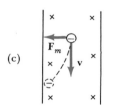

(c)

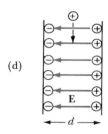

(d)

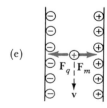

(e)

FIG. 27.6. (a) and (b) A positive charge moving through a magnetic field that points into the paper experiences a magnetic force toward the right wall of the vessel. (c) Negative charges are forced to the left wall. The opposite charges on each wall produce (d) an electric field and (e) an electric force that balances the magnetic force so that trailing charges move straight ahead.

From Eq. (27.1), we know that the magnetic force on the charge has a magnitude

$$F_m = qvB.$$

Substituting these expressions for F_q and F_m into Eq. (27.3), we find that

$$qvB - q\frac{V}{d} = 0, \qquad (27.4)$$

or

$$V = vBd.$$

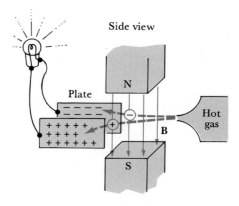

FIG. 27.7. A schematic representation of a magnetohydrodynamic generator.

As we see, the potential difference from one plate to the other depends on the speed v of the moving charges, the strength of the magnetic field B, and the separation of the plates d.

In a magnetohydrodynamic (MHD) generator, gaseous fuel such as pulverized coal enters a combustion chamber and burns there at high temperature (1000–2000 °C) and pressure. If alkali metals, such as potassium, are injected into the burning gas, free electrons and ions are formed. The electrons and ions pass from the combustion chamber through a nozzle and into a magnetic field such as pictured in Fig. 27.7. Opposite-sign charges accumulate on opposite plates at the sides of the moving gas, and the potential difference across these plates provides electric power.

The method we have been describing is also used to measure the speed of blood flow through a vessel. A flow meter produces a magnetic field perpendicular to the vessel. Oppositely charged ions in the blood are forced by the magnetic field to opposite walls of the vessel, thus producing a potential difference V across the walls of the vessel. By measuring V, the magnetic field B, and the diameter d of the vessel, we can substitute in Eq. (27.4) to determine the blood's speed.

EXAMPLE 27.2 Suppose you are an inventor who has derived Eq. (27.4) for the first time. To see if the method leading to this equation has practical promise for measuring blood speed, estimate the potential difference you would expect to measure as blood in an artery passes through the 0.10-T magnetic field of a proposed flow meter.

SOLUTION To solve Eq. (27.4) for potential difference, we need to estimate the speed v of blood in an artery and the diameter d of the artery. The diameter of an artery is approximately 1 cm. The heart pumps about 80 cm³ of blood each second, that is, during each beat. Since the flow rate $Q = vA$ [Eq. (14.2)] where $A = \pi(d/2)^2$ is the cross-sectional area of the vessel, the blood's speed through an artery that carries one-half of the blood pumped each second would be about

$$v = Q/A \simeq \frac{40 \text{ cm}^3/\text{s}}{\pi(1.0 \text{ cm}/2)^2} = 50 \text{ cm/s} = 0.5 \text{ m/s}.$$

Rearranging Eq. (27.4) to solve for the potential difference from one artery wall to the opposite wall and substituting these rough estimates for v, B, and d, we find that

$$V = vBd \simeq (0.5 \text{ m/s})(0.1 \text{ T})(0.01 \text{ m})$$
$$\simeq \underline{5 \times 10^{-4} \text{ V}}.$$

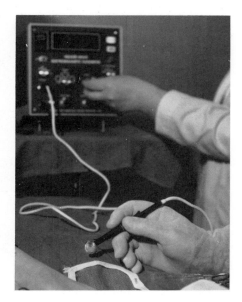

FIG. 27.8. An electromagnetic blood flow meter. The small metal loop at the end of the sensor held in the surgeon's hand fits around an exposed blood vessel and detects blood flow. (Courtesy Carolina Medical Electronics, Inc.)

The voltage looks quite low, but it is measurable. A flow meter, such as described here and shown in Fig. 27.8, is used routinely during heart surgery to monitor blood flow. ∎

27.3 Circular Motion of Charge in a Magnetic Field

You might recall from Chapter 6 that an object moves in a circular path if a centripetal force continuously pulls the object toward the center of a circle. A ball swung at the end of a string is pulled inward by the tension force of the string. The moon as it circles the earth is pulled inward by its gravitational attraction toward the earth. These centripetal forces are exerted perpendicular to the motion of the object and do not change the object's speed. The forces change only the object's direction of motion—instead of flying straight ahead, the object moves in a circle.

Because it is directed perpendicular to the velocity of an electric charge, a magnetic force can act as a centripetal force and cause circular motion. To understand how this happens, lay your physics book on a flat surface. Suppose that a uniform magnetic field points downward into the book from the ceiling. A charge entering the field in a horizontal plane will move in a circle. A top view of the charge's path is pictured in Fig. 27.9.

Suppose, for example, that a positive charge moves along the right edge of the open book toward the top of the page (position *a* in Fig. 27.9). Using the right-hand rule, you can determine the direction of the force on the charge: Your fingers should point down into the book in the direction of **B;** your thumb should point toward the top of the book in the direction of **v;** and the force on a positive charge is in the direction that your palm faces—toward the center of the book. Since the force is perpendicular to the charge's velocity, its speed does not change—only its direction changes. As the charge moves upward, it is deflected slightly to the left by the magnetic force and follows the curved path indicated by the dashed line in Fig. 27.9.

Eventually the charge reaches the top middle of the open book (position *b* in Fig. 27.9). The charge is now moving to the left. Using the right-hand rule again, you will find that the magnetic force again points toward the center of the page. If you do this for every point on the circle shown in Fig. 27.9, you find that the magnetic force always points toward the center of the circle. In Chapter 5 we learned that Newton's second law, when applied to circular motion, was

$$\sum_{\substack{\text{radial} \\ \text{direction}}} F = ma_c,$$

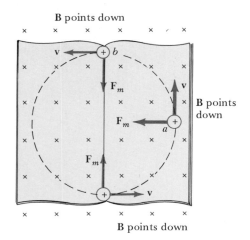

FIG. 27.9. A magnetic field (represented by the crosses) points down into the page. The magnetic force on a charge moving in a circle is shown for several points on the circle.

where the components of all forces acting on the particle in the radial direction are added. The centripetal acceleration a_c caused by these forces has a value v^2/r [Eq. (5.7)], where r is the radius of the circle. In the present situation, the magnetic force provides the centripetal force. Since the angle θ between **v** and **B** is 90°, the magnitude of the force is

$$F_m = qvB \sin 90° = qvB.$$

Substituting in Newton's second law for the force and centripetal acceleration, we find that

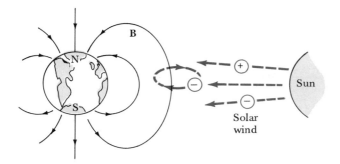

FIG. 27.10. An electron in the solar wind moves in a circle after entering the earth's magnetic field.

$$qvB = m\frac{v^2}{r}. \tag{27.5}$$

Equation (27.5) can be solved for one unknown quantity if the other quantities are known.

EXAMPLE 27.3 An electron from the sun enters the earth's magnetic field high above its surface and undergoes circular motion (Fig. 27.10). Calculate the radius of the circle. At this high elevation, the earth's magnetic field is about 3×10^{-7} T. The electron's speed v is 2×10^8 m/s, the magnitude of its charge q is 1.6×10^{-19} C, and its mass m is 9.11×10^{-31} kg.

SOLUTION We solve Eq. (27.5) for the radius of the circle:

$$r = \frac{mv}{qB} = \frac{(9.11 \times 10^{-31}\,\text{kg})(2 \times 10^8\,\text{m/s})}{(1.6 \times 10^{-19}\,\text{C})(3 \times 10^{-7}\,\text{T})} = 4 \times 10^3\,\text{m} = \underline{4\,\text{km}}. \quad \blacksquare$$

Magnetic Bottles

Example 27.3 leads us to a very interesting phenomenon involving the circular motion of electric charges in magnetic fields. A charged particle can enter a magnetic field with its velocity **v** not completely perpendicular to the field (Fig. 27.11). The perpendicular component of the velocity $\mathbf{v}_\perp$ causes the charge to move in a circle, as we have just seen. However, the component of the velocity that is parallel to the field $\mathbf{v}_\parallel$ is unaffected by the field because the angle between $\mathbf{v}_\parallel$ and **B** is $0°$, and $\sin 0°$ is zero—hence no force. The particle undergoes a helical motion. It rotates in a circle at speed $v_\perp$ while moving along the field at speed $v_\parallel$ (Fig. 27.11).

If the magnitude of the magnetic field is not constant but increases, the particle is in for a real surprise. A field of increasing magnitude is represented by more closely spaced magnetic-field lines (Fig. 27.12a). As the lines come together, the plane in which **v** and **B** are located tilts. For example, at the top of the loop shown in Fig. 27.12a, the **B** field points slightly down, whereas **v** is horizontal. Thus, the plane of **v** and **B** tilts down at the front. Since the magnetic force $\mathbf{F}_m$ is perpendicular to this plane, it points down and slightly back, as shown in Fig. 27.12a. $\mathbf{F}_m$ consists of two components shown in Fig. 27.12b: a radial component F_{radial} pointing toward the center of the circle and a backward component F_{back} that opposes its forward motion. If strong enough, F_{back} can

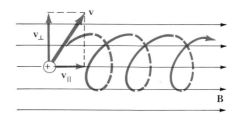

FIG. 27.11. If a charged particle enters a magnetic field with its velocity not completely perpendicular to the field, the charge undergoes a helical motion.

FIG. 27.12. (a) When a charge undergoes helical motion in a converging magnetic field, the magnetic force has (b) a backward component that can reverse the particle's direction.

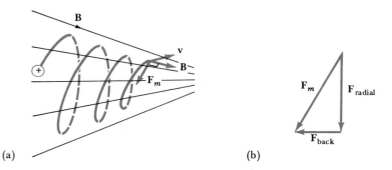

(a)

(b)

actually reverse the particle's direction. The place where the particle's direction is reversed by a converging magnetic field is called a **magnetic mirror** (the particle is reflected as if it hit a mirror).

Magnetic fields can be constructed to confine hot, highly ionized gases, called **plasmas.** The ions are reflected by the "pinched" magnetic field at the ends of the magnetic container—called a **magnetic bottle** (Fig. 27.13). Unfortunately, a magnetic bottle is somewhat leaky. Ions moving along its center axis are not deflected at its ends. Efforts are being made to construct magnetic bottles to confine the hot gas in which nuclear fusion reactions occur (discussed in Chapter 35).

One of the most interesting magnetic bottles occurs high above the earth's atmosphere. The earth is continually "showered" by ions that have escaped from the sun. These showers are especially intense following solar flares. When the ions reach the earth's magnetic field (Fig. 27.14), they are deflected and move in helical paths in two regions at distances of about 3000 km and 16,000 km above the earth's equator (called the Van Allen belts). As the earth's magnetic-field lines converge near the earth's poles, they form a magnetic mirror that reflects the ions back in the opposite direction. The ions are trapped in the magnetic bottle of the earth's own magnetic field.

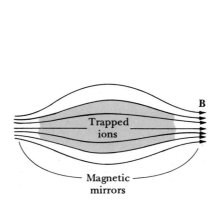

FIG. 27.13. A "magnetic bottle."

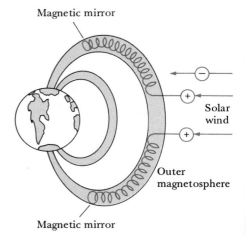

FIG. 27.14. Electrons and protons from the sun are trapped in the earth's magnetic field—a giant magnetic bottle.

Like man-made magnetic bottles, the earth's magnetic bottle leaks, and some of the ions reach the earth's atmosphere near the north and south poles. When these ions collide with atoms in our atmosphere, the atoms are ionized. As the ionized atoms recombine with their missing electrons, light is emitted. The aurora or northern (and southern) lights are produced. A picture of these dramatic lights is shown in Fig. 27.15.

Mass Spectrometers

The circular motion of charged particles in magnetic fields helps us understand natural phenomena such as the production of the earth's northern lights. This circular motion has also been put to practical use by the development of instruments, such as mass spectrometers, that are used to determine the mass of atom-sized particles.

In a **mass spectrometer,** an atom or molecule is first ionized by removal of one of its electrons (Fig. 27.16). The ion is then accelerated across a voltage and passes through a velocity selector (see Problem 8 in this chapter) that allows only those ions moving at a predetermined speed to pass. The ions then enter a magnetic field, where they move in one-half of a circular path. The radius r of the circle can be measured by observing the place where the ions strike a detector after moving halfway around the circle. The mass m of the ion can be calculated using Eq. (27.5):

$$m = \frac{qrB}{v},$$

where q is the charge of the ion, r is the radius of the semicircle, v is the ion's speed, and B is the magnitude of the magnetic field.

27.4 Torque on a Current Loop

Another important application involving magnetic force occurs when electric charge moves through a loop of wire in a magnetic field. The magnetic force acting on these moving charges can exert a torque on the loop that causes it to rotate. An electric motor turns because of such a torque.

To calculate this torque, consider the magnitude of the magnetic force acting on all charges moving along one side of the loop of wire shown in Fig. 27.17. (When calculating the force, it is easier to assume that positive charges move in the direction of the current rather than that negative charges move opposite the current; the result is the same in each case.) Suppose that there are n moving charges in this wire. Each charge q travels perpendicular to the magnetic field. Hence the net magnetic force on the charges and on the wire is

$$F_m = nqvB. \qquad (27.6)$$

If a charge moves the length b of the wire in time t, its speed is

$$v = \frac{b}{t}.$$

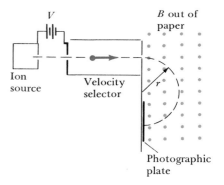

FIG. 27.16. A mass spectrometer.

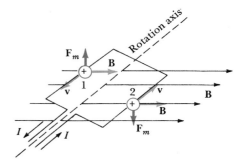

FIG. 27.17. The magnetic force on positive charges moving along the side of a loop of wire causes a torque that will rotate the loop clockwise.

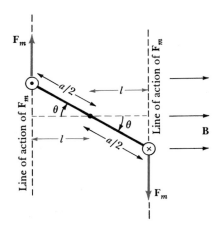

FIG. 27.18. The magnetic force on each side wire of a loop has a moment arm l. The torque due to each force has a magnitude of $\tau = lF_m$.

In time t, all n charges in the wire will pass a cross section at its end. According to Eq. (24.1), the current is

$$I = \frac{nq}{t}.$$

Using the preceding equations to substitute for v in Eq. (27.6), we find that

$$F_m = nqvB = nq\left(\frac{b}{t}\right)B = \left(\frac{nq}{t}\right)bB = IbB.$$

The force is the same magnitude on each side wire shown in Fig. 27.17, but the force on one side wire is directed up, whereas that on the other wire is directed down. Although the net force on the two wires is zero, together they produce a torque on the loop of the wire.

To calculate the torque, consider the end view of the loop shown in Fig. 27.18. If the plane of the loop is oriented at an angle θ with respect to the field, then the moment arm l of the force about an axis along the center of the loop is

$$l = \left(\frac{a}{2}\right)\cos\theta,$$

where a is the width of the loop. The torque due to both wires is calculated using Eq. (6.3):

$$\tau = 2(l)(F_m) = 2\left(\frac{a}{2}\cos\theta\right)(IbB).$$

The inside area A of the loop is the product of its width and length; that is, $A = ab$. Hence, the torque is $\tau = IAB\cos\theta$. For most practical applications, the current travels through a coil with many turns, each of which experiences a torque such as just calculated. If the coil has N turns, the total torque on the coil carrying a current I when in a magnetic field B is

$$\tau = NIAB\cos\theta. \tag{27.7}$$

(Magnetic forces exerted on moving charges at the ends of the loop stretch or compress the loop but cause no torque.)

The torque is greatest when the plane of the loop is oriented parallel to the direction of the magnetic field (Fig. 27.19a). The angle θ is then $0°$, and since $\cos 0° = 1$, the torque is a maximum. When the loop is perpendicular to the

FIG. 27.19. (a) The torque on the loop is maximum when the plane of the loop is parallel to the **B** field. The forces have the greatest moment arm. (b) The torque is zero when the loop is perpendicular to the field, since the moment arms of the forces are zero.

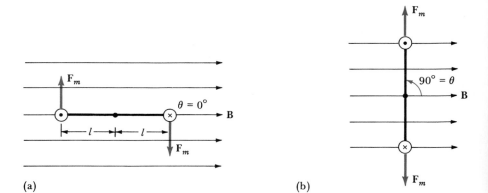

field, as in Fig. 27.19b, two equal and opposite forces are produced, each having a zero moment arm. Thus, no torque is produced. (Note that when θ is 90°, cos 90° = 0, and the torque is zero.)

Although Eq. (27.7) was derived for a rectangular loop, it applies equally to a loop of any shape, as long as the loop is flat.

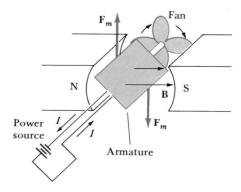

FIG. 27.20. A dc electric motor.

27.5 A Direct-Current Electric Motor

The torque just described is used to turn the shaft of a direct-current electric motor. The motor has a large number of loops of wire, called the **armature,** wound around an iron core. An armature with one loop is shown in Fig. 27.20. In a motor, the armature is placed between the poles of a magnet. When electric current flows through the loops of wire, a magnetic force is exerted on these moving charges. The force causes a torque that turns the armature and its core. The core is attached to a shaft that causes some other object, such as a fan, to rotate.

The motor shown in Fig. 27.20 has one serious problem: When the loop has rotated a little less than one-quarter turn, the torque on the loop becomes counterclockwise rather than clockwise. Look at the side view of a loop shown in Fig. 27.21a. The circle with a cross, $\otimes$, indicates a current moving into the paper along a side wire of the loop. A circle with a dot, $\odot$, indicates a current moving out of the paper on the other side wire of the loop. When in the horizontal position with the currents flowing as shown, a clockwise torque is exerted on the loop. As the loop passes the vertical orientation, the torque becomes counterclockwise. However, if the current is reversed, as in Fig. 27.21b, the torque remains clockwise. Consequently, to make the armature continue to rotate clockwise, the current must change direction each time the loop passes the vertical orientation.

This current reversal is made possible using a commutator (Fig. 27.22). The **commutator** consists of two semicircular rings that are attached to the armature and rotate with it. Sliding contacts or brushes from the power source contact the commutator rings. The current direction is reversed in the middle of each rotation as the contact of the brushes passes from one commutator ring to the next.

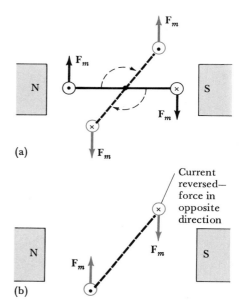

FIG. 27.21. (a) When the loop is horizontal (the solid lines), a clockwise torque is exerted on the loop. As the loop passes the vertical orientation, the torque becomes counterclockwise. (b) If the current direction in the loop is reversed as it passes the vertical, the torque continues to be clockwise.

27.6 Calculating the Magnetic Field Caused by Moving Charges

All the effects we have discussed—magnetic forces, circular motion of charges in a magnetic field, torques on current loops—depend on the presence of a magnetic field. At the start of the chapter we assumed that these fields exist because we saw their effects. But what causes a magnetic field? We know that electric fields are caused by electric charges and that gravitational fields are caused by masses. Is there an analogous object that causes magnetic fields? Some scientists say that there is such an object, called a magnetic monopole, that is analogous to an electric charge. Whereas most magnets have a north and a south pole, a magnetic monopole is a single pole—either north or south. A north pole might be likened to a positive charge and a south pole to a negative charge. Either one supposedly can produce a magnetic field. The search for a magnetic monopole has, to date, been unsuccessful.

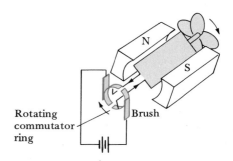

FIG. 27.22. A simple dc motor used to turn a fan.

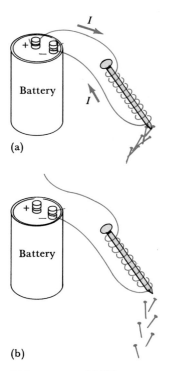

(b)

FIG. 27.23. (a) When a current passes through a wire wrapped around a nail, a magnetic field is produced that allows the nail to become a magnet and attract metal pins. (b) If the current stops, the pins fall, since a magnetic field is no longer being produced.

There are, nonetheless, ways to produce a magnetic field that do not require the existence of a magnetic monopole. For example, putting an electric charge in motion produces a magnetic field. You have probably performed experiments, such as shown in Fig. 27.23, in elementary school that confirm this idea. When a wire is wrapped around a nail and attached to a battery, current flows through the wire. The moving charges produce a magnetic field, and, as a result, the nail acts like a magnet. The nail attracts pins and other objects made of iron when the current is flowing but does not when the current stops.

Two French scientists, Jean Baptiste Biot (1774–1862) and Felix Savart (1791–1841), formulated an equation to calculate the magnetic field caused by moving charges. The equation is difficult to use without the assistance of integral calculus, so we will simply outline the Biot-Savart technique and emphasize some results of its use. Perhaps the most important idea to remember is that moving charges produce magnetic fields. The technique is illustrated with reference to the current moving in a long wire, shown in Fig. 27.24a.

Biot and Savart found experimentally that the magnetic field **B** caused by a charge q moving at a speed v has a magnitude

$$B = \left(\frac{\mu}{4\pi}\right)\frac{qv\sin\theta}{r^2}. \qquad (27.8)$$

The equation is called the **Biot-Savart law.** The quantity r is the distance from the moving charge to the point where the field is being calculated. The quantity μ is called the **permeability** of the medium in which the magnetic field exists. In free space the permeability μ_0 has a value $\mu_0 = 4\pi \times 10^{-7}\,\text{T}\cdot\text{m/A} = 1.257 \times 10^{-6}\,\text{T}\cdot\text{m/A}$; thus,

$$\frac{\mu_0}{4\pi} = 10^{-7}\,\text{T}\cdot\text{m/A}.$$

The permeability of other materials is a complex subject. Even for a particular type of material, such as iron, the permeability depends on its history: Has it been magnetized or demagnetized, and to what degree? Although more will be said about permeability later, we will note that the permeability of iron is about 1000 times that of air. Thus, magnetic fields produced by currents are much stronger in iron than in air.

The angle θ that appears in Eq. (27.8) is the angle between the direction of the charge's velocity **v** and a vector **r** drawn from the moving charge to the field point, as shown in Fig. 27.24a. The direction of **B** is perpendicular to the plane

FIG. 27.24. (a) The magnetic field produced by a moving charge is perpendicular to the planes of **v** and **r**. (b) The magnetic field circles the wire in the direction of your fingers when your right thumb points in the direction of the current.

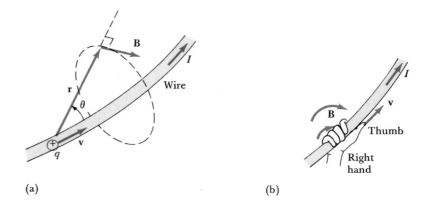

(a)   (b)

containing **v** and **r**. The orientation of **B** relative to that plane is determined using a different right-hand rule than that used to find the force on a moving charge.

Right-hand rule to find direction of B field: Point the thumb of your right hand in the direction of the current—that is, in the direction of a moving, positive charge. The field circles around the current in the same direction as your fingers curl around the current (Fig. 27.24b).

Since most magnetic fields are produced by a very large number of charges moving in a wire, the net magnetic field at a point is the vector sum of the fields caused by each moving charge. Integral calculus must be used to add these fields. Usually the magnitude of the resultant field is expressed in terms of the electric current I causing the field. The direction of the field at different locations is shown by drawing magnetic-field lines. The lines are closely spaced at regions where the field is greatest. Several examples are shown in Fig. 27.25.

EXAMPLE 27.4 A straight wire carries a current of 5 A. At what distance from the wire is the magnetic field equal in magnitude to the earth's magnetic field (5×10^{-5} T)?

SOLUTION The magnetic field at distance a from a straight wire carrying a current I is given by Eq. (27.9), which is listed in Fig. 27.25a:

$$B = \frac{\mu_0 I}{2\pi a}.$$

Rearranging to solve for a, we find that

$$a = \frac{\mu_0 I}{2\pi B} = \frac{(4\pi \times 10^{-7}\,\text{T} \cdot \text{m/A})5\,\text{A}}{2\pi(5 \times 10^{-5}\,\text{T})} = 2 \times 10^{-2}\,\text{m} = \underline{2\,\text{cm}}.$$

Thus, we would not want to put the compass of a ship close to electrical wires. The compass might respond to the magnetic field caused by currents in the wires rather than to the earth's magnetic field. ∎

EXAMPLE 27.5 We wish to observe the effect of a magnetic field on the muscles in a frog's leg. The measurements can be made by placing the frog's leg in a solenoid (Fig. 27.26) that has a diameter of 1.0 cm, is 5.0 cm long, and is made of 200 turns of wire. Calculate the strength of the field inside the solenoid if a current of 4.0 A flows in it. The permeability of the frog's leg (and of all living matter) is about equal to that of air.

SOLUTION The magnetic field at the center of a solenoid is given by Eq. (27.12), shown in Fig. 27.25c:

$$B = \mu_0 \frac{N}{L} I = (4\pi \times 10^{-7}\,\text{T} \cdot \text{m/A}) \times \frac{200}{0.050\,\text{m}}(4.0\,\text{A}) = \underline{2.0 \times 10^{-2}\,\text{T}}.$$

It is unlikely that this field will produce any easily observable effect on the frog's leg. ∎

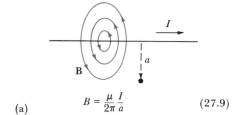

(a)

$$B = \frac{\mu}{2\pi}\frac{I}{a} \qquad (27.9)$$

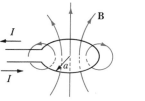

$$B = \frac{\mu I}{2a} \quad \text{At center of loop} \qquad (27.10)$$

$$B = \frac{N\mu I}{2a} \quad \begin{array}{l}\text{At center}\\\text{of loop}\\\text{with } N\\\text{turns}\end{array} \qquad (27.11)$$

(b)

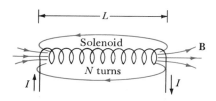

(c) $$B = \mu \frac{N}{L} I \quad \begin{array}{l}\text{Inside}\\\text{solenoid}\end{array} \qquad (27.12)$$

FIG. 27.25. The magnetic fields created by currents in (a) a straight wire, (b) a single loop of wire, and (c) a solenoid.

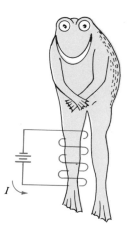

FIG. 27.26. The current through a solenoid creates a magnetic field in a frog's leg.

EXAMPLE 27.6 An electron in the hydrogen atom can be thought of as circling the proton nucleus in a circular orbit of radius 0.53×10^{-10} m. The electron completes one trip around the orbit each 1.5×10^{-16} s. Calculate the magnitude of the magnetic field at the center of the circle where the proton is located.

SOLUTION The electron's motion produces the same effect as if a circular loop of wire carried an electric current around the proton (Fig. 27.25b). The field at the center of the loop is given by Eq. (27.10): $B = \mu_0 I / 2r$. To determine B, we must first calculate the current I due to the electron's motion. Since one electron passes around the loop or orbit each 1.5×10^{-16} s, we can calculate the "current" that is flowing by using Eq. (24.1):

$$I = \frac{q}{t} = \frac{(1.6 \times 10^{-19}\,\text{C})}{(1.5 \times 10^{-16}\,\text{s})} = 1.1 \times 10^{-3}\,\text{A}.$$

This single electron, because of its small orbit and its high speed, produces a current of 1.1 mA.

The magnetic field at the center of the loop is now determined using Eq. (27.10), shown in Fig. 27.25b:

$$B = \frac{\mu_0 I}{2r} = \frac{(4\pi \times 10^{-7}\,\text{T}\cdot\text{m/A})(1.1 \times 10^{-3}\,\text{A})}{2(0.53 \times 10^{-10}\,\text{m})} = \underline{13\,\text{T}}.$$

This is a very strong magnetic field. ∎

27.7 Magnetic Materials

Now that we have examined (1) the force of a magnetic field on a moving charge and (2) the magnetic field produced by moving charges, we can use this information to see how a common bar magnet works. A bar magnet, like all magnetic materials, derives its properties from the magnetic fields created by atoms and the forces (magnetic and others) acting on atoms.

An atom consists of a very small nucleus surrounded by clouds of electrons, which can be thought of as circling the nucleus in elliptical orbits. The motion of each electron is equivalent to a loop of wire carrying one electron. The magnetic field produced by this electron current loop is identical to that produced by a tiny bar magnet (see Fig. 27.27). The electron also produces a magnetic field because it has an intrinsic *magnetic dipole moment* with a north and south pole and is like a very tiny bar magnet. Thus, an electron in an atom creates a magnetic field because of its orbital motion and because of its intrinsic magnetic moment.

In atoms with more than one electron, the fields created by different electrons often cancel each other. The electrons pair off so that the intrinsic magnetic moment of one electron points opposite that of another electron, and their magnetic fields cancel. The field caused by the orbital motion of electrons also cancel since some electron currents move clockwise while others move counterclockwise. If the cancellation is not complete, the atom has a net magnetic dipole moment and produces a magnetic field like that of a small bar magnet. Atoms with nonzero magnetic dipole moments such as iron, aluminum, sodium, oxygen, and nickel are called *paramagnetic* atoms.

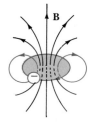

(a)

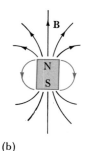

(b)

FIG. 27.27. The field caused by an electron current loop in an atom (a) looks just like that produced by a tiny bar magnet (b).

Normally the paramagnetic atoms in a material are oriented randomly. The magnetic field produced by one atom is cancelled by the field of another atom that is oriented in the opposite direction. The net field produced by all atoms is zero. The iron atoms of the hemoglobin in your blood is a good example. Each iron atom produces a very small magnetic field. But because they are far apart, the iron atoms in the different hemoglobins are oriented randomly. The net magnetic field due to all of these atoms is zero.

However, if a material that is made of these paramagnetic atoms is placed in an externally generated field, the magnetic dipole moments of the atoms have a tendency to align with the external field (like a compass needle aligning with the earth's magnetic field). The magnetic fields caused by the atoms no longer cancel each other completely but add together to produce their own field that adds to the external field. In most materials the paramagnetic atoms contribute little additional field because the thermal agitation by neighboring atoms causes the atoms to remain randomly oriented. For example, the field produced by a unit volume of aluminum atoms is only 2×10^{-5} times the field used to orient the atoms.

In certain materials, such as iron, nickel, cobalt, and gadolinium, the alignment of the atoms is not random. If iron, for example, is placed in an externally generated magnetic field, the atomic magnetic moments of the iron atoms may become aligned so that the net field (the external field plus that caused by the magnetic moments of the iron atoms) is 100 to 1000 times greater than the external field causing the alignment. In these materials, called *ferromagnetic* materials, the magnetic moments of neighboring atoms are locked together and oriented in the same direction by a complicated force that is described by theories of atomic physics.

Even when not in an external magnetic field, the alignment of neighboring atoms in these materials is uniform in small regions called **domains.** In a particular domain the magnetic dipoles of all atoms point in the same direction. Each domain may include 10^{15} to 10^{16} atoms and occupy a space less than a millimeter on a side. In a neighboring domain the magnetic dipoles of the atoms may all point in a different direction. In a piece of iron that is unmagnetized, like a nail, the domains are oriented randomly, and the magnetic field produced by one domain is cancelled by that produced by another.

If an unmagnetized piece of iron is placed in an externally produced magnetic field caused by a loop of wire with current flowing through it, the domains oriented in the direction of the field increase in size, or grow, and those oriented in other directions decrease in size, or shrink. The material is then said to be magnetized, and it produces its own strong magnetic field. In a permanent magnet, the domains remain oriented in the same direction, and the material continues to produce a strong magnetic field even when removed from the external field.

We can now understand why the magnetic permeability μ of a material affects the magnetic field produced by a current. Picture in your mind a solenoid with air inside its coils and another with iron inside. The magnetic field produced by the solenoid with air is caused entirely by the current in the loops of wire, but for a solenoid with iron inside, the field produced by the current loops causes the domains oriented in the direction of the field to grow. Their field adds to that of the coil. If most of the iron magnetic moments become aligned in the direction of the field of the coil, the field of the iron and coil together may be a thousand times greater than if the iron were not present.

This increase in magnetic field caused by ferromagnetic materials has many practical applications. The coating on magnetic tape provides a record of information by the degree to which it becomes permanently magnetized. Airport metal detectors, transformers, electric motors, loudspeakers, and electric generators all depend on the magnetization of ferromagnetic materials.

Summary and Additional Readings

1. **Magnetic force on a moving charge:** If an electric charge q moves with velocity $\mathbf{v}$ through a magnetic field $\mathbf{B}$, a magnetic force $\mathbf{F}_m$ is exerted on the charge. The magnitude of the force is

$$F_m = qvB \sin \theta, \tag{27.1}$$

where θ is the angle between the directions of $\mathbf{v}$ and $\mathbf{B}$. The direction of the magnetic force is perpendicular to both $\mathbf{v}$ and $\mathbf{B}$. The right-hand rule is used to determine the orientation of $\mathbf{F}_m$. Point the fingers of your open right hand in the direction of $\mathbf{B}$ and orient your hand so that the thumb points along $\mathbf{v}$. Your open palm then faces in the direction of the magnetic force on a positive charge. The force on a negative charge is in the opposite direction.

2. **Circular motion in a magnetic field:** When an object of mass m and charge q moves at speed v perpendicular to a uniform magnetic field $\mathbf{B}$, the magnetic force ($F_m = qvB$) acting on the charge causes a centripetal acceleration ($a_c = v^2/r$) that keeps the charge moving in a circle. Newton's second law relating the magnetic force and centripetal acceleration is as follows:

$$qvB = m \frac{v^2}{r}. \tag{27.5}$$

3. **Torque on a current loop:** If a current I flows through a coil of wire with N turns each of area A while the coil is in a magnetic field B, the magnetic force on the charges moving in the wire causes a torque τ on the coil of magnitude

$$\tau = NIAB \cos \theta, \tag{27.7}$$

where θ is the angle between the plane of the coil and the direction of the magnetic field.

4. **Magnetic field produced by moving charges:** Moving electric charges produce magnetic fields. The Biot-Savart law, along with integral calculus, can be used to calculate the field caused by these moving charges. The fields produced by electric currents in wires of different shape are shown in Fig. 27.25.

5. **Magnetic materials:** A material that can be magnetized is made of atoms whose electrons produce magnetic fields because of the orbits and intrinsic magnetic moments of their electrons. The atoms act like tiny bar magnets. In ferromagnetic materials, the atomic bar magnets in a region of the material fall in line; a **domain** is produced. If domains become aligned, the material is magnetized and produces its own large magnetic field.

C. R. Carrigan and David Gubbins, "The Source of the Earth's Magnetic Field," *Scientific American,* pp 118–130, February (1979).

J. R. Nielson, "Hans Christian Oersted—Scientist, Humanist and Teacher," *American Journal of Physics* **7,** 10 (1939).

D. Cohen, "Magnetic Fields of the Human Body," *Physics Today* **28,** 35 (1975).

Questions

1. A beam of electrons is not deflected as it moves through a region of space in which a magnetic field exists. Give two explanations why the electrons might move along a straight path.

2. A beam of electrons moving toward the east is deflected upward by a magnetic field. Determine, using the right-hand rule, the direction in which the magnetic field points. Repeat for an electric field.

3. A box has either an electric field or a magnetic field on the inside. Describe experiments that you might perform to determine which field is present and its orientation.

4. Why are residents of northern Canada shielded less from cosmic rays (mostly protons flying toward the earth from outer space) than are residents of Mexico?

5. A U-shaped wire through which a current flows hangs with the bottom of the U between the poles of an electromagnet (see Fig. 27.28). When the field in the magnet is increased, does the U swing toward the right or the left? Explain.

6. Why does the torque on the current loop of the motor shown in Fig. 27.20 change even though the force on the wires is constant?

7. A magnetic field is said to exist in a region of space. Describe three experiments you could do to confirm this statement.

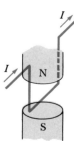

FIG. 27.28

8. An electron enters a solenoid at a small angle relative to the magnetic field inside. Describe the electron's motion.

9. Two parallel wires carry electric current in the same direction. Does the moving charge in one wire cause a magnetic force to be exerted on the moving charge in the other wire? If so, in what direction is the force relative to the wires? Explain. Repeat for currents moving in opposite directions.

10. Why is the magnetic field so much greater at the end of a solenoid that has an iron core than at the end of a solenoid of similar size with an air core?

Problems

27.2 Magnetic Force on a Moving Charge

■ 1. Each of the lettered dots shown in Fig. 27.29 represents a charge of $+1.0 \times 10^{-8}$ C moving at speed 2.0×10^{7} m/s. Calculate the magnitude of the force on each charge and indicate carefully, in a drawing, the direction of the force. A uniform 0.50-T magnetic field points in the positive z direction.

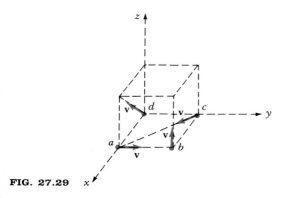

FIG. 27.29

■ 2. A duck accumulates a positive charge of 3.0×10^{-8} C while flying north at speed 18 m/s. The earth's magnetic field at the duck's location has a magnitude of 5.3×10^{-5} T and points in a direction 62° below a horizontal line pointing north. Calculate the magnitude and direction of the magnetic force on the duck.

■ 3. An automobile with a charge of $+2.0$ C is in a (4.0×10^{-5})-T magnetic field that is parallel to the ground and points north. Calculate the magnetic force (magnitude and direction) on the automobile when (a) at rest, (b) moving south at 40 m/s, (c) moving west at 40 m/s, (d) moving 30° east of north at 40 m/s, and (e) moving 37° west of south at 40 m/s.

■ 4. An electron of mass 9.1×10^{-31} kg moves horizontally toward the north at 3.0×10^{7} m/s. Calculate the magnitude and direction of a magnetic field that will cause a magnetic force that balances the electron's weight.

■ 5. A 1000-kg automobile drives west along the equator near Lake George in Uganda. At this location the earth's magnetic field is 3.5×10^{-5} T and points north parallel to the earth's surface. If the automobile carries a charge of -2.0×10^{-6} C charge, how fast must it move so that the magnetic force balances 0.010 percent of its weight?

■ 6. A hydroxide ion (OH$^-$) in a glass of water has an average speed of about 600 m/s. (a) Calculate the electrical force between the hydroxide ion (charge $-e$) and a positive ion (charge $+e$) that is 1.0×10^{-8} m away (about the separation of 30 atoms). (b) Calculate the maximum magnetic force that can act on the ion when in the earth's (5.0×10^{-5})-T magnetic field. (c) On the basis of these two calculations, does it seem likely that the earth's magnetic field has much effect on the biochemistry of the body?

■ 7. A blood flow meter measures a potential difference of 8.0×10^{-5} V across a vessel of diameter 4.0×10^{-3} m. (a) Calculate the magnitude of the electric force acting on an ion of charge 1.6×10^{-19} C. (b) At what speed must the ion move so that the electric force is balanced by a magnetic force caused by a 0.040-T field oriented perpendicular to the flow direction?

■ 8. An electron moves between two parallel plates, as shown in Fig. 27.30. A 480-V/m electric field points from the upper plate toward the lower. A 0.12-T magnetic field points into the paper (represented by the crosses, $\times$, between the plates). (a) Calculate the electric force (magnitude and direction) on an electron between the plates. (b) At what speed and in which direction (right or left) must the electron move so that the magnetic field exerts an opposing force that balances the electric force? (If narrow slits are placed at the entrance and exit of the plates, the device allows charges of only one speed to pass through both slits. Such a device is called a **velocity selector.**)

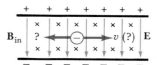

FIG. 27.30

■ 9. An electron moves at speed 8.0×10^{6} m/s toward the right between two parallel plates, such as shown in Fig. 27.30. A 0.12-T magnetic field points out of the paper parallel to the plate surfaces. (a) Calculate the magnitude and direction of the magnetic force on the electron. (b) What should be the magnitude and direction of an electric field caused by charge on the plates to produce an electric force that just balances the magnetic force?

■■ 10. A wire, shown in Fig. 27.31, moves perpendicular to a magnetic field. (a) In which direction are the electrons in the wire

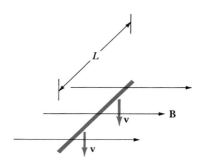

FIG. 27.31

forced to move? (b) After the electrons are forced in one direction, leaving positive charges behind, an electric field and potential difference V are developed from one end of the wire to the other. Electrons no longer move, since the electric and magnetic forces balance. Show that this happens when the potential difference from one end of the wire to the other is

$$V = vLB,$$

where v is the speed of the wire, L is its length, and B is the magnitude of the magnetic field.

■■ **11.** Calculate the potential difference developed across the wings of a Boeing 747 airliner when traveling north at 240 m/s. The wing span is 60 m. The earth's (5.0×10^{-5})-T magnetic field points in a direction 60° below a horizontal line directed north. In solving this problem, follow the outline developed in the previous problem. Note that the magnetic field is not perpendicular to the plane's velocity.

■■ **12.** An alien from a planet in the galaxy M31 (Andromeda) has a ray gun that shoots protons at a speed of 2.0×10^5 m/s. Design a magnetic shield that will deflect the protons away from your body. Using rough *estimates,* show that your magnetic shield will, in fact, protect you. Indicate the orientation of magnetic field in front of you and the direction in which the ions are deflected.

■■ **13.** An electron beam moves toward the east in the tube of a television set. *Estimate* its vertical displacement caused by the earth's magnetic field.

27.3 Circular Motion in a Magnetic Field

14. An electron moves in a circular path perpendicular to the earth's (3.5×10^{-5})-T magnetic field. The radius of the circle is 0.70 m. Calculate the electron's speed.

15. An ion with charge 1.6×10^{-19} C moves at speed 1.0×10^6 m/s into and perpendicular to the 0.30-T magnetic field of a mass spectrometer. After entering the field it moves in a circular path of radius 0.18 m. Calculate (a) the magnitude of the magnetic force acting on the ion and (b) its mass.

■ **16.** An electron and a proton, moving side by side at the same speed, enter a 0.020-T magnetic field. The electron moves in a circular path of radius 7.0 mm. Calculate the radius of the adjacent circle in which the proton moves.

■ **17.** A carbon ion of charge 1.6×10^{-19} C and mass 1.99×10^{-26} kg moves with an oxygen ion of the same charge whose mass is 4/3 times larger than that of the carbon. Both ions move together at speed 2.0×10^5 m/s into and perpendicular to the 0.28-T magnetic field of a mass spectrometer. Calculate the separation of the ions after they move in a semicircular path.

■ **18.** A mass spectrometer has a velocity selector (see Problem 8) that allows ions traveling at only one speed to pass undeflected through slits at the ends. While moving through the velocity selector, the ions pass through a 60,000-V/m electric field and a 0.0500-T magnetic field. The three vectors **v**, **E**, and **B** are mutually perpendicular. (a) Calculate the speed of undeflected ions. (b) After leaving the velocity selector, the ions continue to move in the 0.0500-T magnetic field. Calculate the radius of curvature of a singly charged lithium ion whose mass is 1.16×10^{-26} kg.

■ **19.** (a) Calculate the speed of electrons accelerated through a 28-kV potential difference in a color television set. (b) Calculate the radius of curvature of one of these electrons if moving perpendicular to the earth's (3.5×10^{-5})-T magnetic field at the equa-

tor. (c) If the television set faces east, are the electrons deflected up or down? Explain your answer.

■ **20.** As an electron moves through a bubble chamber (see Section 34.11), it leaves a trail that can be photographed. *Estimate* the velocity of the electron whose path is shown in the lower left spiral of Fig. 34.13 as it moves around the outside loop in a 0.14-T magnetic field. In what direction must the field be pointing? Explain.

■■ **21.** One type of mass spectrometer accelerates ions of charge q, mass m, and initial speed zero through a potential difference V. The ions then enter a magnetic field **B** where they move in a circular path of radius r. Show that the mass of the ions is related to these other quantities by the equation $m = qr^2B^2/2V$.

27.4 Torque on a Current Loop

22. A square coil with 30 turns has sides that are 16 cm long. What current must flow through the wire to produce a maximum torque of 0.60 N · m when the coil is in a 0.30-T magnetic field?

23. A 5.0-A current flows through a 10-turn square coil each side of length 0.12 m. The coil is in a 0.15-T magnetic field. Calculate the torque on the coil (the magnitude and the direction it turns the coil) for each orientation shown in Fig. 27.32.

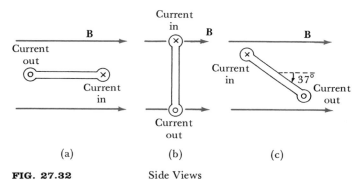

FIG. 27.32   Side Views

24. (a) What is the maximum torque on a 500-turn circular coil of radius 6.0 cm when carrying a 0.30-A current in a 0.10-T magnetic field? (b) How is the coil oriented relative to the field when this maximum torque is exerted on it? (c) What should its orientation be when the torque is reduced by 50 percent?

■ **25.** In Houston, the earth's magnetic field has a magnitude of 5.2×10^{-5} T and points in a direction 57° below a horizontal line pointing north. Determine the magnitude and direction of the magnetic force on a 10-m-long vertical wire carrying a 12-A current straight upward.

■ **26.** An east-west power line at the equator carries 100 A of current toward the east. At this location the earth's magnetic field has a magnitude of 3.5×10^{-5} T and points north. (a) Calculate the magnitude and direction of the magnetic force on a 250-m length of the wire whose mass is 160 kg. (b) Calculate the fraction of the wire's weight that this force equals.

■ **27.** (a) Calculate the magnetic force (magnitude and direction) that acts on segments *a, b,* and *c* of wire shown in Fig. 27.33. Segments *a* and *c* are 2.0 cm long, and segment *b* is 10.0 cm long. A 3.0-A current flows through the wire, which is in a 0.15-T magnetic field. (b) Calculate the torque caused by forces $\mathbf{F}_a$ and $\mathbf{F}_c$ together and the torque caused by force $\mathbf{F}_b$.

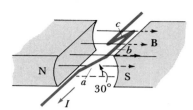

FIG. 27.33

■ **28.** A 500-turn coil of wire is hinged to the top of a table, as shown in Fig. 27.34. Each side of the movable part of the coil has a length of 0.50 m. (a) In which direction should a magnetic field point to help lift the free end of the loop off the table? (b) Calculate the torque caused by a 0.70-T field pointing in the direction described in part (a) when a 0.80-A current flows through the wire and the coil is parallel to the table.

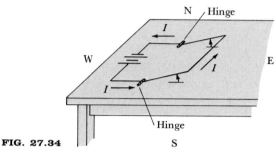

FIG. 27.34

■ **29.** A 20-turn coil, 5.0 cm by 10.0 cm, hangs with the plane of the coil parallel to a 0.12-T magnetic field. (a) Calculate the torque on the coil when a 0.50-A current flows through it. (b) The support for the coil exerts an opposing torque, which increases as the coil is deflected. The opposing torque is calculated using the equation $\tau = -0.016\,\theta$ (in units of N·m), where the angle θ is in units of radians (Fig. 27.35). At approximately what angle does the current torque balance the torque caused by the supporting wire? An ammeter can be calibrated to measure currents in this manner.

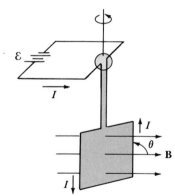

FIG. 27.35

27.5 A Direct-Current Electric Motor

30. An electric motor has a square armature with 500 turns. Each side of the coil is 12 cm long and carries a current of 4.0 A. (a) Calculate the magnetic force on one side of the coil when it is perpendicular to a 1.2-T magnetic field. (b) Calculate the torque on the coil when the armature's surface is in the plane of the field.

31. An electric motor has a circular, 100-turn armature of radius 7.0 cm and carries a current of 50 A when the switch is first turned on. The surface of the coil is parallel to a 0.80-T magnetic field. An 8.0-N·m friction torque initially opposes the rotation of the armature. Calculate the net torque on the armature when the switch is turned on.

32. An electric motor has a circular armature with 250 turns of radius 5.0 cm through which an 8.0-A current flows. The maximum torque developed by the armature as it turns in a magnetic field is 1.9 N·m. Show the orientation of the armature coil in the magnetic field when it produces this torque and calculate the magnitude of the field.

27.6 Calculating the Magnetic Field Caused by Moving Charges

33. An east-west electric power line carries a 500-A current toward the east. The line is 10 m above the earth's surface. Calculate the magnitude and direction of the magnetic field at the earth's surface directly under the wire.

■ **34.** Two long, parallel wires are separated by 2.0 m. Each wire carries a 30-A current, but the currents flow in opposite directions. (a) Calculate the magnitude of the net magnetic field midway between the wires. (b) Calculate the net magnetic field at a point 1.0 m to the side of one wire and 3.0 m from the other wire.

35. During World War II, explosive mines were dropped by the Nazis in the harbors of England. The mines, which lay at the bottom of the harbors, were activated by the changing magnetic field that occurred when a large metal ship passed above them. Small English boats called minesweepers would tow long, current-carrying coils of wire around the harbors. The field created by the coils activated the mines, causing them to explode under the coils rather than under ships. (a) Calculate the current that must flow in one long, straight wire to create a 0.0050-T magnetic field at a depth of 20 m under the water. The magnetic permeability of water is about the same as that of air. (b) How might the field be created using a smaller current?

36. A solenoid of radius 1.0 m with 750 turns and a length of 5.0 m is wound around a pigeon cage. What current must flow so that the solenoid field just cancels the earth's (4.2×10^{-5})-T magnetic field, which is occasionally used by the pigeons to determine the direction they move.

■ **37.** Two concentric solenoids carry currents in opposite directions. The inner solenoid has 50 turns per centimeter, the outer solenoid 20 turns per centimeter. The inner coil carries a current of 0.80 A. (a) Calculate the current needed in the outer coil so that its field cancels the field produced by the inner coil. (b) How large a field is produced by the inner coil?

■ **38.** An excited electron in a hydrogen atom moves with a speed of 1.09×10^6 m/s in a circular orbit whose radius is 2.12×10^{-10} m. (a) Calculate the time required for one trip around the circle. (b) Calculate the current caused by the electron's motion. (c) Calculate the magnetic field at the center of the circular orbit.

■ ■ **39.** Two electrons move past each other along lines separated by 2.0×10^{-8} m. Each electron moves at a speed of 3.0×10^7 m/s. (a) Calculate the magnetic field at one electron caused by the other when the electrons are next to each other. (b) Calculate the magnitude of the magnetic force caused by this field. (c) Calculate the magnitude of the electric force of one electron on the other when separated by 2.0×10^{-8} m.

CHAPTER 28

Induced EMF

Two fundamental ideas dominated our study of magnetism in Chapter 27: (1) A force is exerted on a moving charge as it passes through a magnetic field, and (2) a moving electric charge produces a magnetic field. In this chapter we introduce another important magnetic phenomenon called *electromagnetic induction*, the generation of voltage across the ends of a loop of wire due to a variation of the magnetic field passing through the loop. In addition to describing how electromagnetic induction occurs, we will examine some of the ways it is used—for example, in airport metal detectors and in electric generators and transformers.

28.1 Electromagnetic Induction

A simple example of electromagnetic induction is illustrated in Fig. 28.1. In Fig. 28.1a, a loop of wire is attached to a sensitive voltmeter, which measures the voltage across the ends of the loop. Since there are no batteries or other sources of electromotive force (emf) in the circuit, the potential difference across the ends of the loop is normally zero.

If a bar magnet is moved toward the loop of wire, as in Fig. 28.1b, a voltage is induced across the ends of the wire, and a small current flows around the loop. If the magnet's motion is stopped (Fig. 28.1c), the voltage returns to zero, and the current stops. If the magnet is withdrawn (Fig. 28.1d), a voltage is again produced, and a current flows. However, this time the polarity of the voltage is reversed, and the current flows in the opposite direction. The voltage and current return to zero as the magnet is moved far away (out of sight in Fig. 28.1e).

The most important observation relative to this experiment is that a voltage is produced only when the magnet is moving. The voltage $\mathcal{E}$ produced by moving the magnet relative to the loop of wire is called an *induced emf* and is an example of a process called **electromagnetic induction**—the induction of an emf by a *varying* magnetic field inside a loop of wire. Electromagnetic induction was discovered in 1831 after a seven-year search by the English scientist Michael Faraday. The theoretical description of electromagnetic induction is based on what is called Faraday's law.

586

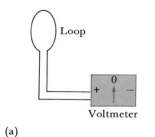

(a)

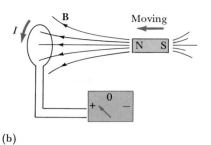

(b)

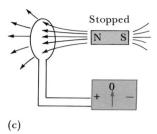

(c)

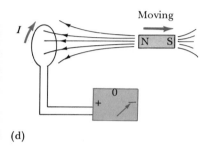

(d)

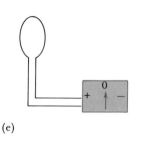

(e)

FIG. 28.1. The voltage across a loop changes as the magnet is moved (b) toward the loop and (d) away from the loop. The voltage is zero when the number of magnetic-field lines through the loop is constant, as in parts (a), (c), and (e).

28.2 Faraday's Law

Before we can examine Faraday's law, we must first define a quantity called magnetic flux. Magnetic flux depends on the magnetic field in a region of space, such as the region inside a loop of wire. Usually the field is depicted by lines, as in Fig. 28.2. If the field is strong, the lines are more closely spaced (Fig. 28.2a); if it is weak, the lines are farther apart (Fig. 28.2b).

If a coil or loop of wire is placed in a magnetic field, some of the field lines pass through the loop. The **magnetic flux** through the loop equals the number of magnetic-field lines that pass through the coil and depends on the magnitude B of the field, on the area A of the loop, and on the orientation of the loop in the field. For example, when the plane of the loop is parallel to the field (Fig. 28.3a), no lines pass through the loop, and the magnetic flux through the loop is zero. If the loop is rotated so that its plane becomes oriented more nearly perpendicular to the field, as in Fig. 28.3b and c, more lines pass through the loop, and the magnetic flux increases.

The **magnetic flux** Φ through a loop is calculated using the equation

$$\Phi = AB \cos \phi, \qquad (28.1)$$

where A is the area inside the loop, B is the magnitude of the magnetic field inside the loop, and ϕ is the angle between the direction of the field and the direction of a line drawn *perpendicular* to the area inside the loop (the dashed line in Fig. 28.3b). When the perpendicular line and the field are parallel (Fig. 28.3c), ϕ is zero, and $\cos 0° = 1$; then the flux through the loop is maximum. When ϕ is 90° as in Fig. 28.3a, $\cos 90° = 0$, and the flux is zero. Magnetic flux is measured in units of tesla-meter^2 (T·m^2).

The magnetic flux through a loop is analogous to the mass of air that blows through a loop held in a wind. The stronger the wind, the greater is the mass of

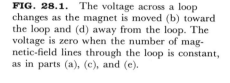

Strong B field

(a)

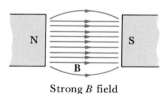

Weak B field

(b)

FIG. 28.2. (a) A large-magnitude magnetic field is represented by closely spaced lines, whereas (b) a weak field has sparsely spaced lines.

Zero flux

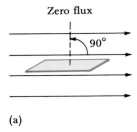

(a)

Intermediate flux

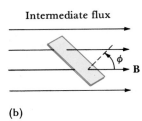

(b)

Maximum flux

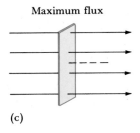

(c)

FIG. 28.3. The magnetic flux through a loop of wire depends on its orientation in a magnetic field.

air blowing through the loop. Also, more air passes through a loop of large area than through one of small area. Finally, the mass of air passing through the loop depends on its orientation. If the loop's surface is parallel to the direction of motion of the wind, the mass of air blows above and below the loop, but not through it—the air flux would be zero. As the loop is rotated perpendicular to the wind's direction, the mass of air blowing through the loop increases.

Magnetic flux plays an important role in electromagnetic induction. Notice that as the bar magnet moves closer to the loop shown in Fig. 28.1b, the flux through the loop increases. Similarly, when the magnet is withdrawn, as in Fig. 28.1d, the flux decreases. In both cases, an emf and current are induced in the coil. However, if the magnet is not moving, the flux does not change, and the emf and current are zero.

Since Faraday's original experiments, a great many other experiments have shown that the emf induced in a loop of wire is equal to the time rate of change of the magnetic flux through the loop. If N loops are wound in a coil, the induced voltage is N times that for one loop. This is called Faraday's law.

Faraday's law: The induced emf (induced voltage) $\mathcal{E}$ across a loop of wire or a coil of wire with N loops (or turns) equals the change of magnetic flux $\Delta\Phi$ through the loops divided by the time Δt needed for that change in flux:

$$\mathcal{E} = -N\frac{\Delta\Phi}{\Delta t} = -N\frac{\Phi - \Phi_0}{t - t_0}. \tag{28.2}$$

The unit of $\mathcal{E}$ is volts if $\Delta\Phi$ is in units of tesla-meter^2 and Δt is in seconds.

In Eq. (28.2), the expression $-N(\Delta\Phi/\Delta t)$ is used to calculate the instantaneous induced emf during a very short time Δt, and the expression $-N(\Phi - \Phi_0)/(t - t_0)$ is used to calculate the average induced emf over a longer time $(t - t_0)$. The negative sign in the equation will be discussed shortly. Regardless of the sign, it is evident from this equation that voltage is induced only while flux is changing and that the voltage is greatest when there is large flux change in a short period of time.

EXAMPLE 28.1 A circular loop of wire with 50 turns ($N = 50$) and a radius of 2.0 cm is moved from outside the poles of a horseshoe magnet to between the poles in 0.10 s (Fig. 28.4). The field between the poles is 0.18 T. (a) Calculate the average induced emf as the loop moves into the magnetic field. (b) How large is the emf if the coil moves out of the field in one-half the time?

SOLUTION (a) The area of the loop is $A = \pi r^2 = \pi(0.020)^2 = 1.3 \times 10^{-3}$ m^2, and a line perpendicular to the plane in which the loop lies is parallel to the field; thus, $\phi = 0$. When outside the field at $t_0 = 0$, the field B_0 through the loop is zero. After moving into the field at $t = 0.10$ s, the field B through the loop increases to 0.18 T. Thus, by combining Eqs. (28.1) and (28.2), we find that the average induced emf across the loop as it moves into the field is

$$\mathcal{E} = -N\frac{\Phi - \Phi_0}{t - t_0} = -\frac{N(AB\cos\phi - AB_0\cos\phi)}{t - t_0}$$

$$= -\frac{NA\cos\phi\,(B - B_0)}{t - t_0} = -\frac{50(1.3 \times 10^{-3}\,\text{m}^2)(\cos 0°)(0.18\,\text{T} - 0)}{0.10\,\text{s} - 0}$$

$$= -0.12\,\text{V}.$$

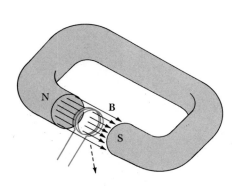

FIG. 28.4

(b) As the coil moves out of the field, the change in flux has the same magnitude as when the coil moved into the field. However, the flux change occurs in one-half the time. Thus, the magnitude of the induced emf will be twice as great ($\varepsilon = +0.24$ V). The polarity is reversed because the initial and final magnetic field values are reversed ($B_0 = 0.18$ T and $B = 0$). ∎

EXAMPLE 28.2 A circular coil of wire with 10 turns ($N = 10$) and a 4.0-cm radius is located between the poles of a large electromagnet that produces a steady magnetic field of 0.50 T. The surface of the loop, which is originally parallel to the field, is rotated in 0.10 s so that its surface is perpendicular to the field. Calculate the average induced emf across the ends of the loop as the coil rotates.

SOLUTION The area of one loop of the coil is $\pi r^2 = \pi(0.040 \text{ m})^2 = 5.0 \times 10^{-3} \text{ m}^2$. The magnetic field B in the region of the coil has a constant value 0.50 T. However, the orientation of the loop changes. The surface of the loop is originally parallel to the magnetic field; thus, a line drawn perpendicular to the loop's surface makes an angle of 90° with the magnetic field ($\phi_0 = 90°$) at $t_0 = 0$. At $t = 0.10$ s, the loop has rotated so that a line perpendicular to its surface is parallel to the field ($\phi = 0°$). By combining Eqs. (28.1) and (28.2), we find that the average induced emf as the coil rotates is

$$\varepsilon = -N\frac{\Phi - \Phi_0}{t - t_0} = -10\frac{AB\cos\phi - AB\cos\phi_0}{t - t_0}$$

$$= -\frac{10AB(\cos\phi - \cos\phi_0)}{t - t_0}$$

$$= -\frac{10(5.0 \times 10^{-3} \text{ m}^2)(0.50 \text{ T})(\cos 0° - \cos 90°)}{(0.10 \text{ s} - 0)}$$

$$= -0.25 \text{ V}.$$

∎

28.3 The Polarity of Induced EMF

We have seen that a changing magnetic flux inside a loop of wire produces an induced voltage. We now turn to the polarity of this induced emf, that is, which end of the loop of wire is at a higher voltage? The polarity of the induced emf has an important effect on the operation of electric motors and on devices called induction coils, which are used in electronic equipment such as television sets and in the electric circuit for the spark plugs of a car.

To determine the polarity of the induced emf, we must first consider the electric current that flows because of the induced emf and the magnetic field created by this current. We learned in Section 27.6 that a current flowing in a loop of wire produces a magnetic field such as the one shown in Fig. 28.5. The direction of the field is determined using the right-hand rule: If the thumb of the right hand points in the direction of the current, then the fingers wrap around the wire in the direction of the field.

When the magnetic flux through a loop changes, a voltage is induced that causes a current in the loop. The induced current produces an "induced" magnetic field inside the loop that *opposes* the change in magnetic flux that caused

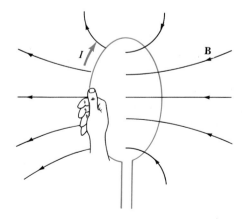

FIG. 28.5. The magnetic field caused by a current in a loop.

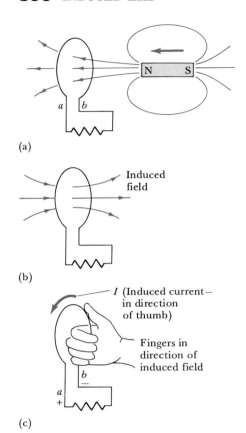

(a)

(b)

(c)

FIG. 28.6. (a) As a magnet moves toward the loop, the flux through the loop increases and (b) induces a field that opposes the increasing flux. (c) The induced field is caused by an induced current that moves from b to a. Thus, a is at a higher voltage than b, since current moves through a source of emf toward higher voltage.

the voltage and current. For example, if the flux through a loop points toward the left and is increasing, then the induced current will produce a flux through the coil that points to the right. However, if the magnetic flux is toward the left and decreasing, the induced current will produce a flux through the coil toward the left in an attempt to prevent the original decrease in flux.

The relation between the changing flux through a coil and the polarity of the induced emf is summarized by **Lenz's law:**

The **polarity** of an induced emf is such that it produces a current whose own magnetic field opposes the change in flux that caused the induced emf.

The negative sign in Faraday's law simply reminds us that the induced emf ε *opposes* the changing flux that caused the emf. The negative sign has no other meaning.

To determine the polarity of the induced emf, we simply determine in which direction the induced field should point so that it opposes the change in flux. Using the right-hand rule, we curl the fingers of the right hand in the direction of the induced field. The thumb then points in the direction of the induced current.

To decide on the polarity of the induced emf, we consider the coil or loop of wire as a voltage source, just as a battery is a voltage source. For a battery, current flows *through* the battery from the negative to the positive terminal and then out into the external circuit. For the coil, the induced current flows around the coil (the voltage source) from the negative to the positive terminal and then out into an external circuit attached to the coil. Thus, *the induced current flows around the coil toward its positive terminal.*

EXAMPLE 28.3 In Fig. 28.6a, a bar magnet moves toward a loop of wire. Which lead of the wire is at a higher voltage, a or b?

SOLUTION The flux of magnetic-field lines passing through the loop from right to left increases as the magnet moves closer to the loop. Hence, the induced current should produce a magnetic field that opposes this increase, and to do so, the induced field should pass through the loop from left to right, as shown in Fig. 28.6b. The right-hand rule shows that the current must flow counterclockwise to produce this induced field (Fig. 28.6c). Since current flows through a source of emf toward the higher-voltage terminal, a must be at higher voltage than b. ∎

EXAMPLE 28.4 A rectangular loop that is 10 cm long on each side is moved into a uniform magnetic field of 1.5 T at a speed of 3.0 m/s. A line normal to the loop's inside area is parallel to the field, as shown in Fig. 28.7. (a) Calculate the induced emf across the ends of the loop as the loop enters the field. (b) Is point a or point b at higher voltage?

SOLUTION (a) In this problem the magnetic field is nonzero in one region inside the coil and zero elsewhere. As more of the loop enters the field, the nonzero region increases. The induced emf is caused by the increasing area ΔA of the region occupied by the field.

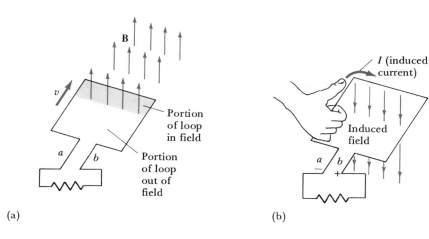

(a)

(b)

In general, for a rectangular loop whose front edge has length L and that moves at a speed $v = \Delta x/\Delta t$, the area increases by

$$\Delta A = L\, \Delta x$$

in a time Δt. Thus, the changing flux per time Δt is

$$\mathcal{E} = -\frac{\Delta\Phi}{\Delta t} = -\frac{(\Delta A)B\cos\phi}{\Delta t} = -\frac{(L\,\Delta x)B\cos\phi}{\Delta t}.$$

Substituting $\Delta x/\Delta t = v$, we find that

$$\mathcal{E} = -LvB\cos\phi.$$

We are given that $L = 0.10$ m, $v = 3.0$ m/s, $B = 1.5$ T, and $\phi = 0°$. Thus,

$$\mathcal{E} = -(0.10\text{ m})(3.0\text{ m/s})(1.5\text{ T})1 = \underline{-0.45\text{ V}}.$$

(b) Because the induced emf is caused by an increasing magnetic flux that points up, the induced field should point down. A downward-induced field is caused by an induced current flowing clockwise from a to b (Fig. 28.7b). Since the induced current flows around the voltage source (the coil) toward the positive terminal, point b must be at higher voltage than point a. ■

28.4 The Electric Generator

An electric generator is an important device whose operation we can understand by using Faraday's law. A homemade version of an electric generator is shown in Fig. 28.8a. Water in a flask is heated by a Bunsen burner. Steam leaves a tube at the top of the flask and strikes the blades of a turbine, causing the turbine to rotate. A loop of wire attached to the turbine also rotates in a steady magnetic field. The changing magnetic flux through the loop causes an induced emf. In the simple power plant shown in Fig. 28.8a, the voltage drives a current through a lightbulb. The power plant of an electric utility company represented schematically in Fig. 28.8b works in much the same manner, but on a much larger scale.

Our concern in this section is with the rotating coil and the induced emf caused by this coil. A top view of a coil with one turn is shown in Fig. 28.9. The

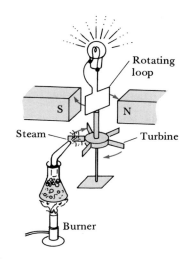

(a)

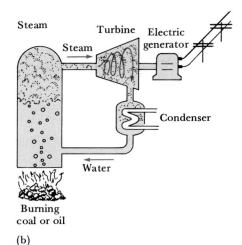

(b)

FIG. 28.8. (a) A homemade version of an electric power plant. (b) A representation of an actual electric power plant.

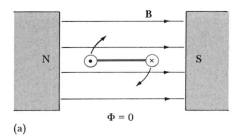

$$\Phi = 0$$

(a)

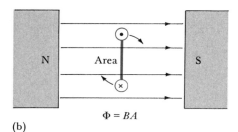

$$\Phi = BA$$

(b)

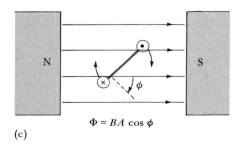

$$\Phi = BA \cos \phi$$

(c)

FIG. 28.9. The flux through a coil depends on its orientation in the field.

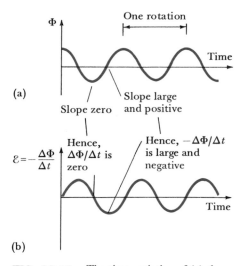

(a)

$$\mathcal{E} = -\frac{\Delta\Phi}{\Delta t}$$

(b)

FIG. 28.10. The time variation of (a) the flux through and (b) the induced voltage across the leads of a coil rotating in a magnetic field.

magnetic flux through the coil depends on the orientation of the coil: If the coil is parallel to the magnetic field (Fig. 28.9a), the field lines pass along the sides of the coil but not through it, and the flux is zero. If the coil's surface is perpendicular to the field direction (Fig. 28.9b), then the flux through the coil is $\Phi = BA$. If a line perpendicular to the coil's surface is oriented at an angle ϕ relative to the field (Fig. 28.9c), then the flux through the coil is

$$\Phi = BA \cos \phi. \tag{28.3}$$

Suppose that the coil is oriented so that $\phi = 0$ at $t = 0$ and that a short time t later the coil, while rotating at constant angular velocity, makes an angle ϕ with the field. The angular velocity ω of the coil is then

$$\omega = \frac{\phi}{t}. \tag{6.1}$$

If the coil rotates through 2π radians in one second, it will turn at a frequency of one rotation per second. If rotating through 4π radians per second, its frequency is two rotations per second. In general, the frequency f of rotation and the angular velocity ω are related by the equation

$$\omega = 2\pi f.$$

Combining the last two equations, we find that

$$\phi = \omega t = 2\pi f t.$$

This expression for ϕ is substituted into Eq. (28.3) to obtain an equation for the changing flux through a rotating coil as a function of time:

$$\Phi = BA \cos 2\pi f t. \tag{28.4}$$

The equation for flux is an example of a sinusoidal function and is plotted against time on the graph in Fig. 28.10a.

Each time the coil makes one-half rotation, the magnitude of the flux through the coil is the same as it was one-half rotation earlier. However, the orientation of the coil has reversed relative to the field. The alternating positive and negative values of flux shown in Fig. 28.10a reflect the changing orientation of the coil relative to the field.

The voltage induced in the coil by the changing flux is given by Eq. (28.2): $\mathcal{E} = -N\,\Delta\Phi/\Delta t$. As the flux changes, so does the voltage induced across its ends. If you are familiar with differential calculus, then you can take the negative time derivative of Φ as given in Eq. (28.4) to determine an equation for $\mathcal{E}$. For those unfamiliar with calculus, we present the result without proof:

$$\mathcal{E} = -\frac{\Delta\Phi}{\Delta t} = (NBA2\pi f)\sin 2\pi f t = \mathcal{E}_0 \sin 2\pi f t. \tag{28.5}$$

The induced emf varies sinusoidally with time. The alternating voltage plotted in Fig. 28.10b changes polarity at the same frequency as the generator's coil rotates. If one terminal is at higher voltage than the other at one instant of time, then one-half turn later the polarity is reversed—the high-voltage terminal has become the low-voltage terminal. The alternating positive and negative voltages shown in Fig. 28.10b reflect the changing polarity of the voltage across the terminals of the generator.

The **peak voltage** $\mathcal{E}_0$ produced by a generator equals the quantity $NBA2\pi f$ that appears in front of the sine function in Eq. (28.5). For example, the electric voltage provided by a power plant for your home has a frequency of 60 Hz and a peak voltage of 170 V. The voltage alternates between $+170$ V and -170 V 60 times each second. The following symbol is used for an alternating current (ac) electric power source that produces an alternating voltage: $-\!\bigcirc\!-$.

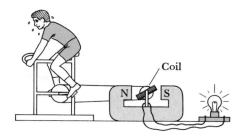

FIG. 28.11. A person produces electricity by rotating a coil in a magnetic field.

EXAMPLE 28.5 A small generator has a coil with 100 loops each of area 80 cm^2. The generator is turned by a gear system attached to the pedals of a stationary bicycle (Fig. 28.11). A student pedals the bicycle so that the generator rotates at a frequency of 20 Hz in a magnetic field of 0.30 T. Find an expression for the voltage produced by the generator.

SOLUTION We are given the following information: $N = 100$, $A = 8.0 \times 10^{-3}$ m^2, $B = 0.30$ T, and $f = 20$ Hz $= 20$ s^{-1} (20 rotations/second). Thus, the generated voltage is

$$\mathcal{E} = (NBA2\pi f) \sin 2\pi f t = [(100)(0.30 \text{ T})(80 \times 10^{-4} \text{ m}^2)(2\pi 20 \text{ s}^{-1})] \sin (40\,\pi t)$$
$$= \underline{30 \sin (40\pi t)} \qquad \text{(in units of volts).}$$

The voltage produced from one terminal of the generator to the other will vary 20 times a second from $+30$ volts to -30 volts, and then back to $+30$ volts. ∎

28.5 Mutual Inductance

Now we encounter a different example of electromagnetic induction. When a current passes through the loops of a coil, a magnetic field is produced. If a second coil is held near the first, some of the magnetic-field lines caused by current in the first coil may pass through the second coil. If the current in the first coil changes, the magnetic field and flux through the second coil will also change, and a voltage will be induced in the second coil because of the changing flux. This process is called **mutual inductance**—a voltage is induced in one coil by a changing current in the other.

The mutual inductance of two coils is defined with reference to Fig. 28.12. A current i_1 passing through coil 1, which has N_1 turns, produces a magnetic field.* Some of the magnetic-field lines produced by coil 1 pass through coil 2 with N_2 turns. By convention, we designate the magnetic flux through each loop of coil 2 by the symbol Φ_{21}. (The subscript 21 refers to the flux through the loops of coil 2 that are caused by the current in coil 1.) The total flux through all N_2 turns of coil 2 is $N_2\Phi_{21}$.

The mutual inductance M of the coils is defined as

$$M = \frac{N_2\Phi_{21}}{i_1}$$

or

$$N_2\Phi_{21} = Mi_1. \tag{28.6}$$

*Varying currents are usually represented by a lowercase i and steady currents by a capital I.

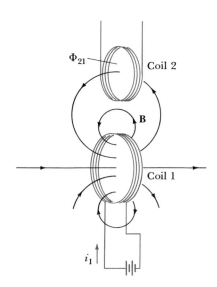

FIG. 28.12. A current in coil 1 produces a magnetic flux Φ_{21} through each loop of coil 2.

The quantity M depends on the size, orientation, and separation of the coils. If they are close to each other, M is greater because a current in coil 1 will produce a greater flux through the loops of coil 2. As we will see later, the value of M also depends on the magnetic permeability of the material in the region between and through the two coils.

If the current in coil 1 varies with time, Eq. (28.6) becomes

$$N_2 \frac{\Delta\Phi_{21}}{\Delta t} = M \frac{\Delta i_1}{\Delta t}.$$

The left side of this equation is just the negative of the emf induced in coil 2 by the changing current in coil 1 ($\mathcal{E} = -N_2 \Delta\Phi/\Delta t$). Consequently,

$$\mathcal{E}_2 = -M \frac{\Delta i_1}{\Delta t}. \tag{28.7}$$

A similar equation applies to coil 1. If the current in coil 2 changes at a rate $\Delta i_2/\Delta t$, then an emf is induced in coil 1:

$$\mathcal{E}_1 = -M \frac{\Delta i_2}{\Delta t}. \tag{28.8}$$

The same M is used in both Eqs. (28.7) and (28.8). The unit of mutual inductance is called a **henry** (H). Rearranging and substituting for the units in Eq. (28.7), we see that

$$1\ \text{H} = \frac{1\ \text{V}}{1\ \text{A/s}}.$$

A 1-H mutual inductance between two coils means that a 1-V potential difference is induced in one coil by a 1-A/s change in current in the other coil.

The unit *henry* was named in honor of the American physicist Joseph Henry (1797–1878), who independently discovered the law of electromagnetic induction, which was named in honor of Michael Faraday. Henry published his findings just a little later than Faraday and had to settle for a unit named in his honor rather than a law.

EXAMPLE 28.6 The spark coil in an automobile ignition system converts an abrupt change in current in one coil into a large induced emf (25,000 V) in a second coil whose leads are across the spark plug. This large voltage produces dielectric breakdown and a spark in the spark plug. If the mutual inductance between the coils is 1.2 H and the current changes from 16 A to zero, during what time must this current change to produce the 25-kV induced emf in the second coil?

SOLUTION Equation (28.7) can be rearranged to determine the time needed for the current in coil 1 to change by $\Delta i_1 = (0 - 16\ \text{A}) = -16\ \text{A}$ to induce a voltage in coil 2 of $\mathcal{E}_2 = 25,000\ \text{V}$. The mutual inductance M between the coils is 1.2 H:

$$\Delta t = -\frac{M(\Delta i_1)}{\mathcal{E}_2} = -\frac{(1.2\ \text{H})(-16\ \text{A})}{(25,000\ \text{V})}$$

$$= 0.77 \times 10^{-3}\,\text{s} = 0.77\,\text{ms}.$$

This abrupt change in current through coil 1, which is connected to the car's battery, occurs as the *breaker points* are opened and closed quickly by a rapidly rotating uneven wheel called the *cam*. The cam and breaker points are located in the distributor, the electrical brain of the car's ignition system. ∎

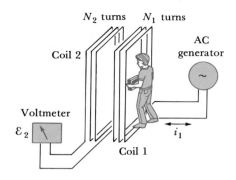

FIG. 28.13. An airport metal detector.

Mutual Inductance and Metal Detectors

A metal detector like those through which you walk at airports can detect metals with high magnetic permeability. The detector could be built to detect a change in the mutual inductance of two large coils, such as shown in Fig. 28.13. An alternating current i_1 in coil 1 produces a magnetic field whose magnitude at each instant of time is given approximately by Eq. (27.11):

$$B = \frac{N_1 \mu i_1}{2r},$$

where N_1 is the number of loops in the first coil, μ is the permeability of the material inside the coil, and r is its radius (the coil is assumed to be circular).

Since the second coil is so near the first, the field through each of its N_2 loops is about the same as that through the loops of the current-carrying coil. Thus, the flux in the second coil caused by the current in the first is approximately:

$$N_2 \Phi_{21} = N_2 BA = N_2 \left(\frac{N_1 \mu i_1}{2r} \right) A,$$

where A is the area inside the second coil. By comparing the above equation to Eq. (28.6), $N_2 \Phi_{21} = M i_1$, we see that the mutual inductance between the coils is

$$M = \frac{N_2 N_1 \mu A}{2r}.$$

The magnitude of the emf induced in coil 2 by the oscillating current in coil 1 is

$$\mathcal{E}_2 = -M \frac{\Delta i_1}{\Delta t}.$$

One way that the airport metal detector could work is by producing an alternating current in the first coil. If the mutual inductance between the first and second coils changes as a person walks through the detector, the induced emf $\mathcal{E}_2$ changes, causing an alarm to go off. Such a change in mutual inductance will occur if a person carries a metal object of high magnetic permeability through the loops of wire.

When a person carrying no metal objects on his or her body walks through the alarm, the mutual inductance is

$$M = \frac{N_2 N_1 \mu_0 A}{2r},$$

since the magnetic permeability of the body is the same as that of air: $\mu_{\text{body}} \simeq \mu_0$. Should a person carry a piece of iron, however, the permeability

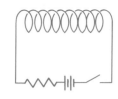

(a)

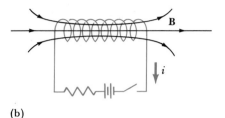

(b)

FIG. 28.14. (a) No current flows through the coil when the switch is open. (b) When the switch is closed, the current produces a magnetic field and flux through the coil.

may change. For example, if the object has a permeability 300 times that of air and occupies 1/300 the area inside the coil, the average permeability of the material in the loop doubles, the mutual inductance between the coil doubles, and the induced emf in coil 2 doubles, causing the alarm to turn on. In the next chapter we will learn how the metal detector can be made even more sensitive.

28.6 Self-Inductance

In the preceding section the induced emf in one coil was caused by a changing current in another coil. A voltage can also be induced in a coil because of a change in its own current. Such a voltage is called a self-induced voltage, or **self-induced emf.**

The circuit shown in Fig. 28.14a illustrates self-inductance. When the switch is open, no current flows and no magnetic field passes inside the coil. When the switch is closed, a current flows and produces a magnetic field in the coil (Fig. 28.14b). The magnetic field and flux through the coil increase from zero (with the switch open) to a finite value (with the switch closed). The changing magnetic field and flux through the coil cause a voltage to be induced in the coil, and this induced emf initiates an induced current that opposes the change in the original current.

The self-induced emf of a coil is defined in terms of the self-inductance L of the coil. Suppose that a current i flowing through the coil shown in Fig. 28.14b produces a magnetic field whose flux through each loop is represented by the symbol Φ. The total flux through the N loops of the coil is $N\Phi$. The **self-inductance** L of the coil is defined by the following equation:

$$L = \frac{N\Phi}{i}.$$

(28.9)

The greater a coil's self-inductance, the greater is the voltage induced in the coil if the current passing through it changes. The unit of self-inductance, like that for mutual inductance, is the henry (H), where 1 H = 1 V/(1 A/s).

The self-inductance of a coil depends on the size and shape of the coil, on the number of turns in it, and also on the magnetic permeability of material inside it. For example, the self-inductance of a coil with an iron core is several hundred times that of a similar coil with an air core. The reason is that the magnetic field caused by the current through the coil aligns the tiny "bar magnets" of the iron atoms, adding their fields to the field produced by the current. As a consequence, there is a much greater flux through a coil that surrounds iron than through one that surrounds air.

Equation (28.9) can be used to determine the induced emf in a coil when the current flowing through it changes. Since $N\Phi = Li$, the changing flux is related to a changing current by the equation

$$N\frac{\Delta\Phi}{\Delta t} = L\frac{\Delta i}{\Delta t}.$$

The left side of this equation is just the negative of the voltage induced in the coil ($\mathcal{E} = -N\,\Delta\Phi/\Delta t$). Consequently,

$$\mathcal{E} = -L\frac{\Delta i}{\Delta t}.$$

(28.10)

The **induced emf** (voltage) can be calculated if the rate of current change and self-inductance of a coil are known. The induced emf is sometimes called a **back emf,** since it opposes the changing current that caused the induced emf.

EXAMPLE 28.7 The self-inductance of a solenoid is 31 mH (Fig. 28.14). What is the average induced emf in the solenoid if the current passing through it changes from zero to 1.5 A in 0.20 s?

SOLUTION The induced emf in the solenoid is determined using Eq. (28.10):

$$\mathcal{E} = -L\frac{\Delta i}{\Delta t} = -(31 \times 10^{-3}\,\text{H})\left(\frac{1.5\,\text{A}}{0.20\,\text{s}}\right) = \underline{-0.23\,\text{V}}.$$

The negative sign implies that the current caused by the induced voltage flows in the opposite direction of the increasing current that induced the voltage. ∎

28.7 Transformers

A **transformer** is a device that uses the mutual inductance between two coils to convert an alternating voltage across one coil into a larger or smaller alternating voltage across the other coil. This increase or decrease in voltage is much like the increase or decrease in force that is caused by the use of a lever; a small force applied to the long end of a lever can produce a large force at the short end, and vice versa. Similarly, a small voltage at the input of a transformer can be converted to a large voltage at its output, and vice versa.

A transformer (Fig. 28.15) consists of two coils of wire wound around a core, or yoke, that is made of a material that can be magnetized. The coil that is attached to the alternating voltage source $\mathcal{E}_1$ is called the **primary coil.** The alternating current flowing through the primary coil produces a fluctuating magnetic field B in the transformer core. This field passes around the core and through the secondary coil. Since the field alternates direction, an induced voltage $\mathcal{E}_2$ is produced in the **secondary coil.** If the secondary coil is attached to a lightbulb, current is forced by the induced voltage through the light's filament, causing it to glow.

Transformers are useful mainly because of the difference between the voltage going into the primary coil and that coming out of the secondary coil. The secondary voltage can be larger than the primary voltage if the secondary has more turns than the primary (a step-up transformer) and smaller if it has fewer turns than the primary (a step-down transformer). The cause of this behavior is outlined below.

The alternating current in the primary coil produces a self-induced voltage of magnitude

$$\mathcal{E}_1 = -N_1\frac{\Delta\Phi}{\Delta t},$$

where Φ is the magnetic flux through each loop of the coil and N_1 is the number of primary loops.

In a good transformer nearly all of the magnetic flux through the loops of the primary coil also passes through the loops of the secondary coil; hence the

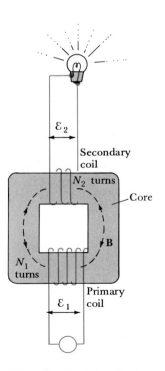

FIG. 28.15. A transformer.

induced emf in this coil is

$$\mathcal{E}_2 = -N_2 \frac{\Delta\Phi}{\Delta t},$$

where N_2 is the number of loops in the second coil and Φ is the same flux that passed through the primary coil.

If we divide the first equation by the second, we find that

$$\frac{\mathcal{E}_1}{\mathcal{E}_2} = \frac{N_1}{N_2}. \tag{28.11}$$

The secondary coil voltage is

$$\mathcal{E}_2 = \frac{N_2}{N_1}\mathcal{E}_1.$$

We see that $\mathcal{E}_2$ depends on the ratio of the number of turns in each coil and also on the voltage in the primary coil.

EXAMPLE 28.8 The 120-V alternating voltage $\mathcal{E}_1$ from a wall socket feeds into the primary coil of a toy electric train transformer, which has 800 turns (N_1). Calculate the number of turns in the secondary coil so that the output voltage $\mathcal{E}_2$ from the transformer is 12 V—enough to run a toy electric train safely.

SOLUTION Equation (28.11) can be rearranged to solve for the number of turns in the secondary coil:

$$N_2 = \frac{\mathcal{E}_2}{\mathcal{E}_1}N_1 = \frac{12\text{ V}}{120\text{ V}}800 = \underline{80}. \qquad \blacksquare$$

Transformers have a number of uses in everyday life. As we saw in the last example, they can reduce voltages to safe levels for toys. Another important use is related to the economics of transmitting electricity from power plants to homes. Suppose that you live in a community with 100,000 people that is located 50 miles from the electric power plant serving the city. The average electric power use rate per person in the United States and Canada for a 24-hour period is about 1000 W. The power plant has to supply $100,000 \times 1000$ W or 1×10^8 W of power to your city continuously (somewhat more during hours of peak use).

Recall from Chapter 24 that electric power used is the product of the voltage difference at its usage point and the current that is flowing:

$$P = VI. \tag{24.9}$$

If the power is delivered at 120 V, then the current to this city must average

$$I = \frac{P}{V} = \frac{1 \times 10^8\text{ W}}{120\text{ V}} = 8 \times 10^5\text{ A}.$$

The wires in a home should carry no more than about 20 A of current. To supply electricity to your city, 8×10^5 A/20 A = 40,000 of these wires must run from the power plant to the city. The cost of the wires alone would prohibit this transmission system. In addition, since such long wires have considerable electrical resistance, the power lost due to resistive heating (I^2R) would be great.

If a way can be found to deliver the electric power at low current, considerable energy is saved and less wire is needed to deliver the power.

Power companies have solved this problem by using a step-up transformer. The electrical generator in a power plant produces electricity at about 12,000 V. The voltage is placed across the primary coil of a large transformer. The secondary coil has about 10 times as many turns as the primary coil, so the voltage is raised to about 120,000 V. Since the voltage is 1000 times greater than the 120 V used in your home, the same power ($P = VI$) can be transmitted by $1/1000$ the current. Only $1/1000$ as much wire is needed to transmit the power, and the heating loss is greatly reduced because the current is reduced. After reaching the city, the voltage is stepped down to lower voltage (120 or 240 V) in local transformers for use in homes.

Summary and Additional Readings

1. **Electromagnetic induction** is the generation of an electrical voltage across a loop of wire by the variation of the magnetic field through the area inside the loop. The **magnetic flux** Φ through the loop is

$$\Phi = AB \cos \phi, \tag{28.1}$$

where A is the loop's area, B is the magnitude of the magnetic field inside the loop, and ϕ is the angle between **B** and a line drawn perpendicular to the loop.

2. **Faraday's law:** The voltage $\mathcal{E}$ induced in a coil with N turns depends on the time rate of change of magnetic flux through the coil:

$$\mathcal{E} = -N\frac{\Delta\Phi}{\Delta t} = -N\frac{\Phi - \Phi_0}{t - t_0}. \tag{28.2}$$

The polarity of the induced emf is such that it produces a current whose own field opposes the change in flux that caused the induced emf.

3. **Electric generator:** An electric generator produces an alternating voltage by rotating a coil with N loops in a steady magnetic field B. The induced emf is

$$\mathcal{E} = (NBA2\pi f)\sin 2\pi ft = \mathcal{E}_0 \sin 2\pi ft, \tag{28.5}$$

where A is the area of each loop, f is the frequency of rotation, and $\mathcal{E}_0$ is the peak voltage produced by the generator.

4. **Mutual inductance:** A mutual inductance exists between two coils if an emf is induced in one coil by a changing current in the other. The **mutual inductance** M between the coils is

$$M = \frac{N_2\Phi_{21}}{i_1}, \tag{28.6}$$

where N_2 is the number of loops in coil 2 and Φ_{21} is the magnetic flux through coil 2 caused by a current i_1 in coil 1. The emf induced in coil 2 by a changing current in coil 1 is

$$\mathcal{E}_2 = -M\frac{\Delta i_1}{\Delta t}. \tag{28.7}$$

5. **Self-inductance:** If the current in a coil changes, the magnetic field and flux produced by the current also change. An emf is induced in the coil by its own changing flux. This emf is called a **self-induced electromotive force.** The **self-inductance** L of a coil indicates the coil's ability to induce a voltage when its current changes. The self-inductance is defined by the equation

$$L = \frac{N\Phi}{i}, \tag{28.9}$$

where Φ is the flux through each loop of a coil with N turns when carrying a current i. The **self-induced emf** due to a changing current is

$$\mathcal{E} = -L\frac{\Delta i}{\Delta t}. \tag{28.10}$$

6. **Transformer:** A transformer is a device that uses the mutual inductance between a primary coil with N_1 turns and a secondary coil with N_2 turns to convert a primary voltage $\mathcal{E}_1$ to a different secondary voltage $\mathcal{E}_2$. The four quantities are related by the equation

$$\frac{\mathcal{E}_1}{N_1} = \frac{\mathcal{E}_2}{N_2}. \tag{28.11}$$

P. E. Glaser, "Solar Power from Satellites," *Physics Today*, **30,** 30 (1977).

J. Barnea, "Geothermal Power," *Scientific American* **226,** 70 (1972).

G. Shiers, "The Induction Coil," *Scientific American* **224,** 80 (1971).

Questions

1. A coil of wire can detect the vibrations or movements of a nearby magnet. How is this accomplished? Describe one possible application of this detection system.

2. Describe three different ways in which the flux through a loop of wire can change, thus inducing a voltage in the loop.

3. An emf can be induced in a coil of wire that is in a magnetic field if any one of three quantities change. List these quantities and give an example showing how a change in the quantity produces an emf.

4. A circular metal plate swings past the north pole of a perma-

nent magnet, as shown in Fig. 28.16. Consider the metal as consisting of a series of rings of increasing radius. (a) Indicate the direction of the current in one ring as the metal swings down from the left. (b) Indicate the direction of the current in the same ring as it swings back up toward the left. Using Lenz's law, justify your answers. Similar currents, called *eddy currents*, are induced in metal sheets exposed to changing magnetic fields.

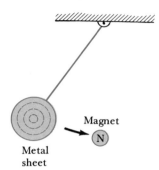

FIG. 28.16

Magnet
N

Metal
sheet

5. A 12-V automobile battery provides the thousands of volts needed to produce a spark across the gap of a spark plug. How might this be done?

6. (a) Show the orientation of a generator coil relative to a constant magnetic field when the *flux* through the coil is maximum. (b) If the coil rotates at constant angular velocity, show the orientation of the coil when the emf generated by the coil is maximum, and (c) when the emf is minimum.

7. A coil connected to a voltmeter can be used to detect alternating currents in other circuits. Such a device can detect currents as small as 5 μA. Carefully explain how one of these devices might work.

8. The induction method of geophysical prospecting depends on the alternating magnetic field produced by an alternating current in a source coil. Refer to Fig. 28.17 and carefully explain how the

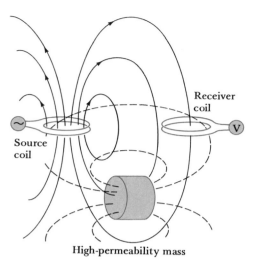

Receiver
coil

Source
coil

High-permeability mass

FIG. 28.17. A geophysical exploration technique involving Faraday's law and mutual inductance.

induced emf in a receiver coil might be affected by a high-permeability mass (iron, for example) under the earth's surface.

9. A respiration detector consists of a coil placed on a person's chest and another placed on the person's back. A current flows through one coil, producing a magnetic flux through the other coil. Why does the flux change as the person breathes, thus inducing a voltage in the other coil?

10. Suppose Lenz's law says that the voltage induced in a coil supports the changing flux through the coil rather than opposing it. If this is true, the current in a coil, once it starts, will continue to increase until the coil melts. Carefully explain in words why this will happen.

Problems

28.1–28.3 Electromagnetic Induction

1. The magnetic field in a region has a magnitude of 0.40 T and points in the positive z direction, as shown in Fig. 28.18.

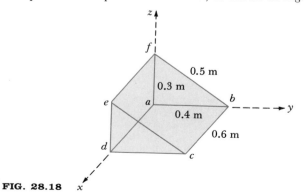

0.5 m

0.3 m

0.4 m

0.6 m

FIG. 28.18

Calculate the magnetic flux through (a) surface *abcd*, (b) surface *bcef*, and (c) surface *adef*. [*Hint:* The calculation in part (b) can be simplified by careful reasoning, but be sure to justify any simplifications you make.]

■ 2. The earth's magnetic field in the southern region of the United States has a magnitude of 5.1×10^{-5} T and points north with a downward inclination of about 60° below the horizontal. *Estimate* the magnetic flux through one side of a car (a) when the car is facing east and (b) when the car is facing north.

■ 3. The earth's magnetic field in British Columbia has a magnitude of 5.8×10^{-5} T and points roughly north with a downward inclination of 72°. Calculate the magnetic flux through a 1.0 m $\times$ 1.8 m window that faces (a) south and (b) west.

■ 4. The earth's magnetic field at one location in the magnetosphere high above the earth's atmosphere is 3.0×10^{-7} T. A coil whose inside area is 5000 m² rotates in the magnetosphere so that the flux varies from the maximum possible value to the minimum possible value in 1.2 s. How many turns must the coil have to develop an average induced emf of 120 V for use by a nearby space platform?

■ 5. A circular coil of 0.30-m diameter with 25 loops rests between the poles of a large electromagnet. The magnet initially produces a 0.50-T field, which passes through the coil perpendicular to its surface. The magnetic field is reduced to zero in 1.5 s. Calculate the magnitude of the induced emf in the coil.

■ 6. A circular coil of wire with 20 turns, each of radius 1.8 cm, is in a 1.0 -T magnetic field. The plane containing the coil is initially perpendicular to the field. (a) Calculate the time needed for a 90° rotation that will produce an average induced emf of 1.0 V. (b) Use a proportion technique to show that the same voltage can be produced if the rotation time is reduced by one-fourth and the radius of the coil is reduced by one-half.

■ 7. A 0.20 m × 0.40 m coil of wire with 12 turns is pulled into a 0.25-T magnetic field at a speed of 1.8 m/s. The 0.20-m side enters first. Calculate the induced emf as the coil enters the field, which is perpendicular to the coil's surface.

■ 8. A person wearing wire-framed glasses rotates her head quickly in the earth's magnetic field. *Estimate* the induced emf in the metal ring encircling one glass lens while the head is being turned. Justify any numbers used in your estimate.

■ 9. The surface of a circular wire loop of radius 9.0 cm is located perpendicular to a uniform 0.35-T magnetic field. The loop is pulled from two sides into a long, thin shape with the two sides next to each other. The change of shape occurs in 0.11 s. Calculate the average induced emf while the loop is being pulled.

■ 10. A pair of rails separated by 1.5 m are connected at one end to form a long, thin U-shape. An axle with wheels on each side rolls along the rails at a speed of 25 m/s. The area of the loop formed by the rails and axle increases as the wheels roll. Calculate the induced voltage across the axle if the earth's magnetic field is 5.4×10^{-5} T and tilts downward 68° below the horizontal.

■ 11. A bar magnet shown in Fig. 28.19 moves toward a single loop of wire of radius 1.4 cm. The average induced emf during the 0.40 s the magnet moves is 1.2×10^{-4} V. (a) Determine the change in the average value of the magnetic field passing through the loop. (b) Is lead *a* or lead *b* at higher voltage? Explain.

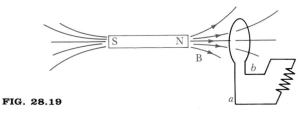

FIG. 28.19

■ 12. The cardboard tube in Fig. 28.20 has two insulated wires wrapped around it. The switch in circuit A is closed, causing

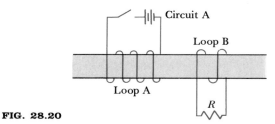

FIG. 28.20

current to flow through loop A. During the short time that the current develops in A, a voltage is induced in loop B. (a) Does the current flow through resistor *R* from left to right or from right to left? Justify your answer, using Lenz's law. (b) In which direction does the current flow through the resistor when the switch is re-opened? Explain.

■ 13. A coil of wire with 10 turns and a radius of 1.5 cm is oriented with the plane of the coil perpendicular to the 0.40-T magnetic field produced by the poles of an electromagnet (see Fig. 28.21). (a) Calculate the average induced emf in the coil when the magnet is turned off and the field decreases to zero in 2.4 s. (b) Is lead *a* or lead *b* at higher voltage? Explain.

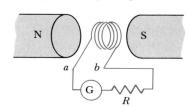

FIG. 28.21

■ 14. The surface of the coil of wire shown in Fig. 28.21 and discussed in Problem 13 is initially oriented parallel to the field with lead *a* on the left. (a) Calculate the average induced emf if the coil is rotated 90° in the counterclockwise direction in 0.050 s. (b) Is lead *a* or lead *b* at a higher voltage? Explain. (c) Calculate the average induced emf if the coil is rotated another 90° in the counterclockwise direction in 0.050 s. (d) Is lead *a* or lead *b* at a higher voltage during the latter 90° rotation? Explain.

■■ 15. A power line 12 m above the ground carries a current of 200 A. (a) Calculate the magnetic field at the ground level below the wire. (b) A hermit builds a 0.50 m × 3.0 m coil with 100 turns and places it below the wire. The coil lies with a 3.0-m side on the ground. Indicate in a drawing the orientation of the coil relative to the power line so that a maximum flux passes through the coil. (c) If the current decreases from 200 A to zero in 1/240 s, calculate the average emf induced in the coil.

■■ 16. A cigarette lighter might be built in principle much like the design shown in Fig. 28.19, with the resistor replaced by a very short gap (0.1 mm) near the wick. (a) *Estimate* the voltage needed across this gap to get dielectric breakdown (a spark) to ignite the fumes from the wick. (b) *Estimate* the shortest time needed to push a small magnet into a coil. (c) As the magnet is pushed toward the coil, the field in the coil increases and causes an induced voltage. If the magnetic flux inside one loop increases by 10^{-6} T·m², how many turns must the coil have to produce the voltage needed to cause a spark?

28.4 The Electric Generator

17. A generator has a 450-turn coil that is 10 cm long and 12 cm wide. The coil rotates at 8 rev/s in a 0.10-T magnetic field. Calculate the generator's peak voltage.

18. You want to make a generator for your bicycle light that will provide an alternating voltage whose peak value is 4.2 V. The generator coil has 55 turns and rotates in a 0.080-T magnetic field. If the coil rotates at 12 rev/s what must the area of the coil be to develop this voltage?

19. A football quarterback, to obtain good grades while strengthening his throwing arm, is asked to vibrate a coil in and out of a 0.18-T magnetic field. The coil has 350 turns, which are 20 cm long and 18 cm wide. At what frequency must the coil

move in and out of the magnetic field to provide a peak voltage of 30 V (assuming the motion is sinusoidal)? The voltage will be used to operate a lightbulb for his studying.

■ 20. A generator has a 100-turn coil that rotates in a 0.30-T magnetic field at a frequency of 80 Hz, causing a peak voltage of 38 V. (a) Calculate the area of each loop of the coil. (b) Write an expression for the voltage as a function of time. (c) Calculate the voltage at 0.0140 s.

■ 21. A 10-Hz generator produces a peak voltage of 40 V. (a) Write an expression for the voltage as a function of time (assuming the voltage is zero at time zero). (b) Calculate the voltage at the following times: 0.025 s, 0.050 s, 0.075 s, and 0.100 s. (c) Plot these values of voltage versus time on a graph and connect the points in a smooth curve. Does the curve resemble a sinusoidal curve?

28.5 Mutual Inductance

22. A heart pacemaker has a small coil in which a voltage is induced by an alternating current in a larger coil placed on a person's chest. The mutual inductance between the two coils is 0.040 H. (a) Calculate the current change per unit time in the chest coil to produce a 0.24-V induced emf in the pacemaker coil. (b) Calculate the induced emf if a steady current of 1.2 A flows in the chest coil.

23. The current in one coil decreases from 4.2 A to zero in 0.80 s. A 0.035-V emf is induced in a neighboring coil. (a) Calculate the mutual inductance between the two coils. (b) Calculate the emf induced in the second coil if the current in the first coil now increases from zero to 1.2 A in 0.050 s. (c) Why is the magnitude of the voltage calculated in part (b) greater than that given in part (a), even though the change in current is less?

24. The mutual inductance between two coils of wire is 0.025 H. (a) Calculate the emf induced in one coil by a 1.2-A current decrease in another coil in a time of 0.036 s. (b) By how much must the current decrease in 0.108 s to produce the same average induced emf? Explain.

■ 25. You are given a 12-V battery, some wire, a switch, and a separate coil of wire. (a) Design a simple circuit that will produce a voltage across the coil even though it is not connected to the battery. (b) Show, using appropriate equations, why your system will work.

■ 26. An airport metal detector consists of two coils whose mutual inductance, when unoccupied, is 8.2 H. An alternating current through one coil induces a 120-mV peak alternating voltage in the second coil. A person wearing an iron medallion walks through the detector. The medallion occupies 1/400 of the space in the second coil and has a magnetic permeability 600 times that of air. Estimate the peak voltage induced as the person passes through the coil. Justify in words your calculation procedure.

■■ 27. You wish to invent a sparker for lighting gas burners. A 1.5-A current from a battery runs through a primary coil. When the switch in this circuit is opened, the current decreases to zero in 0.60 ms, thus inducing a voltage in a secondary coil that causes a spark across a 0.50 mm gap. What must the mutual inductance be between the coils if the dielectric strength of the fumes between the gap is 2.4×10^6 V/m?

■■ 28. A current i flows through a long, straight wire. (a) Write an expression for the magnetic field at a distance a from the wire. (b) If a small coil of radius r with N turns is held perpendicular to the field at distance a from the wire, what is the flux through the coil? (c) Now derive an expression for the mutual inductance be-

tween the wire and coil. (d) If the current in the wire decreases from 15 A to zero in 1/240 s and a coil of 1.2 cm radius with 10 turns is held 15 cm from the wire, what is the average emf induced in the coil?

■■ 29. (a) Two coils are held near each other. A steady current flows through one coil. By using the definition of mutual inductance, show that the voltage induced in the second coil by a change of mutual inductance between the coils is

$$\varepsilon = -i\frac{\Delta M}{\Delta t}.$$

(b) Draw a rough design for a device used to detect metal during food processing that is based on the change in mutual inductance between two coils as given by the preceding equation. Explain in words why it operates as it does—especially why M changes.

28.6 Self-Inductance

30. The current in a circuit changes from 12 A to zero in 0.0040 s, causing a self-induced emf of 105 V. (a) Calculate the self-inductance of the circuit. (b) Calculate the induced emf if the current increases from zero to 4.0 A in 0.0020 s.

31. The current passing through a 0.40-H coil increases from 2.0 A to 10.0 A in 0.12 s. (a) Calculate the voltage induced in the coil. (b) If the current then decreases to 8.0 A in the next 0.020 s, what is the induced emf? Is its polarity the same as in part (a)? Explain.

32. The armature coil of a motor has an inductance of 0.055 H. Calculate the self-induced voltage when the current increases from zero to 8.0 A in 1/240 s. Does the induced voltage aid or oppose the increasing current?

33. The current through the armature of a motor changes from 3.0 A to −3.0 A in 0.0030 s. A 104-V induced emf is produced by the changing current. Calculate the inductance of the armature.

■ 34. The average magnetic field inside a flat coil of area A with N turns around which a current i flows is approximately $B = N\mu_0 i/2\sqrt{A/\pi}$. (a) Determine an expression for the flux through the coil when a current i flows. (b) Determine an expression for the coil's inductance. (c) If the current through a coil of area 45 cm^2 with 20 turns decreases from 4.0 A to 2.5 A in 0.030 s, calculate the induced emf in the coil.

■ 35. (a) Calculate the self-inductance of a 0.10-m-long solenoid with 200 turns, each of 8.0×10^{-5} m^2 area, in which a solid iron core of magnetic permeability 1.0×10^{-3} resides. (b) If a steady current flows through the coil, what is the self-induced emf in the coil? Explain. [*Note:* If the iron core vibrates in and out of the coil, a voltage is induced even though the current is constant. The voltage is caused by the changing flux through the coil as the iron is pushed in and out. The iron core surrounded by the solenoid acts as a vibration detector.]

■■ 36. (a) A steady current flows from a battery through a solenoid. Show, using the definition of inductance and Faraday's law, that if the inductance of the solenoid changes, a back emf is induced whose magnitude is

$$\varepsilon = i\frac{\Delta L}{\Delta t}.$$

(b) The back emf opposes the current and causes it to decrease or increase, depending on whether L increases or decreases. Describe carefully how this solenoid might be used to build a burglar alarm.

28.7 Transformers

37. Calculate the ratio of the secondary to primary turns of a transformer that can step the voltage up from 120 V to 12,000 V to operate a neon sign.

38. A home's electric bell operates on 10 V. Should a step-up or step-down transformer be used to convert the home's 120 V to 10 V? Calculate the ratio of the secondary to primary turns needed for the bell's transformer.

39. A transformer with 100 turns in its secondary coil supplies 8.0 V to the light in a toy. The primary coil is connected across 120 V. (a) How many turns does the primary coil have? (b) If the primary coil is connected to a 240-V source, what voltage appears across the secondary coil?

■ **40.** The power for a city of 50,000 people is to be carried by one large transmission line. The line can carry a maximum current of 180 A. (a) Calculate the voltage at which 1000 W of power per person must be transmitted. (b) Calculate the ratio of secondary to primary turns in a transformer used to step down the voltage to 120 V.

■ **41.** A city of 20,000 people receives 1000 W per person from a pair of transmission lines across which a 120,000-V potential difference exists. The voltage is stepped down to 120 V at a transformer. Calculate (a) the current in the primary coil and (b) the ratio of the number of turns in the primary coil to the number in the secondary coil. (c) How much current would have to flow if the power is transmitted at 120 V instead of 120,000 V?

■ ■ **42.** A transmission line is to carry 1.5×10^8 W of power at 750 kV over a distance of 800 km. Calculate the cross-sectional area of aluminum wires in the transmission line that will cause a 1 percent I^2R power loss.

CHAPTER **29**

Alternating-Current Circuits

We learned in the last chapter that a coil rotating in a magnetic field produces an alternating voltage. At first, one terminal of the alternating-current generator is at a positive voltage relative to the other terminal; when the generator's coil rotates one-half turn, the voltage is reversed. The terminals of the generator are said to change polarity each time the voltage across the terminals is reversed. If a wire is connected from one terminal to the other, electrons in the wire oscillate back and forth, always moving toward the positive terminal. Since the terminals change polarity as the coil rotates, the electrons change direction each time the polarity changes. The back-and-forth motion of the electrons is called an **alternating current** (ac).

About 99 percent of electric power is generated and transmitted by ac circuits. Most appliances in your home operate on alternating current, as do radio, television, high-fidelity, telephone, and all other communications systems.

In our brief discussion of ac circuits, we will consider only a few basic ideas concerning the role of resistors, induction coils, and capacitors in these circuits. A relation between the current through and voltage across each device will be examined. The chapter ends with an analysis of a simple ac circuit containing each of these elements in series.

29.1 AC Generator

An alternating-current (ac) generator is a voltage source whose polarity changes in a repetitive pattern. The generator, symbolized by a circle with a wavy line inside, —⊙—, first has positive voltage on one side and negative on the other, as depicted in Fig. 29.1a. A short time later the positive-voltage side becomes negative, and the negative side positive.

Electronic generators have been built that produce a variety of time-varying voltage patterns, called voltage waveforms. Several examples are shown in Fig. 29.1b–d. Notice that in each case the polarity of the voltage across the generator reverses in a repetitive manner. The change in voltage can be gradual, as in the sinusoidal waveform (Fig. 29.1b), or abrupt, as in the square-wave and sawtooth waveforms in parts (c) and (d). In this chapter we will examine circuits powered only by sinusoidal generators.

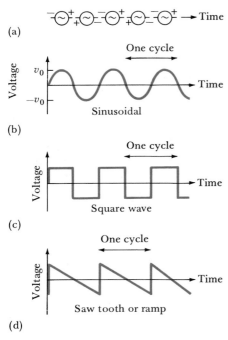

FIG. 29.1. (a) The changing polarity of an alternating voltage. (b)–(d) Different ac voltage waveforms.

A **cycle** or period (shown in Fig. 29.1b–d) is the time needed for the voltage waveform to make one complete vibration pattern. The frequency f of the ac generator is the number of cycles per unit time. The voltage generated by electric power plants in the United States and Canada has a frequency of 60 Hz; the voltage completes 60 cycles in one second, and during each cycle the polarity changes from positive to negative and back to positive again.

The voltage across a sinusoidal generator varies with time, according to the equation

$$v = v_0 \sin (2\pi ft), \qquad (29.1)$$

where v_0 is called the **peak voltage**—that is, the magnitude of the maximum voltage across the generator during each cycle (see Fig. 29.1b). Note that lower-case letters such as v and i are used to represent voltages and currents that change with time. A capital V or I represents constant voltages or currents that do not change with time.

29.2 Alternating Current Through Resistors

The current produced in a resistor when connected to an ac generator (Fig. 29.2) is calculated using Ohm's law: $i = v/R$. Since the voltage varies with time according to Eq. (29.1), we find that the current through the resistor also varies with time:

$$i = \frac{v}{R} = \frac{v_0}{R} \sin (2\pi ft),$$

or

$$i = i_0 \sin (2\pi ft), \qquad (29.2)$$

where $i_0 = v_0/R$ is called the **peak current.**

The ac voltage across the resistor and the current through the resistor are plotted against time in Fig. 29.2b and c, respectively. The current fluctuates in exactly the same pattern as the voltage—first becoming positive, then decreasing and becoming negative, and then completing a cycle by becoming positive again.

What does a negative current mean? If we arbitrarily say that a current moving through the resistor toward the right is positive, then a negative current is one that moves toward the left. The fluctuating current indicates that the electric charge in the resistor first moves in one direction and then in the opposite direction.

EXAMPLE 29.1 The voltage across the wall socket in your home has a peak voltage v_0 of 170 V and a frequency f of 60 Hz. Write an expression for the current through a toaster whose resistance R when hot is 12 Ω.

SOLUTION Substituting $i_0 = v_0/R$ into Eq. (29.2), we find that current varies with time as follows:

$$i = i_0 \sin (2\pi ft) = \frac{v_0}{R} \sin (2\pi ft) = \frac{170 \text{ V}}{12 \text{ }\Omega} \sin (2\pi 60 t)$$

$$= \underline{14 \sin (2\pi 60 t)} \qquad \text{(in units of amperes)}. \qquad \blacksquare$$

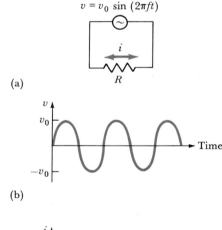

(a)

(b)

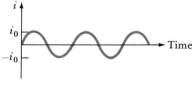

(c)

FIG. 29.2. (a) An ac generator produces an alternating current in a resistor. The variations (b) of voltage with time and (c) of current with time.

The alternating current through a resistor, such as the heating coil in a toaster, produces thermal energy in much the same way as a direct current produces thermal energy, as we see next.

29.3 RMS Voltage and Current

You will recall from past chapters that the word *power* describes a change in energy per unit time. In electricity, power is defined as the electrical energy added to an electric circuit per unit time. If a direct current I flows through a resistor of resistance R, the electrical energy added to the resistor per unit time is

$$P = \frac{\Delta PE_q}{\Delta t} = I^2 R. \tag{24.11}$$

As electrons move through a resistor, they bump into atoms and ions in the resistor, which in turn gain thermal energy, causing the resistor to become hot. The hot resistor transfers thermal energy to its surroundings by radiation, conduction, and/or convection. Eventually, an equilibrium is reached where the resistor gains electrical energy just as fast as the energy is lost by this heat transfer to the surrounding environment.

The same process occurs in a resistor through which an alternating current flows. However, the rate of energy transfer differs from one instant to the next because the current passing through the resistor continually changes. At the instant the current has a value $i = i_0 \sin(2\pi ft)$, the power transferred to the resistor is

$$p = \frac{\Delta PE_q}{\Delta t} = i^2 R = [i_0^2 \sin^2(2\pi ft)]R. \tag{29.3}$$

This expression for power is plotted versus time in Fig. 29.3b. Notice that all values of p are positive because the square of a negative number (the negative value of current in this case) is always positive. Thus, the power (thermal energy generation rate) varies between zero and a maximum of $i_0^2 R$. Usually we are interested only in the average rate of thermal energy production. For example, a slice of bread in a toaster does not respond to the 120 power peaks each second (2 per cycle from the 60-Hz ac voltage); instead, the rate of toasting depends on the average power to the heating elements.

To find the average of the expression on the right of Eq. (29.3), we must take the average of $\sin^2(2\pi ft)$ since it is the only part of this expression that varies with time. The magnitude of the sine of any quantity varies between 0 and 1, as does the magnitude of the sine squared. The average of $\sin^2(2\pi ft)$ over time is, not surprisingly, 1/2.* Referring back again to Eq. (29.3), we see that

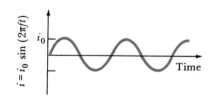

(a)

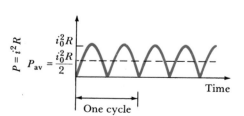

(b)

FIG. 29.3. (a) An oscillating current through a resistor causes electric power absorption that varies with time as shown in part (b).

---

*The average of the square of the sine of a function can be calculated with the aid of the trigonometric identity,

$$\sin^2\theta + \cos^2\theta = 1.$$

The $\sin^2\theta$ and $\cos^2\theta$ functions have exactly the same shape when plotted against θ. The only difference is that one is shifted along the θ axis relative to the other. However, over one cycle of θ, both functions have two full humps, such as shown in Fig. 29.3b, and both have the same average. Thus,

$$(\sin^2\theta)_{av} = (\cos^2\theta)_{av}$$

and

$$(\sin^2\theta + \cos^2\theta)_{av} = (2\sin^2\theta)_{av} = 1, \quad \text{or} \quad (\sin^2\theta)_{av} = \frac{1}{2}.$$

the average power P to a resistor through which an alternating current flows is

$$P = [i_0^2 \sin^2(2\pi ft)]_{av} R = [\sin^2(2\pi ft)]_{av} i_0^2 R = \frac{1}{2} i_0^2 R.$$

Usually we are most interested in the average power used by a device in an electric circuit. Because of this interest in average power, it is convenient to define what is called a root-mean-square (rms) current that relates in a direct way to average power. The **root-mean-square current** I_{rms} is the magnitude of a direct current that produces thermal energy in a resistor at the same rate as the energy is produced by an alternating current of peak value i_0. By equating the expressions for power production by a direct current I_{rms} and by an alternating current of peak value i_0 ($P = I_{rms}^2 R = \frac{1}{2} i_0^2 R$), we find that $I_{rms}^2 = \frac{1}{2} i_0^2$, or

$$I_{rms} = \frac{i_0}{\sqrt{2}} = 0.707 i_0. \qquad \text{(29.4a)}$$

A 0.707-A direct current passing through a resistor produces thermal energy at the same rate as a sinusoidal alternating current of peak value 1.0 A. The rms current is sometimes called the **effective current;** it produces the same effect as the alternating current.

A similar expression can be used to define root-mean-square (rms) or effective voltage. The **root-mean-square voltage** V_{rms} is the magnitude of a dc voltage that produces thermal energy in a resistor at the same rate as an ac voltage of peak value v_0. V_{rms} and v_0 are related by an equation similar to that relating I_{rms} and i_0:

$$V_{rms} = \frac{v_0}{\sqrt{2}} = 0.707 v_0. \qquad \text{(29.4b)}$$

Many ac voltmeters and ammeters are calibrated to read root-mean-square voltages and currents rather than peak values. In fact, most of our future discussion of ac circuits will involve root-mean-square values rather than peak values because they relate directly to power consumption. The same equation ($P = I^2 R = I_{rms}^2 R$) can be used to calculate either ac or dc power used by a resistor. Also, Ohm's law applies equally well for instantaneous voltage and current, peak voltage and current, and rms voltage and current:

$$\frac{v}{i} = \frac{v_0}{i_0} = \frac{V_{rms}}{I_{rms}} = R. \qquad \text{(29.5)}$$

In the rest of this chapter, rms values of current and voltage will be used unless stated otherwise, and the subscript rms will be dropped. This notation for ac circuits is summarized in Table 29.1. Note that $I = 0.707 i_0$, $V = 0.707 v_0$, and $P = I^2 R = \frac{1}{2} i_0^2 R$.

EXAMPLE 29.2 The peak value v_0 of the alternating voltage provided by a power company to your home is 170 V. Calculate the rms value V of this voltage.

SOLUTION We solve Eq. (29.4b) for V as follows:

$$V = 0.707 v_0 = 0.707(170 \text{ V}) = \underline{120 \text{ V}}. \qquad \blacksquare$$

TABLE 29.1 Notation for AC Circuits

	Current	Voltage	Power
Direct current	I	V	P
Alternating current			
Instantaneous value	i	v	p
Peak value	i_0	v_0	p_0
Effective (rms) value	I	V	P

As far as producing thermal energy in a toaster or lightbulb, an alternating voltage of peak value 170 V is equivalent to a constant dc voltage of 120 V. When the leads of an ac voltmeter are placed across the terminals of a wall socket, they measure 120 V. The voltage from the wall socket is almost always referred to as 120 V, since it effectively provides power like a 120-V dc power supply.

EXAMPLE 29.3 A 60-W lightbulb is connected to a 120-V ac power source (the rms voltage V provided to a home by a power company). Calculate the resistance R of the lightbulb.

SOLUTION We are given that $P = 60$ W and $V = 120$ V. The substitution of Eq. (29.5) into Eq. (24.11) results in the expression

$$P = I^2R = (V/R)^2R = V^2/R,$$

or

$$R = V^2/P = (120\text{ V})^2/(60\text{ W}) = \underline{240\ \Omega}\,. \qquad \blacksquare$$

29.4 Apparent Resistance of a Coil

The current through a resistor caused by an alternating voltage is calculated using Ohm's law ($I = V/R$). A similar expression relates the current through and voltage across an induction coil such as those described in Section 28.6. However, the expression for the "resistance" of the coil is quite different from what we would expect!

To get a better understanding of the resistance of a coil to alternating current, we connect a coil of wire to several different voltage sources, as shown in Fig. 29.4. In each case the current through the coil will be measured, and the ratio of the measured voltage and measured current will indicate the coil's resistance to the current. In the future we will use the word *impedance* rather than *resistance* to describe the opposition to current flow through an ac circuit. Thus, for the coil we define its ac impedance as

$$\text{Impedance} = \frac{V}{I}.$$

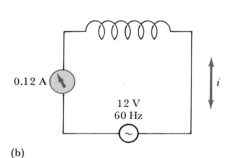

(a)

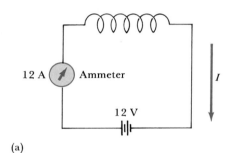

(b)

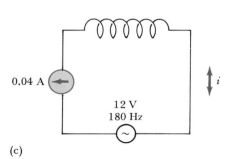

(c)

FIG. 29.4. The current through a coil decreases as the frequency of the voltage source increases.

First, when the coil is connected to a 12-V dc voltage source (Fig. 29.4a), an ammeter measures a current of 12 A. Using Ohm's law, we find that the coil's dc resistance (the word *resistance* is used when direct current flows) is 1 Ω, since

$$R = \frac{V}{I} = \frac{12 \text{ V}}{12 \text{ A}} = 1 \; \Omega \qquad \text{(dc resistance)}.$$

Next the coil is connected to a 12-V, 60-Hz alternating voltage source (Fig. 29.4b). We find, surprisingly, that the alternating current is only 0.12 A, implying an impedance of 100 Ω, not 1 Ω:

$$\text{Impedance} = \frac{V}{I} = \frac{12 \text{ V}}{0.12 \text{ A}} = 100 \; \Omega \qquad \text{(at 60 Hz)}.$$

If the frequency is tripled to 180 Hz (Fig. 29.4c) while the voltage is kept constant, the current is reduced by one-third to 0.04 A, and the impedance is now 300 Ω:

$$\text{Impedance} = \frac{V}{I} = \frac{12 \text{ V}}{0.04 \text{ A}} = 300 \; \Omega \qquad \text{(at 180 Hz)}.$$

Finally, we try one last experiment (not depicted in Fig. 29.4). The coil is stretched so that it is a long, straight wire. Then, when the voltage of the different-frequency ac generators is placed across the stretched coil, they all produce a current of 12 A. The stretched coil's impedance at all frequencies is 1 Ω, the same as its dc resistance:

$$\text{Impedance} = \frac{12 \text{ V}}{12 \text{ A}} = 1 \; \Omega \qquad \text{(stretched coil)}.$$

Evidently, the coil's impedance or opposition to alternating current is a consequence of its shape, since the impedance decreases dramatically when the coil is stretched straight, even though the current travels through the same length of wire.

This puzzling behavior depends on two different types of opposition to current, one related to Ohm's law and the other to Faraday's law. First, consider the Ohm's law resistance. As electrons move through the coil, they bump into atoms and molecules in the wire. This collision-type resistance R is the same for both dc and ac and equals 1 Ω for the example just discussed.

A second type of opposition to current flow becomes dominant as the frequency of the voltage source is increased The frequency-dependent impedance is given the ominous name *inductive reactance* and is represented by the symbol X_L. It has nothing to do with electrons colliding with atoms and ions but is caused instead by a voltage induced in the coil by its own changing current. This frequency-dependent impedance is a result of the coil's self-inductance, a subject discussed in Section 28.6. We found that the voltage ε induced in a coil of inductance L when the current in the coil was changing with time at a rate $\Delta i / \Delta t$ was

$$\varepsilon = -L \frac{\Delta i}{\Delta t}. \qquad \textbf{(28.10)}$$

If the induced voltage ε is large, the opposition to current trying to pass through the coil is also large. We see that the induced voltage and opposition to current depends on two quantities: the inductance L of the coil and the rate of change of current $\Delta i / \Delta t$ through the coil.

Inductance L is a geometrical property of the coil and depends on the number of turns, the inside area of the coil, and the type of material inside the coil. Because a straight wire has no turns, its inductance is essentially zero. Hence, no opposing voltage is induced by an alternating current through a straight wire. On the other hand, a coil with large inductance produces a large induced voltage and offers considerable opposition to a high-frequency alternating current.

The induced voltage that opposes the changing current also depends on $\Delta i/\Delta t$. If a direct current flows through a coil, no opposing voltage is induced because $\Delta i/\Delta t$ is zero—the current is steady and does not change with time. However, a high-frequency alternating current fluctuates back and forth rapidly ($\Delta i/\Delta t$ is large), and the induced voltage opposing the current is also large. The impedance to alternating current caused by the induced voltage in the coil increases in direct proportion to the frequency of the current.

The frequency-dependent impedance, called **inductive reactance** X_L, of a coil is calculated using the equation

$$X_L = 2\pi f L, \tag{29.6}$$

where f is the frequency of the current and L is the self-inductance of the coil. When an alternating voltage source is connected across a coil of inductance L, the rms values of the current and voltage are related by the equation

$$I = \frac{V}{X_L} = \frac{V}{2\pi f L}. \tag{29.7}$$

This equation is analogous to Ohm's law with the reactance of the coil replacing the resistance of the resistor. The peak values of current and voltage are related by a similar equation ($i_0 = v_0/X_L$).

EXAMPLE 29.4 Calculate the current in a coil of inductance 0.050 H when driven by a 120-V, 60-Hz voltage source.

SOLUTION The inductive reactance of the coil is calculated using Eq. (29.6):

$$X_L = 2\pi f L = 2\pi(60 \text{ s}^{-1})(0.050 \text{ H}) = 18.8 \ \Omega.$$

The effective ac current through the coil is

$$I = \frac{V}{X_L} = \frac{120 \text{ V}}{18.8 \ \Omega} = \underline{6.4 \text{ A}}. \qquad \blacksquare$$

EXAMPLE 29.5 A coil, called a "choke," is used to reduce the amplitude of high-frequency signals while allowing low frequencies to pass. (a) Calculate the inductance of a coil that causes a 1200-Ω reactance for a 20-kHz signal. (b) What is the coil's reactance to a 60-Hz signal?

SOLUTION (a) The following information is given: $X_L = 1200 \ \Omega$ when $f = 20 \times 10^3 \text{ Hz} = 20 \times 10^3 \text{ s}^{-1}$. We are to find the value of the coil's inductance L that produces this reactance. Rearranging Eq. (29.6), we find that

$$L = \frac{X_L}{2\pi f} = \frac{1.2 \times 10^3}{2\pi(20 \times 10^3 \text{ s}^{-1})} = \underline{9.5 \text{ mH}}.$$

(b) The same coil offers much less than 1200-Ω resistance to a 60-Hz signal since

$$X_L \ (\text{at } 60 \text{ Hz}) = 2\pi f L = 2\pi(60 \text{ s}^{-1})(9.5 \times 10^{-3} \text{ H}) = \underline{3.6 \ \Omega}.$$

The coil effectively "chokes" high-frequency signals, allowing only lower-frequency signals to pass. ■

In the last chapter we discussed an airport metal detector that depended on the mutual inductance between two coils. The detector can also work using the inductive reactance of a single coil. This type of detector consists of a single coil across which an alternating voltage is connected (Fig. 29.5). An ac ammeter measures the current in the coil. If the current changes, an alarm rings.

When a person walks through the coil, the coil's inductance changes little because the magnetic susceptibility of a person's body is the same as that of air or a vacuum. If the person carries a piece of iron, though, the inductance of the coil changes dramatically. Even a piece of iron that occupies only a tiny fraction of the area inside the coil can double the coil's inductance L. If L doubles, so does the coil's reactance, since $X_L = 2\pi f L$. If X_L doubles, the alternating current through the coil is reduced by half, since $I = V/X_L$. The alarm goes off.

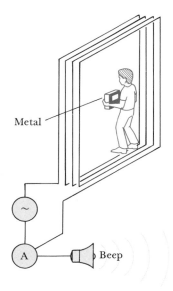

FIG. 29.5. A metal detector detects a change in alternating current if metal is carried through the coil. The reactance of the coil increases, and the current is reduced.

29.5 Apparent Resistance of a Capacitor

In the last section we learned that the reactance—that is, opposition of a coil to the flow of alternating current—depends on the frequency of the current and the inductance of the coil. A capacitor also has a reactance that opposes an alternating current. The capacitive reactance depends on the frequency of the current and the capacitance of the capacitor, as illustrated by the following simple experiments.

A capacitor is first connected to a 9-V dc power source, such as a battery (Fig. 29.6a). At the instant the capacitor is connected, a current flows. Once the capacitor is charged, the current stops. At this point the capacitor is like an open switch that offers infinite resistance to direct current.

Next connect the capacitor to a 9-V, 1000-Hz alternating voltage source (Fig. 29.6b). A small 0.009-A alternating current is measured. The capacitor has an ac impedance (opposition to current flow) of 1000 ohms, since

$$\text{Impedance} = \frac{V}{I} = \frac{9 \text{ V}}{0.009 \text{ A}} = 1000 \ \Omega.$$

If the frequency of the power source is tripled, the current also triples to 0.027 A. Evidently, the impedance has been reduced by one-third:

$$\text{Impedance} = \frac{V}{I} = \frac{9 \text{ V}}{0.027 \text{ A}} = 333 \ \Omega.$$

As the frequency increases, the impedance of the capacitor decreases. This frequency-dependent impedance of a capacitor is called **capacitive reactance.** Like inductive reactance, it has nothing to do with electrons colliding with atoms and ions in a wire but is caused by the voltage across the charged capacitor. As the capacitor accumulates a charge q on one plate and $-q$ on the other plate, a voltage given by

$$V = \frac{q}{C} \tag{26.2}$$

opposes the further accumulation of charge on the plates. If the plates become fully charged, current stops.

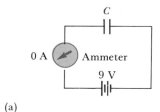

(a)

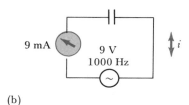

(b)

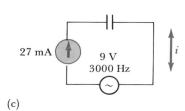

(c)

FIG. 29.6. The apparent resistance of a capacitor decreases as the frequency of the voltage source increases.

Capacitive reactance depends on how fully the capacitor becomes charged during each one-half cycle of an alternating current. At low frequency, more time is available to charge the capacitor during each flow of current in one direction and before flow in the other direction starts, and the capacitor offers a large average resistance at low frequency. At high frequency, however, there is little time for charge to accumulate on the plates before the capacitor discharges as current flows in the opposite direction. As a result, the capacitor offers little resistance to high-frequency current.*

The opposing voltage ($V = q/C$) also depends on the capacitance C of the capacitor. The greater C is, the less the capacitor becomes fully charged during any time period. Hence, a capacitor with large capacitance produces much less resistance to alternating-current flow than does one with small capacitance.

The frequency-dependent impedance (**capacitive reactance X_C**) caused by a capacitor is given by the equation

$$X_C = \frac{1}{2\pi f C},$$ (29.8)

where f is the frequency of the current and C is the capacitance of the capacitor. When an alternating voltage is connected across a capacitor of capacitance C, the rms values of the current and voltage are related by the equation

$$I = \frac{V}{X_C} = \frac{V}{1/(2\pi f C)} = (2\pi f C)V.$$ (29.9)

This equation is analogous to Ohm's law, with the capacitive reactance replacing resistance.

EXAMPLE 29.6 Calculate the capacitive reactance of a 0.030-μF capacitor (a) at 60 Hz and (b) at 90 MHz.

SOLUTION (a) The value of X_C at each frequency is determined using Eq. (29.8):

$$X_C = \frac{1}{2\pi f C} = \frac{1}{2\pi(60 \text{ Hz})(0.030 \times 10^{-6} \text{ F})} = \underline{8.8 \times 10^4 \ \Omega} \qquad \text{(at 60 Hz)}.$$

(b)

$$X_C = \frac{1}{2\pi(90 \times 10^6 \text{ Hz})(0.030 \times 10^{-6} \text{ F})} = \underline{0.059 \ \Omega} \qquad \text{(at 90 MHz)}.$$

A 0.030-μF capacitor offers negligible resistance to an FM radio signal transmitted at 90 MHz. The same capacitor will offer considerable opposition to a 60-Hz noise signal transmitted from fluorescent lights or motors operated by the 60-Hz power used in a home. Consequently, the capacitor reduces the current caused by low-frequency electronic "noise" signals, much of which is at a frequency of 60 Hz, but allows high-frequency radio and television signals to pass. ■

---

*The idea of current flowing "through" a capacitor may bother you. The charge does not actually flow from one plate to the other. The accumulation of charge on the plates allows current to flow in other parts of a circuit attached to a capacitor. An alternating current flows in a circuit with a capacitor even if charge does not cross from one plate to the other.

29.6 Phase Angles of Voltages and Currents in AC Circuits

In the last two sections we have ignored a subtle feature that affects the power used by an ac circuit. This ignored feature is the phase of the current and voltage. Often, current and voltage do not peak at the same times and are said to be *out of phase* with each other. The difference in phase is expressed in terms of a *phase angle*.

Before examining phase angles, let us compare the two different sinusoidal functions shown in Fig. 29.7. The first function (Fig. 29.7a) is a sinusoidal current expressed by the function $i = i_0 \sin(2\pi ft)$. Plotted below the sinusoidal current is $\sin \theta$ versus θ. The two functions, $i_0 \sin(2\pi ft)$ versus t and $\sin \theta$ versus θ, have exactly the same shape. Notice that one cycle of the $\sin \theta$ function corresponds to a change in θ of 2π radians, or 360°. One-half cycle is 180°, and one-quarter cycle is 90°.

Now let us return to voltage and current. In Fig. 29.8a we have plotted a sinusoidal current. Plotted below the current in Fig. 29.8b is a sinusoidal voltage that is in phase with the current—the current and voltage reach their peaks simultaneously.

Other sinusoidal voltage functions are plotted in Fig. 29.8c and d. These voltage functions are different from that shown in part (b) in that they lead or lag the current by what is called a phase angle ϕ. Compare, for example, Fig. 29.8a and c. Notice that at time zero the current shown in part (a) is just starting to rise from zero. However, the voltage shown in part (c) is zero and rising at an earlier time; by time zero it has already reached its positive peak. Since the voltage has already completed one-fourth of the cycle (90°) by time zero, the voltage is said to "lead" the current by 90°. The voltage shown in Fig. 29.8d "lags" the current by 90° since it does not start rising from zero until a time when the current has already completed one-fourth cycle.

In the rest of this section we will talk about the phase angle between the current and the voltage in an ac circuit with one resistor, with one inductor, or with one capacitor. In the next section we consider a circuit with all three elements in the same circuit.

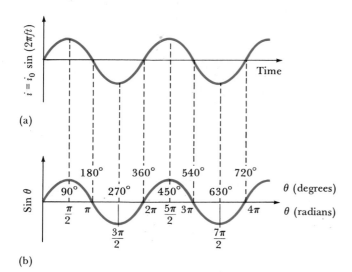

(a)

(b)

FIG. 29.7. (a) A sinusoidal current varies with time in a way that appears identical to (b) the variation of $\sin \theta$ versus θ. One cycle corresponds to 360°.

FIG. 29.8. (a) A sinusoidal current. (b)–(d) A variety of voltages that differ in phase from the current.

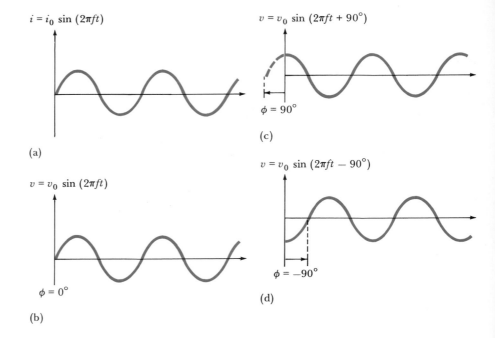

$i = i_0 \sin (2\pi ft)$

(a)

$v = v_0 \sin (2\pi ft + 90°)$

$\phi = 90°$

(c)

$v = v_0 \sin (2\pi ft)$

$\phi = 0°$

(b)

$v = v_0 \sin (2\pi ft - 90°)$

$\phi = -90°$

(d)

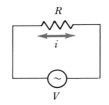

(a)

(b) $i = i_0 \sin (2\pi ft)$ Time

(c) $v = v_0 \sin (2\pi ft)$ Time

FIG. 29.9. (a) An ac circuit with a resistor. (b) The sinusoidal current passing through the resistor is in phase with (c) the voltage that causes the current.

Circuit with One Resistor

For an ac circuit containing a single resistor (Fig. 29.9a), the instantaneous current i and the voltage v are related by Ohm's law:

$$v = Ri = Ri_0 \sin (2\pi ft) = v_0 \sin (2\pi ft),$$

where $v_0 = Ri_0$. *The current and voltage are in phase,* as shown in Fig. 29.9b and c.

Circuit with One Inductor

The phase between the current through an inductor and the voltage across it must be determined carefully. Let us start by applying Kirchhoff's loop rule (Section 25.1) to the inductor circuit shown in Fig. 29.10. As we move clockwise around the circuit, the voltage change across each element is added. We first pass the ac generator, across which there is an unspecified voltage change v. Next we pass through the inductor. The induced emf across the inductor is $-L(\Delta i/\Delta t)$ (Eq. 28.10). Since these are the only two elements in the circuit, the two voltages must add to zero:

$$v + \left(-L\frac{\Delta i}{\Delta t}\right) = 0, \quad \text{or} \quad v = L\frac{\Delta i}{\Delta t}.$$

Notice that the voltage across the coil is proportional to $\Delta i/\Delta t$ and not to i. A sinusoidal current is plotted in Fig. 29.10b. In Fig. 29.10c, the change of current per unit time ($\Delta i/\Delta t$) is plotted. The $\Delta i/\Delta t$ curve leads the current by 90°. Since the generator voltage is proportional to $\Delta i/\Delta t$, it also leads the current by 90°. *The voltage across an inductor leads the current by 90°.*

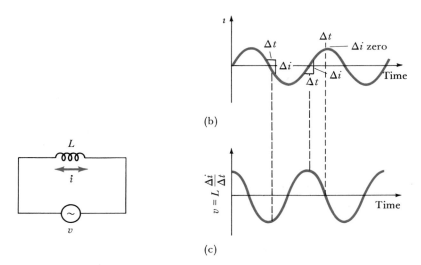

(b)

(a)

(c)

FIG. 29.10. (a) An ac circuit with an inductor. (b) If a sinusoidal current flows, the voltage shown in part (c) must lead the current by 90° so that $v = L(\Delta i/\Delta t)$.

Circuit with One Capacitor

Finally, examine the voltage across the capacitor shown in Fig. 29.11a. The voltage across the ac generator has an unspecified value v. If at a given instant the voltage v across the generator is positive, then the voltage across the capacitor must be negative, and it has a value given by Eq. (26.2): $-q/C$. Applying Kirchhoff's loop rule to the circuit shown in Fig. 29.11, we find that

$$v - \frac{q}{C} = 0. \qquad \textbf{(26.2)}$$

The current flowing through the circuit is related to the charge q on the capacitor plates by the equation

$$i = \frac{\Delta q}{\Delta t}. \qquad \textbf{(24.1)}$$

Solving Eq. (26.2) for q and substituting into Eq. (24.2), we find that

$$i = \frac{\Delta q}{\Delta t} = \frac{\Delta(Cv)}{\Delta t} = C\frac{\Delta v}{\Delta t},$$

where $\Delta v/\Delta t$ is the change in voltage in a short time Δt. If the current has a sinusoidal shape (Fig. 29.11b), then the voltage curve whose change with time $\Delta v/\Delta t$ produces that sinusoidal current curve is as shown in Fig. 29.11c. *The voltage across a capacitor lags the current by 90°.*

29.7 *RLC* Series Circuit

In this section we examine a circuit, called an ***RLC* series circuit,** that contains a resistor, induction coil, and capacitor in series with each other. Our analysis will lead to an understanding of the idea of the electrical impedance of an ac circuit and will develop the background we need to understand electrical resonance and electrical power transfer in ac circuits.

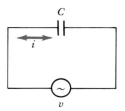

(a)

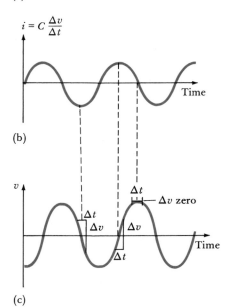

(b)

(c)

FIG. 29.11. (a) An ac circuit with a capacitor. (b) If a sinusoidal current flows, the voltage in part (c) must lag the current by 90° so that $i = C(\Delta v/\Delta t)$.

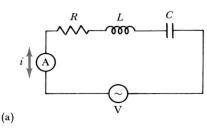

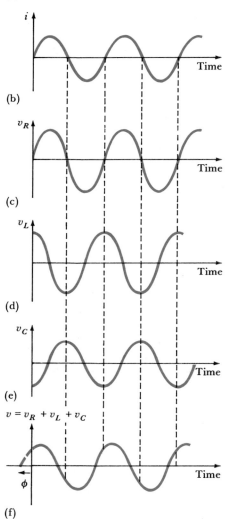

(a)

(b)

(c)

(d)

(e)

$v = v_R + v_L + v_C$

(f)

FIG. 29.12. (a) An *RLC* series circuit. (b) The current flowing in the circuit. (c)–(e) The instantaneous voltages are shown across *R*, *L*, and *C*, respectively. (f) The resultant voltage across the whole circuit.

An *RLC* series circuit is shown in Fig. 29.12a. An ammeter in the circuit measures the current flowing through the circuit. The current in a series circuit is the same at each point in the circuit and is represented by the sinusoidal curve shown in Fig. 29.12b.

We learned in the preceding section that the voltage across each element in the circuit is not necessarily in phase with the current. In fact, if we represent the instantaneous voltages across the resistor, inductor, and capacitor by the symbols v_R, v_L, and v_C, respectively, and the instantaneous current by the symbol i, then

1. v_R is in phase with i;
2. v_L leads the current by 90°;
3. v_C lags the current by 90°.

The net voltage v across all three circuit elements at each instant of time is the sum of the *instantaneous* voltages across each element at that time:

$$v = v_R + v_L + v_C.$$

The net voltage v produced by adding v_R, v_L, and v_C at each instant is shown in Fig. 29.12f. Notice that the voltages v_L and v_C tend to cancel each other and that both are out of phase with v_R. Thus, the amplitude of the net voltage is a complicated function of the amplitudes of v_R, v_L, and v_C, as is the phase of the net voltage. For the example shown in Fig. 29.12, the net voltage shown in part (f) leads the current in part (b) by a phase angle ϕ of about 30° to 40°.

Instead of adding the instantaneous voltages across each element to find the amplitude and phase of the resultant voltage, as in Fig. 29.12, another technique involving vectors is often used. We can represent the voltages across each circuit element by a vector (Fig. 29.13). The length of the vector represents the magnitude of the voltage (the peak voltage), and the direction of the vector represents the phase of the voltage relative to the current.* The positive *x* axis is arbitrarily chosen as the direction of a voltage in phase with the current. Thus, $\mathbf{v}_R$ points in the positive *x* direction (see Fig. 29.13a). The voltage across the inductor coil leads the current by 90° and is represented by a vector $\mathbf{v}_L$ pointing in the positive *y* direction. The voltage across the capacitor lags the current by 90° and is represented by a vector $\mathbf{v}_C$ pointing in the negative *y* direction.

The net or resultant voltage **v** is the vector sum of $\mathbf{v}_R$, $\mathbf{v}_L$, and $\mathbf{v}_C$. To find this sum, let us first add $\mathbf{v}_L$ and $\mathbf{v}_C$ to find $\mathbf{v}_{LC}$, the voltage across the inductor and capacitor:

$$\mathbf{v}_{LC} = \mathbf{v}_L + \mathbf{v}_C.$$

The voltages $\mathbf{v}_L$ and $\mathbf{v}_C$ point in opposite directions; when added, the two vectors tend to cancel each other. The magnitude of the combined voltage of $\mathbf{v}_L$ and $\mathbf{v}_C$ is the difference of their amplitudes:

$$v_{LC} = v_L - v_C.$$

The vector $\mathbf{v}_{LC}$ points up or down depending on whether v_L is greater or less than v_C (see Fig. 29.13b).

Next, $\mathbf{v}_{LC}$ is added to $\mathbf{v}_R$ (Fig. 29.13c). Since the vectors are at right angles, the resultant voltage has a magnitude that is the square root of the sum of the squares of the voltages:

*The vector analysis shown in Fig. 29.13 applies equally well to peak voltages (lower case **v**) and to rms voltage (upper case *V*), as we learn shortly.

$$v = \sqrt{v_R^2 + v_{LC}^2} = \sqrt{v_R^2 + (v_L - v_C)^2}.$$

The phase angle ϕ that **v** makes with the positive x axis is calculated using the tangent of the angle ϕ, shown in Fig. 29.13c:

$$\tan \phi = \frac{v_{LC}}{v_R} = \frac{v_L - v_C}{v_R}.$$

The last two equations apply equally well for the rms voltages across the various circuit elements since $V_{rms} = V = 0.707v$. Substituting rms voltages in these expressions and using the fact that $V_R = IR$, $V_L = IX_L$, and $V_C = IX_C$, we find that

$$V = \sqrt{V_R^2 + (V_L - V_C)^2} = I\sqrt{R^2 + (X_L - X_C)^2} \qquad \textbf{(29.10)}$$

and

$$\tan \phi = \frac{V_L - V_C}{V_R} = \frac{IX_L - IX_C}{IR} = \frac{X_L - X_C}{R}. \qquad \textbf{(29.11)}$$

The quantity under the square root sign in Eq. (29.10) is called the **impedance** of the RLC series circuit and is represented by the symbol Z:

$$Z = \sqrt{R^2 + (X_L - X_C)^2}. \qquad \textbf{(29.12)}$$

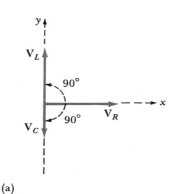

(a)

The impedance of an ac circuit plays a role analogous to the resistance of a dc circuit—both represent the opposition to the flow of current. The big difference between impedance and resistance is that the X_L and X_C parts of impedance are caused by the opposing induced voltage across the inductor coil and the opposing voltage across the charged capacitor. No heating occurs because of these opposing voltages. Thus, the X_L and X_C parts of Z result in no net energy loss due to heating. However, the R part of impedance is caused by collisions of electrons with atoms and ions in a resistor, and it results in thermal energy creation and in heat loss in the electronic circuit.

The equations just developed apply for rms currents and voltages and are summarized as follows.

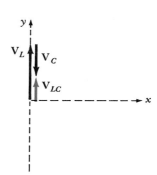

(b)

The rms current through and voltage across an RLC series circuit are related by the equation

$$I = \frac{V}{Z}, \qquad \textbf{(29.13)}$$

where

$$Z = \text{Impedance} = \sqrt{R^2 + (X_L - X_C)^2}. \qquad \textbf{(29.12)}$$

The voltage leads or lags the current by a phase angle ϕ calculated by solving the equation

$$\tan \phi = \frac{X_L - X_C}{R}. \qquad \textbf{(29.11)}$$

The phase angle ϕ is positive if the voltage leads the current and negative if it trails the current. The rms voltages across the individual circuit elements are

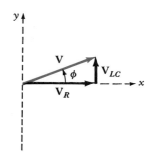

(c)

FIG. 29.13. (a) The rms voltages across a resistor, inductor, and capacitor are represented by vectors. (b) $\mathbf{V}_L$ and $\mathbf{V}_C$ are added to find the voltage across the inductor and capacitor together. (c) $\mathbf{V}_{LC}$ and $\mathbf{V}_R$ are added to find the resultant voltage across the circuit.

$$V_R = IR, \qquad \textbf{(29.5)}$$
$$V_L = IX_L, \qquad \textbf{(29.7)}$$
$$V_C = IX_C. \qquad \textbf{(29.9)}$$

EXAMPLE 29.7 A 0.020-H inductor with 3.0-Ω resistance is connected in series with a 1.2-μF capacitor. A 15-V, 1000-Hz power source is connected across the *RLC* circuit. (a) Calculate the current in the circuit. (b) Calculate the phase angle between the voltage and the current. (c) Calculate the voltage measured across each element. (d) The voltages calculated in part (c) add to more than 15 V. Explain.

SOLUTION (a) The current is given by $I = V/Z$. The circuit's resistance R is 3.0 Ω. To find Z, we must also know the values of X_L and X_C, the circuit's inductive and capacitive reactances. Using Eqs. (29.6) and (29.8), we find that

$$X_L = 2\pi f L = 2\pi(1000 \text{ Hz})(0.020 \text{ H}) = 126 \ \Omega$$

and

$$X_C = \frac{1}{2\pi f C} = \frac{1}{2\pi(1000 \text{ Hz})(1.2 \times 10^{-6} \text{ F})} = 133 \ \Omega.$$

The impedance of the *RLC* series circuit is

$$Z = \sqrt{R^2 + (X_L - X_C)^2} = \sqrt{3^2 + (126 - 133)^2} = \sqrt{9 + 49} = 7.6 \ \Omega.$$

The magnitude of the current through the circuit is

$$I = \frac{V}{Z} = \frac{15 \text{ V}}{7.6 \ \Omega} = \underline{2.0 \text{ A}}.$$

(b) The phase angle between the voltage and the current is calculated using Eq. (29.11):

$$\tan \phi = \frac{X_L - X_C}{R} = \frac{126 - 133}{3} = \frac{-7}{3} = -2.33,$$

or

$$\phi = \underline{-66°}.$$

The voltage lags the current by 66°.

(c) The rms voltages across each element in the circuit are:

$$V_R = IR = (2.0 \text{ A})(3.0 \ \Omega) = \underline{6 \text{ V}},$$
$$V_L = IX_L = (2.0 \text{ A})(126 \ \Omega) = \underline{252 \text{ V}},$$
$$V_C = IX_C = (2.0 \text{ A})(113 \ \Omega) = \underline{266 \text{ V}}.$$

(d) The large voltages across the inductor and capacitor are 180° out of phase with each other and tend to cancel. The net voltage across L and C is their difference: $252 - 266 = -14$ V. This voltage is 90° out of phase with the voltage across the resistor. Together they produce a net voltage

$$V = \sqrt{14^2 + 6^2} = 15 \text{ V}. \quad \blacksquare$$

29.8 Power in AC Circuits

In Section 24.5 we learned that the electric power used by a dc circuit is given by Eq. (24.5), $P = VI$, where V is the voltage across the circuit and I is the current through the circuit. The electric power used in a dc circuit results in the production of thermal energy as current passes through a resistor.

In an ac circuit containing resistors, inductors, and capacitors, the power is used in three ways: (1) Thermal energy is produced in the resistor; (2) a magnetic field builds up and collapses in an inductor; and (3) a capacitor becomes charged and discharged. The last two alternatives result in momentary energy storage or energy release rather than energy loss. For example, as a capacitor is charged, its electrical potential energy increases. During discharge, that energy is returned to the circuit or to the source doing the charging. A similar rise and fall of energy occurs in an inductor (a process not discussed in this text). Averaged over one cycle of alternating-current flow, the inductor and capacitor consume or absorb no net energy.

Since the only energy absorbed by an ac circuit is that due to heating in its resistors, the average power absorbed in an ac circuit is

$$P = V_R I, \tag{24.11}$$

where V_R is the rms voltage across the resistor and I is the rms current through the resistor.

For an RLC series circuit, $V_R = IR$ and $I = V/Z$. Substituting for V_R and I in Eq. (24.11), we find that

$$P = (IR)\left(\frac{V}{Z}\right) = VI\frac{R}{Z},$$

where V is the rms voltage across the whole ac circuit, I is the rms current through the circuit, Z is the circuit's impedance, and R is the resistance of the circuit. We see from Fig. 29.14 that $R/Z = \cos \phi$, where ϕ is the phase angle between the voltage and current. Thus,

$$P = VI \cos \phi, \tag{29.14}$$

where $\cos \phi = R/Z$ is called the **power factor** of the circuit.

The power factor has an important effect on the efficiency of electric power transmission and use. Suppose, for instance, that you have an electric motor that must use electric energy at a rate of 1200 W when connected to the 120-V rms voltage from the wall socket in your home. The wires wound around the armature of the motor cause it to have a self-inductance L and resistance R but little capacitance C. Thus, the capacitive reactance X_C of the motor is zero, but the inductive reactance X_L and the resistance R are not zero. The power factor can be determined by substituting for Z using Eq. (29.12) and setting $X_C = 0$:

$$\cos \phi = \frac{R}{Z} = \frac{R}{\sqrt{R^2 + (X_L - X_C)^2}} = \frac{R}{\sqrt{R^2 + X_L^2}}.$$

Since X_L is not zero, the denominator of the above equation is greater than the numerator, and $\cos \phi$ is less than 1 (for an electric motor, the power factor is typically 0.8). We can now use Eq. (29.14) to determine the current that must flow through the motor (assuming $\cos \phi = 0.8$) to provide it with the desired power:

$$I = \frac{P}{V \cos \phi} = \frac{1200 \text{ W}}{(120 \text{ V})(0.8)} = 12.5 \text{ A}.$$

If the power factor were 1.0, the same power would be provided by a current of only 10 A:

$$I = \frac{1200 \text{ W}}{(120 \text{ V})(1.0)} = 10 \text{ A}.$$

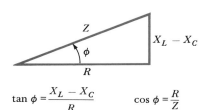

$$\tan \phi = \frac{X_L - X_C}{R} \qquad \cos \phi = \frac{R}{Z}$$

FIG. 29.14. An impedance triangle relating resistance, inductances, impedance, and phase angle.

The electric power company must charge more for electric power because it needs more transmission lines and has greater loss of energy in these lines when delivering extra current to our homes because of the less-than-optimum power factor of our motors and other appliances.

Although fluorescent lightbulbs are much more efficient at producing light than are incandescent lightbulbs, they too have inductive circuits that reduce their power factor. Factories or universities using fluorescent lightbulbs are often charged by the electric power company at a slightly higher rate because of the extra current that must be delivered to provide the needed power for their lights and other appliances that have reduced power factors.

The power factor of a motor or fluorescent lightbulb can be increased to 1.0 by connecting a capacitor in series with the motor or bulb. The same power is provided by less current, as we see in the next example.

EXAMPLE 29.8   An electric motor has an inductance of 0.040 H and a resistance of 20 Ω. (a) Calculate the current through the motor when connected to a 120-V, 60-Hz power source. (b) Calculate the power used by the motor. (c) Calculate the series capacitance that, when added to the motor, increases its power factor to 1.0. (d) What current must now flow through the motor to provide the same power as calculated in part (b)?

SOLUTION   (a) The current depends on the voltage and impedance of the circuit. The impedance depends on the resistance and the inductive and capacitive reactances:

$$R = 20\ \Omega,$$
$$X_L = 2\pi f L = 2\pi(60\ \text{Hz})(0.040\ \text{H}) = 15\ \Omega,$$
$$X_C = 0 \qquad \text{since the circuit has no capacitor.}$$

Thus, the impedance is

$$Z = \sqrt{R^2 + (X_L - X_C)^2} = \sqrt{20^2 + (15 - 0)^2} = 25\ \Omega.$$

The current through the circuit is

$$I = \frac{V}{Z} = \frac{120\ \text{V}}{25\ \Omega} = \underline{4.8\ \text{A}}.$$

(b) The power factor for the circuit is

$$\cos\phi = \frac{R}{Z} = \frac{20\ \Omega}{25\ \Omega} = 0.80.$$

The power used by the circuit is, then,

$$P = IV\cos\phi = (4.8\ \text{A})(120\ \text{V})(0.80) = \underline{460\ \text{W}}.$$

(c) The power factor $\cos\phi = R/Z$ can be increased to 1.0 if a capacitor is added to the circuit whose capacitive reactance X_C equals the inductive reactance X_L of the motor. If $X_C = X_L$, then

$$Z = \sqrt{R^2 + (X_L - X_C)^2} = \sqrt{R^2 + 0} = R,$$

and $\cos\phi = R/Z = R/R = 1.0$. Thus, for this example, we must add a capacitor whose reactance is $X_C = X_L = 15\ \Omega$. The value of the needed capacitance is determined by solving Eq. (29.8) for C:

$$C = \frac{1}{2\pi f X_C} = \frac{1}{2\pi (60 \text{ Hz})(15 \text{ } \Omega)} = 177 \times 10^{-6} \text{ F} = \underline{177 \text{ } \mu\text{F}}.$$

(d) With this capacitance, the motor's impedance is

$$Z = \sqrt{R^2 + (X_L - X_C)^2} = \sqrt{20^2 + (15 - 15)^2} = 20 \text{ } \Omega.$$

The power factor of the motor is

$$\cos \phi = \frac{R}{Z} = \frac{20}{20} = 1.0.$$

If 460 W of power are to be used, the current through the motor capacitor is given by Eq. (29.14):

$$I = \frac{P}{V \cos \phi} = \frac{460 \text{ W}}{(120 \text{ V})(1.0)} = \underline{3.8 \text{ A}}.$$

Instead of delivering 4.8 A, the power company can now deliver 3.8 A and still provide the same power for the motor. ■

29.9 Resonance in *RLC* Series Circuits

The current measured in the *RLC* series circuit shown in Fig. 29.15 is

$$I = \frac{V}{Z},$$

where the circuit's impedance is

$$Z = \sqrt{R^2 + (X_L - X_C)^2}.$$

The circuit's impedance depends on the frequency of the power source, since both X_L and X_C depend on frequency. At one particular frequency, called the **resonant frequency** f_0, the impedance of the circuit is a minimum and the current is a maximum. This current maximum occurs at a frequency when

$$X_L - X_C = 0, \tag{29.15}$$

in which case the impedance is a minimum since

$$Z = \sqrt{R^2 + 0^2} = R.$$

The capacitor and inductor together offer no net opposition to current flow; their effects cancel each other.

To determine an expression for the resonant frequency, we substitute for X_L and X_C into Eq. (29.15) and rearrange:

$$2\pi f_0 L - \frac{1}{2\pi f_0 C} = 0,$$

or

$$f_0 = \frac{1}{2\pi \sqrt{LC}}. \tag{29.16}$$

At this frequency the current in the *RLC* circuit is

$$I = \frac{V}{R} \quad \text{(at resonance)}.$$

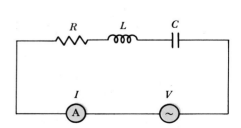

FIG. 29.15. An *RLC* series circuit.

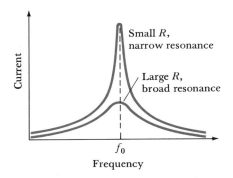

FIG. 29.16. The current is maximum at the resonant frequency of an *RLC* circuit. The breadth of the resonance depends on how large the resistance is.

At frequencies higher than the resonant frequency, $X_L > X_C$ since X_L increases with increasing frequency whereas X_C decreases with increasing frequency. The current

$$I = \frac{V}{Z} = \frac{V}{\sqrt{R^2 + (X_L - X_C)^2}}$$

is less than the current at the resonant frequency since the $(X_L - X_C)^2$ term contributes to the impedance (see Fig. 29.16). The current is also less at frequencies less than f_0. For these frequencies, $X_C > X_L$, and the $(X_L - X_C)^2$ part again adds to the impedance. The impedance is least at resonance, and the current is greatest.

At the resonant frequency, a large ac current oscillates back and forth, first charging the capacitor and then discharging through the inductor. The increasing current in the inductor induces a back emf that sends the charge back toward the capacitor. The charge oscillates back and forth in much the way a person oscillates on a swing.

The resonant curve shown in Fig. 29.16 is broader for circuits with larger resistance (like a swing with lots of friction in its bearings). To construct a circuit with a very narrow resonant curve, the circuit should have large capacitive and inductive reactances and as little resistance as possible. (This circuit oscillates easily, like a swing with frictionless bearings.)

An *RLC* circuit, such as just discussed, is very responsive at its resonant frequency to small input voltages. The tuning circuit on your radio depends on this responsiveness. Usually your radio has a variable capacitor whose capacitance can be changed (Fig. 29.17). You change the capacitor surface area by rotating one set of plates into the region between another stationary set of plates. As the area between plates changes, so does the capacitor's capacitance. To tune the radio for a particular station, we change C so that the resonant frequency of the tuning circuit is at the transmission frequency of that station.

EXAMPLE 29.9   The tuning circuit of a radio consists of a variable capacitor and a 20-mH coil of 8-Ω resistance. (a) Calculate the capacitance needed to tune the circuit to a station broadcasting at 920 kHz. (b) If the radio wave arriving at the antenna produces an ac signal of 2.0×10^{-5} V, how large is the current in the antenna circuit? (c) What is the ac voltage across the coil?

SOLUTION   (a) We solve Eq. (29.16) for the unknown capacitance:

$$C = \frac{1}{(2\pi f_0)^2 L} = \frac{1}{4\pi^2(9.2 \times 10^5 \text{ Hz})^2(20 \times 10^{-3} \text{ H})} = 1.5 \times 10^{-12} \text{ F} = \underline{1.5 \text{ pF}}.$$

With a 1.5-pF capacitor, the circuit will have a resonant frequency of 920 kHz.

(b) At resonance, $X_L - X_C = 0$. Hence, $Z = R$, and the current is

$$I = \frac{V}{R} = \frac{2.0 \times 10^{-5} \text{ V}}{8 \, \Omega} = \underline{0.25 \times 10^{-5} \text{ A}}.$$

(c) The voltage across the inductor is

$$V_L = IX_L = I(2\pi f L) = (0.25 \times 10^{-5} \text{ A})2\pi(9.2 \times 10^5 \text{ Hz})(20 \times 10^{-3} \text{ H})$$
$$= \underline{0.29 \text{ V}}.$$

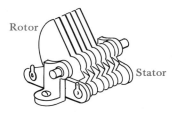

FIG. 29.17.   A variable capacitor.

This fairly large voltage across the inductor provides a relatively large input signal to a radio amplifier. ■

You might wonder how such a small input signal, such as the (2×10^{-5})-V signal discussed in Example 29.9, can produce such a large voltage (0.29 V) across the inductor and across the capacitor. The key to answering this question is the word *resonance*. Let us return to the analogy of a swing. If we continually give the swing a small push at its swinging frequency, it eventually develops large-amplitude oscillations. The same phenomenon occurs at electrical resonance in an *RLC* series circuit. The current oscillates back and forth between the capacitor and the inductor. The frequency of this to-and-fro motion is such that each device gets an optimum current or charge. The external voltage signal gives the vibrating charge a little push each time it returns to the capacitor or inductor for the next swing. Large-amplitude ac current vibrations are caused by many of these small-voltage pushes by the external signal that is being detected.

Summary and Additional Readings

1. **Alternating current:** An alternating voltage source (such as an ac generator) first pushes charge in one direction and then pulls it back in the opposite direction, thus producing an alternating current (ac). The alternating currents, voltages, and power are characterized in terms of their instantaneous values (i, v, and p) that change with time, their peak values (i_0, v_0, and p_0), and their rms or effective values (I, V, and P). The rms values of current and voltage are related to the peak values by the equations $I = 0.707 i_0$ and $V = 0.707 v_0$.

2. **Alternating current through resistors, induction coils, and capacitors:** The rms alternating current I through a resistor, induction coil, or capacitor when an rms voltage V is placed across the device is given by the following equations:

$$\text{Resistor:} \quad I = \frac{V_R}{R}, \tag{29.5}$$

where R is the resistance of the resistor. The instantaneous voltage is in phase with the current.

$$\text{Coil:} \quad I = \frac{V_L}{X_L}, \tag{29.7}$$

where X_L, the **inductive reactance** of the coil, is

$$X_L = 2\pi f L. \tag{29.6}$$

The instantaneous voltage leads the current by 90°.

$$\text{Capacitor:} \quad I = \frac{V_C}{X_C}, \tag{29.9}$$

where X_C, the **capacitive reactance** of the capacitor, is

$$X_C = \frac{1}{2\pi f C}. \tag{29.8}$$

The instantaneous voltage lags the current by 90°.

3. **RLC series circuit:** The rms alternating current I through a series circuit containing a resistor of resistance R, a coil of inductive reactance X_L, and a capacitor of capacitive reactance X_C across which an rms voltage V exists is given by the equation

$$I = \frac{V}{Z}, \tag{29.13}$$

where Z, the **impedance** of the circuit, is

$$Z = \sqrt{R^2 + (X_L - X_C)^2}. \tag{29.12}$$

The voltage leads or lags the current by an angle ϕ that is determined by solving the equation

$$\tan \phi = \frac{X_L - X_C}{R}. \tag{29.11}$$

The voltage leads the current if ϕ is positive and lags the current if ϕ is negative.

4. **Power:** The average power absorbed by an ac circuit is

$$P = VI \cos \phi, \tag{29.14}$$

where the **power factor**, $\cos \phi$, is equal to R/Z.

5. **RLC series resonance:** An *RLC* series circuit has a **resonant frequency** f_0 at which its impedance is a minimum and the alternating current in the circuit is a maximum. Resonance occurs when $X_L = X_C = 0$ and at a frequency

$$f_0 = \frac{1}{2\pi} \sqrt{\frac{1}{LC}}. \tag{29.16}$$

S. A. Hoenig and F. L. Payne, *How to Build and Use Electronic Devices Without Frustration, Panic, and Mountains of Money, or an Engineering Degree*, Little, Brown and Company, Boston (1973).

Scientific American, September (1977). The entire issue deals with microelectronics.

Questions

1. A light has a filament made of a very thin wire that is wrapped in loops. To dim the light, an iron core is inserted into the coil. Why does this dim the light?

2. Explain why a coil of wire placed in series in a circuit allows low-frequency current but not high-frequency current to pass.

3. Explain why a capacitor placed in series in a circuit allows high-frequency current but not low-frequency current to pass.

4. A parallel-plate capacitor is connected in series to a light. An ac voltage source is connected across the two elements. Will the light be bright or dimmer when a material with dielectric constant greater than 1.0 is placed between the capacitor plates? Explain your answer.

5. Under what conditions is the power factor for an *RLC* series circuit 1.0? Draw three simple circuits that have a power factor of zero.

6. A light, a resistor, an inductor, and a capacitor are all connected in series to an ac power source. If an extra inductor is added to the circuit, the light dims, whereas if an extra capacitor is added, the light brightens. Explain why this occurs.

7. An ac generator is connected across an *RLC* series circuit. The frequency of the generator is less than the resonant frequency of the circuit. Does the voltage lead or lag the current? Explain your answer.

8. An *RLC* series circuit has a variable capacitor. If the circuit's resonant frequency is higher than that of a radio station, should the capacitance of the capacitor be increased or decreased to tune the circuit to the radio's frequency? Explain.

Problems

All values of voltage and current given in these problems are rms values unless stated otherwise.

29.1 AC Generator

1. A citizen's band radio transmitter operates at a frequency of 27.1 MHz. The peak voltage across the transmitter is 10 V. Write an expression for the variation of voltage with time. What is the time for one cycle?

■ 2. While exploring for ore, a geophysicist uses an electrical generator that produces a sinusoidal voltage with a peak value of 150 V at a frequency of 20 Hz. (a) Write an expression for the variation of voltage with time. (b) Determine two different times at which the voltage is $+150$ V, two different times at which it is zero, and two different times at which it is -150 V.

■ 3. The voltage across the terminals of an ac generator varies with time according to the equation

$$v = 30 \sin (2\pi 1000t) \text{ V}.$$

(a) What are the peak voltage and frequency of the generator? (b) Calculate the value of the voltage at each of the following times: 0 s, 0.25×10^{-3} s, 0.50×10^{-3} s, 0.75×10^{-3} s, and 1.0×10^{-3} s.

29.2 Alternating Current Through Resistors

4. A 60-Hz ac generator develops a peak voltage of 170 V. The generator is connected to a lightbulb whose resistance when operating is 240 Ω. Calculate an expression for the variation of current through the bulb with time.

■ 5. The generator discussed in Problem 2 is connected to two electrodes placed in the ground 1000 m from each other. The ac current flowing through the ground between the electrodes has a peak value of 250 mA. (a) Calculate the resistance of the ground. (b) Write an expression for the current as a function of time. (c) Calculate the instantaneous current at a time of 0.040 s.

■ 6. A 50-Hz generator with a peak voltage of 310 V (such as used in Europe) is connected to a toaster whose resistance when operating is 12 Ω. (a) Calculate an expression for the current through the toaster as a function of time. (b) Calculate the instantaneous current at the following times: 0 s, 0.005 s, 0.010 s, 0.015 s, and 0.020 s.

29.3 RMS Voltage and Current

7. If you placed an ac voltmeter across the outlets of a wall socket in Europe, you would measure voltage of about 220 V. (a) Calculate the peak voltage. (b) Calculate the rms current and peak current through a toaster whose resistance when operating is 15 Ω. (c) Calculate the power used by the toaster.

8. (a) A number 12 copper wire used for home wiring can safely carry a current of 25 A. Calculate the peak value of the current. (b) Why is the wire rated in terms of rms current rather than peak current?

9. Suppose you accidentally touched wires connected to the 120-V output from a wall socket. (a) Calculate the rms current through your body if the resistance between electrodes is 10,000 Ω. (b) Calculate the peak current. (c) Will any health hazards be caused by this current? (See Fig. 24.2.) [*Do not try this experiment!*]

■ 10. The rms current through a lightbulb when connected to a sinusoidal generator of peak voltage 80 V and frequency 50 Hz is 0.24 A. Calculate (a) the rms voltage, (b) the resistance of the bulb, and (c) the peak current. (d) Write an expression for the variation of the instantaneous current with time.

■ 11. The alternating current through the 15-Ω heating element of a toaster produces power at 980 W. Calculate (a) the rms current through the element, (b) the peak current, (c) the rms voltage across the element, and (d) the peak voltage.

■■ 12. You wish to produce an electric heater by running current through a large resistor. The resistor is connected to a 120-V, 60-Hz power source. (a) Calculate the desired resistance to produce 1000 W of heat. (b) Calculate the rms and peak current through the resistor. (c) If nichrome wire of radius 0.40 mm and resistivity 1.0×10^{-6} Ω · m is to be used, how long should the wire be?

■■ 13. The voltage across your shoulders caused by the changing dipole charge on your heart varies between about 2 mV and -2 mV once each heartbeat. The resistance in the tissue between the shoulders is about 300 Ω. Calculate the peak current. *Estimate* the distance that an ion in the tissue moves in one direction before moving in the opposite direction. The number of ions per unit volume in the tissue is approximately 10^{25} to 10^{26} molecules/m^3. Justify other numbers used in your estimate. [*Hint:* You might consider the typical drift velocity of charges, discussed in Chapter 24.]

29.4 Apparent Resistance of a Coil

14. Inductors are connected, one at a time, to the terminals of a 60-Hz, 12-V power source. Calculate the current through inductors of inductances (a) 0.010 H, (b) 0.10 H, (c) 1.0 H, and (d) 10 H.

15. A 0.35-H inductor is connected across a 12-V ac power source whose frequency can be varied. Calculate the current through the inductor when the frequency is (a) 10 Hz, (b) 100 Hz, (c) 1000 Hz, and (d) 100,000 Hz.

16. The primary coil of a transformer for a toy car racetrack has an inductance of 3.5 H. Calculate the current that flows when the primary coil is connected to a 120-V, 60-Hz power source.

■ 17. The rms current through a coil when connected to an rms voltage of 20 V at 50 Hz is 0.80 A. (a) Calculate the inductance of the coil. (b) Calculate the rms current when the same coil is connected across an rms voltage of 20 V at 200 Hz.

■ 18. A coil has inductive reactance of 6000 Ω when placed in an ac circuit with a 27.1-MHz citizen's band radio signal. What is its reactance to a 60-Hz signal?

■■ 19. A 230-V, 60-Hz voltage is connected across the primary coil of a transformer for an electric welder. The ac voltage across the secondary coil is 30 V, and a current of 250 A flows. Assume that the current is limited by the reactance of the secondary coil. (a) Calculate the inductance of the secondary coil. (b) Calculate the primary to secondary turns ratio.

29.5 Apparent Resistance of a Capacitor

20. A 1000-pF capacitor is connected across a 12-V ac power source whose frequency can be varied. Calculate the capacitor current when the frequency is (a) 10 Hz, (b) 10 kHz, and (c) 10 MHz.

21. Capacitors are connected, one at a time, to the terminals of a 60-Hz, 12-V power source. Calculate the current through capacitors of capacitance (a) 10 pF, (b) 1000 pF, (c) 0.1 μF, and (d) 10 μF.

22. Static-type noise can occur on automobile radios because of ignition sparking in the motor. The noise can be reduced by connecting a 0.50-μF capacitor from the primary coil of the distributor to ground. Some of the high-frequency electrical noise is effectively shorted to ground. Calculate the reactance of the capacitor to 20-kHz noise.

■ 23. A 1-μF and a 2-μF capacitor are connected in series across a 0.50-V, 10-kHz voltage. (a) Calculate the current in the circuit. (b) Repeat the problem for capacitors connected in parallel.

■ 24. (a) The current through a 2.0-μF capacitor when connected to a 20-V generator is 0.16 A. Determine the frequency of the generator. (b) What is the current through a 0.20-H inductor when connected by itself across the same generator?

29.7 RLC Series Circuit

■ 25. A 200-Ω resistor, 1.6-H inductor, and 3.0-μF capacitor are connected in series across a 60-Hz, 120-V power supply. Calculate (a) the inductive reactance of the circuit, (b) its capacitive reactance, (c) its impedance, (d) the current through the circuit, and (e) the phase angle.

■ 26. A 250-Ω resistor, 1.2-H inductor, and 1.8-μF capacitor are connected in series across a 60-Hz, 120-V power supply. Calculate (a) the circuit's impedance, (b) the current, (c) the phase angle of the voltage relative to the current, and (d)–(f) the measured voltages across the resistor, inductor, and capacitor.

■ 27. An ac circuit with a resistor and an inductor has an impedance of 100 Ω at 50 Hz and an impedance of 600 Ω at 500 Hz. Calculate the values of R and L. What is the phase angle between the voltage across and the current through the 50-Hz circuit?

■ 28. An ac circuit with a 1200-Ω resistor and a capacitor of unknown capacitance has an impedance of 1600 Ω at 100 Hz. (a) Calculate the capacitor's capacitance and (b) the circuit's impedance at 500 Hz.

■ 29. A solenoid has an inductance L due to its coiled wire and a resistance R due to the resistivity of the wire. When connected to a 12-V battery, the solenoid draws 4.0 A of current. When connected to a 12-V, 100-Hz power source, the current is 1.8 A. Calculate the values of R and L.

■ 30. A feedthrough capacitor allows alternating current to pass through a circuit but blocks direct current. A 3000-pF capacitor in series with a 5000-Ω resistor is connected to a 10-mV, 9.2-MHz voltage at the input of an amplifier. (a) Calculate the circuit's impedance to the 9.2-MHz input signal and its impedance to direct current. (b) What fraction of the 9.2-MHz impedance is caused by the capacitor?

■ 31. Your neighbor has a citizen's band radio that broadcasts at 27.1 MHz. His communications are being picked up on your television antenna. You can effectively draw his signal to ground by connecting a series capacitor and inductor, which resonate at 27.1 MHz, to ground (see Fig. 29.18). (a) Calculate the impedance of the circuit at 27.1 MHz. (b) Calculate the impedance at 60 MHz, the frequency of the nearest television channel. (c) Explain why the circuit keeps his conversations out of your television but still allows you to have good television reception.

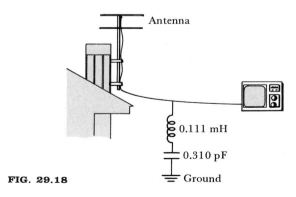

FIG. 29.18

■■ 32. A device for measuring small mechanical movements consists of a 5.0-Ω resistor, 0.20-μF capacitor, and a 0.24-H inductor connected in series across a 10-V ac voltage source whose frequency matches the resonant frequency of the RLC series circuit (726.4 Hz). (a) Calculate the current through the circuit. (b) Calculate the current at the same frequency if the capacitance increases 1.0 percent because a dielectric material is forced between the capacitor plates by a small movement.

■ ■ 33. The high-pass filter circuit shown in Fig. 29.19 can be used to "pass" signals of high frequency while reducing signals of low frequency. The circuit could for example be used to reduce 60-Hz "noise" that enters the amplifier of a radio from its antenna. (a) Calculate a general expression for the alternating current I in terms of V_1, f, C, and R. (b) The output voltage V_2 across the resistor connects to an amplifier circuit in the radio. Show that

$$\frac{V_2}{V_1} = \frac{2\pi f RC}{\sqrt{1 + (2\pi f RC)^2}}.$$

(c) Show that for $2\pi f RC \gg 1$, $V_2/V_1 = 1$ (high frequencies are not attenuated). (d) Show that for $2\pi f RC \ll 1$, $V_2/V_1 \ll 1$ (at low frequencies, the output voltage is greatly attenuated).

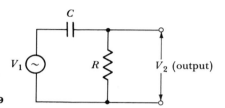

FIG. 29.19

■ ■ 34. A low-pass filter is shown in Fig. 29.20 (see discussion for Problem 33). (a) Calculate the current caused by an input voltage V_1. (b) Show that the ratio of the output and input voltages is

$$\frac{V_2}{V_1} = \frac{1}{\sqrt{1 + (2\pi f RC)^2}}.$$

(c) Calculate the limit of this ratio for very high frequencies and for very low frequencies ($f = 0$).

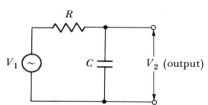

FIG. 29.20

29.8 Power in AC Circuits

35. A 6.0-μF capacitor, a 0.12-H inductor, and an 85-Ω resistor are connected in series across a 10-V, 300-Hz voltage source. Calculate (a) the current in the circuit, (b) the phase angle between the voltage and current, and (c) the power delivered to the circuit.

■ 36. A 240-Ω resistor is connected across a 120-V, 60-Hz voltage source. (a) Calculate the current through the resistor and the electric power used by the resistor. (b) A 0.053-H inductor and 13-μF capacitor are connected in series with the resistor. Calculate the power used by this series circuit when connected to the same power source.

■ 37. A large motor has an armature whose inductance is 0.12 H. The motor has no capacitance. The series resistance R of

the armature's wire is 25 Ω. (a) Calculate the power used by the motor when connected to a 120-V, 60-Hz power source. (b) Calculate the capacitance of a series capacitor that maximizes the power used by the motor. (c) Calculate the power used when this capacitance is added. [*Warning:* Note that both I and ϕ change when the capacitor is added to the circuit.]

■ 38. Your electric lawn mower does not have enough power to cut long grass. To draw more current through the motor and produce more power when it is connected to a 120-V, 60-Hz voltage source, a 540-μF capacitor is connected in series to the mower. The motor's armature has a resistance R of 12 Ω and an inductance of 0.016 H. Calculate the ratio of the power after the capacitor is connected to that before it is connected.

■ 39. A motor with inductance L and resistance R uses 5.0 W of power when connected to a 10-V battery and 400 W of power when connected to a 120-V, 60-Hz power source. Determine the values of R and L.

29.9 Resonance in RLC Series Circuits

40. Television Channel 2 broadcasts at a frequency of about 65 MHz. Calculate the inductance of an inductor in series with a 0.024-pF capacitor such that the circuit resonates at this frequency.

■ 41. A citizen's band radio operator has a receiver circuit with a 0.23-mH inductor and a 0.15-pF capacitor. (a) Calculate the resonant frequency of the receiver circuit. (b) If the circuit also has a resistance R of 9.0 Ω, calculate the current through the receiver when a 0.10-mV input voltage is placed across the circuit at its resonant frequency. (c) Calculate the voltage across the inductor.

■ 42. A radio station broadcasts at a frequency of 9.2 MHz. (a) Calculate the capacitance of a capacitor placed in series with a 0.71-mH inductor such that the circuit resonates at this frequency. (b) What is the wavelength of the radio waves produced by the station? The waves travel at 3.0×10^8 m/s.

■ 43. A 2.0-Ω resistor, 1.9-μF capacitor, and 0.12-H inductor are connected in series. (a) Calculate the resonant frequency of the circuit. (b) The inductor has a metal core that can move in and out of the inductor's coil. Calculate the ratio of the circuit's original impedance at its resonant frequency and its impedance when the inductance is increased by 10 percent because the core has moved. The device is a sensitive detector of movement and vibration.

■ ■ 44. A pressure-measuring device consists of a 0.15-μF capacitor, a 2.0-Ω resistor, and a 0.17-H inductor connected in series across a 6.0-V ac voltage source whose frequency matches the resonant frequency of the RLC series circuit. (a) Calculate the current through the circuit. (b) A pressure increase causes the capacitor plates to be pushed closer together. Calculate the current at the same frequency as used in part (a) if the capacitance increases by 5 percent.

■ ■ 45. An airport metal detector consists of a large coil of wire through which you walk. The inductance of the coil is 83 mH and its resistance due to the resistivity of the wire is 0.50 Ω. (a) Calculate the capacitance of a capacitor that when placed in series with the coil results in a resonant frequency of 200 Hz. (b) Calculate the current that flows when the circuit is connected to a 1.0-V power source oscillating at the resonant frequency of the circuit. (c) A person carries a small piece of metal through the coil so that its inductance increases by 3 percent. Calculate the current in the coil. (d) Calculate the ratio of the current after the inductance change and before the change.

INTERLUDE VI   Impedance Matching

Impedance is a useful concept whenever power transfers from an object that provides the power (the source) to another object that receives the power (the load). The source and load might be the amplifier and loudspeakers of a high-fidelity system, the electrodes and recorder of an electrocardiogram apparatus, or the vibrating air caused by a sound wave and our inner ear where nerves are located to detect sound.

The source and load can each be characterized in terms of their **impedance,** a measure of the resistance they offer to the flow of power. In general, the maximum power transfer from a source to a load occurs when they have the same impedance; their impedances are then said to be **matched.**

As a simple example of power transfer and the effect of impedance on the transfer, imagine a battery that produces an emf ε. All sources of power, including batteries, have an internal resistance that resists the flow of power through the source. In Fig. VI.1, the internal resistance of the battery (its impedance) is represented by R_S, the source resistance. When the battery is connected to a load resistor of resistance R_L, a current flows through the circuit.

The electric power used by the load is given by Eq. (24.11): $P = I^2 R_L$. For the circuit shown in Fig. VI.1, the current is

$$I = \frac{\varepsilon}{R_S + R_L}.$$

Thus, the power used by the load resistor is

$$P = I^2 R_L = \frac{\varepsilon^2 R_L}{(R_S + R_L)^2}$$

and is plotted for different values of R_L in Fig. VI.2. Instead of specifying particular values of R_L, we have shown the value of P for different values of R_L relative to the value of R_S. For example, the value 2 on the abscissa corresponds to $R_L = 2R_S$, the value 3 corresponds to $R_L = 3R_S$, and so forth. We see that the maximum power transfer occurs at 1 when $R_L = R_S$, that is, when the load has the same resistance as the source. The impedances are then matched. For values of R_L not equal to R_S, less power is transferred to the load; the impedances are then **mismatched** (not equal).

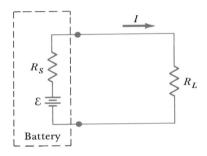

FIG. VI.1. A battery with internal resistance R_S is connected to a load resistor of resistance R_L.

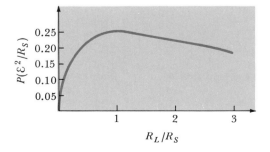

FIG. VI.2. The variation of power to the load resistor is optimum when $R_L/R_S = 1$.

We considered another example of impedance mismatch in Chapter 17 (see Figs. 17.8 and 17.9) in terms of pulses traveling on ropes or strings. If two ropes of different mass per unit length are connected and a pulse is initiated in one rope, the pulse is partially reflected when it reaches the boundary to the second rope. However, if the ropes have the same mass per unit length, their impedances are then matched, and 100 percent of the incident pulse is transmitted to the second rope.

In Section 29.7 we considered another example of impedance and found that a series RLC circuit has an impedance given by Eq. (29.12):

$$Z = \sqrt{R^2 + (X_L - X_C)^2},$$

where R is the resistance of the circuit and X_L and X_C are its inductive and capacitive reactances, respectively. All electrical circuits have impedances given by complicated equations such as Eq. (29.12). The values of R, X_L, and X_C in these equations depend on the arrangement and values of the resistances, inductances, and capacitances in the circuit and on the frequency at which the impedance is calculated. To have maximum power transfer from an electrical source to a load, their resistances must be equal and the reactance parts of their impedance ($X_L - X_C$) must have opposite signs:

$$R_{\text{source}} = R_{\text{load}},$$
$$(X_L - X_C)_{\text{source}} = -(X_L - X_C)_{\text{load}}.$$

Similar expressions for impedance can be developed for mechanical systems that include masses, springs, and friction. In these systems, mass is similar to an electrical inductor, a spring is analogous to an electrical capacitor, and the viscous friction force is similar to an electrical resistor. The analogue between these mechanical and electrical quantities is illustrated in Table VI.1.

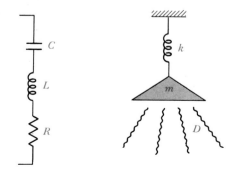

TABLE VI.1 Analogous Quantities and Equations for Electrical and Mechanical Impedance

	Electrical		Mechanical	
	Voltage v		Force F	
	Current i		Velocity v	
Resistance	$v = Ri$	Drag force	$F = Dv$	
Inductance	$v = L(\Delta i/\Delta t)$	Inertia	$F = m(\Delta v/\Delta t)$	
Capacitance	$v = \frac{1}{C}q$, where $I = \Delta q/\Delta t$	Elastic force	$F = kx$, where $v = \Delta x/\Delta t$	
	$V = ZI,$		$\bar{F} = Z\bar{v},$	
	where $Z = \sqrt{R^2 + (X_L - X_C)^2}$		where $Z = \sqrt{D^2 + (X_m - X_k)^2}$	

The human ear is a mechanical system whose impedance can be characterized in terms of the mechanical analogue of Eq. (29.12). The ear has a number of different masses, elastic elements, and sources of viscous friction (see Fig. VI.3). Various hearing disorders of the middle ear can be diagnosed by measuring the ear's acoustic impedance—that is, the impedance of the ear to the absorption of the energy of sound. For example, if the middle ear has a massive adhesion, its inductive-type reactance is greatly increased.

Transformers

Often we wish to transfer power from a source to a load that has a different impedance. If the two are connected directly, an impedance mismatch occurs, and less than optimum power is transferred from the source to the load. This mismatch can be partially corrected using a device called a **transformer,** a two-sided device that has different impedances on each side and is built so that most power entering one side is transferred to the other side. The transformer, when connected between the source and load, matches the source impedance on one side and the load impedance on the other. Power is transferred efficiently from the source to the transformer and from the transformer to the load. Since the power flows efficiently through the transformer, the flow of power from

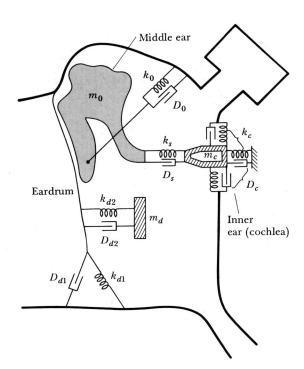

FIG. VI.3. The human ear in terms of masses, springs, and friction devices called dashpots (—⊏—). [Adapted from Marquet et al., *Acta Oto-Rhino-Laryngologica Belgica* **27**, 137 (1973). Reproduced with permission.]

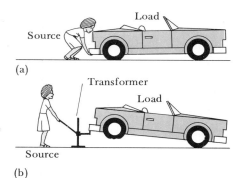

Source — Load

Transformer

Load

Source

(b)

FIG. VI.4. A lever is a transformer that matches the impedance of a person to that of the car.

source to load is greatly improved. The outer and middle ear serve as a transformer that improves the efficiency of transfer of sound from air to the fluid in the inner ear. If sound passed directly from air to the inner ear, the impedance mismatch at the interface would cause most of the sound to be reflected. Instead, the outer ear and eardrum have a relatively low impedance that matches the impedance of air, whereas the bones of the middle ear have a higher impedance that matches the impedance of the fluid in the inner ear. This transformer produces approximately a 30-fold improvement in the transfer of sound to the inner ear.

A lever is another form of transformer that provides efficient transfer of energy from a source to a load. A person trying to lift a car so that a flat tire can be fixed is depicted in Fig. VI.4. Although expending much energy, the person in Fig. VI.4a is unable to lift the car. Using a lever as in part (b), the person lifts the car easily. The lever is a transformer that provides an efficient transfer of energy from the person to the car.

As we have seen, the concept of impedance and impedance matching can be applied to a variety of studies. Whenever power or information or both are to be transferred efficiently between different objects or between different pieces of equipment, efforts must be made to match the impedances of the objects to each other, either by choosing the equipment correctly or by placing impedance-matching transformers between the source and the load.

Questions and Problems

1. Indicate the source, load, and type of energy or power that is transmitted through each of the following transformers: (a) a flyswatter, (b) a megaphone, (c) stair steps.

2. Describe three everyday objects that act as transformers (do not use the transformers mentioned elsewhere in this interlude).

■ 3. A 10-V source of emf has an internal resistance of 50 Ω. (a) Calculate the power dissipated to a load resistor with the following values of resistances: 10 Ω, 30 Ω, 50 Ω, 100 Ω, and 500 Ω. (b) Plot a graph of power used by the load resistor for these different values of resistance and draw a smooth curve through the points.

■ 4. An ac source of electric power operates at a frequency of 300 Hz and has a resistance R of 100 Ω, an inductance L of 0.40 H, and a capacitance C of 10 μF. The source is connected to a load whose inductance is 0.10 H. Calculate the values of R and C for the load so that its impedance matches that of the source.

■ ■ 5. The variation of the resistive part R and of the reactive part $(X_L - X_C)$ of the impedance of a normal ear is shown in Fig. VI.5. (a) Based on the analogy between electrical and mechanical impedance and your knowledge of the frequency dependence of

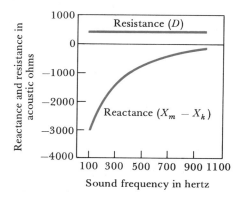

FIG. VI.5. The variation of a normal ear's resistance and reactance with frequency.

X_L and X_C, explain why the ear's elastic constant dominates its impedance at low frequency. (b) Why does the magnitude of the reactive impedance decrease as the sound frequency increases?

■ 6. A long electric power transmission line has a resistance of 10 Ω (Fig. VI.6). Calculate the power consumed by the line and by a 40-Ω load resistance when a 120-V dc voltage is placed across the ends of the line.

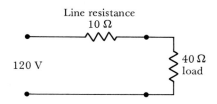

FIG. VI.6

PART VII

Modern Physics

In 1875 a bright 17-year-old student named Max Planck (1858–1947) entered the University of Munich. He was told he should not pursue a career in physics because "all the important discoveries in physics have been made." The laws of mechanics, thermodynamics, electricity, and magnetism accounted for the observed behavior of the universe. Despite this warning, Planck chose physics and eventually became a physics professor. In the last week of 1899 at a meeting of the German Physical Society, Planck presented a paper having to do with the nature of light that ushered in the most exciting thirty-year period in the development of our understanding of the physical universe. It was, according to Albert Einstein, "a marvelous time to be alive."

The experiments and theories in the first thirty years of this century brought us into realms seldom experienced in daily living. Einstein's theory of relativity considered the world of objects moving near the speed of light. Another new principle, called quantum mechanics, offered an extraordinary view of the behavior of electrons and atoms. Revolutionary as these two theories were, they are now well established among the principles of modern physics.

At present, an explanation is being sought for the behavior of atomic nuclei and the particles of which they are made. New words and ideas have emerged—gluon, strangeness, charm, lepton, quark, neutrino. This search is producing an exciting new chapter of progress in physics.

CHAPTER 30

Special Relativity

Albert Einstein (1879–1955) became internationally famous in the fall of 1919. A group of British scientists, after studying a solar eclipse, announced that their study confirmed Einstein's prediction that light passing the sun is bent by the sun's gravitational field. Newspapers around the world carried headlines proclaiming Einstein's new theory of relativity and decrying its complexity and incomprehensibility. The *New York Times* on November 25, 1919, reported that "so complicated has this revolutionary theory proved that even some of the most learned have been confounded."

General relativity, Einstein's 1915 theory that predicted this bending of light, does require considerable mathematical sophistication if used quantitatively. The theory provides a new way of looking at gravitation and has been applied recently to the study of black holes (depicted in Fig. 30.1), extremely dense celestial bodies whose strong gravitational pull allows nothing—not even light—to escape.

Einstein also proposed in 1905 a special theory of relativity, the main subject of this chapter. This less complicated theory provides a description of nature as seen and measured by different observers who move at constant velocity relative to each other. Perhaps the greatest problem encountered in studying special relativity is our willingness to accept some of its predictions—predictions that conflict with our intuition. For example, the hands of a clock will appear to turn more slowly if the clock is moving past an observer than if the clock is stationary in front of the observer. An arrow's measured length is shorter when the arrow moves at high speed than when at rest. An object's mass increases as its speed increases. These paradoxical predictions have been proven correct by a variety of experiments. In most of these experiments, however, the objects under investigation are moving very fast—near the speed of light (3.0×10^8 m/s). At the normal speeds we encounter in our daily lives, these phenomena—variations in time, length, and mass—are so small that we find it hard to observe them. But they do occur.

There is one important relativistic phenomenon whose effects are apparent in our lives each day: Mass is a form of energy and can be converted to other forms of energy, just as other forms of energy can be converted to mass. The sun's radiative energy is provided by the conversion of mass to energy—about 4 million tons each second. Nuclear power plants provide electricity by converting mass to energy.

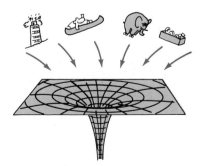

FIG. 30.1. The curved lines of space near a black hole cause objects passing near the black hole to be pulled in by its huge gravitational force. Even light cannot escape.

Our discussion in this chapter will center on two main topics: (1) a determination of the way in which an object's length and a clock's ticks vary as a result of their motion relative to an observer and (2) a reformulation of some of our ideas concerning mass and energy.

30.1 Events and Inertial Reference Frames

Relativity is concerned with differences in the observations made by two observers who examine the same set of events. An **event** is anything that happens—a person's departure and return from a trip or two flashes of light, one at the nose of a rocket and the other at its tail. We will find that if one observer is moving relative to the second observer, the time between the events and their separation in space seem different for each observer. One observer measures a different time for a trip in space than the other observer and measures a different distance for the length of the spaceship.

Before discussing these differences in time intervals and in distances as measured by two observers, we must define carefully the reference frames in which the events are measured. A **reference frame** consists of a clock and a set of coordinate axes. Each observer has a clock that sits at rest relative to the observer and is used to measure the time at which events occur. Each observer also uses a set of coordinate axes that are at rest relative to the observer. These are used to record the location of events and to indicate the size or length of objects. Since one observer moves relative to the other observer, their coordinate systems move relative to each other.

The descriptions of nature by these observers may differ. Suppose that one observer stands beside a train track. Another observer sits in a car of a train that moves at 10 m/s relative to the track. Each observer has a set of coordinate axes at rest relative to himself or herself—one set on the train with its observer and another set beside the track with its observer. If a third person walks down the aisle of the train, the observer on the train may find that the person's speed relative to his or her reference frame is 1 m/s (see Fig. 30.2). However, relative to the observer outside the train, the person's speed is $10 + 1 = 11$ m/s. These velocity measurements differ because of the relative motion of the two observers. (We will find later that the measurements also differ because information about a measurement can be transmitted to an observer no faster than the speed of light.)

Relativity is divided into two broad categories called *special relativity* and *general relativity*. In general relativity no restrictions are placed on the reference frames, whereas in special relativity we can use only reference frames that are not accelerating. In special relativity, for example, we could not use a point on the aisle of an *accelerating* train as the origin of a set of coordinates to describe the location of events occurring either inside or outside the train.

A reference frame that is not accelerating is called an **inertial reference frame.** If a body is placed at rest in an inertial reference frame and if no forces act on it, then it remains at rest. For example, if a piece of dry ice is placed at rest on a horizontal, smooth surface in a train that moves at constant velocity, the dry ice remains at rest, but if the train accelerates, the dry ice slides backward. When accelerating, the train is no longer an inertial reference frame; only when moving at constant velocity is the train an inertial reference frame.

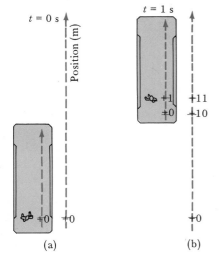

FIG. 30.2. According to an observer on the train, a person walking down the aisle moves 1 m in 1 s, or at a speed of 1 m/s. According to an observer beside the track, the person walking down the aisle moves 11 m in 1 s, or at a speed of 11 m/s.

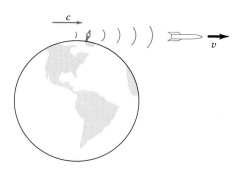

30.2 Postulates of Special Relativity

Einstein formulated the special theory of relativity using only two postulates:

1. The basic laws of nature are the same in all inertial reference frames; that is, basic laws such as Newton's second law ($\Sigma \mathbf{F} = m\mathbf{a}$) should have the same mathematical form in all inertial reference frames.

2. The speed of light in a vacuum is always measured to be 3.0×10^8 m/s and does not depend on whether the light source or the observer that measures the light speed are moving.

While the first postulate seems reasonable, the second seems less so. For example, suppose that an intense pulse of light is shot at a fleeing spaceship from a laser resting on the earth. A stationary observer on the earth measures the light's speed c as 3.0×10^8 m/s (Fig. 30.3a).

Now let us imagine the speed of the light as it passes the spaceship. Suppose that the spaceship moves away from the earth at a speed v that is just a little less than the speed of light. We might expect that the light would have a difficult time reaching the fleeing ship since, according to the earth observer, the light's speed is only $(c - v)$ greater than that of the spaceship. However, an observer on the spaceship measures the light's speed to be $c = 3.0 \times 10^8$ m/s as the light passes the ship (Fig. 30.3b). If a second spaceship approaches the earth at a speed v that is nearly the speed of light, an observer in it also measures the light's speed relative to the spaceship as $c = 3.0 \times 10^8$ m/s, not as $c + v$.

The fact that the speed of light is the same in all reference frames contradicts our intuition. In the last century, some scientists proposed that an invisible substance called ether* filled space and vibrated as light passed through it. (Ether served the same function for a light wave as air does for a sound wave.) Scientists thought that light traveled at speed c through the ether and if an observer moved relative to the ether, light's speed would appear slower or faster than c, depending on whether the observer moved with or against the light.

A famous experiment devised by American physicists A. A. Michelson and E. W. Morley in 1887 showed that the measured speed of light is the same in all reference frames and that it does not depend on the motion of the observer. The experiment resulted in the demise of the ether theory and is the experimental basis for the second postulate of relativity.

Some remarkable consequences result from these two postulates. One of these consequences is called *time dilation*.

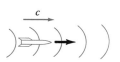

Spaceship observer

(b)

FIG. 30.3. (a) Light leaves the earth in pursuit of a fleeing spaceship. The light's speed relative to an earth observer is c. (b) To an observer on the spaceship moving at speed v, the light's speed is also c and not $c - v$.

30.3 Time Dilation

We usually think of the time between two events as being independent of the motion of an observer. The time between two flashes of light on a passing fire truck appears the same to the driver as to a person standing on the sidewalk nearby. If the truck or a spaceship could move at almost the speed of light, however, the flashes of light from the truck or spaceship would appear less

---

*The ether that supposedly filled space should not be confused with the substance used as an anesthetic.

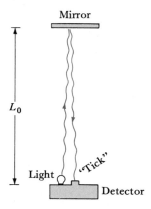

FIG. 30.4. One "tick" of the clock is the time needed for light to travel up to a mirror and back down—a total distance of $2L_0$.

frequent to a person standing on the earth. We say that there is a **time dilation**—a lengthening of the time between events that occur on an object moving relative to the observer. A moving clock runs more slowly than an identical clock sitting at rest in front of the observer. If a clock that has been moving is brought back home, it will have registered fewer ticks than its mate that stayed home. The same is true for the heartbeat and aging of a traveling twin compared to the sibling who remains home! The traveler has fewer beats and ages less.

We now derive an equation that compares the time between two events as measured by two different observers. For making this comparison, we analyze the special situation in which one observer is in the **proper reference frame**—the frame where the place of the events is at rest relative to the observer. An observer on a spaceship in which a light flashes is in the proper reference frame because the light causing the flashes is at rest relative to the observer. The time between events in the proper reference frame is called the **proper time.** Relative to a second observer, such as a person standing on the earth, the events (flashes of light on the spaceship) occur at different places. Thus, the earth is not the proper reference frame for this example, and the time between flashes is delayed for this observer.

To derive an equation relating the time between events as measured in the proper reference frame and in another reference frame, consider the special "clock" on the spaceship shown in Fig. 30.4. An observer on the ship sees a short pulse of light leave the lower surface and travel a distance L_0 to a mirror, which reflects it back to the bottom where it is detected, producing a "tick" sound. We will call the time interval for a round-trip one "tick." Since the light travels a distance $2L_0$ at speed c, the time Δt_0 for one tick as observed by a person on the spaceship is

$$\Delta t_0 = \frac{2L_0}{c}.$$

Suppose a person on the earth could observe the tick as the spaceship passed. Since the clock moves relative to the earth observer, the bottom of the clock is in a different location at the end of a tick than at the beginning (see Fig. 30.5). The light seems to follow a triangular path of length $2L$. Because $2L$ is longer than $2L_0$, the time for one tick seems longer to the earth observer (remember that *light travels at the same speed for both observers*).

The distance L can be determined using the Pythagorean theorem. Note that L is the hypotenuse of a right triangle of height L_0 and base $\frac{1}{2}v \, \Delta t$ (see Fig. 30.5). Thus,

$$L = \sqrt{L_0^2 + \left(\frac{1}{2}v \, \Delta t\right)^2}.$$

The time for one tick according to the earth observer is

$$\Delta t = \frac{2L}{c} = \frac{2\sqrt{L_0^2 + (\frac{1}{2}v \, \Delta t)^2}}{c}.$$

Substituting $L_0 = \frac{1}{2}c \, \Delta t_0$ into the preceding equation and rearranging to solve for Δt, we find that

$$\Delta t = \frac{\Delta t_0}{\left(1 - \dfrac{v^2}{c^2}\right)^{1/2}}. \tag{30.1}$$

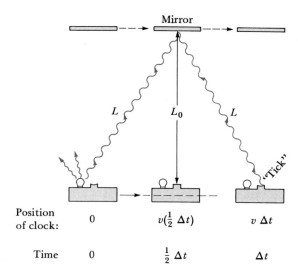

This is the time-dilation equation we have been seeking. The procedure for using this equation is summarized as follows:

Time dilation: Carefully define your events. Then determine in what reference frame the events occur at the same place. This is the **proper reference frame.** The time between the events as measured by an observer in the proper reference frame is the **proper time** Δt_0. The time Δt between the same two events as measured by an observer moving at speed v relative to the proper reference frame is

$$\Delta t = \frac{\Delta t_0}{\left(1 - \dfrac{v^2}{c^2}\right)^{1/2}}. \qquad (30.1)$$

The words "moving at speed v relative to the proper reference frame" are especially important. Recall our example of the spaceship on which a light flashes. The time between flashes as measured by a passenger on the ship is the proper time because the light causing the flashes is at rest relative to the passenger. For an observer on the earth, the flashes occur at different places as the spaceship moves past the earth. The earth is not the proper reference frame. We say that the earth moves at speed v relative to the proper reference frame on the spaceship.

You may feel uncomfortable saying that the earth moves relative to the spaceship, but in relativity we do not attach any special significance to the reference frame on the earth. When using Eq. (30.1), we must first determine which inertial reference frame is the proper one. Then all other reference frames are regarded as moving relative to that frame. The particular reference frame that is the proper frame depends on the events being considered and may differ from one problem to the next.

EXAMPLE 30.1 As a spaceship moves past the earth at a speed of 2.00×10^8 m/s, its light flashes at 10-s intervals, according to the clock of an observer on the spaceship. How much time elapses between flashes as measured by an observer on the earth?

SOLUTION The events are two flashes that are separated in time by 10 s according to an observer on the spaceship. These events occur at the same place on the spaceship (the proper reference frame) and hence

$$\Delta t_0 = 10 \text{ s}.$$

Relative to the spaceship, the earth moves at a speed of 2.00×10^8 m/s. The time Δt between the two events, as measured by the earth observer, is

$$\Delta t = \frac{\Delta t_0}{\sqrt{1 - \dfrac{v^2}{c^2}}} = \frac{10 \text{ s}}{\sqrt{1 - \left(\dfrac{2.00 \times 10^8 \text{ m/s}}{3.00 \times 10^8 \text{ m/s}}\right)^2}} = \underline{13.4 \text{ s}}. \qquad \blacksquare$$

EXAMPLE 30.2 (Muon decay) Direct confirmation of time dilation is provided by an experiment involving particles called *muons*. A muon has about 200 times as much mass as an electron and is produced during nuclear reactions in high-energy particle accelerators. Once produced, muons have a half-life of 2.2×10^{-6} s. This means that if 1000 muons are produced, then 2.2×10^{-6} s later only 500 remain (the other 500 have disintegrated to produce electrons and energy). Suppose that a beam of 1000 muons is produced by a source and emitted at a speed of 0.95c. How far from the source will the muons have traveled before their number is reduced to 500?

SOLUTION The muons have a half-life of 2.2×10^{-6} s in their own proper reference frame (one in which they are produced and decay at the same location). The proper reference frame for the muons when they are flying through the laboratory is a reference frame that moves through the laboratory with them at a speed of 0.95c. The muons at rest *relative to this reference frame* are produced and disintegrate at the same place in this frame as it moves through the laboratory. In this proper reference frame, their half-life is

$$\Delta t_0 = 2.2 \times 10^{-6} \text{ s}.$$

Relative to the proper reference frame, the laboratory seems to move past the muons at a speed of 0.95c. In the laboratory frame, which is considered to move relative to the rest frame of the muons, half the muons live a time Δt given by Eq. (30.1):

$$\Delta t = \frac{\Delta t_0}{\left(1 - \dfrac{v^2}{c^2}\right)^{1/2}} = \frac{2.2 \times 10^{-6} \text{ s}}{\left(1 - \dfrac{0.95^2 c^2}{c^2}\right)^{1/2}} = 7.0 \times 10^{-6} \text{ s}.$$

By the laboratory clock, 50 percent of the muons will remain after 7.0×10^{-6} s. During this time they move a distance (as measured in the laboratory frame) of

$$d = v \, \Delta t = (0.95c)(7.0 \times 10^{-6} \text{ s}) = \underline{2000 \text{ m}}.$$

The distance they travel is more than three times longer than the distance they would travel if their half-life was only 2.2×10^{-6} s in the laboratory frame.

$\blacksquare$

Many experiments have shown that muons and other short-lived particles travel much farther than would be possible during their proper half-lives. Their extended lives, as measured by a laboratory observer, agrees well with the special theory of relativity.

The physiological aging of humans also depends on a person's motion relative to an observer. The rate of a traveler's heartbeat slows relative to an observer on earth. If the traveler is moving near the speed of light, his or her heart may appear to the observer on earth to beat only once in several seconds while to an observer moving with the traveler it appears to beat about once each second. Some unusual family relationships could result from this slowed physiology, as we see in the next example.

EXAMPLE 30.3 Suppose that a 25-year-old mother leaves on a rocket trip. She travels at a speed of $0.99c$ relative to the earth and is gone for 30 years as measured on an earth clock. How old is the mother, in terms of physiological age, when she returns? How old is her daughter, who was 2 when the mother left?

SOLUTION A person's physiological age is meaningful only in his or her own proper reference frame. The earth is the proper reference frame for the daughter. During thirty earth years, the daughter will age thirty years. Since she was 2 years old when the mother left, the daughter will be 32 years old at the end of the mother's trip.

The mother's physiological age change must be determined by a different clock in her own proper reference frame, the spaceship. In that frame she has aged by an unknown amount Δt_0—the time on the spaceship's clock from the start to the finish of the trip. Relative to the spaceship, the earth's clock moves at a speed $v = 0.99c$ and measures a time $\Delta t = 30$ yr for the trip. The mother's age change is calculated using Eq. (30.1):

$$\Delta t_0 = \Delta t \left(1 - \frac{v^2}{c^2}\right)^{1/2} = (30 \text{ yr}) \left[1 - \left(\frac{0.99c}{c}\right)^2\right]^{1/2} = 4.2 \text{ yr.}$$

Since the mother was 25 years old when she left, on returning she is 25 + 4 or 29 years old. The mother is now three years younger than her daughter! ∎

The trip discussed in Example 30.3 is an inappropriate subject for special relativity because a traveler must accelerate and decelerate while making a space trip, and measurements in special relativity are restricted to inertial reference frames—frames that do not accelerate. Over the years careful discussion and argument have led most scientists to believe that such a phenomenon as described in Example 30.3 is possible. A person in an accelerating reference frame would age more slowly than a person in a nonaccelerating frame, and the difference in aging would be the same as that predicted by applying the equations of special relativity.

A test of this idea was performed by J. C. Hafele and Richard Keating in 1971. They transported four very precise cesium atomic clocks around the world on commercial airlines. The clocks were in the air for approximately 45 hours. The time change on these clocks was compared to reference clocks that remained at rest on the earth. The scientists observed a shift in time, a time dilation, that agreed well with the theory of special relativity. Because the airplanes flew at speeds much less than the speed of light, the time dilation was very small—several hundred nanoseconds during 45 hours of flight.

30.4 Length Contraction

Time dilation is linked closely to a phenomenon called **length contraction:** The lengths of objects moving relative to us are shortened. The faster an object moves, the shorter is its length. Length contraction occurs only along the direction of motion. An object's size perpendicular to the direction of motion does not change.

We can derive an expression for the dependence of an object's length on its speed by considering an arrow that moves past a stationary clock (Fig. 30.6). As the arrowhead passes the clock, it trips a switch that starts the clock (Fig. 30.6a). As the arrow's feathers pass the clock, the switch is again tripped, and the clock stops (Fig. 30.6b). The two events—the starting and the stopping of the clock—occurred in the same place in the reference frame of the clock, so the clock is the proper reference frame and its reading Δt_0 is the proper time for this measurement. If a small fly rides on the arrow, the fly would say that the time interval Δt between the events is, according to Eq. (30.1),

$$\overbrace{\Delta t}^{\text{Arrow time}} = \frac{\overbrace{\Delta t_0}^{\text{Stationary clock time}}}{\left(1 - \dfrac{v^2}{c^2}\right)^{1/2}},$$

where v is the speed at which the clock seems to move past a fly sitting on the arrow.

Next, how is the arrow's length affected by its motion relative to an observer? The **proper length** L_0 of the arrow is its length in a coordinate system in which the arrow is stationary. Since the arrow is at rest relative to the fly, the arrow's proper length relative to the fly is L_0. To an observer at rest in the clock's reference frame, the length L of the moving arrow is different. We can relate L to L_0 by noting that the arrow's speed appears the same to an observer on the clock as the clock's speed does to the fly on the arrow. To calculate these speeds, note that the arrow, according to the clock observer, moves a distance L (not a proper length) in a time Δt_0. For the fly, the arrow moves its proper length L_0 past the clock in a time Δt. Thus,

$$v = \overbrace{\frac{L}{\Delta t_0}}^{\text{Clock observer}} = \overbrace{\frac{L_0}{\Delta t}}^{\text{Fly on arrow}}. \tag{30.2}$$

(a)

(b)

Rearranging Eq. (30.2) and substituting Eq. (30.1) for $\Delta t_0/\Delta t$ leads to the following equation, which relates the length L_0 of an object observed in its proper reference frame to its length L in a frame moving at speed v relative to the proper frame:

$$\frac{L}{L_0} = \frac{\Delta t_0}{\Delta t} = \left(1 - \frac{v^2}{c^2}\right)^{1/2}.$$

Length contraction: An object has a length L_0 when measured in a reference frame at rest relative to the object. This frame is the object's **proper reference frame,** and the length L_0 is called the **proper length.** In a reference frame moving at speed v relative to the proper reference frame, the object's length L is measured to be

$$L = L_0 \left(1 - \frac{v^2}{c^2}\right)^{1/2}. \tag{30.3}$$

EXAMPLE 30.4 An arrow flies past a person. When at rest, the arrow is 1.00 m long. Calculate the arrow's length L when the arrow moves (a) at a speed of $0.90c$ and (b) at a speed of 300 m/s.

SOLUTION (a) To a person at rest relative to the arrow, the arrow's length L_0 is 1.00 m. If the arrow moves at speed v past an observer, then we can also say that the observer moves at a speed v relative to the arrow. The arrow is measured by this observer to have a length L (when moving at $v = 0.90c$) given by Eq. (30.3):

$$L = L_0 \left(1 - \frac{v^2}{c^2}\right)^{1/2} = (1.00 \text{ m})\left[1 - \left(\frac{0.90c}{c}\right)^2\right]^{1/2} = \underline{0.44 \text{ m}}.$$

(b) When moving at 300 m/s, the arrow's length is measured to be

$$L = (1.00 \text{ m})\left[1 - \left(\frac{3 \times 10^2 \text{ m/s}}{3 \times 10^8 \text{ m/s}}\right)^2\right]^{1/2}$$

$$= (1.00 \text{ m})[1 - 10^{-12}]^{1/2} \simeq \underline{1.00 \text{ m} - \frac{1}{2}(10^{-12} \text{ m})}.$$

(The binomial expansion has been used for the last approximation.) The arrow appears to be shortened by less than $\frac{1}{100}$ the size of an atom when moving at 300 m/s. Relativistic effects are unobservable at the ordinary speeds of daily life. ∎

30.5 Relativistic Mass

In the past two sections we learned that the time between two events and the length of an object depend on the motion of the observer. In this section, we turn to a third relativistic phenomenon: An object's mass increases as its speed relative to an observer increases. Like the change in time and length, a change in mass is not noticeable at the low speeds of daily living. However, large mass changes occur for objects moving near the speed of light. For example, the mass of an electron increases by a factor of about 40,000 as it is accelerated by intense electric fields to nearly the speed of light at the Stanford Linear Accelerator.

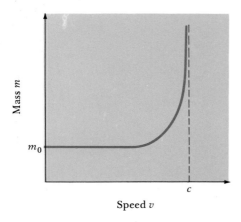

FIG. 30.7. A particle's mass increases dramatically as it approaches the speed of light.

Einstein, by performing a thought experiment, was able to accurately predict the dependence of an object's mass on its speed. The thought experiment involved the collision of two identical objects as viewed by two observers. In order for momentum to be conserved in the reference frame of each observer, the object's mass when moving had to be related to its mass when at rest by the following equation (which we will not derive):

$$m = \frac{m_0}{\left(1 - \dfrac{v^2}{c^2}\right)^{1/2}}, \qquad (30.4)$$

where m_0 is the object's mass when at rest (called its proper mass or **rest mass**) and m is the object's mass when moving at a speed v.

Einstein's relativistic mass equation has been confirmed many times using particle accelerators (discussed in Section 30.6). The variation of an object's mass with speed is shown in Fig. 30.7. At low speeds, an object's mass is nearly constant and equals its rest mass m_0. As the object's speed approaches the speed of light c, the denominator of Eq. (30.4) becomes very small, and the object's mass increases dramatically.

We will see in the next section that the object's increase of mass is a result of energy gained as forces pushing and pulling the object cause it to accelerate to higher speed. However, as its speed approaches the speed of light c, its mass approaches an infinite value. To increase the object's speed to the speed of light would require an infinite force and infinite energy. Since this is not possible, *the speed of an object with nonzero rest mass can never equal or exceed the speed of light.*

EXAMPLE 30.5 An electron is accelerated to a speed of $0.995c$. How much greater is its mass when moving at this speed than its mass when at rest?

SOLUTION Rearranging Eq. (30.4), we find that the ratio of the object's mass when moving at $v = 0.995c$ and its mass m_0 when at rest is

$$\frac{m}{m_0} = \frac{1}{\left(1 - \dfrac{v^2}{c^2}\right)^{1/2}} = \frac{1}{\left[1 - \left(\dfrac{0.995c}{c}\right)^2\right]^{1/2}} = \frac{1}{(1 - 0.990)^{1/2}} = \underline{10.} \quad \blacksquare$$

EXAMPLE 30.6 You leave on a spaceship bound for a planet in the Crab Nebula. If your acceleration from rest is $10g$ (that is, 10 times the acceleration of gravity), how long will you have traveled before your mass increases by 1 percent, as noted by an observer on the earth?

SOLUTION We first use Eq. (30.4) to calculate your speed after your mass has increased by 1 percent:

$$\frac{m}{m_0} = 1.01 = \frac{1}{\left(1 - \dfrac{v^2}{c^2}\right)^{1/2}},$$

or

$$\left(1 - \frac{v^2}{c^2}\right) = \left(\frac{1}{1.01}\right)^2 = 0.980.$$

Hence,

$$v = (1 - 0.980)^{1/2}c = 0.14c = 4.2 \times 10^7 \text{ m/s}.$$

If your acceleration is constant and equal to $10g$, the time needed to reach this speed from rest is given by Eq. (2.9) from kinematics:

$$v - v_0 = at,$$

or

$$t = \frac{v - v_0}{a} = \frac{v - v_0}{10g} = \frac{4.2 \times 10^7 \text{ m/s} - 0}{10(9.8 \text{ m/s}^2)} = 4.3 \times 10^5 \text{ s} = \underline{5.0 \text{ days}}.$$

Your acceleration cannot continue indefinitely at $10g$. If it did, you would soon be moving faster than light, which is impossible. As your speed approaches c, energy originally used to increase your speed is now used primarily to increase your mass. ∎

Relativity and Newtonian Mechanics

The principles of Newtonian mechanics were built on the idea that one can measure time, length, and mass accurately and uniquely when describing the changing motion of an object. Must we reject Newtonian mechanics now that we have learned that measurements of time, length, and mass vary from one observer to another? The answer is yes, if we are talking about objects that move near the speed of light. But at speeds of normal living, the relativistic changes of time, length, and mass are infinitesimal. A person's mass increases by less than a factor of 10^{-13} when the person's speed increases from zero to 100 m/s. Consequently, we can ignore this mass change when using Newtonian mechanics to analyze the person's motion.

As a rough rule of thumb, we can usually ignore relativistic changes in mass, time, and length when the speeds of objects or moving reference frames are less than about one-tenth the speed of light. At this speed ($v = 0.1c$) the factor $[1 - (v^2/c^2)]^{1/2}$ that appears in the equations for time dilation, length contraction, and mass change equals 0.995. Thus, at $v = 0.1c$ there is only a 0.5 percent change in a time interval, length, or mass. At lower speeds the changes are even smaller, and the equations of Newtonian mechanics provide an adequate description of nature.

30.6 Relativistic Energy

A particle accelerator causes the speed of a particle such as an electron or proton to increase by moving the particle across a large voltage (or across a large number of smaller voltages). In the case of an electron, its decrease in electrical potential energy when crossing a voltage V is $\Delta PE_q = -eV$. This decrease is balanced by an increase in kinetic energy. In classical physics, the change in kinetic energy of a particle that starts at rest is

$$\Delta KE = \frac{1}{2}m_0 v^2 - \frac{1}{2}m_0 0^2 = \frac{1}{2}m_0 v^2,$$

where m_0 is the particle's rest mass and is constant. Balancing the particle's kinetic and electrical potential energy changes results in the equation

$$0 = \frac{1}{2}m_0v^2 - eV.$$

Solving for the particle's speed, we find that

$$v = \sqrt{\frac{2eV}{m_0}}.$$

If an electron accelerates across about 10^4 V or less, the preceding equation provides a fairly accurate expression for its speed. For greater voltages, the equation predicts incorrect results. For example, if the electron crosses 10^6 V, its speed is calculated to be 6×10^8 m/s, or twice the speed of light! Electrons never move faster than the speed of light, even when accelerated through voltages much greater than 10^6 V. Evidently, our classical expression for kinetic energy does not apply to particles that move very fast.

Suppose an electron is already traveling near the speed of light when it passes across a large voltage. The decreased electrical potential energy $(-eV)$ must be balanced by increased kinetic energy, yet the electron cannot move much faster because it is already moving near the speed of light. If it experiences a small increase in speed, its mass increases dramatically (see the upward-curved portion of the line near the point $v = c$ in Fig. 30.7). Since the decreased electrical potential energy is balanced by increased mass rather than by increased speed, mass must be a form of energy.

Albert Einstein, using a moderately complicated thought experiment, was able to show that a free particle's total energy is proportional to its relativistic mass:

The **total energy** E of a particle moving at speed v is

$$E = mc^2, \tag{30.5}$$

where

$$m = \frac{m_0}{\left(1 - \dfrac{v^2}{c^2}\right)^{1/2}}$$

and m_0 is the particle's rest mass. If the particle sits at rest, its energy is

$$E_0 = m_0c^2, \tag{30.6}$$

where E_0 is called the particle's **rest-mass energy.**

The total energy in Eq. (30.5) includes both the particle's energy due to its rest mass and the particle's energy due to its motion (kinetic energy). When the particle is at rest, its energy is $E_0 = m_0c^2$. If the particle accelerates to speed v, its energy increases to $E = mc^2$. The difference in the initial energy E_0 and the final energy E is the kinetic energy gained by the particle.

The relativistic **kinetic energy** of a mass moving at speed v is the difference of its total energy E and its rest-mass energy E_0:

$$KE = E - E_0 = mc^2 - m_0c^2, \tag{30.7}$$

where m is the particle's mass when moving at speed v and m_0 is its mass when at rest.

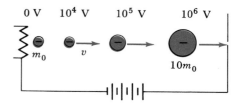

FIG. 30.8. As an electron passes across greater voltages, its mass and energy increase.

As depicted in Fig. 30.8, the increasing kinetic energy of an electron as it moves across a large voltage first appears as an increase in speed; then, as the electron's speed approaches c, continued energy gain results in a noticeable increase in mass.

The relativistic equation for a particle's kinetic energy looks nothing like the equation used in classical physics ($KE = \frac{1}{2}m_0v^2$). We find, though, that if the relativistic expression for mass m [Eq. (30.4)] is substituted into Eq. (30.7) for speeds much less than the speed of light ($v \ll c$), the relativistic equation for kinetic energy reduces to the classical equation. For speeds somewhat less than the speed of light, the classical equation for kinetic energy can be used; for speeds near the speed of light, Eq. (30.7) should be used to calculate a particle's kinetic energy. If the particle's speed is unknown, Eq. (30.7) can be used because it provides the correct answer at all speeds.

EXAMPLE 30.7 Protons are accelerated from rest to a speed of $0.993c$ in a "small" 150-m-diameter booster accelerator that is part of the Fermi National Accelerator Laboratory in Batavia, Illinois (Fig. 30.9). (a) Calculate the rest-mass energy E_0 of a proton whose rest mass is 1.67×0^{-27} kg. (b) Calculate the proton's final energy E as it leaves the booster accelerator. (c) Calculate the increase in kinetic energy of the proton in the booster accelerator.

SOLUTION (a) The proton's initial energy when at rest is determined using Eq. (30.6):

FIG. 30.9. The Fermi National Accelerator Laboratory. (Courtesy of Fermilab.)

$$E_0 = m_0c^2 = (1.67 \times 10^{-27} \text{ kg})(3 \times 10^8 \text{ m/s})^2 = \underline{1.50 \times 10^{-10} \text{ J}}.$$

(b) The proton's energy as it leaves the booster accelerator at a speed of $0.993c$ is determined using Eq. (30.5):

$$E = mc^2 = \frac{m_0c^2}{\left(1 - \dfrac{v^2}{c^2}\right)^{1/2}} = \frac{1.50 \times 10^{-10} \text{ J}}{\left(1 - \dfrac{0.993^2c^2}{c^2}\right)^{1/2}} = \underline{12.70 \times 10^{-10} \text{ J}}.$$

The proton's energy has increased by a factor of 8.5.   (c) The proton's kinetic energy as it leaves the booster accelerator is the difference between its total energy and its rest-mass energy:

$$KE = E - E_0 = 12.70 \times 10^{-10} \text{ J} - 1.50 \times 10^{-10} \text{ J} = \underline{11.20 \times 10^{-10} \text{ J}}. \quad \blacksquare$$

EXAMPLE 30.8   An electron is accelerated from rest across the (2.5×10^4)-V potential difference from the cathode to the screen of a television set. Calculate the percent increase in the electron's mass. An electron's rest mass m_0 is 9.1×10^{-31} kg.

SOLUTION   As the electron moves from the cathode to the screen, its electrical potential energy decreases and its kinetic energy increases by an equal amount. The mass increase can be calculated using the work-energy equation, Eq. (8.10):

$$0 = \Delta KE + \Delta PE_q,$$

where

$$\Delta KE = mc^2 - m_0c^2 = (m - m_0)c^2$$

and

$$\Delta PE_q = qV = (-e)V = (-1.6 \times 10^{-19} \text{ C})(2.5 \times 10^4 \text{ V}) = -4.0 \times 10^{-15} \text{ J}.$$

We find, then, that

$$0 = (m - m_0)c^2 + (-4.0 \times 10^{-15} \text{ J}),$$

or

$$m - m_0 = \frac{4.0 \times 10^{-15} \text{ J}}{c^2} = 4.4 \times 10^{-32} \text{ kg}.$$

The electron's percent increase in mass is

$$\left(\frac{m - m_0}{m_0}\right) 100 = \frac{4.4 \times 10^{-32} \text{ kg}}{9.1 \times 10^{-31} \text{ kg}} \times 100 = \underline{4.9 \text{ percent}}.$$

You probably did not realize that you were increasing the mass of electrons while watching television. (You may even have been increasing your own mass, although for a different reason.) $\blacksquare$

30.7   Equivalence of Mass and Energy

Einstein derived his equation relating mass and energy ($E = mc^2$) using a thought experiment that we will not duplicate here. Nevertheless, a variety of experiments confirm the correctness of the equation. These experiments have convinced physicists that mass, whether stationary or moving, is equivalent to energy.

Electron-Positron Annihilation

A positron is a small particle whose rest mass equals the rest mass of an electron, but it has a charge of $+e$ rather than the $-e$ charge of the electron. Positrons are occasionally produced when high-energy particles from an accelerator collide with the nucleus of an atom. Suppose that a positron created by one of these collisions is near an electron. Because they have opposite electric charges, the positron and electron are pulled together by their electrical attraction. As they collide, a burst of electromagnetic radiation is produced, and the electron and positron cease to exist. Their masses have been converted entirely into the energy of electromagnetic radiation—a form of energy (such as light) that has no rest mass.

Nuclear Fusion

A free proton in the core of the sun collides with and is captured by the nucleus of a nitrogen atom. Electromagnetic radiation sprays outward as the new nucleus is formed. Following this reaction, the mass of the new nucleus is found to be less than the combined masses of the original proton and the nitrogen nucleus. The lost mass has been converted to the energy of electromagnetic radiation (light, ultraviolet radiation, x-rays, and so forth) and into kinetic energy. On the sun this capture of protons by nitrogen nuclei is just one of many nuclear processes that are responsible for producing the radiative energy of the sun. Each second, the sun's mass decreases by about 4×10^9 kg because of these processes, and an equivalent amount of radiative energy is produced. This mass-energy conversion has occurred for over a billion years. Yet because of the sun's huge initial mass (about 2×10^{30} kg), over 99.99 percent of its mass still remains!

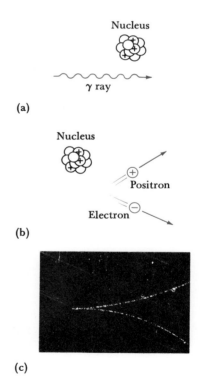

(a)

(b)

(c)

FIG. 30.10. (a) A gamma ray passes a nucleus. (b) It disappears and in its place are an electron and a positron. The gamma ray's energy has been converted to mass. (c) The tracks of an electron and positron in a cloud chamber after their production from a gamma ray (the gamma ray produces no track). The particles move apart because of the force of a magnetic field. (Courtesy of Brookhaven National Laboratory).

Pair Production

In the last two paragraphs we saw that mass could be converted to electromagnetic energy. The reverse can also occur; electromagnetic energy can be converted to mass. The most familiar form of electromagnetic radiation is light; other forms include radio waves, microwaves, infrared radiation, ultraviolet radiation, x-rays, and gamma rays. All these forms of radiation travel at the speed of light in a vacuum and have no rest mass (they do not exist at rest). The types of radiation differ in their frequency and energy, gamma rays being the most energetic form of electromagnetic radiation. A gamma ray may have enough energy to create the mass energy of an electron and a positron. If a gamma ray passes near another atom, it occasionally disappears, and an electron and a positron appear in its place (Fig. 30.10). The nearby atom is unchanged. The energy of the gamma ray has been converted to the mass energy of the two particles, a process called **pair production.**

These are just a few of the many experiments that indicate that energy, in any form, is equivalent to mass and that mass is equivalent to energy.

EXAMPLE 30.9 About 2×10^{10} J of energy are needed to heat and cool your home each year. If the energy could be obtained by converting mass to other forms of energy, how much mass would be needed?

SOLUTION According to Eq. (30.5),

$$m = \frac{E}{c^2} = \frac{2 \times 10^{10} \text{ J}}{(3 \times 10^8 \text{ m/s})^2} = 2.2 \times 10^{-7} \text{ kg} = \underline{0.22 \text{ mg}},$$

approximately the mass of one of the hairs on your head. ∎

30.8 General Relativity

The special theory of relativity allows us to compare observations and measurements made by observers in two reference frames that move at constant velocity relative to each other but that do not accelerate. In these inertial reference frames, the laws of physics are invariant. Newton's second law ($\Sigma \mathbf{F} = m\mathbf{a}$) should work equally well and have the same form for an observer in any inertial reference frame.

Einstein felt that the laws of physics should be invariant in all reference frames—those that are accelerating as well as those that are not. With this requirement in mind, he developed the general theory of relativity, a reformulation of the description of gravitation.

The cornerstone of general relativity is the **principle of equivalence**—the idea that our observations of nature cannot distinguish between an object that is acted on by a gravitational force and an object in an accelerated reference frame. Consider a spaceship at rest or drifting at constant velocity far from any other stars, planets, or objects with mass. A passenger inside the ship would float freely as would his or her belongings. Suppose now that the spaceship's rockets fired, causing the ship to accelerate forward with an acceleration g (9.8 m/s²). If initially floating, the passenger would appear to another passenger to fall to the spaceship's floor with an acceleration g. (Actually, the spaceship's floor rises up to meet the floating passenger.) If the ship's acceleration continued, the passenger would be able to stand or jump on the floor. The passenger could throw a ball in the air and it would return to his or her hand with an acceleration g. The passenger would have exactly the same feeling as if standing on the earth's surface. As we see, the person in the spaceship accelerating at g and a person that stands in a region where the gravitational force causes a gravitational acceleration g both experience equivalent effects.

From the principle of equivalence, Einstein reasoned that the changes in an object's motion caused by gravitational forces could be treated by considering a space that had a curvature. Einstein's space included three spatial coordinates and a time coordinate—a four-dimensional space. A large mass would cause the nearby space to be curved. A small object that accelerates due to the gravitational force of a nearby large mass (such as the earth) could be thought of as moving along a free path in the curved space-time caused by that large mass. Two stones, one larger than the other, would slide along that curved space side by side, just as the large and small stones dropped by Galileo fell side by side from the Leaning Tower of Pisa.

A large mass such as the sun or the black hole depicted in Fig. 30.1 causes Einstein's four-dimensional space to be more distorted than does a smaller mass such as an earth satellite. If an object moves through the curved space near the sun, its path will be noticeably altered. The small distortion of space near the satellite causes a smaller effect on a passing object.

The path of light is also altered by this curved space. Einstein predicted in 1915 that light from distant stars while passing close to the sun would be deflected by an angle of 4.86×10^{-4} degrees (see Fig. 30.11). During a total eclipse of the sun in 1919 astronomers measured such a deflection. Einstein's theory became famous almost overnight.

General relativity solved another problem that had plagued astronomers since the early 1800s. The elliptical path of Mercury around the sun was known to precess, as shown in Fig. 30.12. The precession was so small that the elliptical path required about 3 million years to complete one rotation before returning to its starting orbit. According to Newton's formulation of gravity, the ellipse should not precess. Einstein, however, knew that a planet while moving farther from and closer to the sun on its elliptical orbit experienced a varying gravitational field. The curvature of space along the orbit varied. This varying curvature would cause the planet's orbit to precess, and the precession predicted for Mercury was exactly the precession that had been observed. The orbits of the other planets should also precess. But because they are farther from the sun where the curvature of space is less, their precession is smaller and more difficult to measure.

Another prediction of general relativity, called the *gravitational red shift*, was not confirmed until 1960. According to this theory, gravitation causes time to slow. The stronger the effect of gravitation, the slower time moves. For example, a person working in an office on the first floor of a tall building would seem to an observer on the top floor to age more slowly than the person on the top floor. This is because gravitation is stronger at the bottom than at the top of the building. But don't rush for the first floor of your building: the time difference is only about 10^{-5} s in a lifetime. The gravitational red shift was confirmed in 1960 by R. V. Pound and G. A. Rebka, Jr., at Harvard University. They observed a small reduction in the frequency of vibration of gamma rays emitted from radioactive nuclei on the bottom floor of a laboratory building at Harvard compared to the frequency of gamma rays emitted from the top floor.

Recently, general relativity has been applied to the study of black holes. The space in the immediate neighborhood of these extremely dense objects is so curved (see Fig. 30.1) that nothing, including light, can escape.

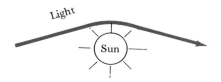

FIG. 30.11. The gravitational force of the sun on light that passes nearby causes the light's path to bend. This bending was first observed in 1919 and brought instant fame to Einstein's general theory of relativity.

FIG. 30.12. General relativity correctly predicts the precession of the elliptical orbit of Mercury around the sun.

Summary and Additional Readings

Our discussion of relativity emphasized two main ideas: (1) a determination of the way an object's appearance varies depending on the motion of an observer relative to the object and (2) a reformulation of some of our ideas concerning energy.

1. **Time dilation:** The time between two events as observed in a reference frame where the events occur at the same location is called the **proper time** Δt_0, and the reference frame is called the **proper reference frame.** When the same events are observed in a reference frame moving at speed v relative to the proper reference frame, the time Δt between the events is

$$\Delta t = \frac{\Delta t_0}{\left(1 - \dfrac{v^2}{c^2}\right)^{1/2}}, \qquad (30.1)$$

where c, the speed of light, equals 3.0×10^8 m/s.

2. **Length contraction:** The **proper length** L_0 of an object is its length as measured in a reference frame in which the object is at rest. Its length L as seen in a reference frame in which the object moves at speed v is

$$L = L_0 \left(1 - \frac{v^2}{c^2}\right)^{1/2}. \qquad (30.3)$$

3. **Mass increase:** An object's mass m increases as its speed relative to an observer increases:

$$m = \frac{m_0}{\left(1 - \dfrac{v^2}{c^2}\right)^{1/2}}, \qquad (30.4)$$

where m_0 is the object's mass when measured at rest (its **rest mass**) and m the object's mass when moving at speed v.

4. **Relativistic energy:** An object's **total energy** E when moving at a speed v is

$$E = mc^2 = \frac{m_0 c^2}{\left(1 - \dfrac{v^2}{c^2}\right)^{1/2}}. \qquad (30.5)$$

The total energy includes both the object's **rest-mass energy** E_0,

$$E_0 = m_0 c^2, \qquad (30.6)$$

and its **kinetic energy** KE,

$$KE = E - E_0 = mc^2 - m_0 c^2. \qquad (30.7)$$

The equivalence of mass and energy, as predicted by Einstein's famous equation $E = mc^2$, has been confirmed by a variety of experiments in which mass is converted to other forms of energy and other forms of energy are converted to mass.

Paul G. Hewitt, *Conceptual Physics,* fifth edition, Little, Brown, Boston (1985). Chapters 34 and 35 provide a very interesting and friendly introduction to special and general relativity.

George Gamow, *Mr. Tompkins in Wonderland,* Cambridge University Press, Cambridge (1939). Describes the way our relativistic world would look if the speed of light were much slower.

The controversy among scientists about the effect of relativity on relative aging (see Example 30.3) is reviewed in a series of articles: J. Bronowski, "The Clock Paradox," *Scientific American* **208**, 134 (1973), and references therein; and G. Builder, "Resolution of the Clock Paradox," *American Journal of Physics* **27**, 656 (1957).

Questions

1. Explain the difference between a proper reference frame and other inertial reference frames in which the time between two events is measured.

2. Explain how it is theoretically possible for a daughter or son to be physiologically older than her or his natural parents.

3. You move toward a star at a speed of $0.9c$. At what speed does light from the star pass you? What if you move away from the star?

4. You pass the earth in a spaceship that moves at $0.9c$ relative to the earth. Do you notice a change in your heartbeat rate? Does an observer on the earth think your heart is beating at the normal rate? Explain both answers.

5. It takes light approximately 10^{10} years to reach the earth from the edge of the observable universe. Would it be possible for a person to travel this distance during a lifetime? Explain.

6. A person holds a meter stick in a spaceship traveling at $0.95c$ past the earth. The person rotates the meter stick so that it is first parallel and then perpendicular to the ship's velocity. Describe its changing appearance to (a) an observer at rest on the earth and (b) to the person in the spaceship.

7. Name several ways in which your life would be different if the speed of light were 20 m/s rather than its actual value.

8. If the speed of light were infinity, how would time dilation, length contraction, and mass increase be affected. Justify your answer carefully.

9. The classical equation for calculating kinetic energy, Eq. (8.4), and the relativistic equation for calculating kinetic energy, Eq. (30.7), appear quite different. Under what conditions is each equation appropriate? Invent a simple example to show that they produce the same result in a velocity region where both equations are appropriate.

Problems

You may need to know the following masses to work some of these problems: $m_{electron} = 9.11 \times 10^{-31}$ kg and $m_{proton} = 1.673 \times 10^{-27}$ kg.

30.3 Time Dilation

These problems should be easier to solve if you first identify the events being considered and the proper reference frame.

1. A particle called Σ^+ lives for 0.80×10^{-10} s in its proper reference frame before breaking apart into two other particles. How long does the particle seem to live according to a laboratory observer when the particle moves past the observer at a speed of 2.4×10^8 m/s?

2. A Σ^+ such as the one discussed in Problem 1 appears to a laboratory observer to live for 1.0×10^{-10} s. How fast is it moving relative to the observer?

3. A person on the earth observes 10 flashes of the light on a passing spaceship in 22 s, whereas the same 10 flashes seem to take 12 s to an observer on the ship. How fast is the ship moving?

4. A spaceship moves away from the earth at a speed of $0.990c$. The pilot looks back and measures the time for the earth to rotate once on its axis. How much time does the pilot measure?

■ 5. A free neutron lives about 1000 s before breaking apart to become an electron and a proton. If a neutron leaves the sun at a speed of $0.999c$, (a) how long does it live according to an earth observer and (b) will the neutron reach Pluto (5.9×10^{12} m from the sun) before breaking apart? Explain your answers.

■ 6. A Σ^- particle lives 1.5×10^{-10} s in its proper reference frame. If traveling at $0.90c$ through a bubble chamber, how far will it move before it disintegrates.

■ 7. A muon that lives 2.2×10^{-6} s in its proper reference frame is created 2400 m above the earth's surface. At what speed must it move to reach the earth's surface at the instant it disintegrates?

■ 8. Suppose the speed of light were 15 m/s. You run a 100-m dash in 10 s according to the timer's clock. How long did the race last according to your clock?

■ 9. Suppose you have roughly 50 more years to live. At what speed must your spaceship move to travel from the earth to the edge of the observable universe (a distance of 10^{10} light-years) before you die?

■ 10. An explorer travels at speed 2.90×10^8 m/s from the earth to a planet of Alpha Centauri, a distance of 4.3 light-years. (a) How long does the trip last according to an earth observer? (b) How much older physiologically is the explorer at the end of the trip?

■■ 11. Suppose that the speed of light is 8.0 m/s. You walk slowly to all of your classes during one semester while a classmate runs at a speed of 7.5 m/s during the time you are walking. *Estimate* your classmate's change in age, as judged by you, and your change in age during that walking time. Indicate how you chose any numbers used in your estimate.

■■ 12. A friend moves at speed v, much less than the speed of light, while on a trip that lasts a time Δt according to you. (a) Use the binomial theorem to show that your change in age minus your friend's change in physiological age during the trip is approximately $\frac{1}{2}(v^2/c^2)\Delta t$. (b) Estimate the difference in aging for a 300-km trip driving at the speed limit v.

30.4 Length Contraction

13. You sit in a spaceship moving past the earth at $0.97c$. Your arm held straight out in front of you seems to you to be 28.0 in long. How long does it appear to an observer on the earth?

14. A javelin hurled by Wonder Woman moves past an earth observer at $0.90c$. Its proper length is 2.7 m. What is its length according to the earth observer?

■ 15. At what speed must a meter stick move past an observer so that it will appear to be 0.50 m long?

■ 16. A billboard is 10 m high and 15 m long according to a person standing in front of it. At what speed must the person drive past in a fast car so that the billboard will appear square?

■ 17. Suppose that the speed of light is 100 m/s and that you are driving a race car at 90 m/s. How much time is required for you to travel 900 m along a track's straightaway, (a) according to a timer on the track and (b) according to your own clock? (c) How long does the straightaway appear to you? (d) Notice that the speed at which the track moves past is your answer to part (c) divided by your answer to part (b). Does this speed agree with the speed as measured by the stationary timer?

■ 18. An observer on a track measures the passage of an extraordinary rocket ship that travels 100 m in 5.0×10^{-7} s. (a) Calculate the ship's speed as determined by the ground observer. (b) How long a distance does the 100 m seem to a man in the rocket ship? (c) What time has elapsed on his watch during this passage? (d) How fast does the earth seem to pass the rocket ship?

30.5 Relativistic Mass

19. Calculate the ratio of an electron's mass and rest mass when moving at the following speeds: (a) 300 m/s; (b) 3.0×10^6 m/s; (c) 3.0×10^7 m/s; (d) 1.0×10^8 m/s; (e) 2.0×10^8 m/s; and (f) 2.9×10^8 m/s.

20. To escape the gravitational pull of the sun, a proton (which is part of the solar wind) must have a speed of at least

6.2×10^5 m/s. By what factor is the proton's mass increased when moving at this speed?

■ 21. At what speed does an object move so that its mass is 1 percent greater than its rest mass? 10 percent greater? Twice its rest mass?

■ 22. A 50-kg space traveler starts at rest and accelerates at $5g$ for 30 days. Calculate the person's mass after 30 days.

■ 23. A pilot and his spaceship of rest mass 1000 kg wish to travel from the earth to planet Scot ML, 30 light-years from the earth. However, the pilot wishes to be only 10 physiological years older when he reaches the planet. (a) At what speed must he travel? (b) What is the mass of his spaceship, according to an earth observer, while making the trip?

■ 24. A person's mass appears to be twice her rest mass when she moves at a certain speed. By what factor must her speed now increase to cause another doubling of her mass?

■■ 25. A rod when at rest is 1.0 m long and has a cross section of 1.0×10^{-4} m^2 and a mass of 1.0 kg. (a) Calculate the rod's density. (b) The rod is now made to move at a speed of $0.995c$ in a direction parallel to its long dimension. Calculate the rod's mass, length, cross section, and density as determined by a stationary observer.

30.6 Relativistic Energy

26. A proton's energy after passing through the accelerator at Fermilab is 500 times its rest energy. Calculate the proton's speed.

27. A 1000-kg rocket starts at rest and accelerates to a speed of $0.90c$. Calculate the kinetic energy needed for this change in speed.

28. Calculate the mass, total energy, rest-mass energy, and kinetic energy of a person with 60-kg rest mass moving at speed $0.95c$.

■ 29. An electron is accelerated from rest across 50,000 V in a machine used to produce x-rays. Calculate the percent increase in the electron's mass.

■ 30. A mass originally moving at $0.90c$ experiences a 5.0 percent increase in speed. By what percent does its kinetic energy increase?

■ 31. An electron is accelerated from rest across a potential difference of 9.0×10^9 V. Calculate the electron's speed (a) using the nonrelativistic kinetic-energy equation and (b) using the relativistic kinetic-energy equation. Which is the correct answer?

■ 32. A proton is given a kinetic energy of 200 GeV (1 GeV = 1.6×10^{-10} J) at the Fermilab. Calculate the ratio of the proton's mass and rest mass.

■ 33. A particle of rest mass m_0 initially moves at speed $0.4c$. (a) If the particle's speed is doubled, calculate the ratio of its final kinetic energy and its initial kinetic energy. (b) If the particle's kinetic energy increases by a factor of 100, by what factor does its speed increase?

■■ 34. A pilot and her spaceship have a mass of 400 kg. The pilot expects to live 50 years and wishes to travel to a galaxy that requires 100 years to reach if she travels at the speed of light. (a) Determine the average speed she must travel to reach the galaxy during her next 50 years. (b) To attain this speed, a certain mass m of matter is consumed and converted to the spaceship's kinetic energy. How much mass is needed? (We will ignore the energy needed to accelerate the fuel that has not yet been consumed.)

■ ■ 35.   The relativistic momentum p of an object of relativistic mass m moving at speed v is $p = mv$. Use this and the relativistic energy equations to verify that $E = (p^2c^2 + m_0^2c^4)^{1/2}$.

30.7   Equivalence of Mass and Energy

36.   Calculate the amount of mass whose rest energy equals the total yearly energy consumption of the world (4×10^{20} J).

37.   Separating a carbon monoxide molecule CO into a carbon and an oxygen atom requires 1.76×10^{-18} J of energy. (a) Calculate the mass equivalent of this energy. (b) Calculate the fraction of the original mass of a CO molecule (4.65×10^{-26} kg) that was converted to energy.

38.   A hydrogen-oxygen fuel cell combines 2 kg of hydrogen with 16 kg of oxygen to form 18 kg of water, thus releasing 2.5×10^5 J of energy. What fraction of the mass has been converted to energy?

■ 39.   Calculate the mass that must be converted to energy during a 70-year lifetime to continually provide electric power for a person at a rate of 1000 W. The production of the electric power from mass is only about 33 percent efficient.

■ 40.   An electric utility company charges a customer 1.5 cents for 10^6 J of electrical energy. At this rate, calculate the cost of a 1-g mass if converted entirely to energy.

■ 41.   A nuclear power plant produces 10^9 W of electric power and 2×10^9 W of waste heat. (a) At what rate must mass be converted to energy in the reactor? (b) What is the total mass converted to energy each year?

■ 42.   *Estimate* the total metabolic energy you use during a day. Calculate the mass equivalent to this lost energy.

■ ■ 43.   (a) A container holding 4 kg of water is heated from $0\,°$C to $60\,°$C. Calculate its change in mass as a result of this increased energy. (b) If the water, initially at $0\,°$C, is converted to ice at $0\,°$C, calculate its change in mass.

■ 44.   (a) Calculate the energy radiated by the sun each second by conversion of 4×10^9 kg of mass to energy. (b) Determine the fraction of this energy intercepted by the earth. The earth is 1.50×10^{11} m from the sun and has a radius of 6.38×10^6 m.

Wave-Particle Duality

We learned in the last chapter that classical theory was inadequate for describing objects moving near the speed of light. Experiments performed early in this century indicated that classical theory was also inadequate for describing both atomic-sized objects and a massless form of energy called electromagnetic radiation. The tiny world of atoms and radiation exhibited a dual behavior that seemed at times like that of a particle and at other times like that of a wave.

Light, which is a form of electromagnetic radiation, exhibits this dual behavior. For example, a beam of light after passing through two slits arrives at the flat surface of a screen as though waves had spread from each slit. These waves produce an interference pattern on the screen, as we learned in Section 21.2. Yet, if observed carefully, the light when it hits the screen consists of individual particlelike bundles of energy called **photons.** The flash of energy released as one of these photons hits a screen might be likened to a broad ocean wave that somehow coalesces into a single large disturbance at part of a beach while the water everywhere else along the beach is calm. How can waves spreading from two slits form individual bundles of energy (photons) at a screen?

Electrons also exhibit a dual behavior. For example, electrons in atoms are found to reside in locations that can best be described using waves—much like the standing waves that we observed on vibrating strings in Section 18.1. Places where the wave's amplitude is large are likely places to find the electron. Yet if one of these electrons gets bumped out of an atom and flies across the tube of a television set, it behaves like a tiny ball of mass as it hits the screen.

Scientists found it necessary to develop a new theory that included both the wave and particlelike aspects of nature. The theory that evolved during the first thirty years of this century, called *quantum mechanics,* is now well established. In this chapter, we consider some of the unusual observations and ideas that led to the formulation of quantum mechanics. We will find that the world of atoms and electromagnetic radiation is very different from our everyday world.

31.1 Electromagnetic Waves

To start our trip through the world of atoms and radiation, let us first consider the wavelike nature of electromagnetic radiation as accepted by most scientists during the latter half of the 1800s. Thomas Young's double-slit interference experiment in 1803 (see Section 20.2) provided the first experimental evidence for a wave theory of light. By the 1860s and 1870s, the Scottish physicist James Clark Maxwell had provided a convincing wave description of light and of other forms of electromagnetic radiation.

According to Maxwell, an **electromagnetic wave** is an electrical and magnetic disturbance that moves through space at the speed of light ($c = 3.0 \times 10^8$ m/s). You probably recognize the names of most of the different types of electromagnetic waves—radio waves, microwaves, infrared waves, light, ultraviolet radiation, x-rays, and gamma rays. The waves differ from each other in their frequency and wavelength and in the way they are produced and interact with matter. The waves are similar in that they all move at the speed of light and consist of moving electric and magnetic fields.

The production of an electromagnetic wave by an oscillating electric charge in an antenna is illustrated in Fig. 31.1. The antenna consists of two vertical metal rods connected to an ac generator that causes charge to oscillate up and down from one rod to the other. In Fig. 31.1a, an excess negative charge resides at the top of the antenna, leaving a positive charge (a deficiency of electrons) at the bottom. In Fig. 31.1b, about one-eighth of a cycle of oscillation later ($\tau/8$), some electrons have moved down from the top rod to the bottom rod, causing the excess charge on each rod to be reduced. As electrons continue to move down, the rods are neutralized (Fig. 31.1c), and eventually the lower rod becomes negatively charged (Fig. 31.1d). One-half cycle of oscillation is completed when the charge distribution shown in Fig. 31.1a is reversed as in Fig. 31.1e. The second half of oscillation involves an upward movement of electrons and a return to the original charge distribution (Fig. 31.1i).

The charge on the antenna produces an electric field directly in front of the antenna. The field fluctuates up and down in phase with the charge distribution on the antenna. In Fig. 31.1a the field points up. When the charge on the antenna is reduced by one-half, as in Fig. 31.1b, the field directly in front of the antenna is reduced by one-half. When the charge distribution is reversed, as in Fig. 31.1e, the field directly in front of the antenna is reversed.

While the field near the antenna oscillates up and down in phase with the oscillating charge distribution on the antenna, the electric field created at earlier

FIG. 31.1. As electrons oscillate up and down on the antenna, an electric field is produced in front of it. As the field moves away at a speed of 3.0×10^8 m/s, the antenna continues to produce new fields.

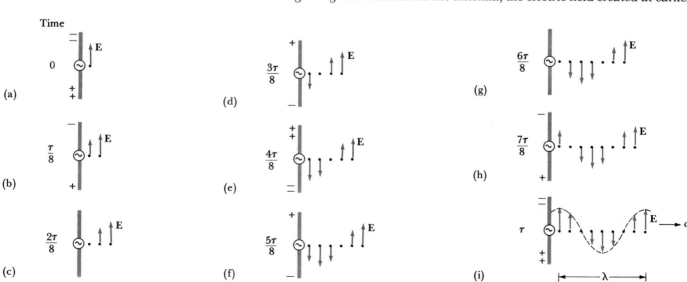

times moves away from the antenna at the speed of light. The symbol **E** traces the position of the original field (produced at time zero) at successive time intervals. A pattern of alternating up-and-down electric field lines leaves the antenna as its charge oscillates up and down. One wavelength of the electric field pattern (Fig. 31.1i) is caused by one cycle of oscillation of charge on the antenna. As the antenna charge continues to oscillate, the wave grows longer and the whole pattern continues to move away from the antenna at the speed of light.

An electric charge some distance from the antenna and in the wave's path is forced by the electric field of the passing wave to vibrate up and down. For example, electrons in the antenna of a car radio are pushed up and down by passing electromagnetic waves generated by a radio station antenna that is miles from the car.

The charges moving in an antenna such as shown in Fig. 31.1 also produce a magnetic field that encircles the antenna rods, as shown in Fig. 31.2. (See Section 27.7 for a discussion of how magnetic fields are produced by moving charges.) If the current moves up the antenna, a magnetic field such as shown in Fig. 31.2a is produced. If the current moves down, the field is reversed (Fig. 31.2b). As the electrons oscillate up and down in the antenna, an alternating magnetic field pattern is produced (Fig. 31.2c). This field also moves away from the antenna at the speed of light. The combined magnetic and electric fields produced by the oscillating charge on the antenna is called an **electromagnetic (EM) wave,** shown in Fig. 31.2d. The magnetic and electric fields are perpendicular to each other and are both perpendicular to the direction in which the wave moves. While the pattern shown in Fig. 31.2d represents the varying fields in one direction, similar patterns can be drawn along other directions since the EM wave spreads in many directions.

31.2 Electromagnetic Spectrum

If we could watch an EM wave as it passed a point in space, the directions of the electric and magnetic fields at that point would alternate back and forth. The frequency f of the wave is the number of complete vibrations per second of the field at a point along the path of the passing wave. The frequency also equals the vibration frequency of the wave's source.* For example, if the electrons in a radio station antenna move up and down at a frequency of 90 MHz, then the electromagnetic waves produced by the antenna have a frequency of 90 MHz.

As with other waves, the wavelength of an electromagnetic wave depends on its frequency and speed. In a vacuum, the wavelength of all electromagnetic waves is given by Eq. (17.8):

$$\lambda = \frac{c}{f},$$

where c, the wave's speed, equals 3.0×10^8 m/s. The different types of waves move at about the same speed in air but at different speeds in other media. (See Table 19.1 for some representative values of the speed of light in different media.)

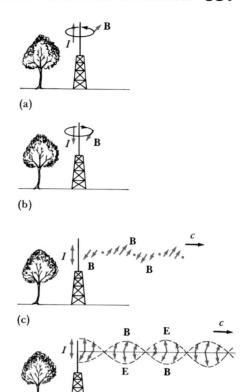

FIG. 31.2. (a) and (b) The magnetic fields produced in front of an antenna by its electric current. (c) A magnetic wave moving at a speed c leaves the antenna. (d) The combined electric and magnetic field moving at a speed c is called an electromagnetic wave.

---

*We assume that the wave's source and the device detecting the wave are at rest relative to each other. If not, the frequency of the wave is shifted because of the Doppler effect.

FIG. 31.3. The electromagnetic spectrum.

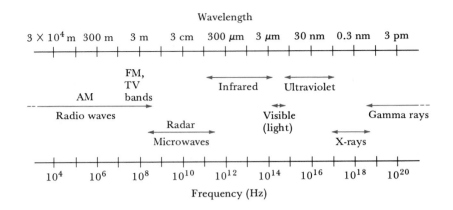

Electromagnetic waves are often classified by frequency in a scheme called the **electromagnetic spectrum** (Fig. 31.3). A particular frequency region in the EM spectrum includes waves produced in a similar manner. Infrared waves are produced by molecular vibrations, whereas radio waves are produced by charges vibrating back and forth in antennas. The boundaries between regions are not sharp.

The low radio-frequency region from 0 to 10^9 Hz consists of waves used mostly for communications—the lower frequencies being used for AM radio and telegraphy and the higher frequencies for FM radio and television communications. Microwaves, the next part of the electromagnetic spectrum, extend from about 10^9 Hz to more than 10^{11} Hz. Microwaves are used for radar tracking, radio navigation, communication, medical diathermy, heating in microwave ovens, drying, and other industrial purposes.

The infrared portion of the spectrum from about 5×10^{11} Hz to 4×10^{14} Hz comes next. Infrared waves are often emitted and absorbed because of the vibrations of atoms and molecules. For example, infrared rays absorbed by your skin cause the amplitude of vibration of molecules in your skin to increase. Your skin is warmed. For this reason infrared waves are sometimes called heat or thermal radiation.

The portion of the electromagnetic spectrum that includes visible light waves starts with red light at about 4×10^{14} Hz and ends with violet light at about 8×10^{14} Hz. This tiny portion of the whole EM spectrum is important to humans because of the sensitivity of our eyes to light. Light waves are caused by the motion of electrons in atoms as they jump between different orbits.

At even higher frequencies, ultraviolet radiation (8×10^{14}–10^{17} Hz) and x-rays (10^{17}–10^{19} Hz) are also produced by the motion of electrons in atoms. Ultraviolet radiation darkens the skin. X-rays are used for examining the interiors of objects that are opaque to light, such as the human body. The spectrum is completed by high-frequency gamma rays (10^{19} Hz and higher) produced by the motion of charged particles in nuclei.

Light waves and many other forms of electromagnetic waves are often described in terms of wavelength instead of frequency. The metric unit used for the wavelength of light waves is the nanometer (1 nm = 10^{-9} m). Red light has a wavelength of about 700 nm, violet light a wavelength of about 400 nm, and x-rays have wavelengths of about 0.1 nm, roughly the radius of an atom.

31.3 Photons

At the beginning of the twentieth century, the wave theory of electromagnetic radiation was well established using only the classical laws of electricity and magnetism. However, one problem still plagued physicists. The wave theory could not explain the frequency dependence of the EM radiation emitted by hot objects. As we see in Fig. 31.4, a hot object emits radiation whose peak intensity shifts toward a higher frequency as the object's temperature increases. The hot coals of a fire emit most of their energy as red light and as high-frequency infrared radiation. Your cooler body emits no light but does emit energy in the lower-frequency infrared region. The intensity of radiation emitted from all objects, hot and cold, approaches zero at high frequency, as shown in Fig. 31.4. Unfortunately, classical theories wrongly predicted just the opposite—that the intensity of emitted radiation should continue to increase with increasing frequency. According to classical theory, more energy would be emitted in the ultraviolet and x-ray regions than in the visible and infrared regions of the EM spectrum. The problem became known as the "ultraviolet catastrophe."

This problem attracted considerable attention in the last decade of the nineteenth century. The German physicist Max Planck finally solved the problem, reporting his new theory in December 1899 for a meeting of the German Physical Society in Berlin. His theory ascribed a new quantum nature to EM radiation—it consisted of tiny bundles of energy called *quanta*. These quanta eventually became known as *photons*.

Planck himself had not sought a new way to describe EM radiation. But by so doing, he was able to predict the correct frequency dependence for this emission from hot bodies. Planck, a conservative physicist, would have preferred to solve the radiation problem using laws of classical electricity, magnetism, and thermodynamics. He later described his proposal of quanta as "an act of despair . . . a theoretical interpretation had to be found at any price."

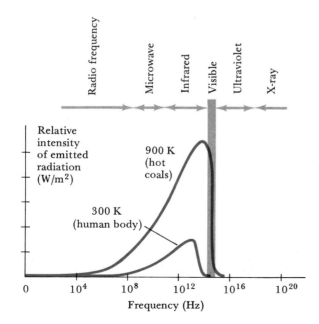

FIG. 31.4. The relative intensity of EM radiation emitted from a hot object and a cooler object is plotted as a function of frequency. The radiation is called blackbody radiation.

This act of despair ushered in the most exciting thirty years in the history of physics and heralded the beginning of the development of quantum mechanics. During this period, much experimental evidence was collected to support the idea that EM radiation behaves, at least in some experiments (see the next section), like photons.

A **photon** is considered to be a massless bundle of electromagnetic energy. The energy E of a single photon depends on its frequency f and is given by the equation

$$E = hf, \tag{31.1}$$

where h, a fundamental constant called Planck's constant, has a value $h = 6.63 \times 10^{-34}$ J·s.

In this definition, the word *massless* is used to distinguish photons from objects that have measurable rest mass, such as electrons and atoms. A photon cannot be captured and weighed on a scale. When a photon is caught, such as happens when light is absorbed by your skin, it ceases to exist. Its energy is converted to some other form of energy—for example, into thermal energy of atoms and molecules in your skin. Photons exist only while moving and do not have rest mass. Their speed is 3.0×10^8 m/s in air or in a vacuum and is usually slower when they are passing through other materials.

Our equation for calculating the energy of a photon depends on frequency, a quantity used to describe vibrations and waves. A discrete bundle of energy called a photon is characterized by a wavelike property. The connection between these bundles of energy and a wave that causes interference patterns is hard to visualize. Scientists in the early 1900s had a difficult time reconciling the idea of photons with the well-accepted wave theory of light. Albert Einstein, for example, used the photon concept in a 1905 paper to explain an important experiment. Yet in 1911 at a conference in Belgium, he told participants, "I insist on the provisional character of this concept which does not seem reconcilable with the experimentally verified consequences of the wave theory."

We will find later that there is no common experience appropriate for picturing electromagnetic radiation. However, it is useful at times to think of photons as moving, localized electromagnetic disturbances such as depicted in Fig. 31.5. In this picture, we represent the light from a flashlight as consisting of individual photons moving past a cross section in the path of the light beam. The brightness of the light depends on the number of photons passing along the light beam each second (about 10^{20} photons per second for the beam of a flashlight).

The energy of these photons in units of joules is very small. A photon of red light has an energy of about 3×10^{-19} J. A more convenient energy unit has been developed to describe photons and other phenomena related to atoms: one **electron volt** (eV) is the magnitude of the change in electrical potential energy that occurs when an electron crosses a potential difference of 1 V $[\Delta PE_q = qV = (1.6 \times 10^{-19}$ C)(1 V) $= 1.6 \times 10^{-19}$ J]. We see that

$$1 \text{ eV} = 1.6 \times 10^{-19} \text{ J}. \tag{31.2}$$

The energy of a photon of red light is about 2 eV. The energies in units of eV associated with other atomic phenomena are listed in Table 31.1.

FIG. 31.5. A flashlight sends out many photons of light. The frequencies f and energies ($E = hf$) of the photons differ.

EXAMPLE 31.1 Calculate the energy of a photon of red light with wavelength 660 nm in units of (a) joules and (b) electron volts. (c) Determine the wavelength of a photon with twice this energy.

SOLUTION (a) Using Eqs. (31.1) and (17.8), we see that

$$hf = h\left(\frac{c}{\lambda}\right) = (6.63 \times 10^{-34} \, \text{J·s})\left(\frac{3.0 \times 10^8 \, \text{m/s}}{660 \times 10^{-9} \, \text{m}}\right) = \underline{3.0 \times 10^{-19} \, \text{J}}.$$

(b) The solution to part (a) in joules is converted to electron volts using Eq. (31.2):

$$3.0 \times 10^{-19} \, \text{J}\left(\frac{1 \, \text{eV}}{1.6 \times 10^{-19} \, \text{J}}\right) = \underline{1.9 \, \text{eV}}.$$

(c) This problem is worked easily using a proportionality method. If $E' = hf' = hc/\lambda'$ is the energy of the photon with twice the energy and $E = hc/\lambda$ is the energy of the photon of wavelength 660 nm, then, after rearranging,

$$hc = E'\lambda' = E\lambda, \quad \text{or} \quad \lambda' = \frac{E\lambda}{E'} = \frac{E\lambda}{(2E)} = \frac{\lambda}{2} = \underline{330 \, \text{nm}}.$$

Photons of shorter wavelength are more energetic. ■

EXAMPLE 31.2 An x-ray photon produced during a medical chest x-ray has an energy of about 40,000 eV. (a) Calculate the photon's wavelength. (b) During a chest x-ray, a person's body typically absorbs x-ray photons whose total energy is about 1.5×10^{-3} J. Calculate the number of x-ray photons absorbed.

SOLUTION (a) The wavelength is determined by combining and rearranging Eqs. (31.1) and (17.8):

$$E = hf = h\left(\frac{c}{\lambda}\right),$$

$$\lambda = \frac{hc}{E} = \frac{(6.63 \times 10^{-34} \, \text{J·s})(3.0 \times 10^8 \, \text{m/s})}{(40,000 \, \text{eV} \times 1.6 \times 10^{-19} \, \text{J/eV})} = 3.1 \times 10^{-11} \, \text{m} = \underline{0.031 \, \text{nm}}.$$

Note that we must convert the energy from electron volts to joules before calculating the wavelength in meters.

(b) The energy of each photon is converted to joules with the assistance of Eq. (31.2):

$$\text{Energy per photon} = 40,000 \, \text{eV}\left(\frac{1.6 \times 10^{-19} \, \text{J}}{1 \, \text{eV}}\right) = 6.4 \times 10^{-15} \, \text{J}.$$

Since the total energy of all x-ray photons absorbed is 1.5×10^{-3} J, the number of photons absorbed is

$$\frac{1.5 \times 10^{-3} \, \text{J}}{6.4 \times 10^{-15} \, \text{J per photon}} = \underline{2.3 \times 10^{11} \, \text{photons}}.$$

■

TABLE 31.1 Energy of Several Atomic Occurrences and Photons in Units of Electron Volts (eV)

Ionizing a hydrogen atom	13.6 eV
Breaking a hydride bond	1 eV
Causing a DNA mutation	3–10 eV
Energy of light photon	2–3 eV
Energy of UV photon	3–300 eV
Energy of x-ray photon	~40,000 eV
Thermal energy of one atom at room temperature	1/40 eV

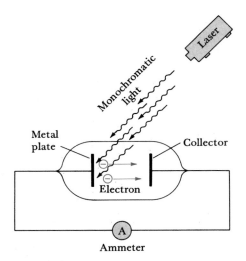

FIG. 31.6. An apparatus used to observe the photoelectric effect.

31.4 Photon Experiments

The photon theory of EM radiation is supported by a variety of experiments. The first success of the photon theory was Planck's explanation of the frequency distribution of electromagnetic radiation that comes from hot objects. Other early successes of the photon theory included the photoelectric effect and the Compton effect.

Photoelectric Effect

The photoelectric effect involves the ejection of electrons from the surface of a metal when it is irradiated with EM radiation. It was first observed experimentally in 1887 by Heinrich Hertz and was explained in terms of the photon theory by Albert Einstein in 1905. In 1921, Einstein was awarded the Nobel Prize for this and other work.

The apparatus for the photoelectric effect is shown in Fig. 31.6. Light of one frequency and wavelength, called *monochromatic light,* strikes a metal plate. If an electron in the metal absorbs enough energy from the light, the electron can escape the metal's surface. The minimum energy needed for an electron to escape, called the metal's **work function** ϕ, is similar to the energy needed by a water molecule to escape the surface of liquid water (to evaporate). The value of the work function differs for various types of metals.

If an electron escapes the metal, it passes across a vacuum and is captured by a collector (another piece of metal). The electron moving from the plate to the collector constitutes what is called a **photoelectric current** (the flow of electrons induced by the absorption of photons). This current, after reaching the collector, continues along a wire back to the original metal plate and is measured by an ammeter along the way.

The wave theory of EM radiation could not explain the photoelectric current. In this theory, the energy of a light wave is spread continuously over the wavefronts of the light (Fig. 31.7a). Calculations have shown that a single electron intercepts only a small fraction of the wave's energy, and considerable time would be needed for an electron to absorb enough energy from the light to escape the metal's surface.

However, when a photoelectric-effect experiment is performed, the current is observed at the instant the light is turned on; there is no delay. This instantaneous current can be explained if one assumes that light consists of photons. A single electron can gain all the energy (hf) of one photon (Fig. 31.7b and c). Part of the energy is used to escape the metal, and any energy remaining after the electron escapes the surface is available as kinetic energy. In terms of energy, the absorption of a photon and the subsequent escape of an electron from the metal's surface are represented by Eq. (8.12):

Initial energy = Final energy,   or

$$\underbrace{hf}_{\substack{\text{Energy of}\\\text{photon}}} = (\underbrace{\phi}_{\substack{\text{To escape}\\\text{metal}}} + \underbrace{\tfrac{1}{2}mv^2}_{\substack{\text{Leftover}\\\text{kinetic energy}}}). \tag{31.3}$$

Energy gained by electron

If the light has a frequency such that hf is less than ϕ, the electron will not gain enough energy from an absorbed photon to escape the surface. Thus, no photoelectric current is observed if low-frequency EM radiation illuminates the metal plate, even if the radiation is intense. The wave theory of light has no explanation for this cutoff.

The work function of a metal can be determined by increasing the frequency of the EM radiation to a value where the photoelectric current starts. The minimum frequency f_0 of the EM radiation needed to start the photoelectric current is such that

$$hf_0 = \phi. \qquad (31.4)$$

At this frequency the electron leaves the surface of the metal plate but has no excess energy. For most metals the photoelectric current is initiated by EM radiation in the visible portion of the EM spectrum. But if the work function of a metal is large, ultraviolet-frequency photons may be needed to initiate the current.

The photon theory of light as presented in 1905 by Einstein did not gain instant support. An American physicist, Robert Millikan, was one of the idea's strongest opponents. He spent ten years experimentally testing the photon idea with the hope of proving it wrong. By 1914 his experiments provided strong experimental support *for* the photon. The experiments also determined very accurately the value of Planck's constant h and the charge e of an electron. Ironically, Millikan was awarded the Nobel Prize in 1923 for his work.

EXAMPLE 31.3 Light strikes the surface of a piece of aluminum whose work function ϕ is 4.20 eV. (a) Calculate the minimum-frequency photon that will supply enough energy for an electron to escape the surface of the aluminum. What is the photon's wavelength? (b) If a 200-nm ultraviolet-wavelength photon is absorbed by an electron in the metal's surface, how much kinetic energy does the electron have when it leaves the metal?

SOLUTION (a) For the electron to gain just enough energy to escape the aluminum surface, the photon's energy hf_0 must equal the energy ϕ needed to remove the electron from the metal, or $hf_0 = \phi$. Thus,

$$f_0 = \frac{\phi}{h} = \frac{(4.20 \text{ eV})(1.6 \times 10^{-19} \text{ J/eV})}{6.63 \times 10^{-34} \text{ J} \cdot \text{s}} = 1.01 \times 10^{15} \text{ s}^{-1} = \underline{1.01 \times 10^{15} \text{ Hz}}.$$

The photon's wavelength is determined by rearranging Eq. (17.8):

$$\lambda = \frac{c}{f} = \frac{3.00 \times 10^8 \text{ m/s}}{1.01 \times 10^{15} \text{ s}^{-1}} = 2.97 \times 10^{-7} \text{ m} = \underline{297 \text{ nm}},$$

a photon in the ultraviolet part of the electromagnetic spectrum.
(b) A shorter-wavelength, 200-nm photon has more energy than needed to knock an electron from the surface. Its energy is determined by combining Eqs. (31.1) and (17.8):

$$E = hf = h\frac{c}{\lambda} = (6.63 \times 10^{-34} \text{ J} \cdot \text{s})\left(\frac{3.0 \times 10^8 \text{ m/s}}{200 \times 10^{-9} \text{ m}}\right)$$

$$= 9.95 \times 10^{-19} \text{ J} = 6.22 \text{ eV}.$$

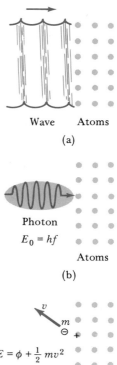

Wave   Atoms
(a)

Photon
$E_0 = hf$

Atoms
(b)

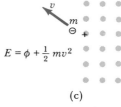

$E = \phi + \frac{1}{2}mv^2$

(c)

FIG. 31.7. (a) If light behaved only as a wave, the energy of the wave is distributed across its wavefronts. As the wavefronts reach a "shore" of atoms in a metal, there is not enough energy given to a single atom to knock one of its electrons out of the metal. However, (b) if the light behaves like photons, a single atom can absorb the entire energy of a photon. (c) For high-frequency photons, the atom gains more than enough energy for an electron to be knocked out of the metal.

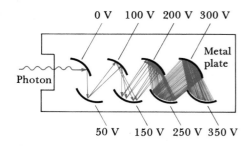

FIG. 31.8. A photomultiplier tube.

Since an electron needs only 4.20 eV of energy to escape the aluminum, the energy that remains after absorbing a photon is in the form of kinetic energy:

$$\frac{1}{2}mv^2 = hf - \phi = 6.22\,\text{eV} - 4.20\,\text{eV} = \underline{2.02\,\text{eV}}. \quad\blacksquare$$

Several devices used to detect light and ultraviolet radiation rely on the photoelectric effect. Photocells in burglar alarms, automatic door openers, and smoke detectors measure the flow of electric charge from a metal surface that is exposed to the light. If the light striking the surface is interrupted, the current leaving the surface stops.

A **photomultiplier tube** is one of the most sensitive light detectors and is found in many scientific instruments that measure the properties of atoms and molecules as indicated by the wavelengths of light they absorb or emit. The tube consists of a series of metal surfaces called *stages* or *dynodes* (Fig. 31.8). A potential difference of about 50 to 100 V is connected across adjacent stages. When a single photon strikes the first stage, a single electron is released, which is accelerated by the voltage toward the second stage. By the time the electron hits the second stage, the electron has gained enough energy to knock several electrons out of its surface. These electrons accelerate across a voltage toward the third stage, where each electron frees several more electrons. This may continue through ten or more stages. At the last stage a huge number of electrons are released. For example, if each electron releases three electrons as it hits the next stage, then the number released at the tenth stage due to the absorption of a single photon at the first stage is $3^9 = 19{,}683$. Thus, one photon can cause a fairly large pulse of electric charge that is easily detected.

The Compton Effect

A. H. Compton (1892–1962), an American physicist from Washington University, provided additional support for the photon theory of EM radiation in 1923. In his experiment Compton irradiated a graphite block with monochromatic x-rays and observed that some of the scattered x-rays had a frequency lower than that of the original x-rays. Just as we would not expect a sound wave to change frequency when scattered from a wall, Compton did not expect x-rays to change frequency when scattered from graphite—that is, if the x-rays were considered to be waves. However, the frequency change could be explained nicely using the photon theory of EM radiation.

To see how the photon theory was used, suppose that an x-ray photon of frequency f_0 collides with a stationary electron in the graphite (Fig. 31.9). During the collision the photon imparts kinetic energy to the electron, causing its speed to increase. The increased kinetic energy of the electron must be balanced by a decrease in the photon's energy from an original value hf_0 to a final value hf, where f is the frequency of the scattered photon. Thus, using the energy-conservation idea again [Eq. (8.12)], we find that

Initial energy = Final energy, or

$$\underbrace{hf_0}_{\substack{\text{Initial}\\\text{photon}\\\text{energy}}} = \underbrace{hf}_{\substack{\text{Final}\\\text{photon}\\\text{energy}}} + \underbrace{\Delta KE_{\text{electron}}.}_{\substack{KE \text{ gained}\\\text{by}\\\text{electron}}}$$

Before collision

Incident
photon

→ ◯ Electron

hf_0

After collision

v ↗

◯

Scattered
photon

hf

FIG. 31.9. The frequency of a photon changes after it is scattered from an electron.

We see that the photon's frequency must decrease as a result of the collision (f is less than f_0, as observed by Compton).

Since the electron's speed has increased, its momentum has also increased. Recall from Chapter 4 that momentum is a conserved quantity; thus, a momentum increase by the electron must be balanced by a momentum decrease of the photon. But the photon must have momentum in order to lose it.

Compton deduced an expression for a photon's momentum by equating the expression for its energy ($E = hf$) to the relativistic expression for the energy of a particle of relativistic mass m ($E = mc^2$). Thus, $mc^2 = hf$, or $mc = hf/c$. The quantity mc represents the momentum of a particle of mass m moving at speed c. The photon moves at speed c and, because of the equivalence of mass and energy, is considered to have relativistic mass $m = E/c^2 = hf/c^2$. Thus, Compton deduced that the momentum of the photon must be

$$\text{Momentum of a photon} = p = mc = \left(\frac{hf}{c^2}\right)c = \frac{hf}{c} = \frac{h}{\lambda}. \qquad (31.5)$$

where $\lambda = c/f$ [Eq. (17.8)] is the photon's wavelength. Compton combined equations for the conservation of momentum and the conservation of energy during the photon-electron collision, and he was able to predict exactly the change in wavelength of x-ray photons that were scattered at different angles by the electrons in the graphite. The experiment and its interpretation provided another important support for the photon theory of electromagnetic radiation.

31.5 The Wave-Particle Nature of EM Radiation

While many experiments imply that EM radiation is made of photons and exhibits a particlelike behavior, other experiments, such as those discussed in Chapter 21, can be explained using a wave theory for EM radiation. What, then, is EM radiation—a wave or a photon?

Our present conception of EM radiation draws from a theory called **quantum electrodynamics.** The dual nature of EM radiation is a consequence of that theory and can be illustrated by Young's double-slit interference experiment discussed in Section 21.2 and depicted in Fig. 31.10a. Coherent light of very low intensity irradiates a pair of slits. After passing through the slits, the light moves toward a photographic plate.

If the plate is exposed to the light for a very short time and if the light intensity is low, the photographic plate becomes exposed as in Fig. 31.10b. Each little spot represents the place where a *photon* has hit the plate and caused it to become exposed. If the light is allowed to expose the plate for a longer time, a pattern of fringes begins to emerge as in Fig. 31.10c and d. In Chapter 21, we explained these fringes as being caused by the interference of light *waves* coming from the two slits. If the light is intense, the fringe pattern is produced rapidly, and the fact that individual photons are causing the pattern is not apparent. However, by reducing the light intensity, we see that photons and not waves are hitting the plate (as in Fig. 31.10b). The photons hit the plate at places where we would expect to see constructive interference of waves.

Based on this and many other experiments, we have evidence that EM radiation has both a wave and a particle (that is, photon) behavior—a **wave-particle duality.** Whenever radiation is detected by an experimental apparatus or interacts with matter, it behaves like a photon. For example, when light

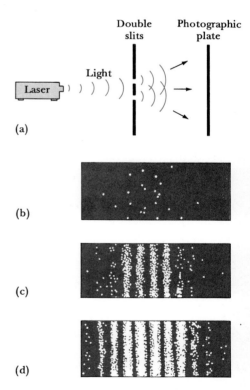

FIG. 31.10. (a) An apparatus for observing the double-slit interference pattern of light. (b), (c), and (d) As more photons hit the photographic plate, a pattern of interference fringes appears.

strikes the plate of a photomultiplier tube or a photographic plate, it behaves like photons. The wavelike behavior of EM radiation determines the probability of detecting photons at different places. When light passes through two slits and then strikes a photographic plate, we must use a wavelike description of the light to decide which are the probable places on the plate for the photons to strike. For the double-slit experiment, photons hit those places on the plate where waves from the slits undergo constructive interference. The dual wave-particle behavior is sometimes summarized as follows:

EM radiation behaves like photons when interacting with matter and behaves like a wave when propagating unobserved through space from a source to a place where it is detected (the wave determines probable places for detecting the photon).

This dual wave-particle behavior of EM radiation differs from the behavior of objects we are used to seeing with our eyes. A baseball, when moving from a bat to the glove of an outfielder, follows a well-defined path and exhibits no wavelike behavior. On the other hand, an ocean wave acts like a wave as it moves toward and collides with a beach; the wave does not strike the beach as discrete bundles of energy at particular locations. The fact that EM radiation exhibits both wave and particle behavior is one of the interesting surprises that physicists have discovered during this century. Perhaps even more surprising has been the discovery that objects with mass also exhibit a dual wave-particle behavior.

31.6 The Wave-Particle Nature of Matter

In 1923 a 31-year-old French graduate student named Louis de Broglie made the unusual proposal that if EM radiation had a dual wave-particle behavior, then objects with mass such as electrons and protons might also exhibit this dual behavior. The idea that an electron might exhibit wave behavior seemed strange. Waves are characterized by their wavelength, frequency, speed, amplitude, and tendency to spread as they move. Waves also undergo interference. None of these properties resembles our usual conception of a particle such as an electron, which we think of as a small, well-defined ball of mass.

Nevertheless, de Broglie thought that particles with mass might also exhibit a wavelike behavior. He even deduced an expression for the wavelength λ that should be associated with the motion of an object of mass m when moving at speed v. First he equated the equations for the momentum of a photon ($p = hf/c = h/\lambda$) and the momentum of a particle of mass m when moving at speed v ($p = mv$) to obtain the equation $mv = h/\lambda$. By rearranging, he obtained an expression for the wavelength (called the **de Broglie wavelength**) associated with the motion of the moving mass:

$$\lambda = \frac{h}{mv}. \tag{31.6}$$

Three years after de Broglie's proposal, two experiments were performed that showed clearly the wave behavior of electrons. Since that time, many other experiments have been performed that show that all tiny masses—electrons, protons, neutrons, atoms, and ions—behave somewhat like waves and that the wavelength associated with their motion is given by Eq. (31.6). Instead of reviewing these experiments, we will again use a modified version of Young's double-slit experiment to illustrate the wave nature of mass particles.

The experiment is illustrated in Fig. 31.11a. A beam of electrons moves at speed v toward a pair of slits. After passing through the slits, the electrons fly toward a fluorescent screen. Each time an electron hits the screen, a flash of light occurs.

If the electrons behave only like particles, we expect to see electrons hitting the screen in two bands (Fig. 31.11b) that lie along a line from the electron source through the slits and to the screen. The pattern that is actually observed is nothing like this. After only a few electrons have hit the screen, the distribution appears quite random, as shown in Fig. 31.11c. As more and more electrons reach the screen, a pattern consisting of alternating bright and dark fringes emerges (Fig. 31.11d and e). This pattern is exactly like the pattern produced when light passes through a pair of slits.

We see that electrons, when they hit the screen, appear as discrete particles; but the places where the electrons hit the screen are places where waves from the two slits undergo constructive interference. The angular deflection to these bright bands where many electrons hit the screen agrees perfectly with calculations using de Broglie's equation for the wavelength of an electron ($\lambda = h/mv$) and the double-slit interference equation developed in Section 21.2:

$$\sin \theta_n = \frac{n\lambda}{d}, \qquad (21.1)$$

where θ is the angular deflection from the slits to the nth interference maximum on the screen, λ is the de Broglie wavelength of the electrons, and d is the slit separation.

The experiment just described is difficult to perform because the wavelengths of electrons are very short, even at low speed, and the slits must be very close to each other to produce a fringe pattern that can be seen. However, the bands are easily observed if electron beams pass through crystals. The space between adjacent rows of atoms serve as closely spaced slits. The actual experiments that confirmed the wavelike nature of electrons used such solid materials as the "slits" to cause interference patterns of the electron waves. These experiments were first performed in 1927 by C. J. Davisson and L. H. Germer of Bell Telephone Laboratories and independently by G. P. Thomson, an English physicist whose father had discovered the electron in 1897.

By 1930 the wave nature of matter had become the heart of a completely new theoretical description of the behavior of atomic-sized masses. The new theory, called *wave mechanics* or *quantum mechanics,* has been fine-tuned for the last fifty years to describe a wide variety of properties of atoms, molecules, and the atomic properties of liquids and solids. In this theory the wave that is calculated for a particle determines the probability of detecting the particle at different locations. For the electrons passing through the double slit, the square of the wave amplitude gives the probability of observing an electron at different places on the fluorescent screen. The electrons hit the screen only at those places where a wave of wavelength $\lambda = h/mv$ undergoes constructive interference after passing through the slits.

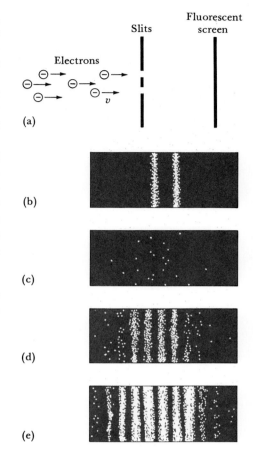

(a)

(b)

(c)

(d)

(e)

FIG. 31.11. (a) An apparatus for observing the double-slit interference pattern of electrons. (b) The pattern of electrons hitting the screen that would be expected if the electrons behaved entirely like particles. (c), (d), and (e) Instead, the electrons seem to be guided to the screen by waves that cause a pattern of interference fringes to evolve as more electrons hit the screen.

Objects with mass exhibit a wave-particle duality just as EM radiation does. The objects behave like particles when interacting with a measuring device or with matter, and they behave like waves when spreading unobserved through space. The wave determines the particle's probability of being detected at different locations.

EXAMPLE 31.4 Calculate the wavelength of the wave associated with an electron of mass 9.11×10^{-31} kg that has been accelerated from rest across a 3000-V potential difference. Ignore any relativistic effects.

SOLUTION To calculate λ, we first need to know the electron's speed. The speed is determined from the energy-conservation equation (see Example 23.7):

$$0 = \Delta KE + \Delta PE_q = \frac{1}{2}mv^2 + (-e)V.$$

Thus,

$$v = \sqrt{\frac{2eV}{m}} = \sqrt{\frac{2(1.6 \times 10^{-19} \text{ C})(3000 \text{ V})}{(9.11 \times 10^{-31} \text{ kg})}} = 3.2 \times 10^7 \text{ m/s}.$$

The electron's wavelength is given by Eq. (31.6):

$$\lambda = \frac{h}{mv} = \frac{6.63 \times 10^{-34} \text{ J} \cdot \text{s}}{(9.11 \times 10^{-31} \text{ kg})(3.2 \times 10^7 \text{ m/s})} = 2.2 \times 10^{-11} \text{ m} = \underline{0.022 \text{ nm}}.$$

This wavelength is about 0.5×10^{-4} times shorter than the wavelength of light. The ability of electron microscopes to view small objects that cannot be seen with light depends on this short wavelength, as we learn in Section 31.8. ∎

You might be wondering why the wave nature of matter is unobserved for large objects such as people and cars. Consider the next example.

EXAMPLE 31.5 Calculate the de Broglie wavelength of a 1-kg ball rolling at a speed of 1 m/s.

SOLUTION

$$\lambda = \frac{h}{mv} = \frac{6.63 \times 10^{-34} \text{ J} \cdot \text{s}}{(1 \text{ kg})(1 \text{ m/s})} = \underline{6.6 \times 10^{-34} \text{ m}}.$$

The ball's wavelength is about 10^{-20} times smaller than the size of the nucleus of an atom. The crests and troughs of the wave are so close to each other that they cannot possibly be used to indicate the probability of finding the bowling ball at one place compared to another. The wave nature of an object is observed only for very small masses about the size of atoms and the particles of which they are made. ∎

31.7 The Uncertainty Principle

We have seen that both EM radiation and objects with mass exhibit a dual wave-particle behavior. This behavior has led to interesting discussions about the limits of our ability to accurately measure the properties of small objects.

The discussion revolves around the idea that a measuring device must affect the quantity being measured.

Suppose, for example, that you wish to measure the temperature of a cup of tea. If a thermometer is placed in the tea, the tea cools a little as the thermometer warms. The final temperature of the thermometer and tea is less than the tea's initial temperature. The measuring device has altered the quantity being measured.

We can usually correct for errors in measurement if we know the characteristics of our measuring device (such as the temperature of the thermometer before entering the tea and its mass and specific heat capacity). There is, however, an intrinsic limit to our ability to make accurate measurements. This limit depends on the dual wave-particle nature of matter.

To see how this limit occurs, consider an experiment in which the position of a small object is determined by scattering electromagnetic radiation from the object. After passing the object, the radiation produces a diffraction pattern on a screen behind the object and indicates its position. A detailed analysis such as performed in Interlude V indicates that the object's uncertainty in position Δx equals roughly the wavelength λ of the radiation used to determine its location:

$$\text{Uncertainty in position} = \Delta x \simeq \lambda.$$

The shorter the wavelength, the better we can determine the object's position.

However, short-wavelength electromagnetic radiation has a disadvantage. Remember that this radiation has a particlelike behavior as well as a wavelike behavior. The electromagnetic "particles" called photons have momentum $p = h/\lambda$ and can transfer momentum to an object that they hit. If a photon has a direct hit, the object moves backward; if the photon has a glancing blow, the object is bumped to the side. The object's sideways momentum after the collision can vary from about zero to $\pm h/\lambda$, where λ is the wavelength of the scattered photon. The uncertainty in the object's x component of momentum after the photon is scattered is

$$\text{Uncertainty in the } x \text{ component of momentum} = \Delta p_x \simeq \frac{h}{\lambda}.$$

To disturb the object's momentum only slightly, long-wavelength radiation should be used.

We are now in a predicament. To locate the object's position most accurately, short-wavelength radiation should be used; but this radiation when scattered from the object creates considerable uncertainty in its momentum. If we use long-wavelength radiation, the object's momentum is only slightly disturbed, but its position is more uncertain. We can express the predicament quantitatively by taking the product of the object's uncertainties in position Δx and momentum Δp_x:

$$\Delta x \, \Delta p_x \simeq (\lambda)\frac{h}{\lambda} = h.$$

The product of these uncertainties equals Planck's constant h and is independent of the type of radiation used for the measurement. This means that no matter what we do, we cannot simultaneously determine an object's exact position and the exact value of its momentum. Similar equations can be derived for the uncertainties of position and momentum along other axes.

A careful derivation of this uncertainty principle, as done in 1927 by the German physicist Werner Heisenberg (1901–1976), shows that the constant h

should be replaced by $h/2\pi$. The Heisenberg uncertainty principle is stated as follows:

Uncertainty principle: It is impossible to know simultaneously with certainty the position and momentum of a particle along any axis. At best, the product of the uncertainties of these two quantities must exceed $h/2\pi$, where $h = 6.63 \times 10^{-34}$ J·s. Thus,

$$\Delta x \, \Delta p_x \geq \frac{h}{2\pi}. \tag{31.7}$$

A similar relation applies to our ability to know the value of a particle's energy and the length of time during which the energy is measured. Suppose that a photon of wavelength λ is used to locate a small object. As we said earlier, the object's position is uncertain by an amount $\Delta x \simeq \lambda$. The photon travels at speed c; hence, it takes a time

$$\Delta t = \frac{\Delta x}{c} \simeq \frac{\lambda}{c}$$

for the photon to pass through a distance the length of the object's position uncertainty. During that time, the photon can transfer all or none of its energy to the object, depending on the way in which it scatters from the object. The object's energy is uncertain by an amount

$$\Delta E \simeq hf = h\frac{c}{\lambda}.$$

The product of these two uncertainties is

$$\Delta E \, \Delta t \simeq h.$$

Heisenberg's more careful derivation of this uncertainty equation showed that the constant h should be replaced by $h/2\pi$.

A measurement on an object requiring a time Δt produces an uncertainty in its energy ΔE that can be no less than is consistent with the equation

$$\Delta E \, \Delta t \geq \frac{h}{2\pi}. \tag{31.8}$$

EXAMPLE 31.6 An electron in an atom is bumped into an orbit farther from the nucleus by a collision with another atom. The electron moves around this "excited" orbit for 10^{-8} s and then falls back to the orbit in which it moved before the collision. Calculate the minimum uncertainty in the energy of the excited orbit.

SOLUTION Since the electron resided in this orbit only 10^{-8} s, this is the maximum time available to measure the energy of an electron when in the orbit. Even if we use the best equipment we could ever devise, the uncertainty in the measurement of the electron's energy would still be equal to or greater than that given by Eq. (31.8):

$$\Delta E \geq \frac{h}{2\pi \, \Delta t} = \frac{6.63 \times 10^{-34} \text{ J·s}}{2\pi(10^{-8} \text{ s})} = \underline{1.1 \times 10^{-26} \text{ J}}, \quad \text{or} \quad 6.9 \times 10^{-8} \text{ eV}.$$

Other factors usually cause greater uncertainty in the measurement of the energy states of atoms. ∎

EXAMPLE 31.7 A free electron in a copper wire moves randomly with a speed of 1000 m/s. Suppose that we can devise an experiment to measure the electron's speed along the x axis with a 10 percent uncertainty. Calculate the uncertainty in the electron's position. The mass of an electron is 9.11×10^{-31} kg.

SOLUTION The electron's speed is uncertain by 10 percent of 1000 m/s, or $\Delta v_x = 100$ m/s. The electron's position uncertainty in the wire is determined using Eq. (31.7) where we substitute $m \, \Delta v_x$ for the momentum uncertainty:

$$\Delta x \geq \frac{h}{2\pi m \, \Delta v_x} = \frac{6.63 \times 10^{-34} \, \text{J} \cdot \text{s}}{2\pi(9.11 \times 10^{-31} \, \text{kg})(100 \, \text{m/s})} = \underline{1.2 \times 10^{-6} \, \text{m}} \, .$$

This position uncertainty is greater than the length of 1000 atoms lined in a row. It is impossible to pinpoint the location of an electron in a wire. In modern physics the free electron is described by a wave that extends over long distances in the wire and whose amplitude squared at each location represents the probability of finding the electron at that location. ∎

31.8 Electron Microscopes

One interesting and useful technological application of the wavelike behavior of matter is the **electron microscope,** a device that views small objects, such as human body cells, by scattering electrons off them. Electrons have considerable advantage over light waves for viewing small objects because of the short wavelength associated with their motion. If the wavelength is shorter, then details of smaller structures can be seen. When we use the best microscope that transmits light, two objects must be farther apart than about 100 nm so that their images are distinct. When we use an electron microscope, however, two objects separated by 0.1 nm form distinct images. The superior resolution of the electron microscope is due to the fact that the scattered electrons have a wavelength ($\lambda = h/mv$) of about 0.005 nm rather than the 500-nm wavelength of light. The resolution of the electron microscope is limited not by the electrons' wavelength but by other factors in the construction of the microscope and in the preparation of the sample being viewed.

A schematic representation of a transmission electron microscope is shown in Fig. 31.12. Electrons are emitted from a hot filament and focused in a parallel beam by a condensing lens that consists of a coil of wire through which a current flows. The current creates a magnetic field that bends and focuses the electron beam, much as glass lenses bend and focus light. After leaving the condensing lens, the electron beam passes through a thin sample (50 nm or less) that is being viewed. The electron beam scattered off the sample is then focused and magnified by two more magnetic lenses—the objective lens and the projection lens. A final image of the sample is formed on a fluorescent screen or on a photographic plate.

The focal length of a magnetic lens depends on the speed of the electrons. To have a uniform focal length for all electrons, the electrons must all move at

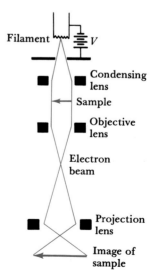

FIG. 31.12. A transmission electron microscope.

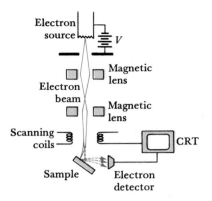

FIG. 31.13. A scanning electron microscope.

about the same speed, which is why the sample must be so thin. If it were thicker, some electrons would be slowed more than others while passing through the sample, and a blurry image would be formed. Thus, sample preparation is a problem when using transmission electron microscopes. In addition, the depth of field is very small in this type of microscope, thus restricting its use to viewing flat, two-dimensional objects.

A scanning electron microscope presents neither of these problems. In this microscope a focused electron beam scans across the specimen along a series of adjacent lines (Fig. 31.13). As the electron beam strikes each point on the specimen, electrons are knocked out of the specimen. These electrons are then collected by a detector. The number of electrons released from each point on the specimen determines the brightness of a spot on the screen of a cathode ray tube (CRT), such as found in a television set. As the microscope electron beam scans across the sample, the spot moves across the CRT. Each point on the specimen corresponds to a point on the screen of the CRT.

If the microscope electron beam strikes the side of an upward projection of the specimen, more electrons are released from the sample and a brighter spot is produced on the CRT screen. In this way the orientation of different parts of the specimen can be determined. If two pictures are taken at different angles, a stereoscopic view of the specimen is seen (Fig. 31.14).

Although the resolution of a scanning electron microscope is only about 10 nm, as opposed to 0.1 nm for the transmission electron microscope, the decreased resolution is balanced by increased depth of field and ease of sample preparation.

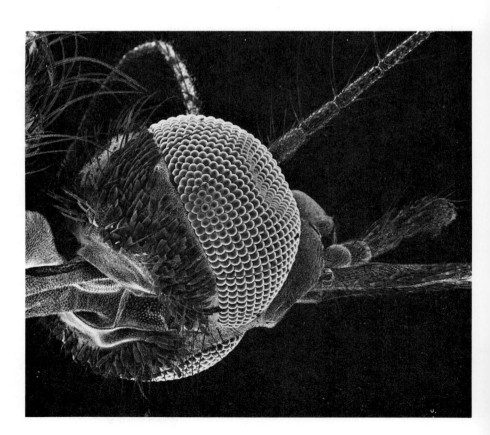

FIG. 31.14. The underside of the head of a female mosquito (*Aedes*) as seen with a scanning electron microscope. Shown are the neck (left), eyes, and proboscis (right), magnified 200 times. (Copyright © David Scharf/Peter Arnold, Inc. All rights reserved.)

Summary and Additional Readings

1. **Wave-particle nature of electromagnetic radiation:** Electromagnetic radiation behaves like a wave when propagating through space and like a particle called a photon when it interacts with matter. An **electromagnetic wave** is an alternating electric and magnetic field that moves through space at a speed of 3.0×10^8 m/s. The various forms of electromagnetic wave, such as radio waves, microwaves, infrared waves, light, ultraviolet radiation, x-rays, and gamma rays, differ from each other in frequency and wavelength. The frequency f, wavelength λ, and speed c of the waves are related by the equation $f\lambda = c$.

A **photon** is a bundle of electromagnetic energy with zero rest mass. A photon's energy depends on its frequency as given by the equation

$$E = hf, \tag{31.1}$$

where h, Planck's constant, equals 6.63×10^{-34} J·s. A photon's momentum is given by the equation

$$p = \frac{h}{\lambda}. \tag{31.5}$$

2. **Wave-particle nature of mass:** Particles with rest mass also exhibit a dual wave-particle behavior. The energy of a particle with mass m is

$$E = mc^2 \tag{30.5}$$

and its momentum is

$$p = mv. \tag{5.1}$$

The particle exhibits a wavelike behavior when propagating through space. The **de Broglie wavelength** associated with the particle's motion when moving at speed v is

$$\lambda = \frac{h}{p} = \frac{h}{mv}. \tag{31.6}$$

The wave is used to determine the probability of observing the particle at different locations.

3. **Uncertainty principle:** It is impossible to know simultaneously with certainty the position and momentum of a particle. At best, the product of the uncertainties of these two quantities must exceed $h/2\pi$, where $h = 6.63 \times 10^{-34}$ J·s. Thus,

$$\Delta x \, \Delta p_x \geq h/2\pi. \tag{31.7}$$

Similarly, a measurement of an object's energy during a time Δt must be uncertain by an amount ΔE consistent with the following equation:

$$\Delta E \, \Delta t \geq h/2\pi. \tag{31.8}$$

H. A. Medicus, "Fifty Years of Matter Waves," *Physics Today* **27**, 38 (1974). A historical account of wave theory.

R. Furth, "The Limits of Measurement," *Scientific American* **183**, 48 (1950).

George Gamow, *Thirty Years That Shook Physics: The Story of Quantum Theory*, Anchor Books, Doubleday and Co., Garden City, N.Y. (1966). An interesting journey through the development of modern physics in the first thirty years of this century.

John Gribbin, *In Search of Schrödinger's Cat: Quantum Physics and Reality*, Bantam Books, New York (1984). A fascinating story of the development of quantum mechanics and its interpretation.

Questions

1. Why are x-ray photons more hazardous than microwave photons?

2. Why do you think that ultraviolet radiation causes a suntan, whereas infrared radiation does not?

3. Why is visible radiation able to remove electrons from sodium but not from zinc?

4. When film is being developed, a special red light is turned on in the developing room, but the regular light is not. Explain.

5. Why is the photon theory supported by the fact that light below a certain frequency, when striking a metal plate, does not cause electrons to be freed from the metal?

6. Light from a relatively powerful laser has lifted and supported the weight of glass spheres that are 20×10^{-6} m in diameter (about the size of a body cell). How is this possible?

7. Is the radiation pressure of light greater on a mirror or on a black surface of the same size? Explain.

8. Why is the wavelike behavior of matter not observed in daily life?

9. Suppose that Planck's constant were 100 J·s rather than 6.63×10^{-34} J·s. Describe several ways in which the world would appear different. Be specific.

Problems

You may need to know the following masses to work some of these problems: $m_{\text{electron}} = 9.11 \times 10^{-31}$ kg, $m_{\text{proton}} = 1.673 \times 10^{-27}$ kg, and $m_{\text{neutron}} = 1.675 \times 10^{-27}$ kg.

31.1 and 31.2 Electromagnetic Waves and Spectrum

1. (a) X-rays used in a medical examination have a wavelength of about 2.5×10^{-11} m. Calculate the frequency of this radiation. (b) A microwave oven produces microwaves of frequency 2.45 GHz. Calculate the wavelength of the microwaves.

2. You change your AM radio from a station broadcasting at frequency 54 kHz to one broadcasting at 140 kHz. Calculate the ratio of the wavelength of the latter station to that of the former.

■ 3. The frequency of microwave generator 1 is 33 percent greater than that of the generator 2. Calculate the ratio of the wavelengths λ_1/λ_2 of the microwaves created by the two generators.

31.3 Photons

4. (a) Calculate the energy in electron volts of a 300-nm ultraviolet photon partly responsible for the production of a suntan. (b) A vibrating carbon monoxide (CO) molecule produces infrared photons of energy 0.37 eV. Calculate the frequency of CO vibration, which is the same as the frequency of the infrared radiation the molecule emits.

5. Suppose the bond in a biologically important molecule is broken by photons of energy 5.0 eV. Calculate the frequency and wavelength of these photons and the region of the electromagnetic spectrum in which they are located.

■ 6. A 1-eV photon's wavelength is 1240 nm. Using a ratio technique, calculate the wavelength of a 5-eV photon.

■ 7. A laser beam of power 3.0×10^{-3} W consists of photons of wavelength 630 nm. Calculate the number of photons passing a cross section of the beam each second.

■ 8. On a bright day with the sun directly overhead, the intensity of light striking the earth is about 1400 W/m². Calculate the number of photons striking a 1.0-m² surface each second, assuming the average photon wavelength is 700 nm.

■ 9. Roughly 5 percent of the power of a 100-W lightbulb is emitted as light, the rest being emitted as heat and longer-wavelength radiation. *Estimate* the number of photons of light coming from a bulb each second. Assume that the wavelength of an average photon appears in the middle of the visible part of the spectrum.

■ 10. For an object to be seen with the unaided eye, the light intensity coming to the eye must be about 5×10^{-12} J/m² · s or greater. Calculate the minimum number of photons that must enter the eye's pupil each second in order for the object to be seen. Assume that the pupil's radius is 0.20 cm and the wavelength of the light is 550 nm.

■ 11. Is the ear better at detecting sound power than the eye is at detecting light power, or vice versa? Explain. [*Note:* Refer to Problem 10 concerning the eye and to Section 18.5 concerning the ear's threshold of audibility.]

■ 12. A small 1.0×10^{-5}-g piece of dust falls in the earth's gravitational field. Calculate the distance it must fall so that its change in gravitational potential energy equals the energy of a 0.10-nm x-ray photon.

■ 13. Calculate the number of 650-nm photons whose energy equals the rest energy of an electron.

■ ■ 14. *Estimate* the number of photons emitted per second from 1.0 cm² of a person's skin if a typical emitted photon has a wavelength of 2000 nm. [*Hint:* See Section 10.3.]

31.4 Photon Experiments

15. The work function of cesium is 1.9 eV. (a) Calculate the lowest-frequency photon that can eject an electron from cesium. (b) Calculate the maximum possible kinetic energy in electron volts of a photoelectron that absorbs a 400-nm photon.

16. Visible light shines on the metal surface of a photocell having a work function of 1.30 eV. The maximum kinetic energy of the electrons leaving the surface is 0.92 eV. Calculate the light's wavelength.

17. Light of wavelength 430 nm strikes a metal surface, releasing electrons with kinetic energy equal to 0.58 eV or less. Calculate the metal's work function.

18. A film becomes exposed when light striking it initiates a complex chemical reaction. A particular type of film does not become exposed if struck by light of wavelength longer than 670 nm. Calculate the minimum energy in electron volts needed to initiate the chemical reaction.

■ 19. Suppose that light of intensity 1.0×10^{-2} W/m² is made of waves rather than photons and that the waves strike a sodium surface with a work function of 2.2 eV. (a) Determine the power in watts incident on the area of a single sodium atom at the metal's surface (the radius of a sodium atom is approximately 1.7×10^{-10} m). (b) How long will it take for an electron in the sodium to accumulate enough energy to escape the surface, assuming it collects all of the light incident on the atom?

■ 20. During photosynthesis, eight photons of 670-nm wavelength can cause the following reaction: $CO_2 + H_2O \rightarrow \frac{1}{6} C_6H_{12}O_6 + O_2$. During respiration, when the human body metabolizes sugar, the reverse reaction releases 4.9 eV of energy per CO_2 molecule. Calculate the ratio of the energy released (respiration) and the energy absorbed (photosynthesis)—a measure of the photosynthetic efficiency.

■ 21. A photomultiplier tube has twelve stages, each of which emits six electrons when struck by an electron released from a previous stage. Calculate the number of electrons emitted after the last stage. Notice that one electron is emitted from the first stage after absorption of a photon, six from the second, and so on.

■ 22. In a Compton-effect scattering experiment, an incident photon's frequency is 2.0×10^{20} Hz; the scattered photon's frequency is 1.4×10^{20} Hz. Calculate the kinetic energy increase of the electron, in units of electron volts, when the photon is scattered.

■ 23. An electron hit by an x-ray photon of energy 5.0×10^4 eV gains 3.0×10^3 eV of energy. Calculate the wavelength of the scattered photon leaving the site of the collision.

■ 24. A laser produces a short pulse of light whose energy equals 0.20 J. The wavelength of the light is 694 nm. (a) How many photons are produced? (b) Calculate the total momentum of the emitted light pulse.

■ ■ 25. A beam of light of wavelength 560 nm is reflected perpendicularly from a mirror. If 10^{20} photons hit the mirror each second, calculate the force of the light exerted on the mirror. [*Hint:* Refer to the impulse-momentum equation from Chapter 4. You may assume that the magnitude of the photons' momenta are unchanged by the collision, but their directions are reversed.]

■ ■ 26. We wish to calculate the net force on the earth due to absorption of light from the sun. (a) Calculate the net area of the earth exposed to sunlight (the earth's radius is 6.38×10^6 m). (b) The solar radiation intensity is 1400 J/m² · s. Calculate the momentum of photons hitting the earth's surface each second. (c) Use the impulse-momentum equation to calculate the average force of this radiation on the earth (F = Change in momentum per unit time).

■ ■ 27. Suppose that we wish to support a 70-kg person by levitating the person on a beam of light. (a) If all of the photons striking the person's bottom surface are absorbed, what must be the power of the light beam, which is made of 500-nm-wavelength photons? (b) *Estimate* the person's temperature change in 1 s. [Small glass beads of about 20-μm diameter can be levitated in this manner, but not humans, as should be apparent from your answer to part (b).]

■ ■ 28. An electron that resides by itself in an open region of space is struck by a photon of light. Using nonrelativistic formulas, show that the electron cannot absorb the photon's energy and simultaneously absorb its momentum. To conserve both energy and momentum, the photon must be absorbed by the electron near another mass, which carries away some of the momentum but little of the energy.

31.6 The Wave-Particle Nature of Matter

29. Calculate the de Broglie wavelengths of an electron and a proton when moving at speed 1.0×10^3 m/s.

30. The United States Golf Association requires that a golf ball have a mass of 45.6 g and reach a speed of 76 m/s when tested on one of its machines. Calculate the de Broglie wavelength of this ball. How does the wavelength compare with the size of a proton (approximately 10^{-15} m in diameter)?

■ **31.** An electron starting at rest crosses a 22-kV potential difference in a television set. Calculate its de Broglie wavelength. (Ignore relativistic effects.)

■ **32.** An electron and a proton have equal kinetic energy (nonrelativistic). Calculate the ratio of the electron's and the proton's de Broglie wavelengths.

■ **33.** A proton and an electron have the same speeds. Calculate the ratio of their de Broglie wavelengths.

■ **34.** A charged particle after passing across a potential difference of 5000 V has a de Broglie wavelength of 1.60×10^{-13} m. Calculate the wavelength of the same particle if it crosses a potential difference that is 30 percent greater. The particle starts at rest.

■ **35.** Calculate the speed of an electron whose de Broglie wavelength is the same as the wavelength of a 50,000-eV x-ray photon.

■ **36.** To study the wave nature of electrons, some physics students propose to accelerate a group of electrons through a voltage so that the de Broglie wavelength of the electrons will equal that of light (about 600 nm). Interference of the electron waves can then be observed easily by passing the electrons through double slits used to study light interference. (a) Calculate the speed of electrons having this wavelength. (b) Calculate the voltage through which the electrons should be accelerated. (c) Does the experiment seem feasible? Explain.

■ **37.** (a) Estimate your de Broglie wavelength as you walk between classes. (b) Suppose that the world were restructured so that Planck's constant h could be much larger than its present value. How large would it have to be in order for your de Broglie wavelength when you move between classes to be 1 m? If this happened, students entering the door of a classroom would be diffracted into a pattern such as shown in Fig. 31.15.

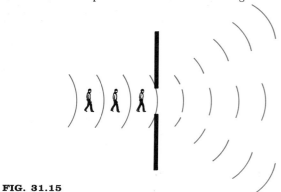

FIG. 31.15

■■ **38.** Show that the de Broglie wavelength of an electron accelerated to nonrelativistic speed through a potential difference V equals $1.23 \times 10^{-9}/\sqrt{V}$ in units of meters.

■■ **39.** Suppose that Planck's constant were 2.0×10^4 J·s. A person running on a track would be guided by his or her de Broglie wave. The most probable places for the person on the track would depend on the amplitude of his or her de Broglie wave at these places. In order for the wave to "fit" around the track's circumference, an integral number of wavelengths must equal the length of the circumference. (a) Calculate the speed that a 50-kg woman must run on a 400-m circular track so that one of her waves will fit around the track. (b) Calculate the speed for two of her waves. A similar phenomenon occurs to electrons in atoms.

■■ **40.** Electrons starting at rest accelerate across a 30-V potential difference. The electrons then pass through a pair of slits and strike a screen. (a) Calculate the slit separation that causes the $n = 1$ interference maximum to have an angular deflection of $10°$. (b) Using the same slits, calculate the angular deflection if the electrons accelerate across a 60-V potential difference.

31.7 The Uncertainty Principle

■ **41.** A proton travels at speed 4.66 ± 0.08 m/s. Calculate the least uncertainty with which its position can be known.

■ **42.** Suppose the speeds of an electron and of a 45.9-g golf ball are each measured with an uncertainty of $\Delta v = 1.0 \times 10^{-3}$ m/s. Calculate the minimum uncertainties in their positions.

■ **43.** A 60-kg runner's instantaneous speed can be determined with an uncertainty of 0.01 m/s. Calculate the minimum uncertainty of the runner's position as limited by the uncertainty principle. Does the uncertainty principle limit the accuracy of times recorded for track races?

■ **44.** A scanning electron microscope can determine a particle's position with an uncertainty of about 1.0×10^{-8} m. Calculate (a) the minimum uncertainty in momentum of the 4.0×10^{-15}-kg nucleus of a cell viewed by the microscope and (b) the minimum uncertainty in its speed.

■ **45.** At one time, scientists thought that an atom's nucleus was made of protons and electrons. However, the electron cannot exist in the nucleus, as is apparent from the uncertainty principle. (a) Calculate the uncertainty in an electron's momentum if restricted to a region the size of a nucleus (10^{-14} m). (b) Assuming that the electron's momentum must be at least as large as its uncertainty, calculate its energy using the relativistic equation $E = \sqrt{p^2 c^2 + m_0^2 c^4}$ (convert your answer to units of electron volts). An electron with such great energy could not reside in the nucleus.

■■ **46.** Students are seldom motionless during a physics lecture. (a) *Estimate* the uncertainty in a person's speed due to wiggling. The number may be small but is not zero. (b) Based on the answer to part (a), estimate the person's uncertainty in momentum. (c) Now suppose that the world were changed, with Planck's constant much larger. Calculate the value of Planck's constant if a person's uncertainty in position while sitting in class is 10 m, given the momentum uncertainty estimated above. With this large position uncertainty, the person might be found almost anywhere in a typical lecture room.

■■ **47.** Your boyfriend or girlfriend walks through a doorway. *Estimate* the least uncertainty with which you think his or her momentum could be measured. Suppose that the world had been created with Planck's constant equal to 1.0×10^{-2} J·s. Would your friend's shape be discernible? Explain.

48. An infrared photon is absorbed by a glucose molecule dissolved in water. The glucose loses this absorbed energy within about 10^{-12} s because of collisions with other molecules. *Estimate* the least uncertainty with which we can measure the energy stored by the molecule during this short time.

■ ■ 49. An elementary particle known as a neutral pion (π^0) lives only 0.8×10^{-16} s before it decays into two photons. The pion's rest mass during its short life is 2.4×10^{-28} kg. (a) Calculate the rest-mass energy of a pion. (b) Estimate the least uncertainty with which we can measure its energy. (c) Calculate the equivalent uncertainty in its mass.

31.8 Electron Microscopes

50. Calculate the wavelength of electrons accelerated across a 40,000-V potential difference. Compare the electrons' wavelength with the wavelength of light.

51. The smallest distance that can be resolved by any microscope under optimum conditions, called its *resolving power,* is approximately equal to the wavelength used in the microscope. A typical electron microscope uses electrons accelerated through a 50,000-V potential difference. Calculate the microscope's theoretical resolving power.

52. In order for an atom to be "seen" with an electron microscope, the wavelength of the electrons should be no bigger than the size of the atom—roughly 0.1 nm. Through what potential difference must the electrons be accelerated to have this wavelength?

CHAPTER 32

The Bohr Model
of the Atom

At the beginning of this century, scientists knew that matter was made of basic building blocks called atoms. The blocks were extremely small (for example, 6×10^{23} helium atoms had a mass of only 4 g, and the same number of carbon atoms had a mass of 12 g). The masses of the various types of atoms differed, as did other of their properties. For example, carbon atoms combined with a variety of atoms to form molecules (CO, CO_2, CH_4, . . .) or solids (graphite and diamond), while helium atoms were inert—they would not combine easily with other atoms. Sodium atoms would combine with chlorine atoms to form a crystalline solid but would not combine with carbon atoms. There was also variation in the wavelengths of light and of other forms of EM radiation emitted and absorbed by different types of atoms. While all atoms of a particular type had identical properties, atoms of different types differed greatly from each other.

Near the beginning of the twentieth century, scientists began to search for a structure of the atom that would account for this variation from one atom to the next. The first atomic theory that could be used to calculate the wavelengths of light emitted and absorbed by an atom was presented by Niels Bohr in 1913 and became known as the Bohr model of the atom. Bohr's model, which applied only to atoms or ions with one electron, was shown after its development to draw on both the particle and the wave behavior of electrons. In 1925 a comprehensive atomic theory, called *quantum mechanics,* was devised. In this theory the dual wave-particle behavior of matter was brought together in one consistent theory.

In this chapter we will examine the method by which Bohr developed his model and the properties of atoms that can be determined using the model. In the next chapter we will examine the quantum mechanical theory of the atom.

32.1 The Nuclear Model of the Atom

Near the end of the nineteenth century, it was well known that matter was made of atoms. Little was known, however, of the structure of the atom. One of the first atomic models, called the "plum-pudding" model (Fig. 32.1), was presented by J. J. Thomson (1856–1940) soon after he discovered the electron in 1897. He

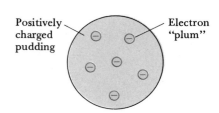

FIG. 32.1 The "plum-pudding" model of the atom.

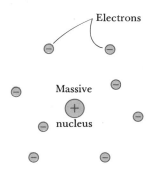

FIG. 32.2 Rutherford's nuclear atom consists of a massive, positively charged nucleus surrounded by electrons with little mass. (Not to scale.)

proposed that negatively charged electrons (the "plums") were embedded in a sphere of uniform positive charge (the "pudding").

In 1911 Ernest Rutherford (1871–1937) and his colleagues in England performed an interesting experiment that led to the development of a different atomic model. Rutherford shot a beam of **alpha (α) particles** (the nuclei of helium atoms—helium atoms with their electrons removed) at a thin foil of gold atoms. Most α particles passed straight through the gold foil. This result was consistent with the plum-pudding model because there were no concentrations of positive charge and mass in the pudding to reflect the α particles. The mass of the negatively charged electrons was about $\frac{1}{7000}$ the mass of the α particles and was too small to deflect an α particle backward.

Rutherford and his colleagues also found that while most α particles passed through the foil undeflected, a few were scattered back toward their source as if they had hit a mass bigger than their own. This result was inconsistent with the plum-pudding model because there was nothing in the "pudding" to cause this back-scattering. This observation led Rutherford to propose the **nuclear model** of the atom. In this model the atom consisted of a tiny, massive core called the *nucleus,* which contained all of the atom's positive charge and most of its mass. Electrons occupied the space outside the core (Fig. 32.2). The electrons outside the nucleus had too little mass to deflect an α particle backward, but the massive nucleus with its large positive charge could cause backscattering during a direct hit by the α particle.

To account for the small fraction of back-scattered α particles, the nucleus could be only about 10^{-14} m in diameter. Since the whole atom occupied a space that was over 10^{-10} m in diameter, the atom seemed to consist mostly of empty space dotted with electrons and a very small nucleus at the center.

To develop a feeling for the small size of the nucleus compared with the region of space occupied by electrons, suppose that your thumbnail were the atom's nucleus. The electrons would then occupy a cross-sectional area a little larger than the size of a football field. Most α particles would be undeflected while traveling through this open space, but when an α particle hit the "thumbnail" nucleus, a dramatic collision would occur.

Although Rutherford's nuclear model of the atom agrees well with our present conception of the atom, it lacks the detail needed to explain many other properties of atomic behavior, such as the wavelengths of light emitted and absorbed by atoms and the reasons why some atoms combine to form molecules and others do not.

32.2 Atomic Spectra

More than thirty years before Rutherford introduced his nuclear model of the atom, scientists had begun to record light spectra emitted by different types of atoms. A **spectrum** consists of the various wavelengths of EM radiation that are emitted (an emission spectrum) or absorbed (an absorption spectrum) by a particular type of atom.

One technique for recording an emission spectrum is illustrated in Fig. 32.3. Atoms in the form of a gas are placed in a tube with glass walls. A large voltage across the metal ends of the tube causes electrons to flow through the gas. As the electrons collide with the atoms in the tube, the atoms gain energy;

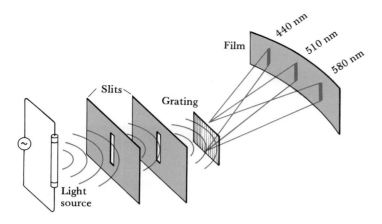

FIG. 32.3 A spectrometer for measuring the wavelength of light coming from a lamp.

they become "excited." The excited atoms quickly lose their extra energy by emitting light and other forms of EM radiation.

The light leaving the tube is collimated as it passes through consecutively placed single slits. After the slits, light passes through a diffraction grating or a prism that bends the light at angles that depend on the wavelength of the light. If a film is placed beyond the grating, it becomes exposed only in narrow bands or lines that correspond to the wavelengths of light emitted from the atoms. A spectrum recorded in this manner is called a **band spectrum** or **line spectrum.**

The line spectra of several types of atoms are shown in Fig. 32.4 and on the color insert facing page 453. The spectrum of each type of atom or molecule is like a fingerprint in that it differs from those of other atoms or molecules. Astronomers can determine the types and relative amounts of atoms that are on distant stars by analyzing their spectra. Geologists and soil scientists use atomic absorption spectroscopy to analyze the types and concentrations of different types of atoms in samples of soil. Using spectral techniques, a blood analyzer can identify the concentrations of 60 different atoms, ions, and molecules in the blood.

During the latter part of the nineteenth century, scientists were looking for one theory that would account for the variety of spectral lines emitted from different atoms. Much of the effort centered on hydrogen, the simplest atom. In 1885 Johann Balmer, a mathematics teacher in a girls' high school in Switzerland, developed (in his spare time) a mathematical equation for calculating the wavelength of a group of these hydrogen lines. The equation became known as **Balmer's equation:**

$$\frac{1}{\lambda} = R\left(\frac{1}{2^2} - \frac{1}{n^2}\right), \qquad \text{where } n = 3, 4, 5, \ldots.$$

The constant R, later called the **Rydberg constant,** has a value of $1.097 \times 10^7 \text{ m}^{-1}$. The wavelength of each line is calculated by substituting an appropriate integer n into the equation.

EXAMPLE 32.1 Use Balmer's equation to calculate the wavelength of a hydrogen emission line that corresponds to the integer $n = 4$.

FIG. 32.4 The emission spectra of several gases (wavelengths are in units of nanometers). (Courtesy of Bausch & Lomb.)

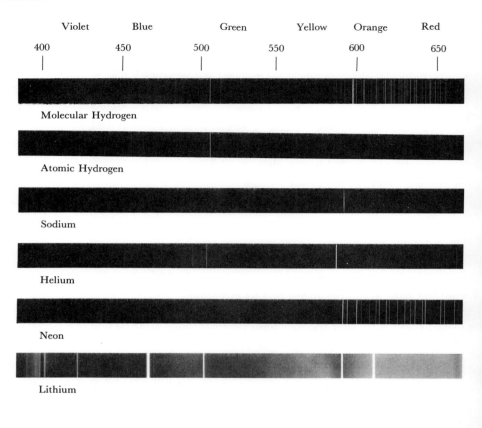

Violet	Blue	Green	Yellow	Orange	Red
400	450	500	550	600	650

Molecular Hydrogen

Atomic Hydrogen

Sodium

Helium

Neon

Lithium

SOLUTION

$$\frac{1}{\lambda} = (1.097 \times 10^7\,\mathrm{m}^{-1})\left(\frac{1}{2^2} - \frac{1}{4^2}\right) = 2.057 \times 10^6\,\mathrm{m}^{-1},$$

or

$$\lambda = \frac{1}{2.057 \times 10^6\,\mathrm{m}^{-1}} = 486.2 \times 10^{-9}\,\mathrm{m} = \underline{486.2\,\mathrm{nm}}.$$

This is one of the wavelengths of light emitted by a hot gas of hydrogen atoms. Light with this wavelength appears blue to the human eye. ∎

In the twentieth century it was found that hydrogen emits electromagnetic radiation in two wavelength series besides those given by Balmer's equation. The wavelengths of these three series of lines emitted by hydrogen atoms are calculated using the following equations:

$$\text{Lyman series:} \qquad \frac{1}{\lambda} = R\left(\frac{1}{1^2} - \frac{1}{n^2}\right), \qquad \text{where } n = 2, 3, \ldots, \quad \textbf{(32.1)}$$

$$\text{Balmer series:} \qquad \frac{1}{\lambda} = R\left(\frac{1}{2^2} - \frac{1}{n^2}\right), \qquad \text{where } n = 3, 4, \ldots, \quad \textbf{(32.2)}$$

$$\text{Paschen series:} \qquad \frac{1}{\lambda} = R\left(\frac{1}{3^2} - \frac{1}{n^2}\right), \qquad \text{where } n = 4, 5, \ldots, \quad \textbf{(32.3)}$$

where $R = 1.097 \times 10^7\,\mathrm{m}^{-1}$ for each series. The **Lyman series** consists mostly of radiation emitted by hydrogen in the ultraviolet region; the **Paschen series** consists of radiation in the infrared region.

The next big step in the history of atomic theory was made by the Danish physicist Niels Bohr. Relying mostly on basic principles, he devised a theory that accounted for all the spectral lines emitted by hydrogen atoms.

32.3 The Bohr Model

Niels Bohr (1885–1962) was a man of many talents (see Fig. 32.5). As a college student he was selected for the All-Danish Soccer Team. He went on to attain a doctorate in physics from the University of Copenhagen and in 1912 spent a few months with Ernest Rutherford at Manchester University in England. Rutherford was so pleased with his new colleague that he told a friend, "This young Dane is the most intelligent chap that I have ever met." In turn, Bohr had a high opinion of Rutherford, who had just developed the nuclear model of the atom. Soon after Bohr returned to Copenhagen in 1913, he extended Rutherford's nuclear model of atoms so that all the spectral emissions of a hydrogen atom could be calculated with great accuracy. In 1922 at age 37 Bohr received the Nobel Prize in physics for his model of the atom.

The Bohr model is restricted to atoms and ions with one electron. Bohr postulated that the electron can move in circular orbits about the atom's nucleus, as pictured in Fig. 32.6, but that only certain radii orbits are allowed. The energy of the atom depends on the radius of the orbit its electron occupies. While the electron was in a particular orbit, Bohr postulated that the atom's energy was stable. The electron could jump from one stable orbit to another, in which case the atom's change in energy was balanced by the emission or absorption of a photon. If the electron moved to an orbit of lower energy, it emitted a photon. In order that energy be conserved, the energy lost by the atom equaled the *energy (hf) of the newly created photon:*

$$E_i - E_f = hf, \tag{32.4}$$

where E_i is the atom's initial energy and E_f is its final energy. If an atom absorbed a photon, its electron moved to an orbit of higher energy.

Using this model of the atom and basic principles of physics, Bohr was able to derive an expression for the radii of the allowed orbits and the energy of the atom when its electron was in these stable orbits. This energy equation, when substituted in Eq. (32.4), could be used to calculate the frequency of photons that hydrogen atoms were expected to emit. When compared to the experimentally observed frequencies of emitted photons, Bohr's predictions agreed perfectly.

Energy of One-Electron Atoms

Much of Bohr's theory relies on principles of classical physics. For example, Bohr knew that when an object moves with speed v in a circular orbit of radius r, it experiences a centripetal acceleration $a_c = v^2/r$ [Eq. (6.7)]. For an electron moving in a circle in a one-electron atom, Bohr assumed that the acceleration was caused by the electrical force of attraction between the electron of charge $-e$ and the nucleus of charge $+Ze$. (The quantity Z, called the **atomic number,**

FIG. 32.5 Niels Bohr and his wife, Copenhagen, 1931. (AIP Niels Bohr Library, Uhlenbeck Collection.)

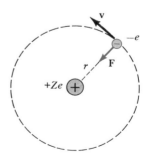

FIG. 32.6 Bohr's model of a one-electron ion. The electrical force that attracts the electron toward the nucleus provides the centripetal force needed to keep it moving in a circular path.

represents the number of protons in the nucleus. For hydrogen, $Z = 1$; for lithium, $Z = 3$, and so forth.) The electrical force keeps the electron in its circular orbit just as the gravitational force of the earth on the moon keeps the moon in its orbit. The magnitude of the electrical force between the electron and the nucleus is given by Coulomb's law [Eq. (22.1)]: $F_q = k(e)(Ze)/r^2$. Substituting the Coulomb force and the centripetal acceleration into **Newton's second law** when written for circular motion ($\Sigma F_{\text{radial}} = ma_c$), we find that

$$\frac{kZe^2}{r^2} = m\frac{v^2}{r}, \tag{32.5}$$

where m is the electron's mass. This equation will be used shortly.

The **atom's energy** E when its electron is in orbit of radius r consists of the kinetic energy of the electron [Eq. (8.4)] plus the electrical potential energy due to the electric charges of the electron and nucleus [use Eq. (23.1) for the case where $r_0 = \infty$]:

$$E = KE + PE_q = \frac{1}{2}mv^2 + \frac{k(Ze)(-e)}{r}$$

$$= \frac{1}{2}mv^2 - \frac{kZe^2}{r}. \tag{32.6}$$

The electrical potential energy is defined to be zero when the electron is infinitely far from the nucleus ($r = \infty$). When the oppositely charged electron and nucleus are closer together, the electrical potential energy is less than zero—hence the negative sign for the PE_q part of Eq. (32.6).

This equation for the atom's energy depends on two unknown quantities: the electron's speed and the radius of its orbit. By combining Eqs. (32.5) and (32.6), we get an expression for the atom's energy that depends only on the radius of the electron orbit. Canceling an r from the denominator of each side of Eq. (32.5), we find that $kZe^2/r = mv^2$. If this expression for mv^2 is substituted in Eq. (32.6), we find that the atom's energy is

$$E = -\frac{kZe^2}{2r}. \tag{32.7}$$

Notice that as the radius of the electron's orbit decreases, the atom's energy lessens (becomes more negative). This is because the atom's electrical potential energy is less when the electron is closer to the nucleus.

We see that according to Eq. (32.7), the atom's energy depends only on the radius of the orbit of its electron. Bohr's next task was to determine the radii of the orbits in which the electron was allowed to move. The energy of these allowed orbits should be such that when an electron jumps from one orbit to another, the change in energy of the atom is balanced by the energy of a photon that is known to be emitted or absorbed by the atom.

After several failures, Bohr found a suitable *condition for determining the allowed orbits.* An electron, according to Bohr, can move in only those orbits in which its angular momentum mvr [Eq. (7.18)] equals a positive integer n times the constant $h/2\pi$, where h is Planck's constant and equals 6.63×10^{-34} J·s:

$$mv_n r_n = n\frac{h}{2\pi} \qquad \text{for } n = 1, 2, 3, 4, \ldots. \tag{32.8}$$

Each possible value of n corresponds to a different allowed orbit. For example, if $n = 3$, the radius r_3 of the third orbit and the electron's speed v_3 when in that orbit must satisfy the condition $mv_3r_3 = 3(h/2\pi)$. When Bohr combined this condition with his other equations, he could calculate the energy of an atom when its electron was in one of these orbits and the correct frequencies of photons emitted from the atom when its electron jumped from one allowed orbit to another.

To calculate the radii of allowed orbits, we solve Eq. (32.8) for v_n and substitute it into Eq. (32.5). We find that

$$r_n = \left(\frac{h^2}{4\pi^2 ke^2 m}\right)\frac{n^2}{Z}.$$

Notice that all the quantities inside the parentheses of the preceding equation are constants ($h = 6.63 \times 10^{-34}$ J·s; $k = 9.0 \times 10^9$ N·m²/C², $e = 1.6 \times 10^{-19}$ C; and $m = 9.11 \times 10^{-31}$ kg). When evaluated, the term in parentheses equals 0.53×10^{-10} m. Thus, the radii of allowed orbits in a one-electron atom or ion are given by the equation

$$r_n = \frac{(0.53 \times 10^{-10})n^2}{Z} \text{ m.} \tag{32.9}$$

EXAMPLE 32.2   Calculate the radii of the smallest three orbits in the hydrogen atom ($Z = 1$). The orbits are pictured schematically in Fig. 32.7.

SOLUTION

$n = 1$: $r_1 = (0.53 \times 10^{-10}$ m$)1 = \underline{0.53 \times 10^{-10}}$ m,

$n = 2$: $r_2 = (0.53 \times 10^{-10}$ m$)4 = \underline{2.12 \times 10^{-10}}$ m,

$n = 3$: $r_3 = (0.53 \times 10^{-10}$ m$)9 = \underline{4.77 \times 10^{-10}}$ m.   ■

If the expression for the radii of the allowed orbits is substituted for r in the equation for the atom's energy [Eq. (32.7)], we find that the energy of the atom when its electron is in an orbit that corresponds to a particular value of n is

$$E_n = -\left(\frac{2\pi^2 k^2 e^4 m}{h^2}\right)\frac{Z^2}{n^2}. \tag{32.10}$$

Once again, all the quantities inside the parentheses are constants. Substituting for these constants, we find that the energy of a one-electron atom or ion when its electron is in the nth orbit is

$$E_n = -\frac{(2.2 \times 10^{-18})Z^2}{n^2} \text{ J.}$$

The energy unit electron volt (eV) is much more convenient for calculations involving atoms. Converting the preceding equation from units of joules to units of electron volts (1 eV $= 1.6 \times 10^{-19}$ J) results in the following equation for the energy of a one-electron atom or ion:

$$E_n = -\frac{13.6Z^2}{n^2} \text{ eV.} \tag{32.11}$$

EXAMPLE 32.3   Calculate the energy of the hydrogen atom ($Z = 1$) when its electron is in orbits $n = 1$, 2, and 3.

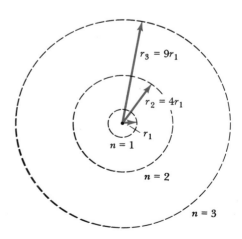

FIG. 32.7   The allowed orbits for an electron in a hydrogen atom ($r_1 = 0.53 \times 10^{-10}$ m).

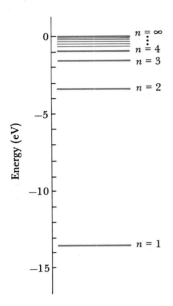

FIG. 32.8 An energy-level diagram for hydrogen.

SOLUTION Using Eq. (32.11) for the energy of an atom having a nucleus with one proton ($Z = 1$), we find that the energies of the atoms with electrons in these different orbits are as follows:

$$n = 1: \quad E_1 = \underline{-13.6 \text{ eV}},$$
$$n = 2: \quad E_2 = -13.6/4 = \underline{-3.4 \text{ eV}},$$
$$n = 3: \quad E_3 = -13.6/9 = \underline{-1.5 \text{ eV}}. \qquad \blacksquare$$

Energy-Level Diagrams

The energy of the atom when its electron is in these different orbits is often represented by means of an **energy-level diagram** (Fig. 32.8). A vertical axis is drawn whose scale is in units of energy. A horizontal line, called an **energy level,** is drawn for each possible value of energy for the atom. Since a hydrogen atom can exist in states with energy -13.6 eV, -3.4 eV, -1.5 eV, and so on, horizontal lines appear at these energies in Fig. 32.8. Usually the value of n corresponding to a particular energy level is placed at its side.

The integer n can assume any positive integral value from one to infinity, and an infinite number of lines could be drawn in the diagram. Usually, however, lines appear for only the first few energy levels. (The rest are crowded into a very small energy region near zero energy and are not shown.)

Ground States and Excited States

The different allowed orbits of a one-electron atom, often called its states, are labeled by the integer n, which is called a **quantum number.** The state with lowest energy (the $n = 1$ state for hydrogen) is called the **ground state.** Higher energy states ($n = 2, 3, 4, \ldots$) are called **excited states.**

In a group of hydrogen atoms, most atoms have their electron in the ground state (the $n = 1$ orbit). As the atoms move about and collide with each other, an electron of one atom is occasionally excited from the ground state to an allowed state of higher energy. The energy gained by the excited atom is balanced by losses of translational kinetic energy of the atoms involved in the collision.

An excited atom remains excited for only about 10^{-8} s; its electron then falls back to the ground state in one step or in a series of steps, and during each step a photon is emitted from the atom.

While one atom is becoming deexcited by emitting a photon, another atom is excited by a collision. At any one time a small fraction of atoms are in excited states. On the sun's surface, for example, approximately six out of every billion hydrogen atoms are in the $n = 2$ state. A smaller fraction are in higher excited states. As the temperature of a group of atoms increases, their collisions become more violent, and a larger fraction of them are bumped by collisions into excited states. Since there are more excited atoms, more photons are being emitted as the atoms become deexcited. Thus, a hot object emits more light than a cool object.

Wavelength of Emitted and Absorbed Photons

We can now use Bohr's equations to calculate the wavelengths (or frequencies) of photons that are emitted and absorbed by one-electron atoms. As an atom emits a photon, an electron in the atom falls from an initial excited state of energy E_{n_i} to a final state of energy E_{n_f}. The energy lost by the atom $(E_{n_i} - E_{n_f})$ equals the energy of the newly created photon (hf), or

$$hf = E_{n_i} - E_{n_f}.$$

Substituting Eq. (17.8) $(f = c/\lambda)$ for the photon's frequency and rearranging, we find that

$$\frac{1}{\lambda} = -\frac{1}{hc}(E_{n_f} - E_{n_i}).$$

If we substitute for E_{n_f} and E_{n_i} using Bohr's equation for the energy of the atom [Eq. (32.10)], we find that

$$\frac{1}{\lambda} = \left(\frac{2\pi^2 k^2 e^4 m}{h^3 c}\right) Z^2 \left(\frac{1}{n_f^2} - \frac{1}{n_i^2}\right).$$

All the quantities inside the first term in parentheses of the preceding equation are constants, and when their values are substituted, the term equals the Rydberg constant R, which was determined empirically by Balmer. Substituting for R, we now have an equation for calculating the *wavelength of a photon emitted by a one-electron atom or ion* whose electron falls from state n_i to state n_f:

$$\frac{1}{\lambda} = RZ^2 \left(\frac{1}{n_f^2} - \frac{1}{n_i^2}\right), \tag{32.12}$$

where R equals $1.097 \times 10^7 \text{ m}^{-1}$ and for hydrogen $Z = 1$. Notice that this equation has the same form as Eqs. (32.1–32.3), which were derived from experimental observations of the wavelengths of light emitted by hydrogen atoms. For example, for the series of spectral lines observed by Balmer, the final electron state had quantum number $n_f = 2$. Light was emitted as an atom fell from an initial higher-energy state with quantum number $n_i = 3$ or more to a final state $n_f = 2$ (see Fig. 32.9). For the Lyman series, the final state was $n_f = 1$ and for the Paschen series the final state of a transition was $n_f = 3$.

Photons can also be *absorbed* by the atoms. But for this to happen, the photon's energy must match the difference between two allowed energy states in the atom, and the lowest state must hold an electron while the upper state must be partially empty. This energy condition is satisfied if the photon's energy $(hf = hc/\lambda)$ equals the difference in the atom's energy $(E_{n_f} - E_{n_i})$ as its electron jumps from an initial orbit n_i to a final orbit n_f.

Bohr had for the first time applied basic physical principles to a model of the atom and had successfully calculated the wavelengths of light emitted and absorbed by an atom. Even with this success, the model had some faults. The Bohr model could not be used to calculate the spectra of atoms or ions with more than one electron. Also, his assumption that the energy of an atom was stable when its electron moved in one of the allowed orbits conflicted with classical theories of electricity and magnetism. When viewed from the side, the electron's orbital motion resembled that of an electron oscillating in an antenna.

FIG. 32.9 The vertical lines represent electron transitions in the hydrogen atom that produce the different spectral series.

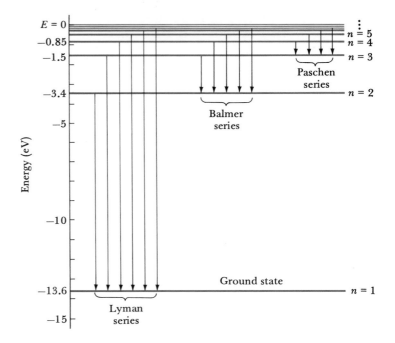

An oscillating charge emits electromagnetic radiation (see Fig. 31.1) and loses energy as it moves back and forth. But if the electron loses energy, its orbit is not stable; its separation from the nucleus would decrease and it would spiral into the nucleus.

It is now known that we cannot picture atoms in the way that Bohr did. Electrons in atoms exhibit a wavelike behavior, and their paths as they move around the nucleus are uncertain. We can determine the probability of finding an electron at different locations relative to the nucleus, but we cannot determine the electron's exact orbit as it moves.

Even though Bohr's model is now regarded as inappropriate, it still serves a useful purpose in helping us visualize many of the interesting properties of atoms.

EXAMPLE 32.4 Two hydrogen atoms ($Z = 1$) are excited into the $n = 4$ state by collisions with other atoms. One atom emits one photon as it falls back to the ground state ($n = 1$). The other atom emits a photon as it falls to the $n = 2$ state and then emits another photon as it falls from the $n = 2$ to the $n = 1$ state. Calculate the wavelengths of each of these three photons.

SOLUTION The electron of the first atom falls from the state with $n_i = 4$ to the state with $n_f = 1$. The wavelength of the photon is determined by solving Eq. (32.12):

$$\frac{1}{\lambda} = (1.097 \times 10^7 \text{ m}^{-1})(1^2)\left(\frac{1}{1^2} - \frac{1}{4^2}\right) = 1.03 \times 10^7 \text{ m}^{-1},$$

or

$$\lambda = 97.2 \times 10^{-9} \text{ m} = \underline{97.2 \text{ nm}}.$$

This is one of the transitions in the Lyman series that produces a photon in the ultraviolet region of the electromagnetic spectrum.

The second atom returns to the ground state in two transitions. First, its electron falls from the state with $n_i = 4$ to the state with $n_f = 2$:

$$\frac{1}{\lambda} = (1.097 \times 10^7 \text{ m}^{-1})(1^2)\left(\frac{1}{2^2} - \frac{1}{4^2}\right) = 2.06 \times 10^6 \text{ m}^{-1},$$

or

$$\lambda = 486 \times 10^{-9} \text{ m} = \underline{486 \text{ nm}}.$$

This is a visible-wavelength photon (it appears bluish-green to the eyes) in the Balmer series. In the second transition the electron falls from the $n_i = 2$ state to the ground state ($n_f = 1$):

$$\frac{1}{\lambda} = (1.097 \times 10^7 \text{ m}^{-1})(1^2)\left(\frac{1}{1^2} - \frac{1}{2^2}\right) = 8.23 \times 10^6 \text{ m}^{-1},$$

or

$$\lambda = 122 \times 10^{-9} \text{ m} = \underline{122 \text{ nm}}.$$

This is an ultraviolet-wavelength photon in the Lyman series. ∎

EXAMPLE 32.5 Calculate (a) the radii and (b) the energy of the three lowest-energy allowed orbits for the electron in the Li^{2+} ion. [Li^{2+} consists of the nucleus of a lithium atom ($Z = 3$) orbited by a single electron.] (c) What are the energy and wavelength of a photon that, when absorbed, causes an electron in Li^{2+} to be excited from the $n = 1$ to the $n = 3$ state?

SOLUTION (a) We determine the radii of the $n = 1$, 2, and 3 orbits using Eq. (32.9) with $Z = 3$:

$$n = 1: \quad r_1 = \frac{0.53 \times 10^{-10} \text{ m}}{3}\, 1^2 = \underline{0.18 \times 10^{-10} \text{ m}},$$

$$n = 2: \quad r_2 = \frac{0.53 \times 10^{-10} \text{ m}}{3}\, 2^2 = \underline{0.71 \times 10^{-10} \text{ m}},$$

$$n = 3: \quad r_3 = \frac{0.53 \times 10^{-10} \text{ m}}{3}\, 3^2 = \underline{1.59 \times 10^{-10} \text{ m}}.$$

(b) Substituting $Z = 3$ into the energy equation [Eq. (32.11)], we find that the energies of the $n = 1$, 2, and 3 orbits are

$$n = 1: \quad E_1 = \frac{-13.6(3^2)}{1^2} = \underline{-122.4 \text{ eV}},$$

$$n = 2: \quad E_2 = \frac{-13.6(3^2)}{2^2} = \underline{-30.6 \text{ eV}},$$

$$n = 3: \quad E_3 = \frac{-13.6(3^2)}{3^2} = \underline{-13.6 \text{ eV}}.$$

(c) The photon's energy must equal the energy ($E_3 - E_1$) needed to excite the electron. Thus, the photon's energy hf is

$$hf = E_3 - E_1 = (-13.6 \text{ eV}) - (-122.4 \text{ eV}) = \underline{108.8 \text{ eV}}.$$

Substituting $f = c/\lambda$ in the preceding equation, rearranging, and converting the energy to units of joules, we find that

$$\lambda = \frac{hc}{E_3 - E_1} = \frac{(6.63 \times 10^{-34}\,\text{J}\cdot\text{s})(3.0 \times 10^8\,\text{m/s})}{(108.8\,\text{eV})(1.6 \times 10^{-19}\,\text{J/eV})}$$
$$= 11.4 \times 10^{-9}\,\text{m} = \underline{11.4\,\text{nm}}. \qquad\blacksquare$$

32.4   De Broglie's Quantum Condition

In developing his atomic model, Bohr had proposed that electrons could move only in those orbits in which their angular momentum equaled an integral multiple of $h/2\pi$; that is, $mvr = n(h/2\pi)$. His justification for this proposal was based on a lengthy analysis of the radiation emitted by one-electron atoms when their electrons were in large orbits with high values of the quantum number n. The fact that this condition helped to produce such a successful theory was perhaps its best justification. The condition on the angular momentum of electrons in orbit about the nucleus was made more concrete ten years later, in 1923, by de Broglie's proposal that an electron of mass m moving at speed v behaves like a wave of wavelength

$$\lambda = \frac{h}{mv}. \qquad (31.5)$$

To see how these two ideas, the quantized angular momentum and the wave nature of electrons, are related, picture in your mind the circular orbit of an electron in a hydrogen atom. We now regard the electron as a wave propagating around the circle. The situation is very similar to the formation of standing waves on a stretched string, such as discussed in Section 18.1. The wave on the string, after reflection from the end of the string, interferes with itself. If the total distance traveled by the wave up the string and back equals an integral number of wavelengths, the interference is constructive and large-amplitude standing waves are formed on the string. For all other wavelengths the interference is destructive, and waves of these other wavelengths cannot resonate on the string.

A similar situation occurs for electron waves moving in orbits about the nucleus of a one-electron atom. If the total distance around the orbit (the circumference $2\pi r$ of a circle) equals an integral number of electron wavelengths, the wave always returns to a particular position on the string with the same amplitude (Fig. 32.10a). The amplitude of vibration at each point on the orbit increases to form a standing-wave vibration—a resonance. If the circumference of the orbit is not an integral number of wavelengths, then destructive interference occurs as the wave makes a trip around the loop (Fig. 32.10b); the wave cannot resonate on these other orbits.

Constructive interference of electron waves on circular orbits of radius r takes place when the circumference $2\pi r$ equals an integral number of wavelengths:

$$2\pi r = n\lambda \qquad \text{for } n = 1, 2, 3, \ldots. \qquad (32.13)$$

When we substitute de Broglie's expression for the wavelength of an electron of mass m moving at speed v ($\lambda = h/mv$), we find that

$$2\pi r = n\frac{h}{mv},$$

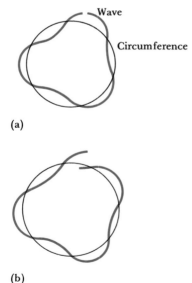

Wave

Circumference

(a)

(b)

FIG. 32.10   (a) The de Broglie wave of the electron returns to its starting position in phase with the wave that started earlier, and constructive interference occurs. (b) The wave is out of phase with itself, and destructive interference occurs.

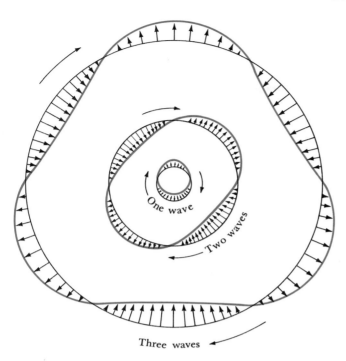

FIG. 32.11 Electron standing-wave patterns in the first three orbits of a hydrogen atom.

One wave

Two waves

Three waves

or

$$mvr = n\frac{h}{2\pi} \quad \text{for } n = 1, 2, 3, \ldots.$$

This is exactly the condition used by Bohr [Eq. (32.8)] to select those orbits on which an electron was allowed to move. The situation is pictured schematically in Fig. 32.11. Notice that the inner orbit with the smallest circumference has room for only one wavelength. The second orbit, longer than the first, holds two waves, and so forth.

De Broglie's justification for Bohr's condition for selecting orbits foreshadowed the role that waves would soon play in the description of atomic phenomena. Two years after de Broglie's wave proposal, a whole new formulation of atomic physics called wave mechanics or quantum mechanics was introduced. The wavelike description of electrons in quantum mechanics was somewhat different from Bohr's orbits, as we learn in the next chapter.

Summary and Additional Readings

1. **Nuclear model of atom:** By scattering α particles from gold atoms in a gold foil, Ernest Rutherford determined that an atom has a nucleus that is small in size (about 10^{-14} m in diameter) but large in mass and is made of protons and neutrons. (Rutherford did not know about the neutrons.) Electrons move around the nucleus in a variety of different orbits whose radii are roughly 10^{-10} m.

2. **Atomic spectra:** Atoms emit and absorb electromagnetic radiation of different wavelengths (called the atom's **spectrum**). The spectrum of each type of atom differs from that of other atoms and depends on the energy of the atom when in different states.

3. **Bohr model of one-electron atom:** Niels Bohr in 1913 developed the first model of the atom that could be used to calculate the spectra of radiation emitted and absorbed by an atom. According to his model, which applies only to one-electron atoms and ions, the electron can move around the nucleus only in certain allowed orbits. The electrical force between the electron and nucleus provides the centripetal force to keep the electron in orbit ($kZe^2/r^2 = mv^2/r$). When the electron is in an allowed orbit, the atom's energy is stable and consists of the electron's kinetic energy and the electrical potential energy due to the charges of the electron and nucleus [$E = \frac{1}{2}mv^2 + k(Ze)(-e)/r^2$]. Bohr postulated

that the angular momentum of an electron when in an allowed orbit must equal an integral multiple of $h/2\pi$; that is, $mvr = n(h/2\pi)$, where $n = 1, 2, 3, \ldots$. Finally, when an electron fell from an initial orbit of higher energy E_{n_i} to a final orbit of lower energy E_{n_f}, a photon was emitted whose energy equaled the difference in energy of the atom:

$$hf = E_{n_i} - E_{n_f}.$$

An electron could jump to an orbit of higher energy either by absorbing a photon whose energy equaled the energy gained by the atom or by a collision with another atom.

By combining the above equations, Bohr was able to derive expressions for the radii of allowed orbits, the energy of the atom when its electron was in these orbits, and the wavelength of radiation emitted as an electron fell from one orbit to another:

$$r_n = \frac{(0.53 \times 10^{-10})n^2}{Z}\,\text{m} \qquad \text{for } n = 1, 2, 3, \ldots, \quad \textbf{(32.9)}$$

$$E_n = -\frac{13.6Z^2}{n^2}\,\text{eV} \qquad \text{for } n = 1, 2, 3, \ldots, \quad \textbf{(32.11)}$$

$$\frac{1}{\lambda} = RZ^2 \left(\frac{1}{n_f^2} - \frac{1}{n_i^2} \right), \qquad \textbf{(32.12)}$$

where Z is the number of protons in the nucleus and R, called the *Rydberg constant*, equals $1.097 \times 10^7 \text{ m}^{-1}$.

4. **De Broglie's electron waves:** Bohr's condition that the angular momentum of an electron in an allowed orbit must equal $n(h/2\pi)$ was given physical meaning by de Broglie's proposal that electrons exhibit a wavelike behavior. Bohr's allowed orbits were found to be those in which an electron wave formed a resonant standing-wave vibration pattern. Resonance occurred if the circumference of the orbit equaled an integral number of de Broglie wavelengths:

$$2\pi r = n\lambda = n(h/mv) \qquad \text{for } n = 1, 2, 3, \ldots. \quad \textbf{(32.13)}$$

E. N. da C. Andrade, "The Birth of the Nuclear Atom," *Scientific American* **195**, 93 (1956).

George Gamow, *Mr. Tompkins Explores the Atom*, Cambridge University Press, New York (1945).

Questions

1. Explain carefully why the back-scattering of α particles from atoms is inconsistent with the plum-pudding model of the atom but is consistent with the nuclear model.

2. How can you determine if the sun contains iron?

3. Write three basic equations that we used to derive the expressions for the allowed radii and the allowed energy of electron states in the Bohr model of the atom.

4. If we know the value of n for the orbit of an electron in a hydrogen atom, we can determine the values of three other quantities related to either the electron's motion or the atom as a whole. Describe briefly these three quantities.

5. Explain carefully why hydrogen gas at room temperature

emits little or no light whereas hydrogen gas on the sun emits considerable light.

6. How can an atom with only one electron have so many spectral lines?

7. A group of hydrogen atoms are in a container at room temperature. Infrared radiation, light, and ultraviolet rays are passed through the tube, but only the ultraviolet rays are absorbed at the wavelengths of the Lyman series. Explain why this is true.

8. Why cannot an electron move in an orbit whose circumference is 1.5 times the electron's de Broglie wavelength?

9. If Planck's constant were approximately 50 percent larger, would atoms be larger or smaller? Explain your answer.

Problems

32.1 The Nuclear Model of the Atom

1. The electron in a hydrogen atom spends most of its time 5.3×10^{-11} m from the proton nucleus, whose radius is about 1.4×10^{-15} m. If each dimension of an atom were increased by the same factor and the proton's radius were increased to the size of a tennis ball, how far from the proton would the electron be?

2. A high-energy α particle scattered from the nucleus of an atom helps determine the size and shape of the nucleus. For best results, the de Broglie wavelength of the α particle should be the same size as the nucleus (approximately 10^{-14} m) or smaller. If the mass of the α particle is 6.7×10^{-27} kg, at what speed must it travel to produce a wavelength of 10^{-14} m?

■ 3. A 10 m × 10 m board has twenty holes in it, each with a 3.0-cm radius. If one million BBs are shot randomly at the board, how many will pass through the holes? This experiment is analo-

gous to shooting α particles at a gold foil. However, in that experiment, α particles pass through in most places and are deflected when they hit the gold nuclei.

■ 4. A single layer of gold atoms lies on a table. The radius of each gold atom is about 1.5×10^{-10} m, and the radius of each gold nucleus is about 6×10^{-14} m. A particle much smaller than the nucleus is shot at the layer of gold atoms. Roughly, what is its chance of hitting a nucleus and being scattered? (The projectile is not affected by the electrons around the atom.)

■ ■ 5. (a) Calculate the mass of a gold foil that is 0.010 cm thick and whose area is 1 cm × 1 cm. The density of gold is 19,300 kg/m³. (b) Calculate the number of gold atoms in the foil if the mass of each atom is 3.27×10^{-25} kg. (c) The radius of a gold nucleus is 6.0×10^{-15} m. Calculate the area of a circle with this radius. (d) Calculate the chance that an α particle passing through the gold foil will hit a gold nucleus. Ignore the α particle's size and assume that all gold nuclei are exposed to it; that is, no gold nuclei are hidden behind other nuclei.

■■ 6. A mass M moving at speed v_0 has a direct elastic collision with a second mass m that is at rest. With the conservation-of-energy and momentum principles, we showed in Chapter 8 that the final velocity of mass M is $v = (M - m)v_0/(M + m)$. Using this result, determine the final velocity of an α particle following a head-on collision with (a) an electron and (b) a gold nucleus, each initially at rest. The alpha particle's velocity before the collision is $0.010c$; $m_a = 6.6 \times 10^{-27}$ kg; $m_{el} = 9.11 \times 10^{-31}$ kg; and $m_{gold} = 3.3 \times 10^{-25}$ kg. (c) Based on your answers, could an α particle be deflected backward by hitting an electron in a gold atom?

32.2 Atomic Spectra

7. Calculate the wavelengths of the first three lines in the Lyman series. In what part of the electromagnetic spectrum do these lines appear?

8. Calculate the longest-wavelength and shortest-wavelength lines in the Paschen series. In what part of the electromagnetic spectrum do these lines appear?

■ 9. Calculate the wavelength, frequency, and photon energy of the line with $n = 5$ in the Balmer series.

■ 10. Calculate the wavelengths, frequencies, and photon energies (in eV) of the first two lines in the Balmer series. In what part of the electromagnetic spectrum do these lines appear?

■ 11. A hypothetical one-electron atom emits radiation in a series of lines whose wavelengths are given by the equation $1/\lambda = 300[1 - (1/n^2)]$ in units of nanometers for $n = 2$, 3, 4, . . . , ∞. Calculate (a) the most energetic photon emitted and (b) the least energetic photon emitted in the series.

■■ 12. An imaginary atom is observed to emit electromagnetic radiation at the following wavelengths: 250 nm, 2250 nm, 6250 nm, 12,250 nm, ... Invent an empirical equation for calculating these wavelengths; that is, determine $\lambda = f(n)$ where f is the unknown function of an integer n, which can have values 1, 2, 3, 4, ...

32.3 The Bohr Model

13. (a) Calculate the radii and energy of the $n = 1$, 2, and 3 orbits in the He$^+$ ion ($Z = 2$). (b) Construct an energy-level diagram for this ion.

14. (a) Calculate the radii and energy of the $n = 1$, 2, and 3 orbits of a sodium ion in which 10 of its 11 electrons have been removed. (b) Construct an energy-level diagram for the ion.

15. A uranium atom ($Z = 92$) has two electrons in an $n = 1$ orbit. *Estimate* the radius of this orbit.

16. *Estimate* the energy needed to remove an electron from (a) the $n = 1$ orbit to the $n = \infty$ orbit of iron ($Z = 26$) and (b) from the $n = 1$ to the $n = \infty$ orbit of hydrogen ($Z = 1$).

■ 17. An electron in a hydrogen atom falls from the $n = 4$ to the $n = 3$ state. Calculate the wavelength of the emitted photon.

■ 18. An electron in a He$^+$ ion falls from the $n = 3$ to the $n = 1$ state. Calculate the wavelength, frequency, and energy of the emitted photon.

■ 19. Calculate the energy, frequency, and wavelength of a photon whose absorption raises the electron in a He$^+$ ion from the $n = 1$ to the $n = 6$ state.

■ 20. A group of hydrogen atoms in a discharge tube emit violet light of wavelength 410 nm. Determine the quantum num-

bers of an atom's initial and final states when undergoing this transition.

■ 21. A helium ion He$^+$ emits an ultraviolet photon of wavelength 164 nm. Determine the quantum numbers of the ion's initial and final states.

■ 22. A gas of hydrogen atoms in a tube is excited by collisions with free electrons. If the maximum excitation energy gained by an atom is 12.5 eV, calculate all of the wavelengths of light emitted from the tube as atoms return to the ground state.

■ 23. Some of the energy states of a hypothetical one-electron atom, in units of eV, are $E_1 = -31.50$, $E_2 = -12.10$, $E_3 = -5.20$, and $E_4 = -3.60$. (a) Draw an energy-level diagram for this atom. (b) Calculate the energy and wavelength of the least energetic photon that can be absorbed by these atoms when initially in their ground state.

■ 24. Some of the energy states of a hypothetical one-electron atom are, in units of eV, $E_1 = -18.72$, $E_2 = -6.36$, $E_3 = -3.70$, $E_4 = -1.74$, $E_5 = -0.96$, . . . , $E_\infty = 0.00$. (a) Draw an energy-level diagram for this atom. (b) An atom in the $n = 1$ state is involved in a collision in which it gains 15.02 eV of energy. In what state is its electron following the collision? (c) After the collision, the electron falls to the next lower state. What is the energy of the photon created by this transition?

■ 25. The average thermal energy due to the random transitional motion of a hydrogen atom at room temperature is $\frac{3}{2}kT$ where k is the Boltzmann constant. Would a typical collision between two hydrogen atoms be likely to transfer enough energy to one of the atoms to raise its electron from the $n = 1$ to the $n = 2$ energy state? Explain your answer. [*Note*: Actually, free hydrogen on the earth is in the molecular form H$_2$. However, the above reasoning still explains why a gas of hydrogen molecules at room temperature is seldom excited so that it can emit light.]

■ 26. A 1000-kg satellite circles the earth once every 24 hours in an orbit of radius 4.23×10^7 m. Assuming the orbit satisfies Bohr's quantum condition, calculate the value of n for this orbit.

■■ 27. An electron in the excited state of an atom usually remains in that state only about 10^{-8} s before falling to a lower state. How many times will an electron in the $n = 3$ state of hydrogen rotate around the orbit before falling to the $n = 2$ or $n = 1$ state?

■■ 28. Calculate the speed and frequency of the revolution of an electron around the first Bohr orbit in hydrogen. According to classical physics, the atom should emit electromagnetic radiation at this frequency. In what portion of the electromagnetic spectrum is this frequency?

■■ 29. Show that the frequency of revolution of an electron around the nucleus of a hydrogen atom is $f = (4\pi^2 k^2 e^4 m/h^3)(1/n^3)$.

■■ 30. (a) Are we justified in using nonrelativistic energy equations in the Bohr theory for hydrogen? (Is the electron's speed less than $0.1c$?) (b) Answer the same question for an inner electron in an $n = 1$ state of uranium ($Z = 92$).

■■ 31. Calculate the ratio of the electric force between a proton and an electron in the ground state of the hydrogen atom and the gravitational force between the two particles. Based on your answer, are we justified in ignoring the gravitational force in the Bohr theory?

32.4 De Broglie's Quantum Condition

■ 32. The speed of an electron when in the second allowed Bohr orbit is 1.09×10^6 m/s. (a) Calculate the de Broglie wavelength of the electron. (b) Calculate the radius of the orbit so that the appropriate electron standing wave is formed around the orbit.

■ 33. (a) Use Eq. (32.5) to calculate the speed of the electron in a hydrogen atom when in the $n = 1$ orbit. The radius of the orbit is 0.529×10^{-10} m. (b) Calculate the electron's de Broglie wavelength. (c) Confirm that the circumference of the orbit equals one de Broglie wavelength.

■ 34. Repeat Problem 33 for the $n = 3$ orbit, whose radius is 4.76×10^{-10} m. Three de Broglie wavelengths should fit around the $n = 3$ orbit.

■■ 35. A 70-kg person runs around a circular track whose radius is 64 m. Calculate the lowest two speeds at which the person can run so that the track's circumference will equal an integral number of the person's de Broglie wavelengths. For this problem assume that Planck's constant equals 1.4×10^5 J·s.

CHAPTER 33

Quantum Mechanics and Atoms

In 1925, twelve years after Bohr developed his model for one-electron atoms, Erwin Schrödinger (1887–1961) and Werner Heisenberg (1901–1976) independently presented different but equivalent forms of a new theory for describing atoms. The theory, called **quantum mechanics,** combines the wave and particle nature of matter in one equation. The solutions to this equation, called *wave functions,* can be used to determine the energy of electrons in atoms and the wavelength and intensity of spectral emissions from the atoms. The theory of quantum mechanics is not restricted to atoms but has also been used to describe the properties of molecules (small and large) and of solid materials such as those used to make transistors and superconductors. In short, quantum mechanics has proved so successful at describing a broad range of phenomena that most physicists regard it as the fundamental theory underlying the behavior of nature at the atomic and molecular level. In this chapter we will describe the results of the quantum theory as applied to atoms.

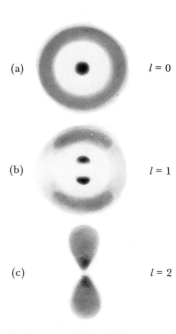

(a) $l = 0$

(b) $l = 1$

(c) $l = 2$

FIG. 33.1. Three different $n = 3$ orbital distributions. The density of the pattern equals the probability of finding an electron at a point.

33.1 Quantum Mechanics

In the approach to quantum mechanics developed by Schrödinger, calculations are performed by solving an equation named in his honor. The solutions to Schrödinger's equation are called **wave functions** and are given the symbol ψ (psi). For atoms, the wave functions are analogous to the orbits in the Bohr theory in that they can be used to determine possible locations for electrons in an atom. In the Bohr theory, however, the electron moves in a well-defined orbit, whereas the path of an electron described by a wave function is not well defined. Instead, the square of the wave function, ψ^2, indicates the *probability* of finding the electron at different locations. This probability (ψ^2) for electrons in atoms is often pictured as in Fig. 33.1, looking something like a cloud; the more dense the cloud is, the greater the probability of finding the electron at that position.

Just as there are many allowed electron orbits in the Bohr theory, many wave functions are solutions of the Schrödinger equation. Each wave function represents a possible state for a particle—a place for it to reside. The energy of

693

a particle depends on the state it occupies. For a one-electron atom the energies of the different states that are calculated using the Schrödinger equation are the same as the energies of electrons when in different Bohr orbits. Whereas the Bohr theory applies only to one-electron atoms, the theory of quantum mechanics has been applied with success to almost every type of atom, to many molecules, to electrons in solids and in biological molecules, and even to particles inside the nucleus of atoms. Unfortunately, the solution of quantum mechanics problems is often complex and usually requires the use of large computer facilities. In the rest of the chapter we will review qualitatively some of the main results of quantum-mechanical calculations with atoms.

33.2 Quantum Numbers for Atomic Wave Functions

In the Bohr model of the one-electron atom, the different allowed orbits for the electron were labeled using a quantum number n. The number could have any integral value from one to infinity. The radius of an orbit and the energy of an electron in that orbit depended on the value of n for the orbit.

A similar situation occurs in quantum mechanics. The wave functions that are solutions of the Schrödinger equation provide a remarkably diverse choice of states, that is, places for the electrons to reside in an atom. In some states the electron remains close to the nucleus, actually swooping through the nucleus on occasion. These so-called inner electrons have little chance of escaping because of the large energy needed to remove them from the nucleus. Other electrons reside in states where they remain much farther from the nucleus and actually flirt more with neighboring atoms and nuclei than with their own nucleus. These outer electrons occasionally join one of the neighboring atoms to produce a negatively charged ion.

The wave functions used to describe these different states are usually labeled using four **quantum numbers,** and the values of the quantum numbers for a particular wave function give a complete determination of the energy and type of orbit of the electron when in that state compared with the electron in other states. (For convenience, we will occasionally use the word *orbit* to indicate an electron's probability distribution.) The quantum numbers can each have different integral or half-integral values; they are described in the following sections along with an indication of the way in which they are related to the electron and its orbit.

Principal Quantum Number n

The **principal quantum number** n is analogous to the quantum number n in Bohr's theory. It can have any positive integral value

$$n = 1, 2, 3, 4, \ldots \qquad (33.1)$$

and is an indicator of the energy and average separation of the orbit from the nucleus. The larger the value of n, the greater the energy of an electron when in that state and the greater the average separation of the electron from the nucleus. Amazingly, the electron energy in hydrogen found from the Schrödinger theory and labeled by the principal quantum number n exactly agrees with the

Bohr result [Eq. (32.10)] even though the orbital shape and description are quite different.

Orbital Angular-Momentum Quantum Number l

The **orbital angular-momentum quantum number** l labels orbits of different shape and also the orbital angular momentum of an electron when in a state. For a particular value of n there are several l-type states of different shape, which are numbered using the following values of l:

$$l = 0, 1, 2, \ldots, n - 1. \tag{33.2}$$

For example, three differently shaped electron distributions are possible for $n = 3$ states (these differently shaped orbits have l values equal to 0, 1, and 2). Five differently shaped orbits are available to electrons in $n = 5$ states (the l values of these states are 0, 1, 2, 3, and 4). An $l = 0$ state is spherically symmetric (Fig. 33.1a); the electron's probability distribution is the same along any direction away from the nucleus. When in states with higher values of l, the electron distribution is not spherically symmetric. Examples of $l = 1$ and 2 states are shown in Fig. 33.1b and c. The angular momentum L of an electron when in a state with quantum number l is $L = \sqrt{l(l + 1)}\, h/2\pi$.

Magnetic Quantum Number m_l

Angular momentum is a vector quantity, and the **magnetic quantum number** m_l indicates the projection of the angular-momentum vector of an electron on a measurement axis, such as the axis of the magnetic field shown in Fig. 33.2. This projection in turn determines the orientation of the electron's orbit when in that state. In general, for a state with orbital angular-momentum quantum number l, $2l + 1$ differently oriented states exist corresponding to the following values of m_l:

$$m_l = 0, \pm 1, \pm 2, \ldots, \pm l. \tag{33.3}$$

For a state with $l = 0$, the angular momentum L of an electron is zero, and the m_l quantum number is also zero. For an $l = 1$ state, however, the angular momentum is not zero, and it can have three different orientations corresponding to the three allowed values of m_l $(-1, 0, +1)$. The orbits shown in Fig. 33.2 represent schematically the three possible orientations of an $l = 1$ electron orbit. For an $l = 2$ state, five orbits of different orientation exist corresponding to the five allowed values of m_l $(-2, -1, 0, +1, +2)$.

The word *magnetic* is chosen for the m_l quantum number because of its importance relative to the magnetic properties of the atom. When an electron moves in an orbit around a nucleus, it creates a magnetic field much like that produced by an electron current in a circular loop of wire. The atom is said to have a magnetic moment because of this orbital motion of the electron and the magnetic field it produces. The m_l quantum number indicates the orientation of the magnetic moment due to the orbital motion of the electron; it is important in a variety of experiments involving the magnetic properties of atoms.

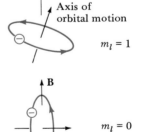

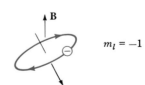

FIG. 33.2. An $l = 1$ electron orbit type can assume three different orientations in a magnetic field.

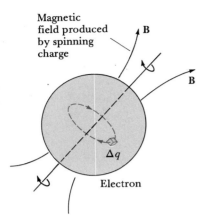

FIG. 33.3. A model of a spinning electron.

Spin Magnetic Quantum Number m_s

The fourth and last quantum number is related not to the motion of the electron about the nucleus (as are n, l, and m_l) but to an intrinsic property of the electron called its *spin angular momentum*. An electron seems to spin about an axis as it moves in orbit about the nucleus, much the way the earth spins about its axis as it orbits the sun. Any spinning or rotating mass has angular momentum. For an electron, the magnitude of its spin angular momentum is $S = \sqrt{s(s + 1)}(h/2\pi)$, where $s = \frac{1}{2}$.

If the electron is pictured as a solid sphere (Fig. 33.3), its electric charge must spin as the mass spins, and each portion of charge Δq moves in a loop and produces an electric current. The current, in turn, produces a magnetic field.* We see that an atom's magnetic field and magnetic moment are caused by two different motions: (1) the orbital motion of its electrons about the nucleus and (2) the spinning motion of the electrons about their own axes. Many careful experiments have shown that the electron's spin magnetic moment can assume only two orientations relative to an externally applied magnetic field—one in the direction of the field and the other opposite the direction of the field (Fig. 33.4). The **spin magnetic quantum number** m_s is used to indicate the orientation of an electron's spin angular momentum and its magnetic moment. The quantum number can have two values,

$$m_s = +\tfrac{1}{2} \quad \text{and} \quad -\tfrac{1}{2}, \tag{33.4}$$

corresponding to the two possible orientations of the electron's spin angular momentum and magnetic moment. The two orientations are sometimes called "up" and "down."

The allowed values for each of the four quantum numbers for electron states in atoms are summarized in Table 33.1.

33.3 Multielectron Atoms

Together, the four quantum numbers introduced in the last section (n, l, m_l, and m_s) are used to label the wave functions for states occupied by electrons in atoms. The symbol $|nlm_lm_s\rangle$ will be used to represent one of these wave functions. For example, $|300-\tfrac{1}{2}\rangle$ represents a wave function with $n = 3$, $l = 0$, $m_l = 0$, and $m_s = -\tfrac{1}{2}$. The probability of finding an electron at different locations around the nucleus when in a state ("orbit") described by this wave function is shown in Fig. 33.1a.

*This model provides an intuitive idea of how an electron's intrinsic magnetic moment is produced. Unfortunately, detailed calculations are not consistent with this model.

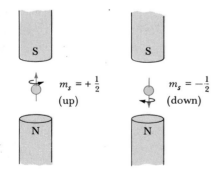

FIG. 33.4. An electron's spin axis can assume only two orientations in a magnetic field.

$m_s = +\tfrac{1}{2}$ (up) $m_s = -\tfrac{1}{2}$ (down)

TABLE 33.1 Quantum Numbers for Electron States in Atoms

Name	Symbol	Allowed Values
Principal	n	1, 2, 3, 4, . . .
Orbital	l	For a given n, l can have the values 0, 1, 2, . . . , $n - 1$.
Magnetic	m_l	For given n and l, m_l can have the values 0, ± 1, ± 2, . . . , $\pm l$.
Spin magnetic	m_s	For each set n, l, and m_l, the spin magnetic quantum number can have two values: $m_s = +\tfrac{1}{2}$ or $-\tfrac{1}{2}$.

Early in the study of the light emitted and absorbed by atoms, a special vernacular was developed for the states of different orbital angular-momentum quantum number l. If an electron fell from an excited state with $l = 0$ to a state of lower energy, the spectral line appeared sharp or well defined. Thus, an $l = 0$ state became known as an _s_ state. Transitions from $l = 1$ states were often the principal bands of light observed, and $l = 1$ states became known as _p_ states. The letter designations for these and other l states are summarized in Table 33.2. This letter designation is used extensively in the rest of this chapter. An $n = 1$ and $l = 0$ state is called a $1s$ state; an $n = 4$ and $l = 2$ state is called a $4d$ state; and so forth (see Table 33.2).

The energy of an electron when in a particular state of the atom depends on the values of n and l but not on the values of m_l and m_s. The smaller the values of n and l, the less the electron's energy. (The energy depends more on n than on l.)

The electrons in atoms usually occupy the lowest energy states that are available. The electron in a hydrogen atom almost always resides in a $1s$ state, the state in which $n = 1$ and $l = 0$. In lithium, a three-electron atom, two of its three electrons are in the $1s$ state. Its third electron normally resides in a $2s$ state. The orbits of the $1s$ states seem to be "filled" by the first two electrons, and the third electron, finding no room, must reside in an orbit of greater energy farther from the nucleus (the $n = 2$, $l = 0$ orbit).

This ability of electrons to exclude other electrons from their orbits is quite remarkable. An electron itself when measured in scattering experiments outside the atom is much smaller than the dimensions of the orbit it occupies when in the atom. Yet this tiny electron manages to keep other electrons out of its orbit.

A principle used to describe this phenomenon was developed in 1925 by the Austrian physicist Wolfgang Pauli (1900–1958).

Pauli exclusion principle: Each electron state $|nlm_lm_s\rangle$ of an atom can hold no more than one electron.

For example, for the $n = 1$, $l = 0$ state, m_l can only be zero and m_s can be $+\frac{1}{2}$ or $-\frac{1}{2}$. Thus, there are two different states with $n = 1$ and $l = 0$; the $|nlm_lm_s\rangle = |100\frac{1}{2}\rangle$ and $|100-\frac{1}{2}\rangle$ states. According to Pauli's exclusion principle, each state can hold one electron; so two electrons can occupy the $n = 1$, $l = 0$ states. These two states are said to form a **subshell**—a group of states with the same values of n and l. If each state in a subshell is occupied by an electron, the subshell is said to be **filled.**

The number of states in a subshell depends on the number of different allowed combinations of m_l and m_s for the given values of n and l. The rules restricting the values of n, l, m_l, and m_s are summarized in Table 33.1.

TABLE 33.2 Letter Designations of Different l States

l value	0	1	2	3	4	5	6
Letter	_s_	_p_	_d_	_f_	_g_	_h_	_i_

EXAMPLE 33.1 Determine the number of states and the quantum number designation of each state for a $3d$ subshell.

SOLUTION For a $3d$ subshell, $n = 3$ and $l = 2$ (see Table 33.2). For an $l = 2$ state, the m_l quantum number can have the values $m_l = -2, -1, 0, 1$, and 2. For each of these five values of m_l, the m_s quantum number can have two values: $m_s = -\frac{1}{2}$ or $+\frac{1}{2}$. Thus, there are ten unique $|nlm_lm_s\rangle$ states for the $3d$ subshell:

$$|32-2\tfrac{1}{2}\rangle, \quad |32-1\tfrac{1}{2}\rangle, \quad |320\tfrac{1}{2}\rangle, \quad |321\tfrac{1}{2}\rangle, \quad |322\tfrac{1}{2}\rangle,$$
$$|32-2-\tfrac{1}{2}\rangle, \quad |32-1-\tfrac{1}{2}\rangle, \quad |320-\tfrac{1}{2}\rangle, \quad |321-\tfrac{1}{2}\rangle, \quad |322-\tfrac{1}{2}\rangle.$$

If one electron resides in each of these states, such as occurs in a copper atom, the atom is said to have a filled $3d$ subshell. ∎

Electron Configurations

The electrons in an atom normally occupy the states with lowest energy. The atom as a whole is then said to be in its **ground state.** An electron configuration for the ground state of an atom indicates the subshells occupied by its electrons and the number of electrons in each subshell. For example, a sodium atom has 11 electrons, distributed as follows: $1s^2 2s^2 2p^6 3s$. The superscript represents the number of electrons in each subshell: $1s^2$ indicates that two electrons occupy the $1s$ subshell, $2p^6$ indicates that six electrons occupy the $2p$ subshell, and so forth. The maximum number of electrons that can be in a subshell and their approximate order of filling are listed in Table 33.3. We see that for sodium, ten electrons are needed to fill the $1s$, $2s$, and $2p$ subshells. The eleventh electron in sodium occupies a state in the next available subshell—the $3s$ subshell. When sodium is ionized to form Na^+, it loses the electron from the $3s$ shell.

33.4   The Periodic Table

Many of the properties of atoms (their ability to form positive or negative ions and to combine with other atoms to form molecules) depend on the number of electrons in the subshell of the atom that is only partially filled and on the type

TABLE 33.3   Atomic Subshells from Lowest to Highest Energy

Subshell	Maximum Number of Electrons in Subshell	Total Number of Electrons in Atom to Fill Subshell
$1s$	2	2
$2s$	2	4
$2p$	6	10
$3s$	2	12
$3p$	6	18
$4s$	2	20
$3d$	10	30
$4p$	6	36
$5s$	2	38
$4d$	10	48
$5p$	6	54
$6s$	2	56
$4f$	14	70
$5d$	10	80
$6p$	6	86
$7s$	2	88
$5f$	14	102
$6d$	10	112

of subshell it is. These "outer" electrons are usually farthest from the nucleus and interact most with other atoms in their vicinity. Also, these outer electrons usually require the least energy to be removed from the atom. When sodium is ionized to form Na^+, it loses the electron that was in the $3s$ subshell before it was ionized.

Groups of atoms with the same number of outer electrons in similar subshells usually exhibit similar properties. This leads naturally to the classification of the atoms as presented in the periodic table of elements (Table 33.4). (The word *element* indicates a substance made of one type of atom.) In the table the different types of atoms are grouped together in columns so that atoms in the same column have similar electron configurations for their outer electrons.

Group I. Atoms in this group all have one outer electron in an s state (a state with $l = 0$). Examples are hydrogen $(1s)$, lithium $(2s)$, sodium $(3s)$, and potassium $(4s)$. This outer electron is easily removed, leaving behind a positive ion: H^+, Li^+, Na^+, and K^+.

TABLE 33.4 Periodic Table of the Elements*

I	II												III	IV	V	VI	VII	0
1 **H** 1.0080																		2 **He** 4.0026
3 **Li** 6.941	4 **Be** 9.0122												5 **B** 10.81	6 **C** 12.011	7 **N** 14.0067	8 **O** 15.9994	9 **F** 18.9984	10 **Ne** 20.179
11 **Na** 22.9898	12 **Mg** 24.305			Transition elements									13 **Al** 26.9815	14 **Si** 28.086	15 **P** 30.9738	16 **S** 32.06	17 **Cl** 35.453	18 **Ar** 39.948
19 **K** 39.102	20 **Ca** 40.08	21 **Sc** 44.956	22 **Ti** 47.90	23 **V** 50.941	24 **Cr** 51.996	25 **Mn** 54.9380	26 **Fe** 55.847	27 **Co** 58.9332	28 **Ni** 58.71	29 **Cu** 63.54	30 **Zn** 65.37	31 **Ga** 69.72	32 **Ge** 72.59	33 **As** 74.9216	34 **Se** 78.96	35 **Br** 79.909	36 **Kr** 83.80
37 **Rb** 85.467	38 **Sr** 87.62	39 **Y** 88.906	40 **Zr** 91.22	41 **Nb** 92.906	42 **Mo** 95.94	43 **Tc** (99)	44 **Ru** 101.07	45 **Rh** 102.906	46 **Pd** 106.4	47 **Ag** 107.870	48 **Cd** 112.40	49 **In** 114.82	50 **Sn** 118.69	51 **Sb** 121.75	52 **Te** 127.60	53 **I** 126.9045	54 **Xe** 131.30
55 **Cs** 132.906	56 **Ba** 137.34	57 **La** 138.906	72 **Hf** 178.49	73 **Ta** 180.948	74 **W** 183.85	75 **Re** 186.2	76 **Os** 190.2	77 **Ir** 192.2	78 **Pt** 195.09	79 **Au** 196.967	80 **Hg** 200.59	81 **Tl** 204.37	82 **Pb** 207.2	83 **Bi** 208.981	84 **Po** (210)	85 **At** (210)	86 **Rn** (222)
87 **Fr** (223)	88 **Ra** 226.03	89 **Ac** 227.028	104 **Rf** (261)	105 **Ha** (262)	106 (263)	107 (262)	108 (265)	109 (266)										

Lanthanide series

58 **Ce** 140.12	59 **Pr** 140.908	60 **Nd** 144.24	61 **Pm** (147)	62 **Sm** 150.4	63 **Eu** 151.96	64 **Gd** 157.25	65 **Tb** 158.925	66 **Dy** 162.50	67 **Ho** 164.930	68 **Er** 167.26	69 **Tm** 168.934	70 **Yb** 173.04	71 **Lu** 174.97

Actinide series

90 **Th** 232.038	91 **Pa** 231.036	92 **U** 238.029	93 **Np** 237.048	94 **Pu** (242)	95 **Am** (243)	96 **Cm** (248)	97 **Bk** (249)	98 **Cf** (249)	99 **Es** (254)	100 **Fm** (257)	101 **Md** (258)	102 **No** (259)	103 **Lr** (260)

*Atomic weights of stable elements are those adopted in 1969 by the International Union of Pure and Applied Chemistry. For those elements having no stable isotope, the mass number of the "most stable" isotope is given in parentheses.

Group II. Atoms in this group have two electrons in an outer s state. Examples are beryllium ($2s^2$), magnesium ($3s^2$), and calcium ($4s^2$). These atoms become ions with a double positive charge when their outer electrons are removed (Be^{2+}, Mg^{2+}, Ca^{2+}).

Group III. Atoms in this group have two outer electrons in s states and a third outer electron in a p state. Examples are boron ($2s^22p$), aluminum ($3s^23p$), and gallium ($4s^24p$).

Group IV–VII. Each atom to the right of the preceding atom in the periodic table has one more p electron than the neighboring atom to the left. Thus, groups IV, V, VI, and VII consist of atoms with outer electron configurations s^2p^2, s^2p^3, s^2p^4, and s^2p^5, respectively. Notice that the group VII atoms are one p electron short of having a filled p subshell. Group VII atoms often capture an extra electron to fill their p subshell because filled subshells are usually very stable. Negative ions, such as F^-, Cl^-, Br^-, and I^-, are formed in this way.

Group 0. All atoms in this group have outer electrons in a filled p subshell, with the exception of helium, which has a $1s^2$ configuration. Examples are neon ($2p^6$), argon ($3p^6$), krypton ($4p^6$), and xenon ($5p^6$). Since filled subshells are very stable, group 0 atoms are less likely to be ionized and are relatively inert.

Transition atoms. This category includes atoms that do not fit neatly into the other groups of the periodic table. The atoms in the first row have 18 electrons in filled $1s$, $2s$, $2p$, $3s$, and $3p$ subshells. Additional electrons occupy the $4s$ and $3d$ subshells; for example, titanium with 22 electrons has its outer electrons in a $4s^23d^2$ configuration, while copper with 29 electrons has its outer electrons in a $4s^13d^{10}$ configuration. When ionized, transition atoms tend to lose their $4s$ electrons rather than their $3d$ electrons. Thus, these transition atoms tend to form positive ions with partially filled $3d$ subshells.

Second-row transition atoms starting with yttrium are atoms whose $5s$ and $4d$ states are being filled. The lanthanides include atoms whose $6s$, $5d$, and $4f$ subshells are partially or totally filled. The actinides include atoms whose $7s$, $6d$, and $5f$ subshells are partially or totally filled.

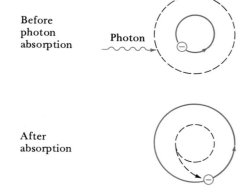

FIG. 33.5. Photon absorption resulting in the excitation of an atom.

33.5 Absorption and Emission of Radiation

Many of the interesting properties of atoms are related to an electron's jump between two different orbits—called an electron **transition.** The absorption of a photon by an atom results in an electron transition from a state of lower energy to one of higher energy, an event called **excitation.** To be absorbed, the photon must have an energy equal to the difference in energy of the atom when in an excited state and when in the ground state. The absorption of photons in the visible- and ultraviolet-wavelength regions usually involves the excitation of an outer electron from the subshell in which it normally resides to a vacant subshell of higher energy (Fig. 33.5). Excitation can also occur as a result of collisions of atoms with each other.

Atoms and molecules can often be identified by their absorption **spectra**—the wavelengths of light they absorb. Absorption spectroscopy is used for purposes such as monitoring the type and concentration of atoms and molecules in samples of blood, air, soil, and water and for measuring the changing concentration of molecules involved in chemical and biochemical reactions.

Photon **emission** occurs when atoms fall from an excited state to a state of lower energy, a process called *deexcitation*. Deexcitation usually involves the transition of a single electron (Fig. 33.6) and occurs within 10^{-8} s after the atom becomes excited. The energy of the photon equals the atom's change in energy.

Fluorescence occurs when an atom that has been excited by absorbing a photon returns to the ground state in two or more steps, where each step involves the emission of a photon (Fig. 33.7). The energy of the emitted photons together equals the energy of the absorbed photon. As a result, each emitted photon has less energy than the absorbed photon that caused excitation.

Fluorescent lightbulbs produce light by fluorescence. A gas inside the bulb is excited by collisions with electrons that have been accelerated by an applied voltage. The excited atoms return to their ground state by emitting ultraviolet-wavelength photons. The ultraviolet-wavelength photons strike a coating of fluorescent material on the inside surface of the bulb, causing these atoms to become excited. These fluorescent atoms return to the ground state in more than one step, and at least one step involves the emission of a photon of visible light.

Phosphorescence is similar to fluorescence except that the intermediate state (E_2 in Fig. 33.7) to which an excited atom falls is a "metastable" ("almost stable") state. The atom when in the **metastable state** remains there somewhat longer than the 10^{-8} s that an atom usually remains in an excited state. The lifetime of an excited state is the time needed for half the atoms to fall from the excited state. For a metastable state the lifetime could be as short as 10^{-5} s or as long as an hour or so, depending on the particular type of phosphorescent atom. A watch that glows in the dark is sometimes made of a phosphorescent material. Some of the light absorbed during the day causes electrons to be excited to the metastable state. Light is emitted by this material long after the exciting light is turned off; the phosphorescent material glows in the dark. Earlier in this century, luminous watches used radioactive materials to give the phosphor energy.

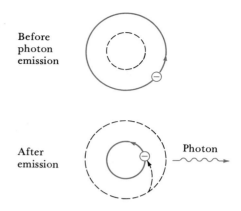

FIG. 33.6. Spontaneous photon emission by deexcitation of an atom.

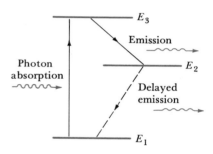

FIG. 33.7. An energy-level diagram for a material made of phosphorescent atoms. The state with energy E_2 is a metastable state. Electrons are excited from E_1 to E_3 and then fall to E_2. Because E_2 is metastable, the fall of the electrons back to E_1 is delayed.

33.6 Stimulated Emission and Lasers

The emission of radiation that was discussed in Section 33.5 is called **spontaneous emission:** An atom spontaneously falls from an excited state to a state of lower energy and emits a photon in the process. No external inducements are needed to encourage the transition.

An excited atom can fall to a state of lower energy by another process called **stimulated emission,** which occurs when a photon passes an excited atom and induces or stimulates the deexcitation of the atom. A new photon is created as the atom falls to a state of lower energy. This photon and the photon that stimulated the deexcitation move off together (Fig. 33.8). Stimulated emission occurs only if the energy lost by the atom as it falls from its excited state equals the energy of the photon that induced the transition. This photon has exactly the same energy and wavelength as the newly created photon. The two photons are also in phase, and their waves add constructively to form a wave of twice the energy as that of the original photon.

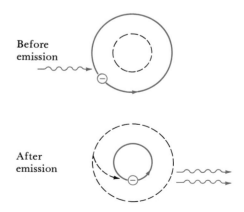

FIG. 33.8. Stimulated emission. A photon stimulates an atom to fall from an excited state, creating a second photon and a deexcited atom.

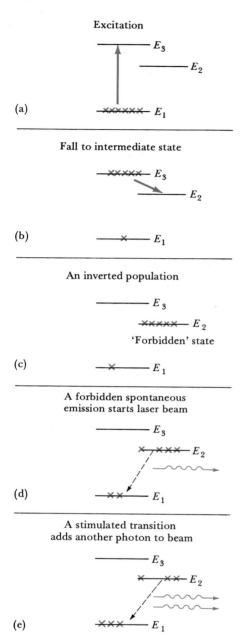

Excitation

(a)

Fall to intermediate state

(b)

An inverted population

'Forbidden' state

(c)

A forbidden spontaneous
emission starts laser beam

(d)

A stimulated transition
adds another photon to beam

(e)

FIG. 33.9. Steps leading to the production of a laser beam.

A **laser** is a device that relies on stimulated emission to produce an intense, narrow beam of light. Lasers can be used to align roads through mountain tunnels, to perform surgery on the eye's retina, to drill holes through metal, to carry thousands of telephone messages along a thin glass fiber, to measure the distance to the moon, to fabricate microelectronic circuit boards, and to detect information compressed on the tiny grooves of a video disc.

A laser consists of a specially chosen group of atoms (or molecules) that can be excited to a metastable state, an apparatus for exciting these atoms, and a container to hold the atoms. Partially or totally reflecting mirrors are usually placed at the two ends of the container. The atoms to be excited may be in a gaseous, liquid, or solid form. Excitation occurs in a variety of ways. For example, in a helium-neon gas laser, such as used in many physics laboratories, electrons are accelerated through a gas of helium and neon atoms. Collisions between the electrons and atoms cause the atoms to become excited. In other lasers, a flash of light may cause excitation. The excitation of an atom to a state of higher energy is represented in Fig. 33.9a by the heavy arrow. Each cross (✕) in this figure represents an atom (or a group of atoms) in the energy state where the cross occurs. Originally, almost all atoms are in the ground state with energy E_1. After excitation, many of these atoms have been excited to state E_3 (Fig. 33.9b).

The excited atoms fall quickly to an intermediate, metastable state of energy E_2; the arrow in part (b) indicates this transition. The atoms have now experienced a population inversion, a situation in which more atoms are in the excited state of energy E_2 than in the ground state of energy E_1 (part c). The transition from the metastable state back to the ground state is called a **forbidden transition** because it occurs spontaneously at a much slower rate than most atomic deexcitation transitions. For this reason the atoms remain much longer in state E_2 than they normally remain in other excited states—the state is "almost stable," or metastable.

The deexcitation of this inverted population begins slowly (compared to the normal deexcitation time). An atom falls spontaneously back to the ground state and emits a photon (part d). This photon, as it passes another atom in the metastable state, can cause it to fall to the ground state by stimulated emission. Two photons of equal energy and in phase with each other leave the second atom (part e). These two photons, in turn, stimulate deexcitation of two more excited atoms. We now have four photons moving together, each of which can stimulate more transitions. Before long, a huge cascade of atoms is returning to the ground state by stimulated emission, and each atom contributes one photon whose energy and phase equals that of the other photons. The process, represented schematically in Fig. 33.10, is called light amplification by the stimulated emission of radiation; hence the acronymn **laser.**

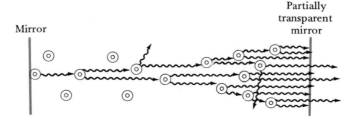

Mirror

Partially
transparent
mirror

FIG. 33.10. The increasing intensity of a laser beam as it passes through a group of excited atoms in a laser.

The deexcitation of the inverted population is not complete after one pass of the light wave through the excited atoms. Mirrors placed at the ends of the laser reflect the light back and forth through the excited atoms many times. If the population remains inverted, the intensity of the light beam grows during each round-trip. However, the mirror at one end of the laser reflects only part of the beam and allows the other part to leave the laser. In a **continuous laser,** an equilibrium is reached where energy lost by the light leaving the laser equals energy gained by the partially reflected light beam as it makes another round-trip through the excited atoms. A continuous narrow beam of light leaves the laser.

A beam of light from a laser is unique for several reasons. (1) The light is **monochromatic;** that is, all the photons have the same wavelength. (2) The photons in the beam are in phase with each other (the light is said to be **coherent**) rather than a group of independent photons whose vibrations are independent of each other. The coherence of a laser beam allows it to be used to transmit information (such as telephone conversations) in much the way that high-frequency radio waves transmit information for television broadcasting. (3) Ordinary sources of light such as tungsten bulbs emit light in all directions, and the light intensity drops as the light spreads after leaving the source. On the other hand, a laser beam is very narrow, being limited only by diffraction at the mirror where it leaves the laser. If the laser is well designed, the angular spread of the beam is approximately λ/D (see Section 21.5 and Interlude V) where λ is the wavelength of the light and D is the diameter of the mirror. Even though the power of the laser may be low, the small spread of the beam causes the light to be relatively intense (Fig. 33.11).

Use of Lasers in Eye Surgery

The fact that a laser beam is very narrow has led to applications requiring a significant amount of energy in very localized places, such as when drilling small holes in metal sheets. Another application is in eye surgery. As you may recall

FIG. 33.11. A narrow laser beam illuminates London's Big Ben from almost three miles away. The laser light on Big Ben is much brighter than the light illuminating the tower from more powerful lamps on the nearby bridge. (United Press International Photo.)

FIG. 33.12. (a) The eye, showing a small retinal detachment. (b) A magnified cross section of a laser weld to hold the retina in place. (Courtesy of International Research and Development Co. Ltd., Newcastle-upon-Tyne, England.)

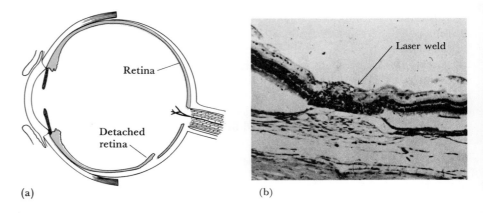

(a)

(b)

from Section 20.4, the retina of the eye contains about 130 million cones and rods, each of which serves as a light detector. The cones and rods connect to nerves that transmit visual information to the brain. If a disease or a severe blow to the head causes part of the retina to detach from the underlying tissue (Fig. 33.12a), the path of nerve impulses to the brain is interrupted, and the affected area becomes blind.

Modern treatment of a detached retina involves welding it to the underlying tissue by using the heat produced by a short laser flash. The cross section of one of these welds is shown in Fig. 33.12b. Lasers are very useful in eye surgery for several reasons. (1) The surgery is noninvasive; other parts of the eye are not invaded by foreign objects such as scalpels, clamps, sponges, or tweezers. (2) The weld spots can be extremely small, an important factor when the detachment occurs in the central part of the retina that is used for seeing fine detail. (3) Welds can be made in a fraction of a second, so the eye does not have to be clamped open. (4) The laser weld causes no pain.

33.7 X-Ray Emission

Usually, the outer electrons in atoms are the only electrons involved in transitions from one state to another. The outer electrons are bumped into excited states when atoms collide with each other, and they jump between different states when photons of light and ultraviolet radiation are absorbed and emitted by atoms. Inner electrons can also change states, but much more energy is needed to excite an inner electron to a vacant state of higher energy. Consequently, these inner electrons usually remain in their orbits close to the nucleus. One opportunity for leaving an inner orbit occurs if an inner electron of an atom is struck by an energetic electron in an x-ray tube.

An x-ray tube, such as shown in Fig. 33.13, has a filament that becomes hot when an electric current flows through it. The extra thermal energy of the atoms and ions in the filament causes electrons to be bumped from the filament. These electrons are accelerated by a large voltage toward a metal anode commonly made of tungsten. In the x-ray tube used to produce x-rays for medical and dental examinations, the voltage from the filament to the anode is typically about 40,000 V. The speed of an electron released from the filament increases to over 10^8 m/s, almost the speed of light, just before it crashes into the anode.

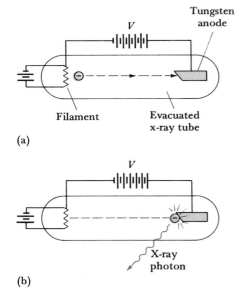

(a)

(b)

FIG. 33.13. (a) An electron starts its trip across an x-ray tube. (b) When the electron crashes into the anode, x-ray photons are created.

The electron, as it strikes the anode, can be stopped by a single dramatic collision or by a series of collisions. During each collision the electron loses kinetic energy, which is balanced by the creation of a photon. The photons produced during these collisions are called **Bremsstrahlung radiation** (from the German word meaning "braking radiation") and are a consequence of the electron's deceleration. Most of the photons are in the x-ray portion of the electromagnetic spectrum.

If the electron stops in one direct hit, it emits one photon whose energy equals the energy gained by the electron as it accelerated across the tube. Thus, the maximum energy of a photon created by a collision is

$$\underbrace{hf_{max}}_{\substack{\text{Photon} \\ \text{energy}}} = \underbrace{eV.}_{\substack{\text{Energy gained as} \\ \text{electron crosses} \\ \text{x-ray tube}}} \qquad (33.5)$$

If the electron stops in a series of collisions, a group of lesser-energy photons is created.

A typical x-ray spectrum produced by these electron collisions is shown in Fig. 33.14. Notice that no photons are produced with frequencies greater than $f_{max} = eV/h$, as predicted by Eq. (33.5). The smooth part of the curve in Fig. 33.14 is caused by less-energetic collisions, which produce Bremsstrahlung radiation.

The pointed lines on the curve in Fig. 33.14, called the **characteristic lines,** result from collisions of the electron flying across the tube with inner electrons of atoms in the anode. If a collision causes the $1s$ electron to be knocked from a tungsten atom, then outer electrons of the atom fall into the vacancy left in the inner subshell. The characteristic lines in Fig. 33.14 are the result of photons emitted by $2p \longrightarrow 1s$ and $3p \longrightarrow 1s$ transitions of tungsten atoms in the anode.

We can estimate the energy needed to knock an inner electron from a many-electron atom by using Bohr's equation for the energy of an electron when in different orbits [Eq. (32.11)]:

$$E_n = -\frac{13.6Z^2}{n^2}\,\text{eV}.$$

Notice that the energy is proportional to Z^2, the number of protons in the nucleus. For tungsten, $Z = 74$, and the $n = 1$ electrons in tungsten should have energy of approximately

$$E_1 \simeq -\frac{13.6(74^2)}{1^2} = -70,000\,\text{eV}.$$

To remove an $n = 1$ electron from tungsten would require about 70,000 eV of energy.

Bohr's equation was developed only for one-electron atoms, and it overestimates the energy needed to remove an $n = 1$ electron from an atom with many electrons. Using Eq. (32.11), we estimate the energy of an $n = 1$ electron in iron ($Z = 26$) to be -9200 eV, whereas the experimentally determined value is -7130 eV. Nevertheless, the equation does provide a rough indication of the great energy needed to dislodge an inner electron from an atom. The energy depends very much on the nuclear charge [the Z^2 dependence in Eq. (32.11)].

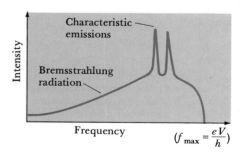

FIG. 33.14. The intensity of radiation emitted by an x-ray tube as a function of frequency.

Summary and Additional Readings

1. **Quantum mechanics** is a theory used to determine the energy, distribution in space, and other properties of particles such as electrons when in different environments. **Schrödinger's equation** is solved for **wave functions,** which, when squared, indicate the probability of finding an electron or some other particle at different locations. The wave functions can also be used to determine the electron's energy when in a state described by a particular wave function and its chance of jumping to another state of the atom.

2. **Quantum numbers:** Usually, many different wave functions are solutions to Schrödinger's equation, and each wave function defines a region for an electron to reside—an electron **state.** For atoms the states are labeled using four quantum numbers, which have integral or half-integral values and indicate various properties about the electron and its orbit.

The **principal quantum number** n has positive integral values

$$n = 1, 2, 3, 4, \ldots \qquad (33.1)$$

and indicates the energy and average distance of the orbit from the nucleus. The larger the value of n, the greater the electron's average separation from the nucleus and the greater its energy.

The **orbital angular momentum quantum number** l indicates the shape of the electron distribution and the angular momentum of an electron. For a particular value of n, states exist for the following values of l:

$$l = 0, 1, 2, 3, \ldots, n - 1, \qquad (33.2)$$
$$s \quad p \quad d \ f.$$

The **orbital magnetic quantum number** m_l indicates the orientation of the orbit and the orientation of its magnetic moment due to its orbital motion. For a particular value of n and l, states exist for the following values of m_l:

$$m_l = 0, \pm 1, \pm 2, \ldots, \pm l \qquad (33.3)$$

The **spin magnetic quantum number** m_s indicates the orientation of the axis of an electron's spinning motion. The two allowed values of m_s are

$$m_s = +\tfrac{1}{2} \text{ (up)} \quad \text{or} \quad -\tfrac{1}{2} \text{ (down)}. \qquad (33.4)$$

3. **Multielectron atoms:** The electrons in an atom usually occupy the states with lowest energy, but subject to the **Pauli exclusion principle:** An electron state of an atom, represented by the symbol $[nlm_lm_s\rangle$, can hold no more than one electron. The lowest-energy electron states are usually those with (1) the smallest value of n and (2) the smallest value of l. An electron configuration indicates the electron distribution among different orbits (for example, $1s^2 2s^2 2p^6 3s^2 3p^6 4s$, and so forth), where the superscript indicates the number of electrons in a subshell with a particular value of n and l. The letters $s, p, \ldots$, indicate the value of l for a subshell (see Table 33.2).

4. **Periodic table:** Atoms whose outer electrons are in similar electron configurations are grouped together in the periodic table. These atoms exhibit similar properties.

5. **Electron transitions:** An atom can **absorb** a photon if the photon's energy equals the difference in energy of the state in which an electron resides and a higher-energy vacant state. An excited atom can emit a photon by falling **spontaneously** to a lower-energy state. An excited atom can also emit a photon by a process called **stimulated emission:** Another photon passes the excited atom and induces or stimulates its deexcitation, thus causing the production of a second photon whose energy and phase equal that of the first photon. The coherent narrow beam of light from a laser is caused by stimulated emission.

6. **X-rays** are produced by the deceleration of fast-moving electrons as they hit the metal anode of an x-ray tube. X-ray photons are also created as outer electrons in atoms fall into a vacant inner orbit.

R. K. Clayton, *Light and Living Matter*, vol. 1: *The Physical Part*, vol. 2: *The Biological Part*, McGraw Hill, New York (1970, 1971, respectively). Excellent presentation of the ways light interacts with living things.

J. F. Hyde, "Let the Elements Teach Periodic Law," *The Physics Teacher* **13**, 538 (1975).

W. Thumm, "Roentgen's Discovery of X-Rays," *The Physics Teacher* **13**, 207 (1975).

Questions

1. Discuss the similarities and differences in the way a hydrogen atom is pictured in Bohr's model and in quantum mechanics.

2. (a) Describe verbally the way in which electron orbits having different values of quantum number n differ from each other. Repeat part (a) only for (b) quantum number l and (c) quantum number m_l. (d) What are the allowed values of m_s and how do these different electron states differ?

3. Explain why each of the following is an unlikely ground-state electron configuration for an atom: (a) $1s^2 2s^2 2p^6 3s^2 4s^2$; (b) $1s^2 2s^2 2p^6 3s^2 4p^3$; (c) $1s^2 2s^2 2p^4 3s^2 4s$; (d) $1s^2 2s^2 2p^6 3s^2 3p^6 4s^2 4d^2$.

4. Why do fluorine and chlorine exhibit similar properties?

5. Why do atoms in the first column of the periodic table exhibit similar properties?

6. Explain how astronomers might determine the different types of atoms that are on the sun and the amount of each type.

7. Can a phosphorescent material absorb visible radiation and emit, after some delay, ultraviolet radiation? Explain.

8. Indicate the difference between spontaneous emission and stimulated emission.

9. What property of a laser beam makes it useful for (a) cutting metal and (b) carrying information?

Problems

33.1 Quantum Mechanics

■ ■ 1. One of the easiest problems involving quantum mechanics is the determination of the wave functions for an electron inside an infinite square-well potential (Fig. 33.15). The electron is prohibited from leaving the well but moves freely inside it. Its allowed wave functions appear exactly like the waves on a string fixed at both ends. For a well of width a, the allowed wavelengths are $\lambda = 2a/n$ where $n = 1, 2, 3, \ldots$ [see Eq. (18.3)]. Using the

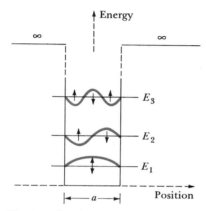

FIG. 33.15. The wave functions for an electron inside an infinitely deep square-well potential look like standing waves on a string fixed at both ends.

expression for the de Broglie wavelength and kinetic energy of an electron (its potential energy in the well is zero), show that the electron's energy when in a state with quantum number n is $E_n = (h^2/8a^2m)n^2$.

33.2 Quantum Numbers for Atomic Wave Functions

2. (a) An electron is in an $n = 4$ orbit. List the possible values of its l quantum number. (b) If the electron happens to be in an $l = 3$ state, list the possible values of its m_l quantum number. (c) List the possible values of its m_s quantum number.

3. (a) An electron in an atom is in a state with $m_l = 3$ and $m_s = +\frac{1}{2}$. What are the smallest possible values of l and n for that state? (b) Repeat for an $m_l = 2$ and $m_s = \frac{1}{2}$ state.

4. (a) An electron resides in an $n = 7$ orbit. List the possible values of its l quantum number. (b) Of these different l states, the electron occupies the $l = 4$ state. List the possible values of the m_l quantum number.

■ 5. Draw schematic orbits and arrows representing the different m_l quantum states for an $l = 2$ atomic electron in a magnetic field.

33.3 Multielectron Atoms

6. List the $|nlm_lm_s\rangle$ states available to an electron in a $4p$ subshell.

7. List the $|nlm_lm_s\rangle$ states available to an electron in a $4f$ subshell.

8. Identify the values of n and l for each of the following subshell designations: $3s$, $2p$, $4d$, $5f$, and $6s$.

■ 9. (a) Determine the electron configuration of sulfur ($Z = 16$). (b) Why are sulfur and oxygen ($Z = 8$) in the same group of the periodic table?

■ 10. (a) Determine the electron configuration of silicon ($Z = 14$). (b) Why are carbon and silicon in the same group of the periodic table?

■ 11. Determine the electron configuration for iron ($Z = 26$).

■ 12. Manganese ($Z = 25$) has two $4s$ electrons. How many $3d$ electrons does it have? Explain your answer.

33.4 The Periodic Table

13. Determine the electron configurations of four elements in group I of the periodic table. Explain why these elements are likely to have similar properties.

■ 14. Determine the electron configurations of three elements in group VI of the periodic table. Explain why these elements are likely to have similar properties.

33.5 Absorption and Emission of Radiation

15. A sodium atom in its ground state absorbs a 589-nm photon when its $3s$ electron jumps to the $4p$ state. Calculate the energy difference of the $3s$ and $4p$ states.

■ 16. A fluorescent molecule, oxyluciferin, absorbs a photon of wavelength 370 nm and emits two photons, one of wavelength 480 nm. Determine the wavelength of the second emitted photon using the fact that the energy of the absorbed photon equals the sum of the energies of the emitted photons.

■ 17. A fluorescent molecule absorbs one photon and then emits two photons, one of wavelength 1610 nm and the other of wavelength 480 nm. Calculate the wavelength of the absorbed photon.

33.6 Stimulated Emission and Lasers

18. (a) A laser pulse emits 2.0 J of energy during 1.0×10^{-9} s. Calculate the power emitted during that short time. (b) Calculate the light intensity (power per area) in the laser beam if its cross-sectional area is 8.0×10^{-6} m^2.

19. A pulsed laser used for welding produces 100 W of power during 10 ms. Calculate the energy delivered to the weld.

■ 20. A pulsed argon laser of 456.5-nm wavelength emits 3.0×10^{-3} J of energy to produce a tiny weld in the retina of a person's eye. How many photons are in the laser pulse?

■ 21. A continuous helium-neon laser emits a beam that diverges at an angle of 4.2×10^{-5} rad. Calculate the width of a cross section of the laser beam after it has traveled (a) 1.0 km and (b) to the moon.

■ ■ 22. A laser used to weld the retina of an eye emits 20 mW of power for 100 ms. The light is focused on a spot 0.1 mm in diameter. Assume that the laser's energy is deposited in a small sheet of water of 0.10-mm diameter and 0.30-mm thickness. (a) Calculate the energy deposited. (b) Calculate the mass of this water. (c) Calculate the increase in temperature of the water (assume that it does not boil and that its heat capacity is 4180 J/kg·C°).

33.7 X-Ray Emission

23. Electrons are accelerated across a 40,000-V potential difference. Calculate the frequency and wavelength of the maximum-energy x-ray photons created when the electrons crash into the anode of an x-ray tube.

24. An x-ray tube emits photons of frequency 1.33×10^{19} Hz or less. Calculate the electrical potential difference across the tube.

25. *Estimate* the energy of a $1s$ electron in gold ($Z = 79$).

■ 26. *Estimate* the minimum voltage across an x-ray tube that results in the removal of $1s$ electrons from molybdenum atoms ($Z = 42$) in the anode of the tube.

■■ 27. The mass of one mole of iron (6.0×10^{23} atoms) is 0.056 kg and iron's density is 7.9×10^3 kg/m^3. (a) Show that the volume of space occupied by one iron atom is 1.2×10^{-29} m^3. (b) Calculate the volume of space occupied by a lithium atom, the metal of least density 0.53×10^3 kg/m^3. The mass of one mole of lithium atoms is 0.0069 kg. (c) Repeat the calculations in part (b) for an osmium atom, the metal of greatest density 22.5×10^3 kg/m^3. The mass of one mole of osmium atoms is about 0.192 kg. (d) Why are these volumes all on the same order of magnitude even though an osmium atom has more than 25 times the number of electrons found in a lithium atom?

■■ 28. An x-ray tube operates at 40,000 V. The beam of electrons flowing from the filament to the anode produces a current of 10 mA. (a) Calculate the rate of heat production in the anode by the collisions of these electrons with the anode. (b) If the specific heat capacity of the 0.10-kg anode is 460 J/kg·C°, calculate its temperature increase in 1.0 s if no cooling water is used.

CHAPTER 34

The Nucleus and Radioactivity

An atom's nucleus occupies only about 10^{-12} percent of the atom's volume. If an atom's size increased so that its cross section equaled that of a football field, the nucleus would be smaller than your thumbnail. Although it is so small, the nucleus contains 99.97 percent of the atom's mass because the constituents of the nucleus, protons and neutrons, are packed closely together and each particle has a mass roughly 2000 times greater than that of an electron, the third particle of which atoms are made. The protons and neutrons in a nucleus are held together by a strong, though incompletely understood, nuclear force. Disturbances of these tightly packed particles in the nucleus may cause reactions in which the nucleus breaks apart, ejects one or a group of its constituent particles, or absorbs a new particle.

In this chapter we will determine the amount of energy released or absorbed during nuclear reactions, with emphasis on nuclei undergoing radioactive decay. We will also learn of several applications of radioactive decay, such as determining the age of ancient objects unearthed by archeologists and the effects that particles emitted during radioactive decay have on biological materials.

34.1 The Nucleus

The protons in a nucleus of an atom each have a charge $+e$ of 1.6×10^{-19} C and a mass m of 1.673×10^{-27} kg. The neutrons found in a nucleus have no net electric charge and a mass of 1.675×10^{-27} kg—slightly greater than the mass of a proton. The protons and neutrons in a nucleus are referred to as **nucleons.**

A nucleus is labeled by indicating its number of protons and neutrons:

Z = Atomic number = Number of protons in nucleus,

N = Neutron number = Number of neutrons in nucleus,

A = Mass number = $Z + N$

= Total number of nucleons (protons and neutrons) in nucleus.

The values of A, Z, and N for several nuclei appear in Table 34.1. The symbol $^{A}_{Z}X$, shown in the first column of the table, combines the information about

TABLE 34.1 Values of A, Z, and N for the Nuclei of Several Atoms

Nucleus	Mass Number $(Z + N)$ A	Atomic Number (Number of Protons) Z	Neutron Number N
$^{1}_{1}\text{H}$	1	1	0
$^{2}_{1}\text{H} = \text{D}$	2	1	1
$^{4}_{2}\text{He}$	4	2	2
$^{6}_{3}\text{Li}$	6	3	3
$^{7}_{3}\text{Li}$	7	3	4
$^{12}_{6}\text{C}$	12	6	6
$^{13}_{6}\text{C}$	13	6	7
$^{14}_{6}\text{C}$	14	6	8
$^{16}_{8}\text{O}$	16	8	8
$^{235}_{92}\text{U}$	235	92	143
$^{238}_{92}\text{U}$	238	92	146

nuclear constituency with the chemical symbol for an atom. For example, the lithium atom, with $Z = 3$, $A = 7$, and chemical symbol Li, is written as $^{7}_{3}\text{Li}$. Although the number of neutrons is not given in this symbol, it can be determined easily by subtracting Z from A.

Providing both the chemical symbol Li and the value of Z is unnecessary, since lithium is the name of an atom with three electrons whose nucleus has three protons, and thus $Z = 3$ for lithium. For this reason we often omit the value of Z when writing the symbol for an atom or nucleus (we write ^{7}Li instead of $^{7}_{3}\text{Li}$).

You will notice in Table 34.1 that several nuclei (such as carbon) have the same number of protons but different numbers of neutrons: $^{12}_{6}\text{C}$, $^{13}_{6}\text{C}$, and $^{14}_{6}\text{C}$. Nuclei with the same number of protons but different numbers of neutrons are called **isotopes**—$^{12}_{6}\text{C}$ is a carbon isotope. Each isotope has the same number of orbital electrons; consequently, their chemical properties are very similar. Because the isotopes have a different number of neutrons in their nuclei, however, the properties of the nuclei may be quite different. For example, the uranium isotope $^{235}_{92}\text{U}$ can take part in the nuclear reactions that provide energy for power plants, while the $^{238}_{92}\text{U}$ isotope cannot.

Isotopes also vary in **natural abundance**—the percentage of an element found in a particular isotopic form. The abundances of the three carbon isotopes mentioned earlier are as follows: $^{12}_{6}\text{C}$ (98.89 percent), $^{13}_{6}\text{C}$ (1.11 percent), and $^{14}_{6}\text{C}$ (10^{-10} percent).

The size of a nucleus can be determined by scattering high-energy electrons, protons, or α particles from the nucleus. Nuclei are found to be roughly spherical in shape (Fig. 34.1) and have radii in meters given by the equation

$$R = 1.2 \times 10^{-15} A^{1/3} \text{ m},$$

where A is the number of nucleons in the nucleus.

34.2 Nuclear Force

Holding a group of protons together within a volume the size of a nucleus is no small task. The protons, having like electrical charges and being so close together, repel each other with an electrical force about 100 million times greater

FIG. 34.1. The nucleus consists of protons ($\oplus$) and neutrons ($\bullet$) held close to each other by a strong nuclear force.

than the force that holds electrons in orbit about the nucleus. Without some strong attractive force to balance this electrical repulsion, nuclei would quickly disintegrate as protons were repelled from the nucleus by their positively charged neighbors.

The attractive force that keeps the protons and neutrons together in a nucleus is called the **strong nuclear force.** A precise mathematical description of this force has not been formulated, but enough is known about the force to realize that it is quite different from the two other fundamental forces we have already studied, the gravitational force and the electrical force. The strong nuclear force is a short-range force: It provides a very strong attraction between two nucleons that are separated by about 5×10^{-15} m or less; but if they are separated farther, the force is essentially zero. By comparison, the gravitational and electrical forces decreases as masses and electric charges are separated farther, but the forces do not decrease to zero until the separation is infinite. The strong nuclear force is the same between any two nucleons (two protons, two neutrons, or a proton and a neutron) and does not depend on their charge. Finally, the magnitude of the strong nuclear force of attraction between two nucleons separated by 5×10^{-15} m or less is much greater than the electrical force and very much greater than the gravitational force between the nucleons. It is the strength of the nuclear force that keeps protons and neutrons bound together tightly in the nucleus.

Because of the short range of the strong nuclear force, a particular proton or neutron in the nucleus is attracted by the strong nuclear force only to its nearest neighbors. Yet a proton feels an electrical repulsion from all other protons in a nucleus. For nuclei with many protons, the electrical repulsion of a proton from all other protons may exceed the strong nuclear attraction of the proton for its neighbors. The nucleus may be unstable and eject one or a group of its particles; the nucleus is said to undergo **radioactive decay.** Surprisingly, the nucleus emits not a proton but an α particle (a small nucleus that consists of two protons and two neutrons) or an electron. The reason for this behavior is discussed in Section 34.5.

34.3   Binding Energy per Nucleon

Discussing the stability of nuclei in terms of the strong nuclear force is a cumbersome problem because of our incomplete understanding of the strong nuclear force and because of the unknown distribution of protons and neutrons in nuclei. We can address the stability of nuclei and many other nuclear phenomena more easily using the ideas of mass defect, binding energy, and binding energy per nucleon.

The mass of every stable nucleus (and even of unstable nuclei) is less than the mass of the nucleons of which it is made. The nucleus is said to have a *mass defect;* it cannot break apart into free protons and neutrons because the nucleus lacks the mass needed for these particles. We can think of the stability of the nucleus as resulting from this mass deficiency. The greater the mass defect, the more stable the nucleus.

The mass defect usually constitutes only a small fraction of the total mass of the nucleus and is calculated conveniently using a unit called the atomic mass unit.

TABLE 34.2   Rest Masses of Several Particles and Atoms

Particle or Atom	Mass		
	kg	u	MeV/c^2
Electron	9.11×10^{-31}	0.000549	0.511
Proton	1.673×10^{-27}	1.007276	938.3
Neutron	1.675×10^{-27}	1.008665	939.6
$_1^1\text{H}$	1.673×10^{-27}	1.007825	938.8
$_2^4\text{He}$	6.647×10^{-27}	4.002604	3,728.4
$_3^7\text{Li}$	11.651×10^{-27}	7.016005	6,535.4
$_{11}^{23}\text{Na}$	38.176×10^{-27}	22.989770	21,414.9

One **atomic mass unit** (u) equals one-twelfth the mass of a $_6^{12}\text{C}$ atom, including the mass of its six electrons. In terms of kilograms,

$$1\text{ u} = 1.660566 \times 10^{-27}\text{ kg.}$$

A list of the masses of several particles and atoms appears in Table 34.2. Appendix C contains a more comprehensive list.

To understand how the mass of a nucleus is used to determine the energy binding it together, we can perform a simple thought experiment using a lithium atom, $_3^7\text{Li}$, whose mass, according to Table 34.2, is 7.016005 u. We pull the atom apart to obtain the separate particles of which it is made: three electrons that had been orbiting the nucleus plus three protons and four neutrons from the nucleus (Fig. 34.2). If we now combine each proton with an electron to form a hydrogen atom, we are left with three hydrogen atoms ($_1^1\text{H}$) and four neutrons ($_0^1 n$).* The mass of the separated parts is

$$3m_{_1^1\text{H}} = 3(1.007825\text{ u}) = 3.023475\text{ u,}$$
$$4m_{_0^1 n} = 4(1.008665\text{ u}) = 4.034660\text{ u,}$$
$$\text{Total mass of constituents} = 7.058135\text{ u.}$$

The total mass of the constituents (7.058135 u) is 0.042130 u greater than the mass of the lithium atom itself (7.016005 u). Almost all of this missing mass is associated with the nucleus and the particles of which it is made. The lithium nucleus lacks 0.042130 u of mass that is needed to spontaneously break apart to form its separate protons and neutrons. Lacking the mass to break apart, the particles must stay together.

*The symbol for a neutron is $_0^1 n$ because its charge is zero and its nucleon number is 1.

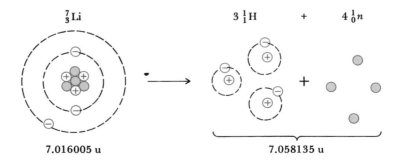

FIG. 34.2.   A $_3^7\text{Li}$ atom has less mass than the particles of which it is made.

7.016005 u

7.058135 u

The "missing mass" or **mass defect** of a nucleus is defined as follows:

$$\text{Mass defect} = \Delta m = [Zm_{^1_1\text{H}} + (A - Z)m_{^1_0 n}] - m_{^A_Z\text{X}}, \tag{34.1}$$

where Z is the number of protons in the nucleus, $m_{^A_Z\text{X}}$ is the mass of the atom whose mass defect is being calculated, $m_{^1_1\text{H}}$ is the mass of a hydrogen atom, and $m_{^1_0 n}$ is the mass of a neutron.

The nuclear **binding energy** (BE) of a nucleus equals the energy that is equivalent to the mass defect Δm. This energy is determined using Einstein's famous equation that relates mass and energy:

$$BE = \Delta mc^2. \tag{34.2}$$

The binding energy represents the total energy needed to separate the atom and its nucleus into hydrogen atoms and neutrons. All but an infinitesimal amount of this energy is needed to pull the nucleus apart.

The binding energy of a nucleus is usually calculated in units of million electron volts (MeV), where

$$1 \text{ MeV} = 10^6 \text{ eV} = 1.602 \times 10^{-13} \text{ J}.$$

The calculation of binding energy from mass defect [Eq. (34.2)] is made easier using a conversion that relates a 1-u mass to its equivalent energy:

$$mc^2 = E,$$

or

$$\begin{aligned}
(1 \text{ u})c^2 &= (1.660566 \times 10^{-27} \text{ kg})(3.0 \times 10^8 \text{ m/s})^2 \\
&= 1.492 \times 10^{-10} \text{ J} \\
&= 931.5 \text{ MeV}.
\end{aligned}$$

Rearranging the last equation, we see that

$$1 \text{ u} = 931.5 \text{ MeV}/c^2. \tag{34.3}$$

EXAMPLE 34.1 Calculate the binding energy of a $^{23}_{11}\text{Na}$ nucleus.

SOLUTION First we calculate the mass defect using Eq. (34.1):

$$\begin{aligned}
\Delta m &= [11\, m_{^1_1\text{H}} + 12 m_{^1_0 n}] - m_{^{23}_{11}\text{Na}} \\
&= 11(1.007825 \text{ u}) + 12(1.008665 \text{ u}) - 22.989770 \text{ u} = 0.200285 \text{ u}.
\end{aligned}$$

Substituting this value for Δm into Eq. (34.2), we find that

$$BE = \Delta mc^2 = \left(0.200285 \text{ u} \times \frac{931.5 \text{ MeV}/c^2}{1 \text{ u}}\right)c^2 = \underline{186.6 \text{ MeV}}.$$

Notice that the conversion $931.5 \text{ MeV}/c^2 = 1$ u has been used to convert the mass defect in atomic mass units to its equivalent mass in MeV/c^2; the c^2 cancelled when we multiplied by c^2 to convert mass to energy. ∎

The stability of a nucleus depends more on the average binding energy of each nucleon than on the total binding energy of the whole nucleus. For sodium, which has 23 nucleons and a total binding energy of 187 MeV, the average binding energy per nucleon is 187 MeV/23 nucleons, or 8.1 MeV/nucleon. The binding energy per nucleon for most stable nuclei ranges from about 6 to 8 MeV/nucleon. To calculate the binding energy per nucleon, divide the total binding energy by the atomic mass number A:

$$\text{Binding energy per nucleon} \equiv \frac{BE}{A}. \tag{34.4}$$

The binding energy per nucleon for stable nuclei is plotted as a function of A in Fig. 34.3. Notice that nuclei with $A \simeq 60$ have the greatest binding energy per nucleon and are also the most stable nuclei, while very small and very large nuclei have less binding energy per nucleon and are less stable. The binding-energy-per-nucleon curve shown in Fig. 34.3 is used to analyze several nuclear processes, such as fission (the splitting of large nuclei) and fusion (the combining of small nuclei). Both processes cause the release of energy because the new nuclei being formed have less mass per nucleon and are more stable than the nuclei from which they were formed.

34.4 Nuclear Reactions

Nuclear reactions, like chemical reactions, involve the conversion or transmutation of nuclear reactants to different nuclei that are products of the reaction. In these reactions two nuclei may interact to form one or more new nuclei, or a single nucleus may divide into two or more nuclei or emit a small particle, thus leaving behind a different nucleus. The nuclear reactions are often written using the atomic symbols rather than special nuclear symbols. Consider the reaction of a proton with a lithium nucleus to form two helium nuclei (Fig. 34.4):

$$^{1}_{1}\text{H} + ^{7}_{3}\text{Li} \longrightarrow ^{4}_{2}\text{He} + ^{4}_{2}\text{He} + 17.3 \text{ MeV}. \tag{34.5}$$

Notice that the proton, the nucleus of a hydrogen atom, is represented by the

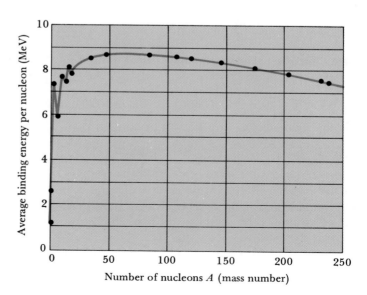

FIG. 34.3. The average binding energy per nucleon is plotted versus the number of nucleons for stable nuclei.

Number of nucleons A (mass number)

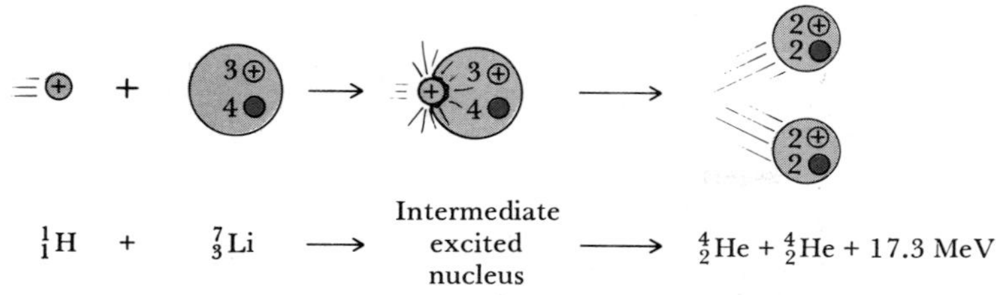

FIG. 34.4. A nuclear reaction in which a proton ($^1_1$H) combines with a lithium nucleus to form two helium nuclei.

$$^1_1\text{H} + {}^7_3\text{Li} \longrightarrow \text{Intermediate excited nucleus} \longrightarrow {}^4_2\text{He} + {}^4_2\text{He} + 17.3 \text{ MeV}$$

symbol for the hydrogen atom ($^1_1$H). The lithium nucleus is represented by $^7_3$Li, the symbol for a lithium atom, and so forth. This convention causes no problem because an equal number of orbital electrons are included on each side of Eq. (34.5)—one for the hydrogen plus three for the lithium equals two each for the two helium atoms. The advantage of using this convention for writing nuclear reactions is that atomic masses found in atomic mass tables can be used to calculate the energy changes during the reaction. Even though atomic masses are used, the energy changes are associated almost entirely with differences in the nuclear masses of the reactants and the products of the reaction.

These nuclear reactions must satisfy several rules, such as those used to balance chemical reactions:

Rule 1: Charge Conservation. The electric charge of the reacting nuclei equals the charge of the products. This condition is satisfied if the sums of the atomic numbers (Z) of the reactants and of the products are equal. Notice, for example, that the atomic numbers for the lithium-proton reaction described by Eq. (34.5) add to four on each side of the equation ($1 + 3 = 2 + 2$).

Rule 2: Baryon Number Conservation. The number of baryons remains unchanged during a nuclear reaction. Baryons are a group of particles found in nuclei that include protons, neutrons, and other exotic particles (introduced in the next chapter; the nuclear reactions discussed in this chapter do not involve these exotic particles). Thus, for now, baryons are conserved if the total number of protons plus neutrons of reacting nuclei equals the number of protons plus neutrons in the product nuclei. This condition is satisfied if the mass numbers (A) are the same on both sides of the equation describing the reaction. For the reaction given by Eq. (34.5), baryons are conserved because $1 + 7 = 4 + 4$.

Rule 3: Energy Conservation. The energy of the reactants including their rest mass energies equals that of the products including their rest mass energies:

$$\Sigma m_{\text{reactants}}\, c^2 = \Sigma m_{\text{products}}\, c^2 + \Delta E.$$

The expression on the left includes the rest mass energy of all the reactants, and the first term on the right is the rest mass energy of all the products. If the mass energies differ, than energy is released by the reaction (positive ΔE) or absorbed (negative ΔE). This energy difference, usually in the form of the kinetic energy of the products or reactants, is sometimes called the Q of the reaction. Rearranging, we find that the Q of the reaction is

$$Q = (\Sigma m_{\text{reactants}} - \Sigma m_{\text{products}})c^2. \tag{34.6}$$

EXAMPLE 34.2 Calculate the Q of the following reaction:

$$^1_1H + {}^7_3Li \longrightarrow {}^4_2He + {}^4_2He.$$

SOLUTION

$$\Sigma m_{\text{reactants}} = m_{^1_1H} + m_{^7_3Li} = 1.007825 \text{ u} + 7.016005 \text{ u} = 8.023830 \text{ u},$$
$$\Sigma m_{\text{products}} = 2m_{^4_2He} = 2(4.002604 \text{ u}) = 8.005208 \text{ u}.$$

Since the products have less mass than the reactants, energy is released during the reaction. Using Eq. (34.6) and the conversion $1 \text{ u} = 931.5 \text{ MeV}/c^2$, we find that the energy released (and the Q of the reaction) is

$$Q = [(8.023830 \text{ u} - 8.005208) \text{ u}]c^2,$$
$$= \left[0.018622 \text{ u} \times \frac{931.5 \text{ MeV}/c^2}{1 \text{ u}} \right]c^2 = \underline{17.3 \text{ MeV}}. \quad\blacksquare$$

EXAMPLE 34.3 Use rules 1 and 2 and the periodic table (Table 33.4) to determine the missing product or reactant for the following nuclear reactions:

(a) $^6_3Li + {}^1_0n \longrightarrow ? + {}^3_1H + \text{energy},$

(b) $^7_4Be + {}^0_{-1}e \longrightarrow ? + \text{energy},$

(c) $^1_0n + {}^{235}_{92}U \longrightarrow {}^{147}_{56}Ba + ? + 3{}^1_0n + \text{energy},$

(d) $^4_2He + ? \longrightarrow {}^{17}_8O + {}^1_1H - \text{energy},$

(e) $^{14}_6C \longrightarrow ? + {}^0_{-1}e + \text{energy}.$

SOLUTION (a) For the values of Z and A to be equal on each side of this reaction, the missing nucleus must have $Z = 2$ and $A = 4$. We see from the periodic table that helium has $Z = 2$. Thus, the missing product is ${}^4_2He$. (b) An electron is represented by the symbol ${}^0_{-1}e$. The value of Z equals -1, since its charge is $-e$, and A equals 0, since it is not a nucleon. The missing nucleus must have $Z = 4 - 1 = 3$ and $A = 7 + 0 = 7$ and is lithium (${}^7_3Li$). (c) You should confirm that the missing nucleus is ${}^{86}_{36}Kr$. (Do not forget to include *three* neutrons in your calculations.) (d) and (e) The missing nucleus in both reactions is ${}^{14}_7N$. $\quad\blacksquare$

34.5 Radioactive Decay

The nuclei of some unstable atoms undergo a process called **radioactive decay** in which a small particle is emitted from a nucleus, resulting in the formation of a new nucleus. Radioactive decay was discovered quite by accident in 1896 by a French physicist, Henri Becquerel (1852–1908). Becquerel had been investigating phosphorescence, the excitation of an atom followed by a delayed emission of light from the atom.

Becquerel's equipment consisted of a photographic plate wrapped carefully in two sheets of heavy black paper so that no light could enter. Becquerel placed some uranium salt crystals on the plate and then set the plate in the sun. When he developed the plate, Becquerel found that it was exposed at the places over which the crystals had been lying. Becquerel thought that the sunlight's interaction with the crystals was somehow involved in producing the radiation that passed through the paper and exposed the plates.

The next day, when Becquerel tried to repeat his experiment, cloudy weather intervened, so he stored his salt sample on top of a box of unexposed photographic plates in a drawer. Several days later, while checking to be sure his plates were still good, he found that they had been exposed in a shape resembling the pile of uranium crystals. Becquerel now realized that the sunlight had nothing to do with this exposure; his plates were being exposed by some unusual emission from the uranium crystals. He had accidentally discovered radioactivity, a discovery for which he received the Nobel Prize in physics in 1903.

We now know that these unusual radioactive emissions are caused by the disintegration or decay of unstable nuclei. The decay usually involves the ejection of a small particle from the "parent" nucleus. A new "daughter" nucleus is left behind. In some cases the newly formed daughter nucleus is unstable and also undergoes radioactive decay.

A nucleus may be unstable for several reasons. If there are too many protons in relation to the number of neutrons, the repulsive electric force between protons exceeds the nuclear force holding the protons together. It would seem that a nucleus could correct this excess-proton instability problem by ejecting one or more protons from the nucleus; yet, surprisingly, we seldom see a radioactive nucleus emit a proton. Instead, a nucleus that spontaneously disintegrates emits two other types of particles: (1) helium nuclei [called **alpha** (α) **particles**], and (2) electrons or positrons [called **beta** (β) **particles**]. A positron has the same mass as an electron and is similar to an electron in all respects, except that its charge is $+e$ rather than $-e$. The positron is called β^+ and the electron β^-. These different decay processes are examined next.

Alpha Decay

Of the small nuclei, helium ($^4_2$He) is one of the most stable. In a large nucleus, small numbers of nucleons tend to group together as helium nuclei (like friends separated in small groups in a large crowd). If the nucleus is unstable, one of these groups may be ejected as a unit—that is, as a helium nucleus. This emission reduces the number of protons in the original nucleus and reduces the electrical repulsive force acting on the protons that remain. Since a helium nucleus also contains two neutrons, the total number of nucleons A in the original nucleus decreases by 4.

Figure 34.5 illustrates the α decay of radium-226 to radon-222 and an α particle. The daughter nucleus and α particle together have less mass than the parent nucleus, and energy is released in the form of kinetic energy.

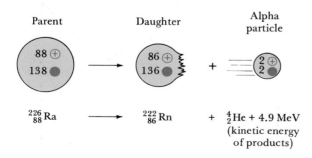

$$^{226}_{88}\text{Ra} \longrightarrow\ ^{222}_{86}\text{Rn}\ +\ ^4_2\text{He} + 4.9\ \text{MeV}$$
(kinetic energy of products)

FIG. 34.5. The alpha decay of a radium-226 to form a radon-222 and an α particle ($^4_2$He).

EXAMPLE 34.4 (a) Calculate the kinetic energy of the product nuclei when $^{212}_{84}\text{Po}$ undergoes alpha decay. (b) Could ^{212}Po decay by emitting a proton from its nucleus? The masses of the nuclei are $m_{212_{\text{Po}}} = 211.9889$; $m_{208_{\text{Pb}}} = 207.9766$, $m_{4_{\text{He}}} = 4.0026$; $m_{211_{\text{Bi}}} = 210.9873$; and $m_{1_{\text{H}}} = 1.0078$.

SOLUTION (a) The alpha decay of polonium-212 is represented by the equation

$$^{212}_{84}\text{Po} \longrightarrow {}^{208}_{82}\text{Pb} + {}^{4}_{2}\text{He} + \text{energy}.$$

The Q of this reaction is given by Eq. (34.6):

$$Q = (m_{212_{\text{Po}}} - m_{208_{\text{Pb}}} - m_{4_{\text{He}}})c^2$$
$$= (211.9889 \text{ u} - 207.9766 \text{ u} - 4.0026 \text{ u})c^2$$
$$= \left(0.0097 \text{ u} \times \frac{931.5 \text{ MeV}/c^2}{1 \text{ u}}\right)c^2 = \underline{9.0 \text{ MeV}}.$$

Of this 9.0 MeV of released energy, 8.5 MeV is converted to the kinetic energy of the α particle (^4He). Most of the remaining 0.5 MeV is converted to the kinetic energy of the recoiling lead-208 nucleus. In addition, a small fraction of the energy may be released as a gamma ray, a high-energy photon.

(b) The decay of polonium-212 by proton emission would be represented by the following reaction:

$$^{212}_{84}\text{Po} \longrightarrow {}^{211}_{83}\text{Bi} + {}^{1}_{1}\text{H}.$$

The total mass of the products together equals 210.9873 u + 1.0078 or 211.9951 u. Since this product mass exceeds the 211.9889-u mass of the original polonium atom, polonium-212 does not have enough mass to spontaneously decay by proton emission from the nucleus. Most decay schemes involving proton emission are prohibited for the same reason. ∎

Beta Decay

Beta decay, another nuclear decay process, occurs by electron or positron emission. For example, a carbon-14 nucleus decays by electron emission (Fig. 34.6a), whereas sodium-21 decays by positron emission (Fig. 34.6b).

Beta decay is unusual in several respects. First, electrons and positrons are not constituents of nuclei, yet they are emitted from many nuclei during radioactive decay. The electron is created in the nucleus when a neutron changes into a proton and electron:

$$^{1}_{0}n \longrightarrow {}^{1}_{1}\text{H} + {}^{0}_{-1}e.$$

The electron is emitted, and the proton is left behind in the nucleus. Positron emission occurs when a proton converts to a neutron and positron ($^{0}_{1}e$):

$$^{1}_{1}\text{H} \longrightarrow {}^{1}_{0}n + {}^{0}_{1}e.$$

The positron is emitted from the nucleus, and the neutron remains behind.

A second intriguing aspect of beta decay concerns the conservation of energy during the reaction. Careful measurements and calculations indicated that energy was not conserved in beta decay reactions. In 1931, Wolfgang Pauli proposed that an as yet undiscovered particle carried away the missing energy.

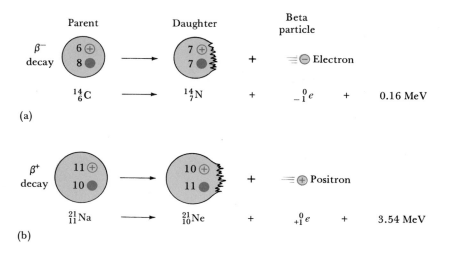

FIG. 34.6. (a) The β^- decay of a carbon-14 nucleus to form nitrogen-14 and an electron ($_{-1}^{0}e$). (b) The β^- decay of a sodium-21 nucleus to form neon-21 and a positron ($_{+1}^{0}e$).

The particle, called a **neutrino,** was postulated to have zero mass and zero charge and to travel at the speed of light. Although this sounds like the description of a photon, there is a subtle difference between a photon and a neutrino. Photons can be absorbed by atoms as their electrons jump between different energy states. These allowed transitions occur only between states that differ in angular-momentum quantum number by $\Delta l = \pm 1$. The photon's angular momentum equals that lost by the atom. On the other hand, a neutrino has only half as much angular momentum as a photon and cannot cause an electron transition in an atom. Neutrinos were discovered 25 years after Pauli proposed their existence. Large numbers of neutrinos are known to continually rain down from above through our bodies. Yet they cause no damage and leave no trail because of their small likelihood of interacting with the atoms in their path.

Beta decay reactions always result in the production of a neutrino ν or antineutrino $\bar{\nu}$ (we will not discuss their differences):

$$_{6}^{14}\text{C} \longrightarrow _{7}^{14}\text{N} + _{-1}^{0}e + \bar{\nu},$$
$$_{11}^{22}\text{Na} \longrightarrow _{10}^{22}\text{Ne} + _{1}^{0}e + \nu.$$

A third unusual feature of beta decay was the need for a new type of force to cause the decay. The force, called the **weak interaction,** was proposed in 1934 by Enrico Fermi (1901–1954). Like the nuclear force, the weak force has a very short range but is much weaker than the strong force that holds nucleons together in nuclei.

Gamma Decay

A number of experiments indicate that protons and neutrons occupy discrete energy states in a nucleus, similar to the energy states occupied by electrons in atoms. If a nucleus becomes excited as a result of a collision with another particle, one of the protons or neutrons in the nucleus may be excited to a higher energy state. When the proton or neutron falls back to its normal state, a **gamma ray photon (γ)** is emitted. These photons are similar to light or x-ray photons, but they have much greater energy.

FIG. 34.7. A beta decay leads to the formation of an excited carbon-12 nucleus (^{12}C*). It settles into its unexcited state by emitting a gamma ray photon.

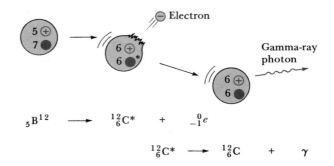

$$_5B^{12} \longrightarrow \, ^{12}_6C* \, + \, _{-1}^{0}e$$

$$^{12}_6C* \longrightarrow \, ^{12}_6C \, + \, \gamma$$

Gamma rays are sometimes emitted from a daughter nucleus after the emission of an α or β particle from the parent nucleus. The daughter may be created with one of its protons or neutrons in an excited state. As the proton or neutron falls into the lowest possible energy state, a gamma ray is emitted. This can happen when boron-12 undergoes beta decay to form carbon-12. An excited $^{12}_6C*$ (the asterisk indicates a nucleus excited to a higher energy state) returns to its lowest energy state by gamma decay (Fig. 34.7).

34.6 Half-Life, Decay Rate, and Exponential Decay

In a sample of radioactive nuclei, each time a nucleus emits an α or a β particle, a different nucleus is formed. The number of the original type of nuclei decreases by one. As time progresses, more and more radioactive decays occur, and the number of radioactive nuclei that remain decreases (see Fig. 34.8).

Half-Life

The graph shown in Fig. 34.8 illustrates an important property that is common to the decay of all types of radioactive nuclei. Notice that the material starts with N_0 radioactive nuclei at time $t = 0$. After a time T, called the **half-life** of the sample, the number of radioactive nuclei has decreased by one-half to $N_0/2$.

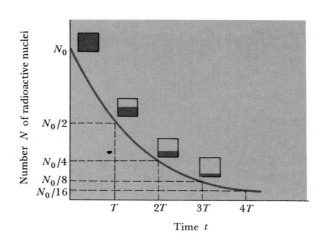

FIG. 34.8. For every time period T the number of radioactive nuclei decreases by one-half.

During the next half-life, the number decreases again by one-half; thus, at time $2T$, $N_0/4$ radioactive nuclei remain. At time $3T$, the number has again been reduced by one-half to $N_0/8$; the fraction of nuclei that remain at time $3T$ is $1/(2^3) = \frac{1}{8}$. By extending this type of reasoning, we find that the fraction of radioactive nuclei that remain after decaying for n half-lives (at time $t = nT$) is $1/(2^n)$. In summary:

The **half-life** T of a particular type of radioactive nucleus is the time required for the number of these nuclei in a given sample to be reduced by one-half. After n half-lives, the fraction of radioactive nuclei that remain is

$$\frac{N}{N_0} = \frac{1}{2^n} \qquad \text{at time } t = nT. \qquad \textbf{(34.7)}$$

The half-lives of several different types of nuclei are listed in Table 34.3.

EXAMPLE 34.5 Carbon-11 with a half-life of 21 min was used in pioneering experiments to trace the absorption of carbon dioxide gas into barley plants. Investigators found that radioactive carbon ^{11}C incorporated in carbon dioxide gas ($^{11}CO_2$) eventually become part of carbohydrate molecules produced by the barley in the presence of the gas. Suppose that transporting the radioactive $^{11}CO_2$ gas from the place where it is made to the barley plants required 2 h 6 min. Calculate the fraction of the original $^{11}CO_2$ molecules that would still be radioactive after this delay.

SOLUTION The delay lasts for 126 min, which equals six 21-min half-lives. Thus, the number of radioactive nuclei has been reduced by the fraction

$$\frac{1}{2^6} = \frac{1}{64} = \underline{0.0156}.$$

When this technique is used to determine the fraction of radioactive nuclei remaining after a certain time, the delay can be any length of time—not just an integral number of half-lives. For example, if the transport of the $^{11}CO_2$ gas had taken 120 min, or $120/21 = 5.7$ half-lives, the fraction remaining would be

$$\frac{1}{2^{5.7}} = 2^{-5.7} = 0.0192. \qquad \blacksquare$$

Decay Rate

In working with radioactive samples, we seldom know or measure directly the number of radioactive nuclei in a sample. Instead, we measure the number of nuclei that decay during a certain period of time. From this measurement we can calculate the number of radioactive nuclei that still remain in the sample. This calculation is possible because the number ΔN of nuclei that decay during a short time Δt is proportional to Δt and to the number N of radioactive nuclei in the sample:

$$\Delta N = -\lambda N \Delta t \qquad \textbf{(34.8)}$$

As we see, N can be calculated if we know or can measure ΔN, Δt, and λ. The proportionality constant λ is called the **decay constant** and has a different value

TABLE 34.3 Half-Lives and Decay Constants of Some Common Nuclei

Isotope	Half-Life	Decay Constant (s^{-1})
$^{87}_{37}Rb$	4.88×10^{10} yr	4.50×10^{-19}
$^{238}_{92}U$	4.5×10^{9} yr	4.9×10^{-18}
$^{40}_{19}K$	1.28×10^{9} yr	1.72×10^{-17}
$^{239}_{94}Pu$	2.44×10^{4} yr	9.00×10^{-13}
$^{14}_{6}C$	5730 yr	3.84×10^{-12}
$^{226}_{88}Ra$	1602 yr	1.37×10^{-11}
$^{137}_{55}Cs$	30.0 yr	7.32×10^{-10}
$^{90}_{38}Sr$	28.1 yr	7.82×10^{-10}
$^{3}_{1}H$	12.4 yr	1.77×10^{-9}
$^{60}_{27}Co$	5.26 yr	4.18×10^{-9}
$^{131}_{53}I$	8.05 day	9.96×10^{-7}
$^{11}_{6}C$	21 min	5.5×10^{-4}

for each *type* of nucleus (see Table 34.3). The greater λ is, the greater the decay rate for that type of nucleus. The negative sign in Eq. (34.8) reflects the fact that the number N of radioactive nuclei *decreases* by an amount ΔN in time Δt.

The ratio of ΔN and Δt is called the decay rate or activity of a sample of radioactive material:

$$\text{Decay rate (activity)} = \frac{\Delta N}{\Delta t} = -\lambda N. \qquad \textbf{(34.9)}$$

Radiation-detection devices such as Geiger counters are used to measure the activity of radioactive samples.

The SI unit for activity is called the **becquerel** (Bq) and equals one disintegration per second: 1 Bq = 1 decay/s. An older unit of activity, the **curie** (Ci), is still in common use. A radioactive sample with an activity of one curie experiences 3.70×10^{10} decays per second. We see that 1 Ci = 3.70×10^{10} Bq. This unusual number for the curie was chosen because it represents roughly the activity of 1 g of pure radium, the radioactive material isolated from tons of uranium ore by Marie and Pierre Curie near the turn of this century. Radioactive tracers used in medicine usually have activities of μCi (1 μCi = 10^{-6} Ci).

Exponential Decay

Using calculus and Eq. (34.8), it can be shown that the number of nuclei varies with time, according to the equation

$$N = N_0 e^{-\lambda t}, \qquad \textbf{(34.10)}$$

where N_0 is the number of radioactive nuclei at time $t = 0$, N is the number at time t, λ is the decay constant of the nucleus, and $e = 2.718 \ldots$ is the base of the natural logarithm system. This equation for N is called an **exponentially decreasing function of time** and is plotted in the graph shown in Fig. 34.8.

We can now determine the number N of radioactive nuclei that remain in a sample at time t compared to the number N_0 at time zero by using either Eq. (34.7) or Eq. (34.10). The former equation, $N/N_0 = 1/(2^n)$, is often easier to use. However, rearrangement of the exponential function ($N = N_0 e^{-\lambda t}$) leads to an important equation for calculating the unknown age of a sample, which we use when we discuss radioactive dating in the next section.

EXAMPLE 34.6 A nuclear reactor has a spent fuel rod that must be stored. The activity of the strontium-90 in the rod when first removed from the reactor is 3.0×10^{16} Bq (8.1×10^5 Ci). (a) Calculate the number of radioactive strontium-90 nuclei in the rod at that time. (b) How many and what fraction of the radioactive strontium-90 nuclei remain after 100 yr?

SOLUTION (a) We are given that the strontium-90 activity at the time of removal of the fuel rod is $(\Delta N/\Delta t) = -3.0 \times 10^{16}$ Bq = -3.0×10^{16} disintegrations/s = -3.0×10^{16} s^{-1}. From Table 34.3, the decay constant λ of strontium-90 is 7.82×10^{-10} s^{-1}. Rearranging Eq. (34.9), we find that the number of radioactive nuclei in the rod when first removed from the reactor is

$$N_0 = \frac{(\Delta N/\Delta t)_{t=0}}{-\lambda} = \frac{(-3.0 \times 10^{16}\ \text{s}^{-1})}{(-7.82 \times 10^{-10}\ \text{s}^{-1})} = \underline{3.8 \times 10^{25}}.$$

(b) The solution to part (a) is the original number N_0 of radioactive nuclei in the sample at time zero when the fuel rod is first removed from the reactor. Later, at time $t = 100$ yr $= 3.15 \times 10^9$ s, the number N of radioactive nuclei that have not yet decayed is determined using Eq. (34.10):

$$N = N_0 e^{-\lambda t} = (3.8 \times 10^{25}) e^{-(7.82 \times 10^{-10}\ \text{s}^{-1})(3.15 \times 10^9\ \text{s})}$$
$$= (3.8 \times 10^{25}) e^{-2.47} = (3.8 \times 10^{25})(0.085)$$
$$= \underline{3.2 \times 10^{24}}.$$

The fraction of nuclei that remain is $N/N_0 = e^{-\lambda t} = e^{-2.47} = \underline{0.085}$.

To check this answer, we can use Eq. (34.7). Notice from Table 34.3 that the half-life of strontium-90 is 28.1 yr. The number of half-lives in 100 yr is $n = (100\ \text{yr}/28.1\ \text{yr}) = 3.56$. Thus, the fraction of nuclei that remain radioactive after 100 yr (3.56 half-lives) is

$$\frac{N}{N_0} = \frac{1}{2^{3.56}} = \frac{1}{11.8} = 0.085,$$

the same as the answer calculated using Eq. (34.10). ∎

Decay Rate and Half-Life

A type of nucleus that has a large decay constant λ decays rapidly and consequently has a short half-life T. The two quantities are related quantitatively by the equation

$$T = \frac{0.693}{\lambda}, \tag{34.11}$$

which we can derive starting with Eq. (34.10). At time $t = T$ (one half-life), the number N of radioactive nuclei is one-half the number N_0 at time zero. Substituting $t = T$ and $N = N_0/2$ into Eq. (34.10), we find that

$$\frac{N_0}{2} = N_0 e^{-\lambda T}.$$

Cancelling the N_0 and taking the natural logarithm of both sides of this equation results in the expression $\ln (1/2) = -\lambda T$. But $\ln 0.5 = -0.693$. Thus, $-0.693 = -\lambda T$, or $T = 0.693/\lambda$, which is Eq. (34.11).

EXAMPLE 34.7   A new material to be used as a radionuclide in medicine is created by bombarding the nuclei of atoms with high-energy protons. One of these new radioactive substances when first produced has an activity of 7.40×10^5 Bq (20 μC). Thirty minutes later, the activity of the same sample has reduced to 6.03×10^5 Bq. (a) Calculate the ratio of the number of radioactive nuclei at 30 min and the number at time zero. (b) Determine the half-life and decay constant of this radioactive substance.

SOLUTION   (a) We first use Eq. (34.9) to determine the ratio of the activities of the substance at $t = 30$ min and $t = 0$:

$$\frac{(\Delta N/\Delta t)_{30}}{(\Delta N/\Delta t)_0} = \frac{-\lambda N_{30}}{-\lambda N_0} = \frac{N_{30}}{N_0} = \frac{6.03 \times 10^5\ \text{s}^{-1}}{7.40 \times 10^5\ \text{s}^{-1}} = \underline{0.815}.$$

By using Eq. (34.9), we have shown that *the ratio of the activities of a radioactive sample at two different times equals the ratio of the number of radioactive nuclei in the sample at those times.* We will use this general result again in the next section.

(b) Next, we substitute $N_{30}/N_0 = 0.815$ into Eq. (34.10) to determine the decay constant of the substance:

$$\frac{N_{30}}{N_0} = e^{-\lambda(30 \text{ min})} = 0.815.$$

Taking the natural logarithm of each side of the preceding equation produces the equation $-\lambda(30 \text{ min}) = \ln 0.815$. Thus

$$\lambda = \frac{\ln 0.815}{-30 \text{ min}} = \frac{-0.204}{-30 \text{ min}} = \underline{0.00679 \text{ min}^{-1}}.$$

The half-life of the sample is given by Eq. (34.11):

$$T = \frac{0.693}{\lambda} = \frac{0.693}{0.00679 \text{ min}^{-1}} = \underline{102 \text{ min}}.$$

The substance is indium-113, used commonly in clinical nuclear medicine.

■

34.7   Radioactive Dating

In Section 34.6, we learned how to calculate the unknown fraction N/N_0 of radioactive nuclei that remain in a sample after a known time of decay. Conversely, archeologists and geologists are interested in determining the unknown age of a radioactive sample from the known fraction N/N_0 of radioactive nuclei that remain in the sample after the time of decay. Their calculation method is based on a rearrangement of Eq. (34.10). First divide both sides of the equation by N_0 and then take the natural logarithm of each side to get

$$\ln \frac{N}{N_0} = \underline{-\lambda t}.$$

Substituting $\lambda = 0.693/T$ (Eq. 34.11) and rearranging, we have an expression for the age of a radioactive sample:

$$t = -\frac{\ln (N/N_0)}{0.693} T. \tag{34.12}$$

In this equation T is the half-life of the decaying nucleus, and t is the sample's age when N radioactive nuclei are left. The sample started at time zero with N_0 radioactive nuclei.

EXAMPLE 34.8   Tritium, an isotope of hydrogen ($^3_1$H), is produced in the atmosphere by the cosmic ray–induced nuclear reaction:

$$^{14}_7\text{N} + ^1_0 n \longrightarrow ^{12}_6\text{C} + ^3_1\text{H}.$$

Tritium undergoes β^- decay with a half-life of 12.4 years. Fresh rainwater contains a small amount of tritium. If isolated, the tritium in the water decays and 12.4 yr later contains only one-half as much tritium. A bottle of wine recovered from a sunken ship contains only 4.0×10^{-4} times the normal amount of tritium. How old is the wine?

SOLUTION Water in the wine at the time the wine was made had the same amount of tritium as an equal volume of fresh rainwater. However, as the wine aged, the tritium underwent beta decay and its concentration decreased. By the time the wine was discovered, the number of tritium nuclei had been reduced by a factor $N/N_0 = 4.0 \times 10^{-4}$. Substituting this and the 12.4-yr half-life of tritium into Eq. (34.11), we find that the wine's age is

$$t = -\frac{\ln{(4.0 \times 10^{-4})}}{0.693}(12.4 \text{ yr}) = -\frac{-7.8}{0.693}(12.4 \text{ yr}) = \underline{140 \text{ yr.}} \quad ■$$

An important type of radioactive dating, called **carbon dating,** is used to determine the age of objects that are less then 40,000 years old. Most of the carbon in our environment consists of the isotope carbon-12, but about 1 in 10^{12} carbon atoms is radioactive carbon-14 with a half-life of 5700 years. The carbon-14 is continuously produced in our atmosphere by collisions of neutrons from the solar wind with nitrogen nuclei in the atmosphere, resulting in the reaction ${}^{1}_{0}n + {}^{14}_{7}N \longrightarrow {}^{14}_{6}C + {}^{1}_{1}H$. At their present concentration, the carbon-14 isotopes disintegrate by radioactive decay just as fast as they are created by the above reaction. An equilibrium is established.

Any plant or animal that metabolizes carbon incorporates about one carbon-14 isotope into its structure for each 10^{12} carbon-12 isotopes that are present. When an animal dies, this is the concentration of radioactive carbon-14 in its bones. Since carbon is no longer absorbed and metabolized after death, the carbon-14 in the bones starts to disintegrate by radioactive decay. After 5700 years the concentration decreases by one-half. Two half-lives, or 11,400 years after death, the concentration dwindles to one-fourth the concentration at death. A measurement of the carbon-14 concentration indicates the age of the bone (see Fig. 34.9).

EXAMPLE 34.9 A bone found by an archeologist contains a small amount of radioactive carbon-14. The radioactive emissions from the bone produce a measured decay rate of 3.3 decays/s. The same mass of fresh cow bone produces 30.8 decays/s. Estimate the age of the sample.

SOLUTION The 30.8 decays/s reflects the activity or decay rate of a sample at death. The bone of unknown age must have had that activity at the animal's death. Using this information and the fact that the decay rate and number of radioactive nuclei are proportional to each other ($\Delta N/\Delta t = -\lambda N$), we can determine the ratio of radioactive carbon now in the bone to that in the bone when the animal died:

$$\frac{(\Delta N/\Delta t)_{\text{now}}}{(\Delta N/\Delta t)_{\text{at death}}} = \frac{3.3 \text{ s}^{-1}}{30.8 \text{ s}^{-1}} = \frac{-\lambda N_{\text{now}}}{-\lambda N_{\text{at death}}},$$

or

$$\frac{N_{\text{now}}}{N_{\text{at death}}} = \frac{N}{N_0} = \frac{3.3}{30.8} = 0.107.$$

FIG. 34.9. An archeologist brushing debris from a layer of butchered bison bones of a Paleo-Indian site near Crawford, Nebraska. The age of the site can be determined by carbon-14 dating of the bones. (© 1976 Tom McHugh/Photo Researchers, Inc.)

Substituting this value into Eq. (34.12), we find the age of the bone:

$$t = -\frac{\ln(N/N_0)}{0.693} T = -\frac{\ln(0.107)}{0.693}(5700 \text{ yr})$$

$$= -\frac{-2.23}{0.693}(5700 \text{ yr}) = \underline{18,400 \text{ yr.}}$$

34.8 Radioactive Decay Series

Many of the daughter nuclei formed by radioactive decay are themselves radioactive. In fact, the decay of the original parent nucleus may start a chain of successive decays called a **decay series**. An example of a decay series is illustrated in Fig. 34.10. First, the alpha decay of $^{238}_{92}$U leads to the formation of $^{234}_{90}$Th. Thorium-234 then undergoes beta decay to form $^{234}_{91}$Pa. The series continues until $^{218}_{84}$Po is formed. Polonium-218 can decay by either alpha or beta emission. The series branches at this point and at several other points as well. Eventually, the stable lead isotope ($^{206}_{82}$Pb) is formed and the series ends.

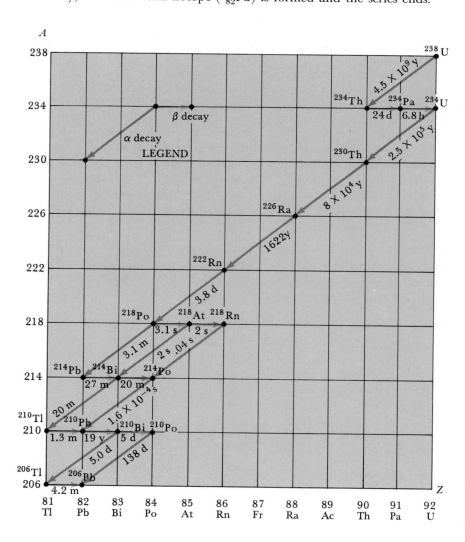

FIG. 34.10. A radioactive decay series beginning with uranium-238 and ending with lead-206.

Radioactive series such as illustrated in Fig. 34.10 replenish our environment with nuclei that would normally have disappeared long ago. For example, radium ($^{226}_{88}$Ra) has a half-life of 1600 years. During the 5×10^9 years of our solar system's existence, the original supply of radium-226 would have been depleted by radioactive decay. However, the supply is continually replenished as uranium-238, with a (4.5×10^9)-year half-life, decays and leads to the production of ^{226}Ra via a series of reactions such as shown in Fig. 34.10.

34.9 Ionizing Radiation

Ionizing radiation consists of photons and moving particles whose energy exceeds the 1–10 eV of energy needed to knock an electron out of an atom or molecule, thus forming an ion. There are a variety of forms of ionizing radiation, such as photons in the ultraviolet, x-ray, and gamma ray regions of the electromagnetic spectrum; α and β particles emitted during the radioactive decay of nuclei; and cosmic rays (high-energy particles) that reach the earth from outer space.

Life has evolved on the earth in a steady background of this radiation, most of it coming from emissions of radioactive nuclei in the earth's crust and from cosmic rays passing down through the earth's atmosphere. The technological use of ionizing radiation *supposedly* began more than 80 years ago (see Fig. 34.11). Today, our exposure from man-made sources of ionizing radiation accounts for about 40 percent of our total exposure (see Table 34.6 in the next section). Before discussing the sources of this radiation and the biological damage it causes, we must first become familiar with the special terminology used to describe ionizing radiation.

Quantities and Units Used to Describe Ionizing Radiation

Five different quantities and their units of measure are used to describe the activity of a source of ionizing radiation and the effect of the radiation on the matter in which the radiation is absorbed. The first quantity, the **decay rate** or

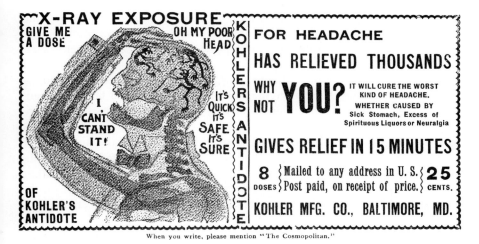

FIG. 34.11. An advertisement from the December 1898 issue of *The Cosmopolitan* magazine claimed that doses of x-rays could be sent anywhere in the United States to cure a headache. X-rays had been discovered just three years earlier by W. C. Roentgen.

activity of a radioactive source, was defined in Section 34.6 and indicates the number of α particles, β particles, or gamma rays emitted by a radioactive material each second. The decay rate is measured in units of becquerel or curies.

A second quantity, called **exposure,** indicates the ionization produced by x-rays and gamma rays as they pass through air. A photon of this radiation as it hits an atom can knock an electron out of the atom, thus forming an ion pair—a free electron and a positive ion. A device called an *ionization chamber* is placed in the path of the radiation. The number of ion pairs produced in air in the chamber indicates the intensity of the radiation. The unit of exposure is the **roentgen** (R), named after Wilhelm Roentgen who discovered x-rays in 1895. One roentgen is the amount of radiation that produces 1.61×10^{15} positive ions in 1 kg of air, or 2.58×10^{-4} coulombs of electric charge per kilogram of air.

The roentgen is defined only for x-rays and gamma rays. Another quantity called **absorbed dose** applies to all types of ionizing radiation and is a better indicator of the effect of radiation on biological material. The **absorbed dose** of radiation is the energy E absorbed by the irradiated part of an object of mass m divided by that mass:

$$\text{Absorbed dose} = \frac{\text{Energy absorbed}}{\text{Mass of absorbing medium}}. \qquad (34.13)$$

The SI unit of absorbed dose is called the **gray** (Gy):

$$1 \text{ Gy} = 1 \text{ J/kg}.$$

Another commonly used absorbed dose unit is called the **rad,** short for <u>r</u>adiation <u>a</u>bsorbed <u>d</u>ose, where

$$1 \text{ rad} = 10^{-2} \text{ J/kg} = 10^{-2} \text{ Gy}. \qquad (34.14)$$

The absorbed dose is not a completely satisfactory indicator of the damage we might expect when a living organism absorbs ionizing radiation. This is because equal absorbed energy of different types of ionizing radiation causes unequal damage. For example, a mass that absorbs 1 rad of α particles experiences about 10 times more damage than the same mass that absorbs 1 rad of x-rays. Alpha particles move more slowly through matter than other forms of radiation and interact with a larger number of atoms per path length because of their double positive charge. On the other hand, x-rays and gamma rays deposit a large fraction of their energy with a small number of atoms. The net damage—the number of bonds broken, the number of ions formed, and the concentration of these ions—is about 10 times greater in the case of the α particles than for x-rays.

Because of this variation in damage caused by different types of radiation, a fourth quantity called the relative biological effectiveness must be introduced. The **relative biological effectiveness** (RBE) indicates the relative damage caused by different types of ionizing radiation. The RBE of α particles is 10 times that of x-rays. For the RBE scale, x-rays whose photons have an energy of 2×10^5 eV are used as a standard. The RBE of some other form of ionizing radiation is the damage caused by a certain amount of absorbed energy of that type of radiation divided by the damage caused by an equal amount of absorbed x-ray energy. The RBE of several different types of radiation are listed in Table 34.4.

This brings us to our last quantity, the **dose equivalent** or simply **dose,** an indicator of the net biological effect of radiation. The dose equivalent of radia-

TABLE 34.4 Relative Biological Effectiveness (RBE) of Different Types of Radiation

Type	RBE
X-rays and gamma rays	1.0
Electrons	1.0–1.7
Slow neutrons	4–5
Protons	10
Alpha particles	10
Heavy ions	20

tion equals the product of the absorbed dose and the relative biological effectiveness of the radiation:

$$\text{Dose or dose equivalent} = (\text{Absorbed dose})(\text{RBE})$$

$$= \left(\frac{\text{Energy absorbed}}{\text{Absorbing mass}}\right)(\text{RBE})$$

The SI unit for dose is the **sievert** (Sv):

$$\text{Dose (in Sv)} = \text{Absorbed dose (in Gy)} \times \text{RBE.}$$

Another common unit for dose is called the **rem** (short for roentgen equivalent man). The rem is related to the rad by the equation

$$\text{Dose (in rem)} = \text{Absorbed dose (in rad)} \times \text{RBE.} \qquad (34.15)$$

We see, using Eq. (34.15) and Table 34.4, that an absorbed dose of 1 rad of α particles (RBE = 10) results in a dose of 10 rem, whereas an absorbed dose of 1 rad of x-rays (RBE = 1) results in a dose of 1 rem.

These various quantities used to describe ionizing radiation and its effects on a medium are summarized in Table 34.5.

EXAMPLE 34.10 During a chest x-ray, you receive a dose of 20 mrem in about 5 kg of body tissue. Calculate the number of x-ray photons absorbed if each x-ray photon has an energy of 50,000 eV.

SOLUTION The RBE of x-rays is 1. Thus, the dose of x-rays in rem equals the absorbed dose in rad. For an absorbed dose of 20 mrad, the energy absorbed per kilogram of exposed tissue is determined using the conversion given by Eq. (34.14):

$$20 \text{ mrad} = (20 \times 10^{-3} \text{ rad})\left(\frac{10^{-2} \text{ J/kg}}{1 \text{ rad}}\right) = 20 \times 10^{-5} \text{ J/kg.}$$

Since 5 kg of tissue receives this dose, the total absorbed energy is

$$(5 \text{ kg})\left(20 \times 10^{-5} \frac{\text{J}}{\text{kg}}\right) = 1.0 \times 10^{-3} \text{ J.}$$

The energy of each x-ray photon in units of joules is determined using the conversion $1 \text{ eV} = 1.6 \times 10^{-19} \text{ J}$:

$$\text{Energy of each photon} = 50{,}000 \frac{\text{eV}}{\text{photon}}\left(\frac{1.6 \times 10^{-19} \text{ J}}{1 \text{ eV}}\right)$$

$$= 8.0 \times 10^{-15} \frac{\text{J}}{\text{photon}}.$$

TABLE 34.5 Summary of Radiation Units

Quantity	Common Unit	Value	SI Unit	Value
Source activity	Ci	$3.7 \times 10^{10} \text{ s}^{-1}$	Bq	1 s^{-1}
Exposure	R	$2.58 \times 10^{-4} \text{ C/kg}$	R	
Absorbed dose	rad	10^{-2} J/kg	Gy	1 J/kg
Dose or dose equivalent	rem	rad $\times$ RBE	Sv	Gy $\times$ RBE

Dividing the total absorbed energy by the energy of each photon gives us the number of absorbed photons:

$$\frac{1.0 \times 10^{-3}\,\text{J}}{8.0 \times 10^{-15}\,\text{J/photon}} = \underline{1.3 \times 10^{11}\ \text{photons.}}$$
∎

34.10 Sources and Effects of Ionizing Radiation

The U.S. Environmental Protection Agency estimated that the average dose of ionizing radiation received by a person in the United States in the year 1980 was about 222 mrem/yr. The average dose for a person in Canada is about the same. This represents the average dose for each kilogram of body mass. The dose varies somewhat for different body parts. Also, the dose varies greatly from one person to another, depending on where a person lives, the number of medical and dental x-rays taken during a year, the person's occupation, and so on. The sources of this radiation, listed in Table 34.6, can be divided into two broad categories: natural and man-made.

Natural Sources

The largest natural source consists of radioactive elements in the earth's crust, such as uranium-238, potassium-40, radon-226 and -228, thorium-232, and the radioactive daughters of these nuclei. These radioactive elements are even incorporated into some building materials such as brick. Also, small amounts of

TABLE 34.6 Estimated Annual Exposure to Ionizing Radiation in 1980 Averaged over the U.S. Population*

Natural Sources	Dose (mrem/yr)
Radioactive elements	
External from surroundings	60
Internal (absorbed in diet and breathing)	25
Cosmic rays	45
	130

Man-made Sources	
Medical and dental	
X-rays	72
Radiopharmaceuticals	14
Nuclear fallout	4
Miscellaneous (TV, fire alarms, clocks, etc.)	2
Occupational	0.8
Nuclear power production	0.04
	92
Total dose/yr	222

*Adapted from "Estimates of Ionizing Radiation Doses in the United States 1960–2000," U.S. Environmental Protection Agency, August (1972).

radioactive radon, an inert atom, diffuse out of the soil into homes and buildings, thus causing exposure to the occupants. Carefully sealing the home's foundation reduces the radon entry. Adequate ventilation also reduces the radon concentration in a home. The earth's radioactive crust and the radioactive atoms leaving the crust contribute a dose of about 60 mrem/yr, or 27 percent of the total average dose absorbed by a U.S. citizen. The dose varies greatly for different locations in the United States and Canada (from 30 mrem/yr in some places to 90 mrem/yr in others). In several inhabited regions of the world the dose due to radioactive elements in rocks, soil, or building materials is considerably higher. In Kerala in India, for example, over 100,000 people receive a background dose from radioactive nuclei of about 1300 mrem/yr. One-sixth of the population of France receive a dose from radioactive granite that ranges from 180 to 350 mrem/yr.

Another natural source results from the ingestion of foods that contain radioactive isotopes. For example, potassium-40, which is present in small concentration in fertilizers containing potassium, finds its way from fields into our food products. These nuclei are absorbed into the body and are called **internal sources**. Every minute, about a million radioactive nuclei in our bodies decay to form different elements. The dose equivalent from this internal radiation is about 25 mrem/y, or 11 percent of the total dose.

Cosmic rays constitute the third natural source. **A cosmic ray** is a charged atomic particle moving at almost the speed of light. Examples of cosmic rays are electrons, positrons, protons, α particles, and other heavier nuclei. The original source of cosmic rays appears to have been huge explosive reactions of supernovas in our galaxy. The mass of cosmic rays in space equals roughly the mass of the stars.

In the upper atmosphere almost 90 percent of all cosmic rays are fast-moving protons. As the protons enter the atmosphere, they collide with other nuclei. About three-quarters of the cosmic rays reaching the earth are muons, particles produced in the atmosphere as a result of these proton-nuclei collisions. At sea level cosmic rays arrive at a rate of about one per square centimeter per minute.

Our exposure from cosmic rays would be well over 100 mrem/yr. However, two-thirds of the rays are diverted away from the earth by the earth's magnetic field. The net dose from cosmic rays is about 45 mrem/yr, or 20 percent of the total dose. The cosmic ray exposure is greater in locations at high elevation—approximately three times greater at an elevation of 2000 m than at sea level.

Man-Made Sources

Our exposure to man-made ionizing radiation results primarily from the use of x-rays and radioactive nuclei as diagnostic tools in medicine and dentistry. Together, these sources account for a dose of 86 mrem/yr—38 percent of our total exposure and more than 90 percent of the man-made exposure. The dose due to medical x-rays is slowly increasing from year to year. The medical use of radioactive nuclei (radionuclides) increased at a rate of about 25 percent per year during the 1970s. The growth rate stopped in the early 1980s but seems to be increasing again.

The second largest source of man-made radiation is the fallout of radioactive nuclei produced in atmospheric testing of nuclear bombs in the 1950s and early 1960s (4 mrem/yr). Miscellaneous sources of radiation (2 mrem/yr) include radiation from television sets, radioactive clock dials, and increased cosmic ray exposure due to air travel. The occupational dose listed in Table 34.6 (0.8 mrem/yr) is associated primarily with extra exposure received by radiologists who give medical and dental x-rays.

In 1980 the world's nuclear power industry accounted for about 0.04 mrem/yr. This includes radiation caused by uranium mining, fuel processing, power plant operation, and waste disposal facilities.

Effects of Ionizing Radiation on Humans

The effects of ionizing radiation on living organisms are divided into two categories: genetic damage and somatic damage. **Genetic damage** occurs when the DNA molecules in the genes of a person's reproductive organs are altered, causing a mutation. These genetic changes are passed on to future generations. **Somatic damage** involves cellular changes caused by ionizing radiation in all other parts of the body except the reproductive organs. Of major concern is the induction of various forms of cancer.

It is difficult to assess the risk of cancer or of other forms of somatic or genetic damage as a result of exposure to low-level ionizing radiation. Populations in regions where the background radiation is somewhat higher than normal such as in Kerala, India, show no apparent effect of their exposure. Yet it is nearly impossible to make an accurate comparison between the 100,000 people living in Kerala and other population groups because of differences in their diet, social habits, and ethnic origin. At present, the attitude of health agencies such as the U.S. Bureau of Radiological Health and the World Health Organization is that small doses of ionizing radiation pose a small risk of genetic and somatic damage. For example, the risk of inducing leukemia that is associated with a single chest x-ray is of the order of 2 in 10 million; for an x-ray examination of the skull it is 1.6 in 1 million. The chance of getting thyroid cancer is approximately 12 in 10 million from a chest x-ray.[*] Approximately three people per one million population die each year as a result of cancer caused by medical and dental x-rays.[†]

Perhaps the most extensive statistical studies of the effects of low-intensity ionizing radiation appear in two reports called BEIR I and BEIR III published in 1972 and 1980, respectively, by the Advisory Committee on the Biological Effects of Ionizing Radiation of the National Academy of Sciences.[‡] The conclusions of BEIR III are complicated, but they lead to an estimate that one cancer death results from the exposure of a population by about 10^4 person rem. This quantity, **person rem,** is obtained by multiplying the total number of people exposed to some dose of radiation by the average dose per person. The

[*]Eric J. Hall, *Radiobiology for the Radiologist,* second edition, p 425, Harper and Row, New York (1978).

[†]R. L. Gotchy, "NRC Estimates of Health Risks Associated with Low-Level Radiation Exposure," 10th Annual National Conference on Radiation Control, HEW Publication (FDA)79-8054, June (1979).

[‡]"The Effects on Populations of Exposure to Low Levels of Ionizing Radiation," Advisory Committee on the Biological Effects of Ionizing Radiation, National Academy of Sciences, National Research Council, Washington, DC 20006.

number of excess somatic cancer deaths that might occur if a large population is exposed to a low-level dose of ionizing radiation can be estimated as illustrated in Example 34.11.

EXAMPLE 34.11 Approximately 2.2×10^6 people living within 50 mi of the Three Mile Island nuclear reactor accident on March 28, 1979, received an average dose of 1 mrem during the two weeks following the accident. Estimate the number of cancer deaths resulting from this radiation.

SOLUTION The collective dose due to the exposure of all these people is $(2.2 \times 10^6 \text{ persons})(1 \times 10^{-3} \text{ rem}) = 2.2 \times 10^3$ person rem. Since approximately one excess cancer death is expected for each 10^4 person rem, the excess number of cancer deaths due to the Three Mile Island accident would be

$$(1 \text{ death}/10^4 \text{ person rem})(2.2 \times 10^3 \text{ person rem}) = \underline{0.2 \text{ deaths.}}$$

By comparison, in a population of 2.2 million persons, the normal death rate due to cancer is approximately 4000 each year. The estimated 0.2 deaths caused by the excess radiation from the Three Mile Island accident could occur years later and would be indistinguishable from the 4000 cancer deaths that occur each year in the absence of this extra radiation. It is, as we see, very difficult to associate a particular cancer death with a single exposure in the past. The statistics for predicting adverse health effects from low-level ionizing radiation are difficult to confirm by observation and consequently are somewhat uncertain. ■

34.11 Radiation-Detection Devices

A variety of devices have been built for detecting ionizing radiation. These devices can be used for making radioactive dating measurements, for calibrating x-ray machines, for detecting particles emitted by radioactive nuclei used in medical diagnosis, for detecting the background radiation at different locations on the earth, and for many other purposes. Most of the devices depend on the fact that ionizing radiation produces ions as it passes through matter.

Geiger Counter

Probably the most widely used radiation detection device, a Geiger counter consists of a gas-filled cylindrical metal tube with a positively charged wire at high voltage ($\sim 10^3$ V) down the center (Fig. 34.12). If a high-energy particle (or photon) enters the window at one end of the tube, the particle's collisions with atoms in the gas inside cause several ions to be formed. The electrons that are freed accelerate toward the positively charged wire and cause other atoms in their path to be ionized. These secondary electrons that are freed strike other atoms, and soon a cascade of electrons rushes toward the wire. When they reach it, they produce a pulse of electric current that passes down the wire into an electric circuit attached across the wire and metal tube. If a loudspeaker is placed in the circuit, a "click" is heard each time a cascade is initiated by the entrance of an ionizing particle or photon into the tube.

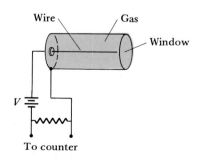

FIG. 34.12. A Geiger tube.

Each pulse of current is called a count. The counting rate is the number of counts per second. Geiger counters are very sensitive to all forms of ionizing radiation but cannot be used to measure high counting rates, nor do they indicate the type of radiation being detected or the energy of the particles or photons in the beam. A Geiger counter is relatively inexpensive and is rugged enough for use as a portable detector.

Scintillation Detector

A scintillation detector is made of a material that emits a flash of light when struck by an energetic particle or photon. A photomultiplier tube (see Section 31.4) near the scintillation material detects the flash of light, thus counting the particle or photon that caused the flash. Scintillation materials usually consist of a special type of plastic, crystals such as sodium iodide, or liquids. Radioactive samples whose age is to be determined are often placed directly in a liquid scintillation material. In this way all radioactive emissions from the sample must pass through the scintillation material and thus produce a flash of light.

Semiconductor Diode Detector

A semiconductor diode is a device that prohibits the flow of electric current in one direction through the diode while allowing it to flow with little resistance in the opposite direction. If an energetic particle passes into the diode, ionization occurs, and a pulse of current can flow in the direction in which current normally does not flow. The diode detects ionizing radiation by a pulse of current.

The three detectors just described are used to indicate the number of high-energy particles or photons entering the device per unit time (the counting rate).

The next three ionizing radiation–detection devices produce a trail along the path followed by individual charged particles.

Photographic Emulsion

If a high-energy charged particle enters a photographic emulsion, the particle ionizes atoms along its path. The chemical changes caused by this ionization become apparent when the photographic emulsion is developed. We see the particle's path as it passed through the emulsion.

Cloud Chamber

A cloud chamber contains a gas vapor cooled to a temperature just below its condensation temperature. Any disturbance of the gas initiates condensation in the region of the disturbance. A high-energy charged particle passing through the cloud causes atoms along its path to be ionized, and condensation is initiated on these ions. The condensation track, which can be photographed, indicates the path of the particle. If a magnetic field exists in the cloud, negative and positive charges are deflected in opposite directions by the magnetic field.

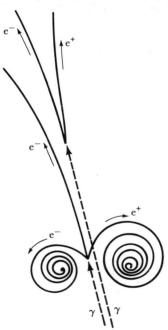

FIG. 34.13. A bubble chamber photo (top) shows electron-positron pairs produced by the annihilation of a gamma ray photon. (Courtesy of the Lawrence Radiation Laboratory, University of California.)

Bubble Chamber

A bubble chamber consists of a liquid that is heated to a temperature near its boiling point. Charged particles passing through the liquid produce a trail of bubbles that can be photographed to show the particle's path (Fig. 34.13).

Summary and Additional Readings

1. **Nucleus.** A nucleus consists of protons and neutrons held together by the strong nuclear force. An atom with Z protons (Z is called the atomic number) and N neutrons is represented by the symbol $^A_Z X$, where the mass number A equals $Z + N$, and X represents the chemical symbol for the element. Nuclei with the same number of protons but different numbers of neutrons are called **isotopes.**

2. **Binding energy per nucleon.** The **mass defect** of a nucleus equals the difference between the mass of its separated parts and the mass of the nucleus itself:

$$\text{Mass defect} = \Delta m = [Z m_{^1\text{H}} + (A - Z) m_{^1_0\text{n}}] - m_{^A_Z X}. \quad (34.1)$$

If a nucleus has less mass than the particles of which it is made, it cannot spontaneously separate into these parts. It is bound together because of a small amount of missing mass (a mass defect). The energy of this mass defect is called the **nuclear binding energy:**

$$\text{Nuclear binding energy} = BE = \Delta m c^2. \quad (34.2)$$

When the binding energy is divided by A, the number of nucleons in the nucleus, we have

$$\text{Binding energy per nucleon} = \frac{BE}{A}. \quad (36.4)$$

When calculating mass defect, we use the atomic mass unit u, where $1 \text{ u} = 1.660566 \times 10^{-27} \text{ kg} = 931.5 \text{ MeV}/c^2$ and $1 \text{ MeV} = 10^6 \text{ eV} = 1.6 \times 10^{-13} \text{ J}$.

3. A **nuclear reaction** results in the transmutation of one or more nuclei into other product nuclei. The reactions must satisfy the rules listed in Section 34.4. The energy released or absorbed during the reaction is called the Q of the reaction:

$$Q = (\Sigma \, m_{\text{reactants}} - \Sigma \, m_{\text{products}}) c^2. \quad (34.6)$$

4. **Radioactive decay.** Unstable nuclei can spontaneously decay by emitting an **alpha particle** (^4_2He), a **beta** (β) **particle** (an energetic electron or positron), or a **gamma ray** (a high-energy photon). The **decay rate,** or **activity,** of a radioactive material is the number of decays ΔN per time Δt and equals

$$\text{Decay rate} \equiv \frac{\Delta N}{\Delta t} = -\lambda N, \quad (34.9)$$

where N is the number of radioactive nuclei in the sample and λ, the **decay constant,** depends on the type of nucleus undergoing radioactive decay. As time progresses and as more of the original type of nucleus decay, the number that remain decreases. The **half-life** T of a particular type of radioactive nucleus is the time required for one-half of the nuclei in a sample to experience radioactive decay. If the sample starts at time zero with N_0 radioactive

nuclei, then at a time $t = nT$ the fraction remaining is

$$\frac{N}{N_0} = \frac{1}{2^n}, \quad (34.7)$$

where n is the number of half-lives during which decay has occurred. The number that remain after time t can also be determined using the **exponentially decreasing function**

$$N = N_0 e^{-\lambda t}. \quad (34.10)$$

The decay constant λ and the half-life T are related by the equation

$$\lambda = \frac{0.693}{T}. \quad (34.11)$$

5. **Radioactive dating.** Equations (34.10) and (34.11) can be rearranged and combined to get an equation for determining the age of a radioactive sample if we can determine the ratio of the number N of radioactive nuclei in the sample at the present time t and the number N_0 in the sample at time zero in the past:

$$t = -\frac{\ln(N/N_0)}{0.693} T. \quad (34.12)$$

6. **Ionizing radiation.** Any particle or photon that has enough energy to ionize an atom or molecule is called ionizing radiation. Several quantities are used to describe this radiation and its effect on living matter. The **activity** or **decay rate** of a radioactive sample has already been defined [Eq. (34.8)] and is given in units of **becquerel** ($1 \text{ Bq} = 1$ decay/s) or in units of **curie** ($1 \text{ Ci} = 3.70 \times 10^{10}$ decays/s). The ionization produced by radiation is measured by the **exposure.** An exposure of one **roentgen** (R) produces 2.58×10^{-4} coulombs of positive ions when passing through one kilogram of dry air. The **absorbed dose** of radiation is the energy absorbed by a medium per unit mass and is measured in units of **gray** ($1 \text{ Gy} = 1 \text{ J/kg}$) or in units of **rad** ($1 \text{ rad} = 10^{-2} \text{ J/kg}$). The **dose equivalent** or simply **dose** is the product of the absorbed dose and the **relative biological effectiveness** (RBE) of the radiation and is measured in **sievert** ($1 \text{ Sv} = 1 \text{ Gy} \times \text{RBE}$) or in units of **rem** ($1 \text{ rem} = 1 \text{ rad} \times \text{RBE}$).

———————

Peter Lindenfeld, "Radioactive Radiations and Their Biological Effects," American Association of Physics Teachers Issue-Oriented Module, College Park, Md. (1984).

Priscilla Laws, "Ionizing Radiation in Health Care," American Association of Physics Teachers Issue-Oriented Module, College Park, Md. (1980).

Eric J. Hall, *Radiobiology for the Radiologist,* second edition, Harper and Row, New York (1978).

Questions

1. *Estimate* the radius of a sphere that would hold all of your mass if its density equaled the density of a nucleus.

2. Determine the chemical elements represented by the symbol X in each of the following: (a) $^{20}_{10}X$; (b) $^{52}_{24}X$; (c) $^{59}_{27}X$; and (d) $^{93}_{41}X$.

3. Explain why very heavy nuclei are unstable.

4. The half-life of radium-226 is 1622 years. Since the earth is about 4 billion years old, essentially all of the radium-226 should have decayed. How can you explain the fact that radium-226 still remains in our environment in moderate abundance?

5. Design an experiment to determine whether O_2 emitted from plants comes from H_2O or from CO_2, the basic input molecules that lead to plant growth.

6. The age of the Dead Sea Scrolls was found by carbon dating. Are they made of stone or wood? Explain.

7. Design an experiment in which radioactive nuclei are used to test the ability of different detergents to remove dirt from clothes.

Problems

34.1 The Nucleus

■ 1. Determine the number of protons, neutrons, and nucleons in each of the following nuclei: $^{9}_{4}$Be, $^{16}_{8}$O, $^{27}_{13}$Al, $^{56}_{26}$Fe, ^{64}Zn, and ^{107}Ag.

■ 2. (a) Calculate the radius of a $^{56}_{26}$Fe nucleus. (b) Do the same for a $^{112}_{48}$Cd nucleus. (c) Even though the cadmium nucleus has twice as many nucleons as iron, its radius is not twice as large. Explain.

■ 3. (a) Calculate the radius of a copper nucleus ($A = 63$). (b) Using the answer to part (a) and the fact that a copper atom has a radius of about 0.1 nm, *estimate* the atom's size if the nucleus grew until it was the size of a tennis ball.

■ 4. (a) *Estimate* the density of a nucleus. (b) Calculate what the earth's radius would be if it had this same density.

■■ 5. *Estimate* the total number of (a) nucleons and (b) electrons in your body. (c) Indicate roughly the volume in cm^3 occupied by these nucleons.

34.2 Nuclear Force

■ 6. *Estimate* the magnitude of the repulsive electrical force between two protons in a helium nucleus.

■ 7. *Estimate* the magnitude of the repulsive electrical force on a proton near the edge of a gold nucleus ($^{197}_{79}$Au) due to all of the other protons in the nucleus. State any assumptions used in your calculation. This is an order-of-magnitude estimate, so do not become bogged down in details.

34.3 Binding Energy per Nucleon

8. Calculate the rest-mass energies of an electron, a proton, and a neutron in units of MeV.

9. Use Fig. 34.3 to *estimate* the total binding energy of $^{197}_{79}$Au.

■ 10. Calculate the total binding energy and the binding energy per nucleon for carbon ($^{12}_{6}$C).

■ 11. Calculate the binding energies per nucleon for $^{238}_{92}$U and $^{120}_{50}$Sn. Based on these numbers, which nucleus is more stable? Explain.

■ 12. Determine the energy that is needed to remove a neutron from $^{7}_{3}$Li to produce $^{6}_{3}$Li plus a free neutron. [*Hint:* Compare the mass of $^{7}_{3}$Li and that of $^{6}_{3}$Li $+ ^{1}_{0}n$.]

■ 13. A 5.5-g marble is dropped from rest in the earth's gravitational field. How far must it fall before its decrease in gravitational potential energy is 938 MeV, the same as the rest-mass energy of a proton?

34.4 Nuclear Reactions

14. Insert the missing symbol in the following reactions.

(a) $^{4}_{2}$He $+ ^{12}_{6}$C $\longrightarrow ^{15}_{7}$N $+$?

(b) $^{2}_{1}$H $+ ^{3}_{1}$H $\longrightarrow ^{4}_{2}$He $+$?

(c) $^{1}_{0}n + ^{235}_{92}$U $\longrightarrow ^{140}_{54}$Xe $+$? $+ 2^{1}_{0}n$

(d) $^{3}_{1}$H $\longrightarrow$? $+ ^{0}_{-1}e$

15. Explain why the following reactions violate one or more of the rules for nuclear reactions.

(a) $^{1}_{0}n + ^{238}_{94}$Pu $\longrightarrow ^{140}_{54}$Xe $+ ^{96}_{40}$Zr $+ 2^{1}_{0}n$

(b) $^{14}_{6}$C $\longrightarrow ^{14}_{7}$N $+ ^{0}_{1}e$

(c) $^{4}_{2}$He $+ ^{27}_{13}$Al $\longrightarrow ^{32}_{15}$P $+ ^{1}_{0}n$

(d) $^{2}_{1}$H $+ ^{3}_{1}$H $\longrightarrow ^{4}_{2}$He $+ ^{1}_{1}$H

■ 16. Explain why the following reaction will not occur spontaneously: $^{4}_{2}$He $\longrightarrow ^{3}_{1}$H $+ ^{1}_{1}$H.

■ 17. The following reaction occurs on the sun: $^{3}_{2}$He $+ ^{4}_{2}$He $\longrightarrow ^{7}_{4}$Be. How much energy is released?

■ 18. Oxygen is produced on stars by the following reaction: $^{12}_{6}$C $+ ^{4}_{2}$He $\longrightarrow ^{16}_{8}$O. How much energy is absorbed or released in units of MeV by the reaction?

■ 19. One part of the carbon-nitrogen cycle that provides energy for the sun is the reaction $^{12}_{6}$C $+ ^{1}_{1}$H $\longrightarrow ^{13}_{7}$N $+ 1.943$ MeV. Using the known masses of ^{12}C and ^{1}H and the results of this reaction, determine the mass of ^{13}N.

■ 20. Determine the missing nucleus in the following reaction and calculate the energy released: $^{232}_{92}$U $\longrightarrow$? $+ ^{4}_{2}$He $+$ energy.

■ 21. Determine the missing nucleus in the following reaction and calculate its mass: ? $\longrightarrow ^{211}_{83}$Bi $+ ^{4}_{2}$He $+ 8.20$ MeV.

34.5 Radioactive Decay

22. The following nuclei produced in a nuclear reactor each undergo radioactive decay. Determine the daughter nucleus formed by each decay reaction. (a) $^{239}_{94}$Pu alpha decay; (b) $^{144}_{58}$Ce β^- decay; (c) $^{129}_{53}$I β^- decay; and (d) $^{65}_{30}$Zn β^+ decay.

23. Potassium-40 ($^{40}_{19}$K) undergoes both β^+ and β^- decay. Determine the daughter nucleus in each case.

■ 24. Radon-222 ($^{222}_{86}$Rn) is released into the air during uranium mining and undergoes alpha decay to form $^{218}_{84}$Po of mass 218.0089 u. Calculate the energy released by the decay reaction. Most of this energy is in the form of α particle kinetic energy.

■ 25. Carbon-11 ($^{11}_{6}$C) undergoes β^+ decay. Determine the product of the decay and the energy released.

■ 26. (a) Calculate the total binding energy of radium-226 ($^{226}_{88}$Ra). (b) Calculate and add together the binding energies of a radon-222 ($^{222}_{86}$Rn) and an α particle. (c) Calculate the difference of these numbers, which equals the energy released during alpha decay of ^{226}Ra.

■ 27. The body contains about 7 mg of radioactive $^{40}_{19}$K, absorbed with foods we eat. Each second, about 2.0×10^3 of these potassium nuclei undergo beta decay (either β^- or β^+). Assuming that all decays are β^+ and that 40 percent of the energy released is absorbed by the body, determine the energy in MeV added to the body each second by ^{40}K decay. What is the rate of energy addition in watts?

■■ 28. A radioactive ^{60}Co nucleus emits a gamma ray of wavelength 0.93×10^{-12} m. If the cobalt was initially at rest, use the conservation-of-momentum equation to calculate its speed following the gamma ray emission.

34.6 Half-Life, Decay Rate, and Exponential Decay

■ 29. Cesium-137, a waste product of nuclear reactors, has a half-life of 30 yr. Use two different methods to calculate the fraction of ^{137}Cs remaining in a reactor fuel rod (a) 120 yr after it is removed from the reactor, (b) 240 yr after, and (c) 1000 yr after.

■ 30. A sample of radioactive technetium-99 of half-life 6 h is to be used in a clinical examination. The sample is delayed 15 h before arriving at the lab for use. Use two methods to determine the fraction that remains.

■ 31. If 120 mCi of radioactive gold-198 with half-life 2.7 days is administered to a patient for radiation therapy, what is the gold-198 activity 3 wk later if none is eliminated from the body by biological means?

■ 32. How many years are required for the amount of krypton-85 ($^{85}_{36}$Kr) in a spent nuclear-reactor fuel rod to be reduced by a factor of $\frac{1}{8}$? $\frac{1}{32}$? $\frac{1}{128}$? The half-life of ^{85}Kr is 10.8 years.

■ 33. A radioactive sample initially undergoes 4.8×10^4 decays/s. Twenty-four hours later its activity is 1.2×10^4 decays/s. Determine the half-life of the radioactive species in the sample.

■ 34. How many years are required for the amount of strontium-90 (^{90}Sr) released from a nuclear explosion in the atmosphere to be reduced by a factor of (a) $\frac{1}{16}$, (b) $\frac{1}{64}$, (c) 0.010.

■ 35. A student nicknamed "Hothead" accidentally swallows 0.10 μg of iodine-131 while pipetting the radioactive material. (a) Calculate the number of ^{131}I atoms swallowed (the mass of ^{131}I is approximately 131 u). (b) Calculate the activity of this material. The half-life of ^{131}I is 8.08 days. (c) What is the mass of radioactive iodine-131 that remains 21 days later if none leaves the body?

■ 36. An unlabeled container of radioactive material has an activity of 90 decays/min. Four days later the activity is 72 decays/min. Calculate the half-life of the material. When will its activity reach 9 decays/min?

■■ 37. A radioactive sample contains two different types of radioactive nuclei: A with half-life 5.0 days and B with half-life 30.0 days. Initially the decay rate of the A-type nucleus is 64 times that of the B-type nucleus. When will their decay rates be equal?

■ 38. To estimate the number of ants in a nest, 100 ants are removed and fed sugar made from radioactive carbon of long half-life. The ants are returned to the nest. Several days later, it is found that of 200 ants tested, only 5 are radioactive. Roughly, how many ants are in the nest? Explain your calculation technique.

■ 39. A small amount of radioactive material placed in 1 cm^3 of water has an activity of 75,000 decays/min. An identical amount of radioactive material is injected into a person's blood system. A short time later, after the material has mixed throughout the blood, a 1-cm^3 sample of blood has an activity of 16 decays/min. Calculate the person's volume of blood.

■ 40. One gram of pure, radioactive radium produces 130 J of heat per hour and has an activity of 1 Ci. Calculate the average energy in electron volts released by each radioactive decay.

34.7 Radioactive Dating

■ 41. A mallet found at an archeological excavation site has $\frac{1}{16}$ the normal carbon-14 decay rate. Use both the $\frac{1}{2}^n$ technique and Eq. (34.12) to calculate the mallet's age.

■ 42. A sample of water from a deep well contains only 30 percent as much tritium as fresh rainwater. How old is the water in the well?

■ 43. The decay rate of ^{14}C from a bone uncovered at a burial site is 12.6 decays/min, whereas the decay rate from a fresh bone of the same mass is 1610 decays/min. Approximately how old is the uncovered bone?

■ 44. The ^{235}U in a rock decays with a half-life of 7.07×10^8 years. A geologist finds that for each ^{235}U now remaining in the rock, 2.6 ^{235}U have already decayed to form daughter nuclei. Calculate the age of the rock.

■■ 45. A tree sample uprooted and buried by the Wisconsin glacier contains 50 g of carbon when it is discovered. (a) If 1 in 10^{12} carbon atoms in a *fresh* tree sample are carbon-14, how many carbon-14 atoms would be in 50 g of carbon from a fresh tree? (b) Calculate the carbon-14 activity of this fresh sample. (c) Determine the age of the buried tree if its 50 g of carbon has an activity of -2.2 s^{-1}

34.8 Radioactive Decay Series

■ 46. A different radioactive series than that shown in Fig. 34.9 begins with $^{232}_{90}$Th and undergoes the following series of decays: $\alpha \ \beta^- \ \beta^- \ \alpha \ \alpha \ \alpha \ \alpha \ \beta^- \ \alpha \ \beta^-$. Determine each nucleus in the series.

■ 47. After a series of α and β decays, plutonium-239 ($^{239}_{94}$Pu) becomes lead-207 ($^{207}_{82}$Pb). Determine the number of α and β particles emitted in the complete decay process. Explain your method for determining these numbers.

34.9 and 34.10 Ionizing Radiation, Sources and Effects of Ionizing Radiation

48. (a) How much energy is absorbed by a 70-kg person receiving a 250-mrad whole-body absorbed dose of radiation? (b) Is it better to absorb 250 mrad of x-rays or 250 mrad of beta rays? Explain. (c) What is the dose in each case?

49. Calculate the dose caused by a 70-mrad absorbed dose of the following types of radiation: (a) x-rays, (b) beta rays, (c) protons, (d) α particles, and (e) heavy ions.

50. (a) X-rays passing through dry air produce 6.68×10^{-10} coulomb of electron charge per kilogram of air. What is the x-ray exposure of the air? (b) The average energy needed to produce one electron ion pair is 33.7 eV. How much energy in joules is deposited per kilogram of dry air by the x-ray exposure in part (a)?

■ 51. A pilot who flies 18 h/wk receives an additional whole-body dose of 0.70 mrem per hour of flight due to increased cosmic radiation. Calculate the pilot's yearly dose due to flying.

52. The yearly whole-body dose caused by radioactive ^{40}K absorbed in our tissues is 17 mrem. (a) Assuming that ^{40}K undergoes beta decay with RBE of 1.4, calculate the absorbed dose in units of rads. (b) How much beta ray energy is absorbed by an 80-kg person? [*Note:* ^{40}K also emits gamma rays, many of which leave the body before being absorbed. Since fatty tissue has low potassium concentration and muscle has higher concentration, gamma ray emissions indicate indirectly the amount of fat in a person.]

■ 53. During an x-ray examination a person receives a dose of 80 mrem in 4.0 kg of tissue. The RBE of x-rays is 1.0. (a) Calculate the total energy absorbed by the body. (b) Calculate the energy in joules of each 40,000-eV x-ray photon. (c) Calculate the number of photons absorbed by the person.

■ 54. A radioactive cloud passes over a city of 5 million people, causing an average dose per person of 10 mrem. *Estimate* the number of excess cancer deaths caused by this exposure.

■ 55. Use the technique of Example 34.10 to *estimate* the excess number of cancer deaths caused by medical and dental x-rays given in the United States each year.

■ 56. Calculate the number of ^{40}K nuclei in the body of an 80-kg person using the information provided in Problem 52 and the fact that ^{40}K has a radioactive half-life of 1.28×10^9 years. Assume that each ^{40}K beta decay results in 1.4 MeV of energy that is deposited in the person's tissue.

■■ 57. While working with ^{14}C, you accidentally swallow through a pipette $10\mu Ci$. (a) If each ^{14}C decay produces a 0.155-MeV β^- particle, which is absorbed by your tissue, calculate the energy in joules absorbed per year (ignore the decrease in ^{14}C activity). (b) If your body mass is 50 kg, calculate your radiation absorbed dose from the ^{14}C for one year, assuming the radiation is distributed uniformly throughout your body.

■■ 58. A body cell of 1.0×10^{-5} m radius and density 1000 kg/m^3 receives a radiation absorbed dose of 1 rad. (a) Calculate the mass of the cell (assume it is shaped like a sphere). (b) Calculate the energy absorbed by the cell. (c) If the average energy needed to produce one positively charged ion is 100 eV, calculate the number of positive ions produced in the cell. (d) Repeat the procedure for the (3.0×10^{-6})-m-radius nucleus of the cell, where its genetic information is stored.

INTERLUDE VII The Exponential Function

We live in a world where growth has become a way of life: Our population grows by a small percentage each year, as does our use of energy, the number of cars we drive, the area of the earth covered by asphalt and concrete, the cost of a loaf of bread, the national debt, and the money in our savings account (assuming there are no withdrawals). If the growth is steady, these quantities can be described by the exponential function. In this interlude we examine the properties of this function and some of the surprising consequences we might expect from exponential growth.*

Growth Rate $\Delta N/\Delta t$

Exponential growth (or decay) occurs when the rate of change of a quantity is proportional to the quantity itself. Suppose that you have some money in a savings account. The money ΔN that is added to your account during a time Δt depends on three quantities: (1) the amount N of money already in the account, (2) the interest rate being paid, and (3) the time Δt that the money is collecting interest. These quantities are related by the equation

$$\Delta N = kN\,\Delta t,$$

or

$$\frac{\Delta N}{\Delta t} = kN, \qquad\qquad \textbf{(VII.1)}$$

where k is a proportionality constant that depends on the interest rate being paid and $\Delta N/\Delta t$ is called the **growth rate**.

The meanings of the proportionality constant k becomes apparent when we rearrange Eq. (VII.1):

$$k = \frac{\Delta N/N}{\Delta t}.$$

We see that k equals the fractional change in N—that is $\Delta N/N$—per time Δt. If the money in a savings account increases by 6 percent per year, then the fractional change in the money during the year is $\Delta N/N = 0.06$ and the proportionality constant

$$k = \frac{\Delta N/N}{\Delta t} = \frac{0.06}{1 \text{ yr}} = 0.06 \text{ yr}^{-1}.$$

Notice that k has dimensions of time^{-1}.

When discussing quantities that grow, it is common to indicate their **percentage growth rate** P. For the money in a savings account, $P = 6$ percent/yr.

---

*Most of the ideas and examples presented here are from a series of articles, referenced at the end of the interlude, written by Professor Albert Bartlett of the Department of Physics at the University of Colorado, Boulder. We would like to thank Professor Bartlett for his permission to use this material.

We relate the percentage growth rate to k by multiplying by 100:

$$P = \text{Percentage growth rate} = 100k. \qquad \textbf{(VII.2)}$$

The Exponential Function

It is quite easy, by means of calculus, to derive from Eq. (VII.1) another equation that can be used to calculate the value of N at some future time t if its value N_0 is known at time zero. The right side of this equation, which we will not derive, is called the **exponential function**:

$$N = N_0 e^{kt}, \qquad \textbf{(VII.3)}$$

where $e = 2.7182\ \ldots$ is the base of the natural logarithm system.

If k is a positive number, exponential growth occurs. If k is a negative

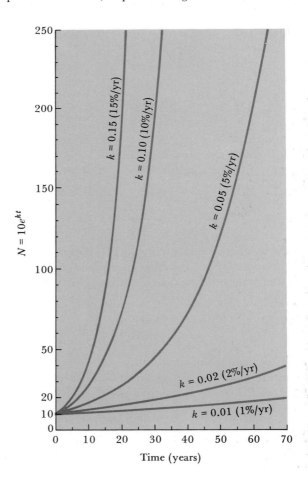

FIG. VII.1. Graphs of five different exponential functions are plotted on linear graph paper. $N_0 = 10$ for each graph, and the value of k is different for each graph.

740

number (or if a negative sign appears in front of a positive-valued k), exponential decay occurs. We studied examples of exponential decay in Sections 34.6 and 34.7 (radioactive decay) and Section 26.6 (the decay of the electric charge on the plates of a capacitor as it discharges through a resistor). In the rest of this interlude we will restrict our attention to examples of exponential growth ($k > 0$).

The general characteristics of exponential growth are illustrated by means of the linear graphs shown in Fig. VII.1, where a steadily growing quantity N is plotted as a function of time for five different values of k. In each curve shown in Fig. VII.1 the value of N at time zero equals 10 ($N_0 = 10$). The curve that increases least rapidly is for $k = 0.01$ yr^{-1} (or a percentage growth rate of 1 percent/yr). The curve that increases most rapidly is for $k = 0.15$ yr^{-1} (a percentage growth rate of 15 percent/yr).

These same graphs are plotted on semilogarithmic graph paper in Fig. VII.2. Note that the exponentially increasing functions appear as straight lines

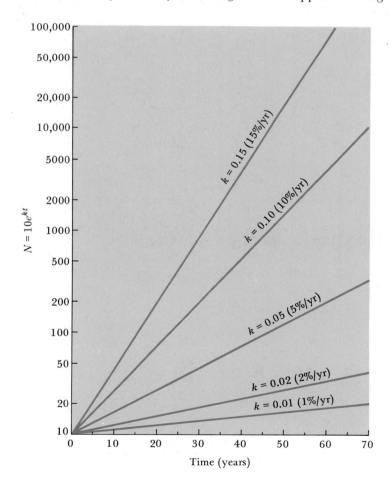

FIG. VII.2. The same exponential functions that were plotted in Fig. VII.1 on linear graph paper are plotted again on semilogarithmic graph paper.

741

on semilog graph paper (see Appendix B). These straight-line graphs can be extended easily to determine the value of an exponentially increasing quantity at some item in the future. (Remember that "exponentially increasing" is synonymous with "steadily growing.") This has been done in Fig. VII.3, where we show the increasing costs of several consumer products that are inflating at a rate of 10 percent/yr. Notice that with 10 percent/yr inflation, the cost of a movie will increase from \$3 at present to \$60 at a time 30 years in the future. An

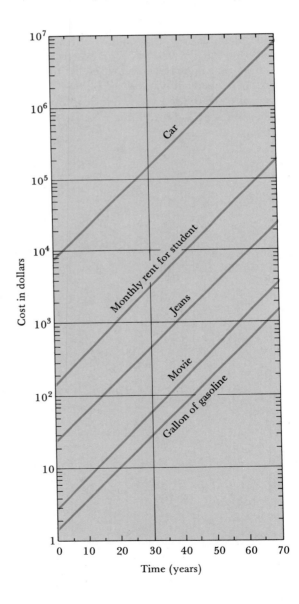

FIG. VII.3. The changing cost of several different consumer products that experience 10 percent/yr inflation are plotted on semilogarithmic graph paper.

economy car's price will increase during the next 30 years from about $8000 to $160,000.

We do not need to rely on graphs alone to estimate the way in which constant growth rate affects the value of a quantity, as we learn next.

Doubling Time

Equation (VII.3) can be used to derive a simple expression for calculating the time needed for a quantity's value to double. Rearranging Eq. (VII.3) and taking the natural logarithm of each side, we find that

$$t = \frac{\ln (N/N_0)}{k}.$$

The growing quantity has a value N_0 at time zero and increases to a value $N = 2N_0$ in a time T, called the **doubling time,** given by the expression

$$T = \frac{\ln (2N_0/N_0)}{k} = \frac{\ln 2}{k} = \frac{0.693}{k}.$$

If we now substitute $k = P/100$ [Eq. (VII.2)] into the above equation, we find that

$$\text{Doubling time} = T = \frac{69.3}{P} \simeq \frac{70}{P}, \qquad \textbf{(VII.4)}$$

where P is the percentage growth rate of the quantity. For example, if the world's population growth rate is 1.9 percent/yr, then the population doubling time is

$$T = \frac{70}{1.9 \text{ yr}^{-1}} = 37 \text{ yr.}$$

If the world's yearly energy use is increasing at a rate of 4 percent/yr, then the energy used per year will double in

$$T = \frac{70}{4 \text{ yr}^{-1}} = 18 \text{ yr.}$$

The units need not be years. For example, if a colony of bacteria grows at a rate of 3.5 percent/min, then the doubling time is

$$T = \frac{70}{3.5 \text{ min}^{-1}} = 20 \text{ min.}$$

The value of an exponentially increasing quantity at some time in the future can be determined using this idea of doubling time. Note, for example, that in one doubling time the original value N_0 of a quantity doubles to become

$2N_0$. In the next doubling time, the quantity increases by another factor of 2 to become $2(2N_0) = 2^2N_0 = 4N_0$. After a third doubling time, the quantity's value becomes $2(2^2N_0) = 2^3N_0 = 8N_0$. In general, in a time $t = nT$, where n is the number of doubling times, the value of a quantity increases to

$$N = 2^nN_0 \qquad \text{in a time } t = nT, \qquad \qquad \textbf{(VII.5)}$$

where N_0 was its original value at the start of that time period. Suppose, for example, that the cost of medical care increases at a rate of 14 percent. By how much would the cost increase in 20 years? Using Eq. (VII.4), we find that the doubling time is $T \simeq 70/(14 \text{ percent/yr}) = 5 \text{ yr}$. The number of doubling times in 20 years is $n = t/T = 20 \text{ yr}/5 \text{ yr} = 4$. Thus, the cost should increase by a factor $N/N_0 = 2^4 = 16$. Equation (VII.5) can be used even if n is not an integer.

Growth in One Human Lifetime

We often hear people talking about the good old days when a cup of coffee cost a nickel, a gallon of gasoline 30 cents, and a nice home $10,000. When we use the exponential function to help us look into the future, though, we may find that the good old days of low prices are now. Equation (VII.3) can be used to derive a simple expression for calculating the factor by which a quantity will increase in the next 70 years (about one human lifetime) if it experiences constant growth. Substituting $t = 70 \text{ yr}$ and $k = P/100$ into Eq. (VII.3) and rearranging, we find that

$$\frac{N}{N_0} = e^{kt} = e^{(P/100)(70)} = e^{0.7P}.$$

But $e^{0.7P} = (e^{0.7})^P = 2^P$. Thus, *in approximately one lifetime of 70 years we can expect an exponentially increasing quantity to increase by a factor*

$$\frac{N}{N_0} = 2^P \qquad \text{(change in 70 yr)}, \qquad \qquad \textbf{(VII.5)}$$

where P is the percentage growth rate per year. If the price of a home increases by 10 percent/yr, then in 70 yr its price will increase by a factor of $2^{10} = 1024$; a home costing $100,000 today would cost $102,400,000 seventy years from now. If the population of a city grows at a rate of 2 percent/yr, then 70 years from now, the city will have to increase its water, sewage, and waste disposal facilities by a factor of $2^2 = 4$. On the other hand, if the city grows by 5 percent/yr, these facilities must be increased by a factor of $2^5 = 32$. Few cities in the United States should experience this type of growth because the percentage growth rate for the country is now about 0.6 percent/yr. In Mexico, the percentage growth rate in 1977 was 3.5 percent/yr. If this growth continues, then in 70 years the population of Mexico will have increased by a factor of $2^{3.5} = 11.3$.

Use During Doubling Period Exceeds All Previous Use

Another interesting property of a quantity that increases exponentially is that the change in the quantity during any doubling period is greater than the sum of the changes in the quantity in all the time preceding that doubling period. Suppose that you have $100 in a savings account and that you are paid 7 percent/yr interest. The doubling time for your money is ten years [Eq. (VII.4)]. The amount of money in the account increases with time as shown in Table VII.1. The money added to the account during the last doubling period ($1600) exceeds the sum of the money added during all earlier doubling periods ($100 + $200 + $400 + $800 = $1500).

For quantities whose yearly rate of use increases exponentially, we can make a similar statement. *The amount of the quantity used during any doubling period exceeds the amount of that quantity used during all the time preceding that doubling period.* For example, the rate of electrical energy use in the United States has increased at about 7 percent/yr (a doubling period of 10 years). If this continues, we will use more electrical energy during the next 10 years than was used in all the previous history of our country!

It is clear that exponential growth cannot continue forever. In the real world we do not have resources that are infinitely large. Unfortunately, we do not realize the scarcity of a resource until it is nearly gone, a condition that arrives abruptly. The idea is illustrated by considering bacteria that grow on a tray. The growth is started at 11:00 A.M. and the number of bacteria double each minute for 1 h, at which time the tray has become completely covered. At 11:58 A.M. the tray is only one-fourth full of bacteria, and considerable room seems still to be available. One minute later, at 11:59 A.M., the bacteria have doubled and the tray is one-half full; there still seems to be enough room. One minute later, at noon, the bacteria double one last time and the tray is completely filled. It seems foolish to wait until the last minute to consider problems associated with growth.

TABLE VII.1 Increase in Money in a Savings Account (7 percent annual interest)

Time	Money in Account	Money Added in Last Doubling Period
0	$ 100	—
10 yr	$ 200	$ 100
20 yr	$ 400	$ 200
30 yr	$ 800	$ 400
40 yr	$1600	$ 800
50 yr	$3200	$1600

Additional Readings

Highly recommended as supplementary reading are a number of articles by Albert A. Bartlett published in *The Physics Teacher*, which are cited below by volume and page.

1. **14** 393 (1976)
2. **14** 485 (1976)
3. **15** 37 (1977)
4. **15** 98 (1977)
5. **15** 225 (1977)
6. **16** 92 (1978)
7. **16** 158 (1978)
8. **17** 23 (1979)

Problems

1. A community decides that its water facilities can be increased by no more than a factor of 2.0 in 24 yr. (a) Calculate the maximum percentage growth rate in population of the community. (b) At this rate, by what factor will the water facilities need to be increased in 70 yr?

2. Suppose that the number of nuclear fission reactions in a

nuclear power plant increased by 0.035 percent every 0.10 s. Calculate (a) the percentage growth rate and (b) the doubling time.

■ 3. The energy in the beam of a laser increases by a factor of 4.0×10^8 in 2.0×10^{-6} s as the beam induces stimulated emission while passing through the laser cavity. Use Eq. (VII.3) to calculate the value of k for this growth process, assuming the growth is exponential.

■ 4. Suppose that the cost today of a bicycle is $150. Use both Eqs. (VII.3) and (VII.5) to calculate the bicycle's cost in 20 yr if the average inflation rate is (a) 5.0 percent/yr and (b) 10.0 percent/yr.

■ 5. If the number of cars registered in the United States is increasing at a rate of 3.9 percent/yr, how many years are required for the number to (a) double; (b) quadruple; (c) increase by a factor of 10 (approximately); and (d) increase by a factor of 100 (approximately)?

■ 6. (a) The world usage of crude oil is increasing at a rate of about 7 percent per year. What is the doubling time? (b) If we are now using 3×10^9 barrels/yr, how many barrels can we expect to use per year in 20 yr (assuming the usage rate continues to grow at 7 percent)? (c) *Estimate* the number of barrels that have been used by the world in all of its history preceding this year.

■ 7. Suppose that the cost of a gallbladder operation has varied as follows: 1965 ($839), 1970 ($1397), 1975 ($2208), 1980 ($3580). (a) Plot these data on semilogarithmic graph paper. Does the cost seem to be increasing exponentially? Explain. (b) Extend your graph to estimate the cost in the year 2000. (c) *Estimate* using your graph the time needed for the cost to double. (d) Use your answer to part (c) to calculate the percentage growth rate.

■ 8. The cost of mailing a letter has varied with time as shown in the following table. (a) Plot these data on semilog graph paper. Does the cost seem to be increasing exponentially? If so, draw a straight line through your data points. (b) Extend your graph to estimate the cost of mailing a letter in the year 2000. (c) *Estimate* using your graph the time needed for the cost to double. (d) Use your result in part (c) to estimate the percentage increase in cost per year.

Year	Cost (cents)	Year	Cost (cents)
1968	6	1978	15
1971	8	1981	18
1974	10	1982	20
1975	13	1985	22

CHAPTER 35

Fission, Fusion, and Particle Physics

In Chapter 34 we learned about the binding energy that holds nuclei together and the radioactive emissions of nuclei that are unstable. Now we will encounter two types of nuclear reactions called *fission* and *fusion*. Fission occurs when a heavy nucleus splits into two smaller nuclei; fusion takes place when two small nuclei combine to form a larger nucleus. Fission reactions release energy for use in nuclear power plants, while fusion reactions release energy that warms the sun and other stars.

Later in the chapter we turn from nuclear physics to a study of the particles found in nuclei and of a large number of other particles that have been discovered in the last fifty years. We will discuss a relatively new theory of particle physics that suggests that protons, neutrons, and many other particles are made of smaller objects called *quarks*.

35.1 Nuclear Fission

Nuclear **fission** occurs when a heavy nucleus, such as $^{235}_{92}\text{U}$, splits to form two daughter nuclei of roughly equal mass. This process is called fission because it resembles the fission or division of a biological cell. Fission does not usually occur spontaneously; instead, it must be induced by collision and absorption of a low-energy neutron into the heavy nucleus (Fig. 35.1). After a short delay, the excited nucleus splits into two pieces. In addition to forming two medium-sized daughter nuclei, fission frees several neutrons. These neutrons may move on to induce fission reactions in other nuclei.

A typical fission reaction involving a uranium nucleus proceeds as follows:

$$^{1}_{0}n + {}^{235}_{92}\text{U} \longrightarrow {}^{236}_{92}\text{U}^* \longrightarrow {}^{141}_{56}\text{Ba} + {}^{92}_{36}\text{Kr} + 3\,{}^{1}_{0}n + \text{energy}. \quad (35.1)$$

The excited $^{236}\text{U}^*$ nucleus that is formed after neutron absorption by ^{235}U can split in many different ways; in all, about 90 different daughter nuclei have been formed by uranium fission. For all of these possible uranium fission reactions we find that, on the average, 2.5 neutrons are produced per fission.

In addition to releasing daughter nuclei and neutrons, fission reactions also release energy; in fact, a single fission reaction releases about 10^7 times more

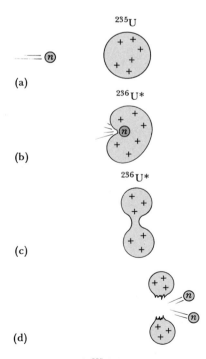

FIG. 35.1. A ^{235}U fission reaction induced by neutron capture.

(a)

(b)

(c)

(d)

energy than does an ordinary chemical reaction, such as the combustion of a molecule of coal. This energy release during fission can be sustained since neutrons released by one reaction can induce further reactions. Energy from controlled fission is now used in nuclear power plants to produce electricity.

The amount of energy released during a fission reaction can be analyzed using the graph of binding energy per nucleon shown in Fig. 35.2 (see Section 34.3 for a review of binding energy). Notice that the binding energy per nucleon for ^{235}U is about 7.6 MeV. Thus, to separate ^{235}U into its individual protons and neutrons, each nucleon needs 7.6 MeV of energy in the form of mass. This mass defect per nucleon—that is, missing mass per nucleon—for ^{235}U is

$$\Delta m = \frac{\Delta E}{c^2} = \frac{7.6\,\text{MeV}\left(\dfrac{1\,\text{u}}{931.5\,\text{MeV}/c^2}\right)}{c^2} = 8.16 \times 10^{-3}\,\text{u}.$$

Each of the two products of fission has about half as many nucleons as uranium; the average of their mass numbers is about 112. According to Fig. 35.2, a nucleus with $A = 112$ has a binding energy per nucleon of about 8.5 MeV, or a mass defect per nucleon of 9.13×10^{-3} u. During a fission reaction, the average mass defect per nucleon increases from about 8.16×10^{-3} u to 9.13×10^{-3} u. Each nucleon loses 0.97×10^{-3} u of mass, equivalent to an energy of 0.9 MeV per nucleon. Since uranium fission involves 235 nucleons, the total energy released by one fission is

$$235 \text{ nucleons}\left(\frac{0.9\,\text{MeV}}{\text{nucleon}}\right) \simeq 200\,\text{MeV}.$$

In comparison, the combustion of one molecule of octane (C_8H_{18}) in the engine of your automobile releases 57×10^{-6} MeV of energy.

35.2 Nuclear Power Plants

Nuclear power plants fueled by the fission reactions of uranium nuclei produce about 10 percent of the energy for human use. This energy comes from approximately 500 nuclear reactors in 36 countries.

FIG. 35.2. The average binding energy per nucleon is 0.9 MeV greater for the products of uranium-235 fission ($A \simeq 112$) than it is for uranium-235. The fission of ^{235}U releases about 0.9 MeV of energy for each of the 235 uranium nucleons.

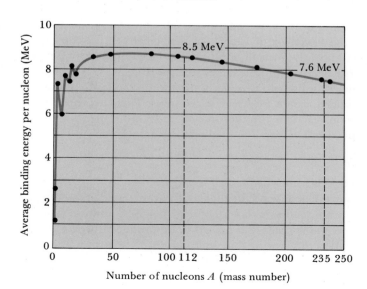

The production of electricity by nuclear fission is similar to the process used when burning oil or coal. In a coal or oil plant, shown schematically in Fig. 35.3a, heat supplied by burning fuel converts water to steam. The steam, under high pressure, passes the blades of a turbine and causes it to rotate. A generator coil attached to the turbine turns in a magnetic field, and electricity is produced. After passing the turbine, the steam is condensed at low pressure by cooling water and returns to the boiler for another trip around the power cycle. A nuclear boiling-water reactor (Fig. 35.3b) works the same way as a fossil-fuel plant except that the heat necessary to boil water is supplied by nuclear fission rather than by combustion of fossil fuel.

In a nuclear pressurized-water reactor, such as the one shown in Fig. 35.3c, fission reactions that occur in the reactor core supply heat to water that is maintained at high pressure so that it does not boil. The hot water is then pumped through a heat exchanger, which transfers heat from the pressurized water to a separate supply of water, thus converting it to steam. The steam then drives a turbine. In this system, the steam used to turn the turbine is isolated from the nuclear fuel, and there is less chance of carrying radioactive nuclei from the core of the reactor to other parts of the power plant.

In the rest of this section we will focus on the reactions that occur in the reactor core and on the mechanisms used to control the heating of water in the core.

Nuclear Fuel

Uranium, the fuel for nuclear reactors, comes in two isotopic forms: $^{238}_{92}U$ (99.3 percent natural abundance) and $^{235}_{92}U$ (0.7 percent natural abundance). While ^{238}U does not undergo fission, ^{235}U does. For most nuclear power plants the ^{235}U concentration must be increased from 0.7 percent to about 3 percent; the uranium is said to be "enriched" in its ^{235}U concentration. At this concentration, at least one of the neutrons produced by a fission reaction has a good chance of causing fission in another ^{235}U nucleus. The fission reactions are sustained, one following the other. At lower ^{235}U concentrations, the rate of reactions may slow down or even stop.

Once enriched, the uranium is placed in ceramic pellets, which retain their solid form even at the high temperature inside the core. Each pellet is about the size of the tip of your little finger. Pellets are stacked end to end in tubes 4 m long (called fuel rods) made of zirconium alloy. About 100 fuel rods are precisely arranged into bundles, called fuel assemblies; a few hundred assemblies provide the fuel for a reactor. A moderate-sized reactor may require 40,000 fuel rods weighing 100 tons or more. All of these fuel rods together are called the reactor **core.**

Water flowing past the fuel rods in the core is heated by contact with the rods, whose average temperature is about 300°C. The water serves three purposes: (1) It is the working fluid that, when heated and converted into steam, drives the turbine; (2) it prevents the reactor core from overheating and melting; and (3) it serves as a moderator for neutrons.

Moderators

A neutron released by a fission reaction normally moves at almost the speed of light. When moving at this speed, its chance of being captured by another ^{235}U

FIG. 35.3. (a) A fossil-fuel electric power plant, (b) a nuclear boiling-water reactor, and (c) a nuclear pressurized-water reactor.

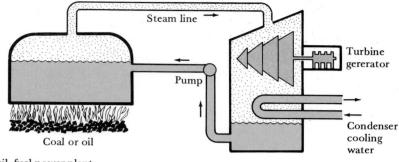

(a) Fossil–fuel power plant

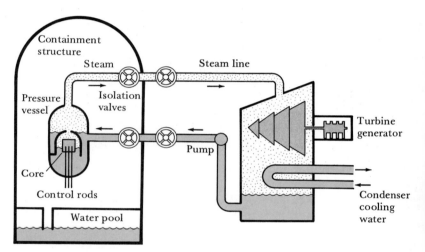

(b) Nuclear boiling–water reactor

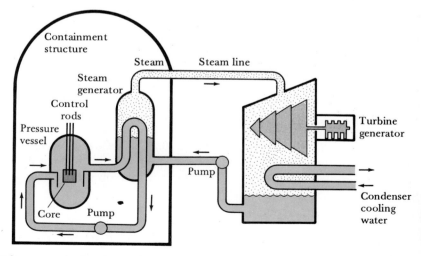

(c) Nuclear pressurized–water reactor

nucleus is small. If the neutron can be slowed, its chance of capture is much better. The function of a moderating material is to slow neutrons. To best accomplish this, the moderating material must have atoms that are approximately the same mass as a neutron. One direct hit between a moving neutron and a proton will completely stop the neutron. (This happens to any moving object that hits head-on another stationary object of equal mass—such as a cue ball hitting another billiard ball.)

Water serves as the moderator for U.S. nuclear power plants. The two nuclei of the hydrogen atoms in water (H_2O) quickly slow neutrons by collisions. Unfortunately, some collisions result in the combining of a proton and neutron to form deuterium ($_1^2H$):

$$_1^1H + {}_0^1n \longrightarrow {}_1^2H.$$

This combination is unfortunate because an absorbed neutron will not produce future fission reactions. The absorption of neutrons by water is one reason why uranium-235 must be enriched to a concentration of 3 percent. At lower concentrations the fission reactions cannot be sustained because too many neutrons are absorbed as they move between ^{235}U nuclei.

Reactors that use water as a moderator are called light-water reactors. The word *light* indicates that the hydrogen nuclei in the water molecules are protons instead of the heavier hydrogen isotope $_1^2H$, deuterium.

Deuterium nuclei also serve as excellent moderators for neutrons and have the extra advantage of absorbing few neutrons. If deuterated water (that is, water in which the protons are replaced by deuterium, D, to form D_2O) is used as the working fluid, the reactor is called a heavy-water reactor. Although expensive, heavy water has an advantage over light water because the fission reactions can be sustained using uranium that has not been enriched. Nuclear reactors in Canada use heavy water.

Chain Reaction

For a reactor core to continue producing a steady supply of energy, each nuclear fission reaction must cause a subsequent fission reaction. Before considering this situation, let us see what happens if each fission reaction releases two neutrons (the average number is actually 2.5) that are absorbed by other ^{235}U nuclei and thus cause two more fission reactions (Fig. 35.4). The new fissions each release two more neutrons for a total of four neutrons. These four neutrons cause four more fusions, which release eight neutrons, and so on. If each neutron produces a fission, the number of nuclei undergoing fission per unit time increases exponentially; the reaction goes out of control, and an explosion may occur. Such uncontrolled explosive reactions occur with uranium enriched to contain about 90 percent ^{235}U, the concentration of ^{235}U in atomic bombs. A fission explosion cannot occur when we use uranium enriched with only 3 percent ^{235}U. At this concentration the majority of the neutrons released by one fission reaction escape the reactor core, are absorbed by a proton, or undergo some other nonfission fate. But even at the low ^{235}U concentration, the reaction rate can increase to the point where the reactor core becomes so hot it melts.

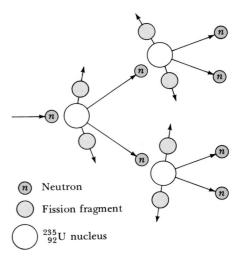

n   Neutron

◯   Fission fragment

◯ $_{92}^{235}U$ nucleus

FIG. 35.4. A chain reaction. Neutrons released by one fission reaction are absorbed by other uranium nuclei and cause them to undergo fission.

Controlling Fission

For each ^{235}U fission, an average of 2.5 neutrons are released. To supply heat at a steady rate, exactly one of these neutrons should cause another fission reaction and the other 1.5 neutrons should not cause fission. With ^{235}U enriched to 3 percent or more, slightly more than one neutron released per fission causes future fission reactions. If uncontrolled, the fission rate would increase slowly and would eventually produce enough heat to melt the core. To prevent this increase, rods made of cadmium are inserted into the core. These rods absorb neutrons and slow the reaction. When the control rods are completely inserted, the reaction stops. If the rods are withdrawn, fewer neutrons are absorbed, and the rate of fission increases.

Spent Fuel Rods

When the fuel rods have been in a reactor for about three years, the ^{235}U concentration decreases so that efficient, sustained reactions are no longer possible. The fuel rods must then be removed. They continue to produce heat long after fission reactions have stoppped because many of the products of fission are radioactive. The krypton-92 and barium-141 formed by the reaction shown in Eq. (35.1) both undergo beta decay with half-lives of 3.0 s and 18 min, respectively. The radioactive decay of these fission products in the fuel rods releases energy and causes the rods to remain hot for days.

A large number of fissionable uranium nuclei still remain in a fuel rod after it is removed from a reactor. The rods could be sent to a fuel-recycling plant to recover and incorporate this material into new fuel rods. At present, fuel-recycling plants are not allowed in the United States. All of the spent fuel is stored in pools of water at the reactor sites.

Nuclear Power Safety

Nuclear power production involves a health hazard, as do most forms of human activity. To put the nuclear power generation hazard in perspective, we will compare it to the hazard of producing electricity by a coal-fired power plant. These two power-generation techniques are chosen for comparison because they are both expected to play an increasing role in future electric-power production as our supplies of oil and gas become depleted.

The economic and health aspects of coal-fired and nuclear-power production have been analyzed in a variety of comprehensive studies.* A study published in 1977 indicated that on the average, 1.7 occupational deaths occur each year in the mining, transportation, and operating of a 1000-MW coal-fired power plant.† In addition, roughly 0.4 to 25 deaths per year result from the emission of sulfur-related pollutants from coal combustion. This number can be reduced by using low-sulfur coal and lime scrubbers to absorb sulfur emissions.

*A Report of Advisory Committee on the Biological Effects of Ionizing Radiations, "Considerations of Health Benefit-Cost Analysis for Activities Involving Ionizing Radiation Exposure and Alternatives," U.S. Environmental Improvement Agency, EPA 520/4-77-03 (1977); Report of the Nuclear Energy Policy Study Group, *Nuclear Power Issues and Choices*, Ballinger, Cambridge, Mass. (1977).
†EPA 520/4-77-03 (1977), cited above.

In summary, for a new coal-fired 1000-MW plant meeting present safety standards, analysis indicates premature deaths for workers and the general public in the range of 2 to 25 per year.* In 1975 the numbers were larger (from 20 to 60) because fewer emission-control devices were used on coal plants.

Major health concerns related to the use of nuclear fuel include the possibility of a major accident and the effects of low-level radiation emitted into the environment. Estimates used to calculate these hazards are even more uncertain than those applied to power generation using coal. Low-level radiation exposure resulting from uranium mining, fuel processing, power plant operation, and waste disposal facilities for a 1000-MW nuclear reactor for one year is expected to cause 0.6 to 1.0 deaths per year. Concerning the hazard from accidents, the WASH-1400 report, a reactor safety study published in 1975,† concluded that the very worst accident involving a core meltdown and the escape of a significant amount of radioactive material has a probability of about 10^{-6} of occurring for each year a reactor operates. This accident would cause the most severe impact on human health if the wind, weather, and population density were in the least favorable conditions (a probability of about 10^{-3} per year of reactor operation that these conditions would prevail during a bad accident). Thus, the worst accident in terms of human fatalities has approximately a $10^{-6} \times 10^{-3} = 10^{-9}$ probability of occurring per year of reactor operation. If such a worst-case accident took place, the result would be about 3000 immediate fatalities and approximately 45,000 cancer fatalities in the 40 years following. To compare this risk to that related to coal-fired power production, the accident risk has been averaged over the number of years estimated to pass between these serious accidents, none of which has ever occurred. The WASH-1400 report projected 0.02 accident-related fatalities per year as a result of the operation of a 1000-MW reactor. This report has been criticized as underestimating the risk; at the extreme upper limit, if all uncertainties are viewed pessimistically, the risk might be as high as ten deaths per year per 1000-MW reactor. In summary, the hazards from low-level radiation and possible accidents at a 1000-MW reactor are estimated to produce approximately 0.6 to 10 fatalities per year, about one-half the estimates for a coal-fired plant.

The risks caused by electric-power production are compared in Table 35.1 to the risks for other types of accidental death. The numbers, except for power production (see the second footnote in the table), represent the expected fatalities each year for a population of 300 million, slightly more than the combined population of the United States and Canada.

35.3 Breeder Reactors

Another type of nuclear reactor, called a breeder reactor, converts nonfissionable ^{238}U with 99.3 percent natural abundance to plutonium-239, a fissionable material. Some of the neutrons emitted by the fission of ^{235}U are absorbed by ^{238}U, and after the following series of reactions, plutonium-239 ($^{239}_{94}Pu$) is formed:

$$^{1}_{0}n + {}^{238}_{92}U \longrightarrow {}^{239}_{92}U^* \longrightarrow {}^{239}_{93}Np + {}^{0}_{-1}e,$$
$$^{239}_{93}Np \longrightarrow {}^{239}_{94}Pu + {}^{0}_{-1}e.$$

*EPA 520/4-77-03 (1977), cited above.

†The Reactor Safety Study, Nuclear Regulatory Commission (1975). This report is often referred to as the Rasmussen Report or the WASH-1400 Report.

TABLE 35.1 Yearly Mortality Risks for a Population of 300 million*

Cause of Death	Numbers of Deaths
One 1000-MW nuclear reactor†	0.6–10
One 1000-MW coal-fired power plant†	2–25
Lightning	200
Medical x-rays	1,000
Electrocution	2,000
Air travel	3,000
Drowning	9,000
Fires and hot substances	10,000
Falls	30,000
Motor vehicles	90,000
Total number of deaths due to all causes	3,000,000

*Adapted from "NRC Estimates of Health Risks Associated with Low-Level Radiation Exposure," 10th Annual Conference on Radiation Control, Bureau of Radiological Health, HEW Publication (FDA) 79-8054 (1979), and from the WASH-1400 report. The numbers, except for the first two, are for a population of 300 million.
†The generation of electricity for a population of 300 million would require about six hundred 1000-MW power plants. If 100 percent of our electric power in the year 2000 were produced by nuclear power plants, the expected number of deaths per year would be from 0.6 to 10 times 600, or 360 to 6000. For coal-fired power plants producing 100 percent of our electricity, the number of deaths per year would be from 1200 to 15,000.

The ^{239}Pu isotope is a fissionable nucleus that releases about three neutrons per fission and can serve as a nuclear fuel for sustained fission reactions.

In a breeder reactor, one neutron released by ^{239}Pu fission produces another fission. A second neutron is absorbed by a ^{238}U nucleus and produces a new ^{239}Pu nucleus. In this way, the concentration of the ^{239}Pu fuel remains constant. If the third neutron is also absorbed by ^{238}U, two ^{239}Pu nuclei are produced for each ^{239}Pu fission. Since more fuel is produced than is consumed, the reactor is said to "breed" nuclear fuel. Breeding fissionable plutonium-239 from nonfissionable uranium-238 increases the supply of fissionable fuel in the earth's crust by a factor of more than 100. Without breeding, only 0.7 percent of natural uranium, the ^{235}U, is fissionable; with breeding, all of the uranium is fissionable.

35.4 Fusion

A proposed technique for producing electric power that should have less impact on the environment than present methods is based on a process called nuclear fusion. **Fusion** occurs when two small nuclei combine or fuse together to form a larger nucleus.

The simplest fusion reaction is the combination of a proton ($^1_1$H) and a neutron ($^1_0 n$) to form a deuterium ($^2_1$H):

$$^1_1\text{H} + {^1_0 n} \longrightarrow {^2_1\text{H}} + \gamma. \tag{35.2}$$

The mass of the deuterium is smaller than the mass of the proton plus neutron. Energy is released by the reaction; part of the energy results in the production of a gamma ray (γ) and the rest is converted to kinetic energy of the deuterium.

EXAMPLE 35.1 Calculate the mass-energy released by the fusion of a neutron and proton.

SOLUTION The energy released equals c^2 times the difference in mass of the reactants and the product of the reaction [Eq. (34.6)]:

$$Q = \text{Energy released} = (m_{^1_1H} + m_{^1_0n} - m_{^2_1H})\,c^2$$
$$= (1.007825\text{ u} + 1.008665\text{ u} - 2.014102\text{ u})\,c^2$$
$$= \left(0.002388\text{ u} \times \frac{931.5\text{ MeV}/c^2}{1\text{ u}}\right)c^2 = \underline{2.22\text{ MeV}}. \qquad \blacksquare$$

Many fusion reactions release relatively large amounts of energy. In fact, fusion provides the energy for the sun and other stars. Since the sun's mass is approximately three-fourths hydrogen, most of the fusion occurring on the sun involves reactions combining protons—the nuclei of hydrogen atoms—with each other and with other particles and nuclei. Helium is also important in these reactions since it accounts for approximately one-fourth of the sun's mass.

An example of a series of fusion reactions that provides some of the sun's energy is the **proton-proton cycle,** summarized as follows:

$$^1_1H + {}^1_1H \longrightarrow {}^2_1H + {}^0_1e + \nu \qquad (0.42\text{ MeV}), \qquad \textbf{(35.3a)}$$
$$^2_1H + {}^1_1H \longrightarrow {}^3_2He + \gamma \qquad (5.49\text{ MeV}), \qquad \textbf{(35.3b)}$$
$$^3_2He + {}^3_2He \longrightarrow {}^4_2He + {}^1_1H + {}^1_1H \qquad (12.86\text{ MeV}). \qquad \textbf{(35.3c)}$$

Net reaction:

$$6\,{}^1_1H \longrightarrow {}^4_2He + 2\,{}^1_1H + 2\,{}^0_1e + 2\nu \qquad (24.68\text{ MeV}). \qquad \textbf{(35.3)}$$

We see that a product of one reaction is involved in the subsequent reaction. Because two heliums (${}^3_2He$) are needed for the last reaction [Eq. (35.3c)], the first two reactions must each occur twice before the last reaction can proceed. Besides helium-4, two neutrinos (ν) and two positrons (${}^0_1e$) are produced by the reaction. The positrons annihilate quickly with electrons to produce an additional 2 MeV of energy.

The net energy released by the proton-proton cycle (including the electron-positron annihilation) is 26.68 MeV. One gram of hydrogen undergoing the reactions summarized by Eq. (35.3) releases enough energy to heat an average home for more than 40 years! Although the abundance of hydrogen in our environment makes these reactions seem like attractive sources of energy, one major obstacle stands in the way of their use. In order for two protons to react as represented by Eq. (35.3a), they must be brought to a separation of about 10^{-14} m or less so that the strong nuclear force between them exceeds their electrical repulsion. This close encounter occurs only if the protons have a large amount of kinetic energy that is converted to electrical potential energy as they approach each other (see Fig. 35.5). Let us calculate the initial kinetic energy needed to bring the protons within 10^{-14} m of each other. From this calculation we will be able to estimate the temperature at which we can expect protons to fuse.

Suppose that the protons are moving straight toward each other with an initial speed v_0 and that they are initially separated by a very large distance. As they draw closer, their speed decreases while their electrical potential energy

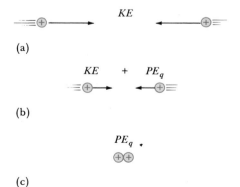

FIG. 35.5. As two protons approach each other, their initial kinetic energy converts to electrical potential energy.

increases. Finally, when separated by about 10^{-14} m or less, the protons become "locked" together by the strong nuclear force. For the protons to get this close, their initial kinetic energy must balance their final electrical potential energy when separated by a distance $R \simeq 10^{-14}$ m:

$$\text{Initial } KE = \text{Final } PE_q$$
$$2 \left(\frac{1}{2} m v_0^2 \right) = \frac{ke^2}{R}.$$

Solving for the initial kinetic energy of one of the protons leads to the expression

$$\frac{1}{2} m v_0^2 = \frac{ke^2}{2R} = \frac{(9 \times 10^9 \, \text{N·m}^2/\text{C}^2)(1.6 \times 10^{-19} \, \text{C})^2}{2(10^{-14} \, \text{m})} = 1.15 \times 10^{-14} \, \text{J}.$$

In Chapter 12 we learned that the temperature of a gas reflects the average random kinetic energy of the atoms in the gas; the temperature T is related to the average kinetic energy by Eq. (12.12):

$$\frac{3}{2} kT = \frac{1}{2} m v^2,$$

where k is Boltzmann's constant (1.38×10^{-23} J/K). For a gas of protons to have sufficient kinetic energy for fusion to occur during an *average* collision, the temperature must be

$$T = \frac{\frac{1}{2} m v^2}{\frac{3}{2} k} = \frac{1.15 \times 10^{-14} \, \text{J}}{\frac{3}{2}(1.38 \times 10^{-23} \, \text{J/K})} = 5.6 \times 10^8 \, \text{K}.$$

The sun's temperature is about 0.15×10^8 K at its core. However, even at this "low" temperature, enough protons have the kinetic energy needed to cause fusion. Obviously, fusion cannot be expected to occur in a home furnace in which the flame temperature is only about 600 K. In fact, even the sun's outer shell, the photosphere, is too cool (5800 K) for fusion to occur.

It is expected that the commercial production of energy by controlled fusion reactions will become a reality, but not in the immediate future. As of 1984, the highest temperature yet attained in an experimental fusion reactor was 0.17×10^8 K, a factor of about 10 too low for a commercial fusion reactor. This temperature was achieved on a device called a *tokamak* in which a magnetic field in a toroid confines the hot fusing gas away from the walls of its container.

The production of energy by fusion has several advantages over energy production by fission. Fusion reactions produce few radioactive isotopes, and those formed have short half-lives. The fusion process can be easily turned off. Also, the earth contains an unlimited supply of fusionable material. Fusion, if it can be made to work, offers a somewhat cleaner source of energy than either coal or nuclear fission.

35.5 Particle Physics

In this and the next two sections we conclude our study of physics by considering one of the most interesting and challenging problems physicists have studied in many years. The significance of this endeavor parallels that of the development of Newtonian mechanics and the atomic theory of quantum mechanics. These great mental achievements of the past have led to many new technologies, such as communications satellites, electronic devices of all sorts, lasers, and various ways of using and providing energy. The technological impact of new theories such as quantum mechanics could not have been predicted or even imagined at the time of their development. When J. J. Thomson discovered the electron in 1897, he could not have envisioned the many modern devices that rely on information and energy transfer by electrons. Scientists working today in the field of elementary-particle physics find themselves in a similar situation. They are trying to understand the nature of the most basic particles, called **elementary particles,** of which our universe is built. They cannot envision what new applications will come of this understanding.

To most of us, electrons, protons, and neutrons seem like unchangeable balls of mass—some with electric charge and others with none. They were supposedly part of the universe when it was born and will supposedly be with us, unchanged, to the end. But scientists now think that these building blocks of nature may not be so basic after all. Physicists currently think that each proton and neutron consists of three smaller, more basic particles called **quarks.** The quarks—not the proton and the neutron—are supposedly the elementary particles, the basic building blocks.

The electron, along with several other particles, still retains its cherished role of being "elementary"—an indivisible, single piece of matter. But even the electron's structure is a puzzle. How can this tiny mass and charge possibly stick together? Its diameter is less than 10^{-15} m. If we think of its mass and electric charge as occupying a sphere of that diameter and draw an imaginary slice through its middle, as depicted in Fig. 35.6a, the electrical repulsive force of one-half of the sphere on the other half is over 200 N (Fig. 35.6b). The strong nuclear force has no effect on the electron, and evidently some other force is needed to hold an electron together. An interesting proposal suggests that an electron is a very small, stable black hole and that the great density of mass inside the electron holds its pieces together, even though they are being pushed apart by their electrical repulsion.

The intense interest in elementary particles in recent years has resulted partly from the discovery of more than 200 new particles that in many ways seem as fundamental as protons and neutrons. All these particles have masses greater than the electron, and many are more massive than protons and neutrons.

A list of several stable particles appears in Table 35.2. A particle is considered stable if it reaches the ripe old age of about 10^{-12} s or older before decaying into other particles. This time is long compared to the life of most of the other 200 particles—about 10^{-23} s. The decay of a particle often leads to the forma-

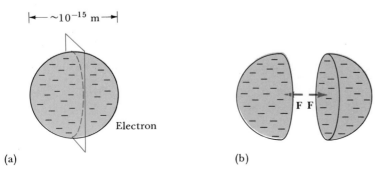

(a)   (b)

FIG. 35.6. (a) An imaginary slice is drawn through an electron. (b) The repulsive electrical force of one half on the other is over 50 lb.

TABLE 35.2 "Stable" Particles

Category	Particle Name	Symbol	Anti-particle	Rest-mass Energy (MeV)	B	S	L_e	L_μ	L_τ	Life-time (s)	Decay Mode
Photon	Photon	γ	Self	0	0	0	0	0	0	Stable	
Leptons	Electron	e^-	e^+	0.51	0	0	1	0	0	Stable	
	Electron neutrino	ν_e	$\bar{\nu}_e$	0	0	0	1	0	0		
	Muon	μ^-	μ^+	105.7	0	0	0	1	0	2.2×10^{-6}	$\mu^- \longrightarrow e^- + \bar{\nu}_e + \nu_\mu$
	Muon neutrino	ν_μ	$\bar{\nu}_\mu$	0	0	0	0	1	0		
	Tau	τ^-	τ^+	1782	0	0	0	0	1	$\sim 10^{-13}$	$\tau^- \longrightarrow e^- + \bar{\nu}_e + \nu_\tau$
	Tau neutrino	ν_τ	$\bar{\nu}_\tau$	0	0	0	0	0	1		
Hadrons											
Mesons	Pion	π^+	π^-	139.6	0	0	0	0	0	2.6×10^{-8}	$\pi^+ \longrightarrow \mu^+ + \nu_\mu$
		π^0	Self	135.0	0	0	0	0	0	0.8×10^{-16}	$\pi^0 \longrightarrow \gamma + \gamma$
	Kaon	K^+	K^-	493.8	0	+1	0	0	0	1.2×10^{-8}	$K^+ \longrightarrow \mu^+ + \nu_\mu$ $\pi^+ + \pi^0$
		K^0_S	$\bar{K}^0_S$	497.8	0	+1	0	0	0	0.9×10^{-10}	$K^0_S \longrightarrow \pi^+ + \pi^-$ $\pi^0 + \pi^0$
		K^0_L	$\bar{K}^0_L$	497.8	0	+1	0	0	0	5.7×10^{-8}	$K^0_L \longrightarrow \pi^\pm + e^\mp + (\bar{\nu}_e)$ $\pi^\pm + \mu^\mp + (\bar{\nu}_\mu)$ $\pi^0 + \pi^0 + \pi^0$
	Eta	η^0	Self	548.8	0	0	0	0	0	$<10^{-16}$	$\eta^0 \longrightarrow \gamma + \gamma$
Baryons	Proton	p	$\bar{p}$	938.3	+1	0	0	0	0	Stable	
	Neutron	n	$\bar{n}$	939.6	+1	0	0	0	0	10^3	$n \longrightarrow p + e^- + \bar{\nu}_e$
	Lambda	Λ^0	$\bar{\Lambda}^0$	1116	+1	-1	0	0	0	2.5×10^{-10}	$\Lambda^0 \longrightarrow p + \pi^-$ $n + \pi^0$
	Sigma	Σ^+	$\bar{\Sigma}^-$	1189	+1	-1	0	0	0	0.8×10^{-10}	$\Sigma^+ \longrightarrow p + \pi^0, n + \pi^+$
		Σ^0	$\bar{\Sigma}^0$	1192	+1	-1	0	0	0	$<10^{-14}$	$\Sigma^0 \longrightarrow \Lambda^0 + \gamma$
		Σ^-	$\bar{\Sigma}^+$	1197	+1	-1	0	0	0	1.5×10^{-10}	$\Sigma^- \longrightarrow n + \pi^-$
	Xi	Ξ^0	$\bar{\Xi}^0$	1315	+1	-2	0	0	0	3×10^{-10}	$\Xi^0 \longrightarrow \Lambda^0 + \pi^0$
		Ξ^-	$\bar{\Xi}^+$	1321	+1	-2	0	0	0	1.7×10^{-10}	$\Xi^- \longrightarrow \Lambda^0 + \pi^-$
	Omega	Ω^-	$\bar{\Omega}^+$	1672	+1	-3	0	0	0	1.3×10^{-10}	$\Omega^- \longrightarrow \Xi^0 + \pi^-,$ $\Lambda^0 + K^-$

tion of another particle, which in turn decays. The decay process eventually stops when a stable particle, such as a proton or electron, is formed. The decay modes of the unstable "stable" particles appear in Table 35.2, along with their lifetimes and a variety of other information about them.

The particles are produced by accelerating protons or electrons to a very high energy. When a moving particle hits some target nucleus, new particles spray out from the collision (Fig. 35.7). The particles produced by the collision hit other target nuclei or decay, thus leading to the formation of new particles.

The different particles produced by the collision can be detected by the trails they leave in bubble chambers, cloud chambers, or nuclear emulsion plates. Bubble-chamber tracks similar to those shown in Fig. 35.8 represent a series of events started by a collision of a moving kaon (K^-) with a stationary proton. The reaction that follows is represented by the equation

$$K^- + p^+ \longrightarrow \Xi^- + K^0 + \pi^+.$$

Charged particles produce tracks in the bubble chamber, whereas particles with no charge, such as the K^0, produce no tracks. From the point of the collision, we see only the short upward track of the Ξ^- and the track toward the right of the π^+.

The Ξ^- lives only about 10^{-10} s before decaying into a Λ^0 and π^-, the latter producing a track directed up toward the right. Both the Λ^0 and the K^0, produced earlier, move undetected through the bubble chamber until they decay, each causing the creation of two new charged particles:

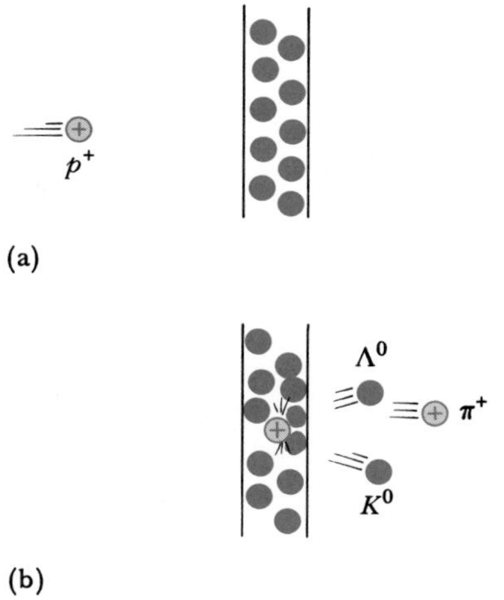

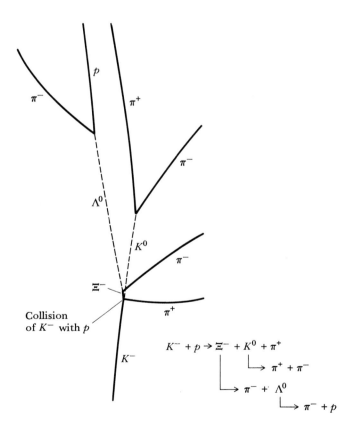

FIG. 35.7. (a) A proton flying at high speed crashes (b) into a target nucleus, causing several new particles to spray out of the target material.

Collision
of K^- with p

$K^- + p \to \Xi^- + K^0 + \pi^+$
$\quad\quad\quad\quad\quad \downarrow \to \pi^+ + \pi^-$
$\quad\quad\quad \downarrow \to \pi^- + \Lambda^0$
$\quad\quad\quad\quad\quad\quad\quad \downarrow \to \pi^- + p$

FIG. 35.8. A K^- hits a proton, resulting in the production of a Ξ^-, K^0, and π^+. The tracks are similar to those produced in a bubble chamber.

$$K^0 \longrightarrow \pi^+ + \pi^-,$$
$$\Lambda^0 \longrightarrow p^+ + \pi^-.$$

The charged products of each decay produce tracks that together look like the letter V. V-tracks appear frequently in these experiments.

Through elaborate analysis of these and many other tracks, it becomes possible to identify the mass and electric charge of the different particles involved in the collisions. The masses are not reported in atomic mass units; instead, we list the rest-mass energy of the particles in units of MeV. For example, a proton of mass 1.0072766 u has a rest-mass energy of 938.3 MeV.

Each particle listed in Table 35.2 has an **antiparticle** that has the same mass but an opposite electric charge. A proton's antiparticle has the same mass as a proton but has charge $-e$; an electron's antiparticle, the positron, has charge $+e$ and the same mass as the electron. Some scientists speculate that, given the great amount of symmetry found in the world, there is probably a part of the universe made of antiparticles. A hydrogen atom in this antiworld would consist of a positron circling a negatively charged antiproton. Antiparticles do not last long in our part of the universe since they quickly join with their oppositely charged particles, causing an annihilation that releases all the mass of each particle in the form of electromagnetic radiation. A more recent theory says that there is no antiworld and that most of the antiparticles were annihilated during the explosive beginning of our universe, the big bang. The excess of particles that were not annihilated by antiparticles constitute the remaining mass of the universe.

The discovery of more than 200 of these particles leaves us with an impression of confusion. By examining a large number of particle collisions, scientists have devised several classification schemes that bring a little order to this chaos. These schemes have led to the development of several new conservation laws for particles.

35.6 Classification Schemes and Conservation Laws

Much of our present understanding of fundamental particles is summarized in classification schemes that group together particles with similar behaviors and properties. One of the most useful schemes groups particles according to the forces they exert on each other. For example, protons and neutrons are subject to the strong nuclear force, whereas electrons are not. In Table 35.2, the **hadron** group consists of particles that exert strong nuclear forces on each other; the proton and neutron are the best known of this group. The **leptons,** including the electron, do not interact strongly with other particles.

The hadrons are subdivided into two groups, the mesons and the baryons.* **Baryons** are distinguished by the fact that their number must be conserved in any nuclear process. The conservation, however, occurs in a very special way. Each baryon listed in Table 35.2 has a baryon number, $B = +1$. The antiparticle of a baryon has $B = -1$. Other particles have baryon number $B = 0$. The **conservation-of-baryon-number law** is stated as follows:

---

*Mesons are supposedly made of two quarks whereas baryons are made of three quarks, elementary particles discussed in the next section.

In any nuclear reaction or decay process, the sum of the baryon numbers of the reactants must equal the sum for the products of the reaction.

The law was developed merely by noting that no exceptions have been observed in any particle-reaction experiment performed so far.

EXAMPLE 35.2 Determine whether each of the following reactions obeys the baryon-number-conservation law:

(a) $p^+ + n \longrightarrow p^+ + n + p^- + p^+$,

(b) $p^+ + n \longrightarrow p^+ + p^+ + p^-$,

(c) $K + p^+ \longrightarrow \Xi^- + K^0 + \pi^+$.

SOLUTION

(a) B: $1 + 1 = 1 + 1 - 1 + 1$.

(b) B: $1 + 1 \neq 1 + 1 - 1$.

(c) B: $0 + 1 = 1 + 0 + 0$.

Reactions (a) and (c) satisfy the law, whereas reaction (b) violates the baryon-number-conservation law. Reaction (b) has never been observed and is said to be forbidden because it violates the baryon-number-conservation law. ∎

On the basis of observation of many particle reactions, other laws have been proposed. A group of particles discovered in the 1950s exhibited what was regarded at that time as strange behavior; these became known as **strange** particles, examples being the K, Λ, and Σ. One example of their "strange" behavior is that they are always produced in pairs. For example, if a π^- meson hits a proton, a K^0 and Λ^0 are produced:

$$\pi^- + p \longrightarrow K^0 + \Lambda^0.$$

According to conservation laws known in the 1950s, a neutron should be able to replace the Λ^0 in the preceding equation, but such a neutron reaction never occurred. Because of this and other strange behavior, another new conservation law was hypothesized, the **conservation-of-strangeness law:**

During any nuclear reaction or decay process involving the strong nuclear force, the strangeness of the reactants must equal that of the reaction products.*

The strangeness number S of the elementary particles listed in Table 35.2 appears in the column labeled S. The strangeness of an antiparticle has the opposite sign to that of the particle.

EXAMPLE 35.3 Determine whether the following reactions satisfy the conservation-of-strangeness law:

(a) $\pi^- + p \longrightarrow K^0 + \Lambda^0$,

(b) $\pi^- + p \longrightarrow K^0 + n$.

*Strangeness does not have to be conserved during radioactive decay reactions caused by the weak nuclear force, which is discussed briefly in the next section.

SOLUTION

(a) S: $0 + 0 = 1 - 1$, allowed.

(b) S: $0 + 0 \neq 1 + 0$, forbidden.

Reaction (b) is forbidden by the conservation-of-strangeness law. ■

In recent years, particles have also been classified in other ways: electron lepton number (L_e), muon lepton number (L_μ), and tau lepton number (L_τ). Most recently, the particles have been postulated as having charm, truth, and beauty; such words reflect the whimsy of physicists and not the actual particle properties.

35.7 Quarks

The classification and conservation laws just described are analogous to the development of the periodic table of elements by Dmitri Mendeleev in 1869. Not until almost 60 years later, after the formulation of quantum mechanics, could scientists explain why certain groups of atoms had similar behavior. Physicists now are searching for a basic theory of elementary particles that will explain why they are classified as baryons, strange, and so forth.

The theory that has gone furthest in explaining these properties of elementary particles is based on a **quark model.** This theory, proposed independently in 1963 by Murray Gell-Mann and George Zweig of the California Institute of Technology, suggests that hadrons are composed of several smaller particles called *quarks*. In the original 1963 theory, three quarks (named "up," "down," and "strange") and their antiquarks were needed to construct all the hadrons then known. Since that time, three other quarks named "charm," "top," and "bottom" have been added to the list.* These quarks have properties such as mass, charge, baryon, number, and strangeness (see Table 35.3), and each quark has an antiquark whose numbers have opposite signs. Baryons are made of three quarks; mesons of two. A particle's net charge, baryon number, strangeness, and so on equal the sum of the numbers for the quarks that compose the particle. The quark compositions of several particles are listed in Table 35.4.

*Top and bottom are also called "truth" and "beauty."

TABLE 35.3 Properties of Quarks and Antiquarks

Name	Quarks					Antiquarks				
	Symbol	Charge	Baryon Number	Strangeness	Charm	Symbol	Charge	Baryon Number	Strangeness	Charm
Up	u	$+\frac{2}{3}e$	$\frac{1}{3}$	0	0	$\bar{u}$	$-\frac{2}{3}e$	$-\frac{1}{3}$	0	0
Down	d	$-\frac{1}{3}e$	$\frac{1}{3}$	0	0	$\bar{d}$	$\frac{1}{3}e$	$-\frac{1}{3}$	0	0
Strange	s	$-\frac{1}{3}e$	$\frac{1}{3}$	-1	0	$\bar{s}$	$\frac{1}{3}e$	$-\frac{1}{3}$	$+1$	0
Charm	c	$+\frac{2}{3}e$	$\frac{1}{3}$	0	1	$\bar{c}$	$-\frac{2}{3}e$	$-\frac{1}{3}$	0	-1
Top (truth)	t	$+\frac{2}{3}e$	$\frac{1}{3}$	0	0	$\bar{t}$	$-\frac{2}{3}e$	$-\frac{1}{3}$	0	0
Bottom (beauty)	b	$-\frac{1}{3}e$	$\frac{1}{3}$	0	0	$\bar{b}$	$\frac{1}{3}e$	$-\frac{1}{3}$	0	0

EXAMPLE 35.4 Show that the charge, baryon number, and strangeness of the proton equal the sum of these numbers for its three quarks (*uud*).

SOLUTION Using Table 35.3, we find that:

$$
\begin{array}{ccc}
u & u & d
\end{array}
$$

$$q = \tfrac{2}{3}e + \tfrac{2}{3}e - \tfrac{1}{3}e = e,$$
$$B = \tfrac{1}{3} + \tfrac{1}{3} + \tfrac{1}{3} = 1,$$
$$S = 0 + 0 + 0 = 0.$$

We see from Table 35.2 that these numbers agree with the values of q, B, and S listed for a proton. ∎

TABLE 35.4 Quark Composition of Several Stable Hadrons

Mesons	Constituent Quarks
π^+	$u\bar{d}$
π^-	$\bar{u}d$
K^+	$u\bar{s}$
K^-	$\bar{u}s$
J/ψ	$c\bar{c}$
Υ	$b\bar{b}$

Baryons	Constituent Quarks
Proton	uud
Neutron	udd
Λ^0	uds
Σ^+	uus
Σ^0	uds
Σ^-	dds
Ξ^0	uss
Ξ^-	dss

In recent years researchers have extended the quark model in an effort to devise a theory for the way quarks interact with each other. One theory in favor at present is called **quantum chromodynamics.** This theory is analogous in many ways to the one used to describe the electrical interactions between charged particles. In electrical theory, particles are said to have electric charge whose value can be $+e$, 0, or $-e$. The concept of electric charge was invented not because electric charge could be seen stuck to a particle but because it was needed to explain the electrical force that one particle exerts on another.

A similar situation exists in particle physics. To explain the fact that quarks must exert very strong forces on each other when held together in a particle, the quarks are said to have a property called *color.* A quark can be either red, blue, or yellow. In quantum chromodynamics, the world **color** does not mean that the quarks actually look red, blue, or yellow. Instead, the color represents a physical property of a quark that causes it to exert a strong force on another quark (just as the electric charge and mass of particles cause them to exert electrical and gravitational forces on each other). This strong force is related in some as yet undetermined way to the strong nuclear force that hadrons exert on each other.

Another analogy between electrical theories and quantum chromodynamics is the ability of charged particles and of quarks to emit and absorb other forms of energy. Particles with electric charge emit and absorb photons; quarks emit and absorb particles called **gluons.** The gluons are supposedly small objects that are exchanged or shared by quarks and bind them together to form protons, neutrons, and the other hadrons. The gluons are the mediators of the strong nuclear force just as photons are responsible for transmitting the electromagnetic force between charged particles.

The idea that the forces between two particles depend on the exchange or sharing of other particles is the dominant theme of recent efforts by physicists to describe the basic forces of nature. Presently, four forces are needed to account for all the observed interactions between particles: (1) the gravitational force, (2) the electromagnetic force, (3) the weak nuclear force, and (4) the strong nuclear force. The electromagnetic and strong nuclear forces are caused by the exchange of photons and gluons, respectively. The gravitational force is thought to occur by exchanging **gravitons,** massless objects that differ from photons and have not yet been observed. The **weak nuclear force,** needed to explain the radioactive decay of nuclei, depends on the exchange of particles called **intermediate vector bosons,** first observed in 1983 at the CERN particle accelerator in Geneva, Switzerland.

Physicists now are trying to develop a **unified theory**—a single theory that includes a description of these four forces and the consequences of their interactions. The weak nuclear force and the electromagnetic force were successfully combined in the early 1970s in what is called an **electroweak theory** by Steven Weinberg, Abdus Salam, and Sheldon Glashow. They were awarded the Nobel Prize in physics for this theory in 1979. Efforts to combine the other forces have led to predictions that are as yet unconfirmed. One of these unified theories predicts that protons are not stable particles but undergo decay. Experiments built to detect this decay have placed a lower limit on their half-life of about 10^{32} years. (At this rate you need not worry about losing by decay many of the 10^{28} protons in your body.)

Knowledge in the field of elementary-particle physics is changing rapidly. In 1968 Abraham Pais, a particle theorist, likened the state of particle physics at that time to a symphony hall some time before the start of the concert.

> On the podium one will see some but not yet all of the musicians. They are tuning up. Short brilliant passages are heard on some instruments; improvisations elsewhere; some wrong notes, too. There is a sense of anticipation for the moment when the symphony starts.*

Now, almost 20 years later, some physicists think that the symphony is ready to start. Our unification has begun. We will soon hear a symphony of photons, gluons, vector bosons, and gravitons dancing between and interacting with the many unusual particles of which our world is made.

*Abraham Pais, "Particles," *Physics Today,* p 28, May (1968).

Summary and Additional Readings

1. **Fission** occurs when a heavy nucleus (such as $^{235}_{92}$U) splits, forming two smaller daughter nuclei. Fission is normally induced by absorption of a neutron. Along with the daughter nuclei, several neutrons are emitted. These neutrons can induce more fissions, thus creating a series of fission reactions. Roughly 200 MeV of energy is released by the fission of one ^{235}U nucleus. This energy can be used to produce electricity in nuclear power plants.

2. **Fusion** is the combination of two small nuclei to form a larger nucleus. Many fusion reactions release a large amount of energy. In order for fusion to occur, the nuclei must move to a distance of about 10^{-14} m or less from each other. This requires large initial kinetic energy that is converted to electrical potential energy as the nuclei with like charges approach each other. Because of the large kinetic energy needed to cause a close approach of the nuclei, fusion usually occurs only at very high temperature, such as in the interior of the sun. Fusion reactors are being developed for controlled release of energy for use in generating electricity.

3. **Particle physics.** At one time the basic building blocks of nature were thought to be electrons, protons, and neutrons. At the present time more than 200 particles have been discovered. Most of these particles have very short lifetimes—10^{-6} s or less. These particles are now classified according to certain properties that seem to be conserved during their reactions: electric charge, lepton number, baryon number, strangeness, charm, and so on.

4. **Quarks.** One of the popular new theories for explaining the properties of particles is based on the idea that many of the particles are constructed from other, more basic particles called quarks. Protons and neutrons are each thought to be made of three quarks, and mesons are made of two quarks. The properties of a particle depend on the properties of the particular quarks of which it is made. The six different quarks (each with an antiquark) are named up, down, strange, charm, top (truth), and bottom (beauty). Each of these six quarks can have a different color, which is a fundamental property (like mass and electric charge) that affects the force quarks exert on each other.

Alvin M. Weinberg, "The Future of Nuclear Energy," *Physics Today,* p 48, March (1981).

"Atmospheric Calculations Suggest a Nuclear Winter," *Physics Today,* p 17, February (1984). A news article with original references concerning the consequences of a nuclear war.

Leon Lederman, "Unraveling the Mysteries of the Atom," *The Physics Teacher,* p 15, January (1982). Discusses particle accelerators and their impact on our knowledge of particles and on other aspects of technological development.

L. M. Brown and L. Hoddeson, "The Birth of Elementary-Particle Physics," *Physics Today,* p 36, April (1982).

Questions

1. Why does a reactor core remain hot long after fission has stopped?

2. Explain carefully how control rods manage to increase or decrease the rate of fission in a nuclear reactor.

3. Suppose that each married couple had three or more children. How would this affect the world's population? Explain the analogy between this situation and the control of nuclear fission by absorbing more or fewer neutrons with the control rods in a reactor core.

4. Discuss the advantages and disadvantages of producing electricity by (a) burning coal, (b) nuclear fission, and (c) nuclear fusion.

5. Why is a high temperature usually needed for fusion to occur?

6. A neutron and a proton will fuse at low temperature to form deuterium, causing the release of energy. Why is this not a suitable way to produce energy?

7. Describe briefly the quark model of elementary particles.

Problems

You may need to use the masses given in Appendix C and the unit conversion $1 \text{ u} = 1.6606 \times 10^{-27} \text{ kg} = 931.5 \text{ MeV}/c^2$ for some of these problems.

35.1 Nuclear Fission

■ 1. Calculate the energy released in the following fission reaction: $^{1}_{0}n + ^{235}_{92}U \longrightarrow ^{141}_{56}Ba + ^{92}_{36}Kr + 3^{1}_{0}n$. The masses of the nuclei are: $m_n = 1.0087 \text{ u}$, $m_U = 235.0439 \text{ u}$, $m_{Ba} = 140.9141 \text{ u}$, and $m_{Kr} = 91.8981 \text{ u}$.

■ 2. Calculate the energy released in the following fission reaction:

$$^{1}_{0}n + ^{235}_{92}U \longrightarrow ^{141}_{55}Cs + ^{92}_{37}Rb + 3^{1}_{0}n,$$

where $m_{Cs} = 140.9356 \text{ u}$ and $m_{Rb} = 91.9191 \text{ u}$.

■ 3. Determine (a) the number of protons and neutrons in the missing fragment of the fission reactions shown below and (b) the mass of that fragment:

$$^{1}_{0}n + ^{235}_{92}U \longrightarrow ? + ^{136}_{54}Xe + 12^{1}_{0}n + 126.5 \text{ MeV}.$$

■ 4. *Estimate* the mass of uranium-235 needed to supply all the electrical energy for the United States in 1985, approximately $1.3 \times 10^{19} \text{ J}$. (Alternatively, students in Canada may perform this estimate for Canada, where about $1.7 \times 10^{18} \text{ J}$ of electrical energy was used in 1985.) Assume that the energy released per fission is 200 MeV and that one-third of the energy produces electricity, the other two-thirds producing waste heat.

■ 5. Convert the 200 MeV per nucleus energy that is released by ^{235}U fission to units of joules per kilogram. By comparison, the energy release of coal is $3.3 \times 10^7 \text{ J/kg}$.

■ 6. (a) Calculate the energy used by a 1200-W hair dryer in 10 min. (b) Approximately how many ^{235}U fissions must occur in a nuclear power plant to provide this energy if 35 percent of the energy released by fission produces electrical energy? The energy released per fission is 200 MeV.

■ 7. (a) Calculate the energy release in MeV of gasoline per molecule of *n*-heptane burned. The molecular mass of *n*-heptane is 100 u, and it releases energy at a rate of $4.8 \times 10^7 \text{ J/kg}$. (b) Calculate the ratio of energy released by one uranium-nucleus fission (approximately 200 MeV) and one *n*-heptane molecule combustion.

■■ 8. *Estimate* the number of ^{235}U nuclei that must undergo fission to provide the energy to lift an ant 1 cm.

■■ 9. An excited $^{236}_{92}U^*$ nucleus decays into $^{144}_{56}Ba$ and $^{92}_{36}Kr$ fragments. Suppose that these fragments are spherical and barely touch just after formation. (a) Calculate the radius of each fragment and their separations. (b) Calculate their electrical potential energy compared with that when they are infinitely far apart.

35.2 Nuclear Power Plants

■ 10. *Estimate* the number of ^{235}U fission reactions per second in a nuclear power plant that produces 1000 MW of electricity and 2000 MW of waste heat.

■ 11. Suppose that a nuclear power plant and a coal-fired power plant both operate at 40 percent efficiency. Calculate the ratio of the mass of coal that must be burned in one day of operating a 1000-MW plant compared to the mass of ^{235}U that must undergo fission in the same plant. The energy released by burning coal is $3.3 \times 10^7 \text{ J/kg}$. The energy released per ^{235}U fission is 200 MeV.

■ 12. The world's uranium supply is approximately 10^9 kg (10^6 tons), 0.7 percent of which is ^{235}U. (a) How much energy is available from the fission of this ^{235}U? (b) The world's energy consumption rate for production of electricity is about 10^{20} J/y. At this rate, how many years would the uranium last if it were used to provide all our electrical energy?

■ 13. A 1000-MW nuclear reactor requires about 1.3 g of ^{235}U per day per MW of electricity generated. Eighty-five percent of this mass is converted to fission products. (a) Calculate the mass of fission products produced in one year by a 1000-MW plant. (b) Calculate the volume of the fission products if their average density is $10{,}000 \text{ kg/m}^3$.

■ 14. When an object of mass m_1 collides elastically and head-on with a stationary object of mass m_2, the final and initial kinetic energies of m_1 are related by the equation $KE_f = [(m_1 - m_2)/(m_1 + m_2)]^2 KE_i$. Use this to show that hydrogen is a better neutron moderator than deuterium. (We ignore the disadvantage that hydrogen absorbs neutrons more than deuterium.)

■ 15. Think of a chain reaction in which each fission, on the average, produces 1.1 future fissions. If at time zero 10 fissions occur, how many occur during the next generation (the second generation)? During the third? During the tenth? During the nth?

■ 16. Suppose that each fission reaction in the core of a reactor produces 1.00005 neutrons that initiate a subsequent fission and that the time between fissions is 5.0×10^{-3} s. By what factor does the rate of fission increase in 60 s?

■■ 17. ^{235}U and ^{238}U can be separated by allowing molecules of UF_6 to diffuse through filters. Since the $^{235}UF_6$ is less massive, it moves faster and becomes more concentrated. Calculate the ratio of the speeds of $^{235}UF_6$ and $^{238}UF_6$ at room temperature.

■■ 18. A 1000-MW nuclear power plant uses about 400 kg of ^{235}U per year. Approximately 0.3 percent of the ^{235}U fission reactions produce a ^{85}Kr nucleus. (a) Calculate the number of ^{85}Kr nuclei produced per year. (b) Calculate the activity in curies of this krypton, whose decay constant is 2.0×10^{-9} s^{-1} (half-life of 10.8 yr). If these ^{85}Kr nuclei were released to the atmosphere, the ^{85}Kr radiation exposure from all nuclear plants would reach about 1 mrem/yr per person by the year 2000.

35.3 Breeder Reactors

■ 19. (a) A ^{239}Pu fission releases 180 MeV of energy per nucleus. Convert this energy yield to joules per kilogram. (b) A breeder reactor converts 1 kg of ^{238}U to 1 kg of ^{239}Pu. If the world supply of ^{238}U is 10^9 kg, calculate the energy available from its conversion to ^{239}Pu and subsequent ^{239}Pu fission. (c) If 10^{20} J of this energy is used each year to produce the world's electricity, how many years will the uranium last?

35.4 Fusion

■ 20. Calculate the energy released by the following fusion reaction: $^{2}_{1}H + {}^{3}_{1}H \longrightarrow {}^{4}_{2}He + {}^{1}_{0}n$.

■ 21. A series of reactions on the sun leads to the combination of three helium nuclei ($^{4}_{2}He$) to form one carbon nucleus ($^{12}_{6}C$). (a) Calculate the net energy released by the reactions. (b) What fraction of the total mass of the three helium nuclei is converted to energy?

■ 22. Deuterium nuclei have an approximately equal chance of reacting in either of the following ways: (a) $^{2}_{1}H + {}^{2}_{1}H \longrightarrow {}^{1}_{0}n + {}^{3}_{2}He$ and (b) $^{2}_{1}H + {}^{2}_{1}H \longrightarrow {}^{1}_{1}H + {}^{3}_{1}H$. Calculate the energy released in each reaction.

■ 23. A series of reactions that might provide power for a nuclear fusion reactor are summarized by the following equation: $6\,{}^{2}_{1}H \longrightarrow 2\,{}^{1}_{1}H + 2\,{}^{1}_{0}n + 2\,{}^{4}_{2}He$. (a) Calculate the net energy released by the reaction. (b) Convert this answer to units of joules per kilogram of deuterium ($^{2}_{1}H$).

■ 24. The synthesis of oxygen on the sun involves the combination of ^{12}C and ^{4}He to form ^{16}O. Does the reaction consume or release energy, and how much?

■ 25. Approximately 4×10^9 kg of mass is converted to energy each second on the sun. If all the energy is a result of the proton-proton series of reactions, including the positron-electron annihilation, calculate the number of proton-proton series reactions occurring each second.

■■ 26. (a) Determine the number of water molecules in 1 kg of water (about 1 qt). One mole of water has a mass of 18 g. (b) For every 67,000 hydrogen nuclei ($^{1}_{1}H$) in water, there is one deuterium ($^{2}_{1}H$) nucleus. Calculate the number of $^{2}_{1}H$ nuclei in the 1 kg of water. (c) The fusion of two deuterium nuclei releases about 4 MeV of energy. Calculate the energy, in units of joules, that could be released by fusion of the deuterons in 1 kg of water. By comparison, burning 1 kg of coal releases 2×10^7 J of energy.

■■ 27. World energy consumption in 1980 was about 2.5×10^{20} J. (a) Calculate the number of deuterium nuclei that would be needed to produce this energy. (b) Calculate the volume

of water of density 1000 kg/m^3 needed to supply this energy. Use the numbers given in the preceding problem.

■■ 28. (a) Calculate the average kinetic energy of a proton at the center of a star whose temperature is 2×10^7 K. (b) If two of these protons have a head-on collision, calculate their closest separation.

35.5 Particle Physics

■ 29. A Ξ^- particle is produced by a high-energy accelerator. Using the decay schemes shown in Table 35.2, develop a sequence of decays that lead to a group of stable particles with lifetimes longer than 10^{-5} s.

■ 30. Work the previous problem, but for the K^+ particle.

■ 31. A K^+ enters a bubble chamber and decays to a μ^+ and ν_μ. Draw a set of lines on a piece of paper representing the tracks you expect to be formed by the K^+ and its decay products and their decay products.

■ 32. Shown in Fig. 35.9 is a set of bubble-chamber tracks caused by the following particles and their reactions: $\Omega^- \longrightarrow \Lambda^0 + K^-$, followed by $\Lambda^0 \longrightarrow p + \pi^-$ and $K^- \longrightarrow \mu^- + \bar{\nu}_\mu$. Redraw the tracks on a separate paper and identify which track is caused by which particle. Also, add dashed lines for particles that cause no tracks.

FIG. 35.9. A hypothetical set of bubble-chamber tracks.

■ 33. A proton and antiproton with negligible kinetic energy annihilate one another to produce two photons. Calculate the energy in electron volts of each photon and its frequency and wavelength.

■ 34. Calculate the wavelength of the lowest-energy photon capable of producing an electron-positron pair. Ignore the question of conserving momentum (a poor assumption).

■■ 35. A Σ^+ baryon with a lifetime of 0.8×10^{-10} s in its proper inertial frame of reference and with a rest-mass energy of 1189 MeV has an equal amount of kinetic energy. (a) Calculate its speed. (b) Calculate its lifetime as determined by a laboratory observer. (c) Calculate the length of its path in a bubble chamber before decay occurs.

35.6 Classification Schemes and Conservation Laws

36. Show that the laws of conservation of charge and baryon number are not violated by the π^+, K^+, Λ^0, and Ξ^0 decay reactions shown in Table 35.2.

37. Show that the following reactions are forbidden by the conservation of charge and/or baryon number laws:

(a) $\Lambda^0 \longrightarrow p + e^- + \nu_e + n$,

(b) $\pi^0 + p \longrightarrow \Sigma^0 + \pi^0$,

(c) $\Omega^- \longrightarrow \Sigma^- + p + \pi^- + e^+ + e^-$,

(d) $\pi^- + p \longrightarrow \bar{K}^0 + \eta^0 + \pi^0$.

38. Use the conservation of charge, baryon number, and strangeness laws to show that the following reactions are not allowed:

(a) $\pi^- + p \longrightarrow \Xi^0 + K^0 + \pi^0$,

(b) $\Lambda^0 + n \longrightarrow p^+ + p^- + 3\pi^+ + 3\pi^-$,

(c) $\pi^- + p \longrightarrow K^+ + \Omega^-$,

(d) $p + p \longrightarrow p + \pi^0 + \Lambda^0 + K^-$.

■ 39. A moving proton strikes another proton that is at rest, causing the reaction $p + p \longrightarrow p + n + \pi^+$. What is the least

kinetic energy of the incident proton for this reaction to occur? Ignore conservation of momentum.

35.7 Quarks

40. Using Tables 35.2 and 35.3, confirm that the charge, baryon number, and strangeness of the following particles equal the sum of the numbers for the quarks of which the particle is made: (a) $n(udd)$; (b) $\Sigma^+ (uus)$; (c) $\Xi^-(dss)$.

41. Repeat Problem 40 for the following particles: (a) $\Sigma^- (dds)$; (b) $\pi^- (\bar{u}d)$; (c) $K^0(d\bar{s})$.

■ 42. A new meson is postulated to have the following quantum numbers: $q = S = +1$ and $B = 0$. What combination of quarks produces these numbers?

APPENDIX A   Mathematics Review

A1. Significant Figures

The uncertainty in the value of a measured quantity can be indicated conveniently by using significant figures. Suppose you use a ruler to measure the length of the cover of a book and find that it is 26.0 cm long. This number has three **significant figures,** the number of digits whose values have been ascertained reliably by your measurement. By convention, the last digit to which a value is assigned may be uncertain by one or two numbers on either side of the assigned value. By stating that the book is 26.0 cm long, we mean that the book's length is 26.0 ± 0.1 cm, or that its length lies between 25.9 and 26.1 cm. We cannot say that the book's length is 26.00 cm unless we can actually measure it with a certainty of ± 0.01 cm. Such certainty is usually not possible using rulers, since the smallest division on most rules is 0.1 cm.

When multiplying or dividing numbers, the number of significant figures in the final answer should equal the number of digits in the least significant number used in the calculation. Suppose, for example, that you wish to calculate the area of a poster that is 5.3 cm high and 11.2 cm wide. Your answer should have only two significant figures since 5.3 has only two significant figures: $A = 5.3 \times 11.2 = 59.36 = 59$ cm^2. To see why we should not retain the .36 in our answer, we must realize that the calculated value of A can range from $5.2 \times 11.1 = 57.72$ cm^2 to $5.4 \times 11.3 = 61.02$ cm. Our answer of 59 cm^2 reflects the fact that the calculated number is uncertain by about ± 1 cm. If we stated our answer as 59.36 cm^2, we would be implying an unjustifiable uncertainty of ± 0.01 cm^2.

When adding or subtracting numbers, the last significant figure of the result is in the same column relative to the decimal point as the column of the last significant figure of the least accurate number used in the calculation. Suppose, for example, that you wish to add the distances listed below:

$$
\begin{array}{r}
1.42 \ \text{m} \\
220.1 \ \ \text{m} \\
\underline{2.357 \ \text{m}} \\
233.877 \ \text{m} \quad \text{or} \quad \underline{233.9 \ \text{m}} \ .
\end{array}
$$

The answer is 233.9 m, not 233.877 m, since the least accurate number, 220.1 m, is uncertain by about ± 0.1 and the answer cannot be more accurate than ± 0.1.

A2. Powers of Ten

Powers of ten are used to write the large and small numbers in science. For instance, the speed of light, 300,000,000 m/s, is written in powers of ten notation as 3×10^8 m/s, where $10^8 = 100,000,000$. Other examples of numbers written in the powers of ten notation are given below:

$$100 = 10 \times 10 = 10^2,$$
$$1000 = 10 \times 10 \times 10 = 10^3,$$
$$5,700,000 = 5.7 \times 10^6,$$
$$49,200 = 4.92 \times 10^4.$$

Numbers less than one can also be written in the powers of ten notation. For instance,

$$0.1 = \frac{1}{10} = 10^{-1},$$

$$0.001 = \frac{1}{1000} = \frac{1}{10^3} = 10^{-3},$$

$$0.0000073 = 7.3 \times 10^{-6},$$

$$0.000000012 = 1.2 \times 10^{-8}.$$

We can see from these examples that

$$\frac{1}{10^n} = 10^{-n}. \qquad \text{(A.1)}$$

In order to multiply and divide numbers that are written in the powers of ten notation, we use the following rules:

$$10^n \times 10^m = 10^{n+m}, \qquad \text{(A.2)}$$

$$\frac{10^n}{10^p} = 10^n \times 10^{-p} = 10^{n-p}. \qquad \text{(A.3)}$$

where n, m, and p can be any number. The following examples use these rules for the case where n, m, and p are integers:

$$10^2 \times 10^4 = 10^6,$$

$$\frac{10^4}{10^2} = 10^4 \times 10^{-2} = 10^2,$$

$$\frac{10^{-2}}{10^{-3}} = 10^{-2} \times 10^3 = 10,$$

$$(5 \times 10^2)(4 \times 10^4) = (5 \times 4)(10^{2+4}) = 20 \times 10^6 = 2 \times 10^7,$$

$$(6.6 \times 10^{-34})(5.0 \times 10^{15}) = (6.6 \times 5.0)(10^{-34+15}) = 33 \times 10^{-19} = 3.3 \times 10^{-18},$$

$$\frac{(6.6 \times 10^{-34})(3 \times 10^8)}{(5 \times 10^{-7})} = \left(\frac{6.6 \times 3}{5}\right)(10^{-34+8+7}) = 4 \times 10^{-19}.$$

The powers of ten notation is preferred in science for a variety of reasons: (1) numbers are more easily read and compared in powers of ten notation than if written with a large number of zeros before or after a decimal point; (2) the numbers can be transcribed from one place to another more easily; (3) calculations are easier and less prone to error; and (4) the notation provides a satisfactory method for indicating the number of significant figures in a number.

A3.   Exponents and Logarithms

The expression a^n means that a should be multiplied by itself n times. For example, $2^3 = 2 \times 2 \times 2$ and $10^4 = 10 \times 10 \times 10 \times 10$. The superscript n is called an **exponent**.

The **logarithm** of a number N is the exponent n to which a number a must be raised in order to equal that number ($a^n = N$). If

$$a^n = N, \quad \text{then} \quad n = \log_a N, \quad \textbf{(A.4)}$$

where a is called the **base number.** Two examples of the use of Eq. (A.4) are given below:

$$10^4 = 10,000 \quad \text{or} \quad 4 = \log_{10} 10,000,$$
$$2^5 = 32 \quad \text{or} \quad 5 = \log_2 32.$$

The decimal system of numbers has the base 10, a common base number for calculations involving logarithms. In fact, when the base number 10 is used, we omit the subscript 10 and write $n = \log N$ which implies that $10^n = N$. This system with base number 10 is called the **common logarithm system.**

Another commonly used system of logarithms has the base number $e = 2.718 \ldots$ and is called the **natural logarithm system.** In this system logarithms are written as $n = \ln N$ which implies that $e^n = N$. A logarithm of a number N in one system can be converted to a logarithm in the other system by using one of the following:

$$\ln N = 2.3026 \log N, \quad \textbf{(A.5)}$$
$$\log N = 0.43429 \ln N. \quad \textbf{(A.6)}$$

We often take logarithms of numbers that are multiplied or divided by each other. The following rules are useful for this purpose:

$$\log AB = \log A + \log B, \quad \textbf{(A.7)}$$
$$\log \frac{A}{B} = \log A - \log B, \quad \textbf{(A.8)}$$
$$\log A^n = n \log A. \quad \textbf{(A.9)}$$

Let us prove the first rule. Suppose that

$$10^x = A \quad \text{and} \quad 10^y = B.$$

Then,

$$x = \log A \qquad y = \log B,$$

and

$$x + y = \log A + \log B.$$

Also,

$$AB = 10^x 10^y = 10^{x+y}.$$

Using the definition of logarithm [Eq. (A.4)], we find that

$$x + y = \log AB.$$

We now have two equations that each have $x + y$ on the left side. Equating the right sides of these equations, we find that

$$\log AB = \log A + \log B,$$

which is the desired result. The other rules can be proved easily in a similar manner.

APPENDIX B Graphing

A **graph** is a diagram that illustrates by means of points and lines a relationship between different quantities whose values are plotted on the graph. Graphical analysis, the drawing and interpreting of graphs, helps us visualize the way a change in one variable affects the values of another variable. Graphical analysis also helps us determine the type of mathematical equation that relates the quantities being graphed.

As an example of graph construction, consider the changing position of a ball that rolls on a smooth, frictionless surface at constant speed. As the ball crosses a starting line, we start a timing clock. We have arranged our experiment so that the ball is at position zero when the time is zero. The observed values of position x and time t for later times are recorded in Table B.1. Notice that the time measurement is so precise that the experimental error in this quantity can be ignored. However, we estimate that the position of the ball can be determined no better than ± 0.1 m. This is an estimate of the uncertainty in the measurement of the ball's position.

Let us now construct a graph of the data recorded in Table B.1.

1. Choose the graph paper: There are many types of graph paper such as linear, semilog, and log-log. We will find that lines passing through our data points have different shapes when plotted on these various types of graph paper. *If the data can be fit by a straight line on one type of graph paper, this allows us to determine the type of equation that relates the variables plotted on that graph.* For example, equations of the form $y = ax + b$ produce straight lines on linear graph paper. Exponential equations of the form $y = y_0 e^{kx}$ produce straight lines on semilog graph paper, and power laws of the form $y = ax^n$ produce straight lines on log-log graph paper. If our data produce a straight line on one of these types of graph paper, we are able to write an equation that relates the variables x and y. For now, we will use only linear graph paper to develop our graphing techniques.

2. Choose the axes: You must decide what variable to put on each axis. You may be told in the instructions for an experiment to plot position versus time. In this case the variable before the "versus" is plotted on the vertical axis (the **ordinate**) and the variable after the versus is plotted on the horizontal axis (the **abscissa**). If you were asked to plot x versus t, the x would go on the vertical axis and the t on the horizontal axis. The variable on the vertical axis is usually called the **dependent variable,** as its values often depend on the value of the variable plotted on the horizontal axis, the **independent variable.** Occasionally you may wish to plot some function of a variable rather than the variable itself (for example, plot x^2 versus t or $1/x$ versus t).

Having chosen the variables for each axis, the axes must be labeled with the name of the variable followed by its unit of measure in parentheses.

3. Choose the scales: The **scale** of a graph is the number of ruled lines on the graph that correspond to a particular change in the value of the variable being plotted on that axis. If plotting time, for instance, we may choose one large division to be 0.25 second. Ten large divisions would be 2.5 seconds.

The scale is chosen by considering the range of values of the variable to be

TABLE B.1 Position of a Ball as a Function of Time

Time in s (± 0.0 s)	Position in m (± 0.1 m)
0.0	0.0
0.5	1:0
1.0	2.3
1.5	3.3
2.0	4.2
2.5	5.3
3.0	6.7

plotted and the total number of divisions on the graph paper. Try to use most of the graph paper. If the values of the time variable range from zero to 3.0 seconds and there are 18 large divisions (18 cm) on the graph paper, we might choose the scale so that five large divisions equal one second. Fifteen large divisions equal three seconds, and the data will be spread over most of the paper (see Fig. B.1). Having chosen the scale, numbers should be placed along this axis to indicate the value of the variable at different places along the axis. A scale must also be chosen for the vertical axis.

4. Plot the data: Having determined the scale and labeled the axes, we next plot the data points on the graph. Notice that the point for the distance measurement at time 2.0 seconds (Table B.1) is a point in Fig. B.1 above the time scale at 2.0 seconds and is horizontally across from the place on the distance scale where $x = 4.2$ m.

It is important to include error bars on your graph. An **error bar** is a line passing through the data point. The length of the error bar indicates the uncertainty in the value of the variable plotted in that direction. The position variable plotted in Fig. B.1, for example, has an uncertainty of ± 0.1 m. As a result, an error bar for a point on the graph is parallel to the position coordinate axis, is 0.2 m long, and extends 0.1 m above and below the data point.

5. Draw a line through your data points: There are a variety of ways to draw lines through the data points. Suppose that your data points can be fit by a straight line. Often, a computer can be used to choose the best straight line, assuming of course that a straight line seems to fit the data. Even with the best straight line, it is unlikely that all of the data points will lie on the line. In fact, none of the data points need lie on the line. A computer chooses the straight line so that the sum of the squares of the deviations of the actual data points from the line is a minimum. This procedure produces a line that passes through the center of the data points, and hopefully most data points lie near the line. Usually an equal number of points will be above the line as below.

If you cannot use a computer for choosing the line, then an "eyeball" procedure must be tried. You will have to draw your line so that it is not too far from any point and so that the points above the line balance the deviations of

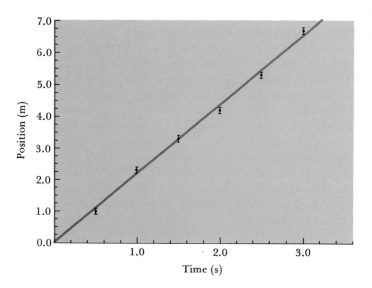

FIG. B.1. A straight line graph on linear graph paper.

points below the line. In the example shown in Fig. B.1, one point lies on the line, two above the line, and three below.

Although your line should pass within the error bars of most points, this is not necessary. The lengths of the error bars are estimates of the uncertainty of the measurements. The line passing through the data points should be within the uncertainty limits of most data points. It is not unusual, though, for the error bars of some data points not to lie on the line. (For a so-called normal distribution of data points, we expect about one out of three error bars to be off the line.)

6. Analyze the graph: If the data points can be fit by a straight line, we can use this line to write an equation that relates the independent variable to the dependent variable. We will review this procedure for each type of graph paper in the following sections.

What if our data points vary in a way that cannot be fit by a straight line? Then we must draw a curved line that moves smoothly through the data points. This curved line will help you develop a feel for the way the data are changing. Does the curved line represent an accelerating or decelerating object? Does it look like an exponentially decreasing or increasing function? The curved line may also help us guess the kind of mathematical equation that relates the independent variable to the dependent variable. This will help us determine whether another type of graph paper should be used to plot the data.

7. Linear graph paper and linear equations: **Linear graph paper** has equal spacing between the lines along each axis. A **linear equation** is of the form

$$y = ax + b, \tag{B.1}$$

where x is the independent variable, y is the dependent variable, and a and b are constants. A linear equation produces a straight line when plotted on linear graph paper. Conversely, if a set of data, when plotted on linear graph paper, can be fit by a straight line, then we know that the dependent variable is related to the independent variable by a linear equation.

The constant a in Eq. (B.1) is called the **slope** of the plotted line. To find the slope, we choose two points on the line. (We do not choose actual data points unless they lie on the line.) It is best to select points that are far from each other; that is, near the ends of the line. The uncertainty in the slope is less when calculated using distant points. The slope is given by the equation

$$\text{slope} = a = \frac{y_2 - y_1}{x_2 - x_1}, \tag{B.2}$$

where (x_1, y_1) and (x_2, y_2) are the values of the two variables at the chosen points on the line. The units of the slope are the units of the y variable divided by the units of the x variable.

The constant b is called the **y intercept** and equals the value of y at the point where the plotted line crosses the y axis (the value of y when $x = 0$).

EXAMPLE B.1 Determine the equation that fits the data plotted in Fig. B.2.

SOLUTION The data produce a straight line on linear graph paper. The dependent variable v must be related to the independent variable t by an equation of the form $v = at + b$. Using Eq. (B.2), we see that

$$a = \frac{v_2 - v_1}{t_2 - t_1} = \frac{(2.22 - 0.80)\text{m/s}}{(0.95 - 0.05)\text{s}} = \frac{1.42 \text{ m/s}}{0.90 \text{ s}} = 1.6 \text{ m/s}^2.$$

FIG. B.2. A linear graph.

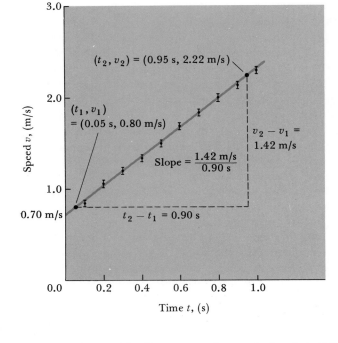

The value of v where the straight line crosses the vertical axis is 0.7 m/s; thus, $b = 0.7$ m/s. The equation that fits this data is

$$v = (1.6 \text{ m/s}^2)t + 0.7 \text{ m/s}.$$ ■

8. Semilog graph paper and exponential functions: **Semilog graph** paper has a linear scale (equal spacing between lines) on the horizontal axis and a logarithmic scale on the vertical axis (Fig. B.3). To make a logarithmic scale, we place a group of equally spaced horizontal marks on the vertical axis. These marks do not represent the numbers 1, 2, 3, 4, 5, . . . as in linear graph paper; instead, they represent the numbers 1, 2, 4, 8, 16, ... Each equally spaced step up the vertical axis represents a doubling in the value of the variable plotted on that axis. **Exponential functions** of the form

$$y = y_0 e^{kx} \tag{B.3}$$

have the property that the magnitude of the change in the y variable, Δy, doubles as the x variable is changed in equal steps, Δx. If we plot the values of y versus x for exponential functions on semilog graph paper, a straight line connects the data points. Conversely, if the data points for the variables y versus x produce a straight line on semilog graph paper, the variables are related by an exponential equation [Eq. (B.3)].

Semilog graph paper comes in different types that vary in the range of values that can be plotted on the logarithmic scale. Each range of 10 is called a **cycle.** A two-cycle log scale can accommodate data that range in value over two powers of ten (for example, from 1–100 or from 10^3–10^5). If the data that you wish to plot on the logarithmic scale range from 2 to 6400, you will need a four-cycle logarithmic scale. The first cycle is for points from 1–10, the second cycle for 10–100, the third cycle for 100–1000, and the fourth for 1000–10,000.

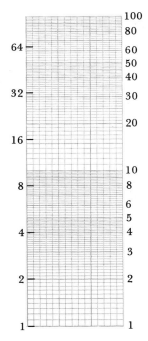

FIG. B.3. On a log scale, the variable doubles in value for equal length steps up the scale.

If the data when plotted on semilog graph paper can be fit by a straight line, you know that the variables are related by an equation like

$$y = y_0 e^{kx}.$$

The constant y_0, the **y intercept,** is the value of y when $x = 0$ (it is also the place where the straight line crosses the y axis). The constant k is determined by choosing two points on the line, (y_1, x_1) and (y_2, x_2). The value of k is determined by taking the natural logarithm of y_2/y_1 and substituting in the equation

$$k = \frac{\ln(y_2/y_1)}{x_2 - x_1}. \tag{B.4}$$

EXAMPLE B.2 Determine an equation that fits the data shown in Fig. B.4. The graph shows the cost of gasoline (y) versus time (t) assuming a 10 percent per year rate of inflation.

SOLUTION Since the data produce a straight line on semilog graph paper, we know that the dependent variable y is related to the independent variable t by an equation of the form $y = y_0 e^{kt}$. We see that when $t = 0$, $y = y_0 e^{k0} = y_0 = \$2.00$. The value of k is determined using Eq. (B.4) and the points ($\$32.00$, 28 years) and ($\$2.00$, 0 years):

$$k = \frac{\ln(32.00/2.00)}{28 \text{ yr} - 0 \text{ yr}} = \frac{2.77}{28 \text{ yr} - 0} = 0.10 \text{ yr}^{-1}.$$

The equation relating y and t is

$$\underline{y = \$2.00 e^{(0.10 \text{ yr}^{-1})t}}. \qquad \blacksquare$$

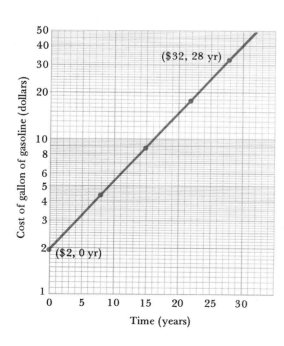

FIG. B.4. A semilog graph.

9. Log-log graph paper and power laws: **Log-log graph** paper has a logarithmic scale on each axis. **Power law equations** of the form $y = ax^n$ produce straight lines on log-log graph paper. The constant n (an exponent) can be determined by choosing two points on the line [(x_1, y_1) and (x_2, y_2)] and substituting the values of x and y for these points into the equation

$$n = \frac{\ln(y_2/y_1)}{\ln(x_2/x_1)}. \tag{B.5}$$

Having found n, the constant a can be determined by substituting the values of n, x_2, and y_2 into the equation

$$a = \frac{y_2}{x_2^n}. \tag{B.6}$$

EXAMPLE B.3 Determine an equation that fits the data shown in Fig. B.5.

SOLUTION Since the data produce a straight line on log-log graph paper, the variables y and t must be related by an equation of the form $y = at^n$. Choosing the points (2 m, 0.5 s) and (200 m, 5 s) that are on the line, we find using Eq. (B.5) that

$$n = \frac{\ln(200 \text{ m}/2 \text{ m})}{\ln(5 \text{ s}/0.5 \text{ s})} = \frac{\ln 100}{\ln 10} = \frac{4.6}{2.3} = 2.$$

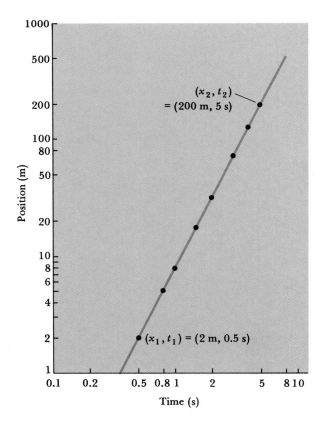

FIG. B.5. A log-log graph.

Substituting $n = 2$, $y_2 = 200$ m, and $t_2 = 5$ s into Eq. (B.6), we find that

$$a = \frac{200 \text{ m}}{(5 \text{ s})^2} = 8 \text{ m/s}^2.$$

The equation relating y and t is

$$y = (8 \text{ m/s}^2)t^2.$$ ∎

APPENDIX C   Selected Isotopes*

(1) Atomic number Z	(2) Element	(3) Symbol	(4) Mass number, A	(5) Atomic mass†	(6) Percent abundance, or decay mode if radioactive	(7) Half-life (if radioactive)
0	(Neutron)	n	1	1.008665	β^-	10.6 min
1	Hydrogen	H	1	1.007825	99.985	
	Deuterium	D	2	2.014102	0.015	
	Tritium	T	3	3.016049	β^-	12.33 yr
2	Helium	He	3	3.016029	0.00014	
			4	4.002604	99.99986	
3	Lithium	Li	6	6.015123	7.5	
			7	7.016005	92.5	
4	Beryllium	Be	7	7.016930	EC, γ	53.3 days
			9	9.012183	100	
5	Boron	B	10	10.012938	19.8	
			11	11.009305	80.2	
6	Carbon	C	11	11.011433	β^+, EC	20.4 min
			12	12.000000	98.89	
			13	13.003355	1.11	
			14	14.003242	β^-	5730 yr
7	Nitrogen	N	13	13.005739	β^+	9.96 min
			14	14.003074	99.63	
			15	15.000109	0.37	
8	Oxygen	O	15	15.003065	β^+, EC	122 s
			16	15.994915	99.76	
			18	17.999159	0.204	
9	Fluorine	F	19	18.998403	100	
10	Neon	Ne	20	19.992439	90.51	
			22	21.991384	9.22	
11	Sodium	Na	22	21.994435	β^+, EC, γ	2.60 yr
			23	22.989770	100	
			24	23.990964	β^-, γ	15.0 h
12	Magnesium	Mg	24	23.985045	78.99	
13	Aluminum	Al	27	26.981541	100	
14	Silicon	Si	28	27.976928	92.23	
			31	30.975364	β^-, γ	2.62 h
15	Phosphorus	P	31	30.973763	100	
			32	31.973908	β^-	14.28 days
16	Sulfur	S	32	31.972072	95.0	
			35	34.969033	β^-	87.4 days
17	Chlorine	Cl	35	34.968853	75.77	
			37	36.965903	24.23	
18	Argon	Ar	40	39.962383	99.60	
19	Potassium	K	39	38.963708	93.26	
			40	39.964000	β^-, EC, γ, β^+	1.28×10^9 yr
20	Calcium	Ca	40	39.962591	96.94	
21	Scandium	Sc	45	44.955914	100	

*Data are taken from C. M. Lederer and V. S. Shirley, eds., *Table of Isotopes,* 7th ed. (New York: John Wiley & Sons, Inc., 1978).
†The masses given in column (5) are those for the neutral atom, including the Z electrons.

(1) Atomic number Z	(2) Element	(3) Symbol	(4) Mass number, A	(5) Atomic mass†	(6) Percent abundance, or decay mode if radioactive	(7) Half-life (if radioactive)
22	Titanium	Ti	48	47.947947	73.7	
23	Vanadium	V	51	50.943963	99.75	
24	Chromium	Cr	52	51.940510	83.79	
25	Manganese	Mn	55	54.938046	100	
26	Iron	Fe	54	53.939612	5.8	
			56	55.934939	91.8	
27	Cobalt	Co	59	58.933198	100	
			60	59.933820	β^-, γ	5.27 yr
28	Nickel	Ni	58	57.935347	68.3	
			60	59.930789	26.1	
29	Copper	Cu	63	62.929599	69.2	
			65	64.927792	30.8	
30	Zinc	Zn	64	63.929145	48.6	
			66	65.926035	27.9	
31	Gallium	Ga	69	68.925581	60.1	
32	Germanium	Ge	72	71.922080	27.4	
			74	73.921179	36.5	
33	Arsenic	As	75	74.921596	100	
34	Selenium	Se	80	79.916521	49.8	
35	Bromine	Br	79	78.918336	50.69	
36	Krypton	Kr	84	83.911506	57.0	
37	Rubidium	Rb	85	84.911800	72.17	
38	Strontium	Sr	86	85.909273	9.8	
			88	87.905625	82.6	
			90	89.907746	β^-	28.8 yr
39	Yttrium	Y	89	88.905856	100	
40	Zirconium	Zr	90	89.904708	51.5	
41	Niobium	Nb	93	92.906378	100	
42	Molybdenum	Mo	98	97.905405	24.1	
43	Technetium	Tc	98	97.907210	β^-, γ	4.2×10^6 yr
44	Ruthenium	Ru	102	101.904348	31.6	
45	Rhodium	Rh	103	102.90550	100	
46	Palladium	Pd	106	105.90348	27.3	
47	Silver	Ag	107	106.905095	51.83	
			109	108.904754	48.17	
48	Cadmium	Cd	114	113.903361	28.7	
49	Indium	In	115	114.90388	95.7; β^-	5.1×10^{14} yr
50	Tin	Sn	120	119.902199	32.4	
51	Antimony	Sb	121	120.903824	57.3	
52	Tellurium	Te	130	129.90623	34.5; β^-	2×10^{21} yr
53	Iodine	I	127	126.904477	100	
			131	130.906118	β^-, γ	8.04 days
54	Xenon	Xe	129	128.90478	26.4	
			131	130.90508	21.2	
			132	131.90415	26.9	
			136	135.90722	8.9	
55	Cesium	Cs	133	132.90543	100	
56	Barium	Ba	137	136.90582	11.2	
			138	137.90524	71.7	
57	Lanthanum	La	139	138.90636	99.911	
58	Cerium	Ce	140	139.90544	88.5	
59	Praseodymium	Pr	141	140.90766	100	
60	Neodymium	Nd	142	141.90773	27.2	
61	Promethium	Pm	145	144.91275	EC, α, γ	17.7 yr

(1) Atomic number Z	(2) Element	(3) Symbol	(4) Mass number, A	(5) Atomic mass†	(6) Percent abundance, or decay mode if radioactive	(7) Half-life (if radioactive)
62	Samarium	Sm	152	151.91974	26.6	
63	Europium	Eu	153	152.92124	52.1	
64	Gadolinium	Gd	158	157.92411	24.8	
65	Terbium	Tb	159	158.92535	100	
66	Dysprosium	Dy	164	163.92918	28.1	
67	Holmium	Ho	165	164.93033	100	
68	Erbium	Er	166	165.93031	33.4	
69	Thulium	Tm	169	168.93423	100	
70	Ytterbium	Yb	174	173.93887	31.6	
71	Lutecium	Lu	175	174.94079	97.39	
72	Hafnium	Hf	180	179.94656	35.2	
73	Tantalum	Ta	181	180.94801	99.988	
74	Tungsten	W	184	183.95095	30.7	
75	Rhenium	Re	187	186.95577	62.60, β^-	4×10^{10} yr
76	Osmium	Os	191	190.96094	β^-, γ	15.4 days
			192	191.96149	41.0	
77	Iridium	Ir	191	190.96060	37.3	
			193	192.96294	62.7	
78	Platinum	Pt	195	194.96479	33.8	
79	Gold	Au	197	196.96656	100	
80	Mercury	Hg	202	201.97063	29.8	
81	Thallium	Tl	205	204.97441	70.5	
82	Lead	Pb	206	205.97446	24.1	
			207	206.97589	22.1	
			208	207.97664	52.3	
			210	209.98418	α, β^-, γ	22.3 yr
			211	210.98874	β^-, γ	36.1 min
			212	211.99188	β^-, γ	10.64 h
			214	213.99980	β^-, γ	26.8 min
83	Bismuth	Bi	209	208.98039	100	
			211	210.98726	α, β^-, γ	2.15 min
84	Polonium	Po	210	209.98286	α, γ	138.38 days
			214	213.99519	α, γ	164 μs
85	Astatine	At	218	218.00870	α, β^-	≈ 2 s
86	Radon	Rn	222	222.017574	α, γ	3.8235 days
87	Francium	Fr	223	223.019734	α, β^-, γ	21.8 min
88	Radium	Ra	226	226.025406	α, γ	1.60×10^3 yr
89	Actinium	Ac	227	227.027751	α, β^-, γ	21.773 yr
90	Thorium	Th	228	228.02873	α, γ	1.9131 yr
			232	232.038054	100, α, γ	1.41×10^{10} yr
91	Protactinium	Pa	231	231.035881	α, γ	3.28×10^4 yr
92	Uranium	U	232	232.03714	α, γ	72 yr
			233	233.039629	α, γ	1.592×10^5 yr
			235	235.043925	0.72; α, γ	7.038×10^8 yr
			236	236.045563	α, γ	2.342×10^7 yr
			238	238.050786	99.275; α, γ	4.468×10^9 yr
			239	239.054291	β^-, γ	23.5 min
93	Neptunium	Np	239	239.052932	β^-, γ	2.35 days
94	Plutonium	Pu	239	239.052158	α, γ	2.41×10^4 yr
95	Americium	Am	243	243.061374	α, γ	7.37×10^3 yr
96	Curium	Cm	245	245.065487	α, γ	8.5×10^3 yr
97	Berkelium	Bk	247	247.07003	α, γ	1.4×10^3 yr
98	Californium	Cf	249	249.074849	α, γ	351 yr
99	Einsteinium	Es	254	254.08802	α, γ, β^-	276 days

(1) Atomic number Z	(2) Element	(3) Symbol	(4) Mass number, A	(5) Atomic mass†	(6) Percent abundance, or decay mode if radioactive	(7) Half-life (if radioactive)
100	Fermium	Fm	253	253.08518	EC, α, γ	3.0 days
101	Mendelevium	Md	255	255.0911	EC, α	27 min
102	Nobelium	No	255	255.0943	EC, α	3.1 min
103	Lawrencium	Lr	257	257.0998	α	$\approx$35 s
104			261	261.1087	α	1.1 min
105			262	262.1138	α	0.7 min
106			263	263.1184	α	0.9 s
107			261	261	α	1–2 ms

Answers to Odd-Numbered Problems

Answers are rounded off to two or more significant figures (see Appendix A1).

Chapter 1

1. 450 m 27° S of W
3. 23 m 24° N of W
5. 9.8 m 2° S of E
7. 200 N 59° above $+x$ axis
9. (a) 200 N 59° above $+x$ axis,
   (b) 200 N 59° above $+x$ axis,
   (c) 250 N 6.8° above $+x$ axis,
   (d) 250 N 6.8° below $-x$ axis
13. 9.8 km W
15. 160 N 37° S of E
17. 0, 15 km; -8.5 km, -18 km
19. -71 N, 71 N; 120 N, 160 N
21. 3.5 m, 2.0 m; 3.1 m, -0.8 m; 3.3 m, -1.5 m
23. 460 N 24° below either the $+x$ or $-x$ axis
25. 8.1 km 30° above $-x$ axis, 36 km 56° below $+x$ axis, 10 km along the $+y$ axis, 20 km along the $-x$ axis
27. 450 m 27° S of W
29. 23 m 24° N of W
31. 200 N 59° above $+x$ axis
33. 23 km 80° N of E
35. (a) 160 N 79° below $-x$ axis,
    (b) 310 N 31° above $-x$ axis,
    (c) 310 N 31° below $+x$ axis
37. 6.3 km 72° N of W
39. 91°

Chapter 2

1. 2.1×10^8 yr
3. 13 m
5. (a) 230 s, (b) 6.9 m/s
7. (a) 82 km/h, (b) 61 km/h 61° S of E
9. (a) 180 s, (b) 2.8 m/s
11. (a) 0.060 km/min,
    (b) -0.140 km/min,
    (c) -0.004 km/min, (d) 7.9 min after it is at the elevation 1.1 km

13. (a) 29 m/s, (b) 11 m/s,
    (c) 62 mi/h, (d) 67 mi/h
15. (a) 1.6×10^4 cm², (b) 17 ft²
17. (a) 0.033 m/s, (b) 0.074 mi/h
19. 6.5 m/s
21. -200 m/s² opposite initial velocity
23. (a) 10.8 m/s², 17.8 m/s², 22.8 m/s², 24.6 m/s², 22.4 m/s², 17.8 m/s², 18.4 m/s², 14.2 m/s²,
    (b) 18.6 m/s²
25. (a) 1.9 m/s, (b) 9.6 m/s,
    (c) 3.9 m/s²
27. 1.2 m/s², 26.8 m/s
29. 55 m/s², 24 m
31. (a) 22 m/s, (b) -2.4 m/s²
33. (a) 0.7 to 0.8 s, (b) -19 to -20 ft/s², (c) ~130 ft
35. (a) 2.85, (b) 136 lb
37. 400 m
39. 120 m
41. 3.8 m/s
43. 5.3 m/s
45. (a) 3.6 s, (b) 29 m/s down
47. 5.0 m
49. 14 m/s
51. 77 m/s
53. (a) 0.089 s, 4.1 s, (b) 2.4 m,
    (c) 110 m, yes
57. (a) ~10 m/s², (b) ~6 m/s²,
    (c) ~3 m/s², (d) ~1 m/s²
59. (a) ~5 m, (b) ~11 m,
    (c) ~14 m, (d) ~5 m, 16 m, 30 m
61. ~30 m/s

Interlude I

1. 4.3 s
3. 7800 N
5. 59 m
7. 8.6 in

Chapter 3

1. 1000 kg
3. (a) 2.3 m/s², (b) 0.20 N
5. 48 N

7. 10,000 N
9. 580 N
11. (a) 37 N, (b) 4.3×10^4 kg
13. (a) 0.094 slug, (b) 1.4 kg,
    (c) 13 N
21. 50 N
23. 2.9×10^4 N up
25. (c) 8.8×10^4 N, (d) 6.9×10^4 N
27. 6.5 m/s²
29. (a) 5.4 s, (b) 11 m/s
31. 2.5
33. 1.2 m/s², 6.9 N
35. 5.3 s
37. 1350 N, 1690 N
39. 3.3 m/s², 65 N
41. 0.65 m/s², 105 N
43. 120 lb
45. 0.84
47. (a) 340 N, (b) 2.0 m/s²
49. 80 N, 0.41
51. (c) 3.5 m/s²
53. 1.4 m/s²
57. 44 percent increase

Chapter 4

1. (a) 6.7 m/s, (b) 2700 m/s
5. (a) 11 kg·m/s, (b) 2.8 kg·m/s
7. (a) 7.5×10^5 N, (b) 2.5×10^5 N
9. 0.28 N
11. (a) 0.080 s, (b) 0.64 m
13. (a) 270 N, (b) 0.040 m/s
15. (a) 7500 N, (c) 0.17 s
17. 0.40 m
19. (a) 0.060 s, (b) 0.45 m
21. (a) -130 kg·m/s, (b) 130 N
23. (b) 1.9×10^4 N, (c) 7.5×10^4 N
25. 8.3 m/s 15° N of E
27. (a) ~150 N·s, (b) ~16 m/s
29. 1.9 m/s opposite cannonball's velocity
31. (a) 0.56 m/s, (b) 18 s
33. ~1 or 2×10^{-14} m/s

35. 1.3 m/s east
37. 8 m/s
39. (a) 12.8 mi/h west, (b) 2260 lb, (c) 1180 lb
41. 13 m/s 42° S of E
43. 12 m/s

Chapter 5

1. 0.87 rad, 2.79 rad, 270°, 23°, 13 rad, 1080°
3. 1600 m
7. 9.4×10^5 rev
9. (a) 1.99×10^{-7} rad/s, (b) 3.0×10^4 m/s
11. (a) 4300 m, (b) 7.1 m/s, (c) 10 rad/s
13. (a) 6.60×10^{15} rev, (b) 2.20×10^6 m/s
15. 0.28 rad/s, 140 m/s
17. (a) 1.6×10^5 m/s, (b) 9.9×10^5 m/s², 1.0×10^5 g
21. 5.3×10^3 N
23. 57 km/h
25. 1.50 rad/s
27. 8.9 m/s
29. (b) 1600 N
31. (b) 80 N, (c) 630 N, (d) 83°
33. 26 m/s
35. 22 percent increase in speed
37. 4.9×10^{26} kg
39. (a) 77.6 kg, (b) 820 N
41. 2.67×10^7 m
43. 1.41 h
45. 200 N
47. (a) r doubles, (b) v increases by 1.41
49. (c) 322 N
51. 0.313 rad/s

Chapter 6

1. (c) −4 N, −32 N, 32.2 N 83° below −x axis
3. (c) 43 N 9.9° above +x axis
5. 120 N, 89 N, 57 N
7. 76.6 N, 64.3 N, 7.8 kg
9. 208 N, 240 N
11. 3250 N 15° above −x axis
13. 1000 N, 500 N, 250 N, 500 N

15. 8.5×10^4 N
17. will break
19. 0, −184 N·m, 240 N·m, −240 N·m
21. −14 N·m, +17 N·m, −3 N·m
23. (a) −153 N·m, −150 N·m, (b) 1.52 m
25. (a) −470 N·m, (b) 250 N
27. 0.33 m from center on side with 10-kg mass
29. 1.7 m
31. 700 N, 5.7 m from left cable
33. 233 N
35. 220 lb, 200 lb
37. 1570 N, 1620 N 15° above horizontal
39. 200 N, 173 N toward right
41. 2600 N
43. 560 N, 250 N
45. 736 N
47. 0.77 m
49. 350 N, 310 N 14° above beam
51. 1960 N, 2060 N 18° above horizontal
53. 1680 N

Chapter 7

1. (a) −0.058 rad/s², (b) 0.99 s
3. 200 m/s², 133 rad/s²
5. (a) 37 m/s², (b) 11 rad/s, (c) 37 rad/s²
7. (a) 0.25 m/s², (b) 0.83 rad/s², (c) 40 s
9. (a) −0.20 rad/s², (b) 30 rad, (c) 9.0 m
11. 1.15×10^{-14} rad/s²
15. 0.040 N
17. (a) 2.5 rad/s², (b) 1.5 s
19. 8.0 kg·m²
21. 4.0 kg·m²
23. 0.71 m from pivot point
25. 2000 N·m
27. 205 N·m
29. 5.9×10^4 N·m
31. 3.5×10^{19} N
33. (d) 17 N, 4.2 m/s²
37. 2.7 kg·m²
39. 12.5 rad/s, 2.5 m/s

41. 1.96 rev/s
43. $\sim 4 \times 10^{-20}$ rad/s, $\sim 5 \times 10^{-11}$ s

Chapter 8

1. (a) 2180 J, (b) 2400 J
3. (a) 4000 J, (b) −4000 J
5. 78 N, 0, −78 N
7. 25° above ground
9. (a) 90 J, (b) 90 J
11. (a) 0.5×10^5 J, (b) 1.5×10^5 J
13. 6.2 m/s
15. yes
17. (a) 1.44×10^5 J, (b) 0.64×10^5 J
19. (a) 6.8×10^5 J, (b) 1.0×10^7 J, (c) 2.0×10^8 J, (d) 80
21. 0.036 J
23. (a) 3.9×10^7 J, (b) 230
25. 3.8×10^6 J
27. (a) $−1.2 \times 10^5$ J, (b) 28 m/s
29. (a) 470 J, (b) 180 J, (c) 290 J
31. (a) 4.0×10^3 J, (b) 6.3 m/s
33. (a) 21.6 J, (b) −16.2 J, (c) −5.4 J
35. (a) −24 J, (b) 4.9 m/s
37. (a) 400 N/m, (b) 4.5 J
39. 16 m/s
41. 1.1×10^5 N/m
43. 1.3×10^4 N
45. (a) 12.5 m/s, (b) 4100 N
47. (b) decreases by 33 percent
49. 3.8 m/s
51. 380 m
53. (a) 1.3×10^5 J, (b) 3.8×10^3 m (2.4 mi)
55. 8.7 m/s
57. 6.4 m/s
59. 0.40
61. 1.6 m/s
63. (a) 0.63 m/s, (b) 0.22 m
65. (a) 0.80 m/s, (b) 0.033 m
67. 1.2 m/s west and 3.2 m/s west
69. 3.9×10^5 W
71. 5.9×10^3 W
73. (a) 4.3×10^5 J, (b) 60 W, (c) 600 W
75. (a) 1.9×10^4 W, (b) 3.9×10^4 W
77. (a) 1.14×10^7 J, (b) 240 g, (c) 210 g

79. (a) 5.5×10^4 kcal, (b) 16 lb

Chapter 9

1. (a) 41.7°C, (b) 103°F
3. (a) 1.9°F, (b) −88°C = 185 K
5. −321°F, 77 K
7. 1.5×10^6 J
9. (a) 2.1×10^4 J, (b) 1.2×10^4 J, (c) 0.23×10^4 J
11. 0.21 m
13. 1.1×10^4 s
15. 14 C°
17. 170 C°
19. (a) 2.5×10^{16} J, (b) 1.1×10^4 C°
21. 38.9°C
23. 6.1 C°
25. 0.078 kg
27. 2.1×10^5 J
29. 0.36 kg
31. 11 g
33. 1.6×10^4 s = 4.3 h
35. (a) 2.3×10^5 J, (b) 0.85×10^5 J
37. 90°C
39. 2.8×10^3 s = 47 min
41. 1.3×10^{-4} kg/s
43. (a) 35 mm Hg, (b) -2.7×10^3 J, (c) −6.5 C°
45. 0.81 m
47. 84.980 m
49. 0.71 m
51. 5750 cm³
53. (a) 7.6×10^{14} m³, (b) 2 m

Chapter 10

1. (a) −470 W, (b) -4.1×10^7 J
3. 173 W
5. 3.7 W
7. (a) -2.7×10^6 J, (b) 8.1 kg
9. −1300 W
11. −370 W
13. (a) −270 W, (b) −130 W
15. 430 W
17. −63 W
19. (a) 297 K, (b) 277 K
21. 28.6 m²
23. 1.026

27. −16 W
29. (a) 8.3×10^{-3} g/s, (b) 60 g
31. (a) −310 W, (b) 1.6×10^{-4} m
33. 84 W
35. (a) −118 W, (b) 214 W
37. (a) −140 W, (b) −116 W, (c) 540 W
39. (a) −29 W, (b) −8700 J, (c) 41°C
41. +5.8 kW
43. 0.55 kW

Chapter 11

1. (b) 1, 5, 10, 10, 5, 1, (c) 0, 1.6 k, 2.3 k, . . .
3. 8.75
5. 0.40
7. (a) (9, 0, 0), (5, 2, 2), . . . , (b) 1680, (c) 1.33
9. (d) two times
11. (a) −609 J/K, (b) 3020 J/K
13. (a) 78 J/K, (b) 1220 J/K, (c) 1290 J/K, (d) 6050 J/K, (e) 100 J/K, (f) 8740 J/K
15. -1.3×10^4 J/K
17. (a) 2.6°C, (b) +8 J/K
19. 1.9×10^4 J/K
21. (a) 15.0°C, (b) 1.6 J/K
23. 1.04 J/K
25. (a) 0.047, (b) 2.1×10^8 J/s
27. 893 K or 620°C, 7.8×10^7 J
29. 0.28
31. (a) 0.65, (b) 1.54×10^9 J/s, (c) 0.54×10^9 J/s, (d) 0.065 C°
33. (a) 8.9, (b) 8900 J, (c) 2500 J
35. (a) 7.0, (b) 7.3×10^4 J, (c) 1.0×10^4 J, (d) 8.3×10^4 J

Chapter 12

1. about eleven times longer
3. (a) 5×10^{-8} m, (b) ∼17 molecules
5. (a) 1300 lb, (b) 220 lb
7. 0.098 m²
9. 4.7 cm
11. 0.91×10^5 N/m², -0.10×10^5 N/m²
13. 2.1×10^5 N up

15. 0.72 cm³
17. 12 liters
19. 0.080 m
21. 0.87 m
23. (a) 9.3×10^{24} molecules/m³, (b) 2.5×10^{25} molecules/m³
25. 18.9
27. 1.9 atm
29. 840 moles
31. 1.8 moles
33. 510 m/s, 0.88 m/s, 5.3×10^{-10} m/s
35. (a) 1.61×10^5 K, (b) 1.00×10^4 K
39. 0.44
41. 0.64
43. (a) 4.8 N, (b) 1.6 N/m²
45. (a) −168 J, (b) +168 J
47. 2000 J, gas expands, T increases
49. (a) 0, (b) +5000 J
51. 9.4×10^3 J
53. (a) −9.0 J, (b) 0, (c) +6.0 J
55. −9 J, 0, 6 J; 19 J, −9 J, −7 J
57. 1.0×10^6 N/m², water flows toward salt
59. (a) 6.4 moles/m³, (b) 9000 N/m²
61. (a) 610 moles/m³, (b) ∼10⁻³ mole, (c) ∼0.2 g

Chapter 13

1. 5400 N, 1200 lb
3. 4×10^4 lb, 20 tons, 2×10^5 N
5. 4.7×10^7 N/m²
7. 630 kg
9. 5500 kg/m³
11. 7.8×10^3 m
13. 5.1×10^5 N/m²
15. 0.14 m
17. 150 N pushing out
21. 1140 kg/m³
23. 0.022 m²
25. (a) 10 N, (b) 1.6×10^{-6} N
27. 4.8×10^3 N
29. 110 N
31. (a) 37.9 N, (b) 38.4 N
33. 172 N
35. (a) 1.5 N, (b) 2.9 m/s²
37. 1960 N

39. 0.0156 m^3
41. $\rho_f = (mg - T)\rho/mg$
43. 4.2×10^{-2} N/m
45. 29 stitches
47. 1.05×10^{-4} kg
49. (a) 5.5×10^{-4} N,
    (b) $(4.4 \times 10^{-2}$ N/m$)h$, (c) 2.7 cm
51. 9.5×10^{-5} m, 6.1 cm
53. (a) 1.5×10^{-4} m,
    (b) 3.0×10^{-5} m,
    (c) 7.4×10^{-6} m,
    (d) 1.5×10^{-6} m,
    (e) 3.0×10^{-7} m
55. 3.0 cm

Chapter 14

1. (a) 140 cm^3/s, (b) 46 cm/s
3. 0.060 m
5. 6.6×10^{-3} cm/s
7. (a) 0.020 m^3/s,
   (b) 1.7×10^5 N/m^2
9. (a) 4.8 m/s, (b) 5.3 m/s,
   (c) 4.95×10^5 N/m^2
11. (a) 406 N/m^2, (b) 1070 N pushing out
13. 46 m/s
15. 6.6×10^{-3} m
17. 9.81×10^4 N/m^2
19. 0.50 cm
21. (a) 2.0×10^3 N/m^2, (b) 5.9 cm
23. 3690 N/m^2
25. (a) decreases by 19 percent,
    (b) increases by 23 percent
27. (a) 1.9 N, (b) 121 N
31. 1040, laminar
33. turbulent
35. (a) 630 N, (b) 1050 N
37. 1.9×10^{-12} N
39. 9.5×10^{-7} m
41. 1.3 m/s
43. (a) 6.5×10^{-3} N,
    (b) 6.5×10^{-3} N,
    (c) 6.7×10^{-4} kg

Chapter 15

1. (a) 180 N, (b) 0.16 m
3. 0.21 m
5. (a) 1.5×10^5 N/m^2, (b) 0.12 m

7. (a) 70 cm, (b) 7.8×10^{-7} m^2,
   5.0×10^{-4} m
9. 3.8×10^{-4} m^2
11. 2.5 cm
13. 1.1×10^{-5} m^2
17. (a) $\Delta L/2$, (b) $2F$, (c) $\Delta L/4$
19. 3.1 cm
21. (a) 1.5×10^5 N, (b) will break
25. 1.7×10^{-4} m
27. $\sim 3 \times 10^9$ m^2
29. 6.4×10^{-7} m
31. (a) -3.2×10^{-6} m^3,
    (b) -6.6×10^{-6} m^3,
    (c) -2.5×10^{-4} m^3,
    (d) -5.6×10^{-4} m^3
33. -3.3×10^{-5} m^3 or -33 cm^3
35. 9 percent more dense
37. 8.0
39. 0.8 J

Interlude III

1. 80 kg
3. (a) 116 nails, (b) 750 N, (c) the mother
7. (a) 1240 lb/in^2, (b) 29 ft
9. 40 N/cm^2

Chapter 16

1. (a) 2.3×10^{-3} s,
   (b) 4.74×10^{13} Hz
3. (a) 20 Hz, (b) 5.0×10^{-5} s
5. 20 N/cm
9. (a) 0.10 m, (b) 14.8 m/s,
   (c) 12.8 m/s
11. (a) 11.3 J, (b) 4.0 m/s,
    (c) 0.052 m, (d) 6.9 m/s
13. double A
15. (a) 0.25, 0.75, (b) 0.71 A
17. 120 m/s
19. 1.2 N/m
21. (a) 22 J, (b) 4300 N/m,
    (c) 9.5 Hz
23. (a) 0.17 s, (b) 4.2 Hz, (c) 8.5 Hz
25. (a) 660 N/m, (b) 3.5×10^{-12} m
27. (a) 2×10^8 N/m, (b) 2×10^{-2} m
29. 32
31. 3.9 Hz
33. 28 m/s

35. (a) 0.25 m
37. 9.75 m/s^2
39. ± 20 percent
41. (a) 5000 J, (b) 4.5 m/s
43. (a) $-\pi \sqrt{L/g}$ ($\Delta g/g$),
    (b) $-\Delta g/2g$, (c) 0.1 percent
45. (a) 0.20 s, (c) $+10$ cm, (d) 0

Chapter 17

1. 17 m, 0.017 m
3. (a) 2.1×10^9 N/m^2, (b) 2980 m/s
5. (b) 820 m
7. 8.1×10^{10} N/m^2, 4.9×10^{10} N/m^2
11. 0.51
13. 7.1 m, out of phase
17. (a) -0.17, (b) $+0.17$
19. (a) 0.017, (b) 0.78
21. 0.0013
27. 47 m, . . .
29. 57 Hz
31. 445 Hz
33. (a) 266 Hz, (b) 236 Hz,
    (c) 265 Hz, (d) 235 Hz
35. 438 Hz, 389 Hz
37. (a) 4.28×10^5 Hz,
    (b) 4.65×10^5 Hz
39. (a) 4.3 m/s, (b) 4.2 m/s
43. 807 nm

Interlude IV

1. $Cg^{1/2}L^{-1/2}$
3. $CL^{-1/2}m^{-1/2}T^{1/2}$

Chapter 18

1. (a) 541 m/s, (b) 784 Hz, . . .
3. 196 Hz
5. (a) 0.066 m, (b) 0.083 m
7. 2.2 Hz
9. 2.25
11. 1.009
13. (a) 425 Hz, 850 Hz, 1275 Hz,
    (b) 212 Hz, 638 Hz, 1063 Hz
15. (a) 350 Hz, (b) 0.12 m
17. (a) 20 Hz, (b) 4.3 m
19. ~ 1000 Hz
21. $CL^{-1}\rho^{-1/2}B^{1/2}$
29. 710 m

31. 40 dB, 100 dB
33. (a) 9 dB, (b) 19 dB, (c) 29 dB
35. 35 dB
37. The intensity can vary by a factor of 2.
39. 86 dB
41. $8.8 \times 10^{-4}\,\text{W/m}^2$, 89 dB
43. $3.5 \times 10^8\,\text{s}$ or 11 yr

Chapter 19

3. 9.7 cm
5. (a) 1.43, (b) 1.30
7. 0.47 m
9. n_2/n_1
11. (a) 45.6°, (b) 1.4
13. (a) 6050 m/s
15. 1.41
17. 20° below horizontal
19. 3.0 m
21. 24.4°
23. 1.48
25. 52°
27. 1.41
29. 90°
31. (a) 29.6°, (b) 30.4°, (c) 56.0°, (d) 49.8°

Chapter 20

5. (c) −30 cm, (d) −6 cm
7. 8.6 cm
9. $(\frac{1}{15})$ s
11. (a) 4.65 m, (b) 8.14 cm
13. (a) 12.4 cm, (b) 372 cm
15. 25 cm
17. −0.16 m
21. (a) 2.10 cm, (b) 2.09 cm, (c) 1.94 cm
23. (a) −150 cm, (b) 33 cm
25. (a) −0.33 diopters, (b) 2.3 diopters
27. 28.6 cm
29. (a) 4.3 cm, (b) 7.0
31. (a) 5.0 cm, (b) 4.9 cm, (c) 4.2 cm, (d) 5.0, 5.1, 5.9
33. (a) 20 cm right of second lens, (b) upright, (c) real
35. (b) ∼2

37. (a) 3.4 cm right of second lens, (b) 2.6 mm tall
39. (a) 82 cm left of diverging lens, (b) 0.20, (c) 5
43. (a) 29 cm left of second lens, (b) 0.015 m
45. (a) 3.60 cm, (b) 5.0
47. (a) 0.515 cm left of objective lens, (b) −280
49. (a) 8.1 cm, (b) 20
51. 1.025 cm left of objective lens, 44.8 cm lens separation
53. (a) −18 cm, upright, virtual, (b) −13 cm, upright, virtual, (c) −6.7 cm, upright, virtual
55. 150 m converging mirror
57. (a) −0.86 m, (b) 0.49 m
59. 30 cm converging mirror
61. (a) −14 m, (b) 0.74

Chapter 21

5. (a) 0.00°, 0.062°, 0.12°, (b) 0.00°, 0.12°, 0.25°
7. (a) 0.28 mm
9. (b) 1.4 m, 2.7 m
11. ∼1.8 cm
13. (a) 25°, (b) 0.73 m
15. 3.3×10^3 lines/cm
17. 450 nm
19. 0.30 m, 1130 Hz
21. 1.5
23. (a) $6.57 \times 10^{-7}\,\text{m}$, $4.56 \times 10^{14}\,\text{Hz}$, (b) $9.93 \times 10^{-7}\,\text{m}$, $3.02 \times 10^{14}\,\text{Hz}$, (c) $1.17 \times 10^8\,\text{m/s}$, away from earth
25. 130 nm, 390 nm
27. 118 nm
29. (b) 320 nm
31. (a) 1.8°, (b) 0.18°, (c) 0.018°
33. 23 m
35. $3.7 \times 10^{-4}\,\text{m}$
37. (a) 2.4 m, (b) 0.48 m, (c) 0.048 m
41. 38.9°
43. (a) 0.125, (b) 0.188

Interlude V

1. $1.8 \times 10^{-6}\,\text{m}$
3. (b) 90°, ∼30°
5. 90°

Chapter 22

1. (a) $-9.6 \times 10^{-14}\,\text{C}$, (b) 9.9×10^7, $-1.6 \times 10^{-11}\,\text{C}$
3. $6.07 \times 10^{-7}\,\text{kg}$
5. 58 N, repulsive
7. (a) $-1.5 \times 10^{-3}\,\text{C}$, (b) $8.5 \times 10^{-15}\,\text{kg}$
9. 1.8×10^{-11}
11. 6.5 N
13. (a) $8.2 \times 10^{-8}\,\text{N}$, (b) $9.0 \times 10^{22}\,\text{m/s}^2$, $2.2 \times 10^6\,\text{m/s}$
15. $8.83 \times 10^8\,\text{m}$
17. 1600 N left
19. between the charges, 41.4 m from the 1.0-C charge
21. $2.5 \times 10^9\,\text{N}$, 44.7° above the $+x$ axis
23. 23 N left
25. 43 N away from charge on opposite corner
27. $4.5 \times 10^{-11}\,\text{N}$
29. $6.4 \times 10^{-4}\,\text{C}$
31. $2.2 \times 10^{-15}\,\text{N}$ up
33. $1.7 \times 10^{-16}\,\text{N}$ left
35. $1.0 \times 10^{-2}\,\text{C}$
37. (a) $3.9 \times 10^{17}\,\text{N/C}$, (b) $6.3 \times 10^{-2}\,\text{N}$
39. 5.0 N/C right
41. $8.2 \times 10^4\,\text{N/C}$ down
43. (a) E = 0, (b) $2.0 \times 10^7\,\text{N/C}$ up
45. $3.0 \times 10^{-27}\,\text{N·m}$
47. $6.8 \times 10^{-36}\,\text{C·m}$

Chapter 23

1. (a) $8.0 \times 10^{-20}\,\text{J}$, (b) $1.0 \times 10^{-21}\,\text{J}$
3. $4.3 \times 10^5\,\text{J}$
5. (a) $-2.3 \times 10^{-14}\,\text{J}$, (b) $3.7 \times 10^6\,\text{m/s}$
7. (a) $4.3 \times 10^{-6}\,\text{N}$ toward plus charge, (b) $3.6 \times 10^{-6}\,\text{N}$ toward plus charge, (c) $3.95 \times 10^{-10}\,\text{J}$, (d) $3.93 \times 10^{-10}\,\text{J}$
9. (a) $+2.5 \times 10^{-15}\,\text{J}$, (b) $-2.5 \times 10^{-15}\,\text{J}$
11. $1.5 \times 10^{11}\,\text{J}$
13. $1.9 \times 10^{10}\,\text{J}$
15. $6.4 \times 10^{-19}\,\text{J}$
17. $1.2 \times 10^{-14}\,\text{m}$

19. 5.5×10^{-3} J
21. 7.3 m/s
23. 20.4 m/s
25. 1.5×10^{-11} C, 9.4×10^7 electrons
27. 2.2×10^5 J
29. 8.8×10^7 m/s
31. 1.41
33. -3.3×10^4 V
35. (a) 3.1×10^6 m/s,
(b) 9.3×10^6 kg·m/s,
(c) 2.3×10^3 m/s
37. (a) 7.5×10^6 J, (b) 6.3×10^3 s
39. (a) 0 V, (b) 7.5×10^4 V
41. 6000 V
43. (a) 2.4×10^4 V/m,
(b) 4.0×10^{-5} m

Chapter 24

1. (a) 3.1×10^{18} electrons/s,
(b) 1.1×10^{22} electrons/h
3. 3.4×10^{22} electrons
5. (a) 3.1×10^{19} electrons,
(b) 9.4×10^{-10} m^3
7. 1.2 mA, 24 mA
9. 1.2 Ω
11. (a) 38 Ω, (b) 0.040 A
13. 2.0 m
15. 1.96
17. (a) 2.0, (b) 0.25, (c) 0.61
19. 10 mV
21. 0.0175 Ω
23. (a) 30.0°C, (b) 3700 Ω
25. 81°C
27. 200 Ω, 0.55 A
29. 2.0×10^4 h
31. 6×10^{10} W
33. (a) 480 W, (b) 0.77
35. 4.2×10^{-5} m^2
37. (a) 0.50, (b) 4.0, (c) 1.6
39. (a) 9×10^{11} kg, (b) 2×10^{19} J
41. 23 A, 1.0 Ω

Chapter 25

1. (b) 0.33 A, 0.67 A, 1.00 A
3. 24 V, 1.6 A, 0.4 A
5. (b) 0.20 A, (d) 0
7. (a) 27 V, (b) -25 V

9. 34 Ω
11. fuse (7.50 A, 56 W), bulb
(0.47 A, 53 W), crock
(1.41 A, 159 W), and toaster
(5.63 A, 634 W)
13. (b) 0.149 A cw, 0.023 A cw, 0.127 A
down, (c) -2.5 V
15. (a) 0.10 A, (b) 50 Ω
17. $\varepsilon = \varepsilon_p/(1 + R_1/R_2)$
19. 0.061 A
21. 300 Ω
23. (a) 60 Ω, (b) 2.0 A, (c) 0.67 A,
1.33 A
25. 56 Ω
27. 15 Ω
29. (a) 23.3 Ω, (b) 4.29 A,
(c) 428 W, (d) 0.95 A, 1.43 A,
1.90 A
31. 30 V, 15 Ω
33. 13.3 W, 6.7 W
35. 135 W
37. (a) 0.50 A, (b) 0.45 A
39. (a) 80 V, (b) 70.6 V

Chapter 26

1. (a) 6.6 m^2, (b) 6.6×10^{-4} m^2
3. (a) 71 pF, (b) 0.071 mm
5. 52 V
7. (a) 880 pF, (b) 0.25 cm,
(c) 0.53 cm
9. on the order of 100 to 1000 V
11. (b) -67 V, (c) -2.2×10^4 V/m
13. 3.0×10^{-5} C
15. (a) 2×10^{-4} C,
(b) 0.67×10^{-4} C, 1.33×10^{-4} C,
(c) 33 V
17. 0.33 m
19. 1.5×10^8 V
21. 6.0×10^4 V
23. (a) 0.16 μF, (b) 1.6×10^{-3} C
25. (a) 360 J, (b) 0.12 C
29. (a) 0.032 J, (b) 0.010 mm
31. (a) 5.0, (b) 3.8×10^{-10} J,
(c) 1.9×10^{-9} J
33. (a) 5.0×10^5 Ω, (b) 4.0 s
35. (a) 4.5 s, (b) 0.33 C
37. 1800 Ω
41. (a) 38 μF, 4.6×10^{-3} C,
(b) 4.1 μF, 4.9×10^{-4} C

43. 2.9×10^{-10} J
45. 0.89 J

Chapter 27

1. (a) 0.10 N in positive x direction,
(b) 0, (c) 0.10 N, (d) 0.071 N
3. (a) 0, (b) 0, (c) 3.2×10^{-3} N
down, (d) 1.6×10^{-3} N up,
(e) 1.9×10^{-3} N down
5. 1.4×10^{12} m/s (unattainable speed
according to relativity—see
Chapter 30)
7. (a) 3.2×10^{-21} N, (b) 0.50 m/s
9. (a) 1.5×10^{-13} N up,
(b) 9.6×10^5 V/m up
11. 0.62 V
15. (a) 4.8×10^{-14} N,
(b) 8.6×10^{-27} kg
17. 5.9 cm
19. (a) 9.9×10^7 m/s, (b) 16 m,
(c) down
23. (a) 0.11 N·m clockwise, (b) 0,
(c) 0.086 N·m counterclockwise
25. 3.4×10^{-3} N west
27. (a) 0.0045 N out of paper, 0.045 N
up, 0.0045 N into paper, (b) 0,
7.8×10^{-4} N·m
29. ~20°
31. 54 N·m
33. 1.0×10^{-5} T north
35. (a) 5.0×10^5 A
37. (a) 2.0 A, (b) 5.0×10^{-3} T
39. (a) 1.2×10^{-3} T,
(b) 5.8×10^{-15} N,
(c) 5.8×10^{-13} N

Chapter 28

1. (a) 0.096 T·m^2, (b) 0.096 T·m^2,
(c) 0
3. (a) 3.2×10^{-5} T·m^2, (b) 0
5. 0.59 V
7. 1.08 V
9. 0.081 V
11. (a) 0.078 T, (b) lead b
13. (a) 1.2×10^{-3} V
15. (a) 3.3×10^{-6} T, (c) 0.12 V
17. 27 V
19. 2.1 Hz
21. (b) 40 V, 0 V, -40 V, 0 V

23. (a) 6.7×10^{-3} H, (b) 0.16 V
27. 0.48 H
31. (a) 27 V, (b) 40 V
33. 0.052 H
35. (a) 0.032 H, (b) 0 V
37. 100
39. (a) 1500, (b) 16 V
41. (a) 170 A, (b) 1000,
    (c) 1.7×10^5 A

Chapter 29

1. 3.7×10^{-8} s
3. (b) 0, 30 V, 0, -30 V, 0
5. (a) 600 Ω, (c) -0.24 A
7. (a) 310 V, (b) 21 A, 15 A,
   (c) 3200 W
9. (a) 0.012 A, (b) 0.017 A
11. (a) 8.1 A, (b) 11.4 A, (c) 120 V,
    (d) 170 V
13. 0.7×10^{-5} A, $\sim 1 \times 10^{-10}$ m
15. (a) 550 mA, (b) 55 mA,
    (c) 5.5 mA, (d) 0.055 mA
17. (a) 0.080 H, (b) 0.20 A
19. (a) 3.2×10^{-4} H, (b) 7.7
21. (a) 4.5×10^{-8} A,
    (b) 4.5×10^{-6} A,
    (c) 4.5×10^{-4} A, (d) 4.5×10^{-2} A
23. (a) 2.1×10^{-2} A,
    (b) 9.4×10^{-2} A
25. (a) 603 Ω, (b) 884 Ω, (c) 345 Ω,
    (d) 0.35 A, (e) $-55°$
27. 80 Ω, 0.19 H, 36°
29. 9.5×10^{-3} H, 3.0 Ω
31. (a) 44 Ω, (b) 3.3×10^4 Ω
35. (a) 0.062 A, (b) 58°, (c) 0.32 W
37. (a) 27 W, (b) 59 μF, (c) 580 W
39. 20 Ω, 0.047 H
41. (a) 27 MHz, (b) 1.1×10^{-5} A,
    (c) 0.43 V
43. (a) 333 Hz, (b) 0.079
45. (a) 7.6 μF, (b) 2.0 A, (c) 0.36 A,
    (d) 0.18

Interlude VI

3. (a) 0.28 W at 10 Ω, 0.47 W at
   30 Ω, . . .

Chapter 30

1. 1.3×10^{-10} s

3. 2.5×10^8 m/s
5. (a) 2.2×10^4 s, (b) yes
7. 2.9×10^8 m/s
9. $(1 - 2.5 \times 10^{-17})^{0.5}$ c
13. 6.8 in
15. 2.6×10^8 m/s
17. (a) 10 s, (b) 4.4 s, (c) 390 m,
    (d) yes
19. (a) 1.00000, (b) 1.00005,
    (c) 1.005, (d) 1.06, (e) 1.34,
    (f) 3.9
21. 4.2×10^7 m/s, 1.25×10^8 m/s,
    and 2.6×10^8 m/s
23. (a) 2.85×10^8 m/s, (b) 3160 kg
25. (a) 1.0×10^4 kg/m^3, (b) 10 kg,
    0.10 m, 1.0×10^{-4} m^2, and
    1.0×10^6 kg/m^3
27. 1.2×10^{10} J
29. 9.8 percent
31. (a) 5.6×10^{10} m/s,
    (b) $\simeq 3 \times 10^8$ m/s
33. (a) 7.3, (b) 2.5
37. (a) 2.0×10^{-35} kg,
    (b) 4.2×10^{-10}
39. 7.4×10^{-5} kg
41. (a) 3.3×10^{-8} kg/s, (b) 1.1 kg/yr
43. (a) 1.1×10^{-11} kg,
    (b) -1.5×10^{-11} kg

Chapter 31

1. (a) 1.2×10^{19} Hz, (b) 0.122 m
3. 0.75
5. 1.2×10^{15} Hz, 250 nm
7. 9.5×10^{15} photons/s
9. 1.4×10^{19} photons/s
11. P(eye) $\simeq 6 \times 10^{-17}$ J/s
13. 2.7×10^5 photons
15. (a) 4.6×10^{14} Hz, (b) 1.2 eV
17. 2.31 eV
19. (a) 9.1×10^{-22} W, (b) 390 s
21. 3.6×10^8
23. 0.026 nm
25. 2.4×10^{-7} N
27. (a) 2.1×10^{11} W, (b) $\sim 10^6$ C°
29. 730 nm, 0.40 nm
31. 8.3×10^{-12} m
33. 5.5×10^{-4}
35. 2.9×10^7 m/s
37. (a) $\sim 10^{-35}$ m, (b) 10 to 100 J·s

39. (a) 1.0 m/s, (b) 2.0 m/s
41. 7.9×10^{-7} m
43. 1.8×10^{-34} m
45. (a) 1.1×10^{-20} kg·m/s,
    (b) 2.0×10^7 eV
49. (a) 2.2×10^{-11} J,
    (b) 1.3×10^{-18} J,
    (c) 1.5×10^{-35} kg
51. 5.5×10^{-12} m

Chapter 32

1. 1200 m
3. 565
5. (a) 1.9×10^{-4} kg, (b) 5.9×10^{20}
   atoms, (c) 1.1×10^{-28} m^2,
   (d) 6.7×10^{-4}
7. 122 nm, 103 nm, 97 nm
9. 434 nm, 6.91×10^{14} Hz, 2.86 eV
11. (a) 5.97×10^{-14} J,
    (b) 4.47×10^{-14} J
13. (a) 0.27×10^{-10} m, -54.4 eV;
    1.06×10^{-10} m, -13.6 eV;
    2.39×10^{-10} m, -6.0 eV
15. 5.8×10^{-13} m
17. 1.88 μm
19. 23.5 nm, 1.28×10^{16} Hz, 52.9 eV
23. (b) 64 nm, . . .
25. not normally
27. 2.4×10^6 times
31. 2.3×10^{39}
33. (a) 2.19×10^6 m/s,
    (b) 3.33×10^{-10} m
35. 5.0 m/s, 10.0 m/s

Chapter 33

15. 3.4×10^{-19} J = 2.1 eV
17. 370 nm
19. 1 J
21. (a) 4.2 cm, (b) 1.6×10^4 m
23. 9.7×10^{18} Hz, 3.1×10^{-11} m
25. $\sim -8 \times 10^4$ eV
27. (a) 1.2×10^{-29} m^3,
    (b) 2.2×10^{-29} m^3,
    (c) 1.4×10^{-29} m^3

Chapter 34

1. 4, 5, 9, . . .
3. (a) 4.8×10^{-15} m

5. (a) $10^{28} - 10^{29}$,
    (b) one-half the number in (a),
    (c) $10^{-10} - 10^{-9}$ cm^3

7. $\simeq 400$ N

9. $\simeq 1600$ MeV

11. 7.57 MeV, 8.50 MeV, tin is more stable

13. 2.8×10^{-9} m

17. 1.59 MeV

19. 13.005739 u

21. 214.99867 u

25. 1.47 MeV

27. 800 MeV, 1.3×10^{-10} W

29. (a) $\frac{1}{16}$, (b) $\frac{1}{256}$, (c) 9.2×10^{-11}

31. 0.55 mCi

33. 12 h

35. (a) 4.6×10^{14}, (b) 4.6×10^8 decays/s $= 0.012$ Ci, (c) 0.017 μg

37. 36.0 days

39. 4700 cm^3

41. 22,900 yr

43. 40,100 yr

45. (a) 2.51×10^{12}, (b) -9.64 s^{-1},

    (c) 12,000 yr

49. (a) 70 mrem, (c) 700 mrem

51. 660 mrem/yr

53. (a) 3.2×10^{-3} J,
    (b) 6.4×10^{-15} J, (c) 5.0×10^{11}

55. 1600

57. (a) 0.29 J/yr, (b) 0.58 rad

Interlude VII

1. (a) 2.9 percent/yr, (b) 7.5

3. 9.6×10^6 s^{-1}

5. (a) 18 yr, (b) 36 yr, (c) 60 yr,
    (d) 120 yr

7. (a) yes, (b) $\sim\$26,000$,
    (c) ~ 7 yr, (d) 10 percent/yr

Chapter 35

1. 199.6 MeV

3. (b) 87.9056 u

5. 8.2×10^{13} J/kg

7. (a) 5.0×10^{-5} MeV/molecule,

    (b) 4.0×10^6

9. (a) 6.3×10^{-15} m, 5.4×10^{-15} m,
    and $r \simeq 11.7 \times 10^{-15}$ m,
    (b) 4.0×10^{-11} J $= 250$ MeV

11. 2.5×10^6

13. (a) 400 kg, (b) 0.04 m^3

15. 11, 12, 24, . . .

17. 1.004

19. (a) 7.3×10^{13} J/kg,
    (b) 7.3×10^{22} J, (c) 730 yr

21. (a) 7.3 MeV, (b) 6.5×10^{-4}

23. (a) 43.2 MeV,
    (b) 3.45×10^{14} J/kg

25. 8.4×10^{37} reactions/s

27. (a) 7.8×10^{32} nuclei,
    (b) 7.9×10^8 m^3

29. end with the following "stable"
    particles:
    p, n, 2e$^-$, $2\bar{\nu}_\mu$, $2\nu_\mu$, $2\nu_e$, and 2γ

33. 938 MeV, 2.26×10^{23} Hz, and
    1.32×10^{-15} m

35. (a) 2.60×10^8 m/s,
    (b) 1.6×10^{-10} s, (c) 4.2 cm

39. 141 MeV

Index

Boldface type is used to indicate the pages on which a term, principle, or law is defined.

Aberrations, **436**–438
Abscissa, **771**
Absolute potential, **507**–508
Absolute pressure, **251**
Absolute temperature scale, 188
Absolute zero, 188
Absorbed dose, of radiation, **728**, 735
Absorption
    photon, 700–701, 706
    of radiation, 700–701, 706
    of wave energy, 359n
Absorption spectra, 678, **701**
ac. *See* Alternating current (ac)
Acceleration, 24–26, 44
    angular, **136**, 138–148
    average, **24**
    centripetal, **103**–105, 113
    constant, linear motion with, 26–33, 44
    graphical analysis of, 38–43, 44
    gravitational, **33**–35, 44, 109–110
    instantaneous, **26**
    linear, 148
    in Newton's second law of motion, 54–56
    tangential, **136**–138
    units of, 25, 56, 57
Accommodation, **425**
Actinides, 700
Action-reaction forces, 59–60, 87
Activity, of radioactive material, **722**, 728, 735
Addition, of vectors, 6–8, 11–15
Adhesion, **288**
Air, thermal expansion of, 201
Air columns, standing waves in, 380–383
Airplane wing, airflow behind, 294
Alpha decay, 717–718, 735
Alpha particles, **678**, 717, 735
Alternating current (ac), 518, **604**, 623
    through capacitors, 611–612, 623
    through coils, 608–611, 623
    through resistors, 605–606, 623
    root-mean-square values for, 606–608

Alternating-current circuits, 604–623
    phase angles in, 613–615
    power in, 618–621
    resonance in, 621–623
    RLC series, 615–618, 621–623
    rms voltage and, 606–608
Alternating-current generator, 592–593, 604–605
Ammeter, **519**, 541–542, 543
Ampere (unit), **517**, 528
Amplitude
    of simple harmonic motion, **336**, 337, 339, 349
    of vibration, **336**, 337, 339, 349
    of wave, **356**, 368
Angular acceleration, **136**, 138–148
    constant, kinematic equations for, 138
    and tangential acceleration, 137
    and torque, 139–140
    units of, 136
Angular deformation, 314
Angular magnification, **427**–430, 442
Angular momentum, **145**–148
    in atoms, 694–696, 706
    orbital, 695, 706
    quantization of, 694–697, 706
    spin, 696, 706
Angular motion, 148
Angular position, 113
    units to describe, 100–101, 113
Angular-position coordinate, 99–**100**, 113
Angular size, **428**, 442
Angular velocity, **102**–103, 112, 113
Antenna, electrical field production by, 656–657
Antinodes, **378**
Antiparticles, **760**
Antiproton, 760
Antiquarks, 762, 764
Archimedes, 283, 399
Archimedes' principle, **282**–286, 290
Arc-length coordinate, **100**
Area, under graph, 42–43
Aristotle, 53
Armature, **577**
Astigmatism, 437–438
Atmosphere (unit), **251**
Atmospheric pressure, 251

Atom(s), **476**
    energy levels of, 681–**684**
    excited states of, **684**–688, 697
    ground state of, **684**, 698, 700–702
    groups of, 699–700
    multielectron, 696–698, 706
    one-electron, 681–688, 689–690
    quantum theory of, 693–706
    states of, 684–688, 693–698
    structure of. *See* Atomic structure
    subshells in, **697**, 706
    transition, 700
Atomic absorption spectroscopy, 678–679
Atomic mass unit, **712**, 735
Atomic number, **681**–682, 709–710, 735
Atomic spectra, **678**–681, 689, 701
Atomic structure, 677–706
    Bohr model of, 681–688, 689–690, 693
    of hydrogen, 681–688, 689–690
    of multielectron atoms, 693–706
    nuclear model of, 677–**678**, 689
    plum-pudding model of, 677–678
Atomic wave functions, **693**–694, 706
    quantum numbers for, 694–696
Audibility, threshold of, **387**
Aurora, 575
Average acceleration, **24**
Average speed, **20**
Average velocity, 20–**21**
Avogadro's number, **255**–257, 270
Axis
    of graph, 771
    of lens, **417**

Back emf, **597**
Balmer, Johann, 679
Balmer's equation, **679**
Balmer series, 680
Band spectrum, **679**
Bar (unit), **251**
Baryon(s), **760**, 762
Baryon number, 715, 760–761
Base number, **770**
Basic quantities, 4
Battery, 515–516
Beat frequency, 362–**363**, 368
Becquerel, Henri, 716–717

790

POWER OF TEN PREFIXES

Prefix	Abbrev.	Value
Exa	E	10^{18}
Peta	P	10^{15}
Tera	T	10^{12}
Giga	G	10^{9}
Mega	M	10^{6}
Kilo	k	10^{3}
Hecto	h	10^{2}
Deka	da	10^{1}
Deci	d	10^{-1}
Centi	c	10^{-2}
Milli	m	10^{-3}
Micro	μ	10^{-6}
Nano	n	10^{-9}
Pico	p	10^{-12}
Femto	f	10^{-15}
Atto	a	10^{-18}

SOME USEFUL FACTS

Area of circle (radius r)	πr^2
Area of sphere (radius r)	$4\pi r^2$
Volume of sphere	$4\pi r^3/3$
Density of dry air (STP)	1.293 kg/m^3
Density of water (4° C)	1000 kg/m^3
Speed of sound in air (0° C)	331.4 m/sec

Trig definitions:

$\sin \theta = $ (opposite side)/(hypotenuse)

$\cos \theta = $ (adjacent side)/(hypotenuse)

$\tan \theta = $ (opposite side)/(adjacent side)

Quadratic equation:

$$0 = ax^2 + bx + c,$$

$$\text{where } x = \frac{-b \pm \sqrt{b^2 - 4ac}}{2a}$$